Sustainable Development Goals Series

The **Sustainable Development Goals Series** is Springer Nature's inaugural cross-imprint book series that addresses and supports the United Nations' seventeen Sustainable Development Goals. The series fosters comprehensive research focused on these global targets and endeavours to address some of society's greatest grand challenges. The SDGs are inherently multidisciplinary, and they bring people working across different fields together and working towards a common goal. In this spirit, the Sustainable Development Goals series is the first at Springer Nature to publish books under both the Springer and Palgrave Macmillan imprints, bringing the strengths of our imprints together.

The Sustainable Development Goals Series is organized into eighteen subseries: one subseries based around each of the seventeen respective Sustainable Development Goals, and an eighteenth subseries, "Connecting the Goals," which serves as a home for volumes addressing multiple goals or studying the SDGs as a whole. Each subseries is guided by an expert Subseries Advisor with years or decades of experience studying and addressing core components of their respective Goal.

The SDG Series has a remit as broad as the SDGs themselves, and contributions are welcome from scientists, academics, policymakers, and researchers working in fields related to any of the seventeen goals. If you are interested in contributing a monograph or curated volume to the series, please contact the Publishers: Zachary Romano [Springer; zachary.romano@springer.com] and Rachael Ballard [Palgrave Macmillan; rachael.ballard@palgrave.com].

More information about this series at https://link.springer.com/bookseries/15486

Udo Schickhoff • R.B. Singh •
Suraj Mal
Editors

Mountain Landscapes in Transition

Effects of Land Use and Climate Change

Springer

Editors
Udo Schickhoff
CEN Center for Earth System
Research and Sustainability
Institute of Geography
University of Hamburg
Hamburg, Germany

R.B. Singh (Deceased)
Department of Geography
Delhi School of Economy
University of Delhi
New Delhi, Delhi, India

Suraj Mal
Department of Geography
Shaheed Bhagat Singh College
University of Delhi
New Delhi, Delhi, India

ISSN 2523-3084 ISSN 2523-3092 (electronic)
Sustainable Development Goals Series
ISBN 978-3-030-70237-3 ISBN 978-3-030-70238-0 (eBook)
https://doi.org/10.1007/978-3-030-70238-0

This Springer imprint is published by the registered company Springer Nature Switzerland AG
The registered company address is: Gewerbestrasse 11, 6330 Cham, Switzerland

To Bruno Messerli

for inspiring leadership and lifetime dedication to the cause of mountains

To R.B. Singh

for having been an outstanding mentor, guide, and life-long advisor to his students and his great contributions to geography

Foreword

Mountains landscapes provide the sustenance of life for humanity in many different ways, with their biodiversity, incredible beauty and culture, and the resources they provide to billions of people. These landscapes are rapidly experiencing change, and we need to pay attention to the signals that mountains provide us. Indeed how humanity manages such changes in landscapes will be an important test of our time, and if we can pass, we know there is hope.

An important first step is to understand the dynamics of the socio-ecological systems in the mountains with our best science. I am pleased that in recent years, much more effort has gone into producing the science needed and that science is well-reflected in this book.

The most striking context of mountains in recent years is rapid change and transformations in mountain societies and ecosystems. We know that mountains are highly vulnerable to climate change, and the chapters in this book document these changes throughout the world. However, there are a variety of change drivers working together that add to the complexity of mountain systems. Outmigration, urbanization, globalization, and connectivity all affect mountain societies. At the same time, utilization and extraction of mountain resources are putting extreme pressure on mountain ecosystems. I am writing during the time of COVID-19, during this short time migration trends have reversed, and the side effects of COVID-19 are adding tremendous additional shocks to mountain systems.

Ultimately, it is up to societies to respond and make changes so that we continue to enjoy the benefits that mountains provide. There are responses needed at the global scale to halt biodiversity loss and slow climate change, and there are responses needed within mountain communities to adapt. In the face of rapid change and sudden shocks, building socio-ecological resilience will be a key. In developing responses, there is a need to understand the potential for mountain landscape conservation and the potential for economic development and to figure out ways in which both conservation goals meet development goals. In order to develop responses at all scales, the book outlines the impacts of climate and other changes on mountain landscapes and people and suggests solutions to move forward.

Mountains play an important role in making connections—essential in today's world. Mountain landscapes themselves connect different types of ecosystems, and people are intimately connected to nature. To understand mountains, there is a need to connect across disciplines. To develop solutions, there is a need to connect with society and connect science with policy. Mountains provide resources to downstream areas and thus play an important role in connecting upstream and downstream landscapes. Mountain landscapes are often shared between countries and thus can play an important role to connect countries. The value of this book is that it will help us build many bridges to make better mountain connections and rise to the challenge of a changing world.

David Molden
Director General
International Centre for Integrated
Mountain Development (ICIMOD)
Kathmandu, Nepal

Preface

Mountains are critical components of the Earth system. Mountains and highlands, inhabited by more than 900 million people, cover almost 25% of the terrestrial surface of the Earth. Mountains considerably influence regional and continental atmospheric circulation as well as water and energy cycles and provide ecosystem services to about half of humanity. Mountains are popularly known as the water towers of the Earth. The supply of water is the key function of mountains as all of the world's major rivers originate in mountains where vast volumes of water are stored in glaciers and in the snow cover and gradually released in the melting season. Over 40% of the global population lives in the watersheds of rivers having its sources in mountain ranges. Other ecosystem services include the provision of energy, forest and agricultural products, minerals, and other natural resources. Moreover, mountains are globally significant as core areas of biodiversity, characterized by higher species richness than adjacent lowlands and high proportions of endemic species. Mountain regions are also centers of cultural diversity, provide ample opportunities for recreation and tourism, and are of spiritual significance. Thus, mountains have a lasting effect on the life of billions living either in mountains or in adjacent lowlands. About one-tenth of the world's population derives their life support directly from mountains.

Being exceptionally fragile and susceptible to global environmental change, the world's mountains have undergone significant modifications in the Anthropocene epoch. Climate and land use change will increasingly threaten the integrity of mountain ecosystems and alter their capacity to provide goods and services for both highland and lowland people. High elevation environments with glaciers, snow, permafrost, water, and a complex altitudinal zonation of vegetation and fauna are widely considered as being most sensitive to climatic changes. The fragility of mountain ecosystems also represents a substantial challenge to sustainable land use and natural resource management. Unsustainable mining, forestry, agricultural practices, and tourism in the context of rapid urbanization and globalization often have drastic consequences, resulting in environmental deterioration and landscape degradation, and their impacts are usually more difficult to correct than in lowland areas.

Increasing temperatures, shrinkage of glaciers, snow cover decline, extreme precipitation events, delayed freezing and early ice melting on rivers and lakes, altitudinal shifts of species, habitat and biodiversity loss, increased soil erosion rates, etc., are among the responses of mountain ecosystems to climate and land use change and their interactive effects. In view of these challenges, the future of mountain communities and their livelihoods is becoming uncertain. Knowing how structures and functions of mountain ecosystems are affected is of fundamental importance taking the significant implications for mountain people as well as hundreds of millions living downstream into account. Understanding the system response is also vital in terms of adaptation and mitigation, for implementing collective and collaborative action and effective strategies of sustainable land use and environmental management. Therefore, an urgent need was felt to study the change in the world's mountains in a holistic perspective.

Attention to mountain issues has grown significantly over the past decades. A first milestone to establish mountains as a research priority and to support sustainable mountain development was the establishment of the UNESCO Man and Biosphere Project 6 (1973) in order to study the "Impact of Human Activities on Mountain and Tundra Ecosystems." Subsequently, applied and interdisciplinary aspects of mountains were studied in detail, supported by the foundation of the International Mountain Society (1980), the International Center for Integrated Mountain Development (1983), and by publishing the journal "Mountain Research and Development" (since 1981). A new commission on "High-Altitude Geoecology" under the International Geographical Union was established in 1968, later titled "Mountain Geoecology and Sustainable Development" and now "Mountain Studies," which invited and popularized mountain studies among the younger community of scientists. The United Nations University's mountain program "High Land Low Land Interactive System" was launched in 1978, which was later renamed as "Mountain Ecology and Sustainable Development Program" under the leadership of the Late Professor Dr. Bruno Messerli and Professor Dr. Jack D. Ives. At the United Nations Conference on Environment and Development (UNCED) held in Rio de Janeiro during 1992, a chapter on sustainable mountain development was included in Agenda 21, thus putting mountains on the global agenda. Mountain-related studies were further popularized and gained global attention with the publications of two landmark books, "Mountains of the World: A Global Priority" and "The Himalayan Dilemma" by the Late Professor Dr. Bruno Messerli and Professor Dr. Jack D. Ives.

Increasing awareness of the importance of mountain research and development resulted in the declaration of the year 2002 as the "International Year of Mountains" and in the designation of December 11 as International Mountain Day. The "Mountain Research Initiative" was also founded in 2002, which is a global scientific promotion and coordination effort toward strengthening the dialog between science and policy. The UN resolution "Sustainable Mountain Development" in 2010 further strengthened the international recognition of the importance of mountain environments and mountain peoples, also reflected in a number of recent pioneering national and global research initiatives such as the Global Observation Research Initiative in Alpine Environments (GLORIA) program, the scientific collaboration network of the World Glacier Monitoring Service (WGMS), or the Global Terrestrial Network for Permafrost (GTN-P) as well as in recent significant internationally coordinated publications including the IPCC "Special Report on the Ocean and Cryosphere in Changing Climate."

The present volume compiles available knowledge of the response of mountain ecosystems to recent climate and land use change addressing key concepts, major drivers, and key processes. After an introductory global review of changing mountain environments in the Anthropocene, the subsequent chapters present case studies from mountains across the world, divided into two parts, viz. (I) climate change and response processes of mountain environments and (II) response processes of mountain environments to land use change. Part I is specifically devoted to climate change impacts substantiated by 19 case studies, while Part II deals with effects of land use change in different mountains of the world exemplified by 12 case studies. We believe the present initiative will be useful to the research and policy-making community and will advance our scientific understanding of change in the world's mountains. We intend to further consolidate the international recognition of the global significance of mountain regions and contribute at the same time to an accelerated implementation of the UN Sustainable Development Goals. The volume might also be useful for teaching in the fields of geography, landscape ecology, environmental studies, hydrology, climatology, and human–environmental interaction.

Hamburg, Germany — Udo Schickhoff
New Delhi, India — R.B. Singh
New Delhi, India — Suraj Mal

Contents

Contributors

Lipy Adhikari International Centre for Integrated Mountain Development, Kathmandu, Nepal

Istak Ahmed Department of Geography and Disaster Management, Tripura University, Tripura, India

Alla Aleksanyan Department of Geobotany and Ecological Physiology, Institute of Botany aft. A. L. Takhtajyan of NAS RA, Yerevan, Armenia

Simon K. Allen Department of Geography, University of Zurich, Zurich, Switzerland

Mangalasseril Mohammad Anees School of Environmental Sciences, Jawaharlal Nehru University, New Delhi, India

Martha E. Apple Department of Biological Sciences, Montana Technological University, Butte, MT, USA

Manohar Arora National Institute of Hydrology, Roorkee, Uttarakhand, India

Manoj Kumar Arora BML Munjal University, Gurgaon, India

Anil Aryal Interdisciplinary Centre for River Basin Environment, University of Yamanashi, Kofu, Japan;
Hydro Energy & Environment Research Center, Kathmandu, Nepal

Mohd Farooq Azam Discipline of Civil Engineering, School of Engineering, Indian Institute of Technology Indore, Simrol, India

Michelle A. Baker Department of Biology and Ecology Center, Utah State University, Logan, UT, USA

Abhishek Banerjee Key Laboratory of Geographic Information Science, Ministry of Education, and School of Geographical Science, Institute of Eco-Chongming, East China Normal University, Shanghai, China

Alina Baranova CEN Center for Earth System Research and Sustainability, Institute of Geography, University of Hamburg, KlimaCampus, Hamburg, Germany

Nabin Bhattarai International Centre for Integrated Mountain Development, Kathmandu, Nepal

Maria Bobrowski Physical Geography, Center for Earth System Research and Sustainability (CEN), Universität Hamburg, Hamburg, Germany; CEN Center for Earth System Research and Sustainability, Institute of Geography, University of Hamburg, Hamburg, Germany

J. Böhner CEN Center for Earth System Research and Sustainability, Institute of Geography, Universität Hamburg, Hamburg, Germany

Karanjot Kaur Brar Center of Advanced Study in Geography (CAS), Department of Geography, Panjab University, Chandigarh, India

Anders Bryn Natural History Museum, University of Oslo, Oslo, Norway; Department of Geography, University of Bergen, Bergen, Norway

B. Bürzle CEN Center for Earth System Research and Sustainability, Institute of Geography, Universität Hamburg, Hamburg, Germany

Vishwa B. S. Chandel Center of Advanced Study in Geography (CAS), Department of Geography, Panjab University, Chandigarh, India

R. P. Chaudhary RECAST Research Centre for Applied Science and Technology, Tribhuvan University, Kirtipur, Nepal

Sunita Chaudhary International Centre for Integrated Mountain Development, Kathmandu, Nepal

Nakul Chettri International Centre for Integrated Mountain Development, Kathmandu, Nepal

Nibedita Das (Pan) Department of Geography and Disaster Management, Tripura University, Tripura, India

Amal Debnath Department of Forestry and Biodiversity, Tripura University, Tripura, India

Jatan Debnath Department of Geography and Disaster Management, Tripura University, Tripura, India

Ignacio J. Diaz-Maroto Department of Agroforestry Engineering, Higher Polytechnic School of Engineering, University of Santiago de Compostela, Lugo, Spain

Maria L. Dvinskaya Sukachev Institute of Forest SB RAS, FRC Krasnoyarsk Science Center SB RAS, Krasnoyarsk, Russia

George Fayvush Department of Geobotany and Ecological Physiology, Institute of Botany aft. A. L. Takhtajyan of NAS RA, Yerevan, Armenia

Lucia Felbauer IGF/ÖAW Institute for Interdisciplinary Mountain Research, Austrian Academy of Sciences, Innsbruck, Austria

Andrea Fischer IGF/ÖAW Institute for Interdisciplinary Mountain Research, Austrian Academy of Sciences, Innsbruck, Austria

Courtney G. Flint Department of Sociology, Social Work, and Anthropology, Utah State University, Logan, UT, USA

Arne Friedmann Institute of Geography, AG Biogeography, University of Augsburg, Augsburg, Germany

Daniel Germain Department of Geography, Université du Québec À Montréal, Montréal, Canada

Jussi Grießinger Institute of Geography, Friedrich-Alexander-University Erlangen-Nürnberg, Erlangen, Germany

Kay Helfricht IGF/ÖAW Institute for Interdisciplinary Mountain Research, Austrian Academy of Sciences, Innsbruck, Austria

Philipp Hochreuther Institute of Geography, Friedrich-Alexander-University Erlangen-Nürnberg, Erlangen, Germany

Helene Hoffmann IGF/ÖAW Institute for Interdisciplinary Mountain Research, Austrian Academy of Sciences, Innsbruck, Austria;
Department of Earth Sciences, University of Cambridge, Cambridge, UK

Sergei T. Im Sukachev Institute of Forest SB RAS, FRC Krasnoyarsk Science Center SB RAS, Krasnoyarsk, Russia;
Siberian Federal University, Krasnoyarsk, Russia;
Reshetnev Siberian State University of Science and Technology, Krasnoyarsk, Russia

Andrina Janicke IGF/ÖAW Institute for Interdisciplinary Mountain Research, Austrian Academy of Sciences, Innsbruck, Austria

Pawan Kumar Joshi School of Environmental Sciences, Jawaharlal Nehru University, New Delhi, India;
Special Center for Disaster Research, Jawaharlal Nehru University, New Delhi, India

Srijana Joshi International Centre for Integrated Mountain Development, Kathmandu, Nepal

Ramchandra Karki CEN Center for Earth System Research and Sustainability, Institute of Geography, University of Hamburg, Hamburg, Germany;
Department of Hydrology and Meteorology, Government of Nepal, Kathmandu, Nepal

Saurabh Kaushik Academy of Scientific and Innovative Research (AcSIR-CSIO), CSIR-CSIO Campus, Chandigarh, India;
CSIR-Central Scientific Instruments Organisation, Chandigarh, India

Yannis Kazoglou Department of Forestry, Wood Sciences & Design, University of Thessaly, Karditsa, Greece

Viacheslav I. Kharuk Sukachev Institute of Forest SB RAS, FRC Krasnoyarsk Science Center SB RAS, Krasnoyarsk, Russia;
Siberian Federal University, Krasnoyarsk, Russia

Oliver Korch Institute of Geography, AG Biogeography, University of Augsburg, Augsburg, Germany

Jörg Löffler Department of Geography, University of Bonn, Bonn, Germany

Manisha Maharjan Department of Environmental Engineering, Kyoto University, Katsura, Nishikyo-ku, 615-8510, Kyoto, Japan;
Hydro Energy & Environment Research Center, Kathmandu, Nepal

Suraj Mal Department of Geography, Shaheed Bhagat Singh College, University of Delhi, Delhi, India;
CEN Center for Earth System Research and Sustainability, Institute of Geography, University of Hamburg, Hamburg, Germany

Philipp Marr Department of Geography and Regional Research, University of Vienna, Vienna, Austria;
Department of Geography, University of Bonn, Bonn, Germany

Alice C. Martin Division of Biological Sciences, University of Montana, Missoula, MT, USA

Wolfgang J. H. Meier Institute of Geography, Friedrich-Alexander-University Erlangen-Nürnberg, Erlangen, Germany

David J. Molden International Centre for Integrated Mountain Development, Kathmandu, Nepal

Dennis J. Moritz Department of Mathematics, University of Montana, Missoula, MT, USA

Geoffrey Mukwada Afromontane Research Unit, Department of Geography, University of the Free State, Bloemfontein, South Africa

Nuwan Gunarathne University of Sri Jayewardenepura, Nugegoda, Sri Lanka

H. Mahendra P. Peiris Postgraduate Institute of Agriculture, University of Peradeniya, Peradeniya, Sri Lanka

Il'ya A. Petrov Sukachev Institute of Forest SB RAS, FRC Krasnoyarsk Science Center SB RAS, Krasnoyarsk, Russia

Kerstin Potthoff School of Landscape Architecture, Norwegian University of Life Sciences, Ås, Norway

Seema Rani Department of Geography, Institute of Science, Banaras Hindu University, Varanasi, Uttar Pradesh, India

Deepika Rawat CEN Center for Earth System Research and Sustainability, Institute of Geography, University of Hamburg, Hamburg, Germany

Macy K. Ricketts Department of Botany, University of Wyoming, Laramie, Wyoming, USA

Vishwambhar Prasad Sati Department of Geography and Resource Management, Mizoram University, Aizawl, Mizoram, India

Udo Schickhoff CEN Center for Earth System Research and Sustainability, Institute of Geography, University of Hamburg, KlimaCampus, Hamburg, Germany

T. Scholten Department of Geosciences, Chair of Soil Science and Geomorphology, University of Tübingen, Tübingen, Germany

Niels Schwab CEN Center for Earth System Research and Sustainability, Institute of Geography, University of Hamburg, Hamburg, Germany

Christopher A. Scott School of Geography & Development, and Udall Center for Studies in Public Policy, University of Arizona, Tucson, AZ, USA

Bandana Shakya International Centre for Integrated Mountain Development, Kathmandu, Nepal

Eklabya Sharma International Centre for Integrated Mountain Development, Kathmandu, Nepal

Manu Raj Sharma University Department of Geography, L. N. Mithila University, Darbhanga, India

Richa Sharma Public Health Foundation of India (PHFI), Gurgoan, India

Aparna Shukla Ministry of Earth Sciences, New Delhi, India

Alexander S. Shushpanov Sukachev Institute of Forest SB RAS, FRC Krasnoyarsk Science Center SB RAS, Krasnoyarsk, Russia;
Reshetnev Siberian State University of Science and Technology, Krasnoyarsk, Russia

R. B. Singh (Deceased) Department of Geography, Delhi School of Economics, University of Delhi, Delhi, India

Tejpal Singh Academy of Scientific and Innovative Research (AcSIR-CSIO), CSIR-CSIO Campus, Chandigarh, India;
CSIR-Central Scientific Instruments Organisation, Chandigarh, India

Sreedharan Sreekesh Centre for the Study of Regional Development, Jawaharlal Nehru University, New Delhi, India

Ludwig Stabile-Caillé Department of Geography, Université du Québec À Montréal, Montréal, Canada

Philipp Stojakowits Institute of Geography, AG Biogeography, University of Augsburg, Augsburg, Germany

Rocky Talchabhadel Texas A&M AgriLife Research, Texas A&M University, El Paso, TX, USA;
SmartPhones for Water Nepal (S4W-Nepal), Lalitpur, Nepal

Michael Vrahnakis Department of Forestry, Wood Sciences & Design, University of Thessaly, Karditsa, Greece

Eva-Maria Wild Faculty of Physics - Isotope Physics, VERA Laboratory, University of Vienna, Vienna, Austria

Stefan Winkler Department of Geography and Geology, University of Würzburg, Würzburg, Germany

Bisma Yousuf University of Jammu, Jammu, India

1 The World's Mountains in the Anthropocene

Udo Schickhoff, Maria Bobrowski, Suraj Mal, Niels Schwab, and R.B. Singh

Abstract

This review summarizes current understanding of drivers for change and of the impact of accelerating global changes on mountains, encompassing effects of climate change and globalization. Mountain regions with complex human–environment systems are known to exhibit a distinct vulnerability to the current fundamental shift in the Earth System driven by human activities. We examine indicators of the mountain cryosphere and hydrosphere, of mountain biodiversity, and of land use and land cover patterns, and show that mountain environments in the Anthropocene are changing on all continents at an unprecedented rate. Rates of climate warming in the world's mountains substantially exceed the global mean, with dramatic effects on cryosphere, hydrosphere, and biosphere. Current climatic changes result in significantly declining snow-covered areas, widespread decreases in area, length, and volume of glaciers and related hydrological changes, and widespread permafrost degradation. Complex adaptations of mountain biota to novel constellations of bioclimatic and other site conditions are reflected in upslope migration and range shifts, treeline dynamics, invasion of non-native species, phenological shifts, and changes in primary production. Changes in mountain biodiversity are associated with modified structure, species composition, and functioning of alpine ecosystems, and compromise ecosystem services. Human systems have been negatively impacted by recent environmental changes, with both inhabitants of mountain regions as well as people living in surrounding lowlands being affected. Simultaneously, accelerating processes of economic globalization cause adaptation strategies in mountain communities as expressed clearly in changing land use systems and mobility patterns, and in increasing marginalization of peripheral mountains and highlands. The current state of the world's mountains clearly indicates that global efforts to date have been insufficient to make significant progress towards implementing the Sustainable Development Goals of the 2030 Agenda for Sustainable Development, adopted by all United Nations member states.

U. Schickhoff (✉) · M. Bobrowski · S. Mal · N. Schwab
CEN Center for Earth System Research and Sustainability, Institute of Geography, University of Hamburg, Hamburg, Germany
e-mail: udo.schickhoff@uni-hamburg.de

S. Mal
Department of Geography, Shaheed Bhagat Singh College, University of Delhi, Delhi, India

R.B. Singh (Deceased)
Department of Geography, Delhi School of Economics, University of Delhi, Delhi, India

U. Schickhoff et al. (eds.), *Mountain Landscapes in Transition*,
Sustainable Development Goals Series, https://doi.org/10.1007/978-3-030-70238-0_1

Keywords

Climate change · Combined mountain agriculture · Cryosphere · Glacier retreat · Globalization · Land use change · Migration · Pastoralism · Permafrost degradation · Range shift · Treeline dynamics

1.1 Introduction

It has been known since Alexander von Humboldt (1769–1859) that the decrease of temperature with increasing elevation in mountains induces vertical climate alterations which are reflected in all climate-dependent landscape elements, especially in the altitudinal zonation of vegetation and land use. From the results of the pioneers and key exponents of geographical high mountain research such as von Humboldt, Carl Troll (1899–1975), and Bruno Messerli (1931–2019) a picture of the natural setting of high mountains and of the interwoven geoecological and human-geographical factor complexes emerged, which has undergone major changes in recent decades. Over the past decades, mountain regions have been subjected to above-average climate warming and significant land use changes. Contemporary climate change and modified land use intensities and land use systems have tremendous effects on mountain landscapes so that the pioneers of high mountain research would hardly recognize certain landscapes on a visit today. These effects are the core theme of this book; they are explored in the following chapters which include compelling examples from around the world.

The significance of mountains for the Earth system (Fig. 1.1) and for a considerable part of the human population is often not rated highly enough. Mountain ecosystems have evolved on every continent, characterized by the complexity of their topography associated with steep environmental gradients, i.e. distinct variations of climatic, edaphic, and other environmental factors over short distances (Schickhoff 2011). Mountains and highlands cover nearly 25% of the terrestrial surface of the Earth (Romeo et al. 2015), 11% of the global land surface are higher than 2000 m above sea level (a.s.l.) (Kapos et al. 2000). Based on topographic ruggedness of the Earth's surface, Körner et al. (2017) calculated an area of 12.5% of the land surface covered by mountains (excluding Antarctica) of which 24% comprise alpine and nival belts. Elias (2020) and Testolin et al. (2020) quantified a comparable land area covered by alpine biomes. As a result of the physiography and diverse topography—major mountain ranges rise prominently above their surroundings—mountains exert a great influence on energy and moisture fluxes and on local and regional airflow patterns up to the large-scale atmospheric circulation. Their influence on airflows, temperature and humidity extends far beyond their geographic boundaries and may be felt for hundreds and thousands of kilometers (Bach and Price 2013).

Mountains provide ecosystem goods and services to more than half of humanity, thus they are of critical importance to people in almost every country of the world (Ives et al. 1997; Schickhoff 2011; Byers et al. 2013). Approximately 13% of the human population derives their life-support directly from mountains (Price 1998; Romeo et al. 2015), including diverse communities of distinct ethno-linguistic and cultural identity. Mountains are essential resource regions for the supply of water, energy, grazing lands, forest and agricultural products, and mineral resources. Many plant and animal species are endemic to mountain regions which are characterized by increased biodiversity relative to the surrounding lowlands (biodiversity hotspots). Mountains are also centres of ethnic, religious and cultural diversity, provide ample opportunities for recreation and tourism, and are of spiritual significance. Water supply is usually considered the key function of mountains for humanity since all of the world's major rivers have their headwaters in mountains, and huge quantities of freshwater are stored as snow and ice as well as in lakes and reservoirs and gradually released to the lowlands. Mountains are often called 'water towers' of the Earth owing to the key role they play for supplying water to billions of people in lowlands used for drinking, domestic use, irrigation,

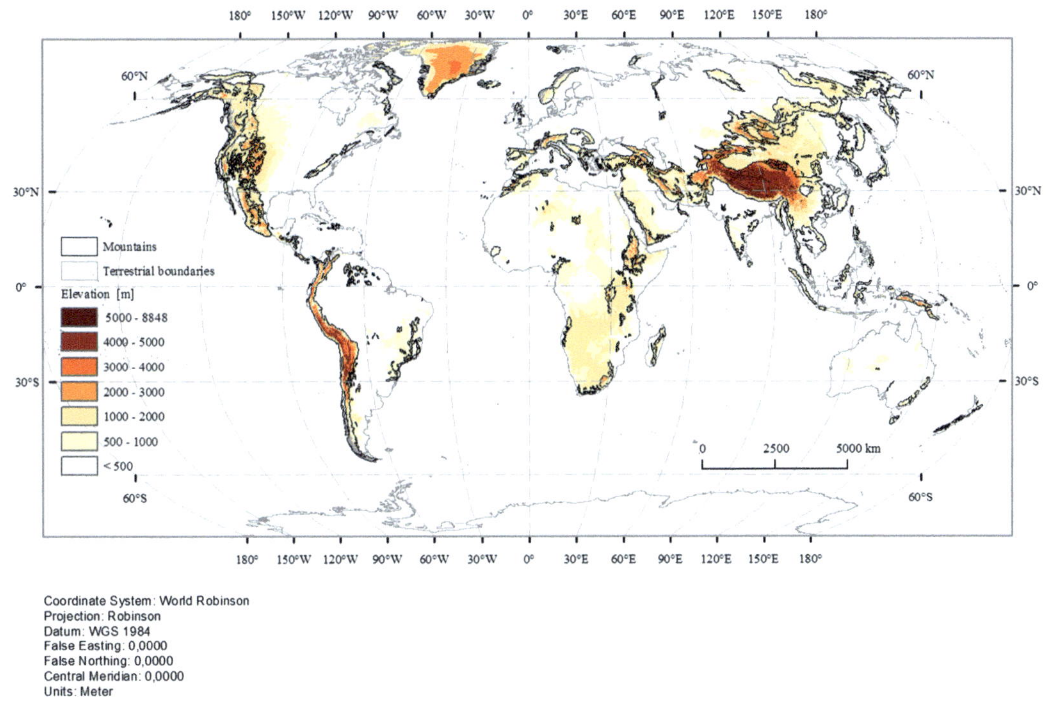

Fig. 1.1 Mountains of the world (background image from https://dds.cr.usgs.gov/srtm/version2_1/SRTM30/)

hydropower, industry, and transportation (Körner et al. 2005; Viviroli et al. 2007; Schickhoff 2011; Byers et al. 2013). Water supply from mountains is essential for life in semiarid and arid regions where the proportion of water generated at higher elevations may be more than 95% as in the basin of the Aral Sea (Messerli 1999). Even in humid regions, 60–80% of the total freshwater available is provided by mountain watersheds. Hydropower from these watersheds provides about one-fifth of the world's total electricity supply (Byers et al. 2013). Water supply from mountains forms the basis for ensuring availability and sustainable management of water and sanitation for billions of people (Goal 6 of the UN Sustainable Development Goals). Integrated water resources management as a global framework covering policies, institutions, management instruments and financing for the comprehensive and collaborative management of water resources has still been implemented at a low level (UN 2020).

Mountains show above-average species richness and comprise many unique biomes that are globally significant as core areas of biodiversity. A quarter of all terrestrial biodiversity is situated in mountains (Körner et al. 2017). Over evolutionary time scales, mountains also have generated high levels of diversity through in situ adaptations and diversification (Badgley et al. 2017; Hoorn et al. 2018). The global hotspots of species diversity, areas with increased levels of species richness and high proportions of endemic species, are predominantly mountainous regions. The particular species richness is related to the topographic complexity and associated high levels of geodiversity, i.e. the small-scale diversity of habitats and site conditions resulting from steep climatic and ecological gradients in fragmented and topographically diverse terrain. The compression of climatic life zones along vertical gradients, spatial isolation, combined with effective reproduction systems, as well as moderate disturbance influences additionally contribute to small-scale extraordinarily high levels of biodiversity. Tropical and subtropical mountain regions in particular are home to highly diverse and species-rich ecosystems constituting the global centres of vascular plant diversity (Körner 2002; Barthlott et al. 2005, 2007). Species diversity includes the most important food staples such as potatoes, maize, wheat, rice, beans or barley which had been domesticated in mountain regions (Brush 1998). Promoting sustainable use of terrestrial ecosystems, reversing land degradation, and halting biodiversity loss are major targets at the heart of Goal 15 of the UN Sustainable Development Goals which need to be supported in particular in mountain regions (UN 2020).

The resource function of mountain regions also contributes substantially to their global significance (Schickhoff 2011). For instance, mountain forests account for more than a quarter of the area of global closed forests (Kapos et al. 2000). They provide diverse goods and services to millions of people including provisioning services (both timber and non-timber forest products such as fuelwood, fodder, grazing resources, medicinal plants, and mushrooms), regulating and supporting as well as cultural services (Price and Butt 2000; Price et al. 2011; Gratzer and Keeton 2017). Mountain forests play a critical role for mountain dwellers and valley communities regarding protection against natural hazards such as landslides, rockfalls, avalanches, and floods as well as for reducing soil erosion and maintaining hydrological cycles. Mountain forests also represent a major carbon sink, and carbon sequestration in those forests is of increasing significance in climate change mitigation. The past two decades have seen a significant increase globally in the extraction of mineral resources from mountains; mines in mountains are the major current source of many of the world's strategic non-ferrous and precious metals (Fox 1997; Jacka 2018), contributing to the fast increasing global material footprint. As mountain regions continue on a path of using natural resources unsustainably, the successful transition to sustainable consumption and production patterns is more essential than it has ever been before (addressed by Goal 12 of the UN Sustainable Development Goals) (UN 2020).

The global significance of mountain regions can only be fully grasped if the focus is on mountain dwellers. Between 2000 and 2012, the global mountain population increased from 789 to 915 million people, and will further increase in the next decades (Romeo et al. 2015). Most mountain populations are nowadays integrated, to varying degrees, economically, socially and politically with lowland communities and the wider world (Funnell and Parish 2001). Nevertheless, mountains are still home to many indigenous peoples, encompassing an amazing diversity of human cultures and communities. For example, 100 different ethnic/caste groups were identified in the 2001 census in the mountainous state of Nepal (Sharma 2008), and more than 700 languages are spoken in mountainous regions of New Guinea (Stepp et al. 2005). This cultural diversity contributes to the attractiveness of mountains that have become key tourism destinations in many parts of the world. The significance of mountains as centres for recreation, adventure, scenic beauty or interaction with local people will increase in coming decades as tourism is the world's largest and fastest growing industry. The large influx of tourists to mountain regions is not without conflicts due to the impacts on fragile high altitude environments and the special spiritual and cultural significance mountains have in many cultures (Price and Kohler 2013; Hamilton 2015).

Mountain ecosystems represent some of the few remaining wilderness areas of the globe, and encompass some of the most intriguing habitats in terms of the particular fascination of high mountain landscapes, with regard to high biodiversity levels and resident biota's special adaptation to the harsh physical environment, as well as in terms of the extraordinary cultural diversity and the sophisticated and complex resource utilization strategies that mountain dwellers have developed over many generations. Mountain ecosystems on the other hand are exceptionally fragile, susceptible to global environmental changes, and less resilient since longer periods of time may be needed for recovery from damage or excessive stress. As elsewhere on the globe, climatic changes and land use changes are the major drivers which are increasingly threatening the integrity of mountain ecosystems, affecting their capacity to provide goods and services.

Mountain regions around the world provide increasing evidence of ongoing impacts of land use change and of climate change on physical and biological systems. High elevation environments with steep relief, complex topography, cryospheric systems (snow, glaciers, permafrost), the compression of ecological vertical gradients and specific human–environmental subsystems are in general considered to be among the most sensitive terrestrial systems to reflect effects of climatic variations and consequences of changes in land use (Huber et al. 2005; Körner et al. 2005; Grabherr et al. 2010; Löffler et al. 2011; Schickhoff 2011, 2016a, b; Gottfried et al. 2012; Grover et al. 2015; Schickhoff et al. 2016a; Pauli and Halloy 2019; Hock et al. 2019; Schickhoff and Mal 2020). Observed changes of glaciers, snow cover, permafrost, hydrological conditions, and of the complex altitudinal zonation of vegetation and fauna indicate a distinct vulnerability, mountains are considered to be at the forefront of climate change impacts (Pihl et al. 2019). Mountain plants and animals, in particular endemic species, are often adapted to relatively narrow ranges of temperature and precipitation, even minor climatic changes can have significant effects (Körner 2003; Grabherr et al. 2010). If the water supply from High Asia is significantly reduced by retreating glaciers, more than half of Asia's population would be adversely affected (cf. Körner et al. 2005; Viviroli et al. 2007). More than a billion people in Asia live in the watersheds of rivers that have their sources in mountains. With regard to physical systems, current global warming has already left distinct traces in the cryosphere and hydrosphere of the world's mountains. It is also a powerful stressor on alpine biota, inducing shifts in phenology, species distributions, community structure as well as other ecosystem changes. As the climate crisis continues unabated, in particular in mountain regions, and as pervasive and catastrophic effects have become obvious, taking urgent action to combat climate change and its impacts and accelerating the transitions needed to achieve the Paris Agreement is the order of the day (Goal 13 of the UN Sustainable Development Goals) (UN 2020).

In many mountain ranges, ongoing alterations of montane and alpine land use systems caused by widespread socio-economic transformation processes are the major underlying driver of the transition of mountain landscapes. From a global perspective, changes in land use affect mountain forests and their ecosystem services in particular. In recent decades, two opposing trends have become apparent in the area covered by forests in mountain regions reflecting general global trends in forest cover: In many countries of the Global South forest cover is further declining, whereas a gradual expansion can be observed in industrialized countries (Schickhoff 2011, 2016b). For both montane and alpine life zones, it needs to be highlighted that the fragility of these high elevation environments poses a tremendous challenge for sustainable land use and natural resource management.

This chapter provides a global overview of the current state of knowledge on the effects of climate change and land use change on mountain landscapes. Presenting examples from major mountain systems around the world, the current knowledge is summarized with respect to climatic changes, impacts on physical systems (changes of snow cover, glaciers, permafrost, and related hydrological processes), biotic responses (phenological shifts, species migrations, range extensions, treeline dynamics, shifts in species composition), and effects of modified land use systems. Understanding how structures and functions of mountain ecosystems are affected by environmental change is a focal point for the mountain research agenda, in particular with regard to the abundance of ecosystem services and the multifunctionality of mountains (cf. Egan and Price 2017; Palomo 2017). At the same time, understanding the effects of environmental change on mountain ecosystems is of vital importance for adaptation planning, both for mountain people and for billions living in lowlands, in order to mitigate implications of climate and land use changes and to enhance the adaptive capacity of mountain socio-ecological systems in response to anticipated future changes. The international recognition of the importance of mountain environments and mountain peoples has increased over recent decades, however, the local and global awareness for the essential role mountain systems play in the geo-biosphere needs to be further supported and increased. Milestones of international efforts to establish mountains as a research priority, to support intergovernmental and nongovernmental processes of advocacy for mountains, and to support sustainable mountain development in general include the establishment of the UNESCO-MAB (Man and Biosphere) project on 'Impact of Human Activities on Mountain and Tundra Ecosystems' in 1973, the United Nations Conference on Environment and Development (UNCED) held in Rio de Janeiro in 1992 (inclusion of a mountain chapter into Agenda 21), the establishment of both the Mountain Forum (a global network of intergovernmental, nongovernmental, scientific, and private-sector organizations and individuals) in 1995 and the Mountain Research Initiative (a global scientific promotion and coordination effort towards strengthening the dialogue between science and policy) in 2002, the International Year of Mountains 2002, and the UN resolution 'Sustainable Mountain Development' in 2010 (Messerli 2012; Price and Kohler 2013; Kohler et al. 2015). Advances in international efforts to increase awareness for the importance of mountain research and development has stimulated scientific interest, reflected in a number of recent pioneering national and global research initiatives such as the GLORIA (Global Observation Research Initiative in Alpine Environments) programme, the scientific collaboration network of the WGMS (World Glacier Monitoring Service) or the Global Terrestrial Network for Permafrost (GTN-P). We strongly endorse further awareness-raising by producing and disseminating mountain-related education and research materials. All efforts towards sustainable mountain development should ideally be embedded in the 2030 Agenda for Sustainable Development, an urgent call for action substantiated by the 17 Sustainable Development Goals (UN 2020).

1.2 Recent Climate Change and Its Effects in Major Mountain Systems of the World

1.2.1 Climatic Changes

1.2.1.1 General Overview

Greenhouse gas emissions which continue to increase are the dominant factor in the observed persistent warming trend for the global mean surface temperature over recent decades and in recent years, with the last five-year period (2015–2019) and the last ten-year period (2010–2019) being the warmest of any equivalent period on record, and with 2015, 2016, 2017, 2018, and 2019 being the five warmest individual years (WMO 2019). July 2019 was the hottest month on record globally. Global warming is currently estimated to be 1.1 °C above pre-industrial values (1850–1900) and 0.2 °C warmer than 2011–2015, with the high latitudes of the Northern Hemisphere, in particular the northern Asian sector, showing the largest increase in mean temperature (Hoegh-Guldberg et al. 2018; WMO 2019). Here, the polar amplification leads to warming rates of more than 2 °C per 50 years, while warming trends and increasing temperature extremes have been generally observed in major mountain systems of the world over the past century (IPCC 2014). Temperature trends in most mountain regions substantially exceed the global mean over recent decades (Fig. 1.2), albeit with distinct patterns of spatial and seasonal differentiations, in particular in terms of vertical gradients. A current warming rate of 0.3–0.4 °C per decade is observed in most mountain regions of the world including western North America, the European Alps, and High Mountain Asia. This rate is significantly higher than the global mean and accelerating (cf. IPCC 2018; Hock et al. 2019; WMO 2019).

A widespread phenomenon is the amplification of warming rates at higher elevations, to be attributed mainly to changes in albedo and downward thermal radiation (Rangwala et al. 2013; Pepin et al. 2015; Hasson et al. 2016; Palazzi et al. 2019). At local and regional scales,

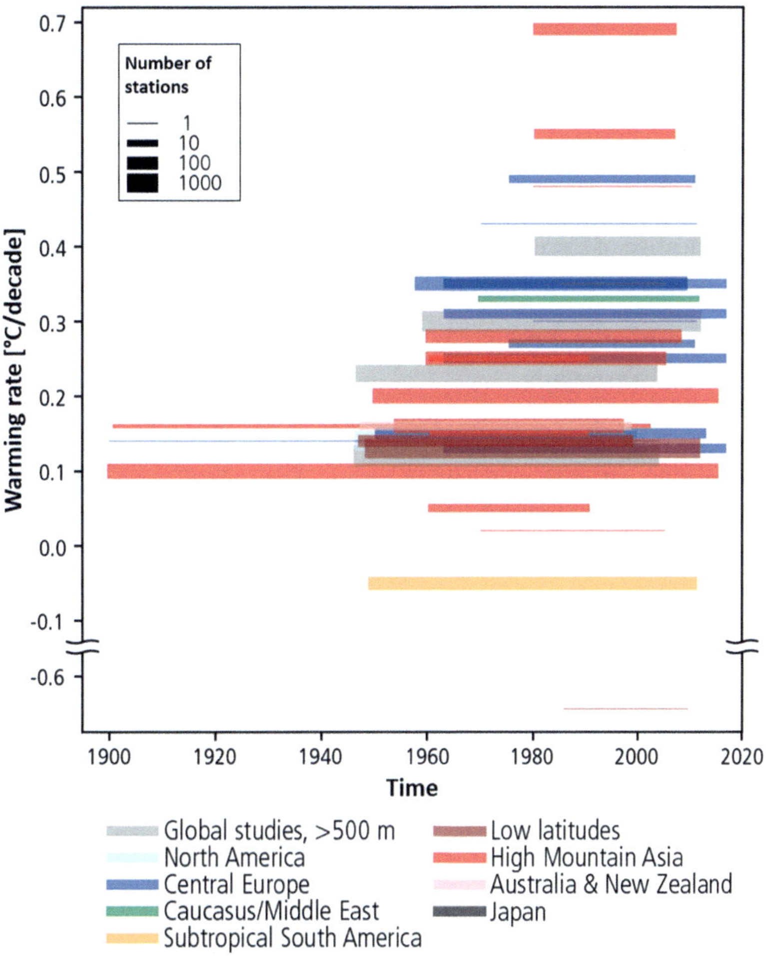

Fig. 1.2 Mean annual surface air temperature in mountain regions; each line refers to a warming rate from one of 40 studies based on multiple observation stations, with line thickness indicating the number of observation stations used. (modified from Hock et al. 2019)

however, evidence for elevation-dependent warming is sometimes contradictory. Obviously, trends in air temperature vary with elevation, but not in a consistent manner. Variations result from the effects of region, season, and selected temperature indicators (cf. Hock et al. 2019). The amplification of warming at higher elevations will increase with higher greenhouse gas emission scenarios, subjecting high elevation environments to comparatively more distinct changes in habitat conditions than lower elevations (Schickhoff et al. 2016a). Regardless of the underlying climate scenario, surface air temperature in mountain regions is projected to further increase at an average rate of at least 0.3 °C per decade until the mid-21st century (IPCC 2018), irreversibly affecting mountain ecosystems and their biodiversity, and impairing their capacity to provide key ecosystem services. This emphasizes the necessity of achieving the climate action target of the UN Sustainable Development Goals (UN 2020).

Compared to temperature changes, precipitation trends in mountain systems of the world are much more heterogeneous. Observations of annual precipitation often do not show significant increases or decreases over the past decades, while snowfall exhibits a more or less consistently decreasing trend, in particular at lower elevations (Hock et al. 2019). All greenhouse gas emission scenarios project a further decrease of snowfall at lower elevations throughout the twenty-first century, thus the rain/snow partitioning will be continuously affected. In contrast, projections of annual precipitation for the next decades show increases in the order of 5–20% for many mountain regions in South and East Asia, East Africa, and temperate Europe; only some mountain regions (the Mediterranean, Southern Andes) will experience a decrease in annual precipitation (Hock et al. 2019). The frequency and intensity of extreme precipitation events is projected to increase in many mountain regions.

1.2.1.2 Regional Overview

Asia[1] and Australasia

Temperature trends in the vast Hindu Kush Himalaya (HKH) are quite representative for many of the extensive mountain systems of Asia. The HKH has experienced warming from 1901 to 1940, cooling from 1940 to 1970, and a strong amplification of warming rates to 0.2 °C per decade over the period 1951–2014 (Fig. 1.3) (Ren et al. 2017; Krishnan et al. 2019a). Without any doubt, the warming trend has accelerated in the past two decades and in recent years (Diodato et al. 2011; Kattel and Yao 2013; Gerlitz et al. 2014; Hasson et al. 2016). At higher elevations, mean annual and mean annual maximum temperatures have been increasing at rates between 0.6 and c. 1 °C per decade over the past 40 years (Shrestha et al. 1999; Liu et al. 2006, 2009; Bhutiyani et al. 2007, 2010; Shrestha and Aryal 2011; Yang et al. 2011). Winter season temperature trends have been generally higher than those of other seasons (Hasson et al. 2016). Extreme warm days and nights show an increasing trend of occurrence in the past decades (nights by 2.54 days per decade), while occurrences of cold days and nights have declined (Hijioka et al. 2014; Krishnan et al. 2019a). In addition to the significant warming the HKH has seen in the past, the climate is projected to change more dramatically in the coming decades, with warming to be at least 0.3 °C higher, and in the NW Himalaya and Karakoram at least 0.7 °C higher than the targeted 1.5 °C as a global mean (Dimri et al. 2018; Krishnan et al. 2019a). Across Asia, the strongest warming of hot extremes is projected to occur in western and central Asia (Hoegh-Guldberg et al. 2018).

Significant and accelerated warming rates were observed over the entire Tibetan Plateau (Hasson et al. 2016; You et al. 2016, 2017; Ren et al. 2017). Yan and Liu (2014) reported a considerably increased warming trend in mean annual temperature of 0.32 °C per decade between 1961 and 2012, overcompensating the global warming slowdown period of 1998–2013 (cf. Ji and Yuan 2020). Current warming rates in Tibet are much higher than previously estimated (cf. Liu and Chen 2000), for the period 1992–2017 a warming rate of 0.47 °C per decade was assessed (Li et al. 2019). Significant warming of winter and annual temperatures are consistently reported from the West and Central Himalaya in India. Over the northwestern subregion, winter temperature has shown an elevated rate of increase (1.4 °C/100 years) compared to the monsoon temperature (0.6 °C/100 years) during the period from 1866 to 2006 (Bhutiyani 2015, 2016). Higher winter season mean temperature trends of up to 2.0 °C were detected for the period 1985–2008 (Bhutiyani et al. 2007, 2010; Shekhar et al. 2010; Dimri and Dash 2012; Singh D et al. 2015; Kumar et al. 2018). Seasonal maximum and minimum temperatures have increased by 2.8 and 1.0 °C, respectively; they show an increasing trend over the Pir Panjal, Shamshawari and Greater Himalayan ranges (Shekhar et al. 2010). Significantly increasing winter, monsoon and annual temperatures are reported from most stations, with the magnitude of warming being higher during recent decades compared to the century average (Bhutiyani et al. 2010; Singh and Kumar 2014; Shafiq et al. 2019; Negi et al. 2020). In Uttarakhand, temperature records of the past 100 years show a notable warming trend, particularly prominent during the last decade and at higher elevations (Mishra 2014; Singh RB et al. 2016).

A recent comprehensive evaluation of temperature trends across Nepal over the period 1980–2016 showed widespread significant warming which is higher for maximum temperature (0.4 °C per decade) than for minimum

[1]The information on mountain systems in Asia compiled in this paper is expanded and updated from Schickhoff & Mal (2020).

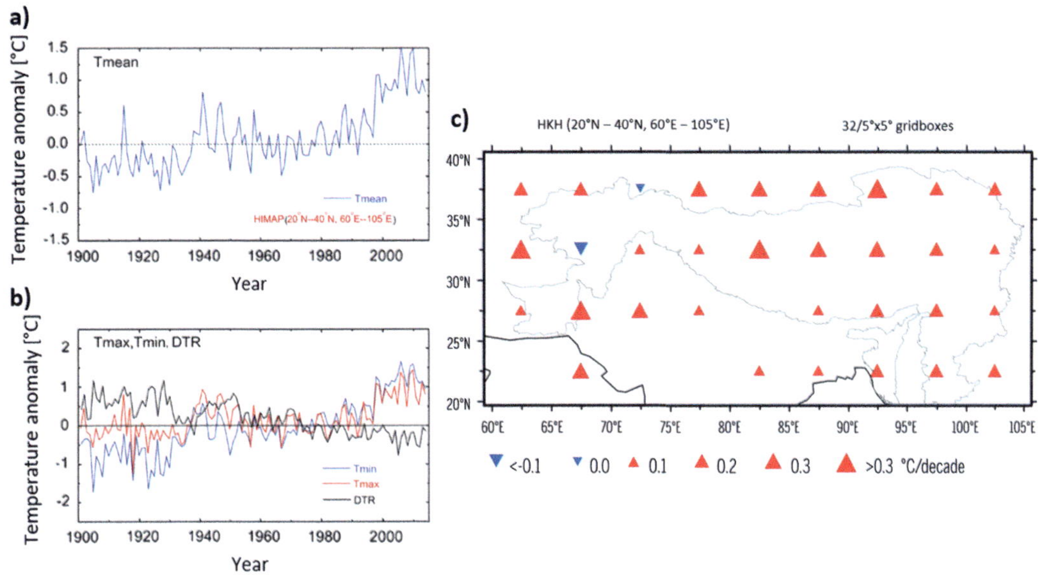

Fig. 1.3 Annual mean temperature anomaly series (°C) for the HKH region between 1901 and 2014 relative to 1961–90 mean values (**a**: Tmean; **b**: Tmax, Tmin, and DTR); **c**: Grid-averaged trends of annual mean temperature in the HKH region since 1901. (modified from Krishnan et al. 2019a)

temperature (0.2 °C per decade), higher in the mountainous region than in valleys and lowlands, and higher in the pre-monsoon season than in the rest of the year (Karki et al. 2019). Shrestha et al. (2019) reported more or less equal magnitudes of warming, with a more pronounced rate of increase after 2005 (see also Dahal et al. 2019). Current mean annual temperature warming rates in Sikkim and Bhutan amount to 0.3–0.4 °C per decade (cf. Hoy et al. 2016; Goswami et al. 2018; Patle et al. 2019), comparable to current warming trends in the eastern Himalaya (Arunachal Pradesh, India) (cf. Yang et al. 2013; Bhagawati et al. 2017). In the western HKH, annual mean temperatures showed a slight increase in recent decades, whereas summer temperatures are slightly decreasing or show rather small magnitude of trends at many climate stations in the Karakoram (Fowler and Archer 2006; Khattak et al. 2011; Bocchiola and Diolaiuti 2013; Raza et al. 2015; Hasson et al. 2017; Waqas and Athar 2019; Latif et al. 2020). In winter and summer, the Karakoram has been near the boundary between large-scale cyclonic and anti-cyclonic trends over recent decades, while the Central Himalaya has been under the influence of an anti-cyclonic trend (Norris et al. 2019). Deviations from the general HKH climate warming pattern are linked to the Karakoram glacier anomaly (see 2.2; Forsythe et al. 2017).

Patterns of elevation-dependent warming have been widely observed in the HKH and in particular on the Tibetan Plateau and surrounding regions (Hasson et al. 2016; Karki et al. 2019; Krishnan et al. 2019a; Dimri et al. 2020). Maximum warming rates have been assessed between 4000 and 5000 m a.s.l., locally even at higher elevations (cf. Gao et al. 2018; Pepin et al. 2019; Rangwala et al. 2020). High resolution temperature trends over the Himalaya for the period since the 1980s show a clear elevational gradient in the pre-monsoon season with maximum values of up to 1.2 °C per decade at higher elevations (Gerlitz et al. 2014; Schickhoff et al. 2015). Thakuri et al. (2019) confirmed elevation-dependent warming based on stations up to 2600 m a.s.l. in Nepal. Higher warming rates at intermediate elevations were reported by Negi et al. (2020) for the NW Himalaya.

Trends in annual precipitation are difficult to derive considering the widespread non-availability of long-term observations and distinct variabilities prevalent in different subregions and seasons (Schickhoff et al. 2016a). Over the last 100-plus years, the trend of annual precipitation in the entire HKH is characterized by a slight decrease (Fig. 1.4) (Ren and Shrestha 2017; Ren et al. 2017; Krishnan et al. 2019a). The marginal reduction in annual precipitation (with concurrent interdecadal variability) over quite a large part of the Indian subcontinent is consistent with a weakening tendency of Indian summer monsoon precipitation, associated with a weakening land-sea thermal gradient, a decline in the number of monsoon depressions and an increase in the number of monsoon break days (Krishnamurthy and Ajayamohan 2010; Kulkarni 2012; Lacombe and McCartney 2014; Roxy and Chaithra 2018; Singh D et al. 2019; Basu et al. 2020). Nevertheless, all global and regional climate models and scenarios project an increase in both the mean and extreme precipitation of the Indian summer monsoon in the twenty-first century, largely due to increased moisture flux from ocean to land (Christensen et al. 2013; Krishnan and Sanjay 2017). Observations in subregions of the HKH over recent decades show either slightly decreasing or slightly increasing trends, but trends are rarely significant. Generally increasing trends for winter precipitation, originating from western disturbances, and positive trends at many stations for summer precipitation (predominantly monsoonal) have been observed in the Karakoram over recent decades (Khattak et al. 2011; Palazzi et al. 2013; Hasson et al. 2017). Increasing trends of winter precipitation at the majority of stations in the NW, W, and Central Himalaya in India are overcompensated by decreasing summer (monsoonal) precipitation rates since the 1960s, resulting in prevailing negative trends of annual precipitation (Sontakke et al. 2008; Bhutiyani et al. 2010; Singh and Mal 2014; Bhutiyani 2016; Shafiq et al. 2019). Decreasing trends of annual precipitation were also observed in Far West Nepal (Wang et al. 2013; Pokharel et al. 2019), while the major remaining parts of Nepal experienced a positive trend of annual precipitation, in particular of monsoonal precipitation, in the period 1979–2016, notably in the years after 2000 (Shrestha et al. 2019; see also Panthi et al. 2015 for the Kali Gandaki River Basin). Further east (Sikkim, Bhutan, Arunachal Pradesh, eastern Himalaya) no significant longer-term trends or slightly positive trends, if any, are observed (Qin et al. 2010; Li et al. 2011; Jain et al. 2013; Hoy and Katel 2019). Annual precipitation on the Tibetan Plateau has slightly increased since the 1960s, although respective trends are not uniform across the entire Plateau region (Hasson et al. 2016; You et al. 2017).

A clear shift in temporal characteristics of precipitation variation has been assessed after 1990 with greater interannual variability and more frequent intense precipitation events and less frequent light precipitation events (Krishnan et al. 2019a). Higher-elevation areas, in particular the Tibetan Plateau, have witnessed a significant increase in annual mean daily precipitation intensity (Ren et al. 2017; Zhan et al. 2017),

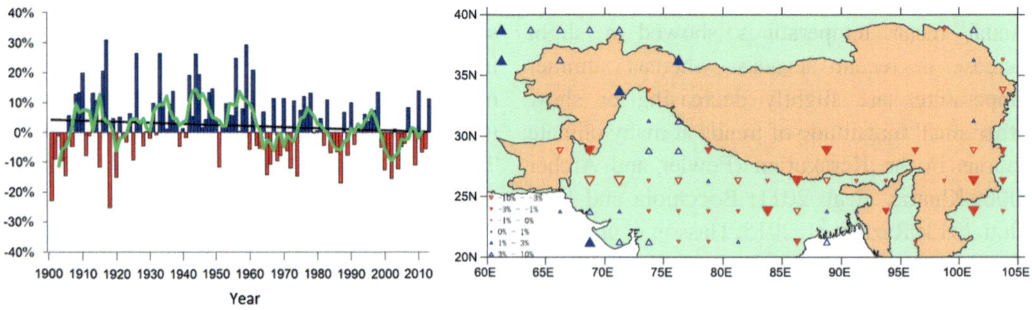

Fig. 1.4 Regional average annual precipitation percentage anomaly (PPA) during 1901–2014 in the HKH region (green line: five-year moving average; black line: linear trend) and spatial distribution of linear trends. (Modified from Krishnan et al. 2019a)

subjecting alpine life zones to additional stress. Over the western Himalaya, Priya et al. (2017) and Krishnan et al. (2019b) identified a rising trend of synoptic-scale western disturbance activity and related precipitation extremes during the recent few decades. For some parts of Nepal, a significant increase of high intensity precipitation extremes was observed during 1970–2012, and at the same time, the number of rainy days is significantly decreasing over the whole of Nepal while the number of consecutive dry days is significantly increasing (Karki et al. 2017).

Significant warming has also characterized surface air temperature trends in other Asian and Australasian mountain systems. Observations in E and NE Asia (China, Taiwan, Korea, Japan) indicate an abrupt increase of summer mean surface air temperature since the mid-1990s (Dong B et al. 2016), with extreme summertime droughts having increased in frequency, severity and duration (Zhang J et al. 2019). Substantial warming rates are to be expected for the coming decades (Hsu and Chen 2002; Lee et al. 2014; Murata et al. 2015). Mountains of southern and eastern Siberia experienced an outstanding 2–3 °C increase of mean annual air temperature over the last three decades (Fedorov et al. 2014; Desyatkin et al. 2015), while the mean winter season temperature in the Siberian Altai has increased by up to 4 °C (Kharlamova et al. 2019). Strong positive temperature trends associated with an increase in summer days and a significant decline in frost days have also been observed in Mongolian mountains (Dashkhuu et al. 2015). High-elevation areas in the Tien Shan and Pamir experienced warming rates of up to 0.5 °C (mean annual air temperature) per decade over recent decades (Chevallier et al. 2014; Deng et al. 2015; Hu et al. 2016). Significant, but slightly lower warming rates were assessed in the Caucasus (Elizbarashvili et al. 2017), Pontic, Zagros and Arabian Mountains (Donat et al. 2014; Ghasemi 2015; Yucel et al. 2015) as well as in the mountains of SE Asia (Supari et al. 2017; Tang 2019). In Australia and New Zealand, mean temperatures have warmed strongly since 1900 (c. +0.9 °C), resulting in warmer, less frosty winters (Mullan et al. 2010; Reisinger et al. 2014). However, a reduced increase of mean temperatures (0.06 °C/decade) has occurred in New Zealand since 1970, while no clear overall pattern can be derived from precipitation variations which are connected with the Southern Oscillation Index (SOI) and the Interdecadal Pacific Oscillation (IPO) (McGlone et al. 2010). Hawai'i has experienced strong warming at higher elevations, with snowfall on Hawai'i's mountain peaks being projected to almost completely disappear by 2100 (Frazier and Brewington 2020).

Europe

In congruence with the global climate response to increasing greenhouse gas concentrations, distinctive long-term temperature trends have been observed in European mountains, with regionally and seasonally different rates of warming. All of Europe has warmed significantly, in particular since the 1960s, with Scandinavia showing strongest winter warming, and SW, Central, and NE Europe particularly high summer warming (Fig. 1.5) (Kovats et al. 2014; EEA 2017). In the European Alps, annual mean temperatures increased by about 2 °C since the late nineteenth century which is a rate more than twice as large as the global or northern hemispheric average (Auer et al. 2007; Brunetti et al. 2009; APCC 2014; Gobiet et al. 2014). Warming rates increased distinctly to c. 0.5 °C per decade since the early 1980s, with the most intense warming since the 1990s, leading to an annual mean temperature increase of more than 1 °C in 25 years (Weber et al. 1997; EEA 2009). In Switzerland, the 1988–2017 summer average was by far the warmest 30-year period over the past 300 years (cf. Fig. 1.5), resulting in more frequent and more intense heatwaves, less frequent cold periods, and an upward shift of the winter zero-degree line by 300–400 m since the 1960s (CH2018 2018). Rottler et al. (2019) detected elevation-based differences in temperature trends during autumn and winter with stronger warming at lower elevations. Precipitation trends are sub-regionally differentiated. In the southern Alps, precipitation trends are small and not significant. Here, Brugnara and Maugeri

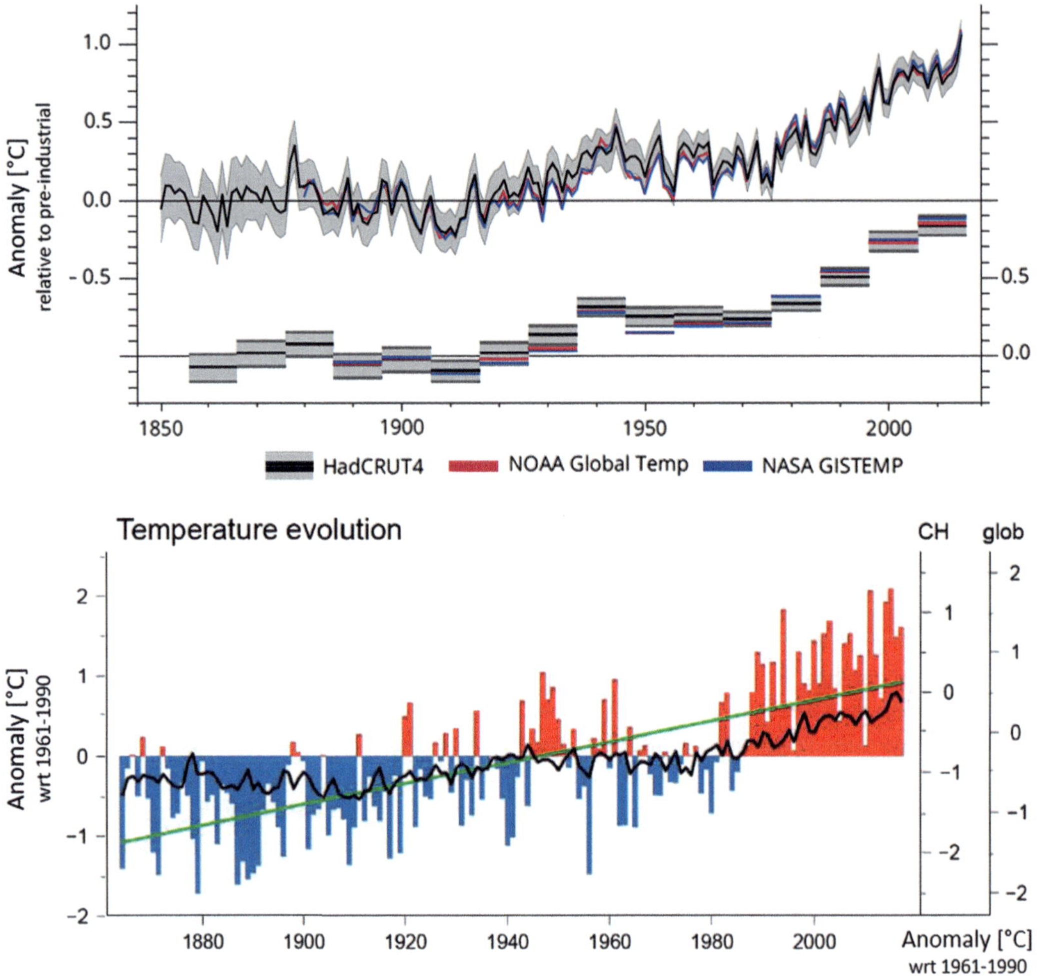

Fig. 1.5 Upper panel: European average temperatures between 1850 and 2015 over land areas relative to the pre industrial period; lower panel: Swiss and global annual mean temperatures, relative to the means for 1961–1990 (left axis) and 1981–2010 (right axes, left: Swiss series (CH) and right: global (glob)); the Swiss mean values are shown as bars, the global values as a black line; the linear trend fit to the Swiss values is shown in green. (Modified from EEA 2017; CH2018 2018)

(2019) assessed a significantly decreasing precipitation frequency over the period 1890–2017, and related this trend to a step-like reduction in cyclonic weather types over central Europe. Considerable and significant precipitation increases, however, were observed in northern Switzerland for the winter season (~20% per 100 years) as well as in the Austrian Alps (a 10 to 15% increase) over the past 150 years (APCC 2014; CH2018 2018). Likewise, the frequency of extreme precipitation events in the Alps increased by about 25% since 1900. In summary, precipitation evolution in the Greater Alpine Region shows significant regional and seasonal differences over the last century, with increases in the NW and decreases in the SE (Auer et al. 2007). Simultaneous to accelerated warming in the next decades, projected changes indicate less precipitation and more severe droughts in summer, and more precipitation in winter (Gobiet et al. 2014). The Carpathians experienced strongest warming in summer seasons, with rates of up to 2.4 °C from 1961 to 2010, and increasing annual precipitation in most of the region, except for the western and southeastern areas (Werners et al. 2014).

Climate observations in the Mediterranean region indicate increasing temperatures and decreasing precipitation, contributing to a progressive and substantial drying of the land surface since 1900. For instance, mean surface air temperature in the Pyrenees increased by 0.21 °C per decade, while precipitation decreased by 2.5% per decade in the period 1950–2010, leading to more frequent and intense droughts (EEA 2017). Warming rates are predicted to be in a similar magnitude in western and eastern Mediterranean mountains over the coming decades, the western mountain ranges such as the Sierra Nevada, the Pyrenees and the Apennines, however, will suffer to a larger extent from decreasing precipitation than the eastern Mediterranean mountains (Dinaric Alps, Balkan, Rhodopes, Pindos) (Nogués-Bravo et al. 2008, 2012). The mean temperature in Scandinavian mountains has increased significantly since the early twentieth century, with particularly warm periods in 1930–1950 and after 1980. From 1964 to 2013, mean annual temperature in the northern Scandinavian mountains increased approximately by 2.0 °C, and winter temperature (January–February) by 3.0 °C, associated with an increasing trend in precipitation (Vuorinen et al. 2017). Significant increases in mean precipitation were also observed in the Norwegian Scandes between 1900 and 2014 (Vanneste et al. 2017). A south-to-north gradient in the magnitude of precipitation increase in the Scandes is projected for the next decades (Christensen et al. 2015).

America

Over most of North America, mean annual temperature has increased over the past century, with higher latitudes of Canada and Alaska experiencing the largest temperature anomalies and warming rates more than double the global rate (Fig. 1.6). Substantial warming has been observed since the 1970s, accompanied by decreases in frost days and cold spells, increases in the occurrence of severe hot events over the USA, and increases in extremely hot seasons in northern Mexico, the USA, and parts of Canada (Vincent and Mekis 2006; Kunkel et al. 2008; Melillo et al. 2014; Romero-Lankao et al. 2014; Bush and Lemmen 2019; Cuervo-Robayo et al. 2020). In western North America, twentieth-century observations show temperature increases over the entire mountain region, from the SW to Alaska, which are higher than the global average and range mostly between 1 and 2 °C, and with minimum temperatures increasing to a greater extent than maximum temperatures (Wagner 2009). Warming rates are considerably higher in winter than in summer, exemplified by mean temperature increase of 3.3 °C in winter, 1.7 °C in spring, 1.5 °C in summer, and 1.7 °C in autumn between 1948 and 2016 in Canada (Fig. 1.7) (Bush and Lemmen 2019). As in Scandinavia and North Asia, a crude south-to-north gradient of increasing warming rates is evident (Kittel et al. 2002), and, as in Asia and Europe, higher elevations show greater temperature increases than lower elevations (Minder et al. 2018). Twentieth-century annual precipitation trends are positive over the Rocky Mountain/Great Basin region, although not always significant, and with seasonally heterogeneous trends (Wagner 2009). Extreme precipitation events have become more frequent and more intense in recent decades (Kunkel et al. 2008).

Across the system of the American Cordilleras, the Alaskan and Yukon subregions have been warming at a faster rate than any other subregion (mean annual temperature increase of up to more than 3 °C in the past 70 years), with considerably more warming in winter than in summer (Chapin et al. 2014; Lader et al. 2016; Zhang X et al. 2019). In the Pacific Coastal and Rocky Mountain ranges of western Canada, precipitation has slightly increased in most seasons. However, a statistically significant decrease in winter precipitation has been observed (Zhang X et al. 2019). Over recent decades (1970–2012), observations in the Pacific Northwest and the northern Rocky Mountains of the USA show accelerated average warming rates of c. 0.2 °C per decade, associated with longer growing

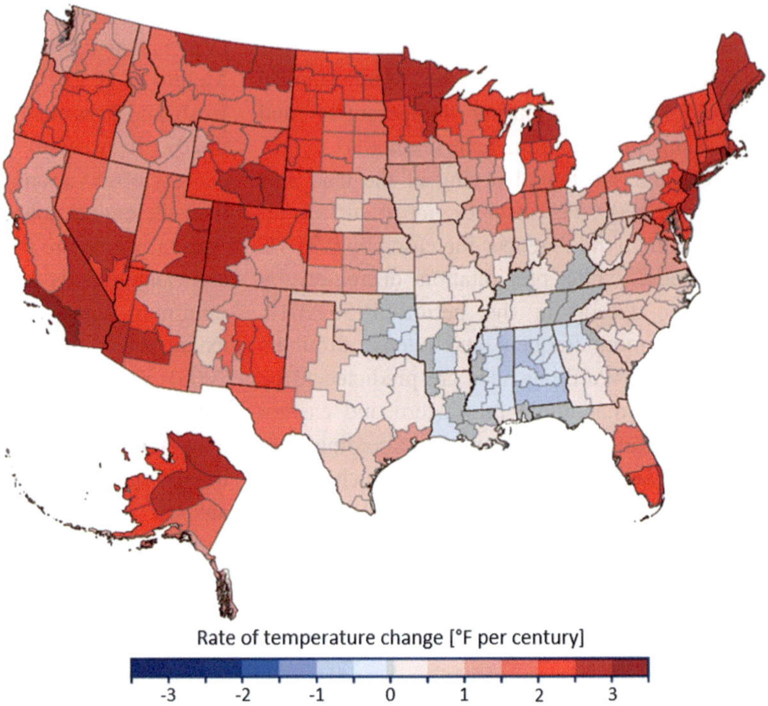

Fig. 1.6 Rate of temperature change in the United States 1901–2015 (modified from https://www.epa.gov)

seasons, increased evapotranspiration across the region, and increased climatic water deficits (Mote et al. 2013; Abatzoglou et al. 2014). In the southern Rocky Mountains and the Sierra Nevada, the decade 2001–2010 was the warmest in the 110-year instrumental record, with temperatures up to 1 °C higher than historic averages, with relatively higher spring and summer warming, fewer cold air outbreaks and more heatwaves, and with spatially varying precipitation trends (decreases in the southern part of the region, with strongest percentage declines during spring and summer, and increases in the northern part) (Hoerling et al. 2013; Garfin et al. 2014). Thus, it will get increasingly difficult to buffer drought effects in the southern mountainous regions of North America.

Significant warming, in the order of up to 1.0 °C since the 1970s, has also been detected throughout Central America and South America (Magrin et al. 2014). The tropical and subtropical Andes are being subjected to significant changes in mean climatic conditions, reflected in a mean temperature increase of about 0.1 °C per decade over the past 70 years (Fig. 1.8) (Bradley et al. 2006; Lavado Casimiro et al. 2013; Vuille 2013; Lopez-Moreno et al. 2016). Significantly positive temperature trends were also confirmed for the Patagonian Andes in the past century (Masiokas et al. 2008). After significant warming during much of the twentieth century, subtropical coastal regions experienced a recent cooling trend, in particular in central and northern Chile, related to the Pacific Decadal Oscillation (Falvey and Garreaud 2009). Higher elevations in the tropical Andes and further south to Central Chile, however, show continued warming of currently c. 0.2 °C per decade (Vuille et al. 2015). Temperatures at higher elevations are obviously now decoupled from the sea surface temperature forcing in the Pacific, which served as a strong predictor for cold or warm periods in the Andes in previous decades (Vuille et al. 2018). Irrespective of this, patterns of elevation-dependent warming have been observed throughout the Andes (e.g. Mora and Willems 2012; Ruiz et al. 2012, Schoolmester et al. 2018).

Precipitation trends are weaker and spatially much more heterogeneous. Stations in the Andes of Ecuador, Peru, and Bolivia showed a trend

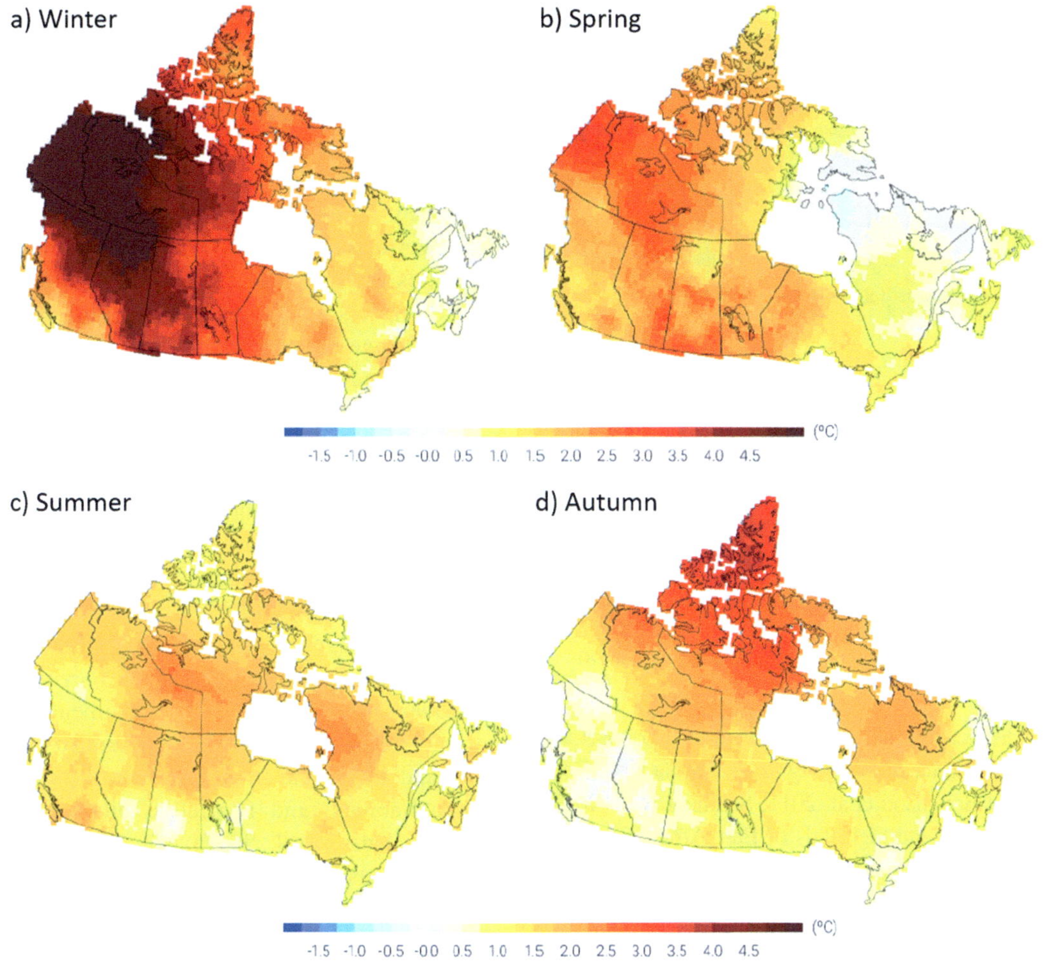

Fig. 1.7 Trends in seasonal temperatures across Canada; observed changes (°C) in seasonal mean temperatures between 1948 and 2016 for the four seasons. (Modified from Bush and Lemmen 2019)

towards increased precipitation north of ~11°S between 1950 and 1994, while most stations located further south showed a precipitation decrease (Vuille et al. 2003), also in Patagonia (Masiokas et al. 2008). However, precipitation trends are not significant over recent decades, and most of the variability in the data appears to be associated with the ENSO (El Nino Southern Oscillation) phenomenon (Lavado Casimiro et al. 2013; Salzmann et al. 2013; Rau et al. 2017). In general, climate anomalies such as ENSO and large-scale ocean-atmospheric indexes have a considerable influence on temperature and precipitation fluctuations in South America.

Africa

Across the continent of Africa, mean annual temperatures have increased by 0.5 °C or more in the past 50–100 years (Fig. 1.9), with minimum temperatures warming more rapidly than maximum temperatures, and temperature anomalies being significantly higher for the period 1995–2010 compared to previous decades (Toulmin 2009; Collins 2011; Niang et al. 2014). Observed and projected temperature rise is comparatively high in NW Africa, in particular in the Atlas Mountains. A very strong warming of about 6 °C is expected here in the course of the twenty-first century while the precipitation trend

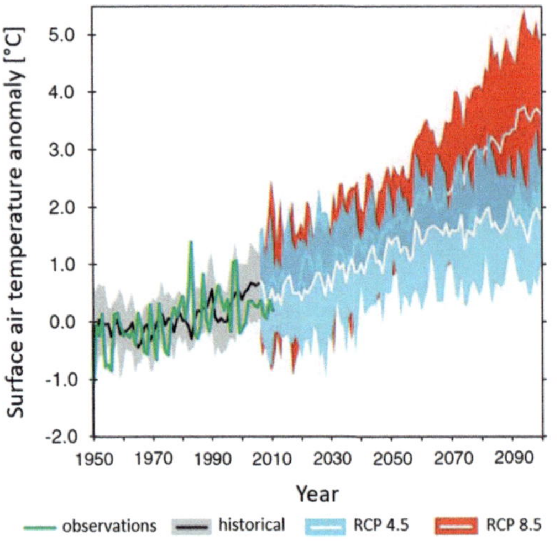

Fig. 1.8 Observed and simulated annual mean air temperature anomalies in the tropical Andes (departures from 1961–1990 mean) derived from station data (green, 1950–2010), historical CMIP5 (grey, 1950–2005), and future CMIP5 scenarios (light blue, RCP 4.5; red, RCP 8.5, 2006–2100). (Modified from Vuille et al. 2018)

is distinctly negative, leading to an earlier onset and longer duration of droughts (Patricola and Cook 2010; Bouchaou et al. 2011; Schilling et al. 2012). Mountains and highlands of East Africa also experienced significant warming over recent decades, up to 1.8 °C since 1950 (Jury and Funk 2013), while long-term precipitation trends are not significant, but rainfall is recently declining in some parts of the region (Anyah and Qiu 2012; Viste et al. 2013; Mengistu et al. 2014; Omondi et al. 2014). A recent increase of warming rates to 0.5 °C per decade was reported for the Rwenzori Mountains in Uganda (Taylor et al. 2006). In Ethiopia, Kenya, and Tanzania, increases in maximum and minimum temperatures are accompanied by increasing trends in warm nights, warm days, warm spell days, and mostly a non-significant change in precipitation indices (Gebrechorkos et al. 2019). Ethiopia's eastern highlands, however, experience significant climate-induced drought and stress on crop and livestock productivity, while large regions of western Ethiopia are becoming wetter (Brown et al. 2017). Most of southern Africa has also experienced significant warming over recent decades (Kruger and Sekele 2013), with marked recent temperature increases in the Drakensberg system (Morris 2017).

1.2.2 Impacts on the Cryosphere and Hydrosphere

1.2.2.1 General Overview

Over recent decades, considerable changes have been observed in cryospheric components (snow, ice, glaciers, permafrost) in mountains of the world that serve as vivid illustrations of mountains being at the forefront of climate change impacts (Hock et al. 2019; Pihl et al. 2019). Changes in cryospheric land conditions potentially induce important albedo feedbacks to the regional and global climate. Climate warming causes cascading effects on cryospheric and related hydrological processes that affect not only mountain catchments but also the lowlands. The cascade of effects extends to human livelihoods, economy, and ecosystems. Widespread changes of the cryosphere and associated changes in water cycle and balance and river discharge regimes have inevitable consequences for erosion rates, sediment and nutrient fluxes, and the biogeochemistry of rivers and lakes, and finally for water quality, aquatic habitats, and respective biotic communities (Huss et al. 2017). Changes of the cryosphere also affect terrestrial communities and ecosystems significantly, for instance, by creating new habitats in glacier forefields, by modifying the length of the growing season and the phenology of plant production and consumers, and by altering soil moisture conditions and nutrient availability. Ultimately, ecosystem functioning is affected due to a novel constellation of site conditions and competitive relationships, and associated changes in species compositions and primary productivity. Water supply from the cryosphere is indispensable for socio-economic systems in both mountains and lowlands. Meltwater from snow and ice is essential for drinking water supplies, irrigated agriculture, mining, hydropower generation, industries, tourism, and other activities (Beniston and Stoffel 2014; Huss et al. 2017).

The snow cover is the largest cryosphere component. Global observations show that climate change has caused a general reduction in

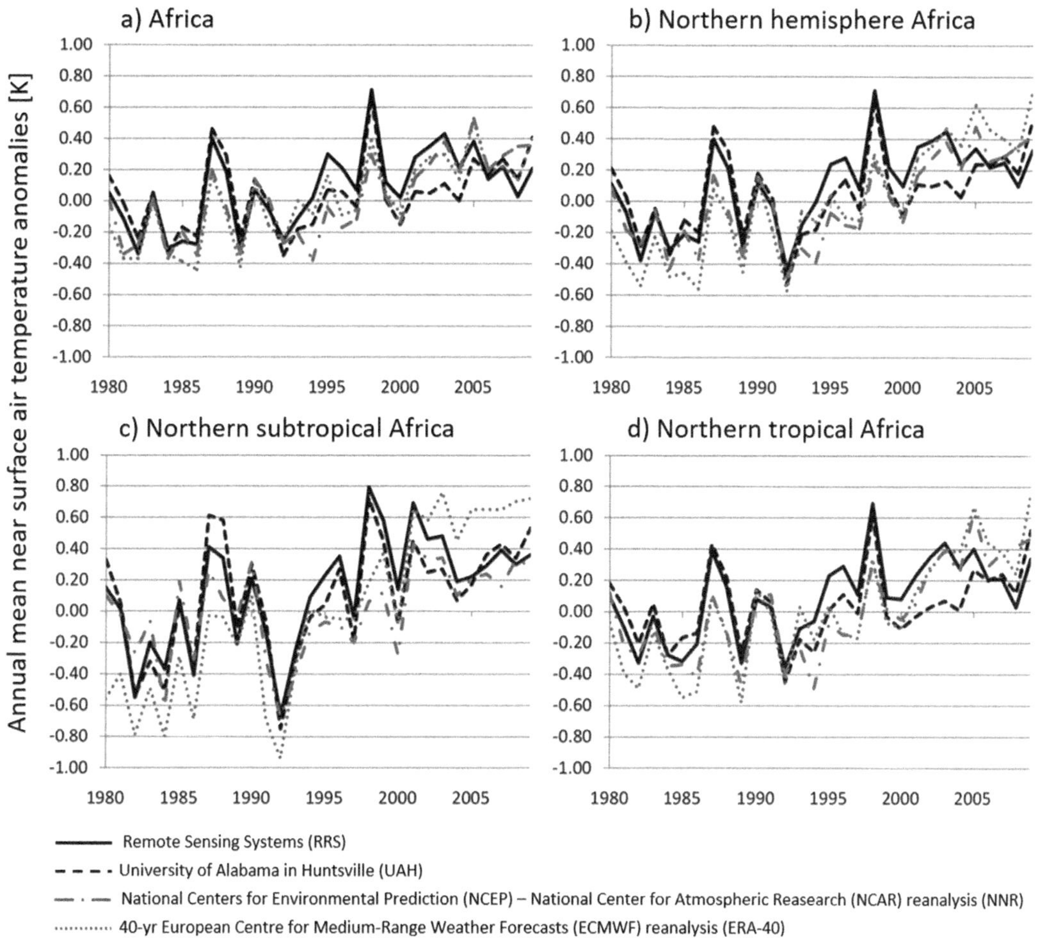

Fig. 1.9 Annual mean near-surface air temperature anomalies (K) between 1979 and 2010 for Africa and selected subregions, with black lines indicating satellite data (solid: RSS; dashed: UAH), and grey lines reanalysis data (dashed-dotted: NNR; dotted: ERA-40). (Modified from Collins 2011)

low-elevation snow cover in recent decades (Fig. 1.10) (Bormann et al. 2018). In nearly all mountain regions around the globe, snow cover duration (SCD) has declined, particularly at lower elevations, with an average decline rate of 5 days per decade (Hock et al. 2019). Snow-covered area (SCA) and snow depth are also decreasing significantly, albeit with high year-to-year variation. Snow cover will further decline in the next decades, a decrease by 10–40% is expected for the period 2031–2050 compared to 1986–2005 (Hock et al. 2019). On the other hand, increased snowfall will occur at higher elevations where the rain/snow partitioning is no longer affected by rising temperatures, and where total winter precipitation is increasing (Kapnick and Delworth 2013). Snow accumulation is critical for water availability in many regions. Such snow-dependent regions are expected to experience increasing stress from the imminent shift towards low snow years within the next three decades and from extreme changes in snow-dominated water resources (Diffenbaugh et al. 2013).

As key indicators and unique demonstration objects of ongoing climate change, glaciers have attracted tremendously increased scientific

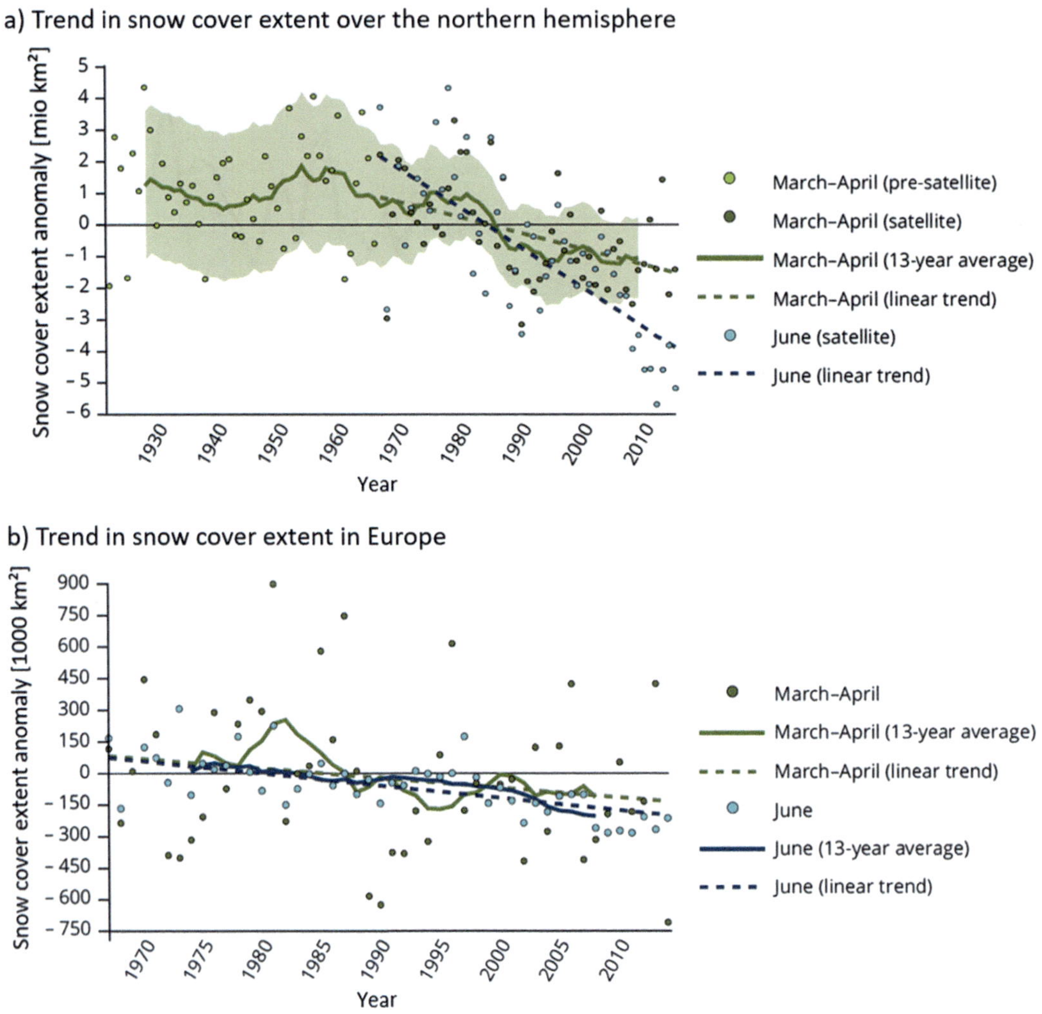

Fig. 1.10 Satellite-derived trends in snow cover extent over the northern hemisphere and Europe 1967–2015; the time series for the northern hemisphere is extended back to 1922 by including reconstructed historical estimates. (Modified from EEA 2017)

interest and accelerating international media attention. Numerous new records of annual mass loss were observed in the past two decades, indicating implications for the water cycle that affect continental-scale water supply and even global-scale sea levels. Glacier mass loss provides a more direct evidence of climate change in remote mountains where meteorological observations are hardly available. Global glacier recession is accelerating (Fig. 1.11), with atmospheric warming considered to be the primary driver, modified by other meteorological variables and internal glacier dynamics (Marzeion et al. 2014; Vuille et al. 2018; Hock et al. 2019). Over the last decades, declines in glacier area, length, and mass have condensed to a globally widely coherent picture of mountain glacier recession, albeit with interannual and regional variations (Zemp et al. 2015). At a global scale, glacier mass loss increased by c. 30% between 1986–2005 and 2006–2015 (Zemp et al. 2019). During the latter period, mountain glaciers lost about 500 kg of mass per square metre per year, a total of 123±24 Gt (billion

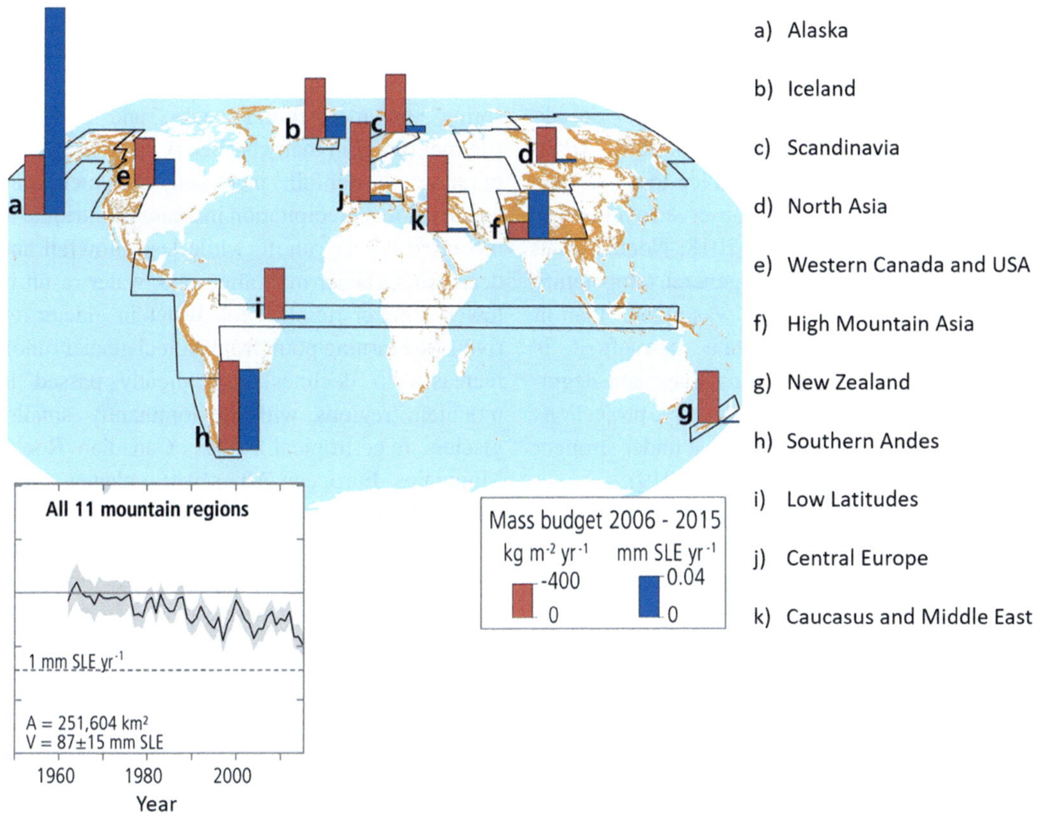

Fig. 1.11 Glacier mass budgets for eleven mountain regions; red and blue bars refer to regional budgets averaged over the period 2006–2015 in units of kg m^{-2} yr^{-1} and mm sea-level equivalent (SLE) per year, respectively. (Modified from Hock et al. 2019)

tonnes) per year (excluding the Arctic and Antarctic) (Hock et al. 2019; Pihl et al. 2019). Most negative glacier mass budgets were observed in the Southern Andes, Caucasus/ Middle East, European Alps and Pyrenees, with total mass loss and corresponding contribution to sea level between 2006 and 2015 being largest in Alaska, followed by the Southern Andes and High Asia (Hock et al. 2019). Notwithstanding the global trend of glacier recession, glaciers in various mountain ranges have shown intermittent re-advances or mass gains due to locally restricted climatic causes or internal glacier dynamics (WGMS 2008). Century-scale projections for mountain glaciers show substantial mass loss by 2100 relative to 2015 in the order of 18% for scenario RCP2.6 and 36% for scenario RCP 8.5 (Hock et al. 2019).

Permafrost is another important component of the cryosphere in high mountain regions, in particular in the Northern Hemisphere. Mountain permafrost accounts for c. 25–30% of the global permafrost occurrence, its distribution is spatially highly heterogeneous (Hock et al. 2019). It significantly influences energy balance, terrain stability-related geophysical hazards, ground and subsurface hydrology, water quality, river sedimentation, and infrastructure. Permafrost degradation due to global warming contributes to mountain slope destabilization and increased mass-movements and related hazards (Haeberli et al. 2017; Patton et al. 2019). As the understanding of permafrost depends on ground and subsurface temperature observations, which are logistically demanding and expensive, it remains largely understudied in many mountain ranges. At a global scale, mountain permafrost warming has been shown to accelerate recently (Fig. 1.12) and to exceed values of the late twentieth century, with an average warming rate of 0.19 °C per decade between 2007 and 2016 (Biskaborn et al. 2019), while general warming, ground-ice loss and permafrost degradation has been observed over longer time periods (e.g. Cao et al. 2018; Noetzli et al. 2018; Mollaret et al. 2019). In general, temperature increase in colder permafrost was greater than in warmer permafrost. Mountain permafrost is expected to undergo increasing thaw and degradation during the twenty-first century, projections reveal increased loss of permafrost under stronger atmospheric warming (Hock et al. 2019).

Changes in the cryosphere have wide-ranging consequences for freshwater availability in both mountain and downstream regions since streamflow timing and magnitude is largely controlled by the meltwater supply from cryospheric components (Rasul and Molden 2019). Runoff from alpine catchments is particularly critical for the water supply in summer months when other water sources in the lowlands are often limited. With regard to climate-cryosphere-hydrosphere interactions in mountain regions, reduced ice and snow cover triggers major shifts in seasonal runoff regimes. In snow and glacier-dominated river basins, recent observations indicate emerging trends of increased average winter runoff, earlier spring snowmelt runoff peaks, and declining summer runoff in many basins. A decreasing ratio of snow to rainfall, increased snowmelt, and local/regional precipitation increases contribute to increased winter runoff, while less snowfall and decreasing glacier melt after peak water result in lower summer runoff. Peak water in glacier-fed rivers (the turning point from annual glacier runoff increases to declines) has already passed in mountain regions with predominantly smaller glaciers (e.g. tropical Andes, Canadian Rocky Mountains, European Alps), while glacier runoff will continue to increase in the next decades in mountain catchments with large ice volumes (northern North America, parts of the HKH region, Central Asia) where peak water will be reached in the late twenty-first century (Huss et al. 2017; Huss and Hock 2018; Hock et al. 2019; Hoelzle et al. 2019).

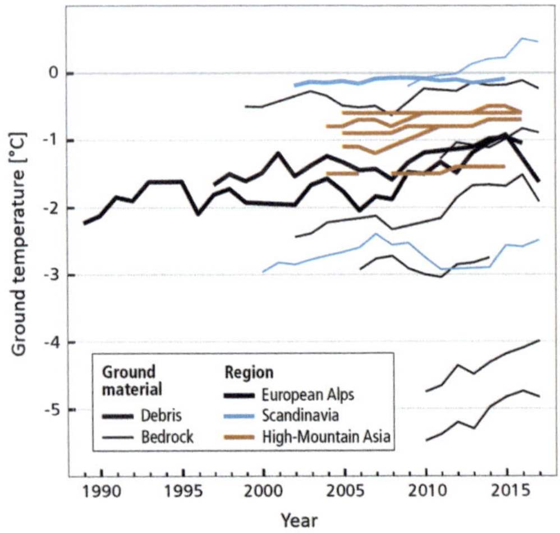

Fig. 1.12 Mean annual ground temperature from boreholes in debris and bedrock in the European Alps, Scandinavia and High Mountain Asia; the depth of measurements is approximately 10 m. (Modified from Hock et al. 2019 after Noetzli et al. 2018)

1.2.2.2 Regional Overview

Asia and Australasia

Although comprehensive observations on snowpack parameters in Asian mountains are still limited, growing and ample evidences from satellite-based global to local studies suggest that the snow cover has significantly declined, particularly since the 1960s (Dietz et al. 2013; Rohrer et al. 2013; Singh et al. 2014; Bolch et al. 2019). The HKH and Tibetan regions show overall negative trends in snow accumulation rates (Bolch et al. 2019). Over the period of 2000–2010, the annual (−1.25%) and seasonal snow-covered area (−1.04 to −0.01%) decreased, except for the autumn season (5.6%) (Gurung et al. 2011). However, westerly dominated basins (Indus basin, NW Himalaya) show increases in winter snow cover (Bolch et al. 2019; but see also Li et al. 2018). Increasing snow-covered area trends in the Karakoram/NW Himalaya contrast with declining trends in the Ganga and Brahmaputra river basins (Singh et al. 2014; Bilal et al. 2019). Declining trends of annual and seasonal snow-covered area were also assessed for southern slopes of NW Himalayan river basins (Jhelum and Shyok to Satluj and Beas), except for winter seasons over 2001–2012 (Sharma et al. 2014). Barman and Bhattacharjya (2015) reported a declining snow-covered area trend in the Brahmaputra river basin, except in winter seasons between 2002 and 2012. A slight decline (0.01% a^{-1}) over the Tibetan Plateau has been observed since the early 2000s (Duo et al. 2014; Li et al. 2018). Based on long-term data (1972–2017), Bormann et al. (2018) found overall declining trends in High Asia, with a slight increase in the Karakoram and in the East Himalaya. In the Siberian region including Kamchatka, the snow-covered area has declined significantly (0.8 × 104 km^2 $a-^1$) over 1970–2012 (Yu et al. 2017). Distinctly declining trends (up to 0.8 × 10^2 km^2 a^{-1}) were assessed for the Pamir, Alay, and Altai over 2000–2015, while the Tien Shan and Kunlun show mixed trends (Dietz et al. 2013; Liu J et al. 2017). In the Tien Shan, negative trends in summer (−0.02% a^{-1}) and winter (−0.1% a^{-1}) contrast with positive trends in spring and autumn (0.1% a^{-1}) (Tang et al. 2017). Another significant decrease in snowpack parameters was detected in the Zagros Mountains and in the Greater Caucasus (Notarnicola 2020).

Snow cover duration, as affected by the precipitation and temperature changes in pre- and post-winters, has decreased in the HKH region, Tien Shan, Kunlun, Altai, and Kamchatka by up to 30 days per decade between 1982 and 2013 (Bulygina et al. 2009; Dietz et al. 2013; Tang et al. 2013; Ye and Cohen 2013; Chen et al. 2016). A large decrease of snow cover duration (4 days a^{-1} between 2000 and 2015) was detected in the Nyainqentanglha Mountains (SE Tibet) (Wang et al. 2017; Notarnicola 2020). By contrast, increases were reported from NE Tibet and some Siberian mountain ranges (Chen et al. 2016). Significantly decreasing snow cover and snow duration is projected for the Southern Alps in New Zealand and alpine regions in Australia (Hennessy et al. 2008; Hendrikx et al. 2012).

Glaciers across Asia have experienced sustained mass loss since the mid-nineteenth century, with accelerated loss in recent decades, except for some of the glaciers in the Karakoram, Pamir, Kunlun, Tien Shan, and Kamchatka which have not changed significantly or, in case of surge-type glaciers, have shown area increases. Recent estimates of total glacier mass change in High Mountain Asia are in the order of −19.0 ±2.5 Gt yr^{-1} for the period 2000–2018, with greatest total mass loss across the Himalayas, Nyainqentanglha, and the Tien Shan and positive mass balance in the western Kunlun Shan and eastern Pamir (Fig. 1.13) (Brun et al. 2017). The average glacier area loss in the entire HKH region was estimated at 0.35% a^{-1} between 1970 and 2000; the rate increased to 0.42% a^{-1} between 2000 and 2010 (Bolch et al. 2019). Simultaneously, the glacier mass balance rate has increased from −0.26 (m w.e.[water equivalent]$^{-1}$) (1970–2000) to −0.37 (m w.e.$^{-1}$) in 2000–2010, with some regional variations and even anomalies (Azam et al. 2018; Bolch et al. 2019). The Imja–Lhotse Shar glacier in the Khumbu region in Nepal showed an exceptionally large loss rate of −1.45±0.52 m w.e. yr^{-1} for 2002–2007, with enhanced ice losses by calving into the Imja Lake (Bolch et al. 2011). There is a strong E-W gradient of glacier retreat, with average glacier area change rates of −0.81% a^{-1} in the eastern Himalaya decreasing to −0.37 and −0.34% a^{-1} in the central and western Himalaya between 2000 and 2010; area loss rates slightly slowed down in the central and western Himalaya, while an increase was observed in the eastern Himalaya during this period (Bolch et al. 2012, 2019; Azam et al. 2018). On the contrary, glacier area changes in the Karakoram show a divergent pattern that is known as the 'Karakoram anomaly' (Hewitt 2005, 2007). Non-surge-type glaciers were relatively stable and surge-type glaciers showed large increases as well as decreases over the past decade (Bhambri et al. 2017; Azam et al. 2018; Bolch et al. 2019). Accordingly, most Karakoram glaciers had a positive mass balance in recent decades (Kääb et al. 2012, 2015; Gardelle et al. 2013; Pratap et al. 2016; Berthier and Brun 2019; Shean et al. 2020). The glacier mass balance anomalies in the HKH region can be explained by contrasting meteorological conditions, reflected in differing energy balances, accumulation regimes and melt dynamics at high elevations (Bonekamp et al. 2019), but the understanding is far from complete (Farinotti et al. 2020). Strong variations in glacier mass balances in High Mountain Asia vividly illustrate that the sensitivity of glaciers to climate change is regionally variable.

The spatial patterns of the terminus change rates of glaciers (>−80 to >80 m a^{-1}) across the HKH correspond to glacier area changes. Over recent decades, glacier terminus recession rates have been assessed to be highest in the eastern Himalaya, while a considerably lower glacier recession is observed in the central and western Himalaya, and partially a surging/advancement (up to 2.5 km) in the Karakoram (Hewitt 2007; Quincey et al. 2015; Mal et al. 2016; Bhambri et al. 2017; Azam et al. 2018). Recently, recession rates of large glaciers in the central and western Himalaya (Gangotri, Milam, Bara Shigri) slowed down (Bhambri et al. 2012; Bhattacharya et al. 2016; Chand et al. 2017; Mal et al. 2019), while the glaciers of the NW Himalaya showed variable, often lower change rates or were relatively stable (Schmidt and Nüsser 2009, 2012; Chand and Sharma 2015; Chudley et al. 2017). Nevertheless, over longer time scales significant glacier retreat and thinning becomes obvious, as exemplified by the Chungpare Glacier at Nanga Parbat (Nüsser and Schmidt 2017). The average glacier area loss rate on the Tibetan Plateau was estimated to be slightly lower (0.27% a^{-1}, with <1.5% of glaciers advanced) compared to the surrounding regions between 1970 and 2009, with higher rates in the SW and SE, and lower rates in the inner, W, NE, E and N parts of the plateau (Bolch et al. 2010b; Wei et al. 2014; Ye et al. 2017). Glacier recession has fragmented larger glaciers into smaller ones, the number of glaciers in Nepal and Bhutan, for instance, increased by 11% and 15% (24% and 23% area loss), respectively, between 1980 and 2010 (Bajracharya et al. 2014a, b). Likewise, a distinct increase in number and area of moraine-dammed glacial lakes was assessed in recent decades, formed due to thinning, flow stagnation and recession of glacier tongues, and fed by glacier meltwater (Fig. 1.14) (Gardelle et al. 2011; Somos-Valenzuela et al. 2014; Zhang et al. 2015; Krause et al. 2019). Hence, glacial lake outburst floods (GLOFs), which have resulted in catastrophic damages and fatalities in the past decades, pose an increasing risk, with the southern Himalaya being a GLOF hotspot region (Fig. 1.15) (Nie et al. 2017; Veh et al. 2019). GLOF frequencies are predicted to increase during the next decades (Harrison et al. 2018). Projections for different RCP scenarios show that much of the glacier ice in High Mountain Asia will disappear towards the end of the century, with potentially serious consequences for regional water management and mountain communities (Kraaijenbrink et al. 2017; Mukherji et al. 2019; Immerzeel et al. 2020). Decreasing water supplies from cryosphere change will affect particularly irrigation-dependent agriculture in the Indo-Gangetic Plains (Biemans et al. 2019) and in arid mountain regions, where local farmers are forced to develop adaptive strategies (Nüsser et al. 2012, 2019a, b; Parveen et al. 2015; Rasul et al. 2020).

Siberian mountains have experienced a substantially high glacier area loss since 2000 (3.4% a^{-1}) compared to the low recession rate since the Little Ice Age (0.29% a^{-1}) (Osipov and Osipova 2014). In Kamchatka, the average glacier area loss rate was 0.33% a^{-1} between 1950 and 2000 (Khromova et al. 2014); it increased substantially to 1.7% a^{-1} in recent years, leading to the disappearance of 46 glaciers between 2000 and 2014 (Lynch et al. 2016). Glacier reductions on the Kamchatka Peninsula range from 10 to 70% over recent decades (Khromova et al. 2019). The area shrinkage of glaciers in the Altai, the Urals and the Tien Shan is also remarkably high (between 0.32 and 0.62% a^{-1}) over the period from the 1950s until recently, associated with respective negative mass balance rates (Shahgedanova et al. 2010; Khromova et al. 2014; Farinotti et al. 2015; Wei et al. 2015; Ganyushkin et al. 2017; Zhang et al. 2017; Barandun et al. 2018). In the Chinese part of the Tien Shan, 182 glaciers vanished in recent decades (Baojuan et al. 2017), some glaciers, however, have shown advances (Shangguan et al. 2015). Even higher recession rates were assessed in the Pamir Alay

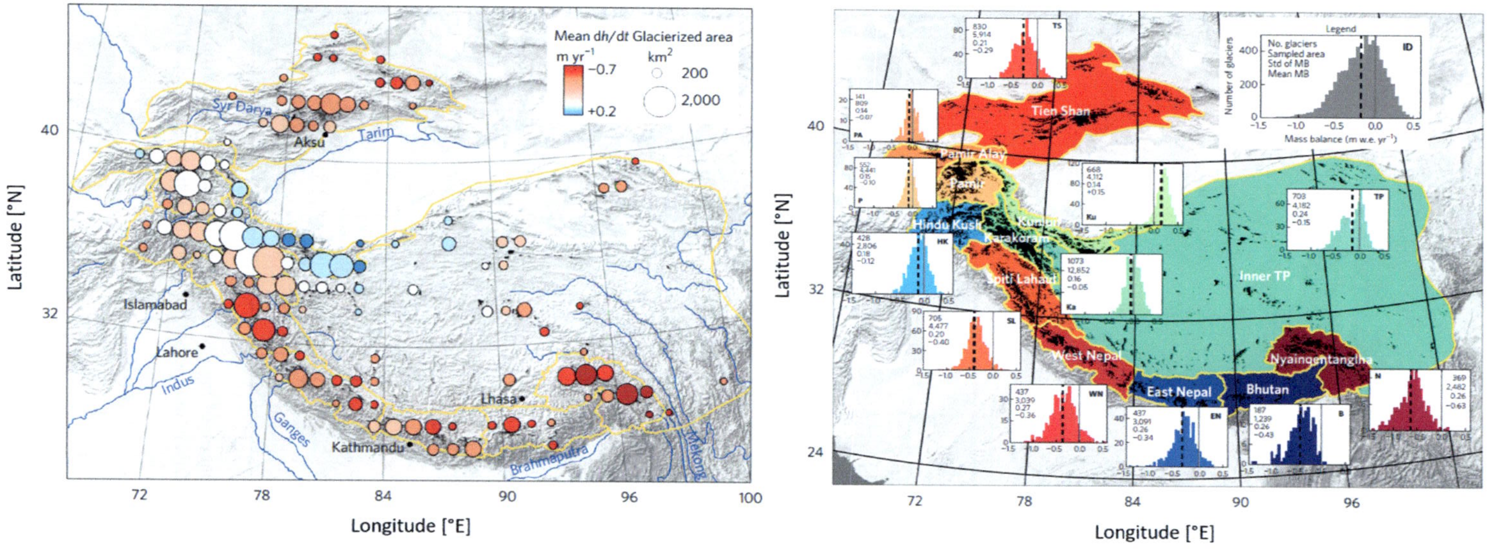

Fig. 1.13 Spatial pattern of glacier elevation changes and mass balance for High Mountain Asia (2000–2016). Left panel: Regional glacier mean elevation change on a 1° × 1° grid. Right Panel: Region-wise distribution of glacier-wide mass balance for every individual glacier (>2 km^2), represented in histograms of the number of glaciers (y-axis) as a function of mass balance (x-axis in m w.e. yr^{-1}); the black dashed line represents the area-weighted mean; numbers denote the total number of individual glaciers, the corresponding total area in km^2, the standard deviation of their mass balances and the area-weighted average mass balance in m w.e. yr^{-1}. (Modified from Brun et al. 2017)

Fig. 1.14 The fast retreat of Himalayan glaciers has resulted in the formation and expansion of meltwater lakes as in the former snout area of Gangapurna glacier (3550 m), Nepal, creating risks from GLOF events. (*Photo* © Udo Schickhoff, September 23, 2013)

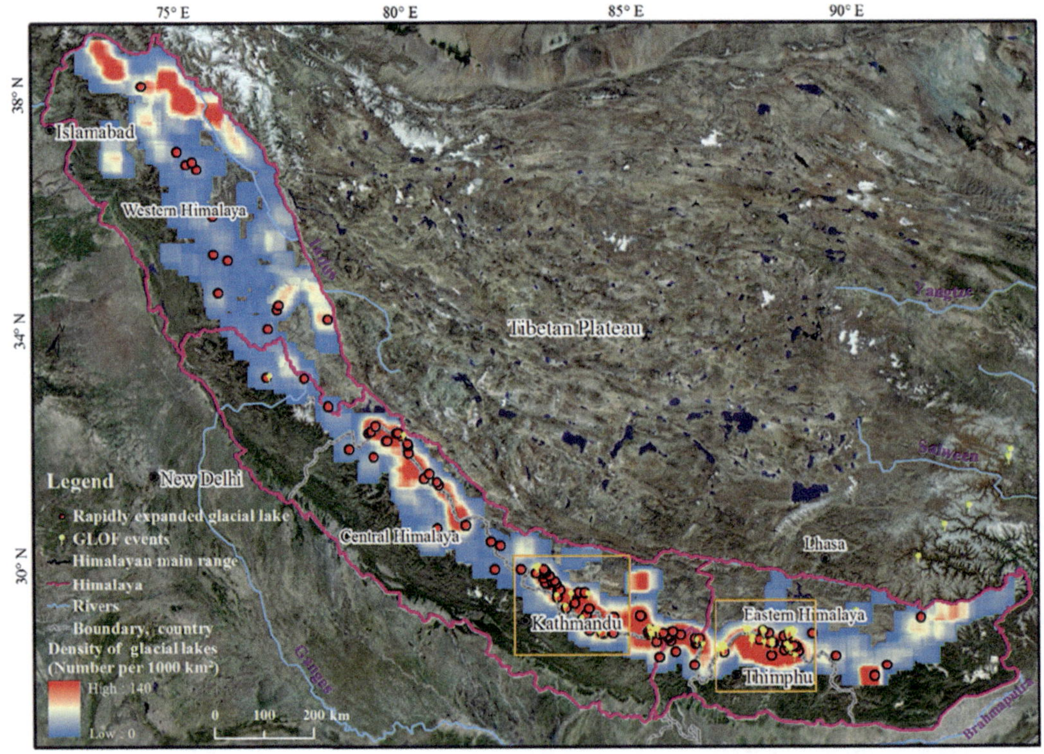

Fig. 1.15 Glacial lakes in the Himalaya in 2015: Spatial distribution of rapidly expanded glacial lakes and historical GLOF events in the Himalaya (potential vulnerable areas of GLOFs in orange boxes). (Modified from Nie et al. 2017)

(0.84% a^{-1} over the period 1978–2001) (Khromova et al. 2014), where a total of 142 glaciers disappeared (Holzer et al. 2016), while some fluctuations are also observed (Bolch et al. 2019). Recent glacier area loss rates in the Caucasus increased to 0.69% a^{-1} between 1986 and 2014 (Tielidze and Wheate 2018).

Tropical glaciers in Australasia show a dramatic recession over recent decades. The glacier areas on Puncak Jaya (4884 m a.s.l.), the highest mountain on the island of New Guinea, were found to decrease by 85% between 1988 and 2015 (Veettil and Wang 2018a), suggesting that these tropical glaciers might disappear before 2050 (Veettil and Kamp 2019). Specific climate conditions may result in exceptional terminus advance of some glaciers, opposed to the global trend. This is the case in New Zealand where several maritime glaciers advanced between 1983 and 2008, including the famous Franz Josef and Fox glaciers, which are steeply inclined and react swiftly and similarly to climate forcing. The glacier advance phase resulted predominantly from discrete periods of reduced air temperature, associated with anomalous southerly winds and low sea surface temperature in the Tasman Sea region (Mackintosh et al. 2017; see also Cullen et al. 2019). Nevertheless, the total ice volume of the Southern Alps for the small and medium glaciers has decreased from 26.6 km^3 in 1977 to 17.9 km^3 in 2018 (a loss of 33%), with accelerating ice loss for the period 1998–2018 (Salinger et al. 2019). Particularly, gentle-sloping, debris-covered glaciers with terminal lakes in the Southern Alps are in decline, as exemplified by the Tasman Glacier which has undergone c. 5 km of retreat into a terminal lake since the early 1980s (Dykes et al. 2011).

Permafrost research in High Mountain Asia is still limited. Nevertheless, there is growing evidence of permafrost warming and degradation. In the extended HKH region, permafrost research has focused on the Tibetan Plateau. It is generally assumed that most permafrost has undergone warming and thaw in recent decades (Zhao et al. 2010; Gruber et al. 2017; Bolch et al. 2019). The Tibetan Plateau is estimated to have the highest decadal permafrost area loss in the northern hemisphere, considerably increasing from 1×10^4 km^2 over the period 1901–2009 to 9×10^4 km^2 between 1979 and 2009 (Guo and Wang 2017). Thermal degradation of permafrost and increasing thickness of the active layer is widespread in Tibet, affecting c. 88% of the permafrost area of the 1960s (Ran et al. 2018). Local studies on the Himalayan South Slope suggest widespread permafrost degradation and the rise of permafrost lower limits by several hundreds of metres since the 1970s (Fukui et al. 2007; Allen et al. 2016). Significant warming and associated degradation of permafrost were also ascertained for Siberian and Mongolian high mountains and the Tien Shan (Marchenko et al. 2007; Sharkhuu et al. 2007; Guo and Wang 2017; Liu G et al. 2017; Biskaborn et al. 2019; Munkhjargal et al. 2020). In New Zealand, a connection between degrading permafrost and the occurrence of rock avalanches and other landslides is suspected (Allen et al. 2011).

Both climate change and anthropogenic activities, especially hydropower projects and irrigation, have significantly affected the hydrology in Asian mountains (river discharge, hydrological budgets) during the past century (Bhutiyani et al. 2008; Xu et al. 2009; Haddeland et al. 2014; Singh S et al. 2016; Scott et al. 2019). River runoff in eastern and Central Asian river basins decreased up to 15% during 1971–2000, even succeeded by the northwestern HKH, Pamir, Kunlun Shan, Qilian Shan, and Caucasus where the runoff decreased by 15–30% during the same period (Haddeland et al. 2014). Hydrological changes that have only been triggered by climate change are difficult to assess in detail due to, inter alia, poor understanding of the role of snow and ice in the regime of catchment basins, interannual variability of meteorological conditions, hardly available long-term series of river discharge, and multiple factors influencing streamflow. Trends may change in space and time within single basins, thus, conclusive evidence of either declining or increasing streamflow trends in the extended HKH region cannot yet be provided (Scott et al. 2019). Nevertheless, several review-based and observational studies on glacier- and snow-fed major basins indicate that river runoff has increased in some basins (Brahmaputra, Salween, Mekong), has no significant change/spatio-temporal mixed responses (Indus, Yangtze), and has decreased in others (Ganges, Yellow River) (Xu et al. 2009; Shrestha and Aryal 2011; Miller et al. 2012; Singh S et al. 2016; Hasson et al. 2017; Scott et al. 2019). Glacierized basins on the Tibetan Plateau show increased discharge, correlated to increased summer and winter temperatures and earlier snowmelt (Ye et al. 2005; Yao et al. 2007; Lin et al. 2008). Modelling studies for the HKH region predict shifts in the timing and magnitude of streamflows, but no significant changes or not more than minor increases in overall annual flows (Immerzeel et al. 2013; Lutz et al. 2014). In general, runoff in catchments with large ice volumes is projected to increase in the next decades indicating later peak water while basins with smaller ice volumes will face a decrease in runoff indicating earlier peak water (Fig. 1.16) (Hock et al. 2019).

The pattern of heterogeneous streamflow responses has been observed in other Asian mountain ranges and basins as well. Contrasts between individual basins become obvious when basins of the HKH region (Indus, Ganges, Brahmaputra) with small melt-to-discharge ratios due to the coincidence of glacier melt season and monsoon season are compared with Central Asian watersheds with a summer-dry climate where glacier melt substantially contributes to

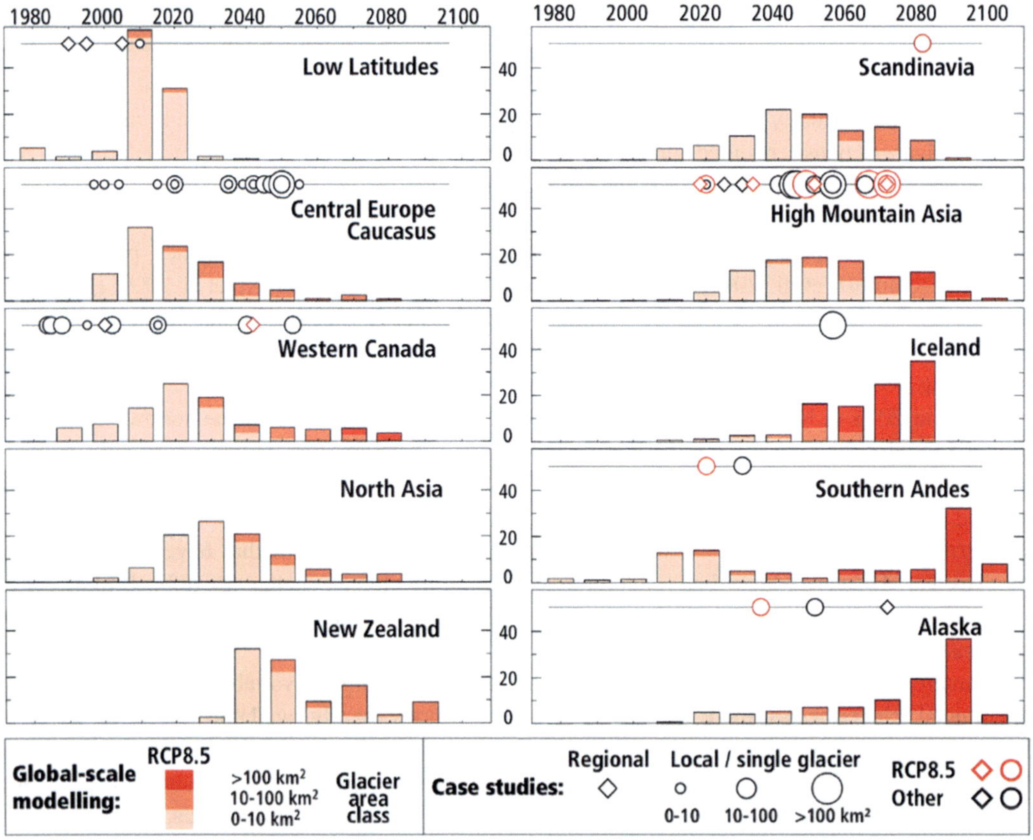

Fig. 1.16 Timing of peak water from glaciers in different regions under the RCP8.5 scenario; shadings of the bars distinguish different glacier sizes indicating a tendency for peak water to occur later for larger glaciers; circles mark timing of peak water from individual case studies, and refer to results from individual glaciers regardless of size or a collection of glaciers covering <150 km^2 in total, while triangles refer to regional-scale results from a collection of glaciers with >150 km^2 glacier coverage (Modified from Hock et al. 2019)

streamflow in July and August (Huss et al. 2017). River discharge in the glacier-dominated Aksu basin (Tien Shan) has increased in summer and winter over the past 50–60 years (Chen et al. 2006; Krysanova et al. 2015; Duethmann et al. 2015), while downstream stations at the main Tarim River show declining trends due to human abstraction of water (Tao et al. 2011). Declining snow cover thickness and duration in the central and western Tien Shan is associated with a decrease in river runoff (Aizen et al. 1997). Increased discharge volumes are reported for the Pamir (Chevallier et al. 2014), also for the northern Caucasus (Rets et al. 2018), and for the Southern Alps of New Zealand (Gawith et al. 2012). Discharge has recently decreased in some Siberian and Mongolian basins (Frolova et al. 2017; Dorjsuren et al. 2018).

Europe

Over recent decades, changes in the mountain cryosphere have already affected landscapes, hydrological regimes, water resources, and infrastructure, with significant downstream impacts in terms of quantity, seasonality, and quality of water (Beniston et al. 2011). Impacts related to climate-cryosphere interactions will continue to cause changes to such an extent that Europe's mountain landscapes will have a completely different visual appearance by the end of the twenty-first century. Seasonal snow lines will shift to much higher elevations, glaciers at low- and mid-range elevations will have disappeared, and even large valley glaciers will be characterized by significant retreat and mass loss (Beniston et al. 2018).

Numerous long-term observations in the European Alps show significantly negative current snow cover trends below 2000 m a.s.l. and negative or no clear trends above 2000 m, while the decadal variability of the snow cover is strong (Fig. 1.17) (Scherrer et al. 2004, 2013; Durand et al. 2009). Recently, Klein et al. (2016) detected a marked decline in all snowpack parameters over the period 1970–2015 irrespective of elevation, with significantly shortened snow cover duration by 8.9 days per decade on average which is largely driven by earlier snowmelt. Marty et al. (2017) provided evidence of a large-scale decline in snow water equivalents, while Schöner et al. (2019) found a clear decrease in mean snow depth over much of the Austrian and Swiss Alps. Similar trends are observed in the Tatra Mountains (Gadek 2014). The existence of a permanent snow cover during summer is very unlikely towards the end of the century, even at the highest elevations in the Alps (Beniston et al. 2018). This has profound implications for the remaining glaciers (Figs. 1.18, 1.19) that have already experienced a substantial mass loss since the nineteenth century and will face an increasing pace of mass loss (Zemp et al. 2015). The ice volume loss in the European Alps is estimated to be c. 50% during the period 1900–2011 (Huss 2012), while the glacier area in Switzerland decreased by 28% between 1973 and 2010 (Fischer et al. 2014), and in Austria by 17% between 1969 and 1998 (APCC 2014), resulting in the disintegration of many glaciers. The reduction in glacier area is even more critical in the case of the small glaciers in southern Europe. In the Pyrenees, Rico et al. (2017) assessed a decline of the glacier area by 88% between 1850 and 2016, with a rapid wastage since the 1980s, confirming the recently accelerated shrinkage trend. Small glaciers in temperate and southern Europe are likely to completely disappear, and even large valley glaciers will have lost much of their current volume by the end of the century (Jouvet et al. 2009; Linsbauer et al. 2013; Zekollari et al. 2014, 2019).

In Norway, Dyrrdal et al. (2013) observed a decrease in snow depth and number of snow days at lower elevations and in regions with warmer winter climate since the early 1960s, only some stations in higher mountain regions show positive trends, in particular in colder regions in the western part of South Norway. Declining snow depths at lower elevations and a shortened duration of snow cover was also assessed in northern Finland and related to large-scale climatic indices (Kivinen and Rasmus 2015). The glacier area in Norway has been reduced by c. 10% between 1960 and the 2000s (Winsvold et al. 2014). While the mass balance of

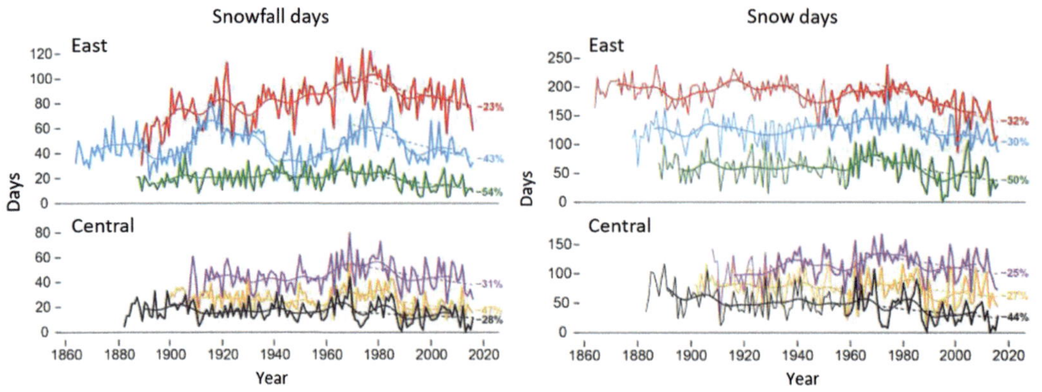

Fig. 1.17 Number of days with snowfall (daily new snow sum ≥ 1 cm) and number of days with snowpack (daily snow height ≥ 1 cm) at Swiss stations; the annual values are shown as a bold line; the thin line represents a 20-year Gaussian smoother. Top: Eastern Switzerland stations (Sils-Maria: red, Elm: blue, Chur: green). Bottom: Central Switzerland stations (Einsiedeln: purple, Meiringen: orange, Luzern: black). The dashed lines and numbers show the linear trends in the period 1970–2016. (Modified from CH2018 2018)

Norwegian glaciers is generally negative in the past 50–60 years with the decade 2001–2010 being the most negative, many maritime glaciers showed intermittent periods of positive mass balance in the late 1980s and 1990s due to higher snow accumulation (Andreassen et al. 2016, 2020), linked to the positive NAO (North Atlantic Oscillation) phase during that period (cf. Bonan et al. 2019). Massive volume losses in the order of 64–81% are predicted for the Scandinavian glaciers for the twenty-first century, some ice caps might lose up to 90% of their current volume, and many glacier tongues will disappear (Beniston et al. 2018).

Direct temperature monitoring and indirect geophysical surveys show accelerated permafrost warming in the Alps and in Scandinavia over recent decades. At monitoring sites in the Alps, the current mean annual ground temperature trend (10–20 m depth) is up to 1.0 °C per decade (Noetzli et al. 2018; Hock et al. 2019). In South Norway, mean ground temperature increase at 6.6–9.0 m depth ranged from ~0.015 to ~0.095 °C a^{-1} between 1999 and 2009 (Isaksen et al. 2011). Increasing permafrost temperatures and observed expansions of active-layer thickness (PERMOS 2016) suggest ongoing permafrost degradation, resulting in an increased frequency of slope instabilities in mountain ranges and to a higher magnitude of mass wasting processes such as rockfalls, rockslides, icefalls, landslides, and debris flows (Stoffel et al. 2014; Patton et al. 2019). Changes in the cryosphere of Europe's mountains will have severe hydrological implications, including a transition of runoff regimes from glacial to nival and from nival to pluvial, as well as shifts in the timing of discharge maxima (Beniston et al. 2018). In glacierized catchments, the glacier melt contribution to runoff will be reduced significantly by the end of the century, with peak discharge occurring 1–2 months earlier in the year (Hanzer et al. 2018). The altered seasonality of high-elevation water availability will have serious consequences for water storage and management in reservoirs for drinking water, irrigation, and hydropower production (Beniston et al. 2018).

America

Alaska has been one of the regions on Earth with highest warming rates over recent decades, with temperature increase being more than twice as high as in the contiguous United States. As a consequence, Alaska experienced a considerable decrease of the snow cover and a significant shrinkage of the ice mass of most of its glaciers, still accounting for 12% of the global ice-covered area outside the Antarctic and Greenland ice sheets (Kienholz et al. 2015). More than 90% of

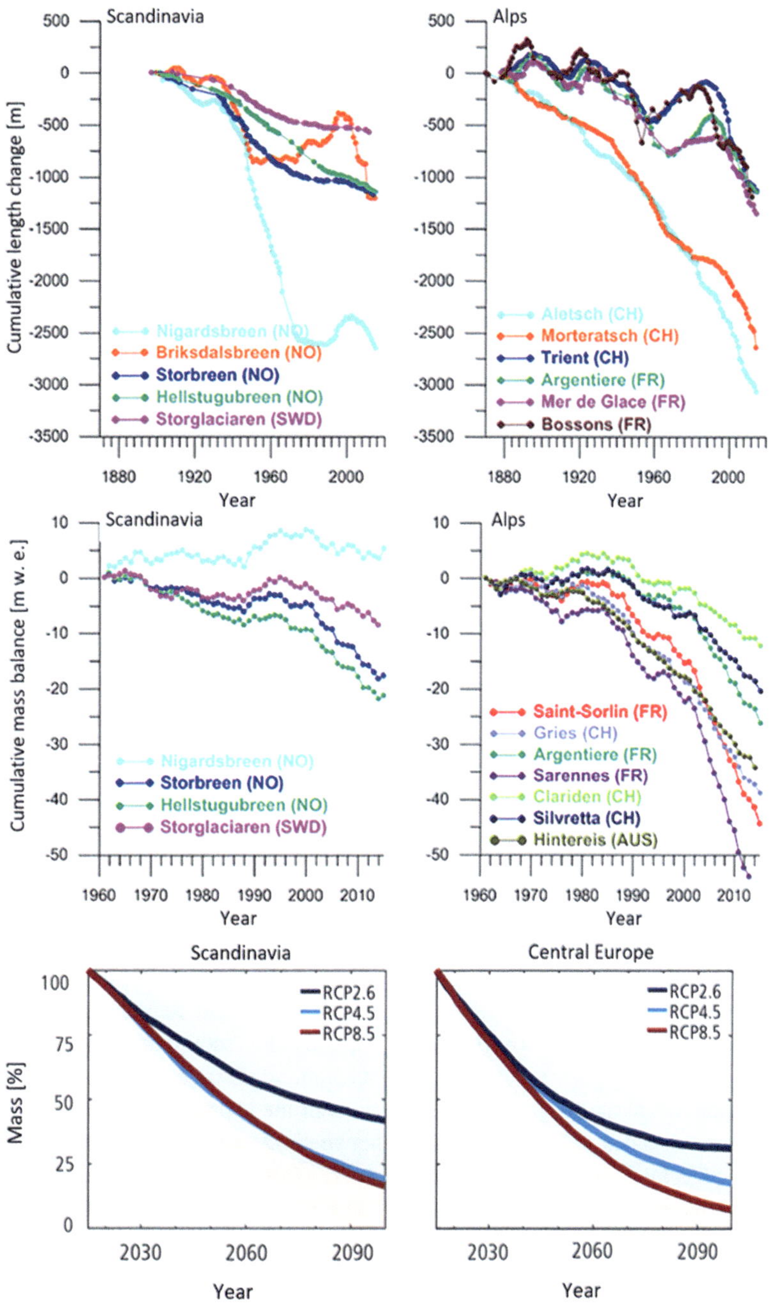

Fig. 1.18 Length and surface mass balance changes documented with in situ measurements for glaciers in Scandinavia and in the European Alps. (Modified from Beniston et al. 2018). Lower panels: Projected glacier mass evolution for Scandinavia and Central Europe between 2015 and 2100 relative to each region's glacier mass in 2015 (100%) based on three RCP emission scenarios (modified from Hock et al. 2019)

Alaska's glaciers are retreating (Thoman and Walsh 2019), as in other regions of North America (Fig. 1.20). Some mountain ranges and individual large glaciers are particularly affected, vividly illustrated by the Chugach Mountains on the south coast of Alaska, where a significant decrease in glaciation was observed. Here, the Columbia Glacier, one of the shrinking tidewater glaciers, is currently in a dramatic retreat. With a loss of about half of its volume since 1957 and 20 km of its length in the past three decades (McNabb and Hock 2014; Carlson et al. 2017a, b), the Columbia Glacier is one of the fastest changing glaciers in the world. Mass losses of

Fig. 1.19 Retreat of Rhone glacier, Switzerland, from the end of the Little Ice Age (1856) to 2005. (Obtained/modified from DFB AG; Copland 2011)

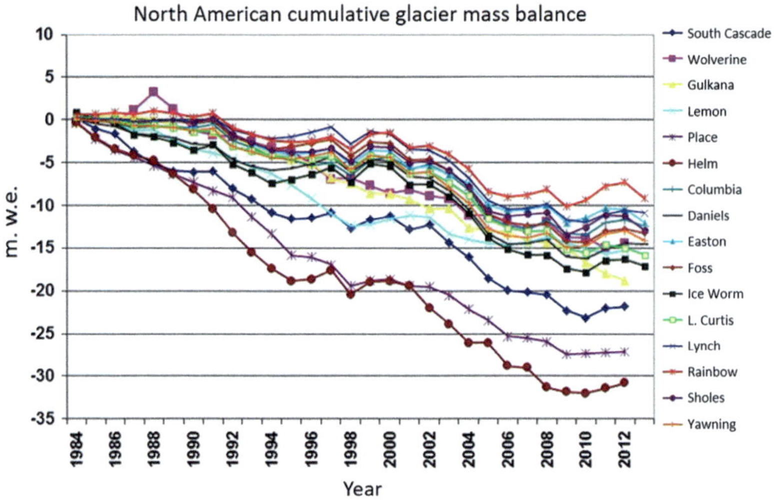

Fig. 1.20 North American cumulative glacier mass balance 1984–2013. (Modified from www.antarcticglaciers.org after M. Pelto)

Alaskan glaciers have been immense (Fig. 1.21). Estimates are in the order of 75±11 Gt per year between 1994 and 2013 (Larsen et al. 2015). Projections suggest continued and substantive glacier retreat and negative mass balances in the coming decades, with volume losses between 32 and 58% by 2100, making Alaskan glaciers large contributors to sea-level rise (Huss and Hock 2015) (cf. Fig. 1.11). The regional equilibrium line altitude is also projected to shift upward by 105 to 225 m, associated with a considerable decrease in snow precipitation (despite an increase in total precipitation), a shift to rain-dominated watersheds at lower elevations, shorter snow seasons, and warming permafrost (McGrath et al. 2017; Littell et al. 2018; Thoman and Walsh 2019).

The effects of widespread warming on the cryosphere such as shorter snow cover duration, earlier spring peak streamflow, thinning glaciers, and thawing permafrost are also evident in the mountain ranges of western Canada and the conterminous United States. These effects are projected to intensify in the coming decades. In the Rocky Mountains of Canada, a spatially coherent pattern of decreasing snow depth and snow cover duration and extent was detected for the period 1950–2013 (Fig. 1.22), with an average decline of the annual snow cover duration of about 4 days per decade, almost entirely due to reductions occurring during the spring season (DeBeer et al. 2016). Mountain glaciers in western Canada are receding at all latitudes, with rates of loss accelerating in the last few decades. While glaciers have exhibited a wide range of local changes from small net advances to complete disappearance, a decline in glacier cover of c. 25% over the past decades was observed in

Fig. 1.21 Top: Retreat of Muir Glacier, Alaska, 1941–2004 (left photo by William O. Field; right photo by Bruce F. Molnia, USGS). Bottom: Retreat of Pedersen Glacier, Alaska, 1917–2005 (left photo by Louis H. Pedersen; right photo by Bruce F. Molnia, USGS); photos obtained from the Glacier Photograph Collection, Boulder, Colorado USA, National Snow and Ice Data Center/World Data Center for Glaciology

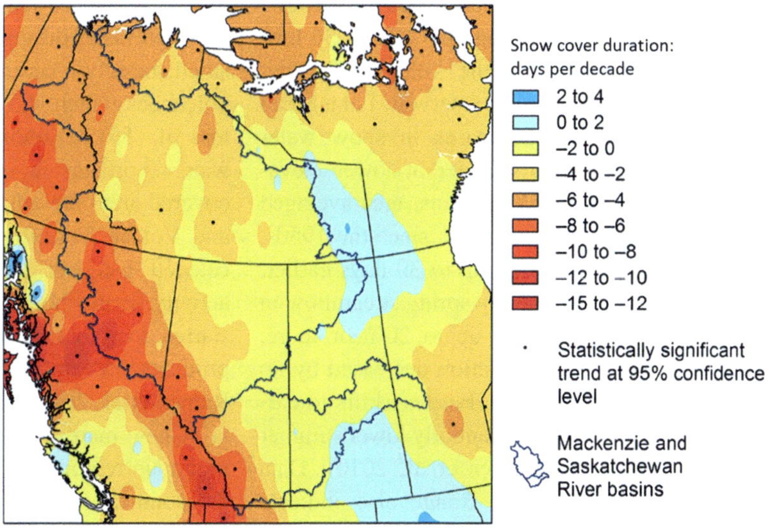

Fig. 1.22 Trends in spring season (February–August) snow cover duration for the period 1972–2013; spatial patterns indicate an above-average decline in the Rocky Mountains. (Modified from DeBeer et al. 2016)

most studies (Bolch et al. 2010a; Tennant et al. 2012; Beedle et al. 2015). A current hotspot of glacier shrinkage is located in the southern Coast Mountains in British Columbia, where the rate of mass loss over the period 2009–2018 was −7.4 ±1.9 Gt per year, about 20% higher than over the period 1985–1999 (Menounos et al. 2019). Projections for 2100 show drastic decreases of glacier area and volume in western Canada, with the volume of glacier ice shrinking by 70±10% relative to 2005, triggering severe hydrological implications and related impacts on aquatic ecosystems, agriculture, forestry, alpine tourism and water quality (Clarke et al. 2015).

The trend towards glacier recession, reduced mountain snowpack and earlier spring snowmelt runoff peaks is also widespread in the western United States, where glaciers cover an area of only 533 km^2, which is only 4% of the glacier area in western Canada (Menounos et al. 2019). Recent estimates suggest a decrease of the glacier and perennial snowfield area by 39% since the mid-twentieth century (Fountain et al. 2017). In Glacier National Park (Montana), only 35% of the Little Ice Age glaciers persisted by 2005 (Martin-Mikle and Fagre 2019). Glaciers in the Pacific Northwest, the most glacierized region in the conterminous United States, have displayed ubiquitous patterns of retreat and long-term negative trends in glacier area, resulting most likely in an immense reduction in late summer discharge volumes (up to 80%) by the end of the century due to post-peak declines in glacier melt and seasonal snowmelt (Frans et al. 2018). Observed declines in snowpack are dramatic, with over 90% of snow monitoring sites with long records across the western United States showing declines, regardless of phase changes in the Pacific Decadal Oscillation (PDO). Snowpack has declined on average by 21% or 36 km^3 since 1915, greater than the volume of water stored in the West's largest reservoir, Lake Mead (Mote et al. 2018). Decreases in snow water equivalent are generally larger at lower elevations. In the Cascade Mountains, area-averaged snowpack decreased by c. 20% since the 1950s, spring snowmelt occurred up to 30 days earlier, the share of late winter/early spring streamflow in annual flow increased by up to 20% or more, while the summer flow fraction decreased by up to 15% (Mote et al. 2014). Further shifts to earlier snowmelts and to substantially lower summer flows are projected (Elsner et al. 2010). Data from the Colorado Front Range also indicate ongoing degradation of mountain permafrost (Leopold et al. 2014).

While southern Sierra Nevada stations at higher elevations showed an upward trend in snow water equivalent over the twentieth century, with increased precipitation more than compensating for the overall warming, massive declines in peak snow water equivalent are projected for the Sierra Nevada and the southern Rocky Mountains for the coming decades (Garfin et al. 2014). Snowpack lows are particularly evident at lower Sierra Nevada elevations. 2015 saw a record low snowpack in the Sierra Nevada (Margulis et al. 2016). The estimated return interval for the 2015 1 April snow water equivalent value was calculated to be 3,100 years, highlighting its exceptional character (Belmecheri et al. 2016). As many watersheds in the Southwest of the United States depend on snowpack to provide the majority of the annual runoff, lower snow water equivalents imply reduced reservoir water storage. Reductions in runoff, streamflow, and soil moisture pose increased risks to the water supplies needed to maintain the Southwest's cities, agriculture, and ecosystems (Garfin et al. 2014). The glaciers of the Sierra Nevada show recently accelerated retreat rates, the absolute ice loss, however, is rather low due to the small glacier mass. Glacier areas have declined by more than half over the past century, and most glaciers will disappear completely from 2070 onwards if the current rate of loss continues (Basagic and Fountain 2011). An even faster disappearance is expected for the small, rapidly receding glaciers on the Mexican volcanoes which showed an overall glacier area loss of 75% between 1973 and 2017, implying water shortages in the surrounding areas. Ice-covered areas are only left on Volcán Citlaltepétl and Volcán Iztaccíhuatl, whereas Volcán Popocatepétl has lost its glaciers due to eruptive activity, even though the glacier shrinkage has started long before the appearance of eruptive products (Veettil and Wang 2018b; Cortés-Ramos et al. 2019).

A new map of snow cover changes in global mountain regions shows the Andes, in particular the southern Andes, as one of the hotspots of negative trends in snow parameters, with the area between Chile and Argentina (latitudes 29 to 42° S) exhibiting an overall snow cover duration decrease between 2500 and 4000 m of −26.6 days, and an earlier last snow day of −21.1 days over the period 2000–2018 (Fig. 1.23) (Notarnicola 2020). Saavedra et al. (2018) observed even more negative snow cover changes, with more pronounced snow loss on the east side of the Andes, and a significant increase in snowline

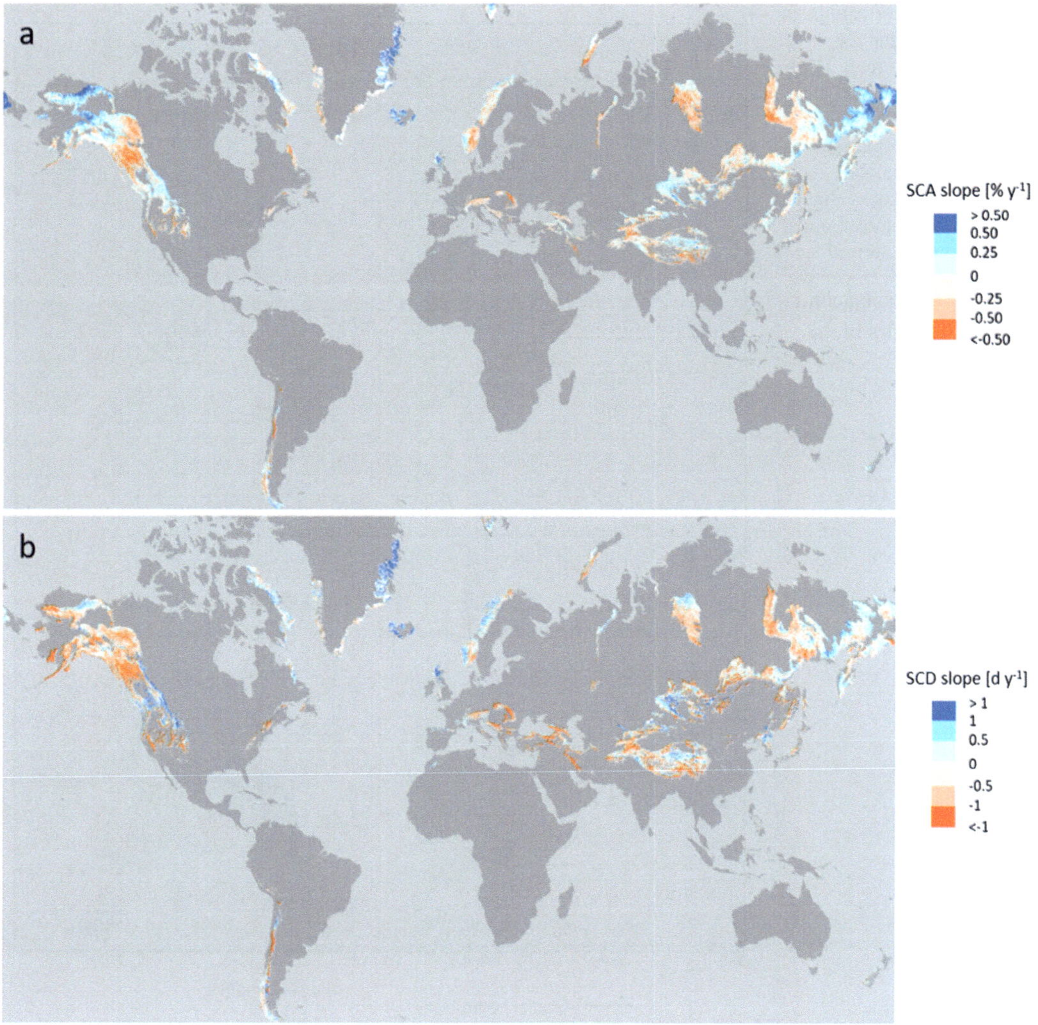

Fig. 1.23 Snow cover changes in global mountain regions shown as spatial distribution of positive and negative Sen's slopes resulting from MODIS products in the period 2000–2018; **a**: Snow-covered area (SCA slopes); **b**: Snow cover duration (SCD slopes). (Modified from Notarnicola 2020)

elevation south of 29–30°. Malmros et al. (2018) obtained similar results indicating adverse impacts on downstream water resource availability to agricultural, densely populated regions in central Chile and Argentina. The tropical Andes exhibit more heterogeneous snow cover trends. Mernild et al. (2017) simulated nonetheless a decrease in the number of snow cover days and in snow cover extent for the period 1979–2014.

Glaciers along the Andes exhibited a large-scale retreat over the past several decades; they are considered to be among the fastest shrinking glaciers on Earth. The recent dramatic recession of Andean glaciers is unprecedented since the maximum glacier extension of the Little Ice Age. Total Andean glacier mass change over the period 2000–2018 is estimated to be -22.9 ± 5.9 Gt yr^{-1}, thus comparable to the glacier mass change in entire High Mountain Asia (see above). The most negative mass balances over this period were assessed in the Patagonian Andes (-0.78 ± 0.25 m w.e. yr^{-1}), followed by the Tropical Andes (-0.42 ± 0.24 m w.e. yr^{-1}), while the Dry Andes showed relatively moderate losses (-0.28 ± 0.18 m w.

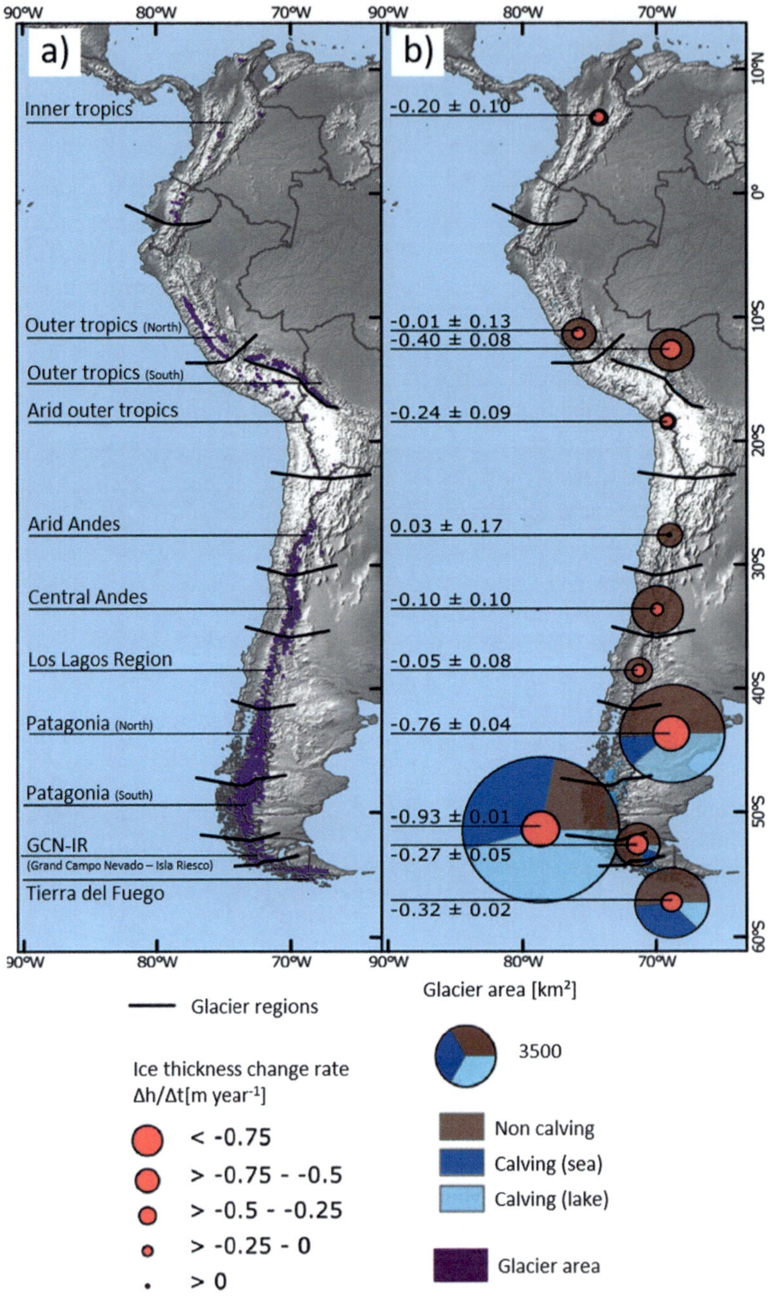

Fig. 1.24 Glacier regions in South America and ice thickness change rates 2000–2011/15; **a**: glacierized areas in purple, with black lines delimiting glacier regions; **b**: inner circles symbolize average ice thickness change rate of glacier regions; outer circles show glacierized area and proportion of respective glacier types. (Modified from Seehaus 2020 after Braun et al. 2019)

e. yr^{-1}) (Fig. 1.24) (Dusaillant et al. 2019). Braun et al. (2019) detected lower values for Andean glaciers and highlighted the massive ice loss of Patagonian icefields. Across the Patagonian Andes, the glacierized area was reduced by c. 20% within the last ~150 years (Meier et al. 2018). Dramatic examples of glacier recession include the Jorge Montt Glacier and the O'Higgins Glacier, the fastest shrinking glaciers in Chile, which lost 20 km and 15 km, respectively, of its length over the twentieth century (Schoolmeester et al. 2018). Accelerated mass loss is recently reported for glaciers of the dry Chilean Andes (Kinnard et al. 2020).

Mass loss from glaciers across the Andes of Colombia, Ecuador, Peru and Bolivia has been substantial, not seldom dramatic in recent decades, with a rather homogeneous pattern of glacier shrinkage and an accelerated retreat rate after 1976, followed by further increases after 2000 and 2013 (Rabatel et al. 2013; Mernild et al. 2015; Seehaus et al. 2019, 2020). Since the 1950s, glacier surface area has decreased to almost zero in Venezuela, which is about to become an ice-free country (Braun and Bezada 2013). In Colombia, the current glacier extent is 36% less than in the mid-1990s, 62% less than in the mid-twentieth century, and almost 90% less than the Little Ice Age maximum extent, and it is predicted that only the largest glaciers on the highest peaks will persist until the second half of this century (Rabatel et al. 2013, 2018). At the Chimborazo volcano in Ecuador, the loss of surface area was 72% between 1962 and 2016 (Schoolmeester et al. 2018). Many glaciers of the tropical Andes show comparatively sensitive and rapid responses to climatic changes, including an enhanced recession during El Niño events. Small glaciers at lower elevations (<5000 m a.s.l.) that do not have a permanent accumulation zone have already completely disappeared or will disappear within the next years/decades (Rabatel et al. 2013; Seehaus et al. 2019). In the Cordillera Blanca in Peru, the world 's most extensively glacier-covered tropical mountain range, glaciers have been rapidly receding as well over the past few decades (Schoolmeester et al. 2018). Projected warming will also result in the loss of permafrost. It is predicted that permafrost areas in the Bolivian Andes will shrink from present day extent by up to 95% under warming projected for the 2050s and by 99% for the 2080s and that almost all of the currently active Bolivian rock glaciers will be lost by the end of the century (Rangecroft et al. 2016).

Projections for the end of the century indicate that the future rise of the equilibrium line altitude may lead to further disappearance of glaciers at inner tropical sites under high emission scenarios, whereas outer tropical glaciers which are more strongly affected by future changes in the hydrologic cycle may persist as smaller glaciers (Vuille et al. 2018). The high ice loss rates of Andean glaciers result in a temporary increase in dry season water supply downstream. Meltwater supplies play a significant role in wetland cover dynamics in the high Andes (Dangles et al. 2017). Peak water, however, has already passed in many glacierized catchments, and, in the long term, dry season river discharge will decrease due to future glacier shrinkage, contributing to emerging water resource crises and environmental hazards for both urban and rural populations relying on glacier-fed streams for agriculture and livelihoods (Thompson et al. 2017; Vuille et al. 2018).

Africa

Snowpack is on the decline in North Africa and thus in accordance with trends in other Mediterranean regions, notwithstanding the fact that the persistence of snow cover is highly variable in space and time (Fayad et al. 2017). In the Atlas Mountains, a statistically significant long-term trend has not been detected yet (Marchane et al. 2015). However, a combination of warming and reduced precipitation, associated with earlier springtime melting, will result in reduced snowpack, adversely affecting the supplies of meltwater for lowland areas in Morocco (García-Ruiz et al. 2011; Marchane et al. 2017). The Drakensberg Range in the Lesotho Highlands is characterized by a very high inter- and intra-annual variability of snow coverage (Wunderle et al. 2016), and at the same time by a decadal trend of declining snow depth and snow cover duration, with much lower values in comparison to the late nineteenth century (Grab et al. 2017).

Climate change impacts on the cryosphere are most obvious in East Africa where the only African mountains are located which have glaciations in their summit regions (Kilimanjaro [5895 m], Mount Kenya [5199 m] and Rwenzori [5109 m]). All these glaciers show an extraordinary recession over the past century, with a loss of more than 80% of the glacier area on all three mountains (Fig. 1.25). Analysed mass and energy fluxes on selected glaciers on Mount Kenya and Kilimanjaro suggest that the

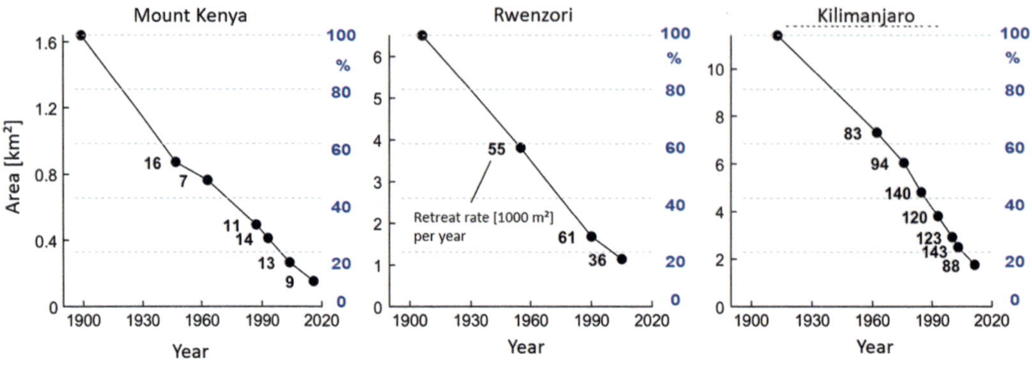

Fig. 1.25 Loss of glacier area on Mt. Kenya, the Rwenzori Mountains, and Mt. Kilimanjaro; absolute values on left y-axes, percentage change, relative to the first available area value on right y-axes; numbers in bold show mean area loss between points in time (in '000 m^2 a^{-1}). (Modified from Prinz and Mölg 2020)

frequency and amount of solid precipitation is the dominant local climatic factor for the mass balance of these glaciers, with decreasing snowfall being interpreted to be a concomitant effect of global warming (Mölg et al. 2009; Prinz et al. 2016). Ice loss is particularly severe on the Lewis Glacier, the largest glacier on Mount Kenya, which has already lost 90% of its area and 95% of its volume since the end of the nineteenth century (Prinz et al. 2018; Chen et al. 2018). 8 glaciers vanished completely until 2004 (Hastenrath 2005). Since 2010, the mass loss of Lewis Glacier has been accelerating due to glacier disintegration, yet another glacier disappeared completely, and Prinz et al. (2018) predict that Mt. Kenya's glaciers will be extinct before 2030, if current retreat rates continue. The loss of ice cover on Kilimanjaro is similarly dramatic (Fig. 1.26), with glaciers having retreated from their former extent of 11.40 km^2 in 1912 to 1.76 km^2 in 2011 (Cullen et al. 2013). About the same magnitude of glacier recession was reported for the Rwenzori Mountains, where only the higher elevated glaciers on Mt. Stanley have shown a slower decrease (Kaser and Osmaston 2002; Mölg et al. 2006). Nevertheless, the Stanley glacier had almost vanished by 2008 (Mumba 2008; Spinage 2012). The scenario of a complete disappearance of all ice at Kilimanjaro and Rwenzori is likely to occur between 2040 and 2060 (Mölg et al. 2003; Cullen et al. 2013). The East African glaciers do not play a major role in the regional water balance, however, they are of great importance for the tourism potential in the respective regions.

1.2.3 Biotic Responses

1.2.3.1 General Overview

High mountain ecosystems and their biodiversity are affected by climate change at an accelerated pace. It is evident from long-term ecological monitoring and large-scale assessments that the high levels of warming to which mountain ecosystems are exposed have resulted in substantial redistributions and losses of habitats and species, and in increased vulnerability to additional stressors such as invasive species or disturbances (Jentsch and Beierkuhnlein 2003; Pauchard et al. 2009; Pauli et al. 2012; Wipf et al. 2013; Alexander et al. 2016; Dainese et al. 2017; Lamprecht et al. 2018; Steinbauer et al. 2018; Pauli and Halloy 2019; Petriccione and Bricca 2019). Climate change effects on temperature, snow, moisture, and nutrient regimes potentially cause alterations in plant physiology and phenology, species interactions, community structure, species distributions, and ecosystem processes (Körner 2003; Winkler et al. 2019), with higher losses of biodiversity and habitats occurring with higher levels of climate warming (Nunez et al. 2019). Respective changes are increasingly observed, the knowledge of the

Fig. 1.26 Retreat of glaciers on Mount Kilimanjaro 1912–2006; 85% of the ice has disappeared during this period. (Modified from Thompson 2010)

alteration of mountain ecosystems, however, is still profoundly deficient, in particular in many of the underresearched mountain ranges in the Global South (Schickhoff and Mal 2020). Species responses to climate change are driven by the capacity to persist in situ by altering fitness-related traits through plastic adjustment or genetic adaptation to novel stresses such as longer growing seasons, increasing temperatures, and less infertile soils. Plants at higher elevations have a low capacity to persist in situ since traits such as slow growth or dwarfism are genotypically determined, and the phenotypic plasticity is constrained under harsh climatic conditions. A greater potential for montane and alpine species to adapt and to survive rapid anthropogenic climate change lies in distributional shifts to track preferred bioclimatic conditions (Schickhoff 2011, 2016a; Pauli and Halloy 2019; Winkler et al. 2019; Winkler 2020). However, clonal, relatively slow dispersal strategies are not uncommon at high elevations which restrict the potential of shifting range limits. Magnitude and rate of climate change as well as induced alterations of abiotic and biotic site conditions will overstretch the adaptive capacity of many species, increased extinction risks and losses of biodiversity are thus inevitable (Thuiller et al. 2005; Moritz and Agudo 2013). To date, only a small percentage of countries is on track to achieve respective national biodiversity targets within the framework of the Convention on Biological Diversity. A fundamental embedding of mountain biodiversity in national biodiversity conservation strategies is necessary in order to better meet the objectives of the UN Sustainable Development Goals (UN 2020).

Changes in species distribution ranges as a complex response to novel constellations of bioclimatic and other site conditions are increasingly observed in mountain regions, with range extensions to higher elevations being considerably overrepresented, compared to range contractions. Species from lower elevations are now colonizing habitats on mountain summits at a rate which is five times faster than half a century ago (Fig. 1.27) (Steinbauer et al. 2018; Pauli and Halloy 2019). Implications of the establishment of these 'neonative' species (Essl et al. 2019) include the gradual transformation of species composition and community structure of resident communities, in particular in the alpine and nival zones. Warmth-demanding and/or less cold-adapted species become more dominant, while strongly cold-adapted high-elevation species are declining in abundance and frequency. Severe area losses of these cryophilic species are expected since upward range shifts are constrained by limited available space (Engler et al. 2011; Elsen and Tingley 2015; Lenoir and Svenning 2015; Freeman et al. 2018). Species-specific migration rates are very different, suggesting that the interaction of multiple internal species-specific traits controls the response to changed climatic conditions (cf. Roux and McGeoch 2008). Asynchronous responses to external driving forces result in 'no-analogue communities' with modified competitive conditions (Williams and Jackson 2007). Novel biocoenoses with modified dominance relationships, competitive conditions and population densities inevitably affect ecosystem functioning, and thus the provision of ecological services and the resilience to disturbances (Pecl et al. 2017).

Treeline ecotones are as well subjected to reinforced dynamics in recent decades, though not yet necessarily reflected in treeline advance. Since the elevational position of natural alpine treelines is primarily caused by heat deficiency (Holtmeier 2009; Körner 2012, 2020), and treeline elevations have responded to climate oscillations throughout the Holocene (Tinner and Theurillat 2003), treelines are often considered to be sensitive indicators of global warming. The findings of observational studies on treeline shifts, however, give evidence of both advancing treelines and insignificant treeline responses (Holtmeier and Broll 2005, 2007, 2017a; Schickhoff et al. 2015). A global meta-analysis of treeline response to climate warming showed advancing treelines at 52% and persistent treelines at 47% of the studied sites (Harsch et al. 2009). A recent meta-analysis across the Northern Hemisphere found almost 90% of treelines ascending over the past century and c. 10% remaining stable while the mean hemispheric shift rate was much lower than expected from climate warming (Lu et al. 2020). Inconsistent responses indicate a highly heterogeneous sensitivity of alpine treelines to the effects of climate warming which is not surprising given the multitude of after-effects of treeline-landscape

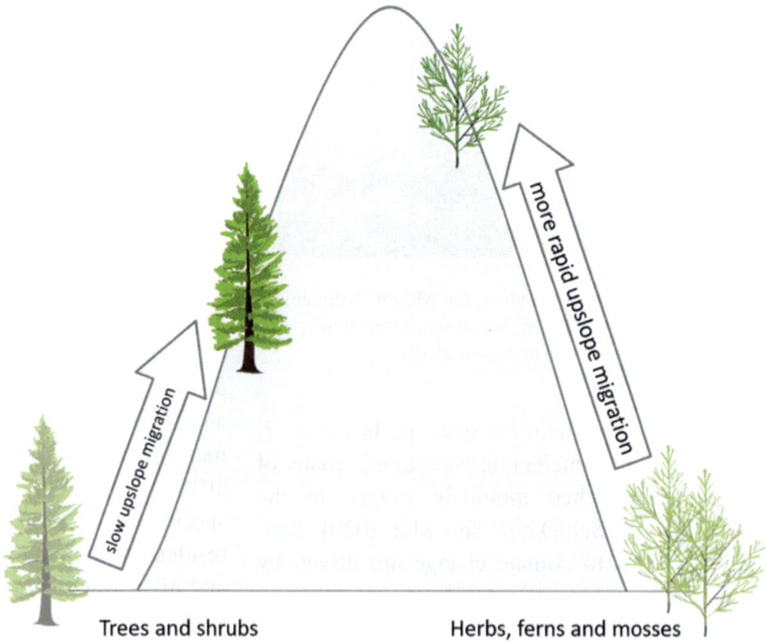

Fig. 1.27 Short-lived plant species exhibit a more rapid upslope migration as compared to long-lived plant species; upward range shifts may be constrained by limited available space. (Modified from Lenoir and Svenning 2013)

history (past climate fluctuations, natural and anthropogenic disturbances) that determine treeline position, spatial patterns and successional stages (Holtmeier and Broll 2017a). For example, human impacts are almost omnipresent at treeline environments in Africa, Asia, and Europe where mountain regions are settled since ancient times, and effects of land use history and dynamics overlap with those of multiple ecological and biophysical factors. Thus, a potential advance of a particular treeline (at local scale) to higher elevations is very difficult to predict. A global comparison of the response variability of different treeline forms revealed a certain correlation between spatial patterns and response dynamics of treeline ecotones, with the majority of diffuse treelines showing an advance (Harsch and Bader 2011; Bader et al. 2020). However, other factors and interrelationships, for instance species-specific traits and response patterns of treeline-forming tree species, may superimpose the response trend of treeline forms (Treml and Veblen 2017). The interactions between climatic changes as regional to global input variables and facilitating, modulating or overriding site factors at the local scale (the complex of abiotic and biotic local site conditions and their interactions and feedback systems including human impact and the entire treeline-landscape history) control current spatial patterns and temporal dynamics in treeline ecotones (Wieser et al. 2014; Elliott 2017; Holtmeier and Broll 2017a; Schickhoff et al. 2020). Lagged changes in treeline positions should not obscure the fact that current warming trends are favourable to growth, development, and reproduction of tree species in many treeline environments. Assuming that alpine treelines would have tracked global warming some day and reached a new steady state at higher elevations, the shrinking of lower and upper alpine/nival life zones would be dramatic. An upslope extension of mountain forests corresponding to a 2.2 K warming is likely to lead to a global loss of c. 24% of the lower alpine zone and of c. 55% of the upper alpine and nival zones (Körner 2012). Large-scale treeline shifts would have serious implications for diversity and function of high elevation ecosystems (Greenwood and Jump 2014).

Mountain endemics and other species with spatially restricted populations will be particularly affected by large magnitudes of climate change, fragmenting populations and reducing vigour and viability of species. In addition, endemic species are particularly vulnerable to genetic swamping due to introgressive hybridization (Gómez et al. 2015). Species in regions with declining precipitation are exposed to a higher risk as well (Engler et al. 2011). Global meta-analyses give impressive evidence of climate change-induced species migrations and range extensions (cf. Tomiolo and Ward 2018). Low dispersal abilities and slow migration rates, however, will prevent many species from keeping pace with the relocation of climatically suitable habitats (Settele et al. 2014). Many alpine plants depend on a certain required day length to become phenologically active (photoperiodism) (Keller and Körner 2003). Many other alpine plants, however, adjust sequences of phenological events to the rise in temperatures and to the advance in the timing of snowmelt. Thus, phenological shifts are considered to be sensitive indications of the response to climate warming. A phenological fingerprint of climate warming has been detected on a global scale, most pronounced at higher latitudes and higher elevations (Peñuelas et al. 2013; Piao et al. 2019).

The response of plants to climate change with regard to fitness and primary productivity has significant feedbacks to the global climate since the terrestrial biosphere plays a key role in the global carbon cycle, i.e. changes in primary production imply changes in the carbon storage of ecosystems. In recent decades, the global net primary production has slightly increased (Settele et al. 2014). Significant biotic responses also include alterations of the dense network of functional relationships, interdependencies and mutual interferences between species. Ecological or biotic interactions ensure a certain degree of self-regulation and resilience of ecosystems. Examples of spatial and temporal decoupling of interacting species, e.g. between herbivores and their food plants, and other mismatches in interactions with potentially adverse effects on ecosystem services are increasingly documented (Valiente-Banuet et al. 2015). Shifts in species composition and community structure are to an increasing extent caused by biological invasions into mountain regions (Pauchard et al. 2009, 2016; Alexander et al. 2016). Invasive species are predominantly climate change winners since they are often thermophilic and very adaptive due to wide ecological amplitudes. The expansion of non-native species further contributes to a reorganization of higher-elevation communities and alters ecological interactions and the provision of ecosystem services. It also contributes to homogenization effects of mountain ecosystems and biota (Jurasinski and Kreyling 2007; Malanson et al. 2011).

The focus in the following regional overview of biotic responses is on terrestrial vascular plants as the main primary producers in high mountain ecosystems and on vegetation as major structural component. Upslope extension of distribution ranges is also evident for numerous animal species and has been documented for many taxonomic groups (Gonzalez et al. 2010; Chen et al. 2011).

1.2.3.2 Regional Overview

Asia and Australasia

Representing four of the 34 global biodiversity hotspots and numerous ecoregions with significant conservation value, the HKH represents a major centre of global biodiversity (Myers et al. 2000; Pandit et al. 2014; Bhattacharjee et al. 2017; Xu et al. 2019). Impacts of climate change such as increasing temperature variability and declining precipitation during the dry season will affect the majority of species, thus threatening biodiversity conservation and the maintenance of mountain ecosystem integrity. A recent assessment based on satellite-derived NDVI datasets indicates that the length of the growing season in the HKH has increased by 4.25 days per decade over the last five decades (Krishnan et al. 2019a), in line with an overall greening trend in NDVI magnitude and an earlier green-up in most parts of the HKH region (Panday and Ghimire 2012; Mishra and Mainali 2017; Baniya et al. 2018). In particular, subnival vegetation above 5000 m has expanded (Fig. 1.28) (Anderson et al. 2020), and will have more space available for future expansion (Keenan and Riley 2018). Respective observations of shifts in species-specific phenological patterns are still limited, but indicate large-scale changes. Several *Rhododendron* and other species were found to currently flower several weeks earlier than in previous decades (Xu et al. 2009; Gaira et al. 2011; Mohandass et al. 2015; Negi and Rawal 2019; see also Yang et al. 2017 and Dorji et al. 2020 for the Tibetan Plateau). Adhikari et al. (2018) reported significant changes in phenological patterns in a treeline ecotone in Uttarakhand over recent decades, with the majority of species showing advanced flowering and extended vegetative phases. Mean date of leafing and flowering in lower elevation forests (Sal, Pine, and Oak forests) in the same region has advanced by 1–2 weeks within a period of 30 years (1985–2015) (Singh and Negi 2016). However, no significant changes over the past century were found for flowering phenology of *Rosaceae* species in the Hengduan Mountains, indicating that phenological responses to climate change are more complex than commonly assumed (Yu et al. 2016).

In the vast HKH region, climate change-induced shifts in species distributions and species composition of communities have been occurring, largely without being noticed or documented yet by science. In a first detailed study on upslope migration, Telwala et al. (2013) provided evidence of warming-driven elevational range shifts in 87% of 124 studied endemic plant species in alpine Sikkim over the last 150 years. Considering shifts of species' upper elevation limits of up to c. 1000 m, present-day plant assemblages and community structures are definitely different from those of the nineteenth century. In recent years, long-term research plots have been established within the Himalayan GLORIA (Global Observation Research Initiative in Alpine Environments) sub-network in order to monitor in detail species

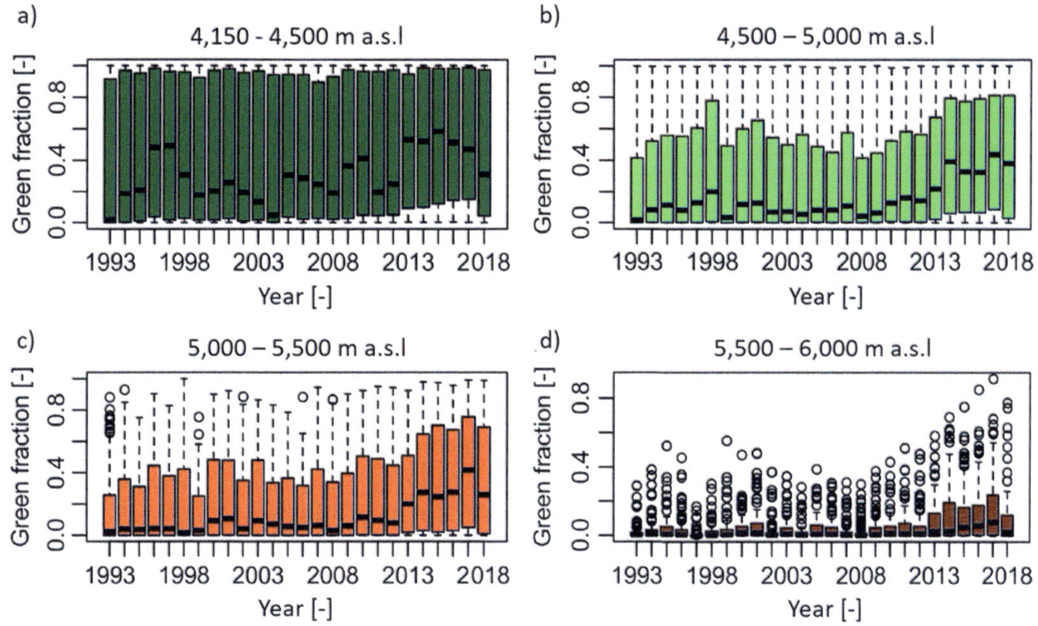

Fig. 1.28 Boxplots of green pixel fraction values for the broader HKH region (1993–2018), showing weakly positive time series trends in vegetated ground, separated by elevational zones; the extent of the boxes represent the 25th and 75th percentiles (quartiles), the bold middle line is the 50th percentile (median), the whiskers are the minimum and maximum values which fall within 1.5 times the interquartile range and the circles represent values beyond this range (outliers). (Modified from Anderson et al. 2020)

distribution patterns (Salick et al. 2014; Sekar et al. 2017). First resampling of plots in the eastern Himalaya (Hengduan Mountains) after seven years yielded the result that alpine plants on high-elevation summits increased in number of species, in frequency and in diversity, and that even Himalayan endemic species showed positive population trends (Salick et al. 2019, 2020). Here, modelling studies also project upslope expansion of distribution ranges (You et al. 2018). Likewise, the total number of vascular plant species on summits in the Kashmir Himalaya increased between 2014 and 2018 (Hamid et al. 2020). Warming-induced upward migration of plants was also observed in Ladakh by Dolezal et al. (2016), who resurveyed outpost populations of subnival plants after ten years and found an extension of the elevational range of 120–180 m. On the Tibetan Plateau, experimental studies highlighted the critical role of soil moisture for plant communities' response to climate warming: Alpine meadows showed increases in net primary productivity, while alpine steppes experienced decreasing productivity, decreasing cover of graminoids and forbs, and rapid species losses due to warming-induced drought conditions (Ganjurjav et al. 2016). The performance of the dominant species in central Tibetan Plateau alpine meadows, the shallow-rooted *Kobresia pygmaea*, in terms of plant cover, reproductive phenology/success and competitiveness was also found to be largely controlled by soil moisture which tends to decrease under climate warming (cf. Dorji et al. 2013, 2018). In line with these results, Lehnert et al. (2016) found that variability in precipitation and soil moisture outweighs overgrazing as the primary driver of recent large-scale vegetation changes on the Tibetan Plateau. Recently deglaciated terrain represents another highly dynamic alpine habitat. The surface area of deglaciated glacier forelands has been increasing considerably due to the ongoing recession of the vast majority of HKH glaciers. Vegetation successions on glacier forelands have not been addressed in greater detail so far. Some preliminary studies are available analysing the colonization of glacier forelands by pioneer species (Mong and Vetaas 2006; Vetaas 2007; Miehe 2015; Bisht et al. 2016).

Treeline dynamics and treeline shifts in the HKH region mostly result from combined effects of land use change and climate change (Schickhoff et al. 2015, 2016b; Shrestha et al. 2015; Suwal et al. 2016). HKH treelines are almost exclusively lowered from their natural elevational position by long-lasting human impacts (anthropogenic treelines). If these treelines are moving upslope, recent land abandonment or declining human impacts are the dominant drivers whereas climate change plays a subordinate role. Substantial treeline advances or shrub encroachments of alpine meadows in recent decades, reported in some studies (e.g. Baker and Moseley 2007; Brandt et al. 2013; Singh et al. 2018), have to be mainly attributed to effects of land use change. Bold statements in remote sensing studies about exceptional short-term climate warming-induced treeline advances (e.g. Mohapatra et al. 2019; Singh et al. 2020) must be viewed with extreme caution, in particular if they are not backed up by field data. Climate change, however, is a more significant driver of near-natural treeline dynamics. Only very few near-natural treeline ecotones have persisted in remote locations, mainly on north-facing slopes where these treelines have to be categorized as krummholz treelines (Schickhoff et al. 2015, 2016b; Schwab et al. 2017; 2020); they show rather low responsiveness to climate warming (see also Liang et al. 2011; Chhetri and Cairns 2015, 2018; Gaire et al. 2017; Sigdel et al. 2020).

Studies at the near-natural treeline in the Rolwaling valley (Nepal Himalaya) showed that the dense, self-sustaining and persistent krummholz belt of *Rhododendron campanulatum* forms a very effective barrier that largely prevents the expected upslope migration of *Abies spectabilis* and *Betula utilis* and other treeline tree species. The site conditions in the krummholz belt, modified by *Rh. campanulatum* itself in particular in terms of light and nutrient deficiencies, lower soil temperatures, and allelopathic effects, severely restrict the competitiveness of other tree species, reflected inter alia in a negative correlation between abundance and density of *Rh. campanulatum* and recruitment of other tree species. The elevational position of the Rolwaling treeline can be regarded as rather stable, suggesting a certain decoupling of treeline dynamics from global warming. However, the sensitivity is clearly evident in terms of stand density, seed-based regeneration and tree growth patterns, while a treeline shift is to be expected in the medium to long term only (decades to centuries) (Schickhoff et al. 2016b, 2020; Schwab et al. 2016, 2017, 2020; Müller et al. 2016; Bürzle et al. 2018).

Increasing stand densification as well as intense tree recruitment within other Himalayan treeline ecotones indicate the potential for future treeline shifts (Lv and Zhang 2012; Gaire et al. 2014, 2017; Shrestha et al. 2015; Wang et al. 2016; Tiwari et al. 2017a; Yadava et al. 2017; Tiwari and Jha 2018; Mainali et al. 2020; Sharma et al. 2020). Moreover, remote sensing studies indicate a general biomass increase in treeline ecotones over recent decades (Rai et al. 2013, 2019), while modelling studies support the concept of climate change-induced upslope range expansion of treeline species (Forrest et al. 2012; Joshi et al. 2012; Zomer et al. 2014; Rashid et al. 2015; Manish et al. 2016; Bobrowski et al. 2017; Lamsal et al. 2017; Chhetri et al. 2018; Gilani et al. 2020). Recent dendroecological studies at HKH treelines indicate enhanced tree growth at some high-elevation sites (Fan et al. 2009; He et al. 2013; Huang et al. 2017; Thapa et al. 2017; Shi et al. 2020), and a widespread strong sensitivity of tree growth to pre-monsoon temperature and humidity conditions (Fig. 1.29) (Dawadi et al. 2013; Liang et al. 2014; Ram and Borgaonkar 2014; Bräuning et al. 2016; Kharal et al. 2017; Panthi et al. 2017; Sohar et al. 2017; Tiwari et al. 2017b; Schwab et al. 2018; Singh SP et al. 2019). Warming-induced higher evapotranspiration and soil moisture deficits during dry spring months adversely affect tree growth in particular on sites which are prone to drought stress. Moisture supply in the pre-monsoon

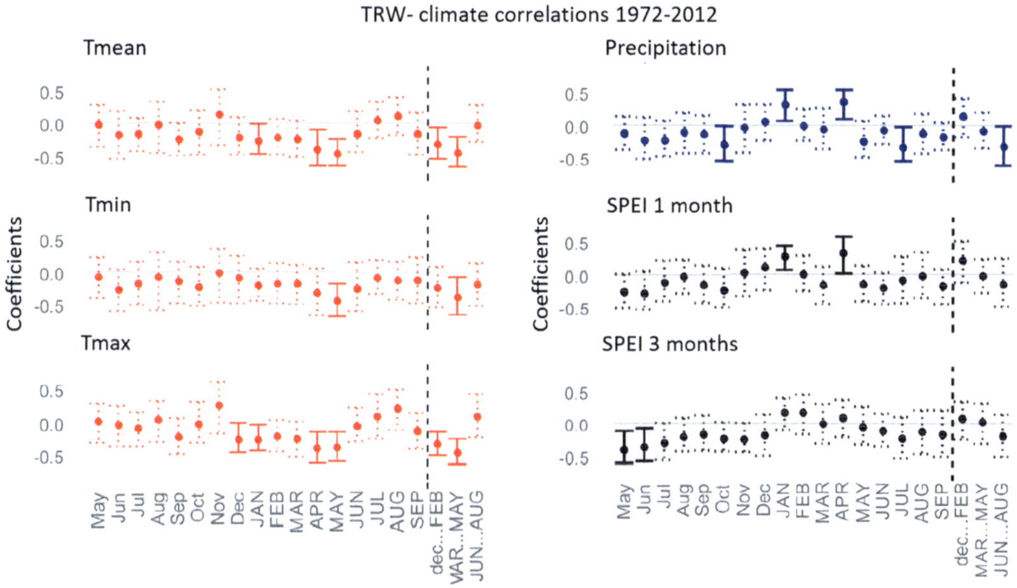

Fig. 1.29 Static correlations (1972–2012) of the tree-ring width chronology with temperature, precipitation and drought indices (SPEI) for current and previous year's months and current year seasons; solid bars indicate significant correlations ($p < 0.05$). (Modified from Schwab et al. 2020)

season might become an effective control of future treeline dynamics (Schickhoff et al. 2016b; Mishra and Mainali 2017; Sigdel et al. 2018; Lyu et al. 2019; Schwab et al. 2020). Correlations between ring width and winter temperatures in treeline ecotones were found to be largely positive (e.g. Bhattacharyya et al. 2006). However, increasing winter temperatures can be detrimental to the growth of *Rhododendron* shrub species above the treeline (Bi et al. 2017).

At elevations below the treeline, climate change is as well a major threat for the integrity of ecosystems (Chakraborty et al. 2018; Chettri et al. 2020). Montane and subalpine forest ecosystems in the HKH region are very critical for biodiversity, watershed protection and livelihoods of forest-dependent communities. Impacts of climate change on species distribution patterns, species composition of forest communities, and ecosystem functioning might degrade the capacity to maintain the provision of ecosystem services. Recently, a considerable upward migration of alien invasive species into the Himalaya was observed (Bajracharya et al. 2015; Negi and Rawal 2019), which is projected to continue (Lamsal et al. 2018). The spread of exotic species such as *Ageratina adenophora* (up to 2800 m) or *Lantana camara* (beyond 1500 m) over vast stretches of lower elevational zones alters species composition and ecosystem services of native plant communities. Higher native species richness obviously facilitates the invasibility of habitats (Bhattarai et al. 2014).

Similar trends of climate change-induced biotic responses, summarized above for the HKH alpine regions, prevail in other Asian and Australasian mountain systems. With regard to phenological changes, results of a meta-analysis across 145 sites in China demonstrated that more than 90% of the spring/summer phenophases time series show earlier trends and 69% of the autumn phenophases records show later trends (Ge et al. 2015). Recent positive trends in vegetation growth and productivity have been detected for the mountain regions of the Chinese landmass (Xu et al. 2014; Fang et al. 2016) and for some Mongolian mountain ranges (Kappas et al. 2020), including partially substantial treeline advances (Du et al. 2018) and upslope expanding distribution ranges (Zong et al. 2016).

Considerable shifts of upper altitudinal limits of mountain plant distributions were assessed in the Central Mountain Range of Taiwan over the last century, in parallel with rising temperatures in the region (Jump et al. 2012). Treeline advance on Mt. Fuji, Japan, is enhanced by climate warming (Sakio and Masuzawa 2012). Encroachment of subalpine bamboo species into alpine meadows, resulting in declining plant species richness, was reported from the Taisetsu Mountains, Hokkaido, northern Japan (Kudo et al. 2011). Warming-induced vegetation dynamics in the Altai and Mongolian mountains and in the Tien Shan and Pamir will be largely controlled by moisture conditions (Dulamsuren et al. 2010a, b; Poulter et al. 2013; Bao et al. 2015; Seim et al. 2016; Yin et al. 2016; Jiang et al. 2017; Dubovyk 2018), which also drives the extent to which forests will be transformed to forest-steppes and steppes in southern Siberian mountains in the next decades (Tchebakova et al. 2016). As temperatures in Inner Asia have increased substantially since the mid-twentieth century, tree growth has declined in many areas of the forest steppe (Dulamsuren et al. 2010b, 2011, 2013; Liang et al. 2016). At treeline elevations in the Chinese and Mongolian Altai, Chen et al. (2012) and Dulamsuren et al. (2014) assessed positive correlations of tree growth and growing season temperature, and no drought-induced growth limitation. Kirdyanov et al. (2012) assessed a densification of formerly open forests and an upslope shift of the treeline of approximately 50 m over the last century in the Putorana Mountains, northern Siberia, corroborating the large-scale treeline advances and tree growth enhancements found over much of Siberia by Esper and Schweingruber (2004), Soja et al. (2007), Kharuk et al. (2010), Petrov et al. (2015), Shevtsova et al. (2020) and others.

In the Russian Altai, a treeline shift of 150 m during the past 50 years was reported, with the rate of upslope movement having accelerated until recently (Gatti et al. 2019). In the North Urals, the upper limits of tree stands with different degrees of canopy closure have risen by about 100 m of elevation since the mid-nineteenth century (Moiseev et al. 2010; Hagedorn et al. 2014). Accelerated forest growth in the treeline ecotone has been detected in the Tien Shan under conditions of rising temperatures and sufficient precipitation (Qi et al. 2015). Elevational belt shifts are expected in the Tien Shan, while shifts of phenological dates have already been observed (Dimeyeva et al. 2015; Imanberdieva et al. 2018). A study of long-term vegetation dynamics of alpine communities in the Caucasus confirmed an upward shift of the upper limit of species distributions and an increasing abundance of species in upper alpine zones (Elumeeva et al. 2013; see also Gigauri et al. 2013). Treeline tree species in the Caucasus (*Betula litwinowii*) expand their range to higher elevations as well, caused by combined effects of land use change (reduced grazing pressure) and climate change (Akatov 2009; Hansen et al. 2018). Climate change-induced migrations will most likely result in northward and upward shifts of subalpine plant species in the mountains around the Iran Plateau (Shamsabad et al. 2018), treeline advances are expected in Pontic Mountains (Kurt et al. 2015).

Preliminary observations in alpine zones of SE Asian mountains point to phenological changes and to subalpine/alpine grasslands affected by shrub and tree encroachment (Hope 2014). Overall trends towards a longer duration of the growing period were detected in different study areas of the Australian and New Zealand Alps (e.g. Thompson and Paull 2017). The average advance of the timing of spring events, based on long-term datasets of c. 350 species, was calculated to be c. 4 days per decade (Chambers et al. 2013). Shifts in species' distributions are predicted for many taxa in the mountains of Australia and New Zealand, with suitable habitats shifting and/or contracting as the climate changes (Cabrelli et al. 2015). A temperature rise of 3 °C may lead to a loss of c. 80% of existing alpine lands in New Zealand and to a loss of up to 50% of vascular plant taxa (Halloy and Mark 2003). The New Zealand *Nothofagus* treelines are relatively unresponsive to recent climate warming, however, and show only little evidence of treeline advance (McGlone et al. 2010; Harsch et al. 2012). Population dynamics at the alpine treeline in SE Australia were found to be responsive to climate change, reflected in a recent short-distance treeline advance, while treeline dynamics is largely controlled by fire (Naccarella et al. 2020). It needs to be highlighted that there are still tremendous knowledge deficits with regard to climate change-induced biotic responses in mountain life zones of Asia and Australasia which need to be reduced in order to develop appropriate management strategies aiming at the maintenance of mountain ecosystem integrity and the continuous provision of essential goods and services.

Europe

Long-term greening trends prevail at higher elevations in representative regions of the European Alps. Significant increases of peak NDVI are widespread over recent decades (Julien et al. 2006; Carlson et al. 2017a, b; Filippa et al. 2019). Accelerated greening of above treeline habitats coincides with a pronounced increase in the amount of snow-free growing degree-days. Remote sensing studies confirm the observed recent colonization of previously glaciated/non-vegetated areas at higher elevations as well as shrub/tree encroachment due to the abandonment of agricultural practices, and highlight the interplay of climate and land use change in controlling greening dynamics in the Alps (Filippa et al. 2019). Reduced human activities also play a major role in recent biomass increases in Scandinavian mountains where regional case studies suggest that climate warming is of subordinate importance (Tømmervik et al. 2019). Growth responses to climate are complex and spatiotemporally unstable (Hofgaard et al. 2019), however, the largely increasing trend in radial growth of trees and in productivity of northern vegetation over recent decades is considered to be climate change-induced (Lopatin et al. 2008; Park et al. 2016). Positive trends in maximum NDVI detected in Arctic mountains are positively correlated with mean summer temperature (Vickers et al. 2016). However, high latitude greening is complex and browning drivers such as extreme winter warming and loss of freeze tolerance, drought stress, thermokarst development, fire, defoliating insects, or rust fungi may temporarily reduce greenness and productivity in different parts of mountain landscapes (Buermann et al. 2014; Phoenix and Bjerke 2016; Tei et al. 2017; Myers-Smith et al. 2020). Detected greening trends in treeline ecotones of the Scandes are attributed to expanding shrub vegetation and densification of previously sparse vegetation cover (Franke et al. 2019). Increased greening was also observed in the Pyrenees as long as productivity of alpine grasslands is not compromised by high stocking rates (Gartzia et al. 2016). Complex interrelationships between climate and land use change determine productivity and biomass in Mediterranean mountains, with the current balance being still towards greening since land abandonment is still buffering the browning drivers (Pausas and Millán 2019; Vicente-Serrano et al. 2020).

The emergence of longer and warmer growing seasons is not only associated with high-elevation plant communities producing more biomass, but also with plants and animals dramatically altering their phenology (Fig. 1.30) (Menzel et al. 2006; Amano et al. 2010; Fu et al. 2014; Garonna et al. 2014). Vitasse et al. (2009) showed for tree species in Pyrenean mountain forests that leaf unfolding is the major driver of extending the growing season with increasing temperature. Spring phenological phases, such as budburst and flowering, occur 20 days earlier at low elevations and 15 days earlier above 1000 m in the Swiss Alps than 50 years before (Defila et al. 2016). Considering the duration of the vegetation period at both elevations, the advance is much more pronounced at high elevation in the Alps (Güsewell et al. 2017). Xie et al. (2017) and Asam et al. (2018) detected correlations with interannual differences in snow cover duration. Climate warming not only affects the timing of phenological events but also the underlying patterns in phenology along environmental gradients. Vitasse et al. (2018) highlighted stronger phenological advance at higher elevations and showed that the elevation-induced shift in the time of leaf-out in four common tree species in the Swiss Alps between low and high elevation has contracted by 35% from the 1960s until

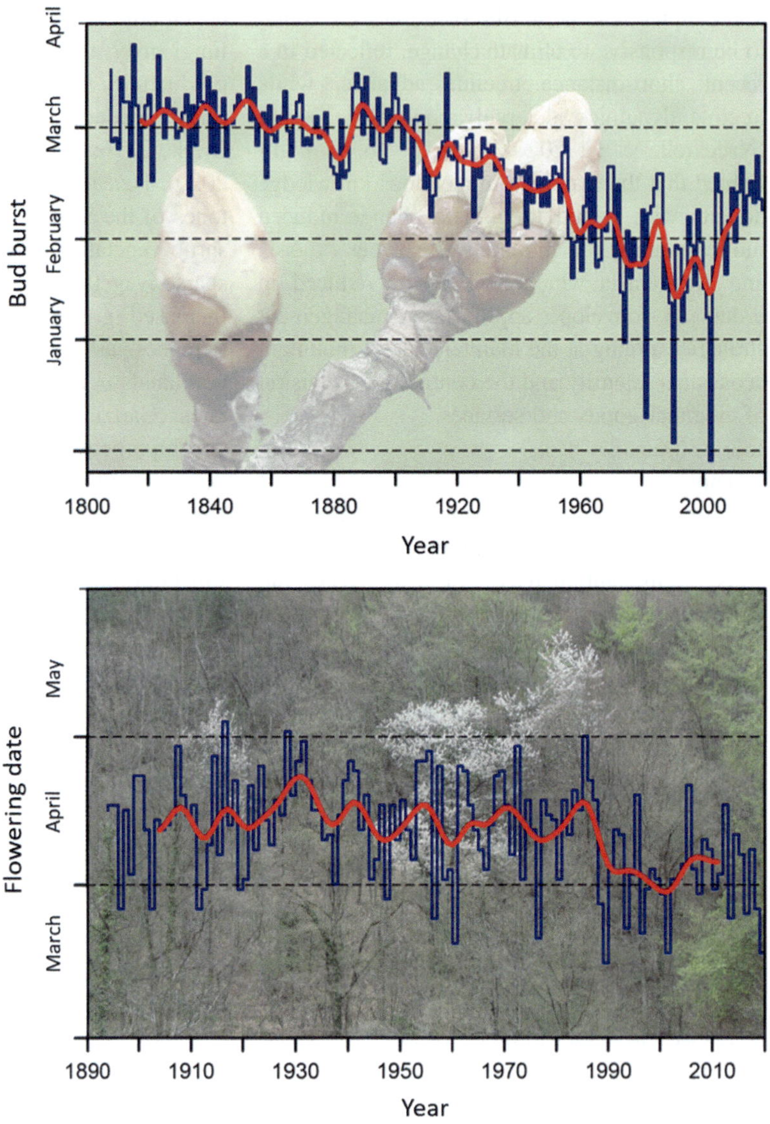

Fig. 1.30 Top: Bud burst of the horse chestnut (*Aesculus hippocastanum*) in Geneva, Switzerland, 1808–2020. Bottom: Flowering of the cherry tree (*Prunus avium*) in Liestal, Switzerland, 1894–2020; the red lines show the respective 20-year weighted average. (Modified from www.meteoschweiz.admin.ch)

today (Fig. 1.31). This increase in the rate of progression of spring leaf-out with elevation is mainly attributed to an increasingly insufficient number of chilling days at low elevations during warmer winters (days with mean temperature of 0–8 °C between November and mean leaf-out date), resulting in less pronounced phenological shifts (see also Asse et al. 2018). Thus, lowland trees are not keeping up with the pace of phenological advance of their conspecifics at higher elevations. The results of Vitasse et al. (2018) are of far-reaching significance in that they suggest that global warming has altered 'Hopkins' bioclimatic law' which specifies the progressive delay in tree leaf-out with increasing latitude, longitude, and elevation. Vandvik et al. (2018) analysed the alteration of this law at other elevation and environmental gradients across Europe and concluded that a change of this law occurs at broader scales, suggesting far-reaching consequences for species, communities, and ecosystems since community composition, trophic interactions, biochemical cycling and the like are affected. A distinct advance in spring phenophases has been observed over much of southern Fennoscandia during recent decades while high mountains areas and northern Fennoscandia showed a delay due to higher winter precipitation and longer snow cover in spring (Pudas et al. 2008; Wielgolaski and Inouye 2013). In the Mediterranean region, warm and dry springs have resulted in advances in flowering, leaf unfolding and fruiting dates, and in lengthening the growing season (Peñuelas et al. 2002; Gordo and Sanz 2010). However, severe drought conditions may reduce the length of the growing season and affect flowering phenology (Spano et al. 2013).

The anticipated shifts in climatic zones in Europe within the next decades (Jylhä et al. 2010) will be associated with further shifts of species distribution ranges and accelerated transformations of montane and alpine vegetation. To date, climate-induced shifts in biodiversity patterns including upward migration of plant species and transformations of plant communities have nowhere been studied in greater detail than in the European Alps. Long-term vegetation monitoring series are available from

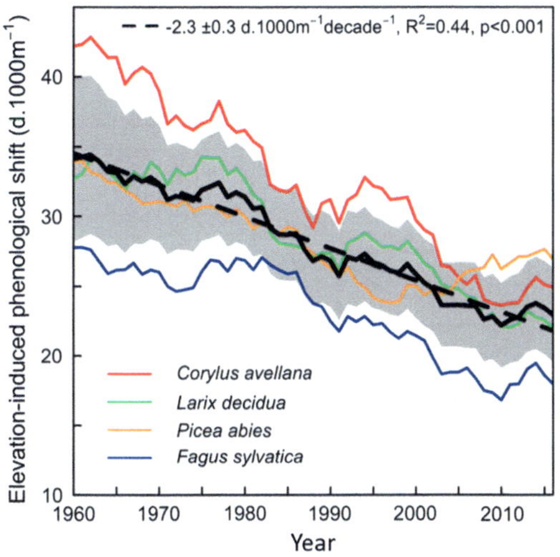

Fig. 1.31 Changes of the elevation-induced phenological shift for four tree species over the period 1960–2016 in Switzerland; eleven-years moving averages are represented (black line) and slope of the linear regression (dashed line) and SD (grey area) across species is also shown. (Modified from Vitasse et al. 2018)

summit areas in the Alps, including detailed surveys dating back to the nineteenth century. First extensive resurveys of summit sites in the alpine-nival ecotone in the 1990s and 2000s provided compelling evidence of increasing vascular plant species richness on most of the summits and a general trend of upward migration in the range of up to more than 100 elevational metres per century, given that appropriate migration corridors are available (Grabherr et al. 1995; Pauli et al. 2001; Bahn and Körner 2003; Grabherr 2003; Walther et al. 2005; Holzinger et al. 2008; Parolo and Rossi 2008; Vittoz et al. 2008a; Stöckli et al. 2011; Wipf et al. 2013). Bergamini et al. (2009) reported a significant upslope migration also for bryophytes (24 m per decade). Based on a large dataset, Frei et al. (2010) confirmed a strong trend towards increasing species richness per summit and found many plant species at an elevation higher than any reported occurrence in the region one century ago. Their results also pointed to a more heterogeneous response at lower range limits, suggesting species-specific differences in response patterns. Increasing species richness of alpine plant communities, albeit without distinct upward shift processes, was reported in a resampling study (1953–2003) by Cannone and Pignatti (2014).

After establishing the GLORIA network in the mountains of Europe, increasingly comprehensive and detailed studies on vegetation dynamics in the alpine-nival ecotone of the Alps have been conducted. It became evident that distinct increases of alpine pioneer species are accompanied by significant declines of subnival-nival plant species, suggesting range contractions at their rear edge (Pauli et al. 2007, 2012). The pattern of expansions of more thermophilic species to higher elevations and concurrent declines of cold-adapted, long-established species of the upper alpine and subnival belt was corroborated in the first pan-European GLORIA resurvey study which substantiated a widespread thermophilization process in alpine vegetation after a period of only seven years (Gottfried et al. 2012). The most compelling evidence of a continent-wide acceleration in the rate of

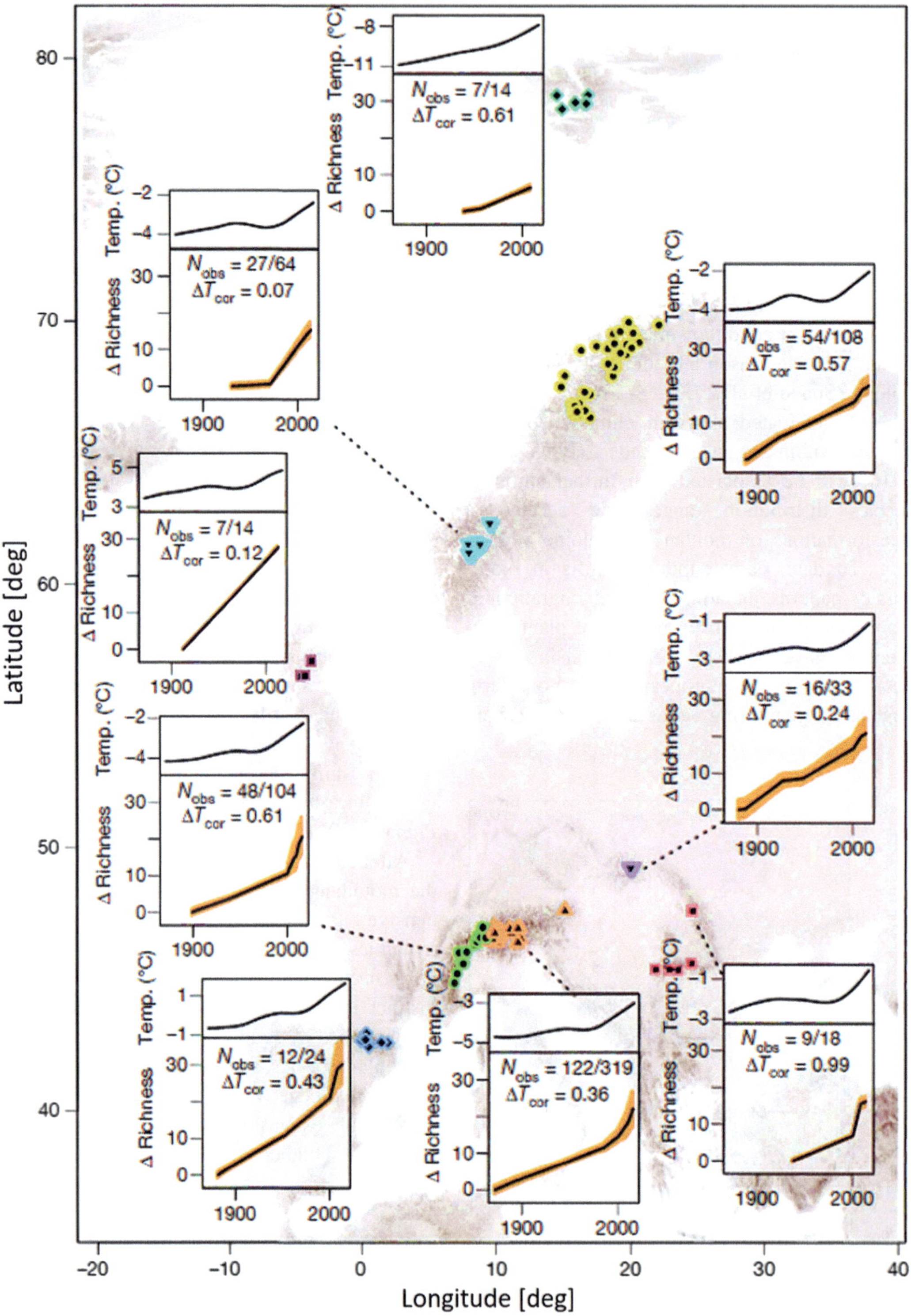

Fig. 1.32 Average species richness change on European mountain summits over time (lower parts of inset panels) compared to mean annual temperature over time (upper part of inset panels); N_{obs}, number of summits/surveys within the mountain region providing data for the panel; correlation between rate of change in species richness and rate of change in temperature (ΔT_{cor}) is positive for all mountain regions. (Modified from Steinbauer et al. 2018)

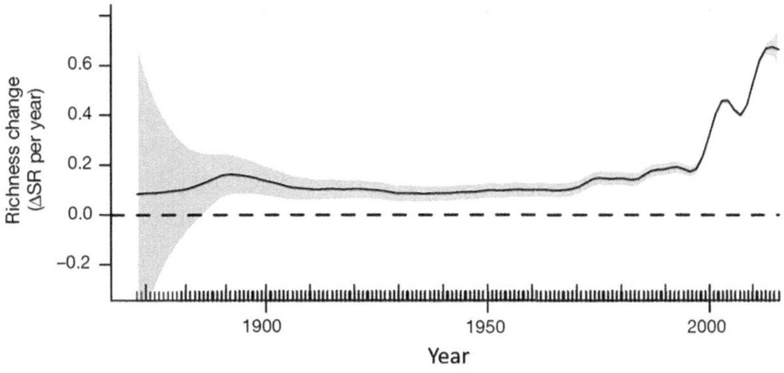

Fig. 1.33 Rate of change in species richness on European mountain summits over time (mean, black line); positive values indicate an increase in species richness on summits and negative values indicate a decrease; rates (ΔSR per year = $(SRt2 - SRt1)/(t2 - t1)$ where SR is species richness and t is time) were averaged across all summits. (Modified from Steinbauer et al. 2018)

increase in plant species richness at high elevations was provided by Steinbauer et al. (2018), who evaluated a dataset of repeated plant surveys from 302 mountain summits across Europe, spanning 145 years of observation. Species enrichment between 2007 and 2016 was five times higher than fifty years ago and found to be strikingly synchronized with accelerated global warming (Figs. 1.32, 1.33).

A recent resurvey of the largest alpine to nival permanent GLORIA plot site in the Alps after two decades showed increasing vascular plant species richness over the entire period while vegetation cover decreased due to the decline of cryophilic species. The increase in richness was reduced in the second decade when disappearance events of more cryophilic species became more numerous, suggesting an accelerating transformation towards more thermophilic and more drought-adapted vegetation (Lamprecht et al. 2018). European GLORIA data also showed larger increases in species richness and higher numbers of newly established species on the warmest slopes of summit zones (east- and south-facing slopes) (Winkler et al. 2016). In the context of increasing maladaptation to warmer habitat conditions and a successive trailing-edge decline of cryophilic species and a leading-edge expansion of more thermophilic species, Steinbauer et al. (2020) highlighted that cryophilic species declines preceded the onset of strong competition pressure from advancing species, suggesting physiological constraints of cold-adapted specialists in adapting to a warmer temperature regime having greater significance than competitive displacement. Another comprehensive resurvey of more than 1500 vegetation plots confirmed that elevational ranges of cold-adapted species tended to contract, while those of thermophilic species which showed a marked increase in species abundance tended to expand (Rumpf et al. 2018). The results of this study suggest that 'losers' of recent range dynamics are overrepresented among high-elevation, cryophilic species with low nutrient demands, and that these species face the risk of displacement by novel, superior competitors that move up faster than they themselves can escape to even higher elevations. The extinction risk of high-elevation plants is alleviated, on the other hand, by topographic complexity and the high diversity of microhabitats, facilitative neighbour interactions, and the longevity of many mountain plants (Scherrer and Körner 2011; Rixen and Wipf 2017; Graae et al. 2018). Nevertheless, long-term warming effects will increase mountain-top extinctions, in particular among endemics, once the accumulated extinction debt will be paid off (Dullinger et al. 2012).

Range shifts, species enrichment on mountain summits, and plant community thermophilization are pan-European phenomena, documented also in Scandinavia (Klanderud and Birks 2003; Kullman 2007a; Odland et al. 2010; Michelsen et al. 2011; Felde et al. 2012), the Carpathians (Czortek et al. 2018; Kobiv 2018), and in some Mediterranean mountain environments (Molero Mesa and Fernández Calzado 2010; Pérez-García et al. 2013; Evangelista et al. 2016; Stanisci et al. 2016; Frate et al. 2018). Grytnes et al. (2014) confirmed the widespread upward shifting of species in their pan-European survey and found elevational shifts in range limits not as clearly related to climatic warming as latitudinal shifts. The thermophilization process on Mediterranean mountain summits is largely characterized by declining species richness, with the loss of high-elevation species, often endemics (Kazakis et al. 2007), outweighing the new appearance of more widespread species. This shift in species composition is attributed to combined effects of increasing temperature and decreasing precipitation in spring and summer (García-Romero et al. 2010; Pauli et al. 2012; Fernández Calzado and Molero Mesa 2013; Jiménez-Alfaro et al. 2014; Giménez-Benavides et al. 2018). Among alpine habitats, snowbeds experience substantial changes and a general homogenization in species composition due to strongly modified snow cover and soil moisture conditions, with the invasion of shrubs and generalists from surrounding grasslands, and increasing species richness and plant cover (Virtanen et al. 2003; Kullman 2007b; Matteodo et al. 2016; Liberati et al. 2019). The ongoing glacier retreat in European mountains extends the surface area of recently deglaciated terrain which is already colonized by first bryophytes and vascular plants after one to three years (Cannone et al. 2008; Burga et al. 2010). Climate warming enhances the establishment of plants on glacier forelands, favouring also other than true pioneer species, and accelerates successional stages. Successful establishment depends in particular on the grain size of the substrate, the associated water capacity, the available gene pool, and on the distance to the seed source (Erschbamer et al. 2008; Erschbamer and Caccianiga 2016; Schumann et al. 2016; Franzén et al. 2019). Compared to vegetation dynamics in glacier forelands 100 years ago, Fickert et al. (2017) assessed an accelerated colonization and more species involved in early colonization. Examples in the Alps (Goldbergkees, Jamtalferner) also show that after 100 years of primary succession roughly 80% of the ground is covered by plants while the number of species (vascular plants) increases to 40–50 per 10 m^2 sample site (Fickert and Grüninger 2018; Fischer et al. 2019).

Enhanced tree growth, intense regeneration and infilling of gaps are common trends in European treeline ecotones (Rolland et al. 1998; Batllori and Gutierrez 2008; Vittoz et al. 2008b; Hofgaard et al. 2009; Holtmeier and Broll 2011; Vitasse et al. 2012; Mathisen et al. 2014; Camarero et al. 2015, 2017; Kaczka et al. 2015; Hedenås et al. 2016; Jochner et al. 2017, 2018; Malfasi and Cannone 2020). Positive climate-growth relationships were also found for shrubs above treeline in most studies, suggesting densification of shrub stands and further expansion (Hallinger et al. 2010; Rundqvist et al. 2011; Francon et al. 2017; Vowles et al. 2017; Weijers et al. 2018). Most alpine treelines have advanced to higher elevations over the past century (Fig. 1.34). Some studies documented substantial treeline shifts, with gains in elevation of 70–100 m or more (Meshinev et al. 2000; Peñuelas and Boada 2003; Cudlin et al. 2017). A recent remote sensing-based study indicated widespread strong treeline advances from the western Pyrenees to the eastern Carpathians over the last 40 years, with eastern European mountains showing the most remarkable changes (Fig. 1.35) (Dinca et al. 2017). In the Swedish Scandes, treeline shifts to even more than 200 m were assessed (Kullman 2007b, 2018, 2019; Kullman and Öberg 2009) as well as upward migration of thermophilic tree species such as *Betula pendula* and *Alnus glutinosa* and of true temperate tree species (*Quercus robur*, *Ulmus glabra*, *Acer platanoides*) into treeline ecotones (Kullman 2008). Many authors refer to correlations of advancing treelines with increases in mean temperatures. The effects of declining land use intensity, however, are certainly often involved, and appear to explain most cases of particularly significant treeline shifts, at least in temperate and southern European mountains (Gehrig-Fasel et al. 2007; Chauchard et al. 2010; Kulakowski et al. 2016; Treml et al. 2016; Cudlin et al. 2017; Kyriazopoulos et al. 2017; Wielgolaski et al. 2017; Wieser et al. 2019). It is evident, for instance, that the cessation of land use has been the most important driver of the large-scale forest expansion at higher elevations in the Alps over the past century (Mietkiewicz et al. 2017). Land use legacies are also considered the major drivers of stand densification processes and treeline advances at anthropogenic Mediterranean treelines (Palombo et al. 2013; Ameztegui et al. 2016; Vitali et al. 2019). Likewise, recruitment patterns in treeline ecotones and treeline advances in northern Europe are not infrequently correlated with impacts of reduced reindeer grazing or other abandoned human disturbances (Bryn 2008; van Bogaert et al. 2011; Aakala et al. 2014; Potthoff 2017).

Biotic responses are pervasive at mid- and lower elevations, though less obvious compared to alpine or treeline environments. General trends include upslope and northward range shifts (Lenoir et al. 2008; Amano et al. 2014), increases of lowland and thermophilic species, and decreases of cold-tolerant species of higher elevations at rear edges of their ranges at lower elevations (Lenoir et al. 2010; De Frenne et al. 2013). Significant upslope shifts over short time periods can be observed in different taxonomic groups as data from the national biodiversity monitoring programme of Switzerland show (Roth et al. 2014). Drought stress and climate-induced disturbances result in vegetation shifts, increasing forest damage and canopy mortality (Martínez-Vilalta and Lloret 2016; Senf et al. 2018). Mountain forests have responded faster over recent decades in terms of shifts in species distribution and plant community composition than lowland forests (Bertrand et al. 2011). A comprehensive analysis in western European temperate and Mediterranean mountains yielded the result of a significant upward shift in species optimum elevation over the twentieth century, averaging 29 m per decade (Lenoir et al. 2008). In Swiss forests, Küchler et al. (2015) detected a strong signal of upslope shift in the understorey

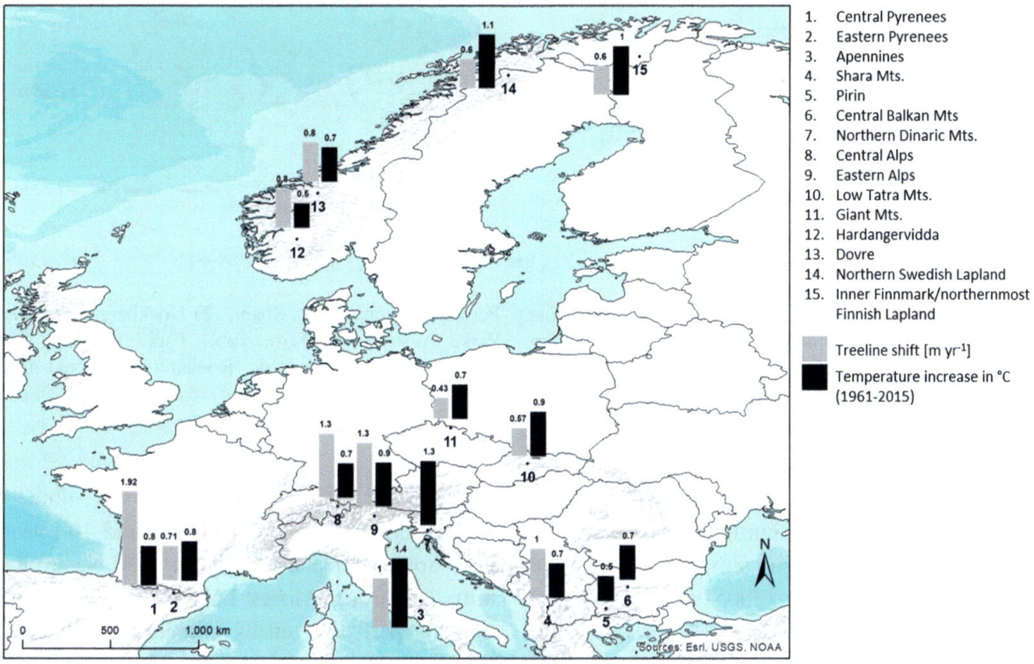

Fig. 1.34 Temperature increase in °C (1961–2015) and treeline shift in m yr^{-1} (different time periods between 1915 and 2015) for selected European mountains (after data in Cudlin et al. 2017)

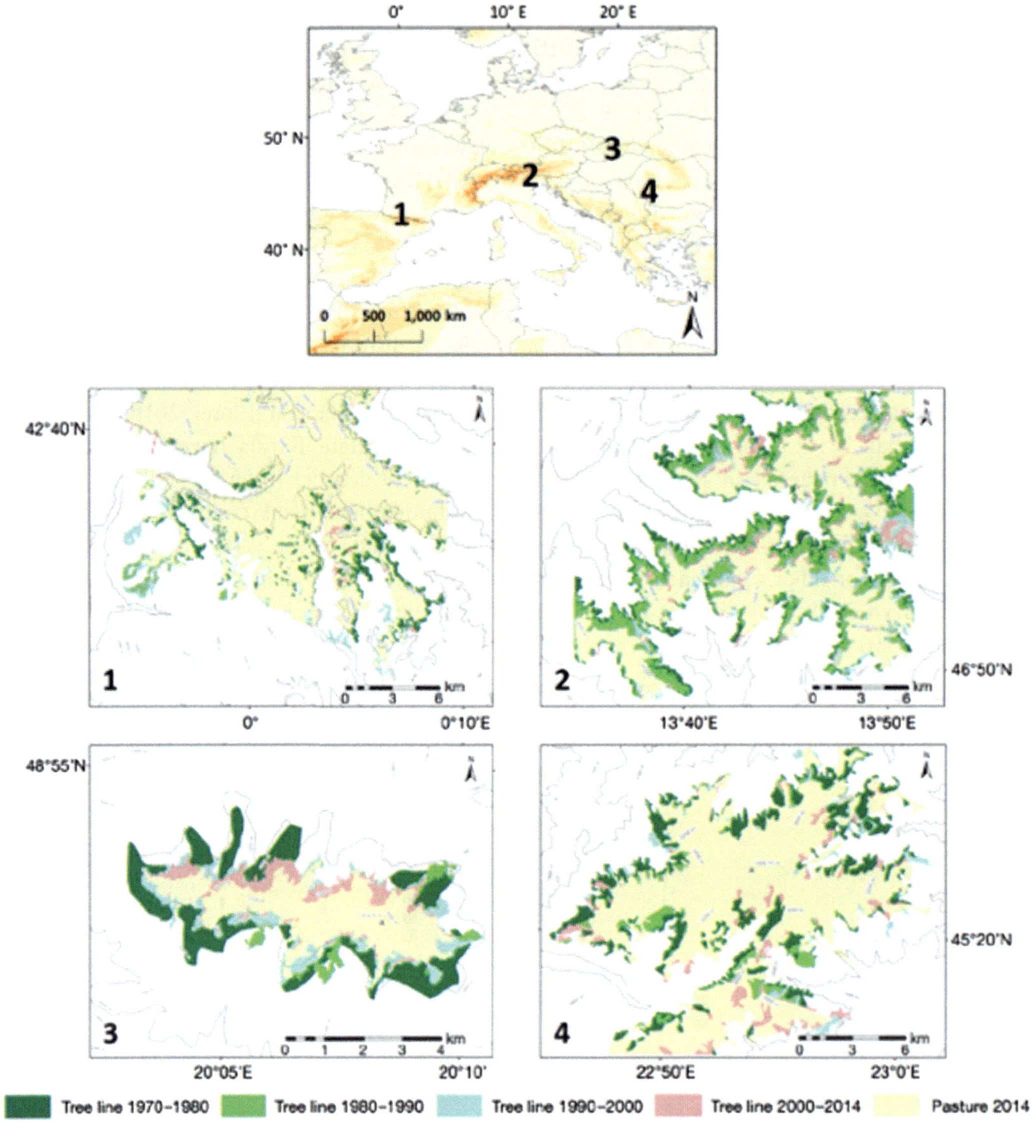

Fig. 1.35 Spatial and temporal dynamics of forest line and cover in 4 European mountain sites obtained from supervised land cover classification of Landsat satellite data between 1970 and 2015; dynamics are represented using a 10 yr time-step interval for (1) Ordesa and Monte Perdido National Park, Spain; (2) Nockberge Biopshere Park, Austria; (3) Low Tatra Park, Slovakia; and (4) Retezat National Park, Romania. (Modified from Dinca et al. 2017)

vegetation of about 10 m per decade since the mid-twentieth century. Significant upslope shifts were observed for single temperature-sensitive species. Dobbertin et al. (2005) resurveyed pine mistletoe (*Viscum album* ssp. *austriacum*) occurrences in pine forests of the European Alps and showed that the current upper limit is roughly 200 m above the limit found 100 years ago. Some evidence suggests that elevational shifts in European forest belts below the treeline are only partly driven by climate warming, and that forest successional changes such as the

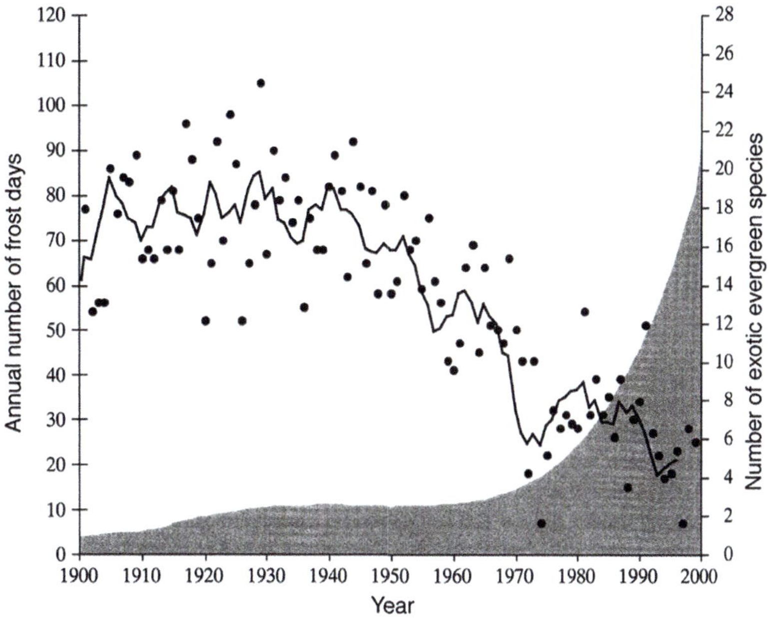

Fig. 1.36 Increase of exotic evergreen species under increasingly milder winter temperatures on the lower south slope of the European Alps; dots: annual number of frost days; solid line: smoothed 5-year averages. (Modified from Walther et al. 2002)

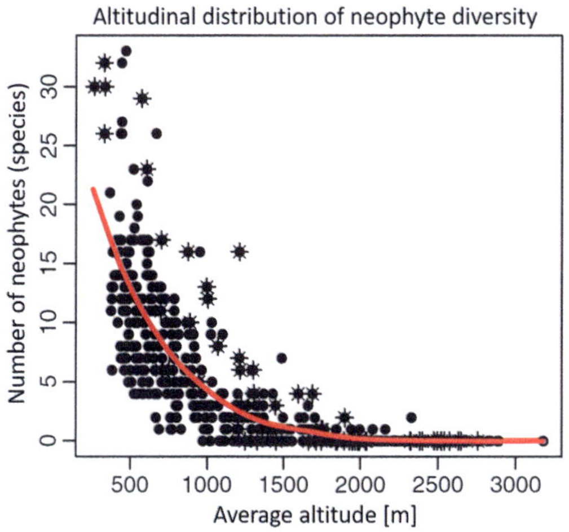

Fig. 1.37 Altitudinal distribution of neophytes in Switzerland based on data from the Swiss Biodiversity Monitoring Programme. (Modified from Nobis 2008)

closure and maturation of forest stands, associated with agricultural land abandonment, play a major role (Bodin et al. 2013).

A well-known example of thermophilization of temperate forests in the southern Alps is the increase in abundance and frequency of indigenous evergreen broadleaved (laurophyllous) species which become increasingly competitive with lengthening of the growing season and decreasing number of frost days (Fig. 1.36) (Walther 2001). Even exotic evergreen species including dwarf palms (*Trachycarpus fortunei*) have succeeded in colonizing these forests, driven by mild winter temperatures and reduced frost occurrence (Walther et al. 2007). Meanwhile, *Trachycarpus fortunei* is regularly recorded and locally established in northern Switzerland and further north (Essl 2019). Shifts in species composition of communities and species richness patterns are increasingly altered by such non-native species invading European mountains. Non-native species affect native species richness and community dissimilarity, resulting in biotic homogenization (Haider et al. 2018). In addition, invasive species affect trophic levels, biotic interaction networks and other ecosystem properties (Gallien et al. 2017). With increasing elevation, however, non-native species decline in probability of occurrence (Fig. 1.37), and their high-elevation range limits do expand, but not rapidly (Becker et al. 2005; Pyšek et al. 2011; Seipel et al. 2016; Siniscalco and Barni 2018). Thermophilic species are prevalent in the alien species pool in the European Alps which has only a small number of mountain specialists (Dainese et al. 2014). In northern European highlands and mountain ranges, an increased risk of non-native plant colonization was assessed, mainly driven by human-mediated dispersal (Wasowicz 2016).

Critical transitions of forest ecosystems in the Alps with potentially severe consequences for ecosystem services may already occur at warming levels of around +2 °C (Elkin et al. 2013; Albrich et al. 2020). Such substantial transitions, for instance, the progressive replacement of cold-temperate ecosystems (*Fagus sylvatica* forests) by Mediterranean ecosystems (*Quercus ilex* forests) from lower elevations during the twentieth century were reported from Mediterranean mountains (Peñuelas and Boada 2003). Rear-edge replacement of Mediterranean fir species (*Abies pinsapo*, *A. cephalonica*) by more drought-resistant pine species (*Pinus halepensis*) also indicate a drought-induced shift in dominance patterns of woodland vegetation (Linares et al. 2009; Sarris et al. 2011). Increasing duration and intensity of drought periods have negative impacts on Mediterranean forests, resulting inter alia in declining tree growth trends, crown condition decline, and increasing tree mortality rates (Carnicer et al. 2011; Linares et al. 2011; Galván et al. 2014).

America

In high latitudes of North America, remote sensing data provide evidence for heterogeneous greenness changes. While the long-term satellite record (1982–2019) in the Arctic indicates greening, the interannual variability in greenness has increased in recent years and browning trends in some regions are increasingly observed (Phoenix and Bjerke 2016; Lara et al. 2018; Frost et al. 2020; Myers-Smith et al. 2020). NDVI analyses in the boreal zone show that areas with productivity decreases have gained predominance in recent decades. While in maritime regions with sufficient precipitation a general greening trend as a response to rapid warming prevails (Ju and Masek 2016), also in alpine treeline ecotones in the Boreal Cordillera ecozone (Bolton et al. 2018), the positive effect of increased temperatures in many dry continental regions is meanwhile offset or even exceeded by the disadvantage of increased evapotranspiration due to temperature rise. The areal fraction exhibiting browning trends in recent years is associated with high winter temperatures and frost drought, fire, or drought limitations (Beck et al. 2011; Beck and Goetz 2012). Tree-ring analyses corroborate drought-induced growth declines in boreal forests of the western Canadian interior (Hogg et al. 2017). While vegetation productivity in high latitude mountain regions still shows a strong dependency on growing season temperature, temperature-induced drought stress has become an important limiting factor in interior mountain regions unless the ongoing warming is accompanied by a significant increase in precipitation (Verbyla and Kurkowski 2019). Dendroecological studies in high-elevation forests and at alpine treelines in Alaska and Yukon point to complex growth responses to continued warming and small-scale differences in climate-growth relationships, with soil moisture often mediating the sensitivity to warm temperatures and affecting productivity (Wilmking et al. 2004; D'Arrigo et al. 2008; Ohse et al. 2012; Wolken et al. 2016; Sherriff et al. 2017; Tei et al. 2017; Dearborn and Danby 2018; Lange et al. 2020). NDVI increases prevail in the Canadian Rocky Mountains (Jiang et al. 2016). However, remote sensing-based studies across the Rocky Mountains and the western US also found water limitation, in particular early summer drought conditions, to impose critical controls on vegetation productivity under continued atmospheric warming (Sloat et al. 2015; Berkelhammer et al. 2017; Berner et al. 2017; Wainwright et al. 2020). In the southwest region of the US, NDVI increases at higher elevations in the southern Rocky Mountains and the Sierra Nevada contrast with drought-induced decreases at lower elevations and in the south of California and the Four Corner States, with recent drought periods accentuating the elevational transition from water-limited to temperature-limited ecosystems (Herrmann et al. 2016; El-Vilaly et al. 2018; Dong et al. 2019). Recent prolonged drought periods facilitated fire severity and extensive tree dieback at low and mid-elevations

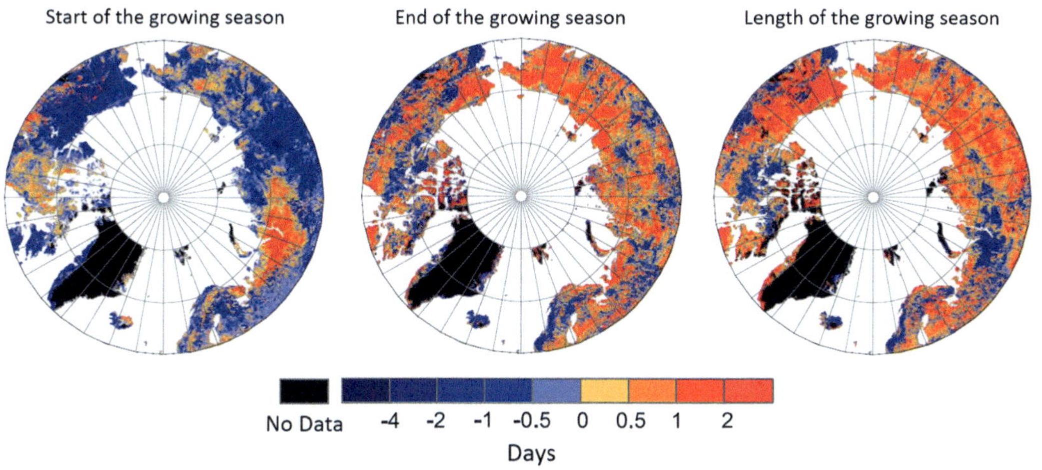

Fig. 1.38 Spatial patterns of the linear trends in SOS (start of the growing season), EOS (end of the growing season) and LOS (length of the growing season) from 2000 to 2010 based on MODIS data; positive values (warm colors) indicate later onset (SOS), later finish (EOS) and longer duration (LOS) of the growing season. (Modified from Zeng et al. 2011)

(Byer and Jin 2017; Potter 2017; Crockett and Westerling 2018).

The thermal potential growing season in temperate and high northern latitudes has lengthened over recent decades (Barichivich et al. 2013; Melaas et al. 2018). This trend is increasing and regionally accelerating. According to MODIS data, the growing season length in the North American Arctic increased by about 14 days between 2000 and 2010, with a significantly earlier start of the growing season of c. 11.5 days (Fig. 1.38) (Zeng et al. 2011). Species-level phenological shifts result in a substantial reshaping of various temporal components of entire plant communities, affecting patterns of temporal overlap among (mutualistic) species and interactions within trophic levels and beyond (phenological mismatch). Notwithstanding the recognition that photoperiod constrains spring plant phenology in alpine regions and the extent to which the growing season can lengthen is limited (Ernakovich et al. 2014), considerable phenological shifts have been assessed at higher elevations. Using a unique long-term record of flowering phenology from the Colorado Rocky Mountains, CaraDonna et al. (2014) showed that the diversity of species-level phenological shifts contributed to altered coflowering patterns within meadow communities, a redistribution of floral abundance across the season, and an expansion of the flowering season by more than one month between 1974 and 2012 (Fig. 1.39). Large shifts in the phenology of rare Colorado Rocky Mountain plants were found by Munson and Sher (2015), who assessed an acceleration of flowering dates by more than 40 days since the late 1800s. With regard to plants of alpine habitats, they found high spring temperatures explaining the accelerated phenology. Correspondingly, flowering initiation in alpine plants of the Colorado Front Range was observed to occur earlier with earlier snowmelt (Inouye and Wielgolaski 2013; Winkler et al. 2018), potentially generating a mismatch in the seasonal timing of interacting organisms, e.g. plants and pollinators (Forrest and Thomson 2011). In the Catalina Mountains of south-central Arizona, precipitation appears to play a much larger role for flowering patterns in spring and summer than further north (Crimmins et al. 2013). Shifts of morphological and physiological phenophases of trees in drier habitats seem to be less pronounced (Hallman and Arnott 2015), despite a considerable lengthening of the growing season (Tang et al. 2015). Climate warming-induced advance in the timing of spring onset is consistent across

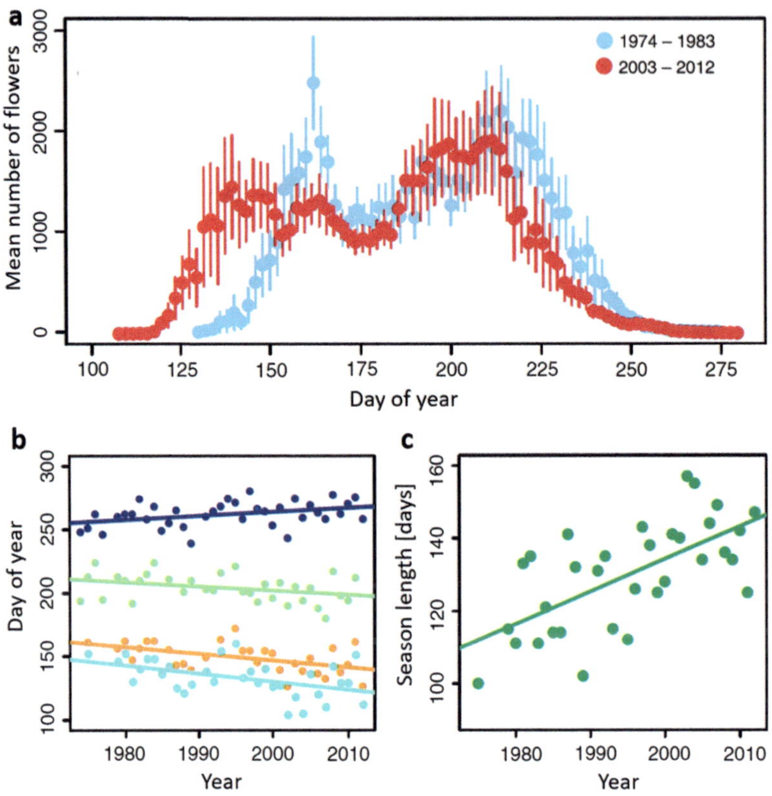

Fig. 1.39 Aggregate community-level shifts in flowering phenology; **a** comparison of the season-wide flowering curves for the first and last 10 years of the dataset; each dot is the 10-y mean number of flowers; **b** phenological shifts through time for first flowering of the community (cyan), last flowering for the community (dark blue), and timing of community-level spring peak (orange) and summer peak (green); each dot represents a community-level phenological measure in 1 y; **c** change in the length of the flowering season; each dot represents the total number of days on which open flowers were present in each year. (Modified from CaraDonna et al. 2014)

the mountain regions of the western and northeastern US (Ault et al. 2011; Schwartz et al. 2013).

As presented for Europe, upward migration of plant species and transformation of montane to alpine plant communities is pervasive across North American mountain ranges as long as the expansion of distribution ranges is not constrained by a decreased water balance and drought stress or other non-thermal drivers (cf. Rapacciuolo et al. 2014). Elmendorf et al. (2015) analysed changes in plant community composition from repeat sampling and experimental warming studies in varied arctic and alpine habitats and found a general increase in the relative abundance of species with a warmer thermal niche. Over vast areas of arctic mountain ranges, climate warming-induced significant changes in plant community composition have occurred (Danby et al. 2011), in accordance with a strong trend towards subarctic forest-tundra ecotone advance which, however, is rarely capable to keep pace with climate change within the twenty-first century (Rees et al. 2020). The velocity of latitudinal tree migration which is predominantly northward is also lower than the velocity of climate warming in temperate and boreal forests in eastern Canada and the eastern US, suggesting a constrained capacity to track climate warming (Boisvert-Marsh et al. 2014; Fei et al. 2017; Sittaro et al. 2017).

Upward range expansion of species, induced or facilitated by climate warming, appears to be a common change pattern across the Rocky

Mountains (Landhäusser et al. 2010; Sproull et al. 2015), while climate change effects on the abundance and distribution of tree species are mediated in particular by ecological disturbances such as wildfires and insect outbreaks (Keane et al. 2018; see also Littell et al. 2013 for the Cascade and Coast ranges). A thermophilization of montane to alpine plant communities is reflected in the results of a resurvey in the Colorado Rocky Mountains (2600 to 4100 m) after 65 years: Zorio et al. (2016) detected significant changes in species composition and dominance, with an upward shift of species' mean elevation of 41 m. Many species from lower elevations, in particular graminoids and shrubs, expanded their ranges into new communities. A study on shrub encroachment into alpine tundra in the Colorado Front Range showed a shrub cover (*Salix planifolia*, *Salix glauca*) expansion by 441% over 62 years (1946–2008) on a 18 ha study site (Formica et al. 2014). Here, data from other long-term monitoring plots (20 + years) showed increasing species and functional diversity (Spasojevic et al. 2013). Most resurvey studies in North American mountain ranges reveal thermophilization processes of plant communities. Examples include shifts in herb community composition in the Klamath-Siskiyou Mountains (California/Oregon) over more than 50 years (Damschen et al. 2010), expansion of subalpine species into alpine plant communities in California's White Mountains over 49 years (Kopp and Cleland 2014), and shifts in plant distributions with elevation in southern California's Santa Rosa Mountains over 30 years (Kelly and Goulden 2008). The average elevation of dominant plant species was found to have shifted upslope by c. 65 m as a consequence of changes in the regional climate in the latter study. Increased dominance of evergreen oaks in foothill woodland and montane hardwood forest of the Sierra Nevada also suggests thermophilization under warmer and drier conditions (Dolanc et al. 2014). Changes in the geographic distributions of species in the US Southwest mountain ranges, strongly associated with observed changes in climate, were highlighted in general by Fleishman et al. (2013). Range shifts are documented for diverse groups of animals as well (Moritz et al. 2008; Forister et al. 2010), including pathogens, thus increasing the risk of forest infestations at higher elevations (Bentz et al. 2010).

Corresponding to recent results from the European Alps, Lesica (2014) found plant species restricted to highest elevations in the Montana Rocky Mountains to decline in abundance, while lower-elevations species expand their range upslope with climatic warming. In accordance with these declines, long-term monitoring (1988–2014) of arctic-alpine and boreal plant species at their southern range limit in the Rocky Mountains revealed overall declining population trends (Lesica and Crone 2017). In the Santa Catalina Mountains of southern Arizona, montane plant species showed significant upward shifts of lower elevation range boundaries and elevational range contractions over the past five decades, attributed to the conditions of decreasing precipitation and increasing temperatures (Brusca et al. 2013). Warming-mediated drought stress is also driving upslope retreat of *Pinus ponderosa* in the Sierra Nevada, where low-elevation ponderosa pine forests have been replaced by montane hardwood forests and annual grasslands (Field et al. 2016). Range shifts in montane forests were reported as well from eastern US and Canadian mountain ranges. Beckage et al. (2008) found a rapid upward movement of the northern hardwood-boreal forest ecotone in the Green Mountains (Vermont) from 1964 to 2004, while Savage and Vellend (2015) detected significantly increasing mean elevations of species distributions (9 m/decade) on Mont Mégantic (southern Québec) between 1970 and 2012 (Fig. 1.40), associated with biotic homogenization at higher elevations.

Upslope elevational range shifts have also been assessed for tree species at alpine treelines. Accordingly, observational studies in many mountain ranges detected a treeline advance. Climatic treelines which still show persistence are expected to shift to higher elevations in the mid- or long term, unless non-thermal site factors do not prevent advances. In particular, limitations to seedling recruitment with warming can

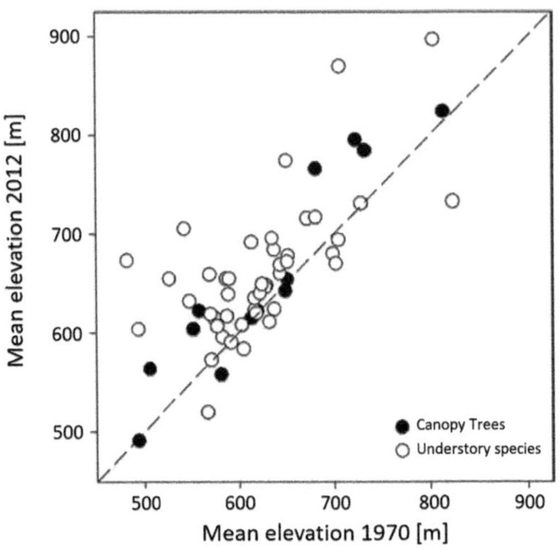

Fig. 1.40 Mean abundance-weighted elevation of species distributions in 1970 and 2012; each point represents a single species (n = 45 understorey species, n = 13 canopy trees); distributions of species above the diagonal 1:1 line increased in mean elevation, and vice versa. (Modified from Savage and Vellend 2015)

constrain the pace of tree range shifts at treelines (Conlisk et al. 2017; Elliott 2017; Kueppers et al. 2017). Rapid upward advance of woody vegetation over the past 60 years (Dial et al. 2007, 2016; Terskaia et al. 2020), and significant increases in treeline elevation and stand density over the past 100-plus years were detected at several boreal-subarctic alpine treelines in Alaska and Yukon (Lloyd and Fastie 2003; Danby and Hik 2007; Stueve et al. 2011). Other alpine treelines at higher latitudes indicate moderate upslope shifts (de Lafontaine and Payette 2012; Trant and Hermanutz 2014), or show ongoing treeline dynamics, for instance by stand infilling, but more or less stagnating elevational positions (Mamet and Kershaw 2012). A recent study, covering nine alpine treeline ecotones in the Canadian Rocky Mountains, revealed a widespread increase in radial growth, establishment frequency, and stand density since the mid-twentieth century, and a concurrent treeline advance at all sites, averaging 40–50 m (Davis et al. 2020). Empirical evidence of increases in tree density and treeline advance since 1950 across a latitudinal gradient of 1100 km in the Rocky Mountains was provided by Elliott and Kipfmueller (2011), Elliott (2012), and Elliott and Petruccelli (2018), with treeline advance ranging between 39 and 140 m. As elsewhere, however, treeline dynamics in the Rocky Mountains is complex, with site- and species-specific responses modifying the general trend of treeline advance (Malanson et al. 2007, 2009; Holtmeier 2009; Elliott 2011; Holtmeier and Broll 2010, 2012, 2017b; Davis and Gedalof 2018; Elliott et al. 2020). Across five mountain ranges of the Great Basin, Smithers et al. (2018) found a mean vertical treeline elevation shift of c. 20 m since 1950, associated with upslope expanding ranges of *Pinus longaeva* and *Pinus flexilis*, whose recruitment and radial growth is controlled by water limitations that complexly interact with temperature (Millar et al. 2015). Millar et al. (2004) documented expansion of subalpine conifers in the central Sierra Nevada, reflected in snowfield and subalpine meadow invasion, branch elongation, and vertical branch release. Here, a resampling-based study revealed a densification of high-elevation forests over the past 75 years with widespread, multiple-species increases in density of young trees, with interactions between water balance and disturbance factors playing a crucial role in future shifts in vegetation composition and structure (Dolanc et al. 2013).

As a result of the colonization from Europe, non-native plant species richness is highest in New World regions, with the US having the highest number of recorded invasive alien species globally (Seipel et al. 2012; Turbelin et al. 2017). In North American mountain ranges, as elsewhere, the abundance of non-native plant species declines with increasing elevation, while their invasibility is facilitated by climate warming. Relative to lowland ecosystems, alpine environments host few non-native plants (Alexander et al. 2016; Malanson 2020). The density of non-native plant species is related to the density of native plant species (Stohlgren et al. 2005), suggesting an increased invasion risk in national parks and other protected areas with high native species richness and high percentage of threatened and endangered plants (Allen et al. 2009). Increasing rates of exotic species introductions are expected in the boreal zone as a result of human activities and climate change (Sanderson et al. 2012). In the Rocky Mountains, dominant exotic species comprise intentionally introduced Eurasian grasses (e.g. *Phleum pratense*, *Poa pratensis*, *Bromus tectorum*, *Bromus inermis*) and herbs (e.g. *Melilotus*, *Medicago*, *Trifolium*, *Verbascum*, *Taraxacum* spp.) which particularly occur along roadways and invade disturbed sites primarily in montane steppes and open forests (Weaver et al. 2001; Pollnac et al. 2012). In the southern Sierra Nevada, non-native species have their main range of elevational occurrence between 1500 and 2000 m, only a few alien species have been ecologically successful invaders in subalpine/alpine ecosystems (Rundel and Keeley 2016). Invasive grasses such as *Bromus tectorum* occur in subalpine forests (Brooks et al. 2016), but mainly invade lower elevations, in particular grazing- and fire-affected sites, causing significant changes in ecosystem composition, structure, and function (Blumler 2011; Grüninger 2015; Millar and Rundel 2016). The distribution of the most common exotic invasive species in California, *Centaurea solstitialis*, is mainly confined to elevations below 1200 m (Pitcairn et al. 2006). Non-native species are a prominent vegetation component on the tropical island of Hawai'i where these species are in an upward niche expansion phase. Exotic species showed a significant upward shift in both their upper and lower elevation limits, by 126.4 and 81.6 m, respectively, between 1970 and 2010 while native species shifted significantly upward in their lower elevation limit only (by 94.1 m), resulting in a drought stress-related range contraction (Koide et al. 2017).

The number of studies on biotic responses to climate change in Central and South American mountain ranges is still comparatively limited. In the Trans-Mexican Volcanic Belt, considerable upward shifts in species distribution ranges are projected, suggesting a high vulnerability of species due to limited habitat space available at higher elevations (Villers-Ruiz and Castañeda-Aguado 2013). Current geographic distributions of temperate/montane pines and oaks in Mexico will most likely decrease significantly (Gómez-Mendoza and Arriaga 2007). Climate change is also threatening montane cloud forests in Mexico. Ponce-Reyes et al. (2012) showed that climatically suitable areas will get lost for more than 90% of protected cloud forests, and that almost three quarters of the entire cloud forests could vanish by 2080. Concurrently, the respective area of suitable habitat for cloud forest species, e.g. small mammals, will be substantially reduced (Lorenzo et al. 2019). Analysing tree species composition in annually censused plots along an altitudinal gradient (70–2800 m) in Costa Rica, Feeley et al. (2013) observed directional compositional shifts, with increased relative abundance of lowland species in 90% of plots caused by disproportionate mortality of highland species. The results point to the significance of successful migrations in order to persist under future warming.

Spatio-temporal patterns of vegetation productivity and phenology along the Andes are highly heterogeneous, affected to a large extent by the moistening trend in the inner tropics and the drying trend in the subtropical Andes, by the precipitation and temperature anomaly patterns associated with ENSO, and by the steep W-E precipitation gradient in the southernmost Andes. South of 9° S, NDVI-based monitoring (1981–2011) alongside the Andes showed positive trends in productivity for temperate forests in Chile and subhumid/humid areas in Peru, Bolivia and Brazil, while arid/semiarid and subhumid vegetation types across Argentina, northern Chile and SE Bolivia showed negative trends (van Leeuwen et al. 2013). A reversal from greening to browning trends around the mid-1990s was assessed by Krishnaswamy et al. (2014). A longer growing season was indicated in southern Chile and southern Argentina. Bianchi et al. (2020) confirmed positive NDVI-temperature relationships over temperate forests in western N Patagonia, while these relationships are weaker east of the Andes and biome-specific. A NDVI analysis in Patagonia covering the period 2001–2016 revealed a greening trend over the western zone, and a drying trend over the eastern zone (Olivares-Contreras et al. 2019).

Tree-ring growth of *Nothofagus pumilio* in northern Patagonia is positively related to growing season temperature and negatively to precipitation at mesic and humid treelines, while at xeric treelines the opposite is observed (Lavergne et al. 2015). A study on the productivity dynamics of high Central Andean peatlands in the semiarid Chilean Altiplano over the past three decades (1986–2017) found more or less stable peatland productivity and a recent regional greening trend over the last seven years (Chávez et al. 2019). In the semiarid region of Chile, Glade et al. (2016) detected negative trends of vegetation productivity below 2000 m and positive trends for higher elevations, associated with an earlier start of the growing period in mountainous ecosystems. On the other hand, high-elevation East Andean ecosystems (>4400 m) in N Argentina and S Bolivia showed decreasing plant productivity over recent decades (radial growth of *Polylepis tarapacana*), attributed to increased aridity (Carilla et al. 2013).

Upward range expansions of species in the Andes under climate warming are predicted (Anderson et al. 2011; Larsen et al. 2011; Ramirez-Villegas et al. 2014), however, only a few observational studies documenting range shifts are available. Nevertheless, the results show more or less consistent patterns of upward species migrations and thermophilization effects throughout elevational gradients, even though wetter biomes and dry biomes may show heterogeneous responses to climate change (Tovar et al. 2013a; Cuesta et al. 2019). In the tropical Andes, Morueta-Holme et al. (2015) revisited the Chimborazo volcano in Ecuador 210 years after an expedition by Alexander von Humboldt and found the limit of plant growth having been strongly pushed upslope (Fig. 1.41). Here, distinct upward shifts in the distribution of vegetation zones are associated with increases in maximum elevation limits of individual plant taxa of >500 m on average. Duque et al. (2015) detected thermophilization effects in N Andean montane forests and adjacent lowlands in NW Colombia, reflected in directionally changing tree communities through time to include relatively more thermophilic species, with compositional shifts occurring primarily via range retractions (high tree mortality at lower elevations). Repeated censuses of forest inventory plots spanning an elevational gradient from 950 to 3400 m in SE Peru showed that most tropical Andean tree genera shifted their mean distributions upslope over the study period (2003/04–2007/08), while the observed mean rate of change was less than predicted from the temperature increases for the region, suggesting a limited ability to respond to increased temperatures and an increased extinction risks with further climate change (Feeley et al. 2011). Widespread thermophilization patterns in Andean forests were confirmed in a recent study based on almost 200 forest plots between 360 and 3360 m spread throughout the tropical and subtropical Andes (Fadrique et al. 2018). The results showed directional shifts in species composition towards having greater relative abundances of species from lower, warmer elevations, while the rates of thermophilization were heterogeneous throughout the elevation gradient, with negative or non-significant rates at highest (treeline) and mid-elevations (cloud base at the transition from montane to cloud forests). A repeated resurvey of permanent plots on four high Andean summits (4040–4740 m) in NW Argentina revealed high rates of plant community turnover and generally decreasing, but temporally fluctuating trends of plant cover, species richness, and diversity, related to the ENSO-influenced short-term temperature and precipitation variability (Carilla et al. 2018). Analysing chronosequences (38 years) in recently deglaciated terrain at high elevations (4700–4900 m) in the Central Andes, Zimmer et al. (2018) observed an overall increase in species richness, abundance, and plant cover and showed that colonization lags behind the velocity of warming and associated glacier retreat, and leads to no-analogue plant communities. As elsewhere, upslope range shifts have also been assessed for diverse groups of animals in the Andes (e.g. Moret et al. 2016; Seimon et al. 2017).

Climate warming-induced treeline dynamics is primarily reflected in tree growth (Lavergne et al. 2015) and increased recruitment above

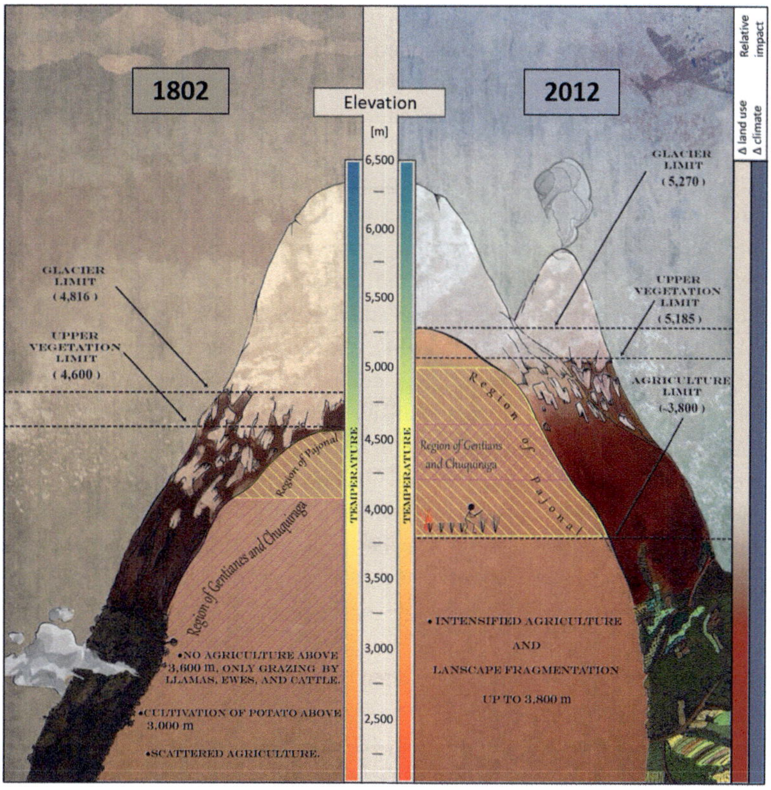

Fig. 1.41 An update of Humboldt's classic study of 1802, showing major changes in overall vegetation limit, average glacier limit, and shifts in topmost vegetation regions on Chimborazo from 1802 to 2012; the major drivers of change, climate, and land use change are represented by the bars to the right: a constant impact of climate change—in particular, increased temperature—the stronger relative impact of land use at the lower sites, mainly through intensified agriculture, and the effect of grass harvesting and local burning. (Modified from Morueta-Holme et al. 2015)

treeline in some places, but not (yet) in distinct treeline shifts. Based on a 42-year span of aerial photographs and high resolution satellite imagery in the high Peruvian Andes, Lutz et al. (2013) found only minor treeline shifts, with migration rates in protected areas being only 2.3% of the rates needed to stay in equilibrium with projected climate by 2100. In the semiarid Peruvian Andes and also in the case of cloud forests in the tropical Andes, initially stationary treelines suggest that other factors (topographic controls, high temperature variation, extreme cold events, water stress, high levels of solar radiation, low seed dispersal, competition with grasses, human impact) override the influence of increasing mean temperatures and may prevent cloud forest tree species from shifting their leading range edges upslope in response to climate warming (Bader et al. 2007; Rehm and Feeley 2015, 2016; Toivonen et al. 2018). Nevertheless, the results of Kintz et al. (2006) and Young et al. (2017) provide landscape-scale evidence of woody plant encroachment, upward treeline shifts, increasing shrubland areas, and increases in the number, size, and connectivity of forest patches at anthropogenic treelines in the Peruvian Andes. At *Nothofagus pumilio* treelines in Patagonia, Fajardo and McIntire (2012) found treelines moving uphill in abrupt pulses until at least 40–70 years ago, but declining tree growth in recent decades. The complexity of treeline dynamics in northern Patagonia was already highlighted by Daniels and Veblen (2004), who stressed the importance of moisture availability for seedling establishment of *Nothofagus pumilio*, and the small-scale differing und unstable relationships of radial growth and seedling demography with climate and ENSO over the late twentieth century (see also Srur et al. 2016). In southern Patagonia, Aravena et al. (2002) found positive correlations between *Nothofagus pumilio* tree growth and temperature at treelines, but a strong influence of local site factors. Srur et al. (2018) corroborated the sensitivity of abrupt *Nothofagus pumilio* treelines to changes in climate variations in the southern Patagonian Andes and found the rate of seedling establishment to be strongly modulated by the interaction between temperature increase and variations in precipitation.

As elsewhere, few non-native plant species have established in higher elevation habitats of the Andes. Alien species are largely restricted to disturbed sites, yet even protected mountain areas have been invaded (Speziale and Ezcurra 2011; Barros and Pickering 2014). Potential impacts of introduced species, e.g. competition for pollination, vary with their density (Muñoz and Cavieres 2008). Currently, the invasive nature of the common gorse (*Ulex europaeus*) causes serious problems in Colombian high Andean forests and paramos. The dense, compact, and homogeneous colonies of this invasive species impoverish or even eliminate native plant communities (Osorio-Castiblanco et al. 2020).

Africa

The increased warming trend across the African continent implies substantial impacts on ecosystems and has triggered similar biotic responses in mountains and highlands as reviewed above for other continents. Remote sensing studies in the Atlas Mountains suggest slightly positive land productivity trends and increases in montane forest cover and density (Del Barrio et al. 2016; Barakat et al. 2018), however, productivity and phenology are strongly controlled by precipitation variability (Otto et al. 2016; Missaoui et al. 2020), and effects of land use changes are pervasive (Mohajane et al. 2018). Positive correlations of radial growth of main tree species and interannual NDVI values in the Ethiopian Highlands suggest that precipitation variability controls landscape-level patterns of vegetation productivity (Siyum et al. 2018). However, increased pressure of human activities often overrides the effects of climatic variables. In the NW Ethiopian Highlands, for instance, monitoring of long-term NDVI changes (2000–2014) revealed a decline in vegetation productivity despite a significant positive trend of annual precipitation (Zewdie et al. 2017). The pattern of positive correlations between rainfall and NDVI and negative correlations between temperature and NDVI is widespread, while the start of the growing season in the highland ecoregions has advanced and the length has extended over recent decades (Workie and Debella 2018; Liou and Mulualem 2019). Significant NDVI declines in dry highland ecoregions suggest an increased risk of land degradation, to be attributed to interacting climate change and land use effects (Gebru et al. 2020). Patterns of vegetation productivity decline are reported for large tracts of land in eastern Africa (Landmann and Dubovyk 2014; Kalisa et al. 2019), largely explained by temperature-induced moisture stress (Krishnaswamy et al. 2014). This does not apply for most of the upper mountain regions of Mt. Kilimanjaro which have undergone a long-term (1982–2011) increase in vegetative signal ('greening up'), to be mainly attributed to vegetation recovery after disastrous fires during the outgoing twentieth century, while the seasonal vegetation activity strongly responds to ENSO and IOD (Indian Ocean Dipole) teleconnections (Torbick et al. 2009; Detsch et al. 2016). Positive trends of recent NDVI values (2002–2017) were also assessed in the Drakensberg Mountains of South Africa (Mukwada and Manatsa 2018).

Very few observational studies on warming-induced changes in plant species distribution patterns and range shifts are available for African mountains and highlands. Modelling studies in the Atlas Mountains suggest that forest species such as *Cedrus atlantica* and *Quercus suber* will disappear from many localities and shift their distribution ranges, which become more contracted and fragmented, to higher elevations (Vessella et al. 2017; Bouahmed et al. 2019). In Algerian mountain forests, fire is considered the most important driver of forest degradation, with fire occurrence being linked to increasing aridity (Djema and Messaoudene 2009). In tropical African highlands, range shifts are mainly driven by anthropogenic pressure and fire as well (Wesche et al. 2000; Wesche 2002), and it is just as difficult to disentangle the role of climate change from the impacts of other drivers. Jacob et al. (2015a) pointed out for treeline environments in tropical African mountain ranges that treeline dynamics cannot be used as a proxy of climate change since treelines are strongly disturbed and have lowered due to high human and livestock pressure. In case studies in the northern Ethiopian highlands and in the Simien Mountains, Jacob et al. (2015b, 2017) provided evidence that treelines tend to shift upslope once anthropogenic pressure is decreasing, suggesting that the strong impact of land use outweighs climate change effects. Notwithstanding, a shift of 150 m of an almost inaccessible *Erica arborea* treeline in the Simien Mountains between 1905 and 2004 indicates involvement of rising temperatures (Jacob et al. 2017). Predicting advances of tropical treelines is, however, a difficult task given the multi-faceted constraints on tree regeneration above the uppermost forest stands (Wesche et al. 2008a).

Nevertheless, climate change and the interaction between climate drivers and land use change have additional effects, causing far-reaching alterations in Africa's mountain ecosystems (Niang et al. 2014). Future suitable habitats of *Juniperus procera*, the endangered and most preferred tree in the northern Ethiopian Highlands, are predicted to shrink by 80–90% (RCP 2.6 and 8.5) by the mid-century (Abrha et al. 2018). Growth patterns of *Juniperus procera* are strongly related to the amount of precipitation, suggesting high sensitivity to future drought periods (Couralet et al. 2007). Studies on *Erica arborea* tree-rings in North Ethiopia showed that tree growth is significantly and positively correlated with minimum temperature in the growing season, but negatively with minimum temperatures in the rainy season in spring (Jacob et al. 2020). In the southern highlands, upward range shifts will most likely create strong potential risks in terms of lowland attrition and range-shift gaps and lead to decreasing population sizes and a higher extinction risk (Kreyling et al. 2010; Kidane et al. 2019). Mekasha et al. (2013) showed that projected warming could significantly affect grassland herbaceous plant communities and that successful migrations of species are essential to mitigate range contraction and habitat losses with range-shift gaps. This also applies to diverse groups of animal species in African highlands (e.g. Raxworthy et al. 2008). Specialized high-alpine giant rosette plants are likely to face very high risk of extinction following climate warming (Chala et al. 2016).

Recurrent fires with climate change-induced higher frequency and intensity have resulted in substantial shrinkage of upper montane forests on Mt. Kilimanjaro, downward shift of the treeline, and in a biotic homogenization between the subalpine and alpine belts (Hemp 2005a, 2009). Increasing isolation of East African mountain ecosystems due to anthropogenic impact increases the threats to diversity and endemism under climate change (Hemp and Hemp 2018). Patterns in plant–pollinator specialization along elevational gradients on Mt. Kilimanjaro suggest that rising temperatures may destabilize pollination networks (Classen et al. 2020). Changes in East African highland ecosystems also include upslope range shifts of malaria vector species. Warmer temperatures at higher elevations facilitate range expansions and the creation of suitable vector habitats in the highlands (Ermert et al. 2012; Kulkarni et al. 2016). Regarding South Africa and Lesotho's mountainous regions of high biodiversity, substantial contractions in species' ranges towards higher elevations are predicted, decreasing the potential regions of occurrence of montane species (Bentley et al. 2019).

Invasive alien species in African highlands sometimes generate conflicts of interest between local communities and governments. On the one hand, they may provide benefits to local people as in the case of Mimosa (*Acacia dealbata*) in the Highlands of Madagascar or Mesquito (*Prosopis juliflora*) in East Africa (Kull et al. 2007; Mwangi and Swallow 2008). On the other, they adversely affect biodiversity and ecosystem services and their control incurs enormous costs across Africa each year (Boy and Witt 2013). At higher elevations, non-native plant species decrease in number and are largely confined to anthropogenic vegetation along roadsides or climbing routes, as exemplified by *Poa annua* on Mt. Kilimanjaro (Hemp 2008).

1.3 Effects of Land Use Changes in Major Mountain Systems of the World

1.3.1 General Overview

Humans have influenced and reshaped much of the world's mountain environments for millennia. In particular, highlands in Africa, Asia and Europe have been subjected to long-lasting land use and anthropogenic landscape transformation (Walsh and Giguet-Covex 2020). For instance, the onset of pastoralism in the Tibetan highlands dates back at least to 8000–9000 years BP (Miehe et al. 2009a, b, 2014, 2019). In many Old World mountain systems, the foundation of permanent settlements and the development of associated land use systems date back at least to the mid-Holocene. In adaptation to the challenges and constraints of harsh high mountain environments, mountain dwellers have developed over many generations sophisticated, complex resource utilization strategies for their sustenance, including a wide spectrum of farming and pastoral practices. Initially, mountain nomadism evolved as a strategy to sustain mountain-related livelihoods, often complementing or replacing subsistence hunting and gathering. It is characterized by animal husbandry as the predominant base for economic and labour activities of mobile communities conducting large-scale seasonal migrations between lowlands and highlands. After the establishment of permanent settlements and village lands, the combination of crop-farming and livestock-keeping evolved as the dominant basis of high mountain agriculture. Pastoral practices in alpine life zones have been increasingly integrated into more complex land use systems including *Alpwirtschaft* (combined or mixed mountain agriculture) and transhumance. However, nomadic pastoralism is still practised in Old World mountain regions, for instance in North and East Africa, Siberia and Mongolia, in the Altai, Tien Shan, Pamir, in Tibet, the HKH region, in the Zagros, and in parts of North and South Europe (Rhoades and Thompson 1975; Grötzbach 1980; Ehlers and Kreutzmann 2000; Kreutzmann 2012; Cunha and Price 2013; Price 2015).

In mountain regions already settled in prehistoric times, combined mountain agriculture has become the most widespread form of traditional land use. The combination of crop cultivation and livestock-keeping reflected the need to incorporate essential natural environmental resources of various altitudinal zones (forests, pastures) and different seasons into the land use system. Developing sophisticated practices of combined mountain agriculture involved interferences in mountain forests which have been increasingly converted to croplands. It also involved encroachments on alpine treelines which have been shifted downslope, often by several hundreds of metres, in order to enlarge alpine grazing lands. However, as long as mountain regions had been sparsely settled, overall impacts remained limited for many generations, and remote mountain ranges probably relatively undisturbed. In previous centuries, mountains provided a degree of isolation from the outside world for their permanent inhabitants and were often characterized by distinct inaccessibility resulting in more or less independent subsistence economies with limited trade and exchange relations with the plains or other mountain regions (Schickhoff 2011).

In some Old World mountain regions, far-reaching transformations of mountain environments are associated with the colonial history. Unlike Europe, where the growing demand for cultivable and pasture land as well as for timber and firewood led to an extensive clearing of mountain forests since the Middle Ages or even much earlier (e.g. in Mediterranean mountains), a significant number of Asian mountains experienced a considerable increase in mountain populations and the concurrent intensification of land use in the course of the past two centuries, encompassing the arrival of colonialism in mountain regions. Nevertheless, cultural landscapes associated with traditional land management also evolved in mountains of Asia over long time periods. In many mountain ranges, however, significant intensifications of agricultural land use took place at a later stage. For instance, rapid landscape transformations in the Himalaya, i.e. large-scale deforestation and substantial changes in the distribution of forests and agricultural lands, occurred only after the British annexation of Himalayan regions in the first half of the nineteenth century. In many Asian mountain ranges, the nineteenth and the twentieth century was a crucial period in the course of cultural landscape evolution and saw a considerable intensification of land use at higher elevations (Schickhoff 1995, 2007, 2011).

During the twentieth century, mountain regions in the Global South were largely characterized by high population growth, poverty, lack of economic opportunities, increased land use pressure, and increased integration into the economy of the lowlands. The primary sector had still been growing in importance, and local mountain farmers were often forced to intensify land use in response to internal drivers, e.g. population growth, and effects of economic globalization, for instance the cultivation of cash crops. Alpine zones were subjected to increased grazing pressure, adversely affecting highland integrity and biodiversity. Heavy grazing implies potentially dramatic losses of biological richness, soil degradation and erosion, and reduced site productivity. Increasing livestock populations, the transformation of traditional pastoral production systems, and inappropriate management practices initiated a general downward spiral in the productivity of many alpine grazing lands and resulted in a loss of biodiversity as well as an increased marginalization of pastoral people (Miller 1997; Schickhoff 2011). At the same time, even the most distant and remote mountain regions were influenced by effects of globalization, and mountains in general have been affected by far-reaching socio-economic transformation processes, notably in the second half of the twentieth century.

In mountain regions of Europe, livelihood diversification has started to gain momentum in the nineteenth century. In the course of the twentieth century, these transformations have eventually led to the extensification of traditional land use and to land abandonment as well as to the concurrent exploitation of mountain environments for tourism, mining, power generation or industrial-scale farming in favourable areas. Traditional forms of agricultural use have been abandoned and mountain farmers were increasingly absorbed in the tourist economy, particularly in winter tourism. The substantial shift from the primary to the tertiary sector has significant environmental implications, e.g. the development of winter mass tourism has neglected many environmental issues. Traditional land use on a moderate level appears to be a key driver for sustaining high levels of biodiversity, both at the ecosystem and landscape scale. Both intensification and abandonment reduce plant species richness relative to traditional land use patterns (Schickhoff 2011). In mountain regions of the Global South, the replacement of farming and herding by the tourism industry as the new economic mainstay has not yet progressed so far as in the European Alps, but the tourism industry has greatly expanded, as evident, for instance, from the mountain tourism in the Nepal Himalaya.

Recently, globalization effects and socio-economic integration into the larger world enhanced modernization trends in mountain agriculture in the Global South. Mountain farmers seek to improve their livelihood by combining alternative farming systems (e.g. agroforestry, cash crops), non-agrarian income (e.g. tourism), and migrant labour remittances, while taking full advantage of the well-established access to lowland markets, provided by the tremendously reinforced road construction. Another intensifying trend is the migration of mountain people from remote locations to surrounding lowlands which could already be observed in the late twentieth century. Impoverished and marginalized mountain people, especially those which are young, energetic and economically active, are increasingly attracted by more diverse and favourable education, job and income opportunities in urban centres of the lowland. Highland-lowland migration, sometimes also stimulated by environmental or political crises (Hugo and Bardsley 2014), often alleviates the population pressure on the scant resource base and leads to a reduced land use intensity at higher elevations. Decreasing population numbers and reduced human pressure may allow ecosystem and biodiversity recovery, where alpine grazing lands had been degraded by previous overuse. It also facilitates the imposition of new forms of land tenure, for instance the establishment of national parks and other protected areas whose number has considerably increased in recent decades. While conservation of most terrestrial ecosystems is not trending towards sustainability, any progress in protecting biodiversity and ecosystems in mountain regions is a vital support for achieving the land degradation-related UN Sustainable Development Goals (UN 2020).

1.3.2 Regional Overview

Asia and Australasia

In the vast HKH region, pastoral strategies are still critically important for sustaining livelihoods of a large human population (Kreutzmann 2012; Dong SK et al. 2016). Livestock grazing in the framework of combined mountain agriculture or by mobile pastoral communities is the predominant land use strategy in the alpine life zone (Fig. 1.42). Alpine grasslands cover more than half of the total land area (including the Tibetan Plateau) and are currently expanding at the expense of snow/glacier cover (cf. Wu et al. 2013; Paudel et al. 2016), thus representing a substantial resource base for animal husbandry. However, as elsewhere, alpine pastoralism is highly susceptible to ongoing social, economic and cultural transformations, resulting in a significant decrease in the importance attached to highland livestock strategies and in a decline of grazing intensity. Labour outmigration is the most important driver of reduced alpine land use intensity. In Nepal, for instance, the migrant population is steadily increasing. Almost 500,000 workers left Nepal in 2014 to work in India, Malaysia, the Gulf countries and other destinations, and remittances have exponentially increased in recent years and already contribute more than 30% to the country's gross domestic product (Fig. 1.43) (Shrestha 2017; Siddiqui et al. 2019). Rural–urban migration within Nepal has also reached high levels and resulted in a largely uncontrolled urbanization process in Kathmandu, leading, inter alia, to severe environmental degradation (Schickhoff 2019). A general decline in pastoral lifestyle and in the number of pastoralists has been assessed for the eastern, central, and western HKH region (Afghanistan might still be an exception), where transformation processes, commercialization of pastoral lands, youth migration and labour shortage, inadequate policy support and institutional arrangements, the decline of trans-Himalayan exchanges (Fig. 1.44), and also the establishment of parks and protected areas aggravate maintaining accustomed pastoral strategies (Nüsser and Gerwin 2008; Bhasin 2011; Schmidt-Vogt and Miehe 2015; Gentle and Thwaites 2016). The livestock sector in the HKH region is characterized by a general decline in the cattle population, while land abandonment and the decrease of traditional agricultural practices due to labour shortage are apparently more pronounced at higher elevations (Chidi 2017; Wang et al. 2019). Whereas the decline in grazing intensity in the Himalaya mainly results from modified pastoral strategies adopted by pastoralists themselves (e.g. Bergmann et al. 2012), reduced high-elevation pasture utilization on the Tibetan Plateau as well as in high mountain ranges of E and S China is caused by external interventions, i.e. state programmes in order to transform the pastoral sector such as resettlement schemes and sedentarization measures aiming at modernization and at reducing grazing pressure and ecological degradation (Ptackova 2012; Hua et al. 2013; Kreutzmann 2013; Qiu 2016).

Over the past few decades, overgrazing by livestock was a major stressor on alpine ecosystems, livestock-environmental interactions had resulted in degradation of alpine grazing lands across the entire HKH region, in particular in drier parts and on the Tibetan Plateau (Harris 2010; Paudel and Andersen 2010; Wu et al. 2013; Baranova et al. 2016; Miehe et al. 2019; Niu et al. 2019; Breckle and Rafiqpoor 2020). In

Fig. 1.42 Pastoralism is declining, but still the predominant land use strategy in the alpine High Asia, exemplified by Ladakhi pastoralists with their goats and sheep at Lake Tso Moriri (4522 m). (*Photo* © Udo Schickhoff, September 23, 2018)

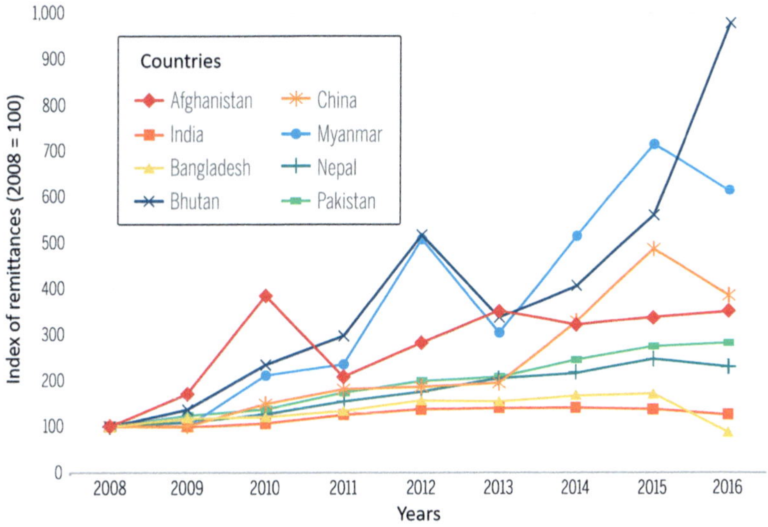

Fig. 1.43 Changes in the index of international remittances received by HKH countries in 2016 (2008 = 100). (Modified from Siddiqui et al. 2019)

quite a few locations, however, local herders have developed effective indigenous rangeland management systems using effective grazing and conservation practices (Dong SK et al. 2007, 2016; Aryal et al. 2014). The current extensification of alpine pastoralism (e.g. Dangwal 2009a) gives grounds for cautious optimism that pasturelands will no longer be grazed beyond their carrying capacity, that formerly degraded rangelands will recover, and livestock grazing will sustain biodiversity and ecosystem services (Cai et al. 2015).

In addition to transformations of high-elevation grasslands, significant land use/land cover changes in the HKH region over recent decades include the conversion of forest to other land uses, mainly farmland, at lower elevations (Wang et al. 2019). However, the (pre)historical dimension of land use/land cover change and deforestation may not be disregarded. As indicated by palaeoecological studies, humans have changed forest environments and transformed forests into replacement communities at least since the mid-Holocene (Miehe et al. 2009a, b;

Fig. 1.44 Desertion of settlements and abandonment of terraced fields after the closure of the Trans-Himalayan trade as exemplified by repeat photographs (1956/2004) of the summer village Milam (3440 m) in Uttarakhand, Indian Himalaya, also symbolizing the recent decrease in land use intensity and the development of a new periphery (upper photo by Bhup Singh Negi; lower photo by Marcus Nüsser; photos courtesy of Marcus Nüsser). (Modified from Nüsser 2006)

Byers 2017), albeit with human interferences and forest clearings having commenced at considerably different times in various Himalayan valleys (Jacobsen and Schickhoff 1995; Beug and Miehe 1999; Schlütz and Zech 2004). It needs to be highlighted that the basic patterns of the present-day cultural landscape in Himalayan valleys are not much different from those of the late nineteenth century (Schickhoff 1995, 2007, 2012). Even though the forests of the Himalaya were considered to be more or less untouched and inexhaustible in pre-colonial times, human impact must have transformed the landscape in many valleys for many centuries, in particular in fertile basins such as Kathmandu or Kashmir Valley which had been inhabited in early times. For instance, the difference between the current upper limit of forests and the potential alpine treeline may be up to 500 m, on south-facing slopes even more, resulting from long-lasting human impact (Miehe 1997; Schickhoff 2005; Miehe et al. 2015; Schickhoff et al. 2015). The expansion of agriculture and trade after the British occupation of Himalayan territories in the first half of the nineteenth century resulted in first significant reductions of forest cover in colonial times. Severe overexploitation of Himalayan forests occurred during the railway building era in the following decades which prompted the constitution of the Imperial Forest Department by the then British India government in 1864. Despite the introduction of 'scientific forestry', unsustainable use in large tracts of mountain forests continued, while the protective influence of silvicultural management was more or less confined to less extensive forest stands, demarcated as 'Reserved Forests' (Schickhoff 1995; Dangwal 2009b). Another phase of massive deforestation arose during World War II and the subsequent struggle for independence.

The first decades of the post-colonial era were characterized by the extensive failure of centralized forest management systems, ultimately resulting in a paradigm shift in forestry (Schickhoff 2014). Continued depletion and degradation of forest resources constituted a threat to rural livelihoods and environmental sustainability and gave rise to the generation of environmental initiatives such as the 'Chipko' movement and to revised forest policies in the 1970s and 1980s, characterized by the introduction of participatory forest management approaches. During this phase, disaster scenarios were fabricated, based on simplified relationships between population growth, deforestation, overgrazing, soil erosion, and floods in the lowland, assuming that the Himalaya was approaching a complete loss of forest cover and catastrophic levels of environmental degradation. Ives and Messerli (1989) clarified that much of this 'Theory of Himalayan Environmental Degradation' is nothing but scaremongering, and encouraged subsequent studies that clearly disproved the theoretical construct (see Ives 2004,

Fig. 1.45 Repeat photographs of Charikot (1989/2010) and Serabesi (1974/2010), Bhimeshwor cluster, indicating the success of community-based forest management in forest restoration. (Nepal Swiss Community Forestry Project; modified from Pokharel et al. 2011; Niraula et al. 2013)

2013). First positive outcomes of participatory and community-based management practices were reflected in an increase of forest areas in c. 25% of all Himalayan districts between 1960 and 1990, while c. 35% reported forest loss (Zurick and Pacheco 2006; Schickhoff 2007). A substantial loss of forest cover was observed in the Karakoram and in the outer Himalayan ranges (Schickhoff 2002, 2006, 2009). In recent decades, decentralized management systems following the 'Community Forestry' approach have been successfully established across the HKH region and have gained relevance for the cultural landscape, in particular in Nepal (Figs. 1.45, 1.46) (Schickhoff 2014). To date, more than 18,500 community forest user groups are managing almost 2 million ha of Nepal's forest, corresponding to c. 30% of the total forest cover (Xu et al. 2019). Remote sensing data show that only 12% of Nepal's districts experienced a loss of forest cover between 1990 and 2013, while 68% showed an increase (Figs. 1.47, 1.48) (Nebelung 2016). Among the national-level forest assessments in Nepal since the 1970s, the latest forest resource assessment 2010–2014 detected the largest forest cover (40%) in Nepal (Fig. 1.49) (DFRS 2015).

In spite of multiple challenges and some limitations and shortcomings such as inequitable benefit-sharing and the exclusion of poor and marginalized groups, the adoption of community-based management approaches has resulted in positive ecological, economic, and social impacts, and most user groups succeeded in regenerating areas of degraded forests and reversing the trend towards forest degradation and deforestation (Gurung et al. 2013; Pathak et al. 2017; Luintel et al. 2018). This also applies to mountain forests in Bhutan, Tibet, India, and to some extent in Myanmar, while Pakistan and Afghanistan are still concerned to achieving a visible impact from community forestry (Xu et al. 2019). On the other hand, the success of community-based forest management should not obscure the fact that forest degradation and deforestation is still an issue at various locations across the HKH (Nüsser 2000; Pandit et al. 2007, 2014; Qasim et al. 2013; Schmidt-Vogt and

Fig. 1.46 Repeat photographs of Dandapakhar, Sindhupalchowk district (1975/2005) and Jiri, Dolakha district (1968/2005), indicating the increase in forest cover caused by the community forestry approach (Nepal Swiss Community Forestry Project. (Modified from Pokharel et al. 2011)

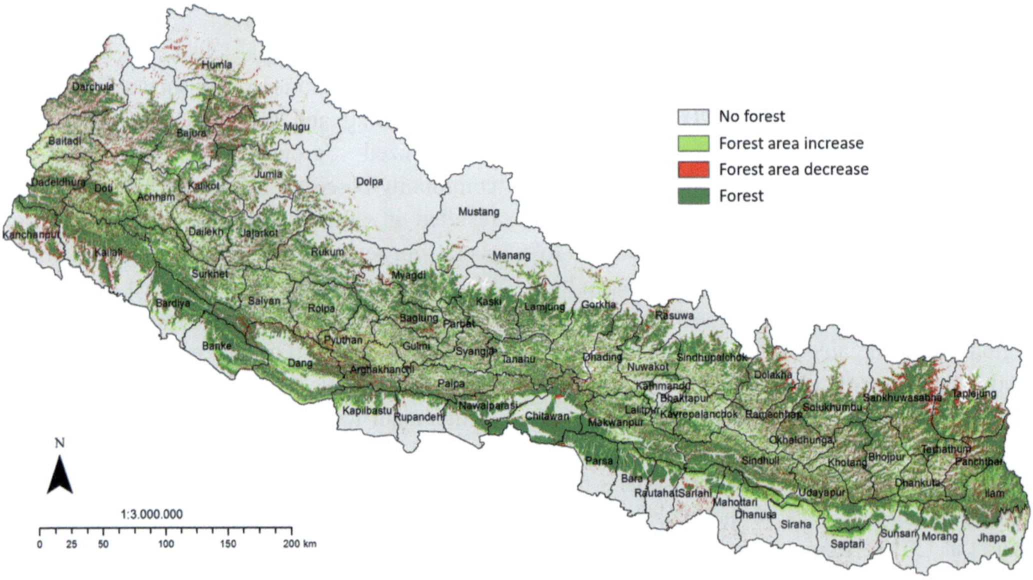

Fig. 1.47 Forest cover change in Nepal 1990–2013. (Modified from Nebelung 2016)

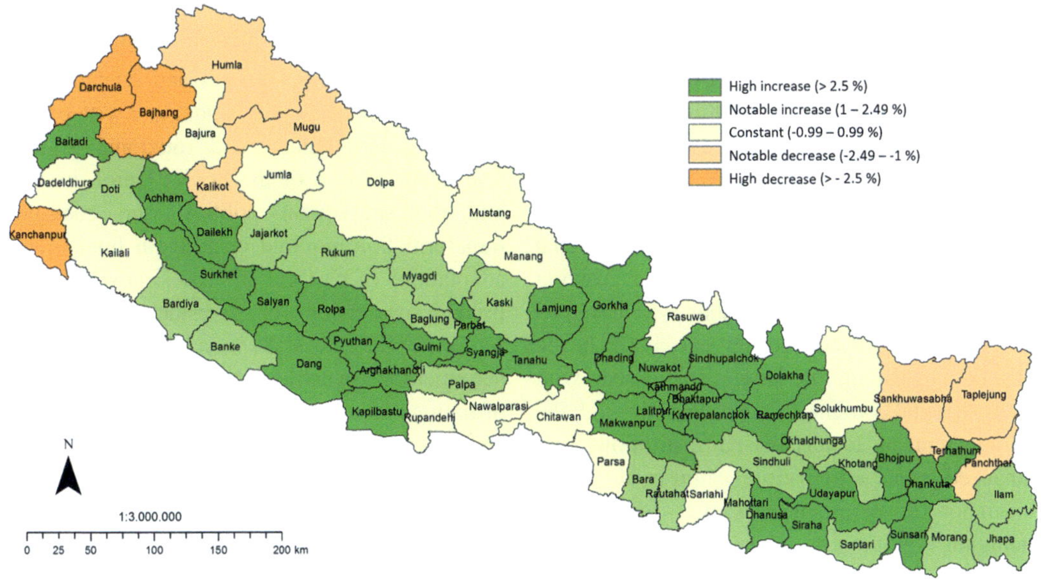

Fig. 1.48 District-wise percentage change of forest cover in Nepal 1990–2013. (Modified from Nebelung 2016)

Fig. 1.49 Extent of forest cover in Nepal 1964–2010/14 according to national-level forest assessments (after data in Paudel et al. 2016; DFRS 2015)

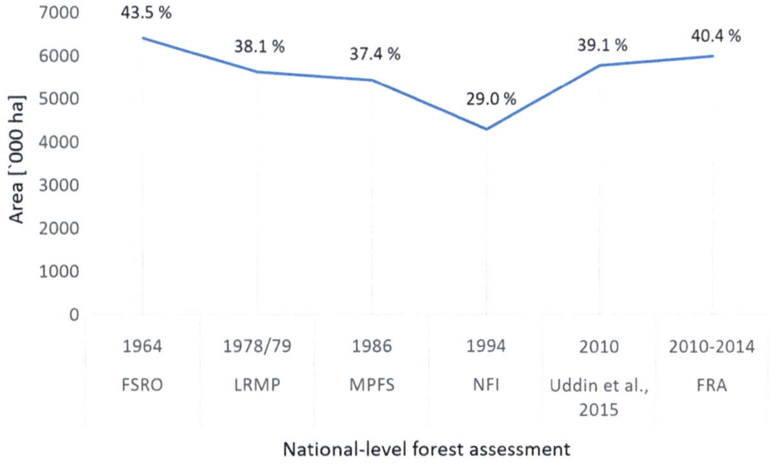

Miehe 2015; Uddin et al. 2015; Garrard et al. 2016; Qamer et al. 2016; Nüsser and Schmidt 2017; Kanade and John 2018; Reddy et al. 2019). It also needs to be highlighted that forest area statistics have little meaning for the qualitative condition of mountain forests. The loss of structural complexity, shifts in species composition, decreasing species richness, erosion of humus horizons and adversely affected ecosystem functions are widespread side effects of forest utilization in recent decades (Schickhoff 2002, 2009, 2012).

As tourism is one of the fastest growing sectors in the world, it has become a significant contributor to the national economy in developing mountain economies. In High Asia, Nepal stands out as a particularly popular destination for international tourism in recent decades, receiving more than one million visitors in 2018. The rapid development of tourism has transformed Nepal's economy, society and environment. While the positive impacts of tourism on local economic growth are widely acknowledged, social and cultural impacts of tourism are viewed critically due to observed changes in local norms, values and behaviour (Shakya 2016). It was feared that the environmental carrying capacity of tourism in the Nepal Himalaya could be exceeded, e.g. the growing demand for firewood and timber was intermittently an object of concern (Byers 2005). In the meantime, tourism is better integrated with environmental conservation, not least through the involvement of locally based institutions and enhanced local participation (Anup et al. 2015). A major development impulse for remote mountain regions is triggered by the expanding rural road network that facilitates the adoption of mobility as an adaptive livelihood strategy. Beazley and Lassoie (2017) recently examined the wide variety of influences on environmental, socio-economic, and sociocultural spheres in the Nepal Himalaya. Human and environmental systems in formerly secluded mountain regions have been tremendously impacted by road construction, as evident from the case of the Karakoram Highway (Kreutzmann 1991; Stellrecht and Winiger 1997; Stellrecht 1998; Schickhoff 2009).

Pastoralism has clearly predominated land use systems at higher elevations in the Pamir, Tien Shan, and Altai, playing a crucial role in Central Asian economies, societies, and cultures since time immemorial. In the former Soviet Central Asian Republics, pastoral traditions and strategies have undergone tremendous changes in the course of the twentieth century, to be attributed to strong external interventions. The first decades of the Soviet era were characterized by forced sedentarization and collectivization campaigns, resulting in a considerable intensification of pastoral land use and its integration into socialist agro-industrial production (Dörre and Borchardt 2012). The pastoral strategy of Soviet times was based on pastoral brigades and herding collectives in the framework of *kolkhozes* (collective farms) and *sovkhozes* (state farms) as well as on permanent high-elevation grazing with short-distance migrations only. This 'detached mountain pastoralism' (Kreutzmann 2011) entailed overuse of grazing resources and related degradation problems that were addressed with pasture irrigation, fertilization, and rotational grazing. The disintegration of the Soviet Union in 1991 and the subsequent political and economic transformation required once again fundamental adaptations of pastoral strategies, now based on private ownership of livestock, subsistence farming, and low state interference in grazing activities. Deindustrialization, the initial decline of national economies, and the disappearance of social securities have led, inter alia, to an increased dependency on grazing land resources. After three decades of post-Soviet transformation, an increased scope and diversity of pasture-related socio-ecological challenges can be observed including conflicts about access to pasture resources, utilization rivalries, insufficient management practices, and degradation processes (Borchardt et al. 2011, 2013; Dörre 2012; Vanselow et al. 2012a), in spite of efforts to decentralize governance and to establish community-based pasture management (Shigaeva et al. 2016). The spatial pattern of pasture degradation has changed in recent decades: Grazing intensity on remote summer pastures at higher elevations has declined due to abandoned seasonal livestock migration (Fig. 1.50), while winter pastures, located close to settlements, have been subjected to more intense grazing pressure with adverse effects on vegetation, plant functional traits, and soils such as lower species richness and diversity, lower biomass, decreased plant height and specific leaf area, lower organic matter content, and higher soil pH values (Akhmadov et al. 2006; Vanselow et al. 2012b; Hoppe et al. 2016a, b, 2018; Mirzabaev et al. 2016; cf. also Liu and Watanabe 2016).

The relative proportion of land covered by mountain forests in Central Asia is rather low. Nevertheless, the natural resources of the forested zones have been an essential component of local land use systems since time immemorial. Forests have been subjected to grazing use and to intensive use of timber and non-timber products (timber, firewood, nuts, fruits, herbs, hay,

Fig. 1.50 The decline in seasonal livestock migration has reduced grazing intensity on remote summer pastures as in Suusamyr valley, Kyrgyzstan, Tien Shan. (*Photo* © Udo Schickhoff, September 26, 2004)

mushrooms, etc.) ever since, resulting in fragmentation, degradation, and transformation. For instance, the extensive walnut forests in the western Tien Shan (Kyrgyzstan) are most likely of anthropogenic origin. Most of these forests replaced mixed juniper-deciduous forests and were established 1,000–1,500 years ago, when fire was used for agricultural purposes and planting of walnut trees was promoted (Beer et al. 2008). The walnut-fruit forests are of high economic value and of essential importance for sustaining the livelihoods of a large population living in the forest area, however, they are characterized by impoverished stand structures, regressive successions, and insufficient regeneration (Borchardt et al. 2010). The deteriorated state of the mountain forests in Central Asia results from the legacy of silviculture practised in the Soviet period and intensified, sometimes unregulated forest utilization in the post-Soviet phase when economic recession increased the pressure on forest resources. Centralized and formal forest management had started with the Russian occupation in the nineteenth century and was strengthened after establishing the planned economy of the USSR. The recent transformation process initiated by the collapse of the Soviet Union and globalization effects have resulted in intensified exploitation and degradation of mountain forests, facilitated by the local population's insecure economic situation, the erosion of managing institutions and institutional weakness with unsustainable and inconsistent management practices, and the appearance of new actors (Schmidt 2005, 2012). Accordingly, the area covered by walnut forests has decreased considerably in recent years (Hardy et al. 2018), adding to the general negative trend of forest cover in the Asian Dryland Belt (Chen et al. 2020).

The history of mobile pastoralists' land use strategies and livelihoods in Mongolian mountain ranges in the twentieth century has many similarities to the former Soviet Central Asian Republics. The system of traditional land use has undergone significant and to some extent dramatic changes, characterized by sedentarization and collectivization during the period of the People's Republic, and by the revival of pastoral nomadism in the early 1990s after the transition to a democracy and market economy. The return of Mongolian nomadism resulted in rapidly growing livestock populations, shifts in herd composition, and widespread degradation of rangelands, also at higher elevations (Fernandez-Gimenez 2002; Janzen 2005; Schickhoff et al.

2007; Zemmrich et al. 2010; Hilker et al. 2014). Reduced livestock mobility, a lack of institutions governing pasture use, and increased poverty among herders are among the challenges to manage rangeland sustainably. The ongoing establishment of community-based rangeland management—over 2000 formally organized herder groups formed since 1999—is a promising institutional innovation which should support implementing strategies towards sustainable pastoral land use (Fernandez-Gimenez et al. 2015). Uncontrolled grazing in mountain forests, fire, and logging are primary drivers of forest degradation and forest depletion and have resulted in substantial annual forest loss in the post-socialist era (Tsogtbaatar 2013). The major industrial sector in Mongolia is mining, accounting for a higher share of the GDP than nomadic animal husbandry. Exploitation of mineral resources has caused severe environmental problems in Mongolia's mountain ranges including devastated rivers and decreasing water resources (Suzuki 2013).

Land use patterns and livelihoods of pastoralists in Russian mountain ranges (Siberia, the Urals) have been affected by the implementation of post-socialist land policy in a similarly fundamental way, subjecting herders to socio-ecological challenges such as unequal allocation of grazing land and localized high grazing pressure (Intigrinova 2010; Istomin and Habeck 2016). In the Caucasus, post-socialist land reforms have reshaped land use patterns meanwhile to that extent that subalpine and alpine zones are currently characterized by outmigration, land abandonment, and increasing recreational activities (Belonovskaya et al. 2016; Gunya 2017). High mountain ranges of Iran have been subjected to intense grazing since ancient times, reflected in the dominance of thorn-cushion formations. Recently, alpine ecosystems are increasingly threatened by reinforced grazing impact, even in protected areas (Noroozi et al. 2008, 2020). Overgrazing has also caused severe pasture degradation in the Pontic Mountains (Curebal et al. 2015). Land use impacts on alpine life zones in New Guinea are considered to be relatively low, exceptions include mining impacts on Mt. Jaya and recently developing ecotourism on Mt. Wilhelm (Hope 2014). However, the mosaic of subalpine forests and grasslands and the fragmentation of the treeline in some mountain areas originated from forest clearings by fire over previous decades and centuries (Hope 2020). Increasingly adverse tourism impacts on the alpine environment have also been assessed on Mt. Kinabalu, Borneo (Latip et al. 2016).

Land use changes in New Zealand's mountain ranges are inextricably linked to the introduction of a large number of non-native species, to which unique island ecosystem biota are particularly vulnerable. New Zealand is one of the most invaded places in the world, many alien species are considered to be invasive pests. Polynesian settlement of New Zealand c. 800 years ago resulted in the clearance of vegetation and in the extinction of 27 bird species, including all moa genera (flightless birds), not least through the introduction of the Pacific rat (*Rattus exulans*) (Bellingham et al. 2010). But only after the late eighteenth century arrival of the Europeans reinforced exploitation of mountain environments (logging, grazing, mining, quarrying) commenced, resulting in large-scale deforestation and substantial landscape transformation (Pawson and Brooking 2013). The period of exploitative pastoralism in montane and alpine grasslands was associated with the depletion of palatable native grasses and herbs that was countered since the 1950s by widespread oversowing with introduced grasses and legumes, leading to the spread of pastoral weeds (Lord 2020). Recently, marginal pastoral high country has been reverted to shrubland and forest. However, indigenous forest, shrubland and grassland vegetation showed a declining trend between 1996/97 and 2012/13, with the exception of subalpine shrubland (Dymond et al. 2017). Numerous non-native plant and animal species have been introduced by the Europeans, some of them such as the brushtail possum (*Trichosurus vulpecula*) and the red deer (*Cervus elaphus*) constitute an important threat due to the damage caused in mountain forests by trampling and browsing (Allan and Lee 2006). As tourism is New Zealand's fastest growing industry, alpine areas are heavily used for sight-seeing, hiking and skiing, placing considerable pressure on higher elevations (Lord 2020). In the Australian Alps, Aboriginal peoples already burned vegetation, however, vegetation physiognomy has undergone more changes during the 200-plus years of Anglo-Australian settlement, inter alia, through the introduction of exotic grasses and weeds. Currently, recurrent disturbance by fire overrides other impacts regarding landscape-scale changes (Collins et al. 2019).

Europe

In many respects, the European Alps can be considered a role model for recent development processes in mountain regions worldwide (Perlik 2019). Over the past 150 years, the Alps have witnessed the process of a profound structural change from an agrarian society to a post-industrial service-based economy, associated with an advanced transformation from a rural to an urban society (Bätzing 2015). Accordingly, land use systems have been reshaped, with modified type and intensity of land use having far-reaching consequences for Alpine landscapes and ecosystems. After-effects of early land use are still visible in the modern Alpine landscape. Neolithic herdsmen already started to expand grazing lands by slash and burn practices in parts of the Alps about 7500 years BP (Conedera et al. 2017). Many centuries of forest clearing have lowered the alpine treeline by 300 m on average, in places by 600 m or more, a process which is of landscape relevance until today. Several waves of increase of the human population and human migration into the Alps, notably between 5,000 and 3,500 and between 1,200 and 700 years BP, entailed the foundation of permanent settlements at higher elevations, leading to widespread human impacts on mountain forests and to large-scale deforestation (Bebi et al. 2017). Complex livelihood systems evolved based on the combination of subsistence agriculture and animal husbandry (*Alpwirtschaft*) in order to make maximum use of resource extraction from multiple altitudinal belts. After the deforestation phase of the Middle Ages, intensive exploitation of mountain forests for energy (in particular for salt processing) and construction materials continued, only slowing down in the aftermath of the Black Death. Renewed population growth and increased demand of wood resources due to the beginning industrialization resulted in another phase of accelerated deforestation across most of the Alps in the late eighteenth and early nineteenth century (Bätzing 2015; Mathieu 2015). Over the centuries the traditional cultural landscape of the Alps had been created, considered to be of high aesthetic value, of vitally contributing to human well-being, and to be the basis for destination marketing of the tourism industry (Schirpke et al. 2019).

In the early nineteenth century, the Alps constituted a less developed region, with Alpine inhabitants facing relative poverty, malnutrition, starvation, and waves of out-migration. The introduction of the potato in the early 1800s and the building of roads and railways in the following decades allowed for some partial mitigation of poverty and hunger. At the same time, the beginning of the industrial revolution led to a successively reduced importance of farming, crafts and mining, prompting the commencement of tourism in the *Belle Epoque* towards the end of the nineteenth century. Livelihood diversification with the decline of traditional farming, the rise of industry and commercial agriculture, and increased economic activities related to tourism has pushed the fundamental structural change that is still unfolding today (Bätzing 2015). The transformation of landscape patterns resulting from the decline of the traditional cultural landscape became most notable after the Second World War, in particular with the initiation of mass tourism in the 1960s and the investment in large-scale winter sports and winter tourist facilities. The former agricultural society has transformed itself into a leisure society (Lichtenberger 1988), not least indicated by the fact that in many Alpine regions income from tourism has become more important than economic returns from farming. Following a stagnation phase (1985–2003), recent trends in Alpine tourism are characterized by the redevelopment of tourism centres and new major projects,

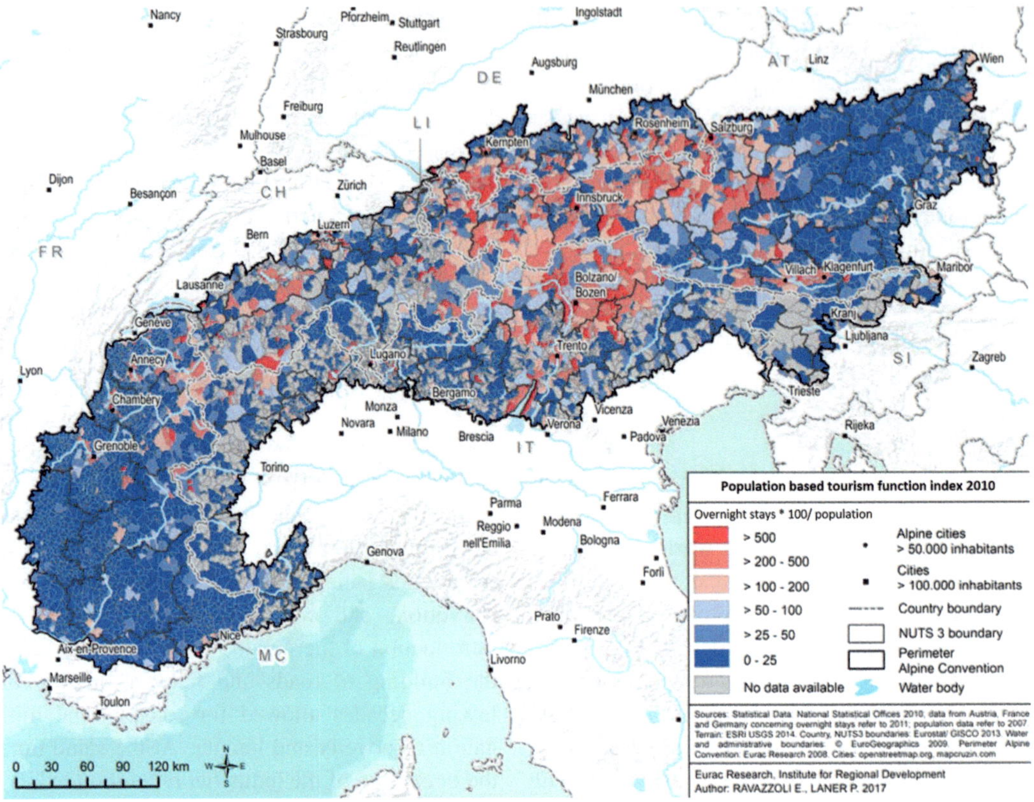

Fig. 1.51 Spatial pattern of tourism intensity in the European Alps, based on the ratio between overnight stays and population. (Modified from Elmi et al. 2018)

associated with a strong centralization in fewer tourism municipalities in more favoured areas with a higher number of touristic beds and overnight stays (Fig. 1.51) (Bätzing 2018). Nowadays, the tourism industry contributes significantly to the Alpine economy, even though the number of jobs directly or indirectly linked to tourism is less than 20%.

Apart from the expansion of touristic infrastructure, the abandonment of agriculturally used areas and the subsequent regeneration of forests has been the essential process of land cover change across the European Alps over the past 150 years. Agricultural land has almost halved between 1850 and 2005, while forest areas have increased by about half and settlement areas quadrupled (Egarter Vigl et al. 2016). In some places, the cessation of land use encompasses as much as 70% of previously used land areas (Tasser et al. 2005). Agriculture in less accessible and marginal areas, in particular on alpine pastures, has tended to become more extensive or has even been abandoned, whereas a trend towards intensifying production can be observed in easily accessible prime locations where much of arable land has been converted to grassland (Tasser et al. 2009; Zimmermann et al. 2010). The observed abandonment of farms is particularly striking in Italy and parts of France and Switzerland (Fig. 1.52). The total number of Alpine farms decreased from 570,000 in 1980 to 260,000 in 2010 (Elmi et al. 2018). The decreasing significance of the agricultural sector is also reflected in the low share of employees in agriculture which was as high as 75% in 1850, but accounts for only 2.5% of total employment in the western Italian Alps and for only 2.3% in the French Alps today (Permanent Secretariat of the Alpine Convention 2015). While mountain agriculture is generally becoming less and less

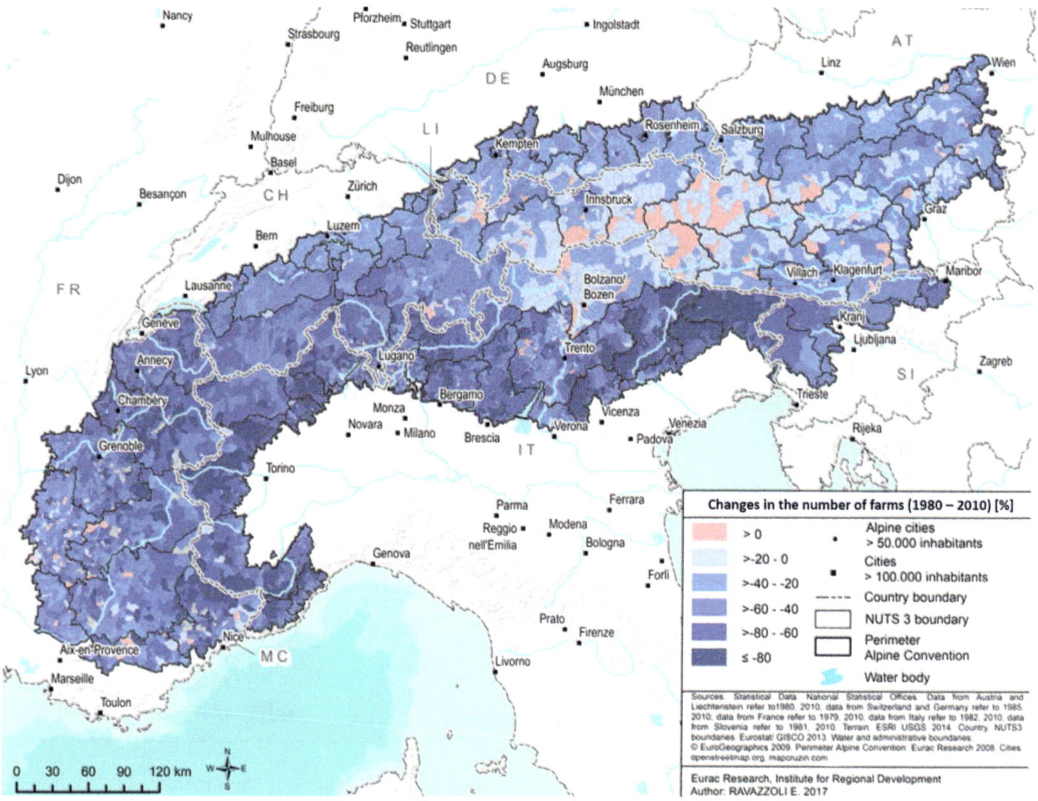

Fig. 1.52 Spatial pattern of the abandonment of farms in the European Alps. (Modified from Elmi et al. 2018)

competitive under economic globalization, it is still highly relevant for maintaining landscape patterns in the Alps. Agriculture still plays a larger role in the northern, German-speaking Alpine countries, facilitated by mountain farming subsidies and the practice of part-time farming (Borsdorf et al. 2015). Forest cover has increased across the entire Alps (Fig. 1.53), with average rates recently accelerating from +3.7% per decade since 1930 to 4.3% per decade since 1990 (Bebi et al. 2017). Secondary forests mainly established on former agricultural land by natural reforestation (Borsdorf and Bender 2007; Tasser et al. 2007). Free succession on abandoned areas inevitably leads to the establishment of new forest areas. Over the past decades, the most rapid increase in forest cover has been observed in the Italian Alps, in the southern Swiss Alps, and in the Austrian province of Salzburg (Bebi et al. 2017). The increase in forestland is a conspicuous effect at landscape scale (Fig. 1.54), associated with a trend towards more monotonous landscapes with reduced structural diversity.

Land abandonment as well as land use intensification results in changes in biodiversity, biogeochemical cycles, climatic and hydrological processes, and related feedback effects on, inter alia, erosion rates, magnitude of floods, snow gliding, and avalanches. Observed biodiversity changes in montane, subalpine and alpine grasslands of the Alps were found to be mainly driven by land management, suggesting that land use change rather than climate change appears to be the most prominent pressure acting on Alpine biodiversity (Vittoz et al. 2009; Dullinger et al. 2020). While land use had a facilitating impact on species and habitat diversity in previous centuries, the transition towards modern high intensity agriculture and the abandoning of land

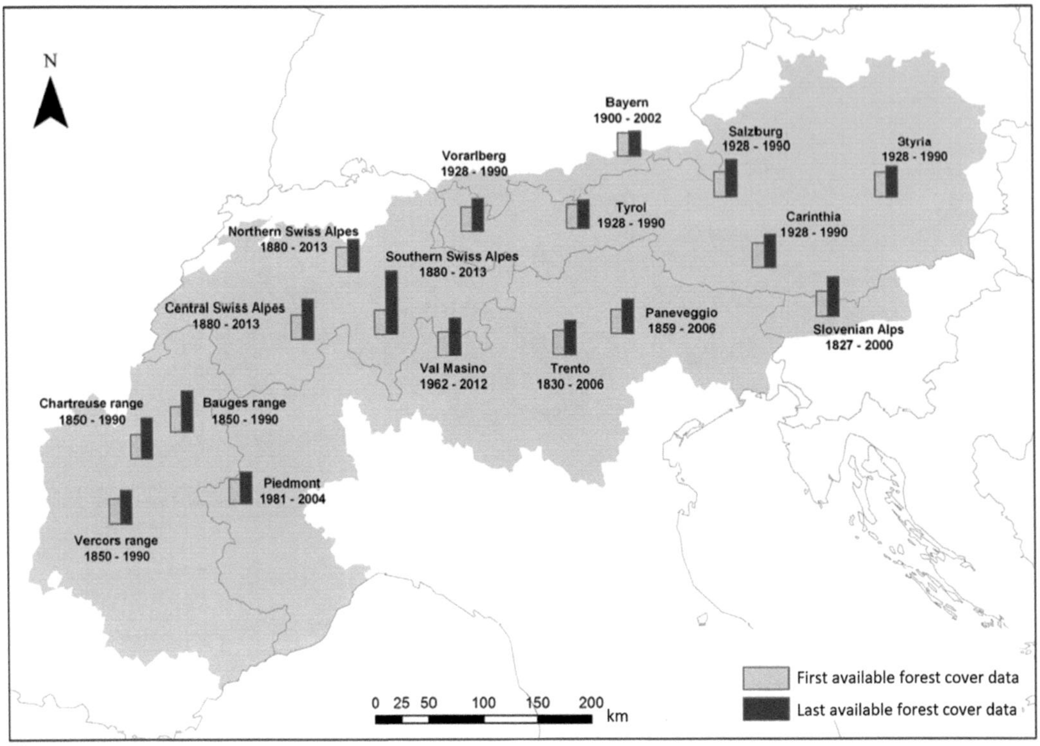

Fig. 1.53 Forest cover changes in different regions of the European Alps, indicating the omnipresent expansion of forest areas. (Modified from Bebi et al. 2017)

use on marginal areas after the Second World War has had the reverse effect (Stöcklin et al. 2007). Resampling of subalpine/alpine grasslands in the northern calcareous Alps revealed a significant long-term decline of plant species richness following land abandonment (Dullinger et al. 2003). On the other hand, high land use intensity has a negative effect on biodiversity on agricultural land (Schmitzberger et al. 2005; Niedrist et al. 2009). It is evident from several studies that both intensification and abandonment change species composition and reduce plant species richness relative to traditional land use patterns (Tasser and Tappeiner 2002; Tasser et al. 2005; Spiegelberger et al. 2006). The loss of biodiversity affects major ecosystem services and ecosystem processes and may lead, in the long term, to decreases in nitrogen mineralization, decomposition rates, nutrient availability, and soil respiration (Tasser et al. 2005). It can be concluded that the goal of sustaining high levels of biodiversity and preserving the diversity of habitats and landscapes can best be achieved by maintaining a wide range of land use types with moderate management intensity (Maurer et al. 2006; Stöcklin et al. 2007; Fischer et al. 2008; Rudmann-Maurer et al. 2008; Strebel and Bühler 2015). Moderate agricultural management intensity also consolidates vegetation cover and soil properties, thus reducing the vulnerability of Alpine ecosystems to landslides, hillslope erosion, and snow gliding processes (Tasser et al. 2003).

The initiation of mass tourism in the Alps, in particular the development of winter sport resorts (Fig. 1.55), has caused severe changes of Alpine landscapes and ecosystems. Winter tourism requires much more extensive technical infrastructures than summer tourism, and ski resorts, ski runs, chairlifts and cableways, and snowmaking facilities are constantly being expanded. Since the 1970s ski runs have been extended to

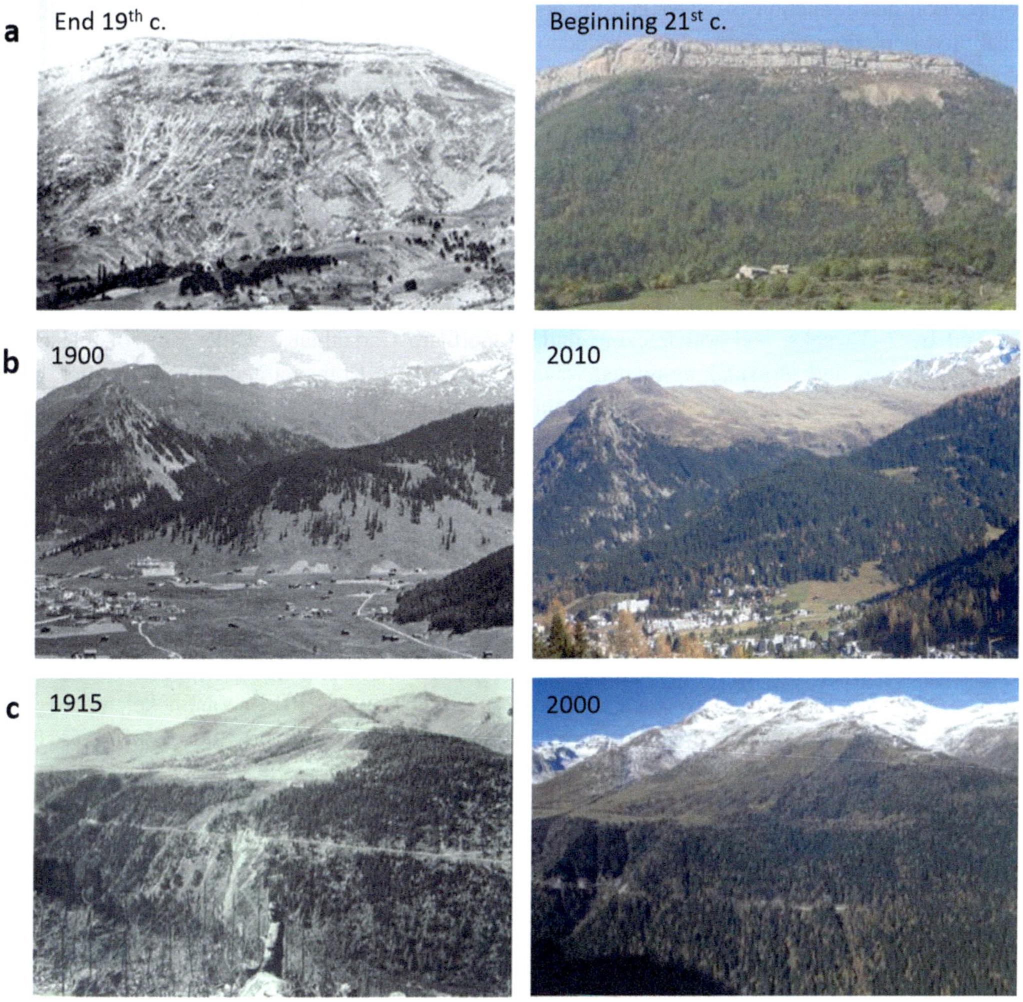

Fig. 1.54 Forest cover changes in the European Alps since the nineteenth century visualized by repeat photographs: **a** Ceüse, southern French Alps; **b** Davos, Central Swiss Alps; **c** Vermiglio, Trentino, Italian Alps. (*Source* Trento Autonomous Province Archive; modified from Bebi et al. 2017)

form wide ski highways, since the 1980s enormous skiing areas have been created, since the 1990s artificial snowmaking has been introduced, and since the 2000s even entire ridge and summit zones in skiing areas are covered with artificial snow, requiring the building of large reservoirs at high elevations (Bätzing 2018). Currently, the Alps capture 43% of total skier visits worldwide and host 80% of the major global ski areas and 38% of the global ski lifts (Vanat 2020). More than 10,000 ski lifts are located in the Alps, covering c. 28,500 km of ski runs that are distributed over ski slopes with high density per massif, pointing to the high pressure exerted by ski activities on mountain territories (Pintaldi et al. 2017). Most winter sports areas in the Alps have caused landscape damage and impairment of ecosystem services that exceeds an acceptable level (Rixen and Rolando 2013; Ringler 2016). The construction of ski runs and skiing has severe impacts on soils in alpine terrain (Fig. 1.56), implying the perturbation of topsoils and the removal of weathered soil horizons as well as subsequent problems such as soil compaction and reduction of water and air permeability, depletion of organic matter, reduction of soil aggregate stability, and nutrient imbalance (Freppaz et al. 2013; Pintaldi et al. 2017; Bacchiocchi et al. 2019). The deterioration of physical, chemical and biological soil properties in turn impairs the establishment and development of plant communities which are also adversely affected by snow compaction and the production of artificial snow. Snowmelt on ski runs is delayed by 2–3 weeks, and soil freezing under compacted snow and snowmaking-related water, salt and ion input are additional stressors that prevent a full recovery of the vegetation (Rixen 2013). Climate warming and the decline in snow cover is an increasing challenge to the winter tourism industry. Austria and Italy bear the highest weather-induced risk of decreasing winter overnight stays related to skiing tourism in Europe (Damm et al. 2017).

Other European mountain systems show many similarities in terms of land use changes over recent decades, but also major differences in historical and political evolution. Integrated in the geo-political context of Eastern Europe, the Carpathians have experienced multiple abrupt shifts in institutions, politics and economics, related to the fall of empires, the collapse of socialism, and the accession of the EU. Recurrent dramatic political, institutional and socio-economic changes have caused several shifts in land management, with land use intensification induced by economic and institutional drivers as well as land abandonment as a result of other socio-demographic and policy changes (Munteanu et al. 2017). Cultural landscapes of the mountain regions have evolved over several thousand years, characterized by small fields, scattered settlements, and large tracts of forests, while larger-scale agriculture was confined to the lowlands. Landscape transformation due to forest clearing for agriculture and for pastures was a dominant process in the Middle Ages up to the nineteenth century. Over the past 200 years, forest cover changes in the Carpathians reflect the turbulent political history, expressed in regionally varying change patterns, while the overall long-term trend indicates an increase in forest areas (Fig. 1.57) (Munteanu et al. 2014): Throughout the nineteenth century until the end of the Austro-Hungarian Empire (1918), forest cover was reported to be stable or slightly decreasing, with a decline in forest cover in the Ukrainian, Romanian and Slovakian Carpathians. The following time periods are characterized by generally increasing forest cover, but regional deviations. During the Interwar and Socialist period, forest cover increases prevailed in the northern Carpathians, while cases of forest loss occurred in Romania and Slovakia. After 1990, forest cover increased with higher mean annual rates, with notable exceptions in parts of the Romanian Carpathians. As in other Carpathian countries, Romania saw a substantial decline of mixed and coniferous forests between 1985 and 2010. Simultaneously, a large-scale successional encroachment of deciduous tree species onto abandoned land has commenced, leading to a net increase in forest cover since the mid-1980s (Griffiths et al. 2014; Vanonckelen and Van Rompaey 2015). Clear-cutting activities (both legal and illegal logging) and widespread natural disturbances, related to an increasing vulnerability of spruce plantations to pests and pathogens, point to regionally highly dynamic forest cover changes. Disturbance patterns in Romanian and Ukrainian forests were attributed to loopholes in national forest laws and illegal harvesting, causing severe damage in valuable old-growth forests (Kuemmerle et al. 2009; Knorn et al. 2013).

In general, forest cover increases in the Carpathians have been synonymous to decreases of agricultural land. Agricultural abandonment on marginal lands and on large tracts of land previously used by state farms accelerated after 1990 due to lack of agricultural subsidies, decreased profitability and migration of labour to western Europe (Kozak 2010; Baumann et al. 2011; Kuemmerle et al. 2011; Bucala-Hrabia 2018). Much cropland has also been converted to grassland. A significant decline of transhumance during the twentieth century has caused considerable forest regrowth at treeline elevations (Shandra et al. 2013; Weisberg et al. 2013). Declining livestock numbers and widespread

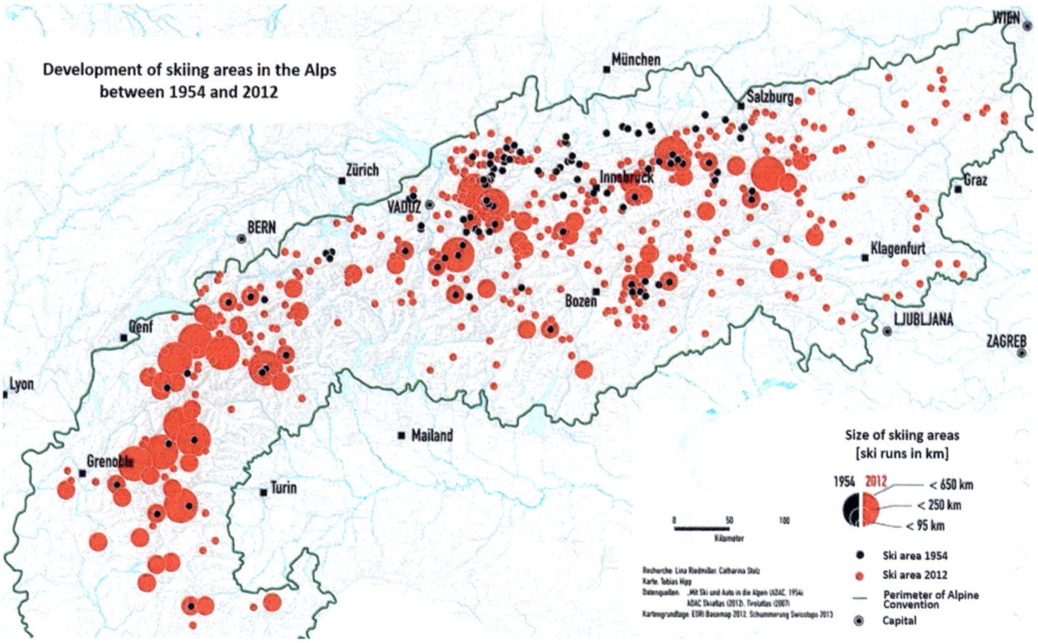

Fig. 1.55 Evolution of the number and size of skiing areas in the European Alps between 1954 (black) and 2012 (red). (Modified from www.alpenverein.de)

Fig. 1.56 Large tracts of land in the European Alps were reshaped in order to develop the ski industry as here above Ischgl, Austria, one of the top winter sport destinations in the Alps. (*Photo* © Udo Schickhoff, August 11, 2011)

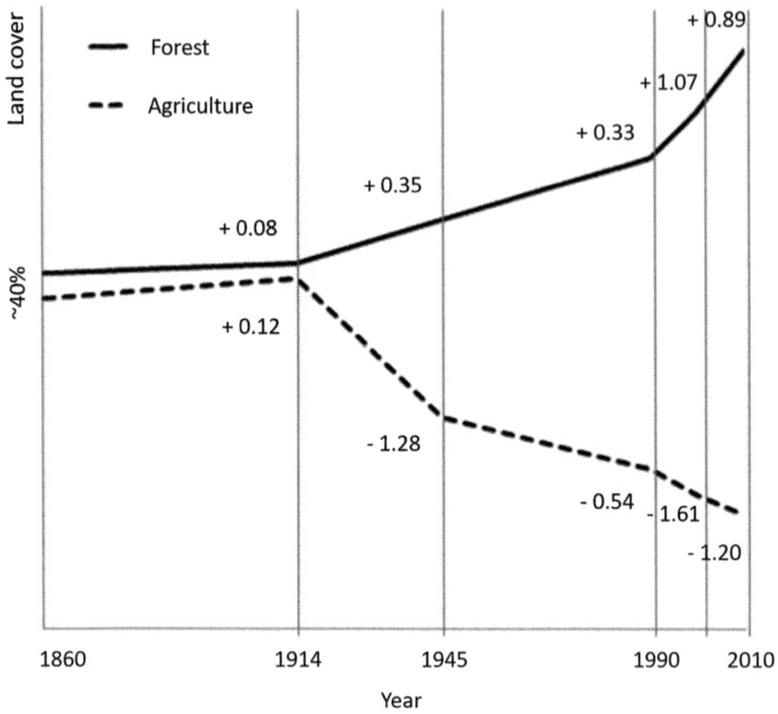

Fig. 1.57 Conceptual reconstruction of the long-term forest and agricultural land cover dynamics in the Carpathians; mean annual rates of land change based on case studies in Munteanu et al. (2014). (Modified from Munteanu et al. 2017)

forest succession have reduced the diversity and area of mountain pastures and meadows, resulting in significantly decreasing species richness and to the entire disappearance of some unique grassland communities (Bezák and Halada 2010; Kricsfalusy 2013). Another significant recent trend in land cover dynamics in the Carpathians is a considerable increase in built-up areas, related to urban sprawl in the lowlands and tourism development and the building of second homes in the mountains (Gerard et al. 2010; Mika 2013).

Mediterranean mountains have one of the longest histories of human intervention, with multiple land use/land cover changes transforming Mediterranean landscapes (Blondel 2006). However, the conspicuous degradation of mountainous environments is arguably a comparatively recent phenomenon, as evident from massive deforestation and soil erosion occurring between 1800 and 1950 (McNeill 1992). Accelerated population growth in the early nineteenth century and improved road connections, accessibility and transport facilities increased the pressure on mountain resources and the exploitation of forests. Deforestation slowed down when large-scale emigration of mountain people began in the late nineteenth century. After the Second World War, Mediterranean mountains have become largely marginal territories, predominantly characterized by rural emigration, abandonment of agricultural land, decline of transhumance of sheep and goats, cessation of grazing pressure, and reforestation of abandoned hill slopes (Papanastasis 2012). Rural depopulation, farmland abandonment and increases in shrubland and forest cover are ubiquitous in Spanish and French Mediterranean mountains including the Pyrenees since the 1950s (Tatoni et al. 2004; Lasanta-Martinez et al. 2005; Chauchard et al. 2007; Ameztegui et al. 2010). The Spanish Pyrenees stand out as one of the European hotspots of forestland increase between 1990 and 2006 (Kuemmerle et al. 2016). In recent decades, winter tourism with the construction of ski resorts has emerged as a land management alternative in the Pyrenees, albeit still with limited territorial impact as compared to the Alps (Lasanta et al. 2013). Similar land transformation processes initiated by rural emigration and manifested by abandoned agricultural land, declining pastoralism and increase in woodland are common and widespread phenomena in mountains of Italy (Torta 2004; Falcucci et al. 2007) and Greece (Papanastasis 2007, 2012). Terraced landscapes are a characteristic anthropogenic imprint on the relief of Mediterranean mountains. Agricultural terraces which are subject to land abandonment and non-maintenance pose an increased risk to gully erosion, terrace failure and landslides which is mitigated, however, in case of colonization by a dense shrub cover or by reforestation (Garcia-Ruiz and Lana-Renault 2011; Tarolli et al. 2014). Land abandonment, cessation of pasture grazing, and increased reforestation induce decreasing availability of habitats for many species of open habitats, but may have beneficial effects for forest-dwelling species (Blondel et al. 2010). While the decline in structural diversity of Mediterranean landscapes may have caused a decrease in floristic species richness in higher successional stages, a recent meta-analysis showed that the overall effect of land abandonment is a slight increase in plant and animal species richness and abundance, albeit with great differences in effect size between taxa, spatial–temporal scales, land uses, landforms, and climate (Plieninger et al. 2014).

Land use in mountains of northern Europe has a long tradition of several thousand years, regardless of harsh climatic conditions and mountain environments being remote and less densely settled. Prehistoric animal husbandry evolved during the Late Neolithic, associated most likely with mountain transhumance (Hjelle et al. 2006). In the Bronze and Iron Age, the use of mountain summer farms became established in southern Norway (Kvamme 1988). The eventful land use history includes a gradual intensification from the seventeenth century onwards, and the full development of Saami reindeer nomadism in the sixteenth and seventeenth centuries in northern Scandinavia where the use of seasonally inhabited farms by farm households has been of limited importance in most inland areas (Moen 2006; Müller-Wille et al. 2006). The high number of mountain summer farms in Norway in the mid-nineteenth century indicates a peak in land use intensity, followed by a strong decline in seasonal farms (Setten and Austrheim 2012). The transition from intensive reindeer herding to more extensive large-scale herding still practised today occurred towards the end of the nineteenth century (Lundmark 2007). Modern land use/land cover changes in Scandinavian mountain landscapes are predominated by forest succession as in most other European mountain regions (Emanuelsson 1987; Hofgaard 1997; Löffler et al. 2004; Bryn 2008; Bryn and Hemsing 2012; Potthoff 2017; Bryn and Potthoff 2018). However, the grazing regime controls establishment of shrubs and trees and treeline expansion to a comparatively greater extent, in particular in northern Scandinavia where semi-domestic reindeer husbandry still exerts a strong influence on mountain ecosystems (Moen 2006). Herd sizes have increased considerably over recent decades (Forbes and Kumpula 2009), and increased reindeer grazing pressure has caused shifts in plant species composition, declines in the cover of lichen heaths, soil erosion, a decline in carrying capacity, and a decrease in productivity, suggesting an overuse of grazing resources at least in some parts of northern Europe (Löffler 2004, 2007; Pape and Löffler 2012). In Finnish Lapland, deteriorating pasture conditions were attributed by the media to intensive Saami reindeer farming and overgrazing. Harkoma and Forbes (2020) highlighted that the underlying causes are more complex and include, inter alia, climate change, regulatory challenges, range restrictions, and other uses of the land such as forestry, infrastructure development, mining, and recreation. Reindeer grazing was also observed to counteract processes of climate-induced encroachment of tall shrubs in tundra (Ims et al. 2013; Bråthen et al. 2017). The recent development in reindeer husbandry is in contrast to the strong decrease of livestock grazing in Norwegian unimproved land since the 1950s (Austrheim et al. 2011). The decrease of livestock grazing is in line with the trend of abandonment of seasonal farming reported from all over Norway (Eiter and Potthoff 2016). Tourism has been part of the mountain economy in Scandinavia since long and has gained significance for local and regional development in recent decades with the decline in extractive industries and agriculture (Fredman and Heberlein 2005). In a warmer future, a northward shift of winter tourism is expected, however, potential ski tourism development zones frequently intersect with established protected areas (Demiroglu et al. 2019; Fredman and Chekalina 2019).

America

With regard to North American mountains and plateaus, a clear distinction must be made between a long period when Native Americans dominated land use and a period of Euro-American dominance of land use that started in the mid-nineteenth century (Vankat 2013). The importance of Native American agriculture increased after about 4,000 years BP with the erection of permanent small settlements. It is often assumed that the impact of Native Americans on mountain environments had been largely insignificant. However, to a certain degree their activity had modified forest extent and composition, created and expanded grasslands, and there is evidence of increased fire frequencies and altered fire regimes at the landscape scale (Denevan 1992; Allen 2002; Roos et al. 2010). Land use effects have clearly reached another dimension since the 1850s when the transition period from Native American to Euro-American dominance of land use ended—three centuries after the first Spanish exploring party had entered the Colorado Plateau. From the Rocky Mountains to the coastal ranges, permanent Euro-American settlements were established, mainly driven by extractive industries such as logging, mining, and grazing, and facilitated by the development of transportation routes, in particular railroads (Wyckoff and Dilsaver 1995; Wildeman and Brock 2000). In the Colorado Front Range, a gold find in 1858 resulted in a gold rush that marked the beginning of permanent Euro-American settlement in the Rocky Mountains (Veblen and Lorenz 1991). In California's mountains, mining emerged as the dominant form of land use after the discovery of gold in 1848 that ignited processes of economic development, settlement, environmental modification, and political adaptation still relevant for California's landscape of today (Dilsaver et al. 2000). In its initial stage, the California gold rush had a comparatively minor effect on landscapes and ecosystems. This changed drastically, however, after the introduction of hydraulic mining, associated with enormous water consumption, the use of dangerous chemicals, the practice of dumping mining debris into mountain rivers, and vast sediment loads. Following the legal ban on hydraulic mining in 1884, the mining industry declined, mining boomtowns became depopulated, and agriculture in the Central Valley became the driving force of California's economy (Alagona et al. 2016). The transformation of the semiarid Central Valley into one of the most productive agricultural regions in the US has required the extraction and diversion of vast amounts of water, primarily from the Sierra Nevada (Ives et al. 1992a). In general, mining was the principal industrial activity that attracted people and brought systematic settlement to the western mountains in Canada and the US from 1850 to 1930 (Harris 1997).

Other significant anthropogenic disturbances of the early phase of Euro-American land use dominance included logging, grazing, and fire management. The nineteenth-century mining booms resulted in heavy demands on timber resources for town construction, fuel, and mine props, and led to large-scale logging of montane forests. In many mining areas, nearly all the timber was cut, causing erosion and soil depletion. Even beyond the immediate mining areas, a large percentage of forests had been logged for fuel and construction purposes towards the end of the nineteenth century, as in the present Rocky Mountain National Park (Veblen and Lorenz 1991). Widespread logging of ponderosa pine (*Pinus ponderosa*) forests commenced in the 1870s when logging became a major industry on the Colorado Plateau, occasionally even at high elevations (Vankat 2013). In the Sierra Nevada, a large timber industry developed to exploit sugar pine (*Pinus lambertiana*) and sequoia

(*Sequoiadendron giganteum*) forests (Alagona et al. 2016). After three decades of heavy logging, the California Board of Forestry estimated in 1886 that already one-third of the Sierra Nevada's timber had been harvested (Beesley 2004). During this deforestation phase, the foothills treeline in the Sierra Nevada was raised by up to 600 m (Dilsaver et al. 2000).

Livestock grazing was the first broad-scale impact on the vegetation of mountains and plateaus in the American West after European colonization. Livestock numbers and grazing impact reached another order of magnitude in the second half of the nineteenth century, when rapid increases in large, commercial ranching operations were supported by the completion of the transcontinental railroad, the final subjugation of the nomadic Native American groups, expanding markets, as well as by the entrepreneurial resource utilization ethic that focused on maximum harvest for maximum profit (Raish 2004). The intensification of the nineteenth-century open-land sheep and cattle ranching also affected higher elevations as evident from severe grazing damage on the Colorado Plateau by high populations of these introduced herbivores (Cole et al. 1997), while seasonal migrations to and pressure on alpine grazing lands remained much less significant compared to Old World mountain regions (Bock et al. 1995). The commercial livestock industry declined at the turn of the century, mainly due to overstocked ranges, droughts, and brutal winters (Huntsinger et al. 2010). As early as 1864, increasing concerns about depleted forests, polluted waterways and degraded rangelands resulted in the designation of California's Yosemite Valley and nearby Mariposa Grove as a nature reserve for conservation and recreation (Beesley 2004). This first groundbreaking success of the American environmental movement was followed eight years later by the establishment of Yellowstone National Park, the world's first national park, and by the establishment of federal forest reserves/national forests around the turn of the century.

The alteration of fire regimes was an important effect of early Euro-American land use that had a lasting impact on vegetation and landscape, still unfolding today (Vankat 2013). Initially, livestock herders set fires to clear vegetation and stimulate forage growth, and many fires were set by mining and other human activities, either accidentally or intentionally. Post-fire stands often showed a shifted dominance of tree species, and post-fire recovery processes still influence recent forest cover changes (Rodman et al. 2019). However, fire frequency decreased in areas where livestock grazing expanded since the shrub and herb layers in open forests and meadows were greatly reduced. This culminated in extensive and effective fire suppression, an important legacy of public land management, becoming widespread in the early twentieth century since fires were then viewed as unnatural events from which vegetation should be protected. For many decades, forests remained unburnt, causing again changes in species composition, structure and dynamics. For instance, pyrophytic species such as the giant sequoia have not been exposed to fire for almost a century and did not regenerate (Harvey et al. 1980). Fire exclusion resulted in an unnatural level of fuel load and increased tree densities in mountain forests, leading to landscape-scale crown fires (Fig. 1.58). In subalpine forests with longer natural fire intervals, fire suppression had less serious implications. In recent decades, fire management practices such as prescribed burning have been developed that are only partially successful in countering the effects of long-term fire exclusion, and bear as well the risk of exceptionally large and intensive crown fires (Fulé and Laughlin 2007; North et al. 2015; Thompson et al. 2018).

Land use management in North American mountain regions remained to be driven by extractive industries during the twentieth century, albeit to a lesser extent, while the environmental movement has strengthened, and tourism, recreation, and residential development have become increasingly important. Mining declined after the global economic depression in the 1930s, but retained its importance as major employer and source of adverse environmental impacts (Fox 1997; Gardner et al. 2013). The migratory sheep

Fig. 1.58 Fire suppression for decades has increased the risk of landscape-scale crown fires, with recent climate change adding to this risk (Yosemite National Park, Sierra Nevada, USA). (*Photo* © Udo Schickhoff, August 23, 2019)

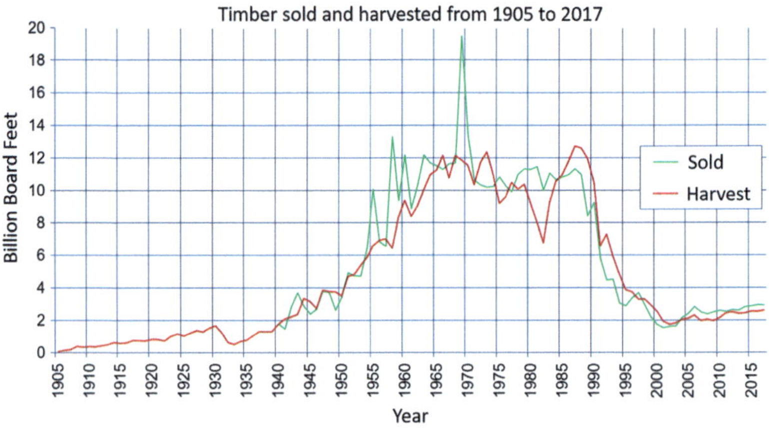

Fig. 1.59 Timber sold and harvested in the national forest system of the United States from 1905 to 2017 (sold data not available before 1940). (modified from www.fs.fed.us)

and cattle industry rapidly declined since the 1930s. The Taylor Grazing Act ended open range grazing in the western US in 1934, and herd movements were further restricted by environmental laws and the addition of more national parks. The percentage of total land area used for grazing in the Rocky Mountains decreased significantly (Cline 2013). Today, transhumance is no longer economically significant in North American mountain regions (Cunha and Price 2013). At lower elevations, however, the influence of the western range livestock industry is still strong and grazing-induced ecological changes have long been debated (Donahue 2005). One of the responses to widespread forest depletion was the establishment of the US Forest Service in 1906, who tried to restore degraded lands, severely limited grazing and regulated logging during the twentieth century (Dilsaver et al. 2000). Henceforth, national forests have been managed under a multiple-use, sustained-yield mandate, combining extractive uses as well as recreation and conservation. The balance among these uses has been spatio-temporally differentiated (Alagona et al. 2016). In the first half of the twentieth century, the focus was on conservative use and resource protection, and national forests were rarely logged (Fig. 1.59). Large-scale timber extractions were resumed in the 1950s, mainly triggered by the postwar housing boom which was fuelled by the growing prosperity of a fast growing population. More stringent environmental legislation, rapid development of plantations (mainly in the Southeast), and foreign producers capturing the US wood supply market were major reasons why the national forest timber harvest plunged to prewar levels in the 1990s (Bosworth and Brown 2007). The sharp decrease in harvest from national forests helped to ensure that the total forest area in North America is currently roughly stable (Masek et al. 2011).

Current land use in North American mountains is characterized by an increasing dominance of conservation, recreation, residential and commercial development, while resource extraction is losing importance. In recent decades, a significant migration to mountain regions can be observed as part of a national population shift to the South and West and from urban to rural areas. Rapid growth of mountain towns and dispersed, landscape-consuming residential development in rural areas reflect emerging land use patterns created by amenity migration, as described for the Colorado and Canadian Rocky Mountains (Riebsame et al. 1996; Leinwand et al. 2010; McNicol and Glorioso 2014) and for the Sierra Nevada (Loeffler and Steinicke 2006). Amenity migrants include semipermanent residents as well as homeworkers and retired persons, establishing permanency in their mountain homes (Moss 2006). Significantly increased housing density has also been the most significant land use change on lands surrounding US national parks in recent decades (Hansen et al. 2014; Resler et al. 2020). At the same time, tourism has received a huge boost and has become an important element in the local economy. Ski area development has dramatically increased (Humphries 2020). The Rocky Mountains are now an international winter tourism destination, giving rise to controversial discussions on further expansion of ski resorts (Childers 2012). The second pillar of the large-scale two-season mass tourism is the increasing nature-based summer tourism, for which the national parks represent a major resource, and which triggers diverse recreation impacts (e.g. Willard et al. 2007). The recent transformation into amenity landscapes is associated with extensive infrastructure networks, a visible expression of contemporary land use in many mountain areas (Alagona et al. 2016).

The historical development of land use in mountain regions of Central and South America exhibits differences to that of North America in the sense that indigenous highland peoples had a much higher population density over many centuries, and had reshaped the environment to a comparatively greater extent. In the Andes, at least thirteen to fourteen thousand years of continuous human occupation had preceded European contact (Erlandson and Braje 2015). Palaeo-Indian hunters burned woodland to expand the game-rich ecotone between forests and the alpine zone and initiated the large-scale deforestation of the highlands. The transformation from woodland to grassland continued to be driven forward when agriculture and pastoralism appeared around 6–7000 BP and land for cultivation and grazing use was needed (Ellenberg 1979; Baied and Wheeler 1993; Gade 1999). This transformation resulted, inter alia, in upper treelines being truly anthropogenic (Miehe and Miehe 2000; Sarmiento and Frolich 2002). Later, advanced civilizations such as the Tiahuanaco and Inca empires developed a highly successful agriculture including the sophisticated management of raised fields and irrigated terraced slopes, and thus remodelled the landscape of whole valley systems. Agriculture extended over a considerable range of altitude from the lowlands to over 5000 m and sustained more than 10 million people in the Central Andes (Grötzbach and Stadel 1997; Borsdorf and Stadel 2015). In late Pre-Columbian times, intensive agriculture was most widespread in the northern and central Andes where, below about 3000 m, maize cultivation resulted in massive landscape transformation, and areas above 4000 m were used for hunting and camelid (llama, alpaca) grazing (Knapp 2007). Grasslands were maintained by clearing, grazing, and burning (Gade 1999). Humans have modified most forest, shrubland, grassland, and wetland vegetation types in the Andes for millennia (Young et al. 2007; Young 2009).

The Spanish conquest (AD 1532) marked the start of profound transformations of Andean society, culture, economy, and environments. Colonial rule aimed at exploiting natural and human resources and at missionizing the indigenous population, focusing on high-yield mining areas, productive agricultural areas, and generally on densely populated regions (Borsdorf and Stadel 2015). The southern Andes were relatively neglected, and have experienced to date a much lower appropriation of land for human use (Hoekstra et al. 2010; but see Inostroza et al. 2016). The Spaniards modified or destroyed traditional community organization, while indigenous agricultural techniques and land use systems largely collapsed (Grötzbach and Stadel 1997). The introduction of land tenure systems, crops, domesticated animals, tools, technologies, institutions, and peoples can be termed an early globalization, involving a variety of impacts on the mountain environment that have been massive in the long term and sometimes substantial or even devastating at a local or regional level (Knapp 2007). In the wake of the Spanish conquest, fundamental socio-economic and administrative changes were introduced, leading to severe societal disruption. The indigenous population started to decline drastically, mainly due to the spread of European diseases or forced labour (Ives et al. 1992b). Declining subsistence needs of a shrinking population resulted in the abandonment of a large number of terraces as well as of raised fields that occupied large areas in highland flats, reflecting that depopulation was associated with disintensification of land use, characteristic of the entire Andes in the 1600s and 1700s (Knapp 2007; see also Butzer and Butzer 1995 for Mexico). Nevertheless, cultivation patterns were continued incorporating a wide range of introduced European crops, which could be used particularly at higher elevations. The most important change in traditional grazing patterns was the replacement of domesticated Andean animals (llamas, alpacas) by European sheep, goats and cattle that contributed most saliently to peasant livelihoods (Gade 1992). Due to the dramatic depopulation of vast tracts of the Andes after the Spanish conquest, large land areas were available for grazing, which was less labour-intensive than traditional farming (Borsdorf and Stadel 2015). Whereas traditional Andean grazing patterns are associated with sustainable production systems, large-scale soil erosion problems and drastic changes in vegetation structure in the post-conquest era are commonly attributed to overgrazing by introduced livestock to which the native vegetation is not adapted (Browman 1974; Millones 1982). Grazing-ecological studies confirm that grazing systems with introduced cattle have a lower efficiency in the use of pastoral space, show a concentration of cattle in fewer places, and have a higher magnitude of environmental impact (Molinillo and Monasterio 2006).

Nevertheless, the colonial period with the population decline of indigenous Andean highland peoples was generally associated with environmental recovery, with the exception of impacts originating from mining activities and from the demand of wood (Denevan 1992; Knapp 2007). The demise of much woodland accelerated since greater quantities of wood were needed for diversified uses including mining activities and charcoal production, controls on wood cutting were far less strict than in Inca times, and forest grazing by introduced livestock caused severe damage. In the course of time, human agency has destroyed over 90% of native Andean forests (Gade 1999). Wood shortages are meanwhile alleviated by ecologically detrimental plantations of exotic eucalyptus and pine species that accounts for a part of the recent increases in woody cover of mid-elevation areas and highland grasslands in the tropical Andes (Balthazar et al. 2015; Aide et al. 2019; but see Restrepo et al. 2015 for the Colombian Andes). At the time of the Spanish American wars of independence in the early nineteenth century, the population started to expand again, followed by an exponential growth since the 1920s that has been attenuated in recent years. The increasing integration into the global system of trade and transfer and the high population growth have resulted in highland resources having been more intensively exploited, and European modification of the environment having accelerated. The

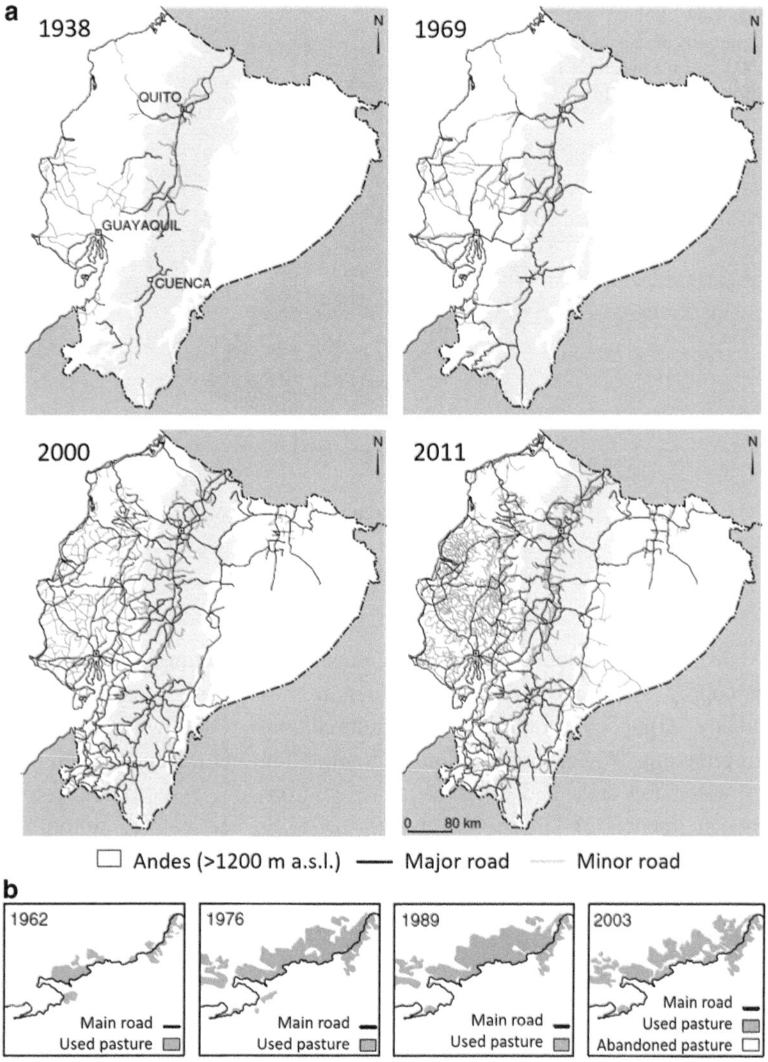

Fig. 1.60 Impacts of expanding rural road networks in the Andes; **a** road networks of Ecuador in 1938, 1969, 2000 and 2011; **b** evolution of land use forms along the main road in the upper Rio San Francisco valley over a 40-year period. (Modified from Peters et al. 2013)

colonization of the highlands was reinforced, the agricultural frontier moved up into the páramo, and the agricultural production intensified, characterized by an increased use of chemical fertilizers and pesticides, indiscriminate use of fires, overgrazing, construction of drainage systems and roads (Monasterio 1980; Hess 1990). Human disturbance still plays a primary role in shaping páramo vegetation patterns, diversity and ecosystem services (Suárez and Medina 2001; Vasquez et al. 2015; Hofstede and Llambi 2020). In general, extensive road construction played a crucial role for land reclamation, with facilitated access to remote valleys supporting deforestation and agricultural intensification (Fig. 1.60) (Peters et al. 2013; Quintero-Gallego et al. 2018).

Intensified use of highland resources applies in particular to mining-related resource extractions. Industrial-scale silver and gold mining in the Andes was already a widespread source of livelihood in Inca times and continued through the colonial and post-colonial periods. One of the largest cities in America at the beginning of the seventeenth century was Potosí, an old silver mining town at 4100 m in present-day Bolivia which had grown to 150,000 inhabitants (Borsdorf and Stadel 2015). With the globalization of economy in recent decades, the exploitation of

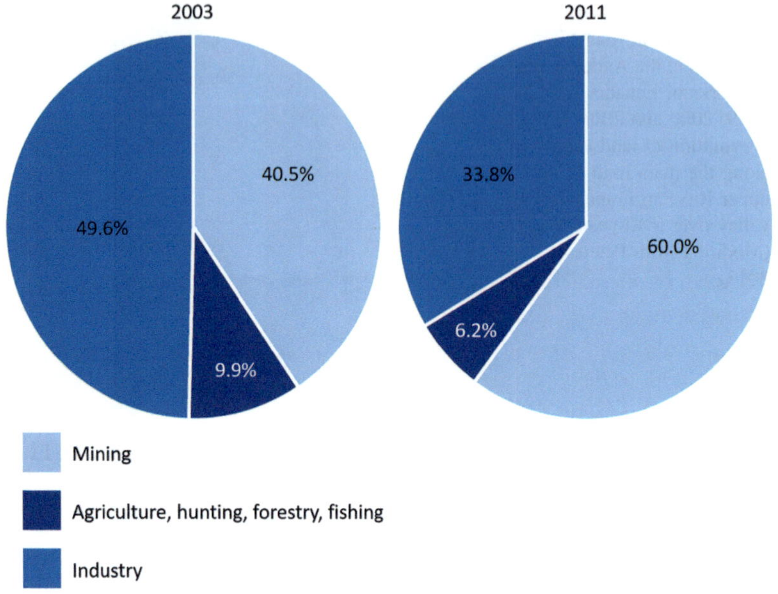

Fig. 1.61 Value share of exports by sector in Chile 2003–2011. (Modified from Simpson et al. 2014)

Andean mineral resources of global interest, such as copper, gold, zinc, tin, and molybdenum, has greatly expanded, controlled by multinational corporations. Water consumption, contamination of water and soils and other negative environmental impacts of large-scale projects are substantial, as vividly illustrated by the case of the open-pit mining project Pascua Lama at an elevation above 4000 m across the border of Argentina and Chile, strongly resisted by rural communities (Romero et al. 2009). This project, aiming at extracting gold, silver and copper, exemplifies the increasing number of conflicts created among enterprises, native ethnic groups, and residents of the lowlands who depend on highland resources such as water and wood withheld from them for the extraction of industrial minerals (Marchant 2010). The huge investment in mining in the Andes has provoked a surge in social mobilization and conflict (Bebbington et al. 2008). Unsustainable highland resource use in the wake of economic globalization is obvious from the fact that Chile, the main producer of copper and associated minerals in the world, concentrates the national mining investment in the Atacama Desert where water resources are extremely scarce and water is even imported from Bolivia (Romero et al. 2009). Mining accounts for an increasing value share of Chile's exports and a significant proportion of GDP, while the importance of the primary sector is further declining (Fig. 1.61).

A common consequence of the deep social, economic, cultural and environmental transformations that have affected the mountain regions in Central and South America was the migration of highland population to the lowland. The decline of traditional mountain economies based on agriculture and livestock had triggered conspicuous migration trends from highland to lowland and from rural to urban in the second half of the twentieth century (Escobar and Beall 1982; Ives et al. 1992b; Lauer 1993; Romero and Rivera 1996; Izquierdo et al. 2018). The huge rural exodus reflected the need to improve livelihoods through getting employment, housing, education, and health services, and resulted in an explosive population growth and slum development in expanding cities. In recent years, these migration trends and the growth of cities have weakened. Globalization and international mobility support in places the emergence of a new rurality, in which international migration patterns stimulate development by increasing the remittance income of households (Yarnall and Price 2010; Borsdorf and Stadel 2015). Recently, páramos and puna grasslands have been increasingly converted to other land uses such as more intensive agriculture and afforestation, involving higher water-demanding trees and crops (Hofstede et al. 2002; Tovar et al. 2013b; Bello et al. 2014).

Spatial patterns of rural settlements are in a process of change as well, exemplified, for instance, by increasing amenity migration, in the case of Santiago de Chile supported by the state's withdrawal from regional planning, deregulation, privatization of the land market and other factors, all of which are linked to globalization (Borsdorf and Hidalgo 2009). Other examples, caused by changing socio-ecological systems at higher elevations, include the concentration of pastoral settlements in the Peruvian puna (Charbonneau 2009) and the expansion of the permanent frontier of agriculture and dwellings to higher elevations in the Ecuadorian páramo (López-Sandoval and Maldonado 2019). The latter case study also illustrates positive conservation outcomes after the establishment of communal governance of natural resources. Unlike the Himalaya (see above), the implementation of community-based management models is less advanced in the Andes and faces diverse challenges, but needs to be promoted in order to contribute to the achievement of sustainability goals (Wilson 2016; Mathez-Stiefel et al. 2017). The view that it is imperative to protect valuable or representative natural and cultural landscapes has increasingly gained ground over recent decades, reflected in a strong increase in protected areas since the 1960s, both in number and surface area (Fig. 1.62). With extraordinary scenic and cultural diversity, the Andes have a tremendous tourism potential. Tourism is considered a key sector of the national economy and an effective strategy to counter poverty and marginalization and to contribute significantly to regional development (Borsdorf and Stadel 2015). In Peru, for example, tourism has become the second most important economic sector after mining, with strongly increasing arrivals of international tourists (more than 4 million in 2017) and annual tourism-induced foreign exchange revenues of more than USD 4 billion (Baumhackl 2019). It is a major challenge for governments and local authorities to alleviate environmental impacts resulting from the spatial concentration of tourist flows, and to make tourism compatible with the ways of living of the local population.

Africa

Since time immemorial African mountains and highlands have been more attractive for human land use than surrounding lowlands since climatic and ecological conditions for agriculture

Fig. 1.62 Number and surface area of protected areas have increased considerably in the Andes in recent decades, exemplified by the Laguna Miscanti (4140 m), Atacama Altiplano, Chile, part of the National Reserve Los Flamencos. (*Photo* © Udo Schickhoff, March 24, 2017)

and for sustaining livelihoods are much more favourable (Grosjean and Messerli 1988; Hurni et al. 1992; Messerli and Winiger 1992). Mountain regions also provided refuge areas for ethnic groups as well as better protection from severe vector-borne human and animal diseases. Accordingly, most of African highlands are areas of large population concentrations, illustrated by the case of Ethiopia where some 10% of the population only is living in areas below 1500 m (Abate 1993; Piguet and Pankhurst 2009). The Atlas Mountains are just another example of sustaining higher population densities than surrounding lowlands by providing economic resources and ecosystem services (Bencherifa 1990; Montanari 2013). Abundant natural resources of varied mountain environments may have facilitated hominid evolution in eastern and southern Africa, while notable human impact dates back to Stone Age populations that had colonized most of East Africa's biomes with very low numbers of individuals by c. 100,000 BP (Spinage 2012). The use of fire, supporting the expansion of savannahs, was the first major impact changing the ecology of Africa, followed much later by pastoralism that, in turn, paved the way for agriculture. The first appearance of goats and sheep in East Africa can be assumed for c. 5,000 BP, coinciding with the terminating African Humid Period. Cattle followed subsequently, spreading slowly from there to southern Africa, while the presence of cattle north of the Sahara is dated back to the eighth-seventh millennia BP (Gifford-Gonzalez 2000, 2017). The introduction of domesticated livestock and the expansion of pastoral communities diversified land use and marked the start of a sequence of significant land cover change (Marchant et al. 2018). After the Bantu expansion in Sub-Saharan Africa and since the Iron Age, impacts from (semi)-permanent settlements, cultivation, pastoralism, and the use of fire became a more widespread and dominating force, resulting in widespread anthropogenic degradation of vegetation and deforestation over the past 2,000–2,500 years (Spinage 2012).

Palaeoecological studies give evidence of human-induced forest clearing, often associated with soil erosion, beginning around 2,000 BP in the Atlas Mountains and in the interlacustrine highlands of East Africa (Lamb et al. 1991; Taylor 1990, 1996; Jolly et al. 1997; Marchant and Taylor 1998; Cheddadi et al. 2015). Large-scale anthropogenic forest destruction appears to have also started at higher elevations in the Rwenzoris and in the Ethiopian Highlands between 1,000 and 2,000 BP, whereas mountains in Kenya and Tanzania retained more forest cover up to modern times (Hamilton 1982; Nyssen et al. 2004; Umer et al. 2007). In East Africa, forest clearings have mainly focused on productive mid-elevation areas. Thus, very little primary forest remained between 1500 and 2500 m, and montane forest vegetation is now largely restricted to protected areas (Marchant et al. 2018). For instance, the submontane forest in the eastern Arc Mountains in Tanzania has lost more than 90% of its mid-Holocene area (Hall et al. 2009). Early deforestation in Ethiopia has been on an exceptionally large scale. The resulting environmental degradation was held partly responsible for the demise of the Axum civilization during the first millennium AD (Butzer 1981).

Over the last several centuries, population growth, migration of peoples, the introduction of new crops and technologies, effects of colonialism, and economic globalization were significant drivers of extensive and pervasive land cover change. Mountain ranges in the Maghreb countries experienced a general decline in forest cover and a matorralisation process, with most of the forests being transformed into various replacement communities including dehesa-like parklands due to high frequency of fires and the intensification of land clearance and grazing pressure (Cheddadi et al. 2015). Reconstructions of human–environment interactions show that the phase of Islamization was associated with population increase and development, including expanded pastoralism, deforestation and agriculture (McGregor et al. 2009). Another phase of agricultural intensification related to colonialism in the Atlas Mountains occurred in the late nineteenth and during the twentieth century, when the mountain ranges and intermontane valleys served as delivery systems for resources for the focal areas of development in the lowlands (Hurni et al. 1992). This function has been maintained in the post-colonial period (in Morocco since 1956), with forests and silvopastoral areas further declining in recent decades, attributed to the interaction of drivers such as drought, fire, soil erosion, and the increasing pressure on resources associated with socio-economic change (Hammi et al. 2010; Chebli et al. 2018; Kouba et al. 2018). Recent transformation processes in mountain livestock farming systems are widening the gap between the utilization of natural resources and the carrying capacity of mountain ranges, in particular in the Middle Atlas where pastoralism always played a predominant role. Socio-ecological changes include the commercialization of pastoralism and a general decline of transhumance, manifested in increasing sedentarization, the decline of traditional institutions regulating herd mobility, reduced pastoral territories and herd mobility, and increased livestock numbers, spatial concentration of herds and grazing season duration (Breuer 2007; El Aich 2018). The ongoing overuse of rangeland resources is a striking contrast to the mountain ranges in the northern Mediterranean basin. Although migration to lowland cities or abroad has a long tradition (in Morocco over the entire post-independence period), and remittance generation has considerably improved living conditions (de Haas 2009; Berriane et al. 2015), unsustainable use of economic resources has not been significantly alleviated. International migration has resulted in increasing agricultural productivity rather than in retreat from agriculture (de Haas 2006; Rössler et al. 2010), while farmers diversify their sources of income with, inter alia, tourist-related activities. Climate change will add a significant challenge to environmental and anthropogenic systems in the Atlas Mountains (Linstädter et al. 2010; Schilling et al. 2012).

The Ethiopian Highlands have been almost entirely reshaped into an anthropogenic agricultural landscape. Favourable natural resources have attracted human settlers ever since, thus deforestation is a very old phenomenon. Many centuries of land resource utilization by a growing population, mainly subsistence agriculture with crop cultivation and animal husbandry, have reduced the original forest cover of c. 80% to below 5%, with the remaining forests located in the southwestern part of the highlands (Hurni et al. 1992). The eventful history explains the spatio-temporal differentiation of land cover changes, with political stability/instability, foreign invasions, population growth, droughts, locusts, repeated famines, and economic prosperity being most important drivers of land use intensity and cultural landscape evolution. Evaluations of historical travel accounts revealed that over the past centuries phases with widespread land degradation alternated with phases of recovery, and that in many parts of the highlands closed forests were already completely absent in the early nineteenth century, in particular within the well-populated elevational belts between 1500 and 2700 m (Ritler 1997, 2003; Munro et al. 2008). A series of historical photographs of 1868 clearly shows that the status of natural resources in northern Ethiopia was already very degraded 150 years ago (Nyssen et al. 2009). After recovery from the major famine and epizootic of 1889–92 and the influenza pandemic of 1918–19, the highlands experienced steady population growth under higher political security over the following decades, resulting in local migration to agriculturally marginal zones where population pressure on land resources (fuelwood, grazing lands, new cultivation areas) increased. Cultivation expanded to steeper slopes and from the long-term mid-elevation settlement zones into lower elevations, while the upper limit of cultivation was shifted to just below the frost line (McCann 1995). An expansion of land use into higher elevations was also observed in the Bale Mountains (Miehe and Miehe 1994; Kidane et al. 2012; Hailemariam et al. 2016). Several studies confirmed a deforestation trend in favour of cultivation over the second half of the twentieth century (Kebrom and Hedlund 2000; Zeleke and Hurni 2001; Bewket 2002), continuing in places to the present day, partly driven by the government policy on land resources and land rights, and by the market-oriented production of high value crops (Lanckriet et al. 2015; Tolessa et al. 2017; Solomon et al. 2018; Strobelt and von Kocemba 2020). It also needs to be highlighted that the population has increased from 6.6 million in 1868 (Nyssen et al. 2009) to 115 million in 2020 (according to UN data), while over 90% of the population's energy requirement is still obtained mainly from biomass (Lemenih and Kassa 2014). Small remnants of the forest climax vegetation only remained in sacred groves around churches and in isolated areas (Wassie et al. 2010; Aerts et al. 2016). Even sacred church forests are threatened by human disturbance (Cardelús et al. 2019).

In general, processes of deforestation, overgrazing, and soil erosion over long time periods have resulted in tremendous land degradation (Nyssen et al. 2004, 2015). The northern highlands appear to be the most severely eroded part of Ethiopia, showing high to extremely high soil loss rates, decreased agricultural productivity (crop and livestock), and increased famine vulnerability (Hurni et al. 1992). Erosion surveys in the late 1960s prompted the initiation of nationwide soil and water conservation programmes and reforestation activities (eucalypt plantations), supported by international development aid after the disastrous drought in the early 1970s (Munro et al. 2008). Meanwhile, positive outcomes of these land rehabilitation programmes, being facilitated by the growing awareness of landholders (Fig. 1.63), are clearly visible (Fig. 1.64), reflected in new eucalypt woodlands, regeneration of indigenous trees and shrubs, and improved soil protection (Bewket 2002; Nyssen et al. 2004, 2009, 2015; de Mûelenaere et al. 2014). Recovery of vegetation was also assessed in subalpine and afro-alpine zones including treeline advance, while a decrease of areas with dense forest has occurred, on the other hand, even in some protected areas (Wondie et al. 2011; Jacob et al. 2017). A promising approach to halt the process of deforestation and forest degradation is participatory forest management which was introduced with pilot projects in the 1990s and found to provide mixed results so far in terms of livelihood and ecological benefits (Ameha et al. 2014, 2016). Even though the depletion of resources continues in places, the positive impact of improved land husbandry shows that land degradation in the Ethiopian Highlands is not principally irreversible (Nyssen et al. 2009). Land rehabilitation programmes should be supported by a rural development policy promoting livelihood strategies that are both environmentally friendly and economically sound. A promising path of rural development could be the shift from the traditionally preferred 'cereal crop-livestock mix' dominated livelihood strategy to one dominated by cash income-based activities such as off-farm business, honey production, poultry, and horticulture (Babulo et al. 2008). To relief pressure from the chronically food-insecure highlands, the government has conducted (much-criticized) resettlement programmes. Rural–urban migration continues to occur at high levels, while international migration flows out of Ethiopia are relatively small (although a much-desired possibility) as are the impacts on the local economy by remittances (Fransen and Kuschminder 2009).

According to historical accounts (compiled in Spinage 2012), mountain regions in Uganda, Rwanda, Kenya and Tanzania have lost much of its forest cover during the past 200 years, often resulting in increased soil erosion and more frequent landslides. In the highlands of Kenya, traditional land use patterns were completely transformed during colonial times, when the colonial rulers established 'white highlands', with white settlers developing export-oriented agriculture on large-scale farms. Following independence in 1963, highland areas were subdivided and transferred to indigenous small-scale farmers, resulting, inter alia, in intensified mixed farming systems, expansion into agriculturally marginal areas, economic marginalization, forest depletion, and land degradation (Hurni et al. 1992). Rapid population growth over recent decades and economic globalization has resulted in substantial agricultural expansion in the Mount Kenya area (Kiteme et al. 2008). Pressure on water and land in the foothills has more recently increased by the expansion of horticultural agribusinesses, while land use in the region remains dominated by small-scale crop and

Fig. 1.63 Land rehabilitation programmes, such as the construction of terraces for soil and water conservation, enjoy firm support from local communities and are actively supported (Tigray, northern Ethiopian Highlands). (*Photo* © Udo Schickhoff, February 21, 2020)

Fig. 1.64 A vivid illustration of successful land rehabilitation at Bolago, northern Ethiopian Highlands: Tree cover has much improved since 1868, afforestation started in the late 1980s. (*Photos* courtesy of Jan Nyssen; modified from Nyssen et al. 2009)

livestock farms, producing both for their own subsistence and for the local markets (Zaehringer et al. 2018). The increase of (increasingly irrigated) cropland is associated with a further decline of small forest patches, bush- and shrubland, but also with enlarged forest plantations. At higher elevations, the adoption of agroforestry systems has increased tree cover, while the land cover of protected areas including Mount Kenya National Park and National Forest remained rather stable over the past 30 years (Eckert et al. 2017). It needs to be highlighted that, even though not free from human impact, almost 100% of the afro-alpine zones in East Africa are under various forms of formal protection (Wesche et al. 2008b; Carbutt 2020).

The forest cover on the slopes of Mt. Kilimanjaro is reported to have been much more extensive in the early nineteenth century (Spinage 2012). However, logging and burning have resulted in significant land cover changes over the twentieth century. Land use pressure has increased due to enormous population growth, with the local population having multiplied 20 times since 1895 (Hemp 2005b). Recurrent fires, mainly started by humans, have played an increasingly destructive role in recent decades (Hemp 2006a). Over the past century, Mt. Kilimanjaro has lost some 300 km^2 of high altitude forests, and the upper closed forest line was lowered by 900 m because of fire (Hemp 2006b). Fire frequency is expected to increase with rising temperatures and decreasing precipitation. In addition to the impact of fire, clear-cutting of montane forests reduced the forest area by 450 km^2 since 1929, resulting in a total loss of the nineteenth-century forest cover of c. 50% (Lambrechts et al. 2002; Hemp 2006b). This forest depletion affects the fog water collection and thus the water balance of the whole mountain. Cutting of trees and illegal logging has been reduced after the introduction of stringent bylaws in 2000 (Kilungu et al. 2019). Current land use changes at lower elevations include the increasing transformation of savannah woodlands into maize fields, the emergence of commercial coffee plantations within the altitudinal zone of the traditional agricultural system of the local Chagga people (Chagga home gardens and grasslands), and the enlargement of forest plantations (Hemp 2006b; Ensslin et al. 2015). Cultivation has expanded to more marginal land down the slopes, associated with the disappearance and extreme fragmentation of bushland and appearance and expansion of settlements (Soini 2005). Logging is insignificant in the upper forest zone, and above the forest belt, grazing and agriculture are non-existent. However, the Kilimanjaro National Park attracts an increasing number of visitors each year, generating increasing human impact on the sensitive alpine zone. Nevertheless, the development of ecotourism is a promising economic alternative for the poverty-stricken, rapidly growing population (Agrawala et al. 2003). While non-agricultural activities and paid employment are becoming increasingly important, considerable entry barriers to remunerable off-farm jobs persist for many households, restricting access to attractive non-farm opportunities (Soini 2005). On the other hand, experiences from Mt. Kenya and the Rwenzori Mountains in Uganda show that alpine tourism has so far failed to meet up to expectations in terms of economic benefits and the promotion of sustainable development, even though a stabilizing effect on the livelihood of rural households is discernible (Neuburger and Steinicke 2012). After the establishment of the Rwenzori Mountains National Park in 1992 (Fig. 1.65), land use restrictions directed the population pressure to the foothills, causing there high population density, unsustainable resource use, and social tensions with adjacent ethnic groups (Steinicke 2011). Poor households in the sub-counties bordering the national park still exhibit a great dependence on forest resources inside the park, which are illegally collected and have a significant impact on reducing income inequalities and making the poor less poor. In order to protect the park, encouraging a pro-poor conservation approach rather than increased law enforcement is required (Tumusiime et al. 2011). In southern Africa, prolonged grazing pressure, originating from prevailing extensive livestock farming mainly practised by commercial farmers, has accelerated soil erosion processes in the Drakensberg mountain region. Lesotho was considered one of the most severely eroded countries in Africa, commonly attributed to overstocking and overgrazing of cattle and sheep on communal lands (Acocks 1988). However, the landscapes of the Drakensberg region have been shaped by multiple factors including legislated disenfranchisement and territorial segregation since the 1800s (Salomon et al. 2012). The highland grasslands have been to some extent converted to cultivation and plantations, while especially the Lesotho Highland basalt grassland is heavily utilized for grazing and subject to severe erosion (Mucina et al. 2006; Brown and du Preez 2020). Recent conservation initiatives including the important Maloti-Drakensberg Transfrontier Park should be accompanied by promoting off-reserve conservation on privately or communally owned land.

Fig. 1.65 The establishment of protected areas in the East African mountain system in recent decades such as the Rwenzori National Park (in 1992) in Uganda has increased land use pressure in the surrounding foothills. (*Photo* © Udo Schickhoff, February 12, 2019)

1.4 Conclusions

An unprecedented dimension of change in the world's mountains is obvious from this review, triggered by global climate change and economic globalization. This novel dimension of change is increasingly well documented in relevant publications (see the comprehensive list of recent references), that allow to identify globally significant trends and processes of transformation, but also regional variations. The dramatic change in magnitude and rate of cryospheric and biotic responses and the rapid pace of implementing adaptation strategies in response to changing socio-economic frame conditions completes the overall picture known as the Great Acceleration which describes accelerating Earth system trends in the Anthropocene. Elevational zones in mountains of the world are experiencing strong levels of temperature increase in the frame of anthropogenic climate change, causing cascading effects on physical, biological, and human systems that, in turn, trigger feedbacks to the climate system. Pervasive cryosphere changes including glacier retreat, snow cover decline, and permafrost degradation increase natural hazard risks, and affect seasonal water supply in river systems, with potentially severe implications for agriculture, hydropower generation, and local water resources availability. Declining water supply from mountains will threaten livelihoods and food security of millions of lowland people, in particular in South and East Asia, and may lead to conflicts over water resources. Biotic responses to climate change such as phenological shifts, changing species distributions, invasion of non-native species, and changes in primary production will modify species composition of communities and thus structure and functioning of ecosystems, affecting the provision of ecosystem services for millions of people in downstream areas. Given the low capacity of alpine plant and animal species to adapt to novel climatic conditions, it must be assumed that loss of species, biodiversity decline, and impairment of ecosystem services will be inevitable. Human systems in the world's mountains are passing through a process of implementing adaptations to an increasing magnitude of impact from climate change and globalization processes. Conforming to the heterogeneity of poverty and marginalization levels within and between mountain regions in the Global South, in emerging markets and in industrialized countries, a wide spectrum of adaptations and responses depending on socio-economic conditions, political guidelines, and environmental changes is discernible. Transformations in mountain agriculture, extractive industries, tourism and other sectors are reflected in land use/land cover changes. In the majority of examined mountain systems in this review, current transformations provide the chance to counter the downward spiral of resource degradation, rural poverty, and livelihood insecurity. From an ecological point of view, the recent trend of reduced land use intensity in alpine zones and of the increase and enlargement of protected areas in mountain regions offers the chance for ecosystem recovery and more efficient biodiversity conservation. However, establishing land use systems in high mountain regions which safeguard livelihood and ecological sustainability remains a considerable task. It needs to be embedded in the overriding priority of mitigating adverse effects of drivers of environmental and socio-economic change in the world's mountains. In order to accelerate the implementation of the UN Sustainable Development Goals, the recognition of the global significance of mountain regions needs to be further consolidated and disseminated.

References

APCC (Austrian Panel on Climate Change) (2014) Österreichischer Sachstandsbericht Klimawandel 2014 (AAR14). APCC-Verlag der Österreichischen Akademie der Wissenschaften, Vienna

Aakala T, Hari P, Dengel S, Newberry SL, Mizunuma T, Grace J (2014) A prominent stepwise advance of the tree line in North-East Finland. J Ecol 102:1582–1591

Abate Y (1993) The society and its environment. In: Ofcansky TP, Berry L (eds) Ethiopia: a country study. Federal Research Division, Library of Congress, Washington, DC, pp 69–141

Abatzoglou JT, Rupp DE, Mote PW (2014) Seasonal climate variability and change in the Pacific Northwest of the United States. J Clim 27:2125–2142

Abrha H, Birhane E, Hagos H, Manaye A (2018) Predicting suitable habitats of endangered *Juniperus procera* tree under climate change in northern Ethiopia. J Sustain For 37:842–853

Acocks JPH (1988) Veld types of South Africa. Memoirs of the botanical survey of South Africa 57. Botanical Research Institute, Pretoria

Adhikari BS, Kumar R, Singh SP (2018) Early snowmelt impact on herb species composition, diversity and phenology in a western Himalayan treeline ecotone. Trop Ecol 59:365–382

Aerts R, Van Overtveld K, November E, Wassie A, Abiyu A et al (2016) Conservation of the Ethiopian church forests: threats, opportunities and implications for their management. Sci Total Environ 551:404–414

Agrawala S, Moehner A, Hemp A, Aalst MV, Hitz S et al (2003) Development and climate change in Tanzania: Focus on Mount Kilimanjaro. OECD, Paris

El Aich A (2018) Changes in livestock farming systems in the Moroccan Atlas Mountains. Open Agric 3:131–137

Aide TM, Grau HR, Graesser J, Andrade-Nuñez MJ, Aráoz E et al (2019) Woody vegetation dynamics in the tropical and subtropical Andes from 2001 to 2014: satellite image interpretation and expert validation. Glob Change Biol 25:2112–2126

Aizen VB, Aizen EM, Melack JM, Dozier J (1997) Climatic and hydrologic changes in the Tien Shan, Central Asia. J Clim 10:1393–1404

Akatov PV (2009) Changes in the upper limits of tree species distribution in the western Caucasus (Belaya river basin) related to recent climate warming. Russ J Ecol 40:33–38

Akhmadov KM, Breckle SW, Breckle U (2006) Effects of grazing on biodiversity, productivity, and soil erosion of alpine pastures in Tajik Mountains. In: Spehn EM, Liberman M, Körner C (eds) Land use change and mountain biodiversity. Taylor & Francis, Boca Raton-London-New York, pp 239–247

Alagona PS, Paulson T, Esch AB, Marter-Kenyon J (2016) Population and land use. In: Mooney H, Zavaleta E (eds) Ecosystems of California. University of California Press, Berkeley, pp 75–94

Albrich K, Rammer W, Seidl R (2020) Climate change causes critical transitions and irreversible alterations of mountain forests. Glob Change Biol 26:4013–4027

Alexander JM, Lembrechts JJ, Cavieres LA, Daehler C, Haider S et al (2016) Plant invasions into mountains and alpine ecosystems: Current status and future challenges. Alp Bot 126:89–103

Allen JA, Brown CS, Stohlgren TJ (2009) Non-native plant invasions of United States national parks. Biol Invasions 11:2195–2207

Allen SK, Cox SC, Owens IF (2011) Rock avalanches and other landslides in the Central southern Alps of New Zealand: a regional study considering possible climate change impacts. Landslides 8:33–48

Allen SK, Fiddes J, Linsbauer A, Randhawa SS, Saklani B, Salzmann N (2016) Permafrost studies in Kullu district, Himachal Pradesh. Curr Sci 11:257–260

Allen CD (2002) Lots of lightning and plenty of people: an ecological history of fire in the upland Southwest. In: Vale TR (ed) Fire, native peoples, and the natural landscape. Island Press, Washington DC, pp 143–194

Allen RB, Lee WG (2006) Biological invasions in New Zealand. Ecological Studies 186, Springer, Berlin

Amano T, Freckleton RP, Queenborough SA, Doxford SW, Smithers RJ, Sparks TH, Sutherland WJ (2014) Links between plant species' spatial and temporal responses to a warming climate. Proc Roy Soc B: Biol Sci 281:20133017

Amano T, Smithers RJ, Sparks TH, Sutherland WJ (2010) A 250-year index of first flowering dates and its response to temperature changes. Proc Roy Soc B: Biol Sci 277:2451–2457

Ameha A, Meilby H, Feyisa GL (2016) Impacts of participatory forest management on species composition and forest structure in Ethiopia. Int J Biodivers Sci Ecosyst Serv Manage 12:139–153

Ameha A, Nielsen OJ, Larsen HO (2014) Impacts of access and benefit sharing on livelihoods and forest: case of participatory forest management in Ethiopia. Ecol Econ 97:162–171

Améztegui A, Brotons L, Coll L (2010) Land-use changes as major drivers of mountain pine (*Pinus uncinata* Ram.) expansion in the Pyrenees. Glob Ecol Biogeogr 19:632–641

Améztegui A, Coll L, Brotons L, Ninot JM (2016) Land-use legacies rather than climate change are driving the recent upward shift of the mountain tree line in the Pyrenees. Glob Ecol Biogeogr 25:263–273

Anderson K, Fawcett D, Cugulliere A, Benford S, Jones D, Leng R (2020) Vegetation expansion in the subnival Hindu Kush Himalaya. Glob Change Biol 26:1608–1625

Anderson EP, Marengo J, Villalba R, Halloy S, Young B et al (2011) Consequences of climate change for ecosystems and ecosystem services in the tropical Andes. In: Herzog SK, Martinez R, Jørgensen PM, Tiessen H (eds) Climate change and biodiversity in the tropical Andes. IAI-SCOPE, Paris, pp 1–18

Andreassen LM, Elvehøy H, Kjøllmoen B, Belart JM (2020) Glacier change in Norway since the 1960s—an overview of mass balance, area, length and surface elevation changes. J Glaciol 66:313–328

Andreassen LM, Elvehøy H, Kjøllmoen B, Engeset RV (2016) Reanalysis of long-term series of glaciological and geodetic mass balance for 10 Norwegian glaciers. Cryosphere 10:535–552

Anup KC, Rijal K, Sapkota RP (2015) Role of ecotourism in environmental conservation and socioeconomic development in Annapurna conservation area, Nepal. Int J Sustain Dev World Ecol 22:251–258

Anyah RO, Qiu W (2012) Characteristic 20th and 21st century precipitation and temperature patterns and changes over the Greater Horn of Africa. Int J Climatol 32:347–363

Aravena JC, Lara A, Wolodarky-Franke A, Villalba R, Cuq E (2002) Tree-ring growth patterns and temperature reconstruction from *Nothofagus pumilio* (*Fagaceae*) forest at the upper tree line of southern Chilean Patagonia. Rev Chil Hist Nat 75:361–376

Aryal S, Maraseni TN, Cockfield G (2014) Sustainability of transhumance grazing systems under socio-economic threats in Langtang, Nepal. J Mountain Sci 11:1023–1034

Asam S, Callegari M, Matiu M, Fiore G, De Gregorio L et al (2018) Relationship between spatiotemporal variations of climate, snow cover and plant phenology over the Alps—an earth observation-based analysis. Remote Sens 10:1757

Asse D, Chuine I, Vitasse Y, Yoccoz NG, Delpierre N et al (2018) Warmer winters reduce the advance of tree spring phenology induced by warmer springs in the Alps. Agric For Meteorol 252:220–230

Auer I, Böhm R, Jurkovic A, Lipa W, Orlik A et al (2007) HISTALP—historical instrumental climatological surface time series of the Greater Alpine Region. Int J Climatol 27:17–46

Ault TR, Macalady AK, Pederson GT, Betancourt JL, Schwartz MD (2011) Northern hemisphere modes of variability and the timing of spring in western North America. J Clim 24:4003–4014

Austrheim G, Solberg EJ, Mysterud A (2011) Spatio-temporal variation in large herbivore pressure in Norway during 1949–1999: has decreased grazing by livestock been countered by increased browsing by cervids? Wildl Biol 17:286–298

Azam MF, Wagnon P, Berthier E, Vincent C, Fujita K, Kargel JS (2018) Review of the status and mass changes of Himalayan-Karakoram glaciers. J Glaciol 64:61–74

Babulo B, Muys B, Nega F, Tollens E, Nyssen J, Deckers J, Mathijs E (2008) Household livelihood strategies and forest dependence in the highlands of Tigray, northern Ethiopia. Agric Syst 98:147–155

Bacchiocchi SC, Zerbe S, Cavieres LA, Wellstein C (2019) Impact of ski piste management on mountain grassland ecosystems in the southern Alps. Sci Total Environ 665:959–967

Bach AJ, Price LW (2013) Mountain climate. In: Price MF, Byers AC, Friend DA, Kohler T, Price LW (eds) Mountain geography. University of California Press, Berkeley-Los Angeles, Physical and human dimensions, pp 41–84

Bader MY, Van Geloof I, Rietkerk M (2007) High solar radiation hinders tree regeneration above the alpine treeline in northern Ecuador. Plant Ecol 191:33–45

Bader MY, Llambí LD, Case BS, Buckley HL, Toivonen JM et al (2020) A global framework for linking alpine-treeline ecotone patterns to underlying processes. Ecography 43:1–24

Badgley C, Smiley TM, Terry R, Davis EB, DeSantis LR et al (2017) Biodiversity and topographic complexity: modern and geohistorical perspectives. Trends Ecol Evol 32:211–226

Bahn M, Körner C (2003) Recent increases in summit flora caused by warming in the Alps. In: Nagy L, Grabherr G, Körner C, Thompson DBA (eds) Alpine biodiversity in Europe. Ecological Studies 167. Springer, Berlin, pp 437–441

Baied CA, Wheeler JC (1993) Evolution of high Andean puna ecosystems: environment, climate, and culture change over the last 12,000 years in the Central Andes. Mt Res Dev 13:145–156

Bajracharya SB, Chaudhary RP, Basnet G (2015) Biodiversity conservation and protected area management in Nepal. In: Miehe G, Pendry CA, Chaudhary RP (eds) Nepal: an introduction to the natural history, ecology and human environment of the Himalayas. Royal Botanic Garden Edinburgh, Edinburgh, pp 473–486

Bajracharya SR, Maharjan SB, Shrestha F, Bajracharya OR, Baidya S (2014a) Glacier status in Nepal and decadal change from 1980 to 2010 based on Landsat data. ICIMOD, Kathmandu

Bajracharya SR, Maharjan SB, Shrestha F (2014b) The status and decadal change of glaciers in Bhutan from the 1980s to 2010 based on satellite data. Ann Glaciol 55:159–166

Baker BB, Moseley RK (2007) Advancing treeline and retreating glaciers: implications for conservation in Yunnan, P.R. China. Arct Antarct Alp Res 39:200–209

Balthazar V, Vanacker V, Molina A, Lambin EF (2015) Impacts of forest cover change on ecosystem services in high Andean mountains. Ecol Ind 48:63–75

Baniya B, Tang Q, Huang Z, Sun S, Techato KA (2018) Spatial and temporal variation of NDVI in response to climate change and the implication for carbon dynamics in Nepal. Forests 9:329

Bao G, Bao Y, Sanjjava A, Qin Z, Zhou Y, Xu G (2015) NDVI-indicated long-term vegetation dynamics in Mongolia and their response to climate change at biome scale. Int J Climatol 35:4293–4306

Baojuan H, Weijun S, Yetang W, Zhongqin L (2017) Glacier shrinkage in the Chinese Tien Shan Mountains from 1959/1972 to 2010/2012. Arct Antarct Alp Res 49:213–225

Barakat A, Khellouk R, El Jazouli A, Touhami F, Nadem S (2018) Monitoring of forest cover dynamics in eastern area of Béni-Mellal Province using ASTER and Sentinel-2A multispectral data. Geol Ecol Landscapes 2:203–215

Barandun M, Huss M, Usubaliev R, Azisov E, Berthier E et al (2018) Multi-decadal mass balance series of three Kyrgyz glaciers inferred from modelling constrained with repeated snow line observations. Cryosphere 12:1899–1919

Baranova A, Schickhoff U, Wang S, Jin M (2016) Mountain pastures of Qilian Shan: plant communities, grazing impact and degradation status (Gansu province, NW China). Hacquetia 15:21–35

Barichivich J, Briffa KR, Myneni RB, Osborn TJ, Melvin TM et al (2013) Large-scale variations in the vegetation growing season and annual cycle of atmospheric CO_2 at high northern latitudes from 1950 to 2011. Glob Change Biol 19:3167–3183

Barman S, Bhattacharjya RK (2015) Change in snow cover area of Brahmaputra river basin and its sensitivity to temperature. Environ Syst Res 4:16

Del Barrio G, Sanjuan ME, Hirche A, Yassin M, Ruiz A et al (2016) Land degradation states and trends in the northwestern Maghreb drylands, 1998–2008. Remote Sensing 8:603

Barros A, Pickering CM (2014) Non-native plant invasion in relation to tourism use of Aconcagua Park, Argentina, the highest protected area in the southern hemisphere. Mt Res Dev 34:13–26

Barthlott W, Hostert A, Kier G, Küper W, Kreft H et al (2007) Geographic patterns of vascular plant diversity at continental to global scales. Erdkunde 61:305–315

Barthlott W, Mutke J, Rafiqpoor D, Kier G, Kreft H (2005) Global centers of vascular plant diversity. Nova Acta Leopoldina NF 92:61–83

Basagic HJ, Fountain AG (2011) Quantifying 20th century glacier change in the Sierra Nevada, California. Arct Antarct Alp Res 43:317–330

Basu S, Mohanty S, Sanyal P (2020) Possible role of warming on Indian summer monsoon precipitation over the North-Central Indian subcontinent. Hydrol Sci J 65:660–670

Batllori E, Gutiérrez E (2008) Regional tree line dynamics in response to global change in the Pyrenees. J Ecol 96:1275–1288

Baumann M, Kuemmerle T, Elbakidze M, Ozdogan M, Radeloff VC et al (2011) Patterns and drivers of post-socialist farmland abandonment in western Ukraine. Land Use Policy 28:552–562

Baumhackl H (2019) Peru "land of the Incas". A tourism destination on the rise. J Tourism Hospitality Manag 7:95–116

Beazley RE, Lassoie JP (2017) Himalayan mobilities: an exploration of the impact of expanding rural road networks on social and ecological systems in the Nepalese Himalaya. Springer, Cham

Bebbington A, Bebbington DH, Bury J, Lingan J, Muñoz JP, Scurrah M (2008) Mining and social movements: struggles over livelihood and rural territorial development in the Andes. World Dev 36:2888–2905

Bebi P, Seidl R, Motta R, Fuhr M, Firm D et al (2017) Changes of forest cover and disturbance regimes in the mountain forests of the Alps. For Ecol Manage 388:43–56

Beck PS, Goetz SJ (2012) Corrigendum: Satellite observations of high northern latitude vegetation productivity changes between 1982 and 2008: ecological variability and regional differences. Environ Res Lett 7:029501

Beck PS, Juday GP, Alix C, Barber VA, Winslow SE et al (2011) Changes in forest productivity across Alaska consistent with biome shift. Ecol Lett 14:373–379

Beckage B, Osborne B, Gavin DG, Pucko C, Siccama T, Perkins T (2008) A rapid upward shift of a forest ecotone during 40 years of warming in the Green Mountains of Vermont. Proc Natl Acad Sci 105:4197–4202

Becker T, Dietz H, Billeter R, Buschmann H, Edwards PJ (2005) Altitudinal distribution of alien plant species in the Swiss Alps. Perspect Plant Ecol Evol Syst 7:173–183

Beedle MJ, Menounos B, Wheate R (2015) Glacier change in the Cariboo Mountains, British Columbia, Canada (1952–2005). Cryosphere 9:65–80

Beer R, Kaiser F, Schmidt K, Ammann B, Carraro G, Grisa E, Tinner W (2008) Vegetation history of the walnut forests in Kyrgyzstan (Central Asia): natural or anthropogenic origin? Quatern Sci Rev 27:621–632

Beesley D (2004) Crow's range: an environmental history of the Sierra Nevada. University of Nevada Press, Reno

Bellingham PJ, Towns DR, Cameron EK, Davis JJ, Wardle DA, Wilmshurst JM, Mulder CP (2010) New Zealand island restoration: seabirds, predators, and the importance of history. N Z J Ecol 34:115–136

Bello JC, Báez M, Gómez MF, Orrego O, Nägele L (2014) Biodiversidad 2014. Estado y tendencias de la biodiversidad continental de Colombia. Instituto Alexander von Humboldt, Bogotá DC, Colombia

Belmecheri S, Babst F, Wahl ER, Stahle DW, Trouet V (2016) Multi-century evaluation of Sierra Nevada snowpack. Nat Clim Chang 6:2–3

Belonovskaya E, Gracheva R, Shorkunov I, Vinogradova V (2016) Grasslands of intermontane basins of central Caucasus: land use legacies and present-day state. Hacquetia 15:37–47

Bencherifa A (1990) Demography and cultural ecology of the Atlas Mountains of Morocco: some new hypotheses. In: Messerli B, Hurni H (eds) African mountains and highlands: problems and perspectives. Walsworth Press, Marceline, pp 369–377

Beniston M, Stoffel M (2014) Assessing the impacts of climatic change on mountain water resources. Sci Total Environ 493:1129–1137

Beniston M, Stoffel M, Hill M (2011) Impacts of climatic change on water and natural hazards in the Alps: can current water governance cope with future challenges? Examples from the European "ACQWA" project. Environ Sci Policy 14:734–743

Beniston M, Farinotti D, Stoffel M, Andreassen LM, Coppola E et al. (2018) The European mountain cryosphere: a review of its current state, trends, and future challenges. Cryosphere 12:759-794

Bentley LK, Robertson MP, Barker NP (2019) Range contraction to a higher elevation: the likely future of the montane vegetation in South Africa and Lesotho. Biodivers Conserv 28:131–153

Bentz BJ, Régnière J, Fettig CJ, Hansen EM, Hayes JL et al (2010) Climate change and bark beetles of the western United States and Canada: direct and indirect effects. Bioscience 60:602–613

Bergamini A, Ungricht S, Hofmann H (2009) An elevational shift of cryophilous bryophytes in the last century—an effect of climate warming? Divers Distrib 15:871–879

Bergmann C, Gerwin M, Nüsser M, Sax WS (2012) State policy and local performance: pasture use and pastoral practices in the Kumaon Himalaya. In: Kreutzmann H (ed) Pastoral practices in High Asia. Springer, Dordrecht, pp 175–194

Berkelhammer M, Stefanescu IC, Joiner J, Anderson L (2017) High sensitivity of gross primary production in the Rocky Mountains to summer rain. Geophys Res Lett 44:3643–3652

Berner LT, Law BE, Hudiburg TW (2017) Water availability limits tree productivity, carbon stocks, and carbon residence time in mature forests across the western US. Biogeosciences 14:365–378

Berriane M, De Haas H, Natter K (2015) Introduction: revisiting Moroccan migrations. J North Afr Stud 20:503–521

Berthier É, Brun F (2019) Karakoram geodetic glacier mass balances between 2008 and 2016: persistence of the anomaly and influence of a large rock avalanche on Siachen Glacier. J Glaciol 65:494–507

Bertrand R, Lenoir J, Piedallu C, Riofrio-Dillon G, De Ruffray P et al (2011) Changes in plant community composition lag behind climate warming in lowland forests. Nature 479:517–520

Beug HJ, Miehe G (1999) Vegetation history and human impact in the eastern Central-Himalaya (Langtang and Helambu, Nepal). Dissertationes Botanicae 318. Cramer, Berlin-Stuttgart

Bewket W (2002) Land cover dynamics since the 1950s in Chemoga watershed, Blue Nile basin, Ethiopia. Mt Res Dev 22:263–269

Bezák P, Halada L (2010) Sustainable management recommendations to reduce the loss of agricultural biodiversity in the mountain regions of NE Slovakia. Mt Res Dev 30:192–204

Bhagawati R, Bhagawati K, Jini D, Alone RA, Singh R et al (2017) Review on climate change and its impact on agriculture of Arunachal Pradesh in the northeastern Himalayan region of India. Nat Environ Pollut Technol 16:535

Bhambri R, Bolch T, Chaujar RK (2012) Frontal recession of Gangotri Glacier, Garhwal Himalayas, from 1965 to 2006, measured through high-resolution remote sensing data. Curr Sci 102:489–494

Bhambri R, Hewitt K, Kawishwar P, Pratap B (2017) Surge-type and surge-modified glaciers in the Karakoram. Sci Rep 7:1–14

Bhasin V (2011) Pastoralists of Himalayas. J Hum Ecol 33:147–177

Bhattacharjee A, Anadon JD, Lohman DJ, Doleck T, Lakhankar T et al (2017) The impact of climate change on biodiversity in Nepal: current knowledge, lacunae, and opportunities. Climate 5:80

Bhattacharya A, Bolch T, Mukherjee K, Pieczonka T, Kropáček J, Buchroithner MF (2016) Overall recession and mass budget of Gangotri Glacier, Garhwal Himalayas, from 1965 to 2015 using remote sensing data. J Glaciol 62:1115–1133

Bhattacharyya A, Shah SK, Chaudhary V (2006) Would tree ring data of *Betula utilis* be potential for the analysis of Himalayan glacial fluctuations? Curr Sci 91:754–761

Bhattarai KR, Måren IE, Subedi SC (2014) Biodiversity and invasibility: distribution patterns of invasive plant species in the Himalayas, Nepal. J Mt Sci 11:688–696

Bhutiyani MR (2015) Climate change in the northwestern Himalayas. In: Joshi R, Kumar K, Palni LMS (eds) Dynamics of climate change and water resources of northwestern Himalaya. Springer, Cham, pp 85–96

Bhutiyani MR, Kale VS, Pawar NJ (2007) Long-term trends in maximum, minimum and mean annual air temperatures across the northwestern Himalaya during the twentieth century. Clim Change 85:159–177

Bhutiyani MR, Kale VS, Pawar NJ (2008) Changing streamflow patterns in the rivers of northwestern

Himalaya: implications of global warming in the 20th century. Curr Sci 95:618–626

Bhutiyani MR, Kale VS, Pawar NJ (2010) Climate change and the precipitation variations in the northwestern Himalaya: 1866–2006. Int J Climatol 30:535–548

Bhutiyani MR (2016) Spatial and temporal variability of climate change in high-altitude regions of NW Himalaya. In: Singh RB, Schickhoff U, Mal S (eds) Climate change and dynamics of glaciers and vegetation in the Himalaya. Springer, Cham, pp 87–101

Bi Y, Xu J, Yang J, Li Z, Gebrekirstos A et al (2017) Ring-widths of the above tree-line shrub *Rhododendron* reveal the change of minimum winter temperature over the past 211 years in southwestern China. Clim Dyn 48:3919–3933

Bianchi E, Villalba R, Solarte A (2020) NDVI spatio-temporal patterns and climatic controls over northern Patagonia. Ecosystems 23:84–97

Biemans H, Siderius C, Lutz AF, Nepal S, Ahmad B et al (2019) Importance of snow and glacier meltwater for agriculture on the Indo-Gangetic plain. Nat Sustain 2:594–601

Bilal H, Chamhuri S, Mokhtar MB, Kanniah KD (2019) Recent snow cover variation in the upper Indus basin of Gilgit Baltistan, Hindukush Karakoram Himalaya. J Mt Sci 16:296–308

Bisht MPS, Rana V, Singh S (2016) Impact of glacial recession on the vegetational cover of Valley of Flowers National Park (a World Heritage Site), Central Himalaya, India. In: Singh RB, Schickhoff U, Mal S (eds) Climate change, glacier response, and vegetation dynamics in the Himalaya. Springer, Cham, pp 377–390

Biskaborn BK, Smith SL, Noetzli J, Matthes H, Vieira G et al (2019) Permafrost is warming at a global scale. Nat Commun 10:264

Blondel J (2006) The 'design'of Mediterranean landscapes: a millennial story of humans and ecological systems during the historic period. Hum Ecol 34:713–729

Blondel J, Aronson J, Bodiou JY, Boeuf G (2010) The Mediterranean region: biological diversity in space and time. Oxford University Press, Oxford

Blumler MA (2011) Invasive species, in geographical perspective. In: Millington A, Blumler M, Schickhoff U (eds) Handbook of biogeography. Sage Publ, London, pp 510–527

Bobrowski M, Gerlitz L, Schickhoff U (2017) Modelling the potential distribution of *Betula utilis* in the Himalaya. Global Ecol Conserv 11:69–83

Bocchiola D, Diolaiuti G (2013) Recent (1980–2009) evidence of climate change in the upper Karakoram, Pakistan. Theoret Appl Climatol 113:611–641

Bock JH, Jolls CL, Lewis AC (1995) The effects of grazing on alpine vegetation: a comparison of the Central Caucasus, Republic of Georgia, with the Colorado Rocky Mountains, USA. Arct Alp Res 27:130–136

Bodin J, Badeau V, Bruno E, Cluzeau C, Moisselin JM, Walther GR, Dupouey JL (2013) Shifts of forest species along an elevational gradient in Southeast France: climate change or stand maturation? J Veg Sci 24:269–283

Van Bogaert R, Haneca K, Hoogesteger J, Jonasson C, De Dapper M, Callaghan TV (2011) A century of tree line changes in sub-arctic Sweden shows local and regional variability and only a minor influence of 20th century climate warming. J Biogeogr 38:907–921

Boisvert-Marsh L, Périé C, De Blois S (2014) Shifting with climate? Evidence for recent changes in tree species distribution at high latitudes. Ecosphere 5:1–33

Bolch T, Kulkarni A, Kääb A, Huggel C, Paul F et al (2012) The state and fate of Himalayan glaciers. Science 336:310–314

Bolch T, Pieczonka T, Benn DI (2011) Multi-decadal mass loss of glaciers in the Everest area (Nepal Himalaya) derived from stereo imagery. Cryosphere 5:349–358

Bolch T, Shea JM, Liu S, Azam FM, Gao Y et al (2019) Status and change of the cryosphere in the extended Hindu Kush Himalaya region. In: Wester P, Mishra A, Mukherji A, Shrestha AB (eds) The Hindu Kush Himalaya assessment. Springer, Cham, pp 209–255

Bolch T, Menounos B, Wheate R (2010a) Landsat-based inventory of glaciers in western Canada, 1985–2005. Remote Sens Environ 114:127–137

Bolch T, Yao T, Kang S, Buchroithner MF, Scherer D et al (2010b) A glacier inventory for the western Nyainqentanglha range and the Nam Co basin, Tibet, and glacier changes 1976–2009. Cryosphere 4:419–433

Bolton DK, Coops NC, Hermosilla T, Wulder MA, White JC (2018) Evidence of vegetation greening at alpine treeline ecotones: three decades of Landsat spectral trends informed by lidar-derived vertical structure. Environ Res Lett 13:084022

Bonan DB, Christian JE, Christianson K (2019) Influence of North Atlantic climate variability on glacier mass balance in Norway, Sweden and Svalbard. J Glaciol 65:580–594

Bonekamp PN, De Kok RJ, Collier E, Immerzeel WW (2019) Contrasting meteorological drivers of the glacier mass balance between the Karakoram and Central Himalaya. Front Earth Sci 7:107

Borchardt P, Oldeland J, Ponsens J, Schickhoff U (2013) Plant functional traits match grazing gradient and vegetation patterns on mountain pastures in SW Kyrgyzstan. Phytocoenologia 43:171–181

Borchardt P, Schickhoff U, Scheitweiler S, Kulikov M (2011) Mountain pastures and grasslands in the SW Tien Shan, Kyrgyzstan—floristic patterns, environmental gradients, phytogeography, and grazing impact. J Mt Sci 8:363–373

Borchardt P, Schmidt M, Schickhoff U (2010) Vegetation patterns in Kyrgyzstan's walnut-fruit forests under the impact of changing forest use in post-Soviet transformation. Erde 141:255–275

Bormann KJ, Brown RD, Derksen C, Painter TH (2018) Estimating snow-cover trends from space. Nat Clim Chang 8:924–928

Borsdorf A, Hidalgo R (2009) Searching for fresh air, tranquillity and rural culture in the mountains: a new lifestyle for Chileans? Die Erde 140:275–292

Borsdorf A, Stadel C (2015) The Andes: a geographical portrait. Springer, Cham

Borsdorf A, Stötter J, Grabherr G, Bender O, Marchant C, Sánchez R (2015) Impacts and risks of global change. In: Grover VI, Borsdorf A, Breuste JH, Tiwari PC, Frangetto FW (eds) Impacts of global change on mountains: responses and adaptation. CRC Press, Boca Raton-London-New York, pp 33–76

Borsdorf A, Bender O (2007) Kulturlandschaftsverlust durch Verbuschung und Verwaldung im subalpinen und hochmontanen Höhenstockwerk: Die Folgen des klimatischen und sozioökonomischen Wandels. In: Geographie Innsbruck, Innsbrucker Geographische Gesellschaft (eds) Alpine kulturlandschaft im wandel. Hugo Penz zum 65. Geburtstag. Innsbrucker Geographische Gesellschaft, Innsbruck, pp 29–50

Bosworth D, Brown H (2007) After the timber wars: community-based stewardship. J Forest 105:271–273

Bouahmed A, Vessella F, Schirone B, Krouchi F, Derridj A (2019) Modeling *Cedrus atlantica* potential distribution in North Africa across time: new putative glacial refugia and future range shifts under climate change. Reg Environ Change 19:1667–1682

Bouchaou L, Tagma T, Boutaleb S, Hssaisoune M, El Morjani ZEA (2011) Climate change and its impacts on groundwater resources in Morocco: the case of the Souss-Massa basin. In: Treidel H, Martin-Bordes JL, Gurdak JJ (eds) Climate change effects on groundwater resources: a global synthesis of findings and recommendations. CRC Press, Boca Raton, FL, pp 129–144

Boy G, Witt A (2013) Invasive alien plants and their management in Africa. CABI Africa, Nairobi

Bradley RS, Vuille M, Diaz HF, Vergara W (2006) Threats to water supplies in the tropical Andes. Science 312:1755–1756

Brandt JS, Haynes MA, Kuemmerle T, Waller DM, Radeloff VC (2013) Regime shift on the roof of the world: alpine meadows converting to shrublands in the southern Himalayas. Biol Cons 158:116–127

Braun MH, Malz P, Sommer C, Farías-Barahona D, Sauter T et al (2019) Constraining glacier elevation and mass changes in South America. Nat Clim Chang 9:130–136

Braun C, Bezada M (2013) The history and disappearance of glaciers in Venezuela. J Lat Am Geogr 85–124

Breckle SW, Rafiqpoor MD (2020) The Hindu Kush/Afghanistan. In: Noroozi J (ed) Plant biogeography and vegetation of high mountains of Central and South-West Asia. Springer, Cham, pp 43–91

Breuer I (2007) Livelihood security and mobility in the High Atlas Mountains. In: Gertel J, Breuer I (eds) Pastoral Morocco. Globalizing scapes of mobility and insecurity. Reichert, Wiesbaden, pp 165–179

Brooks ML, Brown CS, Chambers JC, D'Antonio CM, Keeley JE, Belnap J (2016) Exotic annual *Bromus* invasions: comparisons among species and ecoregions in the western United States. In: Germino MJ, Chambers JC, Brown CS (eds) Exotic brome-grasses in arid and semiarid ecosystems of the western US. Springer, Cham, pp 11–60

Browman DL (1974) Pastoral nomadism in the Andes. Curr Anthropol 15:188–196

Brown ME, Funk C, Pedreros D, Korecha D, Lemma M et al (2017) A climate trend analysis of Ethiopia: examining subseasonal climate impacts on crops and pasture conditions. Clim Change 142:169–182

Brown LR, Du Preez J (2020) Alpine vegetation of temperate mountains of southern Africa. In: Goldstein MI, DellaSala DA (eds) Encyclopedia of the world's biomes, vol 1. Elsevier, Amsterdam, pp 395–404

Brugnara Y, Maugeri M (2019) Daily precipitation variability in the southern Alps since the late 19th century. Int J Climatol 39:3492–3504

Brun F, Berthier E, Wagnon P, Kääb A, Treichler D (2017) A spatially resolved estimate of High Mountain Asia glacier mass balances from 2000 to 2016. Nat Geosci 10:668–673

Brunetti M, Lentini G, Maugeri M, Nanni T, Auer I, Boehm R, Schoener W (2009) Climate variability and change in the Greater Alpine Region over the last two centuries based on multi-variable analysis. Int J Climatol 29:2197–2225

Brusca RC, Wiens JF, Meyer WM, Eble J, Franklin K, Overpeck JT, Moore W (2013) Dramatic response to climate change in the Southwest: Robert Whittaker's 1963 Arizona mountain plant transect revisited. Ecol Evol 3:3307–3319

Brush SSB (1998) Crop diversity in mountain areas and conservation strategy. Revue De Géographie Alpine 86:115–130

Bryn A (2008) Recent forest limit changes in South-East Norway: effects of climate change or regrowth after abandoned utilisation? Norsk Geogr Tidsskr Norw J Geogr 62:251–270

Bryn A, Hemsing LØ (2012) Impacts of land use on the vegetation in three rural landscapes of Norway. Int J Biodivers Sci Ecosyst Serv Manage 8:360–371

Bryn A, Potthoff K (2018) Elevational treeline and forest line dynamics in Norwegian mountain areas—a review. Landscape Ecol 33:1225–1245

Bräuning A, Grießinger J, Hochreuther P, Wernicke J (2016) Dendroecological perspectives on climate change on the southern Tibetan Plateau. In: Singh RB, Schickhoff U, Mal S (eds) Climate change and dynamics of glaciers and vegetation in the Himalaya. Springer, Cham, pp 347–364

Bråthen KA, Ravolainen VT, Stien A, Tveraa T, Ims RA (2017) *Rangifer* management controls a climate-sensitive tundra state transition. Ecol Appl 27:2416–2427

Bucała-Hrabia A (2018) Land use changes and their catchment-scale environmental impact in the Polish western Carpathians during transition from centrally planned to free-market economics. Geogr Pol 91:171–196

Buermann W, Parida B, Jung M, MacDonald GM, Tucker CJ, Reichstein M (2014) Recent shift in Eurasian boreal forest greening response may be associated with warmer and drier summers. Geophys Res Lett 41:1995–2002

Bulygina ON, Razuvaev VN, Korshunova NN (2009) Changes in snow cover over northern Eurasia in the last few decades. Environ Res Lett 4:045026

Burga CA, Krüsi B, Egli M, Wernli M, Elsener S et al (2010) Plant succession and soil development on the foreland of the Morteratsch glacier (Pontresina, Switzerland): straight forward or chaotic? Flora 205:561–576

Bush E, Lemmen DS (eds) (2019) Canada's changing climate report. Government of Canada, Ottawa

Butzer KW (1981) Rise and fall of Axum, Ethiopia: a geo-archaeological interpretation. Am Antiq 46:471–495

Butzer KW, Butzer EK (1995) Transfer of the Mediterranean livestock economy to New Spain: adaptation and ecological consequences. In: Turner BL, Goméz Sal A, González Bernáldez F, Di Castri F (eds) Global land use change: a perspective from the Columbian encounter. CSIC, Madrid, pp 151–193

Byer S, Jin Y (2017) Detecting drought-induced tree mortality in Sierra Nevada forests with time series of satellite data. Remote Sens 9:929

Byers A (2005) Contemporary human impacts on alpine ecosystems in the Sagarmatha (Mt. Everest) National Park, Khumbu, Nepal. Ann Assoc Am Geogr 95:112–140

Byers AC (2017) Khumbu since 1950. Cultural, landscape, and climate change in the Sagarmatha (Mt. Everest) National Park, Khumbu, Nepal. Vajra Books, Kathmandu

Byers AC, Price LW, Price MF (2013) Introduction to mountains. In: Price MF, Byers AC, Friend DA, Kohler T, Price LW (eds) Mountain geography. Physical and human dimensions. University of California Press, Berkeley-Los Angeles, pp 1–10

Bätzing W (2015) Die Alpen—Geschichte und Zukunft einer europäischen Kulturlandschaft. Beck, München

Bätzing W (2018) Die Alpen—Das Verschwinden einer Kulturlandschaft. wbgTHEISS, Darmstadt

Bürzle B, Schickhoff U, Schwab N, Wernicke L, Müller Y et al (2018) Seedling recruitment and facilitation dependence on safe site characteristics in a Himalayan treeline ecotone. Plant Ecol 219:115–132

CH2018 (2018) Climate scenarios for Switzerland. Technical report. National Centre for Climate Services, Zurich

Cabrelli A, Beaumont L, Hughes L (2015) The impacts of climate change on Australian and New Zealand flora and fauna. In: Stow A, Maclean N, Holwell GI (eds) Austral ark: the state of wildlife in Australia and New Zealand. Cambridge University Press, Cambridge, pp 65–82

Cai H, Yang X, Xu X (2015) Human-induced grassland degradation/restoration in the Central Tibetan Plateau: the effects of ecological protection and restoration projects. Ecol Eng 83:112–119

Camarero JJ, García-Ruiz JM, Sangüesa-Barreda G, Galván JD, Alla AQ et al (2015) Recent and intense dynamics in a formerly static Pyrenean treeline. Arct Antarct Alp Res 47:773–783

Camarero JJ, Linares JC, García-Cervigón AI, Batllori E, Martínez I, Gutiérrez E (2017) Back to the future: the responses of alpine treelines to climate warming are constrained by the current ecotone structure. Ecosystems 20:683–700

Cannone N, Diolaiuti G, Guglielmin M, Smiraglia C (2008) Accelerating climate change impacts on alpine glacier forefield ecosystems in the European Alps. Ecol Appl 18:637–648

Cannone N, Pignatti S (2014) Ecological responses of plant species and communities to climate warming: upward shift or range filling processes? Clim Change 123:201–214

Cao B, Zhang T, Peng X, Mu C, Wang Q et al (2018) Thermal characteristics and recent changes of permafrost in the upper reaches of the Heihe river basin, western China. J Geophys Res Atmos 123:7935–7949

CaraDonna PJ, Iler AM, Inouye DW (2014) Shifts in flowering phenology reshape a subalpine plant community. Proc Natl Acad Sci 111:4916–4921

Carbutt C (2020) Nature of alpine ecosystems in tropical mountains of Africa. In: Goldstein MI, DellaSala DA (eds) Encyclopedia of the world's biomes, vol 1. Elsevier, Amsterdam, pp 292–299

Cardelús CL, Woods CL, Bitew Mekonnen A, Dexter S, Scull P, Tsegay BA (2019) Human disturbance impacts the integrity of sacred church forests. Ethiopia. PloS ONE 14:e0212430

Carilla J, Grau HR, Paolini L, Mariano M (2013) Lake fluctuations, plant productivity, and long-term variability in high-elevation tropical Andean ecosystems. Arct Antarct Alp Res 45:179–189

Carilla J, Halloy S, Cuello S, Grau A, Malizia A, Cuesta F (2018) Vegetation trends over eleven years on mountain summits in NW Argentina. Ecol Evol 8:11554–11567

Carlson AE, Kilmer Z, Ziegler LB, Stoner JS, Wiles GC et al. (2017a) Recent retreat of Columbia Glacier, Alaska: millennial context. Geology 45:547–550

Carlson BZ, Corona MC, Dentant C, Bonet R, Thuiller W, Choler P (2017b) Observed long-term greening of alpine vegetation—a case study in the French Alps. Environ Res Lett 12:114006

Carnicer J, Coll M, Ninyerola M, Pons X, Sánchez G, Peñuelas J (2011) Widespread crown condition decline, food web disruption, and amplified tree mortality with increased climate change-type drought. Proc Natl Acad Sci 108:1474–1478

Chakraborty A, Saha S, Sachdeva K, Joshi PK (2018) Vulnerability of forests in the Himalayan region to climate change impacts and anthropogenic disturbances: a systematic review. Reg Environ Change 18:1783–1799

Chala D, Brochmann C, Psomas A, Ehrich D, Gizaw A et al (2016) Good-bye to tropical alpine plant giants under warmer climates? Loss of range and genetic diversity in *Lobelia rhynchopetalum*. Ecol Evol 6:8931–8941

Chambers LE, Altwegg R, Barbraud C, Barnard P, Beaumont LJ et al (2013) Phenological changes in the southern hemisphere. PLoS ONE 8:e75514

Chand P, Sharma MC (2015) Glacier changes in the Ravi basin, north-western Himalaya (India) during the last four decades (1971–2010/13). Global Planet Change 135:133–147

Chand P, Sharma MC, Bhambri R, Sangewar CV, Juyal N (2017) Reconstructing the pattern of the Bara Shigri glacier fluctuation since the end of the Little Ice Age, Chandra valley, north-western Himalaya. Prog Phys Geogr 41:643–675

Chapin FS III, Trainor SF, Cochran P, Huntington H, Markon C, McCammon M, McGuire AD, Serreze M (2014) Alaska. In: Melillo JM, Richmond TC, Yohe GW (eds) Climate change impacts in the United States: the third national climate assessment. U.S Global Change Research Program, Washington DC, pp 514–536

Charbonneau M (2009) Scattered development of settlement and grouping in Andean pastoral societies. Ann De Géog 670:637–658

Chauchard S, Beilhe F, Denis N, Carcaillet C (2010) An increase in the upper tree-limit of silver fir (*Abies alba* Mill.) in the Alps since the mid-20th century: a land-use change phenomenon. For Ecol Manage 259:1406–1415

Chauchard S, Carcaillet C, Guibal F (2007) Patterns of land-use abandonment control tree-recruitment and forest dynamics in Mediterranean mountains. Ecosystems 10:936–948

Chebli Y, Chentouf M, Ozer P, Hornick JL, Cabaraux JF (2018) Forest and silvopastoral cover changes and its drivers in northern Morocco. Appl Geogr 101:23–35

Cheddadi R, Nourelbait M, Bouaissa O, Tabel J, Rhoujjati A et al (2015) A history of human impact on Moroccan mountain landscapes. Afr Archaeol Rev 32:233–248

Chen IC, Hill JK, Ohlemüller R, Roy DB, Thomas CD (2011) Rapid range shifts of species associated with high levels of climate warming. Science 333:1024–1026

Chen X, Liang S, Cao Y (2016) Satellite observed changes in the northern hemisphere snow cover phenology and the associated radiative forcing and feedback between 1982 and 2013. Environ Res Lett 11:084002

Chen J, Ouyang Z, John R, Henebry GM, Groisman PY et al (2020) Social-ecological systems across the Asian Drylands Belt (ADB). In: Gutman G, Chen J, Henebry GM, Kappas M (eds) Landscape dynamics of drylands across Greater Central Asia: people, societies and ecosystems. Springer, Cham, pp 191–225

Chen Y, Takeuchi K, Xu C, Chen Y, Xu Z (2006) Regional climate change and its effects on river runoff in the Tarim basin, China. Hydrol Process 20:2207–2216

Chen AA, Wang NL, Guo ZM, Wu YW, Wu HB (2018) Glacier variations and rising temperature in the Mt. Kenya since the last glacial maximum. J Mt Sci 15:1268–1282

Chen F, Yuan YJ, Wei WS, Fan ZA, Zhang T et al (2012) Climatic response of ring width and maximum latewood density of *Larix sibirica* in the Altay Mountains reveals recent warming trends. Ann For Sci 69:723–733

Chettri N, Shrestha AB, Sharma E (2020) Climate change trends and ecosystem resilience in the Hindu Kush Himalayas. In: Dimri AP, Bookhagen B, Stoffel M, Yasunari T (eds) Himalayan weather and climate and their impact on the environment. Springer, Cham, pp 525–552

Chevallier P, Pouyaud B, Mojaïsky M, Bolgov M, Olsson O, Bauer M, Froebrich J (2014) River flow regime and snow cover of the Pamir Alay (Central Asia) in a changing climate. Hydrol Sci J 59:1491–1506

Chhetri PK, Cairns DM (2015) Contemporary and historic population structure of *Abies spectabilis* at treeline in Barun valley, eastern Nepal Himalaya. J Mt Sci 12:558–570

Chhetri PK, Cairns DM (2018) Low recruitment above treeline indicates treeline stability under changing climate in Dhorpatan Hunting Reserve, western Nepal. Phys Geogr 39:329–342

Chhetri PK, Gaddis KD, Cairns DM (2018) Predicting the suitable habitat of treeline species in the Nepalese Himalayas under climate change. Mt Res Dev 38:153–164

Chidi CL (2017) Patch analysis of cultivated land abandonment in the hills of western Nepal. In: Li A, Deng W, Zhao W (eds) Land cover change and its eco-environmental responses in Nepal. Springer, Singapore, pp 149–162

Childers M (2012) Colorado powder keg: ski resorts and the environmental movement. University Press of Kansas, Lawrence

Christensen JH, Kanikicharla KK, Aldrian E, An SI, Cavalcanti IFA et al. (2013) Climate phenomena and their relevance for future regional climate change. In: IPCC (ed) Climate change 2013: the physical science basis. Contribution of working group I to the fifth assessment report of the Intergovernmental Panel on Climate Change. Cambridge Univ Press, Cambridge-New York, pp 1217–1308

Christensen OB, Kjellström E, Zorita E (2015) Projected change—atmosphere. In: The BACC II Author Team (eds) Second assessment of climate change for the Baltic Sea basin. Springer, Cham, pp 217–233

Chudley TR, Miles ES, Willis IC (2017) Glacier characteristics and retreat between 1991 and 2014 in the Ladakh Range, Jammu and Kashmir. Remote Sens Lett 8:518–527

Chávez RO, Christie DA, Olea M, Anderson TG (2019) A multiscale productivity assessment of high Andean

peatlands across the Chilean Altiplano using 31 years of Landsat imagery. Remote Sens 11:2955
Clarke GK, Jarosch AH, Anslow FS, Radić V, Menounos B (2015) Projected deglaciation of western Canada in the twenty-first century. Nat Geosci 8:372–377
Classen A, Eardley CD, Hemp A, Peters MK, Peters RS, Ssymank A, Steffan-Dewenter I (2020) Specialization of plant–pollinator interactions increases with temperature at Mt Kilimanjaro. Ecol Evol 10:2182–2195
Cline SA (2013) Land use and landscape change in the Rockies: Implications for mountain agriculture. In: Mann S (ed) The future of mountain agriculture. Springer, Berlin, Heidelberg, pp 5–19
Cole KL, Henderson N, Shafer DS (1997) Holocene vegetation and historic grazing impacts at Capitol Reef National Park reconstructed using packrat middens. Great Basin Nat 57:315–326
Collins JM (2011) Temperature variability over Africa. J Clim 24:3649–3666
Collins L, Bennett AF, Leonard SW, Penman TD (2019) Wildfire refugia in forests: severe fire weather and drought mute the influence of topography and fuel age. Glob Change Biol 25:3829–3843
Conedera M, Colombaroli D, Tinner W, Krebs P, Whitlock C (2017) Insights about past forest dynamics as a tool for present and future forest management in Switzerland. For Ecol Manage 388:100–112
Conlisk E, Castanha C, Germino MJ, Veblen TT, Smith JM, Kueppers LM (2017) Declines in low-elevation subalpine tree populations outpace growth in high-elevation populations with warming. J Ecol 105:1347–1357
Copland L (2011) Retreat/advance of glaciers. In: Singh VP, Singh P, Haritashya UK (eds) Encyclopedia of snow, ice and glaciers. Springer, Dordrecht, pp 934–939
Cortés-Ramos J, Delgado-Granados H, Huggel C, Ontiveros-González G (2019) Evolution of the largest glacier in Mexico (Glaciar Norte) since the 50s: factors driving glacier retreat. Geogr Ann Ser B 101:350–373
Couralet C, Sass-Klaassen U, Sahle Y, Sterck FJ, Ayele TB, Bongers FJJM (2007) Dendrochronological investigations on *Juniperus procera* from Ethiopian dry afromontane forests. In: Haneca K, Verheijden A, Beeckman H, Gärtner H, Helle G (eds) Proceedings of the DENDROSYMPOSIUM, 20–22 April, 2006, Tervuren, Belgium. TRACE—Tree Rings in Archaeology, Climatology and Ecology, vol. 5. Forschungszentrum Jülich, Jülich, pp 73–79
Crimmins TM, Crimmins MA, Bertelsen CD (2013) Spring and summer patterns in flowering onset, duration, and constancy across a water-limited gradient. Am J Bot 100:1137–1147
Crockett JL, Westerling AL (2018) Greater temperature and precipitation extremes intensify western US droughts, wildfire severity, and Sierra Nevada tree mortality. J Clim 31:341–354
Cudlin P, Klopčič M, Tognetti R, Máli F, Alados CL et al (2017) Drivers of treeline shift in different European mountains. Climate Res 73:135–150
Cuervo-Robayo AP, Ureta C, Gómez-Albores MA, Meneses-Mosquera AK, Téllez-Valdés O, Martínez-Meyer E (2020) One hundred years of climate change in Mexico. PLoS ONE 15:e0209808
Cuesta F, Llambí LD, Huggel C, Drenkhan F, Gosling WD (2019) New land in the Neotropics: a review of biotic community, ecosystem, and landscape transformations in the face of climate and glacier change. Reg Environ Change 19:1623–1642
Cullen NJ, Gibson PB, Mölg T, Conway JP, Sirguey P, Kingston DG (2019) The influence of weather systems in controlling mass balance in the southern Alps of New Zealand. J Geophys Res Atmos 124:4514–4529
Cullen NJ, Sirguey P, Mölg T, Kaser G, Winkler M, Fitzsimons SJ (2013) A century of ice retreat on Kilimanjaro: the mapping reloaded. Cryosphere 7:419–431
Cunha S, Price LW (2013) Agricultural settlement and land use in mountains. In: Price MF, Byers AC, Friend DA, Kohler T, Price LW (eds) Mountain geography: physical and human dimensions. University of California Press, Berkeley-Los Angeles, pp 301–331
Curebal I, Efe R, Soykan A, Sonmez S (2015) Impacts of anthropogenic factors on land degradation during the Anthropocene in Turkey. J Environ Biol 36:51–58
Czortek P, Eycott AE, Grytnes JA, Delimat A, Kapfer J, Jaroszewicz B (2018) Effects of grazing abandonment and climate change on mountain summits flora: a case study in the Tatra Mts. Plant Ecol 219:261–276
D'Arrigo R, Wilson R, Liepert B, Cherubini P (2008) On the 'divergence problem' in northern forests: a review of the tree-ring evidence and possible causes. Global Planet Change 60:289–305
DFRS (Department of Forest Research and Survey) (2015) State of Nepal's forests. Department of Forest Research and Survey, Kathmandu
Dahal N, Shrestha UB, Tuitui A, Ojha HR (2019) Temporal changes in precipitation and temperature and their implications on the streamflow of Rosi River. Central Nepal. Climate 7:3
Dainese M, Aikio S, Hulme PE, Bertolli A, Prosser F, Marini L (2017) Human disturbance and upward expansion of plants in a warming climate. Nat Clim Chang 7:577–580
Dainese M, Kühn I, Bragazza L (2014) Alien plant species distribution in the European Alps: influence of species' climatic requirements. Biol Invasions 16:815–831
Damm A, Greuell W, Landgren O, Prettenthaler F (2017) Impacts of +2 °C global warming on winter tourism demand in Europe. Clim Serv 7:31–46
Damschen EI, Harrison S, Grace JB (2010) Climate change effects on an endemic-rich edaphic flora: resurveying Robert H. Whittaker's Siskiyou sites (Oregon, USA). Ecology 91:3609–3619

Danby RK, Hik DS (2007) Variability, contingency and rapid change in recent subarctic alpine tree line dynamics. J Ecol 95:352–363
Danby RK, Koh S, Hik DS, Price LW (2011) Four decades of plant community change in the alpine tundra of Southwest Yukon. Canada. Ambio 40:660
Dangles O, Rabatel A, Kraemer M, Zeballos G, Soruco A, Jacobsen D, Anthelme F (2017) Ecosystem sentinels for climate change? Evidence of wetland cover changes over the last 30 years in the tropical Andes. PLoS ONE 12:e0175814
Dangwal DD (2009a) The lost mobility: pastoralism and modernity in Uttarakhand Himalaya (India). Nomadic Peoples 13:84–101
Dangwal DD (2009b) Himalayan degradation. Colonial forestry and environmental change in India. Foundation Books, Delhi
Daniels LD, Veblen TT (2004) Spatiotemporal influences of climate on altitudinal treeline in northern Patagonia. Ecology 85:1284–1296
Dashkhuu D, Kim JP, Chun JA, Lee WS (2015) Long-term trends in daily temperature extremes over Mongolia. Weather Clim Extremes 8:26–33
Davis EL, Brown R, Daniels L, Kavanagh T, Gedalof ZE (2020) Regional variability in the response of alpine treelines to climate change. Clim Change. https://doi.org/10.1007/s10584-020-02743-0
Davis EL, Gedalof ZE (2018) Limited prospects for future alpine treeline advance in the Canadian Rocky Mountains. Glob Change Biol 24:4489–4504
Dawadi B, Liang E, Tian L, Devkota LP, Yao T (2013) Pre-monsoon precipitation signal in tree rings of timberline *Betula utilis* in the Central Himalayas. Quatern Int 283:72–77
DeBeer CM, Wheater HS, Carey SK, Chun KP (2016) Recent climatic, cryospheric, and hydrological changes over the interior of western Canada: a review and synthesis. Hydrol Earth Syst Sci 20:1573
Dearborn KD, Danby RK (2018) Climatic drivers of tree growth at tree line in Southwest Yukon change over time and vary between landscapes. Clim Change 150:211–225
Defila C, Clot B, Jeanneret F, Stöckli R (2016) Phenology in Switzerland since 1808. In: Willemse S, Furger M (eds) From weather observations to atmospheric and climate sciences in Switzerland: celebrating 100 years of the Swiss Society for Meteorology. vdf Hochschulverlag, Zurich, pp 291–306
Demiroglu OC, Lundmark L, Saarinen J, Müller DK (2019) The last resort? Ski tourism and climate change in arctic Sweden. J Tourism Futures 6:91–101
Denevan WM (1992) The pristine myth: the landscape of the Americas in 1492. Ann Assoc Am Geogr 82:369–385
Deng H, Chen Y, Wang H, Zhang S (2015) Climate change with elevation and its potential impact on water resources in the Tianshan Mountains, Central Asia. Global Planet Change 135:28–37
Desyatkin R, Fedorov A, Desyatkin A, Konstantinov P (2015) Air temperature changes and their impact on permafrost ecosystems in eastern Siberia. Therm Sci 19:S351–S360
Detsch F, Otte I, Appelhans T, Hemp A, Nauss T (2016) Seasonal and long-term vegetation dynamics from 1-km GIMMS-based NDVI time series at Mt. Kilimanjaro, Tanzania. Remote Sens Environ 178:70–83
Dial RJ, Berg EE, Timm K, McMahon A, Geck J (2007) Changes in the alpine forest-tundra ecotone commensurate with recent warming in Southcentral Alaska: evidence from orthophotos and field plots. J Geophys Res Biogeosci 112:G04015
Dial RJ, Scott Smeltz T, Sullivan PF, Rinas CL, Timm K et al (2016) Shrubline but not treeline advance matches climate velocity in montane ecosystems of South-Central Alaska. Glob Change Biol 22:1841–1856
Dietz AJ, Kuenzer C, Conrad C (2013) Snow-cover variability in Central Asia between 2000 and 2011 derived from improved MODIS daily snow-cover products. Int J Remote Sens 34:3879–3902
Diffenbaugh NS, Scherer M, Ashfaq M (2013) Response of snow-dependent hydrologic extremes to continued global warming. Nat Clim Chang 3:379–384
Dilsaver LM, Wyckoff W, Preston WL (2000) Fifteen events that have shaped California's human landscape. Calif Geogr 40:1–76
Dimeyeva LA, Sitpayeva GT, Sultanova BM, Ussen K, Islamgulova AF (2015) High-altitude flora and vegetation of Kazakhstan and climate change impacts. In: Öztürk M, Hakeem KR, Faridah-Hanum I, Efe R (eds) Climate change impacts on high-altitude ecosystems. Springer, Cham, pp 1–48
Dimri AP, Choudhary A, Kumar D (2020) Elevation dependent warming over Indian Himalayan region. In: Dimri AP, Bookhagen B, Stoffel M, Yasunari T (eds) Himalayan weather and cimate and their impact on the environment. Springer, Cham, pp 141–156
Dimri AP, Dash SK (2012) Wintertime climatic trends in the western Himalayas. Clim Change 111:775–800
Dimri AP, Kumar D, Choudhary A, Maharana P (2018) Future changes over the Himalayas: mean temperature. Glob Planet Change 162:235–251
Dinca L, Nita MD, Hofgaard A, Alados CL, Broll G et al (2017) Forests dynamics in the montane alpine boundary: a comparative study using satellite imagery and climate data. Climate Res 73:97–110
Diodato N, Bellocchi G, Tartari G (2011) How do Himalayan areas respond to global warming? Int J Climatol 32:975–982
Djema A, Messaoudene M (2009) The Algerian forest: current situation and prospects. In: Palahi M, Birot Y, Bravo F, Gorriz E (eds) Modelling, valuing and managing Mediterranean forest ecosystems for non-timber goods and services. European Forest Institute, Joensuu, pp 17–28
Dobbertin M, Hilker N, Rebetez M, Zimmermann NE, Wohlgemuth T, Rigling A (2005) The upward shift in altitude of pine mistletoe (*Viscum album* ssp. *austriacum*) in Switzerland—the result of climate warming? Int J Biometeorol 50:40–47

Dolanc CR, Safford HD, Dobrowski SZ, Thorne JH (2014) Twentieth century shifts in abundance and composition of vegetation types of the Sierra Nevada, CA, US. Appl Veg Sci 17:442–455

Dolanc CR, Thorne JH, Safford HD (2013) Widespread shifts in the demographic structure of subalpine forests in the Sierra Nevada, California, 1934 to 2007. Glob Ecol Biogeogr 22:264–276

Dolezal J, Dvorsky M, Kopecky M, Liancourt P, Hiiesalu I et al (2016) Vegetation dynamics at the upper elevational limit of vascular plants in Himalaya. Sci Rep 6:24881

Donahue DL (2005) Western grazing: the capture of grass, ground, and government. Environ Law 35:721–806

Donat MG, Peterson TC, Brunet M, King AD, Almazroui M et al (2014) Changes in extreme temperature and precipitation in the Arab region: long-term trends and variability related to ENSO and NAO. Int J Climatol 34:581–592

Dong SK, Lassoie JP, Yan ZL, Sharma E, Shrestha KK, Pariya D (2007) Indigenous rangeland resource management in the mountainous areas of northern Nepal: a case study from the Rasuwa District. Rangeland J 29:149–160

Dong C, MacDonald GM, Willis K, Gillespie TW, Okin GS, Williams AP (2019) Vegetation responses to 2012–2016 drought in northern and southern California. Geophys Res Lett 46:3810–3821

Dong B, Sutton RT, Chen W, Liu X, Lu R, Sun Y (2016) Abrupt summer warming and changes in temperature extremes over Northeast Asia since the mid-1990s: drivers and physical processes. Adv Atmos Sci 33:1005–01023

Dong SK, Shaoliang LY, Yan ZL (2016) Maintaining the human-natural systems of pastoralism in the Himalayas of South Asia and China. In: Dong S, Kassam KAS, Tourrand JF, Boone RB (eds) Building resilience of human-natural systems of pastoralism in the developing world. Springer, Cham, pp 93–135

Dorji T, Hopping KA, Meng F, Wang S, Jiang L, Klein JA (2020) Impacts of climate change on flowering phenology and production in alpine plants: the importance of end of flowering. Agr Ecosyst Environ 291:106795

Dorji T, Hopping KA, Wang S, Piao S, Tarchen T, Klein JA (2018) Grazing and spring snow counteract the effects of warming on an alpine plant community in Tibet through effects on the dominant species. Agric For Meteorol 263:188–197

Dorji T, Totland Ø, Moe SR, Hopping KA, Pan J, Klein JA (2013) Plant functional traits mediate reproductive phenology and success in response to experimental warming and snow addition in Tibet. Glob Change Biol 19:459–472

Dorjsuren B, Yan D, Wang H, Chonokhuu S, Enkhbold A et al (2018) Observed trends of climate and river discharge in Mongolia's Selenga sub-basin of the Lake Baikal basins. Water 10:1436

Du H, Liu J, Li MH, Büntgen U, Yang Y et al (2018) Warming-induced upward migration of the alpine treeline in the Changbai Mountains, Northeast China. Glob Change Biol 24:1256–1266

Dubovyk O (2018) Spatiotemporal assessment of vegetation trends in the post-Soviet Central Asia. In: Egamberdieva D, Öztürk M (eds) Vegetation of central Asia and environs. Springer, Cham, pp 1–13

Duethmann D, Bolch T, Farinotti D, Kriegel D, Vorogushyn S et al (2015) Attribution of streamflow trends in snow and glacier melt-dominated catchments of the Tarim River, Central Asia. Water Resour Res 51:4727–4750

Dulamsuren C, Hauck M, Leuschner HH, Leuschner C (2011) Climate response of tree-ring width in *Larix sibirica* growing in the drought-stressed forest-steppe ecotone of northern Mongolia. Ann For Sci 68:275–282

Dulamsuren C, Khishigjargal M, Leuschner C, Hauck M (2014) Response of tree-ring width to climate warming and selective logging in larch forests of the Mongolian Altai. J Plant Ecol 7:24–38

Dulamsuren C, Wommelsdorf T, Zhao F, Xue Y, Zhumadilov BZ, Leuschner C, Hauck M (2013) Increased summer temperatures reduce the growth and regeneration of *Larix sibirica* in southern boreal forests of eastern Kazakhstan. Ecosystems 16:1536–1549

Dulamsuren C, Hauck M, Khishigjargal M, Leuschner HH, Leuschner C (2010a) Diverging climate trends in Mongolian taiga forests influence growth and regeneration of *Larix sibirica*. Oecologia 163:1091–1102

Dulamsuren C, Hauck M, Leuschner C (2010b) Recent drought stress leads to growth reductions in *Larix sibirica* in the western Khentey, Mongolia. Glob Change Biol 16:3024–3035

Dullinger S, Dirnböck T, Greimler J, Grabherr G (2003) A resampling approach for evaluating effects of pasture abandonment on subalpine plant species diversity. J Veg Sci 14:243–252

Dullinger S, Gattringer A, Thuiller W, Moser D, Zimmermann NE et al (2012) Extinction debt of high-mountain plants under twenty-first-century climate change. Nat Clim Chang 2:619–622

Dullinger I, Gattringer A, Wessely J, Moser D, Plutzar C et al (2020) A socio-ecological model for predicting impacts of land-use and climate change on regional plant diversity in the Austrian Alps. Glob Change Biol 26:2336–2352

Duo C, Xie H, Wang P, Guo J, La J, Qiu Y, Zheng Z (2014) Snow cover variation over the Tibetan Plateau from MODIS and comparison with ground observations. J Appl Remote Sens 8:084690

Duque A, Stevenson PR, Feeley KJ (2015) Thermophilization of adult and juvenile tree communities in the northern tropical Andes. Proc Natl Acad Sci 112:10744–10749

Durand Y, Giraud G, Laternser M, Etchevers P, Mérindol L, Lesaffre B (2009) Reanalysis of 47 years of climate in the French Alps (1958–2005): climatology and

trends for snow cover. J Appl Meteorol Climatol 48:2487–2512
Dussaillant I, Berthier E, Brun F, Masiokas M, Hugonnet R et al (2019) Two decades of glacier mass loss along the Andes. Nat Geosci 12:802–808
Dykes RC, Brook MS, Robertson CM, Fuller IC (2011) Twenty-first century calving retreat of Tasman Glacier, southern Alps, New Zealand. Arct Antarct Alp Res 43:1–10
Dymond JR, Shepherd JD, Newsome PF, Belliss S (2017) Estimating change in areas of indigenous vegetation cover in New Zealand from the New Zealand Land Cover Database (LCDB). N Z J Ecol 41:56–64
Dyrrdal AV, Saloranta T, Skaugen T, Stranden HB (2013) Changes in snow depth in Norway during the period 1961–2010. Hydrol Res 44:169–179
Dörre A, Borchardt P (2012) Changing systems, changing effects—pasture utilization in the post-Soviet transition. Mt Res Dev 32:313–324
Dörre A (2012) Legal arrangements and pasture-related socio-ecological challenges in Kyrgyzstan. In: Kreutzmann H (ed) Pastoral practices in High Asia. Springer, Dordrecht, pp 127–144
EEA (European Environment Agency) (2009) Water resources across Europe—confronting water scarcity and drought. European Union, Luxembourg
EEA (European Environment Agency) (2017) Climate change, impacts, and vulnerability in Europe 2016. EEA Report No 1/2017. European Union, Luxembourg
Eckert S, Kiteme B, Njuguna E, Zaehringer JG (2017) Agricultural expansion and intensification in the foothills of Mount Kenya: a landscape perspective. Remote Sens 9:784
Egan PA, Price MF (2017) Mountain ecosystem services and climate change: a global overview of potential threats and strategies for adaptation. UNESCO Publishing, Paris
Egarter Vigl L, Schirpke U, Tasser E, Tappeiner U (2016) Linking long-term landscape dynamics to the multiple interactions among ecosystem services in the European Alps. Landscape Ecol 31:1903–1918
Ehlers E, Kreutzmann H (2000) High mountain ecology and economy: potential and constraints. In: Ehlers E, Kreutzmann H (eds) High mountain pastoralism in northern Pakistan. Franz Steiner Verlag, Stuttgart, pp 9–36
Eiter S, Potthoff K (2016) Landscape changes in Norwegian mountains: increased and decreased accessibility, and their driving forces. Land Use Policy 54:235–245
El-Vilaly MAS, Didan K, Marsh SE, Van Leeuwen WJ, Crimmins MA, Munoz AB (2018) Vegetation productivity responses to drought on tribal lands in the four corners region of the Southwest USA. Front Earth Sci 12:37–51
Elias SA (2020) Overview of mountains (alpine systems): life at the top. In: Goldstein MI, DellaSala DA (eds) Encyclopedia of the world's biomes, vol 1. Elsevier, Amsterdam, pp 251–264
Elizbarashvili M, Elizbarashvili E, Tatishvili M, Elizbarashvili S, Meskhia R et al (2017) Georgian climate change under global warming conditions. Ann Agrarian Sci 15:17–25
Elkin C, Gutiérrez AG, Leuzinger S, Manusch C, Temperli C, Rasche L, Bugmann H (2013) A 2° C warmer world is not safe for ecosystem services in the European Alps. Glob Change Biol 19:1827–1840
Ellenberg H (1979) Man's influence on tropical mountain ecosystems in South America. J Ecol 67:401–416
Elliott GP (2011) Influences of 20th-century warming at the upper tree line contingent on local-scale interactions: evidence from a latitudinal gradient in the Rocky Mountains, USA. Glob Ecol Biogeogr 20:46–57
Elliott GP (2012) Extrinsic regime shifts drive abrupt changes in regeneration dynamics at upper treeline in the Rocky Mountains, USA. Ecology 93:1614–1625
Elliott GP, Bailey SN, Cardinal SJ (2020) Hotter drought as a disturbance at upper treeline in the southern Rocky Mountains. Ann Am Assoc Geogr. https://doi.org/10.1080/24694452.2020.1805292
Elliott GP, Kipfmueller KF (2011) Multiscale influences of climate on upper treeline dynamics in the southern Rocky Mountains, USA: evidence of intraregional variability and bioclimatic thresholds in response to twentieth-century warming. Ann Assoc Am Geogr 101:1181–1203
Elliott GP, Petruccelli CA (2018) Tree recruitment at the treeline across the Continental Divide in the northern Rocky Mountains, USA: the role of spring snow and autumn climate. Plant Ecolog Divers 11:319–333
Elliott GP (2017) Treeline ecotones. In: Richardson D, Castree N, Goodchild MF, Kobayashi A, Liu W, Marston RA (eds) International encyclopedia of geography: people, the Earth, environment and technology. Wiley Blackwell, https://doi.org/10.1002/9781118786352.wbieg0539
Elmendorf SC, Henry GH, Hollister RD, Fosaa AM, Gould WA et al (2015) Experiment, monitoring, and gradient methods used to infer climate change effects on plant communities yield consistent patterns. Proc Natl Acad Sci 112:448–452
Elmi M, Streifeneder T, Ravazzoli E, Laner P, Petitta M et al (2018) The alps in 25 maps. Permanent Secretariat of the Alpine Convention, Innsbruck-Bolzano
Elsen PR, Tingley MW (2015) Global mountain topography and the fate of montane species under climate change. Nat Clim Chang 5:772–776
Elsner MM, Cuo L, Voisin N, Deems JS, Hamlet AF et al (2010) Implications of 21st century climate change for the hydrology of Washington State. Clim Change 102:225–260
Elumeeva TG, Onipchenko VG, Egorov AV, Khubiev AB, Tekeev DK, Soudzilovskaia NA, Cornelissen JH (2013) Long-term vegetation dynamic in the northwestern Caucasus: which communities are more affected by upward shifts of plant species? Alp Bot 123:77–85

Emanuelsson U (1987) Human influence on vegetation in the Torneträsk area during the last three centuries. Ecol Bull 38:95–111
Engler R, Randin CF, Thuiller W, Dullinger S, Zimmermann NE et al (2011) 21st century climate change threatens mountain flora unequally across Europe. Glob Change Biol 17:2330–2341
Ensslin A, Rutten G, Pommer U, Zimmermann R, Hemp A, Fischer M (2015) Effects of elevation and land use on the biomass of trees, shrubs and herbs at Mount Kilimanjaro. Ecosphere 6:1–15
Erlandson JM, Braje TJ (2015) Stemmed points, the coastal migration theory, and the peopling of the Americas. In: Frachetti MD, Spengler RN (eds) Mobility and ancient society in Asia and the Americas. Springer, Cham, pp 49–58
Ermert V, Fink AH, Morse AP, Paeth H (2012) The impact of regional climate change on malaria risk due to greenhouse forcing and land-use changes in tropical Africa. Environ Health Perspect 120:77–84
Ernakovich JG, Hopping KA, Berdanier AB, Simpson RT, Kachergis EJ, Steltzer H, Wallenstein MD (2014) Predicted responses of arctic and alpine ecosystems to altered seasonality under climate change. Glob Change Biol 20:3256–3269
Erschbamer B, Caccianiga MS (2016) Glacier forelands: lessons of plant population and community development. In: Cánovas F, Lüttge U, Matyssek R (eds) Progress in botany, vol 78. Springer, Cham, pp 259–284
Erschbamer B, Niederfriniger Schlag R, Winkler E (2008) Colonization processes on a central alpine glacier foreland. J Veg Sci 19:855–862
Escobar G, Beall CM (1982) Contemporary patterns of migration in the central andes. Mt Res Dev 2:63–80
Esper J, Schweingruber FH (2004) Large-scale treeline changes recorded in Siberia. Geophys Res Lett 31: L06202
Essl F, Dullinger S, Genovesi P, Hulme PE, Jeschke JM et al (2019) A conceptual framework for range-expanding species that track human-induced environmental change. Bioscience 69:908–919
Essl F (2019) First records of casual occurrences of Chinese windmill palm *Trachycarpus fortunei* (Hook.) H. Wendl. in Austria. BioInvasions Rec 8:471–477
Evangelista A, Frate L, Carranza ML, Attorre F, Pelino G, Stanisci A (2016) Changes in composition, ecology and structure of high-mountain vegetation: a re-visitation study over 42 years. AoB Plants 8:plw004
Fadrique B, Báez S, Duque Á, Malizia A, Blundo C et al (2018) Widespread but heterogeneous responses of Andean forests to climate change. Nature 564:207–212
Fajardo A, McIntire EJ (2012) Reversal of multicentury tree growth improvements and loss of synchrony at mountain tree lines point to changes in key drivers. J Ecol 100:782–794
Falcucci A, Maiorano L, Boitani L (2007) Changes in land-use/land-cover patterns in Italy and their implications for biodiversity conservation. Landscape Ecol 22:617–631
Falvey M, Garreaud RD (2009) Regional cooling in a warming world: recent temperature trends in the Southeast Pacific and along the west coast of subtropical South America (1979–2006). J Geophys Res Atmos 114:D04102
Fan ZX, Bräuning A, Cao KF, Zhu SD (2009) Growth-climate responses of high-elevation conifers in the Central Hengduan Mountains, southwestern China. For Ecol Manage 258:306–313
Fang O, Wang Y, Shao X (2016) The effect of climate on the net primary productivity (NPP) of *Pinus koraiensis* in the Changbai Mountains over the past 50 years. Trees 30:281–294
Farinotti D, Immerzeel WW, De Kok RJ, Quincey DJ, Dehecq A (2020) Manifestations and mechanisms of the Karakoram glacier anomaly. Nat Geosci 13:8–16
Farinotti D, Longuevergne L, Moholdt G, Duethmann D, Mölg T et al (2015) Substantial glacier mass loss in the Tien Shan over the past 50 years. Nat Geosci 8:716–722
Fayad A, Gascoin S, Faour G, López-Moreno JI, Drapeau L, Le Page M, Escadafal R (2017) Snow hydrology in Mediterranean mountain regions: a review. J Hydrol 551:374–396
Fedorov AN, Ivanova RN, Park H, Hiyama T, Iijima Y (2014) Recent air temperature changes in the permafrost landscapes of northeastern Eurasia. Polar Sci 8:114–128
Feeley KJ, Hurtado J, Saatchi S, Silman MR, Clark DB (2013) Compositional shifts in Costa Rican forests due to climate-driven species migrations. Glob Change Biol 19:3472–3480
Feeley KJ, Silman MR, Bush MB, Farfan W, Cabrera KG et al (2011) Upslope migration of Andean trees. J Biogeogr 38:783–791
Fei S, Desprez JM, Potter KM, Jo I, Knott JA, Oswalt CM (2017) Divergence of species responses to climate change. Sci Adv 3:e1603055
Felde VA, Kapfer J, Grytnes JA (2012) Upward shift in elevational plant species ranges in Sikkilsdalen, Central Norway. Ecography 35:922–932
Fernandez-Gimenez ME (2002) Spatial and social boundaries and the paradox of pastoral land tenure: a case study from postsocialist Mongolia. Hum Ecol 30:49–78
Fernandez-Gimenez ME, Baival B, Fassnacht SR, Wilson D (eds)(2015) Building resilience of Mongolian rangelands: a trans-disciplinary research conference, June 9-10, 2015, Ulaanbaatar, Mongolia. Tsogt Print, Ulaanbaatar
Fernández Calzado MR, Molero Mesa J (2013) Changes in the summit flora of a Mediterranean mountain (Sierra Nevada, Spain) as a possible effect of climate change. Lazaroa 34:65–75
Fickert T, Grüninger F (2018) High-speed colonization of bare ground—permanent plot studies on primary succession of plants in recently deglaciated glacier forelands. Land Degrad Dev 29:2668–2680
Fickert T, Grüninger F, Damm B (2017) Klebelsberg revisited: did primary succession of plants in glacier

forelands a century ago differ from today? Alp Bot 127:17–29
Field CB, Chiariello NR, Diffenbaugh NS (2016) Climate change impacts. In: Mooney H, Zavaleta E (eds) Ecosystems of California. University of California Press, Berkeley, pp 251–264
Filippa G, Cremonese E, Galvagno M, Isabellon M, Bayle A et al (2019) Climatic drivers of greening trends in the Alps. Remote Sens 11:2527
Fischer A, Fickert T, Schwaizer G, Patzelt G, Groß G (2019) Vegetation dynamics in alpine glacier forelands tackled from space. Sci Rep 9:1–13
Fischer M, Huss M, Barboux C, Hoelzle M (2014) The new Swiss Glacier Inventory SGI2010: relevance of using high-resolution source data in areas dominated by very small glaciers. Arct Antarct Alp Res 46:933–945
Fischer M, Rudmann-Maurer K, Weyand A, Stöcklin J (2008) Agricultural land use and biodiversity in the Alps. Mt Res Dev 28:148–155
Fleishman E, Belnap J, Cobb N, Enquist CA, Ford K et al (2013) Natural ecosystems. In: Garfin G, Jardine A, Merideth R, Black M, LeRoy S (eds) Assessment of climate change in the Southwest United States. Island Press, Washington, DC, pp 148–167
Forbes BC, Kumpula T (2009) The ecological role and geography of reindeer (*Rangifer tarandus*) in northern Eurasia. Geogr Compass 3:1356–1380
Forister ML, McCall AC, Sanders NJ, Fordyce JA, Thorne JH et al (2010) Compounded effects of climate change and habitat alteration shift patterns of butterfly diversity. Proc Natl Acad Sci 107:2088–2092
Formica A, Farrer EC, Ashton IW, Suding KN (2014) Shrub expansion over the past 62 years in Rocky Mountain alpine tundra: possible causes and consequences. Arct Antarct Alp Res 46:616–631
Forrest JR, Thomson JD (2011) An examination of synchrony between insect emergence and flowering in Rocky Mountain meadows. Ecol Monogr 81:469–491
Forrest JL, Wikramanayake E, Shrestha R, Areendran G, Gyeltshen K et al (2012) Conservation and climate change: assessing the vulnerability of snow leopard habitat to treeline shift in the Himalaya. Biol Cons 150:129–135
Forsythe N, Fowler HJ, Li XF, Blenkinsop S, Pritchard D (2017) Karakoram temperature and glacial melt driven by regional atmospheric circulation variability. Nat Clim Chang 7:664–670
Fountain AG, Glenn B, Basagic HJ IV (2017) The geography of glaciers and perennial snowfields in the American West. Arct Antarct Alp Res 49:391–410
Fowler HJ, Archer DR (2006) Conflicting signals of climatic change in the upper Indus basin. J Clim 19:4276–4293
Fox DJ (1997) Mining in mountains. In: Messerli B, Ives JD (eds) Mountains of the world—a global priority. Parthenon Publishing Group, New York-London, pp 171–198
Francon L, Corona C, Roussel E, Saez JL, Stoffel M (2017) Warm summers and moderate winter precipitation boost *Rhododendron ferrugineum* L. growth in the Taillefer massif (French Alps). Sci Total Environ 586:1020–1031
Franke AK, Feilhauer H, Bräuning A, Rautio P, Braun M (2019) Remotely sensed estimation of vegetation shifts in the polar and alpine tree-line ecotone in Finnish Lapland during the last three decades. For Ecol Manage 454:117668
Frans C, Istanbulluoglu E, Lettenmaier DP, Fountain AG, Riedel J (2018) Glacier recession and the response of summer streamflow in the Pacific Northwest United States, 1960–2099. Water Resour Res 54:6202–6225
Fransen S, Kuschminder K (2009) Migration in Ethiopia: history, current trends and future prospects. Maastricht Graduate School of Governance, Maastricht
Franzén M, Dieker P, Schrader J, Helm A (2019) Rapid plant colonization of the forelands of a vanishing glacier is strongly associated with species traits. Arct Antarct Alp Res 51:366–378
Frate L, Carranza ML, Evangelista A, Stinca A, Schaminée JH, Stanisci A (2018) Climate and land use change impacts on Mediterranean high-mountain vegetation in the Apennines since the 1950s. Plant Ecolog Divers 11:85–96
Frazier AG, Brewington L (2020) Current changes in alpine ecosystems of Pacific Islands. In: Goldstein MI, DellaSala DA (eds) Encyclopedia of the world's biomes, vol 1. Elsevier, Amsterdam, pp 607–619
Fredman P, Chekalina T (2019) Winter recreation trends in the Swedish mountains—challenges and opportunities. In: Pröbstl-Haider U, Richins H, Türk S (eds) Winter tourism: trends and challenges. CABI, Wallingford, pp 183–191
Fredman P, Heberlein TA (2005) Mountain tourism in northern Europe: current patterns and trends. In: Thompson DBA, Price MF, Galbraith CA (eds) Mountains of northern Europe: conservation, management, people and nature. TSO Scotland, Edinburgh, pp 203–212
Freeman BG, Lee-Yaw JA, Sunday JM, Hargreaves AL (2018) Expanding, shifting and shrinking: the impact of global warming on species' elevational distributions. Glob Ecol Biogeogr 27:1268–1276
Frei E, Bodin J, Walther GR (2010) Plant species' range shifts in mountainous areas—all uphill from here? Bot Helv 120:117–128
De Frenne P, Rodríguez-Sánchez F, Coomes DA, Baeten L, Verstraeten G et al (2013) Microclimate moderates plant responses to macroclimate warming. Proc Natl Acad Sci 110:18561–18565
Freppaz M, Filippa G, Corti G, Cocco S, Williams MW, Zanini E (2013) Soil properties on ski-runs. In: Rixen C, Rolando A (eds) The impacts of skiing and related winter recreational activities on mountain environments. Bentham Science Publishers, Bussum, pp 45–64
Frolova NL, Belyakova PA, Grigor'ev VY, Sazonov AA, Zotov LV (2017) Many-year variations of river runoff in the Selenga basin. Water Resour 44:359–371

Frost GV, Bhatt US, Epstein HE, Myers-Smith I, Phoenix GK et al. (2020) Tundra greenness. In: Thoman RL, Richter-Menge J, Druckenmiller ML (eds) Arctic report card 2020. https://doi.org/10.25923/46rm-0w23
Fu YH, Piao S, Op De Beeck M, Cong N, Zhao H et al (2014) Recent spring phenology shifts in western Central Europe based on multiscale observations. Glob Ecol Biogeogr 23:1255–1263
Fukui K, Fujii Y, Ageta Y, Asahi K (2007) Changes in the lower limit of mountain permafrost between 1973 and 2004 in the Khumbu Himal, the Nepal Himalayas. Global Planet Change 55:251–256
Fulé PZ, Laughlin DC (2007) Wildland fire effects on forest structure over an altitudinal gradient, Grand Canyon National Park, USA. J Appl Ecol 44:136–146
Funnell D, Parish R (2001) Mountain environments and communities. Routledge, London-New York
Gade DW (1992) Landscape, system, and identity in the post-conquest Andes. Ann Assoc Am Geogr 82:460–477
Gade DW (1999) Nature and culture in the Andes. University of Wisconsin Press, Madison
Gaira KS, Dhar U, Belwal OK (2011) Potential of herbarium records to sequence phenological pattern: a case study of *Aconitum heterophyllum* in the Himalaya. Biodivers Conserv 20:2201–2210
Gaire NP, Koirala M, Bhuju DR, Borgaonkar HP (2014) Treeline dynamics with climate change at the Central Nepal Himalaya. Climate of the Past 10:1277–1290
Gaire NP, Koirala M, Bhuju DR, Carrer M (2017) Site- and species-specific treeline responses to climatic variability in eastern Nepal Himalaya. Dendrochronologia 41:44–56
Gallien L, Altermatt F, Wiemers M, Schweiger O, Zimmermann NE (2017) Invasive plants threaten the least mobile butterflies in Switzerland. Divers Distrib 23:185–195
Galván JD, Camarero JJ, Ginzler C, Büntgen U (2014) Spatial diversity of recent trends in Mediterranean tree growth. Environ Res Lett 9:084001
Ganjurjav H, Gao Q, Gornish ES, Schwartz MW, Liang Y et al (2016) Differential response of alpine steppe and alpine meadow to climate warming in the Central Qinghai-Tibetan Plateau. Agric For Meteorol 223:233–240
Ganyushkin D, Chistyakov K, Volkov I, Bantcev D, Kunaeva E, Terekhov A (2017) Present glaciers and their dynamics in the arid parts of the Altai Mountains. Geosciences 7:117
Gao Y, Chen F, Lettenmaier DP, Xu J, Xiao L, Li X (2018) Does elevation-dependent warming hold true above 5000 m elevation? Lessons from the Tibetan Plateau. NPJ Clim Atmos Sci 1:19
García-Romero A, Muñoz J, Andrés N, Palacios D (2010) Relationship between climate change and vegetation distribution in the Mediterranean mountains: Manzanares Head valley, Sierra De Guadarrama (Central Spain). Clim Change 100:645–666
García-Ruiz JM, Lana-Renault N (2011) Hydrological and erosive consequences of farmland abandonment in Europe, with special reference to the Mediterranean region—a review. Agr Ecosyst Environ 140:317–338
García-Ruiz JM, López-Moreno JI, Vicente-Serrano SM, Lasanta–Martínez T, Beguería S (2011) Mediterranean water resources in a global change scenario. Earth-Sci Rev 105:121–139
Gardelle J, Arnaud Y, Berthier E (2011) Contrasted evolution of glacial lakes along the Hindu Kush Himalaya mountain range between 1990 and 2009. Global Planet Change 75:47–55
Gardelle J, Berthier E, Arnaud Y, Kääb A (2013) Region-wide glacier mass balances over the Pamir-Karakoram-Himalaya during 1999–2011. Cryosphere 7:1263–1286
Gardner JS, Rhoades RE, Stadel C (2013) People in the mountains. In: Price MF, Byers AC, Friend DA, Kohler T, Price LW (eds) Mountain geography: physical and human dimensions. University of California Press, Berkeley-Los Angeles, pp 267–300
Garfin G, Franco G, Blanco H, Comrie A, Gonzalez P et al (2014) Southwest. In: Melillo JM, Richmond TC, Yohe GW (eds) Climate change impacts in the United States: the third national climate assessment. U.S. Global Change Research Program, Washington DC, pp 462–486
Garonna I, De Jong R, De Wit AJ, Mücher CA, Schmid B, Schaepman ME (2014) Strong contribution of autumn phenology to changes in satellite-derived growing season length estimates across Europe (1982–2011). Glob Change Biol 20:3457–3470
Garrard R, Kohler T, Price MF, Byers AC, Sherpa AR, Maharjan GR (2016) Land use and land cover change in Sagarmatha National Park, a World Heritage Site in the Himalayas of eastern Nepal. Mt Res Dev 36:299–310
Gartzia M, Pérez-Cabello F, Bueno CG, Alados CL (2016) Physiognomic and physiologic changes in mountain grasslands in response to environmental and anthropogenic factors. Appl Geogr 66:1–11
Gatti RC, Callaghan T, Velichevskaya A, Dudko A, Fabbio L, Battipaglia G, Liang J (2019) Accelerating upward treeline shift in the Altai Mountains under last-century climate change. Sci Rep 9:1–13
Gawith D, Kingston DG, McMillan H (2012) The effects of climate change on runoff in the Lindis and Matukituki catchments, Otago, New Zealand. J Hydrol (NZ) 51:121–135
Ge Q, Wang H, Rutishauser T, Dai J (2015) Phenological response to climate change in China: a meta-analysis. Glob Change Biol 21:265–274
Gebrechorkos SH, Hülsmann S, Bernhofer C (2019) Changes in temperature and precipitation extremes in Ethiopia, Kenya, and Tanzania. Int J Climatol 39:18–30
Gebru BM, Lee WK, Khamzina A, Wang SW, Cha S, Song C, Lamchin M (2020) Spatiotemporal multi-index analysis of desertification in dry afromontane

forests of northern Ethiopia. Environ Dev Sustain. https://doi.org/10.1007/s10668-020-00587-3

Gehrig-Fasel J, Guisan A, Zimmermann NE (2007) Tree line shifts in the Swiss Alps: climate change or land abandonment? J Veg Sci 18:571–582

Gentle P, Thwaites R (2016) Transhumant pastoralism in the context of socioeconomic and climate change in the mountains of Nepal. Mt Res Dev 36:173–183

Gerard F, Petit S, Smith G, Thomson A, Brown N et al (2010) Land cover change in Europe between 1950 and 2000 determined employing aerial photography. Prog Phys Geogr 34:183–205

Gerlitz L, Conrad O, Thomas A, Böhner J (2014) Warming patterns over the Tibetan Plateau and adjacent lowlands derived from elevation- and bias-corrected ERA-Interim data. Clim Res 58:235–246

Ghasemi AR (2015) Changes and trends in maximum, minimum and mean temperature series in Iran. Atmos Sci Lett 16:366–372

Gifford-Gonzalez D (2000) Animal disease challenges to the emergence of pastoralism in sub-Saharan Africa. Afr Archaeol Rev 17:95–139

Gifford-Gonzalez D (2017) "Animal disease challenges" fifteen years later: the hypothesis in light of new data. Quatern Int 436:283–293

Gigauri K, Akhalkatsi M, Nakhutsrishvili G, Abdaladze O (2013) Monitoring of vascular plant diversity in a changing climate in the alpine zone of the Central Greater Caucasus. Turk J Bot 37:1104–1114

Gilani H, Goheer MA, Ahmad H, Hussain K (2020) Under predicted climate change: distribution and ecological niche modelling of six native tree species in Gilgit-Baltistan, Pakistan. Ecol Ind 111:106049

Giménez-Benavides L, Escudero A, García-Camacho R, García-Fernández A, Iriondo JM et al (2018) How does climate change affect regeneration of Mediterranean high-mountain plants? An integration and synthesis of current knowledge. Plant Biol 20:50–62

Glade FE, Miranda MD, Meza FJ, Van Leeuwen WJ (2016) Productivity and phenological responses of natural vegetation to present and future inter-annual climate variability across semi-arid river basins in Chile. Environ Monit Assess 188:676

Gobiet A, Kotlarski S, Beniston M, Heinrich G, Rajczak J, Stoffel M (2014) 21st century climate change in the European Alps—a review. Sci Total Environ 493:1138–1151

Gonzalez P, Neilson RP, Lenihan JM, Drapek RJ (2010) Global patterns in the vulnerability of ecosystems to vegetation shifts due to climate change. Glob Ecol Biogeogr 19:755–768

Gordo O, Sanz JJ (2010) Impact of climate change on plant phenology in Mediterranean ecosystems. Glob Change Biol 16:1082–1106

Goswami UP, Bhargav K, Hazra B, Goyal MK (2018) Spatiotemporal and joint probability behavior of temperature extremes over the Himalayan region under changing climate. Theoret Appl Climatol 134:477–498

Gottfried M, Pauli H, Futschik A, Akhalkatsi M, Barančok P et al (2012) Continent-wide response of mountain vegetation to climate change. Nat Clim Chang 2:111–115

Graae BJ, Vandvik V, Armbruster WS, Eiserhardt WL, Svenning JC et al (2018) Stay or go—how topographic complexity influences alpine plant population and community responses to climate change. Perspect Plant Ecol Evol Syst 30:41–50

Grab S, Linde J, De Lemos H (2017) Some attributes of snow occurrence and snowmelt/sublimation rates in the Lesotho Highlands: environmental implications. Water SA 43:333–342

Grabherr G, Gottfried M, Pauli H (2010) Climate change impacts in alpine environments. Geogr Compass 4:1133–1153

Grabherr G (2003) Alpine vegetation dynamics and climate change—a synthesis of long-term studies and observations. In: Nagy L, Grabherr G, Körner C, Thompson DBA (eds) Alpine biodiversity in Europe. Ecological Studies 167. Springer, Berlin, pp 399–409

Grabherr G, Gottfried M, Gruber A, Pauli H (1995) Patterns and current changes in alpine plant diversity. In: Chapin III FS, Körner C (eds) Arctic and alpine biodiversity: patterns, causes and ecosystem consequences. Ecological Studies 113. Springer, Berlin, pp 167–181

Gratzer G, Keeton WS (2017) Mountain forests and sustainable development: the potential for achieving the United Nations' 2030 Agenda. Mt Res Dev 37:246–253

Greenwood S, Jump AS (2014) Consequences of treeline shifts for the diversity and function of high altitude ecosystems. Arct Antarct Alp Res 46:829–840

Griffiths P, Kuemmerle T, Baumann M, Radeloff VC, Abrudan IV et al (2014) Forest disturbances, forest recovery, and changes in forest types across the Carpathian ecoregion from 1985 to 2010 based on Landsat image composites. Remote Sens Environ 151:72–88

Grosjean M, Messerli B (1988) African mountains and highlands: potential and constraints. Mt Res Dev 8:111–122

Grover VI, Borsdorf A, Breuste JH, Tiwari PC, Frangetto FW (eds) (2015) Impact of global changes on mountains. CRC Press, Boca Raton, Responses and adaptations

Gruber S, Fleiner R, Guegan E, Panday P, Schmid MO et al (2017) Review article: inferring permafrost and permafrost thaw in the mountains of the Hindu Kush Himalaya region. Cryosphere 11:81–99

Grytnes JA, Kapfer J, Jurasinski G, Birks HH, Henriksen H et al (2014) Identifying the driving factors behind observed elevational range shifts on European mountains. Glob Ecol Biogeogr 23:876–884

Grötzbach E, Stadel C (1997) Mountain peoples and cultures. In: Messerli B, Ives JD (eds) Mountains of the world—a global priority. Parthenon Publishing Group, New York-London, pp 17–38

Grötzbach E (1980) Die Nutzung der Hochweidestufe als Kriterium einer kulturgeographischen Typisierung von Hochgebirgen. In: Jentsch C, Liedtke H (eds) Höhengrenzen in Hochgebirgen. Arbeiten aus dem Geographischen Institut der Universität des Saarlandes 29. Universität des Saarlandes—Geographisches Institut, Saarbrücken, pp 265–277

Grüninger F (2015) Der ökologische Preis des „Winning of the West". Geogr Rundsch 67:24–31

Gunya A (2017) Land reforms in post-socialist mountain regions and their impact on land use management: a case study from the Caucasus. J Alp Res | Rev De Géog Alpine 105–1. https://doi.org/10.4000/rga.3563

Guo D, Wang H (2017) Simulated historical (1901–2010) changes in the permafrost extent and active layer thickness in the northern hemisphere. J Geophys Res Atmos 122:12285–12295

Gurung A, Bista R, Karki R, Shrestha S, Uprety D, Oh SE (2013) Community-based forest management and its role in improving forest conditions in Nepal. Small-Scale Forest 12:377–388

Gurung DR, Giriraj A, Aung KS, Shrestha BR, Kulkarni AV (2011) Snow-cover mapping and monitoring in the Hindu Kush-Himalayas. ICIMOD, Kathmandu

Gómez JM, González-Megías A, Lorite J, Abdelaziz M, Perfectti F (2015) The silent extinction: climate change and the potential hybridization-mediated extinction of endemic high-mountain plants. Biodivers Conserv 24:1843–1857

Gómez-Mendoza L, Arriaga L (2007) Modeling the effect of climate change on the distribution of oak and pine species of Mexico. Conserv Biol 21:1545–1555

Güsewell S, Furrer R, Gehrig R, Pietragalla B (2017) Changes in temperature sensitivity of spring phenology with recent climate warming in Switzerland are related to shifts of the preseason. Glob Change Biol 23:5189–5202

Gądek B (2014) Climatic sensitivity of the non-glaciated mountains cryosphere (Tatra Mts., Poland and Slovakia). Glob Planet Change 121:1–8

De Haas H (2006) Migration, remittances and regional development in southern Morocco. Geoforum 37:565–580

De Haas H (2009) International migration and regional development in Morocco: a review. J Ethn Migr Stud 35:1571–1593

Haddeland I, Heinke J, Biemans H, Eisner S, Flörke M et al (2014) Global water resources affected by human interventions and climate change. Proc Natl Acad Sci 111:3251–3256

Haeberli W, Schaub Y, Huggel C (2017) Increasing risks related to landslides from degrading permafrost into new lakes in de-glaciating mountain ranges. Geomorphology 293:405–417

Hagedorn F, Shiyatov SG, Mazepa VS, Devi NM, Grigor'ev AA et al (2014) Treeline advances along the Urals mountain range—driven by improved winter conditions? Glob Change Biol 20:3530–3543

Haider S, Kueffer C, Bruelheide H, Seipel T, Alexander JM et al (2018) Mountain roads and non-native species modify elevational patterns of plant diversity. Glob Ecol Biogeogr 27:667–678

Hailemariam SN, Soromessa T, Teketay D (2016) Land use and land cover change in the Bale Mountain eco-region of Ethiopia during 1985 to 2015. Land 5:41

Hall J, Burgess ND, Lovett J, Mbilinyi B, Gereau RE (2009) Conservation implications of deforestation across an elevational gradient in the eastern Arc Mountains, Tanzania. Biol Cons 142:2510–2521

Hallinger M, Manthey M, Wilmking M (2010) Establishing a missing link: warm summers and winter snow cover promote shrub expansion into alpine tundra in Scandinavia. New Phytol 186:890–899

Hallman C, Arnott H (2015) Morphological and physiological phenology of *Pinus longaeva* in the White Mountains of California. Tree-Ring Res 71:1–12

Halloy SR, Mark AF (2003) Climate-change effects on alpine plant biodiversity: a New Zealand perspective on quantifying the threat. Arct Antarct Alp Res 35:248–254

Hamid M, Khuroo AA, Malik AH, Ahmad R, Singh CP, Dolezal J, Haq SM (2020) Early evidence of shifts in alpine summit vegetation: a case study from Kashmir Himalaya. Front Plant Sci 11:421

Hamilton AC (1982) Environmental history of East Africa: a study of the Quaternary. Academic Press, London

Hamilton LS (2015) When the sacred encounters economic development in mountains. George Wright Forum 32:132–140

Hammi S, Simonneaux V, Cordier JB, Genin D, Alifriqui M, Montes N, Auclair L (2010) Can traditional forest management buffer forest depletion? Dynamics of Moroccan High Atlas mountain forests using remote sensing and vegetation analysis. For Ecol Manage 260:1861–1872

Hansen W, Magiera A, Theissen T, Waldhardt R, Otte A (2018) Analysing *Betula litwinowii* encroachment and reforestation in the Kazbegi region, Greater Caucasus, Georgia. J Veg Sci 29:110–123

Hansen AJ, Piekielek N, Davis C, Haas J, Theobald DM (2014) Exposure of US national parks to land use and climate change 1900–2100. Ecol Appl 24:484–502

Hanzer F, Förster K, Nemec J, Strasser U (2018) Projected cryospheric and hydrological impacts of 21st century climate change in the Ötztal Alps (Austria) simulated using a physically based approach. Hydrol Earth Syst Sci 22:1593–1614

Hardy KA, Thevs N, Aliev K, Welp M (2018) Afforestation and reforestation of walnut forests in southern Kyrgyzstan: an economic perspective. Mt Res Dev 38:332–341

Harkoma A, Forbes BC (2020) Traditional reindeer rangeland management and a (human) rights-based approach to food sovereignty. In: Hossain K, Nilsson LM, Herrmann TM (eds) Food security in the High North: contemporary challenges across the circumpolar region. Routledge, Abingdon-New York, pp 34–55

Harris C (1997) The resettlement of British Columbia. University of British Columbia Press, Vancouver

Harris RB (2010) Rangeland degradation on the Qinghai-Tibetan Plateau: a review of the evidence of its magnitude and causes. J Arid Environ 74:1–12
Harrison S, Kargel JS, Huggel C, Reynolds J, Shugar DH et al (2018) Climate change and the global pattern of moraine-dammed glacial lake outburst floods. Cryosphere 12:1195–1209
Harsch MA, Bader MY (2011) Treeline form—a potential key to understanding treeline dynamics. Glob Ecol Biogeogr 20:582–596
Harsch MA, Buxton R, Duncan RP, Hulme PE, Wardle P, Wilmshurst J (2012) Causes of tree line stability: stem growth, recruitment and mortality rates over 15 years at New Zealand *Nothofagus* tree lines. J Biogeogr 39:2061–2071
Harsch MA, Hulme PE, McGlone MS, Duncan RP (2009) Are treelines advancing? A global meta-analysis of treeline response to climate warming. Ecol Lett 12:1040–1049
Harvey HT, Shellhammer HS, Stecker RE (1980) Giant *Sequoia* ecology: fire and reproduction. US Department of the Interior, National Park Service, Washington, DC
Hasson S, Böhner J, Lucarini V (2017) Prevailing climatic trends and runoff response from Hindukush–Karakoram–Himalaya, upper Indus basin. Earth Syst Dyn 8:337–355
Hasson S, Gerlitz L, Schickhoff U, Scholten T, Böhner J (2016) Recent climate change over High Asia. In: Singh RB, Schickhoff U, Mal S (eds) Climate change, glacier response, and vegetation dynamics in the Himalaya. Springer, Cham, pp 29–48
Hastenrath S (2005) The glaciers of Mount Kenya 1899–2004 (Veränderungen der Gletscher am Mount Kenya 1899–2004). Erdkunde 59:120–125
He M, Yang B, Bräuning A (2013) Tree growth-climate relationships of *Juniperus tibetica* along an altitudinal gradient on the southern Tibetan Plateau. Trees 27:429–439
Hedenås H, Christensen P, Svensson J (2016) Changes in vegetation cover and composition in the Swedish mountain region. Environ Monit Assess 188:452
Hemp A (2008) Introduced plants on Kilimanjaro: tourism and its impact. Plant Ecol 197:17–29
Hemp A (2009) Climate change and its impact on the forests of Kilimanjaro. Afr J Ecol 47:3–10
Hemp A, Hemp C (2018) Broken bridges: the isolation of Kilimanjaro's ecosystem. Glob Change Biol 24:3499–3507
Hemp A (2005a) Climate change-driven forest fires marginalize the impact of ice cap wasting on Kilimanjaro. Glob Change Biol 11:1013–1023
Hemp A (2005b) The banana forests of Kilimanjaro: biodiversity and conservation of the Chagga homegardens. Biodivers Conserv 15:1193–1217
Hemp A (2006a) The impact of fire on diversity, structure, and composition of the vegetation on Mt. Kilimanjaro. In: Spehn EM, Liberman M, Körner C (eds) Land use change and mountain biodiversity. Taylor & Francis, Boca Raton-London-New York, pp 51–69
Hemp A (2006b) Vegetation of Kilimanjaro: hidden endemics and missing bamboo. Afr J Ecol 44:305–328
Hendrikx J, Hreinsson EÖ, Clark MP, Mullan AB (2012) The potential impact of climate change on seasonal snow in New Zealand: part I—an analysis using 12 GCMs. Theoret Appl Climatol 110:607–618
Hennessy KJ, Whetton PH, Walsh K, Smith IN, Bathols JM, Hutchinson M, Sharples J (2008) Climate change effects on snow conditions in mainland Australia and adaptation at ski resorts through snowmaking. Climate Res 35:255–270
Herrmann SM, Didan K, Barreto-Munoz A, Crimmins MA (2016) Divergent responses of vegetation cover in southwestern US ecosystems to dry and wet years at different elevations. Environ Res Lett 11:124005
Hess CG (1990) Moving up-moving down": agro-pastoral land-use patterns in the Ecuadorian Paramos. Mt Res Dev 10:333–342
Hewitt K (2005) The Karakoram anomaly? Glacier expansion and the 'elevation effect', Karakoram Himalaya. Mt Res Dev 25:332–340
Hewitt K (2007) Tributary glacier surges: an exceptional concentration at Panmah Glacier, Karakoram Himalaya. J Glaciol 53:181–188
Hijioka Y, Lin E, Pereira JJ, Corlett RT, Cui X et al. (2014) Asia. In: IPCC (ed) Climate change 2014: impacts, adaptation, and vulnerability. Part B: regional aspects. Cambridge University Press, Cambridge-New York, pp 1327–1370
Hilker T, Natsagdorj E, Waring RH, Lyapustin A, Wang Y (2014) Satellite observed widespread decline in Mongolian grasslands largely due to overgrazing. Glob Change Biol 20:418–428
Hjelle KL, Hufthammer AK, Bergsvik KA (2006) Hesitant hunters: a review of the introduction of agriculture in western Norway. Environ Archaeol 11:147–170
Hock R, Rasul G, Adler C, Cáceres B, Gruber S et al. (2019) High mountain areas. In: IPCC (ed) Special report on the ocean and cryopshere in a changing climate. IPCC, Geneva, pp 131–202
Hoegh-Guldberg O, Jacob D, Taylor M, Bindi M, Brown S et al. (2018) Impacts of 1.5 °C global warming on natural and human systems. In: IPCC (ed) Global warming of 1.5 °C. An IPCC special report. IPCC, Geneva, pp 175–311
Hoekstra JM, Molnar JL, Jennings M, Revenga C, Spalding MD et al (2010) The atlas of global conservation: changes, challenges, and opportunities to make a difference. University of California Press, Berkeley
Hoelzle M, Barandun M, Bolch T, Fiddes J, Gafurov A et al (2019) The status and role of the alpine cryosphere in Central Asia. In: Xenarios S, Schmidt-Vogt D, Qadir M, Janusz-Pawletta B, Abdullaev I (eds) The Aral Sea basin. Water for sustainable development in Central Asia, Routledge, Abingdon-New York, pp 100–121

Hoerling MP, Dettinger M, Wolter K, Lukas J, Eischeid J et al (2013) Present weather and climate: evolving conditions. In: Garfin G, Jardine A, Merideth R, Black M, LeRoy S (eds) Assessment of climate change in the Southwest United States. Island Press, Washington, DC, pp 74–100

Hofgaard A (1997) Inter-relationships between treeline position, species diversity, land use and climate change in the Central Scandes Mountains of Norway. Glob Ecol Biogeogr Lett 6:419–429

Hofgaard A, Dalen L, Hytteborn H (2009) Tree recruitment above the treeline and potential for climate-driven treeline change. J Veg Sci 20:1133–1144

Hofgaard A, Ols C, Drobyshev I, Kirchhefer AJ, Sandberg S, Söderström L (2019) Non-stationary response of tree growth to climate trends along the arctic margin. Ecosystems 22:434–451

Hofstede RG, Groenendijk JP, Coppus R, Fehse JC, Sevink J (2002) Impact of pine plantations on soils and vegetation in the Ecuadorian High Andes. Mt Res Dev 22:159–167

Hofstede RGM, Llambi LD (2020) Plant diversity in páramo—neotropical high mountain humid grasslands. In: Goldstein MI, DellaSala DA (eds) Encyclopedia of the world's biomes, vol 1. Elsevier, Amsterdam, pp 362–372

Hogg EH, Michaelian M, Hook TI, Undershultz ME (2017) Recent climatic drying leads to age-independent growth reductions of white spruce stands in western Canada. Glob Change Biol 23:5297–5308

Holtmeier FK, Broll G (2005) Sensitivity and response of northern hemisphere altitudinal and polar treelines to environmental change at landscape and local scales. Glob Ecol Biogeogr 14:395–410

Holtmeier FK, Broll G (2007) Treeline advance - driving processes and adverse factors. Landscape Online 1:1–21

Holtmeier KF, Broll G (2010) Altitudinal and polar treelines in the northern hemisphere—causes and response to climate change. Polarforschung 79:139–153

Holtmeier FK, Broll G (2011) Response of Scots Pine (*Pinus sylvestris*) to warming climate at its altitudinal limit in northernmost subarctic Finland. Arctic 64:269–280

Holtmeier FK, Broll G (2012) Landform influences on treeline patchiness and dynamics in a changing climate. Phys Geogr 33:403–437

Holtmeier FK (2009) Mountain timberlines. Ecology, patchiness, and dynamics. Advances in Global Change Research 36. Springer, Dordrecht

Holtmeier FK, Broll G (2017b) Feedback effects of clonal groups and tree clusters on site conditions at the treeline: implications for treeline dynamics. Clim Res 73:85–96

Holtmeier FK, Broll G (2017a) Treelines—approaches at different scales. Sustainability 9:808

Holzer N, Golletz T, Buchroithner M, Bolch T (2016) Glacier variations in the Trans Alai massif and the Lake Karakul catchment (northeastern Pamir) measured from space. In: Singh RB, Schickhoff U, Mal S (eds) Climate change, glacier response, and vegetation dynamics in the Himalaya. Springer, Cham, pp 139–153

Holzinger B, Hülber K, Camenisch M, Grabherr G (2008) Changes in plant species richness over the last century in the eastern Swiss Alps: elevational gradient, bedrock effects and migration rates. Plant Ecol 195:179–196

Hoorn C, Perrigo A, Antonelli A (2018) Mountains, climate and biodiversity: an introduction. In: Hoorn C, Perrigo A, Antonelli A (eds) Mountains, climate and biodiversity. Wiley-Blackwell, Chichester, pp 1–13

Hope G (2014) The sensitivity of the high mountain ecosystems of New Guinea to climatic change and anthropogenic impact. Arct Antarct Alp Res 46:777–786

Hope G (2020) Current changes in alpine ecosystems of New Guinea. In: Goldstein MI, DellaSala DA (eds) Encyclopedia of the world's biomes, vol 1. Elsevier, Amsterdam, pp 599–606

Hoppe F, Schickhoff U, Oldeland J (2018) Plant species diversity of pastures in the Naryn Oblast (Kyrgyzstan). Die Erde 149:214–226

Hoppe F, Zhusui Kyzy T, Usupbaev A, Schickhoff U (2016a) Rangeland degradation assessment in Kyrgyzstan: vegetation and soils as indicators of grazing pressure in Naryn Oblast. J Mt Sci 13:1567–1583

Hoppe F, Zhusui Kyzy T, Usupbaev A, Schickhoff U (2016b) Contrasting grazing impact on seasonal pastures reflected by plant functional traits: search for patterns in Kyrgyz rangelands. Geo-Öko 37:165–200

Hoy A, Katel O (2019) Status of climate change and implications to ecology and community livelihoods in the Bhutan Himalaya. In: Saikia A, Thapa P (eds) Environmental change in the Himalayan Region. Springer, Cham, pp 23–45

Hoy A, Katel O, Thapa P, Dendup N, Matschullat J (2016) Climatic changes and their impact on socio-economic sectors in the Bhutan Himalayas: an implementation strategy. Reg Environ Change 16:1401–1415

Hsu HH, Chen CT (2002) Observed and projected climate change in Taiwan. Meteorol Atmos Phys 79:87–104

Hu Z, Li Q, Chen X, Teng Z, Chen C, Yin G, Zhang Y (2016) Climate changes in temperature and precipitation extremes in an alpine grassland of Central Asia. Theoret Appl Climatol 126:519–531

Hua XB, Yan JZ, Liu X, Wu YY, Liu LS, Zhang YL (2013) Factors influencing the grazing management styles of settled herders: a case study of Nagqu County, Tibetan Plateau, China. J Mt Sci 10:1074–1084

Huang R, Zhu H, Liu X, Liang E, Grießinger J et al (2017) Does increasing intrinsic water use efficiency (iWUE) stimulate tree growth at natural alpine timberline on the southeastern Tibetan Plateau? Glob Planet Change 148:217–226

Huber UM, Bugmann HKM, Reasoner MA (eds) (2005) Global change and mountain regions. An overview of current knowledge, Springer, Dordrecht

Hugo G, Bardsley DK (2014) Migration and environmental change in Asia. In: Piguet E, Laczko F (eds) People on the move in a changing climate. Springer, Dordrecht, pp 21–48

Humphries HC (2020) Alpine ecosystems in temperate mountains of North America. In: Goldstein MI, DellaSala DA (eds) Encyclopedia of the world's biomes, vol 1. Elsevier, Amsterdam, pp 311–322

Huntsinger L, Forero LC, Sulak A (2010) Transhumance and pastoralist resilience in the western United States. Pastoralism 1:1–15

Hurni H, Bagoora FDK, Laker MC, Mössmer M, Ofwono-Orecho JKW et al (1992) African mountain and highland environments: suitability and susceptibility. In: Stone PB (ed) The state of the world's mountains. Zed Books, London-New Jersey, pp 11–44

Huss M (2012) Extrapolating glacier mass balance to the mountain range scale: the European Alps 1900–2100. Cryosphere 6:713–727

Huss M, Bookhagen B, Huggel C, Jacobsen D, Bradley RS et al (2017) Toward mountains without permanent snow and ice. Earth's Future 5:418–435

Huss M, Hock R (2015) A new model for global glacier change and sea-level rise. Front Earth Sci 3:54

Huss M, Hock R (2018) Global-scale hydrological response to future glacier mass loss. Nat Clim Chang 8:135–140

IPCC (ed) (2014) Climate change 2014: Impacts, adaptation, and vulnerability. Part A: global and sectoral aspects. Cambridge Univ Press, Cambridge-New York

IPCC (ed) (2018) Global warming of 1.5 °C. An IPCC special report. IPCC, Geneva

Imanberdieva N, Imankul B, Severoğlu Z, Altai V, Öztürk M (2018) Potential impacts of climate change on plant diversity of Sary-Chelek Biosphere Reserve in Kyrgyzstan. In: Egamberdieva D, Öztürk M (eds) Vegetation of central Asia and environs. Springer, Cham, pp 349–364

Immerzeel WW, Lutz AF, Andrad M, Bah A, Biemans H et al (2020) Importance and vulnerability of the world's water towers. Nature 577:364–369

Immerzeel WW, Pellicciotti F, Bierkens MFP (2013) Rising river flows throughout the twenty-first century in two Himalayan glacierized watersheds. Nat Geosci 6:742–745

Ims RA, Ehrich E, Forbes BC, Huntley B, Walker DA et al. (2013) Terrestrial ecosystems. In: CAFF (Conservation of Arctic Flora and Fauna) (ed) Arctic biodiversity assessment: status and trends in arctic biodiversity. CAFF, Akureyri, pp 384–440

Inostroza L, Zasada I, König HJ (2016) Last of the wild revisited: assessing spatial patterns of human impact on landscapes in southern Patagonia, Chile. Reg Environ Change 16:2071–2085

Inouye DW, Wielgolaski FE (2013) Phenology at high altitudes. In: Schwartz MD (ed) Phenology: an integrative environmental science. Springer, Dordrecht, pp 249–272

Intigrinova T (2010) Social inequality and risk mitigation in the era of private land: Siberian pastoralists and land use change. Pastoralism – Res Policy Pract 1:178–197

Isaksen K, Ødegård RS, Etzelmüller B, Hilbich C, Hauck C et al (2011) Degrading mountain permafrost in southern Norway: spatial and temporal variability of mean ground temperatures, 1999–2009. Permafrost Periglac Process 22:361–377

Istomin KV, Habeck JO (2016) Permafrost and indigenous land use in the northern Urals: Komi and Nenets reindeer husbandry. Polar Sci 10:278–287

Ives JD (2004) Himalayan perceptions: environmental change and the well-being of mountain peoples. Routledge, London-New York

Ives JD (2013) Sustainable mountain development. Getting the facts right, HimAAS, Lalitpur

Ives JD, Messerli B (1989) The Himalayan dilemma: reconciling development and conservation. Routledge, London-New York

Ives JD, Messerli B, Spiess E (1997) Mountains of the world—a global priority. In: Messerli B, Ives JD (eds) Mountains of the world—a global priority. Parthenon Publishing Group, New York-London, pp 1–15

Ives JD, Ives PAH, Allan NJR, Imkamp C, Watanabe T et al (1992a) Mountains north and south. In: Stone PB (ed) The state of the world's mountains. Zed Books, London-New Jersey, pp 127–184

Ives JD, Ives PAH, Allan NJR, Imkamp C, Watanabe T et al (1992b) The Andes: geoecology of the Andes. In: Stone PB (ed) The state of the world's mountains. Zed Books, London-New Jersey, pp 185–256

Izquierdo AE, Grau HR, Navarro CJ, Casagranda E, Castilla MC, Grau A (2018) Highlands in transition: urbanization, pastoralism, mining, tourism, and wildlife in the Argentinian Puna. Mt Res Dev 38:390–400

Jacka JK (2018) The anthropology of mining: the social and environmental impacts of resource extraction in the mineral age. Annu Rev Anthropol 47:61–77

Jacob M, Frankl A, Hurni H, Lanckriet S, De Ridder M et al (2017) Land cover dynamics in the Simien Mountains (Ethiopia), half a century after establishment of the national park. Reg Environ Change 17:777–787

Jacob M, Frankl A, Beeckman H, Mesfin G, Hendrickx M, Guyassa E, Nyssen J (2015b) North Ethiopian afro-alpine tree line dynamics and forest-cover change since the early 20th century. Land Degrad Dev 26:654–664

Jacob M, Annys S, Frankl A, De Ridder M, Beeckman H, Guyassa E, Nyssen J (2015a) Tree line dynamics in the tropical African highlands—identifying drivers and dynamics. J Veg Sci 26:9–20

Jacob M, De Ridder M, Vandenabeele M, Asfaha T, Nyssen J, Beeckman H (2020) The response of *Erica arborea* L. tree growth to climate variability at the

afro-alpine tropical highlands of North Ethiopia. Forests 11:310
Jacobsen JP, Schickhoff U (1995) Untersuchungen zur Besiedlung und gegenwärtigen Waldnutzung im Hindukush/Karakorum. Erdkunde 49:49–59
Jain SK, Kumar V, Saharia M (2013) Analysis of rainfall and temperature trends in Northeast India. Int J Climatol 33:968–978
Janzen J (2005) Mobile livestock-keeping in Mongolia: present problems, spatial organization, interactions between mobile and sedentary population groups and perspectives for pastoral development. Senri Ethnological Stud 69:69–97
Jentsch A, Beierkuhnlein C (2003) Global climate change and local disturbance regimes as interacting drivers for shifting altitudinal vegetation patterns. Erdkunde 57:216–231
Ji P, Yuan X (2020) Underestimation of the warming trend over the Tibetan Plateau during 1998–2013 by global land data assimilation systems and atmospheric reanalyses. J Meteorol Res 34:88–100
Jiang L, Bao A, Guo H, Ndayisaba F (2017) Vegetation dynamics and responses to climate change and human activities in Central Asia. Sci Total Environ 599:967–980
Jiang R, Xie J, He H, Kuo CC, Zhu J, Yang M (2016) Spatiotemporal variability and predictability of normalized difference vegetation index (NDVI) in Alberta, Canada. Int J Biometeorol 60:1389–1403
Jiménez-Alfaro B, Gavilán RG, Escudero A, Iriondo JM, Fernández-González F (2014) Decline of dry grassland specialists in Mediterranean high-mountain communities influenced by recent climate warming. J Veg Sci 25:1394–1404
Jochner M, Bugmann H, Nötzli M, Bigler C (2017) Among-tree variability and feedback effects result in different growth responses to climate change at the upper treeline in the Swiss Alps. Ecol Evol 7:7937–7953
Jochner M, Bugmann H, Nötzli M, Bigler C (2018) Tree growth responses to changing temperatures across space and time: a fine-scale analysis at the treeline in the Swiss Alps. Trees 32:645–660
Jolly D, Taylor D, Marchant R, Hamilton A, Bonnefille R, Buchet G, Riollet G (1997) Vegetation dynamics in Central Africa since 18,000 yr BP: pollen records from the interlacustrine highlands of Burundi, Rwanda and western Uganda. J Biogeogr 24:492–512
Joshi PK, Rawat A, Narula S, Sinha V (2012) Assessing impact of climate change on forest cover type shifts in western Himalayan eco-region. J For Res 23:75–80
Jouvet G, Huss M, Blatter H, Picasso M, Rappaz J (2009) Numerical simulation of Rhonegletscher from 1874 to 2100. J Comput Phys 228:6426–6439
Ju J, Masek JG (2016) The vegetation greenness trend in Canada and US Alaska from 1984–2012 Landsat data. Remote Sens Environ 176:1–16
Julien Y, Sobrino JA, Verhoef W (2006) Changes in land surface temperatures and NDVI values over Europe between 1982 and 1999. Remote Sens Environ 103:43–55
Jump AS, Huang TJ, Chou CH (2012) Rapid altitudinal migration of mountain plants in Taiwan and its implications for high altitude biodiversity. Ecography 35:204–210
Jurasinski G, Kreyling J (2007) Upward shift of alpine plants increases floristic similarity of mountain summits. J Veg Sci 18:711–718
Jury MR, Funk C (2013) Climatic trends over Ethiopia: regional signals and drivers. Int J Climatol 33:1924–1935
Jylhä K, Tuomenvirta H, Ruosteenoja K, Niemi-Hugaerts H, Keisu K, Karhu JA (2010) Observed and projected future shifts of climatic zones in Europe and their use to visualize climate change information. Weather Clim Soc 2:148–167
Kaczka RJ, Czajka B, Łajczak A (2015) The tree-ring growth responses to climate in the timberline ecotone of Babia Góra Mountain. Geogr Pol 88:163–176
Kalisa W, Igbawua T, Henchiri M, Ali S, Zhang S, Bai Y, Zhang J (2019) Assessment of climate impact on vegetation dynamics over East Africa from 1982 to 2015. Sci Rep 9:1–20
Kanade R, John R (2018) Topographical influence on recent deforestation and degradation in the Sikkim Himalaya in India: implications for conservation of East Himalayan broadleaf forest. Appl Geogr 92:85–93
Kapnick SB, Delworth TL (2013) Controls of global snow under a changed climate. J Clim 26:5537–5562
Kapos V, Rhind J, Edwards M, Price MF, Ravilious C (2000) Developing a map of the world's mountain forests. In: Price MF, Butt N (eds) Forests in sustainable mountain development: a state of knowledge report for 2000. CABI Publications, Wallingford, pp 4–9
Kappas M, Degener J, Klinge M, Vitkovskaya I, Batyrbayeva M (2020) A conceptual framework for ecosystem stewardship based on landscape dynamics: case studies from Kazakhstan and Mongolia. In: Gutman G, Chen J, Henebry GM, Kappas M (eds) Landscape dynamics of drylands across Greater Central Asia: people, societies and ecosystems. Springer, Cham, pp 143–189
Karki R, Hasson S, Gerlitz L, Talchabhadel R, Schickhoff U, Scholten T, Böhner J (2019) Rising mean and extreme near-surface air temperature across Nepal. Int J Climatol 40:2445–2463
Karki R, Schickhoff U, Scholten T, Böhner J (2017) Rising precipitation extremes across Nepal. Climate 5:4
Kaser G, Osmaston H (2002) Tropical glaciers. Cambridge University Press, Cambridge
Kattel DB, Yao T (2013) Recent temperature trends at mountain stations on the southern slope of the Central Himalayas. J Earth Syst Sci 122:215–227
Kazakis G, Ghosn D, Vogiatzakis IN, Papanastasis VP (2007) Vascular plant diversity and climate change in

the alpine zone of the Lefka Ori, Crete. Biodivers Conserv 16:1603–1615

Keane RE, Mahalovich MF, Bollenbacher BL, Manning ME, Loehman RA et al (2018) Effects of climate change on forest vegetation in the northern Rockies. In: Halofsky JE, Peterson DL (eds) Climate change and Rocky Mountain ecosystems. Springer, Cham, pp 59–95

Kebrom T, Hedlund L (2000) Land cover changes between 1958 and 1986 in Kalu District, southern Wello, Ethiopia. Mt Res Dev 20:42–51

Keenan TF, Riley WJ (2018) Greening of the land surface in the world's cold regions consistent with recent warming. Nat Clim Chang 8:825–828

Keller F, Körner C (2003) The role of photoperiodism in alpine plant development. Arct Antarct Alp Res 35:361–368

Kelly AE, Goulden ML (2008) Rapid shifts in plant distribution with recent climate change. Proc Natl Acad Sci 105:11823–11826

Kharal DK, Thapa UK, George SS, Meilby H, Rayamajhi S, Bhuju DR (2017) Tree-climate relations along an elevational transect in Manang Valley, Central Nepal. Dendrochronologia 41:57–64

Kharlamova N, Sukhova M, Chlachula J (2019) Present climate developments in southern Siberia (1963–2017 years). IOP Conf Ser Earth Environ Sci 400:012008

Kharuk VI, Im ST, Dvinskaya ML, Ranson KJ (2010) Climate-induced mountain tree-line evolution in southern Siberia. Scand J For Res 25:446–454

Khattak MS, Babel MS, Sharif M (2011) Hydro-meteorological trends in the upper Indus river basin in Pakistan. Clim Res 46:103–119

Khromova T, Nosenko G, Kutuzov S, Muraviev A, Chernova L (2014) Glacier area changes in northern Eurasia. Environ Res Lett 9:015003

Khromova T, Nosenko G, Nikitin S, Muraviev A, Popova V et al (2019) Changes in the mountain glaciers of continental Russia during the twentieth to twenty-first centuries. Reg Environ Change 19:1229–1247

Kidane Y, Stahlmann R, Beierkuhnlein C (2012) Vegetation dynamics, and land use and land cover change in the Bale Mountains, Ethiopia. Environ Monit Assess 184:7473–7489

Kidane YO, Steinbauer MJ, Beierkuhnlein C (2019) Dead end for endemic plant species? A biodiversity hotspot under pressure. Global Ecol Conserv 19:e00670

Kienholz C, Herreid S, Rich JL, Arendt AA, Hock R, Burgess EW (2015) Derivation and analysis of a complete modern-date glacier inventory for Alaska and Northwest Canada. J Glaciol 61:403–420

Kilungu H, Leemans R, Munishi PK, Nicholls S, Amelung B (2019) Forty years of climate and land-cover change and its effects on tourism resources in Kilimanjaro National Park. Tourism Plann Dev 16:235–253

Kinnard C, Ginot P, Surazakov A, Macdonell S, Nicholson L et al (2020) Mass balance and climate history of a high-altitude glacier, Desert Andes of Chile. Front Earth Sci 8:40

Kintz DB, Young KR, Crews-Meyer KA (2006) Implications of land use/land cover change in the buffer zone of a national park in the tropical Andes. Environ Manage 38:238–252

Kirdyanov AV, Hagedorn F, Knorre AA, Fedotova EV, Vaganov EA et al (2012) 20th century tree-line advance and vegetation changes along an altitudinal transect in the Putorana Mountains, northern Siberia. Boreas 41:56–67

Kiteme BP, Liniger H, Notter B, Wiesmann U, Kohler T (2008) Dimensions of global change in African mountains: the example of Mount Kenya. IHDP Update 2(2008):18–22

Kittel TG, Thornton PE, Royle JA, Chase TN (2002) Climates of the Rocky Mountains: historical and future patterns. In: Baron J (ed) Rocky Mountain futures: an ecological perspective. Island Press, Washington, DC, pp 59–82

Kivinen S, Rasmus S (2015) Observed cold season changes in a Fennoscandian fell area over the past three decades. Ambio 44:214–225

Klanderud K, Birks HJB (2003) Recent increases in species richness and shifts in altitudinal distributions of Norwegian mountain plants. The Holocene 13:1–6

Klein G, Vitasse Y, Rixen C, Marty C, Rebetez M (2016) Shorter snow cover duration since 1970 in the Swiss Alps due to earlier snowmelt more than to later snow onset. Clim Change 139:637–649

Knapp G (2007) The legacy of European colonialism. In: Veblen TT, Young KR, Orme AR (eds) The physical geography of South America. Oxford University Press, Oxford, pp 279–288

Knorn JAN, Kuemmerle T, Radeloff VC, Keeton WS, Gancz V et al (2013) Continued loss of temperate old-growth forests in the Romanian Carpathians despite an increasing protected area network. Environ Conserv 40:182–193

Kobiv Y (2018) Trends in population size of rare plant species in the alpine habitats of the Ukrainian Carpathians under climate change. Diversity 10:62

Kohler T, Balsiger J, Rudaz G, Debarbieux B, Pratt DJ, Maselli D (eds) (2015) Green economy and institutions for sustainable mountain development: from Rio 1992 to Rio 2012 and beyond. Centre for Development and Environment (CDE), Swiss Agency for Development and Cooperation (SDC), University of Geneva and Geographica Bernensia, Bern

Koide D, Yoshida K, Daehler CC, Mueller-Dombois D (2017) An upward elevation shift of native and non-native vascular plants over 40 years on the island of Hawai'i. J Veg Sci 28:939–950

Kopp CW, Cleland EE (2014) Shifts in plant species elevational range limits and abundances observed over nearly five decades in a western North America mountain range. J Veg Sci 25:135–146

Kouba Y, Gartzia M, El Aich A, Alados CL (2018) Deserts do not advance, they are created: land

degradation and desertification in semiarid environments in the Middle Atlas, Morocco. J Arid Environ 158:1–8
Kovats RS, Valentini R, Bouwer LM, Georgopoulou E, Jacob D et al. (2014) Europe. In: IPCC (ed) Climate change 2014: impacts, adaptation, and vulnerability. Part B: regional aspects. Cambridge University Press, Cambridge-New York, pp 1267–1326
Kozak J (2010) Forest cover changes and their drivers in the Polish Carpathian Mountains since 1800. In: Nagendra H, Southworth J (eds) Reforesting landscapes. Linking pattern and process. Springer, Berlin, pp 253–273
Kraaijenbrink PDA, Bierkens MFP, Lutz AF, Immerzeel WW (2017) Impact of a global temperature rise of 1.5° C on Asia's glaciers. Nature 549:257–260
Krause L, Mal S, Karki R, Schickhoff U (2019) Recession of Trakarding glacier and expansion of Tsho Rolpa lake in Nepal Himalaya based on satellite data. Himalayan Geol 40:103–114
Kreutzmann H (1991) The Karakoram Highway: the impact of road construction on mountain societies. Mod Asian Stud 25:711–736
Kreutzmann H (2011) Pastoralism in Central Asian mountain regions. In: Kreutzmann H, Abdulalishoev K, Lu Z, Richter J (eds) Pastoralism and rangeland management in mountain areas in the context of climate and global change. GIZ/BMZ, Bonn, pp 38–63
Kreutzmann H (2012) Pastoral practices in transition: animal husbandry in high Asian contexts. In: Kreutzmann H (ed) Pastoral practices in High Asia. Springer, Dordrecht, pp 1–29
Kreutzmann H (2013) The tragedy of responsibility in High Asia: Modernizing traditional pastoral practices and preserving modernist worldviews. Pastoralism: Research. Policy Pract 3:1–11
Kreyling J, Wana D, Beierkuhnlein C (2010) Potential consequences of climate warming for tropical plant species in high mountains of southern Ethiopia. Divers Distrib 16:593–605
Kricsfalusy VV (2013) Mountain grasslands of high conservation value in the eastern Carpathians: syntaxonomy, biodiversity, protection and management. Thaiszia 23:67–112
Krishnamurthy V, Ajayamohan RS (2010) Composite structure of monsoon low pressure systems and its relation to Indian rainfall. J Clim 23:4285–4305
Krishnan R, Sanjay J (2017) Climate change over India: an interim report. Centre for Climate Change Research, Ministry of Earth Sciences, Govt. of India, Pashan
Krishnan R, Sabin TP, Madhura RK, Vellore RK, Mujumdar M et al. (2019b) Non-monsoonal precipitation response over the western Himalayas to climate change. Clim Dyn 52:4091–4109
Krishnan R, Shrestha AB, Ren G, Rajbhandari R, Saeed S et al. (2019a) Unravelling climate change in the Hindu Kush Himalaya: rapid warming in the mountains and increasing extremes. In: Wester P, Mishra A, Mukherji A, Shrestha AB (eds) The Hindu Kush Himalaya assessment. Springer, Cham, pp 57–96
Krishnaswamy J, John R, Joseph S (2014) Consistent response of vegetation dynamics to recent climate change in tropical mountain regions. Glob Change Biol 20:203–215
Kruger AC, Sekele SS (2013) Trends in extreme temperature indices in South Africa: 1962–2009. Int J Climatol 33:661–676
Krysanova V, Wortmann M, Bolch T, Merz B, Duethmann D et al (2015) Analysis of current trends in climate parameters, river discharge and glaciers in the Aksu river basin (Central Asia). Hydrol Sci J 60:566–590
Kudo G, Amagai Y, Hoshino B, Kaneko M (2011) Invasion of dwarf bamboo into alpine snow-meadows in northern Japan: pattern of expansion and impact on species diversity. Ecol Evol 1:85–96
Kuemmerle T, Chaskovskyy O, Knorn J, Radeloff VC, Kruhlov I, Keeton WS, Hostert P (2009) Forest cover change and illegal logging in the Ukrainian Carpathians in the transition period from 1988 to 2007. Remote Sens Environ 113:1194–1207
Kuemmerle T, Levers C, Erb K, Estel S, Jepsen MR et al (2016) Hotspots of land use change in Europe. Environ Res Lett 11:064020
Kuemmerle T, Olofsson P, Chaskovskyy O, Baumann M, Ostapowicz K et al (2011) Post-Soviet farmland abandonment, forest recovery, and carbon sequestration in western Ukraine. Glob Change Biol 17:1335–1349
Kueppers LM, Conlisk E, Castanha C, Moyes AB, Germino MJ et al (2017) Warming and provenance limit tree recruitment across and beyond the elevation range of subalpine forest. Glob Change Biol 23:2383–2395
Kulakowski D, Barbeito I, Casteller A, Kaczka RJ, Bebi P (2016) Not only temperature: interacting drivers of treeline change in Europe. Geogr Pol 89:7–15
Kulkarni A (2012) Weakening of Indian summer monsoon rainfall in warming environment. Theoret Appl Climatol 109:447–459
Kulkarni MA, Desrochers RE, Kajeguka DC, Kaaya RD, Tomayer A et al (2016) 10 years of environmental change on the slopes of Mount Kilimanjaro and its associated shift in malaria vector distributions. Front Public Health 4:281
Kull CA, Tassin J, Rangan H (2007) Multifunctional, scrubby, and invasive forests? Wattles in the highlands of Madagascar. Mt Res Dev 27:224–231
Kullman L (2008) Thermophilic tree species reinvade subalpine Sweden—early responses to anomalous late Holocene climate warming. Arct Antarct Alp Res 40:104–110
Kullman L (2018) A review and analysis of factual change on the max rise of the Swedish Scandes treeline, in relation to climate change over the past 100 years. J Ecol Nat Resour 2:000150
Kullman L (2019) Early signs of a fundamental subalpine ecosystem shift in the Swedish Scandes—the case of

the pine (*Pinus sylvestris* L.) treeline ecotone. Geo-Öko 40:122–175

Kullman L, Öberg L (2009) Post-Little Ice Age tree line rise and climate warming in the Swedish Scandes: a landscape ecological perspective. J Ecol 97:415–429

Kullman L (2007a) Long-term geobotanical observations of climate change impacts in the Scandes of West-Central Sweden. Nord J Bot 24:445–467

Kullman L (2007b) Modern climate change and shifting ecological states of the subalpine/alpine landscape in the Swedish Scandes. Geo-Öko 28:187–221

Kumar D, Choudhary A, Dimri AP (2018) Regional climate changes over Hindukush-Karakoram-Himalaya region. In: Goel PS, Ravindra R, Chattopadhyay S (eds) Science and geopolitics of the white world. Springer, Cham, pp 143–159

Kunkel KE, Bromirski PD, Brooks HE, Cavazos T, Douglas AV et al (2008) Observed changes in weather and climate extremes. In: Karl TR, Meehl GA, Miller CD, Hassol SJ, Waple AM, Murray WL (eds) Weather and climate extremes in a changing climate. CCSP, Washington DC, pp 35–80

Kurt L, Ketenoglu O, Tug GN, Sekerciler F (2015) Highland vegetation of inner and eastern Anatolia and the effects of global warming. In: Öztürk M, Hakeem KR, Faridah-Hanum I, Efe R (eds) Climate change impacts on high-altitude ecosystems. Springer, Cham, pp 275–288

Kvamme M (1988) Pollen analytical studies of mountain summer farming in western Norway. In: Birks HH, Birks HJB, Kaland PE, Moe D (eds) The cultural landscape—past, present and future. Cambridge University Press, Cambridge, pp 429–443

Kyriazopoulos AP, Skre O, Sarkki S, Wielgolaski FE, Abraham EM, Ficko A (2017) Human-environment dynamics in European treeline ecosystems: a synthesis based on the DPSIR framework. Climate Res 73:17–29

Kääb A, Berthier E, Nuth C, Gardelle J, Arnaud Y (2012) Contrasting patterns of early twenty-first-century glacier mass change in the Himalayas. Nature 488:495–498

Kääb A, Treichler D, Nuth C, Berthier E (2015) Brief communication: contending estimates of 2003–2008 glacier mass balance over the Pamir-Karakoram-Himalaya. Cryosphere 9:557–564

Körner C (2002) Mountain biodiversity, its causes and function: an overview. In: Körner C, Spehn EM (eds) Mountain biodiversity: a global assessment. Parthenon Publishing at CRC Press, London-New York, pp 3–20

Körner C (2003) Alpine plant life. Functional plant ecology of high mountain ecosystems. Springer, Berlin

Körner C (2012) Alpine treelines. Functional ecology of the global high elevation tree limits, Springer, Basel

Körner C (2020) Climatic controls of global high elevation treelines. In: Goldstein MI, DellaSala DA (eds) Encyclopedia of the world's biomes, vol 1. Elsevier, Amsterdam, pp 275–281

Körner C, Jetz W, Paulsen J, Payne D, Rudmann-Maurer K, Spehn EM (2017) A global inventory of mountains for bio-geographical applications. Alp Bot 127:1–15

Körner C, Ohsawa M, Spehn E, Berge E, Bugmann H et al (2005) Mountain systems. In: Hassan R, Scholes R, Ash N (eds) Ecosystems and human well-being: current state and trends, vol 1. Island Press, Washington-Covelo-London, pp 681–716

Küchler M, Küchler H, Bedolla A, Wohlgemuth T (2015) Response of Swiss forests to management and climate change in the last 60 years. Ann For Sci 72:311–320

Lacombe G, McCartney M (2014) Uncovering consistencies in Indian rainfall trends observed over the last half century. Clim Change 123:287–299

Lader R, Bhatt US, Walsh JE, Rupp TS, Bieniek PA (2016) Two-meter temperature and precipitation from atmospheric reanalysis evaluated for Alaska. J Appl Meteorol Climatol 55:901–922

De Lafontaine G, Payette S (2012) How climate and fire disturbances influence contrasted dynamics of *Picea glauca* ecotones at alpine tree lines in atlantic and continental eastern North America. In: Myster RW (ed) Ecotones between forest and grassland. Springer, New York, pp 299–312

Lamb HF, Damblon F, Maxted RW (1991) Human impact on the vegetation of the Middle Atlas, Morocco, during the last 5000 years. J Biogeogr 18:519–532

Lambrechts C, Hemp C, Nnyiti P, Woodley B, Hemp A (2002) Aerial survey of the threats to Mt. Kilimanjaro forests, UNDP, Dar es Salaam

Lamprecht A, Semenchuk PR, Steinbauer K, Winkler M, Pauli H (2018) Climate change leads to accelerated transformation of high-elevation vegetation in the Central Alps. New Phytol 220:447–459

Lamsal P, Kumar L, Aryal A, Atreya K (2018) Invasive alien plant species dynamics in the Himalayan region under climate change. Ambio 47:697–710

Lamsal P, Kumar L, Shabani F, Atreya K (2017) The greening of the Himalayas and Tibetan Plateau under climate change. Global Planet Change 159:77–92

Lanckriet S, Derudder B, Naudts J, Bauer H, Deckers J, Haile M, Nyssen J (2015) A political ecology perspective of land degradation in the North Ethiopian highlands. Land Degrad Dev 26:521–530

Landhäusser SM, Deshaies D, Lieffers VJ (2010) Disturbance facilitates rapid range expansion of aspen into higher elevations of the Rocky Mountains under a warming climate. J Biogeogr 37:68–76

Landmann T, Dubovyk O (2014) Spatial analysis of human-induced vegetation productivity decline over eastern Africa using a decade (2001–2011) of medium resolution MODIS time-series data. Int J Appl Earth Obs Geoinf 33:76–82

Lange J, Carrer M, Pisaric MF, Porter TJ, Seo JW, Trouillier M, Wilmking M (2020) Moisture-driven shift in the climate sensitivity of white spruce xylem anatomical traits is coupled to large-scale oscillation patterns across northern treeline in Northwest North America. Glob Change Biol 26:1842–1856

Lara MJ, Nitze I, Grosse G, Martin P, McGuire AD (2018) Reduced arctic tundra productivity linked with landform and climate change interactions. Sci Rep 8:1–10

Larsen TH, Brehm G, Navarrete H, Franco P, Gomez H et al (2011) Range shifts and extinctions driven by climate change in the tropical Andes: synthesis and directions. In: Herzog SK, Martinez R, Jørgensen PM, Tiessen H (eds) Climate change and biodiversity in the tropical Andes. IAI-SCOPE, Paris, pp 47–67

Larsen CF, Burgess E, Arendt AA, O'neel S, Johnson AJ, Kienholz C (2015) Surface melt dominates Alaska glacier mass balance. Geophys Res Lett 42:5902–5908

Lasanta T, Beltrán O, Vaccaro I (2013) Socioeconomic and territorial impact of the ski industry in the Spanish Pyrenees: mountain development and leisure induced urbanization. Pirineos 168:103–128

Lasanta-Martínez T, Vicente-Serrano SM, Cuadrat-Prats JM (2005) Mountain Mediterranean landscape evolution caused by the abandonment of traditional primary activities: a study of the Spanish Central Pyrenees. Appl Geogr 25:47–65

Latif Y, Yaoming M, Yaseen M, Muhammad S, Wazir MA (2020) Spatial analysis of temperature time series over the upper Indus basin (UIB) Pakistan. Theoret Appl Climatol 139:741–758

Latip NA, Marzukia A, Rais NSM (2016) Conservation and environmental impacts of tourism in Kinabalu Park, Sabah. In: Leng KS, Rahim AA, Weng CN (eds) 1st International conference on society, space & environment, 2–4 November 2016, Ramada Bintang Bali Resort, Bali, Indonesia. School of Humanities, USM, Pulau Pinang, pp 39–46

Lauer W (1993) Human development and environment in the Andes: a geoecological overview. Mt Res Dev 13:157–166

Lavado Casimiro WS, Labat D, Ronchail J, Espinoza JC, Guyot JL (2013) Trends in rainfall and temperature in the Peruvian Amazon-Andes basin over the last 40 years (1965–2007). Hydrol Process 27:2944–2957

Lavergne A, Daux V, Villalba R, Barichivich J (2015) Temporal changes in climatic limitation of tree-growth at upper treeline forests: contrasted responses along the west-to-east humidity gradient in northern Patagonia. Dendrochronologia 36:49–59

Lee JW, Hong SY, Chang EC, Suh MS, Kang HS (2014) Assessment of future climate change over East Asia due to the RCP scenarios downscaled by GRIMs-RMP. Clim Dyn 42:733–747

Van Leeuwen WJ, Hartfield K, Miranda M, Meza FJ (2013) Trends and ENSO/AAO driven variability in NDVI derived productivity and phenology alongside the Andes Mountains. Remote Sens 5:1177–1203

Lehnert LW, Wesche K, Trachte K, Reudenbach C, Bendix J (2016) Climate variability rather than overstocking causes recent large scale cover changes of Tibetan pastures. Sci Rep 6:24367

Leinwand II, Theobald DM, Mitchell J, Knight RL (2010) Landscape dynamics at the public–private interface: a case study in Colorado. Landsc Urban Plan 97:182–193

Lemenih M, Kassa H (2014) Re-greening Ethiopia: history, challenges and lessons. Forests 5:1896–1909

Lenoir J, Gégout JC, Dupouey JL, Bert D, Svenning JC (2010) Forest plant community changes during 1989–2007 in response to climate warming in the Jura Mountains (France and Switzerland). J Veg Sci 21:949–964

Lenoir J, Gégout JC, Marquet PA, De Ruffray P, Brisse H (2008) A significant upward shift in plant species optimum elevation during the 20th century. Science 320:1768–1771

Lenoir J, Svenning JC (2015) Climate-related range shifts—a global multidimensional synthesis and new research directions. Ecography 38:15–28

Lenoir J, Svenning JC (2013) Latitudinal and elevational range shifts under contemporary climate change. In: Levin SA (ed) Encyclopedia of biodiversity. Academic Press, London, pp 599–611

Leopold M, Völkel J, Dethier DP, Williams MW (2014) Changing mountain permafrost from the 1970s to today—comparing two examples from Niwot Ridge, Colorado Front Range, USA. Z. Für Geomorphol Supplementary Issues 58:137–157

Lesica P (2014) Arctic-alpine plants decline over two decades in Glacier National Park, Montana, USA. Arct Antarct Alp Res 46:327–332

Lesica P, Crone EE (2017) Arctic and boreal plant species decline at their southern range limits in the Rocky Mountains. Ecol Lett 20:166–174

Li Z, He Y, Wang C, Wang X, Xin H, Zhang W, Cao W (2011) Spatial and temporal trends of temperature and precipitation during 1960–2008 at the Hengduan Mountains, China. Quatern Int 236:127–142

Li C, Su F, Yang D, Tong K, Meng F, Kan B (2018) Spatiotemporal variation of snow cover over the Tibetan Plateau based on MODIS snow product, 2001–2014. Int J Climatol 38:708–728

Li L, Zhang Y, Qi W, Wang Z, Liu Y, Ding M (2019) No significant shift of warming trend over the last two decades on the mid-south of Tibetan Plateau. Atmosphere 10:416

Liang E, Dawadi B, Pederson N, Eckstein D (2014) Is the growth of birch at the upper timberline in the Himalayas limited by moisture or by temperature? Ecology 95:2453–2465

Liang E, Leuschner C, Dulamsuren C, Wagner B, Hauck M (2016) Global warming-related tree growth decline and mortality on the north-eastern Tibetan plateau. Clim Change 134:163–176

Liang E, Wang Y, Eckstein D, Luo T (2011) Little change in the fir tree-line position on the southeastern Tibetan Plateau after 200 years of warming. New Phytol 190:760–769

Liberati L, Messerli S, Matteodo M, Vittoz P (2019) Contrasting impacts of climate change on the vegetation of windy ridges and snowbeds in the Swiss Alps. Alp Bot 129:95–105

Lichtenberger E (1988) The succession of an agricultural society to a leisure society: the high mountains of Europe. In: Allan NJR, Knapp GW, Stadel C (eds) Human impact on mountains. Rowman & Littlefield, Totowa, pp 401–436

Lin X, Zhang Y, Yao Z, Gong T, Wang H et al (2008) The trend on runoff variations in the Lhasa river basin. J Geog Sci 18:95–106

Linares JC, Camarero JJ, Carreira JA (2009) Interacting effects of changes in climate and forest cover on mortality and growth of the southernmost European fir forests. Glob Ecol Biogeogr 18:485–497

Linares JC, Taïqui L, Camarero JJ (2011) Increasing drought sensitivity and decline of Atlas cedar (*Cedrus atlantica*) in the Moroccan Middle Atlas forests. Forests 2:777–796

Linsbauer A, Paul F, Machguth H, Haeberli W (2013) Comparing three different methods to model scenarios of future glacier change in the Swiss Alps. Ann Glaciol 54:241–253

Linstädter A, Baumann G, Born K, Diekkrüger B, Fritzsche P, Kirscht H, Klose A (2010) Land use and land cover in southern Morocco: managing unpredictable resources and extreme events. In: Speth P, Christoph M, Diekkrüger B (eds) Impacts of global change on the hydrological cycle in West and Northwest Africa. Springer, Berlin-Heidelberg, pp 612–633

Liou YA, Mulualem GM (2019) Spatio-temporal assessment of drought in Ethiopia and the impact of recent intense droughts. Remote Sens 11:1828

Littell JS, Hicke JA, Shafer SL, Capalbo SM, Houston LL, Glick P (2013) Forest ecosystems. Vegetation, disturbance, and economics. In: Dalton MM, Mote PW, Snover AK (eds) Climate change in the Northwest. Island Press, Washington DC, pp 110–148

Littell JS, McAfee SA, Hayward GD (2018) Alaska snowpack response to climate change: statewide snowfall equivalent and snowpack water scenarios. Water 10:668

Liu X, Chen B (2000) Climatic warming in the Tibetan Plateau during recent decades. Int J Climatol 20:1729–1742

Liu X, Cheng Z, Yan L, Yin ZY (2009) Elevation dependency of recent and future minimum surface air temperature trends in the Tibetan Plateau and its surroundings. Global Planet Change 68:164–174

Liu J, Watanabe T (2016) Seasonal pasture use and vegetation cover changes in the Alai Valley, Kyrgyzstan. In: Kreutzmann H, Watanabe T (eds) Mapping transition in the Pamirs. Springer, Cham, pp 113–126

Liu X, Yin ZY, Shao X, Qin N (2006) Temporal trends and variability of daily maximum and minimum, extreme temperature events, and growing season length over the eastern and central Tibetan Plateau during 1961–2003. J Geophys Res Atmos 111: D19109

Liu J, Zhang W, Liu T (2017) Monitoring recent changes in snow cover in Central Asia using improved MODIS snow-cover products. J Arid Land 9:763–777

Liu G, Zhao L, Li R, Wu T, Jiao K, Ping C (2017) Permafrost warming in the context of step-wise climate change in the Tien Shan Mountains, China. Permafrost Periglac Process 28:130–139

Lloyd AH, Fastie CL (2003) Recent changes in treeline forest distribution and structure in interior Alaska. Ecoscience 10:176–185

Loeffler R, Steinicke E (2006) Amenity migration in the high mountain areas of the Sierra Nevada, USA: counterurbanization and consequences. In: Price MF (ed) Global change in mountain regions. Sapiens Publishing, Kirkmahoe, pp 221–223

Lopatin E, Kolström T, Spiecker H (2008) Long-term trends in radial growth of Siberian spruce and Scots pine in Komi Republic (northwestern Russia). Boreal Environ Res 13:539–552

Lord JM (2020) Nature of alpine ecosystems in temperate mountains of New Zealand. In: Goldstein MI, DellaSala DA (eds) Encyclopedia of the world's biomes, vol 1. Elsevier, Amsterdam, pp 335–348

Lorenzo C, Carrillo-Reyes A, Rioja-Paradela T, Sántiz-López E, Bolaños-Citalán J (2019) Projected impact of global warming on the distribution of two pocket mouse species with implications on the conservation of Heteromys nelsoni (Rodentia: Heteromyidae). Rev Biol Trop 67:1210–1219

Lu X, Liang E, Wang Y, Babst F, Camarero JJ (2020) Mountain treelines climb slowly despite rapid climate warming. Glob Ecol Biogeogr. https://doi.org/10.1111/geb.13214

Luintel H, Bluffstone RA, Scheller RM (2018) The effects of the Nepal community forestry program on biodiversity conservation and carbon storage. PLoS ONE 13:e0199526

Lundmark L (2007) Reindeer pastoralism in Sweden 1550–1950. Rangifer Rep 12:9–16

Lutz AF, Immerzeel WW, Shrestha AB, Bierkens MFP (2014) Consistent increase in High Asia's runoff due to increasing glacier melt and precipitation. Nat Clim Chang 4:587–592

Lutz DA, Powell RL, Silman MR (2013) Four decades of Andean timberline migration and implications for biodiversity loss with climate change. PLoS ONE 8: e74496

Lv LX, Zhang QB (2012) Asynchronous recruitment history of *Abies spectabilis* along an altitudinal gradient in the Mt. Everest region. J Plant Ecolss 5:147–156

Lynch CM, Barr ID, Mullan D, Ruffell A (2016) Rapid glacial retreat on the Kamchatka Peninsula during the early 21st century. Cryosphere 10:1809–1821

Lyu L, Zhang QB, Pellatt MG, Büntgen U, Li MH, Cherubini P (2019) Drought limitation on tree growth at the northern hemisphere's highest tree line. Dendrochronologia 53:40–47

López-Moreno JI, Morán-Tejeda E, Vicente-Serrano SM, Bazo J, Azorin-Molina C et al (2016) Recent temperature variability and change in the Altiplano of Bolivia and Peru. Int J Climatol 36:1773–1796

López-Sandoval M, Maldonado P (2019) Change, collective action, and cultural resilience in páramo management in Ecuador. Mt Res Dev 39:R1–R9
Löffler J (2004) Degradation of high mountain ecosystems in northern Europe. J Mt Sci 1:97–115
Löffler J (2007) Reindeer grazing changes diversity patterns in arctic-alpine landscapes in northern Norway. Erde 138:215–233
Löffler J, Anschlag K, Baker B, Finch OD, Diekkrueger B et al (2011) Mountain ecosystem response to global change. Erdkunde 65:189–213
Löffler J, Lundberg A, Rössler O, Bräuning A, Jung G, Pape R, Wundram D (2004) The alpine treeline under changing land use and changing climate: approach and preliminary results from continental Norway. Nor Geogr Tidsskr-Norw J Geogr 58:183–193
Mackintosh AN, Anderson BM, Lorrey AM, Renwick JA, Frei P, Dean SM (2017) Regional cooling caused recent New Zealand glacier advances in a period of global warming. Nat Commun 8:1–13
Magrin GO, Marengo JA, Boulanger JP, Buckeridge MS, Castellanos E et al (2014) Central and South America. In: IPCC (ed) Climate change 2014: impacts, adaptation, and vulnerability. Part B: regional aspects. Cambridge University Press, Cambridge-New York, pp 1499–1566
Mainali K, Shrestha BB, Sharma RK, Adhikari A, Gurarie E, Singer M, Parmesan C (2020) Contrasting responses to climate change at Himalayan treelines revealed by population demographics of two dominant species. Ecol Evol 10:1209–1222
Mal S, Mehta M, Singh RB, Schickhoff U, Bisht MPS (2019) Recession and morphological changes of the debris-covered Milam glacier in Gori Ganga valley, Central Himalaya, India, derived from satellite data. Front Environ Sci 7:42
Mal S, Singh RB, Schickhoff U (2016) Estimating recent glacier changes in Central Himalaya, India, using remote sensing data. In: Singh RB, Schickhoff U, Mal S (eds) Climate change, glacier response, and vegetation dynamics in the Himalaya. Springer, Cham, pp 205–218
Malanson GP, Brown DG, Butler DR, Cairns DM, Fagre DB, Walsh SJ (2009) Ecotone dynamics: invasibility of alpine tundra by tree species from the subalpine forest. In: Butler DR, Malanson GP, Walsh SJ, Fagre DB (eds) The changing alpine treeline: the example of Glacier National Park, MT, USA. Elsevier, Amsterdam, pp 35–61
Malanson GP, Butler DR, Fagre DB, Walsh SJ, Tomback DF et al (2007) Alpine treeline of western North America: linking organism-to-landscape dynamics. Phys Geogr 28:378–396
Malanson GP (2020) Ongoing change in the alpine biome of North America. In: Goldstein MI, DellaSala DA (eds) Encyclopedia of the world's biomes, vol 1. Elsevier, Amsterdam, pp 581–588
Malanson GP, Rose JP, Schroeder PJ, Fagre DB (2011) Contexts for change in alpine tundra. Phys Geogr 32:97–113
Malfasi F, Cannone N (2020) Climate warming persistence triggered tree ingression after shrub encroachment in a high alpine tundra. Ecosystems: https://doi.org/10.1007/s10021-020-00495-7
Malmros JK, Mernild SH, Wilson R, Tagesson T, Fensholt R (2018) Snow cover and snow albedo changes in the Central Andes of Chile and Argentina from daily MODIS observations (2000–2016). Remote Sens Environ 209:240–252
Mamet SD, Kershaw GP (2012) Subarctic and alpine tree line dynamics during the last 400 years in northwestern and central Canada. J Biogeogr 39:855–868
Manish K, Telwala Y, Nautiyal DC, Pandit MK (2016) Modelling the impacts of future climate change on plant communities in the Himalaya: a case study from eastern Himalaya, India. Model Earth Syst Environ 2:92
Marchane A, Jarlan L, Hanich L, Boudhar A, Gascoin S et al (2015) Assessment of daily MODIS snow cover products to monitor snow cover dynamics over the Moroccan Atlas mountain range. Remote Sens Environ 160:72–86
Marchane A, Tramblay Y, Hanich L, Ruelland D, Jarlan L (2017) Climate change impacts on surface water resources in the Rheraya catchment (High Atlas, Morocco). Hydrol Sci J 62:979–995
Marchant C (2010) Paths to sustainable development in the Andes. In: Borsdorf A, Grabherr G, Heinrich K, Scott B, Stötter J (eds) Challenges for mountain regions—tackling complexity. Böhlau, Wien, pp 146–153
Marchant R, Richer S, Boles O, Capitani C, Courtney-Mustaphi CJ et al (2018) Drivers and trajectories of land cover change in East Africa: human and environmental interactions from 6000 years ago to present. Earth Sci Rev 178:322–378
Marchant R, Taylor D (1998) Dynamics of montane forest in Central Africa during the late Holocene: a pollen-based record from western Uganda. The Holocene 8:375–381
Marchenko SS, Gorbunov AP, Romanovsky VE (2007) Permafrost warming in the Tien Shan mountains, Central Asia. Global Planet Change 56:311–327
Margulis SA, Cortés G, Girotto M, Huning LS, Li D, Durand M (2016) Characterizing the extreme 2015 snowpack deficit in the Sierra Nevada (USA) and the implications for drought recovery. Geophys Res Lett 43:6341–6349
Martin-Mikle CJ, Fagre DB (2019) Glacier recession since the Little Ice Age: implications for water storage in a Rocky Mountain landscape. Arct Antarct Alp Res 51:280–289
Marty C, Tilg AM, Jonas T (2017) Recent evidence of large-scale receding snow water equivalents in the European Alps. J Hydrometeorol 18:1021–1031
Martínez-Vilalta J, Lloret F (2016) Drought-induced vegetation shifts in terrestrial ecosystems: the key role of regeneration dynamics. Global Planet Change 144:94–108
Marzeion B, Jarosch AH, Gregory JM (2014) Feedbacks and mechanisms affecting the global sensitivity of glaciers to climate change. Cryosphere 8:59–71

Masek JG, Cohen WB, Leckie D, Wulder MA, Vargas R et al. (2011) Recent rates of forest harvest and conversion in North America. J Geophys Res Biogeosciences 116:G00K03

Masiokas MH, Villalba R, Luckman BH, Lascano ME, Delgado S, Stepanek P (2008) 20th-century glacier recession and regional hydroclimatic changes in northwestern Patagonia. Global Planet Change 60:85–100

Mathez-Stiefel SL, Peralvo M, Báez S, Rist S, Buytaert W et al (2017) Research priorities for the conservation and sustainable governance of Andean forest landscapes. Mt Res Dev 37:323–339

Mathieu J (2015) Die Alpen—Raum, Kultur, Geschichte, Reclam, Stuttgart

Mathisen IE, Mikheeva A, Tutubalina OV, Aune S, Hofgaard A (2014) Fifty years of tree line change in the Khibiny Mountains, Russia: advantages of combined remote sensing and dendroecological approaches. Appl Veg Sci 17:6–16

Matteodo M, Ammann K, Verrecchia EP, Vittoz P (2016) Snowbeds are more affected than other subalpine–alpine plant communities by climate change in the Swiss Alps. Ecol Evol 6:6969–6982

Maurer K, Weyand A, Fischer M, Stöcklin J (2006) Old cultural traditions, in addition to land use and topography, are shaping plant diversity of grasslands in the Alps. Biol Cons 130:438–446

McCann JC (1995) People of the plow. An agricultural history of Ethiopia, 1800–1990. University of Wisconsin Press, London

McGlone M, Walker S, Hay R, Christie J (2010) Climate change, natural systems and their conservation in New Zealand. In: Nottage RAC, Wratt DS, Bornman JF, Jones K (eds) Climate change adaptation in New Zealand. New Zealand Climate Change Centre, Wellington, pp 82–99

McGrath D, Sass L, O'Neel S, Arendt A, Kienholz C (2017) Hypsometric control on glacier mass balance sensitivity in Alaska and Northwest Canada. Earth's Future 5:324–336

McGregor HV, Dupont L, Stuut JBW, Kuhlmann H (2009) Vegetation change, goats, and religion: a 2000-year history of land use in southern Morocco. Quatern Sci Rev 28:1434–1448

McNabb RW, Hock R (2014) Alaska tidewater glacier terminus positions, 1948–2012. J Geophys Res Earth Surf 119:153–167

McNeill JR (1992) The mountains of the Mediterranean world. Cambridge University Press, Cambridge, An environmental history

McNicol BJ, Glorioso RS (2014) Second home leisure landscapes and retirement in the Canadian Rocky Mountain community of Canmore, Alberta. Ann Leisure Res 17:27–49

Meier WJH, Grießinger J, Hochreuther P, Braun MH (2018) An updated multi-temporal glacier inventory for the Patagonian Andes with changes between the Little Ice Age and 2016. Front Earth Sci 6:62

Mekasha A, Nigatu L, Tesfaye K, Duncan AJ (2013) Modeling the response of tropical highland herbaceous grassland species to climate change: the case of the Arsi Mountains of Ethiopia. Biol Cons 168:169–175

Melaas EK, Sulla-Menashe D, Friedl MA (2018) Multidecadal changes and interannual variation in springtime phenology of North American temperate and boreal deciduous forests. Geophys Res Lett 45:2679–2687

Melillo JM, Richmond TC, Yohe GW (eds) (2014) Climate change impacts in the United States: the third national climate assessment. Washington DC, U.S, Global Change Research Program

Mengistu D, Bewket W, Lal R (2014) Recent spatiotemporal temperature and rainfall variability and trends over the upper Blue Nile river basin, Ethiopia. Int J Climatol 34:2278–2292

Menounos B, Hugonnet R, Shean D, Gardner A, Howat I et al (2019) Heterogeneous changes in western North American glaciers linked to decadal variability in zonal wind strength. Geophys Res Lett 46:200–209

Menzel A, Sparks TH, Estrella N, Roy DB (2006) Altered geographic and temporal variability in phenology in response to climate change. Glob Ecol Biogeogr 15:498–504

Mernild SH, Beckerman AP, Yde JC, Hanna E, Malmros JK, Wilson R, Zemp M (2015) Mass loss and imbalance of glaciers along the Andes Cordillera to the sub-Antarctic islands. Glob Planet Change 133:109–119

Mernild SH, Liston GE, Hiemstra CA, Malmros JK, Yde JC, McPhee J (2017) The Andes Cordillera. Part I: snow distribution, properties, and trends (1979–2014). Int J Climatol 37:1680–1698

Meshinev T, Apostolova I, Koleva E (2000) Influence of warming on timberline rising: a case study on *Pinus peuce* Griseb. in Bulgaria. Phytocoenologia 30:431–438

Messerli B (2012) Global change and the world's mountains. Mt Res Dev 32:S1

Messerli B (1999) The global mountain problematique. In: Price MF (ed) Global change in the mountains. Parthenon Publishing Group, New York-London, pp 1–3

Messerli B, Winiger M (1992) Climate, environmental change, and resources of the African mountains from the Mediterranean to the equator. Mt Res Dev 12:315–336

Michelsen O, Syverhuset AO, Pedersen B, Holten JI (2011) The impact of climate change on recent vegetation changes on Dovrefjell, Norway. Diversity 3:91–111

Miehe G, Miehe S (2000) Comparative high mountain research on the treeline ecotone under human impact: Carl Troll's "Asymmetrical zonation of the humid vegetation types of the world" of 1948 reconsidered. Erdkunde 54:34–50

Miehe G, Miehe S, Böhner J, Bäumler R, Ghimire SK et al (2015) Vegetation ecology. In: Miehe G,

Pendry CA, Chaudhary RP (eds) Nepal: an introduction to the natural history, ecology and human environment of the Himalayas. Royal Botanic Garden Edinburgh, Edinburgh, pp 385–472
Miehe G, Miehe S, Böhner J, Kaiser K, Hensen I et al (2014) How old is the human footprint in the world's largest alpine ecosystem? A review of multiproxy records from the Tibetan Plateau from the ecologists' viewpoint. Quatern Sci Rev 86:190–209
Miehe G (2015) Glacial foreland successions. In: Miehe G, Pendry CA, Chaudhary RP (eds) Nepal: an introduction to the natural history, ecology and human environment of the Himalayas. Royal Botanic Garden Edinburgh, Edinburgh, pp 80–90
Miehe G, Schleuss PM, Seeber E, Babel W, Biermann T et al (2019) The *Kobresia pygmaea* ecosystem of the Tibetan highlands—origin, functioning and degradation of the world's largest pastoral alpine ecosystem: *Kobresia* pastures of Tibet. Sci Total Environ 648:754–771
Miehe G (1997) Alpine vegetation types of the Central Himalaya. In: Wielgolaski FE (ed) Polar and alpine tundra. Ecosystems of the World 3. Elsevier, Amsterdam, pp 161–184
Miehe G, Miehe S (1994) Ericaceous forests and heathlands in the Bale Mountains of South Ethiopia. Ecology and man's impact. Warnke Verlag, Reinbek
Miehe G, Miehe S, Kaiser K, Reudenbach C, Behrendes L, Duo L, Schlütz F (2009a) How old is pastoralism in Tibet? An ecological approach to the making of a Tibetan landscape. Palaeogeogr Palaeoclimatol Palaeoecol 276:130–147
Miehe G, Miehe S, Schlütz F (2009b) Early human impact in the forest ecotone of southern High Asia (Hindu Kush, Himalaya). Quat Res 71:255–265
Mietkiewicz N, Kulakowski D, Rogan J, Bebi P (2017) Long-term change in sub-alpine forest cover, tree line and species composition in the Swiss Alps. J Veg Sci 28:951–964
Mika M (2013) Spatial patterns of second homes development in the Polish Carpathians. In: Kozak J, Ostapowicz K, Bytnerowicz A, Wyzga B (eds) The Carpathians: integrating nature and society towards sustainability. Springer, Berlin, pp 497–512
Millar CI, Rundel PW (2016) Subalpine forests. In: Mooney H, Zavaleta E (eds) Ecosystems of California. University of California Press, Berkeley, pp 579–611
Millar CI, Westfall RD, Delany DL, Flint AL, Flint LE (2015) Recruitment patterns and growth of high-elevation pines in response to climatic variability (1883–2013) in the western Great Basin, USA. Can J For Res 45:1299–1312
Millar CI, Westfall RD, Delany DL, King JC, Graumlich LJ (2004) Response of subalpine conifers in the Sierra Nevada, California, USA, to 20th-century warming and decadal climate variability. Arct Antarct Alp Res 36:181–200
Miller JD, Immerzeel WW, Rees G (2012) Climate change impacts on glacier hydrology and river discharge in the Hindu Kush-Himalayas. Mt Res Dev 32:461–468
Miller DJ (1997) Rangelands and pastoral development: an introduction. In: Miller DJ, Craig SR (eds) Rangelands and pastoral development in the Hindu Kush-Himalayas. ICIMOD, Kathmandu, pp 1–5
Millones J (1982) Patterns of land use and associated environmental problems of the Central Andes: an integrated summary. Mt Res Dev 2:49–61
Minder JR, Letcher TW, Liu C (2018) The character and causes of elevation-dependent warming in high-resolution simulations of Rocky Mountain climate change. J Clim 31:2093–2113
Mirzabaev A, Ahmed M, Werner J, Pender J, Louhaichi M (2016) Rangelands of Central Asia: challenges and opportunities. J Arid Land 8:93–108
Mishra A (2014) Changing climate of Uttarakhand, India. J Geol Geosci 3:163
Mishra NB, Mainali KP (2017) Greening and browning of the Himalaya: spatial patterns and the role of climatic change and human drivers. Sci Total Environ 587:326–339
Missaoui K, Gharzouli R, Djellouli Y, Messner F (2020) Phenological behavior of Atlas cedar (*Cedrus atlantica*) forest to snow and precipitation variability in Boutaleb and Babors Mountains, Algeria. Biodiversitas J Biol Divers 21:239–245
Moen J (2006) Land use in the Swedish mountain region: trends and conflicting goals. Int J Biodiver Sci Manage 2:305–314
Mohajane M, Essahlaoui A, Oudija F, Hafyani ME, Hmaidi AE et al (2018) Land use/land cover (LULC) using Landsat data series (MSS, TM, ETM+ and OLI) in Azrou Forest, in the Central Middle Atlas of Morocco. Environments 5:131
Mohandass D, Zhao JL, Xia YM, Campbell MJ, Li QJ (2015) Increasing temperature causes flowering onset time changes of alpine ginger *Roscoea* in the Central Himalayas. J Asia-Pac Biodivers 8:191–198
Mohapatra J, Singh CP, Tripathi OP, Pandya HA (2019) Remote sensing of alpine treeline ecotone dynamics and phenology in Arunachal Pradesh Himalaya. Int J Remote Sens 40:7986–8009
Moiseev PA, Bartysh AA, Nagimov ZY (2010) Climate changes and tree stand dynamics at the upper limit of their growth in the North Ural Mountains. Russ J Ecol 41:486–497
Molero Mesa JM, Fernández Calzado MR (2010) Evolution of the high mountain flora of Sierra Nevada (1837–2009). Acta Bot 157:659–667
Molinillo M, Monasterio M (2006) Vegetation and grazing patterns in Andean environments: a comparison of pastoral systems in Punas and Páramos. In: Spehn EM, Liberman M, Körner C (eds) Land use change and mountain biodiversity. CRC, Boca Raton, pp 137–151
Mollaret C, Hilbich C, Pellet C, Flores-Orozco A, Delaloye R, Hauck C (2019) Mountain permafrost degradation documented through a network of

permanent electrical resistivity tomography sites. Cryosphere 13:2557–2578
Monasterio M (1980) Poblamiento humano y uso de la tierra en los altos Andes de Venezuela. In: Monasterio M (ed) Estudios Ecológicos en los Paramos Andinos. Editorial de la Universidad de los Andes, Mérida, pp 170–198
Mong CE, Vetaas OR (2006) Establishment of *Pinus wallichiana* on a Himalayan glacier foreland: stochastic distribution or safe sites? Arct Antarct Alp Res 38:584–592
Montanari B (2013) The future of agriculture in the High Atlas Mountains of Morocco: the need to integrate traditional ecological knowledge. In: Mann S (ed) The future of mountain agriculture. Springer, Berlin-Heidelberg, pp 51–72
Mora DE, Willems P (2012) Decadal oscillations in rainfall and air temperature in the Paute river basin—southern Andes of Ecuador. Theoret Appl Climatol 108:267–282
Moret P, Arauz MDLA, Gobbi M, Barragán Á (2016) Climate warming effects in the tropical Andes: first evidence for upslope shifts of *Carabidae* (*Coleoptera*) in Ecuador. Insect Conserv Divers 9:342–350
Moritz C, Agudo R (2013) The future of species under climate change: resilience or decline? Science 341:504–508
Moritz C, Patton JL, Conroy CJ, Parra JL, White GC, Beissinger SR (2008) Impact of a century of climate change on small-mammal communities in Yosemite National Park, USA. Science 322:261–264
Morris C (2017) Historical vegetation-environment patterns for assessing the impact of climatic change in the mountains of Lesotho. Afr J Range Forage Sci 34:45–51
Morueta-Holme N, Engemann K, Sandoval-Acuña P, Jonas JD, Segnitz RM, Svenning JC (2015) Strong upslope shifts in Chimborazo's vegetation over two centuries since Humboldt. Proc Natl Acad Sci 112:12741–12745
Moss LA (ed) (2006) The amenity migrants: seeking and sustaining mountains and their cultures. CABI, New York
Mote PW, Abatzoglou JT, Kunkel KE (2013) Climate—variability and change in the past and the future. In: Dalton MM, Mote PW, Snover AK (eds) Climate change in the Northwest. Island Press, Washington, DC, pp 25–40
Mote PW, Li S, Lettenmaier DP, Xiao M, Engel R (2018) Dramatic declines in snowpack in the western US. NPJ Clim Atmos Sci 1:1–6
Mote P, Snover AK, Capalbo S, Eigenbrode SD, Glick et al. (2014) Northwest. In: Melillo JM, Richmond TC, Yohe GC (eds) Climate change impacts in the United States: the third national climate assessment. U.S. Global Change Research Program, Washington, pp 487–513
Mucina L, Hoare DB, Lötter MC, Du Preez PJ, Rutherford MC et al (2006) Grassland biome. In: Mucina L, Rutherford MC (eds) The vegetation of South Africa, Lesotho and Swaziland. SANBI, Pretoria, pp 348–437
Mukherji A, Sinisalo A, Nüsser M, Garrard R, Eriksson M (2019) Contributions of the cryosphere to mountain communities in the Hindu Kush Himalaya: a review. Reg Environ Change 19:1311–1326
Mukwada G, Manatsa D (2018) Spatiotemporal analysis of the effect of climate change on vegetation health in the Drakensberg mountain region of South Africa. Environ Monit Assess 190:358
Mullan B, Stuart SJ, Hadfield MG, Smith MJ (2010) Report on the review of NIWA's ‚seven-station' temperature series. NIWA Information Series No. 78, National Institute of Water and Atmospheric Research (NIWA), Wellington
Mumba M (2008) Unravelling the, 'Mountains of the Moon'. Swara 31:22–26
Munkhjargal M, Yadamsuren G, Yamkhin J, Menzel L (2020) The combination of wildfire and changing climate triggers permafrost degradation in the Khentii Mountains, northern Mongolia. Atmosphere 11:155
Munro RN, Deckers J, Haile M, Grove AT, Poesen J, Nyssen J (2008) Soil landscapes, land cover change and erosion features of the Central Plateau region of Tigrai, Ethiopia: photo-monitoring with an interval of 30 years. CATENA 75:55–64
Munson SM, Sher AA (2015) Long-term shifts in the phenology of rare and endemic Rocky Mountain plants. Am J Bot 102:1268–1276
Munteanu C, Kuemmerle T, Boltiziar M, Butsic V, Gimmi U (2014) Forest and agricultural land change in the Carpathian region - a meta-analysis of long-term patterns and drivers of change. Land Use Policy 38:685–697
Munteanu C, Radeloff V, Griffiths P, Halada L, Kaim D et al (2017) Land change in the Carpathian Region before and after major institutional changes. In: Gutman G, Radeloff V (eds) Land-cover and land-use changes in eastern Europe after the collapse of the Soviet Union in 1991. Springer, Cham, pp 57–90
Murata A, Sasaki H, Kawase H, Nosaka M, Oh'izumi M et al (2015) Projection of future climate change over Japan in ensemble simulations with a high-resolution regional climate model. Sci Online Lett Atmos 11:90–94
Muñoz AA, Cavieres LA (2008) The presence of a showy invasive plant disrupts pollinator service and reproductive output in native alpine species only at high densities. J Ecol 96:459–467
Mwangi E, Swallow B (2008) *Prosopis juliflora* invasion and rural livelihoods in the Lake Baringo area of Kenya. Conserv Soc 6:130–140
Myers N, Mittermeier RA, Mittermeier CG, Da Fonseca GA, Kent J (2000) Biodiversity hotspots for conservation priorities. Nature 403:853–858
Myers-Smith IH, Kerby JT, Phoenix GK, Bjerke JW, Epstein HE et al (2020) Complexity revealed in the greening of the Arctic. Nat Clim Chang 10:106–117
Mölg T, Chiang JC, Gohm A, Cullen NJ (2009) Temporal precipitation variability versus altitude on a tropical high mountain: observations and mesoscale atmospheric modelling. Q J Roy Meteorol Soc J Atmos Sci Appl Meteorol Phys Oceanography 135:1439–1455

Mölg T, Hardy DR, Kaser G (2003) Solar-radiation-maintained glacier recession on Kilimanjaro drawn from combined ice-radiation geometry modeling. J Geophys Res Atmos 108:4731

Mölg T, Rott H, Kaser G, Fischer A, Cullen NJ (2006) Comment on "Recent glacial recession in the Rwenzori Mountains of East Africa due to rising air temperature" by Richard G. Taylor, Lucinda Mileham, Callist Tindimugaya, Abushen Majugu, Andrew Muwanga, and Bob Nakileza. Geophys Res Lett 33: L20404

De Mûelenaere S, Frankl A, Haile M, Poesen J, Deckers J et al (2014) Historical landscape photographs for calibration of Landsat land use/cover in the northern Ethiopian highlands. Land Degrad Dev 25:319–335

Müller M, Schickhoff U, Scholten T, Drollinger S, Böhner J, Chaudhary RP (2016) How do soil properties affect alpine treelines? General principles in a global perspective and novel findings from Rolwaling Himal, Nepal. Prog Phys Geogr 40:135–160

Müller-Wille L, Heinrich D, Lehtola VP, Aikio P, Konstantinov Y, Vladimirova V (2006) Dynamics in human-reindeer relations: reflections on prehistoric, historic and contemporary practices in northernmost Europe. In: Forbes BC, Bölter M, Müller-Wille L, Hukkinen J, Müller F, Gunslay N, Konstantinov Y (eds) Reindeer management in northernmost Europe. Springer, Berlin, pp 27–45

Naccarella A, Morgan JW, Cutler SC, Venn SE (2020) Alpine treeline ecotone stasis in the face of recent climate change and disturbance by fire. PLoS ONE 15: e0231339

Nebelung J (2016) Waldflächenveränderung im Nepal-Himalaya 1990—2013 unter Berücksichtigung der Community Forestry. Eine GIS- und fernerkundungs-basierte Analyse. Unpubl. Dipl. Thesis, Institute of Geography, University of Hamburg, Hamburg

Negi HS, Ganju A, Kanda N, Gusain HS (2020) Climate change and cryospheric response over North-West and Central Himalaya, India. In: Dimri AP, Bookhagen B, Stoffel M, Yasunari T (eds) Himalayan weather and climate and their impact on the environment. Springer, Cham, pp 309–330

Negi GCS, Rawal RS (2019) Himalayan biodiversity in the face of climate change. In: Garkoti SC, Van Bloem SJ, Fulé PZ, Semwal RL (eds) Tropical ecosystems: structure, functions and challenges in the face of global change. Springer, Singapore, pp 263–277

Neuburger M, Steinicke E (2012) Alpine tourism in tropical Africa and sustainable development? Ugandan Rwenzori and Mt. Kenya as case studies. J Sustain Educ 3:1–31

Niang I, Ruppel OC, Abdrabo MA, Essel A, Lennard C et al. (2014) Africa. In: IPCC (ed) Climate change 2014: impacts, adaptation, and vulnerability. Part B: regional aspects. Cambridge University Press, Cambridge-New York, pp 1199–1265

Nie Y, Sheng Y, Liu Q, Liu L, Liu S, Zhang Y, Song C (2017) A regional-scale assessment of Himalayan glacial lake changes using satellite observations from 1990 to 2015. Remote Sens Environ 189:1–13

Niedrist G, Tasser E, Lüth C, Dalla Via J, Tappeiner U (2009) Plant diversity declines with recent land use changes in European Alps. Plant Ecol 202:195

Niraula RR, Gilani H, Pokharel BK, Qamer FM (2013) Measuring impacts of community forestry program through repeat photography and satellite remote sensing in the Dolakha district of Nepal. J Environ Manage 126:20–29

Niu Y, Zhu H, Yang S, Ma S, Zhou J et al (2019) Overgrazing leads to soil cracking that later triggers the severe degradation of alpine meadows on the Tibetan Plateau. Land Degrad Dev 30:1243–1257

Nobis M (2008) Invasive Neophyten auch im Wald? Wald und Holz 8(08):46–49

Noetzli J, Christiansen HH, Deline P, Gugliemin M, Isaksen K et al (2018) Permafrost thermal state. Bull Am Meteor Soc 99:S20–S22

Nogués-Bravo D, Araújo MB, Lasanta T, López-Moreno JI (2008) Climate change in Mediterranean mountains during the 21st century. Ambio 37:280–285

Nogués-Bravo D, López-Moreno JI, Vicente-Serrano SM (2012) Climate change and its impact. In: Vogiatzakis IN (ed) Mediterranean mountain environments. Wiley-Blackwell, Chichester, pp 185–200

Noroozi J, Akhani H, Breckle SW (2008) Biodiversity and phytogeography of the alpine flora of Iran. Biodivers Conserv 17:493–521

Noroozi J, Talebi A, Doostmohammadi M, Bagheri A (2020) The Zagros mountain range. In: Noroozi J (ed) Plant biogeography and vegetation of high mountains of Central and South-West Asia. Springer, Cham, pp 185–214

Norris J, Carvalho LM, Jones C, Cannon F (2019) Deciphering the contrasting climatic trends between the Central Himalaya and Karakoram with 36 years of WRF simulations. Clim Dyn 52:159–180

North MP, Stephens SL, Collins BM, Agee JK, Aplet G, Franklin JF, Fule PZ (2015) Reform forest fire management. Science 349:1280–1281

Notarnicola C (2020) Hotspots of snow cover changes in global mountain regions over 2000–2018. Remote Sens Environ 243:111781

Nunez S, Arets E, Alkemade R, Verwer C, Leemans R (2019) Assessing the impacts of climate change on biodiversity: is below 2° C enough? Clim Change 154:351–365

Nyssen J, Haile M, Naudts J, Munro N, Poesen J et al (2009) Desertification? Northern Ethiopia re-photographed after 140 years. Sci Total Environ 407:2749–2755

Nyssen J, Poesen J, Lanckriet S, Jacob M, Moeyersons J et al (2015) Land degradation in the Ethiopian highlands. In: Billi P (ed) Landscapes and landforms of Ethiopia. Springer, Dordrecht, pp 369–385

Nyssen J, Poesen J, Moeyersons J, Deckers J, Haile M, Lang A (2004) Human impact on the environment in the Ethiopian and Eritrean highlands—a state of the art. Earth Sci Rev 64:273–320

Nüsser M (2000) Change and persistence: contemporary landscape transformation in the Nanga Parbat region, northern Pakistan. Mt Res Dev 20:348–355
Nüsser M (2006) Ressourcennutzung und nachhaltige Entwicklung im Kumaon-Himalaya (Indien). Geogr Rundsch 58:14–22
Nüsser M, Schmidt S (2017) Nanga Parbat revisited: evolution and dynamics of sociohydrological interactions in the northwestern Himalaya. Ann Am Assoc Geogr 107:403–415
Nüsser M, Schmidt S, Dame J (2012) Irrigation and development in the upper Indus basin: characteristics and recent changes of a socio-hydrological system in Central Ladakh, India. Mt Res Dev 32:51–61
Nüsser M, Gerwin M (2008) Diversity, complexity and dynamics: land use patterns in the Central Himalayas of Kumaon, northern India. In: Löffler J, Stadelbauer J (eds) Diversity in mountain systems. Colloquium Geographicum 31. Asgard-Verlag, Sankt Augustin, pp 107–119
Nüsser M, Dame J, Kraus B, Baghel R, Schmidt S (2019a) Socio-hydrology of "artificial glaciers" in Ladakh, India: assessing adaptive strategies in a changing cryosphere. Reg Environ Change 19:1327–1337
Nüsser M, Dame J, Parveen S, Kraus B, Baghel R, Schmidt S (2019b) Cryosphere-fed irrigation networks in the northwestern Himalaya: precarious livelihoods and adaptation strategies under the impact of climate change. Mt Res Dev 39:R1-R11
Odland A, Høitomt T, Olsen SL (2010) Increasing vascular plant richness on 13 high mountain summits in southern Norway since the early 1970s. Arct Antarct Alp Res 42:458–470
Ohse B, Jansen F, Wilmking M (2012) Do limiting factors at Alaskan treelines shift with climatic regimes? Environ Res Lett 7:015505
Olivares-Contreras VA, Mattar C, Gutiérrez AG, Jiménez JC (2019) Warming trends in Patagonian subantarctic forest. Int J Appl Earth Obs Geoinf 76:51–65
Omondi PAO, Awange JL, Forootan E, Ogallo LA, Barakiza R et al (2014) Changes in temperature and precipitation extremes over the Greater Horn of Africa region from 1961 to 2010. Int J Climatol 34:1262–1277
Osipov EY, Osipova OP (2014) Mountain glaciers of Southeast Siberia: current state and changes since the Little Ice Age. Ann Glaciol 55:167–176
Osorio-Castiblanco DF, Peyre G, Saldarriaga JF (2020) Physicochemical analysis and essential oils extraction of the Gorse (*Ulex europaeus*) and French Broom (*Genista monspessulana*), two highly invasive species in the Colombian Andes. Sustainability 12:57
Otto M, Höpfner C, Curio J, Maussion F, Scherer D (2016) Assessing vegetation response to precipitation in Northwest Morocco during the last decade: an application of MODIS NDVI and high resolution reanalysis data. Theoret Appl Climatol 123:23–41
PERMOS (2016) Permafrost in Switzerland 2010/2011 to 2013/2014. Noetzli J, Luethi R, Staub B (eds) Glaciological Report (Permafrost) No. 12–15 of the cryospheric commission of the swiss academy of sciences, Berne. https://doi.org/10.13093/permos-rep-2016-12-15
Palazzi E, Hardenberg J, Provenzale A (2013) Precipitation in the Hindu-Kush Karakoram Himalaya: observations and future scenarios. J Geophys Res Atmos 118:85–100
Palazzi E, Mortarini L, Terzago S, von Hardenberg J (2019) Elevation-dependent warming in global climate model simulations at high spatial resolution. Clim Dyn 52:2685–2702
Palombo C, Chirici G, Marchetti M, Tognetti R (2013) Is land abandonment affecting forest dynamics at high elevation in Mediterranean mountains more than climate change? Plant Biosyst 147:1–11
Palomo I (2017) Climate change impacts on ecosystem services in high mountain areas: a literature review. Mt Res Dev 37:179–187
Panday PK, Ghimire B (2012) Time-series analysis of NDVI from AVHRR data over the Hindu Kush-Himalayan region for the period 1982–2006. Int J Remote Sens 33:6710–6721
Pandit MK, Manish K, Koh LP (2014) Dancing on the roof of the world: ecological transformation of the Himalayan landscape. Bioscience 64:980–992
Pandit MK, Sodhi NS, Koh LP, Bhaskar A, Brook BW (2007) Unreported yet massive deforestation driving loss of endemic biodiversity in Indian Himalaya. Biodivers Conserv 16:153–163
Panthi S, Bräuning A, Zhou ZK, Fan ZX (2017) Tree rings reveal recent intensified spring drought in the Central Himalaya, Nepal. Glob Planet Change 157:26–34
Panthi J, Dahal P, Shrestha ML, Aryal S, Krakauer NY et al (2015) Spatial and temporal variability of rainfall in the Gandaki river basin of Nepal Himalaya. Climate 3:210–226
Papanastasis VP (2007) Land abandonment and old field dynamics in Greece. In: Cramer VA, Hobbs RJ (eds) Old fields: dynamics and restoration of abandoned farmland. Island Press, London, pp 225–246
Papanastasis VP (2012) Land use changes. In: Vogiatzakis IN (ed) Mediterranean mountain environments. Wiley, Chichester, pp 159–184
Pape R, Löffler J (2012) Climate change, land use conflicts, predation and ecological degradation as challenges for reindeer husbandry in northern Europe: what do we really know after half a century of research? Ambio 41:421–434
Park T, Ganguly S, Tømmervik H, Euskirchen ES, Høgda KA et al (2016) Changes in growing season duration and productivity of northern vegetation inferred from long-term remote sensing data. Environ Res Lett 11:084001
Parolo G, Rossi G (2008) Upward migration of vascular plants following a climate warming trend in the Alps. Basic Appl Ecol 9:100–107
Parveen S, Winiger M, Schmidt S, Nüsser M (2015) Irrigation in upper Hunza: evolution of socio-

hydrological interactions in the Karakoram, northern Pakistan. Erdkunde 69–85
Pathak BR, Yi X, Bohara R (2017) Community based forestry in Nepal: status, issues and lessons learned. Int J Sci 6:119–129
Patle GT, Sengdo D, Tapak M (2019) Trends in major climatic parameters and sensitivity of evapotranspiration to climatic parameters in the eastern Himalayan region of Sikkim, India. J Water Clim Change 11:491–502
Patricola CM, Cook KH (2010) Northern African climate at the end of the twenty-first century: an integrated application of regional and global climate models. Clim Dyn 35:193–212
Patton AI, Rathburn SL, Capps DM (2019) Landslide response to climate change in permafrost regions. Geomorphology 340:116–128
Pauchard A, Kueffer C, Dietz H, Daehler CC, Alexander J et al (2009) Ain't no mountain high enough: plant invasions reaching new elevations. Front Ecol Environ 7:479–486
Pauchard A, Milbau A, Albihn A, Alexander J, Burgess T et al (2016) Non-native and native organisms moving into high elevation and high latitude ecosystems in an era of climate change: new challenges for ecology and conservation. Biol Invasions 18:345–353
Paudel KP, Andersen P (2010) Assessing rangeland degradation using multi temporal satellite images and grazing pressure surface model in upper Mustang, Trans Himalaya, Nepal. Remote Sens Environ 114:1845–1855
Paudel B, Zhang YL, Li SC, Liu LS, Wu X, Khanal NR (2016) Review of studies on land use and land cover change in Nepal. J Mt Sci 13:643–660
Pauli H, Gottfried M, Dullinger S, Abdaladze O, Akhalkatsi M et al (2012) Recent plant diversity changes on Europe's mountain summits. Science 336:353–355
Pauli H, Gottfried M, Grabherr G (2001) High summits of the Alps in a changing climate. The oldest observation series on high mountain plant diversity in Europe. In: Walther GR, Burga CA, Edwards PA (eds) "Fingerprints" of climate change—adapted behaviour and shifting species ranges. Springer, New York, pp 139–149
Pauli H, Gottfried M, Reiter K, Klettner C, Grabherr G (2007) Signals of range expansions and contractions of vascular plants in the high Alps: observations (1994–2004) at the GLORIA master site Schrankogel, Tyrol, Austria. Glob Change Biol 13:147–156
Pauli H, Halloy SR (2019) High mountain ecosystems under climate change. In: Oxford research encyclopedia of climate science. https://doi.org/10.1093/acrefore/9780190228620.013.764
Pausas JG, Millán MM (2019) Greening and browning in a climate change hotspot: the Mediterranean basin. Bioscience 69:143–151
Pawson E, Brooking T (eds) (2013) Making a new land: environmental histories of New Zealand. Otago University Press, Dunedin
Pecl GT, Araújo MB, Bell JD, Blanchard J, Bonebrake TC et al. (2017) Biodiversity redistribution under climate change: impacts on ecosystems and human well-being. Science 355:eaai9214
Pepin N, Bradley RS, Diaz HF, Baraer M, Caceres EB et al (2015) Elevation-dependent warming in mountain regions of the world. Nat Clim Chang 5:424–430
Pepin N, Deng H, Zhang H, Zhang F, Kang S, Yao T (2019) An examination of temperature trends at high elevations across the Tibetan Plateau: The use of MODIS LST to understand patterns of elevation-dependent warming. J Geophys Res Atmos 124:5738–5756
Perlik M (2019) The spatial and economic transformation of mountain regions: landscapes as commodities. Routledge, Abingdon-New York
Permanent Secretariat of the Alpine Convention (ed) (2015) Demographic changes in the Alps. Report on the state of the Alps. Alpine Signals—Special Edition 5. Permanent Secretariat of the Alpine Convention, Innsbruck-Bolzano
Peters T, Drobnik T, Meyer H, Rankl M, Richter M et al (2013) Environmental changes affecting the Andes of Ecuador. In: Bendix J, Beck E, Bräuning A, Makeschin F, Mosandl R et al (eds) Ecosystem services, biodiversity and environmental change in a tropical mountain ecosystem of South Ecuador. Springer, Berlin, pp 19–29
Petriccione B, Bricca A (2019) Thirty years of ecological research at the Gran Sasso d'Italia LTER site: climate change in action. Nat Conserv 34:9
Petrov IA, Kharuk VI, Dvinskaya ML, Im ST (2015) Reaction of coniferous trees in the Kuznetsk Alatau alpine forest-tundra ecotone to climate change. Contemp Probl Ecol 8:423–430
Peñuelas J, Boada M (2003) A global change-induced biome shift in the Montseny Mountains (NE-Spain). Glob Change Biol 9:131–140
Peñuelas J, Filella I, Comas P (2002) Changed plant and animal life cycles from 1952 to 2000 in the Mediterranean region. Glob Change Biol 8:531–544
Peñuelas J, Sardans J, Estiarte M, Ogaya R, Carnicer J et al (2013) Evidence of current impact of climate change on life: a walk from genes to the biosphere. Glob Change Biol 19:2303–2338
Phoenix GK, Bjerke JW (2016) Arctic browning: extreme events and trends reversing arctic greening. Glob Change Biol 22:2960–2962
Piao S, Liu Q, Chen A, Janssens IA, Fu Y et al (2019) Plant phenology and global climate change: current progresses and challenges. Glob Change Biol 25:1922–1940
Piguet F, Pankhurst A (2009) Migration, resettlement and displacement in Ethiopia. In: Pankhurst A, Piguet F (eds) Moving people in Ethiopia: development, displacement and the state. Boydell & Brewer, Rochester-New York, pp 1–22
Pihl E, Martin MA, Blome T, Hebden S, Jarzebski MP et al (2019) 10 new insights in climate science 2019. Future Earth & The Earth League, Stockholm

Pintaldi E, Hudek C, Stanchi S, Spiegelberger T, Rivella E, Freppaz M (2017) Sustainable soil management in ski areas: threats and challenges. Sustainability 9:2150
Pitcairn M, Schoenig S, Yacoub R, Gendron J (2006) Yellow starthistle continues its spread in California. Calif Agric 60:83–90
Plieninger T, Hui C, Gaertner M, Huntsinger L (2014) The impact of land abandonment on species richness and abundance in the Mediterranean basin: a meta-analysis. PLoS ONE 9:e98355
Pokharel B, Wang SYS, Meyer J, Marahatta S, Nepal B, Chikamoto Y, Gillies R (2019) The east–west division of changing precipitation in Nepal. Int J Climatol 40:3348–3359
Pokharel BK, Mahat A, Thapa S (2011) Impact of community forestry in Nepal. Kathmandu to Jiri: a photo journey. Nepal Swiss Community Forestry Project, Kathmandu
Pollnac F, Seipel T, Repath C, Rew LJ (2012) Plant invasion at landscape and local scales along roadways in the mountainous region of the Greater Yellowstone Ecosystem. Biol Invasions 14:1753–1763
Ponce-Reyes R, Reynoso-Rosales VH, Watson JE, VanDerWal J, Fuller RA, Pressey RL, Possingham HP (2012) Vulnerability of cloud forest reserves in Mexico to climate change. Nat Clim Chang 2:448–452
Potter CS (2017) Satellite image mapping of tree mortality in the Sierra Nevada region of California from 2013 to 2016. J Biodivers Manage For 6:2
Potthoff K (2017) Spatio-temporal patterns of birch regrowth in a western Norwegian treeline ecotone. Landsc Res 42:63–77
Poulter B, Pederson N, Liu H, Zhu Z, D'Arrigo R et al (2013) Recent trends in Inner Asian forest dynamics to temperature and precipitation indicate high sensitivity to climate change. Agric For Meteorol 178:31–45
Pratap B, Dobhal DP, Bhambri R, Mehta M, Tewari VC (2016) Four decades of glacier mass balance observations in the Indian Himalaya. Reg Environ Change 16:643–658
Price MF (1998) Mountains: globally important ecosystems. Unasylva 195:3–12
Price MF (2015) Mountains. A very short introduction. Oxford University Press, Oxford
Price MF, Butt N (eds) (2000) Forests in sustainable mountain development: a state of knowledge report for 2000. CABI Publishing, Oxon-New York
Price MF, Kohler T (2013) Sustainable mountain development. In: Price MF, Byers AC, Friend DA, Kohler T, Price LW (eds) Mountain geography. University of California Press, Berkeley-Los Angeles, Physical and human dimensions, pp 333–365
Price MF, Gratzer G, Duguma LA, Kohler T, Maselli D, Romeo R (eds) (2011) Mountain forests in a changing world. Realizing values, addressing challenges. FAO-SDC, Rome
Prinz R, Mölg T (2020) Tropische Gletscher: Ostafrika. In: Lozán JL, Breckle SW, Escher-Vetter H, Grassl H, Kasang D, Paul F, Schickhoff U (eds) Warnsignal Klima: Hochgebirge im Wandel. Wissenschaftliche Auswertungen, Hamburg, pp 141–145
Prinz R, Nicholson LI, Mölg T, Gurgiser W, Kaser G (2016) Climatic controls and climate proxy potential of Lewis Glacier. Mt. Kenya. The Cryosphere 10:133–148
Prinz R, Heller A, Ladner M, Nicholson LI, Kaser G (2018) Mapping the loss of Mt. Kenya's glaciers: an example of the challenges of satellite monitoring of very small glaciers. Geosciences 8:174
Priya P, Krishnan R, Mujumdar M, Houze RA (2017) Changing monsoon and midlatitude circulation interactions over the western Himalayas and possible links to occurrences of extreme precipitation. Clim Dyn 49:2351–2364
Ptackova J (2012) Implementation of resettlement programmes amongst pastoralist communities in eastern Tibet. In: Kreutzmann H (ed) Pastoral practices in High Asia. Springer, Dordrecht, pp 217–234
Pudas E, Leppälä M, Tolvanen A, Poikolainen J, Venäläinen A, Kubin E (2008) Trends in phenology of *Betula pubescens* across the boreal zone in Finland. Int J Biometeorol 52:251–259
Pyšek P, Jarošík V, Pergl J, Wild J (2011) Colonization of high altitudes by alien plants over the last two centuries. Proc Natl Acad Sci 108:439–440
Pérez-García N, Font X, Ferré A, Carreras J (2013) Drastic reduction in the potential habitats for alpine and subalpine vegetation in the Pyrenees due to twenty-first-century climate change. Reg Environ Change 13:1157–1169
Qamer FM, Shehzad K, Abbas S, Murthy MSR, Xi C, Gilani H, Bajracharya B (2016) Mapping deforestation and forest degradation patterns in western Himalaya, Pakistan. Remote Sens 8:385
Qasim M, Hubacek K, Termansen M, Fleskens L (2013) Modelling land use change across elevation gradients in district Swat, Pakistan. Reg Environ Change 13:567–581
Qi Z, Liu H, Wu X, Hao Q (2015) Climate-driven speedup of alpine treeline forest growth in the Tianshan Mountains, northwestern China. Glob Change Biol 21:816–826
Qin N, Chen X, Fu G, Zhai J, Xue X (2010) Precipitation and temperature trends for the Southwest China: 1960–2007. Hydrol Process 24:3733–3744
Qiu J (2016) Trouble in Tibet: rapid changes in Tibetan grasslands are threatening Asia's main water supply and the livelihood of nomads. Nature 529:142–146
Quincey DJ, Glasser NF, Cook SJ, Luckman A (2015) Heterogeneity in Karakoram glacier surges. J Geophys Res Earth Surf 120:1288–1300
Quintero-Gallego ME, Quintero-Angel M, Vila-Ortega JJ (2018) Exploring land use/land cover change and drivers in Andean mountains in Colombia: a case in rural Quindío. Sci Total Environ 634:1288–1299
Rabatel A, Ceballos JL, Micheletti N, Jordan E, Braitmeier M et al (2018) Toward an imminent extinction of Colombian glaciers? Geogr Ann Ser B 100:75–95

Rabatel A, Francou B, Soruco A, Gomez J, Ceballos JL et al (2013) Current state of glaciers in the tropical Andes: a multi-century perspective on glacier evolution and climate change. Cryosphere 7:81–102
Rai ID, Bharti R, Adhikari BS, Rawat GS (2013) Structure and functioning of timberline vegetation in the western Himalaya: a case study. In: Wu N, Rawat GS, Joshi S, Ismail M, Sharma E (eds) High-altitude rangelands and their interfaces in the Hindu Kush Himalayas. ICIMOD, Kathmandu, pp 91–107
Rai ID, Singh G, Pandey A, Rawat GS (2019) Ecology of treeline vegetation in western Himalaya: anthropogenic and climatic influences. In: Garkoti SC, Van Bloem SJ, Fulé PZ, Semwal RL (eds) Tropical ecosystems: structure, functions and challenges in the face of global change. Springer, Singapore, pp 173–192
Raish C (2004) Historic and contemporary land use in southwestern grassland ecosystems. In: Finch DM (ed) Assessment of grassland ecosystem conditions in the southwestern United States. General Technical Report RMRS-GTR-135-Vol. 1. USDA Forest Service, Albuquerque, pp 86–119
Ram S, Borgaonkar HP (2014) Tree-ring analysis over western Himalaya and its long-term association with vapor pressure and potential evapotranspiration. Dendrochronologia 32:32–38
Ramirez-Villegas J, Cuesta F, Devenish C, Peralvo M, Jarvis A, Arnillas CA (2014) Using species distributions models for designing conservation strategies of tropical Andean biodiversity under climate change. J Nat Conserv 22:391–404
Ran Y, Li X, Cheng G (2018) Climate warming over the past half century has led to thermal degradation of permafrost on the Qinghai-Tibet Plateau. Cryosphere 12:595–608
Rangecroft S, Suggitt AJ, Anderson K, Harrison S (2016) Future climate warming and changes to mountain permafrost in the Bolivian Andes. Clim Change 137:231–243
Rangwala I, Palazzi E, Miller JR (2020) Projected climate change in the Himalayas during the twenty-first century. In: Dimri AP, Bookhagen B, Stoffel M, Yasunari T (eds) Himalayan weather and climate and their impact on the environment. Springer, Cham, pp 51–71
Rangwala I, Sinsky E, Miller JR (2013) Amplified warming projections for high altitude regions of the northern hemisphere mid-latitudes from CMIP5 models. Environ Res Lett 8:024040
Rapacciuolo G, Maher SP, Schneider AC, Hammond TT, Jabis MD et al (2014) Beyond a warming fingerprint: individualistic biogeographic responses to heterogeneous climate change in California. Glob Change Biol 20:2841–2855
Rashid I, Romshoo SA, Chaturvedi RK, Ravindranath NH, Sukumar R et al (2015) Projected climate change impacts on vegetation distribution over Kashmir Himalayas. Clim Change 132:601–613
Rasul G, Molden D (2019) The global social and economic consequences of mountain cryospheric change. Front Environ Sci 7:91
Rasul G, Pasakhala B, Mishra A, Pant S (2020) Adaptation to mountain cryosphere change: issues and challenges. Climate Dev 12:297–309
Rau P, Bourrel L, Labat D, Melo P, Dewitte B et al (2017) Regionalization of rainfall over the Peruvian Pacific slope and coast. Int J Climatol 37:143–158
Raxworthy CJ, Pearson RG, Rabibisoa N, Rakotondrazafy AM, Ramanamanjato JB et al (2008) Extinction vulnerability of tropical montane endemism from warming and upslope displacement: a preliminary appraisal for the highest massif in Madagascar. Glob Change Biol 14:1703–1720
Raza M, Hussain D, Rasul G, Akbar M, Raza G (2015) Variations of surface temperature and precipitation in Gilgit-Baltistan (GB), Pakistan, from 1955 to 2010. J Biodivers Environ Sci 6:67–73
Reddy CS, Pasha SV, Satish KV, Unnikrishnan A, Chavan SB et al (2019) Quantifying and predicting multi-decadal forest cover changes in Myanmar: a biodiversity hotspot under threat. Biodivers Conserv 28:1129–1149
Rees WG, Hofgaard A, Boudreau S, Cairns DM, Harper K et al (2020) Is subarctic forest advance able to keep pace with climate change? Glob Change Biol 26:3965–3977
Rehm EM, Feeley KJ (2015) The inability of tropical cloud forest species to invade grasslands above treeline during climate change: potential explanations and consequences. Ecography 38:1167–1175
Rehm EM, Feeley KJ (2016) Seedling transplants reveal species-specific responses of high-elevation tropical treeline trees to climate change. Oecologia 181:1233–1242
Reisinger A, Kitching RL, Chiew F, Hughes L, Newton PD et al. (2014) Australasia. In: IPCC (ed) Climate change 2014: impacts, adaptation, and vulnerability. Part B: regional aspects. Cambridge University Press, Cambridge-New York, pp 1371–1438
Ren YY, Ren GY, Sun XB, Shrestha AB, You QL et al (2017) Observed changes in surface air temperature and precipitation in the Hindu Kush Himalayan region over the last 100-plus years. Adv Clim Chang Res 8:148–156
Ren GY, Shrestha AB (2017) Climate change in the Hindu Kush Himalaya. Adv Clim Chang Res 8:137–140
Resler LM, Shao Y, Campbell JB, Michaels A (2020) Land cover and land use change in an emerging national park gateway region: implications for mountain sustainability. In: Sarmiento F, Frolich LM (eds) The Elgar companion to geography, transdisciplinarity and sustainability. Edward Elgar Publishing, Cheltenham, pp 270–292
Restrepo JD, Kettner AJ, Syvitski JP (2015) Recent deforestation causes rapid increase in river sediment load in the Colombian Andes. Anthropocene 10:13–28

Rets EP, Dzhamalov RG, Kireeva MB, Frolova NL, Durmanov IN et al (2018) Recent trends of river runoff in the North Caucasus. Geogr Environ Sustain 11:61–70

Rhoades RE, Thompson SI (1975) Adaptive strategies in alpine environments: beyond ecological particularism. Am Ethnol 2:535–551

Rico I, Izaguirre E, Serrano E, López-Moreno JI (2017) Current glacier area in the Pyrenees: an updated assessment 2016. Pirineos 172:e029

Ringler A (2016) Skigebiete der Alpen: landschaftsökologische Bilanz, Perspektiven für die renaturierung. Jb Ver Schutz Bergwelt 81:29–130

Ritler A (1997) Land use, forests and the landscape of Ethiopia, 1699–1865. Soil Conservation Research Programme Ethiopia, Research Report 38, University of Berne, Berne

Ritler A (2003) Forests, land use and landscape in the Central and northern Ethiopian Highlands, 1865 to 1930. Geographica Bernensia 19, University of Berne, Berne

Rixen C (2013) Skiing and vegetation. In: Rixen C, Rolando A (eds) The impacts of skiing and related winter recreational activities on mountain environments. Bentham Science Publishers, Bussum, pp 65–78

Rixen C, Rolando A (eds) (2013) The impacts of skiing and related winter recreational activities on mountain environments. Bentham Science Publishers, Bussum

Rixen C, Wipf S (2017) Non-equilibrium in alpine plant assemblages: shifts in Europe's summit floras. In: Catalan J, Ninot JM, Mercè Aniz M (eds) High mountain conservation in a changing world. Springer, Cham, pp 285–303

Rodman KC, Veblen TT, Saraceni S, Chapman TB (2019) Wildfire activity and land use drove 20th-century changes in forest cover in the Colorado Front Range. Ecosphere 10:e02594

Rohrer M, Salzmann N, Stoffel M, Kulkarni AV (2013) Missing (in-situ) snow cover data hampers climate change and runoff studies in the Greater Himalayas. Sci Total Environ 468:S60–S70

Rolland C, Petitcolas V, Michalet R (1998) Changes in radial tree-growth for *Picea abies*, *Larix decidua*, *Pinus cembra* and *Pinus uncinata* near the alpine timberline since 1750. Trees 13:40–53

Romeo R, Vita A, Testolin R, Hofer T (2015) Mapping the vulnerability of mountain peoples to food insecurity. FAO, Rome

Romero HI, Smith P, Vasquez A (2009) Global changes and economic globalization in the Andes. Challenges for developing nations. In: Jandl R, Borsdorf A, Van Migroet H, Lackner R, Psenner R (eds) Global change and sustainable development in mountain regions. Innsbruck University Press, Innsbruck, pp 71–92

Romero-Lankao P, Smith JB, Davidson D, Diffenbaugh N, Kinney P et al. (2014) North America. In: IPCC (ed) Climate change 2014: impacts, adaptation, and vulnerability. Part B: regional aspects. Cambridge University Press, Cambridge-New York, pp 1439–1498

Romero H, Rivera A (1996) Global changes and unsustainable development in the Andes of northern Chile. In: Hurni H, Kienholz H, Wanner H, Wiesmann U (eds) Umwelt Mensch Gebirge. Beiträge zur Dynamik von Natur- und Lebensraum. Jahrbuch der Geographischen Gesellschaft Bern 59:103–110

Roos C, Sullivan A III, McNamee C (2010) Paleoecological evidence for indigenous burning in the upland Southwest. In: Dean RM (ed) The archaeology of anthropogenic environments. Southern Illinois University, Carbondale, pp 142–171

Roth T, Plattner M, Amrhein V (2014) Plants, birds and butterflies: short-term responses of species communities to climate warming vary by taxon and with altitude. PLoS ONE 9:e82490

Rottler E, Kormann C, Francke T, Bronstert A (2019) Elevation-dependent warming in the Swiss Alps 1981–2017: features, forcings and feedbacks. Int J Climatol 39:2556–2568

Le Roux PC, McGeoch MA (2008) Rapid range expansion and community reorganization in response to warming. Glob Change Biol 14:2950–2962

Roxy MK, Chaithra ST (2018) Impacts of climate change on the Indian summer monsoon. In: Climate change and water resources in India. Ministry of Environment, Forest and Climate Change (MoEF&CC, (ed) Mishra V, Bhatt JR. Government of India, New Delhi, pp 21–37

Rudmann-Maurer K, Weyand A, Fischer M, Stöcklin J (2008) The role of landuse and natural determinants for grassland vegetation composition in the Swiss Alps. Basic Appl Ecol 9:494–503

Ruiz D, Martinson DG, Vergara W (2012) Trends, stability and stress in the Colombian Central Andes. Clim Change 112:717–732

Rumpf SB, Hülber K, Klonner G, Moser D, Schütz M et al (2018) Range dynamics of mountain plants decrease with elevation. Proc Natl Acad Sci 115:1848–1853

Rundel PW, Keeley JE (2016) Dispersal limitation does not control high elevational distribution of alien plant species in the southern Sierra Nevada, California. Nat Areas J 36:277–287

Rundqvist S, Hedenås H, Sandström A, Emanuelsson U, Eriksson H, Jonasson C, Callaghan TV (2011) Tree and shrub expansion over the past 34 years at the tree-line near Abisko, Sweden. Ambio 40:683–692

Rössler M, Kirscht H, Rademacher C, Platt S, Kemmerling B, Linstädter A (2010) Migration and resource management in the Drâa Valley, southern Morocco. In: Speth P, Christoph M, Diekkrüger B (eds) Impacts of global change on the hydrological cycle in West and Northwest Africa. Springer, Berlin-Heidelberg, pp 634–647

Saavedra FA, Kampf SK, Fassnacht SR, Sibold JS (2018) Changes in Andes snow cover from MODIS data, 2000–2016. Cryosphere 12:1027–1046

Sakio H, Masuzawa T (2012) The advancing timberline on Mt. Fuji: natural recovery or climate change? J Plant Res 125:539–546

Salick J, Fang Z, Hart R (2019) Rapid changes in eastern Himalayan alpine flora with climate change. Am J Bot 106:520–530

Salick J, Ghimire SK, Fang Z, Dema S, Konchar KM (2014) Himalayan alpine vegetation, climate change and mitigation. J Ethnobiol 34:276–293

Salick J, Staver B, Hart R (2020) Indigenous knowledge and dynamics among Himalayan peoples, vegetation, and climate change. In: Welch-Devine M, Sourdril A, Burke BJ (eds) Changing climate, changing worlds. Springer, Cham, pp 55–69

Salinger MJ, Fitzharris BB, Chinn T (2019) Atmospheric circulation and ice volume changes for the small and medium glaciers of New Zealand's southern Alps mountain range 1977–2018. Int J Climatol 39:4274–4287

Salomon M, Bangamwabo V, Everson T, Mutanga O, Fincham R, Allsopp N (2012) Landscapes as libraries: a history of the uKhahlamba Drakensberg from 1818 to 2009. Innov J Appropriate Librarianship Inf Work in South Afr 44:63–80

Salzmann N, Huggel C, Rohrer M, Silverio W, Mark BG, Burns P, Portocarrero C (2013) Glacier changes and climate trends derived from multiple sources in the data scarce Cordillera Vilcanota region, southern Peruvian Andes. Cryosphere 7:103–118

Sanderson LA, McLaughlin JA, Antunes PM (2012) The last great forest: a review of the status of invasive species in the North American boreal forest. Forestry 85:329–340

Sarmiento FO, Frolich LM (2002) Andean cloud forest tree lines. Naturalness, agriculture and the human dimension. Mt Res Dev 22:278–287

Sarris D, Christodoulakis D, Körner C (2011) Impact of recent climatic change on growth of low elevation eastern Mediterranean forest trees. Clim Change 106:203–223

Savage J, Vellend M (2015) Elevational shifts, biotic homogenization and time lags in vegetation change during 40 years of climate warming. Ecography 38:546–555

Scherrer SC, Appenzeller C, Laternser M (2004) Trends in Swiss alpine snow days: the role of local- and large-scale climate variability. Geophys Res Lett 31:L13215

Scherrer D, Körner C (2011) Topographically controlled thermal-habitat differentiation buffers alpine plant diversity against climate warming. J Biogeogr 38:406–416

Scherrer SC, Wüthrich C, Croci-Maspoli M, Weingartner R, Appenzeller C (2013) Snow variability in the Swiss Alps 1864–2009. Int J Climatol 33:3162–3173

Schickhoff U (1995) Himalayan forest-cover changes in historical perspective. A case study in the Kaghan Valley, northern Pakistan. Mt Res Dev 15:3–18

Schickhoff U (2002) Die Degradierung der Gebirgswälder Nordpakistans. Faktoren, Prozesse und Wirkungszusammenhänge in einem regionalen Mensch-Umwelt-System. Erdwissenschaftliche Forschung 41, Steiner Verlag, Stuttgart

Schickhoff U (2005) The upper timberline in the Himalayas, Hindu Kush and Karakorum: a review of geographical and ecological aspects. In: Broll G, Keplin B (eds) Mountain ecosystems. Springer, Berlin, pp 275–354

Schickhoff U (2006) The forests of Hunza Valley—scarce resources under threat. In: Kreutzmann H (ed) Karakoram in transition—The Hunza Valley. Oxford University Press, Oxford-Karachi, pp 123–144

Schickhoff U (2007) Die Gebirgswälder des Himalaya und Karakorum—Sinnbild für Ressourcenübernutzung und Umweltdegradierung? In: Glaser R, Kremb K (eds) Planet Erde—Asien. Wissenschaftliche Buchgesellschaft, Darmstadt, pp 136–149

Schickhoff U (2009) Human impact on high altitude forests in northern Pakistan: degradation processes and root causes. In: Singh RB (ed) Biogeography and biodiversity. Rawat Publ, Jaipur-New Delhi, pp 76–90

Schickhoff U (2011) Dynamics of mountain ecosystems. In: Millington A, Blumler M, Schickhoff U (eds) Handbook of biogeography. Sage Publ, London, pp 313–337

Schickhoff U (2012) Der Himalaya: Wandel eines Gebirgssystems unter dem Einfluss von Klima und Mensch. Berichte der Reinhold-Tüxen-Gesellschaft 24:103–121. Hannover

Schickhoff U (2014) Die Bedeutung gemeinschaftlicher Wald- und Weidenutzung für die Entwicklung der Kulturlandschaft im Himalaya. Berichte der Reinhold-Tüxen-Gesellschaft 26:51–64

Schickhoff U (2016a) Aktuelle Biodiversitätsveränderungen in Hochgebirgen. In: Lozán JL, Breckle SW, Müller R, Rachor E (eds) Warnsignal Klima: die Biodiversität. Wissenschaftliche Auswertungen, Hamburg, pp 107–112

Schickhoff U (2016b) Hochgebirge: Hotspots der Biodiversität im globalen Wandel. In: Schickhoff U (ed) Biogeographie und Biodiversität. Hamburger Symposium Geographie 8, Institut für Geographie der Universität Hamburg, Hamburg, pp 73–97

Schickhoff U (2019) Risikolebensraum Kathmandu (Nepal): Klima- und Umweltveränderungen im Urbanisierungsprozess einer Himalaya-Metropolregion. In: Lozán JL, Breckle SW, Graßl H, Kuttler W, Matzarakis A (eds) Warnsignal Klima: die Städte. Wissenschaftliche Auswertungen, Hamburg, pp 99–105

Schickhoff U, Bobrowski M, Böhner J, Bürzle B, Chaudhary RP et al (2015) Do Himalayan treelines respond to recent climate change? An evaluation of sensitivity indicators. Earth Syst Dyn 6:245–265

Schickhoff U, Bobrowski M, Schwab N (2020) Alpine Waldgrenzen im Klimawandel—Wie sind die heterogenen Reaktionsmuster zu erklären? In: Lozán JL, Breckle SW, Escher-Vetter H, Grassl H, Kasang D, Paul F, Schickhoff U (eds) Warnsignal Klima: Hochgebirge im Wandel. Wissenschaftliche Auswertungen, Hamburg, pp 232–238

Schickhoff U, Mal S (2020) Current changes in alpine ecosystems of Asia. In: Goldstein MI, DellaSala DA (eds) Encyclopedia of the world's biomes, vol 1. Elsevier, Amsterdam, pp 589–598
Schickhoff U, Oyunchimeg D, Jabzan J (2007) Altitudinal gradients of plant species richness as influenced by grazing in Jargalant, Mongolian Altai. In: Gunin PD (ed) Ecosystems of the Inner Asia: issues of research and conservation. Nauka, Moscow, pp 101–113
Schickhoff U, Bobrowski M, Böhner J, Bürzle B, Chaudhary RP et al. (2016b) Climate change and treeline dynamics in the Himalaya. In: Singh RB, Schickhoff U, Mal S (eds) Climate change, glacier response, and vegetation dynamics in the Himalaya. Springer, Cham, pp 271–306
Schickhoff U, Singh RB, Mal S (2016a) Climate change and dynamics of glaciers and vegetation in the Himalaya: an overview. In: Singh RB, Schickhoff U, Mal S (eds) Climate change, glacier response, and vegetation dynamics in the Himalaya. Springer, Cham, pp 1–26
Schilling J, Freier KP, Hertig E, Scheffran J (2012) Climate change, vulnerability and adaptation in North Africa with focus on Morocco. Agr Ecosyst Environ 156:12–26
Schirpke U, Altzinger A, Leitinger G, Tasser E (2019) Change from agricultural to touristic use: effects on the aesthetic value of landscapes over the last 150 years. Landsc Urban Plan 187:23–35
Schlütz F, Zech W (2004) Palynological investigations on vegetation and climate change in the late Quaternary of Lake Rukche area, Gorkha Himal, Central Nepal. Veg Hist Archaeobotany 13:81–90
Schmidt M (2005) Utilisation and management changes in South Kyrgyzstan's mountain forests. J Mt Sci 2:91–104
Schmidt M (2012) Changing human–environment interrelationships in Kyrgyzstan's walnut-fruit forests. For Trees Livelihoods 21:253–266
Schmidt S, Nüsser M (2009) Fluctuations of Raikot glacier during the past 70 years: a case study from the Nanga Parbat massif, northern Pakistan. J Glaciol 55:949–959
Schmidt S, Nüsser M (2012) Changes of high altitude glaciers from 1969 to 2010 in the Trans-Himalayan Kang Yatze massif, Ladakh, Northwest India. Arct Antarct Alp Res 44:107–121
Schmidt-Vogt D, Miehe G (2015) Land use. In: Miehe G, Pendry CA, Chaudhary RP (eds) Nepal: an introduction to the natural history, ecology and human environment of the Himalayas. Royal Botanic Garden Edinburgh, Edinburgh, pp 287–310
Schmitzberger I, Wrbka T, Steurer B, Aschenbrenner G, Peterseil J, Zechmeister HG (2005) How farming styles influence biodiversity maintenance in Austrian agricultural landscapes. Agr Ecosyst Environ 108:274–290
Schoolmeester T, Johansen KS, Alfthan B, Baker E, Hesping M, Verbist K (2018) The Andean glacier and water atlas: the impact of glacier retreat on water resources. UNESCO and GRID-Arendal, Paris-Arendal
Schumann K, Gewolf S, Tackenberg O (2016) Factors affecting primary succession of glacier foreland vegetation in the European Alps. Alp Bot 126:105–117
Schwab N, Janecka K, Kaczka RJ, Böhner J, Chaudhary RP, Scholten T, Schickhoff U (2020) Ecological relationships at a near-natural treeline, Rolwaling Valley, Nepal Himalaya: implications for the sensitivity to climate change. Erdkunde 74:15–44
Schwab N, Kaczka RJ, Janecka K, Böhner J, Chaudhary RP, Scholten T, Schickhoff U (2018) Climate change-induced shift of tree growth sensitivity at a Central Himalayan treeline ecotone. Forests 9:267
Schwab N, Schickhoff U, Bobrowski M, Böhner J, Bürzle B et al (2016) Treeline responsiveness to climate warming: insights from a krummholz treeline in Rolwaling Himal, Nepal. In: Singh RB, Schickhoff U, Mal S (eds) Climate change, glacier response, and vegetation dynamics in the Himalaya. Springer, Cham, pp 307–345
Schwab N, Schickhoff U, Bürzle B, Müller M, Böhner J et al (2017) Implications of tree species-environment relationships for the responsiveness of Himalayan krummholz treelines to climate change. J Mt Sci 14:453–473
Schwartz MD, Ault TR, Betancourt JL (2013) Spring onset variations and trends in the continental USA: past and regional assessment using temperature-based indices. Int J Climatol 33:2917–2922
Schöner W, Koch R, Matulla C, Marty C, Tilg AM (2019) Spatiotemporal patterns of snow depth within the Swiss-Austrian Alps for the past half century (1961 to 2012) and linkages to climate change. Int J Climatol 39:1589–1603
Scott CA, Zhang F, Mukherji A, Immerzeel W, Mustafa D, Bharati L (2019) Water in the Hindu Kush Himalaya. In: Wester P, Mishra A, Mukherji A, Shrestha AB (eds) The Hindu Kush Himalaya assessment. Springer, Cham, pp 257–299
Seehaus T (2020) Die Gletscher der Anden im Klimawandel. In: Lozán JL, Breckle SW, Escher-Vetter H, Grassl H, Kasang D, Paul F, Schickhoff U (eds) Warnsignal Klima: Hochgebirge im Wandel. Wissenschaftliche Auswertungen, Hamburg, pp 146–151
Seehaus T, Malz P, Sommer C, Lippl S, Cochachin A, Braun M (2019) Changes of the tropical glaciers throughout Peru between 2000 and 2016 -mass balance and area fluctuations. Cryosphere 13:2537–2556
Seehaus T, Malz P, Sommer C, Soruco A, Rabatel A, Braun M (2020) Mass balance and area changes of glaciers in the Cordillera Real and Tres Cruces, Bolivia, between 2000 and 2016. J Glaciol 66:124–136
Seim A, Omurova G, Azisov E, Musuraliev K, Aliev K et al (2016) Climate change increases drought stress of Juniper trees in the mountains of Central Asia. PLoS ONE 11:e0153888

Seimon TA, Seimon A, Yager K, Reider K, Delgado A et al (2017) Long-term monitoring of tropical alpine habitat change, Andean anurans, and chytrid fungus in the Cordillera Vilcanota, Peru: results from a decade of study. Ecol Evol 7:1527–1540

Seipel T, Alexander JM, Edwards PJ, Kueffer C (2016) Range limits and population dynamics of non-native plants spreading along elevation gradients. Perspect Plant Ecol Evol Syst 20:46–55

Seipel T, Kueffer C, Rew LJ, Daehler CC, Pauchard A et al (2012) Processes at multiple scales affect richness and similarity of non-native plant species in mountains around the world. Glob Ecol Biogeogr 21:236–246

Sekar KC, Rawal RS, Chaudhery A, Pandey A, Rawat G et al (2017) First GLORIA site in Indian Himalayan region: towards addressing issue of long-term data deficiency in the Himalaya. Nat Acad Sci Lett 40:355–357

Senf C, Pflugmacher D, Zhiqiang Y, Sebald J, Knorn J et al (2018) Canopy mortality has doubled in Europe's temperate forests over the last three decades. Nat Commun 9:1–8

Settele J, Scholes R, Betts R, Bunn S, Leadley P et al. (2014) Terrestrial and inland water systems. In: IPCC (ed) Climate change 2014: impacts, adaptation, and vulnerability. Part A: global and sectoral aspects. Cambridge University Press, Cambridge-New York, pp 271–359

Setten G, Austrheim G (2012) Changes in land use and landscape dynamics in mountains of northern Europe: challenges for science, management and conservation. Int J Biodiver Sci Ecosyst Serv ManageS 8:287–291

Shafiq MU, Rasool R, Ahmed P, Dimri AP (2019) Temperature and precipitation trends in Kashmir Valley, northwestern Himalayas. Theoret Appl Climatol 135:293–304

Shahgedanova M, Nosenko G, Khromova T, Muraveyev A (2010) Glacier shrinkage and climatic change in the Russian Altai from the mid-20th century: an assessment using remote sensing and PRECIS regional climate model. J Geophys Res Atmos 115: D16

Shakya M (2016) Tourism and social capital: case studies from rural Nepal. In: McCool S, Bosak K (eds) Reframing sustainable tourism. Environmental challenges and solutions, vol. 2. Springer, Dordrecht, pp 217–239

Shamsabad MM, Assadi M, Parducci L (2018) Impact of climate change implies the northward shift in distribution of the Irano-Turanian subalpine species complex *Acanthophyllum squarrosum*. J Asia-Pac Biodivers 11:566–572

Shandra O, Weisberg P, Martazinova V (2013) Influences of climate and land use history on forest and timberline dynamics in the Carpathian mountains during the twentieth century. In: Ostapowicz K, Bytnerowicz A, Wyżga B, Kozak J (eds) The Carpathians: integrating nature and society towards sustainability. Springer, Berlin, pp 209–223

Shangguan DH, Bolch T, Ding YJ, Kröhnert M, Pieczonka T, Wetzel HU, Liu SY (2015) Mass changes of southern and northern Inylchek glacier, Central Tian Shan, Kyrgyzstan, during—1975 and 2007 derived from remote sensing data. Cryosphere 9:703–717

Sharkhuu A, Sharkhuu N, Etzelmüller B, Heggem ESF, Nelson FE et al. (2007) Permafrost monitoring in the Hovsgol mountain region, Mongolia. Journal of Geophysical Research: Earth Surface 112:F02S06

Sharma V, Mishra VD, Joshi PK (2014) Topographic controls on spatio-temporal snow cover distribution in Northwest Himalaya. Int J Remote Sens 35:3036–3056

Sharma PK, Tiwari A, Shrestha BB (2020) Changes in regeneration and leaf traits of *Rhododendron campanulatum* along a treeline ecotone in Central Nepal. J Mt Sci 17:602–613

Sharma P (2008) Unravelling the mosaic. Spatial aspects of ethnicity in Nepal. Himal Books, Lalitpur

Shean DE, Bhushan S, Montesano P, Rounce DR, Arendt A, Osmanoglu B (2020) A systematic, regional assessment of High Mountain Asia glacier mass balance. Front Earth Sci 7:363

Shekhar MS, Chand H, Kumar S, Srinivasan K, Ganju A (2010) Climate-change studies in the western Himalaya. Ann Glaciol 51:105–112

Sherriff RL, Miller AE, Muth K, Schriver M, Batzel R (2017) Spruce growth responses to warming vary by ecoregion and ecosystem type near the forest-tundra boundary in South-West Alaska. J Biogeogr 44:1457–1468

Shevtsova I, Heim B, Kruse S, Schröder J, Troeva EI et al (2020) Strong shrub expansion in tundra-taiga, tree infilling in taiga and stable tundra in central Chukotka (north-eastern Siberia) between 2000 and 2017. Environ Res Lett 15:085006

Shi C, Schneider L, Hu Y, Shen M, Sun C et al (2020) Warming-induced unprecedented high-elevation forest growth over the monsoonal Tibetan Plateau. Environ Res Lett 15:054011

Shigaeva J, Hagerman S, Zerriffi H, Hergarten C, Isaeva A, Mamadalieva Z, Foggin M (2016) Decentralizing governance of agropastoral systems in Kyrgyzstan: an assessment of recent pasture reforms. Mt Res Dev 36:91–102

Shrestha AB, Aryal R (2011) Climate change in Nepal and its impact on Himalayan glaciers. Reg Environ Change 11:S65–S77

Shrestha KB, Hofgaard A, Vandvik V (2015) Recent treeline dynamics are similar between dry and mesic areas of Nepal, Central Himalaya. J Plant Ecol 8:347–358

Shrestha UB, Shrestha AM, Aryal S, Shrestha S, Gautam MS, Ojha H (2019) Climate change in Nepal: a comprehensive analysis of instrumental data and people's perceptions. Clim Change 154:315–334

Shrestha AB, Wake CP, Mayewski PA, Dibb JE (1999) Maximum temperature trends in the Himalayas and its vicinity: an analysis based on temperature records

from Nepal for the period 1971–1994. J Clim 12:2775–2786
Shrestha M (2017) Push and pull: a study of international migration from Nepal. Policy Research Working Paper 7965. The World Bank Group, Washington
Siddiqui T, Bhagat RB, Banerjee S, Liu C, Sijapati B et al (2019) Migration in the Hindu Kush Himalaya: drivers, consequences, and governance. In: Wester P, Mishra A, Mukherji A, Shrestha AB (eds)sustainable governance of Andean Hindu Kush Himalaya assessment. Springer, Cham, pp 517–544
Sigdel SR, Liang E, Wang Y, Dawadi B, Camarero JJ (2020) Tree-to-tree interactions slow down Himalayan treeline shifts as inferred from tree spatial patterns. J Biogeogr 47:1816–1826
Sigdel SR, Wang Y, Camarero JJ, Zhu H, Liang E, Peñuelas J (2018) Moisture-mediated responsiveness of treeline shifts to global warming in the Himalayas. Glob Change Biol 24:5549–5559
Simpson M, Aravena E, Deverell J (2014) The future of mining in Chile. CSIRO, Sydney-Santiago
Singh P, Arya V, Negi GCS, Singh SP (2018) Expansion of *Rhododendron campanulatum* krummholz in the treeline ecotone in Tungnath, Garhwal Himalaya. Trop Ecol 59:287–295
Singh RB, Kumar P (2014) Climate change and glacial lake outburst floods in Himachal Himalaya, India. In: Singh M, Singh RB, Hassan MI (eds) Climate change and biodiversity. Springer, Tokyo, pp 3–14
Singh RB, Mal S (2014) Trends and variability of monsoon and other rainfall seasons in western Himalaya, India. Atmos Sci Lett 15:218–226
Singh CP, Mohapatra J, Pandya HA, Gajmer B, Sharma N, Shrestha DG (2020) Evaluating changes in treeline position and land surface phenology in Sikkim Himalaya. Geocarto Int 35:453–469
Singh P, Negi GCS (2016) Impact of climate change on phenological responses of major forest trees of Kumaun Himalaya. ENVIS Bull Himalayan Ecol 24:112–116
Singh SK, Rathore BP, Bahuguna IM (2014) Snow cover variability in the Himalayan-Tibetan region. Int J Climatol 34:446–452
Singh D, Sharma V, Juyal V (2015) Observed linear trend in few surface weather elements over the Northwest Himalayas (NWH) during winter season. J Earth Syst Sci 124:553–565
Singh S, Kumar R, Bhardwaj A, Sam L, Shekhar M et al (2016) Changing climate and glacio-hydrology in Indian Himalayan region: a review. Wiley Interdisc Rev Clim Change 7:393–410
Singh RB, Kumar S, Kumar A (2016) Climate change in Pindari region, Central Himalaya, India. In: Singh RB, Schickhoff U, Mal S (eds) Climate change and dynamics of glaciers and vegetation in the Himalaya. Springer, Cham, pp 117–135
Singh D, Ghosh S, Roxy MK, McDermid S (2019) Indian summer monsoon: extreme events, historical changes, and role of anthropogenic forcings. Wiley Interdisciplinary Reviews: Clim Change 10:e571
Singh SP, Sharma S, Dhyani PP (2019) Himalayan arc and treeline: distribution, climate change responses and ecosystem properties. Biodiversity and Conservation 28:1997–2016
Siniscalco C, Barni E (2018) Are non-native plant species a threat to the Alps? Insights and perspectives. In: Pedrotti F (ed) Climate gradients and biodiversity in mountains of Italy. Springer, Cham, pp 91–107
Sittaro F, Paquette A, Messier C, Nock CA (2017) Tree range expansion in eastern North America fails to keep pace with climate warming at northern range limits. Glob Change Biol 23:3292–3301
Siyum ZG, Ayoade JO, Onilude MA, Feyissa MT (2018) Relationship between space-based vegetation productivity index and radial growth of main tree species in the dry afromontane forest remnants of northern Ethiopia. J Appl Sci Environ Manag 22:1781–1790
Sloat LL, Henderson AN, Lamanna C, Enquist BJ (2015) The effect of the foresummer drought on carbon exchange in subalpine meadows. Ecosystems 18:533–545
Smithers BV, North MP, Millar CI, Latimer AM (2018) Leap frog in slow motion: divergent responses of tree species and life stages to climatic warming in Great Basin subalpine forests. Glob Change Biol 24:e442–e457
Sohar K, Altman J, Lehečková E, Doležal J (2017) Growth-climate relationships of Himalayan conifers along elevational and latitudinal gradients. Int J Climatol 37:2593–2605
Soini E (2005) Land use change patterns and livelihood dynamics on the slopes of Mt. Kilimanjaro, Tanzania. Agric Syst 85:306–323
Soja AJ, Tchebakova NM, French NH, Flannigan MD, Shugart HH et al (2007) Climate-induced boreal forest change: predictions versus current observations. Global Planet Change 56:274–296
Solomon N, Hishe H, Annang T, Pabi O, Asante IK, Birhane E (2018) Forest cover change, key drivers and community perception in Wujig Mahgo Waren forest of northern Ethiopia. Land 7:32
Somos-Valenzuela MA, McKinney DC, Rounce DR, Byers AC (2014) Changes in Imja Tsho in the Mount Everest region of Nepal. Cryosphere 8:1–27
Sontakke NA, Singh N, Singh HN (2008) Instrumental period rainfall series of the Indian region (AD 1813–2005): revised reconstruction, update and analysis. The Holocene 18:1055–1066
Spano D, Snyder RL, Cesaraccio C (2013) Mediterranean phenology. In: Schwartz MD (ed) Phenology: an integrative environmental science. Springer, Dordrecht, pp 173–196
Spasojevic MJ, Bowman WD, Humphries HC, Seastedt TR, Suding KN (2013) Changes in alpine vegetation over 21 years: are patterns across a heterogeneous landscape consistent with predictions? Ecosphere 4:1–18
Speziale KL, Ezcurra C (2011) Patterns of alien plant invasions in northwestern Patagonia, Argentina. J Arid Environ 75:890–897

Spiegelberger T, Matthies D, Müller-Schärer H, Schaffner U (2006) Scale-dependent effects of land use on plant species richness of mountain grassland in the European Alps. Ecography 29:541–548

Spinage CA (2012) African ecology—benchmarks and historical perspectives. Springer, Berlin

Sproull GJ, Quigley MF, Sher A, González E (2015) Long-term changes in composition, diversity and distribution patterns in four herbaceous plant communities along an elevational gradient. J Veg Sci 26:552–563

Srur AM, Villalba R, Rodríguez-Catón M, Amoroso MM, Marcotti E (2016) Establishment of *Nothofagus pumilio* at upper treelines across a precipitation gradient in the northern Patagonian Andes. Arct Antarct Alp Res 48:755–766

Srur AM, Villalba R, Rodríguez-Catón M, Amoroso MM, Marcotti E (2018) Climate and *Nothofagus pumilio* establishment at upper treelines in the Patagonian Andes. Front Earth Sci 6:57

Stanisci A, Frate L, Morra Di Cella U, Pelino G, Petey M, Siniscalco C, Carranza ML (2016) Short-term signals of climate change in Italian summit vegetation: observations at two GLORIA sites. Plant Biosyst 150:227–235

Steinbauer K, Lamprecht A, Semenchuk P, Winkler M, Pauli H (2020) Dieback and expansions: species-specific responses during 20 years of amplified warming in the High Alps. Alp Bot 130:1–11

Steinbauer MJ, Grytnes JA, Jurasinski G, Kulonen A, Lenoir J et al (2018) Accelerated increase in plant species richness on mountain summits is linked to warming. Nature 556:231–234

Steinicke E (2011) Konsequenzen der Nationalparkgründung im Ruwenzori (Uganda). Geogr Rundsch 63:57–63

Stellrecht I (ed) (1998) Karakorum-Hindukush-Himalaya: dynamics of change. Parts I and II, Köppe, Cologne

Stellrecht I, Winiger M (eds) (1997) Perspectives on history and change in the Karakorum, Hindukush, and Himalaya. Köppe, Cologne

Stepp JR, Castaneda H, Cervone S (2005) Mountains and biocultural diversity. Mt Res Dev 25:223–227

Stoffel M, Tiranti D, Huggel C (2014) Climate change impacts on mass movements—case studies from the European Alps. Sci Total Environ 493:1255–1266

Stohlgren TJ, Barnett D, Flather C, Kartesz J, Peterjohn B (2005) Plant species invasions along the latitudinal gradient in the United States. Ecology 86:2298–2309

Strebel N, Bühler C (2015) Recent shifts in plant species suggest opposing land-use changes in alpine pastures. Alp Bot 125:1–9

Strobelt S, von Kocemba M (2020) Mensch-Umwelt-Interaktionen im äthiopischen Hochland. In: Lozán JL, Breckle SW, Escher-Vetter H, Grassl H, Kasang D, Paul F, Schickhoff U (eds) Warnsignal Klima: Hochgebirge im Wandel. Wissenschaftliche Auswertungen, Hamburg, pp 296–302

Stueve KM, Isaacs RE, Tyrrell LE, Densmore RV (2011) Spatial variability of biotic and abiotic tree establishment constraints across a treeline ecotone in the Alaska Range. Ecology 92:496–506

Stöcklin J, Bosshard A, Klaus G, Rudmann-Maurer K, Fischer M (2007) Landnutzung und biologische Vielfalt in den Alpen. Vdf Hochschulverlag, Zürich

Stöckli V, Wipf S, Nilsson C, Rixen C (2011) Using historical plant surveys to track biodiversity on mountain summits. Plant Ecolog Divers 4:415–425

Supari TF, Juneng L, Aldrian E (2017) Observed changes in extreme temperature and precipitation over Indonesia. Int J Climatol 37:1979–1997

Suwal MK, Shrestha KB, Guragain L, Shakya R, Shrestha K, Bhuju DR, Vetaas OR (2016) Land-use change under a warming climate facilitated upslope expansion of Himalayan silver fir (*Abies spectabilis* (D. Don) Spach). Plant Ecol 217:993–1002

Suzuki Y (2013) Conflict between mining development and nomadism in Mongolia. In: Yamamura N, Fujita N, Maekawa A (eds) The Mongolian ecosystem network. Springer, Tokyo, pp 269–294

Suárez E, Medina G (2001) Vegetation structure and soil properties in Ecuadorian páramo grasslands with different histories of burning and grazing. Arct Antarct Alp Res 33:158–164

Tang Z, Wang J, Li H, Yan L (2013) Spatiotemporal changes of snow cover over the Tibetan Plateau based on cloud-removed moderate resolution imaging spectroradiometer fractional snow cover product from 2001 to 2011. J Appl Remote Sens 7:073582

Tang G, Arnone JA III, Verburg PSJ, Jasoni RL, Sun L (2015) Trends and climatic sensitivities of vegetation phenology in semiarid and arid ecosystems in the US Great Basin during 1982–2011. Biogeosciences 12:6985–6997

Tang Z, Wang X, Wang J, Wang X, Li H, Jiang Z (2017) Spatiotemporal variation of snow cover in Tianshan Mountains, Central Asia, based on cloud-free MODIS fractional snow cover product, 2001–2015. Remote Sensing 9:1045

Tang KHD (2019) Climate change in Malaysia: trends, contributors, impacts, mitigation and adaptations. Sci Total Environ 650:1858–1871

Tao H, Gemmer M, Bai Y, Su B, Mao W (2011) Trends of streamflow in the Tarim river basin during the past 50 years: human impact or climate change? J Hydrol 400:1–9

Tarolli P, Preti F, Romano N (2014) Terraced landscapes: from an old best practice to a potential hazard for soil degradation due to land abandonment. Anthropocene 6:10–25

Tasser E, Ruffini FV, Tappeiner U (2009) An integrative approach for analysing landscape dynamics in diverse cultivated and natural mountain areas. Landscape Ecol 24:611–628

Tasser E, Tappeiner U (2002) Impact of land use changes on mountain vegetation. Appl Veg Sci 5:173–184

Tasser E, Tappeiner U, Cernusca A (2005) Ecological effects of land-use changes in the European Alps. In: Huber UM, Bugmann HKM, Reasoner MA (eds) Global change and mountain regions. Springer, Dordrecht, pp 409–420
Tasser E, Walde J, Tappeiner U, Teutsch A, Noggler W (2007) Land-use changes and natural reforestation in the eastern Central Alps. Agr Ecosyst Environ 118:115–129
Tasser E, Mader M, Tappeiner U (2003) Effects of land use in alpine grasslands on the probability of landslides. Basic Appl Ecol 4:271–280
Tatoni T, Médail F, Roche P, Barbero M (2004) The impact of changes in land use on ecological patterns in Provence (Mediterranean France). In: Mazzoleni S, Di Pasquale G, Mulligan M, Di Martino P, Rego F (eds) Recent dynamics of the Mediterranean vegetation and landscape. Wiley, Chichester, pp 105–120
Taylor DM (1990) Late quaternary pollen records from two Ugandan mires: evidence for environmental changes in the Rukiga highlands of Southwest Uganda. Palaeogeogr Palaeoclimatol Palaeoecol 80:283–300
Taylor DM (1996) Mountains. In: Adams WM, Goudie A, Orme AR (eds) The physical geography of Africa. Oxford University Press, Oxford, pp 287–306
Taylor RG, Mileham L, Tindimugaya C, Majugu A, Muwanga A, Nakileza B (2006) Recent glacial recession in the Rwenzori Mountains of East Africa due to rising air temperature. Geophys Res Lett 33: L10402
Tchebakova NM, Parfenova EI, Korets MA, Conard SG (2016) Potential change in forest types and stand heights in Central Siberia in a warming climate. Environ Res Lett 11:035016
Tei S, Sugimoto A, Yonenobu H, Matsuura Y, Osawa A et al (2017) Tree-ring analysis and modeling approaches yield contrary response of circumboreal forest productivity to climate change. Glob Change Biol 23:5179–5188
Telwala Y, Brook BW, Manish K, Pandit MK (2013) Climate-induced elevational range shifts and increase in plant species richness in a Himalayan biodiversity epicentre. PLoS ONE 8:e57103
Tennant C, Menounos B, Wheate R, Clague JJ (2012) Area change of glaciers in the Canadian Rocky Mountains, 1919 to 2006. Cryosphere 6:1541–1552
Terskaia A, Dial RJ, Sullivan PF (2020) Pathways of tundra encroachment by trees and tall shrubs in the western Brooks Range of Alaska. Ecography 43:769–778
Testolin R, Attorre F, Jiménez-Alfaro B (2020) Global distribution and bioclimatic characterization of alpine biomes. Ecography 43:779–788
Thakuri S, Dahal S, Shrestha D, Guyennon N, Romano E, Colombo N, Salerno F (2019) Elevation-dependent warming of maximum air temperature in Nepal during 1976–2015. Atmos Res 228:261–269
Thapa UK, St. George S, Kharal DK, Gaire NP (2017) Tree growth across the Nepal Himalaya during the last four centuries. Prog Phys Geogr 41:478–495
Thoman R, Walsh JE (2019) Alaska's changing environment: documenting Alaska's physical and biological changes through observations. International Arctic Research Center, University of Alaska Fairbanks
Thompson LG (2010) Climate change: the evidence and our options. Behav Analyst 33:153–170
Thompson MP, MacGregor DG, Dunn CJ, Calkin DE, Phipps J (2018) Rethinking the wildland fire management system. J Forest 116:382–390
Thompson LG, Mosley-Thompson E, Davis ME, Porter SE (2017) Ice core records of climate and environmental variability in the tropical Andes of Peru: past, present and future. Rev De Glaciares Y Ecosistemas De Montaña 3:25–40
Thompson JA, Paull DJ (2017) Assessing spatial and temporal patterns in land surface phenology for the Australian Alps (2000–2014). Remote Sens Environ 199:1–13
Thuiller W, Lavorel S, Araújo MB, Sykes MT, Prentice IC (2005) Climate change threats to plant diversity in Europe. Proc Natl Acad Sci 102:8245–8250
Tielidze LG, Wheate RD (2018) The Greater Caucasus glacier inventory (Russia, Georgia and Azerbaijan). Cryosphere 12:81–94
Tinner W, Theurillat JP (2003) Uppermost limit, extent, and fluctuations of the timberline and treeline ecocline in the Swiss Central Alps during the past 11,500 years. Arct Antarct Alp Res 35:158–169
Tiwari A, Jha PK (2018) An overview of treeline response to environmental changes in Nepal Himalaya. Trop Ecol 59:273–285
Tiwari A, Fan ZX, Jump AS, Li SF, Zhou ZK (2017a) Gradual expansion of moisture sensitive *Abies spectabilis* forest in the Trans-Himalayan zone of Central Nepal associated with climate change. Dendrochronologia 41:34–43
Tiwari A, Fan ZX, Jump AS, Zhou ZK (2017b) Warming induced growth decline of Himalayan birch at its lower range edge in a semi-arid region of Trans-Himalaya, Central Nepal. Plant Ecol 218:621–633
Toivonen JM, Gonzales-Inca CA, Bader MY, Ruokolainen K, Kessler M (2018) Elevational shifts in the topographic position of *Polylepis* forest stands in the Andes of southern Peru. Forests 9:7
Tolessa T, Senbeta F, Kidane M (2017) The impact of land use/land cover change on ecosystem services in the central highlands of Ethiopia. Ecosyst Serv 23:47–54
Tomiolo S, Ward D (2018) Species migrations and range shifts: a synthesis of causes and consequences. Perspect Plant Ecol Evol Syst 33:62–77
Torbick N, Ge J, Qi J (2009) Changing surface conditions at Kilimanjaro indicated from multiscale imagery. Mt Res Dev 29:5–13
Torta G (2004) Consequences of rural abandonment in a northern Apennines landscape (Tuscany, Italy). In: Mazzoleni S, Di Pasquale G, Mulligan M, Di

Martino P, Rego F (eds) Recent dynamics of the Mediterranean vegetation and landscape. Wiley, Chichester, pp 157–165
Toulmin C (2009) Climate change in Africa. Zed Books, London-New York
Tovar C, Arnillas CA, Cuesta F, Buytaert W (2013a) Diverging responses of tropical Andean biomes under future climate conditions. PloS ONE 8:e63634
Tovar C, Seijmonsbergen AC, Duivenvoorden JF (2013b) Monitoring land use and land cover change in mountain regions: an example in the Jalca grasslands of the Peruvian Andes. Landscape Urban Planning 112:40–49
Trant AJ, Hermanutz L (2014) Advancing towards novel tree lines? A multispecies approach to recent tree line dynamics in subarctic alpine Labrador, northern Canada. J Biogeogr 41:1115–1125
Treml V, Šenfeldr M, Chuman T, Ponocná T, Demková K (2016) Twentieth century treeline ecotone advance in the Sudetes Mountains (Central Europe) was induced by agricultural land abandonment rather than climate change. J Veg Sci 27:1209–1221
Treml V, Veblen TT (2017) Does tree growth sensitivity to warming trends vary according to treeline form? J Biogeogr 44:1469–1480
Tsogtbaatar J (2013) Deforestation and reforestation of degraded forestland in Mongolia. In: Yamamura N, Fujita N, Maekawa A (eds) The Mongolian ecosystem network. Springer, Tokyo, pp 83–98
Tumusiime DM, Vedeld P, Gombya-Ssembajjwe W (2011) Breaking the law? Illegal livelihoods from a protected area in Uganda. Forest Policy Econ 13:273–283
Turbelin AJ, Malamud BD, Francis RA (2017) Mapping the global state of invasive alien species: patterns of invasion and policy responses. Glob Ecol Biogeogr 26:78–92
Tømmervik H, Bjerke JW, Park T, Hanssen F, Myneni RB (2019) Legacies of historical exploitation of natural resources are more important than summer warming for recent biomass increases in a boreal-arctic transition region. Ecosystems 22:1512–1529
UN (United Nations) (2020) The sustainable development goals report 2020. UN, New York
Uddin K, Chaudhary S, Chettri N, Kotru R, Murthy M et al (2015) The changing land cover and fragmenting forest on the roof of the world: a case study in Nepal's Kailash Sacred Landscape. Landsc Urban Plan 141:1–10
Umer M, Lamb HF, Bonnefille R, Lézine AM, Tiercelin JJ et al (2007) Late Pleistocene and Holocene vegetation history of the Bale mountains, Ethiopia. Quatern Sci Rev 26:2229–2246
Valiente-Banuet A, Aizen MA, Alcántara JM, Arroyo J, Cocucci A et al (2015) Beyond species loss: the extinction of ecological interactions in a changing world. Funct Ecol 29:299–307
Vanat L (2020) International report on snow and mountain tourism. Overview of the key industry figures for ski resorts, April 2020. https://www.vanat.ch/RM-world-report-2020.pdf
Vandvik V, Halbritter AH, Telford RJ (2018) Greening up the mountain. Proc Natl Acad Sci 115:833–835
Vankat JL (2013) Vegetation dynamics on the mountains and plateaus of the American Southwest. Springer, Dordrecht
Vanneste T, Michelsen O, Graae BJ, Kyrkjeeide MO, Holien H et al (2017) Impact of climate change on alpine vegetation of mountain summits in Norway. Ecol Res 32:579–593
Vanonckelen S, Van Rompaey A (2015) Spatiotemporal analysis of the controlling factors of forest cover change in the Romanian Carpathian Mountains. Mt Res Dev 35:338–350
Vanselow KA, Kraudzun T, Samimi C (2012a) Land stewardship in practice: an example from the eastern Pamirs of Tajikistan. In: Squires V (ed) Rangeland stewardship in Central Asia. Springer, Dordrecht, pp 71–90
Vanselow KA, Kraudzun T, Samimi C (2012b) Grazing practices and pasture tenure in the eastern Pamirs. Mt Res Dev 32:324–337
Veblen TT, Lorenz DC (1991) The Colorado Front Range. A century of ecological change. University of Utah Press, Salt Lake City
Veettil BK, Kamp U (2019) Global disappearance of tropical mountain glaciers: observations, causes, and challenges. Geosciences 9:196
Veettil BK, Wang S (2018a) State and fate of the remaining tropical mountain glaciers in Australasia using satellite imagery. J Mt Sci 15:495–503
Veettil BK, Wang S (2018b) An update on recent glacier changes in Mexico using Sentinel-2A data. Geografiska Annaler: Series A, Phys Geogr 100:307–318
Veh G, Korup O, Von Specht S, Roessner S, Walz A (2019) Unchanged frequency of moraine-dammed glacial lake outburst floods in the Himalaya. Nat Clim Chang 9:379–383
Verbyla D, Kurkowski TA (2019) NDVI-climate relationships in high-latitude mountains of Alaska and Yukon Territory. Arct Antarct Alp Res 51:397–411
Vessella F, López-Tirado J, Simeone MC, Schirone B, Hidalgo PJ (2017) A tree species range in the face of climate change: cork oak as a study case for the Mediterranean biome. Eur J Forest Res 136:555–569
Vetaas OR (2007) Global changes and its effect on glaciers and cultural landscapes: historical and future considerations. In: Chaudhary RP, Aase TH, Vetaas OR, Subedi BP (eds) Local effects of global changes in the Himalayas: Manang, Nepal. Tribhuvan University-University of Bergen, Kathmandu-Bergen, pp 23–39
Vicente-Serrano SM, Martín-Hernández N, Reig F, Azorin-Molina C, Zabalza J et al (2020) Vegetation greening in Spain detected from long term data (1981–2015). Int J Remote Sens 41:1709–1740
Vickers H, Høgda KA, Solbø S, Karlsen SR, Tømmervik H, Aanes R, Hansen BB (2016) Changes in greening in the High Arctic: insights from a 30 year AVHRR max NDVI dataset for Svalbard. Environ Res Lett 11:105004

Villers-Ruiz L, Castañeda-Aguado D (2013) Species and plant community reorganization in the trans-Mexican volcanic belt under climate change conditions. J Mt Sci 10:923–931

Vincent LA, Mekis É (2006) Changes in daily and extreme temperature and precipitation indices for Canada over the twentieth century. Atmos Ocean 44:177–193

Virtanen R, Eskelinen A, Gaare E (2003) Long-term changes in alpine plant communities in Norway and Finland. In: Nagy L, Grabherr G, Körner C, Thompson DBA (eds) Alpine biodiversity in Europe. Ecological Studies 167. Springer, Berlin-Heidelberg, pp 411–422

Viste E, Korecha D, Sorteberg A (2013) Recent drought and precipitation tendencies in Ethiopia. Theoret Appl Climatol 112:535–551

Vitali A, Garbarino M, Camarero JJ, Malandra F, Toromani E et al (2019) Pine recolonization dynamics in Mediterranean human-disturbed treeline ecotones. For Ecol Manage 435:28–37

Vitasse Y, Porté AJ, Kremer A, Michalet R, Delzon S (2009) Responses of canopy duration to temperature changes in four temperate tree species: relative contributions of spring and autumn leaf phenology. Oecologia 161:187–198

Vitasse Y, Hoch G, Randin CF, Lenz A, Kollas C, Körner C (2012) Tree recruitment of European tree species at their current upper elevational limits in the Swiss Alps. J Biogeogr 39:1439–1449

Vitasse Y, Signarbieux C, Fu YH (2018) Global warming leads to more uniform spring phenology across elevations. Proc Natl Acad Sci 115:1004–1008

Vittoz P, Bodin J, Ungricht S, Burga CA, Walther GR (2008a) One century of vegetation change on Isla Persa, a nunatak in the Bernina massif in the Swiss Alps. J Veg Sci 19:671–680

Vittoz P, Rulence B, Largey T, Freléchoux F (2008b) Effects of climate and land-use change on the establishment and growth of cembran pine (*Pinus cembra* L.) over the altitudinal treeline ecotone in the Central Swiss Alps. Arctic, Antarct, Alp Res 40:225–232

Vittoz P, Randin C, Dutoit A, Bonnet F, Hegg O (2009) Low impact of climate change on subalpine grasslands in the Swiss northern Alps. Glob Change Biol 15:209–220

Viviroli D, Dürr HH, Messerli B, Meybeck M, Weingartner R (2007) Mountains of the world, water towers for humanity: typology, mapping, and global significance. Water Resour Res 43:7

Vowles T, Gunnarsson B, Molau U, Hickler T, Klemedtsson L, Björk RG (2017) Expansion of deciduous tall shrubs but not evergreen dwarf shrubs inhibited by reindeer in Scandes mountain range. J Ecol 105:1547–1561

Vuille M, Bradley RS, Werner M, Keimig F (2003) 20th century climate change in the tropical Andes: observations and model results. Clim Change 59:75–99

Vuille M, Carey M, Huggel C, Buytaert W, Rabatel A et al (2018) Rapid decline of snow and ice in the tropical Andes—impacts, uncertainties and challenges ahead. Earth Sci Rev 176:195–213

Vuille M, Franquist E, Garreaud R, Lavado Casimiro WS, Cáceres B (2015) Impact of the global warming hiatus on Andean temperature. J Geophys Res: Atmos 120 (9):3745–3757

Vuille M (2013) Climate change and water resources in the tropical Andes. Inter-American Development Bank Technical Note 515, Washington DC

Vuorinen KE, Oksanen L, Oksanen T, Pyykönen A, Olofsson J, Virtanen R (2017) Open tundra persist, but arctic features decline—vegetation changes in the warming Fennoscandian tundra. Glob Change Biol 23:3794–3807

Vásquez DL, Balslev H, Sklenář P (2015) Human impact on tropical-alpine plant diversity in the northern Andes. Biodivers Conserv 24:2673–2683

WGMS (World Glacier Monitoring Service) (2008) Global glacier changes. Facts and figures, WGMS, Zurich

WMO (World Meteorological Organization) (2019) United in science: high-level synthesis report of latest climate science information convened by the science advisory group of the UN Climate Action Summit 2019. WMO, Geneva

Wagner FH (2009) Climate warming and environmental effects in the West: evidence for the twentieth century and implications for the twenty-first. In: Wagner FH (ed) Climate warming in western North America. Evidence and environmental effects. The University of Utah Press, Salt Lake City, pp 143-160

Wainwright HM, Steefel C, Trutner SD, Henderson AN, Nikolopoulos EI et al (2020) Satellite-derived foresummer drought sensitivity of plant productivity in Rocky Mountain headwater catchments: spatial heterogeneity and geological-geomorphological control. Environ Res Lett 15:084018

Walsh K, Giguet-Covex C (2020) A history of human exploitation of alpine regions. In: Goldstein MI, DellaSala DA (eds) Encyclopedia of the world's biomes, vol 1. Elsevier, Amsterdam, pp 555–573

Walther GR, Beißner S, Burga CA (2005) Trends in the upward shift of alpine plants. J Veg Sci 16:541–548

Walther GR (2001) Laurophyllisation—a sign for a changing climate? In: Burga CA, Kratochwil A (eds) Biomonitoring: general and applied aspects on regional and global scales. Springer, Dordrecht, pp 207–223

Walther GR, Post E, Convey P, Menzel A, Parmesan C et al (2002) Ecological responses to recent climate change. Nature 416:389–395

Walther GR, Gritti ES, Berger S, Hickler T, Tang Z, Sykes MT (2007) Palms tracking climate change. Glob Ecol Biogeogr 16:801–809

Wang Y, Pederson N, Ellison AM, Buckley HL, Case BS, Liang E, Camarero JJ (2016) Increased stem density and competition may diminish the positive effects of warming at alpine treeline. Ecology 97:1668–1679

Wang Y, Wu N, Kunze C, Long R, Perlik M (2019) Drivers of change to mountain sustainability in the Hindu Kush Himalaya. In: Wester P, Mishra A, Mukherji A, Shrestha AB (eds) The Hindu Kush Himalaya assessment. Springer, Cham, pp 17–56

Wang X, Wu C, Wang H, Gonsamo A, Liu Z (2017) No evidence of widespread decline of snow cover on the Tibetan Plateau over 2000–2015. Sci Rep 7:1–10

Wang SY, Yoon JH, Gillies RR, Cho C (2013) What caused the winter drought in western Nepal during recent years? J Clim 26:8241–8256

Waqas A, Athar H (2019) Recent decadal variability of daily observed temperatures in Hindukush, Karakoram and Himalaya region in northern Pakistan. Clim Dyn 52:6931–6951

Wasowicz P (2016) Non-native species in the vascular flora of highlands and mountains of Iceland. PeerJ 4: e1559

Wassie A, Sterck FJ, Bongers F (2010) Species and structural diversity of church forests in a fragmented Ethiopian Highland landscape. J Veg Sci 21:938–948

Weaver T, Gustafson D, Lichthardt J (2001) Exotic plants in early and late seral vegetation of fifteen northern Rocky Mountain environments (HTs). W North Am Nat 61:417–427

Weber RO, Talkner P, Auer I, Böhm R, Gajic-Capka M et al (1997) 20th century changes of temperature in the mountain regions of Central Europe. Clim Change 36:327–344

Wei J, Liu S, Guo W, Yao X, Xu J, Bao W, Jiang Z (2014) Surface-area changes of glaciers in the Tibetan Plateau interior area since the 1970s using recent Landsat images and historical maps. Ann Glaciol 55:213–222

Wei JF, Liu SY, Xu JL, Guo WQ, Bao WJ, Shangguan DH, Jiang ZL (2015) Mass loss from glaciers in the Chinese Altai Mountains between 1959 and 2008 revealed based on historical maps, SRTM, and ASTER images. J Mt Sci 12:330–343

Weijers S, Myers-Smith IH, Loeffler J (2018) A warmer and greener cold world: summer warming increases shrub growth in the alpine and high Arctic tundra. Erdkunde 72:63–85

Weisberg PJ, Shandra O, Becker ME (2013) Landscape influences on recent timberline shifts in the Carpathian Mountains: abiotic influences modulate effects of land-use change. Arct Antarct Alp Res 45:404–414

Werners S, Szalai S, Kőpataki E, Csaba Kondor A, Musco E et al (2014) Future imperfect: climate change and adaptation in the Carpathians. GRIDArendal, Arendal

Wesche K, Miehe G, Kaeppeli M (2000) The significance of fire for afroalpine ericaceous vegetation. Mt Res Dev 20:340–347

Wesche K (2002) The high-altitude environment of Mt. Elgon (Uganda, Kenya): climate, vegetation, and the impact of fire. Ecotropical Monographs 2. Society of Tropical Ecology, Bonn

Wesche K, Assefa Y, Von Wehrden H (2008a) Temperate grassland region: equatorial Africa (high altitude). In: Peart B (ed) Compendium of regional templates on the status of temperate grasslands conservation and protection. IUCN World Commission on Protected Areas, Vancouver, pp 41–59

Wesche K, Cierjacks A, Assefa Y, Wagner S, Fetene M, Hensen I (2008b) Recruitment of trees at tropical alpine treelines: *Erica* in Africa versus *Polylepis* in South America. Plant Ecology & Diversity 1:35-46

Wielgolaski FE, Inouye DW (2013) Phenology at high latitudes. In: Schwartz MD (ed) Phenology: an integrative environmental science. Springer, Dordrecht, pp 225–247

Wielgolaski FE, Hofgaard A, Holtmeier FK (2017) Sensitivity to environmental change of the treeline ecotone and its associated biodiversity in European mountains. Climate Res 73:151–166

Wieser G, Holtmeier FK, Smith WK (2014) Treelines in a changing global environment. In: Tausz M, Grulke N (eds) Trees in a changing environment. Springer, Dordrecht, pp 221–263

Wieser G, Oberhuber W, Gruber A (2019) Effects of climate change at treeline: lessons from space-for-time studies, manipulative experiments, and long-term observational records in the Central Austrian Alps. Forests 10:508

Wildeman G, Brock JH (2000) Grazing in the Southwest: history of land use and grazing since 1540. In: Jemison R, Raish C (eds) Livestock management in the American Southwest: ecology, society and economics. Elsevier Science, Amsterdam, pp 1–25

Willard BE, Cooper DJ, Forbes BC (2007) Natural regeneration of alpine tundra vegetation after human trampling: a 42-year data set from Rocky Mountain National Park, Colorado, USA. Arct Antarct Alp Res 39:177–183

Williams JW, Jackson ST (2007) Novel climates, no-analog communities, and ecological surprises. Front Ecol Environ 5:475–482

Wilmking M, Juday GP, Barber VA, Zald HS (2004) Recent climate warming forces contrasting growth responses of white spruce at treeline in Alaska through temperature thresholds. Glob Change Biol 10:1724–1736

Wilson SJ (2016) Communal management as a strategy for restoring cloud forest landscapes in Andean Ecuador. World Development Perspectives 3:47–49

Winkler DE (2020) Contemporary human impacts on alpine ecosystems: the direct and indirect effects of human-induced climate change and land use. In: Goldstein MI, DellaSala DA (eds) Encyclopedia of the world's biomes, vol 1. Elsevier, Amsterdam, pp 574–580

Winkler DE, Butz RJ, Germino MJ, Reinhardt K, Kueppers LM (2018) Snowmelt timing regulates community composition, phenology, and physiological performance of alpine plants. Front Plant Sci 9:1140

Winkler DE, Lubetkin KC, Carrell AA, Jabis MD, Yang Y, Kueppers LM (2019) Responses of alpine plant communities to climate warming. In: Mohan JE (ed) Ecosystem consequences of soil warming. Academic Press, London, pp 297–346

Winkler M, Lamprecht A, Steinbauer K, Hülber K, Theurillat JP et al (2016) The rich sides of mountain summits—a pan-European view on aspect preferences of alpine plants. J Biogeogr 43:2261–2273

Winsvold SH, Andreassen LM, Kienholz C (2014) Glacier area and length changes in Norway from repeat inventories. Cryosphere 8:1885–1903

Wipf S, Stöckli V, Herz K, Rixen C (2013) The oldest monitoring site of the Alps revisited: accelerated increase in plant species richness on Piz Linard summit since 1835. Plant Ecolog Divers 6:447–455

Wolken JM, Mann DH, Grant TA III, Lloyd AH, Rupp TS, Hollingsworth TN (2016) Climate-growth relationships along a black spruce toposequence in interior Alaska. Arct Antarct Alp Res 48:637–652

Wondie M, Schneider W, Melesse AM, Teketay D (2011) Spatial and temporal land cover changes in the Simen Mountains National Park, a World Heritage Site in northwestern Ethiopia. Remote Sens 3:752–766

Workie TG, Debella HJ (2018) Climate change and its effects on vegetation phenology across ecoregions of Ethiopia. Glob Ecol Conserv 13:e00366

Wu N, Rawat GS, Sharma E (2013) High-altitude ecosystem interfaces in the Hindu Kush Himalayan region. In: Wu N, Rawat GS, Joshi S, Ismail M, Sharma E (eds) High-altitude rangelands and their interfaces in the Hindu Kush Himalayas. ICIMOD, Kathmandu, pp 3–14

Wunderle S, Gross T, Hüsler F (2016) Snow extent variability in Lesotho derived from MODIS data (2000–2014). Remote Sens 8:448

Wyckoff W, Dilsaver LM (eds) (1995) The mountainous West: explorations in historical geography. University of Nebraska Press, Lincoln

Xie J, Kneubühler M, Garonna I, Notarnicola C, De Gregorio L et al (2017) Altitude-dependent influence of snow cover on alpine land surface phenology. J Geophys Res Biogeosci 122:1107–1122

Xu G, Zhang H, Chen B, Zhang H, Innes JL et al (2014) Changes in vegetation growth dynamics and relations with climate over China's landmass from 1982 to 2011. Remote Sens 6:3263–3283

Xu J, Grumbine RE, Shrestha A, Eriksson M, Yang X, Wang YUN, Wilkes A (2009) The melting Himalayas: cascading effects of climate change on water, biodiversity, and livelihoods. Conserv Biol 23:520–530

Xu J, Badola R, Chettri N, Chaudhary RP, Zomer R et al (2019) Sustaining biodiversity and ecosystem services in the Hindu Kush Himalaya. In: Wester P, Mishra A, Mukherji A, Shrestha AB (eds) The Hindu Kush Himalaya assessment. Springer, Cham, pp 127–165

Yadava AK, Sharma YK, Dubey B, Singh J, Singh V et al (2017) Altitudinal treeline dynamics of Himalayan pine in western Himalaya, India. Quatern Int 444: 44–52

Yan L, Liu X (2014) Has climatic warming over the Tibetan Plateau paused or continued in recent years? J Earth Ocean Atmos Sci 1:13–28

Yang B, He M, Shishov V, Tychkov I, Vaganov E et al (2017) New perspective on spring vegetation phenology and global climate change based on Tibetan Plateau tree-ring data. Proc Natl Acad Sci 114:6966–6971

Yang J, Tan C, Zhang T (2013) Spatial and temporal variations in air temperature and precipitation in the Chinese Himalayas during the 1971–2007. Int J Climatol 33:2622–2632

Yang X, Zhang T, Qin D, Kang S, Qin X (2011) Characteristics and changes in air temperature and glacier's response on the north slope of Mt. Qomolangma (Mt. Everest). Arct Antarct Alp Res 43:147–160

Yao T, Pu J, Lu A, Wang Y, Yu W (2007) Recent glacial retreat and its impact on hydrological processes on the Tibetan Plateau, China, and surrounding regions. Arct Antarct Alp Res 39:642–650

Yarnall K, Price M (2010) Migration, development and a new rurality in the Valle Alto, Bolivia. J Lat Am Geogr 9:107–124

Ye B, Yang D, Jiao K, Han T, Jin Z, Yang H, Li Z (2005) The Urumqi river source glacier No. 1, Tianshan, China: changes over the past 45 years. Geophys Res Lett 32:21

Ye H, Cohen J (2013) A shorter snowfall season associated with higher air temperatures over northern Eurasia. Environ Res Lett 8:014052

Ye Q, Zong J, Tian L, Cogley JG, Song C, Guo W (2017) Glacier changes on the Tibetan Plateau derived from Landsat imagery: Mid-1970s—2000–13. J Glaciol 63:273–287

Yin G, Hu Z, Chen X, Tiyip T (2016) Vegetation dynamics and its response to climate change in Central Asia. J Arid Land 8:375–388

You J, Qin X, Ranjitkar S, Lougheed SC, Wang M et al (2018) Response to climate change of montane herbaceous plants in the genus *Rhodiola* predicted by ecological niche modelling. Sci Rep 8:1–12

You Q, Min J, Kang S (2016) Rapid warming in the Tibetan Plateau from observations and CMIP5 models in recent decades. Int J Climatol 36:2660–2670

You QL, Ren GY, Zhang YQ, Ren YY, Sun XB et al (2017) An overview of studies of observed climate change in the Hindu Kush Himalayan (HKH) region. Adv Clim Chang Res 8:141–147

Young KR (2009) Andean land use and biodiversity: humanized landscapes in a time of change. Ann Mo Bot Gard 96:492–507

Young KR, León B, Jørgensen PM, Ulloa Ulloa C (2007) Tropical and subtropical landscapes of the Andes. In: Veblen TT, Young KR, Orme AR (eds) The physical geography of South America. Oxford University Press, Oxford, pp 200–216

Young KR, Ponette-González AG, Polk MH, Lipton JK (2017) Snowlines and treelines in the tropical Andes. Ann Am Assoc Geogr 107:429–440

Yu L, Liu T, Zhang S (2017) Temporal and spatial changes in snow cover and the corresponding radiative forcing analysis in Siberia from the 1970s to the 2010s. Advances in Meteorology 2017: ID 9517427
Yu Q, Jia DR, Tian B, Yang YP, Duan YW (2016) Changes of flowering phenology and flower size in rosaceous plants from a biodiversity hotspot in the past century. Sci Rep 6:1–4
Yucel I, Güventürk A, Sen OL (2015) Climate change impacts on snowmelt runoff for mountainous transboundary basins in eastern Turkey. Int J Climatol 35:215–228
Zaehringer JG, Wambugu G, Kiteme B, Eckert S (2018) How do large-scale agricultural investments affect land use and the environment on the western slopes of Mount Kenya? Empirical evidence based on small-scale farmers' perceptions and remote sensing. J Environ Manage 213:79–89
Zekollari H, Fürst JJ, Huybrechts P (2014) Modelling the evolution of Vadret da Morteratsch, Switzerland, since the Little Ice Age and into the future. J Glaciol 60:1155–1168
Zekollari H, Huss M, Farinotti D (2019) Modelling the future evolution of glaciers in the European Alps under the EURO-CORDEX RCM ensemble. Cryosphere 13:1125–1146
Zeleke G, Hurni H (2001) Implications of land use and land cover dynamics for mountain resource degradation in the northwestern Ethiopian highlands. Mt Res Dev 21:184–191
Zemmrich A, Hilbig W, Oyuunchimeg D (2010) Plant communities along an elevation gradient under special consideration of grazing in western Mongolia. Phytocoenologia 40:91–115
Zemp M, Frey H, Gärtner-Roer I, Nussbaumer SU, Hoelzle M et al (2015) Historically unprecedented global glacier decline in the early 21st century. J Glaciol 61:745–762
Zemp M, Huss M, Thibert E, Eckert N, McNabb R et al (2019) Global glacier mass changes and their contributions to sea-level rise from 1961 to 2016. Nature 568:382–386
Zeng H, Jia G, Epstein H (2011) Recent changes in phenology over the northern high latitudes detected from multi-satellite data. Environ Res Lett 6:045508
Zewdie W, Csaplovics E, Inostroza L (2017) Monitoring ecosystem dynamics in northwestern Ethiopia using NDVI and climate variables to assess long term trends in dryland vegetation variability. Appl Geogr 79:167–178
Zhan YJ, Ren GY, Shrestha AB, Rajbhandari R, Ren YY et al (2017) Changes in extreme precipitation events over the Hindu Kush Himalayan region during 1961–2012. Adv Clim Chang Res 8:166–175
Zhang G, Yao T, Xie H, Wang W, Yang W (2015) An inventory of glacial lakes in the Third Pole region and their changes in response to global warming. Global Planet Change 131:148–157
Zhang J, Chen H, Zhang Q (2019) Extreme drought in the recent two decades in northern China resulting from Eurasian warming. Clim Dyn 52:2885–2902
Zhang X, Flato G, Kirchmeier-Young M, Vincent L, Wan H et al. (2019) Changes in temperature and precipitation across Canada. In: Bush E, Lemmen DS (eds) Canada's changing climate report. Government of Canada, Ottawa, pp 112–193
Zhang Y, Enomoto H, Ohata T, Kitabata H, Kadota T, Hirabayashi Y (2017) Glacier mass balance and its potential impacts in the Altai Mountains over the period 1990–2011. J Hydrol 553:662–677
Zhang Y, Liu L, Wang Z, Bai W, Ding M et al. (2019b) Spatial and temporal characteristics of land use and cover changes in the Tibetan Plateau. Chin Sci Bull 64:2865–2875
Zhao L, Wu Q, Marchenko SS, Sharkhuu N (2010) Thermal state of permafrost and active layer in Central Asia during the International Polar Year. Permafrost Periglac Process 21:198–207
Zimmer A, Meneses RI, Rabatel A, Soruco A, Dangles O, Anthelme F (2018) Time lag between glacial retreat and upward migration alters tropical alpine communities. Perspect Plant Ecol Evol Syst 30:89–102
Zimmermann P, Tasser E, Leitinger G, Tappeiner U (2010) Effects of land-use and land-cover pattern on landscape-scale biodiversity in the European Alps. Agr Ecosyst Environ 139:13–22
Zomer RJ, Trabucco A, Metzger MJ, Wang M, Oli KP, Xu J (2014) Projected climate change impacts on spatial distribution of bioclimatic zones and ecoregions within the Kailash Sacred Landscape of China, India, Nepal. Clim Change 125:445–460
Zong S, Xu J, Dege E, Wu Z, He H (2016) Effective seed distribution pattern of an upward shift species in alpine tundra of Changbai Mountains. Chin Geogra Sci 26:48–58
Zorio SD, Williams CF, Aho KA (2016) Sixty-five years of change in montane plant communities in western Colorado, USA. Arct Antarct Alp Res 48:703–722
Zurick D, Pacheco J (2006) Illustrated atlas of the Himalaya. University Press of Kentucky, Lexington

Part I
Climate Change and Response Processes of Mountain Environments

Introduction

This part of the book consists of a collection of 18 case studies from mountains across the world which address climate change itself (past, current and future changes of temperature and precipitation) and its impacts on the cryosphere, hydrosphere, biosphere, and human-environment systems. The case studies illustrate climate change effects on various components of mountain systems and ongoing processes of change, including glaciers, water availability, snow fields, sediment fluxes, vegetation and treeline dynamics, plant species composition, and socio-economic changes, using a variety of methodical approaches including field-based investigations, remote sensing, ground-based observations and modelling techniques. The case studies are arranged in a way that cascading effects of temperature and precipitation changes on the cryosphere, hydrosphere, biosphere, and on socio-economic systems can be retraced.

The first two case studies analyze temperature and precipitation changes in the Himalayan region. An analysis of temperature indices in the western Indian Himalaya (Himachal Pradesh) and its lowland plains (Punjab) over the past six decades (1951–2013) suggests that winters have particularly warmed in the region (M. R. Sharma et al.). In the neighbouring Himalayan state of Uttarakhand declining rainfall over the last century was assessed (S. Mal et al.). Long-term trends of annual and monsoonal precipitation were found to be negative, with a more evident and pronounced decline after the 1960s. Monsoonal precipitation is correlated to the large-scale Southern Oscillation Index (SOI).

Climate change effects on the cryosphere and hydrosphere of the Himalaya are illustrated in the next five case studies. Strongly negative changes in glacier length, surface area and mass balance over the last century across the western Himalayan region have become obvious (S. Kaushik et al.), in line with significant shrinkage of the ice facies of the Satopanth Glacier in the Garhwal Himalaya over a 11-year period (B. Yousuf et al.). Hydrological implications of climatic and cryospheric changes are potentially dramatic. The analysis of historical and future moisture and thermal zones for different seasons under different climate scenarios in the Nepal Himalaya (R. Talchabhadel and R. Karki) reveals the shift of agro-climatological zones, presenting new options and new challenges for agricultural production. Runoff simulation using a hydrological model (SWAT) in the framework of analyzing climate change impacts on water availability in the Thuli Bheri River Basin of Nepal (A. Aryal et al.) indicates considerable uncertainty related to differences in the selected climate models. In the context of climate change-induced modifications of hydrological systems, another case study (J. Grießinger et al.) highlights the significance of local water availability, access to water resources and its sustainable management for securing local livelihoods from irrigation-based land use systems in the Mustang Himalaya (Nepal).

A Holocene time scale is applied in the next three case studies. Palaeoperspectives have a long research tradition in the European Alps. Here, a broad overview is presented of climate and environmental changes, glacier dynamics, natural hazards, and land use changes in the Stubai Valley (Austria) from the early Holocene to the Anthropocene (A. Fischer et al.). Early Holocene glacier retreat had exposed the land surface for the development of human settlements and land uses, which are still being threatened due to time-lag effects on exposed slopes (e.g. slope failures). In Norway, the majority of glaciers are facing overall mass losses, resulting in enhanced paraglacial activity, increased natural risk potential, and changes of glacier foreland ecosystems (Ph. Marr et al.). Widespread glacier retreat adversely affects the high-revenue glacier tourism and hydropower production industries in Norway, however, both natural and socio-economic systems appear to be comparatively rather resilient. The sensitivity of medium- to long-term hillslope sediment fluxes to climate change is in the focus of a case study from the Appalachian Mountains of eastern Canada (D. Germain and L. Stabile-Caillé). The analyzed talus slopes have remained active throughout the Holocene, being characterized by high sediment fluxes and a nonlinear response to climate change, affected by antecedent conditions, geological controls, and climate variability at different spatiotemporal scales.

The subsequent case studies cover a range of biogeographical topics and focus on climate change effects on habitats, plant species, and vegetation dynamics. In the Glacier National Park, Montana (USA), retreating snowfields and glaciers have caused significant changes in the distribution of plant growth forms, plant functional traits and species, with xeromorphy being an important response to water limitation (M. E. Apple et al.). In the treeline ecotone of the Rolwaling Himal (Nepal), analyzed vegetation-environment relationships, plant communities and population densities of tree species indicate that the dense *Rhododendron* krummholz belt largely prevents the upward migration of other tree species and thus constrains the future response of Himalayan krummholz treelines to climate warming (N. Schwab et al.). Modelling the ecological niche of treeline tree species in remote mountain regions faces several challenges as highlighted in the case study on the Himalayan birch which serves as a baseline for projecting the distribution of *Betula utilis* under future climatic conditions (M. Bobrowski). Two case studies from the Altay-Sayan Mountain Region (V. I. Kharuk et al.) analyze tree growth-climate relationships of conifer trees. Results suggest that suitable habitats of the Siberian pine (*Pinus sibirica*) will shrink at middle and low elevations and that the pine will be substituted by drought-resistant larch and softwood species. Drought stress is also a major driver for high recent mortality rates of the Siberian fir (Abies sibirica) which is retreating from its low and middle elevation ranges in the southern Siberian Mountains. An overview of climate change-induced vegetation dynamics is presented in a case study from the Lesser Caucasus (G. Fayvush and A. Aleksanyan). Fundamental changes in vegetation structure and in species composition of plant communities are to be expected, with far-reaching consequences for ecosystem functioning and the provision of ecosystem services.

The final two case studies address climate change effects on human-environment systems. In the Darma Valley (Uttarakhand, Indian Himalaya), the severity of climate change is compared with the perceptions of climate change among local pastoral communities, and the socio-economic dimensions of climate change impacts on local communities and the environment are examined (D. Rawat and U. Schickhoff). A substantial number of pastoralists perceives adverse effects of climatic change and associated impacts on the environment and on their livelihoods. The

Drakensberg Mountains in southern Africa experienced significant warming in recent decades, coupled with increasing frequency of extreme and severe droughts (G. Mukwada). Climate adaptation policies need to be redesigned to ensure secure and sustainable livelihoods of affected local communities and to avert socio-political upheavals in the future. The case studies presented in this part of the present volume illustrate that in line with the UN Sustainable Development Goals urgent action is imperative to combat climate change and its impacts and that it is more important than ever to accelerate the transitions needed to achieve the Paris Agreement.

2 Markers of Climate Change: Analysing Extreme Temperature Indices Over the Himalayan Mountains and Adjoining Punjab Plains

Manu Raj Sharma, Vishwa B. S. Chandel, and Karanjot Kaur Brar

Abstract

Temperature as a central element of weather and climate exerts a strong influence on environment at every spatial and temporal scale. Long-term patterns and trends indicate the direction in which temperature regime has changed in the past and are likely to shape up in the future. Of equal significance is the likelihood of extreme temperature events and their impacts. Climate researchers in the past 20 years have revealed occurrence of significantly longer heat waves in many regions. Large geographical extent and climatic variations produces contrasting patterns of change in temperature extremes in India. A highly diverse landscape of Himalayas has experienced warming in recent decades with significant mean annual and winter warming. Consequently, temperature extremes in such regions are likely to impact the regional development and economy. This paper explores long-term trends, variations, direction and degree of change in temperature extremes over Himachal Pradesh and Punjab. The analysis reveals that warming of the study area has taken place especially during the winter season. The fact which attracts attention is the maximum rate of change over extreme climatic zones especially in the northern Himachal Pradesh and southwestern Punjab Plains. These observed changes may have far-fetching ramification on general climatic regime, environment and human-economic activities.

Keywords

Temperature extremes · Regional development · Winter warming

2.1 Introduction

Climate change on earth is not a rarity (Knoll 2003) but a common phenomenon. The direction, magnitude and patterns of climate have been changing at different spatial and temporal scales ever since the earth originated. However, these deviations were never linear. Some changes were gradual while others were rapid wherein climate shifted at faster rates between the warm and cold phases. Temperature as the most crucial element of weather and climate has a strong influence on earth's environment. The long-term trends and variability of temperature express a general direction of climate change and indicate

M. R. Sharma
University Department of Geography, L. N. Mithila University, Darbhanga, India

V. B. S. Chandel (✉) · K. K. Brar
Center of Advanced Study in Geography (CAS), Department of Geography, Panjab University, Chandigarh, India

U. Schickhoff et al. (eds.), *Mountain Landscapes in Transition*,
Sustainable Development Goals Series, https://doi.org/10.1007/978-3-030-70238-0_2

its expected behaviour in future. One of the facts that have become significant today is associated with the nature and likelihood of extreme temperature events. *"Widespread changes in extreme temperature have been observed over the last 50 years... cold days, cold nights and frost have become less frequent, while hot days, hot nights and heat waves have become more frequent"* (Easterling et al. 2000; IPCC 2007). It is evident that significantly longer heat-waves are increasing (Kawahara and Yamazaki 1999; Zhai and Pan 2003; Ryoo et al. 2004; Batima 2005; Cruz et al. 2006 and Morak et al. 2013; Rohini et al. 2015; Sippel et al. 2016; Mitchell et al. 2016, and King and Karoly 2017). The Southeast Asia is the only region apart from the Amazon Basin where heat extremes are likely to increase strongly (Sillmann et al. 2013). India is also a hot-spot for frequent hot days and multiple-day heat-waves (De and Mukhopadhyay 1998 and Lal 2003) and it is expected that human activities might enhance the adverse impacts of heat waves (Van Oldenborgh et al. 2018).

From the observed records, it is clear that there are wide regional variations in the direction and degree of climatic change (Brohan et al. 2006); it is predicted that some regions may expect extreme aridity (Dai 2013) while others will become relatively wet and hot (Fowler and Wilby 2010). Moreover, such temperature extremes may induce severe hydrological changes (Betts et al. 2018) and enhance the vulnerability of small islands, deltas and coastal cities (Nicholls et al. 2018). Such extremes may also heighten the environmental risk and biodiversity loss (Ford and Perace 2010; Ford and Goldhar 2012; Smith et al. 2018) that will have an exceedingly negative impact on global economy (Pretis et al. 2018). IPCC in its 'Synthesis Report on Climate Change 2014' with a very high confidence level predicted increase in the severity of climate-related extremes, such as heat waves, drought, floods, cyclones and wildfires (IPCC 2014). The nature and impact of temperature extremes in India are likely to vary spatially for it is a geographically and climatically heterogeneous region. Regions like Punjab and Himachal Pradesh that has very sharp spatial temperature gradients and rely on agriculture, horticulture, tourism and hydropower for economic development are expected to suffer from temperature extremes in a big way. Therefore, it is important to understand the behaviour of temperature extremes in such vulnerable regions. With this idea in mind, the chapter attempted to highlight the manifestation, patterns and trends of extreme temperature events in the study region.

2.2 Study Region

The study region is a part of Himalayan system and adjacent Punjab plains in the northwest India. The region can be divided into four broad physiographic zones: Punjab Plains, the Siwalik Hills, the Inner Himalayas (Dhauladhar and Pir Panjal ranges) and the Great Himalayas. Forming roughly a quadrilateral shape between 29°30′N to 33°12′N latitudes and 73°55′E to 79°04′E longitudes, this region has altitude ranging from 200 to 7000 m above mean sea level (Fig. 2.1). Situated between hot desert in the southwest and cold desert in the northeast, the study region spreads over the states of Punjab and Himachal Pradesh. These two states form a physically contiguous geographic unit with strong physiographic, climatic and hydrological links. The study region is climatically pivotal; the Himalayas in the north and east restrict frigid katabatic winds from the Tibetan plateau and the Central Asia. The temperature regime in the southwestern parts of the study area over Punjab Plains and the Thar Desert is critical for the behaviour of southwest monsoons that provide over 60% of total annual rainfall.

The region has conspicuous topographic contrasts that produce several microclimates. The Punjab plains has dry semi-arid, dry sub-humid, moist sub-humid, humid sub-tropical climate while mountains of Himachal Pradesh possess sub-humid tropical climate in south and south-eastern parts to temperate wet and moist climate in the central parts (Fig. 2.1). The northern and eastern mountainous region possess cold semi-arid and cold arid climate. The physiographic and bioclimatic disposition exposes the area to a high

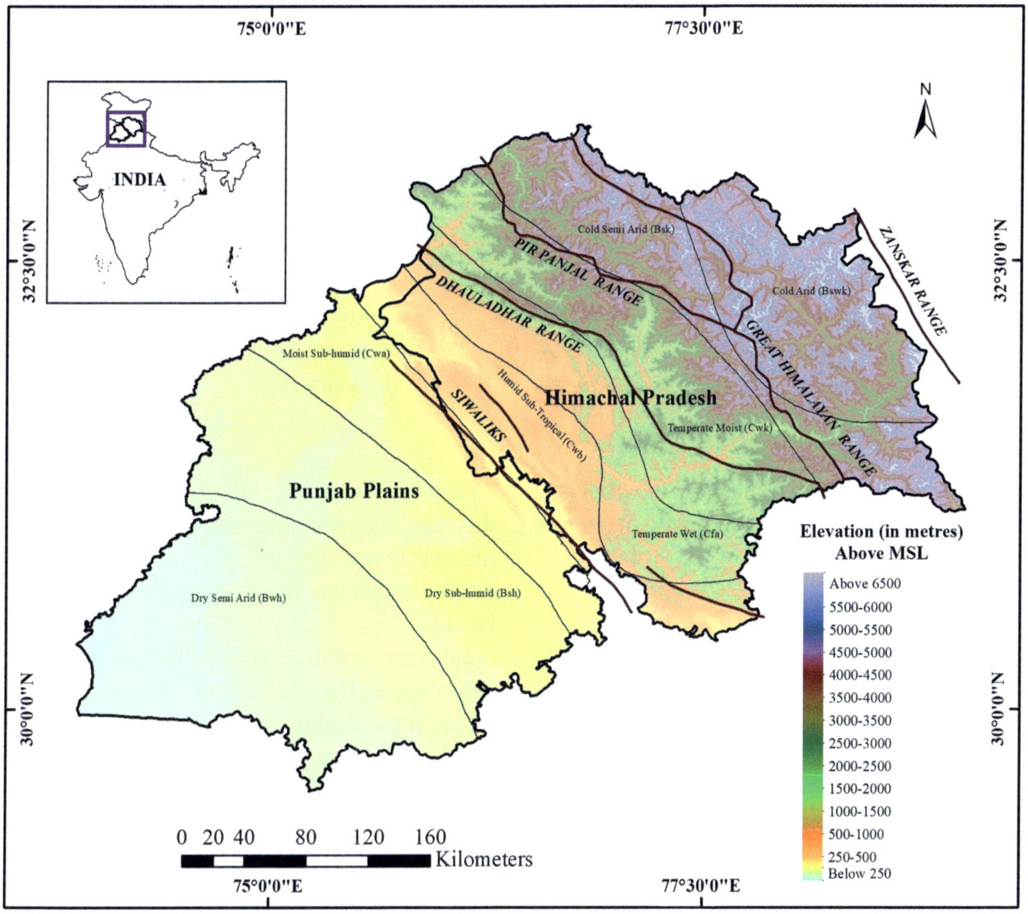

Fig. 2.1 Study Region- Himachal Pradesh and Punjab

degree of climatic variability and extremes. The areas like Dharamshala and Madhopur receive very high annual rainfall (3000 mm) while northeastern Himachal Pradesh and southwestern Punjab receive scanty rainfall. The precipitation is highly variable over space and time; maximum rainfall takes place during July–September with some rainfall induced by western disturbances in January-March. The high mountainous regions receive heavy snowfall during December-February. Similarly, temperature regime over study region is highly diverse. A high annual temperature range of over 60 °C (maximum temperature of 48 °C in Amritsar, Punjab and minimum temperature of −15 °C in Keylong in Lahaul-Spiti district, Himachal Pradesh) makes this region climatically diverse.

2.3 Database and Methodology

2.3.1 Database

This study is based on daily gridded temperature data at 1° × 1° spatial resolution for the period 1951–2013 acquired from India Meteorological Department (IMD). It examines six extreme temperature indices, viz. hottest day, warmest night, coldest day, coldest night, summer days and tropical nights. These indicators were grouped into two categories of frequency indices and intensity indices (Table 2.1). The frequency indices measure absolute number of days wherein a particular extreme event occurs over a specified period of time, i.e. days per unit of

Table 2.1 Indicators for temperature indices

ID		Indicator name	Definitions	Unit
Intensity indices				
1	TXx	Hottest day	Monthly max. value of daily max. temperature	°C
2	TNx	Warmest night	Monthly max. value of daily min. temperature	°C
3	TXn	Coldest day	Monthly min. value of daily max. temperature	°C
4	TNn	Coldest night	Monthly min. value of daily min. temperature	°C
Frequency indices				
5	SU25	Summer Days	Annual count when TX (daily max. temperature) is >25 °C	days
6	TR20	Tropical Nights	Annual count when TN (daily min. temperature) is >20 °C	days

Source Compiled from WMO/CLIVAR List for Climate Extreme Indices (WMO 2009)

time. On the other hand, intensity indices calculate the rarity of events in terms of magnitude, i.e. absolute temperature.

2.3.2 Methodology

The daily gridded data is based on interpolated station wise data. A number of interpolation methods are available to create gridded data. There are many Objective Analysis (OA) techniques such as Barnes (1973), Cressman (1959), Gandin (1965) and Shepard (1968) to interpolate unevenly distributed data that have been applied by researchers on climatic gauge data for conversion into a regular grid data. A comparative analysis of these techniques shows that modified Shepard's method is one of the best OA techniques (New et al. 2000; Kiktev et al. 2003; Caesar et al. 2006). This method is used by IMD as it is easier to implement for preparation of gauge based gridded especially in the data sparse regions. The method is represented as:

$$F(x, y) = \sum_{i=1}^{n} wifi$$

where $\boldsymbol{n}$ is the number of points used to interpolate, $\boldsymbol{fi}$ are the prescribed function values at the points and $\boldsymbol{wi}$ are the weight functions assigned to each point.

Another advantage of using this method is that the datasets processed using Shephard's interpolation technique does not have the effect of data inhomogeneity and therefore can directly be used for analysis. However, temperature indices are sensitive to any change in location, exposure, equipment and observation practice. Therefore, a quality control procedure suggested by Haylock et al. (2006) to remove data errors and inconsistencies was applied on climate dataset using *QC module of 'RclimDex 1.1'* software. The methodology developed by Zhang et al. (2005) was followed to identify presence of any outliers in the dataset. It was checked for every mean value of daily temperature variable whether or not it falls within the range of ±*4 standard deviations.* The mean values outside this range of standard deviation were treated as outliers. An outlier is a value that falls outside a particular range defined as unrealistic. In temperature series, such range is plus or minus four times the standard deviation. The daily temperature value outside this threshold is considered as potentially erroneous. For this study, the direction and rate of change in temperature indices were analysed using least square linear fit. The least-square trends are easy to understand and estimate the uncertainty in the fitted trends that arises from sampling variability. Statistically significant trends were identified at 95% confidence level.

2.4 Results and Discussion

2.4.1 Intensity Indices

Four intensity indices, viz. TXx; TNx; TXn and TNn explain temperature behaviour in absolute terms, i.e. °C. The following significant observations have been recorded about spatial patterns, direction and rate of change.

2.4.1.1 Hottest Day (TXx)

TXx explains monthly maximum value of daily maximum temperature, i.e. the hottest day of each month. The spatial patterns of hottest day temperature (*Max Tmax)* during a year depict peak summer temperature conditions. The southwestern Punjab exhibits maximum TXx (≈45 °C); the values decrease in southwest-northeast direction and attain the lowest value (≈35 °C) in the cold semi-arid region of Himachal Pradesh. The entire Punjab plains has hottest day temperature (TXx) ≥ 44 °C while such values vary between 42 °C in low hilly regions of Punjab to just below 35 °C in northeast and eastern Himachal Pradesh. This clear-cut directional decrease is influenced by variations in altitude and attitude of slopes that controls the daily and seasonal receipt of insolation and radiational heating/cooling. A larger range of hottest day temperature in the mountainous region of northern and eastern Himachal Pradesh is due to micro-climates produced by the topographic variations, factor of continentality and resultant climatic regime.

The study area has recorded increase in the hottest day temperature by approximately 0.4 °C during peak summers since 1950s (Fig. 2.2). The majority of grids covering about 81.82% area has experienced rise in hottest day temperature with significant increase observed over 18.18% grids. The southwestern semi-arid region of Punjab plains has witnessed statistically significant increase of more than 0.30 °C per decade while northern Punjab and adjacent Himachal Pradesh shows an increase of 0.05–0.15 °C/decade. Such a change is less conspicuous in southeastern Punjab and central Himachal Pradesh. While most of the study area has experienced increase in hottest day temperature, the northeastern, eastern, southeastern and southern Himachal Pradesh shows a decline in hottest day temperature although this negative change in TXx is negligible (Fig. 2.3a and b).

2.4.1.2 Warmest Night (TNx)

TNx indicates warmest night temperature conditions during a month. The spatial patterns of TNx are analogues to TXx; Punjab plains have higher night temperature (≥ 28 °C) with warmest nights observed over the dry semi-arid and the dry sub-humid zones. A sharp temperature

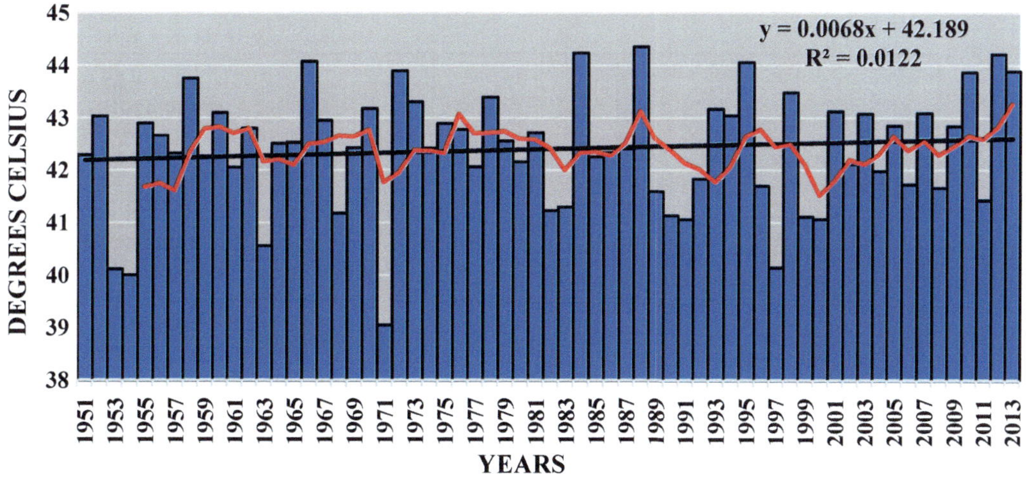

Fig. 2.2 Trends in hottest day (TXx) temperature (1951–2013)

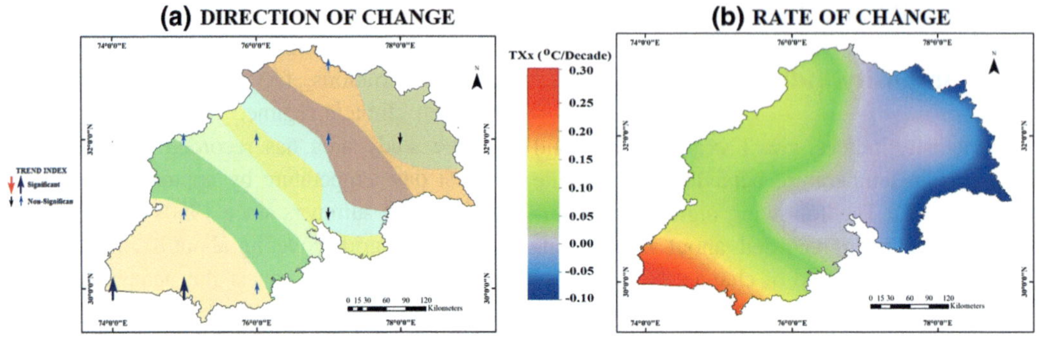

Fig. 2.3 Direction and rate of change in hottest days (TXx)

gradient exists over mountainous region of Himachal Pradesh; TNx values ranges between 25 and 26 °C in temperate wet and moist regions while cold semi-arid region have warmest night temperatures below 24 °C during peak summers with lowest values of 22 °C in extreme eastern parts.

There is an overall decrease of 0.5 °C in warmest night temperatures since 1950s (Fig. 2.4). The region has witnessed a dip in warmest night temperature over 90.91% grids but statistically significant decline is recorded for only 9.09% grids. The most noticeable decrease can be seen over Himachal Pradesh and northern and eastern Punjab. Although, the maximum decline of −0.19 to −0.15 °C per decade is recorded for south and southeastern parts of Himachal Pradesh; this decrease is not statistically significant (Fig. 2.5a and b). Similarly, the northern and western Himachal Pradesh also shows non-significant decrease. The situation is somewhat similar over Punjab where a statistically significant decrease is observed only in the northwestern area. The only area with a miniscule increase in warmest night temperature is the southern and southwestern Punjab.

2.4.1.3 Coldest Day (TXn)

The coldest day during a year defines the lower limit for day time temperatures during peak winters. The spatial pattern based on minimum values of daily maximum temperatures during a year explains the coldest day of the year. The average temperature for coldest day in study area is 13.65 °C with a spatial range of 8–16 °C. The southwestern Punjab experiences maximum

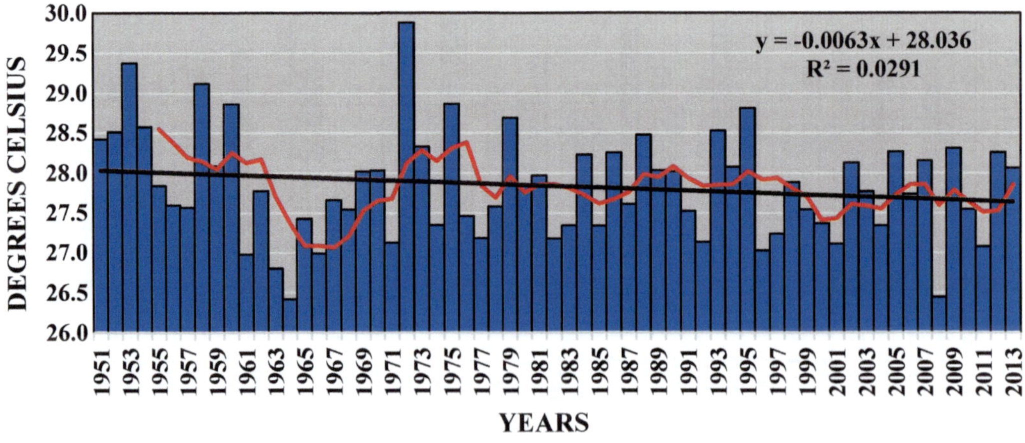

Fig. 2.4 Trends in warmest night (TNx) temperature (1951–2013)

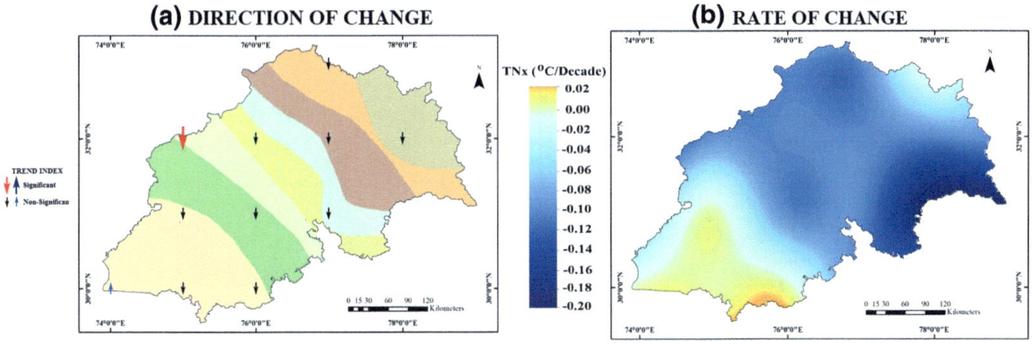

Fig. 2.5 Direction and rate of change in warmest nights (TNx)

temperature during the coldest day that gradually decreases towards the northeast. The temperature gradients during the coldest day are sharper in Himachal Pradesh; TXn values vary between 10 and 12 °C in the middle Himalayan temperate zone whereas such values in the cold semi-arid zone drops to as low as 8 °C over extreme northeast mountains. Overall, the coldest day temperature during peak winters have almost declined by 0.7 °C with noticeable spatial variations in the rate of decline (Fig. 2.6). About 90.91% grids have observed decline in temperature with statistically significant decrease over as many as 36.36% grids (Fig. 2.7a and b).

The most prominent statistically significant decrease is recorded for south and southwestern Punjab (−0.35 to −0.20 °C per decade). For rest of the Punjab plains, the decline is non-significant. Almost the entire Himachal Pradesh except for the northern areas has experienced decrease in coldest day temperatures but rate of change is very small. The only area where coldest day temperature has increased by 0.06 °C to 0.11 °C per decade includes northern high mountainous Himachal Pradesh over Chamba, Lahaul and northern Spiti. Such increase in peak winter day temperature may influence the snowfall pattern.

2.4.1.4 Coldest Night (TNn)

The minimum value of daily minimum temperature in a year represents the coldest winter night; it defines the lower limit for the temperature

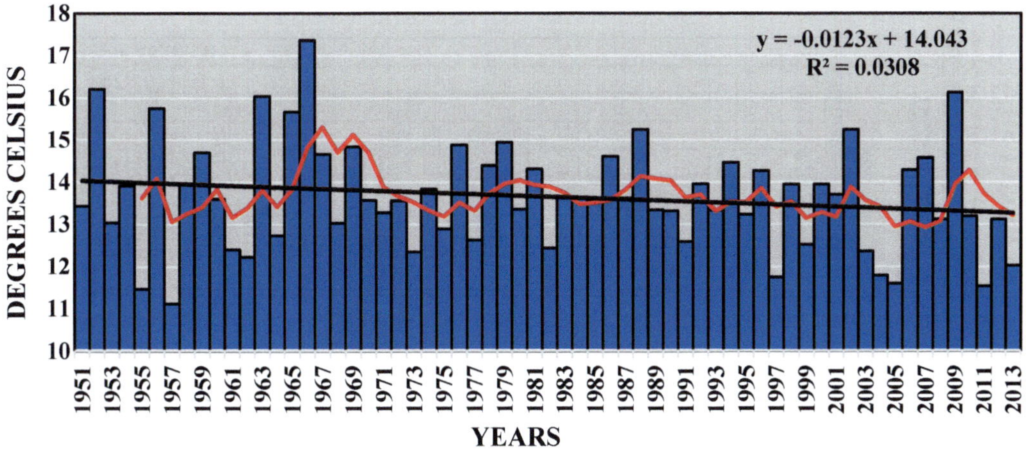

Fig. 2.6 Trends in coldest day (TXn) temperature (1951–2013)

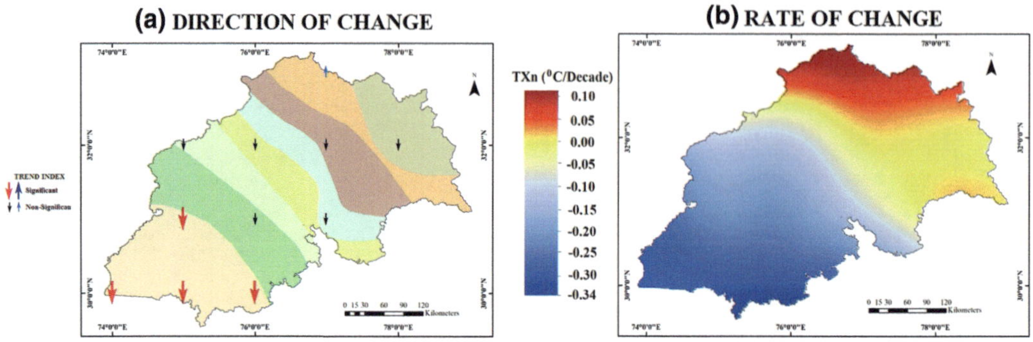

Fig. 2.7 Direction and rate of change in coldest day (TXn)

range. The average coldest night temperature in the region remains at 1.44 °C with spatial variation from 3 °C to just below −1.0 °C. The coldest night temperatures remain around 2.5 °C in northern and southeastern Punjab while minimum values are recorded for north and northeastern Himachal Pradesh where TNn goes well below the freezing point in temperate wet and cold semi-arid zones. There has been a noticeable increase in coldest night temperature by 0.5 °C over study area (Fig. 2.8); the entire grids show increase in temperature with statistically significant rise over 9.09% gridded. A significant increase from 0.16 to 0.20 °C per decade is noticeable over southwestern Punjab (Fig. 2.9a and b) whereas increase is rather small for central and southeastern Punjab. The moist sub-humid and humid sub-tropical zones of Punjab and entire Himachal Pradesh show an increase from 0.10 to 0.15 °C per decade. The uniform direction of change in coldest night temperature suggests warming of peak winters in the study area and the maximum warming is observed to be in the southwestern Punjab plains and northeastern region of high mountains in Himachal Pradesh.

2.4.2 Frequency Indices

Frequency indices analyse the occurrence and behaviours of extreme temperature events in terms of absolute number of days, i.e. number of days per year. The degree and direction of change in these frequency indices is discussed as below.

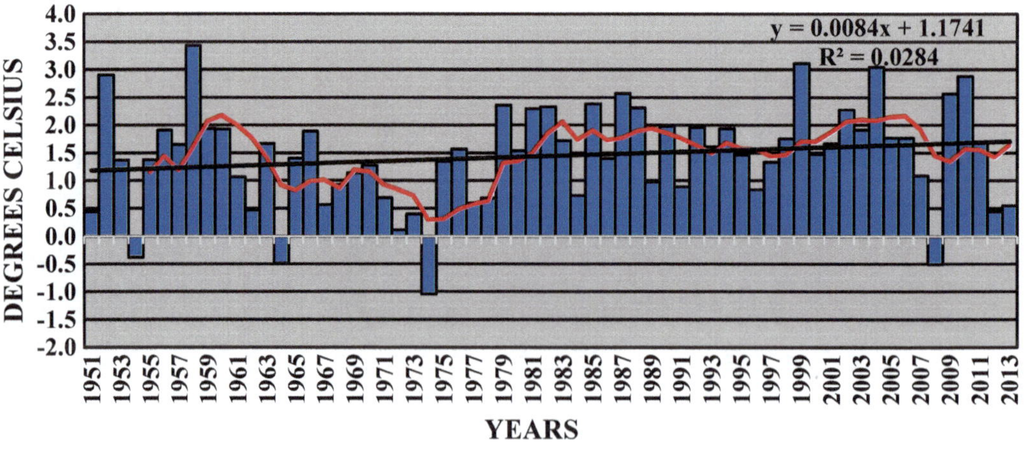

Fig. 2.8 Trends in coldest night (TNn) temperature (1951–2013)

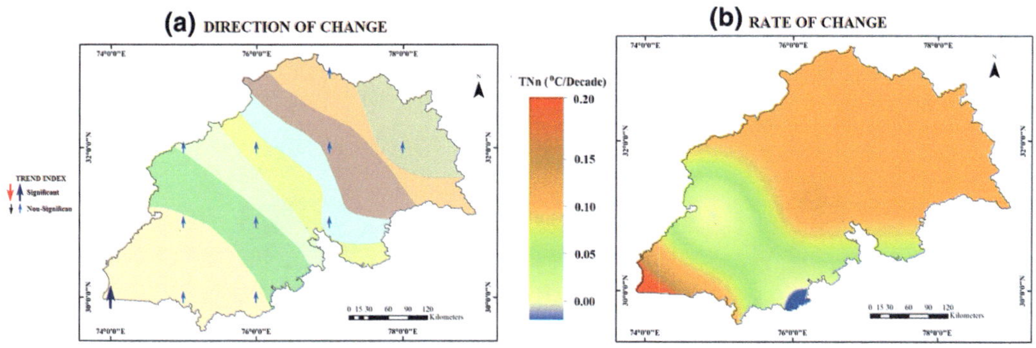

Fig. 2.9 Direction and rate of change in coldest night (TNn)

2.4.2.1 Summer Days (SU25)

SU25 indicates annual count of days when daily maximum temperature remains above 25 °C. This indicator represents the length of warm spell. The study region gets on an average 259 summer days/year with inter-annual variation between 235 and 288 day/year. The occurrence of SU25 is highest (280–290 days/year) in southwestern dry semi-arid Punjab. The frequency decreases to 250 days/year around the Siwalik Hills which further drops to 220–240 days/year over temperate wet and moist regions. For the cold semi-arid zone this duration is less than 200 days/year. The region as a whole shows a slight increase in summer days (Fig. 2.10) that indicates expansion of intra-annual warmer period.

A contrasting pattern of change can be observed over the plains and the hilly regions; most of the Punjab plains show negative/declining summer days while the trend is just the opposite in the Himalayan Mountains (Fig. 2.11a), however the change in statistically insignificant. The southern and central Punjab plains show a slight decrease in summer (Fig. 2.11b) whereas the hilly tracts around the Siwaliks and the high elevation zones of Himachal Pradesh has experienced an increasing trend with maximum increase over northern Himachal Pradesh. The temperate and cold-semi arid regions also exhibit rise in summer days but at much lower and statistically insignificant rates.

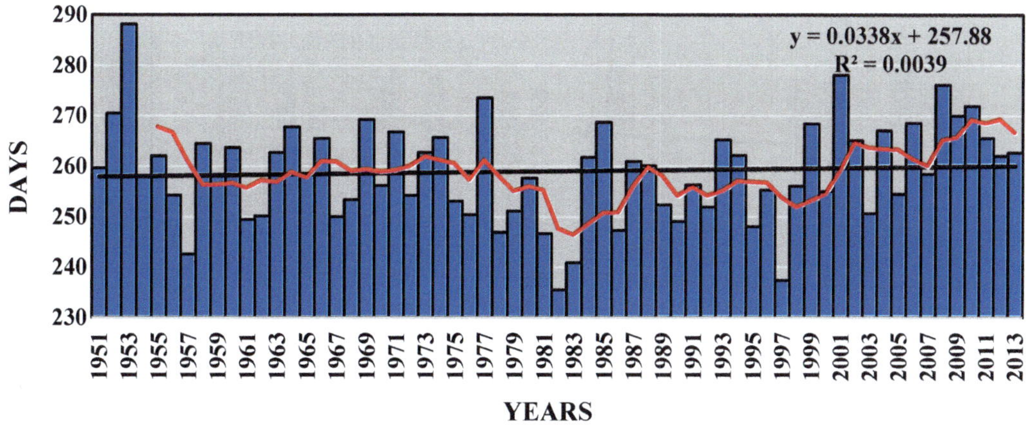

Fig. 2.10 Trends in summer days (1951–2013)

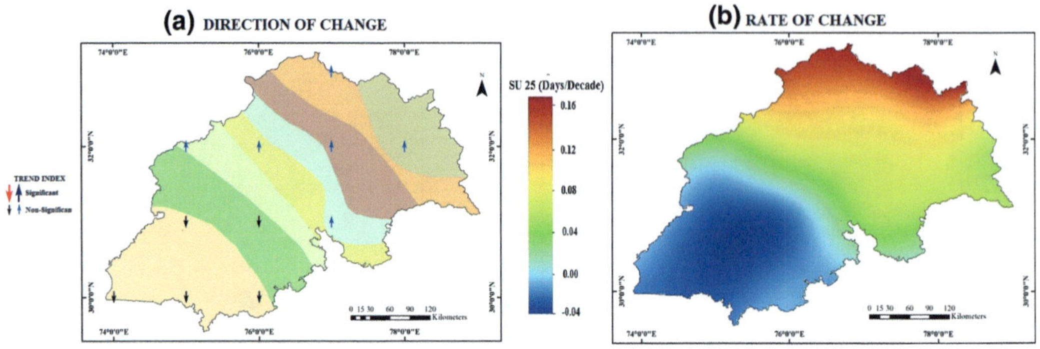

Fig. 2.11 Direction and rate of change in summer days (SU25)

2.4.2.2 Tropical Nights (TR20)

Tropical Nights (TR20) indicates total annual count of days when daily minimum temperature exceeds 20 °C. It indicates the length of warm nights; increase in TR20 values indicates lengthening of warmer nights and vice versa. The study area receives 137 days/year (4–5 months) with TR20 conditions. The count over Punjab plains is about 150–170 days/year which indicates existence of tropical night conditions for almost half of the year. Towards the Siwalik Hills, the proportion is 140–150 days/year while it drops to about 100 days/year over temperate. The high altitude cold semi-arid zone has only 30 days/year when night temperature exceeds 20 °C. An overall increase in tropical nights has been observed in the study area (Fig. 2.12) which implies expansion in warmer night conditions. Such increase is statistically significant in the southwestern and southern Punjab (Fig. 2.13a and b) area while rest of the area shows non-significant increase. The highest rate of increase (1.6 to 1.4 days/decade) in tropical nights is recorded for two contrasting geographical areas. The first zone is the warm semi-arid southwestern and southern Punjab while other area includes the cold semi-arid region in eastern Himachal Pradesh. This overall rise in TR20 frequency over cold semi-arid region and other parts of Himachal Pradesh signifies expansion of tropical night conditions thereby indicating a possible spread of night time warming period.

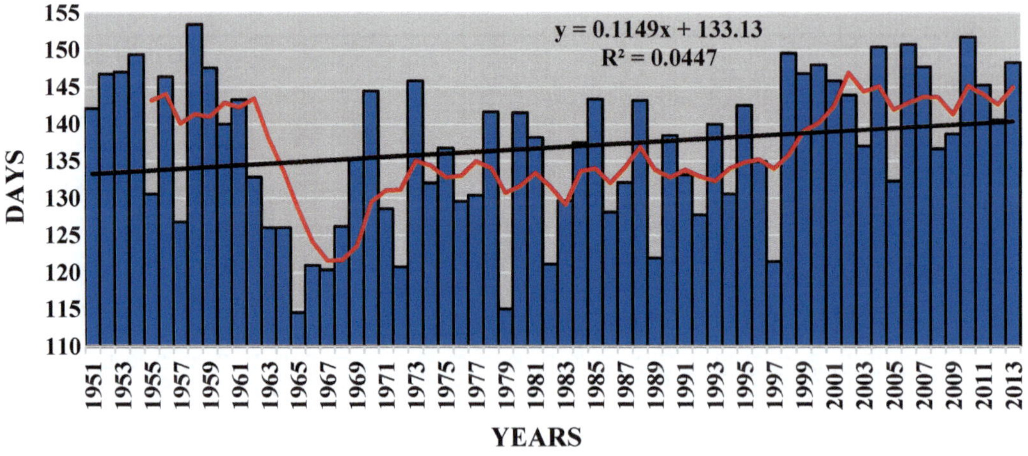

Fig. 2.12 Trends in tropical nights (1951–2013)

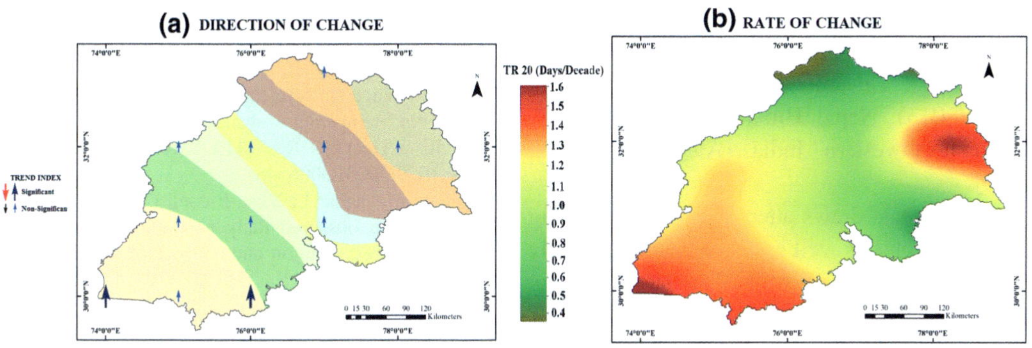

Fig. 2.13 Direction and rate of change in tropical nights (TR 20)

2.5 Conclusions

The temperature regime of study area exhibits noticeable alterations in the long-term temperature conditions. Although, the trends are not very glaring and in some case the change in very small, certain extreme temperature indices hint towards potential warming of the study area. The hottest day temperatures representing peak summers has increased by almost half a degree since 1950 with maximum rise in the southwestern semi-arid Punjab plains. Interestingly the warmest nights representing summer night conditions show an overall decline by 0.5 °C. However, southwestern Punjab is an exception to such trends and overall increase in hottest day and warmest night temperatures in this part of study area implies warming of summer season in semi-arid Punjab plains. On contrary, it appears that winter days over Punjab and large parts of Himachal Pradesh have witnessed cooling. The magnitude of such cooling is relatively higher and statistically significant over Punjab plains especially in southern and southwestern plains. While large parts of Himachal Pradesh do not show much change in coldest-day winter temperatures, the northern mountainous region have experienced warming even during the coldest winter days. It is important to note that the entire study area exhibits increase in the coldest night temperatures which implies that winter nights have definitely become warmer over the time. Such warming is found to be statistically significant over southwestern Punjab. A conspicuous change in peak winter and summer temperature regime is evident. The growing proportion of summer days and tropical nights also suggests overall warming of summer season. Although, Punjab plains do not show much rise in summer days but increased proportion of tropical nights clearly indicate night-time upswing in temperature conditions.

A visible growth in the frequency of summer days is discernible over hilly regions of Himachal Pradesh; the warming of summer season is taking place at higher rates in the temperate and cold semi-arid regions. It is important to note that the rate of warming is noticeable over areas of extreme climatic conditions, i.e. the hot semi-arid southwestern and southern Punjab and the cold semi-arid Himachal Pradesh. The overall scenario points towards warming during peak summers as well as peak winters. The warming of summer days is pronounced over the Himalayan region only whereas summer night warming has taken place for both the Punjab plains and the mountainous region. In addition, the rates of change are much higher for summer night warming than summer days. Such similar results but with varying degree of change has also been either observed or projected in several studies focussed on spatial and inter-annual variability in extreme indices depicting an increase in hottest days, coldest nights and summer days within and around the study region, such as River Satluj basin (Singh et al. 2015); Hindu Kush Himalayas (Kothawale et al. 2010; Wu et al. (2017; Xiu-Bao et al. 2017); Tibetan Plateau Region Rangwala et al. (2013); Nepal Himalayas (Baidya et al. 2008) northwestern India (Alexander et al., 2006).

The fact that draws attentions is the warming of winter nights as well as winter days especially over high Himalayan region. With such trends in place it is likely that winter snowfall patterns may get altered. Similarly, the rise in summer day temperature conditions might accelerate snow ablation leading to upsurge in rivers during early summers. These changes may further introduce hydrological shift with likely repercussions for the Himalayan region as well as the Punjab plains. The manner in which extreme temperature indices might accelerate disaster risk is another serious issue that needs further exploration. It would not be indiscreet to conclude that observed changes in extreme temperature indices hint impending water related issues in the study area. Under such a scenario, it is expected that regional cropping patterns, horticulture productivity, tourism and hydropower generation may face serious distress thereby burdening the regional development.

References

Alexander LV, Zhang X, Peterson TC (2006) Global observed changes in daily climate extremes of temperature and precipitation. J Geophys Res Atmos 111:1042–1063

Baidya SK, Shrestha ML, Sheikh MM (2008) Trends in daily climatic extremes of temperature and precipitation in Nepal. J Hydrol Meteorol 5(1):38–53

Barnes SL (1973) Mesoscale objective map analysis using weighted timeseries observations. NOAA Tech. Memo, ERL NSSL62, National Severe Storms Laboratory, 764 Norman, OK 73069, p 60

Batima P (2005) Potential impacts of climate change and vulnerability and adaptation assessment for livestock sector in Mongolia: Final Report. 2006. https://www.aiaccproject.org

Betts RA, Alfieri L, Caesar J, Feyen L, Gohar L, Koutroulis A, Lewis K, Morfopoulos C, Richardson KJ, Tsanis I, Wyser K (2018) Changes in climate extremes, freshwater availability and vulnerability to food insecurity projected at 1.5 °C and 2 °C global warming with a higher-resolution global climate model. Phil Trans R Soc A376:20160452). https://doi.org/10.1098/rsta.2016.0452

Brohan P, Kennedy JJ, Harris I, Tett SFB, Jones PD (2006) Uncertainty estimates in regional and global observed temperature changes: A new data set from 1850. J Geophys Res 111:D12106. Shttps://doi.org/10.1029/2005JD006548

Caesar J, Alexandar L, Vose Russell (2006) Large-scale changes in observed daily maximum and minimum temperatures: creation and analysis of a new gridded data set. J Geophy Res 111:DO5101. https://doi.org/10.1029/2005JD006280

Cressman GP (1959) An operational objective analysis system. Mon Wea Rev 87(10):367–374

Cruz RVO, Lasco RD, Pulhin JM, Pulhin FB, Garcia KB (2006) Climate change impact on water resources in Pantabangan Watershed, Philippines. AIACC Final Technical Report. 9–107. https://www.aiaccproject.org/FinalReports/final_reports.html

Dai A (2013) Increasing drought under global warming in observations and models. Nature 3:52–58

De US, Mukhopadhyay RK (1998) Severe heat wave over the Indian subcontinent in 1998, in perspective of global climate. Curr Sci 75(12):1308–1311

Easterling DR, Meehl GA, Parmesan SA, Changnon SA, Karl TR, Mearns LO (2000) Climate extremes: Observations, modelling, and impacts. Science 289:2068–2074

Ford J, Goldhar C (2012) Climate change vulnerability and adaptation in resource dependent communities: a case study from West Greenland. Climate Res 54:181–196

Ford JD, Pearce T (2010) What we know, do not know, and need to know about climate change vulnerability in the western Canadian Arctic: a systematic literature review. Environ Res Lett 5. https://doi.org/10.1088/1748-9326/5/1/014008

Fowler HJ, Wilby L (2010) Detecting changes in seasonal precipitation extremes using regional climate model projections: Implications for managing fluvial flood risk. Water Resour 46:W03525

Gandin LS (1965). Objective analysis of meteorological fields. Translated from Russian by R. Hardin. Jerusalem: Israel Program for Sci Transl p 242

Haylock MR, Peterson TC, Alves LM, Ambrizzi T, Anunciação YMT, Baez J, Barros VR, Berlato MA, Bidegain M, Coronel G, Corradi V, Garcia VJ, Grimm AM, Karoly D, Marengo JA, Marino MB, Moncunill DF, Nechet DJ, Rebello Quintana E, Rusticucci M, Santos JL, Trebejo I, Vincent LA (2006) Trends in total and extreme South American rainfall in 1960–2000 and links with sea surface temperature. J Clim 19(8):1490–1512

IPCC (2007) Fourth assessment report. Summary for Policymakers. Cambridge University Press, Cambridge

IPCC (2014) Climate change—impacts, adaptations, and vulnerability. Summary for Policy makers. Cambridge University Press, Cambridge

Kawahara M, Yamazaki N (1999) Long term trend of incidences of extreme high or low temperatures in

Japan. Extended Abstract, Bi-annual meeting of the Meteorological society of Japan (in Japanese)
Kiktev D, Sexton DMH, Alexander L, Folland CK (2003) Comparison of modeled and observed trends in indices of daily climate extremes. J Clim 16:3560–3571
King AD, Karoly, David J (2017) Climate extremes in Europe at 1.5 and 2 degrees of global warming. Environ Res Lett 12:114031. https://doi.org/10.1088/1748-9326/aa8e2c
Knoll AH (2003) Life of a young planet: The first three billion years of evolution on earth. Princeton University Press, Princeton
Kothawale DR, Revadekar JV, Kumar KR (2010) Recent trends in pre-monsoon daily temperature extremes over India. J Earth Syst Sci 119(1):51–65
Lal M (2003) Global climate change: India's Monsoon and its variability. J Environ Stud Policy 6:1–34
Mitchell D, Heaviside C, Vardoulakis S, Huntingford C, Masato G, Guillod BP, Frumhoff P, Bowery A, Wallom D, Allen M (2016) Attributing human mortality during extreme heat waves to anthropogenic climate change. Environ Res Lett 11(7):1–8. https://doi.org/10.1088/1748-9326/11/7/074006
Morak SR, Hegerl GC, Christidis N (2013) Detectable changes in the frequency of temperature extremes. J Clim 26:1561–1574
New MG, Hulme M, Jones PD (2000) Representing twentieth-century space-time climate variability part II: development of 1901–96 monthly grids of terrestrial surface climate. J Clim 13:2217–2238
Nicholls RJ, Brown S, Goodwin P, Wahl T, Lowe J, Solan M, Godbold JA, Ivan DH, Lincke D, Hinkel J, Wolff C, Merkens JL (2018) Stabilization of global temperature at 1.5 °C and 2.0 °C: implications for coastal areas. Philos Trans. Series A, Math Phys Eng Sci 376:2119. https://doi.org/10.1098/rsta.2016.0448
Pretis F, Schwarz M, Tang K, Haustein K, Allen MR (2018) Uncertain impacts on economic growth when stabilizing global temperatures at 15 °C or 2 °C warming. Philos Trans Ser A, Math Phys Eng Sci 376 (2119):20160460. https://www.ncbi.nlm.nih.gov/pubmed/29610370
Rangwala I, Sinsky E, Miller JR (2013) Amplified warming projections for high altitude regions of the northern hemisphere mid-latitudes from CMIP5 models. Environ Res Lett 8(2):279–288
Rohini P, Rajeevan M, Srivastava AK (2015) On the variability and increasing trends of heat waves over India. Nat (Sci Rep) 6:26153. https://doi.org/10.1038/srep26153,2016
Ryoo SB, Jhun KWT, JG (2004) Characteristics of wintertime daily and extreme temperature over south Korea. Int J Climatol 24:145–160
Shepard D (1968) A two-dimensional interpolation function for irregularly-spaced data. In: Proceedings of the 1968 23rd ACM national conference, ACM, pp 517–524
Sillmann J, Kharin VV, Zhang X, Zwiers FW, Bronaugh D (2013) Climate extremes indices in the CMIP5 multimodel ensemble: Part 1. Model evaluation in the present climate. J Geophys Res Atmos 118:1716–1733
Singh D, Gupta RD, Jain SK (2015) Study of daily extreme temperature indices over Sutlej Basin, N-W Himalayan Region India. Global NEST J 17(2):301–311
Sippel S, Otto FEL, Flach M, van Oldenborgh GJ (2016) The role of anthropogenic warming in 2015 Central European heat waves. Bur Am Meteorol Soc 97:S51–S56. https://doi.org/10.1175/BAMS-D-16-0150.1
Smith P, Price J, Molotoks A, Warren R, Malhi Y (2018) Impacts on terrestrial biodiversity of moving from a 2 °C to a 1.5 °C target. Philos Trans R Soc 376:20160456. https://doi.org/10.1098/rsta.2016.0456
Van Oldenborgh GJ, Philip S, Kew S, Van Weele M, Uhe P, Otto F, Singh R, Pai I, Cullen H, Achutarao K (2018) Extreme heat in India and anthropogenic climate change. Hazards Earth Syst Sci 185194:365–381. https://www.nat-hazards-earth-syst-sci.net/18/365/2018/nhess-18-365-2018.pdf
World Meteorological Organisation-WMO (2009) Guidelines on analysis of extremes in a changing climate in support of informed decisions for adaptation. Climate Data and Monitoring, WCDMP-No. 72, Switzerland
Wu J, Ying X, Xue-Jie G (2017) Projected changes in mean and extreme climates over Hindu Kush Himalayan region by 21 CMIP5 models. Adv Clim Change Res 8(3):176–184
Xiu-Bao S, Guo-Yu R, Shrestha AB, Yu-Yu Rm, Qing-Long Y, Yun-Jian Z, Yan X, Rajbhandari R (2017) Changes in extreme temperature events over the Hindu Kush Himalaya during 1961–2015. Adv Clim Chang Res 8 (3):157–165
Zhai P, Pan X (2003) Trends in temperature extremes during 1951–1999 in China. Geophys Res Lett 30 (17):7–10
Zhang X, Hegerl G, Zwiers FW, Kenyon J (2005) Avoiding inhomogeneity in percentile-based indices of temperature extremes. J Clim 18:1641–1651

Spatial Variations and Long-Term Trends (1901–2013) of Rainfall Across Uttarakhand Himalaya, India

3

Suraj Mal, Manohar Arora, Abhishek Banerjee, R.B. Singh, Christopher A. Scott, Simon K. Allen, and Ramchandra Karki

Abstract

Understanding spatial and temporal variations of rainfall is crucial for the ongoing and future socioeconomic and infrastructure developments in the Indian Himalayan region. However, despite such importance, the studies on rainfall distribution, variability and trends are limited and rare in the Uttarakhand Himalayan region. To bridge the gap, this study presents an exploratory analysis of spatial distribution variations and long-term trends of annual and seasonal rainfall in Uttarakhand. At first, spatial variability is investigated, followed by long-term trend analysis using nonparametric Mann–Kendall trend test with trend-free pre-whitening (TFPW) procedure and Theil–Sen's slope estimator. Further, dry and wet periods in Uttarakhand region and their linkages with the large-scale Southern Oscillation Index (SOI) are also explored and investigated.

Annual rainfall is dominated by monsoonal rainfall ($\sim 82\%$), while winter rainfall contribution is only $\sim 9\%$. There is a clear elevational gradient of rainfall in the winter season, but monsoon season does not feature such an elevational pattern. Mostly long-term trends of annual and monsoonal rainfall are negative, which are contributed mainly by a more evident and pronounced decline after the 1960s. Nevertheless, some areas in western and southeastern regions show positive rainfall trends. Likewise, a significant reduction in winter rainfall has been observed in Uttarakhand. Pre-monsoon rainfall experienced mostly negative trends in the central and eastern parts of the study area, while some

The original version of this chapter was revised: For detailed information, please see Correction. The correction to this chapter is available at https://doi.org/10.1007/978-3-030-70238-0_32

S. Mal (✉)
Department of Geography, Shaheed Bhagat Singh College, University of Delhi, Delhi 110017, India
e-mail: suraj.mal@sbs.du.ac.in

M. Arora
National Institute of Hydrology, Roorkee, Uttarakhand 247667, India

A. Banerjee
Key Laboratory of Geographic Information Science, Ministry of Education, and School of Geographical Science, Institute of Eco-Chongming, East China Normal University, Shanghai 200241, China

R.B. Singh (Deceased)
Department of Geography, Delhi School of Economics, University of Delhi, Delhi 110007, India

C. A. Scott
School of Geography & Development, and Udall Center for Studies in Public Policy, University of Arizona, Tucson 85719, AZ, USA

S. K. Allen
Department of Geography, University of Zurich, Zurich CH-8057, Switzerland

R. Karki (Deceased)
CEN Center for Earth System Research and Sustainability, Institute of Geography, University of Hamburg, Hamburg 20146, Germany

U. Schickhoff et al. (eds.), *Mountain Landscapes in Transition*,
Sustainable Development Goals Series, https://doi.org/10.1007/978-3-030-70238-0_3

observatories in western and southeastern indicate increasing trends. Post-monsoon rainfall trends are mostly positive, though statistically insignificant. Monsoon rainfall in Uttarakhand region is highly correlated with large-scale Southern Oscillation Index (SOI), but other seasons' rainfall shows low correlation. These robust findings can be useful for adaptation strategies for different sectors, including agriculture and allied activities.

Keywords

Rainfall trends · Mann–Kendall · Uttarakhand · Himalaya · India

3.1 Introduction

The understanding of the spatial distribution of rainfall and its trends in the Himalayan Mountain regions is crucial due to its major agro-economic and hydro-environmental implications in the densely populated Himalayan Mountain and adjacent lowlands (Sen Roy and Singh 2002; Basistha et al. 2008, 2009; Smadja et al. 2015; Schickhoff et al. 2016; Kumari et al. 2017; Schickhoff and Mal 2020). The Himalayan river basins are inhabited by ~1.3 billion population, and the Ganga basin, including Uttarakhand, is among the most densely populated basins (401 persons/km^2) of the world mountains (Eriksson et al. 2009b). As a result, lower availability of both surface and groundwater (Gosain et al. 2006; Eriksson et al. 2009a; Scott and Sharma 2009) leads to many socioeconomic issues in the Himalayan catchments (Shiva 2009).

The long-term seasonal and annual trends of rainfall are not well explored in the high Himalayan Mountains (Basistha et al. 2009) due to (1) paucity of long-term climatic records, (2) significant data gaps, and (3) poor network of meteorological stations, particularly at <2500 m asl (Duan et al. 2006; Kumari et al. 2017; Schickhoff et al. 2016). Inaccessible terrain and prolonged harsh winters in the Himalayan Mountain regions are the main constraints in maintaining the high altitude observatories (Duan et al. 2006; Palazzi et al. 2013); thus, many of them have been already discontinued (Basistha et al. 2008; Kumari et al. 2017). Consequently, few studies on rainfall trends in the Himalayas have been primarily based on the scarce network of observatories (Table 3.1) or coarse resolution re-analysis and satellite-based data (Palazzi et al. 2013; Mukherjee et al. 2015). However, it is so far understood that the coarser gridded dataset does not represent the localized characteristics of rainfall trends and variability across the highly complex Himalayan Mountain slopes and valleys (Ghosh et al. 2009; Palazzi et al. 2013). Consequently, the studies based on the different spatial resolution gridded data reveal significant dissimilarities at various levels (Goswami et al. 2006; Ghosh et al. 2009). For instance, a comparison of various resolution gridded datasets by Palazzi et al. (2013) reveals contrasting results concerning precipitation trends in the Hindu-Kush Karakoram Himalayan Mountains. A comprehensive review of the major studies based on the stations and gridded data over the Himalayas further supports the regional disparity and contrasting rainfall trends (Table 3.1).

The studies based on station data reveal detailed rainfall trend patterns. For instance, mostly positive trends for annual and seasonal rainfall for most of the stations in the upper Indus Basin in northwestern Himalaya (NWH) are reported (Archer and Fowler 2004). In contrast, a decline in annual, pre-monsoon, and monsoon precipitation is observed, while the winter precipitation mostly increased over the same region (Hasson et al. 2017). However, limited studies on rainfall trends in Western Himalayan (WH) regions show a consistent decreasing trend of monsoon, winter, and annual rainfall (Table 3.1). Nevertheless, inconsistent trends for annual and seasonal rainfall in the Satluj Basin have been observed (Mir et al. 2015). Post-monsoon trends are mostly positive for WH, while some regional anomalies have been found for pre-monsoon rainfall. In the central Himalaya, monsoonal rainfall also shows mostly the declining trends with some local anomalous trends, whereas winter rainfall has considerably decreased.

In contrast, the eastern Himalayan regions show no clear trends of rainfall (Immerzeel 2008;

Table 3.1 Major studies analyzing the rainfall trends in the Himalayan region

Region	Major findings	References
Indus Basin, Northwestern Himalaya	Conflicting precipitation trends	(Archer and Fowler 2004)
	Winter precipitation mainly increased Pre-monsoon, monsoon, and annual precipitation mostly declined Post-monsoon shows mixed trends	(Hasson et al. 2017)
Kashmir Valley, NW and W Himalaya	The decreasing trends in annual, monsoon, and winter rainfall and increasing in pre- and post-monsoon season	(Kumar and Jain 2010)
Jammu and Kashmir and Himachal Pradesh, NW and W Himalaya	The decreasing trends in annual, winter, and monsoon precipitation over the last century	(Bhutiyani et al. 2010; Bhutiyani 2016)
Jammu and Kashmir and Himachal Pradesh, NW and W Himalaya	A decreasing trend of winter rainfall	(Dimri and Dash 2010; Dimri et al. 2016)
Satluj River Basin, Himachal Pradesh, Western Himalaya	The increasing trends for annual and seasonal rainfall except in winter season	(Mir et al. 2015)
Jammu and Kashmir, Himachal Pradesh, and Uttarakhand, Western Himalaya	Annual and seasonal rainfall declined except in post-monsoon season	(Kumar and Jaswal 2016)
Uttarakhand, Western Himalaya	The decreasing trends of annual and summer monsoon rainfall	(Basistha et al. 2009)
	Rainfall has declined in December	(Singh and Mal 2014)
Nepal, Central Himalaya	Winter rainfall has significantly declined	(Adhikari and Devkota 2016)
	Mixed trends of pre-monsoon and monsoon rainfall. Post-monsoon and winter rainfall declined	(Karki et al. 2017)
	Lack of long-term rainfall trend	(Shrestha 2000; Shrestha et al. 2000)
Eastern Himalaya	There is no clear trend of rainfall	(Jain et al. 2013)
	There is no clear trend in precipitation	(Immerzeel 2008)
Assam plains, Eastern Himalaya	Spatially variable trends	(Jhajharia et al. 2012)
Tibet	Spatial variability of precipitation changes. Increase in annual precipitation	(Kuang and Jiao 2016)
Chinese Himalayas	No trend in seasonal totals	(Yang et al. 2013)
The Hindu-Kush and Karakoram Himalayas	Summer monsoon rainfall declined, APHRODITE (–0.0 mm/d^{-1} $yr^{-}1$), and GPCP data (–0.02 mm/d^{-1} $yr^{-}1$). Winter rainfall increased (0.005 mm/d^{-1} $yr^{-}1$) based on CRU data	(Palazzi et al. 2013)
Indian Himalayas	Unclear trends in summer monsoon rainfall in western and central Himalaya, while in eastern Himalaya a negative trend	(Mukherjee et al. 2015)

Jain et al. 2013). The Tibet region, on the contrary, shows a slight increase in annual rainfall, while the seasonal rainfall is overall trendless (Table 3.1). In general, the declining winter rainfall trends are statistically insignificant across large areas of Himalaya, while annual and monsoon rainfalls trends are significant. Interestingly, highly localized and peculiar characteristics of rainfall trends identified only in some station-based studies (Basistha et al. 2009; Dimri and Dash 2012; Hasson et al. 2017; Karki et al. 2017) suggest the need of station-based analysis to identify the local-scale variability for various local-scale impact assessment and adaptation strategies development.

Uttarakhand is a highly vulnerable region for climate change impacts due to its unique geographical location in the dominant summer monsoon regime (Kumari et al. 2017). Primarily, the seasonal distribution of rainfall determines the water availability in the region, thereby significantly influencing the livelihood and wellbeing of downstream communities. The Himalayan regions are prone to extreme precipitation and subsequent floods and landslides due to an influence of orography (Guha-Sapir et al. 2014; Allen et al. 2016). The change in rainfall patterns and trends either due to natural and or anthropogenic forces may thus affect the socioeconomic and environmental settings of the region. Despite such enormous importance, studies on rainfall variability and trends in Uttarakhand are limited with sparse station data for shorter periods, viz. 1902–1980 by (Basistha et al. 2009) and 1957–2007 (Singh and Mal 2014). Therefore, local-scale spatial rainfall variability and long-term trends covering recent decades are mostly unknown over Uttarakhand region.

Given the limited number of rainfall variability and trend studies in the region, the present research focuses on a comprehensive analysis of spatial and temporal characteristics of rainfall for the longer period (1901–2013) across Uttarakhand, based on a number of rainfall observatories with continuous and high-quality data. To our understanding, this is the first study in Uttarakhand region, focusing on rainfall trends for such a long period (1901–2013). In this study, spatial variations of mean seasonal and annual rainfall are analyzed at first, followed by long-term trends analysis. Besides, the dry and wet periods that contributed to overall rainfall trends are explored. Moreover, the linkages of rainfall variability with large-scale circulation features/teleconnections (SOI) are also presented in the end.

3.2 Data and Methods

3.2.1 Study Area

The present study region, i.e., Uttarakhand, is bordered by Uttar Pradesh on the south and southwest, Himachal Pradesh on the west and northwest, Nepal on the east, and Tibet on the north (Fig. 3.1a). Extending between 28°43′ and 31°27′ N latitudes and 77°34′ and 81°02′ E longitudes, the total area of Uttarakhand is $\sim$53,484 km^2. The study area is predominantly characterized by the rugged and high elevation terrain difference, ranging from 184 to 7817 m asl. Broadly, the study region features a north–south elevational gradient, where the Greater Himalayan region (north) is characterized by extremely rugged topography and mountain peaks (e.g., Nanda Devi (7817 m) and Kamet (7756 m, etc.,)), while the southernmost region features the plains (Basistha et al. 2008; Joshi 2004; Kumari et al. 2017) (Fig. 3.1a).

Based on the analysis of SRTM data (https://earthexplorer.usgs.gov), more than 31% of the total geographical area of the state lies >3000 m asl, while between the elevations 1000–2000 m asl and <1000 m asl are $\sim$32.6% and 25.8%, respectively.

The study region is affected by two large-scale weather systems, i.e., southwest monsoon in the summer (JJAS) and western disturbances in the winter (DJF) (Palazzi et al. 2013). The mean

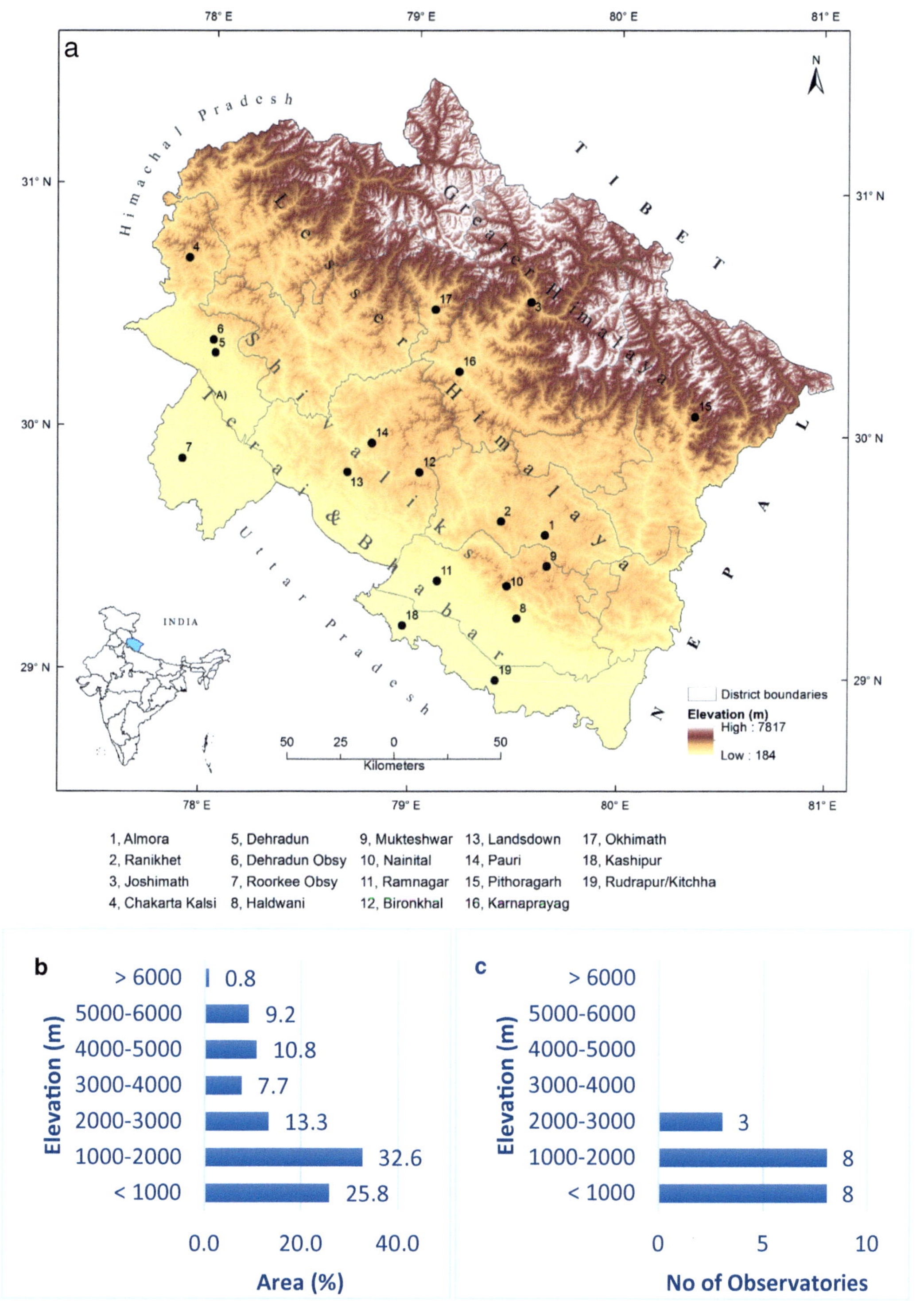

Fig. 3.1 **a** Location of Uttarakhand in India along with the spatial distribution of elevation zones (m) based on Shuttle Radar Topography Mission (SRTM). The solid circles represent locations of rainfall observatories in Uttarakhand; **b** geographical area (%) under different elevation zones based on SRTM, and **c** the number of observatories in different elevation zones

climatology of Uttarakhand (1981–2013) indicates two rainfall seasons, i.e., summer monsoon (JJAS) and winter (DJF) and two relatively drier seasons, i.e., pre- (MAM) and post-monsoon (ON) (Fig. 3.2). The monsoonal system contributes ~81% of annual rainfall, whereas winter rainfall only receives ~9% through the western disturbances. The pre- and post-monsoon seasons receive 7% and 3% of annual rainfall, respectively (Fig. 3.2). Generally, rainfall decreases from south to north under the influence of elevation and local physiography (Basistha et al. 2009; Kumari et al. 2017). The mean monthly temperature (1981–2013) in the study region varies from ~25 °C in the summer season (May–June) to ~11 °C in winter (January) (Fig. 3.2), which indicates the conditions up to 2000 m asl only, since the observed temperature records for higher altitudes contains significant data gaps and hence not used. Generally, the temperature decreases from south to north with increasing elevation (Sah et al. 2005).

3.2.2 Database and Methodology

The monthly rainfall records (1901–2013) available from 128 surface meteorological observatories in Uttarakhand state were collected from the India Meteorological Department (IMD), Pune, India. However, large numbers of rainfall observatories have considerable missing data and thus are inconsistent for a century-long rainfall trend study. Therefore, to identify the observatories with continuous data for 1901–2013, at first visual check was performed by plotting the time series in Excel to detect the data gaps, outliers (extreme values), and negative rainfall (You et al. 2008; Karki et al. 2017; Shrestha et al. 2017). Wherever the available records for an observatory were less than 85% and consist of gaps of more than five consecutive years (Hasson et al. 2017; Karki et al. 2017), those stations were excluded from the study. For the remaining stations, outliers (detected for a few stations) and missing data were filled with a

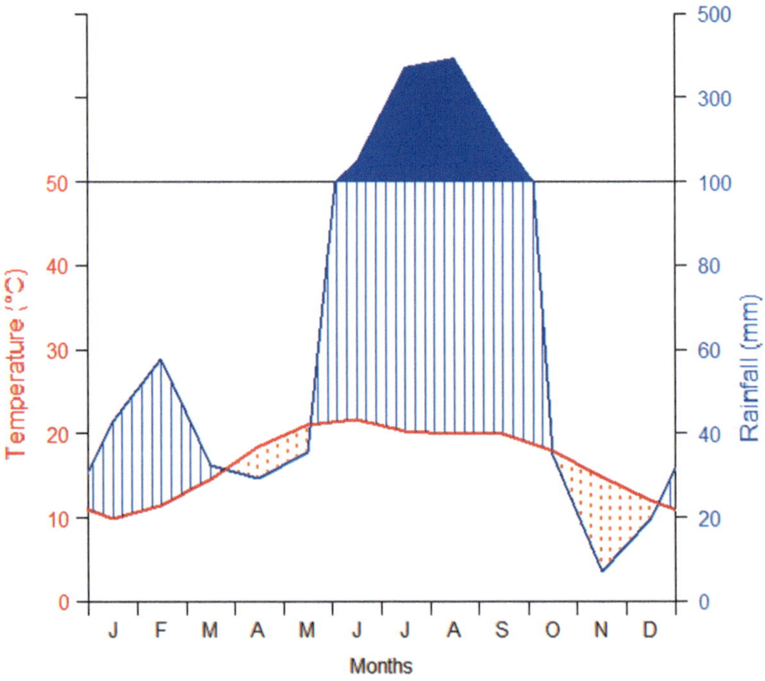

Fig. 3.2 Mean climatology of the study area between 1981 and 2013. The blue line indicates the monthly rainfall, while the red line indicates mean temperature. The blue-shaded regions show rainfall seasons, while the red-shaded regions show drier periods of the year

long-term monthly mean of the respective station. Finally, such a strict quality control resulted in nineteen observatories with consistent rainfall data. Out of the nineteen observatories, thirteen have rainfall records for 1901–1990 and six for 1901–2013. Therefore, we analyzed two trends of the standard periods: (1) 1901–1990 for all the nineteen observatories and (2) 1901–2013 for six observatories (Fig. 3.3) (based on Kumar and Jain 2010). These observatories represent relatively balanced network in terms of spatial coverage, but not in elevation as the climate observations in the Himalayas are biased to lower elevations and valley floors (Fig. 3.1c) (Duan et al. 2006; Kumari et al. 2017). In our study region, the long-term consistent rainfall records for only three high altitude observatories (>2000 m asl) were available (Fig. 3.1a and c).

Further, the all-India and three subregional (Western Uttar Pradesh, Haryana, and Punjab) rainfall time series were obtained from the Indian Institute of Tropical Meteorology (IITM), Pune, India (https://www.tropmet.res.in/). The subregional time series were averaged using the arithmetic mean to prepare a regional time series termed as Northwestern Indian plains (NWIP). Further, the seasonal and annual rainfall data series for all-India, NWIP, and all-Uttarakhand were also prepared to conduct a comparative assessment of variability and trends (Basistha et al. 2009). We further used the Southern Oscillation Index (SOI) data (https://www.bom.gov.au/climate/current/soi2.shtml) to analyze the influence of large-scale circulations/teleconnections on the rainfall of the region (Archer and Fowler 2004; Fowler and Archer 2006; Basistha et al. 2009). The seasonal rainfall analysis has been conducted for four seasons, viz. winters (DJF), pre-monsoon (MAM), monsoon (JJAS), and post-monsoon (ON) for Uttarakhand (Shrestha 2000; Basistha et al. 2009; Kumari et al. 2017).

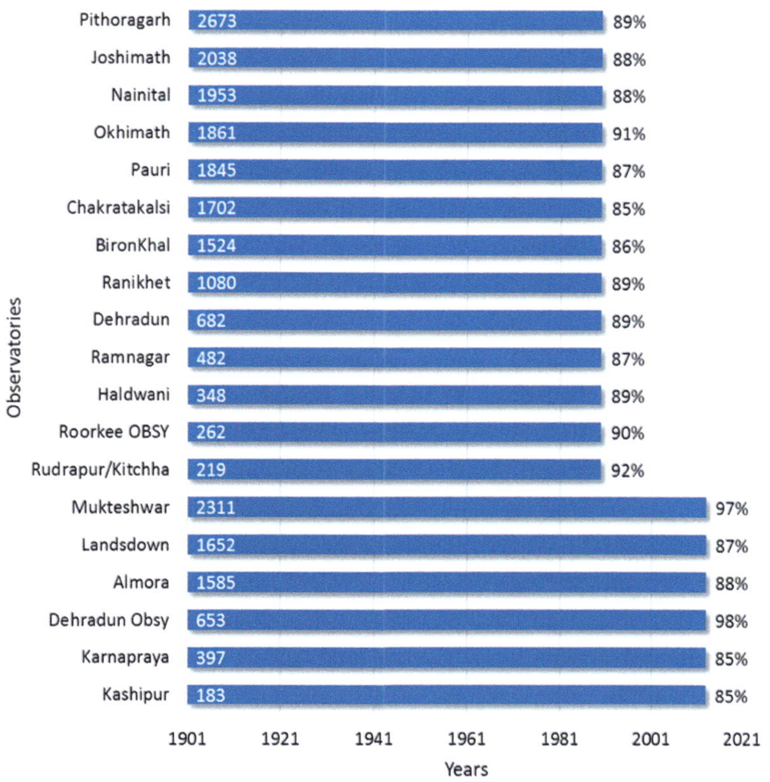

Fig. 3.3 Details of the rainfall observatories used in this study, along with elevation (m) and available rainfall records (%)

3.2.2.1 Trend Analysis

Mann–Kendall (MK) Test and Trend-free Pre-whitening (TFPW)

The statistical significance, as well as the direction of the trends of climatic time series, has been widely estimated based on the nonparametric MK test (Zhang et al. 2009; Kumar and Jain 2010; Kumar et al. 2010; Drápela et al. 2011; Jain and Kumar 2012). It is a robust test and less sensitive to outliers and gaps (Jhajharia et al. 2012; Hasson et al. 2017). It is, however, essential to analyze the serial correlation in the climatic data series and minimize its impact before its application (Yue and Pilon, 2004; Dinpashoh et al. 2014). Previous studies in Uttarakhand and adjacent regions (Basistha et al. 2009; Hasson et al. 2017; Karki et al. 2017) detected serial correlation in some stations and thus suggested the use of the modified MK test and pre-whitening procedures.

Different researchers (Hamed and Ramachandra Rao 1998; Zhang et al. 2000; Yue et al. 2002, 2003) have proposed several pre-whitening procedures. In the study, we use the TFPW (Yue et al. 2003: Yue and Hashino, 2003) (at a 90% confidence level) to minimize the impact of serial correlation in rainfall time series of Uttarakhand to prepare it for the MK test following Karki et al. (2017) and Hasson et al. (2017).

Theil–Sen's (TS) Slope

We apply the TS slope estimator to calculate the magnitude of rainfall trends. The TS test was developed by Theil (1950) and Sen (1968) and has been widely used since then in hydro-climatic data series (Hirsch et al. 1982; Kumar and Jain 2010; Singh and Mal 2014). The TS slope estimator test can be successfully applied to the data series, which is distribution-free and is less affected by data gaps/missing values and outliers (extreme values) (Hasson et al. 2017; Karki et al. 2017).

Inter-annual Variability of Annual and Seasonal Rainfall

To estimate dry and wet (below and above the normal) periods, percentage departures for each time series from respective normal rainfall (long-term mean) have been estimated (Shrestha 2000) as.

$$\text{Rainfall departure} = \frac{Ri - \overline{R}}{\overline{R}} \times 100$$

where Ri is rainfall for ith year, and $\overline{R}$ is the mean rainfall. This analysis has been limited to annual and seasonal rainfall time series for all-India, NWIP, and all-Uttarakhand.

3.3 Results

3.3.1 Spatial Distribution of Rainfall

The spatio-seasonal distribution of rainfall is highly uneven in Uttarakhand (Figs. 3.4 and 3.5). Uttarakhand receives ~1426 mm rainfall annually, of which ~82% (1164 mm) occurs in monsoon season. Therefore, the high contribution of monsoon rainfall dominates the annual rainfall distribution, and hence, both follow similar spatio-seasonal patterns in the study area. The southern foothills of the Himalaya (Shivaliks) and Mid-Himalayan ranges receive a relatively higher amount of monsoon and annual rainfall as compared to southern plains and northern Greater Himalayan regions (Figs. 3.4 and 3.5). Overall, the share of monsoon rainfall to yearly rainfall decreases from the south (>85%) to the northern Greater Himalayan regions (<75%). Some sporadic parts of the northern inner river valleys (e.g., Okhimath) also receive a high amount of rainfall. The relatively lower contribution from the monsoonal rainfall is received in northern-most observatories of the Greater Himalayan region (e.g., Joshimath, 63%). Within the monsoon season, most of the rainfall (62–73% of monsoon rainfall) is concentrated in the two months (July and August). Subsequently, in the post-monsoon season, the rainfall in the region declines (3.6%) with the retreat of the prevailing monsoonal system.

In the winter season, ~9% of annual rainfall is observed, which occur due to the influence of western disturbances in the region (Dimri and Dash 2012; Palazzi et al. 2013). The seasonal

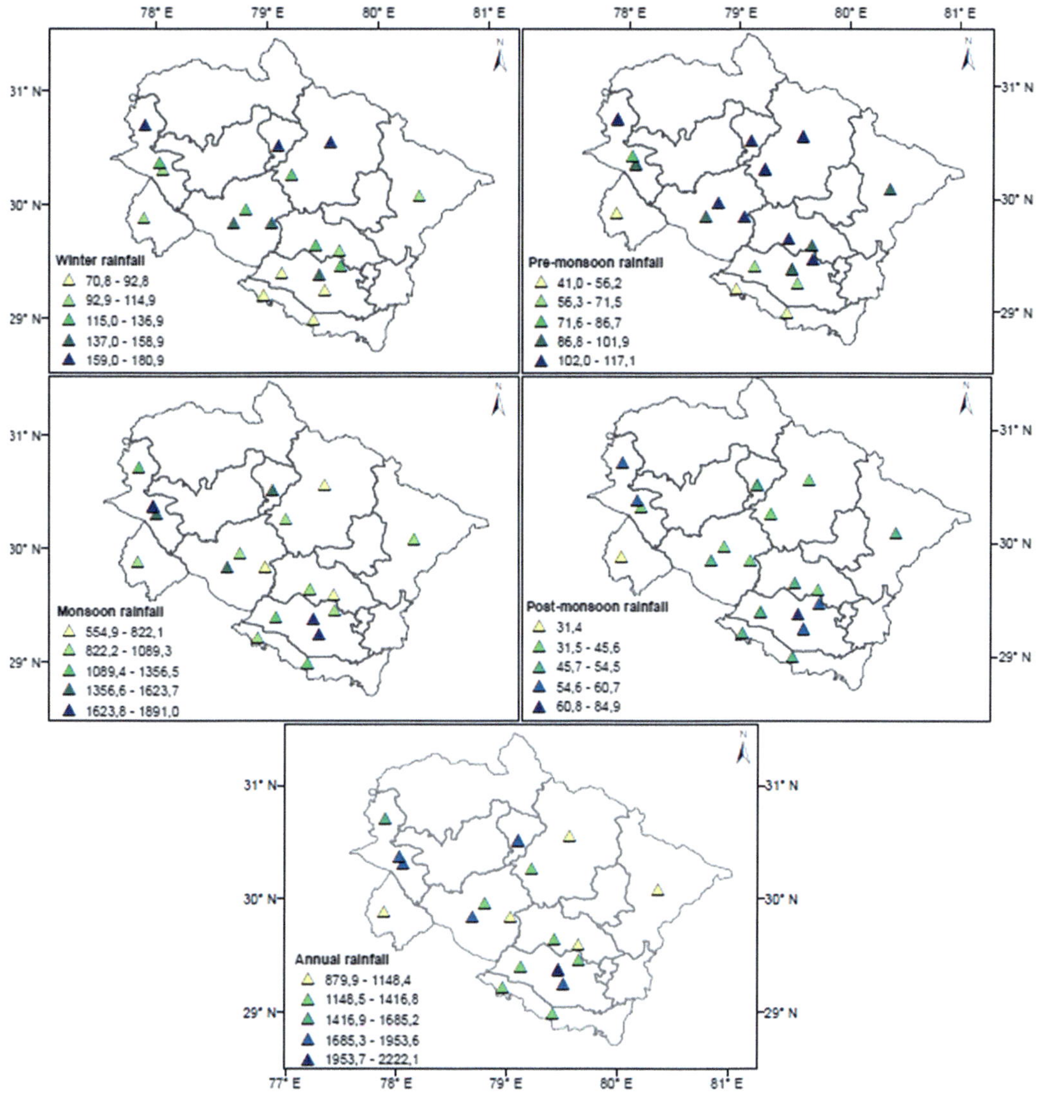

Fig. 3.4 Spatial distribution of mean and annual seasonal rainfall (mm) in Uttarakhand between 1901 and 2013

total and contribution of winter rainfall decrease from the northwest and Greater Himalayan parts (<11%) to the eastern, southern, and lower elevations (>6%) (Figs. 3.4 and 3.5). The southeastern parts of the study area receive the smallest amount of winter rainfall. The winter season is followed by a relatively lower rainfall amount and hence a drier (6% of annual rainfall) pre-monsoon season, which is often an extension of the winter rainfall caused by western disturbances (Dimri and Dash 2012; Dimri et al. 2015).

3.3.2 Spatial Patterns of Rainfall Trends

3.3.2.1 Rainfall Trends for 1901–1990

The rainfall trends for the period of 1901–1990 are based on nineteen observatories (Fig. 3.3). The study shows the mixed spatial patterns of annual rainfall trends across the study area. About 73% of the observatories, mostly concentrated in the eastern sides of the region, indicate declining yearly rainfall trends.

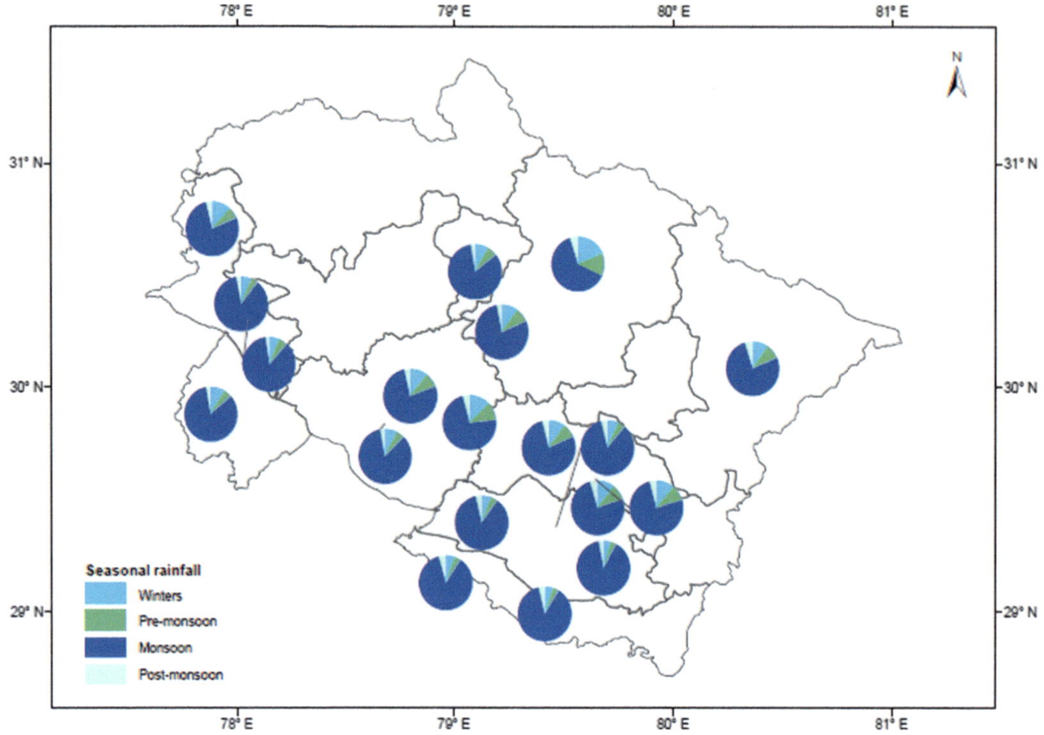

Fig. 3.5 Spatial variation of percentage share of mean seasonal rainfall to mean annual rainfall in Uttarakhand during 1901–2013, indicating a more significant share of monsoonal rainfall in southern observatories, while the northern high elevation observatories receive a relatively high contribution from the winter rainfall

However, statistically significant negative trends are observed only for 37% observatories, primarily located in the mid-Hills and Shivaliks. Statistically significant positive trends are observed for only two observatories located in the northern and northwestern parts of the sutdy region (Fig. 3.6).

The trends of the monsoonal rainfall roughly align with the annual rainfall trends and reveal the spatial concentrations of positive and negative trends. Consistent with spatial patterns of yearly rainfall trends, the eastern part of the study area mainly shows negative trends except for a few cases.

Overall, ~57% of the observatories present negative trends in the monsoon season, where around one-fourth of observed rainfall trends are statistically significant. On the contrary, the western parts reveal the positive trends, where statistically significant trends are observed for two observatories only. In the post-monsoon season, the rainfall features an increasing trend in most of the observatories; though, the trends are mostly statistically insignificant. Central parts of the study area indicate no trends in the post-monsoon season.

In contrast, the winter rainfall has declined over most of the study area. These trends are statistically significant for ~47% of observatories, mostly concentrated mostly in the mid-hills of the study area. Conversely, the rainfall trends in the eastern parts of the study area are statistically insignificant. Likewise, the pre-monsoon rainfall features negative trends for ~63% of the observatories, of which ~16% have statistical significance. The negative trends are most evident in the central and eastern parts, whereas positive trends were largely observed in the western and northern regions.

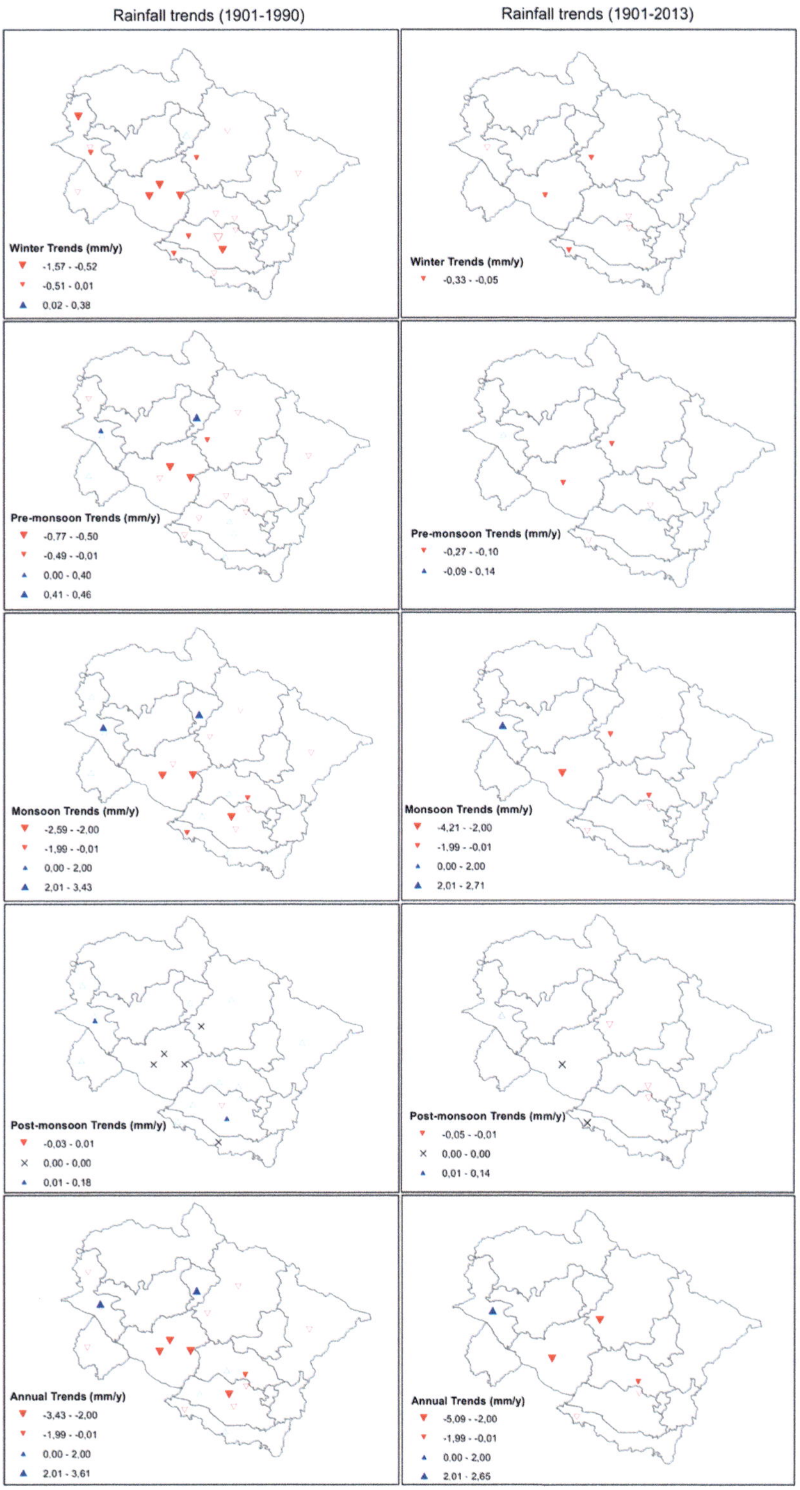

Fig. 3.6 Trends of annual and seasonal rainfall in Uttarakhand during 1901–1990 and 1901–2013 (mm/year) at a 90% confidence level. The positive and negative trends are marked with regular and inverted triangles, respectively. The solid and hollow triangles show statistically significant and insignificant trends, respectively

3.3.2.2 Rainfall Trends for 1901–2013

The rainfall trends for 1901–2013 are estimated for only six observatories because of the limited consistent long-term data availability. The trend patterns from this period are mostly congruent to the trends for the 1901–1990 period, presented in the previous section (Fig. 3.6). Five observatories reveal negative trends of annual rainfall, with statistical significance in three observatories. One observatory on the western side indicates a statistically significant positive trend. Following the yearly rainfall trends, the monsoonal rainfall trends are also negative for five observatories, while only one shows a positive trend. Overall, western parts show positive trends for annual and monsoonal rainfall, while negative in other regions. The post-monsoonal rainfall trends are negative for three observatories and positive for one, while two observatories reveal no trends. The trends in the post-monsoon season are, however, without statistical significance. Winter rainfall trends are negative for all the observatories during 1901–2013, similar to those observed from 1901 to 1990. In the following pre-monsoon season, the rainfall features negative trends for the majority of the observatories, where only two have statistically significant trends (Fig. 3.6).

3.3.2.3 Rainfall Variability

The inter-seasonal rainfall variability of Uttarakhand, as estimated by calculating the percentage departure from the long-term means, is discussed in this section.

The positive (negative) departures from the long-term rainfall means have been considered as wet (dry) periods. Smoothing of inter-annual rainfall variability by seven years moving average allows us to draw a clearer picture of dry and wet periods (Fig. 3.7).

The study shows comparable patterns for the annual and monsoon rainfalls of Uttarakhand. The annual and monsoonal rainfall significantly declined during the post-1960s (Fig. 3.7). The all-Uttarakhand annual rainfall indicates two significant dry (1960–1980 and 1990 to 2000: >10% deficit) and wet (1901–2025 and 1950–1960) periods. The monsoonal rainfall, conversely, indicates a more evident and prolonged dry period post-1960s to 2000 (>15% deficit). The monsoon rainfall also experienced a general surplus before the 1960s, with minor deficit period between 1925 and 1935. There is no apparent cyclic behavior in annual and monsoonal rainfall of Uttarakhand. However, an increasing annual and monsoonal rainfall has been observed after the 2000s (Fig. 3.7c and e). In the post-monsoon season, the rainfall has been

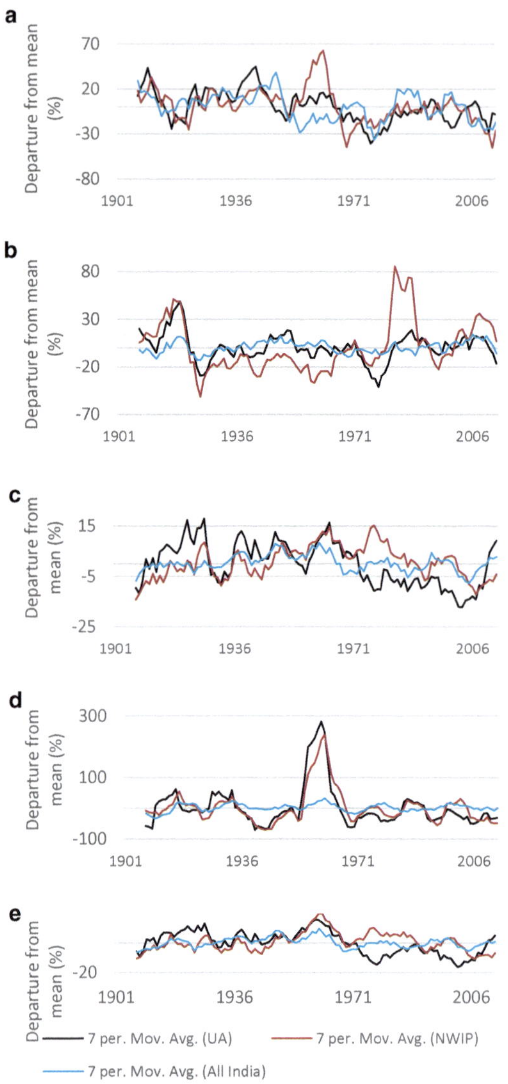

Fig. 3.7 Temporal relationships of percentage departure of all-Uttarakhand rainfall, NWIP rainfall, and all-India rainfall. **a**: winter, **b**: pre-monsoon, **c**: monsoon, **d**: post-monsoon, and **e**: annual

relatively more erratic compared to other seasons (Fig. 3.7d), with a dry period after the 1940s, and a significant wet period between 1945 and 1970.

High rainfall variability has been observed for the winter rainfall when compared to other seasons, except for the post-monsoon (Fig. 3.7a). Overall, the winter rainfall has been in a declining phase, since the 1940s, which weakened in the recent decades. The pre-monsoon rainfall has been relatively stable over Uttarakhand compared to the other seasons (Fig. 3.7b). For this season, one major wet (1910–1920) and one dry period (1970–1980) have been observed. However, the pre-monsoon rainfall, in general, has been in the declining phase between 1950 and 1980s, while a stronger surplus seen over 1980–2005, later following a drier period (Fig. 3.7b).

3.4 Discussion

The results of the present study about seasonality and rainfall trends are broadly consistent with all-Uttarakhand, NWIP, and all-India rainfall as well as with previous studies in the adjacent Himalayan regions (Basistha et al. 2009). It is to note that the rainfall observatories in Uttarakhand region are primarily located at low-mid-elevation (183–2673 m asl) of lower and middle Himalayas (Fig. 3.1a and c). The network of weather observatories in Greater Himalayan parts of the study region and other similar Himalayan regions is rather weak (Duan et al. 2006) due to complex topography, inaccessibility, and prolonged harsh winters, making manual maintenance and data collection from these observatories very difficult and in many cases an impossible task (Basistha et al. 2008; Palazzi et al. 2013; Hasson et al. 2017; Karki et al. 2017; Kumari et al. 2017).

Therefore, the understanding of rainfall and related implications on agro-economic systems, hydrology, and the cryosphere in Greater Himalayan Mountain regions remains a crucial challenge (Immerzeel et al. 2015). Besides, longer duration (>100 years) observations for a large number of stations are reasonably rare in the Himalaya (Basistha et al. 2008; Bhutiyani et al. 2010; Singh and Mal 2014; Bhutiyani 2016; Kumari et al. 2017) as most of the presently functional observatories were installed in the late 1950s (Duan and Yao 2003; Duan et al. 2006; Karki et al. 2017; Kumari et al. 2017).

The study shows that ~82% of the rainfall in Uttarakhand occurs during the monsoon season, which is congruent to the monsoonal rainfall contributions from the all-India and NWIP (Fig. 3.9). Previous studies on the rainfall at all-India (Ghosh et al. 2009; Kumar et al. 2010; Jain and Kumar 2012), all-Uttarakhand (Basistha et al. 2008; Khandelwal et al. 2015; Kumari et al. 2017), and adjacent Himalayan regions levels (Arora et al. 2006; Mir et al. 2015; Karki et al. 2017) confirm the dominance (~80%) of monsoonal rainfall in the region. It is well known that the contribution of monsoon rainfall weakens from the central Himalaya toward the Western Himalaya as the monsoonal system moves from east to west in the region due to the east–west orientation of the Himalayan Mountain (Palazzi et al. 2013; Bharti et al. 2016; Mal et al. 2021). Consequently, western Himalayan areas, e.g., Satluj Basin, receive ~60–70% (Mir et al. 2015), and Chenab, located further west, ~42% of annual rainfall during the monsoon season (Arora et al. 2006). Far northwestern Himalayan Kashmir Valley receives only ~15–25% rainfall in the monsoon season (Kumar and Jain 2010).

In the present study, the monsoon rainfall was observed to decline from Shivaliks and southern slopes of the Mid-Himalayan areas to the northern Greater Himalayan region (Figs. 3.4 and 3.5) but without an evident elevational gradient (Fig. 3.8).

The dominance of valley-based observatories and poor network in the Greater Himalayan region might have resulted in such a weak dependence of monsoon rainfall on the elevation patterns. Although, other studies on the rainfall in western Himalayan states and the Chenab River Basin (Arora et al. 2006; Bharti et al. 2016) also show significant dependence of monsoon rainfall on local physiography. The southern Shivalik and Mid-Himalayan Mountains present the first orographic barrier to the prevailing moisture-laden monsoon winds approaching

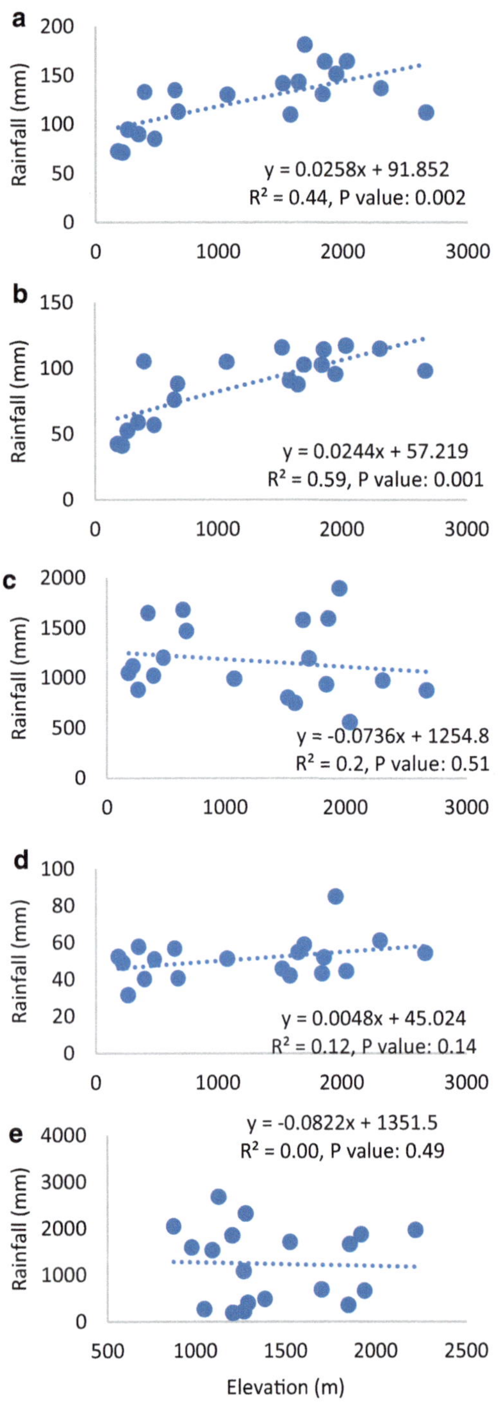

Fig. 3.8 Altitudinal dependence of seasonal rainfall in Uttarakhand (1901–2013). **a**: Winter, **b**: Pre-monsoon, **c**: Monsoon, **d**: Post-monsoon, and **e**: Annual

from the south and force them to precipitate locally in the windward Himalayan slopes (Kulkarni 2012). Consequently, these windward slopes in Shivaliks and Mid-Himalayan ranges (Satluj and Chenab Basins and Nepal Himalaya) observe the high amount of rainfall (Arora et al. 2006; Mir et al. 2015; Karki et al. 2017). Interestingly, some parts of the northern inner valleys (e.g., Okhimath) also observed relatively higher monsoon rainfall, which may be attributed to the transverse orientation of the major river valleys, leading to continuous moisture supply to even deep inner valleys of Greater Himalayan region. Thus, the orientation of valleys and mountain ridges appears to have strong control on rainfall distribution, rather than merely elevation (Kumari et al. 2017).

About 9% of the annual rainfall in Uttarakhand occurs during the winter months, while it is 7% in NWIP and only 3% for all-India (Fig. 3.9). In the present study, the winter rainfall has been observed to decline from northwestern parts to the southeastern parts and from Greater Himalayan regions (north) to the lower plains (south). Such a spatial pattern of the winter rainfall is due to the west–east movement of western disturbances with upper-level westerly circulation, which primarily generate precipitation in the Himalayan Mountains by orographic forcings (Dimri and Niyogi 2013; Dimri et al. 2015, 2016). Consequently, the contribution of winter rainfall declines from the northwestern Himalayan regions to central and further to the eastern Himalayan region as well (Dimri and Dash 2012; Palazzi et al. 2013; Dimri et al. 2015, 2016). According to a study (Kumar and Jain 2010), the Kashmir Valley in northwestern Himalaya receives ~38–48% of the rainfall during winters, while only ~27% in Chenab basin, Himachal Pradesh, positioned east (Arora et al. 2006). The winter rainfall contribution declines further eastward over the Central Himalaya of Nepal, receiving only ~3.5% of the annual rainfall (Karki et al. 2017), indicating a strong west–east gradient of winter rainfall over Himalaya. Nevertheless, it is worth mentioning that present

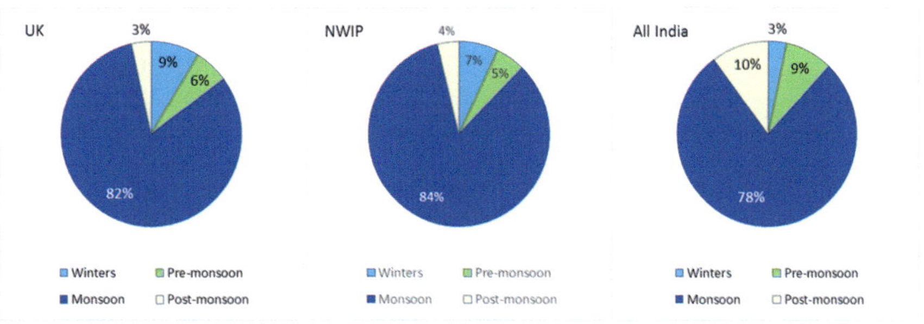

Fig. 3.9 Comparison of % of seasonal rainfall of all-Uttarakhand, NWIP, and all-India (1901–2013)

precipitation measurement methods highly underestimate the solid precipitation that occurs in high elevation areas (Hasson et al. 2017; Karki et al. 2017). The comparison of mean seasonal rainfall for all-Uttarakhand, NWIP, and all-India rainfall is depicted in Fig. 3.9.

The seasonal rainfall of Uttarakhand state is positively correlated with NWIP and all-India rainfall except for the winters (Table 3.2). It indicates that the seasonal rainfall of Uttarakhand is greatly influenced by seasonal rainfall of the NWIP and all-India rainfall, except for the winters. Weak correlation of winter rainfall of Uttarakhand with NWIP and all-India is because the rainfall in the region primarily occurs through the western disturbances, where southern non-mountainous and Shivalik parts of the study area, NWIP, and India are less influenced by this weather system (Dimri et al. 2016). A previous study (Basistha et al. 2009) also indicated a strong relationship of all-Uttarakhand annual and monsoon rainfalls with regional rainfall for the period of 1902–1980.

We further attempted to explore the relationships of annual and seasonal rainfall of Uttarakhand with large-scale phenomena/telecon nections, i.e., SOI (Table 3.3). All-Uttarakhand annual and monsoon rainfalls show high relationships with monsoon and annual SOI. It is so because the monsoon rainfall primarily contributes to the annual rainfall (Bhalme and Jadhav 1984; Shrestha 2000). Previous studies (Bhalme and Jadhav 1984; Shrestha 2000) suggest that Indian and Nepal summer monsoon rainfalls are highly correlated with seasonal SOI. The post-monsoon rainfall of Uttarakhand shows a relatively weaker but positive correlation with seasonal and annual SOI, while the winter and pre-monsoon rainfall shows a negative relationship with seasonal SOI. It is, thus, that the winter rainfall in the study area occurs mostly due to western disturbances and is more related to the source region, while the pre-monsoon rainfall occurs primarily due to the extension of western disturbances (Dimri and Dash 2012; Palazzi et al. 2013; Dimri et al. 2015).

The comparative trends of all-Uttarakhand, NWIP, and all-India rainfall for the period of 1901–1980 and 1901–2013 are presented in Tables 3.4 and 3.5, respectively. There are mixed trends of seasonal and annual rainfall in the study area, with a broadly decreasing trend in the seasonal and annual rainfall except for the post-monsoon rainfall in 1901–1990 (Fig. 3.6). More than half of the trends are generally statistically insignificant, similar to the results from other studies in and adjacent regions (Basistha et al. 2009; Bhutiyani et al. 2010; Kumar and Jain 2010; Singh and Mal 2014; Bhutiyani 2016; Karki et al. 2017). A study (Nuzzo 2014) argues

Table 3.2 Correlation co-coefficients of all-Uttarakhand rainfall with NWIP and all-India rainfall

Regions	Winter	Pre-monsoon	Monsoon	Post-monsoon
NW Indian Plains	−0.03	**0.56**	**0.51**	**0.74**
All India	−0.01	**0.43**	**0.54**	**0.46**

Values in bold are significant at 5% confidence level

Table 3.3 Correlation co-coefficients of all-Uttarakhand rainfall with annual and seasonal SOI

	Annual	Winter	Pre-monsoon	Monsoon	Post-monsoon
SOIDJF	0.01	0.06	**−0.13**	**0.06**	0.13
SOIMAM	**0.31**	−0.11	**−0.13**	**0.32**	**0.24**
SOI JJAS	**0.49**	**−0.23**	**−0.19**	**0.45**	**0.29**
SOION	**0.37**	**−0.15**	−0.29	**0.38**	**0.21**
SOI Annual	**0.40**	−0.14	**−0.25**	**0.41**	**0.29**

Values in bold are significant at 5% confidence level

that the statistical significance is not a robust measure of the statistical validity of the time series trend results; nonetheless, may be of practical use in many sectors (Radziejewski and Kundzewicz 2004; Basistha et al. 2009).

As a result of declining monsoonal rainfall, the declining trend of the annual rainfall of Uttarakhand has been observed for all-Uttarakhand rainfall during 1901–1990 and 1901–2013, whereas these trends are positive for NWIP and all-India rainfall (Tables 3.4 and 3.5) (Naidu et al. 1999; Basistha et al. 2009). The peculiar negative trend of monsoon and annual rainfall for all-Uttarakhand may be linked to the weakening of ISM, as caused by various factors including reduction in the land–ocean thermal contrast, etc. (Kulkarni 2012; Paul et al. 2016; Sandeep et al. 2018; Xu et al. 2018). Therefore, the other nearby Himalayan areas also experienced negative trends of monsoon and annual rainfall (Bhutiyani et al. 2007, 2010; Bhutiyani 2016; Kumar and Jaswal 2016; Karki et al. 2017; Shafiq et al. 2018). As with previous studies, the most prominent negative rainfall trends are observed in the monsoon season (Bhutiyani et al. 2010; Palazzi et al. 2013), consistent with larger-scale decreasing trends observed in monsoon indices, such as the South Asian Summer Monsoon Index (SASMI) (Ma et al. 2019). In contrast, it is to note that the model results, both for historical and projected time horizons, typically reveal positive trends in mean precipitation over this region (Kulkarni et al. 2013; Palazzi et al. 2013; Panday et al. 2015; Maharana et al. 2020). With the vast majority of rainfall occurring during the monsoon months, these contrasting results (model vs. observations) highlight that climate models do not yet adequately capture the complexities of the Himalayan monsoon system and processes influencing its variabilities, such as aerosols and black carbon, snow cover, and influence of orography. Hence, there remains only low confidence in future projections of monsoon timing and intensity for this region (Christensen et al. 2013), which cast uncertainty

Table 3.4 Comparison of rainfall trends of all-Uttarakhand, NWIP, and all-India rainfall (mm/year) during 1901–1990

	Annual	Winter	Pre-monsoon	Monsoon	Post-monsoon
Uttarakhand	−1.3	**−0.4**	−0.1	−0.9	0.2
NW Indian Plains	**1.3**	−0.1	0.0	**1.3**	0.1
All-India	0.4	−0.1	0.0	0.3	0.1

Bold values are significant at 90% confidence level

Table 3.5 Comparison of rainfall trends of all-Uttarakhand, NWIP, and all-India rainfall (mm/year) during 1901–2013

	Annual	Winters	Pre-monsoon	Monsoon	Post-monsoon
Uttarakhand	**−1.6**	**−0.2**	−0.1	**−1.3**	0.1
NW Indian Plains	0.1	**−0.1**	0.0	0.2	0.0
All-India	**0.2**	−0.1	0.0	0.1	0.1

Bold values are significant at 90% confidence level

on the understanding of related hydrological impacts (Maharana et al. 2020).

On the contrary, consistent negative trends of winter rainfall for all the time series were observed, which are consistent with other studies in the Himalayas (Bhutiyani et al. 2007; Dimri and Dash 2012; Bhutiyani 2016; Karki et al. 2017) and linked to the decreasing frequency of western disturbance in the region during the winter months (Cannon et al. 2014; Dimri et al. 2015). The pre-monsoon rainfall decreased in Uttarakhand, while rainfall series in other regions (all-India and NWIP) features no trend. The post-monsoon rainfall trends are slightly stronger and positive when compared with other large-scale time series of NWIP and all-India.

Given the robust and relatively homogenous signal of decreasing rainfall across Uttarakhand observed over the twentieth and early twenty-first century reported in this and previous studies (Basistha et al. 2009; Singh and Mal 2014; Khandelwal et al. 2015; Karki et al. 2017), there can be expected impacts on water availability in the Ganga Basin, with implications particularly for agriculture and allied activities, irrigation, and hydropower sustainability (e.g., as outlined in the Government of Uttarakhand State Action Plan on Climate Change 2014). Furthermore, glaciers are sensitive to changes in precipitation, and the observed decrease in precipitation in winter and monsoon months will strongly affect the accumulation of snow high up in the glaciated catchments (Bolch et al. 2019). This combination of reduced accumulation and observed increasing temperatures is responsible for the generally widespread retreat of Himalayan glaciers (Bolch et al. 2012, 2019; Azam et al. 2018). Consequently, the glaciers' contribution to maintaining the base flow of streams during pre- and post-monsoon, or during unusually dry periods, will be lost, placing further stress on dependent stakeholders and ecosystems (Immerzeel et al. 2010).

While, in general, the decreasing rainfall quantities might suggest a reduced threat of rainfall-induced floods and landslides, but this is inconsistent with increased incidences of such disasters in the region reported over recent decades (Bhambri et al. 2016; Allen et al. 2016). It, to some extent, highlights that many recent disasters have probably been inappropriately linked to changing precipitation. In fact, underlying causes are related more to human settlements and activities occurring in high-risk locations, like floodplains, poor land management (e.g., deforestation and road cutting), land degradation, and expansion of pilgrimage and tourism, as well as other assets within the exposed areas (Pande et al. 2002; Pande 2006; Sati and Gahalaut 2013). Also, it must be noted that such flood and landslide disasters are typically related to extreme meteorological events, which are not generally captured by the monthly data that were analyzed in this study. Events such as the flooding disaster of 2013 of Uttarakhand have been linked to exceptionally heavy rainfall episodes, which have become more likely as a result of atmospheric warming (Singh et al. 2014). Such events are anticipated to increase in frequency and/or intensity in the future (Seneviratne et al. 2012; Panday et al. 2015). Hence, it is possible that while total seasonal or annual rainfall has decreased, extremely intense rainfall events could have increased over the same period. Unfortunately, a lack of daily (or sub-daily) data in this study has prevented a further investigation of this aspect.

3.5 Conclusion

The present study analyzes the spatial variations and long-term trends of annual and seasonal rainfall in the Uttarakhand Himalaya, India, for the period of 1901–2013. We applied a non-parametric MK test with TFPW and TS's slope estimator to analyze the rainfall trends. Further, the wet and dry periods and their linkages of large-scale circulations/teleconnections with rainfall are also explored. Regional consistencies of rainfall variability and trends of all-Uttarakhand rainfall were compared with NWIP and all-India rainfall as well.

The rainfall mostly declined across the study region except in the post-monsoon season. A significant decline in annual and monsoonal rainfall primarily after the 1960s contributed to the overall negative rainfall trend in the study area. Specifically, concerning annual rainfall from 1960 to 1980 and 1990 to 2000 are dry periods, and 1901–1925 and 1950–1960 are wet periods, while the monsoonal rainfall experienced more pronounced and prolonged dry periods between 1960 and 2000. The pre-1960s period is generally a wet period with a minor dry period between 1925 and 1935. Pre-monsoon rainfall in western and southeastern parts of the study area increased, while negative trends are observed for central and eastern regions. Besides, a general decline in winter rainfall post-1940s has contributed to a significant decrease in winter rainfall in Uttarakhand. Annual and monsoon rainfalls in the region are found to be highly correlated with large-scale SOI, but other seasons' rainfall shows low correlation.

This variation of all-Uttarakhand rainfall roughly aligns with other regional and all-India rainfall series with some periodic differences, which needs further investigations. Identifying the causes of declining rainfall is another critical area of research to be conducted in the region as the rainfall variation is resulted by the interaction between synoptic and local topographic features. More importantly, the present study is limited to monthly analysis, but a daily rainfall analysis is urgently needed to identify the trends in droughts, floods, and landslides in the region because such events have very high socioeconomic and ecological impacts, as compared to mean seasonal precipitation. Nevertheless, the present findings of declining and highly variable nature of rainfall have significant implications on the agricultural, hydrometeorological, and social milieu in the region. Studies based on current findings and suitable adaptation strategies development in different sectors are needed to cope with these changing patterns of rainfall to ensure the food and water security of the region.

Acknowledgements The study was conducted as part of the HI-NEX project.

References

Adhikari T, Devkota L (2016) Climate change and hydrological responses in Himalayan basins, Nepal. In: Singh RB, Schickhoff U, Mal S (eds) Climate change, glacier response, and vegetation dynamics in the Himalaya. Springer, Netherlands, pp 65–85. https://doi.org/10.1007/978-3-319-28977-9_4

Allen SK, Rastner P, Arora M et al (2016) Lake outburst and debris flow disaster at Kedarnath, June 2013: hydrometeorological triggering and topographic predisposition. Landslides 13:1479–1491. https://doi.org/10.1007/s10346-015-0584-3

Archer DR, Fowler HJ (2004) Spatial and temporal variations in precipitation in the Upper Indus Basin, global teleconnections and hydrological implications. Hydrol Earth Syst Sci 8:47–61. https://doi.org/10.5194/hess-8-47-2004

Arora M, Singh P, Goel NK, Singh RD (2006) Spatial distribution and seasonal variability of rainfall in a mountainous basin in the Himalayan region. Water Resour Manag 20:489–508. https://doi.org/10.1007/s11269-006-8773-4

Azam MF, Wagnon P, Berthier E et al (2018) Review of the status and mass changes of Himalayan-Karakoram glaciers. J Glaciol. https://doi.org/10.1017/jog.2017.86

Basistha A, Arya DS, Goel NK (2008) Spatial distribution of rainfall in Indian Himalayas—a case study of Uttarakhand Region. Water Resour Manag 22:1325–1346. https://doi.org/10.1007/s11269-007-9228-2

Basistha A, Arya DS, Goel NK (2009) Analysis of historical changes in rainfall in the Indian Himalayas. Int J Climatol. https://doi.org/10.1002/joc.1706

Bhalme HN, Jadhav SK (1984) The Southern Oscillation and its relation to the monsoon rainfall. J Climatol 4:509–520. https://doi.org/10.1002/joc.3370040506

Bhambri R, Mehta M, Dobhal DP et al (2016) Devastation in the Kedarnath (Mandakini) Valley, Garhwal Himalaya, during 16–17 June 2013: a remote sensing and ground-based assessment. Nat Hazards 80:1801–1822. https://doi.org/10.1007/s11069-015-2033-y

Bharti V, Singh C, Ettema J, Turkington TAR (2016) Spatiotemporal characteristics of extreme rainfall events over the Northwest Himalaya using satellite data. Int J Climatol 36:3949–3962. https://doi.org/10.1002/joc.4605

Bhutiyani MR (2016) Spatial and temporal variability of climate change in high-altitude regions of NW Himalaya. In: Singh RB, Schickhoff, Mal S (eds) Climate change, glacier response, and vegetation dynamics in the Himalaya. Springer, Netherlands, pp: 87–101. https://doi.org/10.1007/978-3-319-28977-9_5

Bhutiyani MR, Kale VS, Pawar NJ (2007) Long-term trends in maximum, minimum and mean annual air temperatures across the Northwestern Himalaya during the twentieth century. Clim Change 85:159–177. https://doi.org/10.1007/s10584-006-9196-1

Bhutiyani MR, Kale VS, Pawar NJ (2010) Climate change and the precipitation variations in the north-western Himalaya: 1866–2006. Int J Climatol 30:535–548. https://doi.org/10.1002/joc.1920

Bolch T, Kulkarni A, Kääb A et al (2012) The state and fate of Himalayan glaciers. Science 336:310–314. https://doi.org/10.1126/science.1215828

Bolch T, Shea JM, Liu S et al (2019) Status and change of the cryosphere in the extended Hindu Kush Himalaya region. In: Wester P, Mishra A, Mukherji A, Shrestha AB (eds) The Hindu Kush Himalaya assessment. Springer, Cham, pp 209–255

Cannon F, Carvalho LMV, Jones C, Bookhagen B (2014) Multi-annual variations in winter westerly disturbance activity affecting the Himalaya. Clim Dyn 44:441–455. https://doi.org/10.1007/s00382-014-2248-8

Christensen JH, Krishna Kumar K, Aldrian E et al (2013) Climate phenomena and their relevance for future regional climate change. In: Stocker TF, Qin D, Plattner G-K, et al. (eds) Climate change 2013: the physical science basis. Contribution of working group I to the fifth assessment report of the intergovernmental panel on climate change. Cambridge University Press, Cambridge, United Kingdom and New York

Dimri AP, Dash SK (2010) Winter temperature and precipitation trends in the Siachen Glacier. Curr Sci 98:1620–1625

Dimri AP, Dash SK (2012) Wintertime climatic trends in the western Himalayas. Clim Change 111:775–800. https://doi.org/10.1007/s10584-011-0201-y

Dimri AP, Niyogi D (2013) Regional climate model application at subgrid scale on Indian winter monsoon over the western Himalayas. Int J Climatol 33:2185–2205. https://doi.org/10.1002/joc.3584

Dimri AP, Niyogi D, Barros AP et al (2015) Western Disturbances: A review. Rev Geophys 53:225–246. https://doi.org/10.1002/2014RG000460

Dimri AP, Yasunari T, Kotlia BS et al (2016) Indian winter monsoon: present and past. Earth-Sci Rev 163:297–322. https://doi.org/10.1016/j.earscirev.2016.10.008

Dinpashoh Y, Mirabbasi R, Jhajharia D et al (2014) Effect of short-term and long-term persistence on identification of temporal trends. J Hydrol Eng 19:617–625. https://doi.org/10.1061/(ASCE)HE.1943-5584.0000819

Drápela K, Drápelová I, Drápela DIK (2011) Application of Mann-Kendall test and the Sen's slope estimates for trend detection in deposition data from Bílý Kříž (Beskydy Mts., the Czech Republic) 1997–2010. Beskydy 4:133–146

Duan K, Yao T (2003) Monsoon variability in the Himalayas under the condition of global waming. J Meteorol Soc Japan 81:251–257. https://doi.org/10.2151/jmsj.81.251

Duan K, Yao T, Thompson LG (2006) Response of monsoon precipitation in the Himalayas to global warming. J Geophys Res Atmos 111:1–8. https://doi.org/10.1029/2006JD007084

Eriksson M, Xu J, Shrestha A et al (2009a) The changing Himalayas: impact of climate change on water resources and livelihoods in the greater Himalayas. ICIMOD, Kathmandu, Nepal

Eriksson MXJ, Shrestha AB, Vaidya RA et al (2009b) Perspectives on water and climate change adaptation: The changing Himalayas—Impact of climate change on water resources and livelihoods in the Greater Himalayas. ICIMOD, Kathmandu, Nepal

Fowler HJ, Archer DR (2006) Conflicting signals of climatic change in the upper Indus Basin. J Clim 19:4276–4293. https://doi.org/10.1175/JCLI3860.1

Ghosh S, Luniya V, Gupta A (2009) Trend analysis of Indian summer monsoon rainfall at different spatial scales. Atmos Sci Lett 10:285–290. https://doi.org/10.1002/asl.235

Gosain AK, Rao S, Basuray D (2006) Climate change impact assessment on hydrology of Indian river basins. Curr Sci 90:346–353

Goswami BN, Venugopal V, Sangupta D et al (2006) Increasing trend of extreme rain events over India in a warming environment. Sci 314:1442–1445. https://doi.org/10.1126/science.1132027

Guha-Sapir D, Vos F, Below R (2014) EM-DAT: International Disaster Database. https://www.emdat.be. Université Catholique de Louvain, Brussels, Belgium. [20.11.2014]

Hamed KH, Ramachandra Rao A (1998) A modified Mann-Kendall trend test for autocorrelated data. J Hydrol 204: 182–196. https://doi.org/10.1016/S0022-1694(97)00125-X

Hasson S, Böhner J, Lucarini V (2017) Prevailing climatic trends and runoff response from Hindukush-Karakoram-Himalaya, upper Indus Basin. Earth Syst Dyn 8:337–355. https://doi.org/10.5194/esd-8-337-2017

Hirsch RM, Slack JR, Smith RA (1982) Techniques of trend analysis for monthly water quality data. Water Resour Res 18:107–121. https://doi.org/10.1029/WR018i001p00107

Immerzeel W (2008) Historical trends and future predictions of climate variability in the Brahmaputra basin. Int J Climatol 28:243-254. https://doi.org/10.1002/joc.1528

Immerzeel WW, van Beek LPH, Bierkens MFP (2010) Climate change will affect the asian water towers. Sci 328:1382–1385. https://doi.org/10.1126/science.1183188

Immerzeel WW, Wanders N, Lutz AF et al (2015) Reconciling high-altitude precipitation in the upper Indus basin with glacier mass balances and runoff. Hydrol Earth Syst Sci https://doi.org/10.5194/hess-19-4673-2015

Jain SK, Kumar V (2012) Trend analysis of rainfall and temperature data for India. Curr Sci 102:37–49. https://doi.org/10.1038/nature09063

Jain SK, Kumar V, Saharia M (2013) Analysis of rainfall and temperature trends in northeast India. Int J Climatol 33:968–978. https://doi.org/10.1002/joc.3483

Jhajharia D, Yadav BK, Maske S et al (2012) Identification of trends in rainfall, rainy days and 24h maximum rainfall over subtropical Assam in Northeast India. Comptes Rendus Geosci 344:1–13. https://doi.org/10.1016/j.crte.2011.11.002

Joshi SC (2004) Uttaranchal: environment and development- geoecological overview. Gyanodaya Prakashan, Nainital, Uttaranchal, India

Karki R, Hasson S, Schickhoff U et al (2017) Rising precipitation extremes across Nepal. Climate 5:4. https://doi.org/10.3390/cli5010004

Khandelwal DD, Gupta AK, Chauhan V (2015) Observations of rainfall in Garhwal Himalaya, India during 2008–2013 and its correlation with TRMM data. Curr Sci 108:1146–1150

Kuang X, Jiao JJ (2016) Review on climate change on the Tibetan plateau during the last half century. J Geophys Res 121:3979–4007. https://doi.org/10.1002/2015JD024728

Kulkarni A (2012) Weakening of Indian summer monsoon rainfall in warming environment. Theor Appl Climatol 109:447–459. https://doi.org/10.1007/s00704-012-0591-4

Kulkarni A, Patwardhan S, Kumar KK et al (2013) High-resolution regional climate model PRECIS projected climate change in the Hindu Kush—Himalayan region by using the high-resolution Regional Climate Model PRECIS. Mt Res Dev 33:142–151. https://doi.org/10.1659/MRD-JOURNAL-D-12-00027.1

Kumari M, Singh CK, Bakimchandra O, Basistha A (2017) Geographically weighted regression based quantification of rainfall–topography relationship and rainfall gradient in Central Himalayas. Int J Climatol 37:1299–1309. https://doi.org/10.1002/joc.4777

Kumar V, Jain SK (2010) Trends in seasonal and annual rainfall and rainy days in Kashmir Valley in the last century. Quat Int 212:64–69. https://doi.org/10.1016/j.quaint.2009.08.006

Kumar N, Jaswal AK (2016) Historical temporal variation in precipitation over Western Himalayan Region: 1857–2006. J Mt Sci 13:672–681. https://doi.org/10.1007/s11629-014-3194-y

Kumar V, Jain SK, Singh Y (2010) Analysis of long-term rainfall trends in India. Hydrol Sci J 55:484–496. https://doi.org/10.1080/02626667.2010.481373

Ma H, Zhu Y, HUA W (2019) Interdecadal change in the South Asian summer monsoon rainfall in 2000 and contributions from regional tropical SST. Atmos Ocean Sci Lett 12:399–408. https://doi.org/10.1080/16742834.2019.1648168

Maharana P, Dimri AP, Choudhary A (2020) Future changes in Indian summer monsoon characteristics under 1.5 and 2 °C specific warming levels. Clim Dyn 54:507–523. https://doi.org/10.1007/s00382-019-05012-8

Mal S, Dimri AP, Jeelani G, Allen SK, Scott CA, Arora M, Banerjee A, Lone SA (2021) Determining the quasi monsoon front in the Indian Himalayas. Quaternary International. https://doi.org/10.1016/j.quaint.2021.02.010

Mir RA, Jain SK, Saraf AK (2015) Analysis of current trends in climatic parameters and its effect on discharge of Satluj River basin, western Himalaya. Nat Hazards 79:587–619. https://doi.org/10.1007/s11069-015-1864-x

Mukherjee S, Joshi R, Prasad RC et al (2015) Summer monsoon rainfall trends in the Indian Himalayan region. Theor Appl Climatol 121:789–802. https://doi.org/10.1007/s00704-014-1273-1

Naidu CV, Srinivasa Rao BR, Bhaskar Rao DV (1999) Climatic trends and periodicities of annual rainfall over India. Meteorol Appl 6:395–404. https://doi.org/10.1017/S1350482799001358

Nuzzo R (2014) Scientific method: Statistical errors. Nature 506:150–152. https://doi.org/10.1038/506150a

Palazzi E, Von Hardenberg J, Provenzale A (2013) Precipitation in the Hindu-Kush Karakoram Himalaya: observations and future scenarios. J Geophys Res Atmos 118:85–100. https://doi.org/10.1029/2012JD018697

Panday PK, Thibeault J, Frey KE (2015) Changing temperature and precipitation extremes in the Hindu Kush-Himalayan region: an analysis of CMIP3 and CMIP5 simulations and projections. Int J Climatol 35:3058–3077. https://doi.org/10.1002/joc.4192

Pande RK (2006) Landslide problems in Uttaranchal, India: issues and challenges. Disaster Prev Manag 15:247–255. https://doi.org/10.1108/09653560610659793

Pande A, Joshi RC, Jalal DS (2002) Selected landslide types in the Central Himalaya: their relation to geological structure and anthropogenic activities. Environmen 22:269–287. https://doi.org/10.1023/A:1016536013793

Paul S, Ghosh S, Oglesby R et al (2016) Weakening of Indian summer monsoon rainfall due to changes in land use land cover. Sci Rep. https://doi.org/10.1038/srep32177

Radziejewski M, Kundzewicz ZW (2004) Detectability of changes in hydrological records. Hydrol Sci J 49:39-51. https://doi.org/10.1623/hysj.49.1.39.54002

Sah M, Philip G, Mool PK et al (2005) Uttaranchal Himalaya, India: inventory of glaciers and glacial lakes and the identification of potential glacial lake outburst floods (GLOFs) Affected by Global Warming in the Mountains of Himalayan Region. Wadia Institute of Himalayan Geology (WIHG), International Centre for Integrated Mountain Development (ICIMOD), Asia-Pacific Network for Global Change (APN), Global Change SysTem for Analysis, Research, and Training (START) and United Nation's Environmental Programme (UNEP)

Sandeep S, Ajayamohan RS, Boos WR et al (2018) Decline and poleward shift in Indian summer monsoon synoptic activity in a warming climate. Proc Natl Acad Sci 115:2681 LP–2686. https://doi.org/10.1073/pnas.1709031115

Sati SP, Gahalaut VK (2013) The fury of the floods in the north-west Himalayan region: The Kedarnath tragedy.

Geomatics, Nat Hazards Risk 4:193–201. https://doi.org/10.1080/19475705.2013.827135

Schickhoff U, Mal S (2020) Current changes in Alpine Ecosystems of Asia. In: Goldstein MI, DellaSala DA (eds) Encyclopedia of the World's Biomes 1:589–598. Elsevier. https://doi.org/10.1016/b978-0-12-409548-9.12399-1

Schickhoff U, Singh RB, Mal S (2016) Climate change and dynamics of glaciers and vegetation in the Himalaya: an overview. In: Singh RB, Schickhoff, U, Mal S (ed) Climate change, glacier response, and vegetation dynamics in the Himalaya. Springer, Cham. 1–399. https://doi.org/10.1007/978-3-319-28977-9

Scott CA, Sharma B (2009) Energy supply and the expansion of groundwater irrigation in the Indus-Ganges Basin. Int J River Basin Manag 7:119–124. https://doi.org/10.1080/15715124.2009.9635374

Sen PK (1968) Estimates of the regression coefficient based on Kendall's Tau. J Am Stat Assoc 63:1379–1389. https://doi.org/10.2307/2285891

Sen Roy S, Singh RB (2002) Climate variability, extreme events and agricultural productivity in mountain regions. Oxford and IBH Publication Co. Pvt. Ltd., New Delhi, India

Seneviratne SI, Nicholls N, Easterling D et al (2012) Changes in climate extremes and their impacts on the natural physical environment. In: Field CB, Barros V, Stocker TF et al. (eds) Managing the risks of extreme events and disasters to advance climate change adaptation. A special report of working groups I and II of the intergovernmental panel on climate change (IPCC). Cambridge University Press, Cambridge, UK, and New York, NY, USA, pp 109–230

Shafiq MU, Rasool R, Ahmed P, Dimri AP (2018) Temperature and precipitation trends in Kashmir valley, North Western Himalayas. Theor Appl Climatol 135:293–304. https://doi.org/10.1007/s00704-018-2377-9

Shiva V (2009) Water Wars in India. In: Brauch HG, Spring ÚO, Grin J et al (eds) Facing global environmental change: environmental, Human, energy, food, health and water security concepts. Springer, Berlin, pp 589–592

Shrestha ML (2000) Interannual variation of summer monsoon rainfall over Nepal and its relation to Southern Oscillation Index. Meteorol Atmos Phys 75:21–28. https://doi.org/10.1007/s007030070012

Shrestha AB, Wake CP, Dibb JE, Mayewski PA (2000) Precipitation fluctuations in the Nepal Himalaya and its vicinity and relationship with some large scale. Int J Climatol 327:317–327. https://doi.org/10.1002/(SICI)1097-0088(20000315)20:3<317::AID-JOC476>3.0.CO;2-G

Shrestha AB, Bajracharya SR, Sharma AR et al (2017) Observed trends and changes in daily temperature and precipitation extremes over the Koshi river basin 1975–2010. Int J Climatol 37:1066–1083. https://doi.org/10.1002/joc.4761

Singh RB, Mal S (2014) Trends and variability of monsoon and other rainfall seasons in Western Himalaya, India. Atmos Sci Lett 15:218–226. https://doi.org/10.1002/asl2.494

Singh D, Horton DE, Tsiang M et al (2014) Severe precipitation in Northern India in June 2013: causes, historical context, and changes in probability. In: Herring SC, Hoerling MP, Peterson TC, Stott PA (eds) Explaining extreme events of 2013 from a climate perspective. Bulletin of the American meteorological society, vol. 95, No. 9, Sept 2014. American Meteorological Society

Smadja J, Aubriot O, Puschiasis O et al (2015) Climate change and water resources in the Himalayas. J Alpi Res. https://doi.org/10.4000/rga.2910

State Action Plan on Climate Change (2014) Government of Uttarakhand, India

Theil H (1950) A rank-invariant method of linear and polynomial regression analysis, Part 3. Netherlands Akademie van Wettenschappen, Proceedings 53, 1397–1412.

Xu C, Sano M, Dimri AP et al (2018) Decreasing Indian summer monsoon on the northern Indian subcontinent during the last 180 years: evidence from five tree-ring cellulose oxygen isotope chronologies. Clim Past 14:653–664. https://doi.org/10.5194/cp-14-653-2018

Yang J, Tan C, Zhang T (2013) Spatial and temporal variations in air temperature and precipitation in the Chinese Himalayas during the 1971–2007. Int J Climatol 33:2622–2632. https://doi.org/10.1002/joc.3609

You Q, Kang S, Aguilar E, Yan Y (2008) Changes in daily climate extremes in the eastern and central Tibetan Plateau during 1961–2005. J Geophys Res Atmos 113. https://doi.org/10.1029/2007JD009389

Yue S, Pilon P (2004) A comparison of the power of the test, Mann-Kendall and bootstrap tests for trend detection. Hydrol Sci J 49:21–37. https://doi.org/10.1623/hysj.49.1.21.53996

Yue S, Hashino M (2003) Long term trends of annual and monthly precipitation in Japan. J Am Water Resour Assoc 39:587–596. https://doi.org/10.1111/j.1752-1688.2003.tb03677.x

Yue S, Pilon P, Phinney B, Cavadias G (2002) The influence of autocorrelation on the ability to detect trend in hydrological series. Hydrol Process 16:1807–1829. https://doi.org/10.1002/hyp.1095

Yue S, Pilon P, Phinney B (2003) Canadian streamflow trend detection: Impacts of serial and cross-correlation. Hydrol Sci J 48:51–63. https://doi.org/10.1623/hysj.48.1.51.43478

Zhang X, Vincent LA, Hogg WD, Niitsoo A (2000) Temperature and precipitation trends in Canada during the 20th century. Atmos Ocean 38:395–429. https://doi.org/10.1080/07055900.2000.9649654

Zhang Q, Xu CY, Becker S et al (2009) Trends and abrupt changes of precipitation maxima in the Pearl River basin, China. Atmos Sci Lett 10:132–144. https://doi.org/10.1002/asl.221

4 Spatio-Temporal Heterogeneity in Glaciers Response Across Western Himalaya

Saurabh Kaushik, Pawan Kumar Joshi, Tejpal Singh, and Mohd Farooq Azam

Abstract

The Himalayan cryosphere hosts the largest concentration of glaciers, outside the polar regions, thereby provides extensive resource and opportunity for irrigation, hydro-power generation, and eco-tourism to the people of Central and Southeast Asia. Like any other mountain region, Himalayan glaciers are responding to climate change in terms of length, area, volume, equilibrium line altitude, flow velocity and elevation change. However, the rate of change of these glaciers remains controversial which causes gaps in our understanding of current state of Himalayan glaciers. Through this chapter, we provide a comprehensive picture of glaciers in the Western Himalaya (WH) on the basis of published scientific data and to bridge in some of these gaps. Here, a multi-parametric approach is adopted in order to provide a coherent picture of glacier health in WH. Further, a detailed analysis is carried out at basin scale in order to demonstrate the spatial heterogeneity among glaciers across WH. The chapter also highlights factors responsible for spatial heterogeneneity of glaciers across WH. Additionally, increasing risk of Glacial Lake Outburst Flood (GLOF) to the downstream region is discussed in brief.

Keywords

Himalayan glaciers · Mass balance · Remote sensing · Climate change · Heterogeneity

S. Kaushik · T. Singh
Academy of Scientific and Innovative Research (AcSIR-CSIO), CSIR-CSIO Campus, Chandigarh 160030, India

S. Kaushik · T. Singh
CSIR-Central Scientific Instruments Organisation, Chandigarh 160030, India

P. K. Joshi (✉)
School of Environmental Sciences, Jawaharlal Nehru University, New Delhi 110067, India

P. K. Joshi
Special Center for Disaster Research, Jawaharlal Nehru University, New Delhi 110067, India

M. F. Azam
Discipline of Civil Engineering, School of Engineering, Indian Institute of Technology Indore, Simrol 453552, India

4.1 Introduction

The Hindu-Kush-Himalaya including the Karakoram mountains comprise the largest concentration of glaciers outside the polar region, and thereby it is often referred to as the "*Third Pole*" (Bajracharya and Shrestha 2011; Kaushik et al. 2019a). Out of this the Himalayan-Karakoram (HK) region extends about 2500 km in length from Afghanistan in the west to Yunnan

U. Schickhoff et al. (eds.), *Mountain Landscapes in Transition*,
Sustainable Development Goals Series, https://doi.org/10.1007/978-3-030-70238-0_4

Province in the east through Pakistan, northern India, southern Tibet, Nepal and Bhutan (Azam et al. 2018). The HK region is imperative to the people of South and Central Asia, as it contributes significantly to the hydrology and controls run-off regimes of major perennial river systems, e.g. Indus, Ganga and Brahmaputra (Azam et al. 2018). Therefore, this region is termed as the "water tower of Asia", as a recent estimate reveals that the major river basin (Ganga, Indus and Bharamputra) provides 23.8 km^3 water seasonally. This seasonal run-off act as drought-resilient source of water helps to sustain basic, municipal and industrial need of 221 ± 59 million people (Pritchard 2019). Moreover, HK provides extensive opportunity for irrigation, hydro-power generation and eco-tourism which act as the major drivers of regional economy. The HK region also bears a greater significance in terms of climate change owing to its sensitivity towards changing climate and its capability to preserve the past record of climate change. On the basis of climatic settings, the glaciers of HK can be broadly divided into two categories, i.e. (1) Asian Monsoon type and (2) Westerlies fed (Bolch et al. 2012). Glaciers of central and eastern Himalaya belong to the former category which receive maximum precipitation from Asian Monsoon while glaciers of Karakoram belongs to the later where westerlies provide precipitation in winters (Bolch et al. 2012). Whereas glaciers of Western Himalaya (WH) lie in a transitional zone, resultant they receive precipitation from both climate systems (Azam et al. 2019). This is one of the motivations to carry out the present review of the state of glaciers in WH. The glacierized area for the HK region varies from 36,845 to 50,750 km^2 (Table 4.1). These disparities in the area may be primarily attributed to the differences in adopted methodology (semi-automated or manual), type of data used, scale of mapping, cloud cover, subjectivity in the identification of terminus due to the presence of debris cover and inclusion or not of steep avalanche walls (Azam et al. 2018; Bhambri and Bolch 2009; Kaushik et al. 2019b).

The oldest information of glacier area extents is present in the topographic maps prepared by Survey of India (SOI) in the early 1960s. However, these SOI maps have serious accuracy issues specifically due to the presence of seasonal snow (Bhambri and Bolch, 2009) which has led to an overestimation of glacierized area in some of the studies (e.g. (Kulkarni et al. 2006; Kulkarni et al. 2011). In the recent decade, the proliferation of satellite imageries with increased availability and improved quality (i.e. spatial, spectral and temporal resolutions) has led to fast and cost-effective monitoring of glacier parameters (Kaushik et al. 2019b). Employing various satellite data, numerous studies have estimated glacier parameters at different scales (Bajracharya and Shrestha 2011; Chand et al. 2017; Kaushik et al. 2019a; Kulkarni et al. 2011; Garg et al. 2017; Scherler, Leprince, and Strecker 2008; Shukla and Qadir 2016; Satyabala 2016; Azam et al. 2012; Thakuri et al. 2014; Nuimura et al. 2015; Kulkarni et al. 2007; Kulkarni et al. 2005; Pandey and Venkataraman 2013).

The glaciers of HK region gained global attention after an erroneous statement by Intergovernmental Panel on Climate Change (2007) in its fourth assessment report (AR4) that glaciers of this region will vanish by 2035 (Cogley et al. 2010). However, this error has now been acknowledged and corrected by several later studies (Bolch et al. 2012; Scherler et al. 2011). During the last decade considerable advancement in the knowledge of HK glaciers has taken place mainly due to the increasing interest of scientific community, increase in data availability (satellite and in situ) and development of recent image processing and machine learning algorithms, e.g. (Zhang et al. 2019). The improvement in the satellite image resolution (i.e. spatial, spectral and temporal) along with advancement in Synthetic Aperture Radar (SAR), Interferometric SAR (InSAR), Unmanned Aerial Vehicle (UAV), Light Detection and Ranging System (LiDAR) techniques have made it possible to observe multiple parameters of glaciers which were not possible earlier (Kaushik et al. 2019a). As most of the previous studies (Bajracharya et al. 2014; Kulkarni et al. 2011; Bhambri et al. 2011) could only estimate the snout and length fluctuation in order to assess glacier health, the

Table 4.1 Glacier inventory for Himalaya-Karakoram (HK) region compiled from various studies as referred in the text

Glacierized area (km^2)			
Himalaya	Karakoram	Himalaya-Karakoram	References
33,050	15,400	48,450	(Dyurgerov and Meier 2005)
33,050	15,145	48,195	(Raina 2009)
21,973	21,205	43,178	(Cogley 2011)
20,279	13,646	33,925	(Bajracharya and Shrestha 2011)
22,829	17,946	40,775	(Bolch et al. 2012)
29,000	21,750	50,750	(Kääb et al. 2012)
25,692 ± 700	19,680 ± 1052	45,372	(Pfeffer et al. 2014)
19,460	17,385	36,845	(Nuimura et al. 2015)

current advanced techniques have allowed estimation of multiple parameters (e.g. surface flow velocity (SFV), mass balance, equilibrium line altitude (ELA), elevation change, glacier facies characterization and their relationship with climate) for better assessment. As a consequence, the glaciers of HK region have been investigated extensively by the scientific community. This expansion of knowledge has motivated several reviews (Azam et al. 2018; Bolch et al. 2012; Kaushik et al. 2019a; Bhambri and Bolch 2009; Bhardwaj et al. 2016a; Bhardwaj et al. 2016b) which attempt to provide an up-to-date status of glaciers in HK region and report new emerging areas of scientific research. Yet there are limited studies which reports glacier health taking into account multiple parameter (i.e. area, length, MB, ELA, SFV, etc.) of glaciers to provide a comprehensive picture of their heterogeneity.

The available literature explicitly states that glaciers of HK region exhibit heterogeneous behaviour particularly owing to their geometry (size, slope and curvature), debris cover and precipitation regime (Bolch et al. 2012; Azam et al. 2012; Ali et al. 2017; Wang et al. 2019; Kaushik et al. 2019a; Patel et al. 2018; Scherler et al. 2011; Dehecq et al. 2019; Brun et al. 2017). In general, small glaciers (< 1 km) at lower elevations with thin debris cover are receding faster (Kaushik et al. 2019a; Pandey and Venkataraman 2013; Bhambri et al. 2011; Das and Sharma 2019). A comparative analysis of geographic regions reveals that mean mass loss of High Mountain Asia (HMA) is -0.18 ± 0.04 m w.e.yr^{-1}, whereas Lahual-Spiti and Karakoram show mass budget of -037 ± 0.09 and -0.03 ± 0.07 m w.e.yr^{-1} (Brun et al. 2017). Glaciers of Karakoram and Kunlun region have shown anomalous behaviour as they are stable or advancing (Brun et al. 2017). Moreover, the glacier's response within the same geographical region (e.g. river basin) is quite diverse as each glacier exhibits its own individual behaviour (Scherler et al. 2011; Bolch et al. 2012). Such heterogeneity is the premise of the present study and is further motivated by the recent advancement in knowledge of glaciers, in particular, the HK region. This chapter aims to highlight heterogeneity among glaciers of WH using multi-parametric approach (length, snout, SIV, MB, ELA) and attempt to identify factors responsible for such response. Additionally, review provides comparison of glacier response with other parts of HK region. Moreover, anticipated risk of Glacial Lake Outburst Flood (GLOF) in HK is discussed briefly.

4.2 Western Himalaya (WH)

The Indus River marks the western boundary of the Himalaya and separates it from the adjoining Karakoram Range. The WH extends southeast (approximately 560 km) from the bend of the Indus River in the north-west to the Sutlej River in the southeast (Zurick et al. 2005). The WH region bears great significance in terms of climate change studies as it falls under transitional zone between the Westerlies influence zone and Asian Monsoon influence zone (Ageta and Higuchi 1984; Azam et al. 2019). The region remains an area of interest for the scientific community for climate change studies using various glacier parameters and climate data. This section is subdivided on the basis of major glacier parameters in order to paint a vivid picture of glacier status in the region (Fig. 4.1).

4.2.1 Glacier Area and Length

Survey of India (SOI) and Geological Survey of India (GSI) are the two nodal agencies involved in mapping and monitoring of Himalayan glaciers, and thereby the custodians of the earliest information of glacial extent (in particular, the SOI topographical maps published in 1960s). A brief history of glacier mapping in Indian Himalaya is reported by (Bhambri and Bolch 2009; Kaushik et al. 2019b). Accordingly, the estimation of glacier area and length (snout retreat) is the most widely used parameter to demonstrate glacier response towards on-going

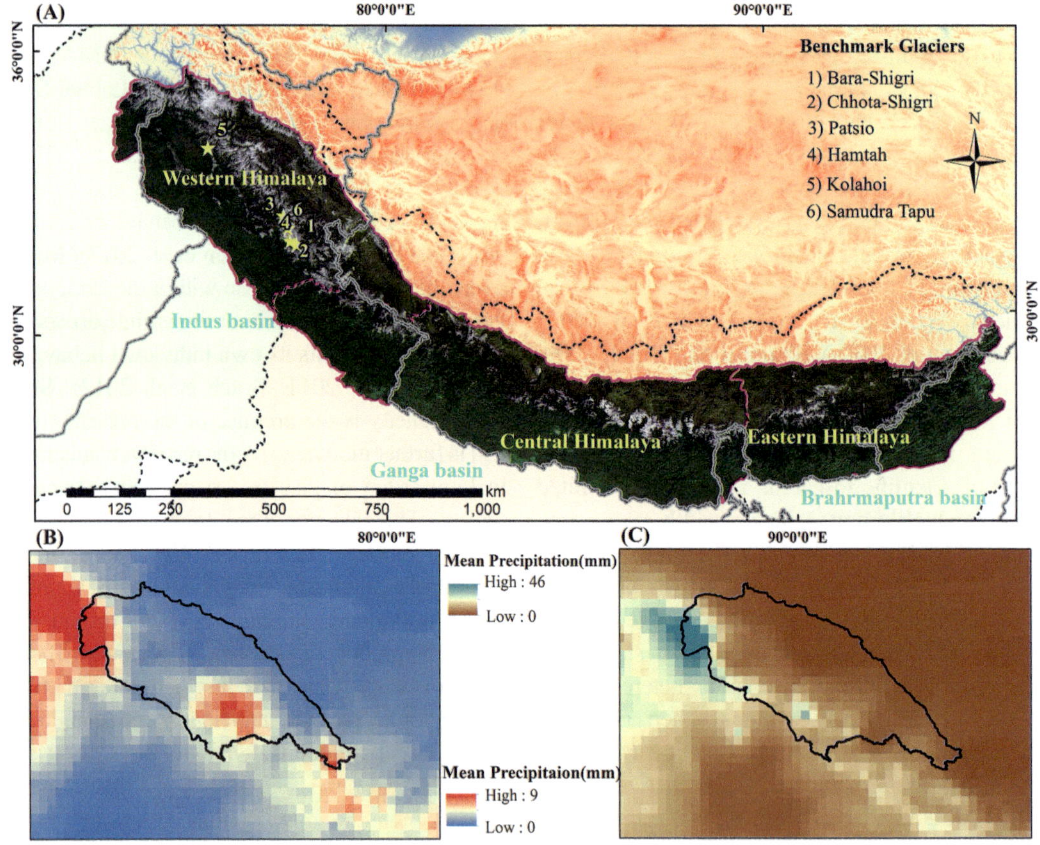

Fig. 4.1 **a** Map of Himalaya showing major river basin and benchmark glaciers of Western Himalaya. **b** Mean precipitation in January 2015 using APHRODITE data. **c** Mean precipitation in July 2015 using APHRODITE data (http://aphrodite.st.hirosaki-u.ac.jp/products.html#)

climate change, as these parameters are comparatively easy to estimate. Studies carried out in the WH with a view to estimate glacier length and area were generally carried out at basin-wide scale with few individual glaciers (Fig. 4.2). Variable rates of glacier loss are quite evident from the compilation of literature which clearly suggest that Himalayan glaciers are receding with heterogeneous rates since post-Little Ice Age (Fig. 4.2). The longest historically recorded glacier fluctuations in WH are available for Bara-Shigri, Chhota Shigri, Samudra Tapu and Hamtah glaciers (Kulkarni et al. 2006; Azam et al. 2014; Chand et al. 2017).

Most of previous studies carried out on single or several glaciers since 1960s use SOI maps and time series remote sensing data along with limited field observation. The highest loss of glacier area (-11.51 km^2 yr^{-1}) in this region is reported (Kulkarni et al. 2011) between 1962 and 2001 which carried out an analysis on 466 glaciers (Table 4.2). Glaciers of Chandra-Bhaga basin were studied extensively at different scales, e.g. (Pandey and Venkataraman 2013; Kulkarni et al. 2011; Kaushik et al. 2019a). However, as a consequence, various studies reported variable rates of deglaciation. Kulkarni et al. (2011) reported much higher rate of deglaciation (-3.64 and -2.79 km^2 yr^{-1}) for Chandra and Bhaga basin, whereas recent studies (Kaushik et al. 2019a; Das and Sharma 2019; Pandey and Venkataraman 2013; Birajdar et al. 2014) reported much lower rates of deglaciation (-0.36, -0.4, -0.19 and -0.31 km^2 yr^{-1}) for the Bhaga basin. In addition, average retreat rate varies from 4.72 ± 0.8 to 20.6 ± 7.6 m yr^{-1}. Similarly, considerable differential rates ($0.017 \pm 0.0001$and -0.14 ± 0.12 km^2 yr^{-1}) of glacier loss is reported for Ravi basin (Chand and Sharma 2015; Chand et al. 2016). Discrepancies among these rates of glacier losses could be attributed to number of glaciers involved, scale of mapping, method of mapping, data used, scene characteristics (e.g. cloud cover) and observation period. They are even further complicated owing to the use of SOI topographic maps as base maps in which seasonal snow was marked in the glacier boundary (Pandey and Venkataraman 2013; Bhambri et al. 2011; Kaushik et al. 2019a) which were reported as

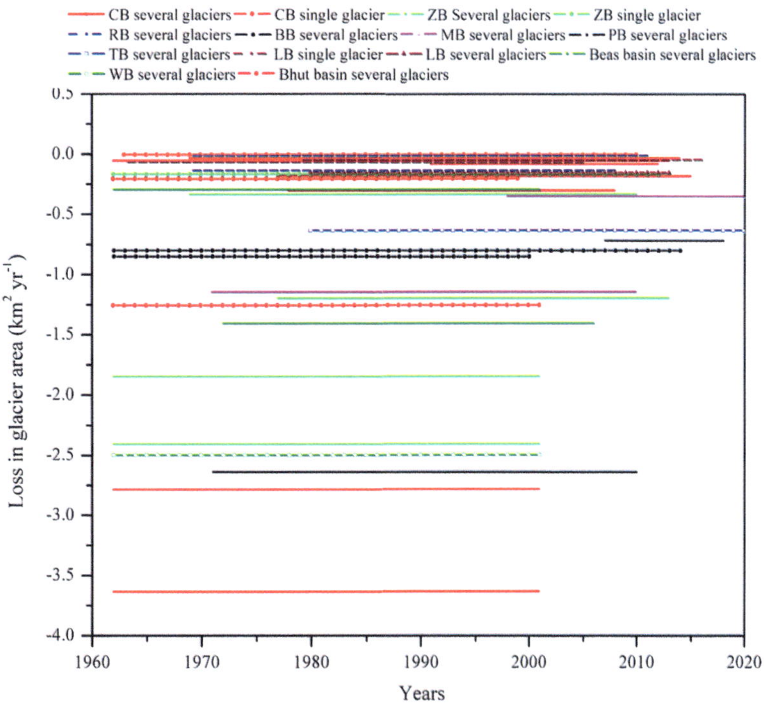

Fig. 4.2 Basin-wise loss of glacier area in the Western Himalaya. (CB Chandra-Bhaga basin; ZB Zanskar basin; RB Ravi basin; BB Baspa basin; MB Miyar basin; PB Parbati basin; TB Trikhund basin; LB Lidder basin; WB Warwan basin)

Table 4.2 Glacier area and length changes studies in the Western Himalaya

No of glacier studied	Data used	Observation period	Initial area km^2	Final area km^2	Area change km^2	Area change km^2 yr^{-1}	Average retreat rate M yr^{-1}	References
Beas Basin								
1	Toposheet, IRS LISS II, LISS III, Landsat TM	1962–2001	48.4	36.9	−11.6	−0.30		(Kulkarni et al. 2005)
224	Landsat MSS/TM/ETM + , IRS LISS III	1972–2006	419	371	−48	−1.41		(Dutta, Ramanathan, and Linda 2012)
Chenab Basin								
1	Toposheet, IRS PAN, IRS LISS III	1962–2000	73	65	−8	−0.21		(Kulkarni et al. 2006)
116	Toposheet, IRS LISS III	1962–2001	696	554	−142.0	−3.64		(Kulkarni et al. 2011)
111	Toposheet, LISS III	1962–2001	363	254	−109.0	−2.79		(Kulkarni et al. 2011)
1	Toposheet, Landsat TM/ETM + , ASTER, IRS LISS III, AWiFS	1963–2010	3.4	3.0	−0.42	−0.01		(Pandey et al. 2011)
15	Landsat MSS/TM, IRS LISS III, AWiFS	1980–2010	377.6 ± 5.7	368.2 ± 14.7	−9.3 ± 0.5	−0.31	−15.5 ± 5.6	(Pandey and Venkataraman 2013)
231	Toposheet, IRS LISS III, ASTER	2001–2011	391.56 ± 3.76	385.17 ± 3.71	−6.39 ± 0.6	−0.639		(Birajdar et al. 2014)
2	Landsat, TM, ETM + , OLI, Map	1962–2013	30.16	26.78	−3.38	−0.06	−9.35 ± 0.7	(Sharma et al. 2016)
3	Landsat TM, ETM, OLI TIRS	1993–2014	158.46 ± 1.2	156.56	−1.9	−0.09	−10.1 ± 2.5	(Garg et al. 2017)
1	Corona, Hexagon, Landsat TM, ETM + , OLI,	1863–2014	–	–	−4 ± 0.6	−0.03 ± 0.004	19.2 + 0.3	(Chand et al. 2017)
238	IRS LISS III, AWiFS, SRTM, Toposheet	2001–2010	751.9 + 43.3	743.36 + 31.1	−8.54	−0.94		(Brahmbhatt et al. 2017)

(continued)

Table 4.2 (continued)

No of glacier studied	Data used	Observation period	Initial area km^2	Final area km^2	Area change km^2	Area change km^2 yr^{-1}	Average retreat rate M yr^{-1}	References
127	Corona, Landsat, TM, ETM + , Sentinel 2 MSI, ASTER	1971–2016	196.0 ± 2.3	181.3 ± 2.3	−14.7 ± 4.3	−0.4		(Das and Sharma 2019)
48	Landsat TM, ETM + , OLI, ASTER, Seninel 2MSI	1979–2017	238.02 ± 9.8	230.76 ± 7.0	−7.26	−0.19	−12.44 ± 3.1	(Kaushik et al. 2019a)
Baspa Basin								
19	Toposheet, IRS LISS III	1962–2001	173	140	−33.0	−0.85		(Kulkarni et al. 2011)
19	Landsat TM, ETM + , OLI, Map	1962–2014	173 ± 15.17	131.7 ± 8.37	−41.1 ± 8.37	−0.8	–	(Gaddam, Kulkarni, and Gupta 2016)
Parbati Basin								
90	Toposheet, IRS LISS III	1962–2001	493	390	−103.0	−2.64		(Kulkarni et al. 2011)
51	Toposheet, IRS LISS III	1998–2009	154.3 ± 0.39	146.3	−8	−0.72		(Kulkarni and Karyakarte 2014)
Miyar Basin								
166	Toposheet, IRS LISS III	1962–2001	568	523	−45.0	−1.15		(Kulkarni et al. 2011)
29	Landsat, TM, ETM, OLI	1989–2014	227 ± 12.8	218 ± 12.8	−9 ± 0.7	−0.36	−9.6 ± 5.2	(Patel et al. 2018)
Warwan Basin								
253	Toposheet, IRS LISS III	1962–2001	847	672	−175.0	−4.49		(Kulkarni et al. 2011)
324	LISSIII, AWiFS, SRTM, Map	1962–2001	946.4 + 50	848.7 + 48.9	−97.7	−2.5		(Brahmbhatt et al. 2017)
Zanskar Basin								
671	Toposheet, IRS LISS III	1962–2001	1023	929	−94.0	−2.41		(Kulkarni et al. 2011)

(continued)

Table 4.2 (continued)

No of glacier studied	Data used	Observation period	Initial area km^2	Final area km^2	Area change km^2	Area change km^2 yr^{-1}	Average retreat rate M yr^{-1}	References
121	Corona, SPOT, Landsat TM/ETM+	1969–2010	96.4	82.6	−13.8	−0.34		(Schmidt and Nüsser 2012)
13	Toposheet, Landsat MSS/TM/ETM + , IRS LISS III, GPS	1962–2001	363.4	291.2	−72.1	−1.85		(Rai, Nathawat, and Mohan 2013)
1	Toposheet, Landsat MSS/TM/ETM + , IRS LISS III, AWiFS	1962–2012	23.82	15.3	−8.5	−0.17		(Ghosh and Pandey 2013)
5	Landsat MSS, TM, ETM + , OLI	1977–2013	291.37 ± 0.02	247.58 ± 0.003	−43.80 ± 0.017	−1.2	−7.8 ± 2.96	(Shukla and Qadir 2016)
Bhut Basin								
189	Toposheet, IRS LISS III	1962–2001	469	420	−49.0	−1.26		(Kulkarni et al. 2011)
Ravi Basin								
157	Landsat ETM +/Aster GDEM/Corona KH 4B/World view/Landsat 8 OLI TRIS	1971–2010	125.9 ± 1.9	120 ± 4.8	−5.9 ± 5.2	−0.14 ± 0.12		(Chand and Sharma 2015)
5	Corona, Landsat, TM, ETM + , OLI, ASTER	1971–2013			−0.73± 0.005	−0.017 ± 0.0001	−4.72 ± 0.8	(Chand et al. 2016)
Tirungkhad Basin								
32	Toposheet, Landsat MSS TM, ETM + , ASTER	1966–2011	112	82	−29.1	−0.64	20.2	(Mir et al. 2014)
Jhelum Basin								
1	Toposheet, IRS LISS III	1963–2005	13.6	10.7	−2.9	−0.07		(Kanth et al. 2011)
45	Landsat TM, EMT + , OLI, ASTER	1996–2014	95.5 ± 1.4	83.6	−11.9 ± 1.4	−0.6	−20.6 ± 7.6	(Ali et al. 2017)

(continued)

Table 4.2 (continued)

No of glacier studied	Data used	Observation period	Initial area km^2	Final area km^2	Area change km^2	Area change km^2 yr^{-1}	Average retreat rate M yr^{-1}	References
9	Landsat MSS,TM, EMT + , OLI, ASTER	1980–2013	29.01	23.79	−5.22	−0.16	−8.07	(Murtaza and Romshoo 2017)
1	Landsat MSS,TM,ETM + , OLI and ASTER	1979–2016	11 ± 0.09	9.26	−1.74 ± 0.02	−0.05		(Mir 2018)
Others								
53	Landsat MSS/TM, ASTER	1976–2003	84.41	77.29	−7.12	−0.26		(Ye et al. 2006)
466	Toposheet, IRS LISS III, LISS IV, GPS	1962–2001	2077	1628	−449	−11.51		(Kulkarni et al. 2007)
–	–	1974–2003	108	100	−8.0	−0.28		(Ye, Yao, and Naruse 2008)
62	ASTER, Landsat TM, ETM + ,	2000–2007	1119.1	1116.92	−2.18	−0.007	−11.95	(Scherler, Bookhagen, and Strecker 2011)

early as 1980 (Vohra 1980). Recent studies (Pandey and Venkataraman 2013; Garg et al. 2017; Mir 2018; Ali et al. 2017; Chand et al. 2016) have reported lower rates of deglaciation compared to previous studies (Fujita and Nuimura 2011; Kulkarni et al. 2007; Kulkarni et al. 2006), particularly, owing to exclusion of SOI topographic maps. Therefore, it is quite clear that errors in SOI topographic maps led to an escalated statement about glacier health in the Himalaya and ensuing furore highlighted limited knowledge and actual fate of Himalayan glaciers (Inman 2010).

It is quite obvious that glaciers of the WH are receding at variable rates; thus, differential loss in glacier area causes fragmentation of large glaciers and ultimately increase in number of glaciers (Kulkarni et al. 2007; Kulkarni et al. 2011). On the basis of available literature, it can be concluded that change in glacier length (e.g. advance or retreat) is a dynamic aspect of ice flow which is primarily governed by glacier geometry, debris cover and regional climate (Fig. 4.2, Table 4.2). Debris cover, climate and geometry of glacier (i.e. slope, extent and aspect) generally govern the glacier response. Small and low elevation glaciers are responding much faster to climate change rather than larger and high elevation glaciers (Scherler et al. 2011; Kaushik et al. 2019a; Patel et al. 2018; Sharma et al. 2016; Bhambri et al. 2011). According to Cogley (2016) unweighted average shrinkage rate for High Mountain Asia (HMA) between 1960 and 2010 is $-0.57\%\ a^{-1}$ and weighted mean shrinkage rate is $-0.40\%\ a^{-1}$ for the same period, whereas the average area shrinkage for HK region is $-0.36\%\ a^{-1}$ (Azam et al. 2018). Thick debris (> 5 cm) cover acts as an insulator and retard the ablation rate, whereas glaciers covered with thin debris cover indicates a higher rate of deglaciation (Banerjee 2017). On the hand, clean ice glaciers are typically characterized by linear mass balance profile (i.e. higher ablation rate towards terminus). Scherler, Bookhagen, and Strecker (2011) explicitly stated higher rate of deglaciation in WH ($\sim 60\ myr^{-1}$), northern central Himalaya and West Kunlun Shan owing to low proportion of debris-covered glaciers. Moreover, several studies (Scherler, Bookhagen, and Strecker 2011; Thakuri et al. 2016; Shukla and Qadir 2016; Garg et al. 2017) have indicated that down-wasting is dominant mode of glacier retreat than area loss particularly in case of debris cover. Thus, there is a general consensus in glaciological community that based on snout monitoring solely imperative conclusions about the overall health of a glacier cannot be made. However, the mass balance of glaciers is a direct and key indicator for assessment of glacier response towards climate change. Further, recent studies (Dehecq et al. 2019; Yao et al. 2012; Garg et al. 2017) suggest a multi-parametric approach in which associated parameters (i.e. ELA, SIV and AAR) and provide a vivid picture of the overall glacier health.

4.2.2 Glacier Mass Balance

Glacier mass balance is difference between mass gain (via snowfall and/or avalanche) and mass loss (i.e. ablation) within a specific hydrological year (Oerlemans 2001). This resultant net gain and loss in glacier mass have a direct implication for water resource management, glacial hazards and climate change studies. Glacier mass balance is generally considered as an un-delayed and direct method to exhibit overall glacier health under the influence of climate change (Oerlemans 2001). Methods of mass balance estimation can be broadly classified in the following categories: glaciological method, geodetic method and glacier mass balance modelling method.

4.2.2.1 Glaciological Method

Glaciological method is generally considered as most accurate method to assess glacier mass balance, although this method is quite difficult, as it involves *in situ* measurements at rugged terrain in harsh climate condition (Pratap et al. 2016; Singh et al. 2018). This method uses stakes and pits placed over representative points (ablation and accumulation) on the glaciers (Pratap et al. 2016). The GSI initiated mass balance studies for Indian Himalaya using glaciological method in 1974 (Gara glacier, Himachal Pradesh). Unfortunately, in terms of mass balance observation, Himalayan range is still limited as there are only 24 glaciers surveyed using glaciological methods (Azam et al. 2018). Out of these 11 glaciers are situated in the WH (Fig. 4.3), Chhota Shigri, Hamtah and Gara are well-studied glaciers of WH in terms of mass balance measurements (Fig. 4.3).

Spatio-temporal variability (−0.11 to −1.43 w.e.yr^{-1}) in mass balance across WH is evident from previous studies (Fig. 4.3; Table 4.3). The highest loss of glacial mass (-1.43 m w.e.yr^{-1}) is reported for Hamtah glacier during 2000-2009, whereas Rulung glacier showed the lowest rate of loss (-0.11 m w.e.yr^{-1}) during 1979-1981. Higher rate of mass loss in Hamtah glacier is attributed to the presence of high debris cover over ablation zone and steep headwalls on the glacier. Azam et al. (2018) classified Hamtah glaciological series under 'dubious category' owing to fact that it is surveyed only in its lower part and contribution of avalanches from headwalls was not included. The mean mass balance of Hamtah glacier (–1.43 m w.e. yr-1) was two and half times more negative than Chhota Shigri (–0.59 m w.e. yr-1) over the common observed period (Azam et al. 2018). From the *in situ* mass balance observation, it is quite clear that rate of mass loss has increased after 1990 and debris cover glaciers are losing mass even with stable front. Azam et al. (2018) classified glaciological observation (Fig. 4.3) as good, fair, excellent and dubious on the basis of several parameters (e.g. geodetic verification of glaciological mass balance, uncertainty analysis, stake density, stake material and accumulation measurements etc.).

4.2.2.2 Geodetic Method

Glaciological method requires enormous amount of man-power and financial support, whereas the geodetic method is cost effective and simpler to estimate mass balance for larger region over a longer duration (Azam et al. 2018; Pratap et al. 2016). Geodetic method involves differencing of digital elevation models (DEMs) derived from stereo remote sensing imageries (Kumar et al. 2018; Berthier et al. 2007). DEM differencing provides change in glacier thickness which is used to determine the mass balance (Pratap et al. 2016; Cogley 2009). However, the accuracy of DEM and uncertainty estimation is of utmost importance in order to draw inferences from results. This method has limitation due to the unavailability of higher resolution DEM and aerial photographs with reference to DEM in public domain.

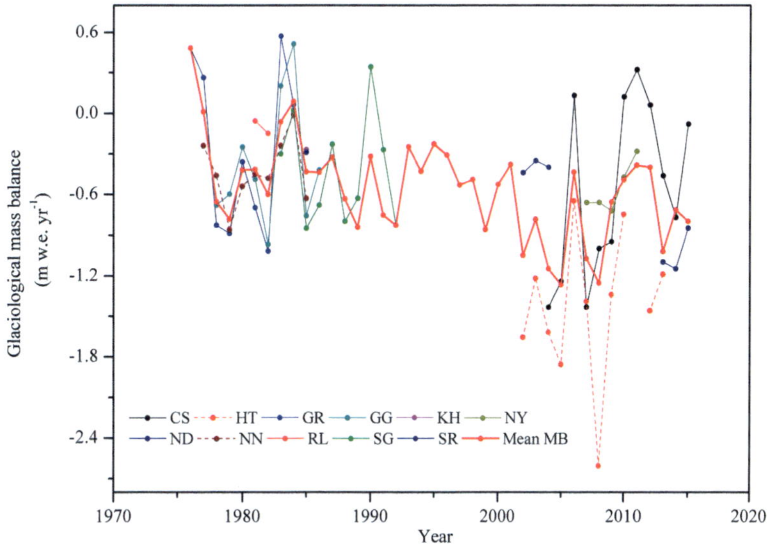

Fig. 4.3 Glaciological mass balance studies in Western Himalaya. Dark red line shows mean mass balance of all 24 glaciers surveyed using glaciological method

Table 4.3 Glaciological mass balance studies in Western Himalayan region

Glacier name (region/country)	Area (km^2)	Debris cover area (%)	Aspect	MB period	Mass balance (m w.e. yr^{-1})	References	Classification by Azam et al 2018
Glaciological mass balance							
Chhota Shigri (CS),	15.5	3.4	N	2002–2014	–0.56 ± 0.40	(Wagnon et al. 2007)	Excellent
Lahaul-Spiti, India						(Azam et al. 2012, 2016)	
Hamtah (HT),	3.2	~70	N	2000–2009	–1.43	GSI (2011)	Dubious
Lahaul-Spiti, India				2010–2012		(Mishra, Kumar, and Singh 2014)	
Gara (GR),	5.2	17	NE	1974–1983	–0.27	(Raina, Kaul, and Singh 1977)	Fair
Baspa Basin, India						(Sangewar and Siddique 2006)	
Gor Garang (GG),	2.0	~60	S	1976–1985	–0.38	(Sangewar and Siddique 2006)	Fair
Baspa Basin, India							
Kolahoi (KH),	11.9	clean	N	1983–1984	–0.27	(Kaul 1986)	Dubious
Jhelum Basin, India							
Naimona'nyi (NN),	7.8	clean	N	2005–2010	–0.56	(Yao et al. 2012)	Fair
Karnali Basin, China							
Naradu (ND),	4.6	~60	N	2000–2003	–0.72	(Koul and Ganjoo 2010)	Fair
Baspa Basin, India				2011–2015			
Neh Nar (NN),	1.3	clean	N	1975–1984	–0.43	GSI (200)	Good
Jhelum Basin, India							
Rulung (RL),	1.1	clean	NE	1979–1981	–0.11	(Srivastava et al. 2001)	
Zanskar Range, India						(Sangewar and Siddique 2006)	Excellent
Shaune Garang (SG),	4.9	24	N	1981–1991	–0.42	GSI (1992);	Fair
Baspa Basin, India						(Sangewar and Siddique 2006)	
Shishram (SR),	9.9	clean	N	1983–1984	–0.29	(Kaul 1986)	Fair
Jhelum Basin, India							

Source (Azam et al. 2018)

Notable studies carried out in WH with view to estimate glacier mass balance using geodetic method are listed in Table 4.4 and shown in Fig. 4.4. Lahaul-Spiti is the most widely studied region in the WH in terms of mass balance observations using geodetic method. Variable rates (−0.37 to −0.70 m w.e. yr^{-1}) are reported for the same region during different periods (Table 4.4). Chhota Shigri is a benchmark glacier of WH (Vincent et al., 2013) with comparable results of mass loss using geodetic (−0.52 m w.e. yr^{-1}; Fig. 4.4) and glaciological methods (−0.56 m w.e. yr^{-1}; Fig. 4.3). The lowest loss (−0.17 m w.e. yr^{-1}) of Chhota Shigri is reported during 1988–2010 which shows an increasing trend of mass loss in recent years due to a combined effect of increasing temperature and decreasing precipitation (Azam et al., 2014). Recently, Kumar et al. (2018) reported mass balance (−0.26 and −0.29 m w.e. yr^{-1}) of Patsio glacier during 2000–2013 for two density functions.

At regional scale, Hindu-Kush-Karakoram-Himalaya (HKKH) showed (−0.21 m w.e. yr^{-1}) mass loss during 2003–2008. A similar rate (−0.20 m w.e. yr^{-1}) is reported for Himalayan region. Recently, a notable contribution is made by Brun et al. (2017) on glacier mass balance observation of 92% of glacierized area of HMA using time series of DEMs over 2000 and 2016 (Fig. 4.4). Except Kunlun region, every other region and glaciers have shown negative mass balance. However, the positive mass balance of Kunlun region may be attributed to its continental setting which leads to marginal influence of East Asian Monsoon (EAM). Azam et al. (2018) reported that conventional glaciological mass balance methods involve error in case of highly debris-covered and avalanche fed glaciers. However, the selection of clean ice glacier at lower elevation as representative of the whole basin will introduce high bias and more negative mass balance of basin. Therefore, remote sensing-based methods are recommended for highly debris-covered glacier and avalanche fed glaciers. However, *in situ* measurements are significant in terms of glacier-climate interactions, glacio-hydrological modelling, ground control and validation of results obtained from remote sensing methods (Azam et al., 2018).

4.2.2.3 Glacier Mass Balance Modelling

In the WH, several studies have exploited modelling approaches to estimate glacier mass balance in which AAR-ELA and temperature index (TI) are quite common (Fig. 4.4). AAR-ELA method is based on the relationship between AAR, ELA and specific mass balance, e.g. Kulkarni, Rathore, and Alex (2004) developed a model (Eq. 1) for mass balance estimation using AAR and specific mass balance relationship. This model is based on a series of *in situ* measurements of Shaune Garang during 1982–1988 and Gor Garang glaciers during 1976–1984.

$$Y = 2.4301 \times X - 1.20187 \qquad (4.1)$$

where Y = specific mass balance in m w.e. and X = AAR

In WH, (Tawde et al. 2017, 2016) reported glacier mass balance (−0.71 and 0.61 m w.e. yr^{-1}) of Chandra basin using AAR and TI method. Here also, the variable rates of mass wastage are due to difference in observation period and number of glaciers studied. Several studies have exploited temperature-index model together with an accumulation model for estimation of glacier mass balance (Azam et al. 2014; Hock 2003). This model computes amount of ablation with positive air temperature and proportionality factor called degree-day-factor which requires daily temperature and precipitation as input. Variable rate mass loss is reported for Chhota Shigri using TI and AM model (−0.30 and −0.68 m w.e. yr^{-1}) during 1969–2012 and 2000–2019 (Table 4.5). However, discrepancies in rates could be attributed to different observation periods. For Chhota Shigri glacier over 2000–2013 period, different methods such as TI and AM (Albedo Methods) method

Table 4.4 Geodetic mass balance observation in Western Himalaya

Region/glacier	Area (km^2)	Period	Mass balance (m w.e. yr^{-1})	Year of observation (periods)	Reference
Western Himalaya					
Lahaul & Spiti (LS1)	915	1999–2004	−0.70 to −0.85	5 (1)	(Berthier et al. 2007)
Lahaul & Spiti (LS2)	2110	1999–2011	−0.45 ± 0.13	12 (1)	(Gardelle et al. 2013)
Lahaul & Spiti (LS3)	1796	2000–2012	−0.52 ± 0.32	12 (1)	(Vijay and Braun 2016)
Chhota Shigri (CS1)	15.5	1988–2010	−0.17 ± 0.09	22 (1)	(Vincent et al. 2013)
Chhota Shigri (CS1)	15.5	1999–2010	−0.44 ± 0.16	11 (1)	(Vincent et al. 2013)
Chhota Shigri (CS3)	15.5	2005–2014	−0.39 ± 0.24	9 (1)	(Azam et al. 2016)
Hamtah (HT)	3.2	1999–2011	−0.45 ± 0.16	12 (1)	(Vincent et al. 2013)
Jammu and Kashmir (JK)	4900	2003–2009	−0.55 ± 0.08	6 (1)	(Kääb et al. 2012)
Lahul & Spiti (LS4)	7960	2000–2016	−0.37 ± 0.09	16(1)	(Brun et al. 2017)
Patsio	2.37	2000–2013	−0.26 ± 0.11(1)* −0.29 ± 0.16(2)	13(1)	(Kumar et al. 2018)
Regional Means					
HKKH	60,100	2003–2008	−0.21 ± 0.05	5 (1)	(Kääb et al. 2012)
PKH	21,900	1999–2008/11	−0.14 ± 0.08	~10 (1)	(Gardelle et al. 2013)
HMA 1	118,200	2003–2009	−0.20 ± 0.10	6 (1)	(Gardner et al. 2013)
HMA 2	100,000	2000–2016	−0.18 ± 0.04	16	(Brun et al. 2017)
Nyainquentanglha	6380	2000–2016	−0.62 ± 0.23	16	(Brun et al. 2017)
Kunlun	9910	2000–2016	+0.14 ± 0.08	16	(Brun et al. 2017)

HKKH = Hindu-Kush-Karakoram-Himalaya; PKH = Pamir-Karakoram-Himalaya; HMA = Himalaya *Mass balance is reported for two different ice density function

have reported variable rate of mass loss (−0.52 and −0.68 m w.e. yr^{-1}) (Brun et al., 2015). Recently, Azam et al. (2019) has reconstructed mean mass balance (−0.30 ± 0.36 m w.e.yr−1) and mean catchment-wide run-off (1.56 ± 0.23 m w.e.yr−1) for Chhota Shigri glacier during 1969–2016. This study exploits glacio-hydrological model which is an ensemble of accumulation module, ablation module and rain module.

4.2.3 Surface Velocity

The preceding sections discuss that the glaciers of Himalayan region have heterogeneous rates of change since the 1970 (Fig. 4.3; 4; 5). The glaciers of some regions, e.g. Lahul-Spiti (−0.37 ± 0.09 m w.e. yr^{-1}) are experiencing mass loss close to global mean mass wastage (−0.42 m w.e. yr^{-1}). However, the link between associated changes in surface flow velocity (SFV) which leads to mass redistribution are

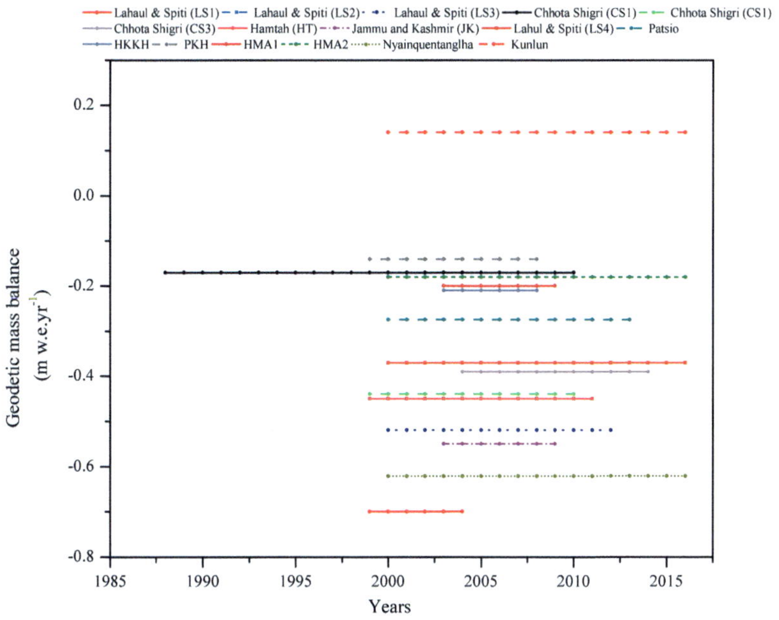

Fig. 4.4 Geodetic mass balance studies in Western Himalaya

Table 4.5 Mass balance modelling studies in the Western Himalaya

Glacier name (region/country)	Location	Area (km^2)	MB period	Mass balance (m w.e. yr^{-1})	Model	Reference
Western Himalaya						
4. Chhota Shigri (CS1),	32°28'N	15.5	1969–2012	–0.30 ± 0.36	TI	(Azam et al. 2014)
Lahaul-Spiti, India	77°52'E					
5. Chhota Shigri (CS2),	32°28'N	15.5	2000–2013	–0.68 ± 0.10	AM	(Brun et al. 2015)
Lahaul-Spiti, India	77°52'E					
Chandra Basin (12 Glaciers)	32°05'N 32°45' 76°15'E–77°50'E	209.91	1999/2000–2008/2009	–0.71 ± 0.34	AAR and TI	(Tawde et al. 2016)
Chandra Basin (146 Glaciers)	32°05'N 32°45' 76°15'E–77°50'E	703.64	1984–2012	−0.61 ± 0.4628	AAR and TI	(Tawde et al. 2017)
Chhota Shigri	32°28'N 77°52'E	15.5	1955–2014	−0.34 ± 0.33	CM	(Engelhardt et al. 2017)
Chhota Shigri	32°28'N 77°52'E	15.5	1989–2017 2003–2014	−0.37 ± 0.51 −0.56 ± 0.72	ELA	(Chandrasekharan et al. 2018)
Chhota Shigri	32°28'N 77°52'E	15.5	1969–2016	−0.30 ± 0.36	GHM	(Azam et al. 2019)

CM, Climate model; uses downscaled gridded climate data from different sources in order to evaluate glacier mass balance; *AM* Accumulation model; *GHM* Glacio-hydrological model poorly understood (Dehecq et al. 2019). It is evident from field measurements that glacier ice flow fluctuates with mass changes at decadal scales. Generally, glacier flow slowed down in regions with negative mass balance (Dehecq et al. 2019). However, no direct link could be established between glacier flow and mass balance at regional scale in Himalaya. Previous studies suggested SFV is of utmost importance in case of debris cover glacier particularly where mass balance information is lacking (Scherler, Bookhagen, and Strecker 2011; Dehecq et al. 2019). Debris-covered glaciers experience downwasting more than frontal retreat, thereby several studies reported direct link of glacier recession with decreasing SFV and formation of proglacial lakes (Thakuri et al. 2016; Garg et al. 2017; Dehecq et al. 2019; Scherler, Bookhagen, and Strecker 2011; Scherler, Leprince, and Strecker 2008; Satyabala 2016; Dobhal, Kumar, and Mundepi 1995). There are limited studies which are carried out to demonstrate spatio-temporal variation in glacier flow velocity across the Himalayan and Karakoram Range. In 2011, Scherler et al. (2011) reported SFV for 289 glaciers across Himalaya and Karakoram range in order to assess the impact of debris cover glacier on glacier terminus. Out of 64 studied glaciers of WH 1.5% glaciers were found to be stagnant with stable glacier front, whereas Hindu-Kush and southern Himalaya has higher percentage of stagnant and stable front (16% and 1.5%). A study by Garg et al. (2017) reported SFV of Sakchum, Chhota Shigri and Bara-Shigri glaciers located in Chandra basin. Results show decreasing trend of SFV for Sakchum (13.5. to 10.63 m yr^{-1}), Chhota Shigri (27.33 to 20.90 m yr^{-1}) and Bara-Shigri (32.50 to 25.30 m yr^{-1}) during 2003/04 to 2013/14. The highest dropdown in SFV is found for Sakchum glacier (21.43%). Moreover, detailed analysis reveals that slowdown in SFV is more prominent in upper ablation zone (UAZ), whereas the accumulation zone (ACZ) exhibits minimum slowdown in SFV. For the same glacier, Dobhal et al. (1995) reported flow rate of ~ 46–52 m yr^{-1} in UAZ during 1987–1989. A study (Azam et al. 2012) made notable contribution, as it attempt to demonstrate how mass wastage is affecting SFV and ice fluxes using field observation. Study reveals negative mass balance of Chhota Shigri glacier (−0.67 ± 0.40 m w.e. yr^{-1}) and decrease in SFV and ice fluxes (24–37%) during 2002–2010. Another study (Tiwari et al. 2014) demonstrated the SFV of Chhota Shigri glacier over the period of 2003–2009 which varies from ~ 20 m yr^{-1} to 40 m yr^{-1}. Moreover, the study reports 10% difference in remotely sensed and field-based SFV. However, recently, remarkable contribution is made by Dehecq et al. (2019) who has reported change in ice flow of all HMA glaciers over the period 2000–2017 using one million pairs of optical satellite images. Trend analysis of study reveals that 9 out of 11 surveyed regions exhibit slowdown concomitant with ice thinning. The WH (i.e. Lahaul and Spiti) exhibits the second largest slowdown (−34.3 ± 4.5% $decade^{-1}$) after Nyainqentanghla (−37.2 ± 1.1% $decade^{-1}$). Karakoram (3.6 ± 1.2% $decade^{-1}$) and West Kunlun (4.0 ± 2.1% $decade^{-1}$) region shows exceptional behaviour with slightly accelerated glacier flow. The study reveals that slowdown in glacier flow was more pronounced during 2005–2008 afterwards more stable condition is observed.

4.3 Comparison of Glacier Fluctuation with Other Parts of Himalaya

Heterogeneous behaviour of the Himalayan glaciers attained global significance as it is imperative to a large population and hydro-economies of south and central Asia in terms of their needs of freshwater resources, hydro-power, eco-tourism, agriculture and regional climate. However, the existing information of Himalayan glaciers is limited in terms of mass balance observation and SFV estimation as most of the studies focussed on area/length change assessment as an indicator of climate change. In Garhwal Himalaya, there are only four (i.e. Dunagiri, Tipra Bank, Chorabari and Dokriani) glaciers which have glaciological mass balance observations (Azam et al. 2018). Dunagiri glacier showed negative mass balance of −1.04 m w.e. yr^{-1} during 1984–1990 (GSI 1991), whereas Tipra Bank showed much lower rate of mass loss −0.14 m w.e. yr^{-1}. Dokriani glacier exhibits negative mass balance (−0.25 m w.e. $yr-^{1}$) during 1992–1995 and 0.39 m w.e.yr^{-1} during 1997–2000 (Gautam and Mukherjee 1992). On the other hand, the Chorabari glacier showed slightly higher rate of mass loss (−0.73 m w.e. yr^{-1}) (Dobhal, Mehta, and Srivastava 2013). In Dudh Koshi basin (i.e. Nepalese Himalaya), four glaciers (AX10, Mera, Pokalde and West Changri Nup) have mass balance records. Fujita et al. (2001) reported mean mass balance of AX10 glacier (−0.69 ± 0.08 m w.e.yr^{-1}) during 1978–1979 and 1995–2010. Mera glacier exhbits negative mass balance (−0.03 ± 0.28 m w.e.yr^{-1}) over the period of 2007–20015 (Wagnon et al. 2012; Sherpa et al. 2017), whereas West Changri Nup glacier showed comparatively higher rate of mass loss (−1.24 ± 0.7 m w.e. yr^{-1}) (Sherpa et al. 2017; Wagnon et al. 2012). Numerous studies (Bhambri et al. 2011; Kulkarni et al. 2011; Bajracharya and Shrestha 2011; Bajracharya, Maharjan, and Shrestha 2014) are carried out in Central Himalaya in order to quantify glacier area/length changes, which reports variable rate of deglaciation. However, higher rates of glacier loss (e.g. −4.79 and −4.26 km^2 yr^{-1}) (Kulkarni et al. 2011 Jin et al. 2005) are reported by studies which utilize SOI topographic map as base map for change assessment. Studies regarding spatio-temporal variation in glacier SFV is very limited, although Satyabala (2016) demonstrates SFV of Gangotri glacier, study explicitly highlight summer speed up in SFV (~ 57% −162%) compared to winter.

Changmekhangpu (India) and Gangju La (Bhutan) are the only representatives from Eastern Himalaya which have *in situ* mass balance records (GSI 2001; Tshering and Fujita (2016). Changmekhangpu (5.6 km^2) showed lower mass wastage of −0.26 m w.e.yr^{-1} during 1979–1986 as compared to Gangju La (0.3 km^2) −1.38 ± 0.38 m w.e.yr^{-1} during 2003–2014. This contrasting response is due to difference in observation period, glacier geometry and debris coverage (e.g. Gangju La is a clean ice, low altitude glacier). However, several studies carried out estimation of glacier area/length changes in Eastern Himalaya (Kulkarni et al. 2011; Basnett, Kulkarni and Bolch 2013; Racoviteanu et al. 2014; Bajracharya and Shrestha 2011). General deglaciation is observed for the region with increasing debris cover and resultant expansion of glacial lakes. The highest number of glaciers (i.e. 817) is studied by Bajracharya, Maharjan, and Shrestha (2014) and the study reports loss in glacier area with −6.52 km^2 yr^{-1}. Another study by Bajracharya, Maharjan and Shrestha (2014; and Racoviteanu et al. 2014) reported loss in glacierized area with 4.85 km2 yr^{-1} during 1962–2000. Recession rate of Sikkim Himalayan glaciers was found to be higher (−3.36 km^2 yr^{-1}) as compared to Nepal Himalaya (−1.44 km^2 yr^{-1}).

Recently, the focus has shifted to monitoring of glacial lake expansion, identification of potential glacial lakes and risk assessment for GLOF (Shukla, Garg, and Srivastava 2018; Aggarwal et al. 2017; Sattar, Goswami, and Kulkarni 2019). Expansion of proglacial lakes and development of supraglacial lakes are more profound in the Eastern Himalayan region as compared to North–West Himalaya (NWH) and Karakoram (Nie et al. 2017). In order to highlight heterogeneity of glacier response, recently, remarkable contribution is made by Brun et al. (2017) and Dehecq et al. (2019) which report mass balance and glacier flow for all HMA glaciers. Total mass change of −0.18 ± 0.04 m w.e. yr^{-1} between 2000 and 2006 is reported (Brun et al. 2017), and the standard deviation of glacier to glacier for regions was of ~ 0.25 m w.e.yr^{-1} which suggest high variable response of glaciers across the HMA.

Existing literature provides strong evidence of increasing temperature, decreasing precipitation which drives deglaciation in the Himalayan region along with reduction in SFV of glacier and expansion of glacial lakes (Azam et al. 2018; Bolch et al. 2012; Dehecq et al. 2019; Nie et al. 2017; Scherler, Bookhagen, and Strecker 2011; Shukla, Garg, and Srivastava 2018; Aggarwal et al. 2017; Bhambri et al. 2011; Kaushik et al. 2019a; Thakuri et al. 2016). In the recent decade, scientific community has widely reported expansion of existing glacial lakes and development of new glacial lakes and identification of potentially dangerous lakes and risk assessment of GLOF. In the HK region, spatially heterogeneous behaviour of glacial lakes is observed, although expansion of such lakes is more prominent in southern slope of central Himalaya (Nie et al. 2017). A study (Nie et al. 2017) reported the presence of 4590 glacial lakes in Himalaya which cover 455.3 ± 72.7 Km^2. Out of these, 118 glacial lakes are identified as potentially vulnerable lakes in terms of GLOF. The GLOFs have emerged as pre-eminent natural hazards in Himalaya which pose serious threat to livelihoods and infrastructure. The past records of GLOF show 15 in Nepal, 6 in the Tibet Autonomous Region of China (with consequences from Nepal) and 5 in Bhutan (Bajracharya et al. 2007). Recent event of GLOF in 2013 occurred in Central Himalaya (i.e. Kedarnath) which caused loss of ~6000 lives and damage of 204 structure (i.e. 96 fully washed out while 108 were partially damaged) (Rafiq et al. 2019). Increasing human settlements along with development activities have further increased vulnerability of GLOFs, which is anticipated to increase significantly in the near future owing to glacial recession. Therefore, the synergistic approach using a combination of techniques of hydrodynamic breach modelling and remote sensing data processing along with consideration of socio-economic factors can provide overall mitigation plan within in the framework of IPCC climate risk concept. Administrative authorities and research centres need to develop an overall strategy to address the possible risk from this anticipated GLOF (Kaushik et al., unpublished data).

4.4 Conclusion

The increasing interest of scientific community and proliferation of satellite imageries have led to huge data generation regarding glacier health in Himalayan region. Yet spatio-temporal information regarding glacier health is very limited in Himalaya as compare other mountain range of the world. The present state of Himalayan glaciers suggests that glaciers are retreating, losing mass and area, ELA shifting upward, decreasing in flow velocity, expansion of glacial lakes and increasing concentration of supraglacial debris. Overall glacier recession corresponds with climate warming and reduced precipitation. Except Karakoram with variable rate and Kulun region which have shown slight but significant positive mass balance. The spatially variable climatic regimes are one of the major reasons the spatially heterogeneous behaviour of glaciers, whereas glacier geometry and debris cover act as controlling factor in glacial response. Further investigation is needed for glacier mass balance and SFV in order to understand changing regime of climate, implication for sea level change and sub-glacial process.

References

Ageta Y, Higuchi K (1984) Estimation of mass balance components of a summer-accumulation type glacier in the Nepal Himalaya. Geografiska Ann: Ser A Phys Geogr 66(3):249–255

Aggarwal Suruchi, Rai SC, Thakur PK, Emmer Adam (2017) Inventory and recently increasing GLOF susceptibility of glacial lakes in Sikkim, Eastern Himalaya. Geomorphology 295:39–54

Ali Iram, Shukla Aparna, Romshoo Shakil A (2017) Assessing linkages between spatial facies changes and dimensional variations of glaciers in the upper Indus Basin, western Himalaya. Geomorphology 284:115–129

Azam MF, Ramanathan AL, Patrick Wagnon C, Vincent A Linda, Berthier E, Sharma P, Mandal A, Angchuk T, Singh VB (2016) Meteorological conditions, seasonal and annual mass balances of Chhota Shigri Glacier, western Himalaya, India. Ann Glaciol 57(71):328–338

Azam Mohd Farooq, Wagnon Patrick, Berthier Etienne, Vincent Christian, Fujita Koji, Kargel Jeffrey S (2018) Review of the status and mass changes of Himalayan-Karakoram glaciers. J Glaciol 64(243):61–74

Azam MF, Wagnon P, Ramanathan A, Vincent C, Sharma P, Arnaud Y, Linda A, Pottakkal JG, Chevallier P, Singh VB (2012) From balance to imbalance: a shift in the dynamic behaviour of Chhota Shigri glacier, western Himalaya, India. J Glaciol 58 (208):315–324

Azam MF, Wagnon P, Vincent C, Ramanathan AL, Kumar N, Srivastava S, Pottakkal JG, Chevallier P (2019) Snow and Ice melt contributions in a highly glacierized catchment of Chhota Shigri Glacier (India) over the last five decades. J Hydrol 574:760–773

Azam MF, Wagnon P, Vincent C, Ramanathan A, Linda A, Singh VB (2014) Reconstruction of the annual mass balance of Chhota Shigri glacier, Western Himalaya, India, since 1969. Ann Glaciol 55(66):69–80

Bajracharya SR, Maharjan SB, Shrestha F (2014) The status and decadal change of glaciers in Bhutan from the 1980s to 2010 based on satellite data. Ann Glaciol 55(66):159–166

Bajracharya SR, Mool PK, Shrestha BR (2007) Impact of climate change on Himalayan glaciers and glacial lakes: case studies on GLOF and associated hazards in Nepal and Bhutan. Int Centre Integr Mt Dev (ICIMOD)

Bajracharya SR, Shrestha BR (2011) "The status of glaciers in the Hindu Kush-Himalayan region." In: International Centre for Integrated Mountain Development (ICIMOD)

Banerjee A (2017) Brief communication: Thinning of debris-covered and debris-free glaciers in a warming climate. The Cryosphere 11(1):133–138. https://doi.org/10.5194/tc-11-133-2017

Basnett S, Kulkarni AV, Bolch T (2013) The influence of debris cover and glacial lakes on the recession of glaciers in Sikkim Himalaya, India. J Glaciol 59 (218):1035–1046

Berthier E, Arnaud Y, Kumar R, Ahmad S, Wagnon P, Chevallier P (2007) Remote sensing estimates of glacier mass balances in the Himachal Pradesh (Western Himalaya, India). Remote Sens Environ 108(3):327–338

Bhambri R, Bolch T (2009) Glacier mapping: a review with special reference to the Indian Himalayas. Prog Phys Geogr 33(5):672–704

Bhambri R, Bolch T, Chaujar RK, Kulshreshtha SC (2011) Glacier changes in the Garhwal Himalaya, India, from 1968 to 2006 based on remote sensing. J Glaciol 57(203):543–556

Bhardwaj A, Sam L, Bhardwaj A, Martín-Torres FJ (2016a) LiDAR remote sensing of the cryosphere: Present applications and future prospects. Remote Sens Environ 177:125–143

Bhardwaj A, Lydia Sam F, Martín-Torres J, Kumar R (2016b) UAVs as remote sensing platform in glaciology: Present applications and future prospects. Remote Sens Environ 175:196–204

Birajdar F, Venkataraman G, Bahuguna I, Samant H (2014) A revised glacier inventory of Bhaga Basin Himachal Pradesh, India: current status and recent glacier variations. ISPRS Ann Photogramm Remote Sens Spat Inf Sci II 8:37–43

Bolch T, Kulkarni A, Kääb A, Huggel C, Frank Paul J, Cogley G, Frey H, Kargel JS, Fujita K, Scheel M (2012) The state and fate of Himalayan glaciers. Science 336(6079):310–314

Brahmbhatt RM, Bahuguna IM, Rathore BP, Kulkarni AV, Shah RD, Rajawat AS, Kargel JS (2017) Significance of glacio-morphological factors in glacier retreat: a case study of part of Chenab basin, Himalaya. J Mt Sci 14(1):128–141

Brun F, Dumont M, Patrick Wagnon E, Berthier MF Azam, Shea JM, Sirguey P, Rabatel A, Ramanathan Al (2015) Seasonal changes in surface albedo of Himalayan glaciers from MODIS data and links with the annual mass balance. The Cryosphere 9(1):341–355

Brun F, Berthier E, Wagnon P, Kääb A, Treichler D (2017) A spatially resolved estimate of High Mountain Asia glacier mass balances from 2000 to 2016. Nat Geosci 10(9):668

Chand P, Sharma MC (2015) Glacier changes in the Ravi basin, North-Western Himalaya (India) during the last four decades (1971–2010/13). Global Planet Change 135:133–147

Chand P, Sharma MC, Bhambri R, Sangewar CV, Juyal N (2017) Reconstructing the pattern of the Bara Shigri Glacier fluctuation since the end of the Little Ice Age, Chandra valley, north-western Himalaya. Prog Phys Geogr: Earth Environ 41(5):643–675. https://doi.org/10.1177/0309133317728017

Chand P, Sharma MC, Prasad RN (2016) "Heterogeneity in fluctuations of glacier with clean ice-covered, debris-covered and proglacial lake in the Upper Ravi Basin, Himachal Himalaya (India), during the past four decades (1971–2013)." In: Climate change, glacier response, and vegetation dynamics in the Himalaya, Springer, pp 155–79

Chandrasekharan A, Ramsankaran R, Pandit A, Rabatel A (2018) Quantification of annual glacier surface mass balance for the Chhota Shigri Glacier, Western Himalayas, India using an Equilibrium-Line Altitude (ELA) based approach. Int J Remote Sens 39 (23):9092–9112. https://doi.org/10.1080/01431161.2018.1506182

Cogley JG (2009) Geodetic and direct mass-balance measurements: comparison and joint analysis. Ann Glaciol 50(50):96–100

Cogley JG (2011) Present and future states of Himalaya and Karakoram glaciers. Ann Glaciol 52(59):69–73

Cogley JG (2016) Glacier shrinkage across High Mountain Asia. Ann Glaciol 57(71):41–49

Cogley JG, Kargel JS, Kaser G, van der Veen CJ (2010) Tracking the source of glacier misinformation. Science 327(5965):522

Das S, Sharma MC (2019) Glacier changes between 1971 and 2016 in the Jankar Chhu Watershed, Lahaul Himalaya, India. J Glaciol 65(249):13–28

Dehecq A, Gourmelen N, Gardner AS, Brun F, Goldberg D, Nienow PW, Berthier E, Vincent C, Wagnon P, Trouvé E (2019) Twenty-first century glacier slowdown driven by mass loss in High Mountain Asia. Nat Geosci 12(1):22

Dobhal DP, Kumar Surender, Mundepi AK (1995) Morphology and glacier dynamics studies in monsoon–arid transition zone: an example from Chhota

Shigri glacier, Himachal Himalaya, India. Curr Sci 68 (9):936–944
Dobhal DP, Mehta M, Srivastava D (2013) Influence of debris cover on terminus retreat and mass changes of Chorabari Glacier, Garhwal region, central Himalaya, India. J Glaciol 59(217):961–971
Dutta S, Ramanathan AL, Linda A (2012) Glacier fluctuation using satellite data in Beas basin, 1972–2006, Himachal Pradesh, India. J Earth Syst Sci 121 (5):1105–1112
Dyurgerov MB, Meier MF (2005) Glaciers and the changing Earth system: a 2004 snapshot, vol 58. Institute of Arctic and Alpine Research, University of Colorado Boulder
Engelhardt M, Ramanathan AL, Eidhammer T, Kumar P, Landgren O, Mandal A, Rasmussen R (2017) Modelling 60 years of glacier mass balance and runoff for Chhota Shigri Glacier, Western Himalaya, Northern India. J Glaciol 63(240):618–628
Fujita K, Kadota T, Rana B, Kayastha RB, Ageta Y (2001) Shrinkage of Glacier AX010 in Shorong region, Nepal Himalayas in the 1990s. Bull Glaciol Res 18:51–54
Fujita K, Nuimura T (2011) Spatially heterogeneous wastage of Himalayan glaciers. Proc Natl Acad Sci 108(34):14011–14014
Gaddam VK, Kulkarni AV, Gupta AK (2016) Estimation of glacial retreat and mass loss in Baspa basin, Western Himalaya. Spatial Information Research 24 (3):257–266
Gardelle J, Berthier E, Arnaud Y, Kaab A (2013) Region-wide glacier mass balances over the Pamir-Karakoram-Himalaya during 1999–2011 (vol 7, pg 1263, 2013). The Cryosphere 7(6):1885–1886
Gardner AS, Geir Moholdt J, Cogley G, Wouters B, Arendt AA, Wahr J, Berthier E, Regine Hock W, Pfeffer T, Kaser G (2013) A reconciled estimate of glacier contributions to sea level rise: 2003 to 2009. Science 340(6134):852–857
Garg PK, Shukla A, Tiwari RK, Jasrotia AS (2017) Assessing the status of glaciers in part of the Chandra basin, Himachal Himalaya: a multiparametric approach. Geomorphology 284:99–114
Gautam CK, Mukherjee BP (1992) "Synthesis of glaciological studies on Tipra Bank Glacier Bhyundar Ganga Basin, district Chamoli." Uttar Pradesh (FS 1980–1988) Geological Survey of India, Northern Region, Lucknow
Ghosh S, Pandey AC (2013) Estimating the variation in glacier area over the last 4 decade and recent mass balance fluctuations over the Pensilungpa Glacier, J&K, India. Glob Perspect Geogr 1(4):58–65
GSI (Geological Survey of India) (1992) Annual general report. Part 8. 125
GSI (Geological Survey of India) (2001) Glaciology of Indian Himalaya. Special Publication no 63
GSI (Geological Survey of India) (2011) Annual general report, Part 8 144
Hock R (2003) Temperature index melt modelling in mountain areas. J Hydrol 282(1–4):104–115
Inman M (2010) The story "Settling the Science on Himalayan Glaciers". Nature
Jin R, Li X, Che T, Lizong W, Mool P (2005) Glacier area changes in the Pumqu river basin, Tibetan Plateau, between the 1970s and 2001. J Glaciol 51(175):607–610
Kääb A, Berthier E, Nuth C, Gardelle J, Arnaud Y (2012) Contrasting patterns of early twenty-first-century glacier mass change in the Himalayas. Nature 488 (7412):495
Kanth TA, Shah AA, ul Hassan Z (2011) "Geomorphologic character & receding trend of Kolahoi Glacier in Kashmir Himalaya." Recent Res Sci Technol 3(9)
Kaul MN (1986) Mass balance of Liddar glaciers. Trans Inst Indian Geogr 8:95–111
Kaushik S, Dharpure JK, Joshi PK, Ramanathan AL, Singh T (2019) "Climate change drives glacier retreat in Bhaga basin located in Himachal Pradesh, India." Geocarto Int 1–20
Kaushik S, Joshi PK, Singh T (2019) "Development of glacier mapping in Indian Himalaya: a review of approaches." Int J Remote Sens 1–28
Koul MN, Ganjoo RK (2010) Impact of inter-and intra-annual variation in weather parameters on mass balance and equilibrium line altitude of Naradu Glacier (Himachal Pradesh), NW Himalaya, India. Clim Change 99(1–2):119–139
Kulkarni AV, Bahuguna IM, Rathore BP, Singh SK, Randhawa SS, Sood RK, Dhar S (2007) "Glacial retreat in Himalaya using Indian Remote Sensing satellite data." Curr Sci 69–74
Kulkarni AV, Dhar S, Rathore, and Rajeev Kalia. BP, Kalia R (2006) Recession of samudra tapu glacier, chandra river basin, Himachal Pradesh. J Indian Soc Remote Sens 34(1):39–46
Kulkarni AV, Karyakarte Y (2014) "Observed changes in Himalayan glaciers." Curr Sci 237–44
Kulkarni AV, Rathore BP, Alex S (2004) Monitoring of glacial mass balance in the Baspa basin using accumulation area ratio method. Curr Sci 86(1):185–190
Kulkarni AV, Rathore BP, Mahajan S, Mathur P (2005) Beas basin, Himachal Pradesh. Curr Sci 88(11):1844
Kulkarni AV, Rathore BP, Singh SK, Bahuguna IM (2011) Understanding changes in the Himalayan cryosphere using remote sensing techniques. Int J Remote Sens 32(3):601–615
Kumar A, Negi HS, Kumar K, Kanda N, Singh KK, Pandit A, Ramsankaran R (2018) "Estimation of recent changes in thickness and mass balance of the Patsio glacier in the Great Himalayan region using geodetic technique and ancillary data." Geocarto Int 1–17
Mir RA (2018) Recent changes of two parts of Kolahoi Glacier and its controlling factors in Kashmir basin, western Himalaya. Remote Sens Appl: Soc Environ 11:265–281
Mir RA, Jain SK, Saraf AK, Goswami A (2014) Glacier changes using satellite data and effect of climate in Tirungkhad basin located in western Himalaya. Geocarto Int 29(3):293–313

Mishra R, Kumar A, Singh D (2014) Long term monitoring of mass balance of Hamtah Glacier, Lahaul and Spiti district, Himachal Pradesh. Geolo Surv India 147(pt 8):230–231

Murtaza KO, Romshoo SA (2017) Recent glacier changes in the Kashmir alpine Himalayas, India. Geocarto Int 32(2):188–205

Nie Y, Sheng Y, Liu Q, Liu L, Liu S, Zhang Y, Song C (2017) A regional-scale assessment of Himalayan glacial lake changes using satellite observations from 1990 to 2015. Remote Sens Environ 189:1–13

Nuimura T, Sakai A, Taniguchi K, Nagai H, Lamsal D, Tsutaki S, Kozawa A, Hoshina Y, Takenaka S, Omiya S (2015) "The gamdam glacier inventory: a quality-controlled inventory of Asian glaciers." Cryosphere 9(3)

Oerlemans J (2001) Glaciers and climate change, Balkema. Rotterdam, Netherlands

Pandey P, Venkataraman G, Shukla SP (2011) Study of retreat of Hamtah glacier, Indian Himalaya, using remote sensing technique. Paper presented at the 2011 IEEE international geoscience and remote sensing symposium

Pandey P, Venkataraman G (2013) Changes in the glaciers of Chandra-Bhaga basin, Himachal Himalaya, India, between 1980 and 2010 measured using remote sensing. Int J Remote Sens 34(15):5584–5597

Patel LK, Sharma P, Fathima TN, Thamban M (2018) Geospatial observations of topographical control over the glacier retreat, Miyar basin, Western Himalaya, India. Environ Earth Sci 77(5):190

Pfeffer WT, Arendt AA, Bliss A, Tobias Bolch J, Cogley G, Gardner AS, Hagen JO, Hock R, Kaser G, Kienholz C (2014) The Randolph Glacier Inventory: a globally complete inventory of glaciers. J Glaciol 60(221):537–552

Pratap B, Dobhal DP, Bhambri R, Mehta M, Tewari VC (2016) Four decades of glacier mass balance observations in the Indian Himalaya. Reg Environ Change 16 (3):643–658

Pritchard HD (2019) Asia's shrinking glaciers protect large populations from drought stress. Nature 569 (7758):649–654. https://doi.org/10.1038/s41586-019-1240-1

Racoviteanu A, Arnaud Y, Williams MW, Manley WF (2014) Spatial patterns in glacier characteristics and area changes from 1962 to 2006 in the Kanchenjunga-Sikkim area, eastern Himalaya. The Cryosphere 9:505–523

Rafiq M, Romshoo SA, Mishra AK, Jalal F (2019) Modelling Chorabari Lake outburst flood, Kedarnath, India. J Mt Sci 16(1):64–76

Rai PK, Nathawat MS, Mohan K (2013) Glacier Retreat in Doda Valley, Zanskar Basin, Jammu & Kashmir, India. Univ J Geosci 1(3):139–149

Raina VK (2009) "Himalayan glaciers: a state-of-art review of glacial studies, glacial retreat and climate change." Himalayan glaciers: a state-of-art review of glacial studies, glacial retreat and climate change

Raina VK, Kaul MK, Singh S (1977) Mass-balance studies of Gara Glacier. J Glaciol 18(80):415–423

Sangewar CV, Siddique NS (2006) Thematic compilation of mass balance data on glaciers in Satluj catchment in Himachal Himalaya. Rec Geol Surv India 141(pt 8):159–161

Sattar A, Goswami A, Kulkarni AV (2019) Hydrodynamic moraine-breach modeling and outburst flood routing-A hazard assessment of the South Lhonak lake, Sikkim. Sci Total Environ 668:362–378

Satyabala SP (2016) Spatiotemporal variations in surface velocity of the Gangotri glacier, Garhwal Himalaya, India: Study using synthetic aperture radar data. Remote Sens Environ 181:151–161

Scherler D, Bookhagen B, Strecker MR (2011) Spatially variable response of Himalayan glaciers to climate change affected by debris cover. Nat Geosci 4(3):156

Scherler D, Leprince S, Strecker MR (2008) Glacier-surface velocities in alpine terrain from optical satellite imagery—accuracy improvement and quality assessment. Remote Sens Environ 112(10):3806–3819

Schmidt S, Nüsser M (2012) Changes of high altitude glaciers from 1969 to 2010 in the Trans-Himalayan Kang Yatze Massif, Ladakh, northwest India. Arctic Antarctic Alpine Res 44(1):107–121

Sharma P, Patel LK, Ravindra R, Singh A, Mahalinganathan K, Thamban M (2016) Role of debris cover to control specific ablation of adjoining Batal and Sutri Dhaka glaciers in Chandra Basin (Himachal Pradesh) during peak ablation season. J Earth Syst Sci 125 (3):459–473

Sherpa SF, Wagnon P, Brun F, Berthier E, Vincent C, Lejeune Y, Arnaud Y, Kayastha RB, Sinisalo A (2017) Contrasted surface mass balances of debris-free glaciers observed between the southern and the inner parts of the Everest region (2007–15). J Glaciol 63 (240):637–651

Shukla A, Garg PK, Srivastava S (2018) Evolution of glacial and high-altitude lakes in the Sikkim, Eastern Himalaya over the past four decades (1975–2017). Frontiers Environ Sci 6:81

Shukla A, Qadir J (2016) Differential response of glaciers with varying debris cover extent: evidence from changing glacier parameters. Int J Remote Sens 37 (11):2453–2479

Singh S, Kumar R, Dimri AP (2018) Mass balance status of Indian Himalayan glaciers: A brief review. Frontiers Environ Sci 6:30

Srivastava D, Sangewar CV, Kaul MK, Jamwal KS (2001) Mass balance of Rulung glacier—a trans Himalayan glacier, Indus basin, Ladakh. GSI Spl Publ 53:41–46

Tawde SA, AV Kulkarni, Bala G (2016) "Estimation of glacier mass balance on a basin scale: an approach based on satellite-derived snowlines and a temperature index model." Curr Sci (00113891) 111(12)

Tawde SA, Kulkarni AV, Bala G (2017) An estimate of glacier mass balance for the Chandra basin, western Himalaya, for the period 1984–2012. Ann Glaciol 58 (75pt2):99–109

Thakuri S, Salerno F, Bolch T, Guyennon N, Tartari G (2016) Factors controlling the accelerated expansion of Imja Lake, Mount Everest region, Nepal. Ann Glaciol 57(71):245–257

Thakuri S, Salerno F, Smiraglia C, Bolch T, C D'Agata, Viviano G, Tartari G (2014) "Tracing glacier changes since the 1960s on the south slope of Mt. Everest (central Southern Himalaya) using optical satellite imagery"

Tiwari RK, Gupta RP, Arora MK (2014) Estimation of surface ice velocity of Chhota-Shigri glacier using sub-pixel ASTER image correlation. Curr Sci 106 (6):853–859

Tshering P, Fujita K (2016) First in situ record of decadal glacier mass balance (2003–2014) from the Bhutan Himalaya. Ann Glaciol 57(71):289–294

Vijay S, Braun M (2016) Elevation change rates of glaciers in the Lahaul-Spiti (Western Himalaya, India) during 2000–2012 and 2012–2013. Remote Sens 8 (12):1038

Vincent C, Ramanathan Al, Patrick Wagnon DP, Dobhal AL, Etienne Berthier P, Sharma YA, Azam MF, Gardelle J (2013) Balanced conditions or slight mass gain of glaciers in the Lahaul and Spiti region (northern India, Himalaya) during the nineties preceded recent mass loss. The Cryosphere 7(2):569–582

Vohra CP (1980) Some problems of glacier inventory in the Himalayas. IAHS Publ 126:67–74

Wagnon P, Linda A, Arnaud Y, Kumar R, Sharma P, Vincent C, Pottakkal JG, Berthier E, Ramanathan A, Hasnain SI (2007) Four years of mass balance on Chhota Shigri Glacier, Himachal Pradesh, India, a new benchmark glacier in the western Himalaya. J Glaciol 53(183):603–611

Wagnon P, Vincent C, Yves Arnaud E, Berthier EV, Stephan Gruber M, Ménégoz AG, Dumont M, Shea JM (2012) Seasonal and annual mass balances of Mera and Pokalde glaciers (Nepal Himalaya) since 2007. The Cryosphere 7(6):1769–1786

Wang R, Liu S, Shangguan D, Radić V, Zhang Y (2019) Spatial heterogeneity in glacier mass-balance sensitivity across high mountain Asia. Water 11(4):776

Yao T, Thompson L, Yang W, Wusheng Y, Gao Y, Guo X, Yang X, Duan K, Zhao H, Baiqing X (2012) Different glacier status with atmospheric circulations in Tibetan Plateau and surroundings. Nat Clim Change 2(9):663

Ye Q, Yao T, Kang S, Chen F, Wang J (2006) Glacier variations in the Naimona'nyi region, western Himalaya, in the last three decades. Ann Glaciol 43:385–389

Ye Q, Yao T, Naruse R (2008) Glacier and lake variations in the Mapam Yumco basin, western Himalaya of the Tibetan Plateau, from 1974 to 2003 using remote-sensing and GIS technologies. J Glaciol 54(188):933–935

Zhang J, Jia L, Menenti M, Guangcheng H (2019) Glacier facies mapping using a machine-learning algorithm: the parlung zangbo basin case study. Remote Sens 11 (4):452

Zurick D, Bajracharya B, Pacheco J, Shrestha BR (2005) Atlas of the Himalaya, ICIMOD

Temporal Variability of the Satopanth Glacier Facies at Sub-pixel Scale, Garhwal Himalaya, India

5

Bisma Yousuf, Aparna Shukla, and Manoj Kumar Arora

Abstract

Growing concern about climate-related glacier change underscores the need to quantify the temporal changes in glacier facies. Classification of glacier facies and assessing their temporal area changes are among the key applications of optical remote sensing in cryosphere. High radiometric resolution (HRR) optical data have an added advantage of overcoming the snow-saturation problem, sometimes observed in medium-to-high-resolution optical sensors. Sub-pixel classification is also known to retrieve the accurate landscape area and addresses the mixed pixel problem in most of the HRR data. Therefore, this paper utilizes the support vector machine (SVM)-based sub-pixel classification approach on bi-temporal HRR data to determine the variability in the surface facies of the Satopanth glacier (SPG), Central Himalaya. Considering the limitations of spectral data in classification, both input Advanced Wide Field Sensor (AWiFS) and reference fine multispectral instrument (MSI) data were aided with the ancillary data like terrain factors, thermal data, band ratios, spectral indices and texture measures. Sub-pixel estimates of SPG facies derived from input AWiFS 2016 image showed good agreement (r >0.7) with their reference MSI-derived estimates. Significant variations were observed in the sub-pixel estimates of SPG facies during the 11-year period (2005–2016). A minimum of ∼2% reduction was observed in fresh and slightly metamorphosed snow (FS) area, whereas ice facies showed maximum shrinkage in area (∼16%). The maximum expansion of ∼8% and ∼7% was observed for supraglacial debris (SGD) and ice-mixed debris (IMD), respectively. Wet-snow (WS) and firn coverages slightly increased by ∼2 and ∼1%, respectively. These changes correspond well with the meteorological data of the SPG obtained from Climate Research Unit Time Series (CRU TS) v.4.01 dataset.

Keywords

AWiFS · Glacier facies · Temporal · Satopanth · Sub-pixel · Sentinel

B. Yousuf
University of Jammu, Jammu, India

A. Shukla (✉)
Ministry of Earth Sciences, New Delhi, India

Manoj KumarArora
BML Munjal University, Gurgaon, India

U. Schickhoff et al. (eds.), *Mountain Landscapes in Transition*,
Sustainable Development Goals Series, https://doi.org/10.1007/978-3-030-70238-0_5

5.1 Introduction

Glaciers around the world possess diverse surface facies that are sensitive to the changes in the annual processes of winter accumulation, subsequent metamorphism and summer melt (Zhou and Zheng 2017). Changes in these facies can affect the glacier's melt characteristics, mass balance, firn-line altitude and global sea level (Wolken et al. 2009). Thus, the temporal assessment of glacier facies is crucial to determine the overall state of the glaciers and their response to the climate change. The advancements in space technology have allowed the rapid and frequent monitoring of glaciers. A wide variety of optical sensors (Terra Advanced Spectral Emission and Thermal Radiometer (ASTER), Landsat Thematic Mapper (TM), Landsat Enhanced Thematic Mapper Plus (ETM +), Landsat 8 Operational Land Imager (OLI), WorldView-2, Satellites Pour l'Observation de la Terre (SPOT) series, Resourcesat Linear Imaging Self-Scanning System (LISS-I/II/III), Resourcesat Advanced Wide Field Sensor (AWiFS) and Sentinel-2A Multispectral Instrument (MSI)) ranging from coarse (56 m) to high (2 m) spatial resolution are available for the extraction of glacier facies (Bhardwaj et al. 2015; Keshri et al. 2009; Paul et al. 2016; Pope and Rees 2014a, b; Shukla and Ali 2016; Shukla and Yousuf 2017; Jawak et al. 2018). Recently, Yousuf et al. (2019) and (2020) recognized that higher radiometric resolution (HRR) data such as AWiFS (10- and 12-bit for Resourcesat-1 and 2, respectively), OLI (12-bit) and MSI (12-bit) are more ideal for the mapping of glacier facies. The study reported that despite the equal potential of AWiFS and MSI data for the accurate characterization of glacier facies, the use of HRR data with coarse-spatial resolution like AWiFS data in the detailed classification of glaciers remains undermined. Moreover, the long archive of AWiFS than OLI and MSI data favors its use for the temporal monitoring of glacier facies.

In order to extract the glacier facies using satellite data, image classification is the standard method (Shukla and Yousuf 2016, 2017; Jawak et al. 2018; Zhang et al. 2019; Yousuf et al. 2020). Depending on the dataset used for facies extraction, the selection of the appropriate classification approach becomes an important consideration to obtain the accurate results. For instance, the HRR AWiFS data is able to capture the spatial heterogeneity in our study glacier bearing six facies, namely fresh and slightly metamorphosed snow (FS), wet-snow (WS), firn, ice, ice-mixed debris (IMD) and supraglacial debris (SGD) (Fig. 5.1a–c). However, due to its coarse-spatial resolution (resampled to 60 m), AWiFS data are fraught with the mixed pixels. Figure 5.1d illustrates an example where the AWiFS pixel showing the firn reflectance actually contains WS, firn and ice on MSI data. Each AWiFS pixel here corresponds to 36 MSI pixels. Hence, if per-pixel classification (PPC) is applied on this AWiFS pixel, it will be assigned to firn only. Contrastively, sub-pixel classification (SPC) techniques will assign the fractions of each facies (WS, firn and ice) to this pixel depending on its constituent spectra. Thus, the application of PPC approach will decrease the classification accuracy here. Whereas, SPC will produce more accurate results by estimating the proportion of each glacier facie within the AWiFS pixel. Considering this, SPC was performed on AWiFS data of the Satopanth glacier (SPG) by the application of support vector machines (SVMs). SVM is a supervised machine learning technique that performs classification by constructing hyperplanes in a multidimensional space separating the dataset into discrete predefined number of classes. SVM details are provided by Kavzoglu and Colkesen (2009). SVMs have been successfully implemented for the PPC/SPC of various land cover types (Huang et al. 2002; Kumar et al. 2010; Liu et al. 2016; Salah 2017). In glaciology, SVMs are exploited for the SPC of snow cover mostly (Zhang et al. 2005; Çiftçi et al. 2017) and rarely for the glacier facies (Yousuf et al. 2020). An exhaustive literature search revealed countable studies documenting the work on snout elevation, geomorphology, field-based and modeled ice thickness/volume, area, length, debris cover, mass balance, velocity and black carbon aerosol measurements, ponds and ice-cliffs, and avalanche contribution to mass balance of the SPG to understand the glacier's behavior toward the climate change (Sah 1991; Heim and Gannser 1939; Nainwal et al. 2007, 2008 and 2016; Bhambri et al. 2011; Scherler et al. 2011; Nair et al. 2013; Panwar et al. 2014; Laha et al. 2017; Mishra et al. 2018; Sharma et al. 2018; Sattar et al. 2019; Remya et al. 2020; Kneib et al. 2021; Panicker et al. 2021). However, to-date, no work has been done on the estimation or variation of glacier facies on the SPG. This is an important aspect to cover as there is a mutual feedback between the glacier facies variations and glacier dynamics. Glacier facies affect and get affected by the changes in glacier dynamics (Ali et al. 2017). So, changes in glacier facies might serve as a reliable indicator of glacier health. Hence, the study aimed to determine the over-decadal change in the SPG facies at sub-pixel scale using SVMs.

5.2 Study Area

The study focuses on the Satopanth glacier (SPG) located between 79°17'35"–79°24'45" E and 30°43'10"–30°46'45" N in the Alaknanda Valley, Chamoli district, Uttarakhand, India (Fig. 5.1a). SPG is a compound valley and east–west flowing glacier, occupying the central region of Indian Himalaya. It is the source of Alaknanda River, a major tributary of the River Ganga. The SPG is ~ 13 km long with its debris-covered tongue spread over 11 km (Source: MSI image dated September 19, 2016). The ablation zone of SPG is thus significantly larger than its accumulation zone. The mean elevation and mean slope of SPG are 4747 m above sea level and 21°, respectively (Source: 30 m Shuttle Radar Topographic Mission (SRTM) Digital Elevation Model (DEM) version 3). The low surface gradient in its debris-covered zone facilitates the formation of supraglacial lakes and ice-cliffs.

The SPG is one among the benchmark Himalayan glaciers which is being continuously monitored in the field (Nainwal et al. 2008, 2016; Laha et al. 2017; Mishra et al. 2018; Sharma et al. 2018). The SPG has experienced continuous recession, area loss, thinning and negative mass balance for different time periods between 1936 and 2017 (Bhambri et al. 2011; Panwar et al. 2014; Nainwal et al. 2016; Remya et al. 2020; Laha et al. 2017). This urges the continuous monitoring of SPG as it may ultimately affect the hydrology of both the Alaknanda and Gangotri Rivers.

5.3 Datasets and Study Techniques

Resourcesat-1 AWiFS images with 56 m spatial resolution (dated August 14, 2005 and July 25, 2016) were exploited to monitor the sub-pixel changes in SPG facies. Sentinel-2A MSI image with 10 m spatial resolution in visible and near-infrared bands (VNIR) and 20 m in shortwave-infrared (SWIR) bands (dated September 19, 2016) was used as reference. Few ancillary layers (Table 5.1) were also included as they assist in the glacier facies characterization (Racoviteanu and Williams 2012; Alifu et al. 2016; Yousuf et al. 2019). Terra ASTER kinetic temperature (KT) product (dated August 19, 2005) and thermal band 10 of Landsat OLI/TIRS (dated September 19, 2016) available at 90 m and 100 m spatial resolutions were used for KT extraction. The meteorological data (mean temperature and precipitation) of SPG were obtained from the Climate Research Unit Time Series (CRU TS) v.4.01 dataset (Harris et al. 2014) to relate the changes in climatic conditions of SPG with its facies changes.

The research methodology involved the following basic steps:

(a) Pre-processing: The raw satellite images were pre-processed for the required geometric, atmospheric (dark object subtraction 4 (DOS 4) approach) and topographic (C-correction approach) corrections (Song et al. 2001; Gupta et al. 2007; Lantzanakis et al. 2017). The OLI/TIRS thermal DN values (band 10) were converted to KT values using emissivity normalization method (Kealy and Hook 1993; Landsat 8 user manual).
(b) Data generation: Two input and one reference datasets were generated at 60 m and

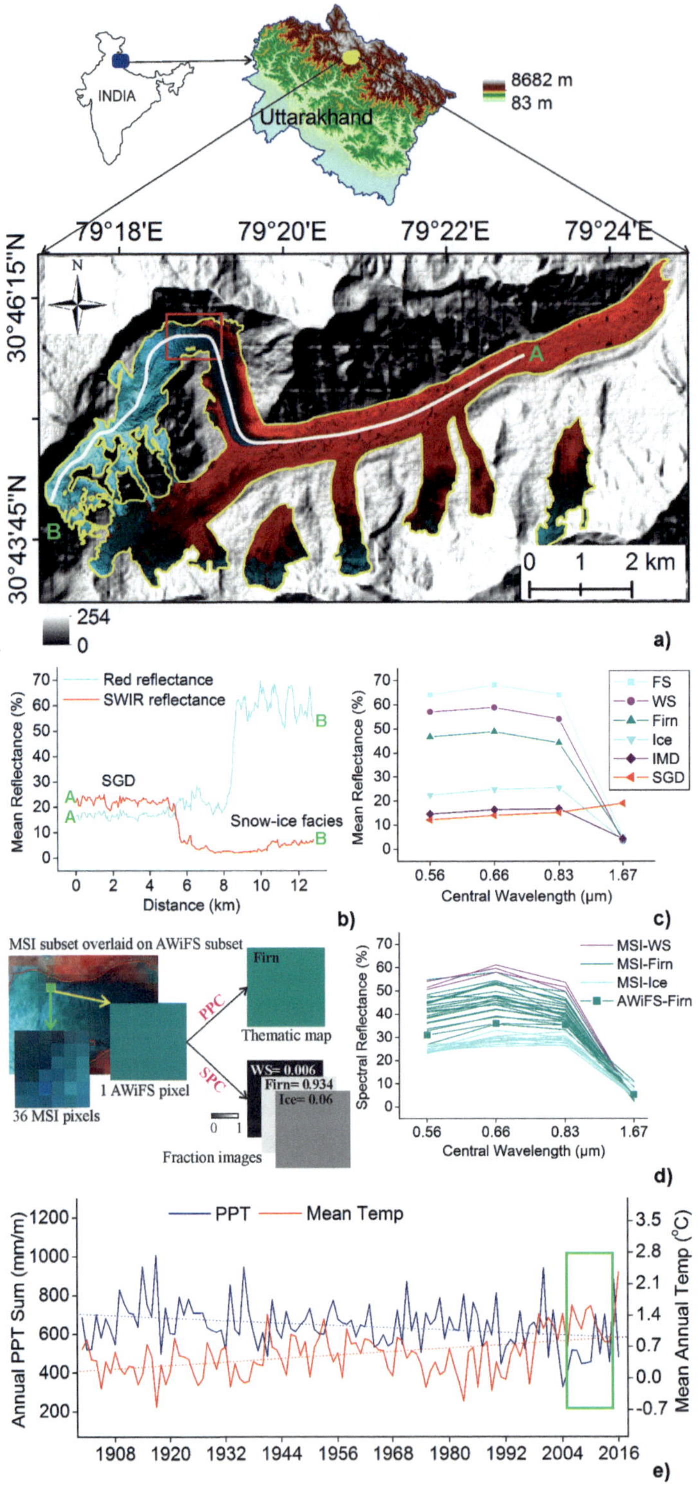
INDIA
Uttarakhand
8682 m
83 m
79°18'E
79°20'E
79°22'E
79°24'E
30°46'15"N
30°43'45"N
N
A
B
0 1 2 km
254
0
a)
Red reflectance
SWIR reflectance
SGD
Snow-ice facies
Mean Reflectance (%)
Distance (km)
b)
FS
WS
Firn
Ice
IMD
SGD
Central Wavelength (μm)
c)
MSI subset overlaid on AWiFS subset
36 MSI pixels
1 AWiFS pixel
PPC
SPC
Firn
Thematic map
WS= 0.006
Firn= 0.934
Ice= 0.06
Fraction images
MSI-WS
MSI-Firn
MSI-Ice
AWiFS-Firn
Spectral Reflectance (%)
d)
PPT
Mean Temp
Annual PPT Sum (mm/m)
Mean Annual Temp (°C)
e)

◀ **Fig. 5.1** Location, heterogeneity and sub-pixel classification process of the study glacier. **a** False color composite of Sentinel-2A MSI image (September 19, 2016) showing the Satopanth glacier (SPG) overlaid on a shaded relief map. Yellow symbol shows its location in the Uttarakhand state on the SRTM DEM version 3 of the Indian Himalaya. **b** Longitudinal spatial profile showing surface heterogeneity across the SPG in the AWiFS red and SWIR reflectances (transect A–B shown as white line in upper panel **a**). **c** AWiFS-derived mean spectral reflectance of identified glacier facies. **d** Illustration of sub-pixel classification (SPC) process by overlaying MSI subset (blended) on AWiFS (resampled to 60 m) subset. These subsets are the zoomed-in views of an area marked by red in upper panel **a**. AWiFS pixel is outlined by yellow and MSI pixels by green. The graph in right panel displays the spectral signatures of facies identified in these pixels. SPC produces fraction images as output while per-pixel classification (PPC) generates a thematic map as output. **e** Meteorological records of the area during 1901–2016. The decreasing precipitation (PPT) and increasing temperature (Temp) trends are highlighted by green for the study period. FS= Fresh and slightly metamorphosed snow; WS= Wet-snow; IMD= Ice-mixed debris; SGD= Supraglacial debris

Table 5.1 List and source of ancillary layers used along with the spectral data for sub-pixel classification of glacier facies

Ancillary data	Source
Elevation	30 m Shuttle Radar Topographic Mission (SRTM) Digital Elevation Model (DEM) version 3
Slope	
KT	OLI/TIRS band 10 and ASTER KT (AST08) product
Green/SWIR ratio	AWiFS and MSI
KT/(green/SWIR) ratio	AWiFS, MSI, OLI/TIRS band 10 and AST08 product
NDSI= (green − SWIR)/(green + SWIR)	AWiFS and MSI
SGI= (NIR − SWIR)/(NIR + SWIR)	AWiFS and MSI
Variance	Second principal component (PC) of principal component analysis (PCA) performed on AWiFS and MSI images
Homogeneity	Second PC of PCA performed on AWiFS and MSI images

10 m spatial resolutions, respectively, for the duration 2005–2016. Each input dataset was a combination of one AWiFS image and a set of nine AWiFS/ASTER/OLI/DEM-derived ancillary layers (Table 5.1). Likewise, the reference dataset consisted of MSI (VNIR and SWIR) bands alongside these nine MSI/OLI/DEM-derived ancillary layers. The relevance of these ancillary layers for the segregation of glacier facies is explained by Yousuf et al. (2019). Each input and reference datasets were segmented into the illuminated and shaded images (Shukla et al. 2010) to avoid misclassification of facies due to the differential illumination.

(c) Facies identification: Six distinct sets of spectral profiles were observed on the SPG surface (Fig. 5.1c), labeled as FS, WS, firn, ice, IMD and SGD depending on their reflectance values and the extracted ancillary information.

(d) SPC: SPC was done using SVMs with radial basis function (RBF) kernel using the pairwise classification strategy. RBF kernel was chosen because it outperforms other kernels in image classification, and involves less numerical difficulties and less hyperparameters than the polynomial kernel (Srivastava and Bhambu 2010). Each of the illuminated (input and reference) and shaded (input and reference) image segments was classified individually. The resultant illuminated and shaded fraction images corresponding to six glacier facies were then merged for each input and reference datasets. Area of each glacier facies was then

Table 5.2 AWiFS-derived fractional area, MSI-derived reference fractional (RF) area and AWiFS-derived over-decadal change for the Satopanth glacier facies

	FS	WS	Firn	Ice	IMD	SGD
AWiFS (2005)-derived area	3.09%	8.96%	7.71%	21.46%	5.79%	53.00%
AWiFS (2016)-derived area	1.29%	11.23%	8.28%	5.80%	12.63%	60.78%
Area change during 2005–2016	**−1.80%**	**2.27%**	**0.57%**	**−15.65%**	**6.84%**	**7.78%**
MSI (2016)-derived area	5.68%	12.73%	3.05%	4.34%	9.73%	64.47%
Deviation between AWiFS (2016)- and MSI (2016)-derived areas	4.39%	1.5%	−5.23%	−1.46%	−2.90%	3.69%
Average deviation ±Standard deviation	**1.25 ±1.54%**					

Facies area is expressed as percentage of the total glaciated area. FS= Fresh and slightly metamorphosed snow; WS= Wet-snow; IMD= Ice-mixed debris; SGD= Supraglacial debris

estimated from these bi-temporal AWiFS-derived fraction and MSI-derived reference fraction (RF) images. AWiFS-derived fractional areas of glacier facies during 2016 were compared with their MSI-derived RF areas for validation. While AWiFS (2005) and AWiFS (2016) fraction images of glacier facies were analyzed to determine the 11-year variability in the glacier facies. Sen's slope estimator was used to determine the magnitude of the trends in precipitation and temperature in terms of percent change over mean (Sen 1968), while the statistical significance was determined by applying the nonparametric Mann-Kendall test (Racoviteanu et al. 2008).

5.4 Results

The comparison between AWiFS (2016)-derived fractional and MSI-derived RF estimates revealed close match between WS, ice, IMD and SGD than FS and firn which showed maximum deviation (Table 5.2 and Fig. 5.2). High correlation (r= 0.79–0.99) and low standard deviation (SD= 0.002–0.024) values were obtained for all the facies, excluding FS where r and SD were observed to be 0.11 and 0.031, respectively (Fig. 5.3). SPG has undergone significant changes over the period of 11 years (Table 5.2 and Fig. 5.2). While FS area has reduced, snow wetness, ice melting, exposed firn, IMD and SGD surfaces have enhanced.

5.5 Discussion

Results illustrate that AWiFS (2016)-derived fractional estimates of SPG facies are close but not identical to the MSI-derived ones (Table 5.2 and Fig. 5.2). The difference could be mainly due to the seasonality effect and partly due to the slight misregistration between the AWiFS and MSI images. This is also emphasized through the absence of correlation between the AWiFS (2016) and MSI-derived estimates of FS (Fig. 5.3). It is possible that solid precipitation might have occurred between July–September 2016 leading to large differences in the FS cover on AWiFS (2016) and MSI images. Glacier facies are highly dynamic in nature; thus, their distribution can change at any point of time as a result of the prevailing meteorological conditions and the glacier phenomena of metamorphism (Yousuf et al. 2019). Therefore, the observed variations among the AWiFS- and MSI-derived estimates of SPG facies may be related to the differences in their image acquisition time. Moreover, the high r and low SD values between most of the facies estimates derived from the AWiFS (2016) and MSI RF images (Fig. 5.3) prove the efficacy of the AWiFS data for facies mapping. The error in our results, expressed as average deviation between the fractional and RF

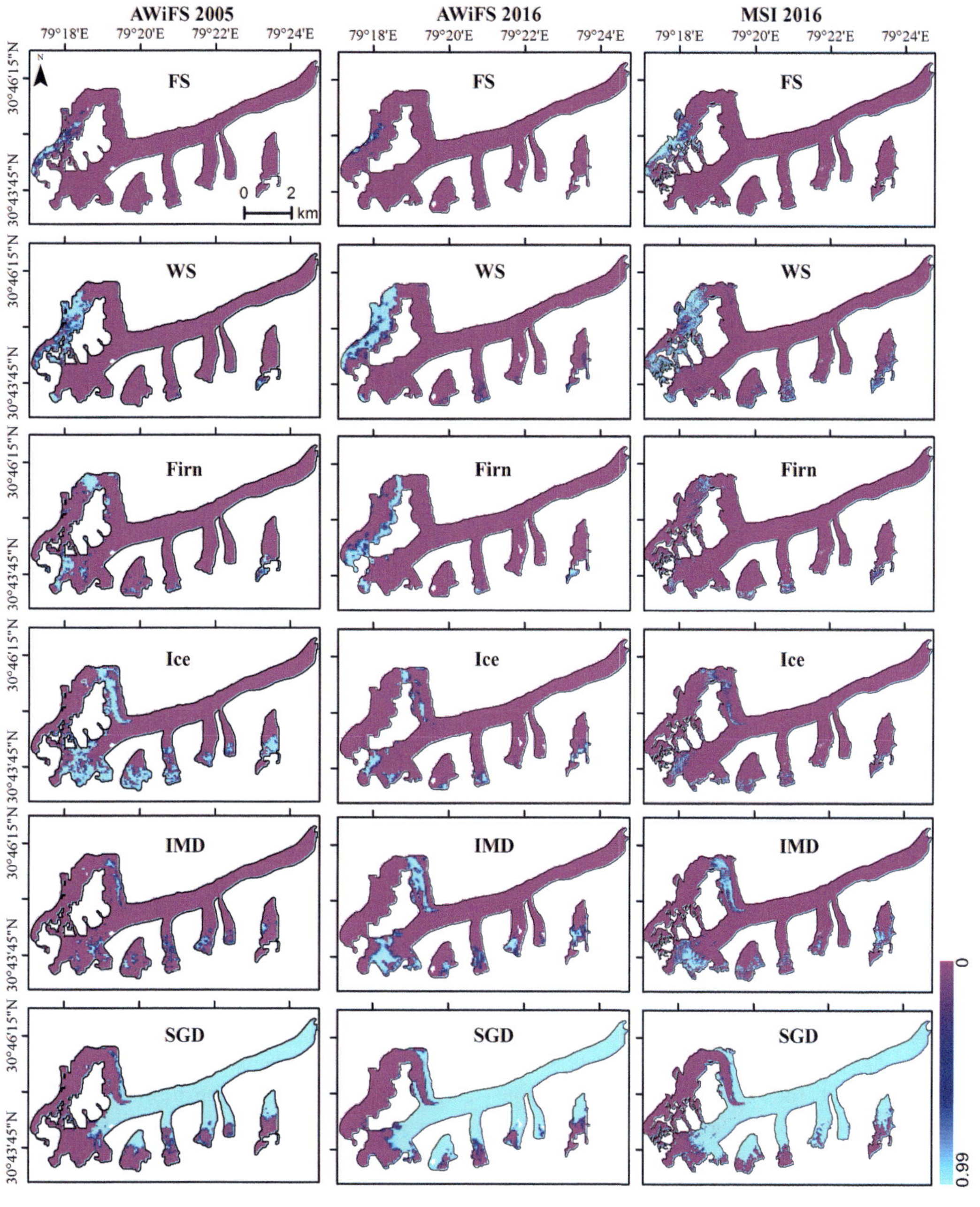

Fig. 5.2 Fraction and reference fraction images of various glacier facies generated from the sub-pixel classification of Satopanth glacier using AWiFS and MSI data acquired during 2005–2016. FS= Fresh and slightly metamorphosed snow; WS= Wet-snow; IMD= Ice-mixed debris, and; SGD= Supraglacial debris

estimates of SPG facies (Table 5.2), was of the order <2%. Our results are valid since the error is less than the acceptable mean error (5–10%) as reported by Painter and Dozier (2004) and Painter et al. (2009), Stroeve et al. (2005), and Czyzowska-Wisniewski et al. (2015) for sub-pixel snow cover mapping. The use of AWiFS for sub-pixel snow cover or per-pixel snow/glacier mapping is already explored in prior studies (Kulkarni et al. 2006 and 2010;

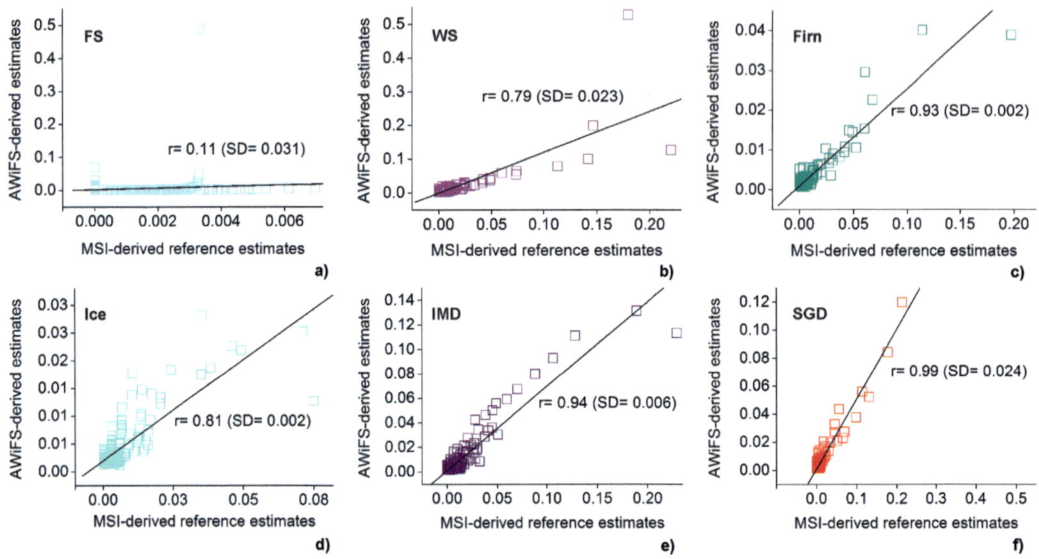

Fig. 5.3 Relationship between input AWiFS-derived and reference MSI-derived sub-pixel estimates of the Satopanth glacier facies. *r= Correlation coefficient; SD= Standard deviation*

Mishra et al. 2009; Negi et al. 2009; Shukla et al. 2010; Arora et al. 2011; Subramaniam et al. 2011; Yousuf et al. 2020). However, the above results reiterate its potential applications in the detailed SPC of glaciers.

Results further indicate noticeable variations in the SPG facies during the 11-year study period (Table 5.2 and Fig. 5.2). All the facies except FS and ice have increased in area. The highest change is observed in the coverage of ice facies (~ −16%) followed by SGD (~ +8%) and IMD (~ +7%). While as, the lowest change is seen in firn (~ +1%) area followed by FS (~ −2%). and WS (~ +2%). Changes in these facies are interlinked and follow the general transition sequence: FS–WS–firn–ice–IMD–SGD. The transition in these facies evolves with the changes in surface temperature and accumulation rate of glacier which are governed by prevailing meteorological conditions. 26% and 38% increasing change in annual mean temperature and precipitation, respectively, in SPG is obtained from CRU TS data (Fig. 5.1a) during 2005–2016, though these results are not statistically significant. Despite the increase in annual precipitation, FS area has reduced which could be probably due to the occurrence of monsoon precipitation in the form of rain. This is also supported by Fujita (2008) who concluded that warming on summer-accumulation type glaciers significantly reduces the snow accumulation. Increased annual precipitation and annual mean temperature would have triggered the snow and ice melt, thus increasing WS and firn coverages while decreasing ice area. The most striking feature is the significant (~16%) reduction in ice coverage which is likely to be associated with the subsequent increase in firn (~1%), IMD (~7%) and SGD (~8%) areas, thus, contributing to the total ice melted. This is because ice melting occurs at different elevations and its extent may change either due to the transformation of firn to ice or ice to IMD/SGD. The meltwater can further speed up the metamorphism of wet-snow (Pomeroy et al. 2006), thus, expanding firn area. These results are consistent with Ali et al. (2017) showing that ice melting predominantly controls the transition of glacier facies. SGD cover on the SPG surface has expanded at an average rate of 0.17 $km^2\ a^{-1}$ during 2005–2016. Increase in the IMD and SGD areas are an indicative of its negative glacier health.

There are no comparative studies to validate the variability of SPG facies over the period

2005–2016. However, Bhambri et al. (2011) reported 1.5% reduction in ice area during 1968-2006. This corroborates the rapid melting of surface ice on the SPG in the recent times. Nainwal et al. (2016) observed that the recession rate of SPG has lowered in the recent decades (11.8 ±3.8 m a^{-1} for 1999–2005 and 4.1 ±0.6 ma^{-1} for 2005–2013). They also reported glacier thinning of 9 ±11 m and 21 ±11 m in the lower and upper ablation zones of SPG for the duration 1962–2013. Laha et al. (2017) estimated a negative mass balance (−2.0 m w.e. a^{-1}) for SPG during 2015. Remya et al. (2020) reported a mass loss rate of −0.55 ± 0.06 m w.e. a^{-1} for the SPG during 2000–2017. Assuming the retreat and thinning rates to be constant after 2013, the slowdown of retreat rates from 2005 onwards and mass loss of SPG can be linked to increase in its SGD cover, as evidenced by other studies (Bolch et al. 2008; Bhambri et al. 2011; Kirkbride and Deline 2013; Scherler et al. 2018). This shows that the variability of SPG facies are in sync with the meteorological data of the study area, and other reported glacier parameters (retreat, ice thickness and mass balance), reflecting the climatic effect on the SPG. The high melting rate of ice exposes more debris on the glacier surface. Decrease in SPG ice with subsequent increase in IMD and SGD are thus linked with the overall response of the glacier. As stated earlier, investigations on SPG have confirmed its degenerating nature, thus, validating our results. This authenticates the usage of spatio-temporal variability of glacier facies as a reliable measure of glacier health.

5.6 Conclusions

The variability in glacier facies is an ideal indicator of climate change. Therefore, the intension of this research was to determine the 11-year variability in the SPG facies using AWiFS data of two time periods (2005 and 2016) and link it with the climate variability. AWiFS data were coupled with ancillary data to avoid misclassification among glacier facies due to their spectral similarity (e.g., between shaded snow and ice facies). SVM-based SPC approach was employed to determine the fractional estimates of glacier facies as it could better represent the intrinsic heterogeneity of the SPG captured with the coarse AWiFS pixels. AWiFS-derived fractional estimates of glacier facies for the year 2016 were compared with their MSI-derived ones for the same year. Nevertheless, it was found that:

i. High r and low SD values between the AWiFS fractional and MSI RF images of glacier facies prove the efficacy of AWiFS data in their detailed SPC.
ii. Owing to the temporal differences among the input and reference data and high dynamicity among glacier facies, RF estimates of glacier facies can never perfectly match their predicted estimates, thus, impacting the SPC accuracy. The deviation in the fractional and RF estimates of SPG facies (particularly FS) is mainly due to the changes in the local meteorological conditions of the area during July–September 2016.
iii. The variability of glacier facies is strongly controlled by the annual precipitation and temperature changes in the area which alter the glacier's accumulation and ablation rates. Rise in the annual mean temperature and the annual precipitation (probably liquid) caused FS shrinkage by ~2% and ice melting by ~16% This might have in turn increased the wetness in snow, thus, expanding the WS and firn areas by ~2% and ~1%, respectively. The melting of surface ice further explains ~7% and ~8% increase in IMD and SGD areas respectively.

This work explores the potential of coarse-resolution data for successful classification of glacier facies at sub-pixel scale. However, the spatial location of constituent facies within the coarse pixels remains still unknown. Several sub-pixel mapping techniques are available to address this issue and generate fine resolution maps using coarse fraction images These maps can enable precise determination of the changes in firn-line altitudes and glacier termini, particularly at basin level.

References

Alifu H, Johnson B, Tateishi R (2016) Delineation of debris-covered glaciers based on a combination of geomorphometric parameters and a TIR/NIR/SWIR Band Ratio. IEEE J Sel Topics Appl Earth Observ Remote Sens 9(2):781–792

Ali I, Shukla A, Romshoo SA (2017) Assessing linkages between spatial facies changes and dimensional variations of glaciers in the upper Indus Basin, western Himalaya. Geomorphology 284:115–129

Arora MK, Shukla A, Gupta RP (2011) Digital image information extraction techniques for snow cover mapping from remote sensing data. In: Singh VP, Singh P, Haritashya UK (eds) Encyclopedia of snow, ice and glaciers. Encyclopedia of earth sciences series. Springer, Dordrecht.https://doi.org/10.1007/978-90-481-2642-2_498

Bhambri R, Bolch T, Chaujar RK (2011) Mapping of debris-covered glaciers in the Garhwal Himalayas using ASTER DEMs and thermal data. Int J Remote Sens 32(23):8095–8119

Bhardwaj A, Joshi PK, Sam L, Singh MK, Singh S, Kumar R (2015) Applicability of landsat 8 data for characterizing glacier facies and supraglacial debris. Int J Appl Earth Obs Geoinf 38:51–64

Bolch T, Buchroithner M, Pieczonka T, Kunert A (2008) Planimetric and volumetric glacier changes in the Khumbu Himal, Nepal, since 1962 using corona, landsat TM and ASTER data. J Glaciol 54(187):592–600

Çiftçi BB, Kuter S, Akyürek Z, Weber GW (2017) Fractional snow cover mapping by artificial neural networks and support vector machines. ISPRS Ann Photogrammetry Remote Sens Spatial Inf Sci 4:179

Czyzowska-Wisniewski EH, van Leeuwen WJ, Hirschboeck KK, Marsh SE, Wisniewski WT (2015) Fractional snow cover estimation in complex alpine-forested environments using an artificial neural network. Remote Sens Environ 156:403–417

Fujita K (2008) Effect of precipitation seasonality on climatic sensitivity of glacier mass balance. Earth Planet Sci Lett 276(1–2):14–19

Gupta RP, Ghosh A, Haritashya UK (2007) Empirical relationship between near-IR reflectance of melting seasonal snow and environmental temperature in a Himalayan basin. Remote Sens Environ 107(3):402–413

Harris I, Jones PD, Osborn TJ, Lister DH (2014) Updated high-resolution grids of monthly climatic observations —the CRU TS3.10 dataset. Int J Climatol 34:623–642

Heim A, Gansser A (1939) Central Himalaya: geological observations of the Swiss Expedition, 1936. Mem Soc Helv Sci Nat 73(1):76–78

Huang C, Davis LS, Townshend JRG (2002) An assessment of support vector machines for land cover classification. Int J Remote Sens 23(4):725–749

Jawak SD, Wankhede SF, Luis AJ (2018) Exploration of glacier surface facies mapping techniques using very high resolution worldview-2 satellite data. Multi Digital Publishing Inst Proc 2(7):339

Kavzoglu T, Colkesen I (2009) A kernel functions analysis for support vector machines for land cover classification. Int J Appl Earth Obs Geoinf 11(5):352–359

Kealy PS, Hook SJ (1993) Separating temperature and emissivity in thermal infrared multispectral scanner data: implications for recovering land surface temperatures. IEEE Trans Geosci Remote Sens 1155–1164

Keshri AK, Shukla A, Gupta RP (2009) ASTER ratio indices for supraglacial terrain mapping. Int J Remote Sens 30(2):519–524

Kirkbride MP, Deline P (2013) The formation of supraglacial debris covers by primary dispersal from transverse englacial debris bands. Earth Surf Processes Land 38(15):1779–1792

Kneib M, Miles ES, Jola S, Buri P, Herreid S, Bhattacharya A, Watson CS, Bolch T, Quincey D, Pellicciotti F (2021) Mapping ice cliffs on debris-covered glaciers using multispectral satellite images. Remote Sens Environ 253:112201

Kulkarni AV, Singh SK, Mathur P, Mishra VD (2006) Algorithm to monitor snow cover using AWiFS data of RESOURCESAT-1 for the Himalayan region. Int J Remote Sens 27(12):2449–2457

Kulkarni AV, Rathore BP, Singh SK (2010) Distribution of seasonal snow cover in central and western Himalaya. Ann Glaciol 51(54):123–128

Kumar A, Saha A, Dadhwal VK (2010) Some issues related with sub-pixel classification using HYSI data from IMS-1 satellite. J Indian Soc Remote Sens 38 (2):203–210

Laha S, Kumari R, Singh S, Mishra A, Sharma T, Banerjee A, Nainwal HC, Shankar R (2017) Evaluating the contribution of avalanching to the mass balance of Himalayan glaciers. Ann Glaciol 58 (75pt2):110–118

Lantzanakis G, Mitraka Z, Chrysoulakis N (2017) Comparison of physically and image based atmospheric correction methods for Sentinel-2 satellite imagery. Perspectives on atmospheric sciences. Springer, Cham, pp 255–261

Liu Q, Trinder J, Turner I (2016) A comparison of sub-pixel mapping methods for coastal areas. ISPRS Ann Photogrammetry Remote Sens Spatial Inf Sci 3(7)

Mishra A, Negi BD, Banerjee A, Nainwal HC, Shankar R (2018) Estimation of ice thickness of the Satopanth Glacier, Central Himalaya using ground penetrating radar. Curr Sci 114(4):785

Nainwal HC, Chaudhary M, Rana N, Negi BD, Negi RS, Juyal N, Singhvi AK (2007) Chronology of the late quaternary glaciation around Badrinath (upper Alaknanda Basin): preliminary observations. Curr Sci 10:90–96

Nainwal HC, Negi BDS, Chaudhary M, Sajwan KS, Gaurav A (2008) Temporal changes in rate of recession: Evidences from Satopanth and Bhagirath Kharak glaciers, Uttarakhand, using Total Station Survey. Curr Sci, 653–660

Nainwal HC, Banerjee A, Shankar R, Semwal P, Sharma T (2016) Shrinkage of Satopanth and BhagirathKharak glaciers, India, from 1936 to 2013. Ann Glaciol 57(71):131–139

Nair VS, Babu SS, Moorthy KK, Sharma AK, Marinoni A, Ajai (2013) Black carbon aerosols over the Himalayas: direct and surface albedo forcing. Tellus B: Chem Phys Meteorol 65(1):19738

Painter TH, Dozier J (2004) Measurements of the hemispherical-directional reflectance of snow at fine spectral and angular resolution. J Geophys Res: Atmos 109(D18)

Painter TH, Rittger K, McKenzie C, Slaughter P, Davis RE, Dozier J (2009) Retrieval of subpixel snow covered area, grain size, and albedo from MODIS. Remote Sens Environ 113(4):868–879

Panicker AS, Sandeep K, Gautam AS, Trimbake HK, Nainwal HC, Beig G, Bisht DS, Das S (2021) Black carbon over a central Himalayan Glacier (Satopanth): Pathways and direct radiative impacts. Sci Total Environ 766:144242

Panwar A, Thapliyal A, Aswal A, Singh D (2014) Recession of Satopanth and Bagirath Kharak glacier, using multi temporal set of data. Int J Innovation Sci Res 9(1):9–15

Paul F, Winsvold SH, Kääb A, Nagler T, Schwaizer G (2016) Glacier remote sensing using sentinel-2. Part II: mapping glacier extents and surface facies, and comparison to landsat 8. Remote Sens 8(7):575

Pomeroy JW, Jones HG, Tranter M, Lilbæk GR (2006) Hydrochemical processes in snow-covered basins. Encycl Hydrol Sci

Pope A, Rees WG (2014a) Impact of spatial, spectral, and radiometric properties of multispectral imagers on glacier surface classification. Remote Sens Environ 141:1–13

Pope A, Rees WG (2014b) Using in situ spectra to explore landsat classification of glacier surfaces. Int J Appl Earth Obs Geoinf 27:42–52

Racoviteanu AE, Arnaud Y, Williams MW, Ordonez J (2008) Decadal changes in glacier parameters in the Cordillera Blanca, Peru, derived from remote sensing. J Glaciol 54(186):499–510

Racoviteanu A, Williams MW (2012) Decision tree and texture analysis for mapping debris-covered glaciers in the Kangchenjunga area, Eastern Himalaya. Remote Sens 4(10):3078–3109

Remya SN, Kulkarni AV, Hassan Syed T, Nainwal HC (2020) Glacier mass loss in the Alaknanda basin, Garhwal Himalaya on a decadal scale. Geocarto Int, 1–19

Sah MP (1991) Some geomorphic observations on Badrinath-Satopanth area, Chamoli District, Garhwal Himalaya. J Him Geol 2(2):185–195

Salah M (2017) A survey of modern classification techniques in remote sensing for improved image classification. J Geomatics 11(1):21

Sattar A, Goswami A, Kulkarni AV, Das P (2019) Glacier-surface velocity derived ice volume and retreat assessment in the dhauliganga basin, central himalaya–A remote sensing and modeling based approach. Front Earth Sci 7:105

Scherler D, Bookhagen B, Strecker MR. Hillslope-glacier coupling (2011) The interplay of topography and glacial dynamics in High Asia. J Geophys Res: Earth Surf 116(F2)

Scherler D, Wulf H, Gorelick N (2018) Global assessment of supraglacial debris-cover extents. Geophys Res Lett 45(21):11–798

Sen PK (1968) Estimates of the regression coefficient based on Kendall's tau. J Am Stat Assoc 63 (324):1379–1389

Sharma T, Semwal P, Shah SS, Nainwal HC, Mishra A (2018) General geomorphological field observations around Satopanth Glacier Area, Garhwal Himalaya, Uttarakhand. IJRAR-Int J Res Anal Rev (IJRAR) 5 (4):493–508

Shukla A, Ali I (2016) A hierarchical knowledge-based classification for glacier terrain mapping: a case study from Kolahoi Glacier, Kashmir Himalaya. Ann Glaciol 57(71):1–10

Shukla A, Yousuf B (2016) Optimization of neural networks for multisource classification in a glaciated terrain. In: Raju NJ (ed) Geostatistical and geospatial approaches for the characterization of natural resources in the environment. Springer, Cham, pp 755–759

Shukla A, Yousuf B (2017) Evaluation of multisource data for glacier terrain mapping: a neural net approach. Geocarto Int 32(5):569–587

Shukla A, Arora MK, Gupta RP (2010) Synergistic approach for mapping debris-covered glaciers using optical–thermal remote sensing data with inputs from geomorphometric parameters. Remote Sens Environ 114(7):1378–1387

Song C, Woodcock CE, Seto KC, Lenney MP, Macomber SA (2001) Classification and change detection using landsat TM data: when and how to correct atmospheric effects? Remote Sens Environ 75 (2):230–244

Srivastava DK, Bhambu L (2010) Data classification using support vector machine. JATIT 12:1–7

Stroeve J, Box JE, Gao F, Liang S, Nolin A, Schaaf C (2005) Accuracy assessment of the MODIS 16-day albedo product for snow: comparisons with Greenland in situ measurements. Remote Sens Environ 94(1):46–60

Subramaniam S, Babu AS, Sivasankar E, Rao VV, Behera G (2011) Snow cover estimation from Resourcesat-1 AWiFS–image processing with an automated approach. Int J Image Process (IJIP) 5(3):298

Wolken GJ, Sharp M, Wang L (2009) Snow and ice facies variability and ice layer formation on Canadian Arctic ice caps, 1999–2005. J Geophys Res: Earth Surf 114 (F3).

Yousuf B, Shukla A, Arora MK, Jasrotia AS (2019) Glacier facies characterization using optical satellite data: impacts of radiometric resolution, seasonality, and surface morphology. Prog Phys Geogr: Earth Environ. https://doi.org/10.1177/0309133319840770

Yousuf B, Shukla A, Arora MK, Bindal A, Jasrotia AS (2020) On drivers of subpixel classification accuracy—an example from glacier facies. IEEE J Sel Top Appl Earth Obs Remote Sens 13:601–608

Zhang H, Zhao J, Jiancheng S (2005) Comparing four sub-pixel algorithms in MODIS snow mapping. In: Proc IGARSS'05, IEEE international, Seoul, South Korea, Jul. 25, 2005, 6:3784–3787

Zhang J, Jia L, Menenti M, Hu G (2019) Glacier facies mapping using a machine-learning algorithm: the ParlungZangbo Basin case study. Remote Sens 11 (4):452

Zhou C, Zheng L (2017) Mapping radar glacier zones and dry snow line in the antarctic peninsula using sentinel-1 images. Remote Sens 9(11):1171

Anticipated Shifting of Thermal and Moisture Boundary Under Changing Climate Across Nepal

6

Rocky Talchabhadel and Ramchandra Karki

Abstract

The zonation of climate helps to understand climate behaviors in different regions better and is widely useful for ecological modeling and climate change impact assessments. Thornthwaite's schemes (original and/or revised) are the most indicated for agricultural application. In this method, the water surplus and deficit in relation to the water need are assessed according to a moisture index (MI). The moisture zonation in a combination with thermal zonation provide a sound knowledge for growing suitable crop/vegetation (s) as per the potential of the zones for maximum harvests. Moist climates have positive MI values; dry climates have negative MI values. We employed a modified Thornthwaite approach based on Thornthwaite (1948) for the classification of MI. Nepal is an agricultural country having more than 60% of people directly engaged in agriculture. The country possesses several microclimates due to the complex topographical settings. Thus, the country is richly endowed with agro-diversity. Expected shifts in such climate boundary under changing climate might be important information in the field of agriculture. In this study, we present comprehensive picture of historical (1960–1990: baseline) and future (2041–2060: F1 and 2061–2080: F2) moisture and thermal zones across Nepal. We used an ensemble of ten selected general circulation models (GCMs) of the fifth phase of the Coupled Model Inter-comparison Project (CMIP5) under two Representative Concentration Pathways (RCPs) scenario *biz.* RCP 4.5 and RCP 8.5. The rate of increment of temperature is comparatively higher in a non-monsoon season than in summer monsoon season (SMS: Jun–Sep). Even though the annual precipitation is anticipated to increase in the future, winter season (WS: Dec–Feb) is expected to be drier by more than 15%. Therefore, the severity of dryness is anticipated to increase in the non-monsoon season because of twofold effects: (1) increased temperature and (2) decreased precipitation. The multi-model ensemble (MME) of selected ten GCMs provides an indication of more than 30%

Ramchandra Karki—Deceased.

R. Talchabhadel (✉)
Texas A&M AgriLife Research, Texas A&M University, 79927, El Paso, TX, USA

R. Talchabhadel
SmartPhones for Water Nepal (S4W-Nepal), Lalitpur 44700, Nepal

R. Karki
Institute of Geography, University of Hamburg, Hamburg, Germany

R. Karki
Department of Hydrology and Meteorology, Government of Nepal, Kathmandu, Nepal

U. Schickhoff et al. (eds.), *Mountain Landscapes in Transition*,
Sustainable Development Goals Series, https://doi.org/10.1007/978-3-030-70238-0_6

reduction of the polar thermal zone and more than 60% increment of the tropical thermal zone in the coming days. It is anticipated that crop production at higher altitudes could increase as a positive impact of climate change under the warming scenario. The diversity of agro-climatological zones allows tropical crops to be grown in highlands during the SMS. The assessment of shifting of moisture and thermal zone can serve as valuable information for exploring the effects of climate change in agricultural practice and irrigational interventions.

Keywords

Moisture zone · Nepal · Representative concentration pathway (RCP) · Thermal zone · Thornthwaite

6.1 Introduction

Climate zonation helps to understand behaviors of climate in different regions and is widely useful for ecological modeling and climate change impact assessments. The climatic characteristics of the region play an important role in assessing crop/vegetation selection and irrigational facilities design (Talchabhadel et al. 2019). Crop yield is mainly influenced by water, climate, and soil. A clear understanding of the supply-demand of water resources is necessary for better food security (Forsythe et al. 2015). Thornthwaite's scheme is based on surplus and deficit of water balance in relation to water needs. Different moisture and thermal zones are identified using the information of climate variables like precipitation and temperature.

Unequivocal phenomena of a warming climate, and abrupt changes in climate variables such as solar radiation, temperature, precipitation, and others, will have serious impacts on the moisture and thermal boundary. Increment of greenhouse gases (GHG) and rapid urbanization are the major causes of existing warming scenarios. The environment, biodiversity (flora and fauna), and human activities are significantly dependent on existing climate conditions and are prone to be affected under changing climate. The warming of the climate system is fast-tracking than our expectations. The required growing degree days (GDD) for the growth and development of plants/insects/animals and available GDD under changing climate affect the selection of crops at different places and different seasons. The GDD and optimum range of temperature and moisture needed are different for different crops, allowing agricultural diversity according to the climate. There are some possibilities that tropical crops to be grown in the summer season of highland under the warming environment (Talchabhadel et al. 2019). To develop adaptation strategies to minimize the impacts, understanding anticipated shifts of climate boundary (especially moisture and thermal zones) for future periods is vital. This study aims to quantify the anticipated shifts in moisture and thermal boundary under changing climate.

Similar to the objective of this study, Talchabhadel and Karki (2019) assessed the climate boundary shifting under changing climate using Köppen–Geiger (KG) scheme. The main prominence of the KG scheme is on temperature limits, whereas Thornthwaite's scheme equally focuses on the effectiveness of precipitation. As crop growth and yield are mainly dependent on water availability on soil (i.e., surpluses or shortages than required), Thornthwaite's scheme is the most indicated for agricultural application. Many researchers [(Aparecido et al. 2016), (Feddema 1994, 2005), (Grundstein 2009), (Guofeng et al. 2016), (Li and Sun 2015), Sun (2015), (Zaman and Rasul 2004) and many others] have employed Thornthwaite's method.

Nayava (1975) and Jha and Karn (2001) categorized the climate boundary across Nepal using Thornthwaite's model. The prior one is limited to only 15 stations for a data period of 5 years during the 1970s, whereas the latter one classified for the administrative districts of Nepal using the meteorological data published during the 1970s in "*Mechidekhi Mahakali*" (HMG 1975). For the recent 30 years (1986–2015), Talchabhadel et al. (2019) updated moisture and thermal zonation using 75 climate stations distributed across the country incorporating a high-resolution spatial interpolation. However, there still exists a research gap exploring the anticipated shift of moisture and thermal zones across the country. We attempted to assess the climate boundary shifting from the baseline period (1960–1990) to two future scenarios (F1: 2041–2060 and F2: 2061–2080). We used bias-corrected and downscaled data from ten different GCMS under two warming scenarios (RCP 4.5 and RCP 8.5). RCP 4.5 is a medium stabilizing warming scenario, whereas RCP 8.5 is a very high emission scenario. We believe this study complements the delineation of agro-climatic potential for different crops under changing climate.

6.2 Study Area

The study domain consists of an entire country Nepal. The country, officially the Federal Democratic Republic of Nepal, is bounded by China in the north and India in the south, east, and west (as shown in Fig. 6.1a). The area of the country is 147,181 km^2. The latitudinal (north–south) extent ranges only 140–250 km, but the elevation varies from 60 m above sea level (asl) in the south to 8848 m asl (Mt. *Sagarmatha*, the highest peak of the world) in the north. The distribution of temperature is strongly dependent on topography. Intra-annual variability of precipitation is very high. Almost 80% of the annual precipitation occurs during SMS. Importantly, precipitation throughout the study domain varies both temporally and spatially. Nayava (1980) highlighted that precipitation in SMS is higher in southeastern parts of the country, whereas northwestern areas experience significant amounts of precipitation in WS. The mean annual precipitation and temperature for the baseline period (1960–1990) are shown in Fig. 6.1b, c, respectively.

6.3 Data and Methods

6.3.1 Data

This study used the freely available WorldClim (http://worldclim.org) gridded data, which has been widely used in many disciplines [(Uddin et al. 2018), (Acharya et al. 2018), (Thapa et al. 2015), (Talchabhadel and Karki 2019)] across Nepal. For a baseline period, this study used monthly climatological WorldClim data in its version 1.4 for 1960–1990 (Hijmans et al. 2005), at a spatial resolution grid of 30 arc second. Our team, Talchabhadel and Karki (2019), analyzed and recommended the performance of WorldClim with observed data (from 247 precipitation and 68 temperature stations) maintained by the Department of Hydrology and Meteorology (DHM), Government of Nepal in different physiographical regions of Nepal. The spatial interpolation was conducted for the observed data, considering the elevation as one of the dominant drivers (Karki et al., 2016).

This study used climate data for future periods (F1 and F2) available from WorldClim, which were downscaled and bias-corrected using the data version 1.4 (Hijmans et al. 2005). WorldClim is a freely available set of global climate layers (gridded climate data). The climate variables available in WorldClim are the monthly minimum and maximum temperature and precipitation for 19 GCMs of the CMIP5. This study considered ten GCMs (biz. ACCESS1-0, BCC-CSM 1-1, CNRM-CM5, HadGEM2-AO, HadGEM2-CC, HadGEM2-ES, MIROC-ESM-CHEM, MIROC-ESM, MIROC5, and MPI-ESM-LR) of the CMIP5.

Table 6.1 shows the detail of the selected ten GCMs. Different GCMs result in different climate variables. Selection of GCMs and understanding their ranges are important while analyzing projected data. In order to increase the confidence in the use of the output of GCMs, validation against the observed historical climate conditions is therefore required. We used a multi-model ensemble (MME) of selected ten GCMs for projected analysis. This study also analyzed moisture and thermal zonation using each GCM

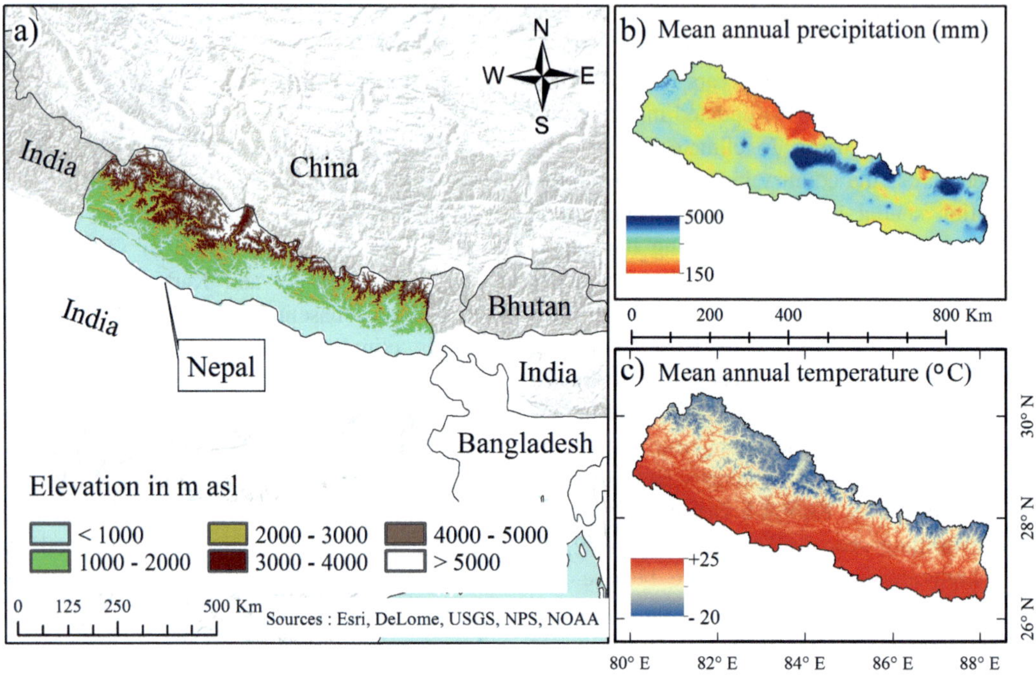

Fig. 6.1 **a** Location of the study area, **b** mean annual precipitation in mm, and **c** mean annual temperature in °C for the baseline period (1960–1990)

to discuss the range of selected GCMs. A moderate (RCP 4.5) and the highest (RCP 8.5) warming scenarios were selected among four RCPs (Vuuren et al. 2011) of GHG emission scenarios for the future periods F1 and F2.

6.3.2 Methods

We followed the methodologies adopted by Talchabhadel et al. (2019) using modified Thornthwaite's scheme. Talchabhadel et al. (2019) used station-wise data from 75 climate stations, whereas this study employed gridded data to calculate the same. The indices for moisture and thermal zonation are calculated on a monthly and annual basis. Moist climates have positive MI values, whereas dry climates have negative MI values. Cold climates have lower potential evapotranspiration (PET) values, and hot climates have larger PET values. Table 6.2 shows the categorization of thermal and moisture zones.

6.4 Results and Discussions

6.4.1 Anticipated Change in Precipitation and Temperature

Figure 6.2 shows the country's average mean monthly precipitation and temperature (minimum, average, and maximum) for the baseline period. The six-month precipitation from May to Oct is almost 90% of the annual precipitation, predominantly with southeastern warm and moist winds from the Bay of Bengal. In general, SMS is hot and humid, whereas WS is dry and cold. During WS, the westerly moisture from the Arabian, Caspian, and the Mediterranean Sea brings few spells of precipitation in the country. If we look into the North–South section of the country, the southern plains are the hottest parts of the country, and the high Himalayas in the north are all-round covered with snow.

Projection in the future shows that different GCMs display a mixed pattern of increasing and

Table 6.1 Detail of selected ten GCMs

SN	Selected CMIP5 model	Institute	Country	Resolution
1	ACCESS1-0	Commonwealth Scientific and Industrial Research Organization (CSIRO) and Bureau of Meteorology	Australia	1.25° x 1.875°
2	BCC-CSM 1-1	Beijing Climate Center (BCC), China Meteorological Administration (CMA), based on National Center for Atmospheric Research (NCAR) CCSM 2.0.1	China	2.8° x 2.81°
3	CNRM-CM5	Centre National de Recherches Me´te´orologiques (CNRM)/National Centre for Meteorological Research	France	1.4° × 1.4°
4	HadGEM2-AO	Met Office Hadley Centre	UK	1.25° × 1.875°
5	HadGEM2-CC	Met Office Hadley Centre	UK	1.25° × 1.875°
6	HadGEM2-ES	Met Office Hadley Centre	UK	1.25° × 1.875°
7	MIROC-ESM-CHEM	University of Tokyo (UoT), National Institute for Environmental Studies (NIES), and Japan Agency for Marine-Earth Science and Technology (JAMSTEC)	Japan	2.8° × 2.81°
8	MIROC-ESM	UoT, NIES, and JAMSTEC	Japan	2.8° × 2.81°
9	MIROC5	UoT, NIES, and JAMSTEC	Japan	1.4° × 1.4°
10	MPI-ESM-LR	Max Planck Institut für Meteorologie/Max Planck Institute (MPI) for Meteorology	Germany	1.865° × 1.875°

ACCESS: Australian Community Climate and Earth System

BCC-CSM: BCC Climate System Model

HadGEM: Hadley Global Environment Model; AO- Atmosphere Ocean, CC- Carbon Cycle, ES-Earth System

MIROC: Model for Interdisciplinary Research On Climate; ESM-Earth system model, CHEM-chemistry coupled

MPI-ESM-LR: MPI-ESM running on Low Resolution grid

decreasing precipitation. The MME, in general, shows an increasing pattern of precipitation in the future. During the future F1, the mean annual precipitation is projected to increase by 3.97% (ranging from −16.26% to 38.9%) under RCP 4.5 and 6.16% (ranging from −18.05% to 50.02%) under RCP 8.5. Similarly, during the future F2, the mean annual precipitation is projected to increase by 7.4% (ranging from −16.47% to 40.75%) under RCP 4.5 and 10.69% (ranging from −17.15% to 58.2%) under RCP 8.5.

All GCMs show the increasing temperature in future periods. During the future F1, the mean annual average temperature is anticipated to increase by 2.22 °C (ranging from 1.23 °C to 3.27 °C) under RCP 4.5 and by 2.74 °C (ranging from 1.54 °C to 4.28 °C) under RCP 8.5. Similarly, during the future F2, the mean annual average temperature is anticipated to increase by 2.77 °C (ranging from 1.58 °C to 4.03 °C) under RCP 4.5 and by 4.03 °C (ranging from 2.42 °C to 5.76 °C) under RCP 8.5. Figure 6.3 shows the country's averaged projected mean monthly precipitation and average temperature for different scenarios. The spread (range) of selected GCMs is higher under RCP 8.5 than RCP 4.5. The MME shows that the projected increment of temperature is lower in SMS than in a non-monsoon season. For instance, during the future F2, the projected increment of average temperature is 3.16 °C during SMS and is 4.46 °C during

Table 6.2 Categories, symbols of different moisture and thermal zones, and their limits. [Adopted from Talchabhadel et al. (2019)]

Based on PET values (thermal zones)		
Categories		PET limits (mm/year)
Tropical		> 1200
Hot		900 to 1200
Warm		600 to 900
Cold		300 to 600
Polar		0 to 300
Based on MI values (moisture zones)		
Categories	Symbol	MI limits (%)
Arid	A	< −50
Semi-arid	SA	−50 to −25
Dry subhumid	SH1	−25 to 0
Moist subhumid	SH2	0 to 25
Humid	H1	25 to 50
	H2	50 to 75
	H3	75 to 100
Per humid	PH1	100 to 200
	PH2	200 to 400
	PH3	> 400

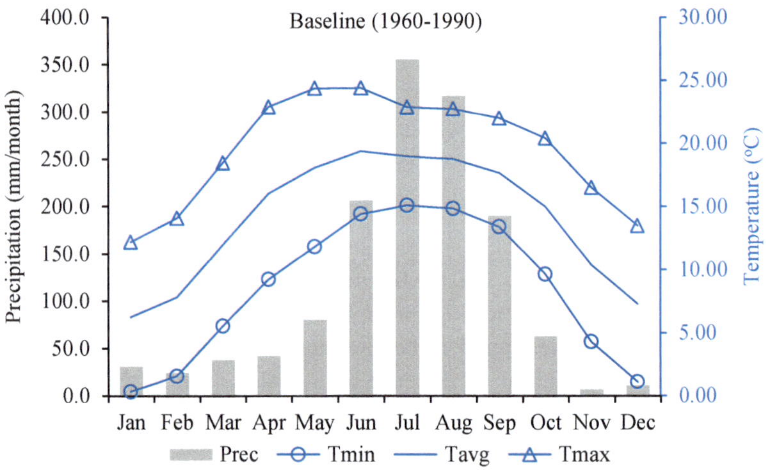

Fig. 6.2 Country's averaged mean monthly precipitation and temperature for the baseline period (1960–1990). Corresponding labels are depicted in their respective colors

the non-monsoon season. In contrast, the % change of precipitation is positive (i.e., increasing) during SMS and is negative (i.e., decreasing) during non-monsoon season. In all future scenarios, the WS precipitation is expected to be drier by more than 15%. Therefore, the severity of dryness is anticipated to increase in the non-monsoon season because of twofold effects: (1) increased temperature and (2) decreased precipitation.

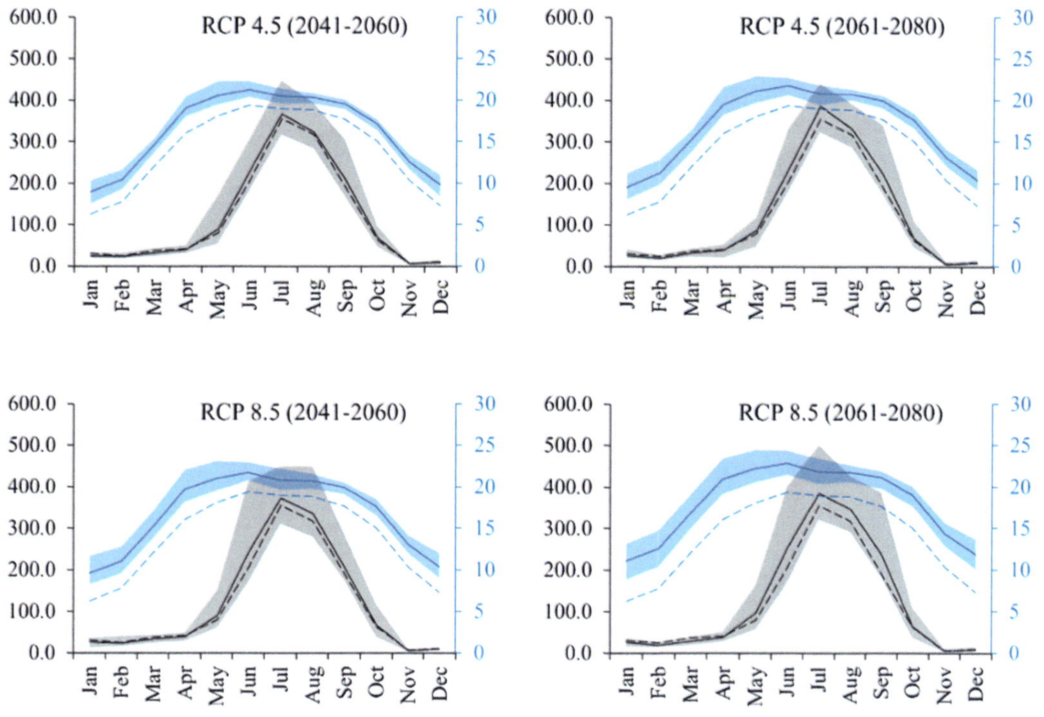

Fig. 6.3 Country's averaged projected mean monthly precipitation and average temperature for the different scenarios. Black color represents precipitation in mm/month, and blue color represents temperature in °C. Shaded regions represent the ranges of selected ten GCMs, whereas solid lines are the MME of selected ten GCMs. The dotted lines represent the values at the baseline period

6.4.2 Anticipated Change in Thermal Zones

Figure 6.4 shows thermal zonation of the country: (a) for the baseline period using observed data maintained by DHM (left panel) and WorldClim data (right panel), (b) for the F1 period using the MME of selected ten GCMs under RCP 4.5 (left panel) and under RCP 8.5 (right panel), and (c) for the F2 period using the MME of selected ten GCMs under RCP 4.5 (left panel) and under RCP 8.5 (right panel). The overall pattern of different thermal zones for the baseline period derived from WorldClim data is almost congruous with that derived from observed data maintained by DHM (Fig. 6.4 a). It ensures the reliability of using WorldClim data for projected analysis under different scenarios (Fig. 6.4 b and c). There exists a prominent increment of the tropical zone and a slight reduction of the polar zone across the country. The quantitative assessment of changes in areal coverage of different zones is summarized in Fig. 6.5 and Table 6.3.

Figure 6.5 shows the total coverage of different thermal zones (expressed in %) for the baseline and future periods using the MME of selected ten GCMs (spatial patterns shown in Fig. 6.4 b and c). Due to the northward shift of the boundary of the tropical thermal zone, the areal coverages of hot and warm thermal zones are reduced. Similarly, due to a noticeable reduction in polar zones, an increment of the cold thermal zone could be found. A closer examination shows the increment of areal coverage of cold thermal zone is not from the shift of boundary from the warm thermal zone but the significant shift of boundary from the polar thermal zone. Table 6.3 shows the areal coverage of different thermal zones during baseline and anticipated % change in different future scenarios with respect to the baseline across the country.

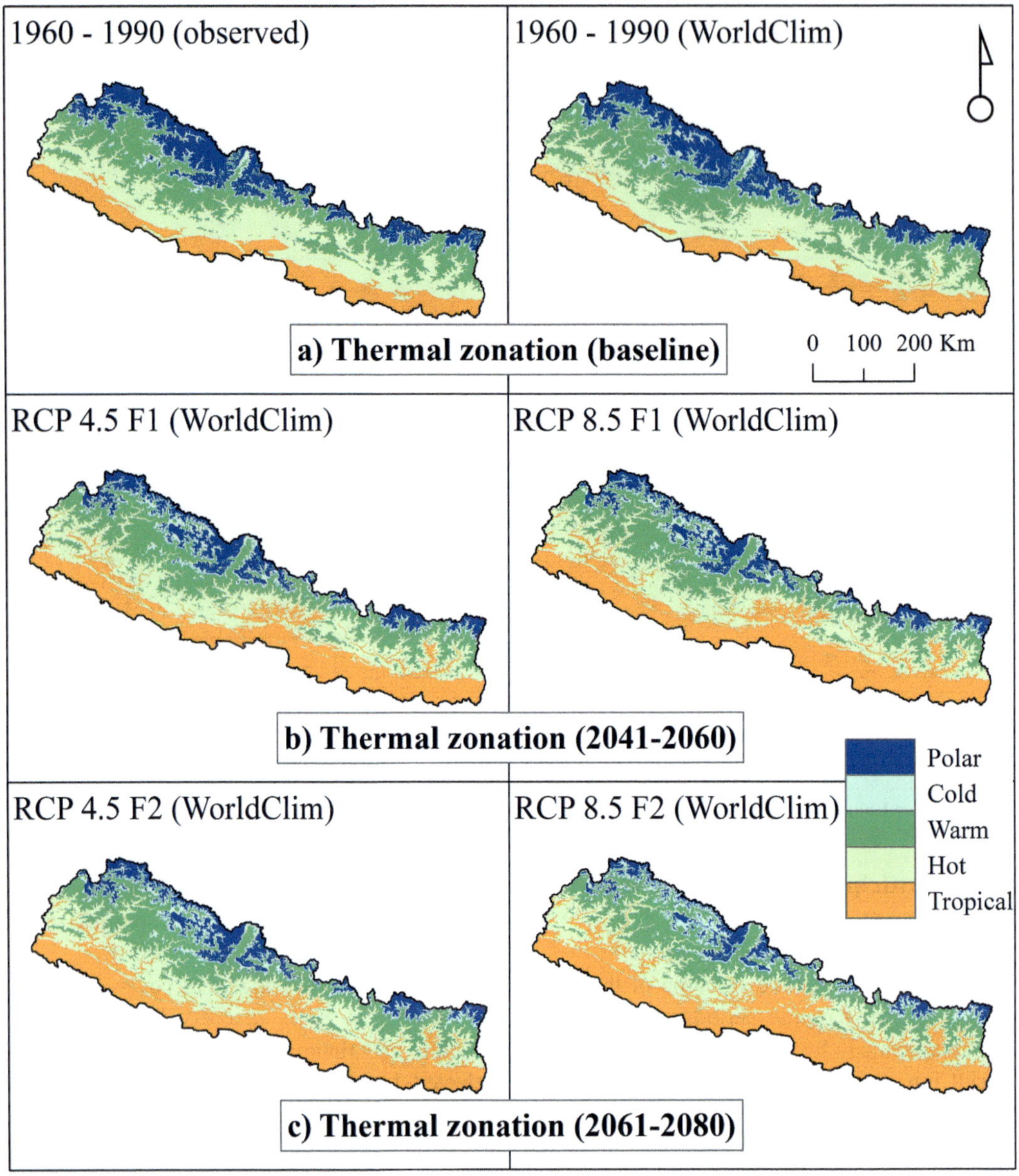

Fig. 6.4 Thermal zonation of Nepal based on modified Thornthwaite's scheme: **a** for the baseline period using observed data (left panel) and WorldClim data (right panel), **b** for the F1 period using the MME of selected ten GCMs under RCP 4.5 (left panel) and under RCP 8.5 (right panel), and **c** for the F2 period using the MME of selected ten GCMs under RCP 4.5 (left panel) and under RCP 8.5 (right panel). The PET limit for the thermal zone represented can be found in Table 6.2

Future scenarios are estimated using the MME of selected ten GCMs, and their ranges are shown. The selected ten GCMs showed a huge variability. For instance, during the future F2, the areal coverage of the polar thermal zone is anticipated to change by −55.6% (ranging from −83.5% to −33.9%), and the tropical thermal zone is projected to change by +104.9% (ranging from +63.5% to +146.4%) under RCP 8.5 with respect to the baseline across the country. In

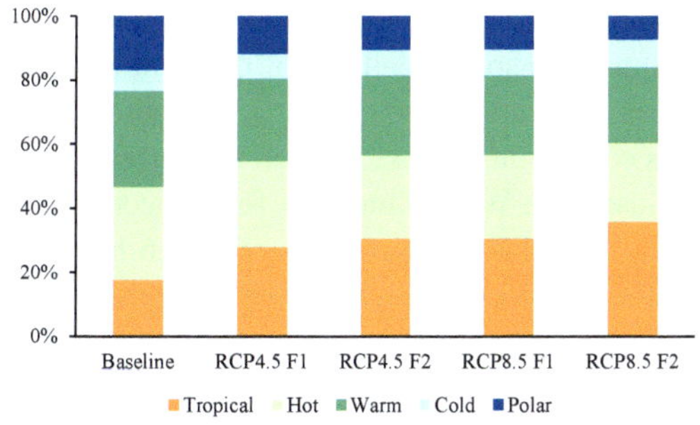

Fig. 6.5 Total coverage of different thermal zones (expressed in %) for the baseline and future periods using the MME of selected ten GCMs

the polar thermal zone is anticipated to reach −37% under RCP 4.5 and −56% under RCP 8.5 with respect to the baseline. During the future F1, the % deviation of the change of areal coverage of the tropical thermal zone is anticipated to reach +60% under RCP 4.5 and +74% under RCP 8.5 with respect to the baseline. Similarly, during the future F2, the % deviation of the change of areal coverage of the tropical thermal zone is anticipated to reach +75% under RCP 4.5 and +105% under RCP 8.5 with respect to the baseline. Due to these two noticeable shifts, the

Table 6.3 Areal coverage of different thermal zones during baseline and anticipated % change in different future scenarios with respect to baseline in Nepal. Future scenarios are estimated using the MME of selected ten GCMs, and their ranges are shown in []

Thermal zones	Baseline	Delta (RCPx.x yy - Baseline)/Baseline in %			
	Area (km^2)	RCP4.5 F1	RCP4.5 F2	RCP8.5 F1	RCP8.5 F2
Polar	24,714.65	−29.8 [− 46.8 to −18.7]	−36.6 [− 62 to −23.2]	−37.8 [− 79.2 to −23.2]	−55.6 [− 83.5 to −35.9]
Cold	9824.46	16.3 [10.5 to 29.6]	18.1 [14.5 to 39.3]	21.2 [14.2 to 37]	27 [21.1 to 31.1]
Warm	44,098.69	−13.8 [− 18.3 to −7.6]	−16.4 [− 18.9 to −9.7]	−17 [− 18.1 to −9.2]	−20.7 [− 22.2 to −15]
Hot	42,787.11	−8.2 [− 13.4 to −1.8]	−10.9 [− 16.9 to −3]	−10.1 [− 13 to −2.3]	−15.9 [− 26.2 to −6.9]
Tropical	25,756.09	59.6 [30 to 87.2]	74.5 [38.4 to 104.9]	74.1 [36.5 to 114.6]	104.9 [63.5 to 146.4]

Values inside [] show the ranges derived from selected ten GCMs.

general, the inter-model variability is higher under RCP 8.5 compared to RCP 4.5 and during the F2 period compared to the F1 period. The MME of selected ten GCMs provides an indication of more than 30% reduction of the polar thermal zone and more than 60% increment of the tropical thermal zone in the coming days.

During the future F1, the % deviation of the change of areal coverage of the polar thermal zone is anticipated to reach −30% under RCP 4.5 and −38% under RCP 8.5 with respect to the baseline. Similarly, during the future F2, the % deviation of the change of areal coverage of areal coverage of the cold thermal zone is anticipated to increase, and the areal coverages of hot and warm climates are anticipated to decrease. Importantly, the thermal zones are mainly governed by the anticipated change in PET (a function of only temperature in the current study). The unequivocal warming climate expects an overall increment in PET and a clear shift toward the torrid climatic condition. A similar analysis for the anticipated shift of moisture zones is conducted in the subsequent section. For the determination of moisture zone, information of anticipated precipitation (as a moisture supply) and PET (as a moisture demand) are exchanged.

6.4.3 Anticipated Change in Moisture Zones

Figure 6.6 shows moisture zonation of the country: (a) for the baseline period using observed data maintained by DHM (left panel) and WorldClim data (right panel), (b) for the F1 period using the MME of selected ten GCMs under RCP 4.5 (left panel) and under RCP 8.5 (right panel), and (c) for the F2 period using the MME of selected ten GCMs under RCP 4.5 (left panel) and under RCP 8.5 (right panel). The overall pattern of different moisture zones for the baseline period derived from WorldClim data is nearly congruous with that derived from observed data maintained by DHM (Fig. 6.6 a). A closer examination indicates a noticeable variation around the higher mountains and in the northwestern region of the country. Talchabhadel and Karki (2019) already reported that the deviation of precipitation between observed and the WorldClim is higher in the high mountain region of the country, and it was also confirmed that the station density in the high mountain is very much limited complying us to rely on the interpolation. Importantly, the difference in the interpolation scheme (although elevation was taken as one of the dominant drivers) between the two datasets could be another reason for the discrepancy because elevation dependency for precipitation is not robust compared to that of temperature.

The amount of moisture is higher for the observed data than the WorldClim in the northwestern region of the country. WS precipitation mainly caused by westerly disturbances contributes almost 6% of the total precipitation in the western region (80–83° longitude) of the country, whereas it contributes only 2.5% of the total precipitation in the eastern region (86‾89° longitude) of the country (Talchabhadel et al. 2018). WorldClim lacks to capture the local phenomena of WS precipitation across the country, therefore, it has underestimated the moisture in the northwestern region of the country. Apart from high mountains and the northwestern region of the country, the spatial patterns of different moisture zones are almost congruous. Without making bias correction in those areas in the current study, we used the WorldClim data for projected analysis under different scenarios (Fig. 6.6 b and c). There is an emergence of arid moisture zone, increment of semi-arid and subhumid moisture zones, and a significant reduction of per humid moisture zones across the country in the coming days. The quantitative assessment of changes in areal coverage of different zones is summarized in Fig. 6.6 and Table 6.4.

Figure 6.7 shows the total coverage of different moisture zones (expressed in %) for the baseline and future periods using the MME of selected ten GCMs (spatial patterns shown in Fig. 6.6 b and c). Table 6.4 shows the areal coverage of different moisture zones during baseline and anticipated % change in different future scenarios with respect to the baseline across the country. Future scenarios are estimated using the MME of selected ten GCMs, and their ranges are shown. The selected ten GCMs show a huge variability. For instance, during the future F2, the areal coverage of semi-arid (SA) moisture zone is anticipated to change by +1540.8% (ranging from −94.7% to +5552.2%) and the dry subhumid is projected to change by +97.3% (ranging from −94.9% to +122.9%) under RCP 8.5 with respect to the baseline across the country. The MME of selected ten GCMs anticipates an emergence of a new moisture zone (i.e., arid) with an area of 42 km^2 under RCP 8.5 during the F2 period. Similarly, a small areal coverage with an area of 641.1 km^2 of semi-arid moisture zone during the baseline is projected to increase significantly in the coming days. It is projected to reach 3047 km^2 under RCP 4.5 and 3996 km^2 under RCP 8.5 during the F1 period. Similarly, it is projected to reach 3744 km^2 under RCP 4.5 and 10,519 km^2 under RCP 8.5 during the F2 period.

During the future F1, the % deviation of the change of areal coverage of per humid (PH1, PH2, and PH3) moisture zone is anticipated to reach −27% under RCP 4.5 and −31% under RCP 8.5 with respect to the baseline. Similarly,

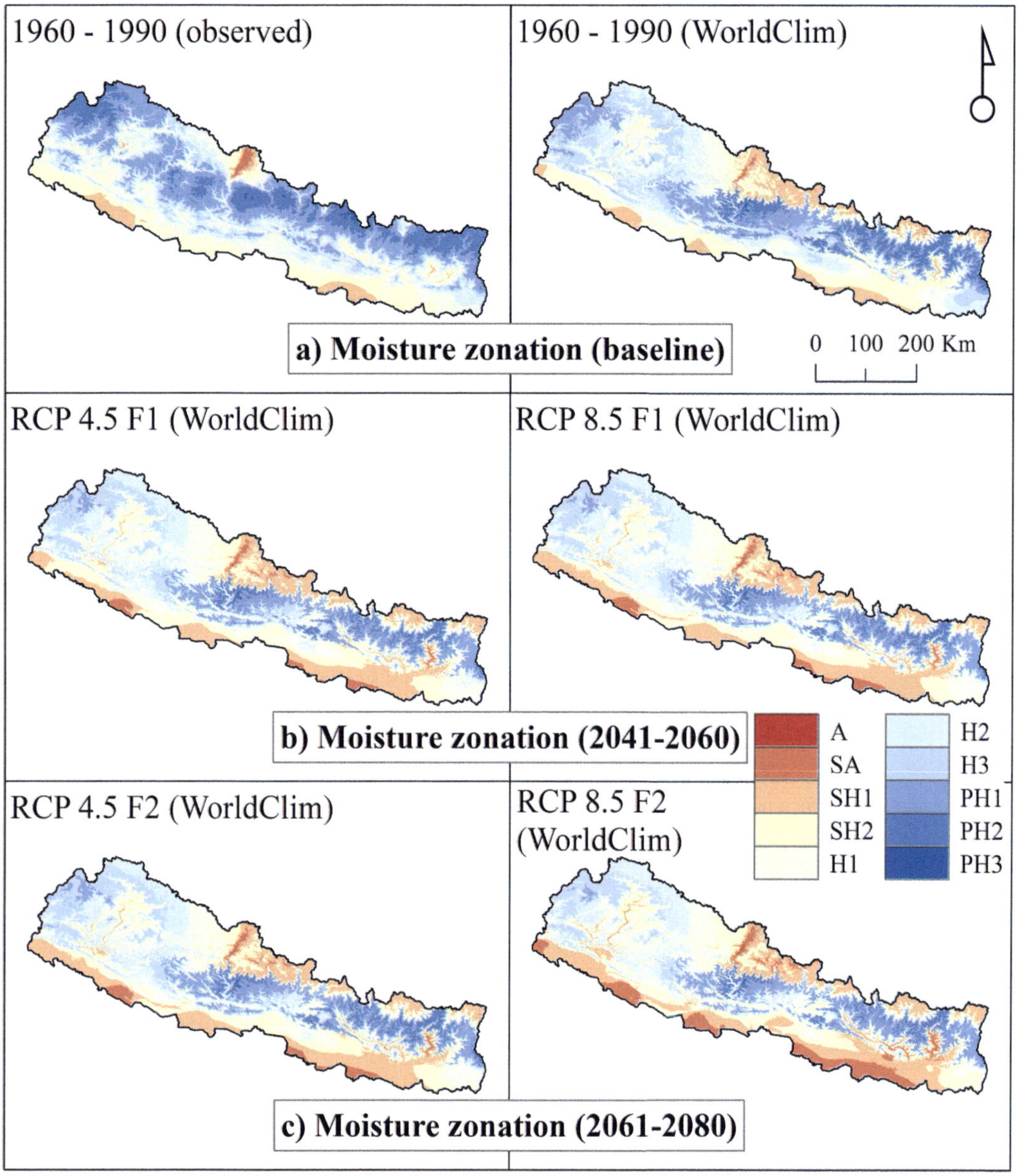

Fig. 6.6 Moisture zonation of Nepal based on modified Thornthwaite's scheme: (a) for the baseline period using observed data (left panel) and WorldClim data (right panel), (b) for the F1 period using the MME of selected ten GCMs under RCP 4.5 (left panel) and under RCP 8.5 (right panel), and (c) for the F2 period using the MME of selected ten GCMs under RCP 4.5 (left panel) and under RCP 8.5 (right panel). The category name and MI limit for the symbol represented can be found in Table 6.2

during the future F2, the % deviation of the change of areal coverage of per humid moisture zone is anticipated to reach −27% under RCP 4.5 and −39% under RCP 8.5 with respect to the baseline. During the future F1, the % deviation of the change of areal coverage of subhumid (SH1 and SH2) moisture zone is anticipated to reach +30% under RCP 4.5 and +35% under RCP 8.5 with respect to the baseline. Similarly, during the future F2, the % deviation of the change of areal

Table 6.4 Areal coverage of different moisture zones during baseline and anticipated % change in different future scenarios with respect to baseline in Nepal. Future scenarios are estimated using the MME of selected ten GCMs, and their ranges are shown in []

Moisture zones	Baseline	Delta (RCPx.x yy - Baseline)/Baseline in %			
	Area (km^2)	RCP4.5 F1	RCP4.5 F2	RCP8.5 F1	RCP8.5 F2
A	0.00	Emergence	Emergence	Emergence	Emergence
SA	641.12	375.4 [− 90.4 to 4623.2]	484 [− 94.1 to 5154.7]	523.4 [− 94.1 to 5662.1]	1540.8 [− 94.7 to 5552.2]
SH1	14,561.36	70.7 [− 90.7 to 112.6]	82.2 [− 93.5 to 123]	87.6 [− 95.4 to 159.7]	97.3 [− 94.9 to 122.9]
SH2	23,755.23	5.3 [− 43 to 34]	1.6 [− 40.2 to 44]	2.8 [− 54.9 to 51.2]	7 [− 55.1 to 47.3]
H1	24,871.17	10.4 [− 34.2 to 25.4]	7.7 [− 34.3 to 27.8]	8.9 [− 49.6 to 27.7]	0.3 [− 45.5 to 18]
H2	26,935.24	−9.8 [− 64.1 to 5.2]	−11.4 [− 73.8 to 3.9]	−10.3 [− 78.6 to 6.1]	−14.3 [− 81 to 6.2]
H3	23,440.69	−20.7 [− 77.4 to −18.7]	−22.6 [− 79.7 to −19.4]	−25.9 [− 79.8 to −23.3]	−38.8 [− 82.3 to −22.5]
PH1	23,478.32	−20.2 [− 54 to 146]	−22.1 [− 59.5 to 148.9]	−23.8 [− 67.7 to 151.7]	−30.2 [− 77.1 to 143.9]
PH2	9372.21	−45.3 [− 94.8 to 106.4]	−40.2 [− 96.3 to 103.8]	−48.3 [− 97.6 to 192.1]	−60.6 [− 98.5 to 172]
PH3	125.67	−49.7 [− 100 to 524.6]	−52.1 [− 100 to 422.8]	−56.9 [− 100 to 692.2]	−77.8 [− 100 to 819.2]

Values inside [] show the range derived from different GCMs.

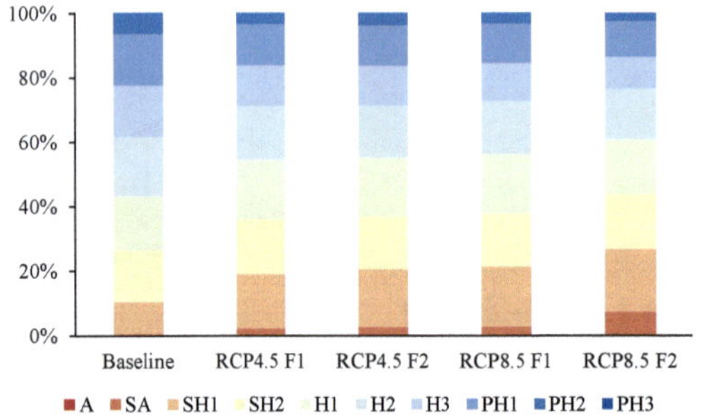

Fig. 6.7 Total coverage of different moisture zones (expressed in %) for the baseline and future periods using the MME of selected ten GCMs. The category name and MI limit for the symbol represented can be found in Table 6.2

coverage of subhumid moisture zone is anticipated to reach +32% under RCP 4.5 and +41% under RCP 8.5 with respect to the baseline.

Figure 6.8 shows the anticipated monthly moisture zonation of the country under RCP 8.5 for the F2 period using the MME of selected ten GCMs. Similar patterns exist in other future scenarios and the baseline period (not shown). In general, the spatial distributions of moisture zonation follow the patterns of monthly precipitation. From the start of December, the country receives few downpours from the northwestern direction as westerly disturbances. The eastern region of the country starts becoming wetter from early April and evolves fully in SMS up to September, resulting in a humid to per humid environment. After the withdrawal of the monsoon around the end of September, the country starts experiencing a cooler and drier environment.

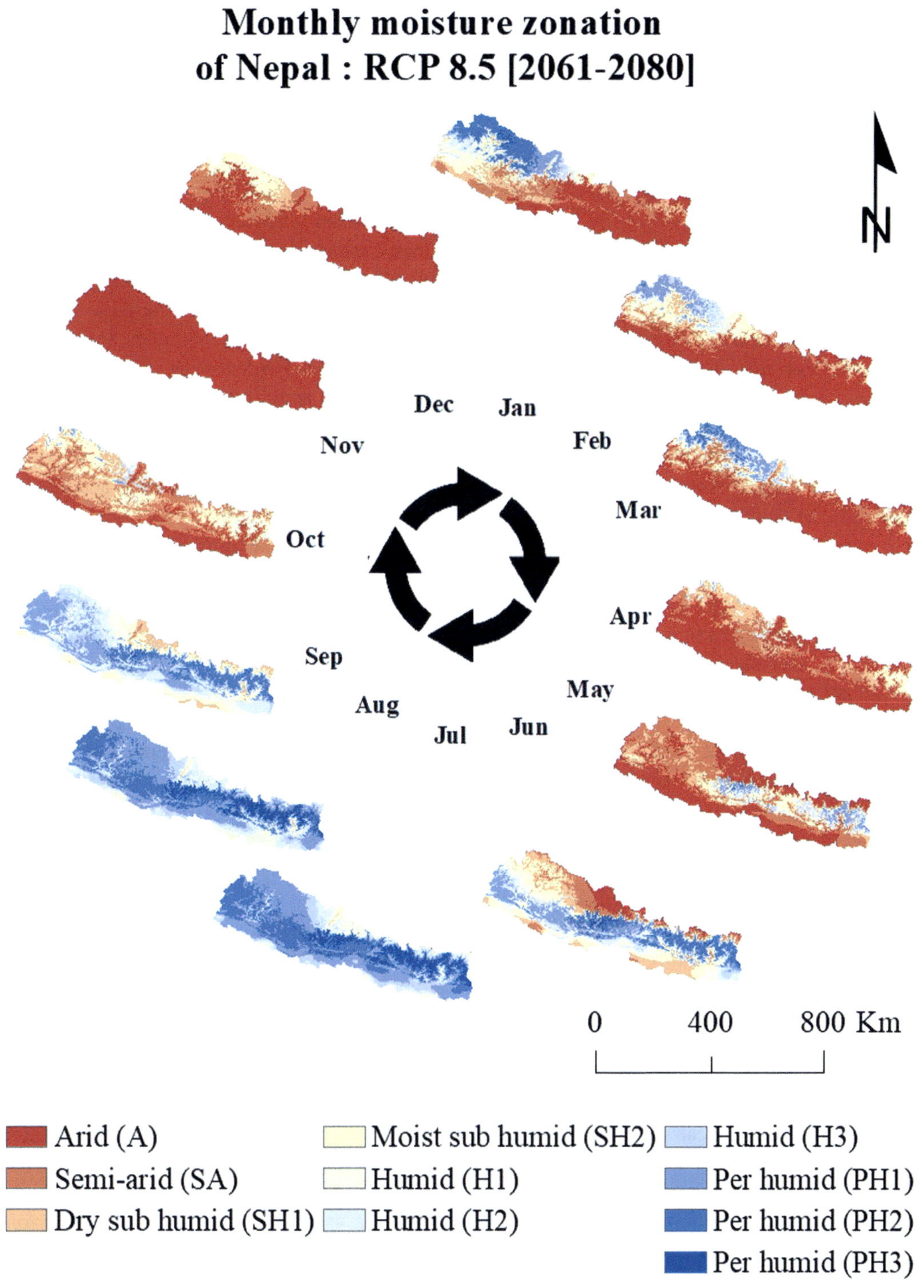

Fig. 6.8 Monthly moisture zonation of Nepal based on modified Thornthwaite's scheme for the F2 period under RCP 8.5 using the MME of selected ten GCMs. The MI limit for the moisture zone can be found in Table 6.2

6.4.4 Discussions

This paper presents the anticipated shifts of thermal and moisture zones across the country under climate change. It is expected that under a warming scenario, crop production at higher altitudes could increase as a positive impact of climate change. The combined information of moisture zonation and thermal zonation provide a better understanding of growing appropriate crop/vegetation (s) as per the potential of the zones for maximum harvests. In general, local farmers follow their typical cropping patterns, cropping rotations over centuries by learning from the experience. They are well known about the crop suitability at their spatial locations during different seasons. The understanding of possible shifts of thermal and moisture zones could help manage the cropping schedule. Soil moisture mainly determines the planting and growing date of different crops. Precipitation, temperature, and soil are key factors influencing moisture availability. Our study could be an initial step toward agro-climatic zonation for different crops under climate change.

The information of soil parameters, crop growing days, and the output of this study (thermal and moisture zonation) may help delineate quantitative agro-climatic potential for different crops. The major crops of Nepal are rice, maize, and wheat. Other crops are millet, potato, barley, buckwheat, vegetables, fruits, pulses, and so on. Our future work includes the preparation of crop suitability mapping for each major crop. The cropping frequencies (one or two times a year) for different crops like rice could be assessed for different thermal and moisture zones. In the southern region of the country, where a tropical thermal zone exists, rice can be produced two times a year, and fruits like mango could easily be cultivated. In contrast, no potential agriculture could be done in the high mountains where the frigid polar environment exists. Cold and temperate crops like barley, buckwheat, millet, potato are suitable in a cold climate, limiting mostly one crop in a year. An irrigation intervention could help reduce the moisture demand for many crops and produce a maximum yield.

6.5 Conclusion

This study presented the anticipated shifts in moisture and thermal boundary across the country under changing climate from the baseline (1960–1990). We selected ten GCMs of CMIP5 under two RCPs for the projected analysis. In general, the inter-model variability is higher under RCP 8.5 compared to RCP 4.5 and during the F2 period compared to the F1 period. The severity of dryness is anticipated to increase in the non-monsoon season because of twofold effects: (1) increased temperature and (2) decreased precipitation. Due to the northward shift of the boundary of the tropical thermal zone, the areal coverages of hot and warm thermal zones are reduced. Similarly, due to a noticeable reduction in polar zones, an increment of the cold thermal zone could be found. The MME of selected ten GCMs provides an indication of more than 30% reduction of the polar thermal zone and more than 60% increment of the tropical thermal zone in the coming days.

There is an emergence of arid moisture zone, increment of semi-arid and subhumid moisture zones, significant reduction of per humid moisture zones across the country in the coming days. A small areal coverage with an area of 641.1 km^2 of semi-arid moisture zone during the baseline is projected to reach 3047 km^2 under RCP 4.5 and 3996 km^2 under RCP 8.5 during the F1 period. Similarly, it is projected to reach 3744 km^2 under RCP 4.5 and 10,519 km^2 under RCP 8.5 during the F2 period. This is due to the combined effect of increased PET (a function of temperature) and decreased precipitation. The intra-annual variability of precipitation and resulting MI is very high. The spatial patterns of the moisture zone are quite different for different seasons. The assessment of shifting of moisture and thermal zone can serve as valuable information exploring the effects of climate change in agricultural practice and irrigational interventions. We believe our results to be useful for different stakeholders especially related to agriculture and crop production.

References

Acharya BK, Cao C, Xu M, Khanal L, Naeem S, Pandit S (2018) Present and Future of Dengue Fever in Nepal: Mapping Climatic Suitability by Ecological Niche Model. Int J Environ Res Public Health 1–15. https://doi.org/10.3390/ijerph15020187

Aparecido LE de O, Rolim G de S, Richetti J, Souza PS de, Johann JA (2016) Köppen, Thornthwaite and Camargo climate classifications for climatic zoning in the State of Paraná, Brazil. Ciência e Agrotecnologia 40(4):405–417. https://doi.org/10.1590/1413-70542016404003916

Feddema JJ (1994) Evaluation of terrestrial climate variability using a moisture index. Publ Climatol XLVII(1)

Feddema JJ (2005) A revised thornthwaite-type global climate classification. Phys Geogr 26(6):442–466. https://doi.org/10.2747/0272-3646.26.6.442

Forsythe N, Blenkinsop S, Fowler HJ (2015) Exploring objective climate classification for the Himalayan arc and adjacent regions using gridded data sources. Earth Syst Dyn 6(1):311–326. https://doi.org/10.5194/esd-6-311-2015

Grundstein A (2009) Evaluation of climate change over the continental United States using a moisture index. Clim Chang 93:103–115. https://doi.org/10.1007/s10584-008-9480-3

Guofeng ZHU, Dahe QIN, Huali T, Yuanfeng LIU, Jiafang LI, Dongdong C, Kai W (2016) Variation of thornthwaite moisture index in hengduan mountains, China. Chinese Geogr Sci 26(5):687–702. https://doi.org/10.1007/s11769-016-0820-3

Hijmans RJ, Cameron SE, Parra JL, Jones PG, Jarvis A (2005) Very high resolution interpolated climate surfaces for global land areas. Int J Climatol 25 (15):1965–1978. https://doi.org/10.1002/joc.1276

HMG (1975) Mechidekhi Mahakali (I-IV Volumes). Department of Information, Ministry of Communication

Jha S, Karn A (2001) Climatic analogues for the adminstrative districts of Nepal. Tribhuvan Univ J, 55–64

Karki R, Talchabhadel R, Aalto J, Baidya SK (2016) New climatic classification of Nepal. Theor Appl Climatol 125(3–4):799–808. https://doi.org/10.1007/s00704-015-1549-0

Li J, Sun X (2015) Valuation of changes of Thornthwaite Moisture Index in Victoria. Aust Geomech 50(3):39–49

Nayava JL (1975) Climates of Nepal. Himal Rev VII: 9–12

Nayava JL (1980) Rainfall in Nepal. Himal Rev 12

Sun X (2015) The impact of climate as expressed by Thornthwaite Moisture Index on residential footing design on expansive soil in Australia. RMIT University

Talchabhadel R, Karki R (2019) Assessing climate boundary shifting under climate change scenarios across Nepal. Environ Monit Assess 191(8):520. https://doi.org/10.1007/s10661-019-7644-4

Talchabhadel R, Karki R, Thapa BR, Maharjan M, Parajuli B (2018) Spatio-temporal variability of extreme precipitation in Nepal. Int J Climatol 38:4296–4313. https://doi.org/10.1002/joc.5669

Talchabhadel R, Karki R, Yadav M, Maharjan M, Aryal A, Thapa BR (2019) Spatial distribution of soil moisture index across Nepal: a step towards sharing climatic information for agricultural sector. Theor Appl Climatol 137(3–4):3089–3102. https://doi.org/10.1007/s00704-019-02801-3

Thapa GJ, Wikramanayake E, Forrest J (2015) Climate-change impacts on the biodiversity of the terai arc landscape and the chitwan-annapurna landscape. Nepal, Kathmandu

Thornthwaite CW (1948) An approach toward a rational classification of climate. Geogr Rev 38(1):55–94. https://doi.org/10.2307/210739

Uddin K, Matin MA, Maharjan S (2018) Assessment of land cover change and its impact on changes in soil erosion risk in Nepal. Sustain 10(12):1–20. https://doi.org/10.3390/su10124715

Vuuren Van DP, Edmonds J, Kainuma M, Riahi K, Nakicenovic N, Smith SJ, Rose SK (2011) The representative concentration pathways: an overview. Clim Chang 109:5–31. https://doi.org/10.1007/s10584-011-0148-z

Zaman QU, Rasul G (2004) Agro-climatic classification of Pakistan. Q Sci Vis 9(1974):59–66

Quantifying Uncertainties in Climate Change Projection and Its Impact on Water Availability in the Thuli Bheri River Basin, Nepal

7

Anil Aryal, Manisha Maharjan, and Rocky Talchabhadel

Abstract

Increase in global mean surface temperature due to greenhouse gases and rapid urbanization has resulted in climate change in both regional and global scale. The Intergovernmental Panel on Climate Change (IPCC) estimated the change in global mean surface temperature in the range of 0.3–0.7 °C for a period of 2016–2035 relative to 1986–2005 under four Representative Concentration Pathways (RCPs). In this study, we aim to quantify the uncertainties associated with projecting the future climate and their impacts on the water availability in the Thuli Bheri River Basin of Nepal. We used Soil and Water Assessment Tool (SWAT) as a hydrological model to simulate the runoff from the basin. Five COordinated Regional Climate Downscaling EXperiment-South Asia (CORDEX-SA) regional climate model (RCM) experiments have been used to analyze the impact of different climate models (CMs) on the future river discharge of the basin. The CMs were bias-corrected using quantile mapping (QM) method. Change in river discharge is evaluated for three future time windows, namely near future (2021–2040), mid-future (2041–2070), and far future (2071–2099). Further, we aim to outline the range of uncertainty arising from different projections of the CMs under the two RCPs 4.5 and 8.5 using the probability density function (PDF). The climate projection analysis indicated a significant increase in temperature in the future. Annual precipitation was projected to change from −4% to 16% under five CMs and two RCPs. The ensemble of the five CMs for both RCPs predicted the change of 9–13% in the future period. This uncertainty in climate projection has impacted water availability in different time periods. It also revealed that the uncertainties due to CMs are significantly higher during the high flow season. The results of this research would be helpful to practitioners, researchers, and decision/policymakers to regulate the issues of water availability in the future.

A. Aryal (✉)
Interdisciplinary Centre for River Basin Environment, University of Yamanashi, 4-3-11 Takeda, Kofu 400-8510, Japan

M. Maharjan
Department of Environmental Engineering, Kyoto University, Katsura, Nishikyo-ku, 615-8510, Kyoto, Japan

R. Talchabhadel
Texas A&M AgriLife Research, Texas A&M University, El Paso, TX 79927, USA

A. Aryal · M. Maharjan
Hydro Energy & Environment Research Center, Kathmandu, Nepal

R. Talchabhadel
Smartphones For Water Nepal (S4W-Nepal), Lalitpur 44700, Nepal

U. Schickhoff et al. (eds.), *Mountain Landscapes in Transition*,
Sustainable Development Goals Series, https://doi.org/10.1007/978-3-030-70238-0_7

Keywords

Climate change · Climate models · Soil and Water Assessment Tool · Thuli Bheri River Basin · Uncertainty

7.1 Introduction

With the onset of the industrial revolution in the early twentieth century, the use of fossil fuel as a source of energy started increasing. Use of fossil fuel, as a major source of energy, has resulted in the emission of harmful air pollutants. In addition, the burning of fossil fuel has embarked in the production of a large number of carbons that are responsible for current global warming. As a result of the global warming, various climate phenomena, such as sea-level rise (IPCC 2013) and loss of glaciers in the Himalayas (NRC 2012), have occurred and are termed as the *climate change*. Since more than three decades, the debate on global warming and climate change has begun. The impact on the availability of freshwater has ignited the discussion toward climate change and global warming. Global change in the average surface temperature and precipitation using multi-model ensemble under RCP 2.6 (low emission scenario) and RCP 8.5 (high emission scenario) shows that average surface temperature will increase by 13 °C under RCP 8.5, whereas average precipitation will increase by 60% relative to the base period 1986–2005 (Allen et al. 2013). The study revealed that increase in surface temperature and average precipitation is more pronounced in the polar while the increase in surface temperature is more prominent in the desert of Africa and China. This will lead to extremities in the weather pattern.

The impact of climate change has a direct and indirect impact on the water availability of freshwater resources (Pandey et al. 2019). Depletion in the freshwater resources has drawn the attention of researchers, scientists, practitioners, planners, and policymakers to improve the understanding of climate change impacts and hydrologic uncertainties for the sustainable management of water resources (Clark et al. 2016). Many researchers have evaluated the climate projections with an ensemble of different CMs for the hydrological analysis using different hydrological models (Chen et al. 2012; Zhang et al. 2016; Kundzewicz et al. 2018). In snow-fed Koshi Basin of Nepal, future runoff is found to be altered by −35 to +51% as the result of increasing temperature and precipitation under the changing climate (Shrestha et al. 2015). Similarly, Mishra et al. (2018) concluded that there is a significant impact of climate change on the freshwater availability in the large Bheri River Basin of Nepal. Also, SWAT has been applied to assess the climate change impacts in different basins of Nepal such as in the Tamakoshi (Aryal et al. 2018), the Bheri (Mishra et al. 2018), the Tamor (Bhatta et al. 2019), and the Mahakali (Pandey et al. 2019).

Uncertainties in future climate projections arise from the boundary and initial condition, emission scenarios, physical feedbacks, carbon cycle ,and structure uncertainty (Knutti 2008; Knutti et al. 2008). These sources of uncertainty are typically classified into three types, including internal variability, inter-model variability, and greenhouse gas emissions scenarios uncertainty. A proper understanding of sources of uncertainty and its impact is necessary to develop a robust and reliable plan for climate change analysis (Dankers et al. 2014; Her et al. 2019). Aryal et al. (2018) quantified the plausible sources of uncertainty in projecting the hydrological responses under various CMs and hydrological models in the Tamakoshi River Basin of Nepal.

This study mainly focuses on uncertainties associated with different CMs (five selected) under different emission scenarios (RCP 4.5 and 8.5) for different future time periods (up to 2100) using only one hydrological model. The future climate projections and its uncertainties are quantified, and their impacts on the water availability in the Thuli River Basin are analyzed.

7.2 Study Area

The Thuli Bheri River Basin with an area of 6888 km^2 (Fig. 7.1), a tributary of large Karnali River, originates in the Chhakra region of Dolpa district. The basin lies in the higher Himalaya geographic region of the country extending from 82° 16'E to 83° 40'E and 28° 41'N to 29° 10'N. The elevation varies from 742 to 7724 m above mean sea level (amsl) with snowcapped mountains covering the 1/4th of the entire area. Snowmelt is the major source of water in this river during winter and spring. Most of the basin is sparsely populated containing little agricultural land which leads to the requirement of lesser water for irrigation. Therefore, this basin is suitable to serve as a water donor basin to the more populated and arable river basins around it. Recently, a water transfer project is under construction to transfer water from the Bheri River to the neighboring Babai River (water deficit basin).

7.3 Data and Methodology

7.3.1 Data Collection

In this study, daily meteorological and hydrological data for the period of 1981–2014 were obtained from the Department of Hydrology and Meteorology (DHM), Government of Nepal (Table 7.1). Because of steep geographic terrain, it is difficult to establish the climate stations in the upper geographic regions of the basin which leads to lesser number of stations. The daily discharge data were acquired at Rimma gauge station.

In this study, the Shuttle Radar Topography Mission (SRTM) Digital Elevation Model (DEM) of 90 m resolution was used. The land use map (Fig. 7.2) of 30 m resolution developed by the International Centre of Integrated Mountain Development (ICIMOD) was used (Uddin et al. 2015). The soil map of 1:1,000,000

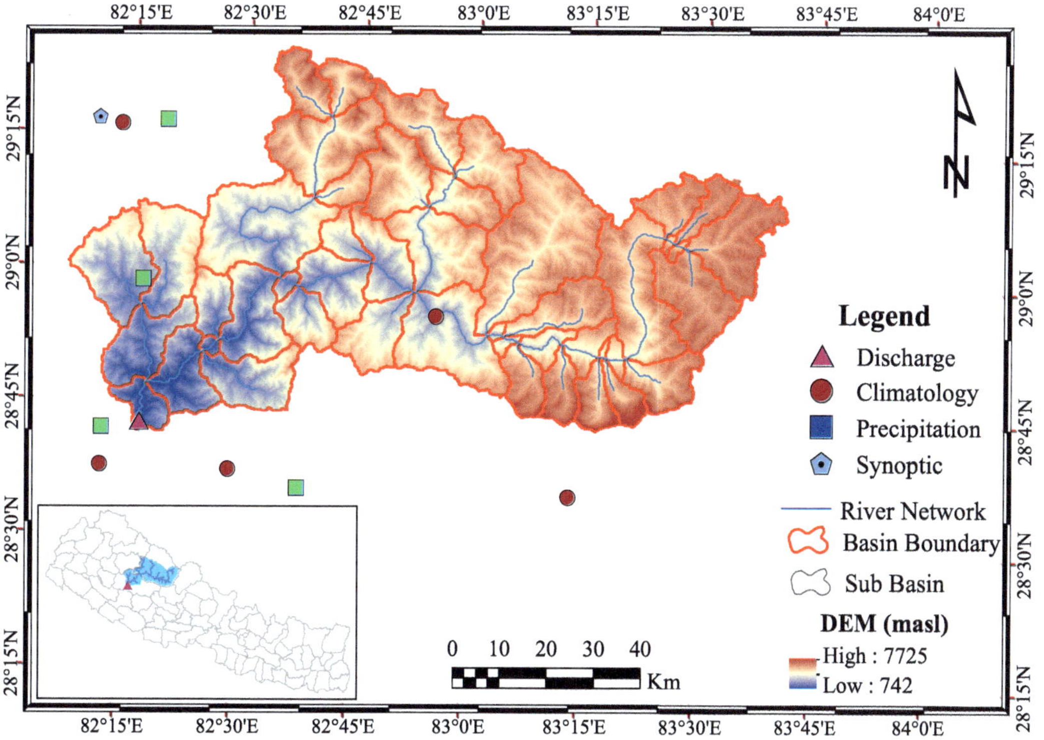

Fig. 7.1 Spatial location of the study area along with hydrometeorological stations

Table 7.1 Detail configuration of the different types of data used in the research

SN	Station index	Station name	Lat. (Deg.)	Long. (Deg.)	Elevation (masl)	Station type
1	303	Jumla	29.28	82.17	2300	Synoptic
2	304	Guthi Chaur	29.28	82.32	3080	Precipitation
3	310	Dipal Gaun	29.27	82.22	2310	Climatology
4	312	Dunai	28.93	82.92	2058	Climatology
5	404	Jajarkot	28.70	82.20	1231	Precipitation
6	418	Maina Gaun	28.98	82.28	2000	Precipitation
7	501	Rukumkot	28.60	82.63	1560	Precipitation
8	513	Chaur Jhari Tar	28.63	82.20	910	Climatology
9	514	Musikot (Rukum)	28.63	82.48	2100	Climatology
10	616	Gurja Khani	28.60	83.22	2530	Climatology
11	265	Rimma	28.71	82.28	550	Discharge

resolution and its properties were acquired from the Soil and Terrain Database (SOTER) for Nepal (https://www.isric.org/projects/soil-and-terrain-soter-database-programme).

7.3.2 Climate Model

The present study includes RCM downscaled using five different global climate model (GCM) forcings (Table 7.2) over the CORDEX-

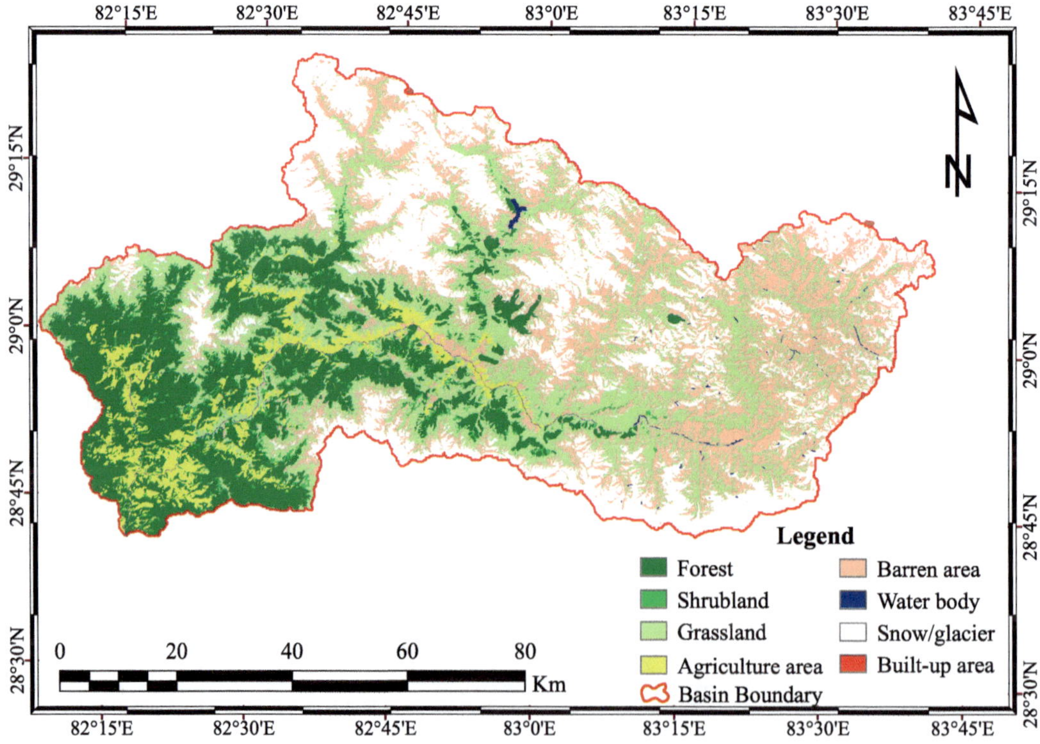

Fig. 7.2 Land use and land cover map for the Thuli Bheri River Basin

Table 7.2 List of CORDEX South Asia RCM experiments used

Experiment name	RCM description	Driving GCM	Contributing institute
CCAM (ACCESS)	Commonwealth Scientific and Industrial Research Organisation (CSIRO), Conformal-Cubic Atmospheric Model	ACCESS1.0	CSIRO Marine and Atmospheric Research, Melbourne, Australia
CCAM (CNRM)		CNRM-CM5	
CCAM (MPI)		MPI-ESM-LR	
CCAM (GFDL)		GFDL-CM3	
CCAM (BCCR)		NorESM-M	

SA domain at a horizontal resolution of 0.44° (~ 50 km). The CMs data were downloaded from Center for Climate Change Research (CCCR), Indian Institute of Tropical Meteorology (IITM), Pune Web site (http://cccr.tropmet.res.in/home/cordexsa_datasets.jsp).

7.3.3 Bias Correction and Uncertainty Quantification

Every CM demonstrates biases (systematic errors) in the projected output because of the limitation in horizontal and vertical (spatial) resolution, complex climate system processes, and improved material science and thermodynamics forms. Such biases are to be corrected assuming that the change in bias behavior is not altered by time for the model. In this study, the QM method is used as the bias correction approach to calibrate the cumulative distribution function (CDF) of the modeled data into the CDF of observed data using transfer function (Gudmundsoon et al. 2012; Chen et al. 2013).

There arises uncertainty in climate change projections due to use of different CMs, emission scenarios, hydrological models, and their parameter uncertainty. These uncertainties are quantified in this study using the boxplot and probability density function (PDF).

7.3.4 Hydrological Modeling

The physically based continuous hydrological model SWAT (Arnold et al. 2012) is used for the simulation of discharge in this study. The SWAT model has been extensively used for both basin scale and small watershed to simulate surface water quantity and quality. The model can also be used to simulate and predict the environmental impacts of soil erosion and sediment yield (Asres and Awulachew 2010; Abdelwahab et al. 2018; Halecki et al. 2018), and land management practices (Briak et al. 2019, Himanshu et al. 2019). Many researchers have widely used SWAT in Nepal to simulate the river discharge (Aryal et al. 2018; Bajracharya et al. 2018; Bhatta et al. 2019).

7.3.5 Calibration and Validation

Calibration and validation of the SWAT model were carried out for the period 1988–2006 and 2007–2013, respectively. The warm-up period of seven years (1981–1987) was used for model calibration. The basin was delineated into 27 sub-basins which was further divided into 358 hydrological response unit (HRUs) using HRU definition threshold of 2%, 5%, and 10% for land use, soil, and slope, respectively. Each sub-basin was divided into ten elevation bands with 500 m elevation band to account the orographic effect on both precipitation and temperature. Sensitivity analysis and calibration were undertaken by SWAT-Cup and its SUFI-2 algorithm (Abbaspour et al. 1997; Abbaspour 2015). The statistical performance of the model was evaluated using the coefficient of determination (R^2), the Nash—Sutcliffe Efficiency index (NSE), and percent bias (PBIAS) (Moriasi et al. 2007).

7.4 Results

7.4.1 Bias Correction of Precipitation

Bias correction has been carried out for all selected CMs. The comparison of the monthly observed precipitation with the raw and bias-corrected precipitation from the ACCESS1-0 CM for the baseline 1981–2005 is shown in Fig. 7.3 (left panel). It demonstrates that bias-corrected monthly precipitation is congruent with the observed data. The right panel in the figure shows the improvement in bias-corrected historical precipitation with R^2 value increasing from 0.36 to 0.48. Thus, the method of bias correction can be applied to correct future data from the CM.

7.4.2 Bias Correction of Temperature

Alike precipitation, the performance of the bias-corrected result for ACCESS1-0 CM was evaluated for the maximum and minimum temperature as shown in Fig. 7.4. The satisfactory results were obtained for other CMs as well. ACCESS1-0 CM showed the underestimation in maximum and minimum temperature.

The result showed the increment in R^2 value from 0.51 to 0.87 for maximum temperature and from 0.94 to 0.96 for minimum temperature after the bias correction (Fig. 7.5). The similar performance of the bias-corrected results was obtained for other CMs as well.

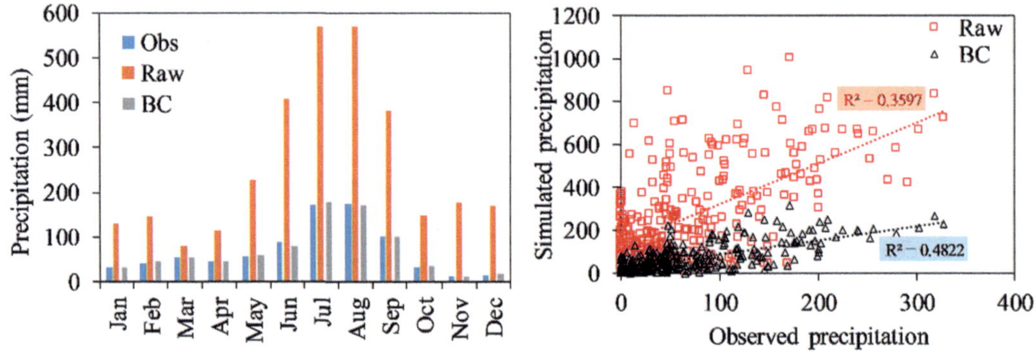

Fig. 7.3 Left panel) comparison of monthly observed, raw, and bias-corrected historical precipitation from ACCESS1-0 for the baseline 1981–2005 and Right panel) improved of R^2 after bias correction

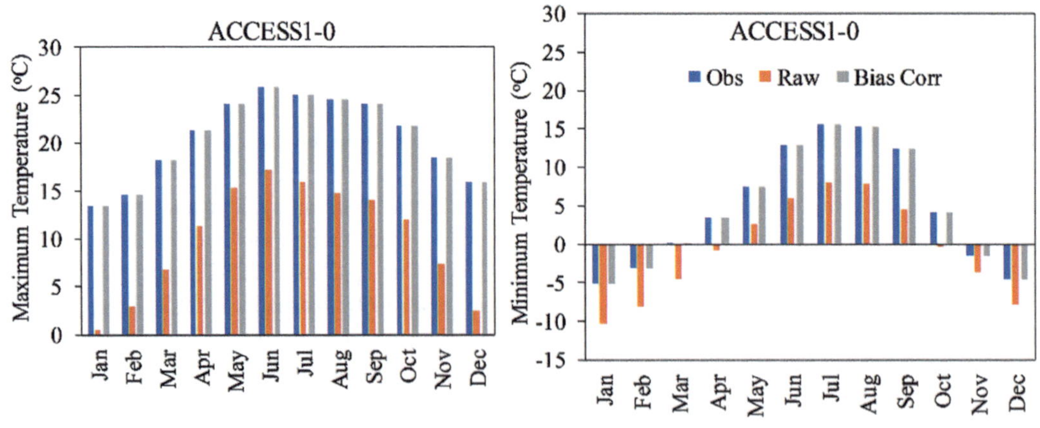

Fig. 7.4 Bias correction of the monthly maximum (left panel) and minimum (right panel) temperature using ACCESS1-0 CM

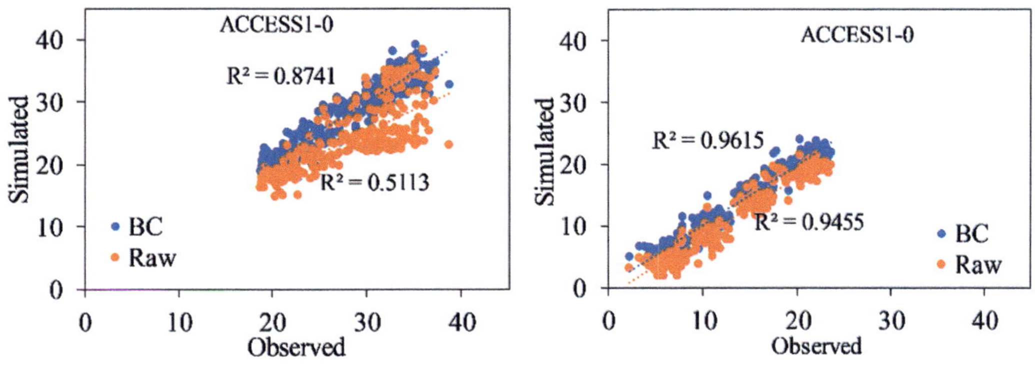

Fig. 7.5 Performance of bias correction of a raw and corrected CM for both maximum (left panel) and minimum (right panel) temperature for ACCESS1-0

7.4.3 Future Annual Precipitation

Annual precipitation projected for the future using different CMs under RCP 4.5 (solid line) and RCP 8.5 (dash line) at stations 304 and 404 is shown in Fig. 7.6. The result shows both increasing and decreasing trends of future precipitation. The combined effects of all CMs show the increasing trend but it differs within CMs and emission scenarios. At station 304, there is less variability in the estimation of future climate from different CMs and two emission scenarios,

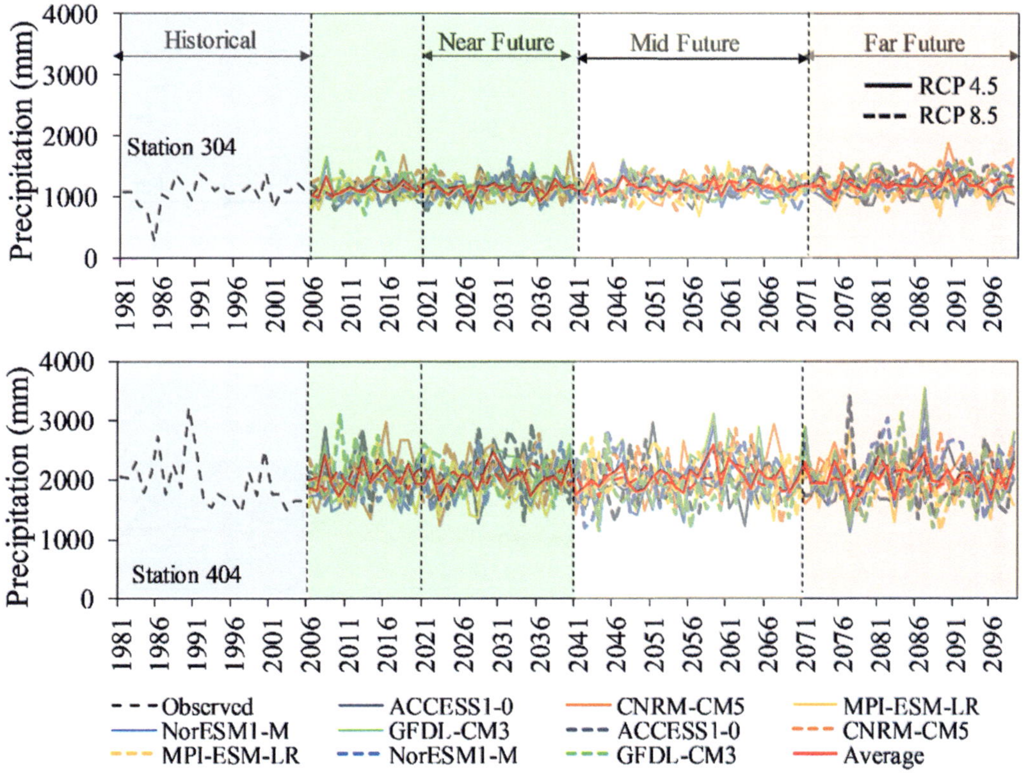

Fig. 7.6 Projection of the annual precipitation under different CMs and two RCPs at precipitation station 304 and 404

but at station 404, more variability is observed in the future precipitation projection. More fluctuations of precipitation range are observed under RCP 8.5 in the far future. Annual precipitation was projected to change from –4 to 16% under five CMs and two RCPs.

7.4.4 Future Annual Temperature

Figure 7.7 shows the future projection of annual maximum and minimum temperature with the observed data. The increasing trend of both maximum and minimum temperatures is observed in the future under RCP 4.5 and 8.5. Less variability in projection is observed in the near future, but the variability becomes higher with time. RCP 8.5 shows more increasing trend than RCP 4.5 with the highest temperature difference being 5–6 °C. The GFDL-CM3 shows the highest increasing trend for both maximum and minimum temperature. The ensemble of five CMs estimated the increase of maximum temperature of +0.021 °C/year under RCP 4.5 and 0.063 °C/year under RCP 8.5. Similarly, the change in minimum temperature is estimated to increase by 0.02 °C/year under RCP 4.5 and 0.058 °C/year under RCP 8.5.

7.4.5 SWAT Model Calibration and Validation

Figure 7.8 shows the comparison of daily observed and simulated discharge for the calibration period (1988–2006) and validation period (2007–2013). In this study, we used 22 parameters for sensitivity analysis, out of which 12 parameters were more sensitive to the observed discharge data. The model performance was evaluated by determining NSE, R^2, and PBIAS. For the

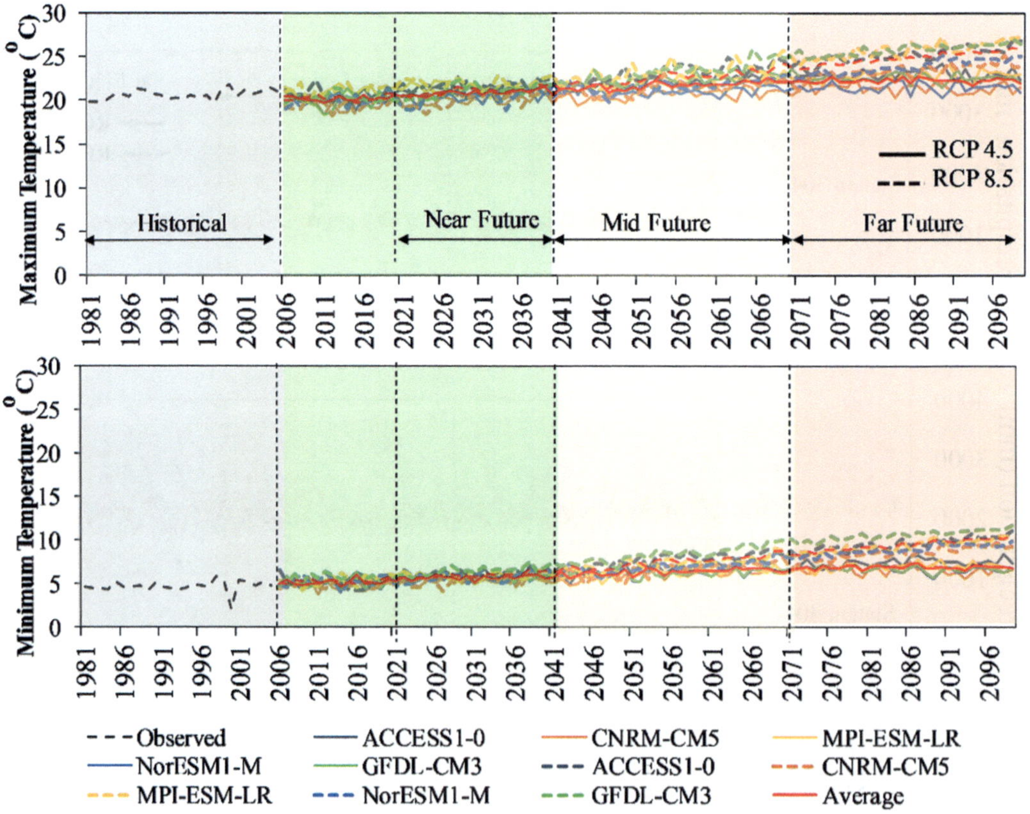

Fig. 7.7 Projection of the annual maximum and minimum temperature under different CMs and two RCPs

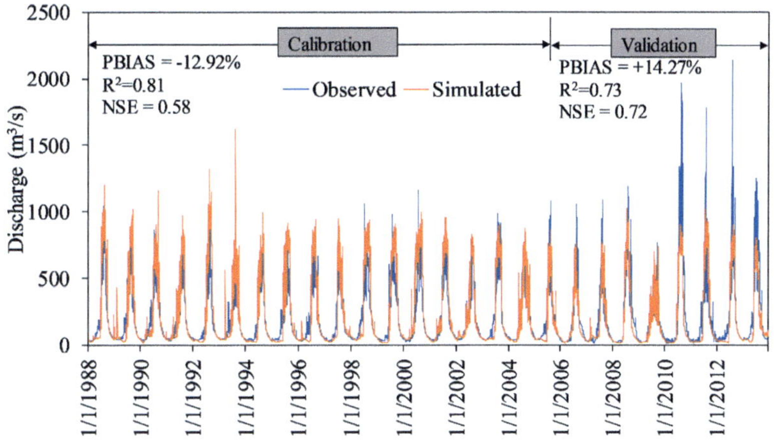

Fig. 7.8 Comparison of daily observed and simulated discharge for the calibration period (1988–2006) and validation period (2007–2013)

calibration period, the performance indicators have values of NSE = 0.58, R^2 = 0.8 and PBIAS = −13.22%. For the validation period, the NSE value is 0.72, R^2 is 0.73, and PBIAS is +14.27%. According to Moriasi et al. (2007), a model's performance can be considered satisfactory at an NSE greater than 0.5, PBIAS within ± 25%, and R^2 greater than 0.6. Thus, the model performance falls under the satisfactory category for calibration and good for the validation period.

7.4.6 Precipitation Uncertainty

With the use of more CMs and different emission scenarios, the uncertainty in the projection becomes wider. In this study, the uncertainty of the projected precipitation is measured through the boxplot as shown in Fig. 7.9. Monthly distribution of the projected precipitation is studied for three future time windows, namely near future (2021–2040), mid-future (2041–2070), and far future (2071–2099) under two RCP scenarios 4.5 and 8.5. The higher range of uncertainty in the projection is observed mainly from June to September. This depicts that uncertainty is higher during peak flows than low flows.

The PDF is plotted to measure the uncertainty in precipitation projection due to different CMs and RCP scenarios (Fig. 7.10). The range of uncertainty increases with the future time period. During the near future, the projected precipitation shows higher deviation of 704 mm by GFDL-CM3 CM under RCP 4.5 while the deviation of 732 mm is more pronounced under NorESM1-M climate under RCP 8.5 against the baseline deviation of 405.5 mm. During the mid-future, the uncertainty in projection ranges from 500 to 1546 mm of precipitation under RCP4.5, while the projection ranges from 574 to 1419 mm under RCP 8.5 in the mid-future. The uncertainty in projection is more exhibited when using the MPI-ESM-LR, range of projection being 839 mm under RCP 4.5, while GFDL-CM3 CM shows more uncertainty under RCP 8.5, range of uncertainty being 778.8 mm. In the far future, the precipitation projection ranges from 541 to 1598 mm, thereby exhibiting the deviation of 929 mm under RCP 4.5 and 948 mm under RCP 8.5 forcing scenario. The uncertainty deviation in the projection of precipitation is shown by all the CMs at all stages of the projection, i.e., near, mid, and far future. Thus, the projections from all the models are equally important in the quantification of the precipitation projection and assessment needs to be more robust.

7.4.7 Temperature Uncertainty

The uncertainty for the projected maximum and minimum temperature is plotted against the observed temperature as shown in Figs. 7.11 and 7.12. The result shows more deviation under RCP 8.5 in comparison with RCP 4.5 for both maximum and minimum temperature. The

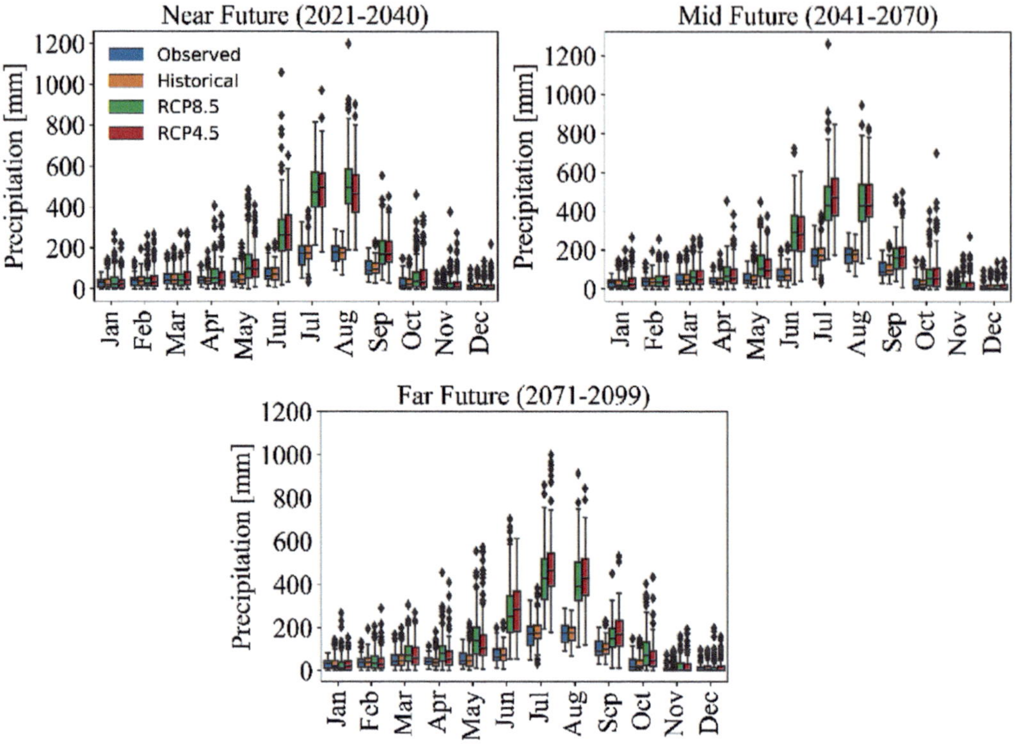

Fig. 7.9 Boxplot of change in monthly projected precipitation for three future time windows (2021–2040, 2041–2070, 2071–2099) under multiple RCP scenarios (4.5 and 8.5)

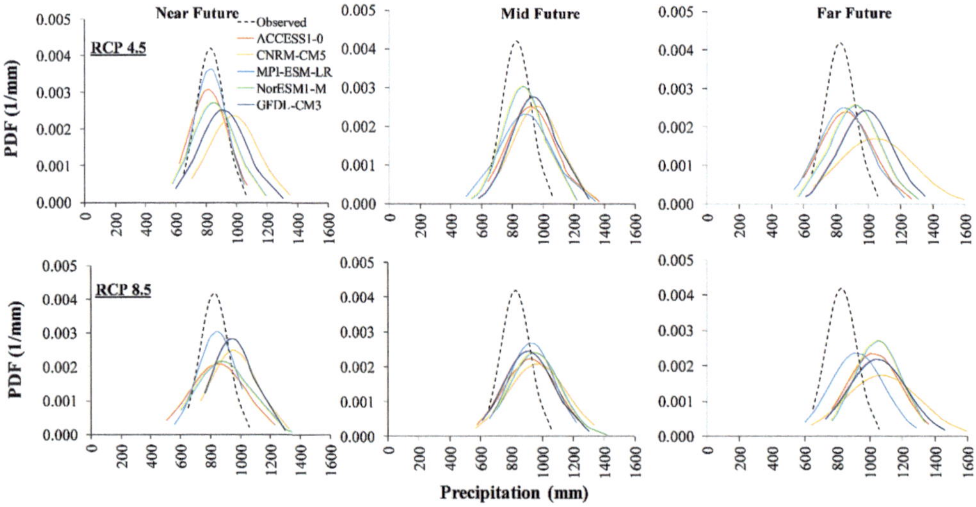

Fig. 7.10 Probability density function (PDF) of observed precipitation against projected precipitation for near, mid, and far future under RCP 4.5 (top panel) and RCP 8.5 (bottom panel)

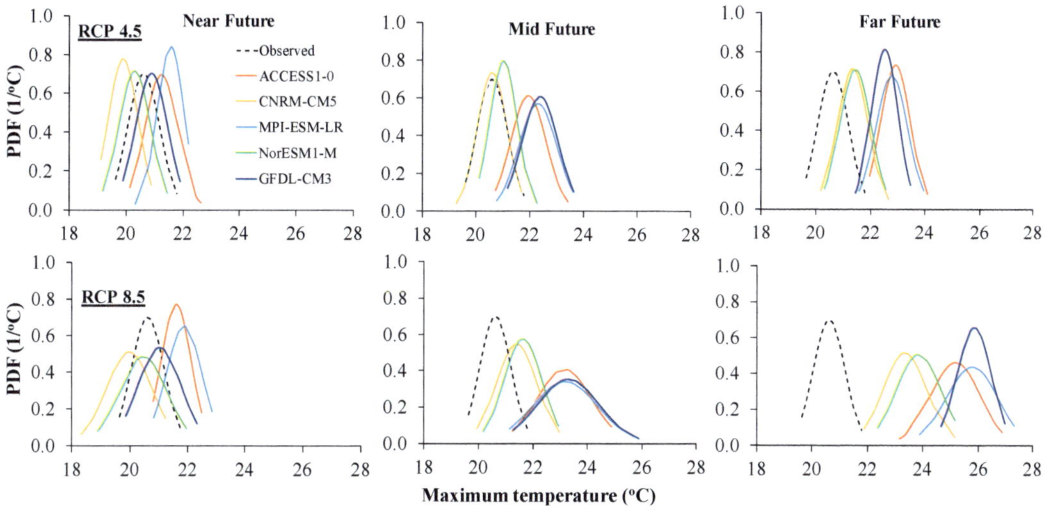

Fig. 7.11 Probability density function (PDF) of observed maximum temperature against the future projection under RCP 4.5 (top panel) and RCP 8.5 (bottom panel)

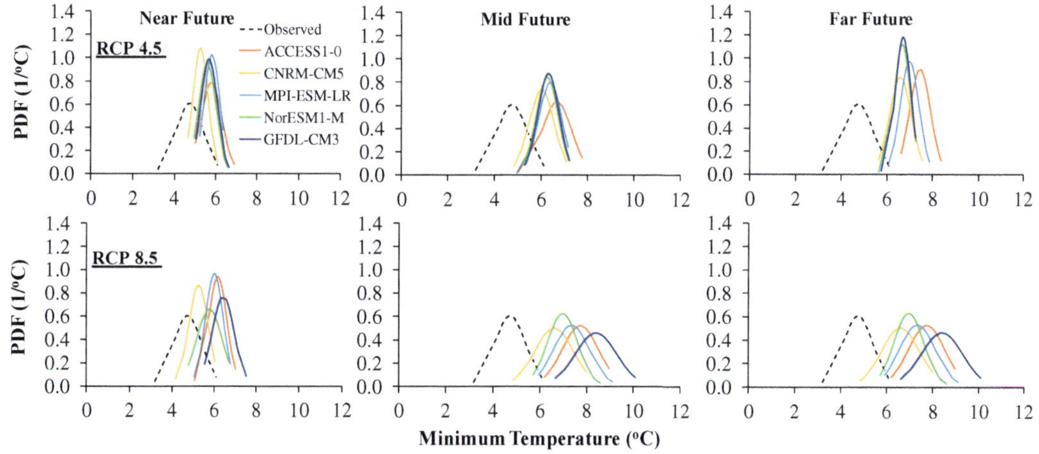

Fig. 7.12 Probability density function (PDF) of observed minimum temperature against the future projection under RCP 4.5 (top panel) and RCP 8.5 (bottom panel)

variability is more pronounced due to ACCESS 1-0 CM under RCP 4.5 for maximum temperature in all future time window while the uncertainty in projection is highly governed by GFDL-CM3 under RCP 8.5. The projection under RCP 8.5 is underestimated when compared to the baseline period. The maximum temperature uncertainty ranges from 21.5 to 24.8 °C during the far future under RCP 4.5 while the uncertainty range exhibits similar phenomena under RCP 8.5 ranging from 20.6 to 24.1 °C for all CMs. The uncertainty curve shows maximum deviation by GFDL-CM3 CM in mid and far future, while the maximum area under the curve is more represented by ACCESS1-0 CM in the near future. Thus, the use of a single model may not be significant in projecting the maximum and minimum temperature.

The minimum temperature range extends from 2.2 to 5.9 °C for the baseline period (1981–2005). The uncertainty in minimum temperature shows more uncertainty in far future under RCP 8.5 ranging from 4.9 to 10.1 °C compared to the minimum temperature ranges from 6.3 to 8.4 °C under RCP 4.5. For the minimum temperature, the uncertainty in projection is more attributed by ACCESS1-0 and CNRM-CM3 CM in far future under RCP 8.5 and attributed by NorESM1-M under RCP 4.5. The projection in minimum temperature shows the variation ranging from 4.2 °C on using MPI-ESM-LR CM to 11.8 °C on using GFDL-CM3 CM showing higher variation against the baseline period.

7.4.8 Discharge Uncertainty

Figure 7.13 displays the monthly observed and future discharge under RCP 4.5 and 8.5. Generally, the high flows show higher uncertainty, followed by mean flows and low flows. The variation in discharge during high flows increased by 51.3% in RCP 4.5 and 45% in RCP 8.5 in the near future, whereas in the mid future, it reached 64.6% in RCP 4.5 and 36.9% in RCP 8.5. Similarly, the variation during high flows increased by 44.1% in RCP 4.5 and 40.1% in RCP 8.5 in the far future. The result demonstrated the increase in discharge from the month of July to October in all three future time windows under both RCPs. The significant decrement in discharge is observed mostly in the month of May. The increasing ratio is slightly decreased in the far future in comparison with the near and mid-future.

The boxplot of monthly discharge is plotted for three future time windows under RCP 4.5 and 8.5 to show the change in discharge due to various RCMs (Fig. 7.14). A higher range of uncertainty is depicted from the month of June to September, i.e., in high flow condition. The mean discharge is generally higher or equal in RCP 4.5 than 8.5. Higher uncertainty range is displayed in the far future for both emission scenarios. The prediction of lowest discharge in the month of July is seen in the far future under RCP 8.5. For example, in the near future, the range of variation of discharge in August is from 420 to 1050 m^3/s in RCP 4.5 and from 389 to 1010 m^3/s in RCP 8.5; whereas in the far future, the discharge varies from 339 to 1005 m^3/s in RCP 4.5 and from 251 to 985 m^3/s in RCP 8.5. The increase in discharge in the wet season depicts enough water availability, but the decrease will exacerbate more problem of water deficit in the basin. Since the Thuli Bheri is the donor river basin, increase in water availability during the wet season will help in water transfer to the neighboring river basins, but the low flow season will face more struggle for the proper irrigation. The uncertainty in the projection of discharge affects the planning of water transfer and supply in the basin and neighboring basins. The use of different CMs results in different scenarios of future discharge; therefore, more studies related to the impact of climate change are essential to meet the future demand for water supply in the basin.

7.5 Discussion

The future projection of precipitation from the ensembles of five CMs under two RCPs demonstrates that there is a wider range of variation in the precipitation amounts in the far future and under RCP 8.5 (Fig. 7.9). Annual precipitation was projected to change from −4% to 16% under five CMs and two RCPs. The precipitation from June to October is likely to increase as projected by all CMs, but in other months, the precipitation has both increasing and decreasing trend. Under RCP 4.5, the projected precipitation varies from 577.6 to 1353.9 mm in the near future; from 500.1 to 1355.5 mm in the mid-future; and from 541.2 to 1598.4 mm in the far future. Under RCP 8.5, the projected precipitation varied from 512.7 to 1413.3 mm in the near future; from 574.1 to 1419.0 mm in the mid future; and from 601.5 to 1598.2 mm in the far future (Fig. 7.10).

The temperature projection showed the increasing trend of both maximum and minimum temperature under different CMs and two RCPs (Fig. 7.7). The ensemble of five CMs estimated

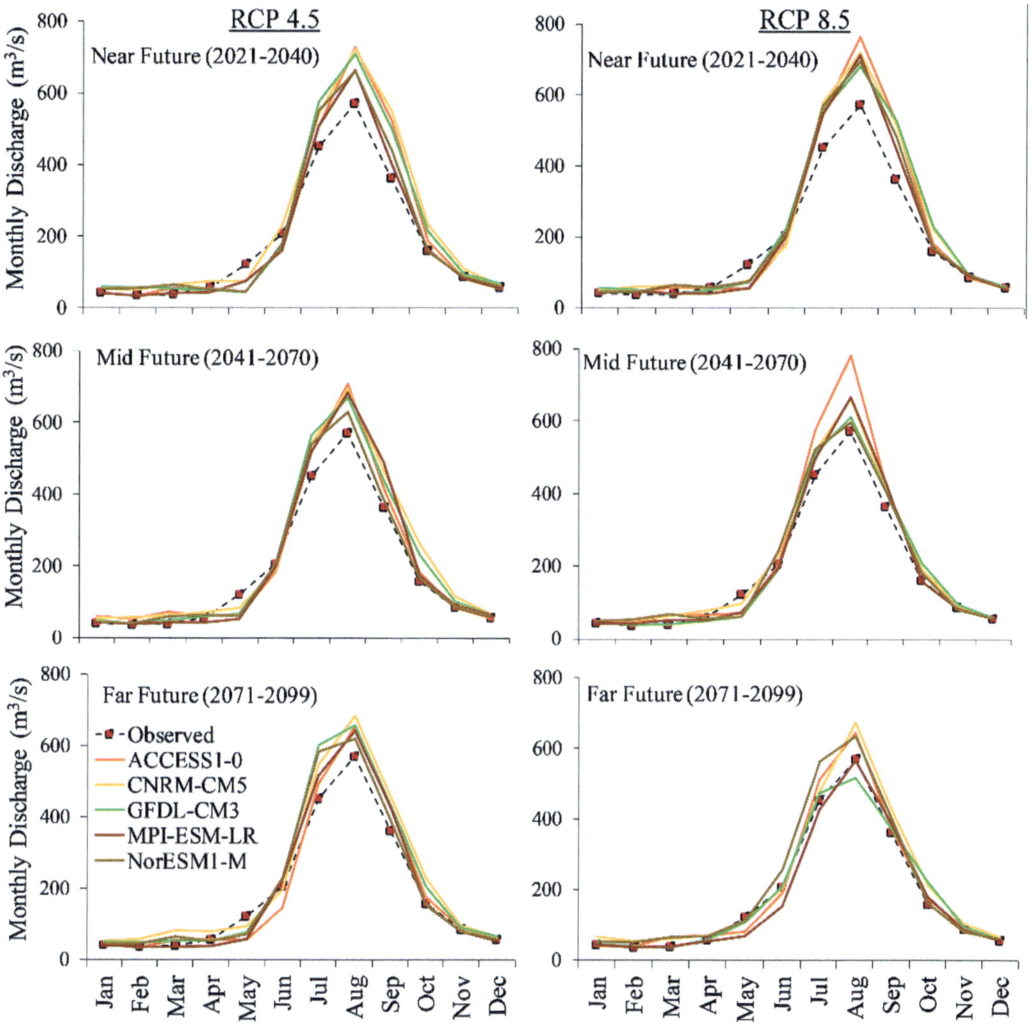

Fig. 7.13 Comparison of observed and future monthly discharge in three future time windows (2021–2040, 2041–2070, 2071–2099) under two RCP scenarios (4.5 and 8.5)

the increase of maximum temperature of +0.021 °C/year under RCP 4.5 and 0.063 °C/year under RCP 8.5. Similarly, the change in minimum temperature is estimated to increase by 0.02 °C/year under RCP 4.5 and 0.058 °C/year under RCP 8.5. This increasing rate is similar to the earlier studies in the extended basin of Bheri (Mishra et al. 2018) and other nearby basins with similar topography (Nepal 2016; Li et al. 2016).

The simulated discharge showed that the changes in mean annual precipitation and temperature have a significant impact on the annual surface runoff. All CMs predicted the increase in surface runoff in the high flow season (Fig. 7.13). In the low flow season, the CMs has predicted both increase and decrease in a runoff in different months. The ensemble of five CMs estimated the increase of 33% under RCP 4.5 and 37.5% under RCP 8.5 in the month of October for the near future. The ensemble of the five CMs for both RCPs predicted the change of 9–13% in the future period. The result showed an increase of discharge from July to March, whereas there is a decrease in discharge from April to Jun. This estimated discharge is consistent with the projected precipitation. Thus, change in precipitation

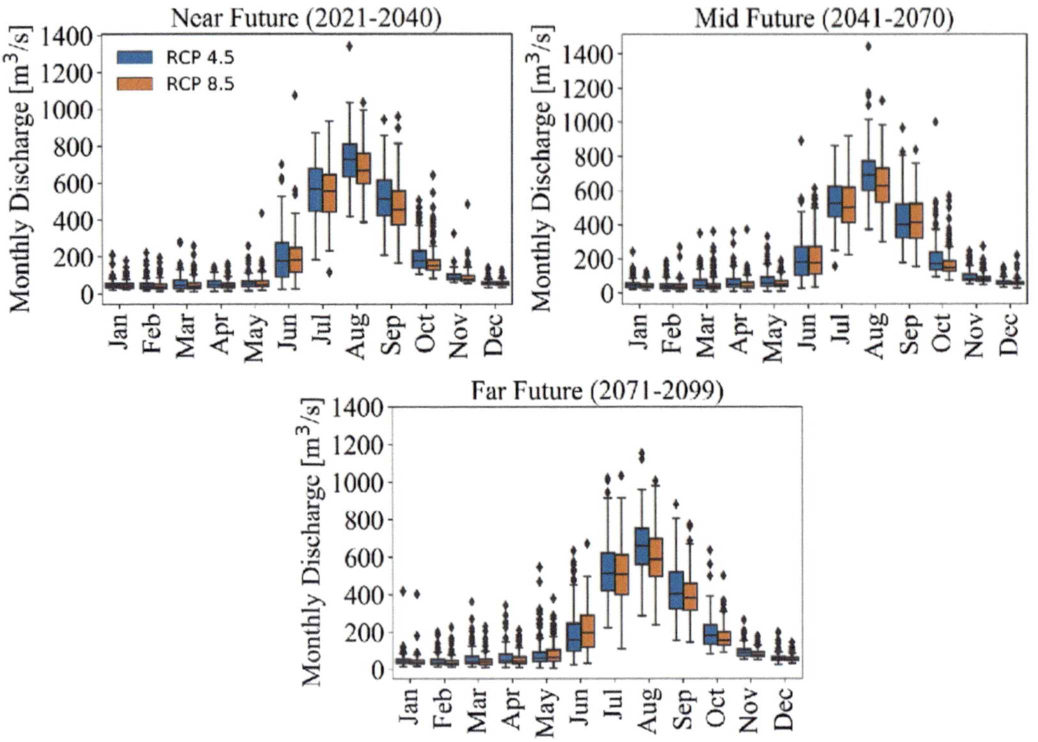

Fig. 7.14 Boxplot of change in monthly projected discharge for three future time windows (2021–2040, 2041–2070, 2071–2099) under multiple RCP scenarios (4.5 and 8.5)

projection has a direct impact on the surface runoff of the basin. This shows the necessity of robust estimation of future precipitation and temperature which will provide a proper simulation of discharge. Immerzeel et al. (2012) evaluated the hydrologic response to future changes in climate for the Langtang Basin in Nepal. The study showed that with an increase in both temperature and precipitation over the next century, there is an increment in evapotranspiration and higher snow and ice melt. This is combined with more snow falling as rain results in a steady decline of the glacier area in the model. Furthermore, the analysis shows that increased precipitation and ice melt will lead to increased streamflow. The seasonal peak in meltwater coincides with the monsoon peak; therefore, no shifts in the hydrograph are expected. Since the Thuli Bheri also lies in the high-altitude zone of Nepal, the similar scenarios occur. Also, the Thuli Bheri is a part of the Bheri River which is considered as a donor river in Western Nepal. The Government of Nepal (GoN) has initiated the water transfer project from Bheri to neighboring Babai River Basin (water deficit basin). Thus, the quantification of minimum flow will be significant to ensure the water availability for the project. Hence, we have also calibrated the SWAT model with the acceptable matching of the base flow with the simulated result. The estimation of future discharge will have a substantial impact in facilitating the water transfer project from Bheri River Basin to the neighboring basin.

Our results show the uncertainty in projected precipitation and temperature which is observed in the discharge as well. The CM-related uncertainty is contributed more during the high flow seasons, where precipitation dominated the river flow regime. Monthly distribution of future precipitation shows the increment in uncertainty mainly from June to September, i.e., high flow season. The CM-driven uncertainty has dominating influence relative to the total uncertainty in the future climate studies (Eisner et al. 2017; Vetter et al. 2017; Buda et al. 2017), all using regional hydrological models in their impact assessments. The range of uncertainty in both precipitation and temperature projections is lesser in the near future and under RCP 4.5 and higher in the far future under RCP 8.5.

A limitation of our study is that we used only data from five CMs to drive the hydrological model and did not consider model parameter-related uncertainty. The CM-driven uncertainty is a serious issue, and further research is necessary to better understand whether (a) this is due to missing or too simplified processes in CMs, e.g., precipitation processes, climate dynamics, and others (b) the complex climate system, or (c) a rigorous model selection process based on a list of agreed performance criteria (Hattermann et al. 2018). The uncertainty in the impacts of river discharge demands for intelligent strategies to adjust water use and management in an uncertain future. The largest changes to the hydrologic system in the future will most likely be due to changes in the timing, location, and intensity of monsoonal activity. Interannual variability of the monsoon strongly affects spatial patterns. Though the result shows the increase in discharge in wet seasons and decrease in the dry season, the interannual variability will affect the water availability spatially in the whole basin. The drier season will become drier which will affect agriculture productivity.

7.6 Conclusion

This study used five CMs for the future climate projections of both precipitation and temperature under RCP scenarios 4.5 and 8.5. QM is used as a bias correction method to reduce the errors in the projected output. The change in future climate projections for three future time windows with the baseline period was analyzed. The uncertainty from using different CMs and emission scenarios is quantified for the future. The physically based hydrological model SWAT was used to analyze the impact of future climate on the hydrology of the river basin.

Future precipitation projection showed that there is variability in projections due to the use of five CMs and RCPs. Both increase and decrease in precipitation are observed in the future. Annual precipitation was likely to change from −4 to 16% under five CMs and two RCPs. Similarly, increasing trend of both maximum and minimum temperature is observed in all future time window and RCPs. The ensemble of five CMs estimated the increase of maximum temperature of +0.021 °C/year under RCP 4.5 and 0.063 °C/year under RCP 8.5. Similarly, the change in minimum temperature is estimated to increase by 0.02 °C/year under RCP 4.5 and 0.058 °C/year under RCP 8.5. The simulated discharge shows that the changes in mean annual precipitation and temperature have a significant impact on the annual surface runoff. All CMs predicted the increase in surface runoff in the high flow season. The CM-related uncertainty in terms of discharge is observed to be more during the high flow seasons, where precipitation dominated the river flow regime. The results showed the intra-annual variability in the future climate projections. The future research needs to assess more CMs to incorporate the uncertainty related to CMs and should consider different models to assess the hydrological model uncertainty. The uncertainty in impacts of climate projections and quantification of river discharge demands for intelligent strategies to adjust water use and management in an uncertain future.

Acknowledgements The authors would like to thank the Department of Hydrology and Meteorology (DHM), Government of Nepal, for the permission to use hydrometeorological data.

References

Abbaspour KC (2015) SWAT-CUP: SWAT calibration and uncertainty programs—a user manual, department of systems analysis, integrated assessment and modelling (SIAM), Eawag. Swiss Federal Institute of Aquatic Science and Technology, Duebendorf, Switzerland, p 100

Abbaspour KC, van Genuchten MT, Schulin R, Schläppi E (1997) A sequential uncertainty domain inverse

procedure for estimating subsurface flow and transport parameters. Water Resour Res 33:1879–1892
Abdelwahab OMM, Ricci GF, Girolamo AMD, Gentile F (2018) Modelling soil erosion in a mediterranean watershed: comparison between SWAT and AnnAGNPS models. Environ Res 166:363–376
Allen SK, Bindoff NL, France FB, Cubasch U, Uk MRA, France OB, Hesselbebjerg J, Denmark C, France PC, Uk MC, Vasconcellos V, Feely, RA (2013) Working group I 2013: the physical science basis
Arnold JG, Moriasi DN, Gassman PW, Abbaspour KC, White MJ, Srinivasan R, Santhi C, Harmel RD, van Griensven A, van Liew MW et al (2012) Swat: model use, calibration, and validation. ASABE 55:1491–1508
Aryal A, Shrestha S, Babel MS (2018) Quantifying the sources of uncertainty in an ensemble of hydrological climate-impact projections. Theor Appl Climatol 135 (1–2):193–209
Asres MT, Awulachew SB (2010) SWAT based runoff and sediment yield modelling: a case study of the Gumera watershed in the Blue Nile basin. Ecohydrol Hydrobiol 10(2–4):191–199
Bajracharya AR, Bajracharya SR, Shrestha AB, Maharjan SB (2018) Climate change impact assessment on the hydrological regime of the Kaligandaki Basin, Nepal. Sci Total Environ 625:837–848
Bhatta B, Shrestha S, Shrestha PK, Talchabhadel R (2019) Evaluation and application of a SWAT model to assess the climate change impact on the hydrology of the Himalayan River Basin. CATENA 181:
Briak H, Mrabet R, Moussadek R, Aboumaria K (2019) Use of a calibrated SWAT model to evaluate the effects of agricultural BMPs on sediments of the Kalaya river basin (North of Morocco). Int Soil Water Conserv Res 7(2):176–183
Buda Su, Huang J, Zeng X, Chao G, Jiang T (2017) Impacts of climate change on streamflow in the Upper Yangtze river basin. Clim Change 141:533–546
Chen H, Xu CY, Guo S (2012) Comparison and evaluation of multiple GCMs, statistical downscaling and hydrological models in the study of climate change impacts on runoff. J Hydrol 434–435:36–45
Chen J, Brissette FP, Chaumont D, Braun M (2013) Finding appropriate bias correction methods in downscaling precipitation for hydrologic impact studies over North America. Water Resour Res 49:4187–4205
Clark MP, Wilby RL, Gutmann ED, Vano JA, Gangopadhyay S, Wood AW, Fowler HJ, Prudhomme C, Arnold JP, Brekke LD (2016) Characterizing uncertainty of the hydrologic impacts of climate change. Curr Clim Change Rep 2(2):55–64
Dankers R, Arnell NW, Clark DB, Falloon PD, Fekete BM, Gosling SN, Heinke J, Kim H, Masaki Y, Satoh Y, Stacke T, Wada Y, Wisser D (2014) First look at changes in flood hazard in the inter-sectoral impact model intercomparison project ensemble. Proc Natl Acad Sci 111(9):3257–3261
Eisner S et al (2017) An ensemble analysis of climate change impacts on streamflow seasonality across 11 large river basins. Clim Change 141:401–417
Gudmundsson L, Bremnes JB, Haugen JE, Engen-Skaugen T (2012) Downscaling RCM precipitation to the station scale using statistical transformations—a comparison of methods. Hydrol Earth System Sci 16:3383–3390
Halecki W, Kruk E, Ryczek M (2018) Loss of topsoil and soil erosion by water in agricultural areas: a multi-criteria approach for various land use scenarios in the Western Carpathians using a SWAT model. Land Use Policy 73:363–372
Hattermann FF, Vetter T, Breuer L, Su B, Daggupati P, Donnelly C, Fekete B, Fl¨orke F, Gosling SN, Hoffmann P, Liersch S, Masaki Y, Motovilov Y, M¨uller C, Samaniego L, Stacke T, Wada Y, Yang T, Krysnaova V (2018) Sources of uncertainty in hydrological climate impact assessment: a cross-scale study. Environ Res Lett 13
Her Y, Yoo SH, Cho J, Hwang S, Jeong J, Seong C (2019) Uncertainty in hydrological analysis of climate change: multi-parameter vs. multi-GCM ensemble predictions. Sci Rep 9(1):1–22
Himanshu SK, Pandey A, Yadav B, Gupta A (2019) Evaluation of best management practices for sediment and nutrient loss control using SWAT model. Soil Tillage Res 192:42–58
Immerzeel WW, van Beek LPH, Konz M, Shrestha AB, Bierkens MFP (2012) Hydrological response to climate change in a glacierized catchment in the Himalayas. Clim Change 110(3):721–736
IPCC (2013) IPCC Ch 13: sea level change
Knutti R (2008) Should we believe model predictions of future climate change? Philos Trans R Soc A 366:4647–4664
Knutti R, Allen MR, Friedlingstein P, Gregory JM, Hegerl GC, Meehl GA, Meinshausen M, Murphy JM, Plattner GK, Raper SCB, Stocker TF, Stott PA, Teng H, Wigley TLM (2008) A review of uncertainties in global temperature projections over the twenty-first century. J Clim 21:2651–2663
Kundzewicz ZW, Krysanova V, Benestad RE, Hov Ø, Piniewski M, Otto IM (2018) Uncertainty in climate change impacts on water resources. Environ Sci Policy 79:1–8
Li H, Xu CY, Beldring S, Tallaksen LM, Jain SK (2016) Water resources under climate change in himalayan basins. Water Resour Manag 30:843–859
Mishra Y, Nakamura T, Babel MS, Ninsawat S, Ochi S (2018) Impact of climate change on water resources of the Bheri River Basin, Nepal. Water (Switzerland) 10 (2):1–21
Moriasi DN, Arnold JG, van Liew MW, Bingner RL, Harmel RD, Veith TL (2007) Model evaluation guidelines for systematic quantification of accuracy in watershed simulations. Trans ASABE 50:885–900

Nepal S (2016) Impacts of climate change on the hydrological regime of the Koshi river basin in the Himalayan region. J Hydro-Environ Res 10:76–89
NRC (2012) Himalayan glaciers: climate change, water resources and water security. National Research Council
Pandey VP, Dhaubanjar S, Bharati L, Thapa BR (2019) Hydrological response of Chamelia watershed in Mahakali Basin to climate change. Sci Total Environ 650:365–383
Shrestha S, Anal AK, Salam PA, Van Der Valk M (2015) Managing water resources under climate uncertainty: examples from Asia, Europe, Latin America, and Australia. Managing water resources under climate uncertainty: examples from Asia, Europe, Latin America, and Australia, (November), 1–438
Uddin K, Shrestha HL, Murthy MS, Bajracharya B, Shrestha B, Gilani H, Pradhan S, Dangol B (2015) Development of 2010 national land cover database for the Nepal. J Environ Manag 148:82–90
Vetter T et al (2017) Evaluation of sources of uncertainty in projected hydrological changes under climate change in 12 large-scale river basins. Clim Change 141:419–433
Zhang Y, You Q, Chen C, Ge J (2016) Impacts of climate change on streamflows under RCP scenarios: A case study in Xin River Basin, China. Atmos Res 178–179:521–534

Decreasing Water Availability as a Threat for Traditional Irrigation-Based Land-Use Systems in the Mustang Himalaya/Nepal

8

Jussi Grießinger, Wolfgang J. H. Meier, and Philipp Hochreuther

Abstract

Mountain ecosystems, which play a major role in the global water cycle, are extremely sensitive to changes in the climate system. Related modifications in the hydrological system affect both socio-economic structures in mountain regions as well as in the downstream lowlands. This is particularly evident in the northern catchment area of the Kali Gandaki River (Mustang/Nepal), an area in the Trans-Himalaya which is part of the greater Ganges catchment. The region is characterized by an extreme gradient in precipitation from a humid to a semi-arid climate and associated strong spatial-temporal differentiation of annual precipitation distribution over a short horizontal distance. While monsoonal summer precipitation in the south bordering High Himalayas contributes significantly to the annual precipitation, extreme winter snowfall indicates a significant influence of the westerly wind circulation on the hydrology of the study area. However, due to the lack of long-term meteorological datasets, the respective significance of these factors for local and regional water resources and the fluctuations and changes they are subject to is still largely unknown. The livelihood of the rural population in the study region depends to a large extent on agriculture for food security as the main source of income. Therefore, local water availability, access to water resources and its sustainable management are essential to secure local livelihoods. Although, locals have adapted their economic and land-use systems over centuries to the challenges in harsh mountain environments, recent and future climate change requires very short-term reactions to enhanced changes within environmental conditions. Of these, the strongly varying water availability in semi-arid environments will have the greatest impact on local communities.

Keywords

Climate change · Glacier retreat · Irrigation-based land-use · Mustang district · Trans-Himalaya

8.1 Introduction

Amongst all environments in the world, mountain ecosystems belong to the most fragile ones (Diaz et al. 2003; Beniston 2003, 2005). This is mainly due to the fact that they are a repository of biodiversity (Spehn et al. 2012; Körner 2004, 2008, 2013, 2014; Körner et al. 2017) and, in

J. Grießinger (✉) · W. J. H. Meier · P. Hochreuther
Institute of Geography,
Friedrich-Alexander-University Erlangen-Nürnberg,
Erlangen, Germany
e-mail: jussi.griessinger@fau.de

U. Schickhoff et al. (eds.), *Mountain Landscapes in Transition*,
Sustainable Development Goals Series, https://doi.org/10.1007/978-3-030-70238-0_8

particular, harbour endangered plant and animal species and ecosystems (Barthlott et al. 2005; Körner 2004). Because of their spatial extent and exposed topographic setting within their adjoining lowlands, high-mountain areas also serve as water towers for a wider surrounding area with adjoining (and often comparably drier) ecosystems (Immerzeel et al. 2010; Viviroli et al. 2011, 2007). Moreover, mountains are inherent of numerous other ecosystem services with corresponding socio-economic linkages. Nowadays, mountains are recognized as important indicators on the impacts of climate and environmental change. This is in the sense that they exhibit dynamics in physical and biological systems that are more directly identifiable than in other geographical regions or ecosystems (Beniston and Stoffel 2014; Loarie et al. 2009; Gottfried et al. 2012; Nogués-Bravo et al. 2007); Diffenbaugh and Giorgi 2012). For high-elevation sites, Pepin and Seidel (2005) reported for the second half of the twentieth century (1951–2000) a median surface warming of +0.13°C/decade. This decadal temperature increase is even more pronounced (+0.65°C/decade), for the last 50 years of their study period (1956–2005). For the last decade 2003–2012, even higher amounts of +0.78°C are observed, clearly indicating a more enhanced and further accelerating temperature increase during the most recent decade (IPCC 2013). In a study solely focusing on the course and effect of elevation-dependent warming (EDW), Pepin et al. (2015; referenced as Mountain Research Initiative EDW Working Group) stated that the warming in mountains is verifiably amplified with elevation. This results in the fact that high-mountain environments are more susceptible to rapid changes in temperature compared to ecosystems at lower elevations. As a consequence, this had and will in the future have direct results on the rate and velocity of coupled changes in and necessary adaptations for mountain ecosystems, especially for cryospheric systems, plant communities and small- to large-scale hydrological catchments.

Besides numerous studies dealing on the effect of a temperature increase on vegetation patterns, dynamics and phenology (Menzel et al. 2006; Rosenzweig et al. 2008; Walther et al. 2002; Yang et al. 2017; Steinbauer et al. 2018; Gottfried et al. 2012; Mayor et al. 2017), focusing on changes in the cryosphere, hydrosphere and the coupled enhancement of morphodynamic activity in mountain regions also increased significantly in recent decades (Fort 2015). Such investigations can help to enable the profound understanding on how mountain environments are affected by, respond and adapt to substantial changes in both, local as well as regional climate (e.g. temperature increase; changes in the amount, timing and variability of precipitation or large-scale regime shifts) (Beniston and Stoffel 2014). Amongst the vast amount of possible impacts on mountain environments, changes in the cryosphere are—in relation to their temporal velocity—amongst the most visible and impressive ones. This is particularly most evident in significant changes in the volume (mass balance) and extent/length of glaciers, icefields and icecaps. Almost all mountain regions of the world exhibited a substantial glacier retreat and mass loss during the recent decades, but also since the Little Ice Age (LIA) (Braun et al. 2019; Bolch et al. 2012; Paul 2011; Sorg et al. 2012; Meier et al. 2018; Yao et al. 2012) that can be directly or indirectly related to a change in temperature, precipitation and/or moisture regimes. Especially for the wider Himalaya region, a profound acceleration in ice loss during the most recent decades is reported, resulting in substantial changes in meltwater supply for the river catchments in South Asia (Maurer et al. 2019; Immerzeel et al. 2010). In addition, it also permafrost as one key component of the terrestrial system in continental high-mountain environments shows an extremely high sensitivity to climate change. For example, large-scale thawing of the alpine permafrost in high-elevation ecosystems is already and will be in the future one key component in the adjustment of the regional hydrological cycles (Haeberli 2013; Wu and Zhang 2008).

In high-mountain areas like the Himalaya-Karakoram-Hindukush region, communities depend on a high degree on the sustained availability of water resources for their livelihood

(Kreutzmann 2000; Parveen et al. 2015). This applies in particular for the semi-arid Trans-Himalayan regions like the Mustang District, where land-use systems are highly triggered by a seasonal water deficit during parts of the vegetation period (Pohle 2001). Therefore, water resources used for irrigation and drinking water are mainly composed of monsoonal (summer) precipitation as well as ice and snow meltwater. Changes in the ratio of snow to rain precipitation, the duration of the snow cover and the ice body have already been partly determined or are expected to do so from climate predictions (Paudel and Andersen 2013). For the northern catchment area of the Kali Gandaki, increasing aridity and water stress are postulated being caused by an increase in the variability of winter precipitation connected to lower snow cover duration (Cannon et al. 2015; Paudel and Andersen 2013; Wipf et al. 2009). In addition, a lower proportion of glacial runoff must be expected, which will be of decisive relevance for local water resources and thus also for the irrigation agriculture, especially during the dry phases in the pre-summer season (Mukherji et al. 2019). Own surveys indicate that water shortage in individual settlements is already so pronounced that the traditional irrigation-based agriculture is no longer secured. Recent economic developments in the field of tourism and the sharp rise in the cultivation of fruit trees are inducing an increase in water demand and consumption, which probably will lead to conflicting demands for various actors.

8.2 Study Area Mustang District

The central-Nepalese Mustang District is enclosed between the main mountain ridges of the Himalayan Arc and the southern fringes of the Tibetan Plateau (Parson et al. 2016) and belongs to the semi-arid geo-ecological zone of the Trans-Himalaya (TH). The Annapurna-Dhaulagiri-Himal (ADH) exhibit various mountain peaks exceeding elevations of 8,000 m a.s.l. (Fig. 8.1). The ADH forms an 62 and 55 km wide, almost closed topographic obstacle at an altitude of 6,000 m a.s.l (Miehe 1984). The orographically forced uplift of moisture saturated air masses during the Indian Summer Monsoon (ISM) season leads to high annual precipitation amounts on the southern slopes of the ADH. In contrast, winter precipitation is mainly caused by westerly disturbances, highly variable peaking during the winter season (Cannon et al. 2015). Subsequently, annual precipitation amounts in the northern part of the Mustang district decreases rapidly due to lee and foehn effects, forming one of the world's sharpest moisture gradients between a humid southern (c.f. meteo station Beni) and a semi-arid northern part (Jomsom) (Fig. 8.1).

Almost 70% of the area of the Mustang District is located above 4,000 m a.s.l. Lower elevations are predominantly restricted to the north-south facing Thakkhola Graben through which the Kali Gandaki river is draining (Fig. 8.1). The transverse valley of the Kali Gandaki connects the Tibetan Plateau with the Indian lowlands and forms—while dividing the ADH—one of the deepest river valleys in the world. The associated diurnal pressure differences between the subtropical Indian lowland and the Tibetan plateau are compensated by a distinctive, extreme upvalley wind highly influencing the regional climatology (Eger et al. 2000). High wind velocities in combination with an extreme solar irradiance lead to generally high evaporation rates within the Mustang district. The temperature conditions are strongly linked to the local topography, whereby the Kali Gandaki valley represents a favourable habitat. Even at elevations exceeding 3,000 m a.sl., the mean annual temperature remains above 6 °C (Fig. 8.2) and can exceed during the warmest quarter of the year 15 °C. The annual 0 °C isotherm is located at an elevation of approximately 4,250 m a.s.l. In higher elevations, especially during the coolest quarter of the year, temperatures are frequently below the negative double-digit range, favouring the appearance of winter-accumulated glaciers.

Due to the geo-ecological framework conditions in the dry northern Himalaya, traditional agriculture is only possible through ingenious irrigation systems situated on small plains, slopes, river terraces and alluvial fans. Besides

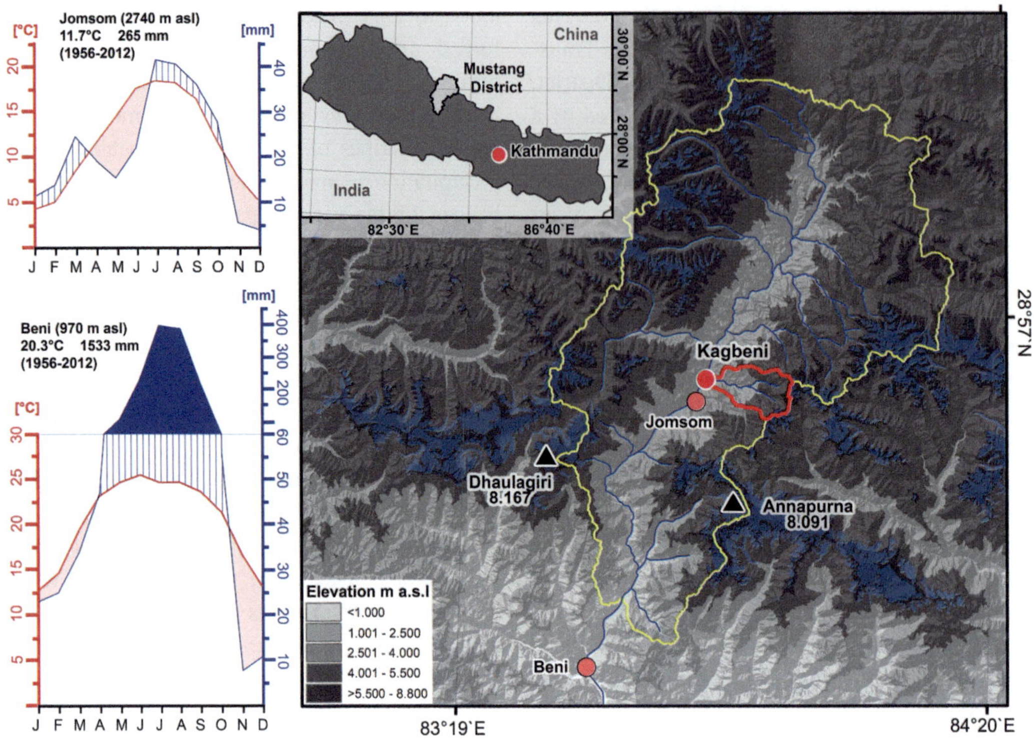

Fig. 8.1 Location of the Mustang District within Nepal and Digital Elevation Model (SRTM, LP DACC NASA Version 3) of the Mustang District (yellow polygon) with the location of the Dzon Chu catchment (red polygon). Climate graphs reflect the gradient between the windward and lee-ward side of the Annapurna-Dhaulagiri Himalaya mountain obstacle during the Indian Summer Monsoon

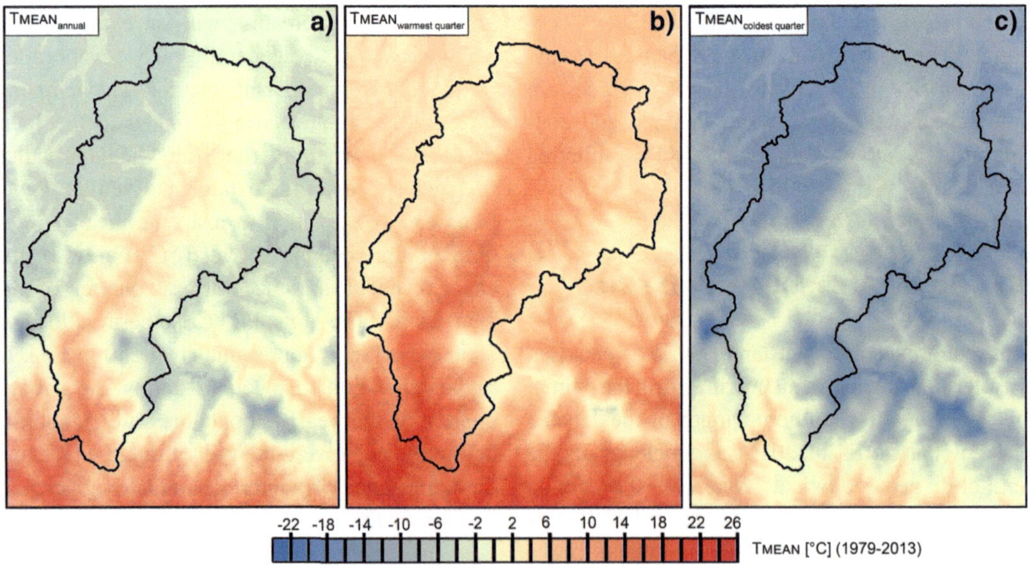

Fig. 8.2 **a** Mean annual temperature, **b** mean temperature of the warmest quarter and **c** mean temperature of the coldest quarter for the Mustang District. Data based on Karger et al. (2017)

land-use on the river terraces of the main river bodies like the Kali Gandaki, adjoining smaller alpine catchments display typical small-scale nested terrace systems highly adapted to the complex hydrological conditions in a high-mountain environment. The necessary constant water supply for crop production is covered by a combination of glacier meltwater, rainfall during the summer monsoon season and water supply from sporadic winterly snow cover and permafrost bodies from the surrounding high-altitude areas. As being the key livelihood service for local people, the irrigation-based agriculture is therefore highly dependent on a continuous water supply from water resources fed by the subalpine to alpine areas above the settlements. However, it can be deduced from this that changes in water supply from the sources mentioned lead to substantial effects on the local irrigation agriculture. Besides the water demand for agriculture, increasing water consumption is caused by a steadily growing number of tourism activity, which is mainly present at the southern area of the Mustang District in Jomsom, Kagbeni and Muktinath (own survey, Lama and Job 2014).

8.3 Evidences of Environmental Change in the Mustang Himalaya

Quantifying the extent and impact of climate change in the Himalayas is due to the extreme topoclimatic variability still a big obstacle. This applies all the more for specific high-elevation regions like the Mustang Himalaya. Caused by the spatial as well as temporal lack of available meteorological data in this region, investigations on the recent course and aggravate statements on future climate changes are substantially impeded. Although numerous studies determined for the Greater Himalaya Region (GHR) a substantial increase in temperature (c.f. Krishnan et al. 2019), it often remains uncertain to which extent the vital rainfall, especially in the semi-arid Trans-Himalaya already had and will change in the future (Xu et al. 2009; Böhner et al. 2015). Verifiably, the cryosphere in the GHR is already affected by a current warming trend, resulting in a rapid area reduction of the majority of Himalayan glaciers (Xu et al. 2009, Bolch et al. 2010; Bolch et al. 2012; Yao et al. 2012). However, a reliable quantification on the impact of climate change in the Mustang Himalaya is still difficult due to spatially wide distributed climate stations with time series frequently spanning only the most recent decade. Since the available climate stations are also mainly located in the valley bottoms, statements for the higher altitudes on climate change and especially changes in precipitation remain difficult. For reliable investigations on the ecological and environmental impacts of climate change, downscaled high-resolution re-analysis climate datasets are therefore indispensable. The recently published climatologies at high resolution for the Earth's land surface areas (CHELSA) dataset offers for the period 1979–2013 a demonstrably well-suited basis to consider regional changes in temperature and precipitation together, especially for complex high-mountain regions like the Himalayas (Karger et al. 2017).

For the Mustang District, analyses of the contained temperature datasets indicate a demonstrable warming trend during the past decades (Fig. 8.3). In the areas north of the main ridge of the Himalayan Arc, a trend towards higher mean annual temperatures (Tmean annual) during the period 1979–2013 are detectable for both, the Kali Gandaki valley and the higher elevations. Throughout the investigation area, the temperature increase exceeds a value of +0.8 K. On closer examination, however, it is noticeable that this trend is even more pronounced in the northern parts of the Mustang District, where especially the higher elevations are subject to a temperature increase of more than +1.0 K.

Maximum values can be seen for the north-western glacierized mountain ranges of the Mustang Himal with an increase of the annual mean temperature of +1.2 K (Fig. 8.3). Interestingly, this substantial temperature increase especially for the higher elevations is even more pronounced while regarding the mean annual minimum temperatures (Tmin annual, Fig. 8.3c)

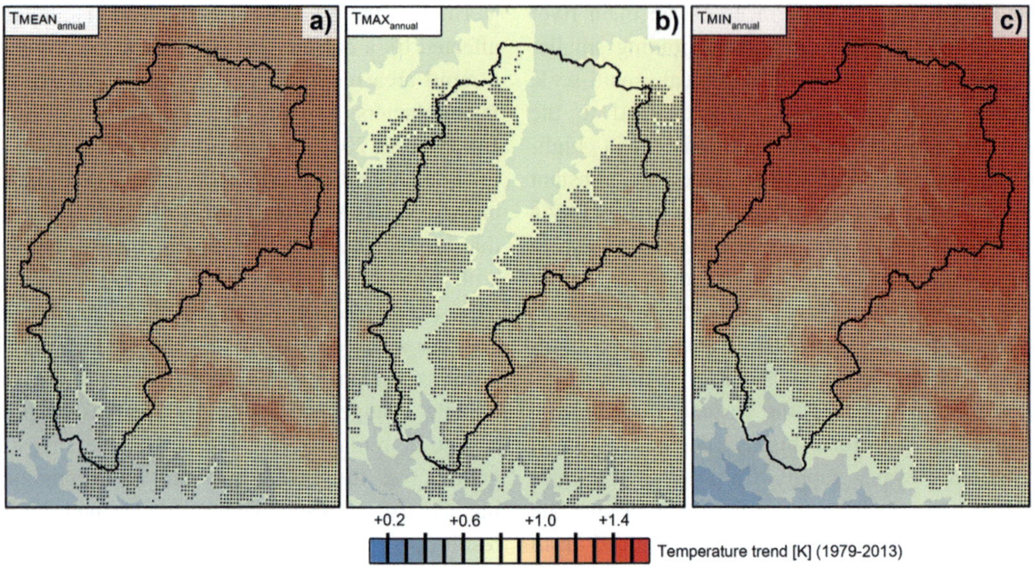

Fig. 8.3 Trends in temperature [in K] for the period 1979-2013 for **a** Mean annual temperature (Tmean annual), **b** mean annual maximum temperature (Tmax annual) and **c** mean annual minimum temperature (Tmean annual). Black dots represent level of significance ($p < 0.01$). Data based on Karger et al. (2017)

and therefore conditions during the winter season. Highly affected areas with temperature increases of more than +1.4 K are indicated especially at the higher elevations in the north-western and north-eastern parts of the Mustang District. Such an increase in temperature is in good accordance with recently reported studies on elevation-dependent warming in high-mountain regions (Pepin et al. 2015). Notably, the trend towards warmer conditions for the Mustang District even exceeds reported values for the Northern Hemisphere (Jones et al. 2016, IPCC 2013).

Along with these changes in temperature, an analysis of precipitation values in the region reveals a slight increase and therefore a positive trend in annual precipitation during the recent four decades (Fig. 8.4). Although this might imply a probable general positive effect for the Mustang District, the maximum increase of +40 mm/year do not generally lead to more humid conditions in the semi-arid investigation area. A major challenge for local agriculture during the year is the high variability in the timing when and how rainfalls occur. Especially during the pre-monsoon season from February to May, the natural variability of rainfall events leads to a low predictability and reliability on rainfall for land-use. Therefore, especially, agriculture in the pre-monsoon season is highly dependent on the constant meltwater supply from glaciers. Additionally, the analysis of a 10-year meteorological dataset from Choser/Upper Mustang indicates a substantial shift to a higher annual variability of rainfall during the last decade, while the total sum in precipitation stays more or less constant (not displayed data by the Department of Hydrology, Nepal). In addition, it is reported by the local population that rainfall events more frequently occur as short-term torrential rains. For the local agriculture, such changes will lead to a increasing demand and use of other available water resources throughout the vegetation period to ensure irrigation, e.g. by meltwater. However, the expressiveness of the shown precipitation increase in the CHELSA dataset must be evaluated with caution. By analysing the standard deviation (SD) of monthly precipitation values during the pre-monsoon season, it becomes evident that especially for the high elevations, the displayed SD exceeds the observed positive trend in annual precipitation.

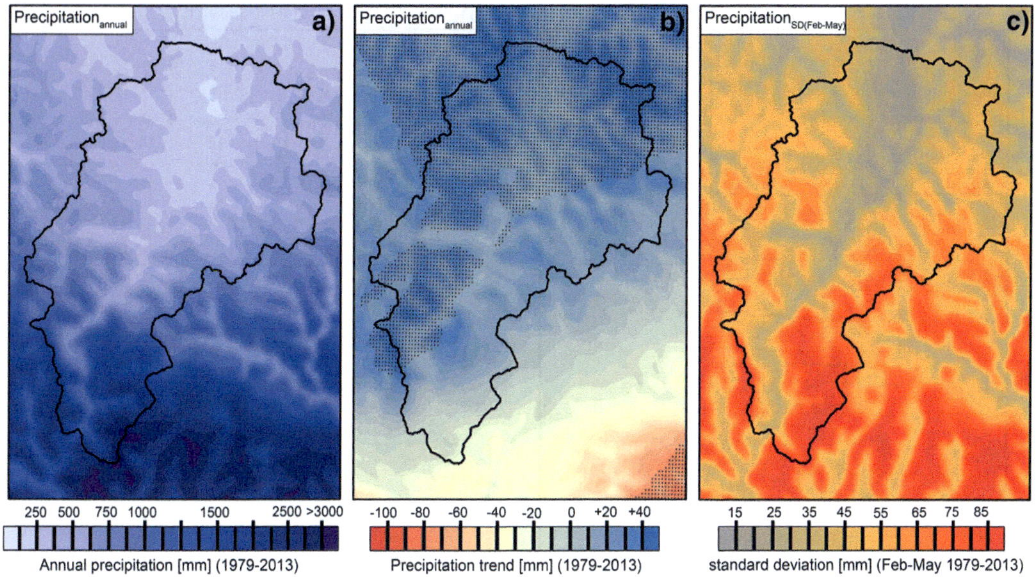

Fig. 8.4 **a** Mean annual rainfall distribution and **b** trends in mean annual rainfall in the Mustang District for the period 1979-2013. Black dots represent the 0.95 level of significance. **c** Standard deviation of the monthly precipitation values for the pre-monsoon period averaged for the period February-May. Data based on Karger et al. (2017)

From the farmers perspective, this implies that the monthly expected rainfall, required for the germination of the seeds may be completely absent, or can be exceeded by several hundred per cent. This unpredictable variability once again underlines the urgent demand for an independent perennial water supply.

In general, continental mountain glaciers are responding to changes in annual temperature increase and variability in solid precipitation highly sensitive (Wang et al. 2019). As already reported for other areas of the Greater Himalaya Region, substantial changes in the extent of the glacierized areas are also apparent for the Mustang district. Although the proportion of the glaciated area is compared to highly glaciated areas of the surrounding mountain massifs of the Himalayan Arc or the Tibetan plateau significantly lower, similar dynamics can be seen. To analyse regional glacier dynamics, Little Ice Age (LIA) glacier extents were digitized for 631 glaciers in the Annapurna-Dhaulagiri-Himal and the northern adjoining Trans-Himalaya landscape on the basis of Sentinel-2 optical satellite imagery. Subsequently, obtained results were compared to the semi-automatic derived glacier extents of 2018 (Landsat-8 and Sentinel-2), using the method described in detail in Meier et al. (2018) (Fig. 8.5a). Uncertainty values were calculated by applying a buffer to the glacier outlines of ± 30 m and ± 15 m, respectively.

As a result, it can be seen that since the end of the LIA the glacierized area in the investigation area was reduced from 1,707 ± 134 to 1,023 ± 62 km^2, clearly indicating a total area loss until 2018 of approximately 680 km^2 (−40%). The larger valley glaciers (>10 km^2) descent primarily from the Annapurna, Dhaulagiri and Peri Himal at an averaged elevation of 7,000 m a.s.l. The relative area loss of these large glaciers, often covered by debris, is by comparison less than for the small glaciers (Fig. 8.5b). Especially the glaciers located in the Trans-Himalaya, e.g. in the Mustang Himal with smaller and lower-elevated catchment areas, display the highest percentage loss in the investigation area. In total, 100 glaciers have already completely disappeared since the LIA, of which

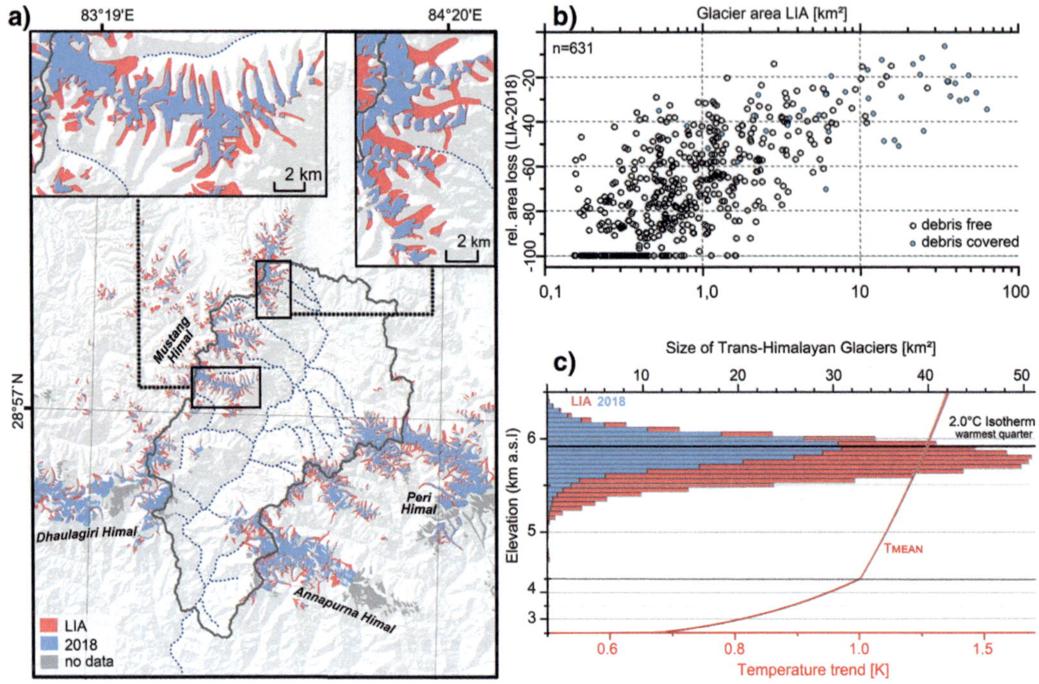

Fig. 8.5 **a** Glacier change and **b** relative change in glacier size between Little Ice Age and 2018. **c** Glacier hypsometry of the Trans-Himalayan Mountain Glaciers and averaged elevation-dependent warming derived from a CHELSA (Karger et al. 2017) trend analysis and SRTM-DEM

95 were located in the semi-arid area north of the ADH. On average, over 300 glaciers covering an initial area of less than 1 km^2 during the LIA have lost almost 80% of their surface area. This striking fact of more rapidly shrinking small glaciers can be detected also on a global scale (Meier et al. 2018; Paul et al. 2004). For 430 glaciers of the Mustang Himal and Trans-Himalaya, the glacier hypsometry was calculated in intervals of 50 m. The most pronounced area loss occurred at elevations between 5,500 and 6,000 m a.s.l. which accounts to almost 85% of the total reduced glacier area (Fig. 8.5c). Moreover, it is clearly evident that a major portion of the currently glacierized area covers a height range of only a few hundred metres. For these rather flat glaciers (averaged elevation range of 350 m), an increase in temperature poses a serious threat, as the accumulation area ratio can change significantly. During the warmest quarter, the 2 °C isotherm rises to almost 6,000 m a.s.l. as deduced from the CHELSA data (c.f. Fig. 8.5c). The temperature trend within the study area clearly represents an elevation-dependent warming and reveals an increase in temperature within the glacierized elevation of +1.25 K. The interplay of a low and highly variable precipitation amount during the major period of snow accumulation and generally increasing temperatures is persistently endangering the glaciers as one of the main local water reservoirs (Bajracharya et al. 2014; Manandhar et al. 2012). The first utilizable optical satellite imagery from the Landsat Mission in 1976, covering the study area, allows a multi-temporal comparison of the glacier extent supplemented by the decadal inventories of Bajracharya et al. (2014). The six-time segments up to 2018 reveal for the glaciers located at the north-western catchment of the Kali Gandaki a steady and ongoing glacier retreat (r = −0.98) without phases of prolonged stagnation or probable glacier re-advances. This demonstrably suggests that the glaciers in the investigation area

are unequilibrated and significantly respond with pronounced area losses to the reported warming trend.

8.4 Potential Impacts of Climate Change on Local Land-Use Systems and Communities

The illustrated substantial changes in climate and corresponding responses of glaciers accompanied with changes in the meltwater supply already endanger the survivability of local communities in the Mustang district. As reported by Sherchan (2019) for the Dhye village in the Upper Mustang, government-controlled re-settlements into areas with perennial water supply (like river terraces of the main tributaries) are already taking place. As the main driving key, climate change induced alterations resulting in a coupled effect of (i) a substantial loss of pasture areas for livestock (Paudel and Andersen 2013) combined with (ii) a massively reduced water availability during the vegetation period for irrigation are postulated. Although a profound scientific evaluation for the reasons of the severe decrease in water supply is lacking, similar challenges can also be observed for settlements further south like Kagbeni and Muktinath in the Dzon Chu catchment (Fig. 8.6).

In comparison with other areas in the Mustang district, the amount of the glaciated areas in the Dzon Chu catchment is significantly lower (c.f. Fig. 8.5). As displayed in Fig. 8.6, the few local glaciers are mainly draining eastwards into the Marsyangdi River (and the Manaslu-Himal). Interestingly, the glacier area feeding the Dzon Chu River amounts to only 2.7 km^2. This contrasts with the relatively high percentage of agricultural areas with more than 4 km^2 surrounding the main settlements of Kagbeni and Muktinath. It must be therefore noted that the river runoff of the Dzon Chu cannot be based solely on meltwater of the glaciated areas. Rather, a complex interaction of meltwater from bare and debris-covered glaciers and visually proven permafrost occurrences (rock glaciers) must be assumed. However, the quantification of the proportion of the latter must remain open within the initial framework of this study.

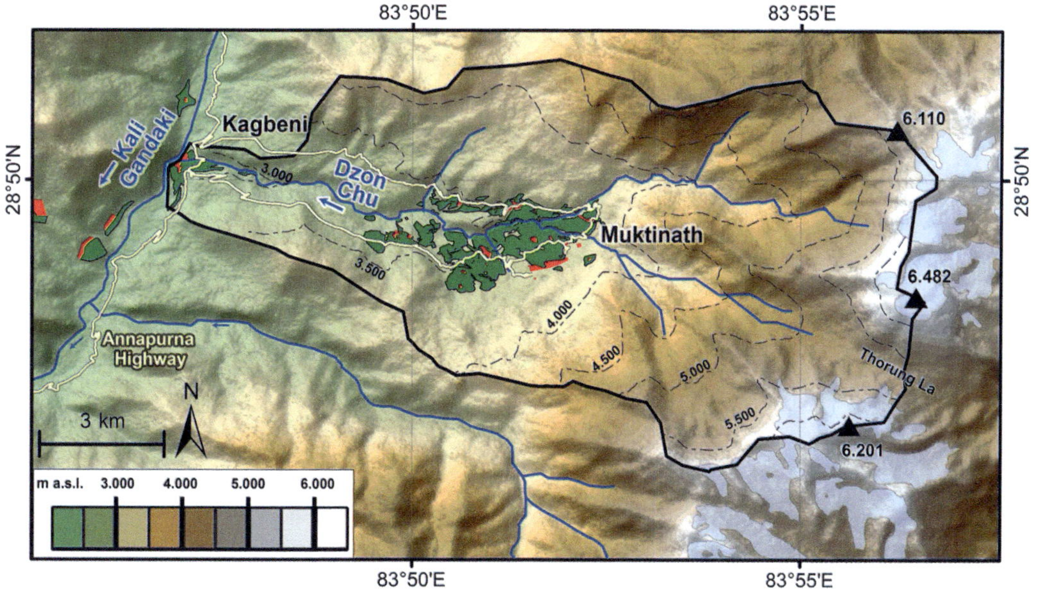

Fig. 8.6 Overview of the Dzon Chu catchment (black outline) including housing areas (red areas) and agriculture areas (green areas) of the main settlements Kagbeni and Muktinath. Blue lines: main river systems and associated tributaries. Glaciated areas are displayed in grey colour

Especially the agricultural areas around the settlement of Kagbeni are an example of a highly customized, small-scale system of fully irrigation-based agriculture within the dry high-alpine environment. An impressive system of 687 small, interlaced irrigated fields with a total area of 0.31 km^2 are pervaded by a complex system of irrigation channels with a total length of around 10 km (Fig. 8.7).

During the recent decades, own surveys as well as investigations by Lama and Job (2014) verify substantial changes in the cultivation of new crop plants, both in Kagbeni as well as in the other surrounding settlements like Muktinath or Ekklebathi. In the period from 1992 to 2018, the areas (mixedly) cultivated with apple trees increased by 20%. With regard to the stated comparatively low water resources fed by glaciers, this is in view of the increased water consumption through apple farming particularly surprising. However, the threat of a potential general water shortage (even outside the vegetation period) is not limited to irrigation farming solely. Simultaneously, the number of new buildings with touristic background (lodges, hotels) drastically increased. While in 1992

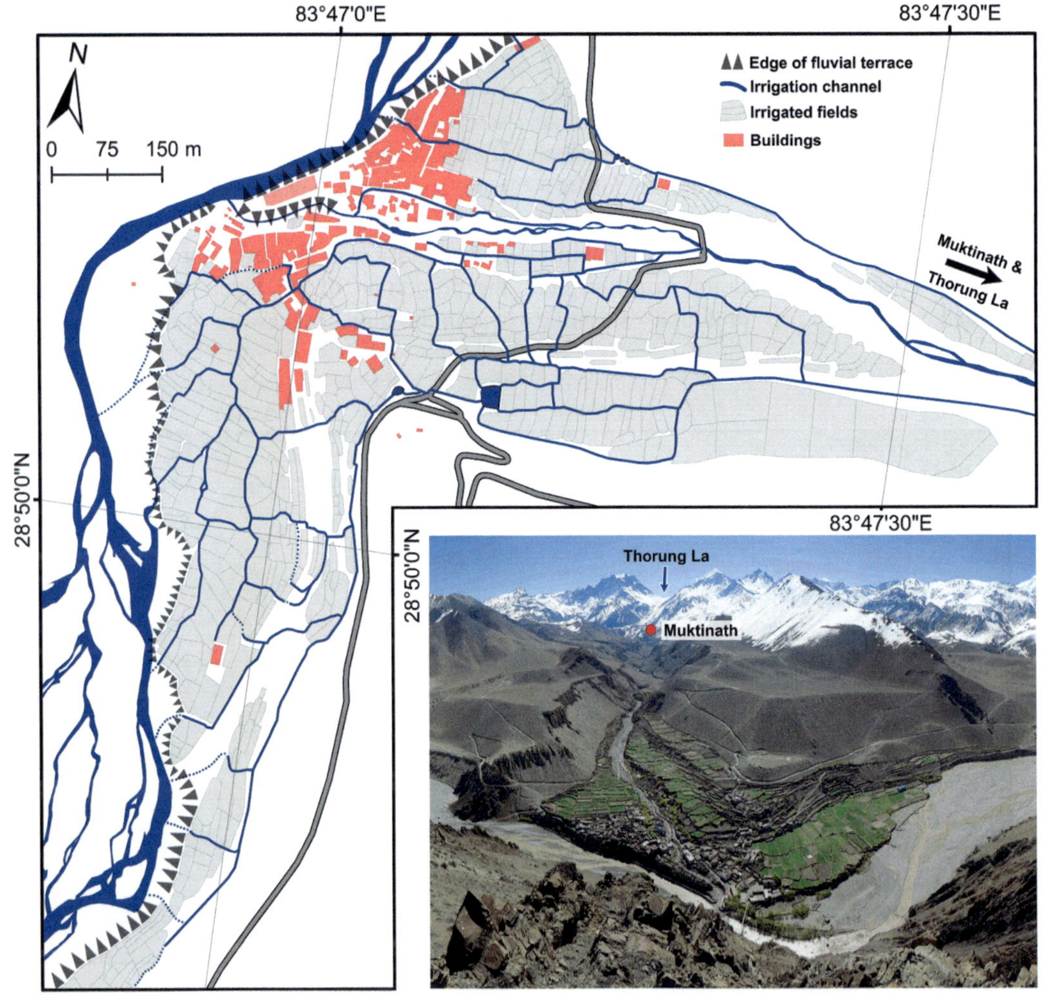

Fig. 8.7 Map showing the agricultural areas in the settlement of Kagbeni, including location of irrigation channels and buildings. Inlay picture is showing an overview from the east on the alluvial fan of the Dzon Chu with the settlement of Kagbeni

Kagbeni only five hotels/lodges existed, a survey of the touristic infrastructure in 2018 already reveal a number of 30 hotels and twelve restaurants.

The underlying reasons are complex and manifold. As indicated by Lama and Job (2014), the number of tourist arrivals in the Kagbeni/Muktinath area increased since 2007 with an average annual growth of 38%. This can be directly linked to the construction and role of the Annapurna highway (Fig. 8.6) leading to a general better accessibility of the Mustang District for tourists as well as pilgrims. As a consequence, the number of tourist infrastructure like hotels and lodges in Kagbeni and Muktinath increased significantly, resulting in an excessive demand on drinking water. In addition, road connectivity intensified the socio-economic linkage between local (agricultural) economy with sales markets situated downstream. As stated by Lama and Job (2014), the traditional, subsistence-based scope of local agriculture have already and probably will further change into a more market-orientated system. This can already be corroborated by own surveys during the recent decade verifying the increasing cultivation of new crops that require more water during germination and/or growth, such as fruit tree plantations (mainly apple) with cover crops or maize. Increasing tourism is accompanied by an intensified demand for meat and therefore a high number of livestock. Over 90% of the households in the village of Kagbeni are engaged in livestock rearing and possess a total of 7,000 cows, yaks, mules and goats (Koirala and Shresta 2017). The meat of the latter is frequently consumed in hotels, representing a rare source of income to the rural population. In consequence, goats account for the largest proportion of livestock population (80%) and are fed in stables with by-products of crop production and in combination with grazing during the warm summer season (Koirala and Shresta 2017; Paudel and Andersen 2013). However, rangeland vegetation is also threatened by a declining trend of snow cover, influencing the soil moisture and water availability for pasture (Paudel and Andersen 2013). This, in turn, is leading to higher demands of agricultural by-products as fodder, e.g. through an enhanced maize production.

8.5 Conclusions

During the recent decades, the Mustang district was like many areas in the Greater Himalaya Region subject to a profound climatic change. In addition to a significant increase in temperature, this is also reflected in a closely related decrease of the glaciated area. The glacier retreat can verifiably be observed since the LIA but is still demonstrably persistent for the past four decades (1976–2018). This results in substantial changes to the numerous (perennial) creeks and rivers originating from glaciers in the higher elevations, which are in turn in large parts of considerable importance for local agriculture. In particular, a steady water supply by glacial meltwater is necessary during phases with highly variable or missing rainfall, e.g. during the spring and/or early summer season when sowing and germination of the crops take place. To that extent, this has already and might in the future lead to enhanced critical implications for the irrigation-based agriculture in the Mustang district. However, the increasing pressure on local water resources in this semi-arid region is for some settlements not solely caused by climate change and its corresponding impacts on local hydrology. Furthermore, a concurrent profound and complex socio-economic change associated with a higher consumption of (drinking) water leads to an increasing competition and overuse of local water resources. This is mainly driven by (i) newly introduced and increasingly cultivated crops (apple, maize), (ii) a substantial increase in water consumption by new infrastructure for the significantly increasing numbers of tourists and pilgrims and (iii) a significantly improved infrastructural connectivity of the region through the recently finished Annapurna highway which connects the Mustang District with the Kathmandu Valley and the Terai. A thorough management between the diverging interest of the irrigation-intensive agriculture, the changes into a new market-orientated agriculture and the parallel increase of a water-consuming tourism infrastructure will be one of the most important challenges for the adaptation to climate change in this region in the upcoming years.

References

Bajracharya SR, Maharjan SB, Shrestha F, Bajracharya OR, Baidya S (2014) Glacier status in Nepal and decadal change from 1980 to 2010 based on landsat data. ICIMOD, Kathmandu

Barthlott W, Mutke J, Rafiqpoor MD, Kier G, Kreft H (2005) Global centers of vascular plant diversity. Nova Acta Leopold 92:61–83

Beniston M (2003) Climatic change in mountain regions: a review of possible impacts. Clim Change 59:5–31

Beniston M (2005) The risks associated with climatic change in mountain regions. In: Huber U, Bugmann H, Reasoner M (eds) Global change and mountain regions: an overview of current knowledge. Springer, Dordrecht, pp 511–520

Beniston M, Stoffel M (2014) Assessing the impacts of climatic change on mountain water resources. Sci Total Environ 493:1129–1137

Böhner J, Miehe G, Miehe S, Nagy L (2015) Climate and weather. In: Miehe G, Pendry CA, Chaudhary R (eds) An introduction to the natural history, ecology and human environment of the Himalayas. Royal Botanical Garden Edinburgh, Nepal

Bolch T, Yao T, Kang S, Buchroithner MF, Scherer D, Maussion F, Huintjes E, Schneider C (2010) A glacier inventory for the western Nyainqêntanglha range and the Nam Co Basin, Tibet, and glacier changes 1976–009. Cryosphere 4(3):419–433

Bolch T, Kulkarni A, Kääb A, Huggel C, Paul F, Cogley JG, Frey H, Kargel JS, Fujita K, Scheel M, Bajracharya S, Stoffel M (2012) The state and fate of Himalayan glaciers. Science 336(6079):310–314

Braun MH, Malz P, Sommer C, Farías D, Sauter T, Cassassa G, Soruco A, Skvarca P, Seehaus T (2019) Constraining glacier elevation and mass changes in South America. Nat Clim Change 9:130–136

Cannon F, Carvalho LMV, Jones C, Bookhagen B (2015) Multi-annual variations in winter westerly disturbance activity affecting the Himalaya. Clim Dyn 44(1–2):441–455

Diaz HF, Grosjean M, Graumlich L (2003) Climate variability and change in high elevation regions: past, present and future. Clim Change 59(1–2):1–4

Diffenbaugh N, Giorgi F (2012) Climate change hotspots in the CMIP5 global climate model ensemble. Clim Change 114(3–4):813–822

Egger J, Bajrachaya S, Egger U, Heinrich R, Reuder J, Shayka P, Wendt H, Wirth V, Giorgi F (2000) Diurnal winds in the Himalayan Kali Gandaki Valley. Part I: Obs Monthly Weather Rev 128(4):1106–1122

Fort M (2015) Impact of climate change on mountain environment dynamics. J Alpine Res 103–2

Gottfired M, Pauli H, Futschik A, Akhalkatsi M, Barancok P, Alonso JLB et al (2012) Continent-wide response of mountain vegetation to climate change. Nat Clim Change 2:111–115

Haeberli W (2013) Mountain permafrost—research frontiers and a special long-term challenge. Cold Reg Sci Technol 96:71–76

Immerzeel WW, Van Beek LPH, Bierkens MFP (2010) Climate change will affect the Asian water towers. Science 328(5984):1382–1385

IPCC (2013) The 5th assessment report of the intergovernmental panel on climate change. Cambridge University Press, Cambridge, p 2013

Jones PD, Parker DE, Osborn TJ, Briffa KD (2016) Global and hemispheric temperature anomalies—land and marine instrumental records. In Trends: a compendium of data on global change. Carbon dioxide information analysis center, Oak Ridge National Laboratory, U.S. Department of Energy, Oak Ridge, Tenn, USA

Kang S, Xu Y, You Q, Flügel WA, Pepin N, Yao T (2010) Review of climate and cryospheric change in the Tibetan Plateau. Environ Res Lett 5:

Karger DN, Conrad O, Böhner J, Kawohl T, Kreft H, Soria-Auza RW, Zimmermann NE, Linder HP, Kessler M (2017) Climatologies at high resolution for the earth's land surface areas. Sci Data 4 (170122):827

Koirala A, Shresta KB (2017) Effects of climate change on livestock population in Mustang District, Nepal. Asian J Agric Dev 14(1):1–13

Körner C (2013) Alpine ecosystems. In: Levin SA (ed) Encyclopedia of biodiversity, vol 1, 2nd edn. Academic Press, Amsterdam, The Netherlands, pp 148–157

Körner C (2014) Mountain ecosystems in a changing environment. eco.mont. J Protected Mt Areas Res 6:71–77

Körner C (2008) Global change affects ecosystems through biodiversity responses. South African J Bot 74:358–358

Körner C (2004) Mountain biodiversity, its causes and function. Ambio, Special Report 13:11–17

Körner C, Jetz W, Paulsen J, Payne D, Rudmann-Maurer K, Spehn EM (2017) A global inventory of mountains for bio-geographical applications. Alpine Bot 127:1–15

Kreutzmann H (2000) Water towers of human kind: approaches and perspectives for research on hydraulic resources in the mountains of South and Central Asia. In: Kreutzmann, H (ed) Sharing water. Irrigation and water management in the Hindukush—Karakoram—Himalaya, Karachi, pp 13–31

Krishnan R, Shrestha AB, Ren G, Rajbhandari R, Saeed S, Sanjay J, Syed M, Vellore R, Xu Y,

You Q, Ren Y (2019) Unravelling climate change in the Hindu Kush Himalaya: rapid warming in the mountains and increasing extremes. In: Wester P, Mishra A, Mukherji A, Shrestha A (eds) The Hindu Kush Himalaya assessment. Springer, Cham

Lama AK, Job H (2014) Protected areas and road development: sustainable development discourses in the Annapurna conservation area, Nepal. Erdkunde 68 (4):229–250

Loarie SR, Duffy PB, Hamilton H, Asner GP, Field CB, Ackerly DD (2009) The velocity of climate change. Nature 462(7276):U1052–U1111

Lutz AF, Immerzeel, WW (2013) Water availability analysis for the upper Indus, Ganges, Brahmaputra, Salween and Mekong river basins. Final Report to ICIMOD, September 2013. FutureWater Report, No. 127. Wageningen, The Netherlands, FutureWater

Manandhar S, Pandey VP, Kazama F (2012) Hydro-climatic trends and people's perceptions: case of Kali Gandaki River Basin, Nepal. Clim Res 54:167–179

Maurer JM, Schaefer JM, Rupper S, Corley A (2019) Acceleration of ice loss across the Himalayas over the past 40 years. Science Advances 5: eaav7266

Mayor JR, Sanders NJ, Classen AT, Bardgett RD, Clément JC, Fajardo A, Lavorel S et al (2017) Elevation alters ecosystem properties across temperate treelines globally. Nature 542:91–95

Meier WJH, Grießinger J, Hochreuther P, Braun MH (2018) An updated multi-temporal glacier inventory for the patagonian andes with changes between the little ice age and 2016. Frontiers in Earth Science 6

Menzel A, Sparks T, Estrella N, Koch E, Aasa A, Ahas R, Alm-Kübler K, Bissolli P, Braslavská O et al (2006) European phenological response to climate change matches the warming pattern. Glob Change Biol 12:1969–1976

Miehe G (1984) Vegetationsgrenzen im extremen und multizonalen Hochgebirge (Zentraler Himalaya). Erdkunde 38(4):268–292

Mountain Research Initiative EDW Working Group., Pepin N, Bradley R. et al. (2015) Elevation-dependent warming in mountain regions of the world. Nature Clim Change 5, 424–430. https://doi.org/10.1038/nclimate2563

Mukherji A, Sinisalo A, Nüsser M, Garrard R, Eriksson M (2019) Contributions of the cryosphere to mountain communities in the Hindu Kush Himalaya: a review. Reg Environ Change 19(5):1311–1326

Nogúes-Bravo D, Araújo MB, Errea MP, Martínez-Rica JP (2007) Exposure of global mountain systems to climate warming during the 21st century. Glob Environ Change 17(3–4):420–428

Parsons AJ, Phillips RJ, Lloyd GE, Law RD, Searle MP, Walshaw RD (2016) Mid-crustal deformation of the Annapurna-Dhaulagiri Himalaya, central Nepal: An atypical example of channel flow during the Himalayan orogeny. Geosphere 12(3):985–1015

Parveen S, Winiger M, Schmidt S, Nüsser M (2015) Irrigation in Upper Hunza: evolution of socio-hydrological interactions in the Karakoram, Northern Pakistan. Erdkunde 69(1):69–85

Paudel KS, Andersen P (2013) Response of rangeland vegetation to snow cover dynamics in Nepal Trans Himalaya. Clim Change 117:149–162

Paul F (2011) Melting glaciers and icecaps. Nat Geosci 4 (2):71–2

Paul F, Kääb A, Maisch M, Kellenberger T, Haeberli W (2004) Rapid disintegration of Alpine glaciers observed with satellite data. Geophys Res Lett 13: L21402

Pepin NC, Seidel DJ (2005) A global comparison of surface and free-air temperatures at high elevations. J Geophys Res 110:D03104

Pohle P (2001) The cultural landscape of Kagbeni—geographical investigations into the present-day land use system. In: Pohle P, Haffner W (eds) Kagbeni—contributions to the village's history and geography. Giessener Geographische Schriften, 77. Gießen

Rangwala I, Miller JR (2012) Climate change in mountains: a review of elevation-dependent warming and its possible causes. Clim Change 114(3–4):527–547

Rosenzweig C, Karoly D, Vicarelli M, Neofotis P, Wu Q, Casassa G, Menzel A et al (2008) Attributing physical and biological impacts to anthropogenic climate change. Nature 453:353–357

Sherchan P (2019) Understanding the nexus of climate change and migration: A case study of Dhye peoples from Upper Mustang. Nepal. Grassroots Journal of Natural Resources 2(1–2):1–19

Sorg A, Bolch T, Stoffel M, Solomina O, Beniston M (2012) Climate change impacts on glaciers and runoff in Tien Shan (Central Asia). Nat Clim Change 2:725–731

Spehn EM, Rudmann-Maurer K, Körner C (2012) Mountain biodiversity. Plant Ecol Divers 4(4):40–43

Steinbauer MB, Grytnes J-A, Jurasinski G, Kulonen A, Lenoir J, Pauli H, Rixen C, Winkler M, Bardy-Durchhalter M, Barni E, Bjorkman AD, Breiner FT, Burg S, Czortek P, Dawes MA, Delimat A, Dullinger S, Erschbamer B, Felde VA, Fernández-Arberas O, Fossheim KF, Gómez-García D, George D, Grindrud ET, Haider S, Haugum SV, Henriksen H, Herreros MJ, Jaroszewicz B, Jaroszynska F, Kanka R, Kapfer J,Klanderud K, Kühn I, Lamprecht A, Matteodo M, Morra di Cella U, Normand S, Odland A, Olsen SL, Palacio S, Petey M, Piscová M, Sedlakova B, Steinbauer K, Stöckli V, Svenning J-C, Teppa G, Theurillat J-P, Vittoz P, Woodin SJ, Zimmermann NE & Wipf S (2018) Accelerated increase in plant species richness on mountain summits is linked to warming. Nature 556 (7700)

Turner AG, Annamalai H (2012) Climate change and the South Asian Monsoon. Nat Clim Change 2:587–595

Viviroli D, Archer DR, Buytaert W, Fowler HJ, Greenwood GB, Hamlet AF, Huang Y, Koboltschnig G, Litaor MI, Lopez-Moreno JI, Lorentz S, Schadler B, Schreier H, Schwaiger K, Vuille M, Woods R (2011) Climate change and mountain water resources: overview Impact of climate change on mountain environment dynamics. Hydrol Earth Syst Sci 15:471–504

Viviroli D, Durr HH, Messerli B, Meybeck M, Weingartner R (2007) Mountains of the world, water towers for humanity: typology, mapping, and global significance. Water Resour Res 43
Walther GR, Post E, Convey P, Menzel A, Parmesan C, Beebee TJC, Fromentin JM et al (2002) Ecological responses to recent climate change. Nature 416:389–395
Wang R, Liu S, Shangguan D, Radic V, Zhang Y (2019) Spatial heterogeneity in Glacier mass-balance sensitivity across high mountain Asia. Water 11
Wipf S, Stoeckli V, Bebi P (2009) Winter climate change in alpine tundra: plant responses to changes in snow depth and snowmelt timing. Clim Change 94:105–121
Wu Q, Zhang T (2008) Recent permafrost warming on the Qinghai-Tibetan Plateau. J Geophys Res 113:D13108
Xu J, Grumbine RE, Shrestha A, Eriksson M, Yang X, Wang Y, Wilkes A (2009) The melting Himalayas: cascading effects of climate change on water, biodiversity and livelihoods. Conserv Biol 23:520–530
Yang B, He M, Shishov V, Tychkov I, Vaganov E, Rossi S, Ljungqvist FC et al (2017) A new perspective on spring vegetation phenology and global climate change based on Tibetan plateau tree-ring data. Proc Natl Acad Sci PNAS 114(27):6966–6971
Yao T, Thompson L, Yang W, Yu W, Gao Y, Guo X, Yang X, Duan K, Zhao H, Xu B, Pu J, Lu A, Xiang Y, Kattel DB, Joswiak D (2012) Different glacier status with atmospheric circulations in Tibetan Plateau and surroundings. Nat Clim Change 2:663–667

Glaciers, Climate and People: Holocene Transitions in the Stubai Valley

9

Andrea Fischer, Lucia Felbauer, Andrina Janicke, Kay Helfricht, Helene Hoffmann, and Eva-Maria Wild

Abstract

The Austrian Stubai Valley starts at the modern transport monument of the Europa bridge of the A13, the lowest motorway crossing of the Alps and ends way back at the ice-covered peaks of the main Alpine ridge. The glaciers released the valley floor of today's main villages during the Early Holocene, but natural processes still are major macro-drivers of the valley's economic development. The steepness of the slopes necessitates warning systems, technical barriers to prevent avalanches and mudflows, as well as land use planning. These are the major strategies for coping with the omnipresent natural hazards, which have shaped the valley landscape for centuries. The article presents a broad overview of glacier development and also compiles a wealth of existing studies on past and present processes from the Early Holocene to the Anthropocene. The synopsis reveals that the effects of climate change and extreme events cannot be anticipated or discussed without a profound debate of cultural practices in the various societies and that a story of transitions underlies the nearly continuous land use in the area during the last millennia.

Keywords

Climate warming · Global change · Mountain region · Stubai · Holocene

9.1 Introduction

Mountain regions are under pressure from global climate change—yet a considerable part of humanity relies on ecosystem services provided by mountain regions. The United Nations Organization (UNO) accounts for the crucial role of mountains and their vulnerability by mentioning mountains explicitly in its 2030 Agenda (UNO 2015). Mountains are acknowledged as important sources of water, energy, biodiversity and key resources, such as minerals, forest and agricultural products and recreation. Research on related topics is an explicit aim of related actions, building on past mountain research (e.g. Messerli 2012; Price et al. 2013).

A. Fischer (✉) · L. Felbauer · A. Janicke · K. Helfricht · H. Hoffmann
IGF/ÖAW Institute for Interdisciplinary Mountain Research, Austrian Academy of Sciences, Technikerstrasse 21 a, Innsbruck, Austria
e-mail: andrea.fischer@oeaw.ac.at

E.-M. Wild
Faculty of Physics - Isotope Physics, VERA Laboratory, University of Vienna, Währinger Straße 17, Vienna, Austria

H. Hoffmann
Department of Earth Sciences, University of Cambridge, Cambridge, UK

U. Schickhoff et al. (eds.), *Mountain Landscapes in Transition*,
Sustainable Development Goals Series, https://doi.org/10.1007/978-3-030-70238-0_9

Human impact on environmental transformations is a central topic of Anthropocene research (e.g. Whitehead 2014). In addition to the current focus on the global framework of transformation towards a sustainable society, mountain research had and still has a second focus on the detection and attribution of transitions. Research on transitions includes prominent features like geomorphological processes as well as sociocultural transitions.

Mountains generally are recognized as sentinels of change, with higher change rates than in other regions (IPCC 2013). The Alps thus play a key role for research to develop process understanding and models, as the shifting baselines can be tackled most precisely on the basis of a rich pool of natural scientific and socio-economic data.

From the perspective of system science, complex human-environment system changes cannot be attributed easily to either external forcings or internal feedbacks. A considerable part of the forcings includes chaotic dynamic subsystems, for example, short-time-scale local weather phenomena or multicausal mass movements. A general rule like that increases in precipitation come along with an increasing number of extreme events can present great variability in observations in small catchments, so that the observation period needed for valid statistics might be longer than the existing dataset. Nevertheless, the study of past processes and practices in a small region can deliver highly important insights not only for system switches, but also on vulnerability and resilience of such a complex system.

This study compiles various recorded states of the man-environment system in the Stubai valley during the Holocene from a transdisciplinary perspective. The reason for doing so is not the development of a generally valid unified theory of all kinds of interaction between man and environment, but rather to step back behind disciplinary framings and defining small subsystems to 'glimpse the dappled world of mottled objects' (Cartwright 1999) as a first step towards an open transdisciplinary scientific debate on potential future system states and their implications. As a general hypothesis, transitions are considered multicausal and non-linear responses of man-environment systems. Necessarily, for this first step, framing is avoided as far as possible. This includes avoiding an a priori classification of 'causing' and 'resulting processes'.

The study area is defined to be within the present political district of the four communities in Stubai valley, part of the district Innsbruck Land since 1869 AD. Historically, this area is roughly the same as the Gericht Stubai (after 1326 AD) located initially in Telfes and, from about 1690 AD onwards, in Mieders (including the neighbouring community of Ellbögen).

9.2 Study Area

The Stubai valley close to Innsbruck offers an excellent playground for studying Holocene transitions and transformations[1] up to now. The valley is located in Tyrol, Austria, and has a long tradition of research, as its start is only about 10 km south of the University of Innsbruck. The Stubai valley, first mentioned between 994 and 1005 AD (Töchterle 1991), hosts the five communities of Schönberg, Telfes, Mieders, Fulpmes and Neustift (Fig. 9.1). Their physical and socio-economic structure is diverse and also has been in the past. This ranges from the glacier ski resort of Neustift with 1,000,000 arrivals per season to the traditional metal industry village of Fulpmes, economic bases reflected in both villages by a higher rate of non-Austrians in the working population (Fig. 9.1).

The valley is 34 km long (north-east oriented) and covers an area of 317.7 km^2 and an altitude range from 662 to 3488 m a.s.l. Glacial history of the area has been subject to very early studies, with the stadials Egesen and Fernau named after type localities in Stubai (e.g. Kinzl 1929, 1949). Human activities in the valley have been traced back to 8250 ± 40 BP (Schäfer et al. 2016). The traditional Brenner road as lowest pass over the

[1]In this study, the term transformation is used for changes caused by an intentional and active intervention, other changes are termed transitions.

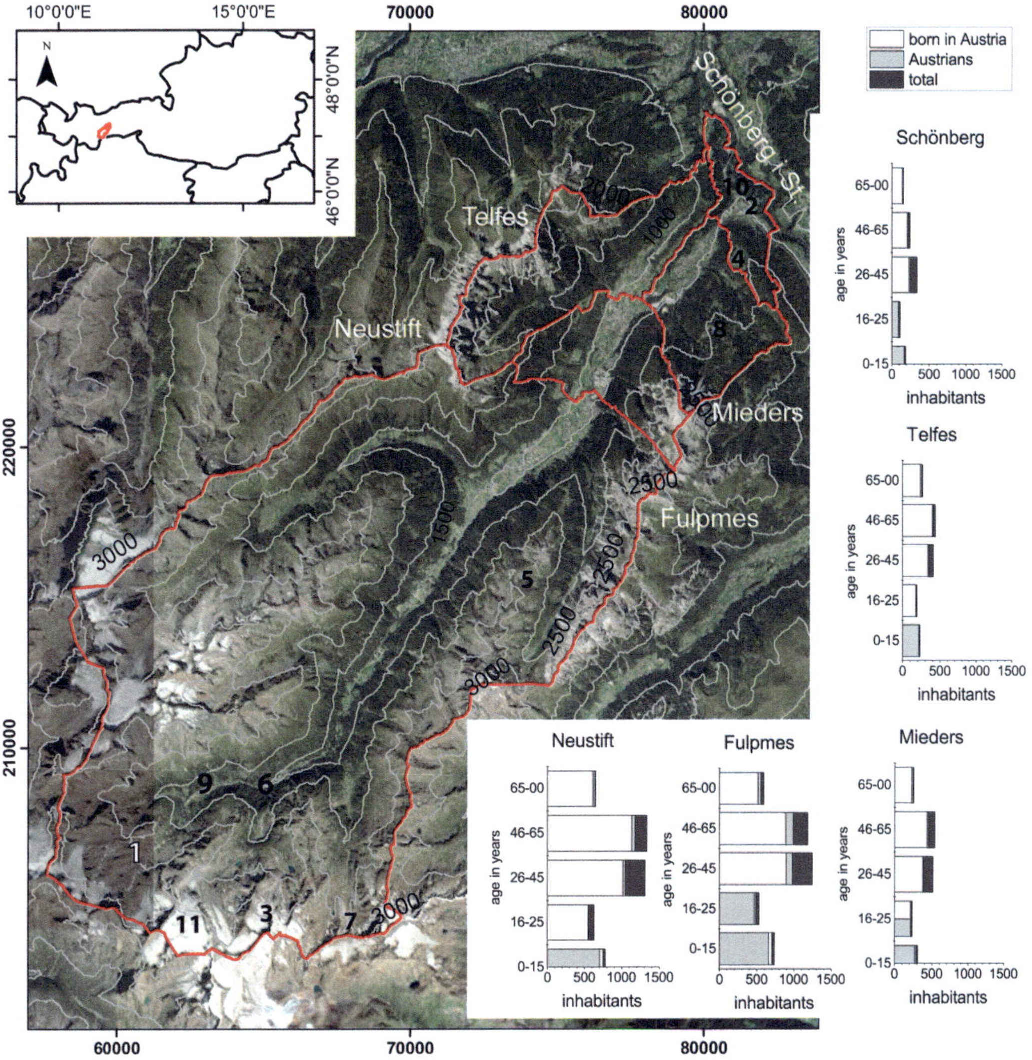

Fig. 9.1 Map of stubai valley with the five communities: Schönberg im Stubaital, Telfes, Mieders, Fulpmes and Neustift im Stubaital, with the respective age distribution of inhabitants by age classes and nationality. 1. Dresdner Hütte, Oberfernau, 2. Europabrücke, 3. Freigerferner, 4. Gleins, 5. Gratzergrübl, 6. Grawa, 7. Grüblferner, 8. Koppeneck, 9. Mutterberg, 10. Stephansbrücke, 11. Sulzenauferner. Orthophoto 2015, DEM for the calculation of contour lines of elevation (grey lines) and borders (red lines) provided by the federal government of Tyrol (data.tirol.gv.at), population data: Statistik Austria

Alps crosses Schönberg at the entrance to the Stubai valley, and today the Europa bridge of the A13 motorway offers direct access to the Stubai valley. More recent research covers the topics of tourism (Kariel et al. 1993, 1989; Scharr and Steinicke 2011; Fischer 2014), climate (Patzelt 2016; Ilyashuk et al. 2015; Feng et al. 2019), ecosystems (Frank et al. 2015; Leitinger et al. 2015; Bahn et al. 2014; Fondevilla et al. 2016) and glaciers (Fischer et al. 2015). Intensive research was done on snow management and glacier covers (e.g. Fischer et al. 2016). The LSTER site Tyrolean Alps with LTER master site Stubai fosters not only ecological, but also trans- and interdisciplinary research.

9.3 Material and Methods

In this section, only the specific methods applied in this study are described in short, without a broader summary of the state of the art in the field. A more extensive treatment of methods and the respective state can be found in the cited papers and references herein.

Glacier extent has been mapped manually in the glacier inventories based on orthophotos for 1969 and 1998 and LiDAR data in 2006 (Fischer et al. 2015). Newest data was mapped from the federal laser scan DEMs in 2017, tackling glacier margins using hillshade and volume change. The glacier extent during past maximum states (Little Ice Age, Egesen, Daun, Gschnitz) is mapped by tracing the remnants of moraines from the LiDAR DEM. The late glacial glacier extent was mapped from moraines visible in the laser scan DEMs and DTMs from 2006, including basic ideas of Senarclens-Grancy (1938), Kinzl (1929, 1949) and personal communications with Gernot Patzelt (Patzelt 2019). Rock glacier extent has been mapped manually by Krainer and Ribis (2011) based on field surveys, orthophotos and LiDAR DEMs. LiDAR DEMS have a very high spatial resolution (~1 m depending on elevation) and high vertical accuracy (~cm to dm depending on slope), so that these are helpful in mapping human traces in landscape, as past roads or paths.

The long-term instrumental climate data of the station Innsbruck University has been homogenized (Auer et al. 2007) to correct for the bias caused by changes in observation method, observer and relocations of the station. Climate information from the Oberfernau bog can be deduced by radiocarbon dating of organic material and palynological analysis (Patzelt 2016).

Archeological findings as artefacts (Krösbacher 2004) or fireplaces (Patzelt 2013) are an empirical evidence for human presence in the area. Later on, historical documents and statistics (e.g. Staffler 1842) give information on the demography and socioeconomy. Latest statistical data is collected by the institute of Statistics Austria. In contrast to the first maps of the study area which had the purpose to support travellers, the federal maps from 1817 onwards had also a military strategic aim and described roads, settlements and population as well as horses (for military use) very detailed. Data on past transport of toll goods are found in historical collections of toll books (Stolz 1955). Modern traffic data is collected by national and federal statistical offices.

Two chronologies of natural disasters have been compiled in recent decades from different perspectives: Fliri (1998) compiled events from the perspective of climatology, Jäger (2010) from the historical perspective, including a view on the general developments and socio-economic situation.

9.4 Results

9.4.1 Glaciers and Rock Glaciers

Glaciological research in the Stubai valley started later than in the neighbouring Ötz valley, but still became very famous during the area of ice age research (*Diluvialforschung*). Systematic measurements of glacier fluctuations started with the glacier survey of the Austrian Alpine Club in 1891 (Fischer et al. 2018). In 2017, the 52 glaciers in the Stubai valley cover an area of 18.8 km^2. This is only 41% of the glacier area at the end of the Little Ice Age (Fischer et al. 2015, Figs. 9.2 and 9.3). The GI3 glacier volume calculated for all glaciers in the Stubai valley is 0.88 km^3, which corresponds to a mean ice thickness of 39 m and a share of 6% of the estimated total glacier volume for Austria (Helfricht et al. 2019).

After the disintegration of the large glaciers of the last ice age, the mild interstadial conditions were interrupted by the cold reversals of Older (Gschnitz and Daun stadials) and Younger Dryas (Egesen stadial), when glaciers covered a major portion of the Stubai valley (Table 9.1).

The recent data from 2017 exhibits an increasing annual area loss (which is calculated for latest periods only, as advances in the 1920s and 1980s and long measurement periods hamper a straightforward interpretation of annual rates).

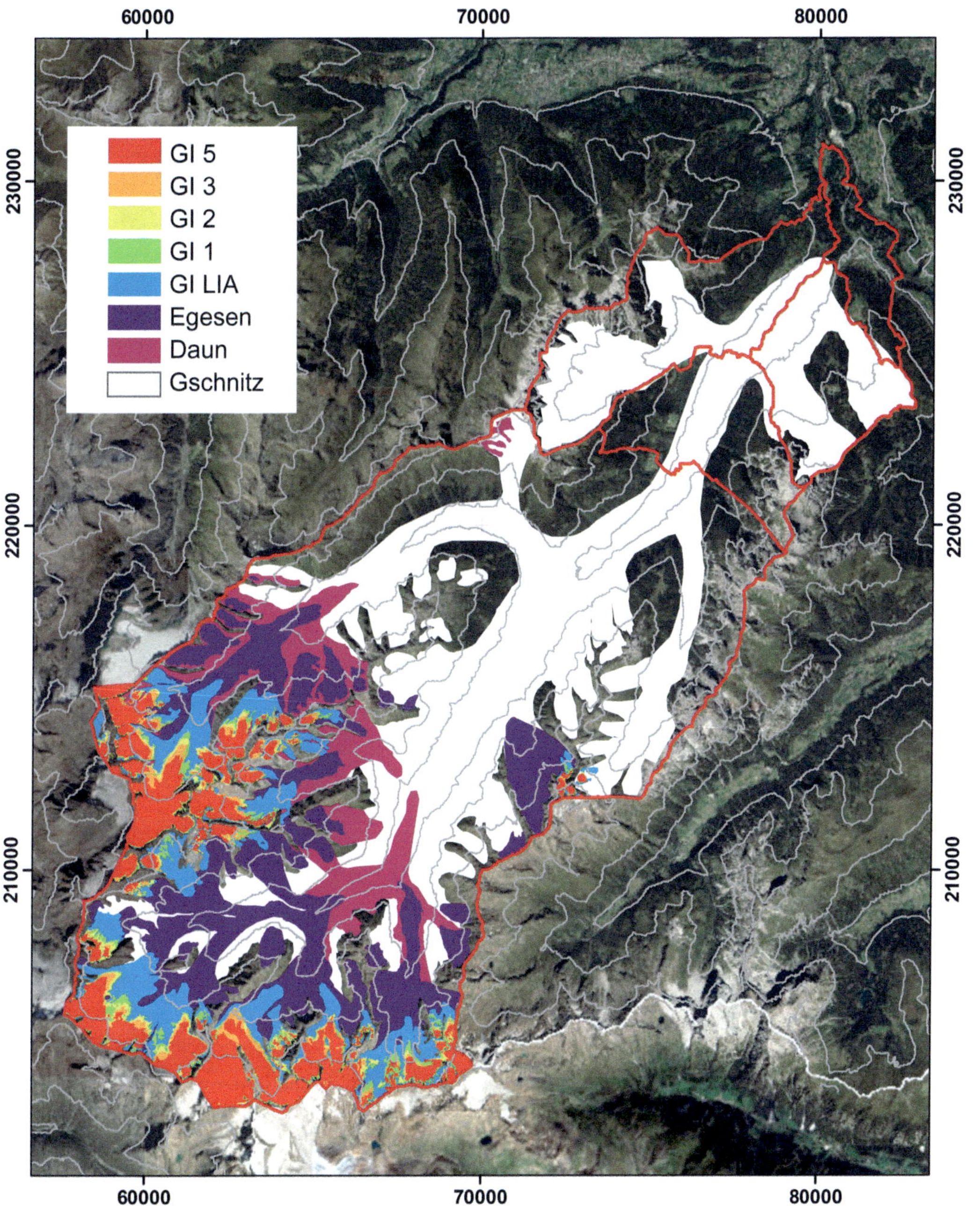

Fig. 9.2 Glacier areas for the different time periods in Table 9.1. The Orthophoto 2015 and the DEM for the calculation of contour lines of elevation and borders provided by the federal government of Tyrol (data.tirol.gv.at)

Current warming leads not only to rapid losses of glacier length and area, but also to changes in the geomorphology of the glacier, which increases its sensitivity to climate change (Fig. 9.4). In the periglacial area, new lakes form. Currently, the largest one at Sulzenauferner still has contact with the dead ice from the recession of Sulzenauferner above a steep ridge. A partial release of water has been reported for the extreme precipitation event of 10 August

Fig. 9.3 Dresdner hut (founded in 1875 as first Alpine hut in the Stubai valley) as drawn by Gatt (1878) with Fernauferner and Schaufelferner. In 2018 (photo: Andrea Fischer), glacier tongues have receded to higher elevations, with parts of Schaufelferner still visible. Within the former moraines, ski slopes (with snow-making facilities) have been constructed. In 2018, the hut has 140 beds

Table 9.1 Glacier cover in the Stubai valley for different stages during the Holocene. The dates for Egesen, Daun and Gschnitz stadials are taken from the Auer et al. (2014). Ivy-Ochs et al. (2008) differ slightly in the time frame, mainly for the Daun stadial (<14.7 kyr). In the light of the local variability of advances, as evident from modelling results of Seguinot et al. (2018), timing in the Stubai valley might differ by an unknown period in the absence of direct datings

Glacier inventory	Approx. time	Glacier area	Area loss	Area loss	Annual area loss	Portion of valley
		km^2	km^2	%	%	%
GI5	2017 AD	18.840	3.985	17	1.59	6
GI3	2006 AD	22.826	1.886	8	0.95	7
GI2	1998 AD	24.711	3.592	13	0.44	8
GI1	1969 AD	28.303	17.416	38		9
LIA	1850 AD	45.719	35.369	44		14
Egesen	10,500 BC	81.088	16.061	17		26
Daun	13,000 BC	97.149	110.939	53		31
Gschnitz	14,500 BC	208.088				66

2017, damaging infrastructure at the Sulzenau hut.

In 2011 Krainer and Ribis mapped 114 rock glaciers in the Stubai valley with a total area of 7.0 km^2. Only 21 of them are still active, covering 2.8 km^2. Most rock glaciers (60) can be classified as fossil, covering approximately the same area as active rock glaciers (2.7 km^2). There has been a scientific debate on thawing permafrost as one of the causes for debris flows during an extreme precipitation event in the Mischbach basin (Reiskopf 2018), with the result that permafrost in the basin in significant amount is only likely in the uppermost parts of the basin. Nevertheless, the existence of permafrost cannot be ruled out in the light of observed ice build-up in early summer at a talus slope on the valley floor south of Neustift (Wakonigg 1996; Punz et al. 2005).

Even if the rock glacier area today is much smaller than the glacier area, the significance of these features for human land use cannot be neglected, even apart from being a potential source for natural disasters. For example, the rock glacier in Gratzergrübl/Pinis valley (Fig. 9.5) has favoured the use of alpine pastures

Fig. 9.4 Glaciers in Stubai not only lose area, but also disintegrate, forming geomorphological features like circular crevasses and increasing debris cover. In the periglacial areas, new lakes form, here at **a** Freigerferner **b** Sulzenauferner **c** Grüblferner and Aperer Feuersteinferner (*photos* Andrea Fischer)

by providing easy construction material for shelters and fences as well as water. Nowadays, the pasture buildings are no longer used, the last ones were abandoned during the twentieth century. In addition to water resources, the generally changing cultural practices of land use might have played a role as one of the reasons to start cultivating land in the vicinity of prehistorical and historical mining activities at Glücksgrat. A vicinity of glacierized landforms and prehistorical meadows has been also documented by Patzelt (2013).

9.4.2 Climate Change

The longest instrumental record of climate change in the vicinity of the Stubai valley was initiated at the University of Innsbruck by Franz Zallinger, starting as early as 1777. Later on, the Central Institute for Meteorology and Geodynamics established a weather station in Neustift/Milders (2004), the Hydrographical Survey of the Federal Government in Tyrol established stations in Schönberg in 1895 AD, Neustift Volderau in 1947 AD, Telfes in 1958

Fig. 9.5 Past huts of the alpine pasture in Gratzergrübl is located directly at the headwall of a now inactive rock glacier. Today, the active rock glacier on top does not feed a spring at the location of the hut. The pasture is confined by late glacial lateral moraines, which facilitated the construction of stone fences. Orthophoto 2015, DEM for the calculation of contour lines of elevation provided by the federal government of Tyrol (data.tirol.gv.at), *photos* Andrea Fischer

AD and at the Dresdner hut in 1978 AD. The runoff gauges at the river Ruetz in Krössbach dating from 1990 AD and Fulpmes established in 1974 AD record the variability of the glacier-fed river Ruetz. Groundwater level is recorded at four stations in Neustift (https://ehyd.gv.at/). At the LTER sites in the valley close to Neustift and at Kaserstattalm, additional stations are maintained for ecological studies (https://deims.org/node/6262).

The homogenized instrumental data at Innsbruck (Auer et al. 2007; https://www.zamg.ac.at/histalp/) shows strongest temperature increases after about 1980 during summer months (Fig. 9.6). In general, variabilities in seasonal temperatures are higher than for annual mean temperature, and variability in precipitation is higher than in the temperature records. Therefore, temperature trends are more straightforward to tackle statistically, whereas long-term precipitation changes lack quantification. Snow height records also show high variability (Fischer 2014), without a clear trend in either height or duration of snow cover at the Dresdner hut station.

A general question is the impact of climate on human activities in a specific area. For weather events like heavy precipitation or extreme snowfall events, the vulnerability of the people living in the area strongly depends on cultural practices and land use, which changed significantly even during the last century. The probability of a statistical ensemble of extreme weather events occurring is clearly linked to general atmospheric circulation patterns and thus to climate and climate change. The smaller the scale and the sample, the less clear is the link. For example, the frequency of flood events during the cooler, but very variable LIA shows no clear connection to the temperature changes within

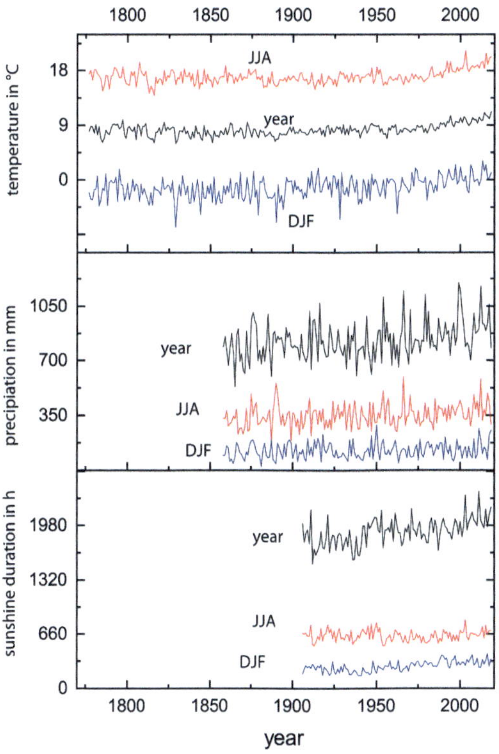

Fig. 9.6 Instrumental climate data (seasonal means of air temperature, sums of precipitation and sunshine duration) measured at the station at Innsbruck University. Year… annual mean/sum, JJA… mean/sum for the summer months June, July and August, DJF …. mean/sum for the winter months December, January and February (*Source* www.zamg.ac.at/HISTALP)

LIA (Auer et al. 2014). For a natural disaster to occur that affects local society and economy, a number of conditions and phenomena must come together: a weather event, plus, for example, direct or indirect impact on important infrastructure and a lack of resilience towards the impact, for example when damages cannot be repaired due to missing resources. In any case, small-scale weather events like extreme precipitation during thunderstorms or heavy snowfall are not very well represented even in the current measurement system, so that geo- and historical chronologies often provide better information for analysing natural disasters and their impacts. For example, the summer of 2003 was clearly extreme in terms of summer temperatures, but no direct damages have been reported as a result of these meteorological extremes. The debris flows of 2017 have occurred in a basin without meteorological measurements and are recorded in runoff data and reports only. Apart from large-scale events like droughts or extreme precipitation resulting from stationary atmospheric flows (blockings), events with a smaller spatial range often occurred and still occur unrecorded even by modern measurement systems. This is a major scaling problem, not only in hydrology, but also for mass movements or avalanche activity, and a tremendous gap of knowledge for coping with future climate change in mountain regions. Only much stronger interdisciplinary effort can help to increase our knowledge of the past to have an idea on baseline and variability for reserving necessary resources in time.

Apart from instrumental data, the Stubai valley offers rich data and potential for additional proxy datings of climate and climate changes. For example, the Oberfernau bog close to today's Dresdner hut contains one of the most complete and long chronologies of glacial and biotic response to climate change. Located next to the LIA lateral moraine of Fernauferner (Fig. 9.7), glacial sediment deposits and peat layers document warm and cold periods for at least from 8000 BC onwards (Patzelt 2016). The large logs found after erosion by an extreme precipitation event in 2017 date back to 7570–7440 BC (VERA-6546) and are located directly above the lake sediment that forms the lowest part of the profile analysed by Patzelt (2016).

9.4.3 Population and Settlement History

Evidence for human activities in the Stubai valley exists for the major part of the Holocene (Fig. 9.8 and references therein). Mesolithic and Neolithic artefacts have been found in the inner and higher part of the valley. Later and younger findings concentrate in the outermost part of the valley and document great continuity of human activities there. Pre-Roman and Romanic place names are found all over the Stubai valley (Fig. 9.8). Mining activities took place at high

Fig. 9.7 Oberfernau bog with the tree eroded (white square) during the extreme precipitation event in August 2017. At the base of the bog, the light grey sediment accumulated during the existence of the lake which drained after the rock dam leaked, so that upper parts of the bog might even show a longer record of organic material. Today the timberline is a bit lower than the bog, which could be attributed to sheep grazing in the area

elevations throughout the valley (Srbik 1929), but were not very productive in historical times. Nevertheless, they may have played a role in earlier times for the development of a hammer mill industry. Place names like *Silbergasse, Schmelzgasse, Schmelzhüttengasse* and *Arzgruben* hint at mining activities. Great amounts of charcoal needed for smelting and forging was also produced in the valley.

The cultivation of Grawa pasture (Fig. 9.1), located south of Tschangelair pasture (a Romanic name, older spelling: Schöngeleir) was dated to 420–560 AD (95.4%; VERA 6480). A charcoal layer with a similar age (545–742 AD) was found in Schaufelferner ice cave. For Gleins, the cultivation of the pasture can be dated back to 990–1160 AD (95.4%, VERA-6068), which is consistent with historical documents.

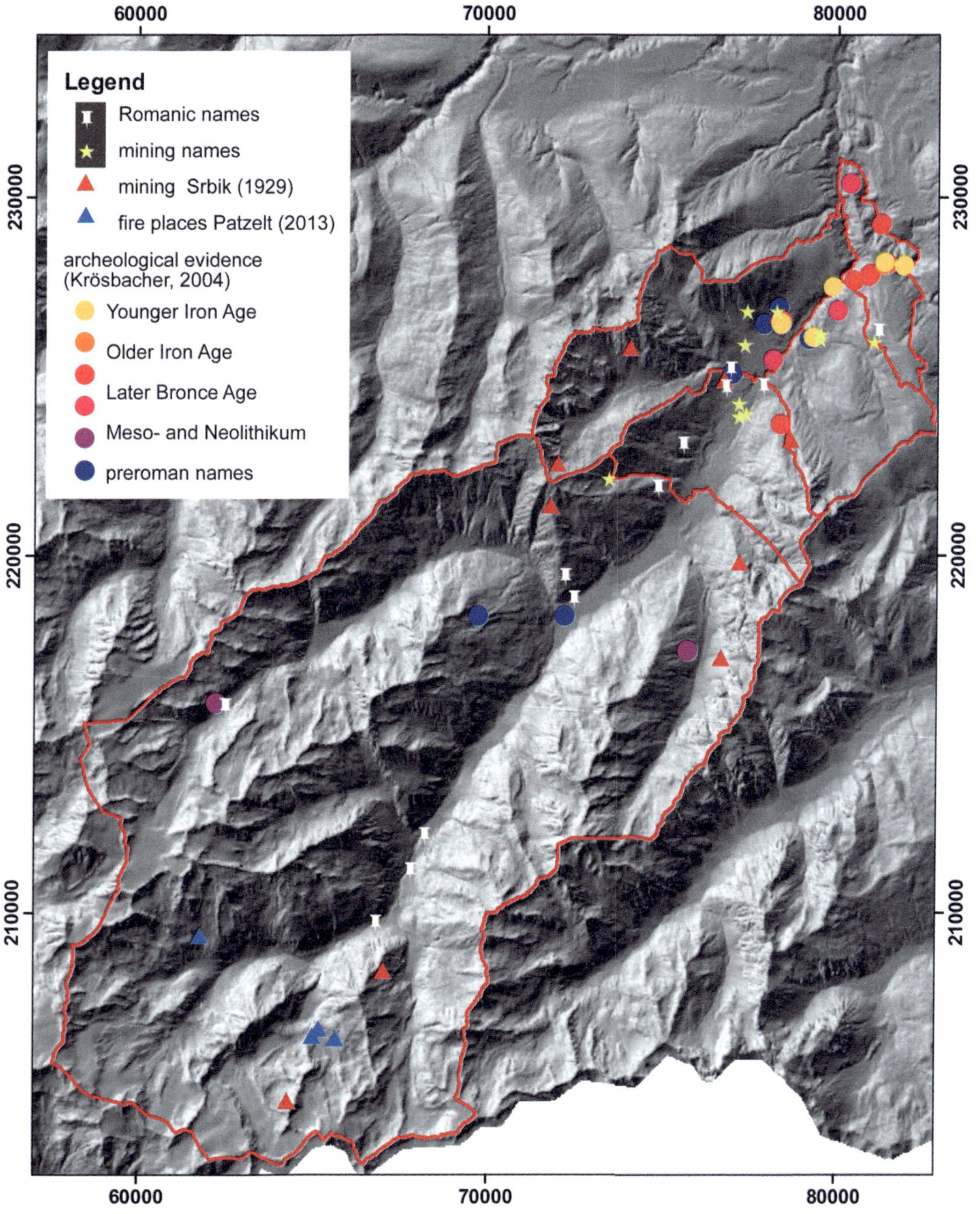

Fig. 9.8 Archeological evidence (summarized by Krösbacher 2004), historical mines (Srbik 1929), place names related to mining activities, fireplaces related to pastures (Patzelt 2013) and Romanic names (Töchterle 1991). Hillshade and borders provided by the federal government of Tyrol (data.tirol.gv.at)

Population data (Table 9.2) show a general increase in population in all communities of the valley, most pronounced from the 1970s onwards. Strongest recent increases are in Neustift and Mieders. Neustift is the only community where the population shrank in the first period (until 1869).

Table 9.2 Population dynamics in the communities of the Stubai valley (*1839: Staffler 1842, all other years: Statistik Austria)

Population	Fulpmes	Mieders	Neustift	Schönberg	Telfes
1839*	954	482	1357	283	519
1869 (31.12.)	1028	448	1241	273	502
1880 (31.12.)	1102	399	1265	264	468
1890 (31.12.)	1079	395	1217	247	474
1900 (31.12.)	1159	390	1238	273	444
1910 (31.12.)	1384	449	1344	454	458
1923 (7.3.)	1577	440	1372	492	506
1934 (22.3.)	1624	492	1646	430	559
1939 (17.5.)	1648	503	1805	406	552
1951 (1.6.)	2068	697	2018	556	686
1961 (21.3.)	2282	587	2195	590	649
1971 (12.5.)	2582	702	2794	668	987
1981 (12.5.)	2973	952	3307	782	1069
1991 (15.5.)	3611	1163	3791	916	1192
2001 (15.5.)	3895	1592	4328	1000	1369
2011 (31.10)	4183	1752	4557	1007	1476

9.4.4 Traffic

Roads are a physical sign in the landscape of interaction between people and provide historical information in maps and toll books. For prehistoric travel routes, artefacts can provide clues. Rich data are available for the outermost part of the Stubai valley, as part of the Brenner Pass road is in the community of Schönberg. Radiocarbon datings of fireplaces at the ridges of Koppeneck (560–400 BC, 78.5%, VERA-6211) and Gleins (60–240 AD, 95.4%, VERA 6143) can very likely be seen in the context of the early Brenner route. Both places provide drinking water and ponds. The summit of Koppeneck also offers a preview of the further route to Innsbruck.

Potentially the first paved road was built under Claudius in 46/47 AD and was improved in 192–215 AD under Septimus Severus and Decius (250 AD), with milestones in Schönberg and Unterberg.

We can assume a fairly continuous history of improvements at this steepest part of the Brenner road. A new, less steep route installed in 1582 and 1584 was damaged in 1776 by a debris flow. In 1842 the Stephansbrücke was opened over the river Ruetz. In 1867 AD, the railway to Bolzano was introduced, followed in 1904 by the Stubaitalbahn connecting Fulpmes and the industry there with Innsbruck main station. In 1963 the Europa bridge of the A13 motorway was opened, with a direct exit to the Stubai valley.

From an economic point of view, traffic brought some prosperity directly related to the transportation of goods: For the steep parts of the road, additional horses were needed that could be borrowed in Schönberg and Unterberg. Accommodation, catering and maintenance stations brought revenue and still do. Recently, some businesses along the Brenner road, mainly accommodation facilities (Schönberger Hof and others), closed down, as travellers tend to avoid the negative impacts of traffic (noise).

In 1734, most of the transport took place in the winter months and in early summer, when horses were available (Stolz 1955). Available data (Table 9.3) suggest an almost continuous increase in traffic and the volume of goods transported over the Brenner, apart from reported decreases caused by wars (Stolz 1955) or recent economic slowdowns. The modern roads seem to have reached the limits of capacity, construction work on a railway tunnel is under way.

Not only the quantity, but also the type of transported goods varies considerably (Fig. 9.9). In 1627, wine and vinegar are the main goods transported over the Brenner pass, at least among products subject to tolls (timber, for example, was exempt from toll at that time and thus not recorded). Some types of goods from the breakdown for 2015, such as furniture or waste, are missing from the lists of goods for 1627, potentially not only because they have not been subject to tolls, but because it made no sense to transport these types of goods within the past economic system. For 1779, Stolz (1955) lists metal products, tobacco, sugar and leather as additional product groups. Although there is a slight chance that toll-free products went unrecorded, limited transport capacity and cheaper manpower will have meant a preference for locally produced items, for example furniture or even energy, of course at much lower levels of consumption than today.

The road itself changes with adaptation to the geomorphological situation, but also to vehicle requirements. In recent centuries, the steepness of roads was softened, the width increased and the road surface improved. This increased bot costs and effort for construction and maintenance as well as the potential vulnerability. The road segment between Unterberg and Schönberg, Stephansbrücke and Europabrücke illustrates the effort undertaken in the twentieth century, in contrast to earlier routes. The actual primary Brenner road (opened in the 1840s) runs along the steeper eastern face (Fig. 9.10).

The changes in the roads close to Gleins exhibit various mechanisms of transition/transformation of anthropogenic systems (Fig. 9.11). Today Gleins is a part of Schönberg and the modern road only leads there. Historically, Gleins also had strong links with Mieders and belonged to the parish of Mieders. Older roads thus lead to Mieders as well as Schönberg. The modern road is built on the eastern slope, whereas all older roads are on northwestern or southwestern slopes. On top of the ridge, close to today's community boundaries, an old path leads towards the Brenner. In contrast to today's route, this path is steeper but less endangered by debris flows, rockfall or avalanches. With foehn winds a frequent occurrence, winter snow is deposited on the lee side, with erosion and very low snow heights directly on the ridge at the old route. When the Brenner road was blocked by debris flows in September 2017 and October 2018, the old path remained undamaged. Being too steep for modern traffic, the old path fulfilled a requirement which had been very important in earlier history: A number of springs and small ponds offer water for travellers and pack animals. A good view of the area might be of strategic advance at the beginning of the road. Older routes connect a number of remnants of charcoal stacks with the communities in the valley, indicating economic exchange with industries there. As this type of business collapsed once cheaper coal could be imported by railway, roads were also closed down.

Travel times shrank rapidly during the last century. The Baedeker travel guide of 1888 (Baedeker 1888) lists different types of vehicles travelling twice a day to Fulpmes (4.5 h) or a hike of 3.5 h to the same location along the river Ruetz. The further path through the villages is described in great detail (cross the river and the forest, turn around a rock …) to Neustift (1.5 h), Volderau (1.25 h), Ranalt (1.25 h) and 2.25 h to Mutterbergalm. As touristic infrastructure is recommended, we can assume that travellers did take breaks over overnights during this trip. In Baedeker (1912), the travel Innsbruck/Mutterberg sounds less adventurous, as the Stubaitalbahn railway could be used to travel to Fulpmes in 1 h (same travel time as in 2019). Travelling by railway is described as spectacular. From Fulpmes the guide recommends taking a vehicle to reach Neustift within 1 h, Volderau within another 1.5 h and 0.25 h to Ranalt. This still was the end of the road—until in the 1970s, the road to the glacier ski resort was built. In 2019, the quickest bus takes about 1 h from Innsbruck main station to Mutterberg, where the cable car can take you up to the Dresdener hut in about 10 min. In recent years, traffic jams have increased travel times by car.

Table 9.3 Transport of goods on the Brenner route at different times

Year	Toll station	Cargo		Cargo	
		tons		tons	
1300			Lueg	3000	Stolz (1955)
1410–1430			Lueg	3600	Stolz (1955)
1500			Lueg	4500	Stolz (1955)
1550			Lueg	5400	Stolz (1955)
1627	Schönberg	3000	Lueg	5000	Stolz (1955)
1734			Lueg	12,500	Stolz (1955)
1960	A13 and railway			3,140,000	Land Tirol (2000)
1980	A13 and railway			15,470,000	Land Tirol (2000)
2000	A13 and railway			35,500,000	Land Tirol (2001)
2016	A13 and railway			46,900,000	Land Tirol (2018)

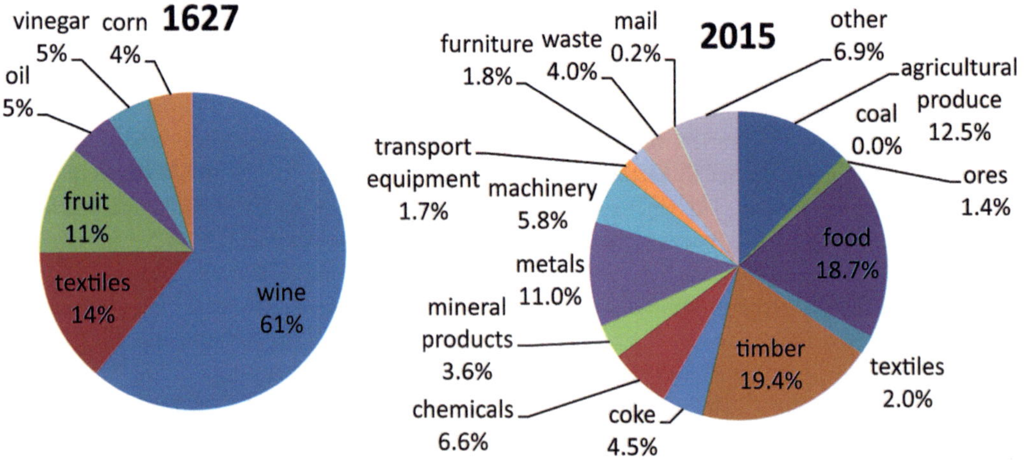

Fig. 9.9 Relative proportion of goods transported over the Brenner, as evident from toll documents in 1627 and in 2015 (CAFT15, 2016). The main type of textiles in 1627 was silk

9.4.5 Land Use

Land use in Stubai is changing constantly. 211.777 km^2 or two-thirds of the total area of Stubai (317.660 km^2) is protected. Forests and alpine pastures are major forms of land use, with settlements and transport infrastructure accounting for only 2% of the total area. The ski resorts (Schlick 2000 in Fulpmes, Serles lifts in Mieders, Elfer lifts and Stubai glacier skiing in Neustift) cover 5.58% of the area. Although the classification system of Graf (1880) cannot be compared directly, and his data includes the village of Ellbögen, the area used for settlements was much smaller in 1873 (Tables 9.3 and 9.4).

Increased areas for settlements and infrastructure also increase the risk that a geomorphological event causes damage. Figure 9.12 shows the changes since the early nineteenth century. New settlements have been built in the

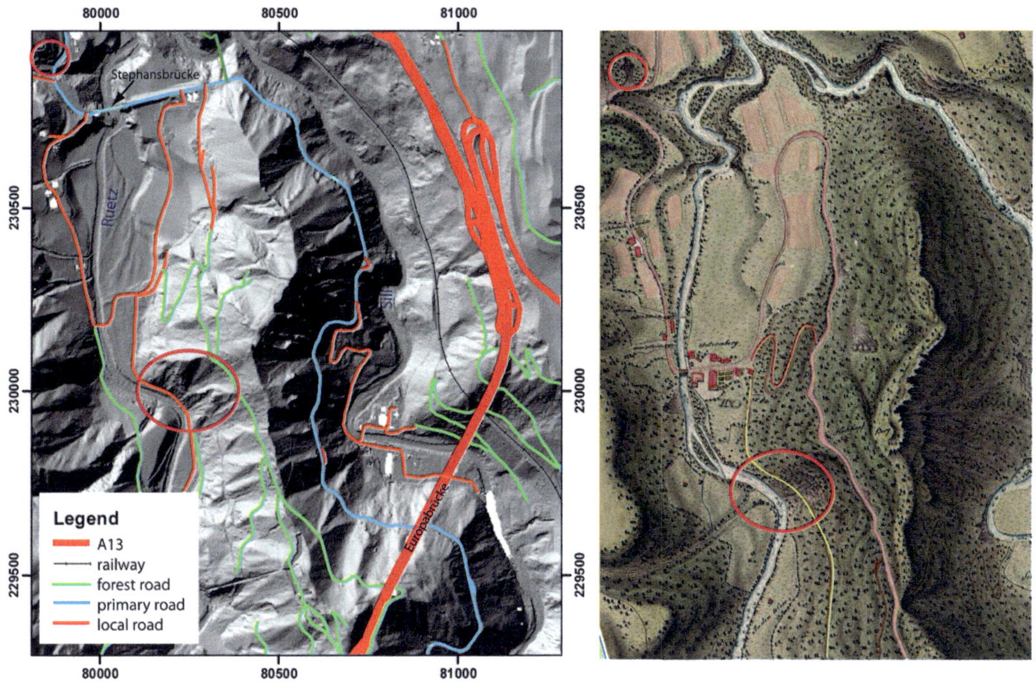

Fig. 9.10 The steepest section of the Brenner route was continuously adapted in response to mass movements (red circles) and the requirements of transport (left: modern traffic routes with older roads used in part as forest roads, right: three generations of roads in the historical map Schönbergstraße, 1830)

areas of Schlickerbach, Seibach and Margarethenbach rivers, which are documented as subject to floods, as is also reflected in the bed of the river Ruetz formed by the talus cones of these rivers. Today, safety measures and infrastructures are in operation in the river beds.

9.4.6 Natural Disasters

Both, Fliri (1998) and Jäger (2010) list a great number of natural disasters in the Stubai valley, mainly floods on all scales, diseases and avalanches during the Little Ice Age. The latest larger events affecting the total valley were floods on 19.07. and 25.08.1987 (Fliri 1998). After that, small-scale events like small-scale debris floods affected roads, houses and agricultural land. Parts of both the Brenner and the Stubaital road are closed for several days each year due to the threat of avalanches or mass movements. In these cases, the Brenner motorway can usually be used to reach the outer parts of the Stubai valley. During events as described by Reiskopf (2018) most urgent affairs can be handled by helicopter transport during road closure.

Fairy tales/oral history tell of debris flows destroying Medraz and Mieders, with both villages relocated afterwards. Currently, there is no known scientific evidence of these events.

Most interesting from a perspective of environmental history are the damages caused by the Schlickerbach river in Fulpmes. As the river was needed as a source of energy for the early industrial work on iron (*Hammerwerke*), many of the buildings are located very close to steep parts of the river. At times of higher discharges, this meant great damage. Even after electricity replaced water as source of energy, the buildings remained in place. Today, the companies have moved to an industrial area, but the buildings are

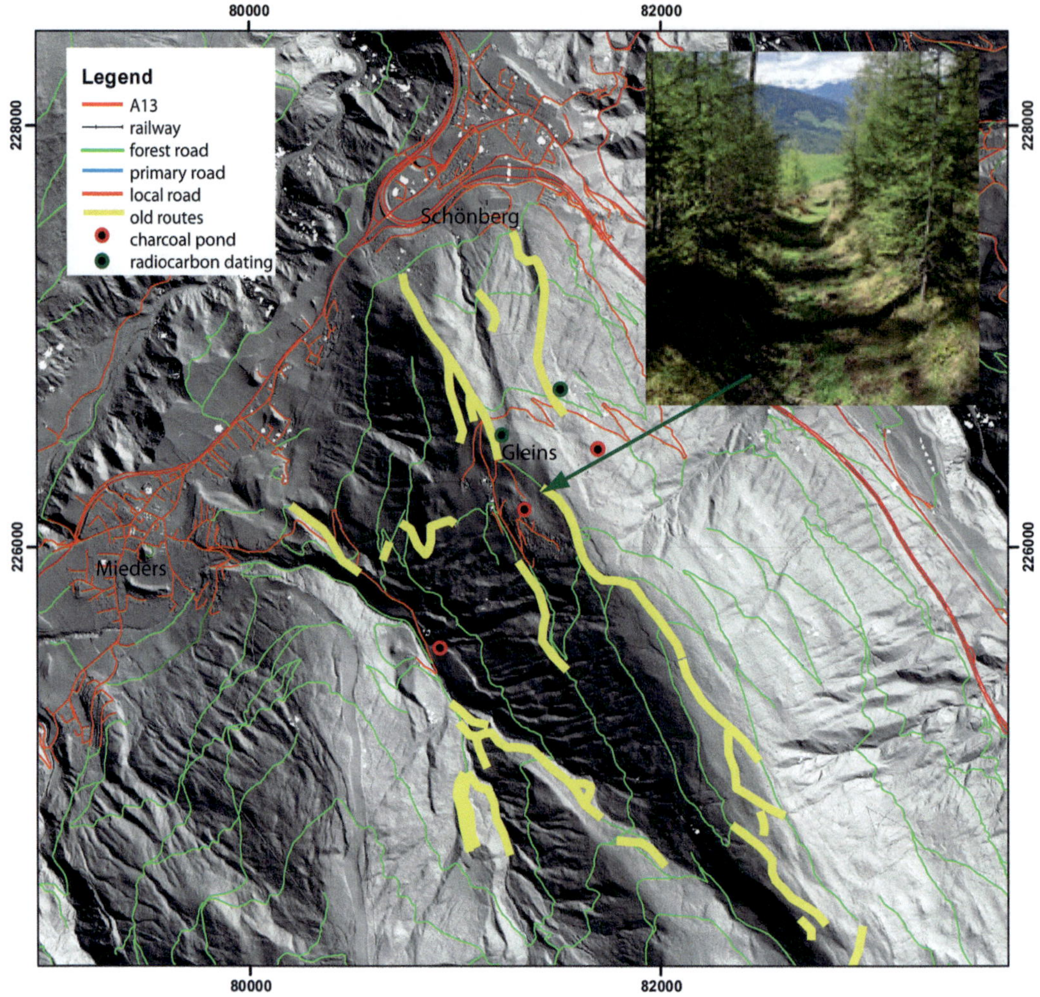

Fig. 9.11 Road system at Gleins with older routes/paths in yellow, mapped in the field and from the laser DEM 2006. Insert: example of an old path with steep sides, suited for the transport of goods with horses

still in place and used as residences. So the historical reason for building close to river, including the benefits, has gone, but the risk is still there.

Generally, changes in the anthropogenic part of the system (land use, settlements and protection measures) can be considered to have a greater impact on the occurrence of disasters than changes in the natural part of the system, such as the size of glaciers and firn cover and the formation of paraglacial lakes.

9.5 Discussion and Conclusion: Past, Present and Future Transitions

This article compiles a number of disciplinary data on past and present transitions and transformations in a small and well-investigated region of the Austrian Alps, with the long-term aim to proceed towards an inter- and transdisciplinary understanding of the past to shape our transformation to a sustainable future.

Table 9.4 Land use in the Stubai valley in 2017 and 1873

	2017 km^2	2017(%)		1873 km^2	1873(%)
Arable land	0.48	0.2	Arable land	7.47	2.1
Alpine and other pastures	217.45	68.5	Alpine and other pastures	78.34	21.9
Meadows	14.94	4.7	Meadows	46.12	12.9
Forest	107.93	34.0	Forest	87.25	24.4
Settlements	3.63	1.1	Settlements	0.34	0.1
Transport infrastructure	2.82	0.9	Unproductive	138.51	38.7
Ski resorts	17.62	5.5			
Total area	317.66		Including the nearby village of Ellbögen	358.03	
(Katalog Landnutzung Tirol 2017)			(Graf 1880)		

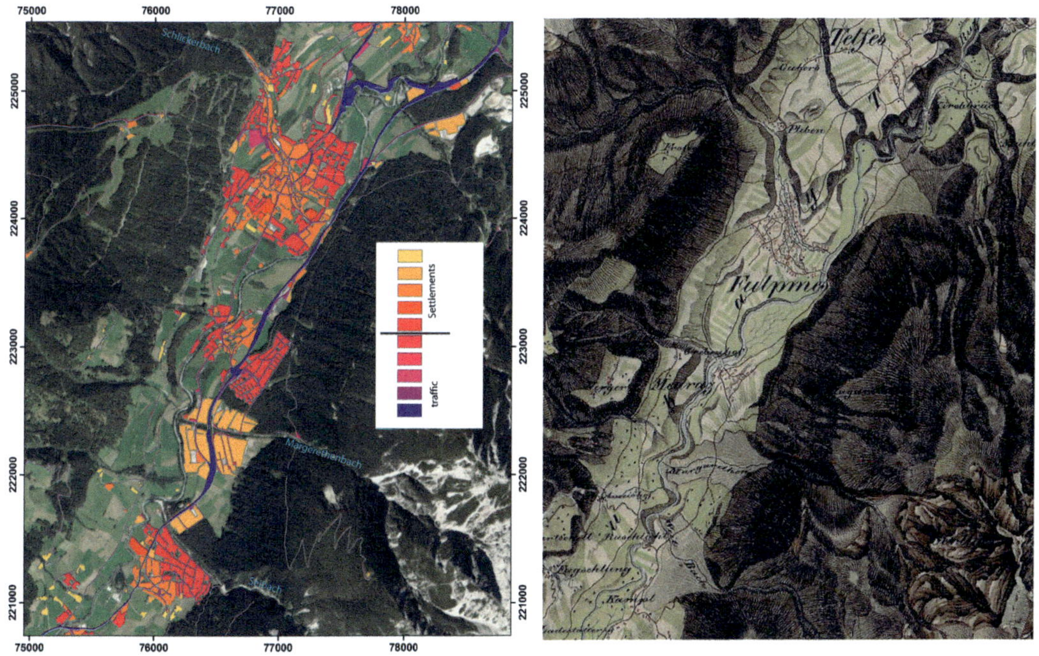

Fig. 9.12 The land used for vulnerable infrastructures like settlements and roads is significantly larger in 2017 than in the early nineteenth century as mapped in the first federal survey (Franziszeische Landesaufnahme, 1816–1821)

What can be confirmed from this compilation is a continuous, but not steady, transition of the landscape as a result of geomorphological processes. Anthropogenic impact and climate change triggering the retreat of glaciers and the succession of biota. A key site here is the Oberfernau bog.

The persistence of change is nothing new and could be tracked backed to Heraklit's *panta rhei*, or the general understanding of a system as a dynamic equilibrium. Recently, we experienced a strong retreat of glaciers, but also rapid succession of biota in the Austrian Alps (e.g. Fischer et al. 2015; Steinbauer et al. 2018). The next decades will show if succession is quick enough to stabilize the paraglacial terrain when glaciers shrink further even under current climatic conditions, but more quickly in case of additional warming. Glacial runoff will decrease as the melting area shrinks. Settlements have grown to take up most of the available land using numerous protective measures in small basins. For any further expansion, individuals, the public and insurance companies will have to agree how much risk (and related costs) they will accept. The development of traffic is interrelated with global or at least European economic developments. As road capacities seem to be exhausted, control measures gain in importance. For the future, a total collapse of traffic is as possible as strict regulations in combinations with tolls. A very short period in history of free traffic seems to come to an end.

Developments generally seem to be path-dependent, in a way that areas, practices or paths in operation tend to be used further, even if the context changes.

A major interest for future transformations is to avoid tipping points, i.e. points of no return to a prior state, which includes not only tipping points of the natural system like triggers of debris flows, but also the ability of humans to cope. This leads to issues of the political organization of society, for example, the relation of individuals (and individual losses) to the collective—and its ability and willingness to compensate for individual losses or even protect from losses if possible.

What can also be concluded from the compilation is a long continuity in human presence in the region, at lower as well as higher elevations. This is consistent with the findings of Kutschera et al. (2014), with evidences of pasturing and mining activities from the Bronze Age onwards (Festi et al. 2014; von Scheffer et al. 2019). Exact comparisons of prehistorical human activities with climate changes are plagued by an inaccuracy in dating (Walser and Lambers 2012), so that the direct impact of climate, weather or geomorphological events on local changes in the socio-economic system is unclear for the past.

The analysis of anthropogenic imprints as indicators of human presence and activity has enormous potential for tackling processes of both the Anthroposphere and the Geosphere in same location. The first has attracted growing attention in archaeology recently (e.g. Sevara et al. 2018). A more extensive combination of geodata analysis with radiocarbon datings of the various paths indicating high and continuous use of a route would help to qualify the periods of use.

The impression that a great number of loose ends remain in the analysis of the data compiled for this study results from basically two factors: first, the scales and system sizes of all the presented data vary (Fig. 9.13). Second, disciplinary framing of both research questions and results, and the related lack of a general theoretical framework hampers efforts at presenting a synopsis from the compilation.

Comparing two different ways of colonialization in the Alps, for example, by large monasteries between around 1000 AD and the free Walser people in the seventeenth century show that different types of societies act in different ways and have different potentials and limits. Current efforts towards a sustainable society to confine climate warming might need new forms of societal organization. Careful research and an open public debate are needed to define values and aims of the present society. For example, a reduction in traffic could reduce greenhouse gas emissions, return some value creation to the regions, but would minimize tourism.

An active transformation of socio-economic structures should aim to avoid tipping points in natural systems which cannot be compensated by socio-economic measures. This does not mean an a priori rejection of development in a region, turning it into a wilderness, or of its further development into a leisure park (Bätzing 2015), or of continuing a regional socio-economic system in a global environment as today. The decision for one of these options will be made, hopefully intentionally, during a political process which should not exclude scientific investigations.

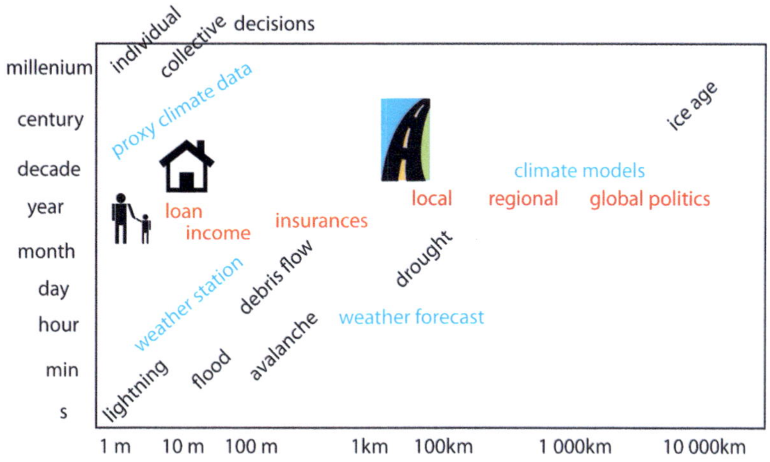

Fig. 9.13 Events with direct impact on humans and their infrastructure happen on a very different scale than can be covered by models. Direct measurements often do not happen directly at the location of an event

Essential in shaping the future of mountain environments will be a political decision that will determine the vulnerability of mountain populations on an individual and a collective basis. The role of science could be to point out potential vulnerabilities and impacts of decisions. To have a better idea of which impacts can occur in future, we urgently need interdisciplinary palaeo-environmental multiproxy studies at localities/benchmark research sites to fit all parts of the natural system puzzle together, in combination with social sciences to set the scenarios of a framework for a future society.

Acknowledgements We thank the Federal Government of Tyrol for providing the geodata via the Open Government Data (OGD) Portal.

Authors Contributions AF designed the study, did the analysis, wrote the text and compiled the Figures. LF and AJ helped with mapping the late glacial stadials, KH mapped GI 5, H.H. did the radiocarbon dating of the charcoal layer in Schaufelferner, and E.-M. W. did the radiocarbon dating of all other samples.

References

Anonymous (1825) Das Thal Stubei und dessen Bewohner. In: Beiträge zur Geschichte, Statistik, Naturkunde und Kunst von Tirol und Vorarlberg, Innsbruck, vol 1, pp 166–246. http://zeitschrift.tiroler-landesmuseen.at/index.php?mybuch=Beitraege_GS_Bd1_Jg1825&mypage=170

Auer I, Böhm R, Jurkovic A, Lipa W, Orlik A, Potzmann R, Schöner W, Ungersböck M, Matulla C, Briffa K, Jones PD, Efthymiadis D, Brunetti M, Nanni T, Maugeri M, Mercalli L, Mestre O, Moisselin J-M, Begert M, Müller-Westermeier G, Kveton V, Bochnicek O, Stastny P, Lapin M, Szalai S, Szentimrey T, Cegnar T, Dolinar M, Gajic-Capka M, Zaninovic K, Majstorovic Z, Nieplova E (2007) HISTALP—historical instrumental climatological surface time series of the greater Alpine region 1760–2003. Int J Climatol 27:17–46

Auer I, Foelsche U, Böhm R, Chimani B, Haimberger L, Kerschner H, Koinig KA, Nicolussi K, Spötl C (2014) Vergangene Klimaänderung in Österreich. Österreichischer Sachstandsbericht Klimawandel 2014 (AAR14). Austrian Panel on Climate Change (APCC), Verlag der Österreichischen Akademie der Wissenschaften, Wien, pp 227–300

Baedeker K (1888) Südbaiern. Tirol und Salzburg. Handbuch für Reisende, Karl Baedeker, Leipzig

Baedeker K (1912) Südbayern. Tirol und Salzburg. Handbuch für Reisende, Karl Baedeker, Leipzig

Bahn M, Reichstein MM, Dukes JS, Smith MD, McDowell NG (2014) Climate–biosphere interactions in a more extreme world. New Phytol 202:356–359

Bätzing W (2015) Zwischen Wildnis und Freizeitpark. Rotpunktverlag, Zürich

CAFT 15 (2016) Erhebung alpenquerender Güter/Brenner. https://www.bmvit.gv.at/verkehr/gesamtverkehr/statistik/aqgv_15/download_caft15/caft15_stra_brenner.pdf

Cartwright N (1999) The dappled world. Cambridge Univ Press, Cambridge

Feng Z, Bohleber P, Ebser S, Ringena L, Schmidt M, Kersting A, Hopkins P, Hoffmann H, Fischer A, Aeschbach W, Oberthaler MK (2019) Dating glacier ice of the last millennium by quantum technology. Proc Nat Acad Sci USA 116:8781–8786

Festi D, Putzer A, Oeggl K (2014) Mid and late holocene land-use changes in the Ötztal Alps, territory of the neolithic iceman "Ötzi." Quatern Int 353:17–33
Fischer A (2014) Snow flakes and fates: what hope is there for Alpine tourism? In: Brebbia CA, Pineda FD, Favro S (eds) Sustainable tourism VI. WIT Press, Southampton, pp 293–305
Fischer A, Seiser B, Stocker Waldhuber M, Mitterer C, Abermann J (2015) Tracing glacier changes in Austria from the little ice age to the present using a lidar-based high-resolution glacier inventory in Austria. Cryosphere 9:753–766
Fischer A, Helfricht K, Stocker-Waldhuber M (2016) Local reduction of decadal glacier thickness loss through mass balance management in ski resorts. Cryosphere 10:2941–2952
Fischer A, Patzelt G, Achrainer M, Groß G, Lieb GK, Kellerer-Pirklbauer A, Bendler G (2018) Gletscher im Wandel: 125 Jahre Gletschermessdienst des Alpenvereins. Springer Spektrum, Berlin, Heidelberg
Fliri F (1998) Naturchronik von Tirol. Beiträge zur Klimatographie von Tirol, Wagner, Innsbruck
Fondevilla C, Àngels Colomer M, Fillat F, Tappeiner U (2016) Using a new PDP modelling approach for land-use and land-cover change predictions: a case study in the Stubai Valley (Central Alps). Ecol Model 322:101–114
Frank D, Reichstein M, Bahn M, Thonicke K, Frank D, Mahecha MD, Smith P, Van der Velde M, Vicca S, Babst F, Beer C, Buchmann N, Canadell JG, Ciais P, Cramer W, Ibrom A, Miglietta F, Poulter B, Ramming A, Seneviratne SI, Walz A, Wattenbach M, Zavala MA, Zscheischler J (2015) Effects of climate extremes on the terrestrial carbon cycle: concepts, processes and potential future impacts. Glob. Change Biol. 21:2861–2880
Gatt F (1878) Dresdener Hütte im Stubai. In: Amthor E (ed) Der Alpenfreund 11, Gera
Gleeson EH, Wymann von Dach S, Flint CG, Greenwood GB, Price MF, Balsiger J, Nolin A, Vanacker V (2016) Mountains of our future earth: defining priorities for mountain research—a synthesis from the 2015 Perth III conference. Mt Res Dev 36:537–548
Graf L (1880) Statistik der Alpen von Deutsch-Tirol. Wagnersche Universitätsbuchhandlung, Innsbruck
Gurung AB, Dach S, Price MF, Aspinall R, Balsiger J, Baron JS, Sharma E, Greenwood G, Kohler T (2012) Global change and the world's mountains—research needs and emerging themes for sustainable development. Mt Res Dev 32:47–54
Helfricht K, Huss M, Fischer A, Otto J-C (2019) Calibrated ice thickness estimate for all glaciers in Austria. Frontiers Earth Sci 7:1–15. https://doi.org/10.3389/feart.2019.00068
Ilyashuk EA, Ilyashuk BP, Tylmann W, Koinig KA, Psenner R (2015) Biodiversity dynamics of chironomid midges in high-altitude lakes of the Alps over the past two millennia. Insect Conserv Divers 8:547–561
Ivy-Ochs S, Kerschner H, Reuther A, Preusser F, Heine K, Maisch M, Kubik PW, Schlüchter C (2008) Chronology of the last glacial cycle in the European Alps. J Quaternary Sci 23:559–573
Jäger G (2010) Schwarzer Himmel-kalte Erde-weißer Tod. Wagner, Innsbruck
Kariel HG (1989) Socio-cultural impacts of tourism in the Austrian Alps. Mt Res Dev 9:59–70
Kariel HG (1993) Tourism and society in four Austrian alpine communities. Geo J 31:449–456
Karl K, Ribis M (2011) Blockgletscherinventar Tirol. Mitteilungsblatt des hydrographischen Dienstes in Österreich, vol 87, Wien, pp 67–88
Katalog Landnutzung Tirol (2017) Datenquelle: Land Tirol—data.tirol.gv.at, downloaded from https://www.data.gv.at/katalog/dataset/0eaa80ce-5156-4043-aeab-77f2b24b76b5
Katalog Orthophoto Tirol Datenquelle: Land Tirol—data.tirol.gv.at, downloaded from https://www.data.gv.at/katalog/dataset/35691b6c-9ed7-4517-b4b3-688b0569729a
Katalog DEM Tirol Datenquelle: Land Tirol—data.tirol.gv.at, downloaded from https://www.data.gv.at/katalog/dataset/0454f5f3-1d8c-464e-847d-541901eb021a
Kinzl H (1929) Beiträge zur Geschichte der Gletscherschwankungen in den Ostalpen. Z. Gletscherkde 17:66–121
Kinzl H (1949) Formenkundliche Beobachtungen im Vorfeld der Alpengletscher. Veröff. Mus. Ferdinandeum 26:61–82
Krösbacher R (2004) Fundtopographie des nördlichen Wipptales, Diploma Thesis, University of Innsbruck
Kutschera W, Patzelt G, Wild EM, Haas-Jettmar B, Kofler W, Lippert A, Oeggl K, Pak E, Priller A, Steier P, Wahlmüller-Oeggl N, Zanesco A (2014) Evidence for early human presence at high altitudes in the Ötztal Alps. Radiocarbon 56:923–947
Franziszeische Landesaufnahme (1816–1821) https://maps.tirol.gv.at/HIK/TIRIS
Leitinger G, Ruggenthaler R, Hammerle A, Lavorel S, Schirpke U, Clement JC, Lamarque P, Obojes N, Tappeiner U (2015) Impact of droughts on water provision in managed alpine grasslands in two climatically different regions of the Alps. Ecohydrology 8:1600–1613
Messerli B (2012) Global change and the world's mountains. Mt Res Dev 32:55–63
Patzelt G (2013) Datierung von Feuerstellen in prähistorischen Hirtenhütten im Waldgrenzbereich ostalpiner Gebirgsgruppen. Praearchos 4: 38–43
Patzelt G (2016) Das Bunte Moor in der Oberfernau (Stubaier Alpen, Tirol)—eine neu bearbeitete Schlüsselstelle für die Kenntnis der nacheiszeitlichen Gletscherschwankungen der Ostalpen—Jahrbuch der geologischen Bundesanstalt vol 156, pp 97–107
Patzelt G (2019) Gletscher: Klimazeugen von der Eiszeit bis zur Gegenwart. Hatje Cantz, Berlin
Price M, Byers A, Friend D, Kohler T, Price L (eds) (2013) Mountain geography: physical and human dimensions. University of California Press, Berkeley

Punz W, Sieghardt H, Maier R, Engenhart M, Christian E (2005) Kaltlöcher im Ostalpenraum. Verh Der Zoologisch-Botanischen Ges Österr 142:27–45
Reiskopf B (2018) Ereignisdokumentation ausgewählter Murenereignisse im Stubaital—Analyse und Interpretation. Diplomarbeit/Masterarbeit—Institut für alpine Naturgefahren (IAN), BOKU-Universität für Bodenkultur, Wien
Schäfer D, Bertola S, Pawlik A, Geitner C, Waroszewski J, Bussemer S (2016) The landscape-archaeological Ullafelsen project (Tyrol, Austria). Preistoria Alpina 48:29–38
Scharr K, Steinicke E (eds) (2011) Tourismus und Gletscherschigebiete in Tirol. Eine vergleichende geographische Analyse. Innsbruck Univ Press, Innsbruck
Schönbergstraße (1830) https://maps.tirol.gv.at/HIK/
Seguinot J, Ivy-Ochs S, Jouvet G, Huss M, Funk M, Preusser F (2018) Modelling last glacial cycle ice dynamics in the Alps. Cryosphere 12:3265–3285
Senarclens-Grancy W (1938) Die Gliederung der stadialen Moränen im Stubaital. Jahrb der Geologischen Bundesanst 88:23–24
Severa C, Verhoeven F, Doneus M, Draganits E (2018) Surfaces from the visual past: recovering high-resolution terrain data from historic aerial imagery for multitemporal landscape analysis. J Archaeol Method Theory 25:611–642
Srbik RV (1929) Bergbau in Tirol und Vorarlberg in Vergangenheit und Gegenwart. Berichte des naturwissenschaftlich-medizinischen Vereines Innsbruck vol 41, pp 118–279
Staffler JJ (1842) Tirol und Vorarlberg, topographisch, mit geschichtlichen Bemerkungen. Felician Rauch, Innsbruck
Steinbauer MJ, Grytnes JA, Jurasinski G, Kulonen A, Lenoir J, Pauli H et al (2018) Accelerated increase in plant species richness on mountain summits is linked to warming. Nature 556:231–234
Stolz O (1955) Deutsche Zolltarife des Mittelalters und der Neuzeit. Steiner, Wiesbaden
Land Tirol (2001) Verkehr in Tirol—Bericht 2000. Sachgebiet Verkehrsplanung. Amt der Tiroler Landesregierung, Verkehr und Strasse, Innsbruck. https://www.tirol.gv.at/fileadmin/themen/verkehr/service/publikationen/downloads/VB_2000_netz.pdf
Land Tirol (2018) Verkehr in Tirol—Bericht 2017. Sachgebiet Verkehrsplanung. Amt der Tiroler Landesregierung, Verkehr und Strasse, Innsbruck. https://www.tirol.gv.at/fileadmin/themen/verkehr/verkehrsplanung/downloads/verkehrsberichte/VB_2017_web.pdf
Töchterle KH (1991) Stubai. Tyrolia, Innsbruck-Wien
United Nations (2015) Transforming our world: The 2030 agenda for sustainable development, A/RES/70/1, sustainabledevelopment.un.org, https://sustainabledevelopment.un.org/index.php?page=view&type=400&nr=2125&menu=1515, downloaded 29.07.2019
Von Scheffer C, Lange A, De Vleeschouwer F, Schrautzer J, Unkel I (2019) 6200 years of human activities and environmental change in the northern central Alps. E&G Quaternary Sci J 68:13–28
Wakonigg H (1996) Unterkühlte Schutthalden. In: Beiträge zur Permafrostforschung in Österreich. Arb Aus d Inst f Geogr d Univ Graz 33:209–223
Walser C, Lambers K (2012) Human activity in the Silvretta massif and climatic developments throughout the holocene. Landscape archaeology. Berlin, 6. 8 Jun 2012. Jun 2012. In: Bebermaier W et al (ed) Landscape archaeology: proceedings of the international conference held in Berlin, 6th–8th June 2012. Berlin: Exzellenzcluster, pp 55–62
Whitehead M (2014) Environmental transformations. A Geography of the Anthropocene, Routledge, London
IPCC (2013) Climate change 2013: The physical science basis. In: Stocker TF et al (eds) Contribution of working group I to the fifth assessment report of the intergovernmental panel on climate change. Cambridge Univ Press, Cambridge-New York
Zemp M, Haeberli W, Hoelzle M, Paul F (2006) Alpine glaciers to disappear within decades? Geophys Res Lett 33:13504–13508

10 Environmental and Socio-Economic Consequences of Recent Mountain Glacier Fluctuations in Norway

Philipp Marr, Stefan Winkler, and Jörg Löffler

Abstract

Mountain glaciers currently experience significant mass losses and frontal retreat at the global scale. Because mountain glaciers generally respond sensitively to climate and are differently affected by climate variations at the regional scale, they may significantly and specifically impact their natural and human environment. Norway has the largest glacier mass in continental Europe and its glaciers are generally well-studied and monitored. Norway may, therefore, provide valuable insights into both causes and consequences of recent glacier fluctuations. In this chapter, the Holocene glacier history of Norway is presented with special focus on glacier fluctuations since the beginning of the twentieth century CE. In line with global patterns, the majority of Norwegian glaciers are facing overall mass losses which are predicted to accelerate in the future. Glacier retreat has an environmental impact by enhanced paraglacial activity, increased natural risk potential, and changes of glacier foreland ecosystems. The socio-economic consequences of mountain glacier changes in Norway are most relevant in the high-revenue glacier tourism and hydropower production industries. It appears that the natural and socio-economic systems in Norway are rather resilient to the anticipated changes and consequences of recent and future glacier fluctuations in comparison with other mountain regions worldwide.

Keywords

Climate change · Glacier retreat · Environmental change · Socio-economic impacts · Norway

10.1 Introduction—The Importance of Mountain Glaciers in a Global Context

Mountain glaciers are currently experiencing significant mass loss and frontal retreat almost without exception at the global scale (Zemp et al. 2017). This development has intensified since the beginning of the twenty-first century CE (Zemp et al. 2015) and is often interpreted as the consequence of the current climate change (IPCC 2014). There is consensus that mountain glaciers

P. Marr
Department of Geography and Regional Research, University of Vienna, Vienna, Austria

P. Marr (✉) · J. Löffler
Department of Geography, University of Bonn, Bonn, Germany
e-mail: marr@uni-bonn.de; philipp.marr@univie.ac.at

S. Winkler
Department of Geography and Geology, University of Würzburg, Würzburg, Germany

U. Schickhoff et al. (eds.), *Mountain Landscapes in Transition*, Sustainable Development Goals Series, https://doi.org/10.1007/978-3-030-70238-0_10

constitute key indicators of short- to long-term variability of climatic conditions (Zemp et al. 2008; Winkler et al. 2010; Beniston et al. 2018). The mass of mountain glaciers responds sensitively to fluctuations of individual meteorological parameters and underlying atmospheric circulation patterns. It reveals a dependency on summer air temperatures but also on factors such as winter precipitation, sea surface temperatures, and changed general airflow patterns in regionally specific modes and patterns (Mutz et al. 2016). Due to their limited sizes compared to polar ice sheets, mountain glaciers exhibit a much shorter time delay in their response to any changes in climatic conditions. Any impact of the mountain glaciers in response to the current climate change on the surrounding natural and human environment is, therefore, often instantaneous. In the wake of current discussions on causes and consequences of climate change, the importance of glaciers as part of the alpine cryosphere is mirrored by an increased attention of scientists, policymakers and various groups of stakeholders worldwide (e.g. Alean 2010; Barry and Gan 2011; Solomina et al. 2016; Beniston et al. 2018). This is mostly due to the anticipated significant consequences for sustainable development of human societies in affected mountain regions because of the current glacier shrinkage (Kovats et al. 2014).

10.2 Mountain Glaciers in Norway

Glaciers in Norway, which represent the largest glacier area and volume in continental Europe (Beniston et al. 2018), are comparably well monitored. Since the 1960s high-quality mass balance data are available complementing earlier annual length change records dating back to the early twentieth century CE (Andreassen et al. 2005, 2015, 2016; NVE 2019). According to the latest glacier inventory (Andreassen and Winsvold 2012), the glacier area in Norway totals 2692 ± 81 km^2, the equivalent of ~ 0.7% of the land area in mainland Norway. 1252 glaciers (1575 glacier units) comprising 57% of the glacier area are located in South Norway, 1282 glaciers (1568 glacier units) and the remaining 43% glacier area in North Norway. The frequent occurrence of larger ice caps and plateau glaciers with multiple outlets (e.g. Jostedalsbreen, Svartisen, Folgefonna) accounts for the differentiation between glaciers and glacier units. The significant increase in the number of individual glaciers compared to previous inventories (Østrem and Ziegler 1969; Østrem et al. 1973, 1988) is the result of utilising modern, remote sensing-based methodology (Andreassen and Winsvold 2012). By contrast, a comparison of glacier area and length change based on these inventories reveal a loss of c. 326 km^2 (equals 11%) between the penultimate and the most recent inventory (Winsvold et al. 2014; details see below).

Norway's glaciers have recently undergone an overall significant mass loss and are predicted to continuously experience substantial mass losses or even disappear by the end of the twenty-first century CE (Nesje et al. 2008a; IPCC 2014; Mutz et al. 2016). In this scenario, manifold impacts are to be expected, such as the disturbance of the geomorphological process systems by paraglacial activity (Winkler 2019), the modification of the near-glacier ecosystems (Matthews and Vater 2015; Hill et al. 2018), and the challenge of performing tourism and recreational activities of locally significant economic value (Furunes and Mykletun 2012). Additionally, future changes of Norwegian glaciers are crucial in the context of hydropower production which is responsible for 99% of the electric power in Norway and c. 50% of all glacier units are located in catchments regulated for hydropower production (Andreassen and Winsvold 2012).

10.3 Pre-Recent History of Mountain Glaciers in Norway

Glacier growth and decay have had severe impacts on environmental changes in Norway during the Quaternary. Whereas the ice-margins during the last glaciation are quite well constrained, other aspects as the past ice-thickness are still debated (Mangerud 2004; Marr and Löffler 2017; Marr et al. 2018, 2019b). However, our knowledge about pre-recent glaciation history steadily increased towards the present The glacial dynamics in the currently glaciated mountain regions of Norway were potentially still affected by the remnants of the Scandinavian ice-sheet for some time by the termination of the Younger Dryas at around 11.7 cal. ka BP. Individual glaciers that can be characterised as precursors of late Holocene and modern mountain glaciers exhibited first advances as early as during the early Holocene 'Preboreal Oscillation' at around 11.1 cal. ka BP (e.g. Nordre Folgefonna; Bakke et al. 2005a). Few other early Holocene glacier advances are reported from South Norway (Nesje et al. 2008a; Nesje 2009) as well as from North Norway (Bakke et al. 2005b; Jansen et al. 2016). These early Holocene glacier advances have demonstrably been linked to meltwater outburst into the North Atlantic. Thus, influencing the thermohaline circulation causing unstable climatic conditions (Prasad et al. 2006; Hoek and Bos 2007).

During the Holocene Thermal Maximum, almost all available glacier records from Norway lack evidence of glacial activity, and a widespread (in South Norway regionally complete) glacier disappearance is assumed (Nesje et al. 2008a). Those few glaciers possibly having survived this period must have been very small and restricted to continental and high-elevation locations, preferably in North Norway. Bakke et al. (2010) present evidence that glaciers may have existed in the Okstindan Mountains near the Arctic Circle during the entire Holocene. This is, however, likely a local exception because both at nearby Svartisen (Jansen et al. 2016) and in northernmost Norway (Lyngen-Peninsula—Bakke et al. 2005b; Langfjordjøkulen—Wittmeier et al. 2015) prolonged periods without signs of glacier activity characterise the Holocene Thermal Maximum. Among other things, the latter lead to the stabilisation of periglacial and related landforms in parts of south-western Norway (Marr et al. 2019a) and seems to have commenced at the latest immediately after the Greenlandic 8.2 ka BP-event (possibly earlier in some regions, Bakke et al. 2005a). Although some internal climatic variability has to be expected, it confirms the 'classic' concept of a mid-Holocene 'postglacial climatic optimum' subsequently followed by a re-formation of glaciers ('Neoglaciation'). Generally, the spatial extent of the glaciated area in Norway during the Holocene Thermal Maximum has been less than ever since.

First signs of re-formation of previously absent glaciers are detected at about 6.0 cal. ka BP (Nesje et al. 2000, 2008a) and can be classified as the onset of the regional 'Neoglaciation'. In many regions, however, this re-establishment of glaciers started later as a consequence of a cold period around 4.0 cal. ka BP which saw a general rise in glacier activity (Bakke et al. 2010; Wittmeier et al. 2015; Jansen et al. 2016). For instance, neoglacial activity seems to have generally started later in more maritime locations. During the late Holocene neoglacial glacier activity persistently increased in magnitude and a higher level of glacier fluctuations commenced around 2.0 cal. ka BP. It was, nevertheless, interrupted by periods of lower activity related to warmer climate conditions (Matthews and Dresser 2008) but finally reached its culmination during the 'Little Ice Age' (LIA, Nesje 2009). Unlike in the European Alps or other regions around the globe where multiple prominent advances within the LIA occurred (Winkler 2002; Grove 2004; Solomina et al. 2016), the LIA at Jostedalsbreen and most other Norwegian glacier regions constitutes a single prominent glacier advance. It was followed by an initially slow, albeit frequently interrupted retreat. The LIA in Norway has had a comparatively late onset at the end of the seventeenth century CE (Grove 2001). There is a fairly uniform image of the timing of the LIA-maximum at Jostedalsbreen between 1740 and c. 1760 (Bickerton and Matthews 1993; Winkler 2002). The timing of the LIA-maximum in other Norwegian mountain regions to the mid- to late eighteenth century CE (Jotunheimen; Matthews 2005; Breheimen; Winkler et al. 2003, Svartisen and Okstindan; Winkler 2003) is comparable, but in some cases, it occurred later (e.g. Folgefonna during the late 1870s; Nussbaumer et al. 2011). The LIA constituted the most extensive neoglacial advance in Norway and its potential climatic causes have been investigated in detail (Nesje and Dahl 2003; Nesje et al. 2008a, b; Imhof et al. 2011). This research yielded that increased winter precipitation must have played a major role. The retreat from the LIA-maximum frontal positions continued into the twentieth century CE. This means that not before the mid-twentieth century glaciers in Norway can be expected to have adjusted to concurrent climatic conditions following the significant LIA glacier advances.

10.4 Mountain Glacier Fluctuations in Norway—From the Twentieth Century to Present

10.4.1 The General Scene

Short advances in some parts of Norway were recorded during the first three decades of the twentieth century CE (Bogen et al. 1989; Andreassen et al. 2005). At Jostedalsbreen, the mass balance perturbations responsible for two advances culminating around 1910 and 1930 have not been strong and/or long enough to cause larger outlet glaciers with longer terminus response times to advance (Winkler 1996). Meteorological data reveals that the first advance was mainly triggered by above-average winter precipitation, the second advance by substantially below-average summer air temperatures. There were parallel glacier advances at Svartisen and Okstindan during the early twentieth century coinciding with periods of high winter precipitation. During the 1930s and 1940s, a widespread and commonly substantial glacier retreat linked to warm summers and overall above-average air temperatures commenced in almost entire Norway at this time (Nesje et al. 1995; Winkler 2002; Andreassen et al. 2005; Winsvold et al. 2014). During the late 1940s and early 1950s, the glacier fronts of the sensitively responding short outlets of Jostedalsbreen stabilised and the strong retreat terminated which was followed by more or less stationary frontal positions experiencing only minor oscillations. As a result of longer terminus response times (Nesje 1989; Winkler 1996) and additional local factors (e.g. calving over proglacial lakes), larger outlet glaciers of Jostedalsbreen concluded their strong retreat 20 or more years later. In other mountain glaciers of Norway, especially those in more maritime settings, the fast and substantial mass loss during the mid-twentieth century simultaneously slowed down (Andreassen et al. 2005), probably indicating final adjustment after termination of the LIA.

As an exception at the global scale (Zemp et al. 2008), a comparable short but strong advance in response to a series of consecutive positive mass budget years was recorded at Jostedalsbreen during the 1990s. Even if the unprecedented mass increase recorded at glaciers in western South Norway had to be adjusted to somewhat lower values after careful re-analysis of the long-term mass balance series (Andreassen et al. 2016), the mass balance perturbation was significant and the advance was the strongest in the region since the LIA-maximum advance (Winkler et al. 1997, 2009; 'Briksdalsbre event', Nesje and Matthews 2011). The climatic driver for this positive mass balance perturbation was predominately increasing winter precipitation connected to a positive NAO (North Atlantic Oscillation)-index indicating strong zonal circulation. A similar situation to what is assumed responsible for the main LIA-advance (Pohjola and Rogers 1997; Chinn et al. 2005; Nesje 2005; Steiner et al. 2008).

Since about 2000, most glaciers in Norway, regardless if they participated in the above-mentioned advance or not, experienced a significant mass deficit and related frontal retreat accelerating towards today (Andreassen et al. 2016; NVE 2019). Although single years showed positive mass balances for some, mostly maritime glaciers (NVE 2019), the current mass loss needs to be identified as fairly uniform trend and is linked to an acceleration of the mean global earth surface temperatures increase of $\sim$ 0.6 °C during the twentieth century CE and may further increase since the beginning the twenty-first century (Imhof et al. 2011; Zemp et al. 2015; Andreassen et al. 2016). Below we present some examples for the recent glacier fluctuations in Norway since the second half of the twentieth century. Apart from the availability of long-term glaciological data records, our selection considers the spatial variability by subdividing Norway's glaciers into three main regions (North, Central, South; Fig. 10.1) following Winsvold et al. (2014). Additionally, South Norway is subdivided into maritime and continental glacier subregions (Fig. 10.1, Table 10.1) to address a more detailed view as supported by previous studies (Winkler and Nesje 2009; Winkler et al. 2010).

10.4.2 North Norway (Regions North and Central)

Although mountain glaciers in Norway show similar trends with respect to their mass balance and length changes over the past few decades and in particular since c. 2000 CE, some differences occurred (Fig. 10.2). Apart from local factors like glacier size, glacier geometry, altitudinal distribution, or aspect individually affecting glaciers, regional patterns reflecting spatial diversity can be identified. The regional subdivision of glaciers in mainland Norway (see above) is helpful in this context. This subdivision does, perhaps surprisingly at first, not follow the traditional geographical division into North and South Norway but separates glaciers in North Norway around the Arctic Circle ('Central' region; e.g. Svartisen, Okstindan mountains) from those in the northernmost part of Norway ('North', e.g. Øksfjordjøkelen, Lyngen-Peninsula). On basis of both LIA glacier chronologies and recent glacier fluctuations, this is fully justified. In their recent comparison, Winsvold et al. (2014) highlighted that whereas the highest retreat rates since the mid-twentieth century are recorded within the northernmost parts of region 'North', lowest rates have been recorded in the 'Central' region (see Table 10.2).

In their study comprising the entire Lyngen-Peninsula, Stokes et al. (2018) conclude that a steady reduction in glacier area since the LIA until the end of the 1980s was subsequently paused until c. 2000 CE but followed by accelerated recession until today. The most recent recession is linked to a +0.5 °C increase in air temperature that exceeds the previous summer air temperature rise following the LIA-maximum extension (cf. Andreassen 2000). The above-mentioned pause within the overall glacier recession during the 1990s is explained by above-average winter precipitation (Stokes et al. 2018). The length record of Steindalsbreen (Fig. 10.2, note multi-year breaks in the 1980s and 1990s) does, however, not indicate any considerable mass balance surplus leading to a significant standstill or readvance as it occurred in parts of South Norway and the 'Central' region (cf. Hausberg and Andreassen 2009). By contrast, Langfjordjøkelen with the longest mass balance record in northernmost Norway shows consistently negative net balances since

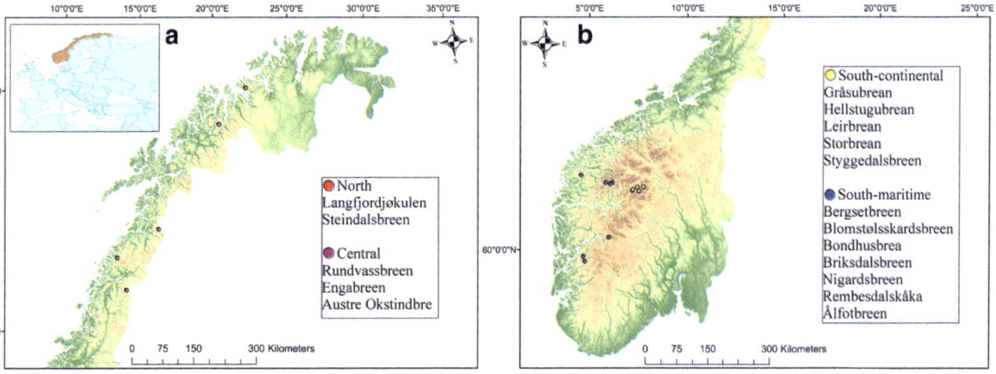

Fig. 10.1 Detailed maps of the location of the selected glaciers in Norway. **a** Glaciers of the northern and central subregions are shown. **b** Glaciers of the South maritime and South continental subregions are shown. (*Source* https://www.kartverket.no, last access: 02.07.2019)

Table 10.1 Glaciological key characteristics for glaciers displayed on Figs. 10.2 and 10.3 taken from the current glacier inventory (Andreassen and Winsvold 2012). Glaciers are grouped into regions as described in the text

Glacier	Area (km^2)	Elevation (m a. s.l.)	Aspect	Slope (degrees)	Morphology
(North)					
Langfjordjøkulen	3.46	313–1039	SE	13	Outlet (Langfjordjøkulen)
Steindalsbreen	5.14	474–1504	E	15	Valley Glacier
(Central)					
Austre Okstindbre	14.14	772–1784	N	9	Outlet (Okstindbreen)
Engabreen	36.02	14–1581	NW	6	Outlet (Svartisen ice cap/Vestisen)
Rundvassbreen	11.11	838–1419	N	5	Outlet (Blåmannsisen ice cap)
(South maritime)					
Bergsetbreen	11.15	854–1957	SE	11	Outlet (Jostedalsbreen ice cap)
Blomstølsskardsbreen	23.11	1033–1638	SW	5	Outlet (Søndre Folgefonna ice cap)
Bondhusbrea	10.91	533–1637	N	5	Outlet (Søndre Folgefonna ice cap)
Briksdalsbreen	11.73	349–1917	NW	9	Outlet (Jostedalsbreen ice cap)
Nigardsbreen	42.02	345–1946	SE	8	Outlet (Jostedalsbreen ice cap)
Rembesdalskåka	17.33	1038–1860	W	4	Outlet (Hardangerjøkulen ice cap)
Ålfotbreen	3.99	899–1384	NE	10	Outlet (Ålfotbreen)
(South continental)					
Gråsubrean	2.17	1860–2399	NW	10	Cirque glacier
Hellstugubrean	2.81	1494–2212	NE	13	Valley glacier
Leirbrean	4.76	1513–2089	NW	12	Outlet (Smørstabbrean)
Storbrean	5.22	1398–2079	NE	14	Valley glacier
Styggedalsbreen	2.02	1280–2253	N	23	Cirque Glacier

measurements commenced (NVE 2019) and a mass loss stronger than for any other glacier in Norway's observation network (cf. Andreassen et al. 2012).

Engabreen is the only glacier of the 'Central' region that matches the long-term annual mass balance records from South Norway. It shows a remarkable similar course in its cumulative mass balance to maritime South Norway (e.g. Nigardsbreen; Andreassen et al. 2016) and its length changes are equally comparable. Short-term measurements at other glaciers of Svartisen, Blåmannsisen (Rundvassbreen), and Okstindan (Kjøllmoen 2017a) indicate that Engabreen's behaviour is quite representative. The similarity of glacier fluctuations of the

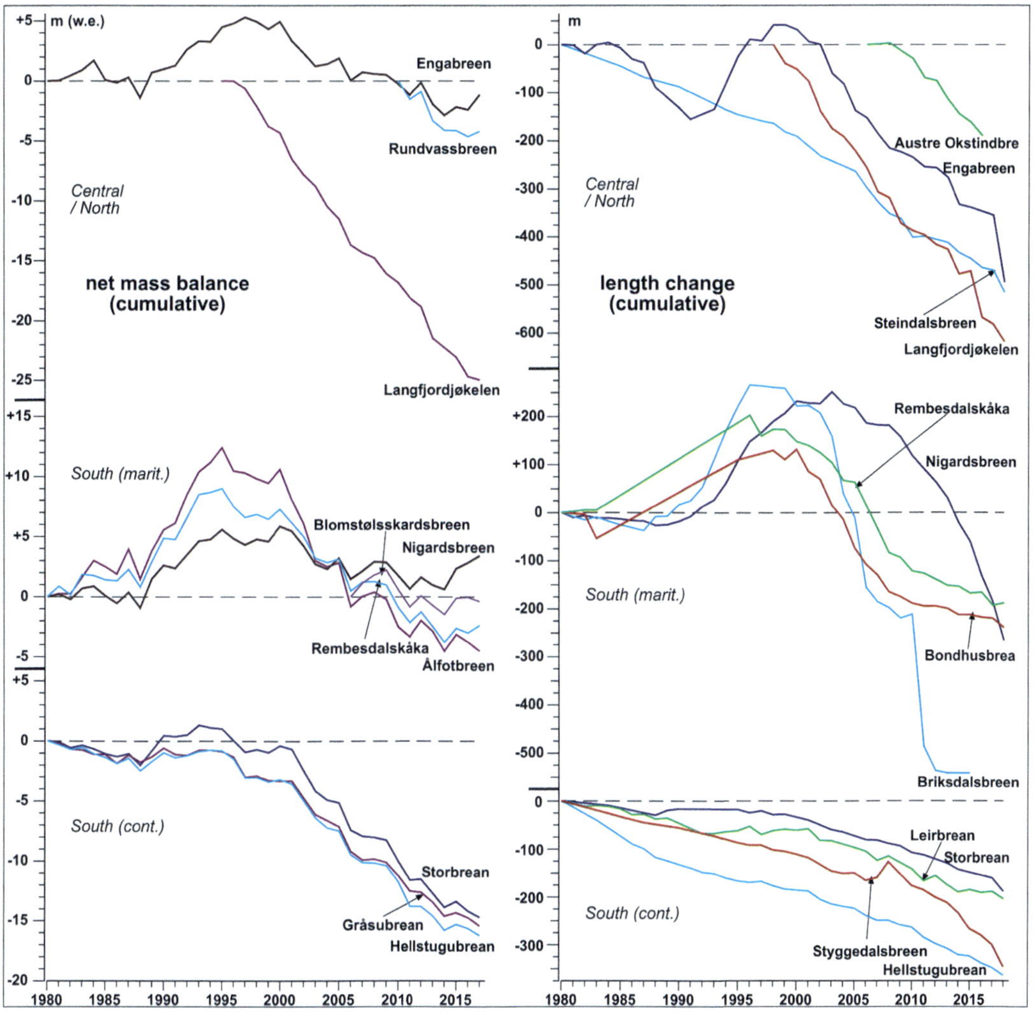

Fig. 10.2 Cumulative glacier net balance and length changes for selected glaciers in Norway since 1980 CE. The changes are clustered corresponding to four main glacier regions as outlined in the text. For both types of glaciological data displayed here a considerable bias needs to be noted because continuous

mostly coast-near glaciers of the 'Central' region and glaciers in maritime South Norway has previously been recognised regarding the timing of the LIA-maximum (Winkler 2003). This suggests that climatic drivers may have comparable influence upon glacier response despite considerable latitudinal distance (Winkler 2002). The latter has been demonstrated by Mutz et al. (2016) who predict very similar responses of Nigardsbreen and Engabreen for the 2000–2100 CE time period based on different emission scenarios.

10.4.3 South Norway (Regions South Maritime and Continental)

The good glaciological database for South Norway allows detailed analysis of both climatic triggers and related interactions of glacier variability as well as relevant factors influencing different glacier response. This justifies a separation of 'South maritime' and 'South continental' regions while Winsvold et al. (2014) integrate all glaciers in South Norway in one

Fig. 10.3 Visual comparison of three selected glacier tongues from outlets of Jostedalsbreen in maritime South Norway [Bergsetbreen: 22nd July 1999 **a**—4th August 2018 **b**; Briksdalsbreen: 5th September 1998 **c**—29th July 2018 **d**; Nigardsbreen: 28th August 2001 **e**—1st August 2018 **f**]. It displays morphological changes that occurred since the termination of the recent advance at the end of the twentieth century CE. It is obvious why due to almost collapsing lowermost glacier tongues the length change records at several of the shorter outlets had to be discontinued (e.g. at Bergsetbreen and Briksdalsbreen). All photos S.Winkler

region (see above). This differentiation allows highlighting differences between recent changes of maritime and continental glaciers. Over the entire period investigated (1947–2006) glaciers in maritime South Norway showed higher overall retreat, but glaciers in more continental regions like Jotunheimen retreated more or less consistently since the mid-twentieth century CE and did not participate in the substantial mass increase during the late 1980s and 1990s, typical for maritime glaciers (Nesje et al. 1995; Winkler et al. 1997; Andreassen et al. 2005, 2016; Paul et al. 2011; Fig. 10.2).

The reason for the above-mentioned mass gain of maritime glaciers in South Norway during the late twentieth century CE was increasing winter precipitation connected to a positive NAO-index regime indicating strong zonal circulation (see 10.3). This advance ceased, however, at about 2000 and the subsequent retreat

Table 10.2 Recent glacier area and length change for glacier regions in Norway for three different time periods based on different glacier inventories (data taken from Winsvold et al. 2014)

Region[a]	Δ GI 1/2[b] (km²)	Δ GI 1/2 (%)	Δ GI 1/2[c] (Ø km²/m)	Δ GI 2/3 (km²)	Δ GI 2/3 (%)	Δ GI 2/3 (Ø km²/m)	Δ GI 1/3 (km²)	**Δ GI 1/3 (%)**	**Δ GI 1/3** (Ø km²/m)
North									
Area change	−87.4	−19.5	−0.141	+17.7	+4.7	+0.023	**−76.4**	**−16.5**	**−0.116**
Length change			−254			−82			**−357**
Central									
Area change	−83.2	−10.8	−0.145	+56.9	+8.3	+0.092	**−31.8**	**−4.0**	**−0.048**
Length change			−221			−22			**−204**
South									
Area change	−18.8	−3.2	−0.038	−41.8	−7.1	−0.073	**−218.0**	**−12.6**	**−0.156**
Length change			−129			−68			**−221**
Norway (total)									
Area change	−189.4	−10.5	−0.112	+32.9	+2.0	+0.017	**−326.1**	**−10.9**	**−0.12**
Length change			−199			−55			**−241**

[a]Glacier regions as defined by Winsvold et al. (2014), see text
[b]GI 1 = Glacier inventory based on aerial photography 1947–1985; GI 2 = Glacier inventory based on remote sensing 1988–1997; GI 3 = Glacier inventory based on remote sensing 1999–2006. The spatially different dates of the underlying surveys explain apparent discrepancies (see Winsvold et al. 2014 for details)
[c]Average for glacier units included in the study of Winsvold et al. (2014) and calculated based on actual differences in years between GI at individual glaciers. This explains any apparent discrepancies

accelerated dramatically during the following few years. This partially caused a morphological collapse of small, steep glacier tongues and forced an abandonment of several length record series (NVE 2019; Fig. 10.3). High air temperatures during the ablation seasons since 1996 and especially since 2001 seem mainly responsible for the continuing retreat (Nesje 2005; Winkler et al. 2009). Additionally, a prolongation of the ablation season into autumn affecting the critical transitional period from rain to snow likely exhibits an important influence (Winkler and Nesje 2009) and winter snow accumulation has accordingly decreased at average since about 2000 (Nesje and Matthews 2011).

An interesting observation during the initial phase of the current strong retreat at maritime glaciers was an apparent lack of the delay expected based on previously experienced individual terminus response times in favour of immediate retreat by excessive ablation at the lower tongues (Winkler and Nesje 2009). High summer temperatures seem to have (temporarily?) disturbed the previous mode of dynamic response and lead to the above-mentioned virtual collapse of nearly stagnant lower glacier tongues. The sensitivity of maritime glaciers with high mass turnovers may, however, cause such situations as exemplified from climatically comparative maritime glaciers in New Zealand currently experiencing similar developments with comparable implications, for example, for glacier tourism (see 10.8; Purdie 2013; Purdie et al. 2015; Winkler 2015; Stewart et al. 2016). In maritime

South Norway, glaciers seem affected by a shift towards a more continental and ablation-season influenced glaciological regime, i.e. stronger impact of (high) summer air temperatures within the complex mass balance system (Winkler et al. 2009). This may explain why the length change records at outlets of Jostedalsbreen (NVE 2019) currently are well aligned with the global trend of accelerated glacier recession (Zemp et al. 2015) despite the fact that the mass balance record of Nigardsbreen shows an overall positive trend since 2006 (Fig. 10.2). The most steep and low-lying outlets do not (yet) benefit from the slight mass increase that has obviously mainly affected the high altitudinal parts of the Jostedalsbreen ice cap.

In Jotunheimen with its dominating mountain-type glaciers, the decrease in total glacier area between the 1960s (as reported in Østrem and Ziegler 1969) and 2003 was 12% (Andreassen et al. 2008) and comparable to the 9% area decrease between the 1960s and 2006 at Jostedalsbreen (Paul et al. 2011), the latter highly dependent on glacier size. This is considerably less than the 28% area decrease in Seiland and Øksfjord in northernmost Norway (Andreassen and Winsvold 2012), but more than at Svartisen where virtually no change was recorded for that particular time period (Paul and Andreassen 2009; cf. Figure 10.2). Although a comparable total area loss between the 1960s and 2003/6 was measured for Jotunheimen and Jostedalsbreen, corresponding data for the entire second half of the twentieth century shows a higher area loss in most of 'South maritime' Norway's subregions than in Jotunheimen (Winsvold et al. 2014). The above-mentioned mass increase during the late 1980s and 1990s seems not to have influenced the data, because single years of positive mass balance in Jotunheimen did not create a mass balance perturbation large enough for any advance. Furthermore, only some glacier in its west and west-central part of this region slowed down their retreat concurrently (Winkler et al. 2009). A large proportion of the documented overall mass loss in maritime South Norway occurred during the mid-twentieth century and the recent advance was not able to compensate for it, in particular if compared to the consistent but more moderate retreat in Jotunheimen. This confirms the suggested less sensitive response of the continental glaciers in South Norway located at higher elevations to climatic variability (Rasmussen 2004; Winkler et al. 2010). This is, furthermore, reflected by the calculated area loss of 27% for glaciers in Jotunheimen since the LIA-maximum until the 1980s, which is significantly less than, for example, in the European Alps or New Zealand (cf. Baumann and Winkler 2010). That continentality rather than latitude influences the sensitivity of Norwegian glaciers in their response to climate variability is meanwhile consensus (Engelhardt et al. 2015).

10.5 Predictions and Scenarios for Mountain Glaciers in Norway

All available predictions about the future development of Norwegian glaciers point towards a continuation of the current trend of accelerated mass loss. According to Nesje et al. (2008a), a predicted summer air temperature increase of 2.3 °C until 2070–2100 CE will result in an equilibrium line altitude rise of 260 $\pm$ 50 m and 34% reduction of the glacier area by 2100. Although also a 16% increase of winter precipitation is predicted, higher air temperatures might obviously shorten the accumulation season and likely offset any theoretical positive effect (cf. Winkler and Haakensen 1999). This is confirmed by Mutz et al. (2016) utilising a statistical modelling approach to obtain predictions of less winter snow accumulation by the end of the twenty-first century (combined with increased summer ablation). Laumann and Nesje (2014) employ a dynamic glacier model for Spørteggbreen (western Breheimen, maritime South Norway) and demonstrate that a small increase in winter precipitation will not compensate for increase in summer ablation. It has to be noted that while future simulations for more continental glaciers yield relatively uniform mass losses regardless of the various climate change scenarios (cf. IPCC 2014) applied (e.g. 30% volume loss of Storbreen by 2050; Andreassen et al. 2006), the results for maritime glaciers range considerably (Andreassen et al. 2006; Mutz et al. 2016). Depending on their size and elevational distribution individual glaciers will be affected differently, but 98% of all individual glaciers in Norway are at risk to disappear according to these scenarios, especially the smaller ones and those at low elevation (Nesje et al. 2008a). Nesje et al. (2008a) highlight that in comparison with the fast disappearance of glacier at the onset of the Holocene Thermal Maximum (see 10.3), these predictions seem realistic in the light that the reconstructed mid-Holocene summer temperature increase of 2–3 °C is comparable to what needs to be expected for the mid-twenty-first century.

10.6 Geomorphological Implications of Mountain Glacier Retreat in Norway

10.6.1 Paraglacial Processes and Glacier-Related Geomorphological Hazards in Norway

The expected ongoing trend of global glacier retreat is generally supposed to have wide-ranging consequences for the recently deglaciated glacier forelands and their surroundings. The transition of formerly glaciated to deglaciated terrain exposes landscapes to unstable conditions which are prone to rapid and extensive modification (Ballantyne 2002). These changes in slope systems are part of the so-called ‘paraglacial process system’ introduced by Church and Ryder (1972) describing geomorphological processes occurring on recently deglaciated terrain and conditioned by glacier retreat. In comparison with mountain regions like the European Alps, the Andes, and the Himalayas, only relatively few studies on recent paraglacial processes and glacier-related hazards in context of modern glacier retreat have been conducted in Norway. This is not surprising as this reflects the lower magnitude of paraglacial processes connected to the recent glacier retreat and glacier-related hazards potentially affecting settlements and vital infrastructure. A number of reasons can be brought forward for explanation, the glaciated mountain regions of Norway are all geologically ‘old’, the neotectonic activity is mostly negligible, and despite being overprinted by multiple glaciations many parts of Norway still resemble preglacial land surfaces of moderate relief (Klemsdal and Sjulsen 1988; Etzelmüller et al. 2007). Due to this specific gross morphology influencing the topography of many Holocene glacier forelands, there is a widespread lack of glacial landforms such as huge and steep lateral moraines typically vulnerable to instability and high erosion by gullying (Winkler 2009). In some regions, the glacier forelands are remarkably flat (e.g. Jostedalsbreen, Winkler 2019) and the slopes of many typical glacial valleys are shaped in weathering-resistant bedrock that is either exposed or has only a thin cover of till and other loose sediments. The small dimension of Holocene glaciers in those valleys is unlikely to have a considerable impact on the stability of these slopes, and all major mass movement events that occurred in Norway in historical times cannot be connected to any recent glacier retreat (see below).

Nevertheless, paraglacial processes and post-depositional erosion may inflict some minor to moderate modification on the recently deglaciated glacier forelands alongside geomorphological processes affecting surrounding slopes (Laute and Beylich 2013, 2014). Ballantyne (1995) studies debris cone formation on the foreland of Bergsetbreen in Jostedalen and found that due to exhaustion of entrainable sediment on the upper slopes their formation already had ceased within 100 to 200 years of deglaciation. An example of an exception in the same region is the inner foreland of Fåbergstølsbreen (Fig. 10.4). The sediment-covered northern slope has recently undergone significant gullying, a process still active today but only affecting the less frequently used access path to the glacier snout (Ballantyne and Benn 1994; Curry 1999, 2000; Curry and Ballantyne 1999). Glacial sediment is eroded from the upper slopes and deposited as debris cones at their base by debris flows and snow avalanche activity. Because the gullies usually stabilise if the (limited) sediment becomes exhausted, paraglacial reworking on the steep slope will cease relatively fast. In most other glacier forelands talus cones or alluvial fans as well as avalanche deposits do not substantially influence their general character and significantly alter the local geomorphological process system. The regionally specific characteristics of modern glacier forelands in Norway are contrasting in several aspects with typical glaciated valley systems sensu Benn et al. (2005) are responsible for low to barely moderate paraglacial activity and very limited impact on any infrastructure in place.

Mass movements have occurred in Norway throughout the Holocene following deglaciation and there is no evidence of recently increased rock-slope failure activity. A higher frequency of mass movements and rock-slope failures has, however, been postulated following local deglaciation and for few periods during the Holocene caused by climatic fluctuations or extreme local weather events (cf, Blikra et al. 1989, 2006; Bøe et al. 2004; Kalsnes et al. 2016; Hermanns et al. 2017; Matthews et al. 2018). In the context of the current glacier retreat in Norway, no such events have been reported so far, but the anticipated future warming for Norway (NCCS 2017) may lead to enhanced paraglacial-related rock-slope failures and mass movement activity (Mercier 2008). In comparison with predictions developed for other mountain regions at the global scale, the above-mentioned specific topographic conditions and geological history needs to be taken into account alongside with the glaciological fact that most glaciers in mainland

Fig. 10.4 Images of paraglacial slope modification at Fåbergstølsbreen, an outlet of Jostedalsbreen: **a** Oblique aerial view on the debris-mantled slope of the northern inner foreland affected by gully erosion with the outer foreland visible in the foreground (paraglacial processes at this location have been studied in detail by Ballantyne and Benn 1994; Curry 1999, 2000; Curry and Ballantyne 1999); **b** Slope close to the retreating modern glacier tongue in the innermost glacier foreland; **c** View from the innermost glacier foreland downvalley depicting the northern valley slope (left) affected by paraglacial modification contrasting with the southern slope of exposed bedrock with little paraglacial activity. All photos: S. Winkler

Norway are warm-based. Only few mostly small glaciers are known to be polythermal and located within a permafrost environment. An example is high-elevation glaciers in Jotunheimen (e.g. Gråsubrean) characterised by ice-cored moraines (Østrem et al. 1988; Andreassen and Winsvold 2012; Matthews et al. 2014; Winkler et al. 2019). They will not be effected by glacier retreat but potentially melt of their ice-core as part of permafrost degradation has to be expected. Due to their morphology and location, this process is not considered to pose any future hazards. Although rock-slope failures, rock-avalanches and landslides are generally considered to be important natural hazards in Norway (Blikra et al. 2006; Kalsnes et al. 2016), the current glacier retreat actually is, by contrast to permafrost degradation and long-term slope adjustment following deglaciation and glacio-isostatic rebound, not (yet) considered an important factor. No indications of such geomorphological processes have been observed and Holocene glacier forelands are usually neither inhabited nor do they contain important infrastructure except few installations for hydropower production and tourism.

10.6.2 Glacier Lake Outburst Floods and Potential Glacier-Related Hazards in Norway

The development of ice-marginal lakes or proglacial lakes is often observed consequences of glacier retreat (Quincey et al. 2007; Barry and Gan 2011; Benn et al. 2012). They can act as dams having hazardous potential due to outburst floods or glacier ice falling into the lake which can create hazardous displacement waves (Evans and Clague 1994; Hambrey and Alean 2004; Benn and Evans 2010). Norway is no exception and some glacier lake outburst floods (GLOFs) related to subglacially or englacially stored meltwater have been observed in historic and more recent times (Jackson and Regulina 2014). The recent inventory of glacier-related hazards in Norway (Jackson and Regulina 2014) lists a total of 69 glaciers (76 glacier units) as potentially dangerous. This number includes, however, glaciers where mountaineering accidents or minor ice falls at glacier termini occurred. Only 12 glaciers may cause GLOFs in the near future and additional 8 glaciers with future potential for GLOFs have been identified. Given the high overall number of glaciers in Norway and its comparatively large glaciated area this number seems low and GLOFs as well as other glacier-related hazards seem to pose low risks. Glacier retreat since the LIA-maximum has substantially reduced the potential for GLOFs, ice avalanches, and other glacier induced hazards. Despite the low risk, there are areas in Norway where GLOFs occurred. One of the most active GLOF areas in Norway during the last 20 years is located at Rundvassbreen in central Norway. There, the first recorded GLOF occurred in September 2001 when a glacier dammed lake subglacially drained into the Siso hydropower plant reservoir, positively influencing its productivity. The water level of the drained lake dropped about 50 m during 35 h. This GLOF is linked to glacier retreat and prolonged thinning of the glacier until it lost its ability to dam the water (Engeset et al. 2005). Following the 2001 event, a number of additional events occurred (Jackson and Regulina 2014; Kjøllmoen 2017b). The most dangerous GLOFs were reported from Rembesdalskåka. Since the eighteenth century CE late summer GLOFs, sometimes with catastrophic impacts, were described which lead to the construction of an artificial lake drainage, later used for hydropower production (see also Kjøllmoen 2018 for a more recent GLOF). Likewise, these events are related to extensive thinning of the glacier (Jackson and Regulina 2014). But even if few other examples, as Flatbreen (Breien et al. 2008) show that GLOFs may well occur locally in Norway, it is not considered as major general threat. In the wake of current glacier retreat and their related geometric changes, however, similar events at other glaciers cannot be excluded and the current risk assessment may have to be updated. The predicted retreat of glaciers in Norway will potentially increase the number of proglacial lakes as overdeepened basins in glacial valleys become subsequently deglaciated and filled up with meltwater (Bogen et al. 2015). But because these lakes are not dammed by potentially unstable natural dams like ice-cored moraines but are constraint to existing (bedrock) basins, GLOFs have usually not considered important and widespread future hazards. As a result of the well-maintained monitoring programme in Norway and the overall low number of glacier-related hazards, the current glacier retreat exhibits a less significant contribution to future natural hazards in mountain regions compared to the global scale.

For the Norwegian Mountains, an observed increase of runoff in winter and spring and the projected earlier snowmelt will overall result in relatively small changes of total annual runoff (NCCS 2017; see 10.8). Glacial meltwater will increase towards the mid-twenty-first century CE, but subsequently meltwater-induced floods are expected to decrease over time. Given that many glaciated areas of Norway will experience increased overall precipitation and especially extreme precipitation events are predicted to occur at shorter intervals in the near future, situations where locally high glacier meltwater runoff and extreme precipitation events occur simultaneously and cause severe flooding (as the 1979 flooding event in Jostedalen, Faugli 1987) may become more frequent in the near future before the glacier area has significantly been reduced by 2100. Sediment availability will increase due to exposure of firstly non-vegetated terrain in the course of glacier retreat and paraglacial activity (see 10.6.1). Consequently, more sediment will potentially be transported into the rivers which could fill up drainage pipes and culverts (Kalsnes et al. 2016), damage hydropower plant infrastructure and increase sediment infilling rates of reservoirs. This effect will, however, be buffered in those cases where existing and future proglacial lakes act as sediment traps (Bogen et al. 2015). In the long run, increasing vegetation cover compensates this development by stabilising slopes and further reduce sediment yield (Stoffel and Huggel 2012).

An increased occurrence of mass movements such as rock-slope failures and landslides is also expected, though there is a high uncertainty within these projections (Kalsnes et al. 2016; NCCS 2017). Landslides are more likely to occur following an intensive short-term rainfall event or prolonged wet period up to 15 days (cf. Kalsnes et al. 2016). During the past few years, some very local extreme rainfall events have caused significant damage to the infrastructure (e.g. 2017 in Utvik/Nordfjord in western Norway), but most catchments have not been glaciated, meaning that increased glacier melt has not contributed to these events. Spatial and temporal variabilities are expected in the future, with more intense landslide periods during spring due to snowmelt and rain in south-east Norway, and during autumn due to intense rainfall events in west Norway (Kalsnes et al. 2016).

10.7 The Impact of Mountain Glacier Retreat on the Ecosystems in Norway

Retreating mountain glaciers successively expose new terrain in glacier forelands which may, therefore, be utilised as an experimental field laboratory for investigating geo-ecological succession of microorganisms, plants and anthropods, simply because no biological legacy exists (Fægri 1934; Matthews 1992; Walker and del Moral 2003; Matthews and Vater 2015; Hill et al. 2018; Fig. 10.5). The colonisation is governed by multiple factors which are subject to changes in relation to the progressed terrain age (Matthews 1992). Ecological changes in glacier forefields can be explored and linked to increasing terrain age since deglaciation and increasing distance to the modern glacier terminus. This pattern allows establishing chronosequences where the spatial properties of one point in time represent change through time (cf. Matthews and Vater 2015).

The colonisation of redwood ants is thought to be of key importance in glacier forefield ecosystems as their nest mounds impact local nutrient cycles and can offer habitats for myrmecophiles (cf. Hill et al. 2018). Additionally, as dominating predators ants can influence

Fig. 10.5 Images of recently deglaciated terrain at the margins of Storjuvbrean/Jotunheimen (c. 1380 m a.s.l.) and Kjenndalsbreen/Jostedalsbreen (c. 200 m a.s.l.). In both cases the terrain is located inside a terminal moraine system formed around 2000 CE (see Winkler and Matthews 2010 for details): **a** Recent glacier front of Storjuvbrean seen from inside the 2000 moraine system (19.07.2018); **b** *Oxyria digyna* (mountain sorrel) on the terrain shown on **a**. *Poa alpinae-Oxyria digyna* communities are considered the regional pioneer vegetation on glacier forelands in southern Norway (see Robbins and Matthews 2009); **c** Eastern inner slope of Kjenndalen seen from the 2000 CE glacier front position. The vegetation trimline resulting from that recent advance is obvious (30.07.2018); **d** Newly established vegetation inside the 2000 CE moraine system at Kjenndalsbreen. Both willowherb (*Epilobium* spec.) and elder scrubs (*Alnus* spec.) colonised relatively short time after glacier retreat, partly supported by the low elevation of the glacier forelands. All Photos: S. Winkler

the insect community, disperse plant seeds, enrich the soil with nutrients around their nests and can positively and negatively affect tree growth (cf. Hill et al. 2018). In Jostedalsbreen forelands Hill et al. (2018) found that first colonisation of the redwood occurs 50–80 years following deglaciation. A key component of nest establishment is the sufficient presence of biological resources (*Betula pubescens*) (Robbins and Matthews 2010), for example the nest mound height is related to the number of trees 5 m around it (Hill et al. 2018).

As a result of their study on the primary succession of invertebrates on different glacier forelands at Jostedalsbreen and in Jotunheimen, Vater and Matthews (2015) showed that the addition and persistence model of succession fits their findings best (see also Vater and Matthews 2013). It is mostly driven by individualistic behaviour of mobile species with low dependence on the succession of vegetation and explains the difference to models of primary plant succession on the forelands. Elevation of the glacier forelands is the most important factor other than local ones and in general forelands in the higher alpine zone experience a 2-stage succession with stronger persistence of pioneer taxa into mature stages of succession than lower glacier forelands in the subalpine and boreal zones; the latter showing a 3-step succession and lower proportions of the pioneer elements at mature stages (Vater and Matthews 2015). Anyway a high number of pioneer taxa has been observed to colonise deglaciated terrain within 20 years (see also Vater 2012 for details) and as regionally influential and important factor elevation is a strongly climate-determined one, ongoing climate change may primary trigger changes altitudinal rises of existing boundaries of vegetation zones and potentially resulting modes of succession. This prediction can well be transferred to the colonisation of vegetation as, for example, Matthews (1979a, b) already highlighted that the complexity and course of vegetation succession on the Storbrean glacier foreland is closely linked to the factor elevation. A similar difference regarding a 2- versus a 3-stage succession depending on elevation proposed by Vater and Matthews (2015) for invertebrates had already been outlined by Matthews and Whittaker (1987) for vegetation succession. Robbins and Matthews (2014) detect that differences in the change of ecological factors partly depend on the different altitudinal zones of their 39 investigated glaciers forelands at Jostedalsbreen and in Jotunheimen. They share, however, a common decrease of mean ecological indicator values for light, reaction, and nitrogen. But, in both the lower subalpine and boreal zones, the decrease of indicator values for pH and productivity to levels found on mature sites is completed after c. 70 years. It takes somewhat longer at higher elevations. Interestingly, the ecological indicator values for moisture did not show any indicative patterns during succession which points towards moisture being predominately locally controlled and independent of the chronosequence for the deglaciated terrain. Except for the high-alpine zone the ecological indicator values for light significantly decreased during subsequent succession stages due to woodland-canopy becoming established below c. 1000 m a.s.l. within 70 years and dwarf-scrub and snowbed vegetation at intermediate elevations between 1000 and 1600 m a.s.l. within 250 years (Robbins and Matthews 2010). The herbaceous pioneer vegetation currently being able to become persistent in the high-alpine zone above 1600 m a.s.l. may become under stress at their lower boundary as climatic conditions promoting vegetation communities of later successional stages are expected to rise in elevation according to future climate scenarios. Today's dominating patterns of vegetation succession on glacier forelands may become affected accordingly and upper boundaries or transitional zones of individual species as well as communities increase in elevation.

10.8 The Impact of Mountain Glacier Retreat on the Socio-Economy in Norway

Recent changes in the cryosphere become a vital determinant for any glacier-related aspects of the tourism industry in Norway. Norwegian glaciers are of particular interest for tourists since more than 100 years (Aall and Høyer 2005) and constitute an integrated part of the Norwegian mountain landscape, thus also contribute to its overall attractiveness. Glacier tourism first peaked between the end of the eighteenth century CE up to the 1930s and ceased completely during World War II. Subsequently, glacier tourism remained low until the 1960s which coincided with shrinking glacier area and length in almost all parts of Norway (Aall and Høyer 2005). Because typically convex-shaped glacier tongues characteristic for advancing or stationary glaciers appear visually more attractive, this development may not purely be accidental. During the last decades of the twentieth century, glacier tourisms subsequently experienced a gradual increase simultaneously with the recent glacier mass increase and readvance in some regions, for example in Jostedalen in maritime South Norway (see 10.3). Regional factors have, however, to be considered for explaining local boosts in glacier-related tourism. In the wake of establishing Jostedalsbreen National park in 1991, three national park centres were established around Jostedalsbreen (Norsk Bremuseet/Fjærland in 1991, Breheimsenteret/Jostedalen and Jostedalsbreen Nasjonalparksenteret/Oppstryn in 1993). The exhibitions focus on local glaciers and the centres partly promote and offer booking services for glacier activities (e.g. glacier guiding; Fig. 10.6).

Whereas glaciers retreated and decreased in size from about 2000, the number of visitors did at first neither follow this trend as during the mid-twentieth century (Aall and Høyer 2005) nor have the authors observed any general decrease of tourist numbers at glacier-related tourist hotspots. This is interesting because the current glacier retreat and partly disintegration of lower glacier tongues successively limited their accessibility for organising safe glacier walks and increased the risks involved with any activity on or near the glacier. As a result, glacier guiding subsequently stopped at a number of glaciers (e.g. Briksdalsbreen) after it initially has been started shortly before or around 2000, especially around Jostedalsbreen. Upvalley retreat of glacier tongues and related morphological changes create limitations for maintaining access infrastructure, especially for inexperienced day-trip tourists, and further complicates the supply of the special glacier hike equipment (Furunes and Mykletun 2012). This caused a concentration of touristic glacier activities at few glaciers where safe glacier tours seemed still possible, for example at Nigardsbreen (Furunes and Mykletun 2012). But even at these glaciers the ongoing retreat and morphological change of glacier tongues pose a threat for continuation of glacier guiding suitable for inexperienced visitors (i.e. the majority of tourists; Fig. 10.6).

Glacier melting is considered as the main factor driving the decrease of tourists participating and operators in glacier tourism activities by 30% from 2003 to 2009 (Furunes and Mykletun 2012). This demands a diversification of activities to other nature-based and adventurous

Fig. 10.6 Glacier guiding at Nigardsbreen back in August 2008. Access to the starting point of the guided tours from the parking lot (far distance) was by boat and a short walk allowing multiple departures per day of tours of different duration and difficulty. The most popular tours ('family tours') lasted two to three hours in total and were suitable for untrained tourists and children over the age of 6 years. All safety equipment was provided and boots could be hired if required. In 2019, only longer and more difficult tours for tourist with some experience will be offered, mainly because access to the glacier has become difficult due to its considerable frontal retreat (Løset 2019). Because it will be the first year without the popular 'family tours' at Nigardsbreen, one may only speculate about the local economic impact, especially because alternatives of easy glacier-guided tours do not exist

activities and could be a measure for glacier tour operators to cope with the changes of glaciers in Norway (Furunes and Mykletun 2012). However, Saarinen and Tervo (2006) and Furunes and Mykletun (2012) report that nature-based tour operators are somewhat sceptical about climate change in general and its effect on their business which is reflected in limited adaptation measures.

A comparison with glacier tourism at the West Coast of New Zealand reveals some interesting opportunities. Those glaciers experienced a similar development of advance/growth during the late twentieth century CE followed by a current strong retreat/disintegration of glacier tongues (Chinn et al. 2005; Winkler 2015). Local communities and tour operators are equally relying on easy glacier access and related offers for tourist activities. Recent studies on the impact of climate/glacier change on tourism have revealed that viewing the glaciers was a significant travel motive to an extent that some tourists claimed to use their 'last chance' (Stewart et al. 2016). The public has undisputed become much more aware of glaciers as indicators for current climate change during the past two decades (Zängl and Hamberger 2004; Lozán et al. 2015). In Norway, the opening of the Ulltveit-Moen senter for klimaviten (at the Norsk Bremuseet) in 2007 and Klimaparken 2469 at Juvflye in Jotunheimen in 2012 demonstrates that tourist-focused educational offers have taken up an increased awareness of tourists on that topic. A 'climate change tourism' could evolve as tourist become aware of disappearing glaciers (Bauer 2011; Furunes and Mykletun 2012). A problematic but inevitable consequence would, however, be an increase of pressure on the few glaciers remaining accessible for tourists (Hay and Elliot 2008). Some tourists may avoid these localities due to overcrowding (Hall 2006), but it would hardly affect the increasing number of organised groups (cruise ship day trips or multi-day coach tours) visiting such localities. Despite the majority of tourists experience the glacier environment without taking part in physical activities on the glacier itself (Furunes and Mykletun 2012), the rather high-priced touristic glacier activities remain an important resource of income for the local tourism industry. Experience from New Zealand demonstrates that local tour operators may have a high-adaptive capacity under rapidly changing environmental conditions (e.g. compensating deteriorating accessibility by helicopter-hiking (Purdie 2013; Stewart et al. 2016). Even if some strategies cannot be transferred to Norway due to conservational and legislative differences regarding vehicle/operations within National Parks, significant recent touristic investments like the Skylift cable car in Loen that opened in 2017 providing a unique view of the Jostedalsbreen ice cap can be characterised as adaption to locally deteriorating glacier access and visibility. Anticipated changes in touristic glaciers-related activity offers will likely create opportunities for new, non-glacier-related activities that may potentially compensate for any job losses within mountain glacier tourism in Norway. The key strategy for a sustainable adaption is, however, based on reliable predictions of the demand of future glacier tourists and the flexibility of future markets for glacier tourism.

Changes to glacier-related tourism are perhaps the most obvious impact of recent glacier wastage in Norway. But the future role of glaciers within the hydrological system of glaciated mountains is another important topic for determining any socio-economic impact. Predicting future streamflow including glaciated catchments serves not only mitigating potential flood risk for settlements and critical infrastructure in Norway's mountainous regions but is also essential for the important hydropower production utilising significant proportions of glacier meltwater by reservoirs and power stations in glaciated catchments (Fig. 10.7). A number of studies on the projection of future streamflow in Norway have been conducted that all benefit from a detailed and reliable database (cf. Fleig 2013). Regional simulations of future flood levels in Norway indicate that a predicted increased in frequency and magnitude of extreme rain events together with a general increase of precipitation in many parts of Norway will cause higher flood levels. Particularly, this is predicted for western South Norway and maritime parts of North Norway (Lawrence 2016). The coincidence of (earlier and/or more sudden) snowmelt with extreme rain events and an enhanced effect within regions with precipitation maxima in autumn and winter is highlighted as a potential future flood risk. The expected increase in glacier runoff during the mid-twenty-first century CE caused by future retreat and downwastage is not explicitly mentioned as a factor increasing future flood risk. The seasonality of flood regimes is projected to change from dominating spring/summer flood regimes in mountain regions (in particular eastern and northern ones) to a autumn/winter flood regime alongside an increasing importance of rainfall as flood generating process subsequently replacing snowmelt (Vormoor et al. 2015, 2016).

The comparison of modelled changes in streamflow for a future climate (2071–2100 CE) with the reference period (1961–1990 CE) conducted between representatively selected glaciated and nearby non-glaciated catchments (Lappegard et al. 2006) showed, however, considerable uncertainties related to different climate scenarios utilised as basis for the projections. These climate scenarios result in different dominating circulations patterns and due to the specific topography of Norway, a west- versus east-dominance of circulation/airflow will exhibit significant impact and complicates any detailed predictions. The overall projected changes in mean annual streamflow showed to be moderate but seasonal changes may be large with implications for hydropower production. As long as glaciers remain present in today's glaciated catchments, the summer streamflow will increase by 15–70% solely due to enhanced glacier melt

Fig. 10.7 Top **a** Styggevatn reservoir in upper Jostedalen in western Norway. Although an exception with a glacier Austdalsbreen (still) calving into the reservoir, many reservoirs in Norway have inflow from glaciated catchments at variable percentages. **b** Pelton wheel and water nozzles of the Jostedalen hydropower plant build deep into the bedrock of the valley side. The water utilised for electricity production from different sources (among those Styggevatnet) is transported to the power plant in artificial tunnels to minimise the environmental impact. **Bottom a** Dam of the Styggevatnet reservoir in upper Jostedalen in western Norway. **b** Central (maintenance) hall of the Jostedalen hydropower plant build deep into the bedrock on the valley side

whereas non-glaciated catchments may experience a 20–60% reduction. Once the glacier will have disappeared, the summer streamflow in northern and eastern Norway will generally decrease between 30 and 75% (Lappegard et al. 2006). Only in western Norway the streamflow will increase compared to the reference period due to expected increased precipitation. In those parts of Norway where today glaciated catchments buffer dry summers (e.g. central South Norway east of the main watershed). Summer droughts may be more severe once glaciers will mostly have disappeared towards the second half of the twenty-first century. An earlier spring flood (earlier snowmelt) and higher evaporation may locally cause problems for agricultural businesses and adaption strategies (e.g. irrigation, change of cultivation methods and crops) need to be considered.

By contrast, the Norwegian energy market and hydropower production will most likely generally benefit from future climate change and resulting streamflow changes. As Beisland et al. (2015) point out, the predicted general increase in precipitation will allow a higher hydropower production and expected changes in the ratio of snow vs. rain in favour of the latter will level inflow-depending reservoir filling and improve availability of water for energy production. The expected higher streamflow in winter (mild weather periods causing snowmelt and rain events are expected to cause occasional winter floods) and an earlier spring flood agrees better with the seasonality of energy consumption than more recent conditions (Lappegard et al. 2006; Golombek et al. 2012). Different regional developments need, however, to be observed. The inflow curves for reservoirs in maritime South and North Norway will be shallower than in the rest of Norway and the increase of inflow is expected to peak in the mid-twenty-first century rather than at this end, the latter the above-mentioned effect of surplus glacier runoff during their predicted peak mass loss (Beisland et al. 2015; Fig. 10.8).

Once the glacier area has been significantly reduced, the inflow in North Norway, currently dominated by the Svartisen ice cap will be slightly reduced. In western South Norway, the same trend is expected around Jostedalsbreen but compensated by generally increased precipitation. But the outlook for the Norwegian hydropower market remains overall positive (Beisland et al. 2015) with an expected increase in power production by 8% thanks to increasing inflow and high reservoir capacity (Golombek et al. 2012). Net export of hydropower-generated power from all Nordic countries is predicted to rise and where other countries in South and Central Europe may experience negative impacts of climate change for their hydropower production (Golombek et al. 2012). Hydropower production in Norway is an example for regional socio-economic benefit from the current climate change. Challenges at regional and local scale are, however, to be expected and as a result, awareness is promoted (Steen 2016) and hydropower is included with any considerations about strategies of adapting to climate change (NOU 2010).

10.9 Conclusion

Mountain glaciers in Norway exhibit no exception from the current global trend of significant mass loss and frontal retreat generally related to climate variability. This trend is expected to continue and accelerate during the twenty-first century. Individual glaciers will be affected differently depending on their specific characteristics, but 98% of all individual glaciers in Norway are at risk to disappear. Both, current and predicted glacier shrinkage have important implications on different aspects of the natural and societal environment in Norway's mountain regions and beyond. The anticipated future increase of mass movement activity and natural risk of rock-slope failures in Norway's mountains are primarily linked to permafrost degradation, higher frequency of extreme climatic events, and the predicted climate change in general. Given the high number of individual glaciers in Norway, the potential for glacier

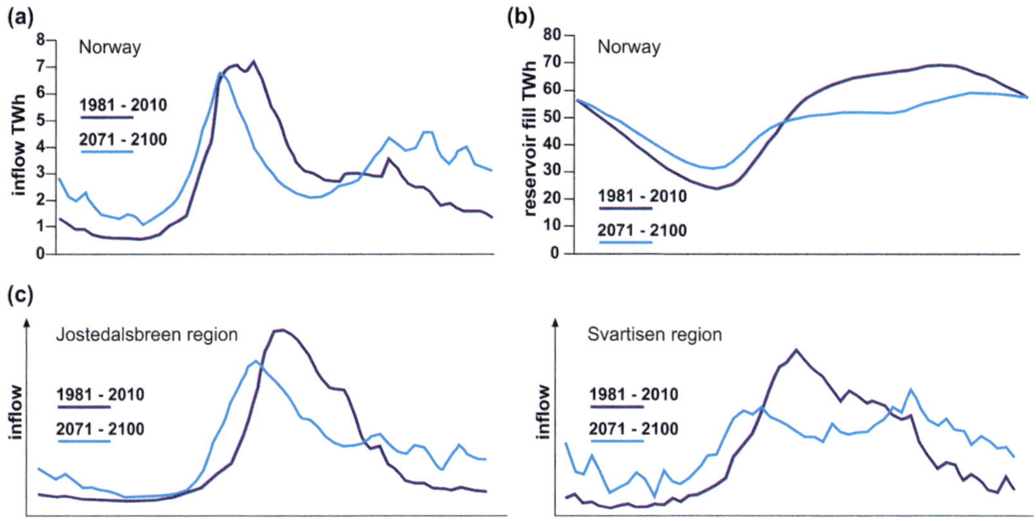

Fig. 10.8 **a** Average yearly inflow into Norwegian reservoirs calculated/predicted for the 30-year-periods 1981–2010 and 2071–2100; **b** Average weekly infill of Norwegian reservoirs calculated/predicted for the 30-year-periods 1981–2010 and 2071–2100; **c** Average inflow profile for glaciated catchments (Nigardsbrevatet–Jostedalsbreen region; Berget–Svartisen region) depicting the proportional changes calculated/predicted for the 30-year-periods 1981–2010 and 2071–2100 (all figures modified after Beisland et al. 2015)

outburst floods and other glacier-related hazards does not pose major threats. It appears that the anticipated changes related to recent glacier fluctuations in Norway will have limited consequences on the natural and socio-economic system. In sharp contrast to the Andes and the Himalayas, the current glacier retreat in Norway will only generate moderate geomorphological hazards and ecological modifications. The anticipated impact on touristic glacier-related activity may create opportunities for new, non-glacier-related activities that may compensate for any job losses within the glacier tourism industry in Norway. The runoff from glacier melt is expected to increase during the mid-twenty-first century before many mountain catchments may eventually become glacier-free towards 2100. Most predictions see only minor changes of the annual runoff, and some parts of Norway will experience higher precipitation, especially in autumn and winter. This increased winter runoff and the projected earlier snowmelt will likely overall result in a more levelled annual runoff. Higher inflow to the hydropower reservoirs during periods of high demand for electricity (winter), meaning that hydropower production in Norway may benefit from projected climate changes in form of increased productivity. Despite all these significant environmental and socio-economic consequences of recent mountain glaciers fluctuations in Norway, it appears that the Norwegian mountain system is by comparison more resilient to possible future climate variability than other high-mountain systems elsewhere.

References

Aall C, Høyer KG (2005) Tourism and climate change adaptation—the Norwegian case. In: Hall CM, Higham J (eds) (2005) Tourism, recreation and climate change. Channelview Press, London, pp 209–223

Alean J (2010) Gletscher der alpen. Haupt, Bern

Andreassen LM (2000) Regional change of glaciers in northern Norway. Norwegian water resources and energy directorate (NVE), Oslo, Rapport 2000/1

Andreassen LM, Winsvold SH (2012) Inventory of norwegian glaciers. Norwegian water resources and energy directorate (NVE), Oslo, Rapport 38–2012

Andreassen LM, Elvehøy H, Kjøllmoen B, Engeset RV, Haakensen N (2005) Glacier mass balance and length variations in Norway. Ann Glaciol 42:317–325

Andreassen LM, Elvehøy H, Jóhannesson T, Oerlemans J, Beldring S, Van den Broeke, MR (2006) Modelling the climate sensitivity of Storbreen and Engabreen, Norway. Norwegian water resources and energy directorate (NVE), Oslo, Rapport 2006/03

Andreassen LM, Paul F, Kääb A, Hausberg JE (2008) Landsat-derived glacier inventory for Jotunheimen, Norway, and deduced glacier changes since the 1930s. Cryosphere 2:131–145

Andreassen LM, Kjøllmoen B, Rasmussen A, Melvold K, Nordli Ø (2012) Langfjordjøkelen, a rapidly shrinking glacier in northern Norway. J Glaciol 58:581–593

Andreassen LM, Huss M, Melvold K, Elvehøy H, Winsvold SH (2015) Ice thickness measurements and volume estimates for glaciers in Norway. J Glaciol 61:763–775

Andreassen LM, Elvehøy H, Kjøllmoen B, Engeset RV (2016) Reanalysis of long-term series of glaciological and geodetic mass balance for 10 Norwegian glaciers. Cryosphere 10:535–552

Bakke J, Dahl SO, Nesje A (2005) Lateglacial and early Holocene palaeoclimatic reconstruction based on glacier fluctuations and equilibrium-line altitudes at northern Folgefonna, Hardanger, western Norway. J Quaternary Sci 20:179–198

Bakke J, Dahl SO, Paasche Ø, Løvlie R, Nesje A (2005) Glacier fluctuations, equilibrium-line altitudes and palaeoclimate in Lyngen, northern Norway, during the lateglacial and Holocene. Holocene 15:518–540

Bakke J, Dahl SO, Paasche Ø, Simonsen JR, Kvisvik B, Bakke K, Nesje A (2010) A complete record of Holocene glacier activity at Austre Okstindbreen, northern Norway: an integrated approach. Quaternary Sci Rev 29:1246–1262

Ballantyne CK (1995) Paraglacial debris-cone formation on recently deglaciated terrain, western Norway. Holocene 5:25–33

Ballantyne CK (2002) Paraglacial geomorphology. Quaternary Sci Rev 21:1935–2017

Ballantyne CK, Benn DI (1994) Paraglacial slope adjustment and resedimentation following glacier retreat, Fåbergstølsdalen, Norway. Arctic Alpine Res 26:255–269

Barry RG, Gan TY (2011) The global cryosphere: past, present, future. University Press, Cambridge

Bauer C (2011) Climate change and alpine summer tourism—chances and strategies in vent and obergurgl. Paper presented at Managing Alpine Future II, Innsbruck

Baumann S, Winkler S (2010) Parameterization of glacier inventory data from Jotunheimen/Norway in comparison to the European Alps and the Southern Alps of New Zealand. Erdkunde 64:155–177

Beisland CS, Birkelund H, Endresen H, Haddeland I, Vik MA (2015) Et væravhengig kraftsystem—og et klima i endring. Norwegian water resources and energy directorate (NVE), Oslo, Rapport 85–2015

Beniston M, Farinotti D, Stoffel M, Andreassen LM, Coppola E, Eckert N, Fantini A, Giacona F, Hauck C, Huss M, Huwald H, Lehning M, López-Moreno JI, Magnusson J, Marty C, Morán-Tejéda E, Morin S, Naaim M, Provenzale A, Rabatel A, Six D, Stötter J, Strasser U, Terzago S, Vincent C (2018) The European mountain cryosphere: a review of its current state, trends, and future challenges. Cryosphere 12:759–794

Benn DI, Evans DJA (2010) Glaciers and glaciation, 2nd edn. Hodder, London

Benn DI, Kirkbride MP, Owen LA, Brazier V (2005) Glaciated valley systems. In: Evans DJA (ed) (2005) Glacial landsystems. Hodder Arnold, London, pp 372–406

Benn DI, Bolch T, Hands K, Gulley J, Luckman A, Nicholson LI, Quincey D, Thompson S, Toumi R, Wiseman S (2012) Response of debris-covered glaciers in the Mount Everest region to recent warming, and implications for outburst flood hazards. Earth-Sci Rev 114:156–174

Bickerton RW, Matthews JA (1993) Little ice age' variations of outlet glaciers from the Jostedalsbreen ice-cap, southern Norway: a regional lichenometric-dating study of ice-marginal moraine sequences and their climatic significance. J Quaternary Sci 8:45–66

Blikra LH, Hole PA, Rye N (1989) Skred i Norge—Hurtige massebevegelser og avsetningstyper i alpine områder, Indre Nordfjord. Trondheim: Norges Geologiske Undersøkelse, Skrifter 92

Blikra LH, Longva O, Braathen A, Anda E, Dehls JF, Stalsberg K (2006) Rock slope failures in Norwegian fjord areas: examples, spatial distribution and temporal pattern. In: Evans SG, Mugnozza GS, Strom A, Hermanns RL (eds) (2006) Landslides from massive rock slope failure. Springer, Dordrecht, pp 475–496

Bøe R, Longva O, Lepland A, Blikra LH, Sønstegaard E, Haflidason H, Bryn P, Lien R (2004) Postglacial mass movements and their causes in fjords and lakes in western Norway. Norw J Geol 84:35–55

Bogen J, Wold B, Østrem G (1989) Historic glacier variations in Scandinavia. In: Oerlemans J (ed) (1989) Glacier fluctuations and climatic change. Reidel, Dordrecht, pp 109–128

Bogen J, Xu M, Kennie P (2015) The impact of proglacial lakes on downstream sediment delivery in Norway. Earth Surf Proc Land 40:942–952

Breien H, De Blasio F, Elvehøi A, Høeg K (2008) Erosion and morphology of a debris flow caused by a glacial lake outburst flood, western Norway. Landslides 5:271–280

IPCC (2014) Climate change 2013: the physical science basis. Contribution of working group I to the 5th Assessment report of the Intergovernmental Panel on Climate Change. Cambridge

Chinn TJH, Winkler S, Salinger MJ, Haakensen N (2005) Recent glacier advances in Norway and New Zealand—a comparison for their glaciological and meteorological causes. Geogr Ann A 87:141–157

Church M, Ryder JM (1972) Paraglacial sedimentation: a consideration of fluvial processes conditioned by glaciation. Geol Soc Am Bull 83:3059–3071
Curry AM (1999) Paraglacial modification of slope form. Earth Surf Proc Land 24:1213–1228
Curry AM (2000) Observations on the distribution of paraglacial reworking of glacigenic drift in western Norway. Norsk Geogr Tiddskr 54:139–147
Curry AM, Ballantyne CK (1999) Paraglacial modification of glacigenic sediment. Geogr Ann A 81:409–419
Engelhardt M, Schuler TV, Andreassen LM (2015) Sensitivities of glacier mass balance and runoff to climate perturbations in Norway. Ann Glaciol 56:71–88
Engeset RV, Schuler TV, Jackson M (2005) Analysis of the first jökulhlaup at Blåmannsisen, northern Norway, and implications for future events. Ann Glaciol 42:35–41
Etzelmüller B, Romstad B, Fjellanger J (2007) Automatic regional classification of topography in Norway. Norw J Geol 87:167–180
Evans SG, Clague JJ (1994) Recent climatic change and catastrophic geomorphic processes in mountain environments. Geomorphology 10:107–128
Fægri K (1934) Über die längenänderungen einiger gletscher des Jostedalsbre und die dadurch bedingten Pflanzensukzessionen. Bergen: Bergens Museums Årbok 1933—Naturvidenskapelig rekke No. 7
Faugli PE (1987) FoU i Jostedøla. Norwegian water resources and energy directorate (NVE), Oslo, Publikasjoner V 6
Fleig AK (2013) Norwegian hydrological reference dataset for climate change studies. Norwegian water resources and energy directorate (NVE), Oslo, Rapport 02–2013
Furunes T, Mykletun RJ (2012) Frozen adventure at risk? A 7-year follow-up study of Norwegian glacier tourism. Scand J Hosp Tour 12:324–348
Golombek R, Kittelsen SAC, Haddeland I (2012) Climate change: impacts on electricity markets in Western Europe. Clim Change 113:357–370
Grove JM (2001) The initiation of the "little ice age" in regions round the North Atlantic. Climatic Change 46:53–82
Grove JM (2004) The little ice age. Routledge, London
Hall CM (2006) The geography of tourism and recreation: environment. Place and Space, Routledge, London
Hambrey MJ, Alean J (2004) Glaciers, 2nd edn. University Press, Cambridge
Hausberg JE, Andreassen LM (2009) Satellitbasert brekartlegging I Lyngen. Norwegian water resources and energy directorate (NVE), Oslo, Rapport 2009/7
Hay J, Elliot T (2008) New Zealand's glaciers: key national and global assets for science and society. In: Orlove B, Wiegandt E, Luckman B (eds) (2008) Darkening peaks: glacier retreat, science, and society. University of California Press, Berkeley, pp 185–195
Hermanns RL, Schleier M, Böhme M, Blikra LH, Gosse J, Ivy-Ochs S, Hilger P (2017) Rock-avalanche activity in W and S Norway peaks after the retreat of the Scandinavian ice sheet. In: Mikoš M, Vilimek V, Yin Y (eds) (2017) Advancing culture of living with landslides (WLF 2017). Springer, Cham, pp 331–338
Hill JL, Vater AE, Geary AP, Matthews JA (2018) Chronosequences of ant-nest mounds from glacier forelands of Jostedalsbreen, southern Norway: insights into the distribution, succession and geo-ecology of red wood ants (Formica lugubris and F. aquilonia). Holocene 28:1113–1130
Hoek WZ, Bos JAA (2007) Early Holocene climate oscillations—causes and consequences. Quaternary Sci Rev 26:1901–1906
Imhof P, Nesje A, Nussbaumer SU (2011) Climate and glacier fluctuations at Jostedalsbreen and Folgefonna, southwestern Norway and in the western Alps from the 'Little ice age' until the present: The influence of the North Atlantic Oscillation. Holocene 22:235–247
Jackson M, Regulina G (2014) Inventory of glacier-related hazardous events in Norway. Norwegian water resources and energy directorate (NVE), Oslo, Rapport 83–2014
Jansen HL, Riis Simonsen J, Dahl SO, Bakke J, Ringkjøb Nielsen P (2016) Holocene glacier and climate fluctuations of the maritime ice cap Høgtuvbreen, northern Norway. Holocene 26:736–755
Kalsnes B, Nadim F, Hermanns RL, Hygen HO, Petkovic G, Dolva BK, Berg H, Høgvold DO (2016) Landslide risk management in Norway. In: Ho K, Lacasse S, Picarelli L (eds) (2016) Slope safety preparedness for impact of climate change. CRC Press, Boca Raton, pp 215–251
Kjøllmoen B (2017a) Homogenisering av korte massebalanseserier I Norge. Norwegian water resources and energy directorate (NVE), Oslo, Rapport 33/2017
Kjøllmoen B (2017b) Glaciological investigations in Norway 2016. Norwegian water resources and energy directorate (NVE), Oslo, Rapport 76/2017
Kjøllmoen B (2018) Glaciological investigations in Norway 2017. Norwegian water resources and energy directorate (NVE), Oslo, Rapport 82/2018
Klemsdal T, Sjulsen E (1988) The Norwegian macrolandforms: definition, distribution and system of evolution. Norsk Geogr Tidsskr 42:133–147
Kovats RS, Valentini R, Bouwer LM, Georgopoulou E, Jacob D, Martin E, Rounsevell M, Soussana J-F (2014) Europe. In: Climate Change 2014: Impacts, Adaptation, and Vulnerability. Part B: Regional Aspects. Contribution of Working Group II to the Fifth Assessment Report of the Intergovernmental Panel on Climate Change. Cambridge University Press, Cambridge, pp 1267–1326.
Lappegard G, Beldring S, Roald LA, Engen-Skaugen T, Førland EJ (2006) Projection of future streamflow in glaciated and non-glaciated catchments in Norway. Norwegian Water Resources and Energy Directorate (NVE), Oslo, Oppdragsrapport A pp 9–2006
Laumann T, Nesje A (2014) Spørteggbreen, western Norway, in the past, present and future: Simulations with a two-dimensional dynamical glacier model. Holocene 24:842–852

Laute K, Beylich AA (2013) Holocene hillslope development in glacially formed valley systems in Nordfjord, western Norway. Geomorphology 188:12–30
Laute K, Beylich AA (2014) Morphometric and meteorological controls on recent snow avalanche distribution and activity at hillslopes in steep mountain valleys in western Norway. Geomorphology 218:16–34
Lawrence D (2016) Klimaendring og framtidige flommer i Norge. Norwegian Water Resources and Energy Directorate (NVE), Oslo, Rapport pp 81–2016
Løset O (2019) Turistar blir straffa for rekordsommar i fjor. https://www.nrk.no/sognogfjordane/turistar-blir-straffa-for-rekordsommar-i-fjor-1.14591188. Accessed 20 June 2019
Lozán JL, Grassl H, Kasang D, Nolz D, Escher-Vetter H (2015) Warnsignal klima: das eis der erde. Universität Hamburg (Wissenschaftliche Auswertungen)
Mangerud J (2004) Ice sheets limits in Norway and on the Norwegian continental shelf. In: Ehlers J, Gibbard PL (eds) (2004) Quaternary glaciations extent and chronology. Elsevier, Amsterdam, pp 271–294
Marr P, Löffler J (2017) Establishing a multi-proxy approach to alpine blockfield evolution in south-central Norway. AUC Geogr 52:219–236
Marr P, Winkler S, Löffler J (2018) Investigations on blockfields and related landforms at Blåhø (Southern Norway) using Schmidt-hammer exposure-age dating: palaeoclimatic and morphodynamic implications. Geogr Ann A 100:285–306
Marr P, Winkler S, Löffler J (2019) Schmidt-hammer exposure-age dating (SHD) performed on periglacial and related landforms in Opplendskedalen, Geirangerfjellet, Norway: Implications for mid- and late-Holocene climate variability. Holocene 29:97–109
Marr P, Winkler S, Binnie SA, Löffler J (2019b) ^{10}Be-based exploration of the timing of deglaciation in two selected areas of southern Norway. E&G Quaternary Sci J 69:1–12. (accepted)
Matthews JA (1979) The vegetation of the Storbreen Gletschervorfeld, Jotunheimen, Norway. I. Introduction and approaches involving classification. J Biogeogr 6:17–47
Matthews JA (1979) The vegetation of the Storbreen Gletschervorfeld, Jotunheimen, Norway. II. Approaches involving ordination and general conclusions. J Biogeogr 6:133–167
Matthews JA (1992) The ecology of recently-deglaciated terrain: a geoecological approach to glacier forelands. Cambridge University Press, Cambridge
Matthews JA (2005) 'Little ice age' glacier variations in Jotunheimen, southern Norway: a study in regionally controlled lichenometric dating of recessional moraines with implications for climate and lichen growth rates. Holocene 15:1–19
Matthews JA, Dresser P-Q (2008) Holocene glacier variation chronology of the Smørstabbtindan massif, Jotunheimen, southern Norway, and the recognition of century- to millennial-scale European Neoglacial events. Holocene 18:181–201
Matthews JA, Vater AE (2015) Pioneer zone geo-ecological change: observations from a chronosequence on the Storbreen glacier foreland, Jotunheimen, southern Norway. CATENA 135:219–230
Matthews JA, Whittaker RJ (1987) Vegetation succession on the Storbreen glacier foreland, Jotunheimen, Norway: a review. Arctic Alpine Res 19:385–395
Matthews JA, Winkler S, Wilson P (2014) Age and origin of ice-cored moraines in Jotunheimen and Breheimen, Southern Norway: Insights from Schmidt-hammer exposure-age dating. Geogr Ann A 96:531–548
Matthews JA, Winkler S, Wilson P, Tomkins MD, Dortch JM, Mourne RW, Hill JL, Owen G, Vater AE (2018) Small rock-slope failures conditioned by Holocene permafrost degradation: a new approach and conceptual model based on Schmidt-hammer exposure-age dating in Jotunheimen, southern Norway. Boreas 47:1144–1169
Mercier D (2008) Paraglacial and paraperiglacial landsystems: concepts, temporal scales and spatial distribution. Géomorphologie 4:223–234
Mutz S, Paeth H, Winkler S (2016) Modelling of future mass balance changes of Norwegian glaciers by application of a dynamical-statistical model. Clim Dynam 46:1581–1597
Nesje A (1989) Glacier-front variations at the outlet glaciers from Jostedalsbreen and climate in the Jostedalsbre region of western Norway in the period 1901–1980. Norsk Geogr Tidsskr 43:3–17
Nesje A (2005) Briksdalsbreen in western Norway: AD 1900–2004 frontal fluctuations as a combined effect of variations in winter precipitation and summer temperature. Holocene 15:1245–1252
Nesje A (2009) Late pleistocene and Holocene alpine glacier fluctuation in Scandinavia. Quaternary Sci Rev 28:2119–2136
Nesje A, Dahl SO (2003) The 'Little ice age'—only temperature? Holocene 13:139–145
Nesje A, Matthews JA (2011) The Briksdalsbre event: a winter precipitation-induced decadal-scale glacial advance in southern Norway in the AD 1990s and its implications. Holocene 22:249–261
Nesje A, Johannesen T, Birks HJB (1995) Briksdalsbreen, western Norway: climatic effects on the terminal response of a temperate glacier between AD 1901 and 1994. Holocene 5:343–347
Nesje A, Dahl SO, Andersson C, Matthews JA (2000) The lacustrine sequence in Sygneskardvatnet, western Norway: a continuous, high-resolution record of the Jostedalsbreen ice cap during the Holocene. Quaternary Sci Rev 19:1047–1065
Nesje A, Bakke J, Dahl SO, Lie Ø, Matthews JA (2008a) Norwegian mountain glaciers in the past, present and future. Global Planet Change 60:10–27
Nesje A, Dahl SO, Thun T, Nordli Ø (2008b) The "Little ice age" glacial expansion in western Scandinavia: summer temperature or winter precipitation? Clim Dynam 30:789–801
NOU (2010) Tilpassing til eit klima i endring. Oslo, Noregs Offentlege Utgreiingar 2010:10

Nussbaumer SU, Nesje A, Zumbühl HJ (2011) Historical glacier fluctuations of Jostedalsbreen and folgefonna (southern Norway) reassessed by new pictorial and written evidence. Holocene 21:455–471

NVE (2019) Updated glacier data base—https://www.nve.no/hydrologi/bre/bredata/ (last accessed 30.01.2019)

NCCS (2017) Climate in Norway 2100—a knowledge base for climate adaptations, NCCS report no. 1/2017, Norwegian Centre for Climate Services, Oslo

Østrem G, Ziegler T (1969) Atlas over breer i Sør-Norge. Meddelelse nr 20 fra Hydrologisk avdeling, NVE

Østrem G, Haakensen N, Melander O (1973) Atlas over breer i Nord-Skandinavia. Oslo: NVE, Meddelelser fra Hydrologisk Avdeling p 22

Østrem G, Dale Selvig K, Tandberg K (1988) Atlas over breer i Sør-Norge, Oslo: NVE, Meddelelser fra Hydrologisk Avdeling p 61

Paul F, Andreassen LM (2009) A new glacier inventory for the Svartisen region, Norway, from Landsat ETM+ data: challenges and change assessment. J Glaciol 55:607–618

Paul F, Andreassen LM, Winsvold SH (2011) A new glacial inventory for the Jostedalsbreen region, Norway, from landsat TM scenes of 2006 and changes since 1966. Ann Glaciol 52:153–162

Pohjola VA, Rogers JC (1997) Atmospheric circulation and variations in Scandinavian glacier mass balance. Quaternary Res 47:29–36

Prasad S, Brauer A, Rein B, Negendank JFW (2006) Rapid climate change during the early Holocene in western Europe and Greenland. Holocene 16:153–158

Purdie H (2013) Glacier retreat and tourism: insights from New Zealand. Mt Res Dev 33:463–472

Purdie H, Gomez C, Espiner S (2015) Glacier recession and the changing rockfall hazard: Implications for glacier tourism. New Zeal Geogr 71:189–202

Quincey D, Richardson SD, Luckman A, Lucas RM, Reynolds JM, Hambrey MJ, Glasser NF (2007) Early recognition of glacial lake hazards in the Himalaya using remote sensing datasets. Global Planet Change 56:137–152

Rasmussen LA (2004) Altitude variation of glacier mass balance in Scandinavia. Geophys Res Lett 31:L13401

Robbins JA, Matthews JA (2009) Pioneer vegetation on glacier forelands in southern Norway: emerging communities? J Veg Sci 20:889–902

Robbins JA, Matthews JA (2010) Regional variation in successional trajectories and rates of vegetation change on glacier forelands in South-Central Norway. Arct Antarct Alp Res 42:351–361

Robbins JA, Matthews JA (2014) Use of ecological indicator values to investigate successional change in boreal to high-alpine glacier-foreland chronosequences, southern Norway. Holocene 24:1453–1464

Saarinen J, Tervo K (2006) Perceptions and adaptation strategies of the tourism industry to climate change: the case of finnish nature-based tourism entrepreneurs. Int J Innov Sustain Dev 1:214–228

Solomina ON, Bradley RS, Hodgson DA, Ivy-Ochs S, Jomelli V, Mackintosh AN, Nesje A, Owen LA, Wanner H, Wiles GC, Young NE (2016) Glacier fluctuations during the past 2000 years. Quaternary Sci Rev 149:61–90

Steen R (2016) Klimatilpasning i energiforsyningen 2009–2016—Hvor står vi nå? Norwegian water resources and energy directorate (NVE), Oslo, Rapport pp 76–216

Steiner D, Pauling A, Nussbaumer SU, Nesje A, Luterbacher J, Wanner H, Zumbühl HJ (2008) Sensitivity of European glaciers to precipitation and temperature —two case studies. Climatic Change 90:413–441

Stewart EJ, Wilson J, Espiner S, Purdie H, Lemieux C, Dawson J (2016) Implications of climate change for glacier tourism. Tourism Geogr 18:377–398

Stoffel M, Huggel C (2012) Effects of climate change on mass movements in mountain environments. Prog Phys Geog 36:421–439

Stokes CR, Andreassen LM, Champion MR, Corner GD (2018) Widespread and accelerating glacier retreat on the Lyngen peninsula, northern Norway, since their 'little ice age' maximum. J Glaciol 64:100–118

Vater AE (2012) Insect and arachnid colonization on the Storbreen glacier foreland, Jotunheimen, Norway: persistence of taxa suggests an alternative model of succession. Holocene 22:1123–1133

Vater AE, Matthews JA (2013) Testing the 'addition and persistence model' of invertebrate succession in a subalpine glacier-foreland chronosequence: fåbergstølsbreen, southern Norway. Holocene 23:1151–1162

Vater AE, Matthews JA (2015) Succession of pitfall-trapped insects and arachnids on eight Norwegian glacier forelands along an altitudinal gradient: patterns and models. Holocene 25:108–129

Vormoor K, Lawrence D, Heistermann M, Bronstert A (2015) Climate change impacts on the seasonality and generation processes of floods—projections and uncertainties for catchments with mixed snowmelt/rainfall regimes. Hydrol Earth Syst Sc 19:913–931

Vormoor K, Lawrence D, Schlichting L, Wilson D, Wong WK (2016) Evidence for changes in the magnitude and frequency of observed rainfall versus snowmelt driven floods in Norway. J Hydrol 538:33–48

Walker LR, del Moral R (2003) Primary succession and ecosystem rehabilitation. Cambridge University Press, Cambridge

Winkler S (1996) Front variations of outlet glaciers from Jostedalsbreen, western Norway, during the twentieth century. Norg Geol Unders B 431:33–47

Winkler S (2002) Von der 'kleinen eiszeit' zum 'globalen gletscherrückzug'—eignen sich gletscher als klimazeugen? Akademie der Wissenschaften und der Literatur, Abhandlungen der Mathematisch-naturwissenschaftlichen Klasse, p 3

Winkler S (2003) A new interpretation of the date of the 'little ice age' maximum at Svartisen and Okstindan, northern Norway. Holocene 13:83–95

Winkler S (2009) Gletscher und ihre landschaften. Wissenschaftliche Buchgesellschaft/Primus, Darmstadt

Winkler S (2015) Die gegenwärtige situation der gletscher auf neuseeland. In: Lozán JL, Grassl H, Kasang D, Nolz D, Escher-Vetter H (eds) (2015) Warnsignal klima: das eis der erde. Universität Hamburg (Wissenschaftliche Auswertungen), pp 123–129

Winkler S (2019) Terminal moraine formation processes and geomorphology of glacier forelands at selected outlet glaciers of Jostedalsbreen, South Norway. In: Beylich AA (ed) (2019) Landscapes and landforms of norway. Springer, Dordrecht

Winkler S, Haakensen N (1999) Kritische überprüfung der möglichkeit zur prognose des gletscherverhaltens auf grundlage von modellierungen—dargestellt anhand von regionalen beispielen aus Norwegen. Petermann Geogr Mitt 143:291–304

Winkler S, Matthews JA (2010) Observations on terminal moraine-ridge formation during recent advances of southern Norwegian glaciers. Geomorphology 116:87–106

Winkler S, Nesje A (2009) Perturbation of climatic response at maritime glaciers? Erdkunde 63:229–244

Winkler S, Haakensen N, Nesje A, Rye N (1997) Glaziale dynamik in westnorwegen—ablauf und ursachen des aktuellen gletschervorstoßes am jostedalsbreen. Petermann Geogr Mitt 141:43–63

Winkler S, Matthews JA, Shakesby RA, Dresser PQ (2003) Glacier variations in Breheimen, southern Norway: dating little ice age moraine sequences at seven low-altitude glaciers. J Quaternary Sci 18:395–413

Winkler S, Elvehøy H, Nesje A (2009) Glacier fluctuations of Jostedalsbreen, western Norway, during the past 20 years: the sensitive response of maritime mountain glaciers. Holocene 19:389–408

Winkler S, Chinn T, Gärtner-Roer I, Nussbaumer SU, Zemp M, Zumbühl HJ (2010) An introduction to mountain glaciers as climate indicators with spatial and temporal diversity. Erdkunde 64:97–118

Winkler S, Donner A, Suntrup gen. Tintrup A (2019) Periglacial landforms in Jotunheimen, central southern Norway, and their altitudinal distribution. In: Beylich AA (ed): Landscapes and landforms of norway. Springer, Dordrecht

Winsvold SH, Andreassen LM, Kienholz C (2014) Glacier area and length changes from repeat inventories. Cryosphere 8:1885–1903

Wittmeier HE, Bakke J, Vasskog K, Trachsel M (2015) Reconstructing holocene glacial activity at landfjordjøkulen, arctic norway, using multi-proxy fingerprinting of distal glacier-fed lake sediments. Quaternary Sci Rev 114:78–99

Zängl W, Hamberger S (2004) Gletscher im treibhaus: eine fotographische zeitreise in die alpine eiswelt. Steinfurt, Tecklenborg

Zemp M, Roer I, Kääb A, Hoelzle M, Paul F, Haeberli W (2008) Global glaciers changes: facts and figures. WGMS/UNEP (Zürich)

Zemp M, Frey H, Gärtner-Roer I, Nussbaumer SU, Hoelzle M, Paul F, Haeberli W, Denzinger F, Ahlstrøm AP, Anderson B, Bajracharya S, Baroni C, Braun LN, Cáceres BE, Casassa G, Cobos G, Dávila LR, Delgado Granados H, Demuth MN, Espizua L, Fischer A, Fujita K, Gadek B, Ghazanfar A, Hagen JO, Holmlund P, Karimi N, Li Z, Pelto M, Pitte P, Popovnin VV, Portocarrero CA, Prinz R, Sangewar CV, Severskiy I, Sigurðsson O, Soruco A, Usubaliev R, Vincent C (2015) Historically unprecedented global glacier decline in the early 21st century. J Glaciol 61:745–762

Zemp M, Nussbaumer SU, Gärtner-Roer I, Huber J, Machguth H, Paul F, Hoelzle M (2017) Global glacier change bulletin no. 2 (2014–2015). ICSU(WDS)/IUGG (IACS)/UNEP/UNESCO/WMO, WGMS (Zürich)

11 Paraglacial Timescale and Sediment Fluxes for Hillslope Land Systems in the Northern Appalachian Mountains of Eastern Canada

Daniel Germain and Ludwig Stabile-Caillé

Abstract

The Appalachian Mountains of Eastern Canada are prone to several mass-wasting processes related to the geology and the nearby presence of large water bodies that influence the climate. Superimposed on this rugged terrain is the impacts of ongoing climate change, which may increase the magnitude, frequency, and duration of an array of hillslope phenomena. In this regard, the quantification of sediment fluxes at various spatiotemporal scales is prerequisite to reducing the exposure of infrastructure and communities, as well as to better understanding the mountain landscape evolution. Here, we report the quantitative modeling of sediment fluxes of several hillslope processes, mainly based on radiocarbon dating, which in turn improves understanding of how sediment has been eroded and transported through these mountain catchments since deglaciation. The results show a variable pattern of paraglacial effects at local and regional scales, highlighting the importance of ecological and hydro-climatic conditions in controlling the duration of glacially conditioned sedimentary stock exhaustion, and therefore the delay of paraglacial responses by geomorphic land systems. Current active scree slopes under the cold-temperate climate are characterized by sedimentation rates slightly lower than those calculated for the periglacial period following deglaciation, and even the sporadic remobilization of the primary stock by alluvial fan dynamics appears to be significant, testifying to a duration of paraglacial processes of more than 10,000 years.

Keywords

Paraglacial · Sediment flux · Hillslope processes · Gaspé Peninsula

11.1 Introduction

Climate changes at various spatiotemporal scales are expected to impact the biosphere, hydrosphere, and cryosphere of the Earth system, and, as a result, the general sediment availability (Knight and Harrison 2013; Lane et al. 2017). These changes are likely to influence patterns of erosion, transport, and deposition related to a given geosystem—the dynamic combination of biotic, abiotic, and anthropogenic factors—across defined landscape components. Quantitative estimates or analyses of denudation rates, source-to-sink fluxes, and sedimentary budgets

D. Germain (✉) · L. Stabile-Caillé
Department of Geography, Université du Québec À Montréal, Montréal, Canada
e-mail: germain.daniel@uqam.ca

U. Schickhoff et al. (eds.), *Mountain Landscapes in Transition*,
Sustainable Development Goals Series, https://doi.org/10.1007/978-3-030-70238-0_11

(Brown et al. 2009) are therefore useful to better understand the ongoing transformation of mountain landscapes and their sensitivity to climate change and disturbance regimes. In this regard, mountain environments appear to be useful natural laboratories within which to study sediment fluxes related to mass-wasting processes, given their sharp climatic and biotic altitudinal gradients, as well as their major role in terrestrial Earth surface dynamics (Knight and Harrison 2014). Indeed, Antonelli et al. (2018) have recently demonstrated, for example, the link between geological and climatic influences on mountain biodiversity. In recently deglaciated mid-latitude mountains, now evolving under a cold-temperate climate, most of the scree slopes are relicts (e.g., Curry and Morris 2004; Hinchliffe and Ballantyne 2009) due to the exponential decrease in sediment availability, known as paraglacial exhaustion (Ballantyne 2002a, b) or paraglacial sediment cascade (Cossard and Fort 2008). Consequently, most active scree slopes currently remain limited to upland environments where physical weathering dominates (Goudie 2004).

Under current global warming, increased temperatures are expected to result in latitudinal and altitudinal tree line expansions (Payette 2007; Moen et al. 2004; Liang et al. 2011). The meta-analysis conducted by Harsch et al. (2009) for 166 sites worldwide has indeed shown a progression of forest fronts in 52% of cases, versus only 1% of sites characterized by a recession. However, several authors have already emphasized the importance and interest of including geomorphic processes in studying the altitudinal limit of trees (see the reviews of Holtmeier and Broll 2005; Whitesides and Butler 2011). Indeed, hillslope processes tend to respond quickly to climate change and other natural and anthropogenic disturbances, as has been observed during past climatic variations, such as the Neoglacial and Little Ice Age periods, both characterized by an increased frequency and intensity of mass-wasting processes (Mathews et al. 1997; Curry 2000; McCarroll et al. 2002). Several authors have attempted to develop frequency–magnitude relationships for different mass-wasting processes from historical events and various environmental archives (van Steijn et al. 2002; Stoffel et al. 2005; Jakob and Friele 2010), highlighting once again the interest of long-term reconstructions for hazard and risk assessment, as well as to improve understanding of the evolution of terrestrial eco-geosystems (Bogaart et al. 2003; Bebi et al. 2009; Pawlik 2013).

On scree slopes, where long-term studies are especially difficult and rare due to the high sediment availability and restricted timeframes of lichenometric and dendrogeomorphic approaches (van Steijn 2002; Germain and Hétu 2016), extreme events might reveal significant stratigraphic opportunities to explore past hillslope dynamics over a longer timeframe. Indeed, heavy rainfall at local or regional scales has the potential to trigger low-frequency, high-magnitude events capable of erosion and transportation beyond the steady-state equilibrium of the system, often corresponding to, for example, the formation of deeply incised channels and soil erosion. These morphological impacts then make it possible to explore several stratigraphic sections and potentially to reconstruct past dynamics related to storm activity and sediment fluxes (Noren et al. 2002; Beylich and Sandberg 2005). In the Chic-Chocs Range (Gaspé Peninsula) of the northern Appalachian Mountains of eastern Canada, increased climatic variability (freezing rain, rain-on-snow, ice-crust formation) in the 1980-90s has resulted in an increased frequency of large snow avalanches (Germain et al. 2009) despite the absence of any significant warming trend, once again indicating that geomorphic processes do not respond linearly to climate change. It is then particularly appropriate to focus on sediment connectivity through investigations that emphasize the nature, frequency, magnitude, and spatial extent of hillslope processes, which will in turn provide better knowledge of both structural and functional landscape components (Fryer et al. 2013; Bracken et al. 2015; Wohl et al. 2019).

The main purpose of this chapter is to provide a quantitative example of medium- to long-term sediment fluxes originating from various hillslope processes evolving into contact with the forest. As previously reported by several authors for the northern Gaspé Peninsula (cf. Germain and Hétu 2016) and other mountainous environments (Blikra and Nemec 1998; Berthling and Etzelmüller 2007), the frequency and magnitude of these mass-wasting processes are extremely variable in time and space. However, the observations and measurements made at a regional scale over several active scree slopes, debris flow cones, and alluvial fans can provide valuable information about the sensitivity of hillslope sediment fluxes to climate change. The specific objectives are: (1) to qualitatively and quantitatively describe the mass-wasting processes related to the studied scree slopes, debris flow cones, and alluvial fans; (2) to evaluate the sedimentation rates based on radiocarbon dates; (3) to estimate the sediment fluxes related to the construction of lobate rock glacier derived from scree accumulation under periglacial conditions shortly after deglaciation; and (4) to discuss the relative importance of sediment fluxes in a cold-temperate climate over the past centuries (climate-weathering-mass movement process link) through comparison to the periglacial phase following deglaciation (paraglacial evolution).

11.2 Geographical Settings

The study region is part of the Chic-Choc Mountains, which are located at the northeastern extremity of the folded sedimentary Appalachian system (Fig. 11.1). The topography is dominated by three levels of plateau, from the highlands (1100 m a.s.l.) to the coast. These plateaus are likely remnants of a Pre-Quaternary erosion surface deeply dissected by glacial valleys (Peulvast et al. 1996; Jutras and Schroeder 1999), particularly in the northern part of the Gaspé Peninsula, where the local topography is characterized by a steeply cliffed coastline ($\sim$400 m a.s.l.).

The presence of the Gulf of St. Lawrence and the continental effect of the northern USA impose diverse and variable climatic influences due to the alternating high- and low-pressure systems associated with westerly atmospheric

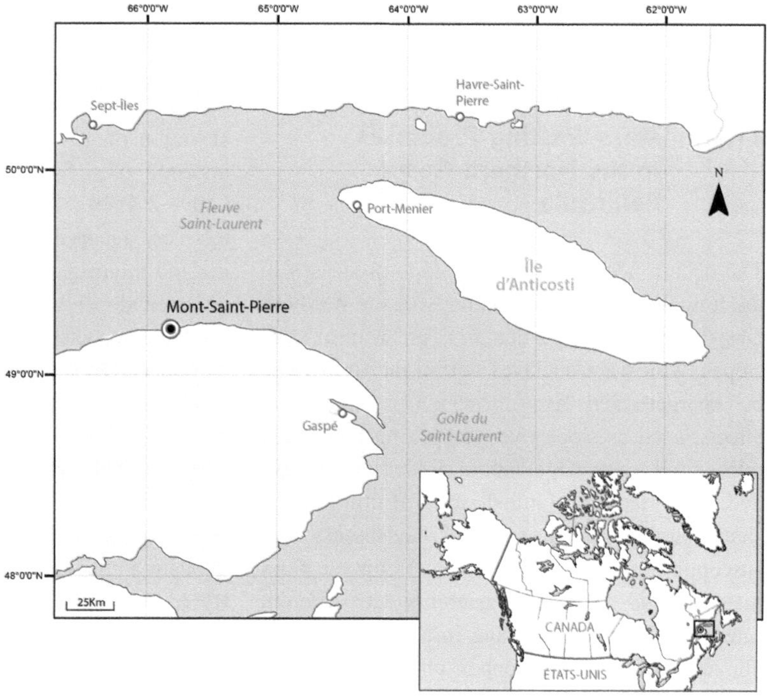

Fig. 11.1 Location of the study area in the Gaspé Peninsula, eastern Canada. The sites are located in the Mont Saint-Pierre valley and the surrounding valleys

circulation (Gagnon 1970). The winter season is cold (−13.8 °C in January), long (from mid-October to the end of April), and snowy (average of 330 cm). In summer, the growing season is warm (+20.0 °C in July) and long enough to allow for the development of a closed forest canopy in low-elevation coastal valleys, from sugar maples (*Acer saccharum* Marsh) and yellow birch (*Betula alleghaniensis* Britton) at lower levels to mixed stands of balsam fir (*Abies balsamea* (L.) Mill.) and yellow birch at higher elevations near the rock walls and the upper portions of scree slopes ($\sim$400 m a.s.l.). Spring and autumn are prone to freeze–thaw cycles, which average 43 days per year (Trenhaile and Rudakas 1981; Fortin et al. 2011). Finally, superimposed upon this regional climatic pattern are climatic phenomena, such as the North Atlantic Oscillation (NAO) and El Nino Southern Oscillation (ENSO), which usually cause mild, dry winters during the negative NAO phase and El Nino years. However, the rapid succession of contrasting meteorological contexts on a daily basis creates dynamic situations favorable to the occurrence of various mass-wasting processes, such as snow avalanches, debris flows, frost-coated clast flows, and niveo-aeolian sedimentation (Hétu and Vandelac 1989).

11.2.1 Mass-Wasting Processes in the Northern Gaspé Peninsula

The occurrence of several hillslope processes in the low-elevation coastal valleys of the northern Gaspé Peninsula, particularly on active talus slopes, which are enclaves within the forest, can be characterized as follows: (1) paraglacial effects, such as slope over-steepening and stress release following deglaciation unloading; (2) folded, contorted, and faulted strata of thin-bedded shales, argilites, siltstones, and sandstones, very susceptible to weathering; and (3) climatic fluctuations and weather variability at different timescales. Indeed, evolving in a cold-temperate climate (Dfb class of Köppen climate classification), the rockwalls located in the upper part of the slopes undergo several freeze–thaw cycles, resulting in a substantial production of small size debris (average of 7 cm long) with a low sphericity index (average of 0.3) (Hétu and Vandelac 1989; Hétu et al. 1994; Hétu and Gray 2000a). Environmental monitoring and punctual field observations made over the last 30 years, in addition to some published and other unpublished data resulting from doctoral and masters theses, have revealed that individual particles tend to accumulate near the rockwalls (talus shift of 15 cm per year) until mobilized by cold season dynamics, which account for 95% of the yearly sediment budget on the active scree slopes (Germain and Hétu 2016). According to Germain and Hétu (2016), the quantitative measurements of sediment fluxes, although rough estimates, demonstrate the prominent role of frost-coated clast flows on these active fine-grained scree slopes, as well as the sensitivity of these landforms and hillslope processes to ongoing climate change.

On a longer timescale, the evolution of scree slopes in the study area can be summarized as follows:

The deglaciation of the low-elevation U-shaped glacial valleys in the northern Gaspé Peninsula began 15,000 calendar years before present (cal. years BP) and was followed by the postglacial Goldthwait Sea, which reached an elevation of up to 50 m, corresponding to a sea invasion of a few kilometers inland (Hétu and Gray 2000b). The subsequent glacio-isostatic recovery favored the fashioning of several marine and fluvio-glacial terraces. Hillslope processes on scree slopes were certainly active at that time, but their impacts were more significant during the Younger Dryas, where lobate rock glaciers resulting from scree accumulations were built on marine and fluvio-glacial terraces. These rock glaciers likely remained active until later than 9700 cal. years BP, under a periglacial regime (Hétu et al. 2003), and paleoecological proxies indicate a cold climate until circa 8000 cal. years BP (de Vernal et al. 1993; Marcoux and Richard 1995; Richard and Larouche 1994; Sawada et al. 1999). Based on pollen analysis, the afforestation of the plateau overlooking the talus slopes began around 12,200 cal. years BP, and the closed forest was established around 10,000 cal. years BP (Labelle and Richard 1984). Conversely, the forest colonization of slopes was much slower because of the intense geomorphic activity, where stratigraphic data show that sediment fluxes related to mass-wasting processes have reached the base of the slopes later than 7,500 and even later than 4,700 cal. years BP in some valleys (Hétu and Gray 2000a).

Since that time and until today, the battle between active scree development and forest colonisation has persisted throughout the Holocene, certainly in concomitance with climatic oscillations and disturbance regimes, as revealed by dendroecological studies of the recent period (Hétu 1990; Lafortune et al. 1997; Germain et al. 2005). Indeed, at the end of the Little Ice Age (circa 1850), the upper limit of the forest edge on scree slopes was several meters (10 to 100 m) lower than it currently is (Hétu 1990; Lafortune et al. 1997), indicating greater geomorphic activity. Between 1850 and 1950, forest vegetation reached higher up the scree slopes due to reduced geomorphic activity on slopes and sedimentation rates. After 1950, the ecological impacts of snow avalanches, for example, allowed frost-coated clast flows to enter further into the forest edge, burying forest vegetation and generally lowering the altitudinal tree limit on scree slopes (Germain and Hétu 2016).

At present, slope exposure is an extremely important determinant of the nature of geomorphic processes acting on talus slopes (Stabile-Caillé 2019). Indeed, west-facing slopes are exposed to the prevailing winds from the northwest, but also to insolation. The active portion of these scree slopes is therefore generally snow-free and subject to frost-coated clast flows (Fig. 11.2), the main geomorphic process causing forest recession on the concerned slopes (Hétu et al. 1994; Lafortune et al. 1997; Germain and Hétu 2016). In contrast, the east-facing slopes are sheltered from the prevailing winds and are generally covered with a significant snowpack during the winter. The dominant hillslope processes on these slopes are then snow-related, namely avalanches (Gratton et al. 2019), snow creep, and debris flows (Fig. 11.2; Jobin 2019). Along the coast, niveo-aeolian processes dominate due to the exposure of the slopes to strong winds from the St. Lawrence Estuary (Hétu 1992).

11.3 Methodology

Based on a research program with the overall objective of better understanding sediment routing and evolution in cold-temperate and forested mountain environments from deglaciation to the present day, several sites were visited and monitored on a yearly basis in several low-elevation coastal valleys of the northern Gaspé Peninsula for at least the past decade. These include active scree slopes, debris flow cones, and alluvial fans (Fig. 11.2).

In July 2014, Hurricane Arthur hit the Gaspé Peninsula, leaving approximately 60 mm of rain in less than 24 h. Rainfall amounts were likely much higher in certain places due to the orographic effect, but this is unconfirmed because of the scarcity of meteorological stations in the area. However, subsequent field visits revealed the high-magnitude geomorphic activity generated by this heavy rainfall episode on several talus slopes. Indeed, the intensity of the hillslope processes was such that, at many sites, channels of several meters deep were incised (Fig. 11.2), significant volumes of sediment were transported, and the displacement of large blocks was recorded. In addition, considering that all of these slopes are evolving into contact with the forest cover, major ecological impacts (e.g., broken, uprooted, and buried trees) were also recorded and clearly visible. Many roads were cut off and several culverts were partially or completely destroyed, confirming the high-intensity nature of these geomorphic events.

In order to improve estimates of medium- to long-term sedimentation rates in various hillslope environments and landforms at a regional scale, the results presented here are mainly based on radiocarbon dating related to the exceptional stratigraphic opportunities that resulted from the extreme geomorphic events of July 2014.

Fig. 11.2 Photographs of investigated sites: **a** West-facing scree slope; the presence of narrow, shallow channels related to debris flow and frost-coated clast flow activity is visible. On the upper part of the slope, the channels tend to be rectilinear, whereas on the lower part of the slope they start to meander between islands of shrubs and trees. See Hétu et al. (1994) and van Steijn et al. (1995) for more details about frost-coated clast flows. **b** Incised channel in alluvial fan in a forested environment following the Hurricane Arthur in July 2014. **c** East-facing debris flow tracks showing a limited rockwall on the upper part of the slope. **d** The close-up shows one of the several meters deep incisions of a debris flow track after the significant rainfall event related to Hurricane Arthur

11.3.1 *Calculation of the H_0/H_i Ratio*

The H_0/H_i ratio represents the relative height of the scree (H_0) with respect to that of the entire slope (H_i). Therefore, a value of near zero indicates a rockwall without or with little accumulation at its foot, whereas a value of close to one indicates a nearly completely buried rockwall. A massive rockwall is, in general, associated with greater geomorphic activity, since the particles falling from the rockwall have more potential energy on average (Jomelli and Francou 2000). The H_0/H_i ratio is then expected to represent the maturity of the slope, since a higher value corresponds to a more stable slope and therefore to less direct geomorphic activity (Jomelli and Francou 2000; Hinchliffe and Ballantyne 2009; Curry and Morris 2004).

Because of the diversity of sites and landforms investigated here (active scree slopes, debris flow cones, and alluvial fans), and the related and previously studied hillslope processes in the area, such as frost-coated clast flows (Hétu et al. 1994; van Steijn et al. 1995; Lafortune et al. 1997; Germain and Hétu 2016), debris flows (Jacob 2001; Jobin 2019), and hyperconcentrated flows (Caron-Fournier 2009; Ouellet and Germain 2014), the H_0/H_i ratio was calculated.

11.3.2 Radiocarbon Dating

In the field, incised channels on scree slopes, debris flow cones, and alluvial fans were surveyed to sample buried wood and organic matter. Although organic horizons show relative stability for a given period of time, buried wood still provides benchmarks for calculating sedimentation rates. Most of this wood comes from broken and fallen branches and trees, subsequently buried by mass-wasting processes that allowed them to be preserved. The position of each sample was recorded with a Garmin Oregon 700 GPS, and radiocarbon dating was done at the André E. Lalonde AMS Laboratory at the University of Ottawa, Canada, with calibration performed using OxCal v4.2.4 (Bronk Ramsey 2009). Because the majority of the ages fall within a section of the calibration curve affected by the so-called Seuss Effect, which is a flat portion of the calibration curve caused by the burning of fossil fuels, several calibrated ages unfortunately are of low precision. However, in all cases, sedimentation rates were calculated based on the median calibrated age, which also corresponds to the highest statistical probability.

11.3.3 Estimation of Rock Glacier Formation Volume and Sediment Flux

As previously mentioned, long-term studies on scree slopes are especially difficult due mainly to the rarity of datable material. As such, the presence of rock glaciers at the foot of several scree slopes offers an interesting and rare opportunity to evaluate sediment fluxes, at least for the periglacial period following deglaciation. Indeed, rock glacier volumes can provide a rough estimate of rockwall recession rates during their development (Ballantyne 1984; Humlum 2000; Berthling and Etzelmüller 2007). Calculated volumes were corrected using a porosity of 30%. The retreat rates were then estimated by dividing the rock glacier volumes by the area of the rockwalls. The retreat values were converted into mm ka^{-1} by considering the total duration of the periglacial period, of approximately 3250 years (i.e., from 10,500 to 7250 cal. years BP) (Hétu and Gray 2000a; Germain and Hétu 2016). The mapping and morphometric measurements were done by photo-interpretation from a mosaic of orthophotos at a scale of 1:20,000 and with a resolution of 30 cm (Ministère de l'Énergie and des Ressources Naturelles 2016). It should be noted that the results are rough estimates given the resolution of available orthophotos, but nevertheless provide an order of magnitude for comparative purposes.

11.4 Results

11.4.1 Morphometric Characteristics of Hillslope Processes

The investigated talus slopes, which include active scree slopes, debris flow cones, and alluvial fans, show significant discrepancies. Along the coast, scree slopes have an average length of 69 ± 16 m and a rockwall height of 98 ± 18 m. In the coastal valleys, the active section of the scree slopes is more prominent, with an average length of 208 ± 34 m, but the rockwall height is usually less, on average 41 ± 10 m. However, in both cases the measurements are highly variable, particularly for the active scree slope sections, ranging between 50 and 450 m.

The difference in H_0/H_i ratios is statistically significant between coastal talus slopes and those located in the valleys. On the coast, the talus slopes have a more massive rockwall and a lower H_0/H_i ratio (Fig. 11.3). These data suggest a less advanced stage of the coastal slopes, whose natural evolution is also disturbed by the national road 132 located at the foot of the slopes. Because of human intervention, notably regularly removing the sediments downslope, these coastal talus slopes are constantly unbalanced, since slope inclination does not correspond to the angle of repose, which increases geomorphic activity

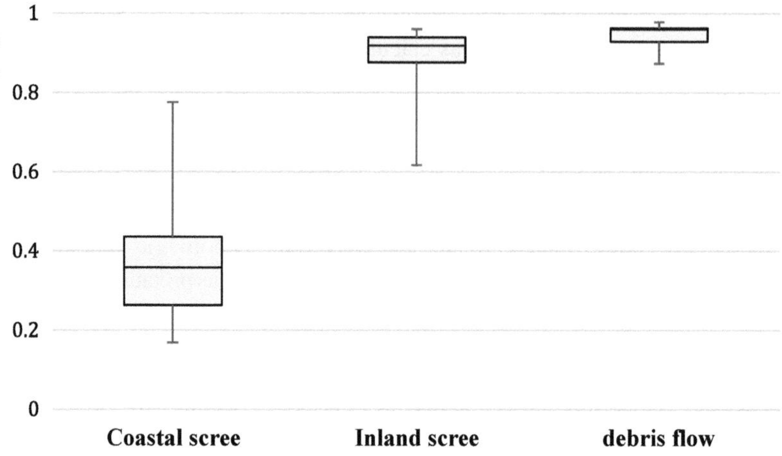

Fig. 11.3 H_0/H_i ratio for active coastal and inland scree slopes, and debris flow tracks. Boxplots represent the minimum and maximum, as well as the 25th and 75th percentile, and the horizontal line is the median

related to mass-wasting processes (Statham 1976; Kirkby and Statham 1975). The H_0/H_i ratio of debris flow cones is less variable than inland scree slopes (Fig. 11.3), but both hillslope systems are characterized by a very high ratio, testifying to a more advanced stage of maturity. The threshold used to distinguish a massive rockwall is generally a H_0/H_i value of 0.5 or greater (Francou and Manté 1990; Jomelli and Francou 2000; Hinchliffe and Ballantyne 2009). Even the minimum value measured for scree slopes (0.66) is greater than this threshold.

Except for scree slopes on the coast, which are all north-facing, the investigated talus slopes in the valleys display the following exposition: 7% north-facing, 23% south-facing, 25% east-facing, and 45% west-facing slopes. As mentioned earlier, most of these active scree slopes in the low-elevation coastal valleys are located on west-facing slopes. Interestingly, the east-facing slopes are mainly dominated by debris flows, and therefore show a smaller rockwall in the upper segment of the slope. Indeed, the geomorphically active sections appear to be smaller when compared to the previously described scree slopes. The H_0/H_i ratio is not statistically significantly different between these two hillslope systems, yet the mean and median is higher for debris flows. Alluvial fans were not reported in Fig. 11.3, given that all investigated fans were characterized by the absence of a rockwall in the upper part of their forested catchment.

11.4.2 Mid- to Long-Term Sediment Fluxes

Table 11.1 details radiocarbon dating results. Twelve samples were taken from active scree slopes, nine from debris flow tracks and cones, and seven from alluvial fans. All samples were of buried wood found in several stratigraphic sections related to the formation of incised channels resulting from the heavy rain of July 2014. The oldest radiocarbon date obtained is from the year 988, from a debris flow cone, and the deepest piece of wood sampled was from five meters below the surface, from the upper part of a scree slope.

On scree slopes, the highest probability of a calibrated age ranging between 1770 and 1984 corresponded to samples from between 55 and 500 cm deep. Five samples were dated to the 1770s, four to the 1800s, and three to the second half on the twentieth century. The highest sedimentation rates (13.16 and 5.15 mm yr^{-1}) were also recorded for the recent period, after 1979 and 1984, respectively. The mean and median values were 2.24 and 0.73 mm yr^{-1}, respectively (Fig. 11.4).

For debris flow cones, the calibrated dates are more variable, ranging from 988 to 1880 and correspond to buried wood from between 70 and 183 cm deep. Except for the oldest date, of 988, all other dates correspond to the Little Ice Age period (circa 1550–1895), with mean and median

Table 11.1 Radiocarbon dates, calibrated ages, and sedimentation rates for scree slopes, debris flow cones, and alluvial fans

	Lab code	Sample	Material	Radiocarbon date (Yr. BP)	Calendar date (Yr. AD/BC)	Median calendar date (Yr. AD/BC)	Depth from the Surface (cm)	Accumulation Rate (mm yr^{-1})
Scree slopes	UOC-5090	T17A	Wood	111 ± 24	1805–1935	1838	130	0.726
	UOC-5091	T17B	Wood	181 ± 23	1662–1954	1770	70	0.028
	UOC-5112	T21B	Wood	31 ± 28	1696–1955	1895	100	0.574
	UOC-5092	RA	Wood	199 ± 23	1653–1954	1779	130	0.546
	UOC-5093	RB	Wood	179 ± 23	1662–1954	1769	100	0.403
	UOC-5094	RC	Wood	78 ± 28	1691–1921	1848	180	1.065
	UOC-5095	RD	Wood	141 ± 27	1669–1954	1807	135	0.643
	UOC-5096	RE	Wood	Modern ± 23	1983–1986	1984	170	5.152
	UOC-5097	RF	Wood	196 ± 24	1654–1954	1776	210	0.871
	UOC-5098	RG1	Wood	Modern ± 29	1954–1957	1955	90	1.452
	UOC-5099	RG2	Wood	182 ± 23	1661–1954	1770	145	0.587
	UOC-5100	RH	Wood	Modern ± 22	1979–1980	1979	500	13.158
Debris flow cones	UOC-5107	M1A	Wood	96 ± 23	1691–1925	1842	380	2.171
	UOC-5108	M1B	Wood	334 ± 23	1480–1640	1565	500	1.106
	UOC-5109	M1C	Wood	225 ± 22	1678–1954	1782	147	0.626
	UOC-5110	M5A	Wood	68 ± 22	1695–1955	1880	125	0.947
	UOC-5111	M5B	Wood	147 ± 24	1668–1954	1798	200	0.913
	UOC-5103	C1B1	Wood	103 ± 23	1686–1927	1840	290	1.638
	UOC-5104	C1B2	Wood	158 ± 24	1666–1954	1770	100	0.405
	UOC-5105	C1B3	Wood	1062 ± 25	948–1022	988	340	0.330
	UOC-5106	C1B4	Wood	139 ± 24	1670–1954	1810	55	0.266
Alluvial fans	UOC-5085	7AB1	Wood	684 ± 26	1271–1310	1293	242	0.334
	UOC-5086	7AB2	Wood	209 ± 23	1681–1954	1782	122	0.519
	UOC-5087	7AB3	Wood	879 ± 23	1147–1220	1169	80	0.094
	UOC-5088	7AB4	Wood	197 ± 22	1654–1954	1778	152	0.636
	UOC-5089	7AB5	Wood	276 ± 23	1620–1665	1628	183	0.470
	UOC-5101	M8A	Wood	514 ± 23	1399–1442	1420	380	0.637
	UOC-5102	M8B	Wood	Modern ± 24	1955–1956	1956	350	5.738

sedimentation rates of 0.93 mm yr^{-1} and 0.46 mm yr^{-1} respectively (Fig. 4). As for the H_0/H_i ratios, debris flow environments are less variable in this respect than scree slopes and alluvial fans.

The buried wood found between 90 and 210 cm deep in alluvial fans were dated to between 1169 and 1956 cal. years. Three dates (1169, 1293, and 1420) are older than the Little Ice Age, while three others (1628, 1778, 1782) fall during this period. Finally, the recent date of 1956 also corresponds to the highest sedimentation rate (5.74 mm yr^{-1}) calculated for alluvial fans, which is also significantly different from the mean (1.20 mm yr^{-1}) and the median (0.12 mm yr^{-1}) values (Fig. 11.4).

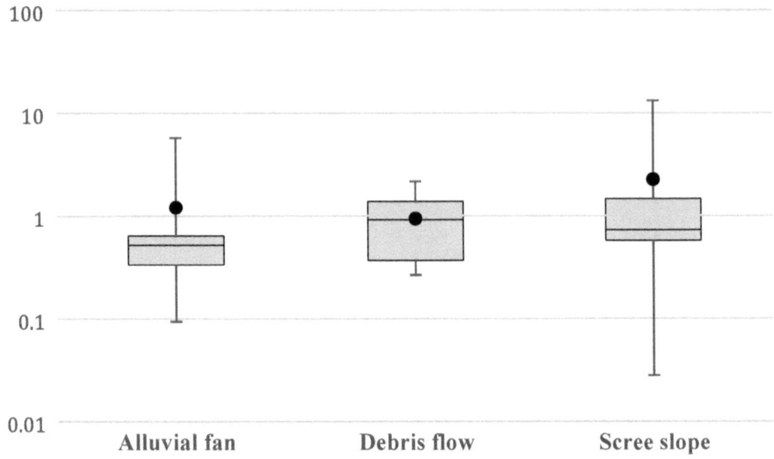

Fig. 11.4 Sedimentation rates for scree slopes, debris flow cones, and alluvial fans. Boxplots represent the minimum and maximum, as well as the 25th and 75th percentile, the horizontal line is the median, and the black circle is the mean

11.4.3 Sediment Volumes and Fluxes Related to Rock Glacier Formation

The sedimentation rate of the supposed most active period under periglacial climate and paraglacial conditions immediately following deglaciation was estimated using the volume of rock glaciers identified at the foot of seven scree slopes. The volumes of these rock glaciers are between 400,000 and almost three million cubic meters (Table 11.2). On average, the rockwalls have retreated approximately 30 m to allow a sufficient volume of sediment to construct these rock glaciers between 10,500 and 7250 cal. years BP. These values correspond to an average rockwall retreat of more than 8,000 mm ka^{-1}, which in turn gives an average sedimentary flux of over 400 $m^3\ yr^{-1}$. Depending on the size of the rock glaciers, sediment fluxes range between 164 and 873 $m^3\ yr^{-1}$ (Table 11.2), corresponding to sedimentation rates of over 3 mm yr^{-1}, with a maximum value of 6.9 mm yr^{-1}.

Because of the variably dominant nature of hillslope processes on west-facing (frost-coasted clast flows) and east-facing slopes (snow avalanches and debris flows), an analysis of variance was performed, but did not support any significant effect of slope aspect on calculated sediment fluxes for rock glacier construction during the periglacial period (p-value = 0.81).

11.5 Discussion

11.5.1 Discrepancies Between Mass-Wasting Processes

The north-facing coastal scree slopes are characterized by a low H_0/H_i ratio (<0.5), due to human disturbances related to the national road 132 at the base of these slopes. Because the Quebec Ministry of Transport periodically removes colluvium near the road, these talus slopes are still evolving as an open system. Moreover, because they are very exposed to strong winds from the St. Lawrence River, a significant amount of sediment is transported by wind from the rockwalls (Hétu 1992; Hétu and Gray 2000a; Germain and Hétu 2016). In the Scottish Highlands, Ballantyne (1998) also concluded that wind plays a significant role in the removal of unstable material from rockwalls. It is therefore likely that wind is important in reshaping the coastal scree slopes and thus contributes to maintain a lower H_0/H_i ratio compared to that of inland scree slopes.

In the low-elevation coastal valleys, talus slopes maintain significant geomorphic activity despite their advanced stage of evolution due to their high H_0/H_i ratio, particularly compared to other scree slopes in cold-temperate climates (Hinchliffe and Ballantyne 1999; Ballantyne and

Table 11.2 Rock glacier volumes, sediment fluxes, and related rockwall retreat rates for the periglacial period from 10,500 to 7250 cal. years BP

	Aspect	Rock glacier			Rockwall	
		Volume (m^3)	Sediment flux[a] (m^3 yr^{-1})	Accumulation rate[b] (mm yr^{-1})	Retreat (m)	Retreat rate (m ka^{-1})
1	W	1 919 119	590	6.80	19	5.71
2	W	2 836 821	873	6.90	56	17.12
3	E	1 166 500	359	4.26	33	10.26
4	W	422 162	130	3.07	8	2.38
5	E	1 733 047	533	5.02	31	9.52
6	E	903 327	278	3.44	29	8.91
7	E	531 898	164	3.57	30	5.25
Mean		1 358 982	418	4.72	29	8.45

[a]Sediment flux calculated from the rock glacier volume and the duration of the periglacial period
[b]Accumulation rate based on morphometric measurements of the rock glacier and calculated sediment flux

Eckford 1984; Curry and Morris 2004; Curry and Black 2003). This might be related to the diversity of mass-wasting processes, which are capable of remobilizing the debris accumulated near the rockwall further downslope. These hillslope processes are of sufficient magnitude in time and space to locally disrupt the treeline, as previously reported in the area (Hétu 1990; Lafortune et al. 1997; Germain and Hétu 2016). For example, frost-coasted clast flows, a process recognized and described for the first time in the northern Gaspé Peninsula (Hétu et al. 1994), are considered to be the most important sedimentary transport mechanism on south- and west-facing slopes (Lafortune et al. 1997; Hétu and Gray 2000a; Germain and Hétu 2016). On east-facing slopes, debris flows appear to be responsible for most of the sediment transfer, given that snow avalanches are usually devoid of geomorphic impact. In this regard, Fig. 11.5 shows a decrease in the active area of the scree slopes and debris flow cones as the rockwall decreases in size, testifying once again to the usefulness of the H_0/H_i ratio to evaluate the potential for sediment release. Unfortunately, no significant correlation was found between rockwall area and sedimentation rate given the variability of mass-wasting processes, as well as the position of the radiocarbon dated wood on talus slopes.

Finally, where rockwalls are absent, the presence of channels with permanent or intermittent flow ensures discontinuous and sporadic

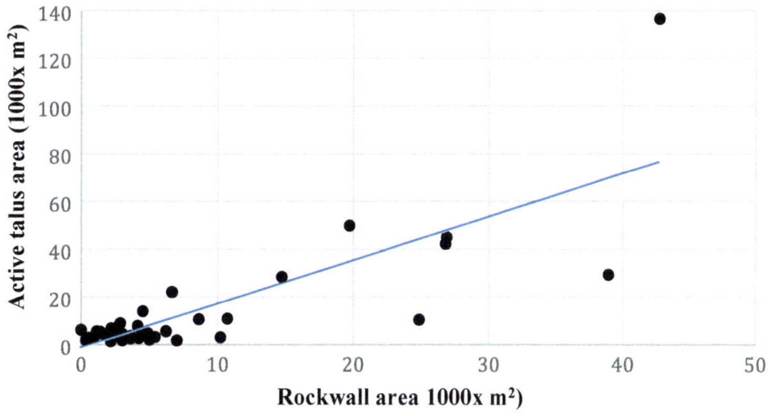

Fig. 11.5 Linear regression model of the relationship between the area of the active section of talus slope and the rockwall area for 43 talus slopes in low-elevation coastal valleys of the Gaspé Peninsula

sedimentary transport on alluvial fans (Ouellet and Germain 2013).

11.5.2 Long-Term Evolution of Talus Slopes

The term paraglacial refers to the statement of Church and Ryder (1972, p. 3059), wherein paraglacial processes are considered to be '*non-glacial processes that are directly influenced by glaciation.*' Since then, the concept has been substantially reviewed and improved (Ballantyne 2002a, b; Cossart and Fort 2008; Mercier 2008; Knight and Harrison 2018). However, it remains based on the assumption that the retreat of the ice caps related to the last glaciation exposed the landscape to paraglacial processes, such as glacial debuttressing, isostatic instability, and high sediment availability, among others, providing an ideal context for high geomorphic activity. It therefore follows that the construction of talus-derived rock glaciers is particularly well-representative of the paraglacial period. The calculated rockwall retreat rates based on rock glacier volumes are on average 8.45 m ka^{-1}, one to two orders of magnitude higher than current rates reported in the scientific literature (Curry and Morris 2004; Ballantyne and Harris 1994). However, the obtained values are on the same order of magnitude ($\sim$1 to 10 m ka^{-1}) when compared to the most active Holocene period (Humlum 2000; Barsch 2012; Berthling and Etzelmüller 2007) and are based on the same methodological approach. Although these are rough estimates, these results nevertheless indicate a high intensity of geomorphic activity during this period of very rapid rock glacier and talus slope formation, as soon as rockwalls were exposed following deglaciation (Hétu and Gray 2000a; Germain and Hétu 2016).

As reported by Hétu and Gray (2000a), it is unlikely that rock glaciers were still active after 7250 cal. years BP due to global warming. The evolution of talus slopes after this period becomes very complex, since there is now significant interaction with the surrounding forest cover. Indeed, for an isolated land system (e.g., scree slope, debris flow cone, alluvial fan), the paraglacial period ends at the exhaustion of the available sediment, which usually follows an exponential decline. However in the northern Gaspé Peninsula, the very susceptible lithology to mechanical weathering, and the diversity of geomorphic processes related to contrasting meteorological conditions have maintained paraglacial activity locally for more than 10,000 years. The sedimentation rates calculated for the construction of rock glaciers (4.72 mm yr^{-1}) versus those of recent scree slope dynamics (2.24 mm yr^{-1}) show smaller differences than those reported in the scientific literature (cf. André 1997). Indeed, Lafortune et al. (1997) reported an even higher sedimentation rate on a scree slope in the study area based on a tree-ring approach, testifying to the extended duration of paraglacial landscape relaxation. Of course, the geomorphic activity, as well as the altitudinal forest limit, on these scree slopes has fluctuated throughout the Holocene, illustrating the complex geomorphic response and dynamic equilibrium between mass-wasting processes, vegetation cover, and hydroclimatic conditions. For example, since the middle of the twentieth century, geomorphic processes on scree slopes have increasingly fragmented the treeline despite the warming trend (Lafortune et al. 1997; Germain and Hétu 2016). However, a warmer and drier climate generally tends to favor an upward movement of the treeline and a decrease in mass-wasting activity, with the opposite true in colder and wetter climates.

At present, the coastal valleys of the northern Gaspé Peninsula have a diachronous pattern of paraglacial influences (Martin and Germain 2016). On the west-facing slopes, the rockwalls of the upper part of the slopes are still active, while on the east-facing slopes, the rockwalls have been almost completely eliminated. The scree slopes located on west-facing slopes appear to be primary paraglacial stock, for which the sediment flux is mainly related to frost-coated clast flow activity occurring in the winter (Germain and Hétu 2016). On east-facing slopes, the sediment flux on debris flow cones appears to be more sporadic in time and space, and less significant (0.93 mm yr^{-1}) than that on scree slopes. Although related to a smaller rockwall, these debris flow land systems should not be characterized as limited by sediment availability given the rapid rockwall dismantling. On the other hand, alluvial fans are secondary sources, since the sediment is not transported directly from the rockwall retreat, but rather from sediment stored on forested talus slopes and elsewhere in small watersheds (Ballantyne 2002b). However, hyperconcetrated flows on these fans can efficiently (1.20 mm yr^{-1}) rework glacigenic sediment stored in primary paraglacial stocks, and these catchments appear, as was the case for debris flow cones, to be transport- rather than sediment-limited (Schrott et al. 2002; Schlunegger et al. 2009). Finally, these results show the high sensitivity of these environments to climate variability at different scales, especially given specific local conditions, maintaining high sedimentation rates from deglaciation to the present day.

11.6 Conclusion

Due to their geographical, climatic, and geomorphological context, talus slopes in the northern Gaspé Peninsula developed rapidly following deglaciation and have remained active throughout the Holocene (Germain and Hétu 2016), which is not frequently reported for cold-temperate climates (Hétu and Gray 2000a; Curry and Morris 2004; Hinchliffe and Ballantyne 2009), except for highlands above the altitudinal tree limit. Our results show that the response of these land systems to climate change is nonlinear, is affected by antecedent conditions, geological controls, and climate variability at different spatiotemporal scales (Knigth and Harrison 2014), and is characterized by high sediment fluxes (e.g., Meigs et al. 2006; Warburton 2007; Brown et al. 2009; Beylich et al. 2011) testifying to the sensitivity of these geomorphological environments to climate change (Knigth and Harrison 2013). However, mid-latitude mountains are also foreseen to undergo significant transformations in the coming decades and century, hence the imperative to better understand the past and ongoing modification of these environments that provide a range of ecosystem services (Price et al. 2004). Mountain systems should be regarded as the result of a long and complex evolutionary process, focusing particularly on the ecological and hydroclimatic conditions that control the duration of glacially conditioned sedimentary stock exhaustion and the related paraglacial responses of different land systems. Given that paraglacial refers to an unstable geomorphological state, vulnerable to erosion processes (Martin and Germain 2016), it is useful in predicting mountain geohazards (Korup and Tweed 2007; Korup and Clague 2009), also showing the interest of capturing the longer timescale and larger spatial-scale behavior of such mountain systems under current global warming.

References

Andre M-F (1997) Holocene rockwall retreat in Svalbard: a triple-rate evolution. Earth Surf Processes 22:423–440

Antonelli A, Kissling WD, Flantua SGA, Bermudez MA, Mulch A, Mueller-Riehl AN, Kreft H, Linder HP, Badgley C, Fjeldsa J, Frtiz SA, Rahbek C, Herman F, Hooghiemstra H, Hoorn C (2018) Geological and climatic influences on mountain biodiversity. Nat Geosci 11:718–725

Ballantyne CK (1998) Aeolian deposits on a Scottish mountain summit: characteristics, provenance, history and significance. Earth Surf Processes 23:625–641

Ballantyne CK (2002) Paraglacial geomorphology. Quaternary Sci Rev 21:1935–2017

Ballantyne CK (2002) A general model of paraglacial landscape response. Holocene 12:371–376

Ballantyne CK, Eckford JD (1984) Characteristics and evolution of two relict talus slopes in Scotland. Scot Geogr Mag 100:20–33

Ballantyne CK, Harris C (1994) The periglaciation of great Britain. Cambridge University Press

Barsch D (2012) Rockglaciers: indicators for the present and former geoecology in high mountain environments. Springer Science & Business Media

Bebi P, Kulakowski D, Rixen C (2009) Snow avalanche distrbances in forest ecosystems—state of research and implications for management. Forest Ecol Manag 257:1883–1892

Berthling I, Etzelmüller B (2007) Holocene rockwall retreat and the estimation of rock glacier age, Prins Karls Forland, Svalbard. Geogr Ann A 89:83–93

Beylich AA, Sandberg O (2005) Geomorphic effects of the extreme rainfall event of 20–21 July, 2004 in the Latnjavagge catchment, northern Swedish Lapland. Geogr Ann A 87:409–419
Beylich AA, Lamoureux SF, Decaulne A (2011) Developing frameworks for studies on sedimentary fluxes and budgets in changing cold environments. Quaest Geogr 30:5–18
Blikra LH, Nemec W (1998) Postglacial colluvium in western Norway: depositional processes, facies and paleoclimatic record. Sedimentology 45:909–959
Bogaart PW, Van Balen RT, Kasse C, Vandenberghe J (2003) Process-based modelling of fluvial system response to rapid climate change II. application to the river maas (The Netherlands) during the last glacial-interglacial transition. Quaternary Sci Rev 22:2097–2110
Bracken LJ, Turnbull L, Wainwright J, Bogaart P (2015) Sediment connectivity: a framework for understanding sediment transfer at multiple scales. Earth Surf Processes 40:177–188
Bronk Ramsey C (2009) Bayesian analysis of radiocarbon dates. Radiocarbon 51:337–360
Brown AG, Carey C, Erkens G, Fuchs M, Hoffmann T, Macaire J-J, Moldenhauer K-M, Walling DE (2009) From sedimentary records to sediment budget: multiple approaches to catchment sediment flux. Geomorphology 108:35–47
Caron-Fournier É (2009) Stratigraphie, sedimentology et dynamique d'un petit cône alluvial holocène du nord de la Gaspésie. Université du Québec à Rimouski. Unpublished Master thesis p 117
Church M, Ryder JM (1972) Paraglacial sedimentation: a consideration of fluvial processes conditioned by glaciation. Geol Soc A Bull 83:3059–3072
Cossard É, Fort M (2008) Sediment release and storage in early deglaciated areas: towards an application of the exhaustion model from the case of massif des écrins (French Alps) since the little ice age. Norsk Geogr Tidsskr 62:115–131
Curry AM (2000) Holocene reworking of drift-mantled hillslopes in glen Docherty, Northwest highlands, Scotland. Holocene 10:509–518
Curry AM, Black R (2003) Structure, sedimentology and evolution of rockfall talus Mynydd Du, south Wales. Proc Geol Ass 114:49–64
Curry AM, Morris CJ (2004) Lateglacial and Holocene talus slope development and rockwall retreat on Mynydd Du, UK. Geomorphology 58:85–106
de Vernal A, Guiot J, Turon J-L (1993) Late and postglacial environments of the Gulf St. lawrence: marine and terrestrial palynological evidence. Géogra Phys Quatern 47:167–180
Fortin G, Hétu B, Germain D (2011) Climat hivernal et régime avalancheux dans les corridors routiers de la Gaspésie septentrionale (Québec, Canada). Climatologie 8:9–26
Francou B, Manté C (1990) Analysis of the segmentation in the profile of Alpine talus slopes. Permafrost Periglac 1:53–60
Fryirs K (2013) (Dis)Connectivity in catchment sediment cascades: a fresh look at the sediment delivery problem. Earth Surf Processes 38:30–46
Gagnon RM (1970) Le climat des chic-chocs. Ministère des Richesses Naturelles du Québec, Service de la Météorologie, Rapport M.P, p 36
Germain D, Hétu B (2016) Hillslope processes and related sediment fluxes on a fine-grained scree slope of Eastern Canada. In: Beylich AA, Dixon JC, Zwolinski Z (eds) Source-to-Sink fluxes in undisturbed cold environments. Cambrisge University Press, pp 79–95
Germain D, Ouellet M-A (2013) Subaerial sediment-water flows on hillslopes: essential research questions and classification challenges. Prog Phys Geog 37:813–833
Germain D, Filion L, Hétu B (2005) Snow avalanche activity after fire and logging disturbances, northern Gaspé Peninsula, Québec, Canada. Can J Earth Sci 42:2103–2116
Germain D, Filion L, Hétu B (2009) Snow avalanche regime and climatic conditions in the chic-choc range, eastern Canada. Climatic Change 92:141–167
Goudie AS (2004) Scree. In: Goudie AS (dir.) Encyclopedia of geomorphology. London, Routledge
Gratton M, Germain D, Boucher É (2019) Meteorological triggering scenarios of tree-ring-based snow avalanche occurrence on scree slopes in a maritime climate. Eastern Canada. Phys Geogr pp 1–18
Harsch MA, Hulme PE, McGlone MS, Duncan RP (2009) Are treelines advancing? A global meta-analysis of treeline response to climate warming. Ecol Lett 12:1040–1049
Hétu B (1990) Évolution récente d'un talus d'éboulis en milieu forestier, Gaspésie, Québec. Géogr Phys Quatern 44:199–215
Hétu B (1992) Coarse cliff-top aeolian sedimentation in Northern Gaspésie, Québec (Canada). Earth Surf Processes 17:95–108
Hétu B, Gray JT (2000) Effects of environmental change on scree development throughout the postglacial period in the Chic-Choc Mountains in the northern Gaspé Peninsula, Québec. Geomorphology 32:335–355
Hétu B, Gray JT (2000) Les étapes de la déglaciation dans le nord de la Gaspésie (Québec, Canada): les marges glaciaires des Dryas ancien et récent. Géogr Phys Quatern 54:5–40
Hétu B, Vandelac P (1989) La dynamique des éboulis schisteux au cours de l'hiver, Gaspésie septentrionale, Québec. Géogr Phy Quatern 43:389–406
Hétu B, van Steijn H, Vandelac P (1994) Les coulées de pierres glacées: un nouveau type de coulées de pierraille sur les talus d'éboulis. Géogr Phys Quatern 48:3–22
Hétu B, Gray JT, Gangloff P, Archambault B (2003) Postglacial talus-derived rock glaciers in the Gaspé Peninsula, Québec (Canada). Proceedings 8th International Conference Permafrost, Zurich, Switzerland, July 20–25, 2003. pp 389–394
Hinchliffe S, Ballantyne CK (1999) Talus accumulation and rockwall retreat, Trotternish, Isle of Skye, Scotland. Scot Geogr Mag 115:53–70

Hinchliffe S, Ballantyne CK (2009) Talus structure and evolution on sandstone mountains in NW Scotland. Holocene 19:477–486
Holmeier FK, Broll G (2005) Sensitivity and response of northern hemisphere altitudinal and polar treelines to environmental change at landscape and local scales. Global Ecol Biogeogr 14:395–410
Humlum O (2000) The geomorphic significance of rock glaciers: estimates of rock glacier debris volumes and headwall recession rates in West Greenland. Geomorphology 35:41–67
Jacob N (2001) Fréquence, intensité et déclenchement des coulées de débris en milieu forestier, Gaspésie septentrionale, Québec. Université Laval, Unpublished Master thesis. p 76
Jakob M, Friele P (2010) Frequency and magnitude of debris flows on Cheekye River, British Columbia. Geomorphology 114:382–395
Jobin A (2019) Impacts géomorphologiques de la tempête post-tropicale de juillet 2014 sur un versant raide de la vallée de Mont-Saint-Pierre, Gaspésie, Québec. Université du Québec à Montréal, Unpublished Master thesis. p 90
Jomelli V, Francou B (2000) Comparing the characteristics of rockfall talus and snow avalanche landforms in an Alpine environment using a new methodological approach: Massif des Ecrins, French Alps. Geomorphology 35:181–192
Jutras P, Schroeder J (1999) Geomorphology of an exhumed carboniferous paleosurface in the southern Gaspé Peninsula, Québec: paleoenvironmental and tectonic implications. Géogr Phys Quatern 53:249–263
Kirkby MJ, Statham I (1975) Surface stone movement and scree formation. J Geol 83:349–362
Knigth J, Harrison S (2013) The impact of climate change on terrestrial Earth surface systems. Nat Clim Change 3:24–29
Knigth J, Harrison S (2014) Mountain glacial and paraglacial environmnets under global climate change: lessons from the past, future directions and policy implications. Geogr Ann A 96:245–264
Knigth J, Harrison S (2018) Transience in cascading paraglacial systems. Land Degrad Dev. pp 1–11
Korup O, Clague JJ (2009) Natural hazards, extreme events, and mountain topography. Quaternary Sci Rev 28:977–990
Korup O, Tweed F (2007) Ice, moraine, and landslide dams in mountainous terrain. Quaternary Sci Rev 26:3406–3422
Labelle C, Richard PJH (1984) Histoire postglaciaire de la végétation dans la région de Mont-Saint-Pierre, Gaspésie, Québec. Géogr Phys Quatern 38:257–274
Lafortune M, Filion L, Hétu B (1997) Dynamique d'un front forestier sur un talus d'éboulis actif en climat tempéré froid (Gaspésie, Québec). Géogr Phys Quatern 51:67–80
Lane SN, Bakker M, Gabbud C, Micheletti N, Saugy J-N (2017) Sediment export, transient landscape response and catchment-scale connectivity following rapid climate warming and Alpine glacier recession. Geomorphology 277:210–227
Liang E, Lu X, Ren P, Li X, Zhu L, Eckstein D (2011) Annual increments of juniper dwarf shrubs above the tree line on the central Tibetan Plateau: a useful climatic proxy. Ann Bot-London 109:721–728
Marcoux N, Richard PJH (1995) Végétation et fluctuations climatiques postglaciaires sur la côte septentrionale gaspésienne, Québec. Can J Earth Sci 32:79–96
Martin J-P, Germain D (2016) Late-glacial and Holocene evolution as a driver of diversity and complexity of the northeastern North American alpine landscapes: a synthesis. Can J Earth Sci 53:494–505
Matthews JA, Dahl S-O, Berrisford MS, Nesje A, Quentin Dresser P, Dumayne-Peaty L (1997) A preliminary history of Holocene colluvial (debris-flow) activity, Leirdalen, Jotunheimen, Norway. J Quaternary Sci 12:117–129
McCarroll D, Shakesby RA, Matthews JA (2002) Enhanced rockfall activity during the little ice age: further lichenometric evidence from a Norwegian talus. Permafrost Periglac 12:157–164
Meigs A, Krugh WC, Davis K, Bank G (2006) Ultra-rapid landscape response and sediment yield following glacier retreat, Icy Bay, southern Alaska. Geomorphology 78:207–221
Mercier D (2008) Paraglacial and paraperiglacial landsystems: concepts, temporal scales and spatial distribution. Géomorphologie 223–234
Ministère de l'Énergie et des Ressources Naturelles (2016) Le modèle numérique d'altitude (MNA) à l'échelle de 1/20 000—fiche technique. Direction générale de l'information géographique, Gouvernement du Québec
Moen J, Aune K, Edenius L, Angerbjörn A (2004) Potential effects of climate change on treeline position in the Swedish mountains. Ecol Soc 9:16
Noren AJ, Bierman PR, Steig EJ, Lini A, Southon J (2002) Millenial-scale storminess variability in the northeastern United States during the Holocene epoch. Nature 419:821–824
Ouellet M-A, Germain D (2014) Hyperconcentrated flows on a forested alluvial fan of Eastern Canada: geomorphic characteristics, return period, and triggering scenarios. Earth Surf Processes 39:1876–1887
Pawlik L (2013) The role of trees in the geomorphic system of forested hillslopes—a review. Earth Sci Rev 126:250–265
Payette S (2007) Contrasted dynamics of northern Labrador tree lines caused by climate change and migrational lag. Ecology 88:770–780
Peulvast JP, Bouchard M, Jolicoeur S, Pierre G, Schroeder J (1996) Palaeolandforms and morphotectonic evolution around the Baie des Chaleurs (eastern Canada). Geomorphology 16:5–32
Price M, Jansky L, Iatsenia AA (2004) Key issues for mountain areas. United Nations University Press, Tokyo, Japan

Richard PJH, Larouche A (1994) Histoire postglaciaire de la végétation et du climat dans la région de Rimouski, Québec. Paléo-Québec 22:49–111

Sawada M, Gajewski K, de Vernal A, Richard PJH (1999) Comparaison of marine and terrestrial Holocene climatic reconstructions from northeastern North America. Holocene 9:267–277

Schlunegger F, Badoux A, McArdell BW, Gwerder C, Schnydrig D, Rieke-Zapp D, Molnar P (2009) Limits of sediment transfer in an alpine debris-flow catchment, Illgraben, Switzerland. Quat Sci Rev 28:1097–1105

Schrott L, Niederheide A, Hankammer M, Hufschmidt G, Dikau R (2002) Sediment storage in a mountain catchment: geomorphic coupling and temporal variability (Reintal, Bavarian Alps, Germany). Z. Geomorph N.F. Suppl.-Bd. 127:175–196

Stabile-Caillé L (2019) Dynamique des talus d'éboulis en milieu forestier de la Gaspésie septentrionale: processus et évolution au cours de l'Holocène. Université du Québec à Montréal, Unpublished Master thesis. p 106

Statham I (1976) A scree slope rockfall model. Earth Surf Processes 1:43–62

Stoffel M, Schneuwly D, Bollschweiler M, Lièvre I, Delaloye R, Myint M, Monbaron M (2005) Analyzing rockfall activity (1600–2002) in a protection forest—a case study using dendrogeomorphology. Geomorphology 68:224–241

Trenhaile AS, Rudakas PA (1981) Freeze-thaw and shore platform development in Gaspé, Québec. Géogr Phys Quatern 35:171–181

van Steijn H (2002) Long-term landform evolution: evidence from talus studies. Earth Surf Processes 27:1189–1199

van Steijn H, Bertran P, Francou B, Hétu B, Texier J-P (1995) Models for the genetic and environmental interpretation of stratified slope deposits: review. Permafrost Perigl 6:125–146

van Steijn H, Boelhouwers J, Harris S, Hétu B (2002) Recent research on the nature, origin and climatic relations of blocky and stratified slope deposits. Prog Phys Geog 26:551–575

Warburton J (2007) Sediment budgets and rates of sediment transfer across cold environments in Europe: a commentary. Geogr Ann A 89:95–100

Whiteside CJ, Butler DR (2011) Adequacies and deficiencies of alpine and subalpine treeline studies in the national parks of the western USA. Prog Phys Geogr 35:19–42

Wohl E, Brierley G, Cadol D, Coulthard TJ, Covino T, Fryirs K, Grant G, Hilton RG, Lane SN, Magiligan FJ, Meitzen KM, Passalacqua P, Poeppl RE, Rathum SL, Sklar LS (2019) Connectivity as an emergent property of geomorphic systems. Earth Surf Processes 44:4–26

12 Distance from Retreating Snowfields Influences Alpine Plant Functional Traits at Glacier National Park, Montana

Martha E. Apple, Macy K. Ricketts, Alice C. Martin, and Dennis J. Moritz

Abstract

The snowfields and glaciers of Glacier National Park, Montana, USA, are retreating due to climate change. This presents alpine plants with changes in habitat and hydrology as the extent of snowfield plant habitat diminishes. We established georeferenced transects at the formerly permanent snowfields of Siyeh Pass, Piegan Pass, and at the Clements Mountain Moraine in the Lewis Mountain Range of Glacier National Park for long-term monitoring of plant functional traits and species distribution. Field observations, taxonomic literature, and measurements of collected leaves provided data to calculate community weighted trait means (CWTM) of quantitative functional traits and the relative percent cover (RPC) of qualitative functional traits. The total percent cover of plants increased significantly with distance from the snow. Raunkiaer plant growth forms differed significantly as there was a greater abundance of cryptophytes with subterranean overwintering buds near the snow but a greater abundance of woody chamaephytes and phanerophytes away from the snow. The significantly lower CWTM of specific leaf area (SLA, mm^2/mg dry weight) away from the water-rich snowfield edge suggests xeromorphy as a response to water limitation. Rhizomes may be an important colonizing mechanism for habitat exposed by retreating snow and ice, as the RPC of rhizomatous species was significantly greater near the snow and since rhizomes are clonal and carry vegetative and reproductive buds. The distribution of plant functional traits and species can be used to predict responses of alpine plants to the disappearance of snowfields and glaciers.

Keywords

Alpine plants · Snowfields · Functional traits · Climate change

M. E. Apple (✉)
Department of Biological Sciences, Montana Technological University, Butte, MT 59701, USA
e-mail: mapple@mtech.edu

M. K. Ricketts
Department of Botany, University of Wyoming, Laramie, Wyoming 82071, USA

A. C. Martin
Division of Biological Sciences, University of Montana, Missoula, MT 59801, USA

D. J. Moritz
Department of Mathematics, University of Montana, Missoula, MT 59801, USA

U. Schickhoff et al. (eds.), *Mountain Landscapes in Transition*,
Sustainable Development Goals Series, https://doi.org/10.1007/978-3-030-70238-0_12

12.1 Introduction

Alpine snowfields can be vast, permanent, and in possession of extensive perimeters which constitute very important habitats for alpine plants. Alpine snowfields retract inward as they melt during the brief summer growing season. This seasonal retraction supplies meltwater (Vitasse et al. 2016) for alpine plants and also supplies space for them to grow when overwintering alpine plants extend their shoots aboveground as the snow melts. Globally, the area occupied by the Koppen classification of alpine tundra has decreased (Diaz and Eischeid 2007) and climate change brings the concomitant retreat of snowfields and glaciers.

This phenomenon is pronounced at Glacier National Park in Montana, USA (Hall and Fagre 2003). Snowfields at Glacier National Park that used to persist throughout the year and that were designated as permanent are now diminishing. With this retreat comes a reduction in snowfield edge habitat, which is especially important because alpine plants are sensitive to climate change (Gottfried et al. 2012; Pauli et al. 2012; Lesica 2014; Hotaling et al. 2017; Lamprecht et al. 2018). The perimeters of the snowfields and glaciers, and therefore, the extent of the edge habitats will be diminished. The alpine plants that inhabit snowfield edges may not only be subjected to habitat loss but also to other effects of climate change such as increased temperatures and changes in precipitation regimes. Newly exposed ground which has the potential to be colonized but which lacks the supply of meltwater appears with the retreat of snowfields and glaciers. This newly exposed ground may eventually be colonized by current snowfield plant species. In this chapter, we consider a study of the distribution of snowfield plant species and their functional traits, which are characteristics of plants that interact with the environment, along a gradient of distance from snowfield edges at Glacier National Park.

Glacier National Park is characterized by extensively glaciated alpine landforms. It is a biologically and culturally significant national park adjacent to Waterton Lakes National Park in Alberta, Canada. Together the two parks form Waterton Glacier International Peace Park, which became a World Heritage Site in 1995. The effects of climate change on snowfields and glaciers at the park (Pederson et al. 2010) are of major interest (Carey 2007; Fagre et al. 2017), in part because of their influence on alpine plants. Although rare arctic-alpine plant species extend southward into Glacier National Park, their populations are declining in response to the increased temperatures associated with climate change (Lesica and McCune 2004).

Currently, some of the snowfields at Glacier National Park are still fairly vast. Their dimensions and characteristics vary, though. Three representative and current snowfields that provide extensive edge habitat are the Clements Mountain Moraine snowfield, which sits between the foot of Clements Mountain and a sinuous glacial moraine. This snowfield was a glacier until the 1930s, and it can be reached along a short trail from the Logan Pass Visitor Center, which receives many visitors each season. The other two are further afield but still accessible on foot. They are the large Siyeh Pass snowfield, (Fig. 12.1), which covers a steep hill extending downward from Siyeh Pass near the Sexton Glacier, and the somewhat smaller but still steep Piegan Pass snowfield, which is adjacent to the Continental Divide Trail.

A gradient exists with distance from the edge of an alpine snowfield. Plants closer to the edge may have a shorter growing season but may complete their growth and flowering in a shorter time than plants farther from the edge, but which may have a longer time each year in which they are not covered by snow and during which they can actively grow aboveground. The edge habitat can be complex and rife with microhabitats that include rocks and other substrata near snowfields, steep topographic and environmental gradients, channels of snowmelt water, variations in the steepness of slopes, and periglacial patterned ground (Scherrer and Körner 2011; Apple et al. 2019).

The functional traits of alpine plants allow them to live in harsh alpine environments, where

Fig. 12.1 Siyeh pass snowfield with a pool of melted snow at Glacier National Park on July 25, 2018

they are subjected to extreme cold, fluctuations in temperature, herbivory, intense sunlight, and high winds. Functional traits of plants vary with environmental gradients and can be defined as structural and physiological characteristics that govern interactions of plants with the environment. (Cornwall and Ackerly 2009). Examples include but are not limited to leaf morphology, photosynthetic pathways, stature, longevity, height, the ability to form clones, types of overwintering structures, root architecture, and the presence of belowground symbioses with nitrogen-fixing bacteria and/or with mycorrhizal fungi. Other functional traits are flower color and morphology, pollination syndromes, and phenology. Individual traits as well as suites of traits can be important determinants of where a particular plant can live. Snowfield plants have passed through the environmental filter of the conditions posed by the snowfield edge (Venn et al. 2011; 2014). Therefore, they have the functional traits necessary for life in the snowfield edge habitat. Interestingly, plants with very different growth forms and leaf morphologies can be adjacent to each other in the snowfield edge habitat, suggesting that suites of traits and multiple strategies exist for life on the snowfield's edge.

Snowfield plants may have a strong tendency to spread clonally by rhizomatous growth and may have an abundance of vegetative and floral buds that can expand quickly upon the seasonal retreat of the snowfield. Clonality also takes the form of bulbs or corms that divide underground and of bulbils that form in the usual position of seeds on the herbaceous *Polygonum viviparum* and *Festuca viviparoidea*, a grass (Lesica 2012). Clonal growth allows the plants to colonize newly exposed ground upon the retreat of snowfields, and it also allows plants to spread without relying on sexual reproduction. Clonal plants can act as stabilizers by virtue of their long and perhaps indefinite life spans (Grabherr 2003; de Whitte and Stöcklin 2010). Clonal plants can live at elevations of 6100 m (Dvorsky et al. 2016).

The availability of water from melted snow varies with distance from snowfields. Xeromorphic, or drought-tolerant, traits become increasingly important where water availability is limited. Xeromorphic belowground traits include substantial taproots that can store water, water-seeking fine roots, adventitious roots for anchorage and exploration of new environments, and the ability to survive the frozen extent of winter and dry periods as bulbs, corms, or other underground structures. Aboveground xeromorphic traits consist of overall morphological features such as cushion and shrub growth forms as well as smaller scale traits that include thick bark and modified leaves. Cushion plants can actually form their own microhabitats to shield themselves from drought and extreme temperatures (Cavieres et al. 2007).

Leaf morphology is important in xeromorphy, which is important in drought tolerance. Xeromorphic leaves can be succulent, evergreen, have thick cuticles, and rolled inward in response to drought. They can also be small or divided, with low surface to volume ratios that limit transpiration (De Micco and Aronne 2012). Photosystem-damaging heat can be dissipated by dissected leaves (Creese et al. 2010; Buchner et al. 2015). Water influences leaf expansion (Pantin et al. 2011), and with low water availability, the specific leaf area, SLA, (mm^2/mg^{-1}) is generally lower. Leaves with lower SLA have greater density than those with higher SLA. Reducing SLA in response to drought is a phenotypic adjustment (Wellstein et al. 2017). SLA is a useful measure of leaf density, which is in turn useful in determining plant responses to variations in the availability of water (Poorter et al. 2009).

Interspecific variation in functional traits can occur with distance from the snowfield and with season. For example, leaves may be less dense if expansion occurred near the edge of the snowfield and/or early in the growing season with an abundance of meltwater. In contrast, leaves of the same species may have greater density and xeromorphy if expansion occurred away from the supply of melting snow or later in the growing season if the meltwater supply was diminished.

Community weighted traits means (CWTMs) are important tools in understanding the responses of plant functional traits along environmental gradients (Choler 2005; Mouillot et al. 2013), including spatiotemporal change in alpine snowfield habitats and community responses to climate change (McGill et al. 2006; Venn et al. 2011, 2014; Apple et al. 2019). Studies of species distributions as well as of CWTMs can be used to explore responses of alpine plants to changes in snowfields at Glacier National Park.

Because of the sensitivity of alpine plants to climate change and because of the disappearance of snowfields and glaciers in response to climate change (Pederson et al. 2010), we established monitoring sites at the edges of snowfields at Glacier National Park in order to obtain baseline data on snowfield plant and functional trait distribution. These monitoring sites consist of georeferenced transects for the collection of data on the distribution of snowfield plant species and their functional traits. The transects were established with the goal of long-term monitoring tin order to provide reference points and to increase our understanding of the responses of snowfield and other alpine plants to habitat change due to climate change. We are interested in continued monitoring to determine which plants, if any, colonize the ground exposed by the retreat of snowfields and which functional traits these plants might have. Monitoring is essential for understanding the responses of alpine plants to shifts in habitat wrought by climate change and instrumental in generating data for predictive models of the fate of these plants when their icy homes have disappeared (Fagre et al. 2017; Valles et al. 2015, 2017).

12.2 Study Area

12.2.1 Site Descriptions of Snowfields at Glacier National Park

The georeferenced snowfield monitoring sites were established in 2012 and 2014 in the alpine zone of the Lewis Mountain Range, which is east of the Continental Divide at Glacier National Park. Here, rare arctic-alpine and other snowfield plants grow near extensive snowfields at Piegan Pass, Siyeh Pass, and at Logan Pass on the Clements Mountain Moraine. The Piegan Pass snowfield slopes downward to where its leading edge (2307 m) intersects the Continental Divide Trail. Krummholz trees are found within 40 m of the leading edge, which has a northern aspect. Since the Piegan Pass snowfield slopes downhill, its elevation varies but the midpoint of the lateral edge is at approximately (2335 m). The lateral edge has a western aspect and intersects periglacial patterned ground. The Siyeh Pass snowfield slopes downward to where its north facing leading edge (2362 m) supplies a puddle of meltwater which leads to an almost flat fell field. The lateral edge of the Siyeh Pass snowfield varies with elevation but at 2415 m, it has a lateral moraine ascending and then descending toward the western aspect. Close to Logan Pass, the Clements Mountain Moraine snowfield is vast but has a sampling site at 2195 m. The Clements Mountain Moraine Snowfield ends in the moraine, which has a western aspect and a 35° slope extending 17 m from the snowfield's lateral edge to the top of the moraine.

12.3 Materials and Methods

12.3.1 Transects and Sampling

At Piegan and Siyeh Pass snowfields, we established 50 m transects extending perpendicularly from the lateral and leading edges of the snowfields. The 0 m point marks the edge of the snowfield at the time of establishment. Zero points at the time of establishment were georeferenced, marked with cairns, and photographed. The Piegan Pass lateral transect (PPSL) is at 2335 m (+48.72037°N, −113.688049°W), and the Piegan Pass leading transect (PPST) is at 2307 m (+48.72037°N, −113.68746°W). The Siyeh Pass lateral transect (SPSL) is at 2415 m (+48.718202° N, −113.627278°W), and the Siyeh Pass leading transect (SPST) is at 2362 m (+48.719139°N, −113.627179°W). The Clements Mountain Moraine transect (CMM) at 2195 m (+48.413720°N, −113.435544°W) is shorter than the other two, because there are only 17 m between the snowfield's edge and the summit of the moraine.

Beginning at the edges and at 5 m intervals along the 50 m transects at the Piegan and Siyeh Pass snowfields, 1 m X 1 m quadrats were placed immediately above and below each transect line (n = 2 quadrats/interval/transect, n = 22/transect). Each interval had an effective sampling area of 1 m × 2 m, although the quadrats were quantified separately for statistical purposes. At the Mount Clements Moraine transect, two quadrats were placed at 5 m intervals with an additional quadrat at the crest of the moraine at three parallel 17 m (n = 2 quadrats/interval/transect, n = 12/transect). All quadrats were photographed. The total percent cover of plants, percent cover of individual species, species presence, and species richness were recorded for each quadrat. Step-pointing was used in 5 cm X 5 cm areas at one-meter intervals along the transects to determine the distribution of monocots, dicots, gymnosperms, bryophytes, and lichens and of snow, rock, scree, and soil. These data are from 2014, and the sites were revisited in 2016 and 2018.

12.3.2 Plant Functional Traits

Fully expanded leaves (n = 20) of each visible species were collected from mid-stem or rosette at each interval from the Clements Mountain Moraine, Siyeh Pass, and Piegan Pass transects in early August 2014. The leaves were photographed with a Zeiss/Sony or Canon camera and measured for width, length, area, and perimeter with Image-J image analysis software (Schneider et al. 2012), then dried at 70 °C for 48 h and weighed. Circularity and specific leaf area (SLA, mm^2/mg dry weight) were calculated for each leaf (Pérez-Harguindeguy et al. 2013). Height–frequency distributions were determined at 5 cm height increments for all plants in each quadrat.

Community weighted trait means (CWTM) for quantitative traits were calculated by multiplying the relative percent cover by the measured or calculated trait value (Garnier et al. 2007). Relative percent cover was calculated by multiplying the percent cover of individual species in a quadrat by total vascular plant cover in the quadrat.

Qualitative functional traits were determined by observations and from taxonomic literature (Clawson et al. 2004; Cripps and Eddington 2005; Lesica 2002, 2012; Markham 2009) and included taxonomic classification, rarity, evergreen or deciduous leaves, leaf shape, growth habit, presence of mycorrhizal and N-fixing symbioses, root architecture, clonality and mechanisms thereof, and Raunkiaer classification of the position of persistent buds (Raunkiaer 1934). Qualitative functional traits were assigned a value of one if present and a zero if not. The relative percent cover values (RPC) for all species with a particular qualitative trait were added and divided by the total vascular plant cover for the quadrat.

12.3.3 Statistics

Analysis of variance (Anova) and linear regressions were used to test for significant differences ($p \leq 0.05$) in the CWTM of quantitative traits and in the RPC for qualitative traits and species distribution along the length of the snowfield transects at Siyeh Pass, Piegan Pass, and at the Clements Mountain Moraine. Means and standard error values were calculated for each 1 m X 1 m quadrat at each 5 m interval. Linear regressions used values from each interval along the transects. Statistical analyses were conducted with JMP Software (SAS Institute, North Carolina, USA).

12.4 Results

12.4.1 The Distribution of Snowfield Plant Species and Their Functional Traits Along the Gradient of Distance from Snowfields

The distribution of alpine plant species differed significantly ($p < 0.05$) along the gradient of distance from the snowfield edges at the Siyeh Pass, Piegan Pass, and Clements Mountain Moraine snowfields of Glacier National Park, as did the community weighted trait means (CWTM) of quantitative functional traits and the relative percent cover (RPC) of qualitative functional traits (Tables 12.1, 12.2, 12.3). Quantitative morphometric traits of leaves, leaf morphology, growth form, plant height, mechanisms of clonality, belowground functional traits, and species distribution varied significantly between the edges of the snowfields and the distal ends of the transects.

Table 12.1 Relative percent cover presented as means and standard errors of vascular plant species at the proximal and distal ends of the Siyeh pass, Piegan pass, and Clements Mountain Moraine snowfield transects

Species	Proximal	Distal	P Value	R^2
Abies lasiocarpa	0.0 ± 0.0	**15.69 ± 15.69**	**0.0348**	0.2061
Anemone lithophila	5.8 ± 2.9	0.0 ± 0.0	0.1900	0.1527
Antennaria alpina	0.0 ± 0.0	2.5 ± 2.5	0.4990	0.2305
Aquilegia jonesii	0.0 ± 0.0	**22.0 ± 21.0**	**0.0077**	0.2303
Arenaria capillaris	0.0 ± 0.0	2.85 ± 1.47	0.8220	0.1122
Arnica alpinum	0.9 ± 0.04	0.0 ± 0.0	0.2141	0.1610
Astragalus bourgovii	0.0 ± 0.0	0.25 ± 0.1	0.6323	0.1918
Boechera lemmonii	**2.22 ± 2.22**	0.0 ± 0.0	**0.0255**	0.2160
Carex paysonis	**17.78 ± 11.44**	0.0 ± 0.0	**0.0455**	0.2594
Castilleja rhexifolia	0.0 ± 0.0	**1.5 ± 0.03**	**0.0094**	0.2248

(continued)

Table 12.1 (continued)

Species	Proximal	Distal	P Value	R^2
Cerastium beeringianum	0.0 ± 0.0	4.5 + 0.20	0.1805	0.2340
Claytonia lanceolata	**5.00 ± 5.00**	0.0 ± 0.0	**0.0241**	0.2045
Claytonia megarhiza	0.0 ± 0.0	2.5 ± 2.5	0.5582	0.1976
Crepis nana	0.0 ± 0.0	0.2 ± 0.2	0.5670	0.1826
Draba macounii	0.0 ± 0.0	0.8 ± 0.6	0.0606	0.1751
Dryas octopetala	0.0 ± 0.0	**15.6 ± 15.6**	**0.0257**	0.2155
Epilobium anagallidifolium	**20.71 ± 5.85**	0.0 ± 0.0	**0.0001**	0.8357
Erigeron lanatus	0.0 ± 0.0	**3.2 ± 2.5**	**0.0228**	0.1988
Erigeron peregrinus	0.0 ± 0.0	3.0 ± 3.0	0.7303	0.2325
Hieracium triste	1.0 ± 0.5	0.0 ± 0.0	0.0823	0.2723
Luzula hitchcockii	0.0 ± 0.0	1.2 ± 1.2	0.1843	0.2173
Luzula spicata	0.0 ± 0.0	**1.2 ± 0.6**	**0.0311**	0.4528
Micranthes lyallii	0.0 ± 0.0	0.5 ± 0.5	0.8510	0.2007
Minuartia obtusiloba	0.0 ± 0.0	2.0 ± 2.0	0.5531	0.1540
Oxyria digyna	**15.71 ± 6.96**	0.0 ± 0.0	**0.0001**	0.3882
Penstemon ellipticus	2.0 ± 2.0	2.0 ± 2.0	0.1817	0.0958
Phacelia hastata	0.0 ± 0.0	0.3 ± 0.3	0.7024	0.2337
Phyllodoce empetriformis	0.0 ± 0.0	2.5 ± 2.5	0.5879	0.2234
Poa alpina	1.4 ± 1.4	2.5 ± 2.5	0.1427	0.1480
Polygonum viviparum	0.0 ± 0.0	**1.02 ± 0.8**	**0.0311**	0.1648
Potentilla diversifolia	0.0 ± 0.0	6.5 ± 5.5	0.0502	0.1971
Ranunculus eschscholtzii	**1.42 ± 1.42**	0.0 ± 0.0	**0.0120**	0.3108
Salix arctica	0.0 ± 0.0	11.0 ± 10.0	0.1409	0.2574
Sedum lanceolatum	0.0 ± 0.0	0.98 ± 0.98	0.7879	0.1540
Senecio cymbalarides	1.42 ± 1.42	3.1 ± 2.0	0.0835	0.2163
Senecio fremontii	2.85 ± 2.85	2.5 ± 2.5	0.2123	0.2085
Sibbaldia procumbens	**6.3 ± 6.3**	1.2 ± 0.8	**0.0023**	0.2676
Silene acaulis	1.8 ± 1,8	2.0 ± 2.0	0.2254	0.1206
Smelowskia calycina	0.0 ± 0.0	**3.4 ± 1.4**	**0.0387**	0.1562
Solidago multiradiata	0.0 ± 0.0	0.0 ± 0.0	0.3225	0.1830

12.4.2 The Distribution of Snowfield Plant Species and Their Functional Traits Close to the Snow

Close to the snow, the CWTM of specific leaf area (SLA mm^2/mg^{-1}) was significantly greater, meaning that leaves close to the snowfield were less dense (Table 12.3). Leaf area (mm^2) and width (mm) were also significantly greater, suggesting greater leaf expansion with higher water availability close to the snow. The relative percent cover (RPC) was significantly greater for species with orbicular, arrow-shaped, and basal leaves and no plants taller than 5 cm were found within 5 m of the snowfields (Table 12.2). The RPC of Raunkiaer cryptophytes, which have overwintering structures such as bulbs, was

Table 12.2 Relative percent cover presented as means and standard errors of qualitative vascular plant traits at the proximal and distal ends of the Siyeh pass, Piegan pass, and Clements Mountain Moraine snowfield transects

Trait	Proximal	Distal	P Value	R^2
Species richness	3.6 ± 1.22	**8 ± 2.84**	**0.0004**	0.2988
Total % cover	5.7 ± 3.29	**51.4 ± 16.61**	**0.0002**	0.3104
Rare species	0.0 ± 0.0	12.3 ± 11.3	0.1675	0.1470
Phanerophyte	0.0 ± 0.0	**15.69 ± 15.69**	**0.0255**	0.1938
Chamaephyte	0.0 ± 0.0	**28.33 ± 12.08**	**0.0056**	0.2575
Hemicryptophyte	59.29 ± 15.07	54.14 ± 15.47	0.2834	0.1801
Cryptophyte	**20.71 ± 5.85**	0.67 ± 0.67	**0.0004**	0.4917
Gymnosperms	0 ± 0	**20.00 ± 20.00**	**0.0226**	0.1920
Angiosperms	100 ± 0	93.33 ± 6.66	0.2025	0.2170
Monocots	30.00 ± 9.21	7.00 ± 4.45	**0.0001**	0.3532
Dicots	55.79 ± 15.97	74.59 ± 16.14	0.4343	0.2300
Cushion	6.42 ± 4.8	**11.39 ± 8.61**	**0.0404**	0.1871
Mat	0.0 ± 0.0	**27.87 ± 13.19**	**0.0042**	0.2924
Rosette	22.22 ± 22.22	4.02 ± 1.86	0.0825	0.1694
Leaves: Simple	75.00 ± 19.36	85.81 ± 10.42	0.4972	0.1112
Evergreen	0.0 ± 0.0	**29.51 ± 17.48**	**0.0009**	0.2957
Lobed or divided	**15.04 ± 2.18**	0.160 ± 0.0	**0.0232**	0.2448
Basal	**17.94 ± 7.38**	5.22 ± 2.66	**0.0137**	0.2156
Oblong	5.71 ± 5.71	11.06 ± 6.69	0.1571	0.2349
Orbicular	**17.5 ± 0.5**	0.0 ± 0.0	**0.0012**	0.3831
Arrow	**1.42 ± 1.42**	0.0 ± 0.0	**0.0279**	0.2159
Entire	70.71 ± 18.43	67.29 ± 12.62	0.3052	0.1863
Scalloped	1.6 ± 1.6	3.63 ± 3.63	**0.5163**	0.2258
Clonality	39.15 ± 26.20	36.64 ± 26.28	0.1064	0.1880
Stolons	0.0 ± 0.0	0.007 ± 0.006	0.2843	0.3043
Rhizomes	**60.63 ± 17.54**	42.09 ± 14.55	**0.0016**	0.4222
Adventitious roots	0.0 ± 0.0	**3.70 ± 1.83**	**0.0178**	0.3903
Tap roots	2.00 ± 2.00	10.00 ± 10.00	0.6719	0.2307
Substantial roots	12.00 ± 4.00	**66.00 ± 15.00**	**0.0009**	0.4572
Woody roots	0.0 ± 0.0	**21.75 ± 15.34**	**0.0077**	0.2317
Fibrous roots	3.6 ± 2.7	1.10 ± 1.1	0.6340	0.1994
N-fixing *D. octopetala*	0.0 ± 0.0	**15.6 ± 15.6**	**0.0257**	0.2155
N-fixing fabaceae	0.0 ± 0.0	0.0 ± 0.0	0.6020	0.1918
Mycorrhizae	34.29 ± 12.48	**79.15 ± 6.05**	**0.0136**	0.4066
VAM	52.00 ± 8.00	60.00 ± 20.00	0.4499	0.2126
Ectomycorrhizae	0.0 ± 0.0	**36.78 ± 16.38**	**0.0001**	0.3618
Height: 0 − 5 cm	4.5 ± 2.0	3.0 ± 2.0	0.4391	0.2051
5 − 10 cm	30.0 ± 15.0	29.0 ± 13.32	0.6560	0.1547
10 − 15 cm	0.0 ± 0.0	**6.7 ± 3.0**	**0.0141**	0.2019
25 − 30 cm	0.0 ± 0.0	0.47 ± 0.47	0.2566	0.2386
50 − 55 cm	0.0 ± 0.0	20.00 ± 20.00	0.0990	0.2337

Table 12.3 Community weighted trait means (CWTM) presented as means and standard errors of quantitative leaf functional traits at the proximal and distal ends of the Siyeh pass, Piegan pass, and Clements Mountain snowfields

CWTM	Proximal	Distal	P Value	R^2
Dry weight (mg)	0.88 ± 0.38	0.99 ± 0.40	0.2823	0.0313
Area (mm^2)	**27.50 ± 11.07**	13.07 ± 2.91	**0.01831**	0.0500
SLA (mm^2/mg)	**4.38 ± 1.53**	1.45 ± 0.14	**0.0016**	0.1293
Length (mm)	20.87 ± 6.36	19.95 ± 4.02	0.1836	0.0752
Width (mm)	**21.83 ± 7.35**	11.07 ± 1.95	**0.0365**	0.0735
Perimeter (mm)	83.04 ± 27.67	61.85 ± 7.40	0.0638	0.1155
Circularity	0.12 ± 0.03	0.095 ± 0.01	0.5176	0.1848

significantly greater. Rhizomes and cryptophytes were the predominant means of clonality. The species with significantly greater RPC close to the snow were *Boechera lemmonii* (Brassicaceae), *Carex paysonis* (Cyperaceae), *Epilobium anagallidifolium* (Onagraceae), *Oxyria digyna* (Polygonaceae), and *Ranunculus eschscholtzii* (Ranunculaceae), (Table 12.1). Of these five species, the sedge, *C. paysonis*, is the sole monocot and it is strongly rhizomatous and mat-forming. The remaining four are small herbaceous dicots with simple leaves. *E. anagallidifolium* and *O. digyna* are strongly rhizomatous (Fig. 12.2).

12.4.3 The Distribution of Snowfield Plant Species and their Functional Traits Away from the Snow

Species richness and percent cover increased significantly with distance from the snow. The RPC of rare arctic-alpine species increased. Rare arctic-alpine plants were most abundant at Siyeh Pass, where they were found on both the lateral and leading transects (Fig. 12.3). At Piegan Pass, they grew only at the distal end of the lateral transect and they were not found at the Clements Mountain Moraine. The RPC was

Fig. 12.2 Red-leaved *Oxyria digyna* and other snowfield plants emerging near the edge of the Mount Clements Moraine snowfield

Fig. 12.3 Rare arctic-alpine *Papaver pygmaeum* blooms away from the Siyeh pass snowfield

significantly greater for species with evergreen and/or scalloped-edged leaves. Plant morphology and stature also differed since the dwarf shrubs (chamaephytes), mat-forming plants, and phanerophytes (trees) had significantly greater relative percent covers, although there was no significant difference in the RPC of hemicryptophytes (herbs) (Fig. 12.4). Here species distribution coincides with functional trait distribution, since the dominant chamaephyte species, *Dryas octopetala*, (Rosaceae) has evergreen scalloped leaves and the dominant

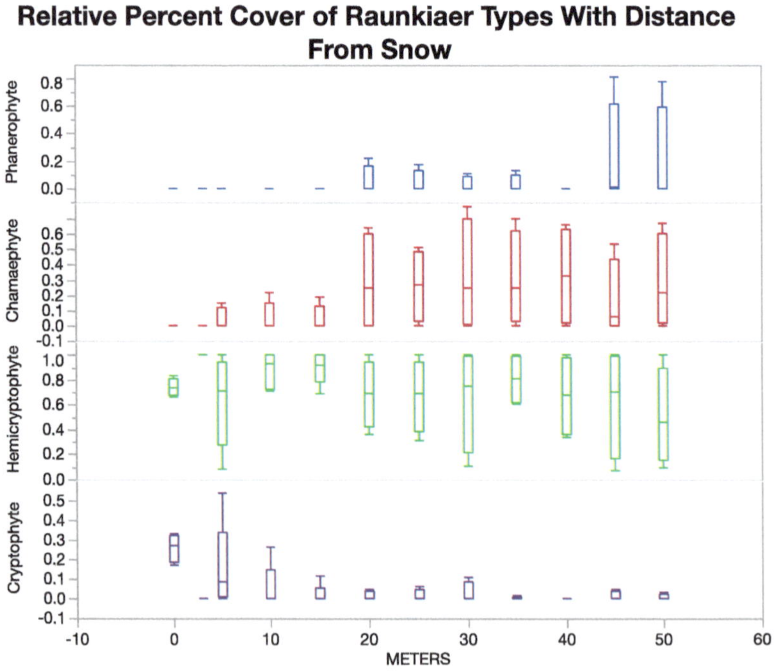

Fig. 12.4 Relative percent cover of Raunkiaer types changed with distance from the snow at the Clements Mountain Moraine, Siyeh pass, and Piegan pass transects

phanerophyte, *Abies lasiocarpa*, (Pinaceae) has evergreen needles. Plants in the 10–15 cm height range had a significantly greater RPC. There were no plants in the intermediate heights of 30–35 cm or 40–45 cm, and the tallest plants were the 50–55 cm krummholz subalpine firs, *Abies lasiocarpa* (Pinaceae) that grew toward the distal end of the PPSL transect. In contrast to the predominantly rhizomatous or cryptophytic clonality close to the snow, significantly more clonal plants were adventitiously rooted or viviparous.

The RPC of species with woody roots, taproots, and root crowns was significantly higher. Mycorrhizal species had a higher RPC, as did the ectomycorrhizal species *A. lasiocarpa* (Subalpine Fir), and the dwarf shrubs, *Dryas octopetala* (Rosaceae) and *Salix arctica* (Salicaceae). *D. octopetala* is symbiotic with nitrogen-fixing *Frankia sp.* and had a higher RPC, Symbiotic nitrogen-fixing Fabaceae did not differ significantly with distance (Tables 12.1 and 12.2).

The RPC was greater for *A. lasiocarpa*, *Aquilegia jonesii (*Ranunculaceae), *Castilleja rhexifolia* (Orobanchaceae), *Dryas octopetala* (Rosaceae), *Erigeron lanatus* (Asteraceae), *Luzula spicata* (Juncaceae), *Polygonum viviparum* (Polygonaceae), and *Smelowskia calycina* (Brassicaceae). Of these, *A. jonesii* and *E. lanatus* are rare arctic-alpine species. The root systems of most of these species are fairly robust, as *A. lasiocarpa* has extensive woody roots, *A. jonesii* and *P. diversifolia* arise from caudices, *E. lanatus* has a taproot, and *S. calycina* arises from branched rootcrowns. The remaining two, *L. spicata* and the viviparous *P. viviparum*, have rhizomes.

12.4.4 Differences in the Distribution of Snowfield Plant Species and their Functional Traits Among Snowfields

Quantitative and qualitative traits and species differed among snowfields. Species presence differed with snowfield and not all species were found at all snowfields. *Oxyria digyna* was found at the Piegan Pass and Clements Mountain Moraine but not at Siyeh Pass, while *Aquilegia jonesii* was found at Siyeh and Piegan Passes but not at the Clements Mountain Moraine.

At Siyeh Pass, The RPC of the rare arctic-alpine species, *A. jonesii* and *E. lanatus* was significantly greater away from the snow, with greater abundance on the lateral transect. The RPC was significantly greater with distance for two common species, *Arenaria capilaris* (Caryophyllaceae) and *Erigeron perigrinus* (Asteraceae). Close to the snow at the leading edge transect, the RPC of the cryptophytic, corm-producing *Claytonia lanceolata* (Portulacaceae) was significantly greater, while the RPC of oblong leaves, upright stems, and 10–15 cm tall plants increased significantly with distance from the snow.

At Piegan Pass, the CWTM for leaf dry weight, area, width, and perimeter were greater close to the snow, as were the RPC for monocots, graminoid leaves, rosettes, herbaceousness, and for *Carex paysonis* (Cyperaceae). Close to the snow at the lateral transect, the CWTM was significantly greater for leaf dry weight, SLA, area, length, width, perimeter, and circularity. Away from the snow, the RPC of scalloped-leaved species and of *Abies lasiocarpa* (Pinaceae), *Polygonum viviparum* (Portulacaceae) *Gentiana calycosa* (Gentianaceae), and *Silene acaulis* (Caryophyllaceae) increased significantly. On the leading transect, the RPC of *C. paysonis* was significantly greater while away from the snow *A. lasiocarpa* had a significantly greater RPC. Species richness did not increase significantly, possibly due to the dominance of *A. lasiocarpa* toward the distal end of the leading transect.

At Clements Mountain Moraine, there were no significant differences in total percent cover, species richness, and SLA. Close to the snow, the RPC of lobed and orbicular leaves and of *Oxyria digyna* (Polygonaceae) were significantly greater. Away from the snow, the RPC of *Penstemon ellipticus* (Plantaginaceae) was significantly greater.

Step-pointing highlighted the differences among snowfields (Table 12.4). At the lateral

Table 12.4 Plant types, lichens, and surface along snowfield transects at Glacier National park

Distance (m)	SPSL	SPST	PPSL	PPST	CMM
0	Rock	Snow	Snow	Snow	Rock
1	Rock	Monocot	Scree	Scree	Bryophyte
2	Scree	Soil, Dicot	Rock	Monocot	Scree
3	Scree	Monocot	Scree	Monocot	Scree
4	Scree	Scree	Scree	Monocot	Scree
5	Scree	Monocot	Scree	Scree	Scree
6	Scree	Scree	Scree	Scree	Scree
7	Scree	Scree	Soil	Scree	Monocot
8	Scree	Dicot	Scree	Scree	Monocot
9	Scree	Scree	Scree	Monocot	Scree
10	Scree	Monocot	Soil	Scree	Scree
11	Scree	Scree	Scree	Rock	Scree
12	Scree	Scree	Monocot	Dicot	Scree
13	Scree	Dicot	Monocot	Dicot	Scree
14	Scree	Scree	Monocot, Dicot	Scree	Scree
15	Scree	Scree	Scree	Monocot, Dicot	Scree
16	Scree	Scree	Monocot	Lichen	Scree
17	Scree	Scree	Scree	Lichen	
18	Scree	Scree	Monocot	Soil	
19	Scree	Scree	Monocot	Monocot	
20	Scree	Scree	Scree, Monocot	Soil	
21	Scree	Rock	Dicot	Dicot	
22	Scree	Scree	Dicot	Gymnosperm	
23	Dicot	Scree	Scree, Dicot	Gymnosperm	
24	Scree	Scree	Dicot	Soil	
25	Scree	Scree	Lichen	Monocot	
26	Scree	Rock	Lichen	Dicot	
27	Rock	Dicot	Scree	Dicot	
28	Scree	Dicot	Dicot	Dicot	
29	Scree	Scree	Soil	Soil	
30	Scree	Scree	Soil	Dicot	
31	Scree	Scree	Bryophyte	Dicot	
32	Scree	Scree	Dicot	Monocot	
33	Scree	Scree	Dicot	Dicot	
34	Scree	Scree	Soil	Monocot	
35	Scree	Scree	Dicot	Monocot	
36	Dicot	Scree	Monocot	Monocot	
37	Rock	Scree	Monocot	Dicot	
38	Scree	Monocot	Dicot	Monocot	

(continued)

Table 12.4 (continued)

Distance (m)	SPSL	SPST	PPSL	PPST	CMM
39	Dicot	Scree	Dicot	Soil	
40	Scree	Dicot	Dicot	Bryophyte	
41	Scree	Monocot	Dicot	Dicot	
42	Scree	Dicot	Rock	Monocot	
43	Scree	Monocot	Lichen	Scree, Lichen	
44	Scree	Monocot	Scree	Soil	
45	Scree	Monocot	Scree, Monocot	Rock, Lichen	
46	Dicot	Monocot	Monocot	Litter	
47	Scree	Monocot	Scree	Litter	
48	Scree	Soil	Scree	Rock, Lichen	
49	Scree	Litter	Scree	Rock, Lichen	
50	Rock	Monocot	Rock, Dicot	Rock, Lichen	

Fig. 12.5 *Silene acaulis*, a pink cushion plant, and *Salix arctica*, a dwarf shrub at Siyeh pass

transect of Siyeh Pass, (SPSL) scree was prevalent, no plants were intersected until meter 23 of 50 and only 4 of 50 points had plants. The leading edge of Siyeh Pass (SPST) was markedly different, with snow, soil, scree, monocots, and dicots in the first five meters and 19 of 50 points had plants (Fig. 12.5). At Piegan Pass (PPSL), no plants were intersected on the lateral transect until meter twelve, and 24 of 50 points had plants. On the Piegan Pass leading edge (PPST), monocots were found at meters three through five, gymnosperms at meters 22 and 23, epilithic lichens at meters 43, 45, 48, 49, and 50, and 28 of the 50 points had plants. At the Clements Mountain Moraine (CMM), scree was prevalent, monocots were found at meters seven and eight and dicots were present but not intersected.

12.5 Discussion of Snowfield Plant Species and Functional Traits

12.5.1 Context and Significance

Snowfield edges provide a water-rich habitat for alpine plants, but both snowfields and alpine plants are sensitive to climate change (Fagre et al. 2017; Gottfried et al. 2012; Pauli et al. 2012). The linkage between snowfields and alpine plants is an important aspect of alpine ecology, especially since the bioclimatic envelope of alpine plants varies with distance from the snow (Byrne et al. 2014). Differences in functional traits are inherently tied to species distribution and to community composition (Venn et al. 2014) because functional traits allow a given species to live in a particular habitat if not excluded by an environmental filter such as timing of snowmelt (Venn et al. 2011) as in the case of abundant graminoid species near late-lying snow patches in the Cairngorm Mountains of Scotland (Valles et al. 2015).

At Glacier National Park, species and functional traits of alpine plants differed significantly along the environmental gradient of distance from the snowfield edges, which may act as a filter for species distribution. Since the snowfield edge habitat will diminish in area with the ongoing and future retreat of the snowfields, it may be that the species and traits of plants that are currently at a distance from the snowfields will inhabit the area vacated by the retreat of the snowfields. What was formerly edge habitat may become unsuitable for the current edge species but suitable for the currently distant species provided that can and do migrate inward by seed dispersal and seedling establishment and/or by clonal mechanisms.

While clonal plants grew along the transects, the mechanism of clonality differed significantly with distance from snowfields. Near the snow, geophytic and rhizomatous species had significantly higher RPC. These clonal species can conceivably move inward to follow the retreating edge. The geophytic, or cryptophytic, *Claytonia lanceolata* reproduces vegetatively from carbon-storing corms and its appearance aboveground follows the edges of retreating snowfields. Rhizomes mobilized for rapid growth during the brief summer can follow the front of the melting snow. The vegetative and floral buds of rhizomes as well as their stores of carbohydrates allow rhizomes to be instrumental in colonization (Klimešová et al. 2012). The rhizomatous *Epilobium anagalidifolium* and *Oxyria digyna* were found near the snowfield edges, which have low percent cover and are ostensibly open to colonization. Snowfield retreat may result in a wave of colonization by rhizomatous species. Clonality away from the snow was dominated by adventitiously rooted species, where the RPC of the adventitiously rooted, anchored, mat-forming dwarf shrubs *Dryas octopetala* and *Salix arctica* was significantly higher.

Nutrient dynamics of snowfield habitats may be greatly influenced by the significant increase in ectomycorrhizae and N-fixing symbioses with distance from the snow. *Salix arctica* (the arctic willow, Salicaceae) and *D. octopetala* (Mountain Avens, Rosaceae) are ectomycorrhizal, and *D. octopetala* is symbiotic with N-fixing *Frankia sp.* (Clawson et al. 2004; Markham 2009). These dwarf shrubs may influence and be influenced by the interactions of *Frankia* and ectomycorrhizae on crucial soil nutrients (Monson et al. 2006). Patterns of plant cover and species distribution may be perpetuated by differences in nutrient dynamics wrought by underground symbioses. Soil microbes may play a key role in determining the distribution of plant species and functional traits (King et al. 2013).

Environmental conditions influence leaf development in alpine plants (Li et al. 2015). The CWTM for leaf area, width, and perimeter and for specific leaf area (SLA) was significantly less with distance from the snow. The smaller, denser leaves were increasingly xeromorphic when farther from their supply of melted snow (Gutiérrez-Girón and Gavilán 2013). It depends somewhat on where the edge of the snow and the water supply was when the leaves expanded. Other gradients besides distance from the snow can influence leaf development, as in the 35° slope ascending from the Clements Mountain Moraine snowfield.

Raunkiaer categories (Raunkiaer 1934) varied significantly with distance from the snow. Ephemeral or early season cryptophytes had a higher RPC close to the snow. They may not have been visible by late July and August when the snowfields became accessible on foot, suggesting a higher RPC than what we recorded. While the RPC of hemicryptophytes (with buds near the snow) did not vary significantly, the RPC of cushion plants (a type of hemicryptophyte) was significantly greater away from the snow. The temperature of cushion plants is decoupled from the external environment (Larcher and Wagner 2010), thus affording some thermal protection in cold mountain environments. Leaf temperatures of *Silene acaulis*, a predominant cushion plant of the GNP snowfields, can exceed ambient temperature by over 20 °C on sunny days with little wind (Gauslaa 1984; Neunerr et al. 2000). Cushion plants can overheat (Körner 2003) and alpine plants can sustain photosystem damage in response to high temperatures, which is important in the context of climate change (Buchner et al. 2015).

The RPC of dwarf shrubs (chaemophytes) and trees (phanerophytes) increased significantly with distance from the snow. Outlier trees at the ends of the Piegan Pass transects are potential seed sources for colonization and establishment of trees on land exposed by the retreat of snowfields and on land currently devoid of trees but populated by cryptophytes, hemicryptophytes, and chaemophytes. This potential influx of trees can contribute to changes in alpine habitats (Resler et al. 2005; Malanson et al. 2007). Changes in the abundance and distribution of dwarf shrubs with environmental change are likely to have major implications in alpine communities (Dawes et al. 2011).

Although many alpine plants live for decades and in some cases centuries (Forbis and Doak 2004), changes in distribution can still take place relatively quickly on the decadal and sub-decadal scales (Kullman 2007; Gottfried et al. 2012; Pauli et al. 2012). Local variability of alpine species may prove to be a valuable buffer against climate change (Malanson et al. 2012). Knowledge of the current distributions of plant species and functional traits with respect to distance from snowfields may prove instrumental in predicting which plant species and with which traits will colonize the current snowfield edge habitat once the snowfields retreat or vanish (Erschbamer et al. 2008; Valles et al. 2017). Previous research has shown that on an alpine glacial foreland, colonization is dispersal-limited, species richness increased, pioneer species decreased, and plant longevity was linked to size (Erschbamer and Retter 2004; Erschbamer et al. 2008). Clonality and clonal mechanisms, along with life spans, of pioneer species on newly exposed ground will likely influence species distribution (de Witte and Stöcklin 2010). Mycorrhizae may not be necessary for initial colonization, but it may be important in later stages, as plants colonizing glacial forefronts in the Cascade Mountains of Washington were predominantly nonmycorrhizal and the proportion of mycorrhizal plants increased in the more distal, established communities (Casares et al. 2005).

12.5.2 Long-Term Monitoring of Alpine Ecosystems

The effects of snowfield retreat on snowfield plants can be understood in part via long-term monitoring of alpine ecosystems (Gottfried et al. 2012; Pauli et al. 2012; Mark et al. 2015; Hotaling et al. 2017). Diverse methods and scales of monitoring exist (Strachan et al. 2016), and monitoring was instrumental in determining the transformation of Scandes Mountain snowfield communities into alpine grasslands in west-central Sweden (Kullman 2007). While *in-situ* monitoring of plant-specific parameters remains integral to understand change in alpine ecosystems, it can now be readily combined with sensor networks for use in detecting environmental parameters and interpreting their effects on alpine plants. *In-situ* cameras can be (CaraDonna et al. 2014) used to study the phenology of alpine plants in the context of seasonal snowmelt. Phenological changes are important bioindicators of climate change. The georeferenced snowfield plots at Glacier National Park constitute a snowfield plant observatory and were established with the goal of understanding responses of the vulnerable snowfield plants to the retreat of snowfields. While mountains are often inherently isolated, this and other studies of the responses of alpine ecosystems to climate change can be used to synthesize our understanding of and constructive responses to the different aspects of climate change.

Fig. 12.6 Mount Clements Moraine and snowfield

12.6 Conclusions

Snowfield plant species have the functional traits necessary for life in the alpine environment. This baseline study represents the establishment of long-term monitoring sites that were initiated because alpine plants, which includes the subcategory of snowfield plants, are susceptible to the effects of climate change and because the snowfields (Fig. 12.6) and glaciers of Glacier National Park and of other alpine regions of the world are likewise susceptible to the effects of climate change and are therefore diminishing in area. We found that the snowfield plant species at the Glacier National Park snowfields were significantly more likely to be rhizomatous near the snowfields, which may prove instrumental in colonizing new area uncovered by the retreat of the snowfields. Leaves that expanded nearer to the snow had significantly higher specific leaf areas, which means that they are less dense, albeit because of expansion with greater availability of water. Conversely, leaves away from the snow had a lower specific leaf area and were likely to have expanded with a reduced availability of water. Xeromorphic, or drought-tolerant, plants with traits such as cushion morphology, low specific leaf area, substantial roots, and evergreen leaves had significantly greater relative percent covers with distance from the snowfields' edges. Data from this study can be used in predictive models of the responses of snowfield plants to climate change.

Acknowledgements We acknowledge the support of the Rocky Mountain-Cooperative Ecosystem Study Unit, (RM-CESU), the Crown of the Continent Research Learning Center at Glacier National Park, Montana Technological University, and the field assistance of Nicky Ouellet, Rene Ouellet, Lindsay Carlson, James Gallagher, and Charlie Apple.

References

Apple M, Ricketts M, Martin A (2019) Plant functional traits and microbes vary with position on striped periglacial patterned ground at Glacier National Park. Montana J Geo Sci 29(7):1127–2114

Buchner O, Stoll M, Karader M, Kranner I, Neuner G (2015) Application of heat stress in situ demonstrates a protective role of irradiation on photosynthetic performance in alpine plants. Plant, Cell, Env 38 (4):812–826

Byrne J, Fagre D, MacDonald R, Muhlfield C (2014) Climate change and the Rocky Mountains. In: Grover V et al. (eds) Impact of global changes on mountains: responses and adaptation. CRC Press, pp 432–463

CaraDonna P, Iler A, Inouye D (2014) Shifts in flowering phenology reshape a subalpine community. Proc Nat Acad Sci 111(13):4916–4921

Carey M (2007) The history of ice: how glaciers became an endangered species. Env Hist 12(3):497–527

Casares E, Trappe J, Jumpponen A (2005) Mycorrhiza-plant colonization patterns on a subalpine glacier

forefront as a model system of primary succession. Mycorrhiza 15(6):405–416
Cavieres L, Badano E, Sierra-Almeida A, Molina-Montenegro M (2007) Microclimatic modifications of cushion plants and their consequences for seedling survival of native and non-native herbaceous species in the high andes of central chile. Arc Antarc Alp Res 39(2):229–236
Clawson M, Bourret A, Benson D (2004) Assessing the phylogeny of *Frankia*-actinorrhizal plant nitrogen-fixing root nodule symbioses with *Frankia* 16S rRNA and glutamine synthetase gene sequences. Molec Phylogen Evol 31:131–138
Choler P (2005) Consistent shifts in alpine plant traits along a mesotopographical gradient. Arc Ant Alp Res 37(4):444–453
Cornwall W, Ackerly D (2009) Community assembly and shifts in plant trait distributions across an environmental gradient in coastal California. Ecol Monogr 79 (1):109–126
Cripps C, Eddington L (2005) Distribution of mycorrhizal types among alpine vascular plant families on the beartooth plateau, Rocky Mountains, USA, in reference to larger-scale patterns in arctic-alpine habitats. Arc Ant Alp Res 37:177–188
Creese C, Lee A, Sack L (2010) Drivers of morphological diversity and distribution in the Hawaiian fern flora: trait associations with size, growth form, and environment. Am J Bot 98(6):955–966
Dawes M, Hagedorn F, Zumbrunn T, Handa I, Hattenschwiler S, Wipf S, Rixen C (2011) Growth and community responses of alpine dwarf shrubs to in-situ CO_2 enrichment and soil warming. New Phytol 191:806–818
De Micco V, Aronne G (2012) Morpho-anatomical traits for plant adaptation to drought. In: Aroca R (ed) Plant responses to drought stress. Springer, Berlin, pp 37–61
de Witte L, Stöcklin J (2010) Longevity of clonal plants: why it matters and how to measure it. Ann Bot 106 (6):859–870
Diaz H, Eischeid J (2007) Disappearing 'alpine tundra' Köppen climatic type in the western United States. Geophys Res Lett 34:L18707
Dvorský M, Chlumská Z, Altman J et al (2016) Gardening in the zone of death: an experimental assessment of the absolute elevation limit of vascular plants. Sci Rep 6:24440
Erschbamer B, Retter V (2004) How long can glacier foreland species live? Flora 199:500–504
Erschbamer B, Niederfriniger R, Winkler E (2008) Colonization processes on a central Alpine glacier foreland. J Veg Sci 19(6):855–862
Fagre D, McKeon L, Dick A, Fountain A (2017) Glacier margin time series (1966, 1998, 2005, 2015) of the named glaciers of Glacier National Park, MT, USA. US Geological Survey. https://doi.org/10.5066/F7P26WB1
Forbis T, Doak D (2004) Seedling establishment and life history trade-offs in alpine plants. Am J Bot 91:1147–1153
Garnier E, Lavorel S, Ansquer P, Castro H, Cruz P, Dolezal J et al (2007) Assessing the effects of land-use change on plant traits, communities and ecosystem functioning in grasslands: a standardized methodology and lessons from an application to 11 European sites. Ann Bot 99:967–985
Gauslaa Y (1984) Heat resistance and energy budget in different Scandinavian plants. Holarctic Ecol 7:1–78
Grabherr G (2003) Alpine vegetation dynamics and climate change—a synthesis of long-term studies and observations. In: Nagy L, Grabherr G, Körner C, Thompson D (eds) Alpine biodiversity in Europe. Ecological studies vol 167. Springer, Berlin, pp 399–409
Gottfried M, Pauli H et al (2012) Continent-wide response of mountain vegetation to climate change. Nat Clim Change 2:111–115
Gutiérrez-Girón A, Gavilán R (2013) Plant functional strategies and environmental constraints in mediterranean high mountain grasslands in central Spain. Plant Ecol Divers 6:435–446
Hall M, Fagre D (2003) Modeled Climate-induced glacier change in Glacier National Park, 1850–2100. Biosci 53(2):131–140
Hotaling S, Hood E, Hamilton T (2017) Microbial ecology of mountain glacier ecosystems: biodiversity, ecological connections, and implications of a warming climate. Env Micro 19(8):2935–2948
Kenzo T, Tanaka-Oda A, Mastuura Y, Hinzman L (2016) Morphological and physicochemical traits of leaves of different life-forms of various broadleaf woody plants in interior Alaska. Can J For Res 46:1475–1482
King A, Farrer E, Suding K, Schmidt S (2013) Co-occurrence patterns of plants and soil bacteria in the high-alpine subnival zone track environmental harshness. Fron Micr 4:239
Klimešová J, Doležal J, Prach K, Košnar J (2012) Clonal growth forms in arctic plants and their habitat preferences: a study from Petuniabukta. Spitsbergen Pol Pol Res 33(4):421–442
Körner C (2003) Alpine plant life. Functional plant ecology of high mountain ecosystems. Springer, Berlin
Kullman L (2007) Long-term geobotanical observations of climate change impacts in the Scandes of west-central Sweden. nor J Bot 24:445–467
Lamprecht A, Semenchuk P, Steinbauer K, Winkler M, Pauli H (2018) Climate change leads to accelerated transformation of high-elevation vegetation in the central Alps. New Phytol 220(2):447–459
Larcher W, Wagner J (2010) Temperatures in the life zones of the Tyrolean Alps. Sitzungsberichte Abt I 213:31–51
Lesica P (2002) Flora of Glacier National Park. Oregon State University Press, Corvallis
Lesica P (2012) Manual of montana vascular plants. Botanical Research Institute of Texas, Austin
Lesica P (2014) Arctic-alpine plants decline over two decades in Glacier National Park, Montana, USA. Arc Ant Alp Res 46(2):327–332
Lesica P, McCune B (2004) Decline of arctic-alpine plants at the southern margin of their range following a decade of climatic warming. J Veg Sci 15:679–690

Li H, Nicotra A, Danghui X, Guozhen D (2015) Habitat-specific responses of leaf traits to soil water conditions in species from a novel alpine swamp meadow community. Cons Phys 3(1):1–8
Malanson G, Bengtson L, Fagre D (2012) Geomorphic determinants of species composition of alpine tundra, Glacier National Park, USA. Arc Ant Alp Res 44 (2):197–209
Malanson G, Butler D, Fagre D et al (2007) Alpine treeline of western north America: linking organism-to-landscape dynamics. Phys Geog 28(5):378–396
Markham R (2009) Does *Dryas integrifolia* fix nitrogen? Bot 87(11):1106–1109
Mark A, Korsten A, Urrutia Guevara D et al (2015) Ecological responses to 52 years of experimental snow manipulation in high-alpine cushionfield, old man range, south-central New Zealand. Arc Ant Alp Res 47(4):751–772
Massicotte H, Melville L, Peterson R, Luoma D (1998) Anatomical aspects of field ectomycorrhizas on Polygonum viviparum (Polygonaceae) and Kobresia bellardii (Cyperaceae). Mycorrhiza 7(6):287–292
McGill B, Enquist B, Weiher E, Westoby M (2006) Rebuilding community ecology from functional traits. Trend Eco Evo 21(4):178–185
Monson R, Rosenstiel T, Forbis T, Lipson D, Jaeger C III (2006) Nitrogen and carbon storage in alpine plants. Int Comp Bio 46(1):35–48
Mouillot D, Graham N, Villéger S, Mason N, Bellwood D (2013) A functional approach reveals community responses to disturbances. Trends Ecol Evol 28(3):167–177
Neuner G, Buchner O, Braun V (2000) Short-term changes in heat tolerance in the alpine cushion plant silene acaulis ssp. Excapa [All.]. J Braun Differ Altitudes Plant Bio 2:677–683
Pantin F, Simonneau T, Rolland G, Dauzat M, Muller B (2011) Control of leaf expansion: a developmental switch from metabolics to hydraulics. Plant Phys 158:803–815
Pauli H, Gottfried M, Dullinger S et al (2012) Recent plant diversity changes on Europe's mountain summits. Science 336(6079):353–355
Pederson G, Graumlich L, Fagre D, Kipfer T, Muhlfeld C (2010) A century of climate and ecosystem change in Western Montana: what do temperature trends portend? Clim Change 98(1):133–154
Pérez-Harguindeguy N, Diaz S, Garnier E, Lavorel S et al (2013) New handbook for standardized measurement of plant functional traits worldwide. Aus J Bot 61:167–234
Poorter H, Niinemets U, Poorter L, Wright I, Villar R (2009) Tansley review: causes and consequences of variation in leaf mass per area (LMA): a meta-analysis. New Phytol 182:565–588
Raunkiær C (1934) The life forms of plants and statistical plant geography. Oxford University Press, Oxford
Resler L, Butler D, Malanson G (2005) Topographic shelter and conifer establishment and mortality in an alpine environment, Glacier National Park. Montana Phys Geog 26(2):112–125
Scherrer D, Körner C (2011) Topographically controlled thermal-habitat differentiation buffers alpine plant diversity against climate warming. J Biogeo 38:406–416
Schneider C, Rasband W, Eliceiri K (2012) NIH image to imageJ: 25 years of image analysis. Nat Met 9(7):671–675
Strachan S, Kelsey E, Brown R et al (2016) Filling the data gaps in mountain climate observatories through advanced technology, refined instrument siting, and a focus on gradients. Mount Res Dev 36(4):518–527
Valles D, Apple M, Dick J, Andrews C, Gutiérrez-Girón A, Pauli H (2015) Modeling plant functional traits and elevation in the cairngorm mountains of Scotland. Paper Presented Int Conf Mod Sim Vis Met, Las Vegas, NV 15:3–9
Valles D, Apple M, Andrews C (2017) Visual simulations correlate plant functional trait distribution with elevation and temperature in the cairngorm mountains of Scotland. Paper presented at the 2017 international conference computer science computer Intelligence. Las Vegas, NV https://doi.org/10.1109/CSCI.2017.220
Venn S, Green K, Pickering C, Morgan J (2011) Using plant functional traits to explain community composition across a strong environmental filter in Australian alpine snowpatches. Plant Eco 212:1491–1499
Venn S, Pickering C, Green K (2014) Spatial and temporal functional changes in alpine summit vegetation are driven by increases in shrubs and graminoids. AoB Plants 6:plu008. https://doi.org/10.1093/aobpla/plu008
Vitasse Y, Rebetez M, Filippa G, Cremonese E, Klein G, Rixen C (2016) 'Hearing' alpine plants growing after snowmelt: ultrasonic snow sensors provide long-term series of alpine plant phenology. Int J Biomet 61 (2):349–361
Wellstein C, Poschlod P, Gohlke A et al (2017) Effects of extreme drought on specific leaf area of grassland species: a meta-analysis of experimental studies in temperate and sub-mediterranean systems. Glob Change Bio 23:2473–3248

13 Environmental Drivers of Species Composition and Tree Species Density of a Near-Natural Central Himalayan Treeline Ecotone: Consequences for the Response to Climate Change

Niels Schwab, Birgit Bürzle, Jürgen Böhner, Ram Prasad Chaudhary, Thomas Scholten, and Udo Schickhoff

Abstract

Climate warming is expected to facilitate alpine treeline advance to higher elevations. However, empirical studies in diverse mountain ranges give evidence of both advancing alpine treelines and rather insignificant responses. In this context, we aim at analysing environmental drivers of species composition and tree species density in the near-natural treeline ecotone in Rolwaling Himal, Nepal, in order to infer the sensitivity and responsiveness to climate warming. We differentiated plant communities and analysed population densities of tree species along the treeline ecotone from closed forest stands via the krummholz belt to alpine dwarf shrub heaths (3700–4300 m). We determined vegetation–environment–soil relationships, i.e. the effects of changing environmental conditions (e.g. nutrient and thermal deficits, plant interactions) on plant communities and stand structures across the ecotone by means of multivariate statistics. In particular, we focus on explaining the high competitiveness of *Rhododendron campanulatum* forming a dense krummholz belt and on its relation to climate change. We identified five plant communities, belonging to two different classes. Soil temperature, nitrogen supply and availability, and soil moisture content mainly differentiate species composition of the identified communities. Results indicate that trees in the ecotone show species-specific responses to the influence of site conditions, and that juvenile and adult tree responses are modulated by environmental constraints in differing intensity. In general, the analysed vegetation–environment relationships in the treeline ecotone suggest that the dense *Rhododendron* krummholz belt largely prevents the upward migration of other tree species and thus constrains the future response of Himalayan krummholz treelines to climate warming.

This chapter was prepared by merging, modifying and completing the previously published papers Bürzle et al. (2017), Schwab et al. (2017) and Schwab (2018).

N. Schwab (✉) · B. Bürzle · J. Böhner · U. Schickhoff
CEN Center for Earth System Research and Sustainability, Institute of Geography, Universität Hamburg, Hamburg, Germany
e-mail: niels.schwab@uni-hamburg.de

R. P. Chaudhary
RECAST Research Centre for Applied Science and Technology, Tribhuvan University, Kirtipur, Nepal

T. Scholten
Department of Geosciences, Chair of Soil Science and Geomorphology, University of Tübingen, Tübingen, Germany

U. Schickhoff et al. (eds.), *Mountain Landscapes in Transition*,
Sustainable Development Goals Series, https://doi.org/10.1007/978-3-030-70238-0_13

Keywords

Alpine vegetation · Central Himalaya · Climate warming · Multivariate analyses · Nepal · *Rhododendron campanulatum* · Rolwaling Himal · Species–environment relationships · Subalpine forest · Vegetation–environment relationships

13.1 Introduction

In general, the upper limit of tree life depends on the heat balance. At a global scale, low air and soil temperatures during growing season determine the position of natural alpine treelines (e.g. Troll 1973; Stevens and Fox 1991; Holtmeier 2009; Körner 2012). Thus, climate warming is expected to cause treelines to advance to higher elevations. Treelines fluctuated repeatedly as a result of climate changes during the Holocene era (Reasoner and Tinner 2009; Schickhoff et al. 2016a). However, recent empirical studies in diverse global mountain ranges showed both advancing alpine treelines and rather insignificant responses (Harsch et al. 2009). Such contrasting responses are not always sufficiently understood. At local scales, specific constellations of various abiotic factors and biotic interactions govern the elevational treeline position and the response to climatic controls (Case and Duncan 2014; Wieser et al. 2014; Weiss et al. 2015; Müller et al. 2016a). Inconsistent and sometimes contradictory responses to climate warming, observed in empirical studies, must be attributed to the local-scale complexity of interacting site factors (Schickhoff et al. 2015). At anthropogenic treelines, land abandonment is often the dominant driver of treeline advance (e.g. Durak et al. 2015), making it difficult to disentangle effects of land use and climate change (Schickhoff 2011; Schwab et al. 2017; Schwab 2018).

Treeline ecotones in the Himalaya are subjected to above-average warming rates. In line with global warming trends in other high mountain regions, several studies have observed above-average current warming trends for the Himalayan region. Warming trends of the annual mean surface air temperature of up to 1.5 °C were detected over the Tibetan Plateau and the Himalaya during the period of 1991–2012 (IPCC 2013, 2014; Mountain Research Initiative EDW Working Group 2015; Schickhoff et al. 2016b; Krishnan et al. 2019). Maximum values were found for high elevations and during winter and pre-monsoon seasons (Liu and Chen 2000; Bhutiyani et al. 2007; Krishnan et al. 2019). For the Rolwaling Valley, the target area of this study, monthly temperature trends in the order of 0.7 °C per decade were assessed in winter and pre-monsoon seasons (Gerlitz et al. 2014). Trend analyses of precipitation amounts in the Himalaya do not exhibit a consistent pattern in past decades. Sub-regions and seasons vary strongly, and long-term observations are underrepresented (IPCC 2013; Schickhoff et al. 2016b; Krishnan et al. 2019). The regional average intensity of annual precipitation and of annual mean daily precipitation has increased since the 1960s (Zhan et al. 2017; Krishnan et al. 2019). Some studies, however, have detected negative trends of winter and pre-monsoon precipitation over the western and central Himalaya (Duan et al. 2006; Bhutiyani et al. 2010; Jain et al. 2013). Wang et al. (2013) have reported an enhanced frequency of winter and pre-monsoon drought events for western Nepal.

Recent climatic changes and related environmental conditions will inevitably affect species composition and community patterns, stand structure and tree recruitment in Himalayan treeline ecotones. Vegetation will be modified to a regionally differentiated extent and at a point in time when ecological and phytosociological knowledge of Himalayan vegetation types is still rather deficient (Schickhoff et al. 2015; Schickhoff et al. 2016a; Bürzle et al. 2017; Schwab et al. 2017).

In fact, detailed phytosociological studies in the Himalayan region are rare, and the knowledge of vegetation–environment–soil relationships is still very limited, in particular in treeline ecotones (cf. Bürzle et al. 2017; Müller et al.

2016a). Miehe (1990) provided the most comprehensive account of subalpine and alpine vegetation types to date containing extensive floristic and ecological information, which is based solely on vegetation sampling. Detailed information on treeline floristic, structural and spatial patterns as well as on human impact are provided in the overviews of Schickhoff (2005) and Miehe et al. (2015).

Himalayan ecosystems are highly sensitive and vulnerable to climate change effects, with multifaceted interactions and diverse response patterns (Shrestha et al. 2012; Telwala et al. 2013; Ferrarini et al. 2014; Salick et al. 2014; KC and Ghimire 2015; Padma Alekhya et al. 2015; Schickhoff et al. 2016a). Schickhoff (2005) and Schickhoff et al. (2015) summarized the available knowledge of geographical and ecological aspects of Himalayan treelines and pointed out that treeline ecological conditions and processes, such as regeneration, carbon balance, frost, drought, snow cover, wind, soil physical and chemical conditions and others are still largely unexplored. Recent reviews of the sensitivity and response of Himalayan treeline ecotones to climate change emphasize the low responsiveness of krummholz treelines, but also highlight intense recruitment of treeline trees within the treeline ecotone and beyond, suggesting a high potential for future treeline advance (Schickhoff et al. 2015; Schickhoff et al. 2016a). Treeline shifts are reported in studies that consider uppermost seedling positions as synonymous with treeline advance (e.g. Gaire et al. 2014). However, occurrence of seedlings does not necessarily mean effective regeneration and treeline advance given the generally low survival rate of seedlings after germination and during critical later life stages (Graumlich et al. 2005; Schickhoff et al. 2016a).

Detailed knowledge of vegetation–environment–soil relationships is among the basic requirements for a better understanding of treeline response patterns to region-wide climate warming inputs. Local and landscape-scale studies are highly needed to understand how the change of soil and other environmental conditions along the elevational gradient in treeline ecotones (e.g. nutrient and thermal deficits, plant interactions) is correlated to modified population densities of tree species and to species compositions of plant communities (Bürzle et al. 2017; Schwab et al. 2017).

To reduce the aforementioned research deficits, we present a comprehensive evaluation of vegetation–environment relations, based on previous studies (Bürzle et al. 2017; Schwab et al. 2017). We hypothesized that tree populations of different life stages and plant communities show modified environmental relationships along the elevational gradient. We further hypothesized that nutrient availability and moisture supply as well as thermal conditions are crucial site factors governing the response of species and communities.

We aim at (i) analysing the relationships between environmental conditions and densities of adult as well as juvenile tree species populations and plant communities. We focus in particular on (ii) identifying crucial site factors for the high competitiveness of the krummholz belt species *Rhododendron campanulatum* and (iii) assess the susceptibility of these site factors to climate change.

13.2 Material and Methods

13.2.1 Study Area

We conducted the studies on the north-facing slope of the Rolwaling Valley (27°52′N; 86°25′E), located in Dolakha District (Province 3), east-central Nepal and part of the Gaurishankar Conservation Area. The investigated slopes cover the entire treeline ecotone from upper subalpine closed forests via a dense *Rhododendron campanulatum* krummholz belt to alpine dwarf shrub heaths with small and stunted tree species individuals, encompassing an elevational range from 3745 to 4300 m a.s.l. (Fig. 13.1). Upper subalpine closed forests are primarily composed of *Acer caudatum*, *Abies spectabilis* as well as *Betula utilis* and constitute upper limits of tall and upright tree growth. The climate of the study site is considered continental and temperate, with a dry winter and a warm summer and distinct spatial variability of temperature and precipitation (Böhner et al. 2015; Karki et al. 2016; Weidinger et al. 2018). The monsoon season accounts for approximately 80% of the total annual precipitation (Karki et al. 2017). The warming trend of the study area corresponds to general Himalayan trends (Gerlitz et al. 2014). Soils in the treeline ecotone are classified as podzols (Müller et al. 2017). The study area has a remote location. Consequently, the Rolwaling Himal (mountain) treeline at the north-facing slope exhibits a near-natural state and represents a climatic treeline (Schwab et al. 2016). Thus, given that land-use effects have disturbed most Himalayan treeline sites, the study slopes provide a unique research opportunity to detect a climate change signal when assessing treeline dynamics (cf. Schickhoff et al. 2015, 2016a; Bürzle et al. 2017; Schwab et al. 2017; Schwab 2018).

13.2.2 Data Collection

Vegetation

The studied site comprises three north-facing slopes, which are named NE1, NE2 (both northeast exposure) and NW (northwest exposure) as per their predominant exposition (Bürzle et al. 2017; Schwab 2018). We stratified these slopes according to elevational differences and changes in vegetation in elevational zones A (lower section of upper subalpine closed forests) to D (alpine dwarf shrub heaths, cf. Figure 13.1). In total, 91 square plots of 20 m × 20 m were randomly selected due to the homogeneity of physical features and vegetation structure and sampled from April to September 2013 and from July to October 2014. In each plot, height and cover of the separate vegetation layers were recorded, followed by detailed inventories of all

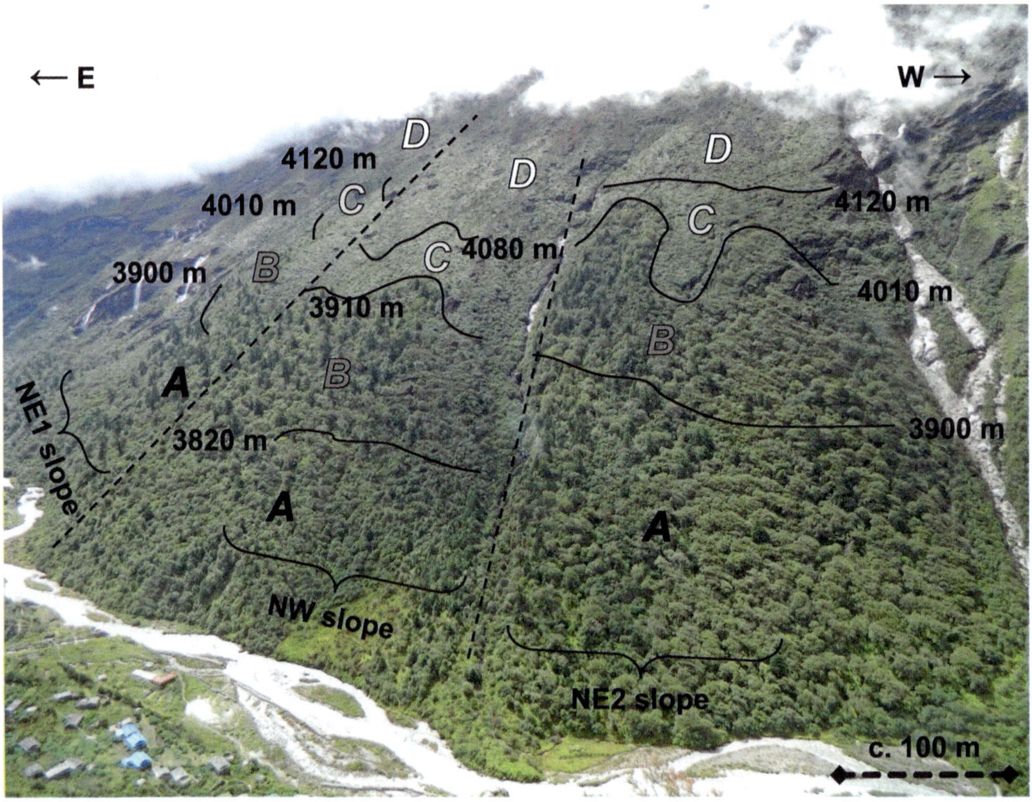

Fig. 13.1 Stratification of the study area by elevation and aspect (NE, NW). Zones A and B represent the lower and upper sections of the upper subalpine forest, C labels the krummholz belt and D indicates the alpine dwarf shrub heath (18 September 2014)

vascular plants. We estimated species cover using the established cover-abundance scale of Braun-Blanquet (1964). In addition, we counted all individuals of all occurring tree species at 50 of the aforementioned plots and termed individuals with dbh $\geq$ 7 cm 'adult trees', whilst smaller ones were categorized as 'juvenile individuals' or 'recruits' (cf. Bürzle et al. 2017; Schwab et al. 2017). Unknown specimens were collected and identified at the herbarium of the Botanische Staatssammlung München (M). In addition, we used determination keys of eFloras (https://www.efloras.org; accessed January 2015), the Flora of Nepal (Watson et al. 2011), and the Flora of Bhutan (Grierson and Long 1984–2001).

Soil, Topography and Climate

Soil samples from all plots were analysed via standard methods at the University of Tübingen, Laboratory for Soil Science and Geoecology, and at the University of Hamburg, Institute of Geography (cf. Müller et al. 2016b, 2017; Bürzle et al. 2017; Schwab et al. 2017). To capture the variation in topography between and within plots, several topographic variables characterizing aspect, slope, curvature, ground cover and microrelief were determined (Schwab et al. 2017). In addition to soil temperature and soil moisture (see Müller et al. 2016b), we recorded air temperatures from April 2013 to June 2014 through mobile climate stations which were installed in the lower and upper parts of the ecotone (Gerlitz et al. 2016).

13.2.3 Data Analyses

Plant Community Classification

For plant community classification, we applied standard phytosociological methods such as cluster analyses, table arrangement with iterative re-sorting of relevés, determination of diagnostic species by calculating the phi-coefficient and definition of character species (cf. Bürzle et al. 2017 for a comprehensive description of the methods). In view of the deficit state of knowledge, we consider the proposed classes as provisional, symbolizing the higher rank in comparison with the subordinated groups (Bürzle et al. 2017). Please refer to Bürzle et al. (2017) for nomenclature of plants and taxa.

Vegetation–Environment Relationships

We assessed significant differences in site conditions between phytosociological units with one-way analysis of variance (ANOVA) and post-hoc tests. We visualized compositional pattern of phytosociological units by means of detrended correspondence analysis (DCA). Corresponding scatter plots were restricted to two dimensions and showed the diversity in species composition along the first two axes, measured in standard deviation units of species turnover (SD) (Gauch 1982; Kent 2012). Visualized compositional patterns were interpreted applying post-hoc correlations of the first two ordination axes (with highest eigenvalue) with environmental variables (cf. Bürzle et al. 2017).

We analysed tree species population density–environment relationships through redundancy analyses (RDA) with backward elimination of explanatory variables as well as through variation partitioning (Legendre and Legendre 2012). To preselect important variables and apply variation partitioning, the explanatory dataset was split into three groups: soil, topography and climate variables. To avoid high multicollinearity, we removed within-group correlations of $|r| > 0.7$ by excluding variables. Missing values in variables (mostly soil temperature and soil moisture) would have restricted the multivariate analyses of tree species density to a minor part of the entire dataset. Hence, we applied multivariate imputation by chained equations (MICE; Van Buuren 2012) based on a random forest classification (Doove et al. 2014) to estimate the missing values based on the relationships between the variables (cf. Schwab et al. 2017). Please refer to Bürzle et al. (2017) and Schwab et al. (2017) for detailed descriptions of data analyses including applied software packages.

13.3 Results

13.3.1 Plant Communities and Environmental Conditions

We differentiated five plant communities, belonging to two classes, comprising two and three groups, respectively. Class 1, the *Betula utilis-Abies spectabilis* forests, contains the *Synotis alata-Abies spectabilis*, *Ribes glaciale-Abies spectabilis* and *Boschniakia himalaica-Rhododendron campanulatum* communities whilst class 2, the *Potentilla fruticosa-Rhododendron anthopogon* dwarf shrub heaths, comprises *Pedicularis cf. microcalyx-Rhododendron anthopogon* and *Anaphalis royleana-Rhododendron anthopogon* communities. The differentiation of the communities of the Rolwaling study site is discussed in detail in Bürzle et al. (2017).

The *Betula utilis-Abies spectabilis* forests (class 1) are developed on upper montane and subalpine N/NE slopes up to an elevation of c. 4100 m a.s.l. These forest stands are primarily composed of *Abies spectabilis* and *Betula utilis*, with *Rhododendron campanulatum* and *Sorbus microphylla* forming a second tree layer (*Synotis alata-Abies spectabilis* and *Ribes glaciale-Abies spectabilis* communities) (Fig. 13.2). At its upper distribution, closed forest stands give way to *Rhododendron campanulatum* krummholz forming an extensive belt between 3900 and 4000 m a.s.l. (NW-exposed), and 4000 and 4100 m a.s.l. (NE-exposed) (*Boschniakia himalaica-Rhododendron campanulatum* community). The layer (mean cover 85%) of gnarled and stunted krummholz trees attains a mean height of about three metres, which gradually decreases upslope. The stems form a dense and largely impenetrable thicket dominated by *Rhododendron campanulatum* with few interspersed, multi-stemmed *Sorbus microphylla* individuals. The name-giving, highly diagnostic taxon *Boschniakia himalaica* is parasitic on *Rhododendron* species (Miehe 1990). Shrub and herb layers have low productivity, indicated by small height (2 m and 0.1 m, respectively) as well as low cover (17.5% and 25%, respectively) and numbers of vascular plant species. As the DCA (Fig. 13.3) and the ANOVA (Table 13.1) suggest, altitude and mean soil temperature appear to be the decisive ecological factors for differences in species composition of the three syntaxa of this class (Bürzle et al. 2017).

In general, the three forest communities are developed on podzol soils with an extremely low pH value of c. 3.0 (Table 13.1) (Müller et al. 2017). The sandy fraction dominates the particle size distribution of the soils (sandy loams); its percentage increases slightly towards higher altitudes (Table 13.1), implying a deterioration of nutrient storage and water-holding capacity. Reduced water retention capacity in turn results in decreasing soil moisture along the elevational gradient (Table 13.1). Slight differences in soil pH do not play a role for the differentiation of species composition, as mean values are extremely low throughout the entire treeline ecotone with slightly higher values in the *Synotis alata-Abies spectabilis* community (Table 13.1) (Bürzle et al. 2017; Müller et al. 2017). Soil C:N ratios increase with elevation, most notably from the subalpine forest to krummholz and the alpine tundra, indicating increasing nutrient (N, P) shortage especially in the alpine tundra (Müller et al. 2017).

In detail, environmental conditions of each community differ specifically: the *Synotis alata-Abies spectabilis* community (community 1.1) occurs on moderately steep slopes (33° on average) between 3700 and 3900 m a.s.l., i.e. in the lower subalpine range of the *Betula utilis-Abies spectabilis* forests. The environmental conditions of this syntaxon show comparatively favourable growth conditions with higher soil temperature sums and soil moisture as well as better nutrient supply (nitrogen and potassium contents, C:N ratio) (cf. Table 13.1, Fig. 13.3, discussion and Bürzle et al. 2017).

The *Ribes glaciale-Abies spectabilis* community (community 1.2) of tall mixed forest stands occupies steep slopes (mean of 37°) in the upper subalpine range of the treeline ecotone (3800 to

Fig. 13.2 Photographs from the investigated treeline ecotone. **a** View across the ecotone with mixed subalpine forest in front followed upward by *Rhododendron campanulatum* krummholz belt and reddish flowering alpine *Rhododendron* dwarf shrub heath (*Anaphalis royleana-Rhododendron anthopogon* community); **b** elevational zone D featuring the *Anaphalis royleana-Rhododendron anthopogon* community; **c** *Rhododendron campanulatum* dominating the tree species density at the upper part of elevational zone B; **d** subalpine forest at elevational Zone A showing the *Synotis alata-Abies spectabilis* community (a: 4 October 2014; b 17 September 2014; c: 4 August 2013; d: 20 July 2013)

4050 m a.s.l.). The constellation of site factors (e.g. soil temperatures, nutrient availability) and resulting growth conditions give evidence of an intermediate position of the *Ribes glaciale-Abies spectabilis* community within the vegetation–environment pattern of the study site. Soil temperature during summer months is significantly lower compared to the *Synotis alata-Abies spectabilis* community, but higher (insignificant) than in the adjoining krummholz belt (Table 13.1). The same holds true for C:N ratio and content of potassium. However, nitrogen supply in the *Ribes glaciale-Abies spectabilis* community is lowest (insignificant) within the three forest communities, but still higher than in the *Rhododendron* dwarf shrub communities (Table 13.1) (cf. Bürzle et al. 2017).

The *Boschniakia himalaica-Rhododendron campanulatum* community (krummholz belt, community 1.3) occupies steep slopes (mean of 35°) between the treeline and the dwarf shrub heaths. The community has to cope with further deteriorating growth conditions along the elevational gradient. As the ANOVA revealed (Table 13.1), annual mean soil temperature and connected growing degree days are lower compared to the downslope communities. At the same time, C:N ratio widens, indicating lower

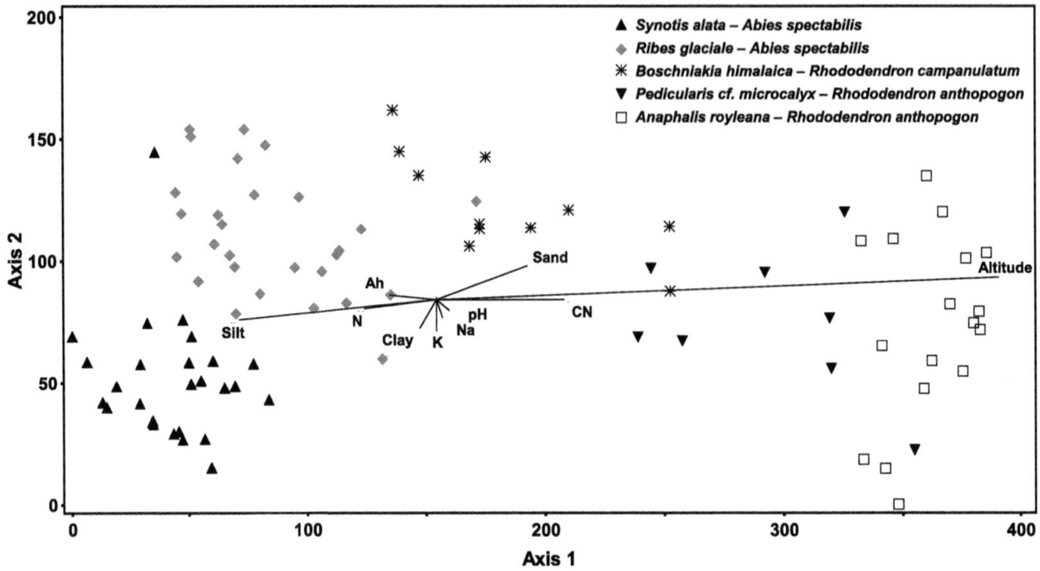

Fig. 13.3 Detrended correspondence analysis (DCA) with post-hoc correlation of environmental parameters. Relevés (N = 91) were classified into five plant communities ((modified from Bürzle et al. 2017)

decomposition rates and lower availability of nitrogen. Mean pH value marginally decreases to 2.8 (cf. Bürzle et al. 2017).

Potentilla fruticosa-Rhododendron anthopogon dwarf shrub heaths (class 2) occur above the treeline in an elevational range between 3950 and 4300 m a.s.l. The *Rhododendron* dwarf thickets are distributed on podzols with pH values of c. 3.0 and formed by *Rhododendron anthopogon*, *Rhododendron setosum* and *Rhododendron lepidotum*. As mentioned above, growth conditions deteriorate along the elevational gradient (see also Müller et al. 2016a,2017). Thus, sites of *Rhododendron* dwarf shrub communities are characterized by lower soil temperatures, wider C:N ratios and lower nitrogen contents compared to the forest communities of class 1 (Table 13.1) (cf. Bürzle et al. 2017).

The dense *Pedicularis* cf. *microcalyx-Rhododendron anthopogon* community (community 2.1) is established on steep slopes (37° on average) immediately above the krummholz belt at elevations between 3950 and 4150 m a.s.l. Growth conditions are less favourable with reduced mean soil temperatures (insignificant), base saturation (insignificant) and nitrogen contents (insignificant) (cf. Table 13.1) in comparison with communities at lower elevation. The dwarf shrub heaths reach a mean height of 0.8 m and a mean cover of 80%. Different *Rhododendron* species (*Rhododendron anthopogon*, *Rhododendron lepidotum*, dwarf individuals of *Rhododendron campanulatum*) dominate the community, with interspersed *Bistorta vaccinifolia* shrubs and *Sorbus microphylla* shrubs or small trees. The herb layer is less developed (mean height 0.1 m, mean cover 45%) and composed of fern species and species of the genera *Kobresia* and *Calamagrostis* (cf. Bürzle et al. 2017).

The open *Anaphalis royleana-Rhododendron anthopogon* community (community 2.2) occurs in the uppermost zone of the treeline ecotone, extending over an elevational range between 4100 and 4300 m a.s.l. (Figs. 13.1 and 13.2). Site conditions are the most unfavourable along the elevational gradient (see DCA, Fig. 13.3) e.g. in terms of mean soil temperatures during the vegetation period, number of growing degree days, soil moisture and thickness of the Ah-horizon (Table 13.1). The herbaceous layer has a mean cover of 50% and is not as sparse as in the *Pedicularis* cf. *microcalyx-Rhododendron anthopogon* community. The *Rhododendron*

Table 13.1 Mean values (Mean) and standard deviation (SD) of environmental parameters of plant communities. Parameters of communities with different letters in column 'Sig.' are significantly different at $p < 0.05$ (one-way-ANOVA, post-hoc pairwise t-test with Holm correction) ((modified from Bürzle et al. 2017; Müller et al. 2016a, b, 2017)

Communities parameters	*Synotis alata—A. spectabilis*				*Ribes glaciale—A. spectabilis*				*Boschniakia himalaica—R. campanulatum*				*Pedicularis* cf. *microcalyx—R. anthopogon*				*Anaphalis royleana—R. anthopogon*			
	N	Mean	SD	Sig	N	Mean	SD	Sig	N	Mean	SD	Sig	N	Mean	SD	Sig	N	Mean	SD	Sig
Landscape parameters																				
Altitude (m.a.s.l.)	27	3815.5	44.1	a	28	3928.5	58.0	b	11	4051.4	60.1	c	8	4098.4	72.8	c	17	4215.1	43.3	d
Inclination (°)	27	33.2	4.4	a	28	36.9	3.5	b	11	35.1	3.1	ab	8	37.1	3.6	ab	17	33.4	3.3	a
Exposition (°)	27	41.4	63.0		28	50.9	62.1		11	91.7	128.4		8	75.3	114.2		17	38.0	17.0	
Soil parameters																				
pH (KCl)	27	3.1	0.4	ab	28	3.0	0.3	a	11	2.8	0.3	a	8	3.0	0.2	ab	17	3.2	0.2	b
Ct (%)	27	29.6	12.3		28	25.9	13.6		11	34.4	12.4		8	26.6	13.7		17	23.6	11.8	
Nt (%)	27	1.5	0.6	a	28	1.3	0.6	ab	11	1.4	0.4	ab	8	1.1	0.5	ab	17	1.0	0.3	b
C:N ratio	27	19.6	3.2	a	28	20.4	2.6	a	11	24.1	4.6	b	8	23.9	3.1	b	17	23.9	4.8	b
CEC (µmolc/g)	27	188.0	135.1		28	188.6	116.4		11	234.3	117.4		8	183.7	120.3		17	151.0	68.0	
Base saturation (%)	27	61.1	22.7		28	57.0	24.6		11	66.8	19.7		8	40.5	17.7		17	67.0	20.5	
Exchange acidity (µmolc/g)	27	67.1	39.7	ab	28	68.2	43.2	ab	11	75.5	45.3	ab	8	106.0	53.2	a	17	45.2	18.6	b
Na (µmolc/g)	27	0.3	0.2		28	0.2	0.3		11	0.2	0.1		8	0.2	0.2		17	0.3	0.5	
K (µmolc/g)	27	10.8	7.5		28	8.2	5.3		11	8.9	4.4		8	7.8	4.3		17	10.5	5.4	
Mg (µmolc/g)	27	32.0	25.6		28	29.7	23.3		11	42.3	23.8		8	19.7	12.2		17	30.0	20.4	
Ca (µmolc/g)	27	109.6	119.3		28	107.1	99.1		11	142.0	122.8		8	58.3	71.9		17	86.8	60.8	
Mn (µmolc/g)	27	5.4	12.7	ab	28	3.7	7.3	ab	11	0.9	0.9	ab	8	0.5	0.3	a	17	4.4	4.8	b
Fe (µmolc/g)	27	7.7	8.9		28	7.3	7.5		11	11.4	10.0		8	9.3	2.7		17	7.4	5.3	
Al (µmolc/g)	27	35.4	29.8	ab	28	32.7	26.9	ab	11	34.7	25.5	ab	8	49.6	21.2	a	17	24.2	15.9	b
Ah (cm)	27	8.3	7.1	a	28	9.2	6.2	a	11	9.3	5.4	a	8	11.1	10.9	a	17	2.8	2.4	b
Organic (%)	27	51.7	21.6		28	44.3	23.4		11	59.2	21.4		8	46.3	22.9		17	41.3	20.0	
Sand (%)	23	56.9	16.9		25	61.1	14.0		9	68.6	10.7		8	69.5	6.2		17	68.0	5.9	

(continued)

Table 13.1 (continued)

Communities parameters	*Synotis alata*—*A. spectabilis*				*Ribes glaciale*—*A. spectabilis*				*Boschniakia himalaica*—*R. campanulatum*				*Pedicularis* cf. *microcalyx*—*R. anthopogon*				*Anaphalis royleana*—*R. anthopogon*			
	N	Mean	SD	Sig	N	Mean	SD	Sig	N	Mean	SD	Sig	N	Mean	SD	Sig	N	Mean	SD	Sig
Silt (%)	23	28.0	6.6	a	25	26.0	5.3	a	9	22.1	6.7	ab	8	22.1	5.7	ab	17	19.0	4.1	b
Clay (%)	23	15.0	11.3		25	12.9	11.0		9	9.3	5.2		8	8.4	3.6		17	13.0	5.4	
Climate parameters																				
Mean soil temp. (year) [°C]	6	3.5	0.6	ab	8	3.6	0.4	b	3	2.7	0.4	abc	6	2.5	0.7	ac	3	2.5	0.8	c
Mean soil temp. (ON) [°C]	7	3.1	0.7		10	3.3	0.6		3	2.8	0.4		6	2.6	0.8		3	2.7	0.2	
Mean soil temp. (DJF) [°C]	7	−2.1	1.4		8	−0,9	0.7		3	−1.9	0.8		6	−2.9	1.5		3	−2.2	0.9	
Mean soil temp. (MAM) [°C]	6	1,3	0.6		8	1,3	0.8		3	0.3	0.2		6	0.3	0.5		3	0.0	0.4	
Mean soil temp. (JJAS) [°C]	7	9,5	0.2	a	10	8.8	0.4	b	3	7.8	0.6	ab	6	8.2	0.7	ab	3	7.8	0.7	ab
Growing degree days	7	169.6	8.3	a	8	169.3	15.0	ab	3	141.7	4.5	c	6	146.8	11.9	bc	3	137.7	10.7	ac
Mean Soil Moisture (year) [pF]	7	2.1	0.5		10	2.3	0.6		3	2.3	0.2		7	2.3	0.4		3	2.6	0.4	

shrub layer has a mean cover of 60% and attains a height of less than 50 cm, interspersed by single higher shrub individuals (*Potentilla fruticosa, Sorbus microphylla*). The shrub layer is dominated by *Bistorta vaccinifolia, Rhododendron setosum* and *Potentilla fruticosa*. The comparatively rich herb layer involves the comparatively high species richness mentioned above, grass taxa are especially frequent (genera *Calamagrostis* and *Kobresia* and others) (cf. Bürzle et al. 2017).

13.3.2 Explanation of Variation in Tree Species Density

Adult and juvenile tree species densities varied greatly along the elevational gradient (Fig. 13.4). The density of *Abies spectabilis* was smaller in the lower section of the subalpine forest (zone A) than in the upper section (zone B). The most distinct difference between these zones was the higher proportion of *Rhododendron campanulatum* in zone B compared to zone A (Fig. 13.2). Closed forests gave way to a dense and nearly impenetrable *Rhododendron campanulatum* krummholz belt (zone C) at approximately 3910 m a.s.l. (NW slope) and 4010 m a.s.l. (NE slopes). There, *Rhododendron campanulatum* dominated the tree species composition. Tree species individuals were rare above the krummholz belt at the alpine section of the transect (zone D), which was mainly composed of *Rhododendron* dwarf shrubs (*Rhododendron anthopogon, Rhododendron lepidotum, Rhododendron setosum*) (Fig. 13.2). We found juvenile *Rhododendron campanulatum* and *Sorbus microphylla* individuals in rather high abundance in this alpine dwarf shrub heath. Nevertheless, no tree-sized individual of these species occurred there (cf. Figure 13.4).

Results from RDA showed that the environmental variables explained 77% of variability in adult tree density distribution of the treeline ecotone (cf. Schwab et al. 2017). The soil group of variables was the most important independent predictor of species distribution (28% explained variability), whilst climatic and topographic variation was of secondary (7%) and tertiary importance (6%; Fig. 13.5a). We found 25% shared variation of adult tree density explained by soil and/or climate variables (Fig. 13.5a) (cf. Schwab et al. 2017).

In case of juvenile tree species density, results from RDA showed that variables could explain 66% of variability. As in adult tree density, the soil group of variables was the most important independent predictor of species distribution (24% explained variability), whilst climatic and topographic variation was of secondary (19%) and tertiary importance (14%; Fig. 13.5b). We found small amounts of shared variation of juvenile tree density explained by soil and/or climate and/or topographic variables. In comparison with partitioning of variation of adult tree density, the juvenile stand density variation partitioning revealed a more balanced distribution of explained variance per variable group (cf. Figure 13.5a, b). Whilst topography was of minor importance in case of adult trees, its share in total explained variation of juvenile density was substantial (cf. Schwab et al. 2017).

13.4 Discussion

13.4.1 Species Composition Along the Elevational Gradient

Plant communities and tree species composition correspond largely to patterns at other West and Central Himalayan treeline sites (cf. Bürzle et al. 2017; cf. Schwab et al. 2017). However, a closed belt of pure *Betula utilis* stands above the mixed forest stands of *Abies spectabilis-Betula utilis* stands, as described in Schickhoff (1993, 2005), Miehe (1990) and Miehe et al. (2015) for shady slopes in the western and central Himalaya, is not developed in the Rolwaling Valley. In Rolwaling, subalpine mixed forests directly merge into the *Rhododendron campanulatum* krummholz belt (*Boschniakia himalaica-Rhododendron campanulatum* community) without an intermediate *Betula utilis* belt. It remains unclear if this difference is a matter of succession status, a consequence of the exceptional near-natural state

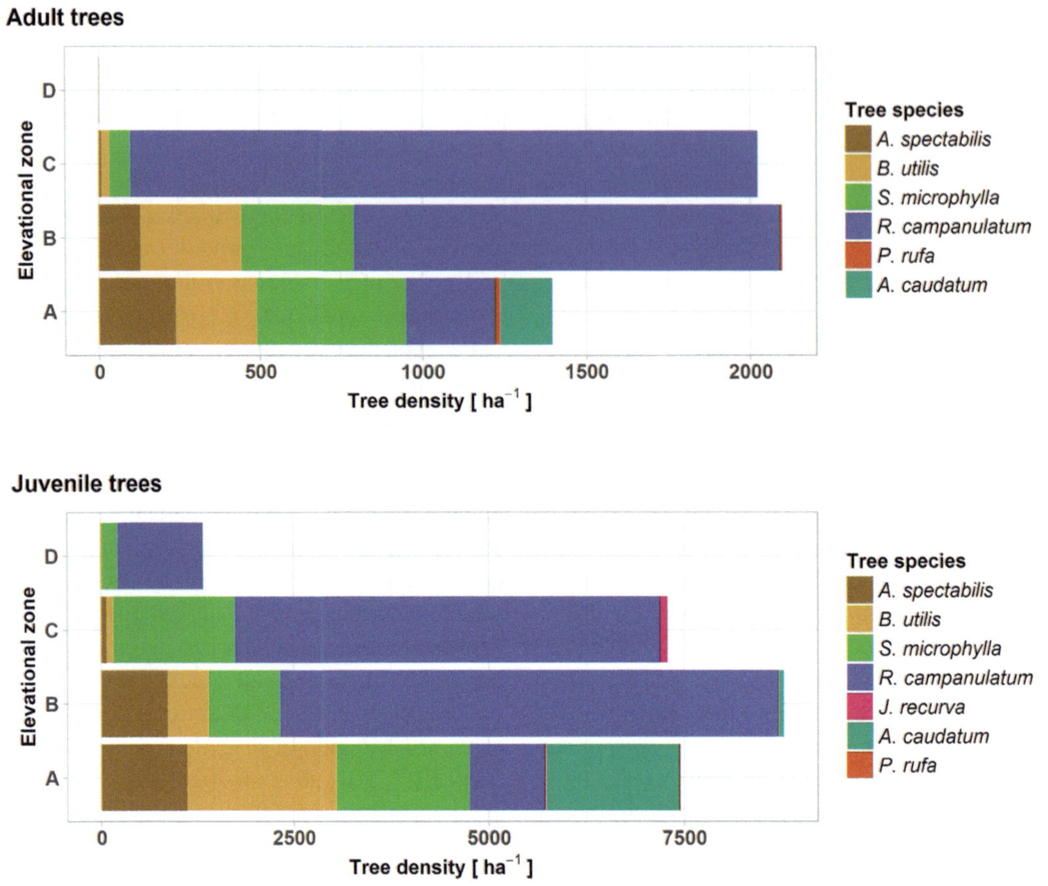

Fig. 13.4 Tree species population density of adult individuals (≥ 7 cm dbh) and juvenile individuals (<7 cm dbh) across the elevational gradient

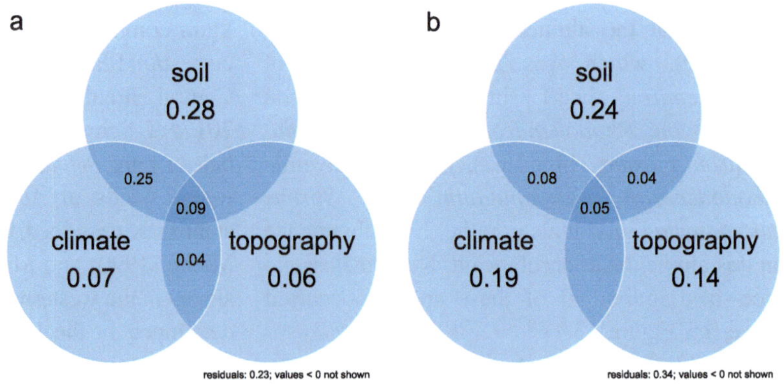

Fig. 13.5 Variation partitioning of tree density using the matrices of soil variables, climate variables and topographic variables (**a**: adult trees, **b**: juveniles)

of the investigated treeline, or related to the decreasing competitiveness of *Betula utilis* along the climatic gradient from the continental NW-Himalaya to the more oceanic SE-Himalaya (cf. Schickhoff 2005; Schwab et al. 2016; Bürzle et al. 2017).

13.4.2 Vegetation–Environment Relationships

Our study reveals distinct relationships of communities with various soil physical and chemical properties. Relevant drivers for community differentiation are soil temperature, nitrogen content and availability, as well as soil nutrient storage and soil water-holding capacity. Annual mean and growing season mean soil temperatures generally decrease from closed forest to the krummholz zone and dwarf shrub thickets (Müller et al. 2016a) and concurrently along the elevational sequence of communities in Rolwaling. The spatial pattern of nitrogen content and availability, soil nutrient storage and soil water retention capacity represents an elevational gradient of decreasing soil fertility, parallel to the overall decrease of temperatures in the same direction (Bürzle et al. 2017). However, the effect of dense tree canopies preventing soil heat flux and radiative warming of soil temperatures was observed on the NE-exposed slope during certain periods of the year, resulting in temporarily higher soil temperatures under dwarf shrub thickets compared to the krummholz zone (Müller et al. 2016a).

As variation partitioning results indicate, species-specific variation in adult and juvenile tree species density along the treeline ecotone depends, to varying extents, on soil, climatic and topographic variables. Variables with a strong influence on the ordinations reflect significant differences in mean values between zones with differing densities of juvenile and adult *Rhododendron campanulatum*. Low temperature and poor nutrient availability characterizes the ecological niche that *Rhododendron campanulatum* occupies (Schwab et al. 2017). Juvenile population density of all species shows a stronger dependence on the climatic variable of temperature and on microtopographic features in comparison with older trees (Schwab et al. 2017), which supports previous assumptions of differences between juvenile and adult tree establishment, occurrence and growth (e.g. Smith et al. 2003; Wieser et al. 2014). In addition to such life stage-specific differences, Bürzle et al. (2018) found species-specific microhabitat preferences for the establishment of seedlings at our study site. They reported that *Abies spectabilis* seedlings grow on ground that is covered mainly by litter, whilst *Betula utilis* and *Rhododendron campanulatum* seedlings are established on bryophyte mats. However, they found little variation in microsite cover among older seedlings, similar to the regeneration of other subalpine tree species in other treeline environments (Jones and del Moral 2005; Maher and Germino 2006).

Soil nutrient cycle studies in Rolwaling Valley revealed significantly decreasing nutrient availability (N, K, Mg, P) along the elevational gradient from subalpine forests to krummholz and alpine dwarf shrub thickets (Müller et al. 2016a, b, 2017; Drollinger et al. 2017). In addition to climate and microtopography variables, manganese content and the C:N ratio as proxy for nutrient availability explained a substantial proportion of variation in tree species density along the gradient (Schwab et al. 2017). On the level of plant communities, elevational gradients of soil nutrient contents show either significant differences between single plant communities in the ANOVA (Mn, Al) or significant correlation with DCA axes (Fe, Na, K). Nitrogen is the only nutrient variable significant in both analyses (Table 13.1). Moreover, all soil nutrient contents show relatively wide standard deviations. Differences in the detectability of soil nutrient gradients between a priori defined elevational vegetation zones and the elevational sequence of plant communities have to be attributed to the mosaic-like pattern of respective habitat patches in the treeline ecotone. Since alpine soils show pronounced small-scale heterogeneity in terms of fertility and other properties in general (Nagy and Grabherr 2009; Bäumler 2015), further investigations are necessary to detect micro-scale relationships between plant species composition and soil properties (Bürzle et al. 2017).

Nitrogen is the most limiting factor to alpine plant productivity, whereas the supply of plant available nitrogen is primarily determined by soil temperature (Larcher 2003; Baptist and

Aranjuelo 2012; Hawkesford et al. 2012). Nevertheless, plant communities and tree densities by themselves are not nitrogen limited, as species composition is adapted to the prevailing nutrient supply, and nutrient enrichment creates modified communities (Körner 2003). Lower nitrogen availability (wider C:N ratio) prevails in the krummholz and dwarf shrub communities (Table 13.1). The dense cover of ericaceous trees (*Rhododendron* krummholz belt) and shrubs (*Rhododendron* dwarf shrubs), associated with a low leaf litter quality (wide C:N ratio, high lignin content) (Körner 2003), may contribute to this nutrient deficiency. Thick layers of recalcitrant *Rhododendron* litter are associated with low N mineralization rates and N availability (Maithani et al. 1998). Moreover, *Rhododendron* leaves contain high amounts of polyphenols (Hegnauer and Hegnauer 1966; Fan et al. 1999), which are an important C source for microorganisms in forest ecosystems (Schimel et al. 1996; Souto et al. 2000; DeLuca et al. 2002). High concentration of polyphenols may lead to the binding of N into complex polyphenolic compounds, and further to an increase in NH_4^+ and NO_3 immobilization (Northup et al. 1995; Schimel et al. 1996; Bradley et al. 2000; DeLuca et al. 2002) and thus to a decrease of nitrogen availability (Bürzle et al. 2017).

Analyses of soil moisture indicated a decreasing trend along the elevational gradient and detected seasons with soil water scarcity (cf. Müller et al. 2016a; b). Despite positive correlations of natural regeneration of tree species with soil moisture and temperature at Himalayan treeline ecotones (cf. Schickhoff et al. 2015), the soil moisture trend did not contribute significantly to explained variation of tree species density (Schwab et al. 2017). Nevertheless, soil moisture and soil water-holding capacity complements the complex pattern of environmental factors to which the species composition is adapted. For instance, the *Synotis alata-Abies spectabilis* community (lower section of the closed subalpine forest) shows highest annual soil moisture values (Table 13.1) (Bürzle et al. 2017). This is reflected by the highly diagnostic species *Clintonia udensis*, which is considered an indicator species for humid soil conditions (Miehe 1990; Bürzle et al. 2017). In addition to temperature, several studies found soil moisture and/or precipitation to be a relevant site factor at treelines for tree seed germination and tree seedling establishment (Bürzle et al. 2018) and tree growth performance, e.g. tree-ring increment (e.g. Gaire et al. 2014, 2017; Tiwari et al. 2017; Schwab et al. 2018). However, soil moisture was not related to population density. *Rhododendron campanulatum* shows higher water use efficiency compared to other species at same elevations, and the most distinct increase in water use efficiency with increasing elevation (De Lillis et al. 2004). Accordingly, *Rhododendron campanulatum* is obviously better adapted than competing tree species to drier conditions which might occur under high irradiance in the *Rhododendron* krummholz belt and at the alpine dwarf shrub heath and in drier seasons (Schwab et al. 2017).

Lowest soil moisture values were detected in the uppermost part of the *Rhododendron* dwarf shrub heath and the *Anaphalis royleana-Rhododendron anthopogon* community, respectively, representing the front edge of the ecotone (Müller et al. 2016b; Bürzle et al. 2017; Schwab et al. 2017). This is remarkable, as plant water supply is often improved at higher altitudes due to reduced evapotranspiration (Körner 2003) as long as increasing wind speed does not overcompensate a decrease in evaporation (cf. Holtmeier 2009). In spite of increasing solar radiation, evaporation often decreases along the elevational gradient, conditioned by temperature, solar intensity, atmospheric pressure, soil moisture, the degree of saturation of the air and wind (Nullet and Juvik 1994; Nagy and Grabherr 2009; Kuhn 2012; Bach and Price 2013). In our study area, low topsoil moisture might be mainly caused by overall low water-holding capacities, according to a high proportion of sand within the soil texture (Müller et al. 2016b). In addition, the soil depth of the *Anaphalis royleana-Rhododendron anthopogon* community is low and the profile shows only weakly developed, thin Ah-horizons (Table 13.1) (Bürzle et al. 2017; Müller et al. 2017). Dehydration of topsoil has strong effects on plant nutrition by interrupting topsoil nutrient cycling and thus reducing nutrient availability (Marschner and Rengel 2012). Effects of this 'drought-enhanced nutrient shortage' on plant life and species distribution are even more appreciable than moisture stress by itself (Körner 2003).

In addition to nitrogen deficiency and rather low soil moisture, we hypothesize the impoverished flora of the *Boschniakia himalaica-Rhododendron campanulatum* community (*Rhododendron* krummholz belt), and the extremely low productivity of the herbaceous layer can be attributed to allelopathic effects of *Rhododendron campanulatum* (Bürzle et al. 2017). Aside from allelopathic effects and competition for nutrients, plant–plant interference in the evergreen, dense *Rhododendron* stands involves a strong competition for light (Bürzle et al. 2017).

Deteriorating site conditions with increasing elevation as assessed in soil studies in the Rolwaling Valley (Müller et al. 2016a; b) can be confirmed for the elevational sequence of plant communities and changing densities of tree species populations, especially in terms of soil temperature, nitrogen supply and availability, and soil moisture (Bürzle et al. 2017; Schwab et al. 2017; Müller et al. 2017). These factors are crucial for differences in species composition along the steep environmental gradient of the treeline ecotone. Multifaceted interrelations between environmental factors make it difficult to rank them in order of their importance for variations in the ecotone's vegetation. Apart from parent material and litter fall, nitrogen supply, for instance, is influenced by temperature and soil moisture. Moisture deficits in plants, in turn, can be strengthened by low soil temperatures (Tranquillini 1982; Larcher 2003). Low temperatures also affect life processes of plants directly by freezing stress and low temperature limit of net photosynthesis (Sakai and Larcher 1987; Larcher 2003; Taiz et al. 2015). Most likely, reduced availability of both nitrogen and manganese at the elevated sites leads to lowered competitive strength of tree species. As an exception, the *Rhododendron* species and ericaceous perennials in general are low-nutrient users that grow in soils that are poor in most essential elements and consequently unsuitable to other species; in fact, they are even favoured by poor soil conditions (Cox 1990; Ristvey et al. 2007). The high root-to-shoot ratio of *Rhododendron campanulatum* and its evergreen nature further accentuate these competitive advantages due to efficient nutrient storage, increased photosynthetic capacity in transitional seasons and allelopathic effects of polyphenol-rich *Rhododendron* litter (Schwab et al. 2017, 2018). Low N and P availabilities in alpine tundra soils correspond to lower litter input and a decline in litter mineralization in this altitudinal zone resulting in small accumulation of SOM (Müller et al. 2017). It can be concluded that altitudinal variations in plant communities themselves govern nutrient cycling through the input of C, N and P into soils by differences in leaf fall (Müller et al. 2017).

Notwithstanding the random selection of plots, the spatial dependence of the species data is most likely not only related to functional correlation with the environmental variables. The spatial configuration of species may be the result of neutral processes of various population and community dynamics leading to spatial autocorrelation, not considered by here presented RDA and variation partitioning results. Schwab et al. (2017) showed that the dominance of *Rhododendron campanulatum* controls the broad scale spatial pattern, indicating the huge contrast in tree species composition between the lower and upper elevational zones. Environmental variables explain considerable amounts of variation in spatial patterns, with higher amounts explained for adult compared to juvenile population density spatial patterns. Neutral processes of community and population dynamics could explain remaining spatial variation (Schwab et al. 2017). These include variation in species demography due to competition (ecological drift) and propagule dispersion (random dispersal) creating spatial autocorrelation in response variables (Legendre and Legendre 2012). Especially finer scaled spatial patterns and the juvenile population density pattern contain proportions of variation assigned to neutral processes such as genetically fixed plant characteristics, mainly according to recruitment strategies. This finding reveals that in addition to microtopographic features, nutrient availability and temperature, flowering phenology, seed maturation including seed size and seed number, seed dispersal, seed germination and seedling establishment need to be considered to explain variation of tree species density. The sampled tree species propagate in different ways: For instance, *Betula utilis* and *Rhododendron campanulatum* use anemochorous dispersal paths. In addition, we observed frequent clonal propagation of *Betula utilis*, whilst *Abies spectabilis* relies rather on zoochorous paths. These different dispersal strategies might result in species-specific responses to climate change. In addition to dispersal, seed quality and quantity might play a role (Cuevas 2000; Dullinger et al. 2004; Holtmeier 2009; Batllori et al. 2010; Kroiss and Hille Ris Lambers 2015; Brodersen et al. 2019). In addition, also plant–plant interaction (especially competition and facilitation) lead to variations in tree species density. Future studies are needed to analyse these species-specific properties in order to determine their influence on tree species density and interaction with upslope migration potential (Schwab et al. 2017).

13.4.3 Responsiveness of Community Patterns and Tree Species Density to Climate Warming

Accumulated knowledge of climate change-induced alteration of Himalayan ecosystems in terms of plant cover, plant functional type dominance, species distributions, species compositions and community structure is still very deficient. The sensitivity of species compositions and community structure in treeline ecotones is likewise largely unknown. For the Rolwaling Valley, our data clearly show the significance of soil temperature, soil moisture and nitrogen (supply and availability) for species distribution along the elevational gradient. Since these factors are directly (soil temperature) or indirectly (soil moisture, nitrogen) affected by climatic conditions, we presume that species compositions and community structure in the treeline ecotone will change and new niches are created with ongoing climate warming (Schickhoff et al. 2015, 2016a, b; Bürzle et al. 2017).

Increasing temperatures might imply enhanced nitrogen supply. Low soil temperatures are the main constraint for nitrogen availability, which controls alpine plant productivity (Larcher 2003; Baptist and Aranjuelo 2012; Hawkesford et al. 2012). Given the above-average warming in Himalayan treeline ecotones including the Rolwaling Valley, an increase in plant height and cover within plant communities and a shift in dominance patterns owing to an earlier start of the growing season are to be expected (Bürzle et al. 2017).

Moreover, climate warming involves changes in precipitation patterns, leading to alterations in snow cover and modified soil moisture. No clear trend could be detected for precipitation patterns in the greater Himalayan region; observations show more decreasing than increasing precipitation trends (Hasson et al. 2016; Schickhoff et al. 2016a). As soil moisture was identified as one of the controlling factors for tree regeneration and stand structural patterns (Schickhoff et al. 2015; Müller et al. 2016b; Schwab et al. 2017; Drollinger et al. 2017) as well as for species composition (Bürzle et al. 2017), we anticipate shifts in community structure and species composition due to changing precipitation and soil moisture supply.

Even if climate warming will change site conditions in the treeline ecotone and create new niches, the general constraints of low-temperatures and low-nutrient availability remain. *Abies spectabilis* and *Betula utilis* could potentially establish populations above the krummholz belt, given their generally intense regeneration, and comparatively high mean annual temperatures at the Rolwaling Himal treeline (Müller et al. 2016a, b). Regardless of slight increases in mean temperatures, site conditions will probably continue to be more beneficial for *Rhododendron campanulatum* and *Rhododendron* dwarf shrub species in the krummholz belt and alpine dwarf shrub zone, potentially resulting in an upward migration of the *Rhododendron campanulatum* belt into the alpine dwarf shrub zone (Schwab et al. 2017; Schwab 2018).

Since the *Rhododendron campanulatum* population of the krummholz belt is firmly established and feedback mechanisms maintain its predominance, this thicket will most likely continue to constitute an insurmountable barrier for seedlings and saplings of *Betula utilis* and *Abies spectabilis* even under warmer conditions (Schwab et al. 2020). It can be assumed that the small numbers of *Abies spectabilis* and *Betula utilis* individuals that were found in the krummholz belt and the alpine dwarf shrub heath are related to microtopographic features similarly to observed establishment patterns in Taiwanese *Abies* treelines (Greenwood et al. 2015). The few individuals of other tree species might consolidate towards established populations at locations above the contemporary krummholz belt position only in the long term. Persistent low-nutrient availability and allelopathic effects will most likely prevent juvenile tree individuals from growing into mature, fruiting trees within or above the current krummholz belt. A substantial establishment in the dwarf shrub heaths above the krummholz belt is similarly unlikely, as comparable soil and environmental conditions prevail. However, the few *Abies spectabilis* and *Betula utilis* individuals in and above the krummholz belt illustrate a certain potential to survive and to reach tree dimensions (Schwab et al. 2017, 2020; Schwab 2018).

The results suggest that a short-term shift of the Rolwaling Himal treeline to higher elevations is rather unlikely. A rather low responsiveness is assumed for near-natural Himalayan treelines and krummholz treelines in general (Harsch and Bader 2011; Chhetri and Cairns 2015; Schickhoff et al. 2016a). However, changes in population structures and species compositions will be inevitable under novel constellations of site conditions (Schwab et al. 2017; Schwab 2018). Temperature and water supply are critical drivers for seedling establishment, and thus, microhabitats that buffer climatic extremes may become more important under climate change conditions in the Himalaya. Therefore, the availability of safe sites will play a major role in future treeline advance (Bürzle et al. 2018).

In general, ongoing climatic changes will already have triggered shifts in species distributions and abundances in the Himalaya, widely without having been noticed or documented by science. For instance, Telwala et al. (2013) provided evidence of warming-driven elevational range shifts in 87% of 124 studied endemic plant species in the alpine zone of Sikkim over the last 150 years. Rana et al. (2017) observed an upward migration tendency of *Rhododendron campanulatum* in a treeline ecotone in central Nepal. Upward range shifts of up to 150 m were detected for the shrub *Myricaria elegans* in the NW Himalaya by Dolezal et al. (2016), stressing that plant species responses to ongoing climate change will not be unidirectional upward range shifts but rather multidimensional, species-specific and spatially variable. Thus, present-day plant assemblages and community structures are definitely different from those of the nineteenth century (Bürzle et al. 2017).

13.5 Conclusion

Our study provides detailed insights into species compositions of plant communities and vegetation–environment relationships in the treeline ecotone in the Rolwaling Valley, an area not previously studied. The study expands on the still very limited knowledge of Himalayan subalpine-alpine vegetation with regard to floristic diversity, ecology and syntaxonomy. When classifying the ecotone with phytosociological methods, we had to deal with rank-less communities and a provisional status of classes; i.e. phytosociological studies in the Himalaya are still pioneer studies. Differences in species composition of subalpine forests and dwarf shrub communities between the Rolwaling Valley and other Himalayan study areas point to the need of further detailed local studies (Bürzle et al. 2017; Schwab et al. 2017).

Community patterns and forest stand densities are distinctly correlated to a deterioration of growth conditions along the elevational gradient, reflected in an increasingly unfavourable constellation of microtopography, soil temperature, soil moisture and soil fertility. Differentiations of species compositions are understood as resulting from complex and changing effects of interrelated site factors along the elevational gradient. Our findings corroborate assumptions that local treeline elevation, species composition and tree density are not defined by thermal deficits alone. The feedback mechanisms of the *Rhododendron campanulatum* krummholz belt constrain the treeline response to climate warming through retarding or inhibiting upward migration of other tree species (Bürzle et al. 2017; Schwab et al. 2017, 2020). Further, altitudinal variations in stand structures themselves govern nutrient cycling through the input of C, N and P into soils by differences in leaf fall (Müller et al. 2017).

The sensitivity of treeline ecotones to climate change should be further assessed by analysing species- and community-specific responses to the constellation of environmental site factors, with a special focus on non-thermal site factors. Differences in variables explaining adult and juvenile population densities point to the need to investigate different life stages and their relation to abiotic and biotic conditions on different temporal scales. Spatial pattern analyses reveal that dispersal mechanisms and biotic interactions should be considered in future studies on treelines. In general, for a thorough monitoring of Himalayan treeline vegetation requires much more local and landscape-scale studies in order to detect climate change-induced shifts in vegetation patterns and to assess the vulnerability of treeline flora and vegetation (Bürzle et al. 2017; Schwab et al. 2017).

Acknowledgements We are grateful to Tenzing and Lakpa Sherpa from Beding who provided lodging and support during field data collection. We thank Ram Bahadur, Simon Drollinger, Helge Heyken, Nina Kiese, Yanina Katharina Müller, Hanna Wanli, Ronja Wedegärtner and Lina Marie Wernicke for assistance in the field. We are obliged to Lena Geiger and Matthias Tetzlaff for providing HemiView data and to Lars Gerlitz for providing climate data. We thank Michael Müller for providing and discussing soil data and the late Ramchandra Karki for discussions on climate data. We acknowledge Bijay Raj Subedi, Madan K. Suwal, Yadu Sapkota and Chandra Kanta Subedi for great support in logistics and administrative issues. B. Bürzle was funded by Studienstiftung des deutschen Volkes. We are indebted to the German Research Foundation for funding (DFG, SCHI 436/14-1, BO 1333/4-1, SCHO 739/14-1), to Nepalese authorities for research permits and to the community in Rolwaling for friendly cooperation and hospitality.

References

Bach A, Price LW (2013) Mountain climate. In: Price MF, Byers AC, Friend DA, Kohler T, Price LW (eds) Mountain geography: physical and human dimensions. University of California Press, Berkeley, pp 41–84

Baptist F, Aranjuelo I (2012) Interaction of carbon and nitrogen metabolisms in alpine plants. In: Lütz C (ed) Plants in alpine regions—cell physiology of adaption and survival strategies. Springer, Wien, Austria, pp 121–134

Batllori E, Camarero JJ, Gutiérrez E (2010) Current regeneration patterns at the tree line in the Pyrenees indicate similar recruitment processes irrespective of the past disturbance regime. J Biogeogr 37:1938–1950. https://doi.org/10.1111/j.1365-2699.2010.02348.x

Bäumler R (2015) Soils. In: Miehe G, Pendry C, Chaudhary RP (eds) Nepal: an introduction to the natural history, ecology and human environment in the Himalayas. Royal Botanic Garden Edinburgh, Edinburgh, United Kingdom, pp 125–134

Bhutiyani MR, Kale VS, Pawar NJ (2007) Long-term trends in maximum, minimum and mean annual air temperatures across the Northwestern Himalaya during the twentieth century. Climatic Change 85:159–177. https://doi.org/10.1007/s10584-006-9196-1

Bhutiyani MR, Kale VS, Pawar NJ (2010) Climate change and the precipitation variations in the Northwestern Himalaya: 1866–2006. Int J Climatol 30:535–548. https://doi.org/10.1002/joc.1920

Böhner J, Miehe G, Miehe S, Nagy L (2015) Climate and weather. In: Miehe G, Pendry C, Chaudhary RP (eds) Nepal: an introduction to the natural history, ecology and human environment in the Himalayas. Royal Botanic Garden Edinburgh, Edinburgh, United Kingdom, pp 23–90

Bradley RL, Titus BD, Preston CP (2000) Changes to mineral N cycling and microbial communities in black spruce humus after additions of $(NH_4)_2SO_4$ and condensed tannins extracted from *Kalmia angustifolia*

and balsam fir. Soil Biol Biochem 32:1227–1240. https://doi.org/10.1016/S0038-0717(00)00039-0
Braun-Blanquet J (1964) Pflanzensoziologie. Springer, Wien
Brodersen CR, Germino MJ, Johnson DM, Reinhardt K, Smith WK, Resler LM, Bader MY, Sala A, Kueppers LM, Broll G, Cairns DM, Holtmeier F-K, Wieser G (2019) Seedling survival at timberline is critical to conifer mountain forest elevation and extent. Front For Glob Change 2:9. https://doi.org/10.3389/ffgc.2019.00009
Bürzle B, Schickhoff U, Schickhoff U, Schwab N, Oldeland J, Müller M, Böhner J, Chaudhary RP, Scholten T, Dickoré WB (2017) Phytosociology and ecology of treeline ecotone vegetation in Rolwaling Himal. Nepal. Phytocoenologia 47:197–220. https://doi.org/10.1127/phyto/2017/0130
Bürzle B, Schickhoff U, Schwab N, Wernicke LM, Müller YK, Böhner J, Chaudhary RP, Scholten T, Oldeland J (2018) Seedling recruitment and facilitation dependence on safe site characteristics in a Himalayan treeline ecotone. Plant Ecol 219:115–132. https://doi.org/10.1007/s11258-017-0782-2
Case BS, Duncan RP (2014) A novel framework for disentangling the scale-dependent influences of abiotic factors on alpine treeline position. Ecography 37:838–851. https://doi.org/10.1111/ecog.00280
Chhetri PK, Cairns DM (2015) Contemporary and historic population structure of *Abies spectabilis* at treeline in Barun valley, eastern Nepal Himalaya. J Mt Sci 12:558–570. https://doi.org/10.1007/s11629-015-3454-5
Cox PA (1990) The larger Rhododendron species. Timber Press, Portland, OR, USA
Cuevas JG (2000) Tree recruitment at the *Nothofagus pumilio* alpine timberline in Tierra del Fuego, Chile. J Ecol 88:840–855. https://doi.org/10.1046/j.1365-2745.2000.00497.x
De Lillis M, Matteucci G, Valentini R (2004) Carbon assimilation, nitrogen, and photochemical efficiency of different Himalayan tree species along an altitudinal gradient. Photosynthetica 42:597–605. https://doi.org/10.1007/S11099-005-0019-9
DeLuca T, Nilsson M-C, Zackrisson O (2002) Nitrogen mineralization and phenol accumulation along a fire chronosequence in northern Sweden. Oecologia 133:206–214. https://doi.org/10.1007/s00442-002-1025-2
Dolezal J, Leheckova E, Sohar K, Dvorsky M, Kopecky M, Chlumska Z, Wild J, Altman J (2016) Annual and intra-annual growth dynamics of *Myricaria elegans* shrubs in arid Himalaya. Trees 30:761–773. https://doi.org/10.1007/s00468-015-1318-9
Doove LL, Van Buuren S, Dusseldorp E (2014) Recursive partitioning for missing data imputation in the presence of interaction effects. Comput Stat Data Anal 72:92–104. https://doi.org/10.1016/j.csda.2013.10.025
Drollinger S, Müller M, Kobl T, Schwab N, Böhner J, Schickhoff U, Scholten T (2017) Decreasing nutrient concentrations in soils and trees with increasing elevation across a treeline ecotone in Rolwaling Himal. Nepal J Mt Sci 14:843–858. https://doi.org/10.1007/s11629-016-4228-4
Duan K, Yao T, Thompson LG (2006) Response of monsoon precipitation in the Himalayas to global warming. J Geophys Res Atmos 111:D19110. https://doi.org/10.1029/2006JD007084
Dullinger S, Dirnböck T, Grabherr G (2004) Modelling climate change-driven treeline shifts: relative effects of temperature increase, dispersal and invasibility. J Ecol 92:241–252. https://doi.org/10.1111/j.0022-0477.2004.00872.x
Durak T, Żywiec M, Kapusta P, Holeksa J (2015) Impact of land use and climate changes on expansion of woody species on subalpine meadows in the Eastern Carpathians. For Ecol Manag 339:127–135. https://doi.org/10.1016/j.foreco.2014.12.014
Fan CQ, Yang GJ, Zhao W, Ding BY, Qin GW (1999) Phenolic components from *Rhododendron latoucheae*: part 5 in the series "Chemical studies on Ericaceae plants." Chin Chem Lett 10:567–570
Ferrarini A, Rossi G, Mondoni A, Orsenigo S (2014) Prediction of climate warming impacts on plant species could be more complex than expected. Evidence from a case study in the Himalaya. Ecol Complex 20:307–314. https://doi.org/10.1016/j.ecocom.2014.02.003
Gaire NP, Koirala M, Bhuju DR, Borgaonkar HP (2014) Treeline dynamics with climate change at the central Nepal Himalaya. Clim Past 10:1277–1290. https://doi.org/10.5194/cp-10-1277-2014
Gaire NP, Koirala M, Bhuju DR, Carrer M (2017) Site- and species-specific treeline responses to climatic variability in eastern Nepal Himalaya. Dendrochronologia 41:44–56. https://doi.org/10.1016/j.dendro.2016.03.001
Gauch HG (1982) Multivariate analysis in community ecology. Cambridge University Press, New York, US
Gerlitz L, Bechtel B, Böhner J, Bobrowski M, Bürzle B, Müller M, Scholten T, Schickhoff U, Schwab N, Weidinger J (2016) Analytic comparison of temperature lapse rates and precipitation gradients in a Himalayan treeline environment: implications for statistical downscaling. In: Singh RB, Schickhoff U, Mal S (eds) Climate change, Glacier response, and vegetation dynamics in the Himalaya. Springer International Publishing, Cham, Switzerland, pp 49–64
Gerlitz L, Conrad O, Thomas A, Böhner J (2014) Warming patterns over the Tibetan plateau and adjacent lowlands derived from elevation- and bias-corrected ERA-Interim data. Clim Res 58:235–246. https://doi.org/10.3354/cr01193
Graumlich LJ, Waggoner LA, Bunn AG (2005) Detecting global change at alpine treeline: coupling paleoecology with contemporary studies. In: Huber UM, Bugmann HKM, Reasoner MA (eds) Global change and mountain regions: an overview of current knowledge. Springer, Dordrecht, The Netherlands, pp 501–508
Greenwood S, Chen J-C, Chen C-T, Jump AS (2015) Temperature and sheltering determine patterns of

seedling establishment in an advancing subtropical treeline. J Veg Sci 26:711–721. https://doi.org/10.1111/jvs.12269
Grierson AJC, Long DG (1984–2001) Flora of Bhutan. Including a record of plants from Sikkim, vol 1–3. Royal Botanic Garden, Edinburgh, United Kingdom
Harsch MA, Bader MY (2011) Treeline form—a potential key to understanding treeline dynamics. Glob Ecol Biogeogr 20:582–596. https://doi.org/10.1111/j.1466-8238.2010.00622.x
Harsch MA, Hulme PE, McGlone MS, Duncan RP (2009) Are treelines advancing? A global meta-analysis of treeline response to climate warming. Ecol Lett 12:1040–1049. https://doi.org/10.1111/j.1461-0248.2009.01355.x
Hasson S, Gerlitz L, Schickhoff U, Scholten T, Böhner J (2016) Recent climate change over High Asia. In: Singh RB, Schickhoff U, Mal S (eds) Climate change, glacier response, and vegetation dynamics in the Himalaya. Springer International Publishing, Cham, Switzerland, pp 29–48
Hawkesford M, Horst W, Kichey T, Lambers H, Schjoerring J, Møller IS, White P (2012) Functions of macronutrients. In: Marschner P (ed) Marschner's mineral nutrition of higher plants. Elsevier/Academic Press, London, United Kingdom, pp 135–189
Hegnauer R, Hegnauer R (1966) Dicotyledoneae: *Daphniphyllaceae - Lythraceae*. Birkhäuser, Basel, Switzerland
Holtmeier F-K (2009) Mountain timberlines. Springer, Dordrecht, The Netherlands
IPCC (2013) Climate change 2013: the physical science basis. Contribution of working group I to the fifth assessment report of the Intergovernmental Panel on climate change. In: Stocker TF, Qin D, Plattner G-K, Tignor M, Allen SK, Boschung J, Nauels A, Xia Y, Bex V, Midgley PM (eds) Cambridge University Press, Cambridge, United Kingdom and New York, USA
IPCC (2014) Climate change 2014: impacts, adaptation, and vulnerability. Part A: global and sectoral aspects. Contribution of working group II to the fifth assessment report of the intergovernmental panel on climate change. In: Field CB, Barros VR, Dokken DJ, Mach KJ, Mastrandrea MD, Bilir TE, Chatterjee M, Ebi KL, Estrada YO, Genova RC, Girma B, Kissel ES, Levy AN, MacCracken S, Mastrandrea PR, White LL (eds). Cambridge University Press, Cambridge, United Kingdom and New York, NY, USA
Jain SK, Kumar V, Saharia M (2013) Analysis of rainfall and temperature trends in northeast India. Int J Climatol 33:968–978. https://doi.org/10.1002/joc.3483
Jones CC, del Moral R (2005) Effects of microsite conditions on seedling establishment on the foreland of Coleman Glacier, Washington. J Veg Sci 16:293–300. https://doi.org/10.1111/j.1654-1103.2005.tb02367.x
Karki R, Hasson S, Gerlitz L, Schickhoff U, Scholten T, Böhner J (2017) Quantifying the added value of convection-permitting climate simulations in complex terrain: a systematic evaluation of WRF over the Himalayas. Earth Syst Dynam 8:507–528. https://doi.org/10.5194/esd-8-507-2017
Karki R, Talchabhadel R, Aalto J, Baidya SK (2016) New climatic classification of Nepal. Theor Appl Climatol 125:799–808. https://doi.org/10.1007/s00704-015-1549-0
KC A, Ghimire A (2015) High-altitude plants in era of climate change: a case of Nepal Himalayas. In: Öztürk M, Hakeem KR, Faridah-Hanum I, Efe R (eds) Climate change impacts on high-altitude ecosystems. Springer International Publishing, Cham, Switzerland, pp 177–187
Kent M (2012) Vegetation description and data analysis: a practical approach, 2nd edn. Hoboken, NJ, Wiley-Blackwell, Chichester, West Sussex, UK
Körner C (2003) Alpine plant life: functional plant ecology of high mountain ecosystems. Springer, Berlin Heidelberg, Germany
Körner C (2012) Alpine treelines: functional ecology of the global high elevation tree limits. Springer, Basel, Switzerland
Krishnan R, Shrestha AB, Ren G, Rajbhandari R, Saeed S, Sanjay J, Syed MdA, Vellore R, Xu Y, You Q, Ren Y (2019) Unravelling climate change in the Hindu Kush Himalaya: rapid warming in the mountains and increasing extremes. In: Wester P, Mishra A, Mukherji A, Shrestha AB (eds) The Hindu Kush Himalaya assessment: mountains, climate change, sustainability and people. Springer International Publishing, Cham, pp 57–97
Kroiss SJ, Hille Ris Lambers J (2015) Recruitment limitation of long-lived conifers: implications for climate change responses. Ecology 96:1286–1297. https://doi.org/10.1890/14-0595.1
Kuhn M (2012) Rain and snow at high elevation. In: Lütz C (ed) Plants in alpine regions—cell physiology of adaption and survival strategies. Springer, Wien, Austria, pp 1–10
Larcher W (2003) Physiological plant ecology: ecophysiology and stress physiology of functional groups. Springer, Berlin Heidelberg, Germany
Legendre P, Legendre L (2012) Numerical ecology. Elsevier, Amsterdam, The Netherlands
Liu X, Chen B (2000) Climatic warming in the Tibetan plateau during recent decades. Int J Climatol 20:1729–1742. https://doi.org/10.1002/1097-0088(20001130)20:14%3c1729::AID-JOC556%3e3.0.CO;2-Y
Maher EL, Germino MJ (2006) Microsite differentiation among conifer species during seedling establishment at alpine treeline. Ecoscience 13:334–341
Maithani K, Arunachalam A, Tripathi RS, Pandey HN (1998) Influence of leaf litter quality on N mineralization in soils of subtropical humid forest regrowths. Biol Fertil Soils 27:44–50. https://doi.org/10.1007/s003740050398
Marschner P, Rengel Z (2012) Nutrient avialability in soils. In: Marschner H, Marschner P (eds) Marschner's mineral nutrition of higher plants.

Elsevier/Academic Press, London, United Kingdom, pp 315–330
Miehe G (1990) Langtang Himal: Flora und Vegetation als Klimazeiger und -zeugen im Himalaya. Cramer, Berlin
Miehe G, Miehe S, Böhner J, Ghimire SK, Bhattarai K, Chaudhary RP, Subedi M, Jha PK, Pendry C (2015) Vegetation ecology. In: Miehe G, Pendry C, Chaudhary RP (eds) Nepal: an introduction to the natural history, ecology and human environment in the Himalayas. Royal Botanic Garden Edinburgh, Edinburgh, United Kingdom, pp 385–472
Mountain research initiative EDW working group (2015) Elevation-dependent warming in mountain regions of the world. Nat Clim Change 5:424–430. https://doi.org/10.1038/nclimate2563
Müller M, Schickhoff U, Scholten T, Drollinger S, Böhner J, Chaudhary RP (2016) How do soil properties affect alpine treelines? General principles in a global perspective and novel findings from Rolwaling Himal. Nepal. Prog Phys Geogr 40:135–160. https://doi.org/10.1177/0309133315615802
Müller M, Schwab N, Schickhoff U, Böhner J, Scholten T (2016) Soil temperature and soil moisture patterns in a Himalayan alpine treeline ecotone. Arct Antarct Alp Res 48:501–521. https://doi.org/10.1657/AAAR0016-004
Müller M, Oelmann Y, Schickhoff U, Böhner J, Scholten T (2017) Himalayan treeline soil and foliar C:N: P stoichiometry indicate nutrient shortage with elevation. Geoderma 291:21–32. https://doi.org/10.1016/j.geoderma.2016.12.015
Nagy L, Grabherr G (2009) The biology of alpine habitats. Oxford University Press, Oxford
Northup RR, Yu Z, Dahlgren RA, Vogt KA (1995) Polyphenol control of nitrogen release from pine litter. Nature 377:227–229. https://doi.org/10.1038/377227a0
Nullet D, Juvik JO (1994) Generalised mountain evaporation profiles for tropical and subtropical latitudes. Singap J Trop Geogr 15:17–24. https://doi.org/10.1111/j.1467-9493.1994.tb00242.x
Padma Alekhya VVL, Pujar GS, Jha CS, Dadhwal VK (2015) Simulation of vegetation dynamics in Himalaya using dynamic global vegetation model. Trop Ecol 56:219–231
Rana P, Bhuju DR, Koirala M, Boonchird C (2017) Dendroecological studies of *Rhododendron campanulatum* D. Don along the elevational gradient of Manaslu conservation area. Nepal Himalaya Pak J Bot 49:1749–1755
Reasoner M, Tinner W (2009) Holocene treeline fluctuations. In: Gornitz V (ed) Encyclopedia of paleoclimatology and ancient environments. Springer Verlag, Heidelberg, pp 442–446
Ristvey AG, Lea-Cox JD, Ross DS (2007) Nitrogen and phosphorus uptake efficiency and partitioning of container-grown azalea during spring growth. J Am Soc Hortic Sci 132:563–571
Sakai A, Larcher W (1987) Frost survival of plants: responses and adaptation to freezing stress. Springer, Berlin Heidelberg
Salick J, Ghimire SK, Fang Z, Dema S, Konchar KM (2014) Himalayan alpine vegetation, climate change and mitigation. J Ethnobiol 34:276–293. https://doi.org/10.2993/0278-0771-34.3.276
Schickhoff U (2011) Dynamics of mountain ecosystems. In: Millington AC, Blumler MA, Schickhoff U (eds) The SAGE handbook of biogeography. SAGE, London, United Kingdom, pp 313–337
Schickhoff U (2005) The upper timberline in the Himalayas, Hindu Kush and Karakorum: a review of geographical and ecological aspects. In: Broll G, Keplin B (eds) Mountain ecosystems. Studies in Treeline Ecology, Springer, Berlin, Germany, pp 275–354
Schickhoff U (1993) Das Kaghan-Tal im Westhimalaya (Pakistan). Bonner Geogr Abh 87. Dümmler, Bonn
Schickhoff U, Bobrowski M, Böhner J, Bürzle B, Chaudhary RP, Gerlitz L, Heyken H, Lange J, Müller M, Scholten T, Schwab N, Wedegärtner R (2015) Do Himalayan treelines respond to recent climate change? An evaluation of sensitivity indicators. Earth Syst Dyn 6:245–265. https://doi.org/10.5194/esd-6-245-2015
Schickhoff U, Bobrowski M, Böhner J, Bürzle B, Chaudhary RP, Gerlitz L, Lange J, Müller M, Scholten T, Schwab N (2016a) Climate change and treeline dynamics in the Himalaya. In: Singh RB, Schickhoff U, Mal S (eds) Climate change, glacier response, and vegetation dynamics in the Himalaya. Springer International Publishing, Cham, Switzerland, pp 271–306. https://doi.org/10.1007/978-3-319-28977-9_15
Schickhoff U, Singh RB, Mal S (2016) Climate change and dynamics of glaciers and vegetation in the Himalaya: an overview. In: Singh RB, Schickhoff U, Mal S (eds) Climate change, Glacier response, and vegetation dynamics in the Himalaya. Springer International Publishing, Cham, Switzerland, pp 1–26
Schimel JP, Cleve KV, Cates RG, Clausen TP, Reichardt PB (1996) Effects of balsam poplar (*Populus balsamifera*) tannins and low molecular weight phenolics on microbial activity in taiga floodplain soil: implications for changes in N cycling during succession. Can J Bot 74:84–90
Schwab N (2018) Sensitivity and response of alpine treelines to climate change—insights from a Krummholz treeline in Rolwaling Himal, Nepal. PhD thesis, Universität Hamburg, Hamburg, Germany
Schwab N, Janecka K, Kaczka RJ, Böhner J, Chaudhary RP, Scholten T, Schickhoff U (2020) Ecological relationships at a near-natural treeline, Rolwaling valley, Nepal Himalaya: implications for the sensitivity to climate change. Erdkunde 74:15–44. https://doi.org/10.3112/erdkunde.2020.01.02
Schwab N, Kaczka RJ, Janecka K, Böhner J, Chaudhary RP, Scholten T, Schickhoff U (2018) Climate change-induced shift of tree growth sensitivity at a

central Himalayan treeline ecotone. Forests 9:267. https://doi.org/10.3390/f9050267

Schwab N, Schickhoff U, Bürzle B, Müller M, Böhner J, Chaudhary RP, Scholten T, Oldeland J (2017) Implications of tree species—environment relationships for the responsiveness of Himalayan krummholz treelines to climate change. J Mt Sci 14:453–473. https://doi.org/10.1007/s11629-016-4257-z

Schwab N, Schickhoff U, Müller M, Gerlitz L, Bürzle B, Böhner J, Chaudhary RP, Scholten T (2016) Treeline responsiveness to climate warming: insights from a krummholz treeline in Rolwaling Himal, Nepal. In: Singh RB, Schickhoff U, Mal S (eds) Climate change, Glacier response, and vegetation dynamics in the Himalaya. Springer International Publishing, Cham, Switzerland, pp 307–345

Shrestha UB, Gautam S, Bawa KS (2012) Widespread climate change in the Himalayas and associated changes in local ecosystems. PLoS ONE 7:e36741. https://doi.org/10.1371/journal.pone.0036741

Smith WK, Germino MJ, Hancock TE, Johnson DM (2003) Another perspective on altitudinal limits of alpine timberlines. Tree Physiol 23:1101–1112

Souto XC, Chiapusio G, Pellissier F (2000) Relationships between phenolics and soil microorganisms in spruce forests: significance for natural regeneration. J Chem Ecol 26:2025–2034. https://doi.org/10.1023/A:1005504029243

Stevens GC, Fox JF (1991) The causes of treeline. Annu Rev Ecol Syst 22:177–191. https://doi.org/10.1146/annurev.es.22.110191.001141

Taiz L, Zeiger E, Møller IM, Murphy AS (eds) (2015) Plant physiology and development. Sinauer, Sunderland, Massachusetts

Telwala Y, Brook BW, Manish K, Pandit MK (2013) Climate-induced elevational range shifts and increase in plant species richness in a Himalayan biodiversity epicentre. PLoS ONE 8:e57103. https://doi.org/10.1371/journal.pone.0057103

Tiwari A, Fan Z-X, Jump AS, Li S-F, Zhou Z-K (2017) Gradual expansion of moisture sensitive *Abies spectabilis* forest in the trans-Himalayan zone of central Nepal associated with climate change. Dendrochronologia 41:34–43. https://doi.org/10.1016/j.dendro.2016.01.006

Tranquillini W (1982) Frost-drought and its ecological significance. In: Lange OL, Nobel PS, Osmond CB, Ziegler H (eds) Physiological plant ecology II: water relations and carbon assimilation. Springer, Berlin, pp 379–400

Troll C (1973) The upper timberlines in different climatic zones. Arct Alp Res 5:A3–A18. https://doi.org/10.2307/1550148

Van Buuren S (2012) Flexible imputation of missing data. CRC Press, Boca Raton, FL, USA

Wang S-Y, Yoon J-H, Gillies RR, Cho C (2013) What caused the winter drought in western Nepal during recent years? J Clim 26:8241–8256. https://doi.org/10.1175/JCLI-D-12-00800.1

Watson MF, Akiyama S, Ikeda H, Pendry CA, Rajbhandari KR, Shrestha KK (eds) (2011) Flora of Nepal: Magnoliaceae to Rosaceae. Royal Botanic Garden Edinburgh, Edinburgh, United Kingdom

Weidinger J, Gerlitz L, Bechtel B, Böhner J (2018) Statistical modelling of snow cover dynamics in the central Himalaya region, Nepal. Clim Res 75:181–199. https://doi.org/10.3354/cr01518

Weiss DJ, Malanson GP, Walsh SJ (2015) Multiscale relationships between alpine treeline elevation and hypothesized environmental controls in the western United States. Ann Assoc Am Geogr 105:437–453. https://doi.org/10.1080/00045608.2015.1015096

Wieser G, Holtmeier F-K, Smith WK (2014) Treelines in a changing global environment. In: Tausz M, Grulke N (eds) Trees in a changing environment. Springer, Dordrecht, The Netherlands, pp 221–263

Zhan Y-J, Ren G-Y, Shrestha AB, Rajbhandari R, Ren Y-Y, Sanjay J, Xu Y, Sun X-B, You Q-L, Wang S (2017) Changes in extreme precipitation events over the Hindu Kush Himalayan region during 1961–2012. Advan Clim Change Res 8:166–175. https://doi.org/10.1016/j.accre.2017.08.002

14 Modelling the Ecological Niche of a Treeline Tree Species (*Betula utilis*) in the Himalayas—A Methodological Overview

Maria Bobrowski

Abstract

Mountains are fascinating habitats, characterized by steep ecological vertical gradients and corresponding altitudinal vegetation zonation. Alpine treelines as upper boundaries of more or less contiguous tree stands are the most conspicuous vegetation limits; they have always attracted great research interest. Globally, alpine treeline elevations in the mountains are caused by heat deficiency. At landscape and local scales, however, multiple interactions of influencing factors and mechanisms determine treeline position, spatial pattern and dynamics. In the course of climate change, it is postulated that treelines will shift to higher elevations. To be able to quantify potential shifts, an analysis of the underlying factors and a correct modelling of the treeline ecotone under current climatic conditions are of great importance. For this purpose, statistical models are used to calculate the ecological niche of species based on climatic factors. These models serve as a baseline for models that project the distribution under future climatic conditions. The Himalayas are the largest mountain range in the world, yet they are often under-represented in the scientific literature. This holds particularly true in relation to modelling studies. Modelling treeline species in remote high-altitude regions faces several challenges, especially the availability of occurrence data and high-quality environmental variables. This book chapter summarizes recent results modelling the ecological niche of the Himalayan birch (*Betula utilis*) under present climatic conditions in the Himalayan mountain system. *B. utilis* represents a favourable target species for modelling studies, since it is widespread as a treeline-forming species along the entire Himalayan arch. Due to less distinctive habitat requirements and high adaptation potential, it is gaining importance as a pioneer tree species for possible succession developments at treelines under future climate conditions. In a synergistic approach, a detailed study on comparing the underlying climatic, topographical and plant phenological factors was undertaken to model the potential and the actual distribution of the focal species. The present results provide a new starting point for further investigations aimed at modelling the distribution of the species under past or future climate scenarios. Simultaneously, the presented approaches can also be transferred to other treeline species in high mountains.

This book chapter consists of excerpts of a previously published dissertation (Bobrowski 2018). All maps were created in ArcGIS (ESRI 2018).

M. Bobrowski (✉)
Physical Geography, Center for Earth System Research and Sustainability (CEN), Universität Hamburg, Bundesstr. 55, 20146 Hamburg, Germany
e-mail: maria.bobrowski@uni-hamburg.de

U. Schickhoff et al. (eds.), *Mountain Landscapes in Transition*,
Sustainable Development Goals Series, https://doi.org/10.1007/978-3-030-70238-0_14

14.1 Introduction

Since treeline elevations are characterized by low temperatures, high-elevation climatic treelines can be considered sensitive indicators of past and recent climate change and variability at local and global scales (Kullman 1998; Holtmeier 2009; Smith et al. 2009; Körner 2012). During recent decades, investigation of climate change-driven treeline dynamics has generated considerable research interest, and results have been widely reported from various treelines around the world (e.g. Randin et al. 2009; Harsch et al. 2009; Paulsen and Körner 2014; Schibalski et al. 2014; Schickhoff et al. 2015, 2016a, b). Since high mountain environments are subjected to above-average warming rates, treeline dynamics under future climate change scenarios are of particular interest in this respect (Schickhoff 2011; IPCC 2014). Global average mean temperature has increased by +0.85 °C between 1880 and 2012 (IPCC 2013). For the Himalayan mountain system, it is hypothesized that the climate is changing at a faster rate than the global average (Shrestha et al. 2012; Schickhoff et al. 2016b). Since 1989, temperature increases during winter months of up to +0.8 °C per decade have been determined in the eastern Himalayas (Gerlitz et al. 2014), whereas pre-monsoon season temperature increases of up to +1.0 °C per decade have been found for higher elevations along the entire Himalayan arc (Schickhoff et al. 2015). Shrestha et al. (2012) found an extended growing season by 4.7 days at average during a 25-year period, with seasonal and regional variations. For the previous century, decreases in annual precipitation (up to 20%) have been identified for the western, but not for the eastern Himalayas (Jain et al. 2013; Schickhoff et al. 2016b). It is to be expected that an increase in temperature and coherently evapotranspiration, combined with a decrease in precipitation, results in amplified drought stress, primarily in the pre-monsoon season (Schickhoff et al. 2015).

One popular hypothesis is that, due to changing regional climatic conditions, ranges of subalpine and alpine species as well as treelines shift upwards along altitudinal, thermally defined gradients (Gottfried et al. 2012; Pauli et al. 2012; Wieser et al. 2014). Furthermore, it is postulated that species respond by altered seasonal phenology (Hughes 2000; Smith et al. 2012; Anadon-Rosell et al. 2014; Ernakovich et al. 2014; Hart et al. 2014), while some species are threatened by extinction or are already extinct (Parmesan 2006; Pauli et al. 2012; Alexander et al. 2015; Cotto et al. 2017). Treelines are regarded as particularly responsive to changing temperature regimes, and initial effects of future climate-induced range shifts are expected for species in high-altitude treeline ecotones.

Modelling the distributional range of treeline species and predicting changes under future climate scenarios have become an increasingly applied component in investigations of high-altitude treelines (e.g. Dullinger et al. 2004; Thuiller et al. 2005; Parolo et al. 2008). In contrast to other mountains of the world, the Himalayan region has largely been neglected and is clearly under-represented in the scientific literature on climate change-induced species range shifts (Schickhoff 2005; Miehe et al. 2007; Telwala et al. 2013; Dutta et al. 2014; Schickhoff et al. 2015). A number of studies that aim to predict species' distribution or forecast species range shifts under climate change scenarios is limited (e.g. Kumar 2012; Menon et al. 2012; Ranjitkar et al. 2014 on *Rhododendron* spp.; Menon et al. 2010 on *Gymnocladus assamicus*; Jaryan et al. 2013 on *Sapium sebiferum*; Gajurel et al. 2014 on *Taxus wallichiana*; Ranjitkar et al. 2014 on *Oxybaphus himalaicus* and *Boerhavia diffusa;* Shrestha and Bawa 2014 on *Ophiocordyceps chinensis*), although a couple of modelling studies were published in the last three years.

Moreover, high-altitude treeline studies in the Himalayas have investigated coniferous tree species (e.g. *Abies*, *Juniperus*, *Pinus*) and broadleaved evergreen tree species (e.g. *Rhododendron*) while deciduous tree species (e.g. *Betula*) have remained largely out of focus. Presently, only a few researchers have addressed the problem of modelling distribution ranges of deciduous treeline species[1] in the Himalayas. Conducted studies on *B. utilis* were mainly local scale studies (e.g. Huo et al. 2010: SW China; Singh et al. 2013: Indian Himalaya, Uttarakhand; Wang et al. 2017: Tibetan Plateau). Recently published large-scale research contained serious shortcomings with regard to a number of *Betula* occurrences, climatic predictor variables and resolution (Lamsal et al. 2017; Chhetri et al. 2018; Mohapatra et al. 2019; Hamid et al. 2019).

The genus *Betula* is known to inhabit a considerably wide ecological niche in the northern hemisphere and can be found in high-altitude and high-latitude treeline ecotones (Truong et al. 2007; Holtmeier 2009; Speed et al. 2011). Alpine treelines with *Betula* as conspicuous treeline species can be found in Russia in the Urals (*B. litwinowii*; Hansen et al. 2018), in Kamchatka (*B. ermanii*; Krestov et al. 2008) and Japan (*B. ermanii*, *B. platyphylla*, *B. maximowicziana*; Koike et al. 2003; Yasaka 2005). The target species, *Betula utilis*, is widespread in Himalayan alpine treelines (Schickhoff 2005; Ashburner and McAllister 2013), and some authors consider *B. utilis* as an indicator species for climate-driven treeline dynamics (e.g. Liang et al. 2014). *B. utilis* shows many characteristics of a pioneer species, for instance, a high degree of adaptability to altered environmental and climatic conditions. *B. utilis* is able to rejuvenate readily under changed light and soil conditions, and facilitates natural reforestation processes and forest edge closure, since it promotes humus accumulation in the course of natural succession.

The treeline-forming species *B. utilis* as a target species provides considerable study organism because (a) underlying environmental factors of the species distribution have not adequately been described and (b) improved accuracy in modelling the current distribution is a precondition for more precisely modelling potential range expansions of treelines under climate change conditions (Schickhoff et al. 2015). The latter applies in particular to a pioneer species such as *B. utilis*, characterized by high adaptability to changing environments. To date, modelling the ecological niche of *B. utilis* covering the entire Himalayan mountain region has remained a major research deficit.

14.2 Challenges and Limitations of Ecological Niche Models in High-Altitude Regions

In order to investigate high-altitude treeline dynamics, modelling techniques have become an indispensable method to predict species distributions under current climate conditions, to hindcast distributions under past climate conditions and to forecast changed distributional ranges under future climate scenarios (e.g. Dullinger et al. 2004; Thuiller et al. 2005; Parolo et al. 2008; Schorr et al. 2012). Inherently, the accuracy of models under climate change scenarios depends on their accuracy under current climate conditions, the importance of which should not be underestimated (Bobrowski et al. 2021).

Modelling ecological niches across vast distribution ranges in remote, high mountain regions like the Himalayas remains a challenging task. Challenges include, first and foremost, the lack of species occurrence data and fine-scale environmental information of sufficiently high quality (i.e. environmental variables).

In many cases, presence–absence data are not available, and presence-only data are often derived from databases of natural history museums and herbaria, which contain occurrences sampled by numerous researchers and with different techniques, intensities and periods of time (Soberón and Peterson 2004). Moreover, sampling records often cluster near the centre of climatic conditions under which the species occurs (Loiselle et al. 2008). This leads to species documentations that do not cover the entire range of suitable habitat conditions for respective species. Such geographic sampling bias can lead to sampling bias in environmental space, which represents a major problem for modelling (Veloz 2009; Anderson and Gonzalez 2011). This holds particularly true for sampling treeline species in

[1]For reasons of readability, the terms 'treeline' and 'treeline ecotone' will be used synonymous in the presented chapter.

remote areas like the Himalayan region. Due to the lower accessibility of treeline sites, the number of available sampling plots is sparse, which demonstrates a reciprocal effect on model prediction performance (Aráujo et al. 2005).

This also applies to environmental variables, as most modelling studies use climatic variables for predicting the distribution range of the species. In topographically complex areas like the Himalayas, climate stations are quite rare due to rough terrain and complicated accessibility. This in turn leads to a poor data basis for calculating climate data sets compared to other more accessible terrains. In addition, climate stations are prevailingly located near settlements at lower elevations, where climatic conditions are most suitable for habitation, livestock farming and agriculture. Those climate stations are not representative of climatic conditions at higher elevations. Besides the data basis, the calculation method and bias correction also influence the quality of climate data sets.

The choice of environmental variables used to model species distributions may result in different distribution maps for the same species (Luoto et al. 2007). If important local abiotic or biotic factors that influence the actual species' distribution (i.e. the realized niche) are disregarded, predictions will represent the potential distribution (i.e. the existing fundamental niche), since climate is not the exclusive factor determining habitat suitability (Thuiller 2004). This will gain extraordinary importance since an improved understanding and modelling capacity of the current distribution constitutes a precondition for modelling treeline dynamics under climate change scenarios.

14.2.1 Modelling Treeline Dynamics Under Climate Change

Global and regional treeline responsiveness to climate change is highly complex and influenced by a variety of abiotic and biotic factors and their interrelations. Given the fact that the elevational position of treelines is attributed to prevailing thermal conditions, worldwide treeline ecotones constitute sensitive indicators to changing climate conditions (Körner 2012).

During the Holocene, treeline fluctuations were caused by climate variability. Evidence was found for the upslope movement during warmer periods and recession during cooler periods (Alps: Schwörer et al. 2014, Himalaya: Schickhoff et al. 2016a). More specifically, after the Pleistocene–Holocene transition period (11.7 kyr. BP) the treeline position in the Himalayas was situated several hundred metres higher than today. The highest elevational positions of treelines in the early Holocene can be attributed to warm and moist climate conditions resulting from a reinforced Asian monsoon regime (Schickhoff et al. 2016a). In the mid-Holocene (5 kyr BP), treeline positions shifted to somewhat lower elevations due to decreasing temperatures (Schickhoff et al. 2016a). In recent millennia, human impact has become the dominant driver of treeline elevational positions. In the Himalayas, Holocene treeline history was not uniform due to regional and local particularities.

Under future climate change scenarios, treeline positions are postulated to advance to higher elevations. To date, treeline responses do not show consistent patterns at global and local scales (Dullinger et al. 2004; Harsch et al. 2009; Körner 2012; Schickhoff et al. 2015). The degree of treelines' susceptibility to being significantly affected by changing climate depends on treeline type and form (Schickhoff et al. 2015). Climatic treelines are highly susceptible to climate warming (Holtmeier and Broll 2007; Körner 2012), whereas orographic treelines do not show significant changes (Schickhoff et al. 2016a). In terms of their responsiveness to climate warming, anthropogenic treelines can be compared to climatic treelines. Based on treeline types, four treeline forms with different responsiveness patterns can be distinguished (i.e. diffuse, abrupt, island and krummholz treeline forms; Harsch and Bader 2011). Only diffuse treelines exhibit a strong response signal, whereas the other forms remain rather unreactive in terms of elevational shifts.

In the Himalayas, explicit differences can be found between treelines on north- and south-facing slopes. Whereas human impact transformed treelines on south-facing slopes to a large extent, north-facing slopes have a limited use potential, and near-natural treeline ecotones, including an intact krummholz belt, still exist (Schickhoff et al. 2015). Krummholz treelines usually show a lagged response to climate warming, and treeline shifts may occur only in the long term (Schickhoff et al. 2015, 2016a, b). It is assumed that the strong competition within the krummholz belt and the occurrence of dense dwarf scrub heaths located above hinders the upward migration of tree species (Schickhoff et al. 2015). However, stand densification and prolific regeneration within the treeline ecotone indicate beneficial preconditions for treeline advance in the future (Schickhoff et al. 2015).

The analyses of treelines responding to changed climatic conditions and differentiation of small- and broad-scale underlying mechanisms and factors remain a pending question. Recent studies have focused on ecological, dendroecological, forest-ecological and remote sensing aspects (Liang et al. 2011; Bharti et al. 2011; Rai et al. 2013; Shrestha et al. 2014; Müller et al. 2016a, b; Schwab et al. 2016; Gaire et al. 2017; Bürzle et al. 2018). Increasingly, modelling approaches have been applied to gain a better understanding of treeline dynamics, particularly to investigate the underlying process-based relationships and identify potential range shifts of species in response to changed climatic conditions and altered land-use regimes (Dullinger et al. 2004; Wallentin et al. 2008; Paulsen and Körner 2014; Schickhoff et al. 2015, 2016a, b). However, the results of recent modelling studies (Lamsal et al. 2017; Chhetri et al. 2018; Mohapatra et al. 2019; Hamid et al. 2019) should be critically examined, since the input parameters (i.e. occurrence and climate data) may have been used without critical examination of potential errors. A particular challenge for modelling studies involves the lack of natural treeline sites since the treeline position is almost everywhere depressed to lower altitudes due to human impact (Miehe et al. 2015; Schickhoff et al. 2015, 2016a, b). At anthropogenically depressed treelines, it is a challenge to disentangle the climatic signal and anthropogenic land-use impacts as the driver behind treeline dynamics. Upslope shifts of treeline species in the short term might be attributed to changes in land-use regimes rather than to climate change. For near-natural treelines, it is postulated that changes in their elevational position will be a medium- to long-term process (Schickhoff et al. 2015, 2016a, b).

When comparing the vegetation map of Scheinfurth (1957) with the modelled potential distribution of *Betula utilis* under current climate conditions deviations were revealed (Bobrowski et al. 2017), this is possibly attributed to land-use change rather than changed climatic conditions (Fig. 14.1). Even if the effects of climate warming trigger upslope shifts of *B. utilis* in the coming decades (Schickhoff et al. 2015), this might be primarily attributed to the cessation of land use. In the reanalysis, the same pattern can be observed for the entire Himalayan Mountains, where similar discrepancies between the Schweinfurth map (1957) and predictions of the modelled actual distribution using climatic, topographic and phenological data (Bobrowski et al. 2018) were found (Fig. 14.1). However, it should be mentioned, that the real distribution of *B. utilis* is presumably smaller than shown in this map, since climate and topography are not the only factors, driving distribution ranges of treeline species. Nevertheless, the additional consideration of phenological traits leads to more precise modelling results, compared to solely climate and topography-based models (Bobrowski et al. 2018).

Furthermore, changing species' distributional patterns and phenology are responses to recent climate change that will modify the structure, composition and position of the treeline in the Himalayan mountain system. Remotely sensed data of plant phenological seasonal variations can be used to track changes in vegetation phenology (Beck et al. 2007) since shifts in seasonal phenological events are among the first responses at plant and ecosystem levels to climate change (Badeck et al. 2004). Shifts of flowering dates have been reported for *Rhododendron* species (Xu et al. 2009), and earlier green-up data resulting in an extension of the growing season

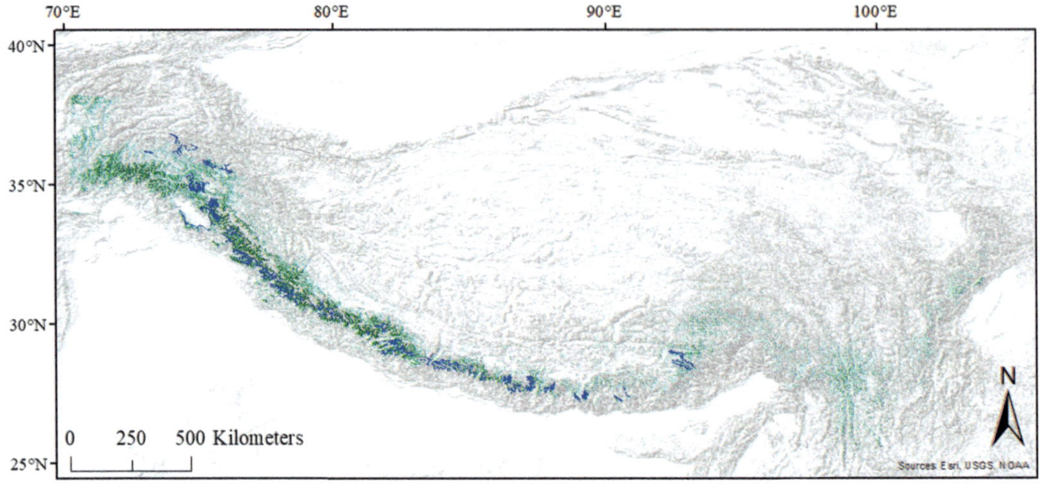

Fig. 14.1 Reanalysis of the modelled ecological niche of *Betula utilis* based on climatic, topographic and phenological data (modified after Bobrowski et al. 2017, 2018) and *B. utilis* forests according to the vegetation map of Schweinfurth (1957), whereas blue areas represent *Betula utilis* dominated forest after Schweinfurth, 1957 and green areas represent the current modelled distribution of *B. utilis* (Bobrowski et al. 2017, 2018)

(Panday and Ghimire 2012; Shrestha et al. 2012) have been reported for the Himalayas. However, responses to above-average warming rates projected for the twenty-first century will most likely be associated with biodiversity loss and a decrease of ecosystem functions (Schickhoff et al. 2016a).

Therefore, investigations of underlying climatic factors and the quantification of changing plant phenological traits provide the basis for efficient nature conservation management, expansion of protected areas, and appropriate habitat restoration strategies. The recent results constitute a stepping stone for further investigations of treeline dynamics in the Himalayan mountain system by incorporating remotely sensed variables.

14.2.2 Limitations of Ecological Niche Models and Potentials of Remote Sensing Data

The investigation of factors driving the current distribution of treeline species is a *conditio sine qua non* of factors behind treeline dynamics. In order to obtain meaningful modelling results, a synthesis of various ecology-related disciplines would be desirable. Since the availability and quality of input parameters determine model performance, complete high-resolution long-term data is advisable. Besides species occurrence data, information ranging from plant-specific characteristics and responsiveness to changing climatic conditions and inter- and intra-specific competition to succession experiments would enhance modelling procedures.

However, in reality, things are different. By using correlative modelling approaches, limitations and errors may occur at any step of the procedure. Far-reaching consequences can be traced back to the input parameters. The model is only as precise as the quality and relevance of the biotic and abiotic parameters used to build the model for the targeted species. In an extensive literature review, He et al. (2015) presented numerous applications of remotely sensed data for modelling species' distributions. They demonstrated the adaptability of remote sensing products for modelling marine and terrestrial biota, and how they can be customized in accordance with specific research questions.

A major source of uncertainty can be traced back to presence-only species occurrence data instead of presence–absence data. They are often derived from databases of natural history museums and herbaria, whereby sampling techniques, intensities and periods of time may differ (Soberón and Peterson 2004). Sampling bias in geographic space leads to sampling bias in environmental space, which must certainly be considered problematic (Veloz 2009). Spatial filtering (i.e. only one point per 1 × 1 km grid cell) of the occurrence points was applied to decrease sampling bias and spatial autocorrelation. No assurance can be provided regarding afflicted biases of the museum- and literature-based occurrences.

In many studies, the investigation and evaluation of input parameters are neglected. The impact of the results' implementation may have severe consequences (Bobrowski and Schickhoff 2017). Abiotic and biotic data derived from remote sensing may open up new opportunities for analysing and modelling species' distributions since they provide response and predictor variables.

The benefits of remotely sensed data in deriving tree species occurrences could be highlighted. Almost 80% (Bobrowski et al. 2018) and 55% (Bobrowski and Schickhoff 2017; Bobrowski et al. 2017) of the *Betula utilis* occurrence points were extracted from freely available satellite imagery (Google 2015) (Fig. 14.2). The potential of remote sensing data for future studies lies in the generation of presence and absence data sets, which are highly required in ENMs (Fithian et al. 2015). Due to unique biophysical properties, hyperspectral sensors can detect subtle differences in reflectance based on unique plant chemistries, which is beneficial for identifying plant species (Buermann et al. 2008). Not only noticeable vegetation structures like treeline ecotones can be distinguished, but technologies may also be applicable for detecting rare or invasive species at the plant species level (He et al. 2015 and references therein). Another advantage is the possibility to incorporate biotic interactions into the models, which are often disregarded due to data limitations (Kissling et al. 2012; Dormann et al. 2018). The inclusion of biotic interactions of tree species treeline species associated with *B. utilis* would be beneficial for modelling possible future range shifts.

Often, standardized statistically derived parameters do not fully reflect the species' physiological needs and habitat requirements, and therefore lead to poor modelling results. As illustrated in Bobrowski and Schickhoff (2017), evaluating and comparing the performance of climate data sets remains a challenging task, and it may be worthwhile to compare different climate data sets (e.g. Chelsa: Karger et al. 2016; Worldclim: Hijmans et al. 2005). In Bobrowski et al. (2018) thermal metrics derived from MODIS land surface temperatures (LST) (Bechtel 2015) were tested, which may be beneficially incorporated into further treeline studies in remote mountainous regions, as they provide freely accessible, complete and long-term data. The main advantages of LST-related variables are continuous observations without interpolation and geographical bias, and therefore with less uncertainty (He et al. 2015). Recent studies have revealed how LST data could improve species modelling studies (e.g. Buermann et al. 2008; Bisrat et al. 2012; Still et al. 2014). These parameters offer numerous possibilities, such as tailored predictors in high resolution. As time series data of vegetation characteristics (i.e. phenological metrics) are becoming more and more readily available, changing habitat suitability can be estimated and incorporated into model approaches. In this way, knowledge can be generated that is particularly important for modelling spatial expansion of invasive species, extinction risk assessment, and range shifts under future climate change (He et al. 2015 and references therein). In mountainous areas, the resolution of climate data (i.e. 1 × 1 km) is often too coarse for models to distinguish between north- and south-facing slopes. With high-resolution remote sensing data, however, the heterogeneity of the terrain can be taken into account, leading to more precise modelling results.

The results underline the relevance of additional remotely sensed environmental variables for reducing the gap between the potential and actual distribution of *B. utilis* (Fig. 14.3). It becomes apparent that the core distribution of *B. utilis* was predicted in the western part of the

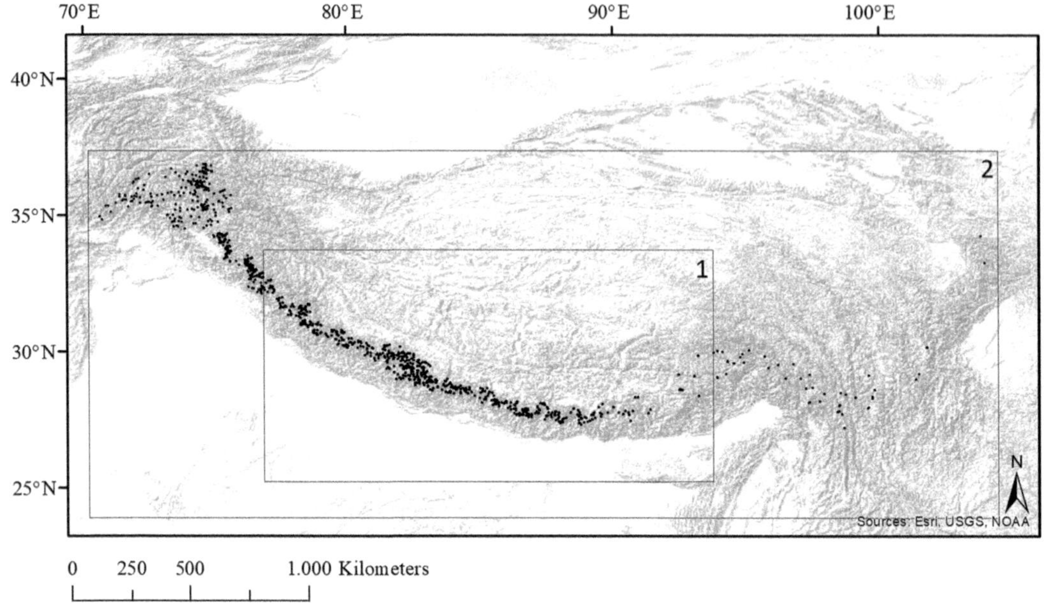

Fig. 14.2 Occurrences of Betula *utilis* used in the modelling approaches: (1) 590 occurrences (Bobrowski et al. 2017; Bobrowski and Schickhoff 2017), of which c. 55% were extracted from satellite images via GoogleEarth; (2) 1041 occurrences (Bobrowski et al. 2018), of which c. 80% were extracted from satellite images via GoogleEarth

Himalayan mountain system, whereas only the LST model predicted a principal distribution in the central part of the mountains. All models showed a uniform distribution along the Himalayan arc. The habitat predicted by CLIMATE tends to be wider in range compared to the other predictions.

The model solely based on climate predictor variables (CLIMATE) roughly met the lower limit of occurrences compared to the CLIMATE + TOPO + PHENO model, but overpredicted the uppermost limits of *B. utilis* (G_P in Fig. 14.3a) Overall, the broadleaved deciduous treeline could not be distinguished from other vegetation formations. The same picture emerged for LST models (G_P in Fig. 14.3b).

By contrast, in the CLIMATE + TOPO + PHENO and LST + TOPO + PHENO models, remotely sensed predictors supplemented information on, e.g. topographical barriers (TOPO; USGS 2004) and distinction between phenological divergent vegetation formations (PHENO; LP DAAC 2012), leading to the modelled actual distribution G_O of *B. utilis*, respectively (Fig. 14.3a, b). However, the actual distribution of *B. utilis* might be smaller than predicted, since topoclimate variables and phenological traits are not the only factors determining habitat suitability (for a detailed discussion see Bobrowski et al. 2018).

Concluding, that although the available data derived from remote sensing technology is rather short term, the modelling results for *Betula utilis* using a combination of statistically derived and remote sensing data may serve as a baseline for future studies. Restrictions in the practical applicability arise from the fact that high-resolution satellite imagery is still often very expensive. On the other hand, the free of cost imagery and software is already available and will become more customary in the future (Petorelli et al. 2016). Recent results showed that, even with freely available data, model performances could be improved, indicating the potential for future modelling studies (Bobrowski et al. 2018). Airborne technology is a continually expanding field, and high-resolution remotely

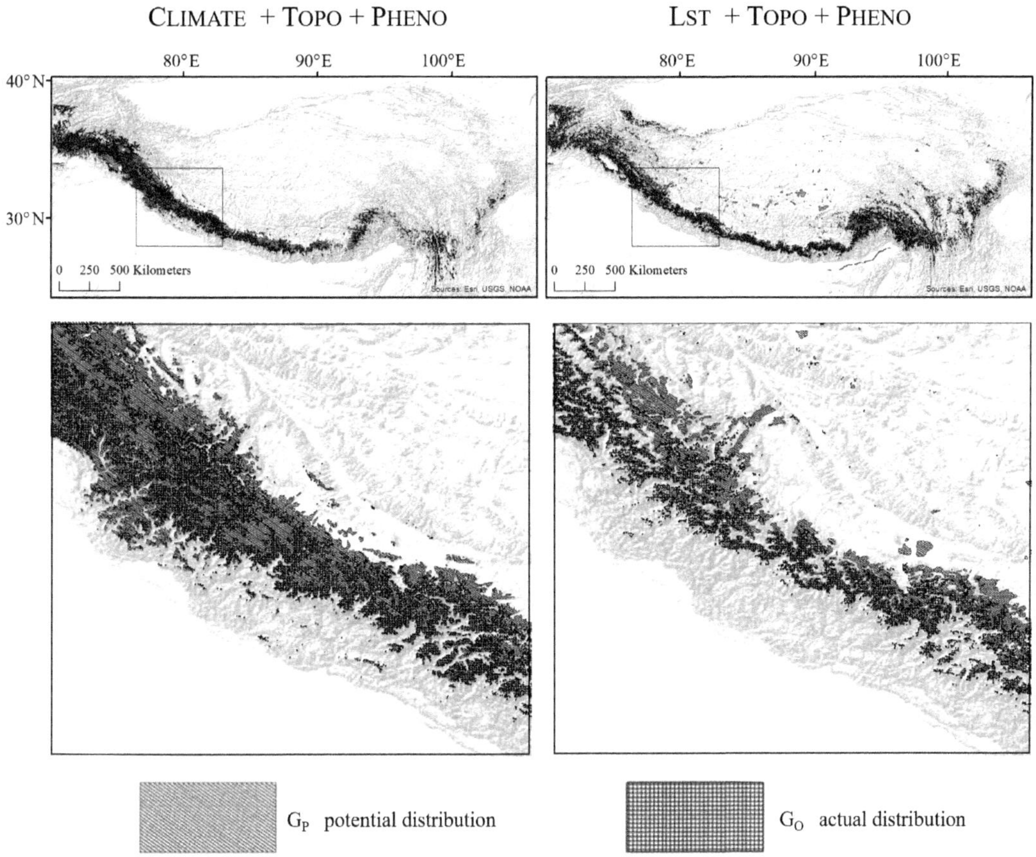

Fig. 14.3 Reanalysis of the model predictions for modelling the ecological niche of *Betula utilis*. G_P the potential distribution (grey) was modelled solely based on climate-related variables **a** CLIMATE and **b** LST. G_O the actual distribution (black) was modelled with additional remotely sensed variables like topography and phenological traits **a** CLIMATE + TOPO + PHENO and **b** LST + TOPO + PHENO. See Bobrowski et al. 2018 for further details

sensed data will provide more insights into spatial patterns and underlying factors in future modelling studies.

14.3 Conclusions

The aim of this chapter was to summarize possible pitfalls and challenges involved in modelling the ecological niches of *Betula utilis* in remote, high-elevation treeline ecotones along the Himalayan mountain range. Moreover successful approaches were presented, such as compiling *B. utilis* occurrences in the treeline ecotone along the Himalayan arc primarily from freely available satellite images.

The results of the consecutive studies provide a comprehensive analysis of the underlying environmental factors (climatic patterns, topography and phenological traits) determining the ecological niche of *B. utilis* in the Himalayan region under current climate conditions. Potential suitable habitats of the species were successfully predicted as a function of climatic variables that characterize current climatic conditions at treeline locations (Bobrowski et al. 2017). It can be concluded, that ecological niche modelling presents a valuable predictive tool for analysing the distribution of treeline species when the existing complexity of remote high-altitude regions is denoted in climate input variables.

It needs to be highlighted that global climate data sets should not be used to model ecological niches without critically scrutinizing the origin of climate data and the computation method of the climate data set, and without being aware of potentially afflicted limitations (Bobrowski and Schickhoff 2017). The obtained results could be particularly misleading, when modelling ecological niches in heterogeneous landscapes like the Himalayan region, emphasizing the use of high resolution (<1 km^2) local climate data sets for future modelling studies.

By expanding the solely climate-based approach with freely available remotely sensed variables modelling the actual distribution of the species was attempted (Bobrowski et al. 2018). The inclusion of variables characterizing spatial variation in environmental variables, such as remotely sensed vegetation indices, provided key inputs. *Betula* is a conspicuous broadleaved deciduous tree species at treelines, allowing for a clear separation on the basis of phenological traits from adjacent vegetation types (evergreen coniferous and evergreen broadleaved species). The incorporation of remote sensing data led to a more refined modelled distribution since, based on the real information of the Earth's surface, as they account for non-climatic dimensions (i.e. anthropogenic impacts), leading in turn to a more realistic actual distribution. Although the model predictions are in general agreement with several vegetation maps, the actual distribution might be smaller than indicated by the models. The results point to the need for further investigations of microclimatic conditions with parameters related to soil properties or solar radiation, as well as investigations of biotic interactions or dispersal limitations.

The presented synergetic modelling approach can be transferred to any other species in conspicuous vegetation formations, such as treeline ecotones, and the applied environmental predictors are transferable without severe modification due to global coverage. Further research may include additional remotely sensed metrics, such as solar radiation, precipitation amounts and snow cover.

Transferability to other deciduous treelines remains to be examined, but promising results and insights can be expected. The obtained insights may serve as a role model for other climatic treelines, and especially for other *Betula* treelines in mountain regions. The findings may serve as a baseline for further investigations of treeline dynamics under future climate change scenarios in regions with limited data availability.

References

Alexander JM, Diez JM, Levine JM (2015) Novel competitors shape species' responses to climate change. Nature 525:515–518

Anadon-Rosell A, Rixen C, Cherubini P, Wipf S, Hagedorn F, Dawes MA (2014) Growth and phenology of three dwarf shrub species in a six-year soil warming experiment at the alpine treeline. PLoS One 9:e100577

Anderson RP, Gonzalez I Jr (2011) Species-specific tuning increases robustness to sampling bias in models of species distributions: an implementation with Maxent. Ecol Model 222:2796–2811

Araújo M, Pearson RG, Thuiller W, Erhard M (2005) Validation of species—climate impact models under climate change. Glob Change Biol 11:1504–1513

Ashburner K, McAllister HA (2013) The genus *Betula*—a taxonomic revision of birches. Botanical Magazine Monograph 5. Royal Botanic Gardens, Kew

Bechtel B (2015) A new global climatology of annual land surface temperature. Remote Sens 7:2850–2870

Beck PSA, Jönsson P, Høgda KA, Karlsen SR, Eklundh L, Skidmore AK (2007) A ground-validated NDVI dataset for monitoring vegetation dynamics and mapping phenology in Fennoscandia and the Kola Peninsula. Int J Remote Sens 28:4311–4330

Badeck FW, Bondeau A, Böttcher K, Doktor D, Lucht W, Schaber J, Sitch S (2004) Responses of spring phenology to climate change. New Phytol 162:295–309

Bharti RR, Rai ID, Adhikari BS, Rawat GS (2011) Timberline change detection using topographic mal and satellite imagery: a critique. Trop Ecol 52:133–137

Bisrat SA, White MA, Beard KH, Richard Cutler D (2012) Predicting the distribution potential of an invasive frog using remotely sensed data in Hawaii. Divers Distrib 18:648–660

Bobrowski M (2018) Modelling the ecological niche of a treeline species (*Betula utilis*) in the Himalayan region. PhD thesis, Universität Hamburg, Germany

Bobrowski M, Gerlitz L, Schickhoff U (2017) Modelling the potential distribution of *Betula utilis* in the Himalaya. Glob Ecol Conserv 11:69–83

Bobrowski M, Schickhoff U (2017) Why input matters: selection of climate data sets for modelling the

potential distribution of a treeline species in the Himalayan region. Ecol Model 359:92–102
Bobrowski M, Bechtel B, Böhner J, Oldeland J, Weidinger J, Schickhoff U (2018) Application of thermal and phenological land surface parameters for improving ecological niche models of *Betula utilis* in the Himalayan region. Remote Sens 10:814
Bobrowski M, Weidinger J, Schickhoff U (2021) Is new always better? Frontiers in global climate datasets for modeling treeline species in the himalayas. Atmos 12:543
Buermann W, Saatchi S, Smith TB, Zutta BR, Chaves JA, Milá B, Graha CH (2008) Predicting species distributions across the Amazonian and Andean regions using remote sensing data. J Biogeogr 35:1160–1176
Bürzle B, Schickhoff U, Schwab N, Wernicke LM, Müller YK, Böhner J, Chaudhary RP, Scholten T, Oldeland J (2018) Seedling recruitment and facilitation dependence on safe site characteristics in a Himalayan treeline ecotone. Plant Ecol 219:115–132
Chhetri P, Gaddis K, Cairns D (2018) Predicting the suitable habitat of treeline species in the Nepalese Himalayas under climate change. Mt Rest Dev 38:153–163
Cotto O, Wessely J, Georges D, Klonner G, Schmid M, Dullinger S, Thuiller W, Guillaume F (2017) A dynamic eco-evolutionary model predicts slow response of alpine plants to climate warming. Nat Commun 8:1–9
Dormann CF, Bobrowski M, Dehling M, Harris DJ, Hartig F, Lischke H, Moretti MD, Pagel J, Pinkert S, Schleuning M, Schmidt S, Sheppard C, Steinbauer MJ, Zeuss D, Kraan C (2018) Biotic interactions in species distribution modelling: ten questions to guide interpretation and avoid false conclusions. Global Ecol Biogeogr 27:1004–1016
Dullinger S, Dirnböck T, Grabherr G (2004) Modelling climate-change driven treeline shifts: relative effects of temperature increase, dispersal and invasibility. J Ecol 92:241–252
Dutta PK, Dutta BK, Das AK, Sundriyal DRC (2014) Alpine timberline research gap in Himalaya: a literature review. Indian For 140:419–427
Ernakovich JG, Hopping KA, Berdanier AB, Simpson RT, Kachergis EJ, Steltzer H, Wallenstein MD (2014) Predicted responses of arctic and alpine ecosystems to altered seasonality under climate change. Glob Change Biol 20:3256–3269
ESRI (2018) ArcGIS desktop: release 10.5.1. Environmental Systems Research Institute, Redlands, CA.
Fithian W, Elith J, Hastie T, Keith DA (2015) Bias correction in species distribution models: pooling survey and collection data for multiple species. Methods Ecol Evol 6:424–438
Gaire NP, Koirala M, Bhuju DR, Carrer M (2017) Site- and species-specific treeline responses to climatic variability in eastern Nepal Himalaya. Dendrochronologica 41:44–56
Gajurel JP, Werth S, Shrestha KK, Scheidegger C (2014) Species distribution modeling of *Taxus wallichiana* (Himalayan Yew) in Nepal Himalaya. Asian J Conserv Biol 3:127–134
Gerlitz L, Conrad O, Böhner J (2014) Warming patterns over the Tibetan Plateau and adjacent lowlands derived from elevation- and bias-corrected ERA-Interim data. Clim Res 58:235–246
Google Earth, ver. 7.1.1.1888, Google 2015
Gottfried MM, Pauli H, Futschik A, Akhalkatsi M, Barančok P, Benito A, José L, Coldea G, Dick J, Erschbamer B, Fernández Calzado MR, Kazakis G, Krajči J, Larsson P, Mallaun M, Michelsen O, Moiseev D, Moiseev P, Molau U, Merzouki A, Nagy L, Nakhutsrishvili G, Pedersen B, Pelino G, Puscas M, Rossi G, Stanisci A, Theurillat JP, Tomaselli M, Villar L, Vittoz P, Vogiatzakis I, Grabherr G (2012) Continent-wide response of mountain vegetation to climate change. Nat Clim Change 2:111–115
Hamid M, Khuroo A, Charles B, Khuroo R, Singh CP, Aravind NA (2019) Impact of climate change on the distribution range and niche dynamics of Himalayan birch, a typical treeline species in Himalayas. Biodivers Conserv 28:2345–2370
Hansen W, Magiera A, Theissen T, Waldhardt R, Otte A, Rocchini D (2018) Analysing *Betula litwinowii* encroachment and reforestation in the Kazbegi region, Greater Caucasus, Georgia. J Veg Sci 29:110–123
Harsch MA, Hulme PE, McGlone MS, Duncan RP (2009) Are treelines advancing? A global meta-analysis of treeline response to climate warming. Ecol Lett 12:1040–1049
Harsch MA, Bader MY (2011) Treeline form—a potential key to understanding treeline dynamics. Glob Ecol Biogeogr 20:582–596
Hart R, Salick J, Ranjitkar S, Xu J (2014) Herbarium specimens show contrasting phenological responses to Himalayan climate. Proc Natl Acad Sci USA 111:10615–10619
He KS, Bradley BA, Cord AF, Rocchini D, Tuanmu MN, Schmidtlein S, Turner W, Wegmann M, Pettorelli N (2015) Will remote sensing shape the next generation of species distribution models? Remote Sens Ecol Conserv 1:4–18
Hijmans RJ, Cameron SE, Parra JL, Jones PG, Jarvis A (2005) Very high resolution interpolated climate surfaces for global land areas. Int J Climatol 25:1965–1978
Holtmeier FK, Broll G (2007) Treeline advance—driving processes and adverse factors. L O 1:1–33
Holtmeier FK (2009) Mountain timberlines—ecology, patchiness and dynamics. In: Advances in global change research. Springer, Berlin, p 36
Hughes L (2000) Biological consequences of global warming: is the signal already apparent? Trends in Ecol Evol 15:56–61
Huo C, Cheng G, Lu X, Fan J (2010) Simulating the effects of climate change on forest dynamics on Gongga Mountain, Southwest China. J For Res 15:176–185
IPCC (2013) Summary for policymakers. Climate change 2013: the physical science basis. In: Stocker TF,

Qin D, Plattner GK, Tignor M, Allen SK, Boschung J, Nauels A, Xia Y, Bex V, Midgley PM (eds) Contribution of working group I to the fifth assessment report of the intergovernmental panel on climate change. Cambridge University Press, Cambridge, New York, pp 1–30

IPCC (2014) Climate change 2014: synthesis report. In: Core Writing Team, Pachauri RK, Meyer LA (eds) Contribution of working groups I, II and III to the fifth assessment report of the intergovernmental panel on climate change international panel on climate change, Switzerland, Geneva

Jain SK, Kumar V, Saharia M (2013) Analysis of rainfall and temperature trends in northeast India. Int J Climatol 33:968–978

Jaryan V, Datta A, Uniyal SK, Kumar A, Gupta RC, Singh RD (2013) Modelling potential distribution of Sapium sebiferum—an invasive tree species in western Himalaya. Curr Sci India 105:1282–1288

Karger DN, Conrad O, Böhner J, Kawohl T, Kreft H, Soria-Auza RW, Zimmermann N, Linder HP, Kessler M (2016) Climatologies at high resolution for the earth land surface areas. arXiv:1607.00217 [physics]

Kissling WD, Dormann CF, Groeneveld J, Hickler T, Kühn I, McInerny GJ, Montoya JM, Römermann C, Schiffers K, Schurr FM, Singer A, Svenning J, Zimmermann NE, O'Hara RB (2012) Towards novel approaches to modelling biotic interactions in multispecies assemblages at large spatial extents. J Biogeogr 39:2163–2178

Koike T, Kitao M, Quoreshi AM, Matsuura Y (2003) Growth characteristics of root-shoot relations of three birch seedlings raised under different water regimes. Plant Soil 255:303–310

Körner C (2012) Alpine treelines—functional ecology of the global high elevation tree limits. Springer, Berlin

Krestov P, Omelko A, Nakamura Y (2008) Vegetation and natural habitats of Kamchatka. Berichte der Reinhold-Tüxen-Gesellschaft 20, Hannover

Kullman L (1998) Tree-limits and montane forests in the Swedish Scandes: sensitive biomonitors of climate change and variability. Ambio 27:312–321

Kumar P (2012) Assessment of impact of climate change on Rhododendrons in Sikkim Himalayas using Maxent modelling: limitations and challenges. Biodiv Conserv 21:1251–1266

Lamsal P, Kumar L, Shabani F, Atreya K (2017) The greening of the Himalayas and Tibetan Plateau under climate change. Global Planet Change 159:77–92

Liang E, Wang Y, Eckstein D, Luo T (2011) Little change in the fir tree-line position on the Southeastern Tibetean Plateau after 200 years of warming. New Phytol 190:760–769

Liang E, Dawadi B, Pederson N, Eckstein D (2014) Is the growth of birch at the upper timberline in the Himalayas limited by moisture or by temperature? Ecology 95:2453–2465

Loiselle BA, Jørgensen PM, Consiglio T, Jiménez I, Blake JG, Lohmann LG, Montiel OM (2008) Predicting species distributions from herbarium collections: does climate bias in collection sampling influence model outcomes? J Biogeogr 35:105–116

LP DAAC (2012) NASA Land processes distributed active archive center, USGS/Earth Resources Observation and Science (EROS) Center. https://lpdaac.usgs.gov/data_access/data_pool

Luoto M, Virkkala R, Heikkinen RK (2007) The role of land cover in bioclimatic models depends on spatial resolution. Glob Ecol Biogeogr 16:34–42

Menon S, Choudhury BI, Khan ML, Peterson AT (2010) Ecological niche modelling and local knowledge predict new populations of *Gymnocladus assamicus*, a critically endangered tree species. Endanger Species Res 11:175–181

Menon S, Khan ML, Paul A, Peterson AT (2012) Rhododendron species in the eastern Himalayas: new approaches to understanding rare plant species distributions, vol 38. Grand Valley State University, Biology Department. Peer Reviewed Publications, pp 78–84

Miehe G, Miehe S, Vogel J, Co S, La D (2007) Highest treeline in the Northern Hemisphere found in Southern Tibet. Mt Res Dev 27:169–173

Miehe G, Pendry CA, Chaudhary R (2015) Nepal: an introduction to the natural history, ecology and human environment of the Himalayas. Royal Botanic Garden Edinburgh, Edinburgh

Mohapatra J, Singh CP, Hamid M, Verma A, Chandra S, Gajmer B, Khuroo A, Kumar A, Nautiyal M, Sharma N, Pandya H (2019) Modelling *Betula utilis* distribution in response to climate-warming scenarios in Hindu-Kush Himalaya using random forest. Biodiv Conserv 28:2295–2317

Müller M, Schickhoff U, Scholten T, Drollinger S, Böhner J, Chaudhary RP (2016a) How do soil properties affect alpine treelines? General principles in a global perspective and novel findings from Rolwaling Himal, Nepal. Prog Phys Geog 40:1–26

Müller M, Schwab N, Schickhoff U, Böhner J, Scholten T (2016b) Soil temperature and soil moisture patterns in a Himalayan alpine treeline ecotone. Arct Antarct Alp Res 48:501–521

Panday PK, Ghimire B (2012) Time-series analysis of NDVI from AVHRR data over the Hindu Kush-Himalayan region for the period 1982–2006. Int J Remote Sens 33:6710–6721

Parmesan C (2006) Ecological and evolutionary responses to recent climate change. Annu Rev Ecol Evol S 637–669

Parolo G, Rossi G, Ferrarini A (2008) Toward improved species niche modelling: *Arnica montana* in the Alps as a case study. J Appl Ecol 45:1410–1418

Pauli H, Gottfried M, Dullinger S, Abdaladze O, Akhalkatsi M, Benito Alonso JL, Coldea G, Dick J, Erschbamer B, Fernández Calzado R, Ghosn D, Holten JI, Kanka R, Kazakis G, Kollár J, Larsson P, Moiseev P, Moiseev D, Molau U, Molero Mesa J, Nagy L, Pelino G, Puşcaş M, Rossi G, Stanisci A, Syverhuset AO, Theurillat JP, Tomaselli M,

Unterluggauer P, Villar L, Vittoz P, Grabherr G (2012) Recent plant diversity changes on Europe's mountain summits. Science 336:353–355

Paulsen J, Körner C (2014) A climate-based model to predict potential treeline position around the globe. Alpine Bot 124:1–12

Pettorelli N, Wegmann M, Skidmore A, Mücher S, Dawson TP, Fernandez M, Lucas R, Schaepman ME, Wang T, O'Connor B, Jongman RH, Kempeneers P, Sonnenschein R, Leidner AK, Böhm M, He KS, Nagendra H, Dubois G, Fatoyinbo T, Hansen MC, Paganini M, de Klerk HM, Asner GP, Kerr JT, Estes AB, Schmeller DS, Heiden U, Rocchini D, Pereira HM, Turak E, Fernandez N, Lausch A, Cho MA, Alcaraz-Segura D, McGeoch MA, Turner W, Mueller A, St-Louis V, Penner J, Vihervaara P, Belward A, Reyers B, Geller GN (2016) Framing the concept of satellite remote sensing essential biodiversity variables: challenges and future directions. Remote Sens Ecol Conserv 2:122–131

Rai ID, Bharti RR, Adhikari BS, Rawat GS (2013) Structure and functioning of timberline vegetation in the Western Himalaya: a case study. In: Ning W, Rawat GS, Joshi S, Ismail M, Sharma E (eds) High-altitude rangelands and their interfaces in the Hindu Kush Himalayas. ICIMOD Nepal, Kathmandu, pp 91–106

Randin CF, Engler R, Normand S, Zappa M, Zimmermann NE, Pearman PB, Vittoz P, Thuiller W, Guisan A (2009) Climate change and plant distribution: local models predict high-elevation persistence. Glob Change Biol 15:1557–1569

Ranjitkar S, Kindt R, Sujakhu NM, Hart R, Guo W, Yang X, Shrestha KK, Xu J, Luedeling E (2014) Separation of the bioclimatic spaces of Himalayan tree Rhododendron species predicted by ensemble suitability models. Global Ecol Conserv 1:2–12

Schibalski A, Lehtonen A, Schröder B (2014) Climate change shifts environmental space and limits transferability of treeline models. Ecography 37:321–335

Schickhoff U (2005) The upper timberline in the Himalaya, Hindu Kush and Karakorum: a review of geographical and ecological aspects. In: Broll G, Keplin B (eds) Mountain ecosystems. Studies in treeline ecology, Springer, Berlin-Heidelberg, pp 275–354

Schickhoff U (2011) Dynamics of mountain ecosystems. In: Millington A, Blumler M, Schickhoff U (eds) Handbook of biogeography. Sage Publications, London, pp 313–337

Schickhoff U, Bobrowski M, Böhner J, Bürzle B, Chaudari RP, Gerlitz L, Heyken H, Lange J, Müller M, Scholten T, Schwab N, Wedegärtner R (2015) Do Himalayan treelines respond to recent climate change? An evaluation of sensitivity indicators. Earth Sys Dynam 6:245–265

Schickhoff U, Bobrowski M, Böhner J, Bürzle B, Chaudhary RP, Gerlitz L, Lange J, Müller M, Scholten T, Schwab N (2016) Climate change and treeline dynamics in the Himalaya. In: Singh RB, Schickhoff U, Mal S (eds) Climate change, glacier response, and vegetation dynamics in the Himalaya. Springer, Switzerland, pp 271–306

Schickhoff U, Singh RB, Mal S (2016) Climate change and dynamics of glaciers and vegetation in the Himalaya: an overview. In: Singh RB, Schickhoff U, Mal S (eds) Climate change, glacier response, and vegetation dynamics in the Himalaya. Springer, Switzerland, pp 1–26

Schorr G, Holstein N, Pearman PB, Guisan A, Kadereit JW (2012) Integrating species distribution models (SDMs) and phylogeography for two species of Alpine Primula. Ecol Evol 2:1260–1277

Schwab N, Schickhoff U, Bobrowski M, Böhner J, Bürzle B, Chaudhary RP, Gerlitz L, Müller M, Scholten T (2016) Treeline responsiveness to climate warming: insights from a krummholz treeline in Rolwaling Himal, Nepal. In: Singh RB, Schickhoff U, Mal S (eds) Climate change, glacier response, and vegetation dynamics in the Himalaya. Springer, Switzerland, pp 307–345

Schweinfurth U (1957) Die horizontale und vertikale Verbreitung der Vegetation im Himalaya. Bonner Geographische Abhandlungen 20, Dümmlers, Bonn

Schwörer C, Kaltenrieder P, Glur L, Berlinger M, Elbert J, Frei S, Gilli A, Hafner A, Anselmetti FS, Grosjean M, Tinner W (2014) Holocene climate, fire and vegetation dynamics at the treeline in the North-western Swiss Alps. Veg Hist Archaeobot 23:479–496

Shrestha UB, Gautam S, Bawa KS (2012) Widespread climate change in the Himalayas and associated changes in local ecosystems. PLoS One 7:e36741

Shrestha KB, Hofgaard A, Vandvik V (2014) Recent treeline dynamics are similar between dry and mesic areas of Nepal, Central Himalaya. J Plant Ecol 8:347–358

Shrestha UB, Bawa KS (2014) Impact of climate change on potential distribution of Chinese Caterpillar Fungus (*Ophiocordyceps chinensis*) in Nepal Himalaya. PLoS One 9:e106405

Singh CP, Panigrahy S, Parihar JS, Dharaiya N (2013) Modeling environmental niche of Himalayan birch and remote sensing based vicarious validation. Trop Ecol 54:321–329

Smith J, William K, Germino MJ, Johnson DM, Reinhardt K (2009) The altitude of alpine treeline: a bellwether of climate change effects. Bot Rev 75:163–190

Smith J, Sconiers W, Spasojevic M, Ashton I, Suding K (2012) Phenological changes in alpine plants in response to increased snowpack, temperature, and nitrogen. Arct Antarct Alp Res 44:135–142

Soberon J, Peterson AT (2004) Biodiversity informatics: managing and applying primary biodiversity data. Philos Trans R Soc Lond B Biol Sci 359:689–698

Speed JDM, Austrheim G, Hester AJ, Mysterud A (2011) Growth limitation of mountain birch caused by sheep browsing at the altitudinal treeline. For Ecol Manage 261:1344–1352

Still CJ, Pau S, Edwards EJ (2014) Land surface skin temperature captures thermal environments of C3 and C4 grasses. Global Ecol Biogeogr 23:286–296

Telwala Y, Brook BW, Manish K, Pandit MK (2013) Climate-induced elevational range shifts and increase in plant species richness in a Himalayan biodiversity epicentre. PLoS One 2:e57103

Thuiller W (2004) Patterns and uncertainties of species' range shifts under climate change. Glob Change Biol 10:2220–2227

Thuiller W, Lavorel S, Araujo MB (2005) Niche properties and geographical extent as predictors of species sensitivity to climate change. Global Ecol Biogeogr 14:347–357

Truong C, Palmé AE, Felber F (2007) Recent invasion of the mountain birch *Betula pubescens* ssp. *tortuosa* above the treeline due to climate change: genetic and ecological study in Northern Sweden. J Evol Biol 20:369–380

USGS (2004) Shuttle radar topography mission, 1 arc second scene SRTM_u03_n008e004, Unfilled Unfinished 2.0, Global Land Cover Facility, University of Maryland, College Park, Maryland, Feb 2000

Veloz SD (2009) Spatially autocorrelated sampling falsely inflates measures of accuracy for presence-only niche models. J Biogeogr 36:2290–2299

Wallentin G, Tappeiner U, Strobl J, Tasser E (2008) Understanding alpine tree line dynamics: an individual-based model. Ecol Model 218:235–246

Wang X, Gao Q, Wang C, Yu M (2017) Spatiotemporal patterns of vegetation phenology change and relationships with climate in the two transects of East China. Global Ecol Conserv 10:206–219

Wieser G, Holtmeier FK, Smith WK (2014) Treelines in a changing global environment. In: Tausz M, Grulke N (eds) Trees in a changing environment. Springer, Dordrecht, pp 221–263

Xu J, Grumbine RE, Shrestha A, Eriksson M, Yang X, Wang YUN, Wilkes A (2009) The melting Himalayas: cascading effects of climate change on water, biodiversity, and livelihoods. Conserv Biol 23:520–530

Yasaka M (2005) The pollen production and dispersal of *Betula platyphylla* var. *japonica* and *B. ermanii*. J Jap Forestry Soc 87

Conifer Growth During Warming Hiatus in the Altay-Sayan Mountain Region, Siberia

15

Viacheslav I. Kharuk, Sergei T. Im, and Il'ya A. Petrov

Abstract

"Warming hiatus" occurred in the Altay-Sayan Mountain Region, Siberia, in c. 1997–2014. We analyzed evergreen conifer (EGC: *Pinus sibirica* du Tour and *Abies sibirica* Ledeb. mainly) stands area (satellite data) and trees (*Pinus sibirica, Larix sibirica* Ledeb.) growth response to climate variables before and during the hiatus. During the hiatus, the EGC area increased in highlands (+30%), whereas at lower elevations (<1000 m a.s.l.), the area decreased (−7%). In highlands, the EGC area changes correlated with summer air temperature mainly, whereas at lower elevations, the changes correlated with drought index SPEI. EGC mortality (Siberian pine and fir mainly) in lowland was caused by the synergy of water stress (inciting factor) and bark-beetle attacks (contributing factor). Within alpine forest–tundra ecotone (2000–2280 m), the larch growth index (GI) was limited by air temperature, whereas the Siberian pine GI was also sensitive to precipitation, root zone moisture content (RZM) and sunshine duration. Warming led to transformation of krummholz Siberian pine into vertical form, whereas larch had vertical forms before warming. Within high elevation belt (1200–2000 m), the Siberian pine growth index (GI) permanently increases since warming onset; the GI positively responded to June–July temperatures and negatively responded to moistening parameters (precipitation, root zone moisture content, and SPEI). At middle elevation, the Siberian pine GI curve has a breakpoint (c. 1983) followed by GI depression. After the breakpoint, the GI correlation with air temperature switched from positive to negative. At the same time, positive correlations between the GI and "moisture parameters" (precipitation, RZM, SPEI) increased. Under projected climate change scenario, Siberian pine will shrink its habitat at middle and low elevations with substitution by drought-resistant larch and softwoods species.

Keywords

Growth increment · Warming hiatus · Warming impact · Tree growth · Tree mortality · Conifer mortality · Water stress · Alpine forest–tundra ecotone

The original version of this chapter was revised: For detailed information, please see Correction. The correction to this chapter is available at https://doi.org/10.1007/978-3-030-70238-0_32

V. I. Kharuk (✉) · S. T. Im · I. A. Petrov
Sukachev Institute of Forest SB RAS, FRC Krasnoyarsk Science Center SB RAS, Krasnoyarsk, Russia

V. I. Kharuk · S. T. Im
Siberian Federal University, Krasnoyarsk, Russia

S. T. Im
Reshetnev Siberian State University of Science and Technology, Krasnoyarsk, Russia

U. Schickhoff et al. (eds.), *Mountain Landscapes in Transition*, Sustainable Development Goals Series, https://doi.org/10.1007/978-3-030-70238-0_15

15.1 Introduction

The effects of climate change on coniferous forests, both positive and negative, were observed throughout the boreal zone (e.g., Andregg et al. 2013; Allen et al. 2015). Conifer decline (*Picea ajanensis* Fisch., *Abies nephrolepis* (Trautv.) Maxim.) was noticed in the Russian Far East (Man'ko et al. 1998) and in Siberia (*Abies sibirica* Ledeb., *Pinus sibirica* du Tour) (Kharuk et al. 2013a, 2017b, c, 2018). Spruce (*Picea abies* L.) mortality in the European part of Russia and Eastern and Western Europe is associated with the deterioration of water condition (Chuprov 2008; Sarnatczkii 2012; Haynes et al. 2014; Kharuk et al. 2015). Along with moisture-sensitive spruce, mortality of drought-tolerant *Pinus sylvestris* L. has been observed on the southern range of this species in the Ukraine and Belarus (Luferov and Kovalishin 2017). Extensive conifer mortality has been reported in the forests of the USA and Canada (Millar and Stephenson 2015; Kolb et al. 2016). Alongside conifers, deciduous trees (*Populus tremuloides* Michx., *Betula pendula* Roth) also suffer from the increased drought (Michaelian et al. 2011; Kharuk et al. 2013b; Hogg et al. 2017). Climate warming also leads to mass insect attacks—bark-beetles, xylophages, needle-eating pests (Allen et al. 2010, 2015; Kharuk et al. 2017b, c). In particular, climate changes promote Siberian silkmoth (*Dendrolimus sibiricus* Tschetv.) northward migration (Kharuk et al. 2017d). Currently, one of the main factors of forests degradation is earlier dormant species, such as *Polygraphus proximus* Blandford which attacks have led to the mass mortality of *Abies sibirica* Ledeb. in Siberia (Krivets et al. 2015; Kharuk et al. 2017b). In the North American forests, the synergy of water stress and insects attacks resulted in tree mortality in the area of 25 million ha (Coleman et al. 2014; Millar and Stephenson 2015).

Alongside negative impacts, climate change led to the upward and northward treeline shifts (Devi et al. 2008; Kharuk et al. 2010, 2017a; Petrov et al. 2015). Warming promotes "dark needle conifer" (*Abies sibirica, Pinus sibirica, Picea obovata* Ledeb.) migration into the zone of larch dominance (Kharuk et al. 2005), and evergreen conifer (EGC) density increase in some ecoregions (He et al. 2017). These positive impacts referred to "accelerating warming" period (c. 1970s 2014 late 1990s) mainly. Meanwhile, controversial data are reported for the followed "hiatus" period, i.e., warming anomaly observed in 1998–2013, when air warming rate fell below long-term average warming rate (Hartmann et al. 2013; Medhaug et al. 2017).

This study aims at the analysis of the EGC (mainly *Pinus sibirica* and *Abies sibirica*) area dynamics in different elevational belts of the Altay-Sayan region (ASR) during warming hiatus. The ASR is one of the priority ecoregions in the Asian continent (Fig. 15.1). The mountainous terrain of the ASR shapes considerable eco-climatic gradients which make mountain forests a sensitive indicator of climate changes. These forests should experience noticeable distributional and compositional dynamics driven by changes in air temperature, water regime, and growing season length.

We hypothesized different EGC response to warming in different elevation belts with the modification effect of relief features. We are seeking the answers to the following questions: (a) What is the dynamics of the EGC area in different elevation belts? (b) How do relief features (exposure, slope steepness) modify the EGC area dynamics? (c) How did the tree growth respond to the climatic variables?

15.2 Materials and Methods

15.2.1 Study Area

The Altai-Sayan Region (Fig. 15.1) has a total area of ~85 million ha. The forested area is about 39 million ha, including ~7.7 million ha of evergreen conifers (MODIS satellite derived estimates). Mountain relief prevails on the territory with the absolute height up to 4330 m a.s.l. The main tree species are Siberian pine (*Pinus*

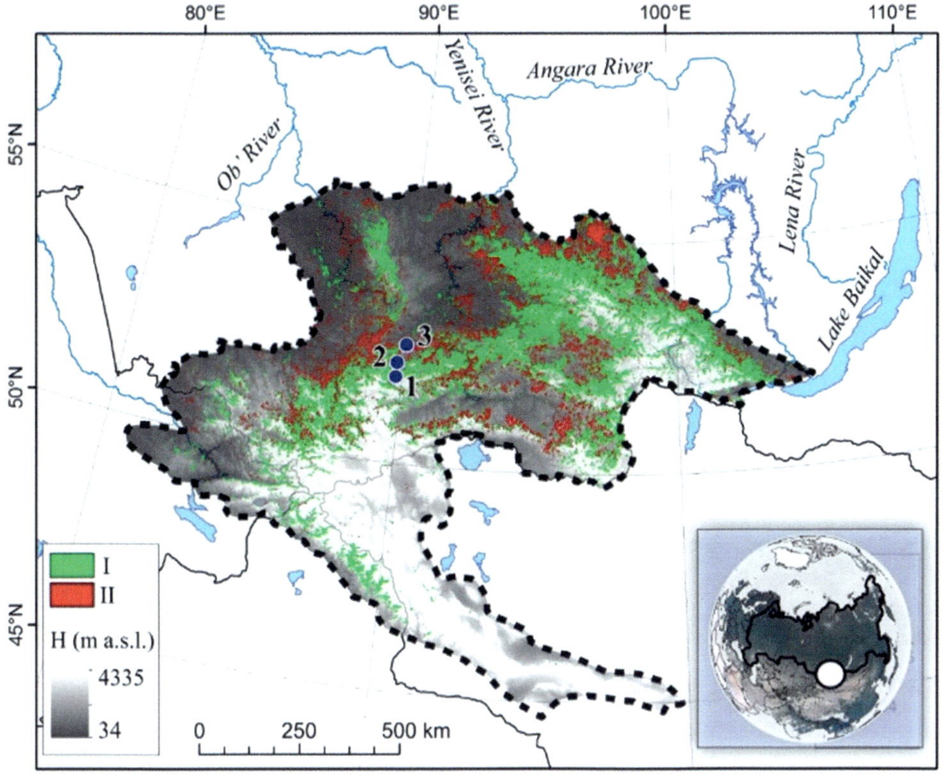

Fig. 15.1 Sketch map of the Altay-Sayan Region. Positive (*I*) and negative (*II*) trends in EGC areas are indicated by *green* and *red*, respectively. *1, 2, 3*—test sites within alpine forest–tundra ecotone, high and middle elevation belts. *H*—elevation above sea level

sibirica), fir (*Abies sibirica*), spruce (*Picea obovata*), pine (*Pinus sylvestris*), and larch (*Larix sibirica*) (Fig. 15.2).

The low elevation belt of the northern megaslope is composed by mixed forests of pine, larch, birch (*Betula sp.*), and aspen (*Populus tremula* L.). At higher elevation (800–900 m), these forests are replaced by "dark coniferous taiga" (composed by *Pinus sibirica* and *Abies sibirica*). This taiga covers elevations up to 1700–1800 m, where it gradually turns into Siberian pine or larch woodlands. Alpine tundra occupies elevations above 2000–2200 m. On the southern megaslope, the mountain forest steppe (with larch domination) prevails at elevations up to 1200–1500 m and is then replaced by the belt of mixed larch and Siberian pine forests up to 1800–2100 m.

The climate is sharply continental with cold winters and cool summers. The average temperatures are −15 °C ... −18 °C in January, +10 °C ... +14 °C in July (at the foothills around +19 °C ... +20 °C). The amount of precipitation on the windward slopes reaches 1200–2500 mm. The averages (2000–2017) for the region temperature and rainfall were −2 °C (summer +15 °C, winter −21 °C) and 495 mm (summer—215 mm, winter—50 mm), respectively.

15.2.2 Materials

The dynamics of the EGC stand area was analyzed using MODIS products (on-ground resolution 500 m, period of 2001–2013; product MCD12Q1: https://lpdaac.usgs.gov/dataset_discovery/modis/

Fig. 15.2 High-elevation West Sayan Mountain taiga forests

modis_products_table/mcd12q1; Friedl et al. 2010). The burned areas were excluded from the analysis with the available MCD64A1 data (https://modis-fire.umd.edu). DEM GMTED2010 with on-ground resolution 250 m and elevation error 28 m (https://lta.cr.usgs.gov/GMTED2010) was used for geospatial analysis.

Dependences of the EGC area and growth index (GI) were analyzed with the main eco-climatic variables: air temperature, precipitation, vapor pressure deficit (VPD), drought index SPEI, root zone moisture content (RZM), sum of active temperatures ($t \geq +5$ °C), and growth period length (the number of days with $t \geq +5$ °C). According to Rossi et al. (2008), in cold regions, conifer cambium activity starts at temperatures of about +4 … +6 °C. Temperature and precipitation data were drawn from the CRU TS 4.01 dataset (https://crudata.uea.ac.uk/cru/data/hrg; Harris and Jones 2017). The values of SPEI were taken from the website https://spei.csic.es (spatial resolution 0.5° × 0.5°). SPEI is determined by the difference between precipitation and potential evapotranspiration (Vicente-Serrano et al. 2010). The root zone moisture content and the number of days with the specified temperature were calculated with MERRA2 data (0.5° × 0.625° resolution, https://gmao.gsfc.nasa.gov/reanalysis/MERRA-2; Gelaro et al. 2017). The wood samples for dendrochronological analysis were taken using the increment borer at "breast height" level (1.3 m) or root collar during field studies in 2014–2018.

15.2.3 Methods

The maps of EGC density trends were generated based on the MCD12Q1 maps (the differences between the average values in 2011–2013 and 2001–2003). In the specified maps, according to the IGBP classification, lands dominated by woody vegetation with a percent cover >60% are referred to evergreen forests. Along with absolute, the normalized areas A_i were assumed as the ratio of an absolute area B_i to C_i area of i-elements of relief (Kharuk et al. 2010). The analysis algorithm of the EGC maps included the following stages. (1) Creation of a burned area mask based on MCD64A1 data. (2) Filtration of burned areas. (3) Cutting the selected fragment of the territory and transforming it into Albers equal-area projection. (4) Formation of binary layers (EGC and background). (5) Assessment of maps for the initial (2001–2003) and final (2011–2013) periods assuming that EGC pixel is registered on the maps for the initial and final periods only if it was observed simultaneously on all images of the considered period. (6) Assessment of changes in the EGC areas between the initial and final periods. (7) Calculation of area changes distribution for the relief features (elevation, exposure, slope steepness). (8) Calculation of the EGC stands spatial density for each year using the method of focal statistics with averaging within a sliding window of 5 pixels. (9) Generation of a multilayer composite covering the entire analyzed period. (10) Calculation of linear trends for each pixel. The raster of trend lines slope coefficients, the significance levels of trends, and the Pearson correlation coefficient were calculated. (11) Calculation of areas with statistically significant ($p < 0.05$) trends of the EGC density. (12) Calculation of zones with statistically significant ($p < 0.05$) trends in climate variables. (13) Calculation of trends distribution for the relief features. The algorithm is implemented using ESRI ArcGIS and Python script. In each elevation belt, test sites (TS) were established. TS description includes forest type, species composition, and number of trees, including their height and diameter, and the soil cover and soil and topographic features (direction, steepness, slope convexity/concavity, and altitude above sea level). Samples for dendrochronological analysis were randomly taken in the area of ~0.5 ha around TS with elevation range of about 10 m a.s.l. Within alpine forest–tundra ecotone (2280–2000 m a.s.l.), *Pinus sibirica* and *Larix sibirica* trees were sampled for the dendrochronological analysis at root collar level ($N = 20$ and $N = 13$, respectively). At high elevation (1200–2000 m) and middle (800–1200 m) elevations, *Pinus sibirica* only were sampled at dbh (1.3 m) height ($N = 28$ and $N = 46$, respectively). LINTAB 3 platform with precision 0.01 mm was used to measure wood cores (Rinn 1996). As a result, absolute individual tree-ring chronologies (in millimeters) were obtained. TSAP and COFECHA were used to assess the quality of crossdating and measurement accuracy (Holmes 1983). To eliminate the age trend, we applied the standardization procedure that converts the time series of the annual rings width to the time series of unitless indices with a defined mean of 1.0 and a relatively constant variance (Speer 2010).

15.3 Results

15.3.1 Climate Variables Dynamics

At both low and high elevations, the "warming hiatus" was observed from c.1997 till 2014 (Fig. 15.3a, e). At high elevation, summer temperature during the hiatus increased by about +1.0 °C in comparison with the "pre-warming" period (Fig. 15.3a); the growing season increased for about three days. After the precipitation increase in 1970s–80s, a strong negative trend of minimal precipitation values has been observed (Fig. 15.3d). At low elevations, precipitation trends were not observed (Fig. 15.3b).

15.3.2 EGC Area Dynamics in the Altai-Sayan Region

The total EGC area within the ASR increased by ~+20% during the hiatus. Meanwhile, at lower elevations (<1000 m a.s.l.), the area decreased (−7%), whereas in the highlands, the area increased by +30% (Fig. 15.4a). Negative and positive EGC trends covered 8 and 17% of the total EGC area, respectively (Fig. 15.1).

Along the elevation gradient, the EGC area changes (ΔS) switched from minus to plus at about 800 m a.s.l. (Fig. 15.4a). Maximums of ΔS decrease and increase located at 650 and 1450 m a.s.l, respectively. There is a similarity between the EGC and SPEI area distributions. Thus, forest mortality observed mainly within negative SPEI areas, whereas the EGC area increases corresponded to positive SPEI trends (Fig. 15.4b).

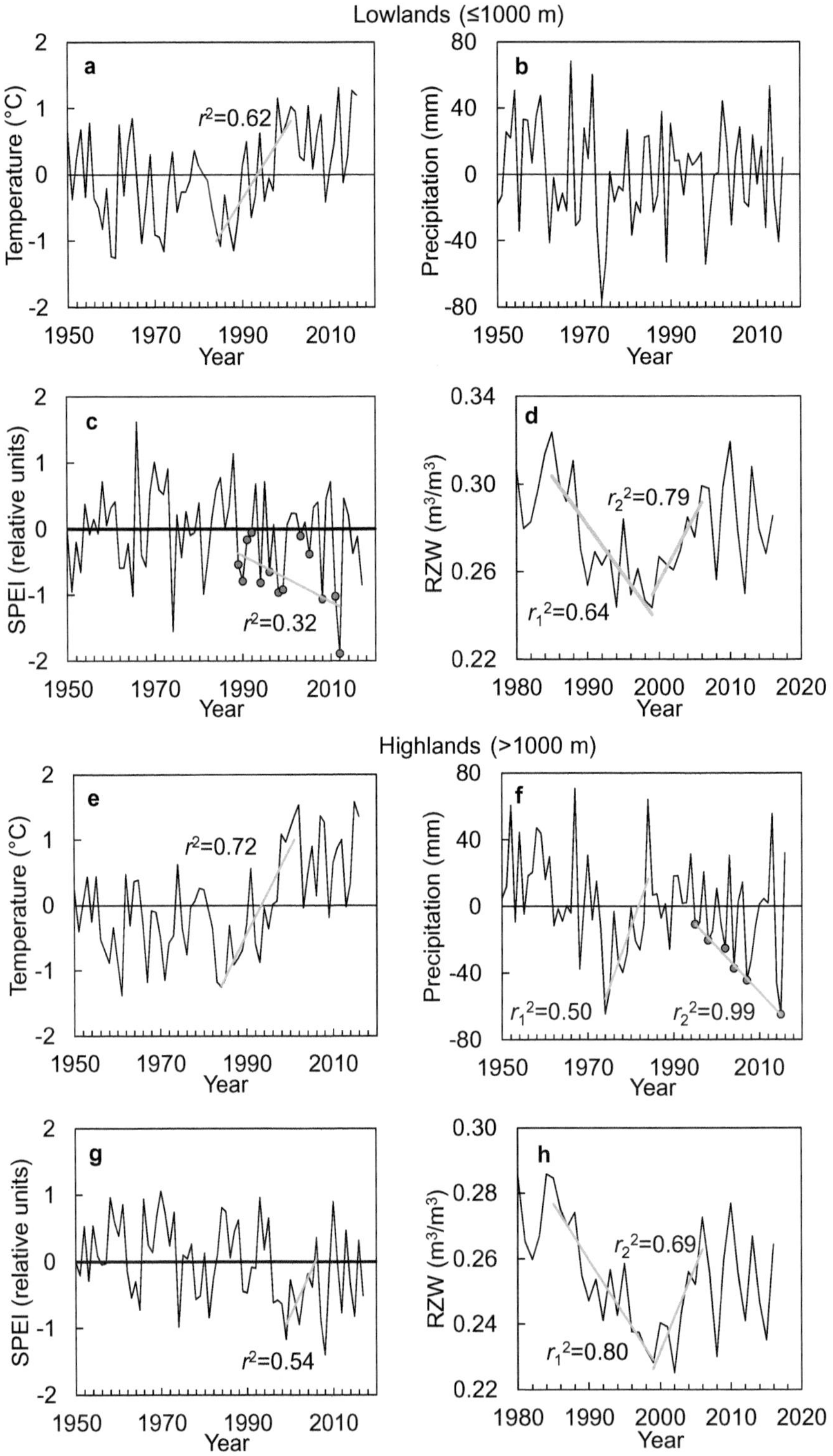
Lowlands (≤1000 m)
a
r^2=0.62
Temperature (°C)
Year
b
Precipitation (mm)
Year
c
r^2=0.32
SPEI (relative units)
Year
d
r_2^2=0.79
r_1^2=0.64
RZW (m³/m³)
Year
Highlands (>1000 m)
e
r^2=0.72
Temperature (°C)
Year
f
r_1^2=0.50
r_2^2=0.99
Precipitation (mm)
Year
g
r^2=0.54
SPEI (relative units)
Year
h
r_2^2=0.69
r_1^2=0.80
RZW (m³/m³)
Year

◀ **Fig. 15.3** Dynamics of eco-climatic variables within the ASR for low and high elevations. **a**, **e** summer temperature anomaly (base period 1950–2016), **b**, **f** precipitation anomaly, **d**, **g** drought index SPEI, **e**, **h** RZM (root zone moisture content)

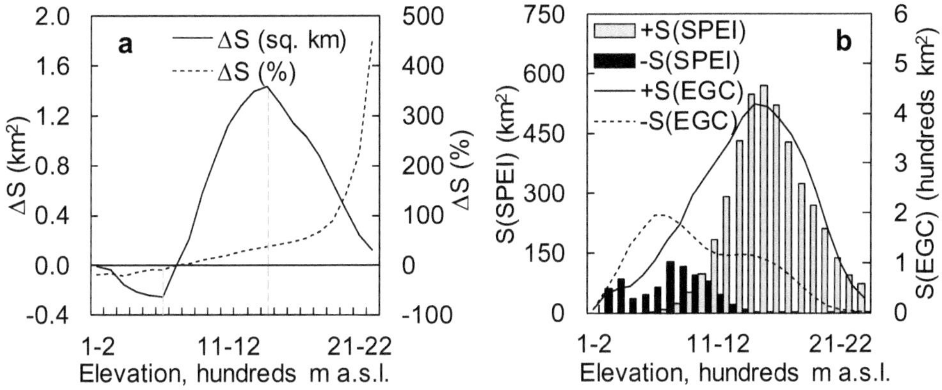

Fig. 15.4 **a** EGC stands area changes (in square km and in %) along elevation; maximums are indicated by vertical lines; **b** EGC and drought index SPEI area trends (positive and negative) along elevation

Slope steepness and exposure have a modified impact on the EGC area changes Thus, the maximal EGC absolute area increase is located at slopes with about 10°; the relative area (in %) is increasing with slope steepness increase (Fig. 15.5a). As for exposure, the EGC area decreased on the western slopes and increased on the northern ones (Fig. 15.5b, c). In the highlands, the EGC area changes positively correlated with total summer precipitation ($r = 0.67 \pm 0.08$; $p < 0.05$) and mean summer SPEI ($r = 0.64 \pm 0.07$; $p < 0.05$). In the lowlands, the EGC area changes also positively correlated with the mean summer temperature ($r = 0.64 \pm 0.06$; $p < 0.05$), whereas correlations with SPEI are negative ($r = -0.64 \pm 0.07$; $p < 0.05$).

15.3.3 Tree Growth Index Dynamics Within Different Elevation Belts

Trees growth index dynamics was different along the elevation gradient (Fig. 15.6). Let us consider these differences with respect to climate variables.

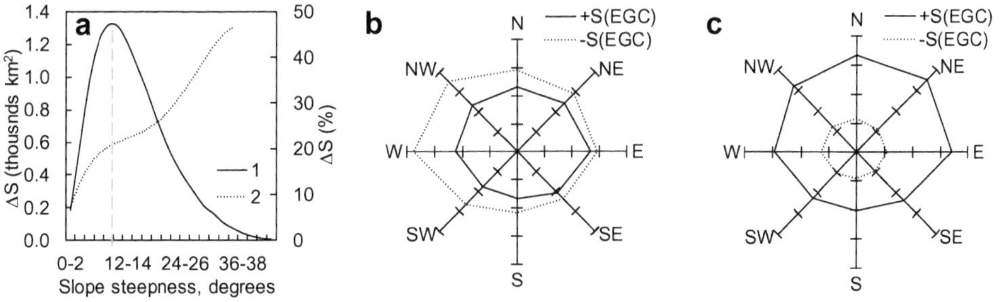

Fig. 15.5 **a** EGC area changes (ΔS) relative to slope steepness. **b**, **c** EGC area trends dependence on exposure for elevations, **b** ≤1000 m.a.s.l and **c** >1000 m a.s.l. respectively

15.3.3.1 Alpine Forest–Tundra Ecotone (2000–2280 m)

Alpine forest–tundra ecotone is formed by *Pinus sibirica* and *Larix sibirica* species.

Both species show a minor GI since 1970s with followed depression in the mid of 1980s and a strong GI increase since the late 1990s (Fig. 15.6a). The latter period coincided with the transformation of krummholz Siberian pine into the upright form (Fig. 15.7). Larch was growing upright during the entire observed period.

The GI of both species positively correlated with air temperature (Fig. 15.8a). The GI of Siberian pine also showed positive correlation with moisture parameters (precipitation and root zone moisture content, Fig. 15.8b, c) and negative one with sunshine duration (Fig. 15.8d). For larch, those correlations are insignificant.

15.3.3.2 High (1200–2000 m) and Middle Elevation (800–1200 m) Belts

At high and middle elevations, stands were composed by Siberian pine (dominant species) with larch as a minor component; because of that, the larch GI was not considered.

The Siberian pine GI was permanently increasing since late 1960s and that increase correlated with June temperature. Moisture parameters (precipitation, SPEI, RZM content) have a negative effect on the GI (Fig. 15.9a, b). At middle elevations, GI increase followed by growth depression with a breakpoint in c. 1983 (Fig. 15.6b, c). In late 1990s, trees split into "decliners" and "survivors" cohorts (Fig. 15.6c). At the breakpoint, the GI correlations switched from positive to negative with air temperature and from negative to positive with precipitation and SPEI; this switch was observed for both tree cohorts (except for precipitation for "survivors") (Fig. 15.9). Trees from "decliners" cohort eventually die back (Fig. 15.10).

15.3.3.3 GI Dynamics of Old-Growth Trees Within Refugium

Finally, we considered the GI dynamics of old-growth Siberian pine trees from the refugium, i.e., the zone where trees survive during LIA. The "refugee's line" is located, dependent on relief features, at about 1800–2000 m a.s.l. (Fig. 15.11).

The elevation difference between old-growth trees and regeneration lines varied within 30–50 m. Tree establishment was poor during the period of the GI decrease (i.e., during air temperature decrease); establishment of a new generation occurred since the beginning of the GI increase (and, accordingly, to air temperature increase). The GI of old-growth trees decreased during the LIA period until the mid of the nineteenth century with following GI increase until the mid of 1940s (Fig. 15.12). Next period of the GI increase coincided with warming in 1980s. The GI of old-growth trees correlated with summer air temperature ($r^2 = 0.4$; analyzed period since the beginning of instrumental observations, 1930). No correlation with precipitation was found.

15.4 Discussion

In the Altay-Sayan Mountain Region despite of the warming hiatus, the total EGC area increased by +20%. Meanwhile, the EGC area changes were opposite in high and low elevations: in the highlands (>1000 m a.s.l.), conifer area increase was +30%, whereas in the lower elevations, area decrease observed (−7%). Similarly, observations in Southern Tibet, China, showed that while *Abies georgei* population is expanding and upper limit of this species has advanced upslope, the lowest limit has retreated upslope (Wong et al. 2010; Shen et al. 2016).

The EGC area increase in the highlands could be attributed to the higher mean air temperature (+1.0 °C) during the hiatus than in the pre-warming period (1950–1970) and longer growing season period (+3 days). The Siberian pine growth index in the highlands also permanently increased since warming onset and during the hiatus (Fig. 15.6b). Thus, during the hiatus, the EGC stand's area in the highlands increased due to improved thermal conditions on the background of sufficient precipitation. Similarly, He et al. (2017) found that evergreen conifer forest

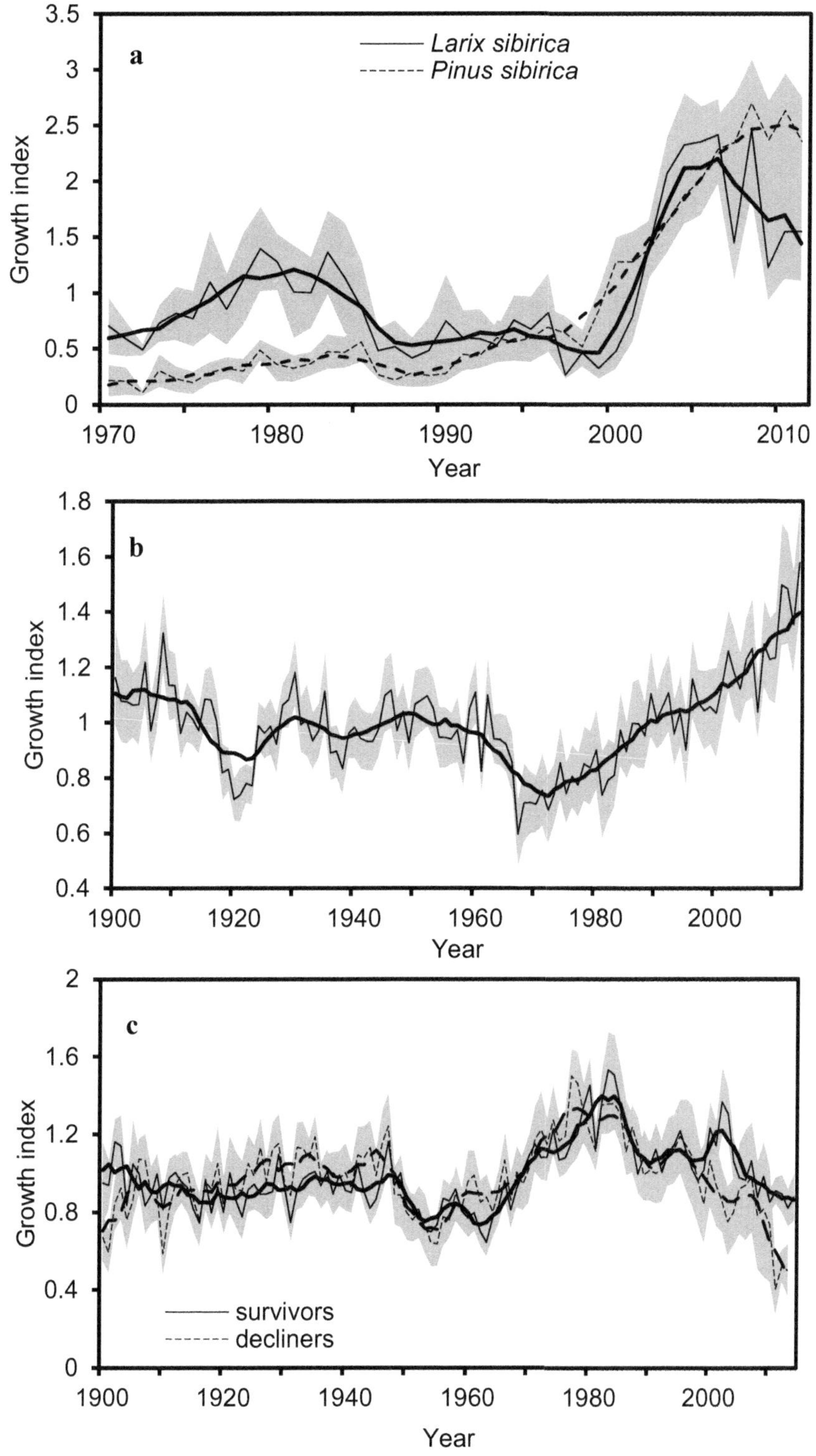

Fig. 15.6 Growth index (GI) dynamics of Siberian pine (N = 20) and larch (N = 13) in forest–tundra ecotone (**a** elevation 2030–2280 m a.s.l.), and Siberian pine at high (**b** 1200–2000 m a.s.l; N = 28) and middle (**c** 800–1200 m; N = 22 for survivors, N = 24 for decliners) elevations. *Dense line*: data filtered by 11 yr. window. *Gray background*: $p > 0.05$

Fig. 15.7 *Pinus sibirica* krummholz transformed into the upright form

expansion in Western Siberia continued during the warming hiatus. Meanwhile, the highlands remained the zone of excessive moistening, which was indicated by the GI negative correlations with precipitation, root zone moisture content, and SPEI (Fig. 15.9b).

The highest relative conifer area increase observed within the alpine forest–tundra ecotone (Fig. 15.4a). On the contrary to high-elevation stands, Siberian pine within ecotone is sensitive, alongside to temperature, to the water supply, which is indicated by positive correlation with precipitation and root zone moisture content (Fig. 15.8b, c).

It is noteworthy that within adjacent high-elevation closed stands, Siberian pine growth does not depend neither on the precipitation nor the RZM content. This phenomenon is caused by poor snow accumulation due to low closure vegetation cover (Fig. 15.7). Therefore, winter winds (with mean wind speed of about 4 m/s) blow off snow from the forest–tundra ecotone. Consequently, seedlings are located within the sites of snow accumulation, e.g., in microdepressions, behind stones or within shrubs. That explains the GI dependence on the RZM content in the beginning of vegetation period (Fig. 15.8 c). Growth of more drought-resistant larch depended on the air temperature only. Larch is also more cold-resultant species; due to that, the larch regeneration line is located about 10 m higher in comparison with the Siberian pine one. It is worth noting that *Pinus sibirica* growth negatively correlated with sunshine duration (Feb–April), whereas larch did not (Fig. 15.8d). The effect of SD impact should be attributed to the evergreen pattern of Siberian pine: it is known that needles increased twig's surface area about 150–300 times which led to about two orders evaporation increase, Larch escaped that

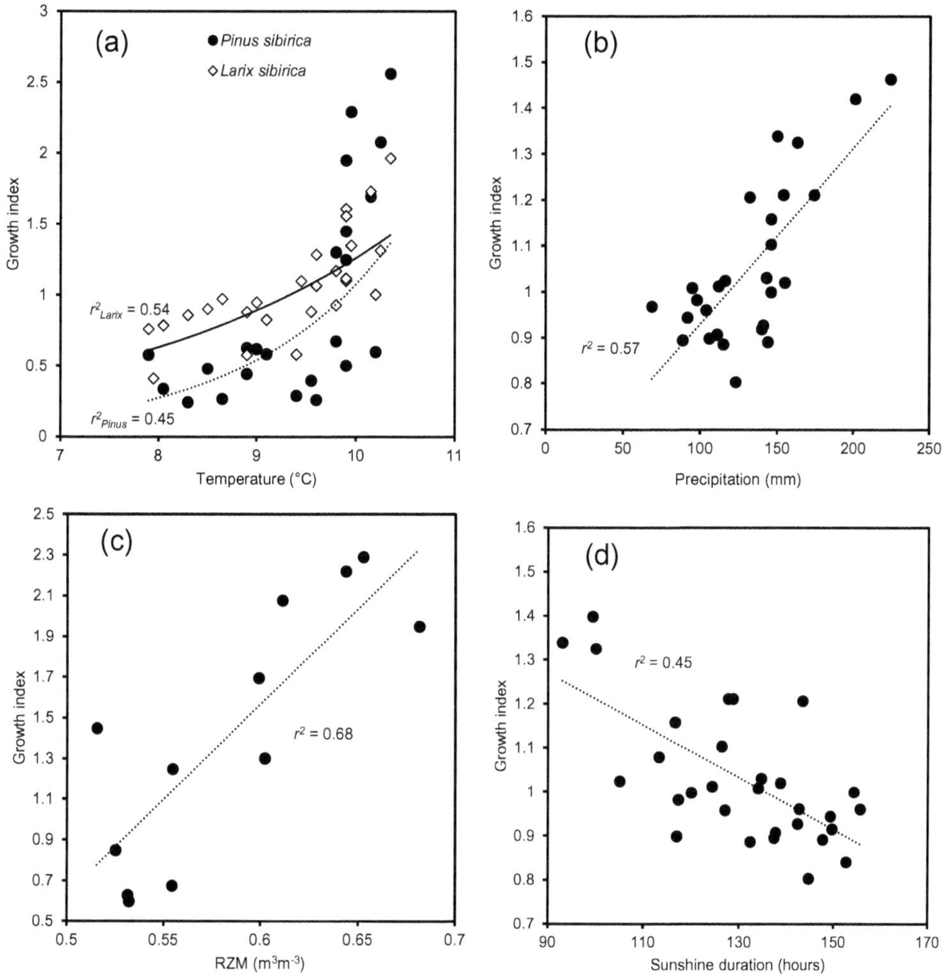

Fig. 15.8 Trees GI correlations with eco-climatic parameters within the alpine ecotone. **a** Siberian pine and larch GI correlations with air temperature (June–July 1983–2011); **b–d** Siberian pine correlations with precipitation (March–June 1982–2009), root zone moisture content (June 1996–2007) and sunshine duration (February–April 1978–2011)

effect due to its deciduous pattern. Alongside the GI increase, warming results in the fascinated phenomenon of Siberian pine krummholz transformation into upright forms (Fig. 15.7). Larch trees were vertical mainly even before warming. Due to its higher cold resistance, this species formed an upper regeneration line, whereas Siberian pine line is located about 10–20 m below. Let us note that the conifer area increase within the alpine ecotone and high elevations attributed mainly to the growth of pre-existing small trees because the observation period is too short for the area expansion due to the new trees establishment. For example, warming-driving regeneration line upward shift estimated as 0.35–0.8 m/yr. (Kharuk et al. 2010, 2017a), or about 10–15 m for the observation period; that difference is hardly detectable by MODIS sensor.

At the middle elevation, the conifer area decrease (−7%) coincided with the moistening decrease and a strong drought episode (as indicated by SPEI values; Figs. 15.3c and 15.4).

The Siberian pine GI positive response to warming at middle elevations switched to the GI depression with a breakpoint in 1983. After the breakpoint, growth limitation by temperature

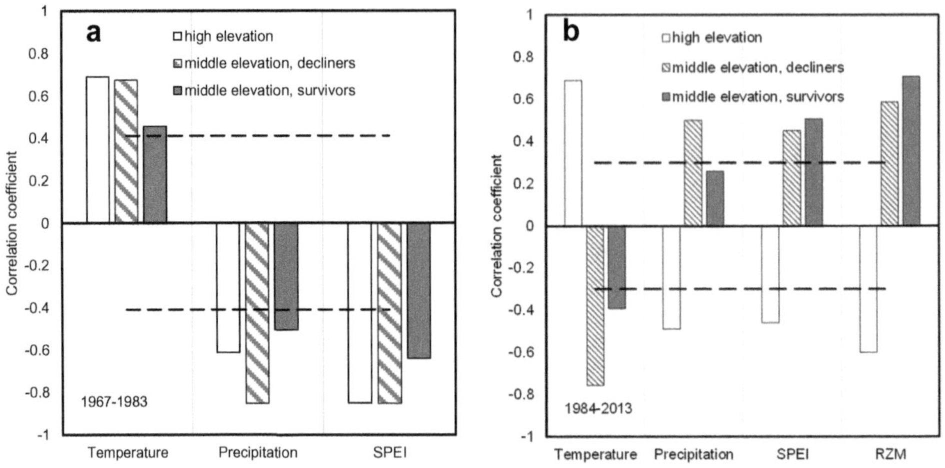

Fig. 15.9 Correlation of the GI of trees at high and middle ("survivors" and "decliners") elevations with climate variables **a** before and **b** after growth breakpoint. Variables: June temperature, May–August precipitation, May–August SPEI, July–August RZM (root zone moisture content; available since 1980). *Dashed lines* indicate $p < 0.05$ level. *Note*: SPEI decreases corresponded to drought increase

switched to limitation by moisture (Fig. 15.9). Thus, growth depression was caused by water stress via elevating air temperature. Similar switch (from temperature to moisture limitation) was described for *Pinus mugo* in high Alps (Churakova et al. 2016). Notably, the described GI decreasing pattern (Fig. 15.6c) is typical for declining stands and may precede stands mortality (Cailleret et al. 2017).

The main species that experienced mortality, as shown earlier (Kharuk et al. 2013a,2017b, c), were Siberian pine and fir (*Abies sibirica*). Meanwhile, there are no reports on the *Pinus sylvestris* or notable spruce (*Picea obovata*) mortality within the ASR, although the latter species mortality was described in Western Siberia lowlands. As for fir mortality, it was caused by synergy impact of bark-beetles and water stress (Kharuk et al. 2016). Similarly, increased water deficit together with pest attacks caused Douglas fir (*Pseudotsuga menziesii* (Mirb.) Franco) growth decrease in Western US forests (Restaino et al. 2016).

Projected air temperature increase (IPCC 2014) will lead to further Siberian pine and fir area reduction in the lowlands, and area increase in the highlands, as well as trees migration into the alpine tundra. Moreover, warming will also adversely affect wildfire regime; thus, both fire frequency and burned area in Siberia have increased in recent decades (Kharuk and Ponomarev 2017). An additional factor of tree mortality is an activation of primary pests such as Siberian silkmoth (*Dendrolimus sibiricus*), which extended its range northward and caused huge forest mortality (about 800 thousand ha) in the mid-taiga zone (Kharuk et al. 2017d).

15.5 Conclusions

1. Warming hiatus in the ASR caused a general increase in the EGC area (+20% mean). However, in the lowlands (<1000 m a.s.l.), the EGC area decreased (−7%) with significant (+30%) increase at high elevations (>1000 m). The EGC area increase in the highlands correlated with air temperature mainly. Tree mortality in the lowlands was caused by the increased water stress via elevated temperature in synergy with bark-beetles attacks.

Fig. 15.10 Siberian pine trees **a** and stands **b** (gray color) mortality at middle elevations

Fig. 15.11 Old-growth Siberian pines in the refugium (c. 1900 m)

2. Elevated temperatures facilitated *Pinus sibirica* growth index (GI) since 1970s within all elevation belts. Within the zone of sufficient precipitation (highlands), the GI demonstrates permanent increase since warming onset. Meanwhile, within middle elevation belt (i.e., "Siberian pine—softwoods" transition), the GI curve has a breakpoint (c. 1983) with subsequent GI depression. After the breakpoint, the GI correlation switched from positive to negative with air temperature. Along with that, positive correlations between the GI and "moisture parameters" (precipitation, RZM, SPEI) have arisen.
3. Within the alpine forest–tundra ecotone, warming leads to transformation of prostrate Siberian pine into the vertical form. Alongside air temperature, growth of Siberian pine was also limited by moistening.
4. Under the projected climate change scenario, Siberian pine will shrink its habitat at middle and low elevations with substitution by drought-resistant larch and softwoods species.

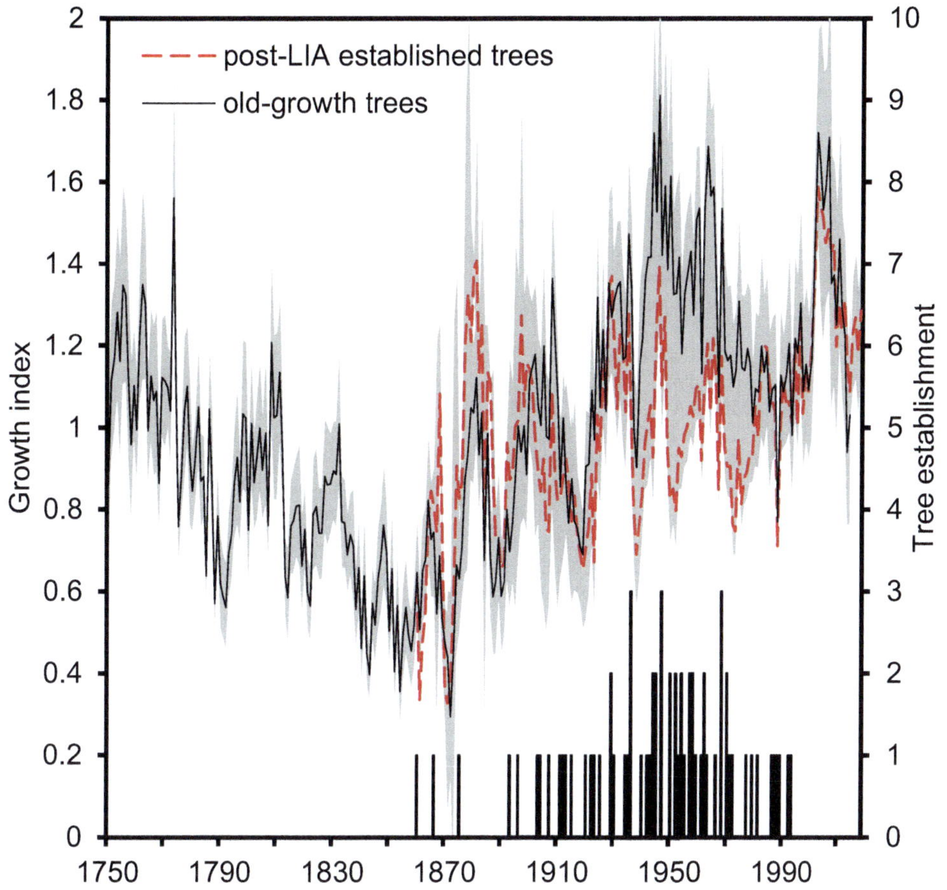

Fig. 15.12 GI dynamics of old-growth trees within the refugium (established before LIA; $N = 20$) and trees established since post-LIA warming ($N = 70$). Data of tree establishments indicated by *columns*

Acknowledgements The research was funded by Russian Fund of Basic Research, Krasnoyarsk Territory and Krasnoyarsk Regional Fund of Science, project numbers 18-45-240003 and 18-05-00432.

References

Allen CD, Macalady AK, Chenchouni H et al (2010) A global overview of drought and heat-induced tree mortality reveals emerging climate change risks for forests. For Ecol Manage 259(4):660–684. https://doi.org/10.1016/j.foreco.2009.09.001

Allen CD, Breshears DD, McDowell NG (2015) On underestimation of global vulnerability to tree mortality and forest die-off from hotter drought in the Anthropocene. Ecosphere 6(8):129. https://doi.org/10.1890/ES15-00203.1

Andregg LDL, Andregg WRL, Berry JA (2013) Not all droughts are created equal: translating meteorological drought into woody plant mortality. Tree Physiol Rev 33(7):701–712. https://doi.org/10.1093/treephys/tpt044

Cailleret M, Jansen S, Robert EMR et al (2017) A synthesis of radial growth patterns preceding tree mortality. Glob Change Biol 23(4):1675–1690. https://doi.org/10.1111/gcb.13535

Chuprov NP (2008) About the problem of spruce decay in the European North of Russia. Russ Forest 1:24–26 (in Russian)

Churakova SO, Saurer M, Bryukhanova MV et al (2016) Site-specific water-use strategies of mountain pine and larch to cope with recent climate change. Tree Physiol 36:942–953. https://doi.org/10.1093/treephys/tpw060

Coleman TW, Jones MI, Courtial B et al (2014) Impact of the first recorded outbreak of the Douglas-fir tussock moth, Orgyia pseudotsugata, in southern California

and the extent of its distribution in the Pacific Southwest region. For Ecol Manage 329:295–305. https://doi.org/10.1016/j.foreco.2014.06.027
Devi N, Hagedorn F, Moiseev P et al (2008) Expanding forests and changing growth forms of Siberian larch at the Polar Urals treeline during the 20th century. Glob Change Biol 14(7):1581–1591. https://doi.org/10.1111/j.1365-2486.2008.01583.x
Friedl MA, Sulla-Menashe D, Tan B et al (2010) MODIS Collection 5 global land cover: Algorithm refinements and characterization of new datasets. Remote Sens Environ 114:168–182. https://doi.org/10.1016/j.rse.2009.08.016
Gelaro R, McCarty W, Suárez MJ et al (2017) The modern-era retrospective analysis for research and applications, version 2 (MERRA-2). J Clim 30:5419–5454. https://doi.org/10.1175/JCLI-D-16-0758.1
Harris IC, Jones PD (2017) CRU TS4.01: University of East Anglia Climatic Research Unit; Climatic Research Unit (CRU) Time-Series (TS) version 4.01 of high-resolution gridded data of month-by-month variation in climate (Jan 1901–Dec 2016). Centre for Environmental Data Analysis, 04 Dec 2017. https://doi.org/10.5285/58a8802721c94c66ae45c3baa4d814d0
Hartmann DL, Klein Tank AMG, Rusticucci M et al (2013) Observations: atmosphere and Surface. In: Stocker TF, Qin D et al (eds) Climate change 2013: the physical science basis. Contribution of working group I to the fifth assessment report of the intergovernmental panel on climate change. Cambridge University Press, Cambridge, United Kingdom and New York, NY, USA
Haynes KJ, Allstadt A, Klimetzek D (2014) Forest defoliator outbreaks under climate change: effects on the frequency and severity of outbreaks of five pine insect pests. Global Change Biol 20:2004–2018. https://doi.org/10.1111/gcb.12506
He Y, Huang J, Shugart HH et al (2017) Unexpected evergreen expansion in the Siberian forest under warming hiatus. J Clim 30:5021–5039. https://doi.org/10.1175/JCLI-D-16-0196.1
Holmes RL (1983) Computer-assisted quality control in tree-ring dating and measurement. Tree-Ring Bull 43:69–78
Hogg EH, Michaelian M, Hook TI, Undershultz ME (2017) Recent climatic drying leads to age-independent growth reductions of white spruce stands in western Canada. Glob Change Biol 23(12):5297–5308. https://doi.org/10.1111/gcb.13795
IPCC (2014) Climate change 2014: impacts, adaptation, and vulnerability. In: Field CB, Barros VR et al (eds) A contribution of working group II to the fifth assessment report of the intergovernmental panel on climate change. World Meteorological Organization, Geneva, Switzerland, p 190. Available online: https://www.ipcc.ch/report/ar5/wg2/. Accessed on 19 May 2017
Kharuk VI, Ponomarev EI (2017) Spatiotemporal characteristics of wildfire frequency and relative area burned in larch-dominated forests of Central Siberia. Russ J Ecol 48(6):507–512
Kharuk VI, Dvinskaya ML, Ranson KJ, Im ST (2005) Expansion of evergreen conifers to the larch-dominated zone and climatic trends. Russ J Ecol 36 (3):164–170. https://doi.org/10.1007/s11184-005-0055-5
Kharuk VI, Ranson KJ, Im ST, Vdovin AS (2010) Spatial distribution and temporal dynamics of high-elevation forest stands in southern Siberia. Glob Ecol Biogeogr 19:822–830. https://doi.org/10.1111/j.1466-8238.2010.00555.x
Kharuk VI, Im ST, Oskorbin PA et al (2013) Siberian pine decline and mortality in southern Siberian Mountains. For Ecol Manage 310:312–320. https://doi.org/10.1016/j.foreco.2013.08.042
Kharuk VI, Ranson KJ, Oskorbin PA, Im ST, Dvinskaya ML (2013) Climate induced birch mortality in Trans-Baikal lake region, Siberia. For Ecol Manage 289:385–392. https://doi.org/10.1016/j.foreco.2012.10.024
Kharuk VI, Im ST, Dvinskaya ML et al (2015) Climate-induced mortality of spruce stands in Belarus. Environ Res Lett 10:125006. https://doi.org/10.1088/1748-9326/10/12/125006
Kharuk VI, Demidko DA, Fedotova EV et al (2016) Spatial and temporal dynamics of Siberian silk moth large-scale outbreak in dark-needle coniferous tree stands in Altai. Contemp Probl Ecol 9:711–720. https://doi.org/10.1134/S199542551606007X
Kharuk VI, Im ST, Dvinskaya ML et al (2017a) Tree wave migration across an elevation gradient in the Altai Mountains. J Mt Sci 14(3):442–452.https://doi.org/10.1007/s11629-016-4286-7
Kharuk VI, Im ST, Petrov IA et al (2017b) Fir decline and mortality in the southern Siberian Mountains. Reg Environ Change 17:803–812.https://doi.org/10.1007/s10113-016-1073-5
Kharuk VI, Im ST, Petrov IA et al (2017c) Climate-induced mortality of Siberian pine and fir in the Lake Baikal Watershed, Siberia. For Ecol Manage 384:191–199.https://doi.org/10.1016/j.foreco.2016.10.050
Kharuk VI, Im ST, Ranson KJ, Yagunov MN (2017d) Climate-induced northerly expansion of Siberian Silkmoth range. Forests 8(8):301. Available online: https://www.mdpi.com/1999-4907/8/8/301/pdf. Accessed on 5 June 2018. https://doi.org/10.3390/f8080301
Kharuk VI, Im ST, Petrov IA (2018) Warming hiatus and evergreen conifers in Altay-Sayan Region, Siberia. J Mountain Sci 15(12):2579–2589. https://doi.org/10.1007/s11629-018-5071-610.1007/s11629-016-4286-7
Kolb TE, Fettig CJ, Ayres MP et al (2016) Observed and anticipated impacts of drought on forests insects and diseases in the United States. For Ecol Manage 380:321–324. https://doi.org/10.1016/j.foreco.2016.04.051

Krivecz SA, Kerchev IA, Bisirova EM et al (2015) Expansion of four-eyed fir bark beetle Polygraphus proximus blandf. (coleoptera, curculionidae: colytinae) in Siberia. Issues Saint-Petersburg For Tech Acad 211:33–45 (in Russian)

Luferov AO, Kovalishin VR (2017) Problem of mortality of pine stands on the territory of Belarus and Ukraine woodlands. Materials of the fifth international conference-meeting "Preservation of forest genetic resources". 2–7 Oct 2017, Gomel, Belarus. Belarus Institute of forest NAN, "Kolordryg" publishing. pp 119–120. (in Russian)

Man'ko YI, Gladkova GA, Butovets GN, Norihiza Kamibayasi (1998) Monitoring of fir-spruce forests decay in the Central Sikhote-Alin. Russ For Sci 1:3–16. (in Russian)

Medhaug I, Stolpe MB, Fischer EM, Knutti R (2017) Reconciling controversies about the global warming hiatus. Nature 545:41–47. https://doi.org/10.1038/nature22315

Michaelian M, Hogg EH, Hall RJ, Arsenault E (2011) Massive mortality of aspen following severe drought along the southern edge of the Canadian boreal forest. Global Change Biol 17(6):2084–2094. https://doi.org/10.1111/j.1365-2486.2010.02357.x

Millar CI, Stephenson NL (2015) Temperate forest health in an era of emerging megadisturbance. Science 349 (6250):823–826. https://doi.org/10.1126/science.aaa9933

Petrov IA, Kharuk VI, Dvinskaya ML, Im ST (2015) Reaction of coniferous trees in the Kuznetsk Alatau Alpine Forest-Tundra Ecotone to climate change. Contemp Probl Ecol 8(4):423–430. https://doi.org/10.1134/S1995425515040137

Restaino CM, Peterson DL, Littell J (2016) Increased water deficit decreases Douglas fir growth throughout western US forests. Proc Natl Acad Sci 113(34):9557–9562. https://doi.org/10.1073/pnas.1602384113

Rinn F (1996) Tsap V 3.6. Reference manual: computer program for tree-ring analysis and presentation. Heidelberg, Germany

Rossi S, Deslauriers A, Griçar J et al (2008) Critical temperatures for xylogenesis in conifers of cold climates. Glob Ecol Biogeogr 17:696–707. https://doi.org/10.1111/j.1466-8238.2008.00417.x

Sarnatczkii VV (2012) Zonal-typological patterns of periodic large-scale spruce decay in Belarus. In: Proceedings of BGTU. Forest estate, pp 274–276. (in Russian)

Shen ZQ, Lu J, Hua M, Fang JP (2016) Spatial pattern analysis and associations of different growth stages of populations of Abies georgei var. smithii in Southeast Tibet, China. J Mt Sci 13(12):2170–2181. https://doi.org/10.1007/s11629-016-3849-y

Speer JH (2010) Fundamentals of tree-ring research. University of Arizona Press, p 368

Vicente-Serrano SM, Beguería S, López-Moreno JI (2010) A multi-scalar drought index sensitive to global warming: the standardized precipitation evapotranspiration index—SPEI. J Clim 23:1696–1718. https://doi.org/10.1175/2009JCLI2909.1

Wong MHG, Duan CQ, Long YC, Luo Y, Xie GQ (2010) How will the distribution and size of subalpine Abies georgei forest respond to climate change? A study in Northwest Yunnan, China. Phys Geography 31 (4):319–335. https://doi.org/10.2747/0272-3646.31.4.319

Climate-Induced Fir (*Abies sibirica* Ledeb.) Mortality in the Siberian Mountains

16

Viacheslav I. Kharuk, Sergei T. Im, Il'ya A. Petrov, Alexander S. Shushpanov, and Maria L. Dvinskaya

Abstract

The mortality of *Abies sibirica* Ledeb. trees and stands in the Siberian Mountains was analyzed. Fir trees' growth index (GI) response to warming was two-phased. Since warming onset, the GI was increasing, whereas further air temperature increase caused the GI depression via water stress. Since the GI breakpoint (c. 1983–84), the GI dependence on the moisture increased. Distributions of dead and alive stands with respect to the sum of positive temperatures ($\Sigma(t > 0\ °C)$), precipitation, root zone moisture and drought index SPEI were different. Mortality of fir stands was strongly increasing with the $\Sigma(t > 0\ °C)$ increase, and it was decreasing with precipitation and root zone moisture increase and atmospheric drought decrease. Stands' mortality was predisposed by poorer soil moisture within those stands location. Those stands also had initially lower GPP in comparison with the "survived" stands. With respect to relief features, mortality was located mostly on the southeastern slopes, and it was decreasing with elevation increase. Water-stressed fir trees were attacked by bark-beetle *Polygraphus proximus* Blandford, the bark-beetle that was not observed earlier within the *Abies sibirica* range. In synergy with water stress, that has led to stands mortality since 2000s. Fir mortality is unprecedented, covered over 5% of fir range and continues to increase. Thus, *Abies sibirica* is retreating from its low and middle elevation range in the Southern Siberian Mountains.

Keywords

Tree mortality · Conifer decline · Climate change · Pest attacks · Bark-beetles · *Polygraphus proximus* · Drought · *Abies sibirica* · Siberian fir · Siberian taiga

16.1 Introduction

Siberian taiga forests formed by fir (*Abies sibirica* Ledeb.) and Siberian pine (*Pinus sibirica* du Tour) experienced an increasing decline and mortality in recent decades. Conifer stands' vigor is decreasing in the Baikal, Kuznetsk Alatau and Sayan

The original version of this chapter was revised: For detailed information, please see Correction. The correction to this chapter is available at https://doi.org/10.1007/978-3-030-70238-0_32

V. I. Kharuk (✉) · S. T. Im · I. A. Petrov · A. S. Shushpanov · M. L. Dvinskaya
Sukachev Institute of Forest SB RAS, FRC Krasnoyarsk Science Center SB RAS, Krasnoyarsk, Russia
e-mail: v7sib@mail.ru

V. I. Kharuk · S. T. Im
Siberian Federal University, Krasnoyarsk, Russia

S. T. Im · A. S. Shushpanov
Reshetnev Siberian State University of Science and Technology, Krasnoyarsk, Russia

U. Schickhoff et al. (eds.), *Mountain Landscapes in Transition*,
Sustainable Development Goals Series, https://doi.org/10.1007/978-3-030-70238-0_16

Mountains (Kharuk et al. 2013, 2016, 2019). Conifer mortality increase is also observed within the other parts of "cold forests" zone. Mortality in the stands formed by *Pinus ponderosa* Engel spreads over an area of several million ha in North America (Logan et al. 2003; Millar and Stephenson 2015). Mass spruce (*Picea abies* L.) mortality occurs in the Eastern and Western European forests (Yousefpour et al. 2010; Martínez-Vilalta et al. 2012), as well as in Belarus and the European part of Russia (Sazonov et al. 2013; Kharuk et al. 2016). Climate-driven redistribution of tree species is predicted for the whole boreal zone, including reduction of fir and Siberian pine range in the southern taiga (Anderegg et al. 2013; Kharuk et al. 2016). Adverse climatic impact on the coniferous species is increasing due to the synergy with a pest impact (Raffa et al. 2008; Kolb et al. 2016; Kharuk et al. 2016). Warming activated an insect attack on Siberian taiga forests, including both "traditional" species (Siberian silk moth, *Dendrolimus sibiricus* Tschetv) and those not observed in this area earlier (*Polygraphus proximus* Blandf.). Thus, an outbreak of the Siberian silk moth in the Middle Siberia (2014–18) caused decline and mortality of the stands over 800 thousand ha. Moreover, this pest outbreak migrated northward of its traditional range (Kharuk et al. 2018a). The other species, bark-beetle *Polygraphus proximus* Blandford, have become the most dangerous pest for *Abies sibirica* trees. Currently, its habitat covers the most part of fir range in the southern taiga (Krivetz et al. 2015).

We aim to analyze the dynamics and causes of mortality of Siberian fir (*Abies sibirica*) trees and stands within the Siberian Mountains.

We are seeking answers to the following questions:

1. What are the temporal and spatial patterns of fir stands' mortality?
2. How do relief features modify mortality spatial pattern?
3. How do climate-driven warming and moisture availability changes affect fir growth and mortality?

16.2 Materials and Methods

16.2.1 Study Area

The study area included fir range in the Siberian Mountains (with key-site in the Eastern Sayan Mountains).

16.2.1.1 Southern Siberian Mountains

The highest elevations within the study area reached above 2500 m, although the typical elevation is from 800 to 2000–2200 m. The lowlands are located at elevation of 300–800 m and are formed by narrow ridges stretching into the direction of the foothill plains. The climate is continental with negative annual temperatures. The average January temperatures are about −18 °C in the foothills up to −34 °C within the intermountain depressions. The average July temperature is about +15 °C in the mountain and about +18 °C in the foothills. The precipitation is strongly influenced by mountainous terrain and varies from 100–200 to 1500–2500 mm/year with minimal values on the eastern rain shadow slopes and in the intermountain depressions. The upper tree limit varies from 2300–2400 in the south to 1200–1600 m a.s.l. in the north. Soils are typically shallow rocky or mountain-podzolic. Stands in the humid areas are formed mostly by fir (*Abies sibirica* Ledeb.) and Siberian pine (*Pinus sibirica* du Tour) with an admixture of spruce (*Picea obovata* Ledeb.), larch (*Larix sibirica* Ledeb.) and Scotch pine (*Pinus sylvestris* L.). Deciduous species (birch, *Betula sp.* and aspen, *Populus tremula* L.) are regularly found as an admixture to coniferous within the lower elevation belt mainly.

16.2.1.2 The Eastern Sayan Mountains Key-Site

This site is typical for "dark needle conifer" taiga (i.e., composed by fir, Siberian pine and spruce). The maximal elevation within the key-site is about 800 m a.s.l. Dark needle conifers occupy 500–800 m a.s.l. elevations mostly. Stands are

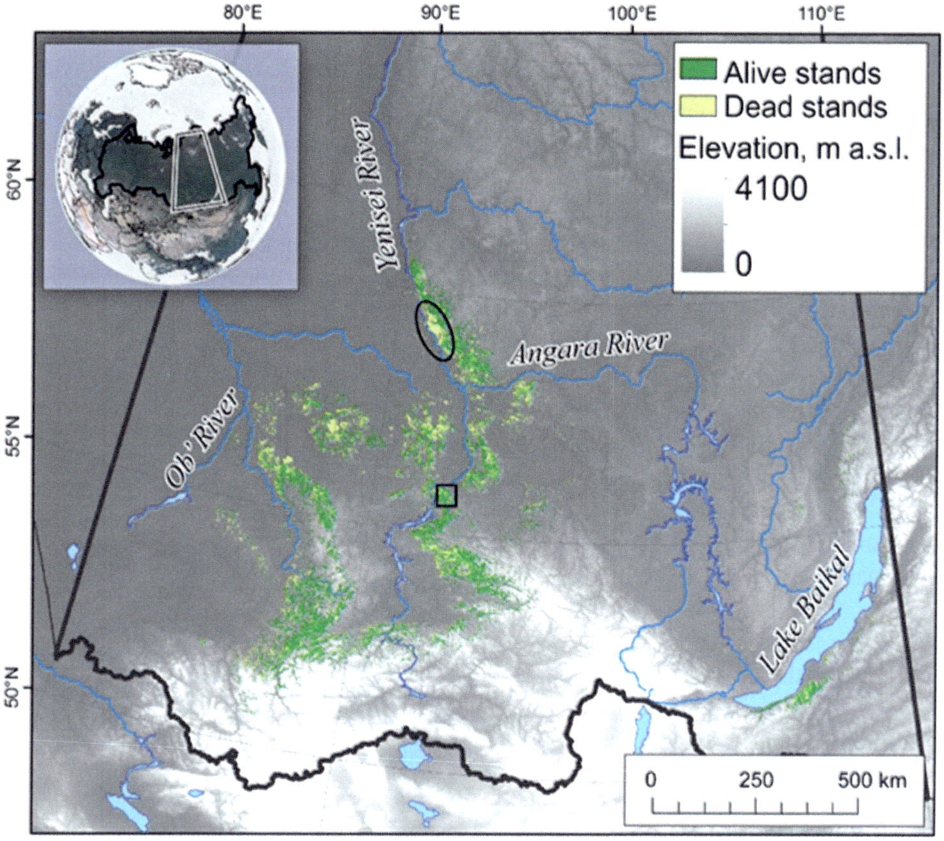

Fig. 16.1 Study area within Siberia. *Green* and *yellow* colors indicate alive and dead fir stands. The key-site in the Eastern Sayan Mountains indicated by the *square*. Stands killed by Siberian silk moth are indicated by the *oval*. The analyzed territory (which included >95% of fir range) is delineated by the *bold line*. The *background* shows elevation above sea level

formed mainly by fir (about 75%); spruce is a minor (about 1–2%) component located along creeks; the rest of the territory is occupied by Siberian pine. At lower elevations (200–500 m a. s.l.), forests are composed mainly by *Pinus sylvestris* L., *Populus tremula* L., *Betula sp.* and *Larix sibirica* Ledeb. The climate is continental with average temperatures +16 °C in summer and minus 17 °C in winter. The annual precipitation is 680 mm (Fig. 16.1).

16.2.2 Materials

The study was based on remote sensing, ground surveys, dendrochronology and climate data. The values of precipitation, temperature and root zone moisture (RZM) were extracted from the CRU TS 4.02 databases (https://www.cru.uea.ac.uk/; resolution 0.5° × 0.5°; Harris and Jones 2017) and MERRA2 (https://gmao.gsfc.nasa.gov/reanalysis/MERRA-2; 0.625° × 0.5°; Gelaro et al. 2017). The Standardized Precipitation Evapotranspiration Index (SPEI; the difference between precipitation values and potential evapotranspiration; Vicente-Serrano et al. 2010) was calculated using the data of https://sac.csic.es/spei (resolution 0.5° × 0.5°).

We used Landsat-based forest cover loss product (loss year), which is the part of the Global Forest Change 2000–2017 v.1.5 dataset created in the University of Maryland (Hansen et al. 2013) to investigate fir mortality dynamics.

Fir stands were identified according to the map of forest species obtained from VEGA-PRO portal (https://pro-vega.ru/eng/maps; based on the MODIS data; spatial resolution 230 m). We analyzed stands with tree coverage $\geq 60\%$ of dominated species.

The burned areas were excluded from the analysis with the available MCD64A1 data (https://modis-fire.umd.edu). DEM SRTM 1-Arc second with on-ground resolution 30 m (downloaded from https://earthexplorer.usgs.gov) was used for geospatial analysis.

A time series of the Landsat (5, 7, 8) scenes (https://glovis.usgs.gov) were used to analyze fir mortality dynamics within the Eastern Sayan key-site. A high-res data (WorldView, GeoEye; 0.41–0.46 m; www.google.com/maps; www.bing.com/maps) were used for verification.

16.2.3 Methods

Fir stands were obtained from forest cover loss product by masking it by fir stands according to the map of forest species (https://pro-vega.ru/eng/maps). Next, the area of dead fir stands was related to topography and eco-climatic variables. Mortality was calculated as a ratio of the dead fir stand area to the total fir stand area for the same year, in percent.

From the analysis of the distribution of dead fir stands with respect to elevation, a relatively mountainous area was identified as a territory with the elevation >300 m a.s.l.

Spatial data were processed using ESRI ArcGIS and Erdas Imagine software, and statistical analysis done using Statsoft Statistica.

16.2.3.1 The Eastern Sayan Key-Site

Maps Generation

Time series of alive and dead stands' maps were generated based on Landsat scenes for the period since 1999 (before forest mortality) until 2017. The scenes were topographically corrected (Riano et al. 2003). The analysis included the following steps:

1. Generating a mask of "dark needle" stands based on the Landsat scene acquired in 1999. 24 samples were used (1444 $\pm$ 217 pixels in each sample). Accuracy of classifications was estimated using KHAT(κ)-statistics. Omission and commission errors were 8 and 13%; with $\kappa = 0.79$, it corresponds to the good level of accuracy.
2. Normalized difference infrared index was applied to identify declined stands (NDII):

$$\text{NDII} = (\text{NIR} - \text{SWIR})/(\text{NIR} + \text{SWIR}), \tag{16.1}$$

where NIR is a digital value from the near-infrared band (851–879 nm), and SWIR is a digital value from the shortwave infrared band (1566–1651 nm). NDII ranges ± 1; this index is efficient for estimation of vegetation vigor (Gu et al. 2007). Applicability of the NDII to identify dead stands was investigated based on the interpretation of the Landsat-8 scene (2017). Test points ($N = 40$) with relatively persistent spectral brightness (stony and water surfaces, urban areas) were used to calibrate other Landsat scenes obtained by different sensors and time (Landsat-5/TM, Landsat-7/ETM+). Linear regressions were calculated based on the obtained data, which related NDII values from the master Landsat scene with NDII values of other scenes. Classifications accuracy was moderate; it was estimated using a set of test sites ($N = 255$ with size about 50 pixels).

3. The dead stands spatial distribution was analyzed with respect to exposure (45° sectors) and slope steepness (1° resolution). To remove bias caused by non-uniform distribution of the relief features, the data were normalized:

$$K_{c(i)} = \left(A_{c(i)f}/A_{c(i)I}\right) * 100, \tag{16.2}$$

where $K_{c(i)}$ is the normalized area (%), $c(i)$ is the i-th category of the relief feature c, $A_{c(i)f}$ is the area of the given class f within i-th category of the relief feature c and $A_{c(i)I}$ is the total area of the i-th category of the relief feature c.

16.2.3.2 Gross Primary Productivity Calculations

Gross primary productivity (GPP) values were obtained from MODIS/Terra (eight days composites with 500 m pixel size; period 2000–2018; https://lpdaac.usgs.gov/products/mod17a2hv006/). GPP calculations are based on the equation:

$$GPP = \varepsilon * APAR = \varepsilon * NDVI * PAR\left(g\,m^{-2}\right), \quad (16.3)$$

where PAR is photosynthetic active radiation, ε is PAR use efficiency, and APAR stands for absorbed PAR (Running et al. 2004). For the Eastern Sayan key-site, mean summer GPP dynamics of "survived" and "declined" stands was analyzed (polygon size 500 × 500 m, polygons number $N = 19$ and 20, correspondingly). GPP values were normalized against the "background" data, i.e., the whole analyzed area ($S = 15{,}000$ ha).

16.2.3.3 Fieldwork

During in situ studies, temporary test plots (TP) were established within the key-site ($R = 9.8$ m). Within each TP relief characteristics (exposure, slope steepness, elevation), forest type, canopy closure, tree mortality, species composition, ground cover and soil type) were described and tree height and diameter were determined. Samples for dendrochronological analysis were randomly taken within the territory of ~0.5 ha (centered on the TP) at 1.3 m height by an increment borer.

16.2.3.4 Dendrochronological Analysis

Dendrochronological analysis was performed based on a representative sample ($N = 166$) of fir trees. The measurements were carried out on the platform LINTAB 3 with an accuracy of 0.01 mm. As a result, absolute individual chronologies were obtained for each tree (in mm). The TSAP and COFECHA programs were used to check the quality of cross-dating (Holmes 1983; Rinn 1996). The average interserial correlation coefficient was 0.48 for a cohort of living trees and 0.44 for declining ones. The expressed population signal (EPS) for both cohorts is 0.98. To eliminate the age trend, standardization was applied using the ARSTAN program, which converts tree-ring widths to time series of dimensionless indices (growth index, GI) with an average of 1.0 and relatively constant dispersion (Speer 2010).

16.3 Results

16.3.1 *The Eastern Sayan Mountains Key-Site*

The stands were composed by fir and Siberian pine with crown closure of 0.4–0.6. The average heights and diameters were 16 m and 20 cm for fir and 17.5 m and 36.0 cm for Siberian pine, respectively. The average age of fir and Siberian pine was 85 and 115 years, respectively. Soil drainage was characterized as moderate and good. The majority of fir trees were declined or dead (Figs. 16.2 and 16.3). The age difference between survived and dead trees was insignificant. Siberian pine trees mortality was low (<5%).

16.3.2 Stand Mortality and GPP Dynamics

Mortality in fir stands has been observed since 2005, strongly increasing in 2014–17; the total area of dead stands is about 75% of the "dark coniferous forest" area (Figs. 16.4 and 16.5). Actually, these were fir stands, whereas Siberian pine and spruce stands were not killed (Figs. 16.2 and 16.3). The mortality phenomenon was preceded and coincided with periodic droughts and aridity increase (i.e., drought index SPEI decrease), and root zone moisture (RZM) decrease (Fig. 16.6).

The GPP of "declined" stands was lower than "survived" ones even before fir mortality, increasing since mass stands' mortality (c. 2008; Fig. 16.7).

Fig. 16.2 Fir tree attacked by *Polygraphus proximus*. The resin streams are tree response to the bark-beetles invasion

16.3.3 Fir Trees' Growth Index Dynamics

The growth index (GI) of fir trees has been increasing since the mid-1960s with a subsequent depression after the breakpoint in c. 1983–84 yrs. A minor GI observed around 2010 was followed by a sharp GI drop and trees division into "decliners" and "survivors" cohorts. The GI of the trees sampled within the treeline (elevation about 1400 m) showed a relatively steady GI increase (Fig. 16.8). The tree mortality pattern (on-ground data) was similar to the satellite-derived data of stands' mortality (Figs. 16.5 and 16.8).

Before the GI breakpoint, the GI of both tree cohorts showed a weak correlation with air temperatures, whereas after the GI breakpoint, a significant correlation with May temperature occurred when drought and root zone moisture were increasing (Figs. 16.8 and 16.9a, c). After the breakpoint, the GI of both survivors and decliners correlated with drought index SPEI, whereas sensitivity to RZM was demonstrated by decliners only (Fig. 16.9b, d).

Maximal GI dependence of "decliners" on SPEI ($r^2 = 0.74$) and RZM ($r^2 = 0.56$) was observed during 1997–2010. In the period of mass tree mortality (since 2010), the GI correlations of both cohorts with climatic variables become insignificant.

16.3.4 Stands' Mortality Within the Fir Range in Siberia

Within the fir range, the distribution of both dead and alive stands is bi-modal with the main maximum at low elevations (200–300 m. a.s.l.; Fig. 16.10a). These elevations are typical for the West Siberian Plain. Thus, the following analysis will be focused on the elevations higher than 300 m, i.e., the mountain area.

There was an increase of fir mortality since the beginning of 2000s for both mountain and total fir range (Fig. 16.10a, b). Annual fir mortality was increasing with saturation in 2010 (Fig. 16.10c).

Fir mortality depends on the moisture parameters (SPEI, precipitation, and RZM) and the sum of positive temperatures (Fig. 16.11a–c). Fir mortality is strongly increasing with $\sum(t > 0\ °C)$ increase and decreasing with RZM increase (Fig. 16.11c, d). Median values of $\sum(t > 0\ °C)$ for dead and alive stands are significantly different (1720 °C and 1620 °C, respectively; $p < 0.05$).

Dead stands' area (relative) is strongly decreasing at elevation higher than ~1100 m a. s.l.; medians of dead and live stands distribution differ significantly (830 and 965 m, respectively; Fig. 16.12a; Table 16.1).

Mortality is strongly decreasing with elevation and slope steepness increase (Fig. 16.12b, c). Alive stands are located on the northern slopes, whereas mortality is observed mainly on the southern ones (Fig. 16.12d).

Fig. 16.3 Fir stands killed by *Polygraphus proximus* attacks

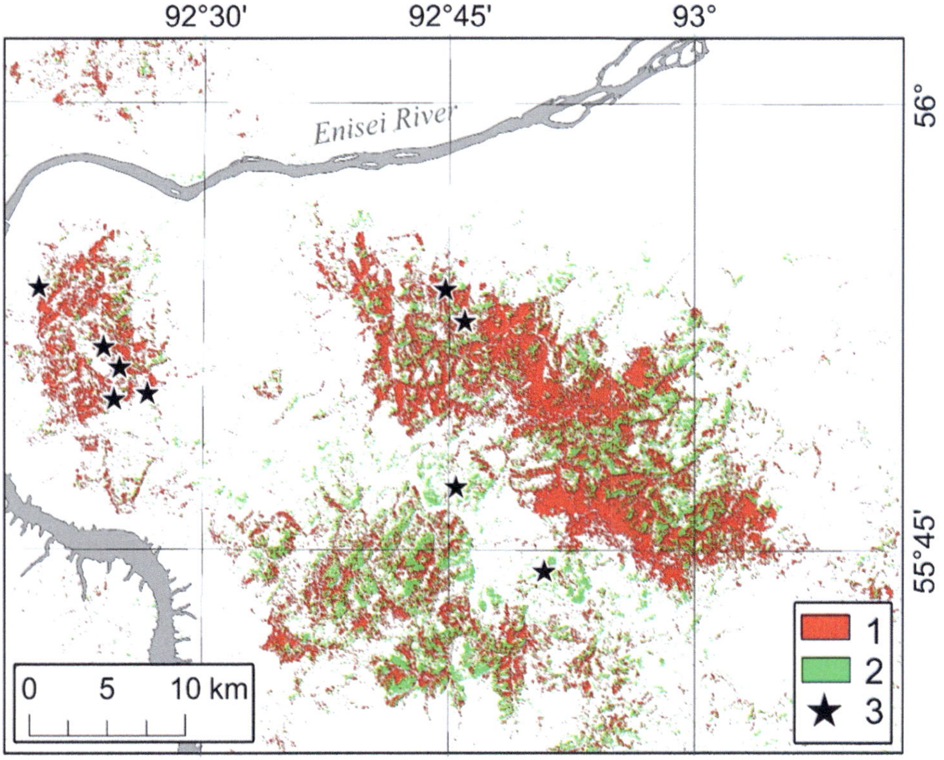

Fig. 16.4 Eastern Sayan Mountains key-site. *1*, *2*—dead and alive fir stands, respectively; *3*—radial increment sample sites

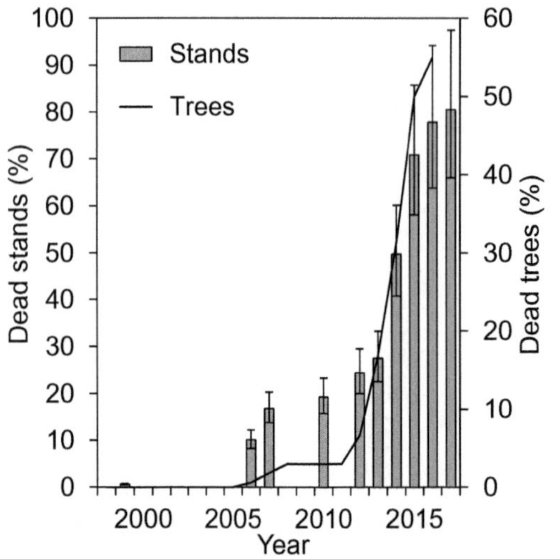

Fig. 16.5 Stands (satellite data) and trees (dendrochronology data) mortality dynamics.

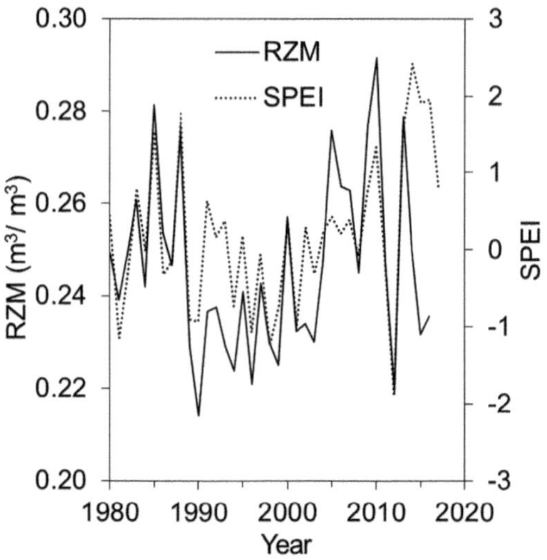

Fig. 16.6 Root zone moisture (RZM) and drought index SPEI dynamics

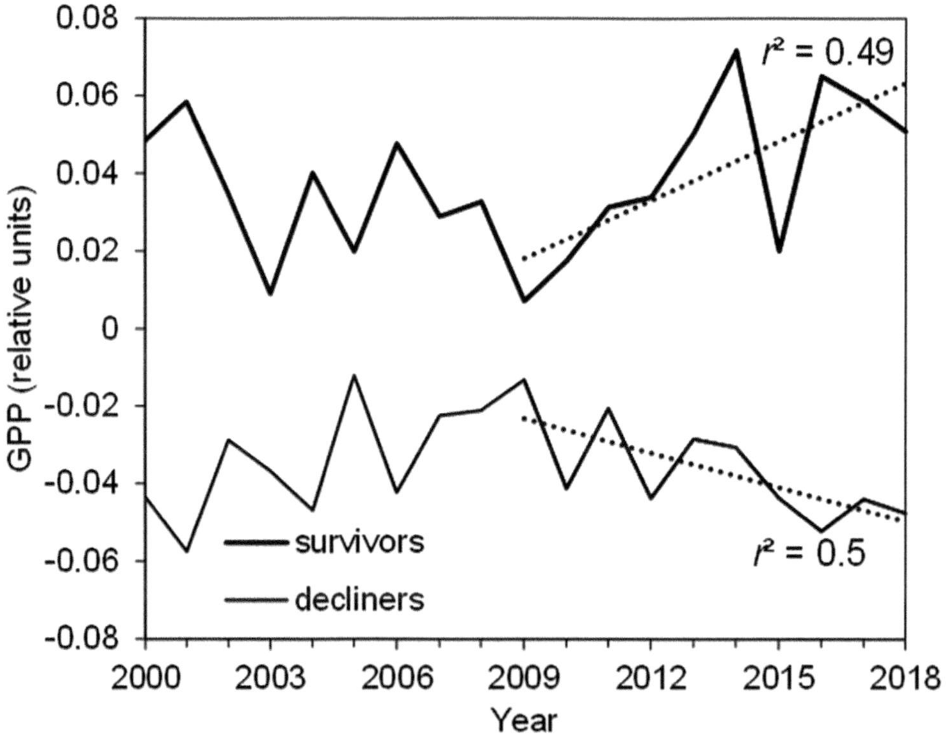

Fig. 16.7 GPP dynamics of "survived" and "declined" stands in the Eastern Sayan key-site. GPP presented as a relative to the mean "background stands" GPP (number of polygons, $N = 580$)

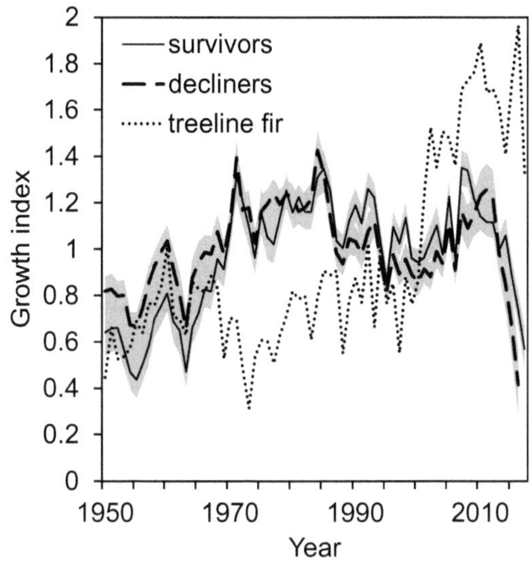

Fig. 16.8 Growth index of *Abies sibirica* "survivors" ($N = 80$), "decliners" ($N = 86$) and "treeline" ($N = 32$) fir cohorts. Confidence level ($p < 0.05$) indicated by the gray background

The location of dead and alive stands is significantly different with respect to elevation, a sum of positive temperatures, atmospheric and soil humidity (indicated by SPEI and RZM), and summer precipitation (Table 16.1).

16.4 Discussion

The unprecedented mass fir stands' mortality in the Siberian Mountains was induced by drought increase via elevated temperatures. Water-stressed fir trees were attacked by bark-beetle *Polygraphus proximus* Blandford, the species that was not observed earlier within *Abies sibirica* range.

Fir trees' mortality was preceded by the GI depression since the mid of 1980s, whereas since warming onset (c. 1970) increasing air temperature at the beginning of the growing season stimulated tree growth. The "breakpoint" of the fir GI curve (c. 1983–84) corresponded to the threshold after which the GI stimulation by temperature switched to depression via water stress. A similar growth trajectory is described also for Siberian pine and larch (*Larix sibirica, L. dahurica*) that experience limitation by moisture, whereas within areas with sufficient precipitation, the GI depression was not observed (Kharuk et al. 2017, 2018a, b).

The cohort of "decliners" was more sensitive to root zone moisture and possessed initially lower GPP in comparison with "survivors" (Fig. 16.7, 16.9a, c). These observations indicated "decliners" pre-disposition to mortality by initially worse growth conditions.

The fir growth negatively reacted to the high air temperature in the beginning (May) of the growth period (Fig. 16.9a, c). That effect was a consequence of temperature enhanced evaporation while water supply from still-frozen soil was limited; as result, needles desiccation (one-year-old mostly) occurred. In the years with anomalously high spring temperatures, it leads to fir crowns reddening over vast areas.

Fir experienced maximal water stress in the first decade of the twenty-first century; that coincided with the first findings of *Polygraphus proximus* bark-beetle, an aggressive Siberian fir pest (Krivetz et al. 2015). Earlier this species was known in the Russian Far East forests only (Krivolutskaya 1983). Ground surveys showed that all dead and declining fir trees were attacked by *Polygraphus proximus*. Eventually, that leads to about 5% stands' mortality within the fir range (with up to 75% within some areas) (Fig. 16.5 and 16.10c). It is known that water-stressed trees decrease protective substances synthesis such as phenols and terpenes (Kolb et al. 2016; Sangüesa-Barreda et al. 2015). Along with that, the soluble carbohydrates concentration in plant tissues is increasing which increases its nutritional value for pests (Liu et al. 2011). An additional factor that promotes pest reproduction was the growing season increase by 7–8 days within the study area.

Bark-beetles attacks of physiologically stressed trees resulted in quick stands' mortality; thus, during 2013–2015, mortality increased from 30 to 70% (Fig. 16.5). A similar phenomenon was described for the Sierra Nevada forests, USA, where mortality of drought-weakened coniferous trees increased by 40–50% during 2015–16 (Pile et al. 2018). Notably, that fir regeneration under

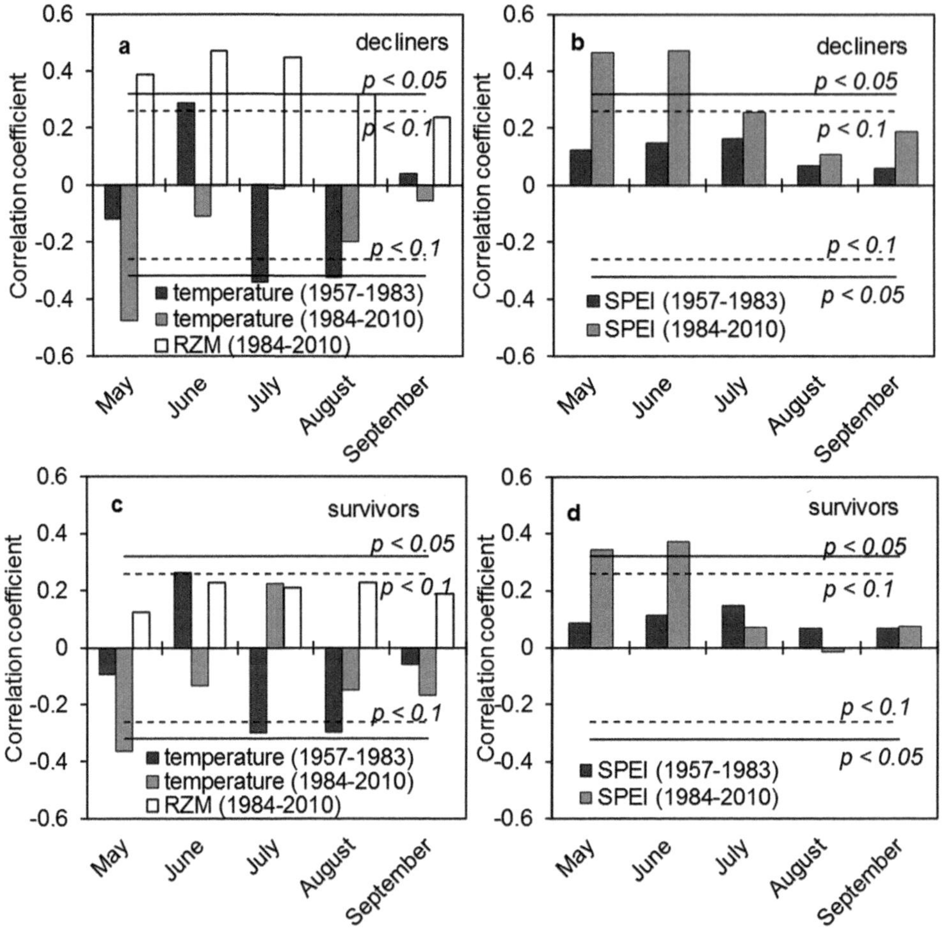

Fig. 16.9 **c, d** "Survivor" and **a, b** "decliner" trees' growth index correlations with **a, c** air temperature, **b, d** drought index SPEI and **a, c** root zone moisture (RZM) before and after the GI breakpoint (1983–84 yr.). The confidence levels are shown by solid and dashed lines

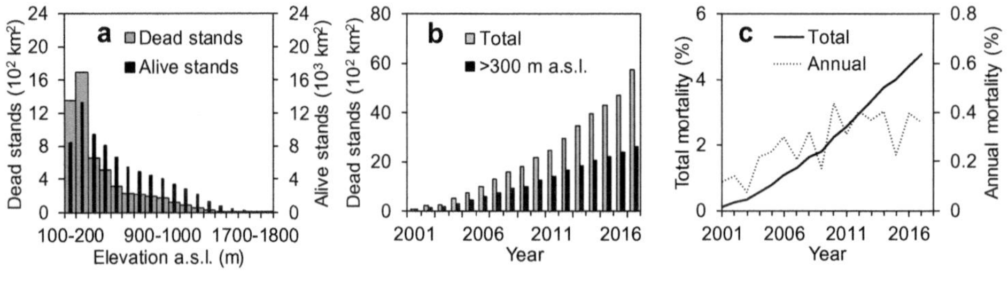

Fig. 16.10 **a** Dead and alive stands' distribution with respect to elevations; **b** dead stands' area dynamics within fir range total and elevations > 300 m a.s.l.; **c** fir mortality dynamics within elevations > 300 m a.s.l. Note: mortality was calculated based on the previous year value

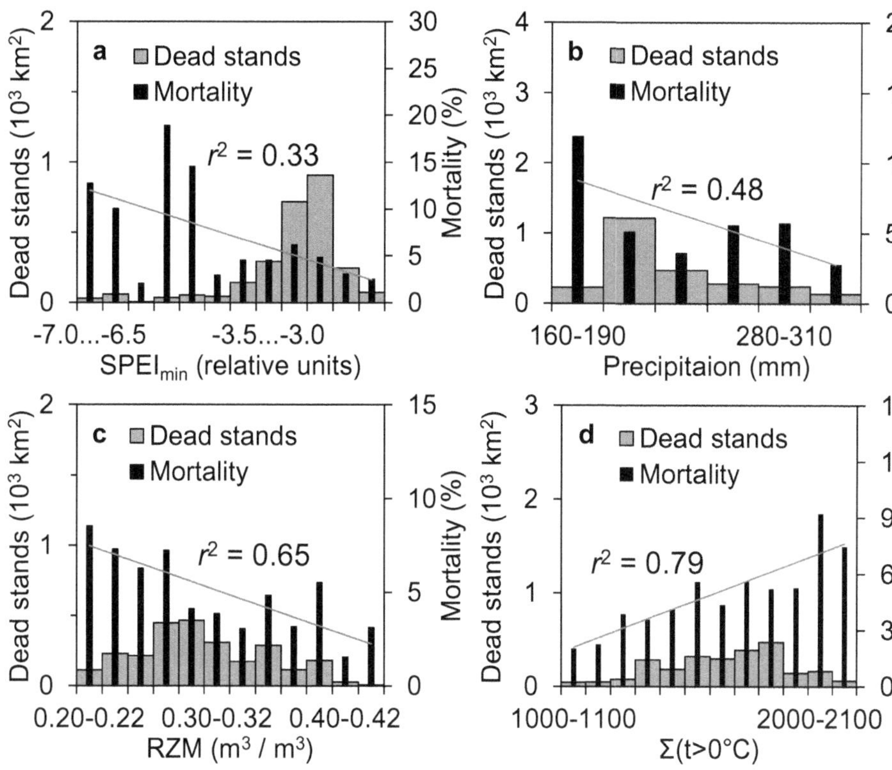

Fig. 16.11 Fir dead stands' area and mortality versus **a** drought index SPEI summer minimum, **b** summer precipitation, **c** mean summer root zone moisture (RZM) and **d** sum of positive air temperatures ($\sum(t > 0\ °C)$)

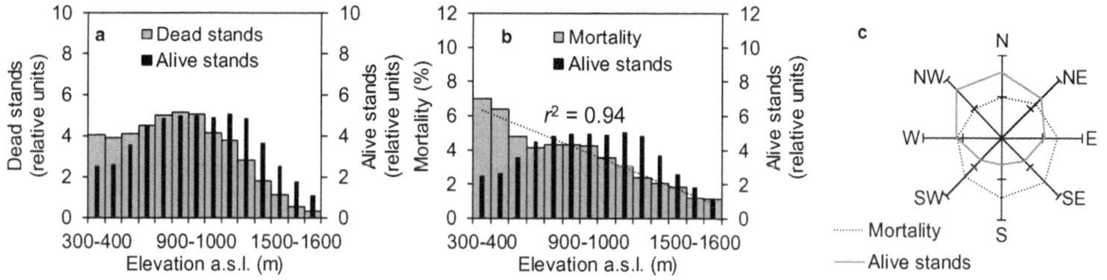

Fig. 16.12 **a** Fir dead and alive stands' distribution with respect to elevation; fir mortality dependence on **b** elevation, and **c** azimuth

Table 16.1 Quantiles of dead and alive stands' location with respect to eco-climatic parameters

Parameter range	Dead stands			Alive stands		
	5%	Median	95%	5%	Median	95%
Elevation, m a.s.l	355	830	1390	405	965	1560
$\Sigma(t > 0\ °C)$	1225	1720	2195	1095	1620	2025
SPEI	−12.4	−5.8	−3.9	−11.1	−5.7	-3.6
RZM (m^3/m^3)	0.21	0.29	0.39	0.22	0.30	0.40
Mean JJA PRE (mm)	180	215	310	195	220	325

the dead canopy was mostly healthy which is probably due to upper canopy drought mitigation. Healthy regeneration is a potential for fir recovering under a favorable climatic scenario. Meanwhile, models predicted drought increase in the southern taiga zone (Pachauri and Meyer 2014).

Within the fir range in the Siberian Mountains, dead and alive stands location significantly different with respect to eco-climatic variables: elevation, a sum of positive temperatures, atmospheric and soil humidity (indicated by SPEI and RZM) and summer precipitation (Table 16.1). Dead stands area strongly ($r^2 = 0.94$) decreasing with elevation increase, and the median of its location (830 m) significantly differs from that one of alive stands (965 m) (Fig. 16.12a; Table 16.1). Alive stands found on the northern slopes mostly, whereas mortality observed mainly on the southeastern ones (Fig. 16.12d). The shift from expected maximal mortality on the southern slopes is due to lower precipitation on the eastern rain-shadow slopes.

Fir mortality is strongly increasing with $\sum(t > 0\ °C)$ increase; dead stands located in areas with a higher $\sum(t > 0\ °C)$ (1,720 °C in comparison 1,620 °C for alive stands) and, consequently, a higher water stress. With moisture regime improvement (root zone moisture and precipitation increase, and atmospheric drought decrease), mortality is decreasing (Fig. 16.11). Similarly, Stephenson et al. (2019) found that *Abies concolor* Gordon mortality in the US forests was dominated by water stress.

The upper boundary of 90% fir mortality was located at 1390 m (Table 16.1); this value corresponded to the reported elevation limit of *Polygraphus proximus* habitat in the Southern Siberian Mountains (1400 m a.s.l.). At that elevation, $\sum(t > 0\ °C)$ is about 1350 °C which could be considered as a thermal limit proxy of *Polygraphus proximus* activity. On the other side, fir trees at higher elevation experience less water stress and, consequently, possess a higher resistance to pest attacks; thus, this issue needs more studies.

Within mixed *Abies sibirica- Pinus sibirica* stands, Siberian pine experienced less mortality, although both species are precipitation sensitive. The factors of higher Siberian pine drought resistance are the deeper root zone (2.0–3.0 m vs 0.5–1.0 m for fir) and lower leaf area index (LAI), which is an important water balance parameter. There are no reliable LAI data for these species. However, similar North American species, *Pseudotsuga menzieslii* Mirb and *Pinus strobus* L., have LAI 9.3 and 4.5, respectively (Thomas and Winner 2000; Guiterman et al. 2012), that could be a proxy of *Abies sibirica* and *Pinus sibirica* LAI values. It is also important that Siberian pine experienced only native pest's attacks (such as *Ips sexdentatus* Boern) and did not face pests similar to *Polygraphus proximus* that is considered as an invader (Krivetz et al. 2015). Since that pest species was first found, its synergy with water stress caused fir mortality on about 600 thousand ha (Fig. 16.1). A similar bark-beetle *Dendroctonus ponderosae* Hopkins, species also from *Scolytinae* subfamily, caused catastrophic mortality of drought-weakened North American conifers over 25 million ha (Millar and Stephenson 2015).

The majority of fir stands' mortality in the Siberian Mountains associated with bark-beetles, with exception to stands killed by Siberian silk moth (*Dendrolimus sibiricus;* Fig. 16.1). It is known that leaf-eating outbreaking insects kill trees independently of their stress. Meanwhile, latest catastrophic Siberian silk moth outbreak was also induced by consecutive warm years with drought events which stimulating pest population. Eventfully, silk moth outbreak killed over 800 thousand ha of forest (predominantly stands formed by Siberian pine). Moreover, that outbreak crossed outbreaks in historical northern boundary and moved about 50 km northward (Kharuk et al. 2018a).

Finally, it is noteworthy that both "survivors" and "decliners" growth index trajectories are demonstrating the decreasing pattern which indicates a chronic stress (Cailleret et al. 2017). In the case of likely climate aridity increase in the southern Siberia (Pachauri and Meyer 2014), the observed GI depression may indicate a potential "survivors" mortality within the southern range of *Abies sibirica*.

The results of this study show that fir mortality is increasing which indicated *Abies sibirica* retreat from its low and middle elevation ranges in the Southern Siberian Mountains.

16.5 Conclusions

1. Fir trees response to warming was two-phased. Since warming onset (1970) the fir GI was increasing, whereas further temperature elevation caused GI depression via water stress. Since the GI breakpoint (c. 1983–84), trees growth showed a higher dependence on the moisture availability.
2. Water-stressed fir trees were attacked by bark-beetle *Polygraphus proximus* Blandford, the species that was not observed earlier within *Abies sibirica* range. The synergy of drought and pest attacks had led to tree mortality since the beginning of 2000. The tree cohort of "decliners" was predisposed to mortality due to location within worse growth conditions, including lower soil moisture.
3. Dead and alive stand distributions differ with respect to the sum of positive temperatures (1720 °C and 1620 °C, correspondingly). Fir mortality is increasing with $\Sigma(t > 0\ °C)$ and decreasing with precipitation and root zone moisture and SPEI increase.
4. Stands' mortality is observed on the south-eastern slopes mostly and is decreasing with the elevation increase.
5. Stands' mortality spreads over 5% of fir range in the Siberian Mountains and continues to increase. Thus, *Abies sibirica* is retreating from its low- and middle elevation range in the Southern Siberian Mountains.

Acknowledgements The research was funded by Russian Fund of Basic Research, Krasnoyarsk Territory and Krasnoyarsk Regional Fund of Science, project numbers 18-45-240003 and 18-05-00432.

References

Anderegg LDL, Anderegg WRL, Berry JA (2013) Tree Physiology review: Not all droughts are created equal: translating metheorological drought into woody plant mortality. Tree Physiol 33(7):701–712. https://doi.org/10.1093/treephys/tpt044

Cailleret M et al (2017) A global synthesis of radial growth patterns preceding tree mortality. Global Change Biol 23:1675–1690. https://doi.org/10.1111/gcb.13535

Gelaro R et al (2017) The modern-era retrospective analysis for research and applications, version 2 (MERRA-2). J Clim 30:5419–5454. https://doi.org/10.1175/JCLI-D-16-0758.1

Gu Y et al (2007) A five-year analysis of MODIS NDVI and NDWI for grassland drought assessment over the Central Great Plains of the United States. Geophys Res Lett 34(6):L06407. https://doi.org/10.1029/2006GL029127

Guiterman C, Seymour RS, Weiskittel AR (2012) Long-term thinning effects on the leaf area of Pinus strobus L. as estimated from litterfall and individual-tree allometric models. Forest Sci 58(1):85–93. https://doi.org/10.5849/forsci.10-002

Hansen MC et al (2013) High-resolution global maps of 21st-century forest cover change. Science 342:850–53. https://earthenginepartners.appspot.com/science-2013-global-forest

Harris IC, Jones PD (2017) CRU TS4.01: University of East Anglia Climatic Research Unit; Climatic Research Unit (CRU) Time-Series (TS) version 4.01 of high-resolution gridded data of month-by-month variation in climate (Jan 1901–Dec 2016). Centre for Environmental Data Analysis, 04 Dec 2017. https://doi.org/10.5285/58a8802721c94c66ae45c3baa4d814d0

Holmes RL (1983) Computer-assisted quality control in tree-ring dating and measurement. Tree-Ring Bull 44:69–75

Kharuk VI et al (2016) Decline of dark coniferous stands in Baikal Region. Contemp Probl Ecol 9(5):617–625. https://doi.org/10.1134/S1995425516050073

Kharuk VI et al (2017) Fir decline and mortality in the Southern Siberian Mountains. Reg Environ Change 17(3):803–812. https://doi.org/10.1007/s10113-016-1073-5

Kharuk VI et al (2013) Siberian Pine decline and mortality in Southern Siberian mountains. For Ecol Manage 310:312–320. Doi: 10.1016%2Fj.foreco.2016.10.050

Kharuk VI, Im ST, Yagunov MN (2018a) Migration of the Northern Boundary of the Siberian Silkmoth Habitat. Contemp Probl Ecol 11(1):26–34https://doi.org/10.1134/S1995425518010055

Kharuk VI, Im ST, Petrov IA (2018b) Warming hiatus and evergreen conifers in Altay-Sayan Region Siberia. J Mt Sci 15(12):2579–2589.https://doi.org/10.1007/s11629-018-5071-610.1007/s11629-016-4286-7

Kharuk VI et al (2019) Fir (*Abies sibirica* Ledeb.) mortality in mountain forests of Eastern Sayan Ridge, Siberia. Contemp Probl Ecol (accepted)

Kolb TE et al (2016) Observed and anticipated impacts of drought on forests insects and diseases in the United States. For Ecol Manage 380:321–324. https://doi.org/10.1016/j.foreco.2016.04.051

Krivetz SA et al (2015) Distribution of the Ussuri polygraph Polygraphus proximus Blandf. (Coleoptera, Curculionidae: Scolytinae) in Siberia. News St. Petersburg For Acad 211:190–211

Krivolutskaya GO (1983) Ecological and geographical characteristics of bark beetle fauna (Coleoptera, Scolytidae) in North Asia. Entomol Rev 62(2):287–301

Liu C et al (2011) Effect of drought on pigments, osmotic adjustment and antioxidant enzymes in six woody plant species in karst habitats of southwestern China. Environ Exper Bot 71:174–183. https://doi.org/10.1016/j.envexpbot.2010.11.012

Logan JA, Regniere J, Powell JA (2003) Assessing the impacts of global warming on forest pest dynamics. Front Ecol Environ 1:130–137. https://doi.org/10.1890/1540-9295(2003)001[0130:ATIOGW]2.0.CO;2

Martínez-Vilalta J, Lloret F, Breshears DD (2012) Drought-induced forest decline: causes, scope and implications. Biol Lett 2012 8(5):689–691. Doi: 10.1098%2Frsbl.2011.1059

Millar CI, Stephenson NL (2015) Temperate forest health in an era of emerging megadisturbance. Science 349 (6250):823–826. https://doi.org/10.1126/science.aaa9933

Pachauri RK, Meyer LA (eds) (2014) IPCC, 2014: climate change 2014: synthesis report. In: Contribution of working groups I, II and III to the fifth assessment report of the intergovernmental panel on climate change. IPCC, Geneva, Switzerland

Pile LS, Meyer MD, Rojas R, Roe O (2018) Characterizing tree mortality after extreme drought and insect outbreaks in the southern Sierra Nevada. In: Kirschman JE (comp.) Proceedings of the 19th biennial southern silvicultural research conference. e-General Technical Report SRS-234. U.S. Department of Agriculture Forest Service, Southern Research Station, Asheville, NC, pp 89–96

Raffa KF et al (2008) Cross-scale drivers of natural disturbances prone to anthropogenic amplification: the dynamics of bark beetle eruptions. Bioscience 58:501–517. https://doi.org/10.1641/B580607

Riano D et al (2003) Assessment of different topographic corrections in Landsat-TM data for mapping vegetation types. IEEE Trans Geosci Remote Sens 41:1056–1061. https://doi.org/10.1109/TGRS.2003.811693

Rinn F (1996) TSAP V 3.6 reference manual: computer program for tree-ring analysis and presentation. Heidelberg, Germany

Running SW et al (2004) A continuous satellite-derived measure of global terrestrial primary production. Bioscience 54: 547–560.https://doi.org/10.1641/0006-3568(2004)054[0547:ACSMOG]2.0.CO;2

Sangüesa-Barreda G, Linares JC, Camarero JJ (2015) Reduced growth sensitivity to climate in bark-beetle infested Aleppo pines: connecting climatic and biotic drivers of forest dieback. For Ecol Manage 357:126–137. https://doi.org/10.1016/j.foreco.2015.08.017

Sazonov AA et al (2013) The problem of large-scale drying of spruce forests in Belarus and ways of solution. Forest Hunt A Sci Product Pract J Forest Workers 7:10–15 (in Russian)

Speer JH (2010) Fundamentals of tree-ring research. University of Arizona Press, Tucson

Stephenson NL et al (2019) Which trees die during drought? The key role of insect host-tree selection. J Ecol. https://besjournals.onlinelibrary.wiley.com/. Accessed on 25 Apr 2019. https://doi.org/10.1111/1365-2745.13176

Thomas SC, Winner WE (2000) Leaf area index of an old-growth Douglas-fir forest estimated from direct structural measurements in the canopy. Can J For Res 30:1922–1930. https://doi.org/10.1139/x00-121

Vicente-Serrano SM, Beguería S, López-Moreno JI (2010) A multiscalar drought index sensitive to global warming. The standardized precipitation evapotranspiration index—SPEI. J Clim 23:1696–1718. https://doi.org/10.1175/2009JCLI2909.1

Yousefpour R, Hanewinkel M, Le Moguédec G (2010) Evaluating the suitability of management strategies of pure Norway spruce forests in the black forest area of Southwest Germany for adaptation to or mitigation of climate change. Environ Manage 45(2):387–402. https://doi.org/10.1007/s00267-009-9409-2

17 Climate Change and Dynamics of Vegetation in the Lesser Caucasus: An Overview

George Fayvush and Alla Aleksanyan

Abstract

This review addresses potential changes of natural ecosystems of the Lesser Caucasus due to predicted climate change. Using methods of ecological modeling we show not only possible changes of ecosystems, but also changes of distribution of component species. Particular attention is paid to wetlands and two rare ecosystems of Lesser Caucasus, which have very limited local distribution and occupy very small areas.

Keywords

Climate change · Natural ecosystems · Lesser Caucasus · Ecological modeling

17.1 Introduction

Mountain ecosystems in connection with various cardinal orographic processes are unstable and because of this, very vulnerable to any additional influencing factors.

One of these factors is climate change, which has a serious impact on natural ecosystems and social life of population in mountainous regions. This is especially visible in mountainous systems like the Lesser Caucasus, which is one of the centers of development of human civilization and accordingly has intensive anthropogenic impact on nature in general and on natural ecosystems in particular. Mountain ecosystems are an important production resource for the population of this area, 50% of which resides in the countryside and for centuries has been using the goods and services provided by ecosystems (The Fifth National Report to Convention on Biological Diversity 2014).

The Lesser Caucasus, which is mainly located in the northeastern part of the Armenian Highlands, it is the core part of the so-called Transcaucasian Highlands (Fig. 17.1). From the orographic and physico-geographical points of view, the Transcaucasian Highlands form the northern edge of the system of folded-blocky mountains of the Armenian Highland. Unlike the Greater Caucasus, the Lesser Caucasus is not a single, distinct watershed ridge. It is a system of coulisse-spaced ridges that merge with the mountain formations of the inner parts of the Armenian Highland and adjacent high areas.

Since early geological epochs the land surface of Armenia, and the surrounding Armenian plateau, has been mountainous, with further mountain building occurring during the Cenozoic era (particularly after the Miocene). These complex

G. Fayvush · A. Aleksanyan (✉)
Department of Geobotany and Ecological Physiology, Institute of Botany aft. A. L. Takhtajyan of NAS RA, Yerevan, Armenia

U. Schickhoff et al. (eds.), *Mountain Landscapes in Transition*,
Sustainable Development Goals Series, https://doi.org/10.1007/978-3-030-70238-0_17

Fig. 17.1 Lesser Caucasus

tectonic shifts have resulted in a country dominated by a series of mountain massifs and valleys as well as in extensive volcanic activity. Climatic changes over the last million years also have left their mark on the country, with evidence of two glacial periods (Riss and Wurm) preserved on almost all mountains over 3000 m a.s.l. (Aslanyan 1958, 1985).

Climate change in mountainous regions first of all leads to a change in the distribution of altitudinal belts. And this in turn leads to changes in distributions of species and respective changes of mountainous ecosystems.

17.2 Database and Methodology

17.2.1 Climate

A wide range of climatic zones is distinguished within the Lesser Caucasus. Accurate and detailed data on climate change in the modern period are available for the central part of the Lesser Caucasus, namely for Armenia. The main reference data for this study have been taken from long-term meteorological information from 44 meteorological stations located throughout the country in different altitudinal belts, received from the governmental non-commercial organization "Armenian State Hydrometeorological and Monitoring Service." The territory of the Lesser Caucasus shows a pronounced vertical succession of six basic climate types—from dry subtropical up to severe alpine (Figs. 17.2 and 17.3). The average annual temperature ranges from −8 °C in high-altitude mountainous regions (2500 m a.s.l. and higher) to 12–14 °C in low-traced valleys. In the lowlands, the average air temperature in July and August reaches 24–26 °C, and in the alpine belt, the temperature does not exceed 10 °C. January is the coldest month with an average temperature of −6.7 °C. The absolute minimum temperature is −42 °C. The overall climate is best characterized as dry continental, in some areas with an annual rhythm more or less similar to the Mediterranean climate regime. The average annual precipitation in this area is 592 mm. The most arid regions are the Ararat valley and the Meghri region with annual precipitations of 200–250 mm. The highest annual precipitation, 800–1000 mm, is observed in high-altitude mountain regions. Most precipitation falls in the spring. In the northern part of Armenia, humidity comes from the Black Sea in the west, in the southern part from the Caspian Sea in the east, while the central part lies in the rain shadow of mountain ridges and is the driest area (Baghdasaryan 1958; Third National Communication on Climate Change 2015; Fourth National Communication on Climate Change 2020).

Prediction of climate change for different scenarios of greenhouse gas emissions is given in the Third and Fourth National Communications on Climate Change (2015, 2020). Climate change in Armenia is assessed using the CCSM4 model in accordance with the IPCC recommended RCP8.5 and RCP6.0 scenarios for CO_2 emissions. Therefore, as per the RCP6.0 scenario (equivalent to the SRES B2 scenario) CO_2 concentration will be 670 ppm by 2100 and it will be 936 ppm according to the RCP8.5 scenario (equivalent to the SRES A2 scenario). Future change forecasts for ambient air temperature and precipitation have been developed up until 2100. The results indicate that the temperature will continue to increase in all seasons of the year. However, according to the RCP8.5 scenario, starting from the mid-twenty-first century (2041–2100), the temperature will rise at a more rapid rate. According to the RCP8.5 scenario, it is very likely that, by 2100, the average annual temperature in Armenia will be 10.2 °C, which exceeds the baseline (1961–1990) by 4.7 °C. Evaluation results for precipitation change show that, according to the RCP8.5 scenario, there might be 16.3% increase in annual precipitation in Armenia by the mid-twenty-first century. There will be no changes in precipitation according to the RCP6.0 scenario. However, according to both scenarios for the summer months, there is an expected significant decrease in precipitation in all three periods: in 2011–2040 summer precipitation is expected to decrease by about 23% compared to the baseline (1961–1990) period. At the same time, the changes in temperature and distribution of precipitation will be very uneven both during the seasons and in separate regions of Armenia (Third National Communication on Climate Change 2015; Fourth National Communication on Climate Change 2020).

17.2.2 Vegetation

Forecast of vegetation changes in Lesser Caucasus is developed based on our personal observations using the «Holdridge Life Zones» scheme (Holdridge 1966), and for some rare ecosystems species distribution models were used: generalized boosted regression models, Breiman and Cutler's random forests for classification and regression, multiple adaptive regression splines, and maximum entropy.

17.3 Climate Change

According to data of meteorologists and climatologists, changes in annual ambient temperature and precipitation in Armenia have been assessed for various time periods; the results were used in preparations for First and Second National Communications (1998, 2010). Particularly in

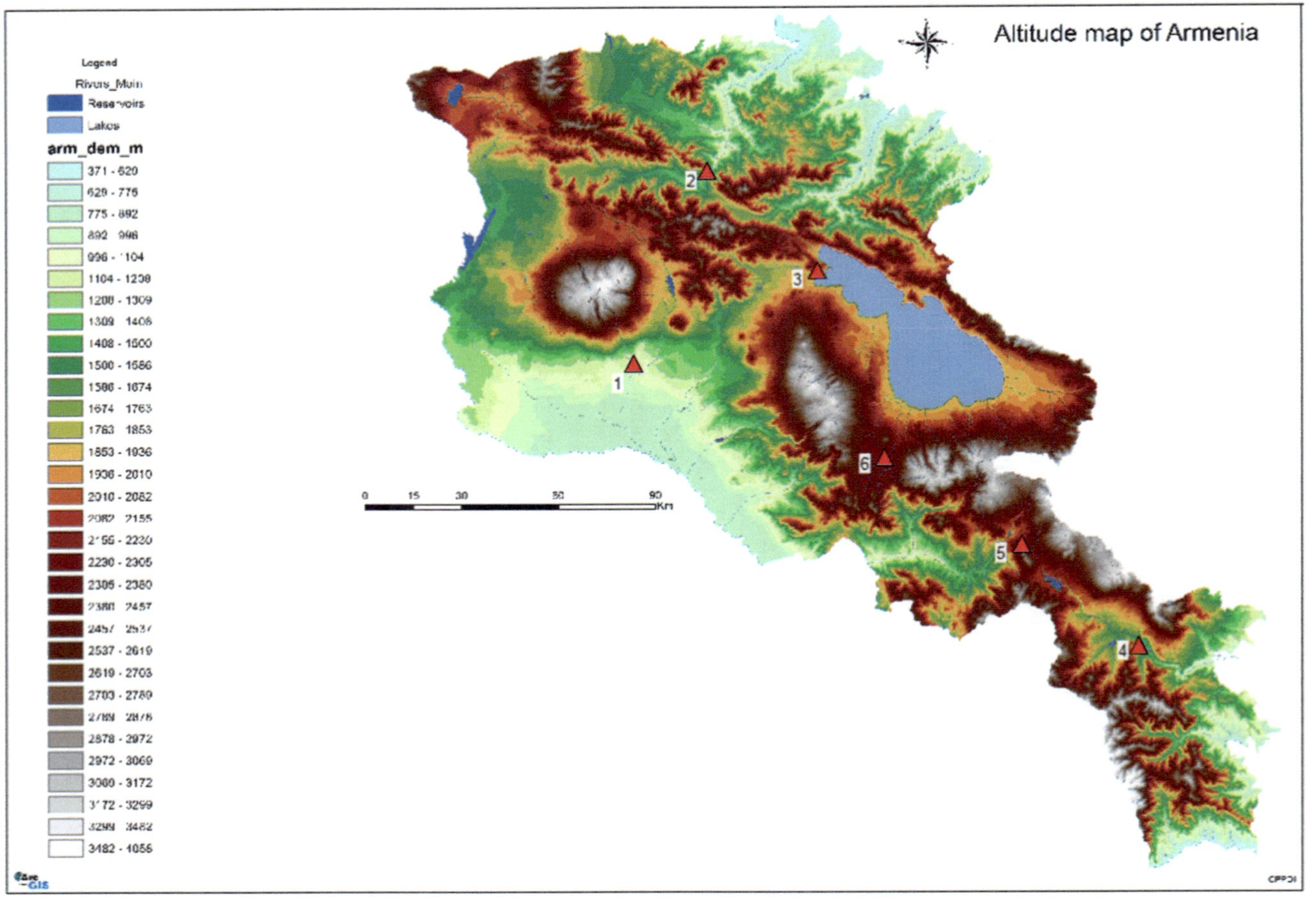

Fig. 17.2 Altitudinal map of central part of the Lesser Caucasus (Armenia) and location of meteorological stations for which climatic diagrams were prepared. (1—Yerevan, 2—Vanadzor, 3—Sevan, 4—Goris, 5—Gorhayk, 6—Janykh)

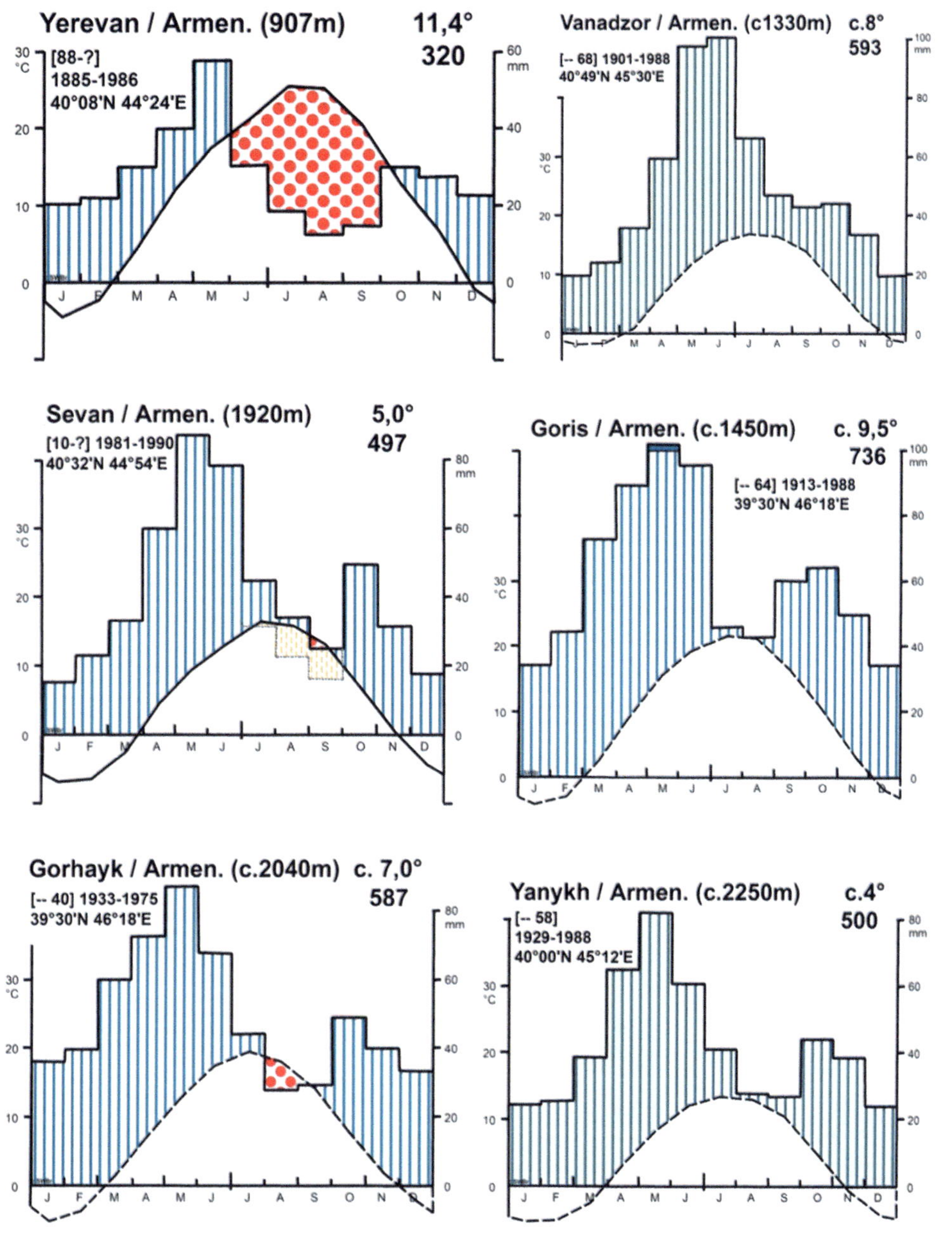

Fig. 17.3 Climatic diagrams for some regions of Armenia (1970–2010)

the first national communication it is said: The mountain ecosystems are vulnerable to global climate change and are the bio-indicators of these changes. The sum-up of historical, archeological, palaeo-botanic, palaeo-palynological data certifies that significant changes in the ecosystems of Armenia have taken place in the last three millenniums and were connected to global warming and climate aridization. During that period, the forest areas have significantly reduced, the semi-desert and steppe vegetation belts have expanded, and the alpine vegetation belt has reduced, etc. Based on this, maps were prepared of the distribution of the main natural ecosystems in the past and their distribution in future (Fig. 17.4).

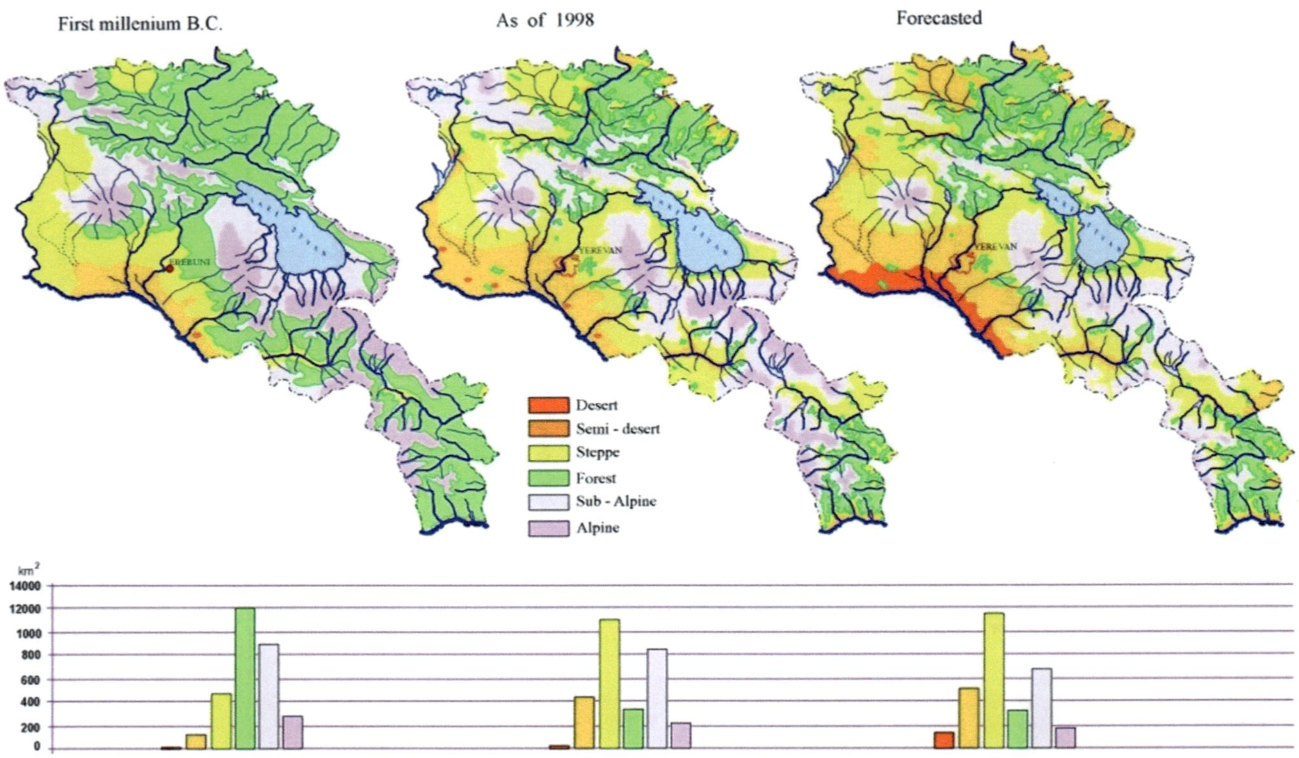

Fig. 17.4 Distribution of main ecosystems in the central part of the Lesser Caucasus (Armenia). (In the past—2000 years ago, current situation, forecasted situation to 2100)

These results show that, in recent decades, there has been a significant temperature increase. In the period of 1929–1996, the annual mean temperature increased by 0.4 °C; in 1929–2007 by 0.85 °C; in 1929–2012 by 1.03 °C. The comparison of changes in the assessment of precipitation amounts for different periods demonstrates that precipitation continues to decline. Observations showed that, in 1935–1996, there was a 6% decrease in annual precipitation, while in 1935–2012 it was close to a 10% decline. The spatial distribution of changes in precipitation amounts is fairly irregular. Over the last 80 years, the climate in the northeastern and central (Ararat Valley) regions of the country has turned arid, while precipitation has increased in the southern and northwestern regions, as well as in the western part of the Lake Sevan basin.

17.4 Results and Discussion

17.4.1 Climate Change and Vegetation Dynamics

All the main Caucasian ecosystems (besides humid subtropics) are represented in the Lesser Caucasus—deserts and semi-deserts, steppes, meadow-steppes, forests and open woodlands, subalpine and alpine vegetation as well as intrazonal ecosystems.

Current distribution of natural ecosystems in altitudinal belts and in territory of Armenia was accounted on the basis of our own long-term studies, as well as based on literature data of the vegetation of Armenia (Magakyan 1941; Takhtadjan 1941, etc.).

Currently, when predicting changes in vegetation with an expected climate change, first of all, researchers pay attention to a possible change in the zonal and altitudinal distribution boundaries of individual types of vegetation. Back in 1976, Turmanina (1976) suggested that the displacement of these boundaries is most likely where the conditions are extreme and air temperature is a limiting factor. For example, a shift to the north of woody vegetation of European Russia and Western Siberia is supposed (Rumyantsev et al. 2013). Interesting studies were conducted in the alpine zone in the North Caucasus (Russia) (Soudzilovskaia et al. 2013). It was shown that plant functional traits can be used as predictors of vegetation response to climate warming, accounting in the test ecosystem (the species-rich alpine belt) for 59% of variability in the per-species abundance relation to temperature. In this mountain belt, traits that promote conservative leaf water economy (higher leaf mass per area, thicker leaves) and large investments in belowground reserves to support next year's shoot buds (root carbon content) were the best predictors of the species increase in abundance along with temperature increase. This finding demonstrates that plant functional traits constitute a highly useful concept for forecasting changes in plant communities, and their associated ecosystem services, in response to climate change. As a result of these studies, it was shown that an increase in the average temperature of 0.6 °C over the past 25 years caused an increase in the optimum for the studied plants by 100 m. It is also indicated that the change in the amount of precipitation did not affect the studied parameters of plants. Most likely, this is due to the fact that in the alpine zone in both the Greater and the Lesser Caucasus, the amount of precipitation usually exceeds the needs of the plants living here. In addition, it should be noted that in the Greater Caucasus, alpine plants have a reserve of altitudes; that is, when glaciers melt, it is possible to migrate up the slope to reach their optimal climatic conditions (without taking into account edaphic conditions). In the Lesser Caucasus, in most cases, this is not possible for most alpine plants. Of course, it should be borne in mind that many species have a fairly wide ecological amplitude and high adaptive potential, and they will be able to adapt to new conditions. But at the same time, quite naturally, there will be a change in plant communities. Therefore, we used the "Holdridge Life Zones" scheme (Holdridge 1966) to predict the dynamics of vegetation in Armenia and the main ecosystems of wetlands, where both temperature and rainfall are taken into account. And when forecasting changes in individual rare ecosystems, other factors were also taken into account (soil conditions, steepness and exposure of slopes, etc.).

According to the results of our forecasts over the twenty-first century, the following changes in ecosystems may occur (Fayvush and Aleksanyan 2016).

Alpine meadows. Prediction of changes of bioclimatic conditions shows that the general direction of condition changes will not be in the direction of subalpine meadows, as expected, but in the direction of subalpine tall-grasses and expansion of wetlands.

Subalpine meadows. The transition is predicted to meadow-steppes, possibly extension of forest ecosystems on the territory of current meadows. In forest regions probably will occur raising of upper limit of the forest and in non-forest regions—transition to meadow-steppe ecosystems. It has to be noticed that alpine and subalpine meadows are the most vulnerable natural ecosystems in Armenia.

Forests. In the humid forests of the middle belt probably probably processes of will begin processes of xerophytization, thinning and penetration of plants of the steppes, arid woodlands, and shibliak. Some xerophytization of wet forests will move it into the humid forests. Modern forests of subalpine zone with time will be replaced to common humid forests, there will occur rising of upper limit of forest vegetation with a corresponding shift of subalpine crooked forests and park forests.

Meadow-steppes. The transition of these ecosystems to steppes is expected, in some cases (when the amount of precipitation will be increased), the formation of subalpine tall-grasses, and sometimes the extension to the territory of modern forest ecosystems will be possible.

Steppes. The general direction of ecosystem changes is xerophytization. The modern dry steppes can be replaced by semi-deserts, the areas of traganth steppes will be expanded. Current relatively mesophile steppe ecosystems can be replaced by drier sub-types.

Semi-desert. In the most of cases, is assumed the conservation of semi-desert vegetation, with an extension of phryganoid zone. Also is expected expansion of areas of desert ecosystems, such as solonchaks and saline deserts.

Shibliak and arid woodlands. In general, the conditions of these ecosystems will conserve and even slightly will increase, but natural regeneration of trees and shrubs can be worsen, and eventually these ecosystems, especially in the lower mountain belt can be replaced by phryganoids.

Petrophilous ecosystems and wetlands are intrazonal, and their vulnerability depends on their altitudinal and geographical locations.

Due to the predicted decrease of precipitation and especially their uneven distribution during the seasons (the absolute maximum will be in spring) wetland ecosystems will be especially vulnerable. According to our forecasts with climate change in the Lesser Caucasus the following changes may occur in these ecosystems (Fig. 17.5).

Changes of precipitation and seasonal regime will first of all lead to changes of rivers and water flows. In particular, there will be replacements of one ecosystem with another. Permanent non-tidal, fast, turbulent watercourses will be replaced with permanent non-tidal, smooth-flowing watercourses.

The increase of temperature first of all will affect the trophic level of lakes. Nowadays basically all oligotrophic lakes of Armenia are located above 3000 m and the increase of temperature will lead to intensification of eutrophication processes and their transition to permanent mesotrophic lakes.

Consequently, as a result of the same processes, mesotrophic lakes will change to eutrophic and even dystrophic lakes. Even now the similar processes can be observed in the lakes of Lori plateau.

Current ecosystem	Forecasted climate change	**Forecasted ecosystem**
 Permanent non-tidal, fast, turbulent water courses (a)	Decrease of precipitation and change in their regime	 Permanent non-tidal, smooth-flowing water courses (b)
 Permanent non-tidal, smooth-flowing water courses (c)	Decrease of precipitation and change in their regime	 Temporary running waters (d)
 Permanent oligotrophic lakes (e)	Increasing temperature	 Permanent mesotrophic lakes (f)
 Permanent mesotrophic lakes (g)	Increasing temperature	 Permanent eutrophic lakes (h)
 Permanent eutrophic lakes (i)	Increasing temperature	 Permanent dystrophic lakes (j)

Fig. 17.5 Forecasted change in wetland's ecosystems

  Permanent mesotrophic lakes (k-1, k-2)	Decrease of precipitation	 Temporary lakes, ponds and pools (l)  Communities with *Typha latifolia* dominance (m)
 Salt marshes with *Juncus acutus* dominance (n)	Decrease of precipitation and increase of temperature	 Ecosystems with *Cynodon dactylon* dominance (o)

Fig. 17.5 (continued)

As a result of the lakes stagnation, increase of the temperature and changes of precipitation regime changes in riparian zone ecosystems can cause transition of inland surface water ecosystems to marshes and wetlands. Besides, many lakes can become temporary lakes due to the decrease in precipitation. Salt marshes of lower belts as a result of reduced precipitation can make the transition to grasslands particularly to the continental inland salt steppes.

17.4.2 Change in Rare Ecosystems

Besides of general predictions for the main important types of vegetation of the Lesser Caucasus, we carried out special investigation and forecast of changes in rare ecosystems of Armenia.

As an example below is given forecast of changes of two very rare ecosystems of the Lesser Caucasus—Steppe scrub with

Asphodeline taurica and Plane Grove under climate change (Aleksanyan 2017).

First ecosystem is located in southern slopes of Shirak range on 2150 m above sea level. It occupies a small area but includes 11 rare species of the Red Book of Armenia. The dominant species here is *Asphodeline taurica.* Based on ecological modeling for all rare species favorable and unfavorable areas for growth are separated. According to our results, under predicted climate change, this ecosystem will enlarge its distribution, but the structure and compostion can undergo significant changes. For example, climate change can seriously harm following species: *Allium rupestre, Tragopogon armeniacus* and *Asperula affinis.*

Second ecosystem is Plane Grove, which is located in the basin of Tsav river, 650–750 m above sea level. The dominant species here is oriental plane (*Platanus orientalis*). This area is unique for this ecosystem in the Caucasus. According to ecological modeling results 6 mesophilous species (*Carex pendula, Euonymus velutina, Platanus orientalis, Pteridium tauricum, Pyrus raddeana*, and *Ranunculus cicutarius*) in future will have a reduction of favorable growth areas. For xerophyte rare species like *Lathyrus cassius, Medicago arabica, Nonea rosea, Lens ervoides, Thlaspi umbellatum* ecological conditions will be favorable, and for some other species (*Calendula persica, Galanthus artjuschenkoae, Lathyrus sylvestris, Trifolium angustifolium*) favorable growth areas will even enlarge.

17.5 Conclusion

As a result of the research and ecological modeling, it was shown that the main part of natural ecosystems of studied area are very vulnerable due to climate change. Very serious changes are expected both in the distribution of plant species and ecosystems as a whole. This should take into account that that even if climate change will contribute to expansion of the areas occupied by different ecosystems based on different adaptation potential of individual species the changes of their distribution will be very uneven and the structure of ecosystems will change greatly.

Because of the different adaptation potential of individual species, the change in their distribution will be very uneven and as a result the structure of ecosystems will change strongly. It means that in future we can and we will have the same ecosystems (forests, meadows, steppes, and others) but they will be completely different from current ecosystems called with the same names.

Acknowledgements We are very grateful to Prof. S.W. Breckle for preparation of climatic diagrams for different regions of Armenia.

References

Aleksanyan T (2017) Estimation of climate impact on some rare ecosystems of Armenia. PhD thesis, Yerevan, p 23

Aslanyan A (1958) Regional geology of Armenia. Haypetrat, Yerevan (in Russian)

Aslanyan A (1985) On the age of the relief of Armenia (main points of the evolution). Probl Holocene Geol Yerevan 14–19 (in Russian)

Bagdasaryan A (1958) Climate of Armenian SSR. Yerevan (in Russian)

Fayvush GM, Aleksanyan AS (2016) Climate change as threat to plant diversity of Armenia. Takhtadjania 3:112–126

First National Communication on Climate Change (1998) Yerevan, p 81

Fourth National Communication on Climate Change (2020) Yerevan UNDP Armenia, 213 p

Holdridge LR (1966) The life zone system. Adansonia 199–203

Magakyan A (1941) Vegetation of Armenian SSR. Moscow-Leningrad (in Russian)

Rumjantsev VY, Malkhazova SM, Leonova NB, Soldatov MS (2013) Forecast forecast of possible changes in the zonal borders of vegetation in European Russia and Western Siberia due to global warming. Siberian Ecol J 4:449–458 (in Russian)

Second National Communication on Climate Change (2010) Yerevan, p 134

Soudzilovskaia NA, Elumeeva TG, Onipchenko VG, Shidakov II, Salpagarova FS, Khubiev AB, Tekeev DK, Cornelissen JHC (2013) Functional traits predict relationship between plant abundance dynamic and long-term climate warming. PNAS 110 (45):18180–18184

Takhtadjan A (1941) Phyto-geographical review of Armenia. Proc Inst Botany Armenian Branch USSR Acad Sci 2:3–156 (in Russian)

The Fifth National Report to Convention on Biological Diversity (2014) Yerevan, p 106
Third National Communication on Climate Change (2015) Yerevan, UNDP Armenia, p 129
Turmanina VI (1976) Phytoindication of climate fluctuations. Landscape indication of natural processes. Proc MOIP 15:64–70 (in Russian)

Changing Climate Scenario in High Altitude Regions: Comparison of Observed Trends and Perceptions of Agro-Pastoralists in Darma Valley, Uttarakhand, India

18

Deepika Rawat and Udo Schickhoff

Abstract

Climate change in the Himalayan region has serious implications for livelihood support systems and overall human well-being. In Darma valley (Uttarakhand, India), local people, known as Bhotias, use alpine pasturelands (bugyals) for transhumant grazing of their livestock. In the region, transhumant pastoralism has seen a tremendous decline due to various environmental and socio-economic changes. In the context of present climatic conditions, the article addresses the extent of climate change in the region, the perceptions of climate change among local pastoral communities and socio-economic dimensions of climate change impacts on the local communities and environment. A substantial step of the study is to investigate the local community's perceptions of climate change which affects their motivation to engage in alternative livelihood options. Socio-economic data on impacts of climate change were correlated with meteorological data (CRU TS climate data v4.02) and remote sensing-based indicators such as NDSI (Normalized Difference Snow Index) in order to validate the impacts of changing climate as experienced by the local community. The results suggest that a substantial number of pastoralists perceived climatic change and its associated impact on the environment and on their livelihoods.

Keywords

Transhumant pastoralism · Bhotia community · Socio-economic changes · Mountain livelihoods · Himalaya

18.1 Introduction

Climate change is an urgent and serious problem for human civilization. Globally, warming of approximately 1 °C above pre-industrial levels has been observed (IPCC 2018). It is also projected that with the current rate warming is expected to reach 1.5 °C between 2030 and 2050 (IPCC 2018). In the Himalayan region, an increasing trend of annual mean surface air temperatures has been observed between 1901 and 2014 (Ren et al. 2017), resulting inter alia in a considerable rate of recession of the majority of high altitude Himalayan glaciers (Bolch et al. 2019), and in earlier snowmelt and shorter winter seasons at higher altitudes. A significant amount of total annual precipitation is now received in the form of rain instead of snow (Sharma et al. 2009; Zomer et al. 2016; Bhutiyani 2016; Bolch et al. 2019; Mal et al. 2019). Other climatic

D. Rawat (✉) · U. Schickhoff
CEN Center for Earth System Research and Sustainability, Institute of Geography, University of Hamburg, Hamburg, Germany

U. Schickhoff et al. (eds.), *Mountain Landscapes in Transition*,
Sustainable Development Goals Series, https://doi.org/10.1007/978-3-030-70238-0_18

change impacts in the Himalaya include water scarcity, altitudinal shifts of plant and animal species, changes in species compositions, changing productivity of pasturelands and agricultural lands, the emergence of insects and pests, and severe socio-economic implications (IPCC 2014).

Climate change has a wide array of socio-economic implications for the local population and their livelihood practices. Changing climatic conditions facilitate changes in traditional agricultural and pastoral practices of rural communities which need to cope with globalization effects, in particular with emerging consequences of high levels of poverty, rapid urbanization, population growth, low levels of economic development, and social transitions such as predominant male outmigration (O'Brien and Leichenko 2000; Sharma et al. 2009; Ogra and Badola 2015; Rautela and Karki 2015; Yi et al. 2007). It has been observed that climate change has adversely affected transhumant pastoralism in various socio-economic ways across the Himalayan region, such as in India (Negi et al. 2017; Sati 2015; Rautela and Karki 2015), Nepal (Aryal et al. 2014; Gentle and Thwaites 2016), Bhutan (Namgay et al. 2014), Gilgit-Baltistan Province of Pakistan (Joshi et al. 2013a), Tibetan Plateau (Wu and Yan 2002) and the extended Hindukush Himalayan region (Xu et al. 2009; Dong et al. 2010; Chaudhary and Bawa 2011) Transhumant pastoral communities who seasonally move with their livestock between fixed summer and winter pastures are particularly affected since they depend on alpine pasturelands which are among the most fragile ecosystems and most vulnerable to climate change (Sati 2015). The practice of transhumant pastoralism is based on a high dependence on natural resources and involves maintaining an ecologically balanced relationship between pastures, livestock, and people (Bhasin 2011). Pastoralists' seasonal migration is directly related to the timing of rainfall, snowfall, water availability, nutritive grass production and therefore, it is highly sensitive to any change in climate (Aryal et al. 2014). Pasture productivity is highly correlated with temperature and precipitation trends (Williams and Albertson 2006; Wilkes 2008; Eriksson et al. 2009). In general, warmer temperatures enhance pasture productivity as long as humidity and soil moisture is sufficient to compensate for higher evapotranspiration (Luo et al. 2004).

In mountain regions of Pakistan, reduced livestock productivity and diminishing pastoral practices was a result of increased occurrences of prolonged droughts which has led to degradation of pastures by reducing the vegetation cover (Afzal et al. 2008). In such circumstances, the absence of prior conditioning often leads to reduced cattle performance and livestock deaths, ultimately leading to economic losses (Mader 2003). Climate change also facilitates an upward shift of the treeline and encroachment of woody vegetation on alpine meadows, thus affecting principal alpine habitats and grazing land distribution (Schickhoff et al. 2015). Acute shortage of fodder is forcing pastoral communities to abandon traditional livestock-based livelihood systems and engage themselves in other forms of non-agrarian incomes (Wu et al. 2014; Berhanu and Beyene 2015). The practice of transhumant pastoralism, in general, is declining due to various climatic and non-climatic factors, as reported in other mountain regions of the world (Maikhuri et al. 2001; Intigrinova 2005; Afzal et al. 2008; Banerjee 2009; Kerven et al. 2012; Namgay et al. 2014; Schickhoff and Mal 2020). Studies show that the high dependence of pastoral communities on alpine pasturelands has affected different aspects of their lives such as livestock management, pastoral production and their overall socio-economic development (Joshi et al. 2013b; Wu and Yan 2002). Vasquez et al. (2010) and Gentle and Thwaites (2016) observed an increasing presence of exotic invasive weeds, potentially inducing outbreak of livestock diseases and potentially leading to lesser numbers of livestock, declining livestock productivity, and changes in seasonal migration patterns.

In the Himalayan state of Uttarakhand, complex climate change impacts have been observed at regional and local scales. The time period 2007–2012 was the warmest in all thirteen districts of Uttarakhand (Mishra 2014, 2017), associated with glacier retreat and erratic patterns of precipitation (Bhutiyani et al. 2007; Mal and Singh 2014; Dobhal and Pratap 2015; Mal et al. 2019). In some villages in Uttarakhand, farmers have shifted to less water-intensive crops (such as cabbage, carrots, maize, and pumpkin) and to other sources of livelihood due to climate variability and water stress (Kelkar et al. 2008). High altitude villages of Niti and Mana valleys of Chamoli District, and of Chaudas, Darma and Byans valley in Pithoragarh District in Uttarakhand have experienced a shortage of water resources used by livestock over the last 15–20 years, enforcing pastoralists to reduce the livestock population as well as abandon their agricultural practices (Negi et al. 2017). Another study by Rautela and Karki (2015) in Johar, Byans, Niti and Bhagirathi valley highlighted that scarcity of fodder for the livestock has ultimately forced local people to purchase fodder from markets. Due to this extra financial pressure, most of the people have abandoned their pastoral livelihoods and have shifted to other sources of income.

Observations by Negi (2007) in Johar Valley highlighted a decline in the sheep population due to governing environmental and socio-economic factors, followed by a significant decline in the traditional art of weaving. Economic uncertainties, along with other push factors of demographic, political, environmental and social concerns have resulted in accelerated labor outmigration from villages to urban areas across Himalayan regions (Hoermann et al. 2010; IPCC 2014; Siddiqui et al. 2019) as also reported from the above mentioned valleys (Rautela and Karki 2015; Negi et al. 2017). In recent decades, labor outmigration has been a general trend in the Himalayan regions as a result of globalization and constraints posed by subsistence agriculture, especially the younger generation outmigrates in search of better livelihood opportunities (Hoermann and Kollmair 2009; Schickhoff and Mal 2020). Climate change is an additional stressor which is likely to influence the rate of outmigration (Hoermann and Kollmair 2009; Banerjee et al. 2011; Mueller et al. 2014; Namgay et al. 2014; Gentle and Thwaites 2016; Siddiqui et al. 2019). In Nepal (Gentle and Thwaites 2016) and Bhutan (Namgay et al. 2014), climate-driven socio-economic changes have modified demography as mostly male members out-migrate, leading to a higher workload on women and to a shift to non-pastoral livelihoods. In Pakistan, Mueller et al. (2014) found a positive relationship between adverse effects of increasing heat stress on farming and long-term migration of men. Another study by Banerjee et al. (2011) reports that labor migration is a viable strategy for adaptation to severe droughts and floods among mountain communities in India, Nepal, China, and Pakistan. In Uttarakhand, decreasing agricultural productivity has been a major reason for the accelerated migration of people to urban areas (Hoermann et al. 2010; Tiwari and Joshi 2016). Thus, outmigration can be perceived as an adaptive strategy of vulnerable pastoral communities to minimize economic hardship under globalization and climate change conditions (Wu et al. 2014; Hoermann et al. 2010).

In this regard, a deepened understanding of the local community's perception of climate change is needed which shapes their adaptation strategies (Vedwan and Rhoades 2001; Adger et al. 2009; Weber 2010; Chaudhary and Bawa 2011; Loria and Bhardwaj 2016). There have been many studies addressing local perceptions of climate change and its adverse impacts, the vulnerability of local communities, and their adaptations across Himalayan regions (for example, Chaudhary and Bawa 2011; Chaudhary et al. 2011; Gentle and Maraseni 2012; Shrestha et al. 2012; Macchi et al. 2015; Vidya et al. 2015; Loria and Bhardwaj 2016; Ndungu and Bhardwaj 2015; Aryal et al. 2016). A few of these studies have substantiated their results with quantitative data, and have found them quite consistent (for example, Sujakhu et al. 2016; Aryal et al. 2016). The *Bhotias* in Darma valley are subjected to adverse socio-economic changes, facilitated by ongoing fast-paced urbanization, predominant male outmigration, and various environmental concerns. In the context of the current climate change regime, it is necessary to understand the effects of changing climate on the pastoral economy and the obvious need to adopt alternative sustainable livelihood strategies. To date, there has been no proper documentation on how transhumant pastoralists in the higher Himalayan regions of Uttarakhand have perceived climate change and socio-economic changes mainly because of its nearly inaccessible and remote location, especially in Darma valley. Therefore, this study aims to address the demand for more coherent research on examining (i) temperature trends, rainfall trends and snow availability; (ii) local community's perception regarding key climatic variables and their associated impacts; and (iii) comparison of meteorological trends with local community's perception of climate change; which is crucial for designing, planning and proper implementation of effective adaptation and mitigation strategies.

18.2 Methods

18.2.1 Study Area

The Darma valley lies in Kailash Sacred Landscape region, situated in the north of the Panchachuli mountain, between 29° and 31° North latitude and 79° and 81° East longitude in Dhauliganga watershed, Dharchula subdivision, Pithoragarh District, Kumaon region, Uttarakhand, India (Figs. 18.1 and 18.2a, b), bordering Tibet in the North and Nepal in the East. It consists of 12 villages namely, Sela, Chal, Nagling, Baling, Dugtu, Dantu, Go, Philam, Bon, Tidang, Marcha and Sipu. Having a striking topography and subtropical, temperate and alpine vegetation, the entire valley lies at an altitude between 7500 and 14,000 ft adjacent to Johar valley in the West and Byans valley in the East. Rainfall throughout these regions is significantly variable.

The valley is inhabited by an indigenous tribe, the *Bhotias* (sub-tribes: *Rang*) who are semi-nomadic agro-pastoralists, also locally known as *Darmya* or *Darmani*. Earlier, along with seasonal migration their main source of livelihood was cross-border trade with Tibet until the disruption caused by the Indo-China War in 1962. Products such as sugar, grain, and wool from India were exchanged for borax, wool, and salt from Tibet (Bergmann et al. 2008, 2011). After the distress caused by the war, the importance of agriculture, forest, and pasture use became even more of significant worth for their livelihood security (Bergmann et al. 2008). These migratory households practice mixed mountain agriculture, i.e., a mix of animal husbandry, crop farming, forest and pasture use (Ehlers and Kreutzmann 2000; Bergmann 2016) linked through seasonal migration (Nüsser 2006). The main livestock types include sheep, goat, cow/oxen, yak and yak hybrids locally known as *Jhubbu* (male) and *Jumma* (female) (Fig. 18.2g, h). At the beginning of winter season (October–November), they migrate to warmer regions in lower locations near Dharchula and they return to Darma valley in March–April to spend the summers with their livestock. Their seasonal migratory pattern is fully dependent on resource availability (nutritional fodder and water) for their livestock and extends over several altitudinal zones of the Kumaon Himalaya (Nüsser 2006). At different altitudinal belts, environmental conditions such as the duration of the snow cover or the onset of vegetation growth necessitate certain potentials and limitations for pastoral land use by the transhumant community (Bergmann et al. 2008).

Transhumant pastoralism also enables them to collect rare medicinal herbs and to utilize agricultural land in the summer villages to cultivate high altitude crops and vegetables such as buckwheat (*Fagopyrum esculentum*), barley, cabbage and potatoes (Fig. 18.2c–e, h). Due to the decline in the practice of transhumance in this region, collection of wild medicinal plants mainly Caterpillar Fungus (scientific name: *Cordyceps sinensis*; locally known as *Keedajadi* or *Yarshagumba*) and cultivation of medicinal plants like Kutki (*Picrorhiza scrophulariiflora*), Atish (*Aconitum heterophyllum*), Hathazari (*Dactylorhiza hatagirea*), Kuth (*Saussurea costus*), Chippi (*Pleurospermum angelicoides*), Jambu (*Allium stracheyii*), Kaala Jeera (*Carum carvi*), etc. as a form of subsistence has become prevalent in the study area, representing a considerable share of the household income (Fig. 18.2e, f).

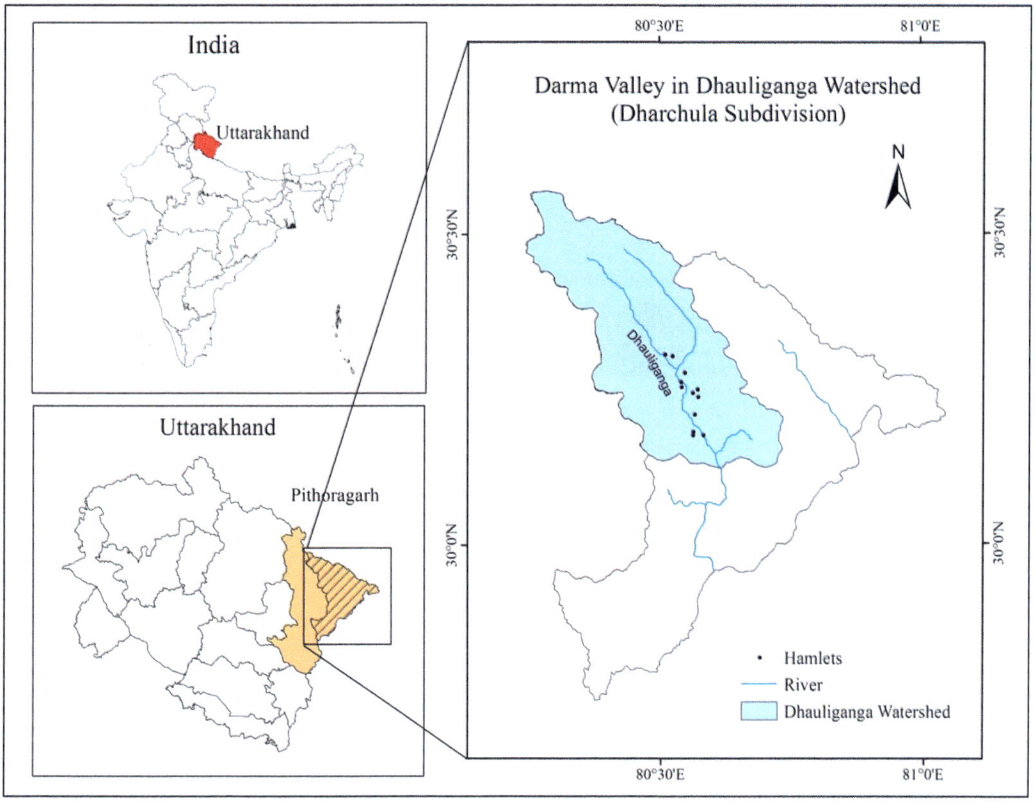

Fig. 18.1 Location of Darma valley in Dhauliganga watershed, Pithoragarh district, Uttarakhand. *Source* Own design

18.2.2 Data Collection and Analysis

For this study, the annual and seasonal temperature and precipitation trends were computed and analyzed through linear regression analysis for the time period between 1975 and 2016 with the CRU TS climate data v4.02 (0.5° × 0.5°, 1968–2016). It is a high-resolution gridded dataset for multiple variables on a 0.5° × 0.5° or finer grid developed by the Climatic Research Unit (University of East Anglia). The seasons were classified as hot summer or pre-monsoon (March, April, May), Monsoon (June, July, August, September), Post-monsoon (October, November) and Winter (December, January, February) to conveniently analyze temperature and precipitation trends. This seasonal classification is also used in other studies in Uttarakhand (Mal et al. 2019).

Data analysis for climatic data (Mann–Kendall Test and linear regression analysis) and information gathered through primary survey was computed and analyzed using XLSTAT and SPSS. In Mann–Kendall test, if p-value is less than the significance level α (alpha) = 0.05, H0 is rejected which means that there is no trend in the tested time series while accepting H0 indicates that no trend has been found in the time series. If Null Hypothesis (H0) is rejected it means the result is statistically significant. Moreover, in Mann–Kendall test, Kendall's tau represents a measure of correlation which has values between +1 and −1 where positive correlation indicates that both variables increase together as opposed to a negative correlation which indicates an increase in one variable and decrease in another. For both temperature and precipitation data, the null hypothesis is tested at 95% confidence level.

Fig. 18.2 **a** A view of Panchachuli range from Dugtu village, Darma valley; **b** Open pastureland near Panchachuli base camp; **c** and **d** Pastoral herders grazing sheep and goats in open fields in Sipu village, Darma valley; **e** Caterpillar Fungus (locally known as *Keedajadi* or *Yarshagumba*); **f** A group of Caterpillar fungus collectors; **g** and **h** Farmers ploughing the fields with Yak hybrids in Nagling and Dugtu village, respectively

To delineate the trend of the presence of snow in the study area over the period of last 28 years (1990–2018), Landsat satellite imageries were pre-processed and Normalized Difference Snow Index (NDSI) values were computed in SAGA-GIS 7.0. Subsequently, the NDSI values were reclassified and NDSI maps were generated in ArcGIS 10.6.1. Landsat TM, ETM+ and OLI imageries for the years 1990, 2001, 2011 and 2018 were acquired from the United States Geological Survey (USGS) Landsat Missions website www.earthexplorer.org for the snow-clad months of March (for 1990, 2001 and 2011) and April (for 2018). NDSI uses green spectral bands (high reflectance of the snow) and short-wave infra-red (SWIR) (low reflectance). For calculating NDSI, the equation used was: NDSI = (Green − SWIR)/(Green + SWIR). According to Hall et al. (1995), for an effective threshold of snow mapping, NDSI values above 0.4 typically indicate the presence of snow usually represented in light color shades (close to white). The time period 1990–2018 is chosen because of the absence of SWIR band in previous Landsat MSS (1972–1983) imageries.

An extensive field survey, based on qualitative research approaches (historical transect, semi-structured interviews, key informant interviews, focused group discussions, etc.), was also conducted in 8 villages of the study area, stratified by different altitudes. The main aim was to investigate the socio-economic characteristics of the affected community, their perceptions related to changing climate and its impacts on their livelihood and environment, and their local response strategies. A sample size of 200 respondents (household heads) was selected based on stratified random sampling representative of around 750 households (District Census Handbook of Pithoragarh, Census of India 2011). Initially, a pilot survey was done to pretest the questionnaire which was later followed by actual primary data collection. Focused Group Discussions (FGDs) were conducted in 4 villages with a group of 6–10 participants based on their availability (Fig. 18.3a–c). 10 Key Informant Interviews (KIIs) were held with village heads and a few government officials (Fig. 18.3d). Furthermore, the primary information gathered on the people's perception of climatic variability and trends was later correlated with the meteorological data and satellite imagery analysis in order to verify whether perceived climate changes are actually taking place and to reappraise adaptive strategies of the agro-pastoral community in the study area.

18.3 Results and Discussion

18.3.1 Precipitation and Temperature Trends

The results of the trend analysis for precipitation and temperature (Table 18.1) show that in Darma valley, annual, pre-monsoon and monsoon precipitation have significantly declined as compared to the post-monsoon and winter precipitation which showed no significant trend (Fig. 18.4). These results are consistent with studies from other high altitude regions in Uttarakhand such as Milam glacier (Mal et al. 2019), and Bhagirathi and Saraswati-Alaknanda basins (Bhambri et al. 2011b). Also, the decreasing trend of monsoon precipitation and overall annual precipitation is also observed in other north-western Himalayan regions (Bhutiyani 2016).

The overall temperature trends in the study area are significantly increasing (Figs. 18.5, 18.6). The mean maximum temperature (in annual, pre-monsoon and post-monsoon) showed an increasing trend while no significant trend is observed in monsoon and winter months between 1975 and 2016. An increase in pre-monsoon maximum temperature in the study area is consistent with a significant high warming trend (0.086 °C/yr) observed in Darchula district of Nepal (DHM 2017) which lies in the closest proximity of the study area. Moreover, there is an overall seasonal and annual rise in the mean minimum temperature for the same time period. These results are consistent with the study conducted by Mishra (2014, 2017) in all 13 districts of Uttarakhand between 1911 and 2012, which showed an increase in annual mean temperatures

Fig. 18.3 **a** FGD with mixed group of respondents (household heads) in Dugtu village; **b** FGD in Sipu village; **c** FGD with male household heads in Philam village; and **d** KII with village heads in Dugtu village

in Pithoragarh, where the present study area lies. Considerable warming trend in mean annual temperature increase (between 1968 and 2016) is also observed in nearby Himalayan regions in India (Bhutiyani et al. 2007, 2009; Zomer et al. 2014; Mal et al. 2019; Shafiq et al. 2018) and in Nepal (Qi et al. 2013; Kattel and Yao 2013; DHM 2017). At higher altitudes, the temperature increase has changed precipitation patterns (from snow to rain), causing cascading effects on the snow cover and glacial mass and leading to recession of glaciers (Sharma et al. 2009; Zomer et al. 2016; Bhutiyani 2016; Bolch et al. 2019; Mal et al. 2019). Thus, such changes in temperature and precipitation patterns can have severe socio-economic implications on the life and livelihood of the local community.

18.3.2 Snow Cover Availability

In Darma valley, there has been a decline in the snow-covered area from 881.25 km^2 in 1990 to 664.41 km^2 in 2018 (March–April). When analyzing the values of the snow-covered areas in the valley, it is apparent that the share of land covered by snow slightly increased from 1990 to 2011 and then eventually decreased by 2018 (Fig. 18.7). This corresponds to a drop of 216.84 km^2 and a decline of 24.6% in total of the snow-covered area. For the representation of snow-covered areas (Fig. 18.8), only four Landsat imageries could be used due to the unavailability of cloud-free satellite imageries for the selected months (March and early April) in the selected time period. To substantiate our results, reference from other studies has been taken.

Table 18.1 Mann–Kendall test for precipitation and temperature trends (1975–2016)

Varible	Mean Kendall Statistic (S)	Kendall's Tau	Var (S)	*p*-value (two-tailed test)	Alpha	Sen's slope	Test interpretation	Trend
Precipitation								
Annual	−245	−0.285	8514.333	0.008	0.05	−4.505	Reject H0	Significant declining trend
Pre-monsoon	−118	−0.137	8518.333	0.021	0.05	−1.138	Reject H0	Significant declining trend
Monsoon	−181	8514.333	8514.333	0.049	0.05	−3.329	Reject H0	Significant declining trend
Post-monsoon	−81	−0.094	8514.333	0.388	0.05	−0.3	Accept H0	No significant trend
Winter	−68	−0.079	8513.333	0.461	0.05	−0.352	Accept H0	No significant trend
Minimum temperature								
Annual	599	0.697	8510.333	<0.0001	0.05	0.049	Reject H0	Significant declining trend
Pre-monsoon	433	0.505	8506.333	<0.0001	0.05	0.053	Reject H0	Significant declining trend
Monsoon	660	0.771	8504.667	<0.0001	0.05	0.047	Reject H0	Significant declining trend
Post-monsoon	435	0.508	8503.667	<0.0001	0.05	0.06	Reject H0	Significant declining trend
Winter	370	0.434	8492.667	<0.0001	0.05	0.04	Reject H0	Significant declining trend
Maximum temperature								
Annual	257	0.3	8507.667	0.005	0.05	0.02	Reject H0	Significant declining trend
Pre-monsoon	239	0.28	8501	0.01	0.05	0.035	Reject H0	Significant declining trend
Monsoon	138	0.161	8504.667	0.135	0.05	0.01	Accept H0	No significant trend
Post-monsoon	184	0.215	8504.667	0.046	0.05	0.021	Reject H0	Significant declining trend
Winter	169	0.198	8497	0.067	0.05	0.025	Accept H0	No significant trend

Significance level (%): 5; Confidence level (%): 95
Source Own study

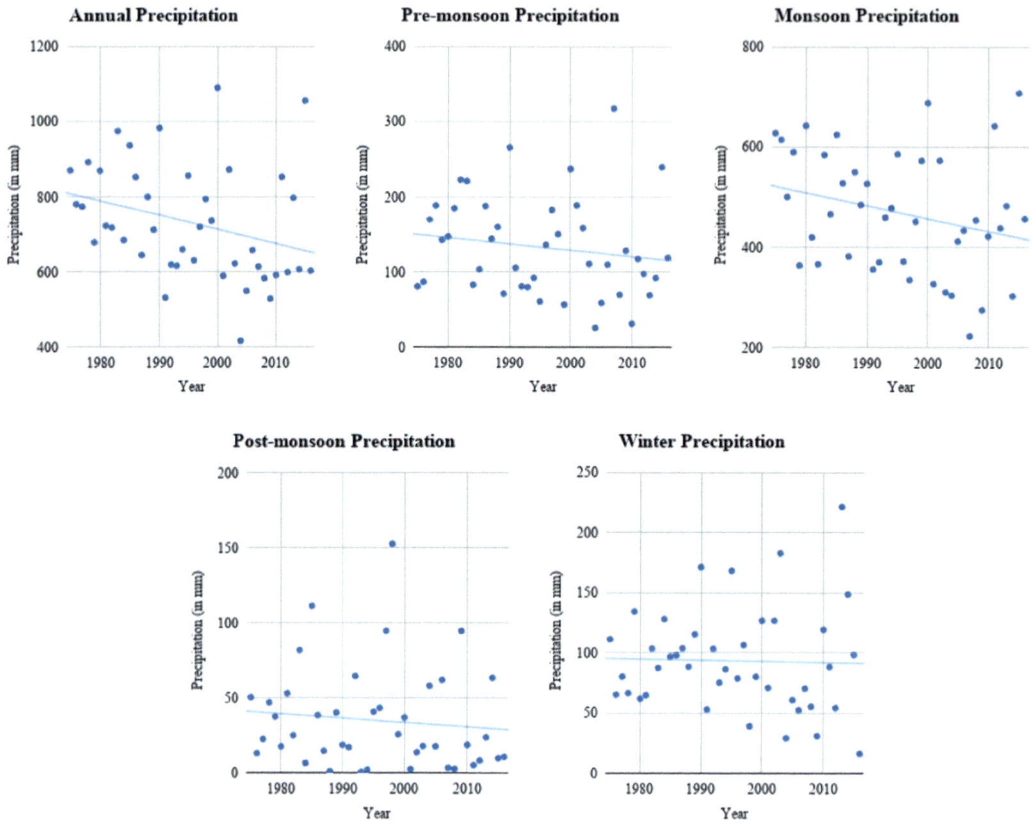

Fig. 18.4 Precipitation trends in Darma valley (1975–2016)

In our analysis, although winter precipitation shows no significant trend, the NDSI analysis shows a considerable decline in the snow-covered area. This can be explained by reinforced warming and melting processes. Studies on glacier flow and recession (Sharma et al. 2009; Zomer et al. 2016; Bhutiyani 2016; Bolch et al. 2019; Mal et al. 2019) reported change of snow to rain due to warming temperatures in the high altitude Himalayan regions, causing reduced snow cover, melting of snow, and loss of glacial mass. Bhutiyani et al. (2009) and Bhutiyani (2016) reported reduced snowfall duration and less snow cover due to increased warming in pre- and post-monsoon seasons in higher altitudes of western Himalaya. A study conducted by Mal et al. (2019) on Milam glacier in the Gori Ganga valley, Pithoragarh district, Uttarakhand, which is close to the present study area, observed a slower recession rate (21.1 ± 1.7 m a-1) between the time period of 2001 and 2017. Over time, other glaciers in Uttarakhand have also shown recessions, such as Satopanth (Nainwal et al. 2016) Bhagirathi, and Gangotri glaciers (Bhambri et al. 2011a, b, 2012).

18.3.3 People's Perceptions Towards Climate Variable

The majority of the respondents perceived changes in precipitation, snowfall, and temperature (Table 18.2). 82% of the respondents perceived increasing change in annual mean temperature whereas 78 and 85% of the respondents perceived increasing changes in average temperature in summer and winter months, respectively. This corresponds to the assessed overall rise in temperature in the study area. Moreover, the perception of respondents about

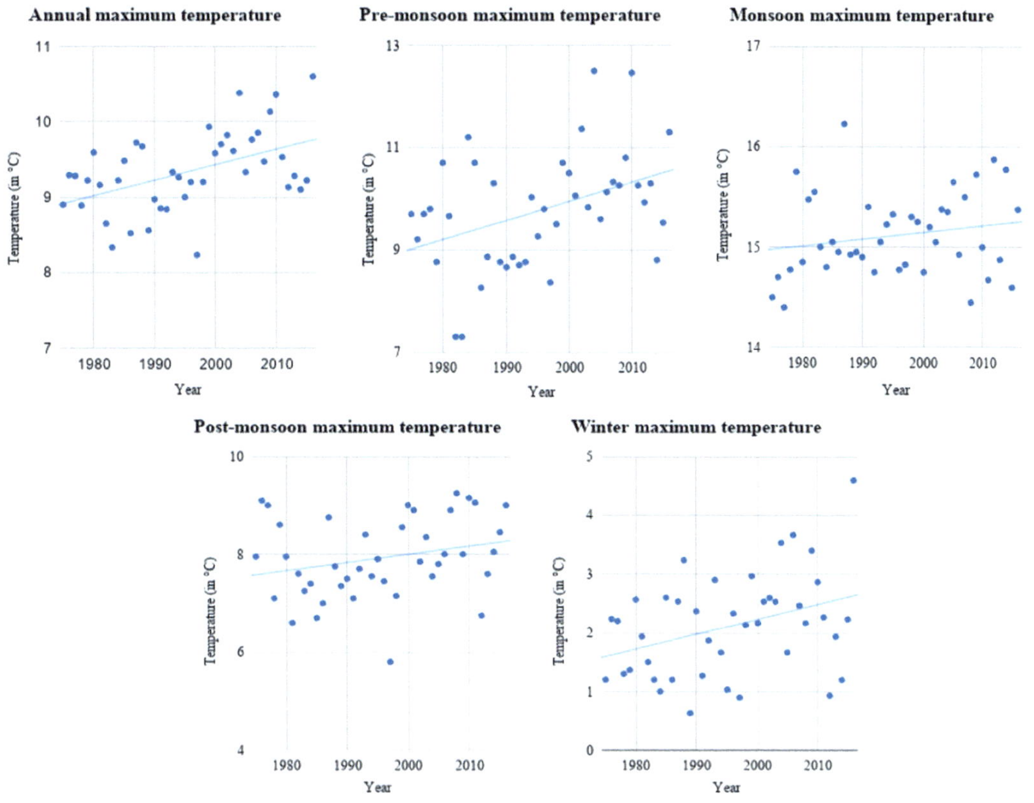

Fig. 18.5 Maximum temperature trends in Darma valley (1975–2016)

precipitation patterns was also in line with the actual trends except for winter months. The meteorological evidence shows that there has been no significant trend in the pattern of winter precipitation over the study period whereas 64% of respondents perceived a decline in precipitation. 69% of the respondents perceived a decrease in overall precipitation while 50.5% of the respondents perceived a change in precipitation in monsoon months. These observations of the respondents about change in temperature and precipitation were reconcilable with the actual observed trends. In addition, the respondents also perceived a declining trend in snowfall availability or occurrence.

Furthermore, various impacts of climate change as perceived by the respondents are depicted in Figs. 18.9 and 18.10. For the majority of respondents, the perceived changes in climatic variables are expressed mainly in the pattern of higher temperatures in summer months, warmer temperatures in winter months, less amount of snow in winter months, unpredictable rainfall patterns, rainfall scarcity, less number of rainy days, increased heat stress and higher intensity of extreme events like flash-floods, landslides, etc. An overall decline in precipitation has been perceived by respondent.

Additionally, during questionnaire survey, FGDs and KIIs respondents reported other perceived climate change impacts such as drying up of water sources, presence of invasive species in pasturelands, lesser availability of wild medicinal plants, increase in livestock diseases (foot-and-mouth disease), change in flowering and maturing season, decline in crop productivity, rapid melting of snow in the pasturelands, less nutritious forage, encroachment of shrub species, etc. As mentioned by the respondents, these changes have affected the socio-economic aspects of the pastoral communities. Implications include a decline in livestock production, diminishing

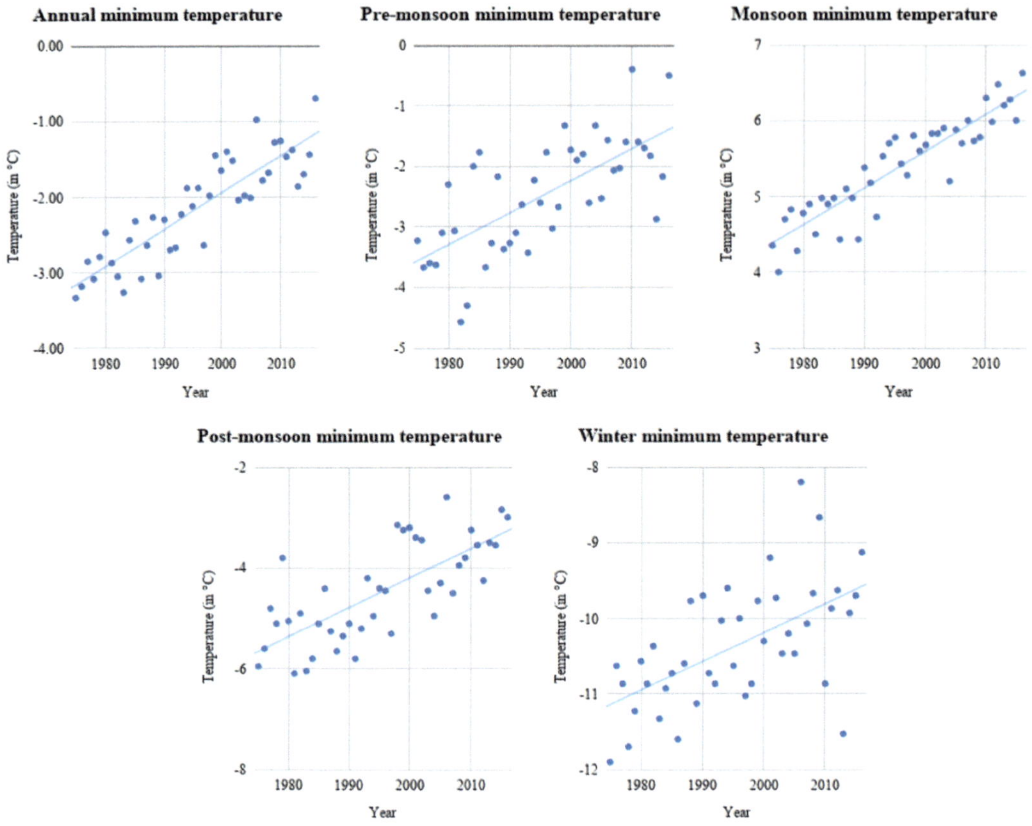

Fig. 18.6 Minimum temperature trends in Darma valley (1975–2016)

Fig. 18.7 Trend in the snow-covered area in the study area between 1990 and 2018

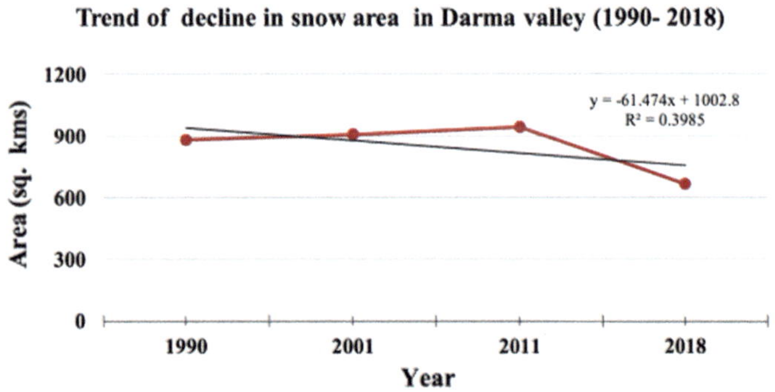

prices of pasture products, financial uncertainty, land abandonment due to low agricultural productivity, food insecurity, change in dietary habits, high dependence on the market for food and fodder, etc. Environmental and socio-economic changes have consequently resulted in a decline in the overall practice of transhumance while also leading to the male-dominated rural exodus and increased pressure on women, and social tensions within the community.

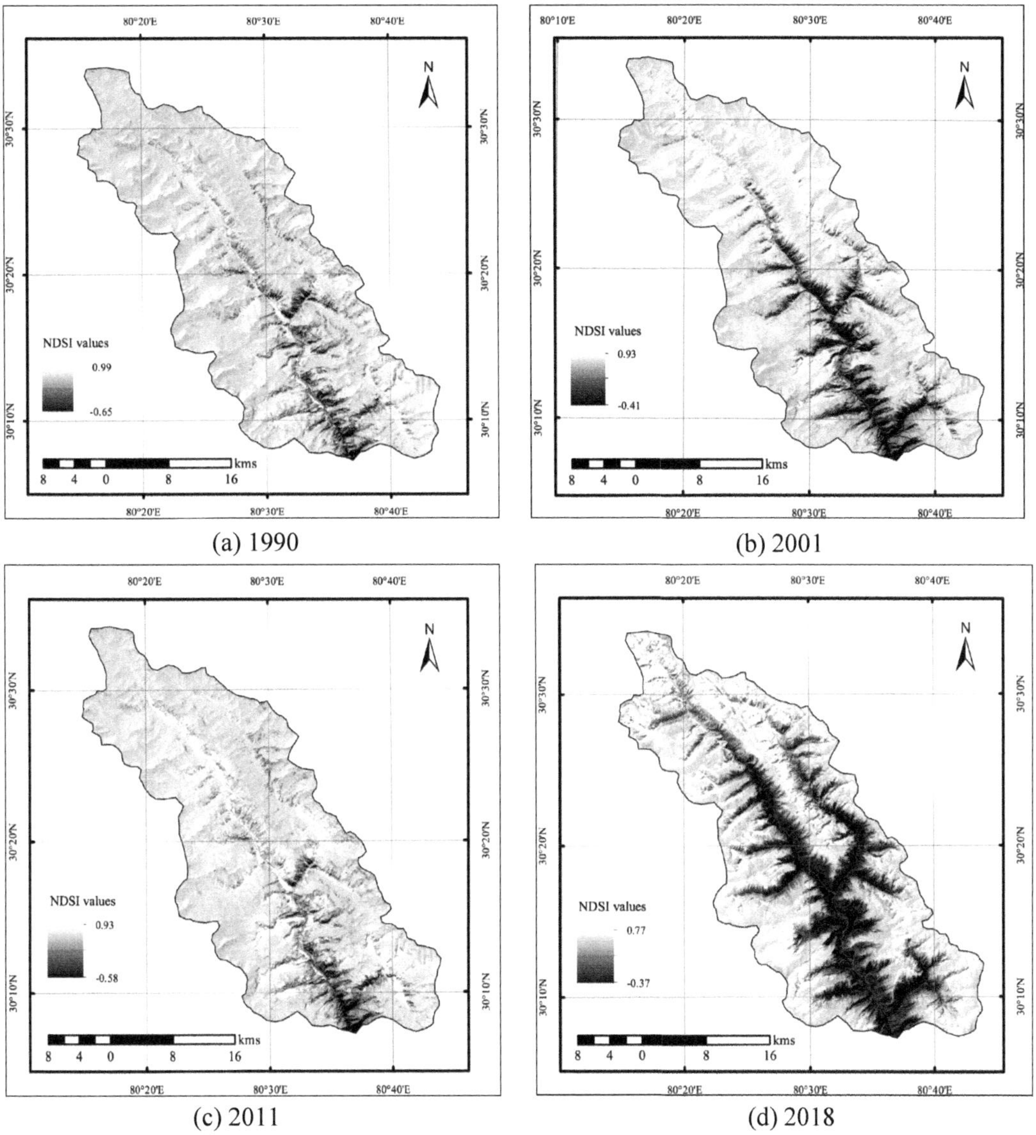

Fig. 18.8 NDSI snow cover maps for the years 1990, 2001, 2011 and 2018

18.3.4 Comparison of Observed Trends and People's Perceptions

The results show that the majority of respondents have experienced changes in climate in Darma valley. Their perceptions in relation to the pre-monsoon and winter (maximum and minimum) temperature correspond to the assessed changes in temperatures based on the CRU TS meteorological dataset. Perceived changes in precipitation for the winter months, however, could not be substantiated by the meteorological evidence which does not show a significant trend in the pattern of winter precipitation over the study period. However, people's response to adverse changes in precipitation patterns has affected the cropping patterns of the transhumant community

Table 18.2 Perception of transhumant pastoralists towards climatic variables

	Variable	Increasing	Decreasing	Unpredictable	No change	Don't know
1	Annual mean temperature	164 (82)	3 (1.5)	27 (13.5)	3 (1.5)	3 (1.5)
2	Temperature in summer (pre-monsoon) months	156 (78)	13 (6.5)	29 (14.5)	0 (0)	2 (1)
3	Temperature in winter months	170 (85)	2 (1)	17 (8.5)	7 (3.5)	4 (2)
4	Annual precipitation	3 (1.5)	138 (69)	55 (27.5)	4 (2)	0 (0)
5	Monsoon precipitation	9 (4.5)	101 (50.5)	82 (41)	2 (1)	6 (3)
6	Winter precipitation	5 (2.5)	128 (64)	56 (28)	3 (1.5)	8 (4)
7	Snowfall availability/occurrence	3 (1.5)	154 (77)	37 (18.5)	3 (1.5)	3 (1.5)

Note $n = 200$; the brackets indicate percentage of respondents

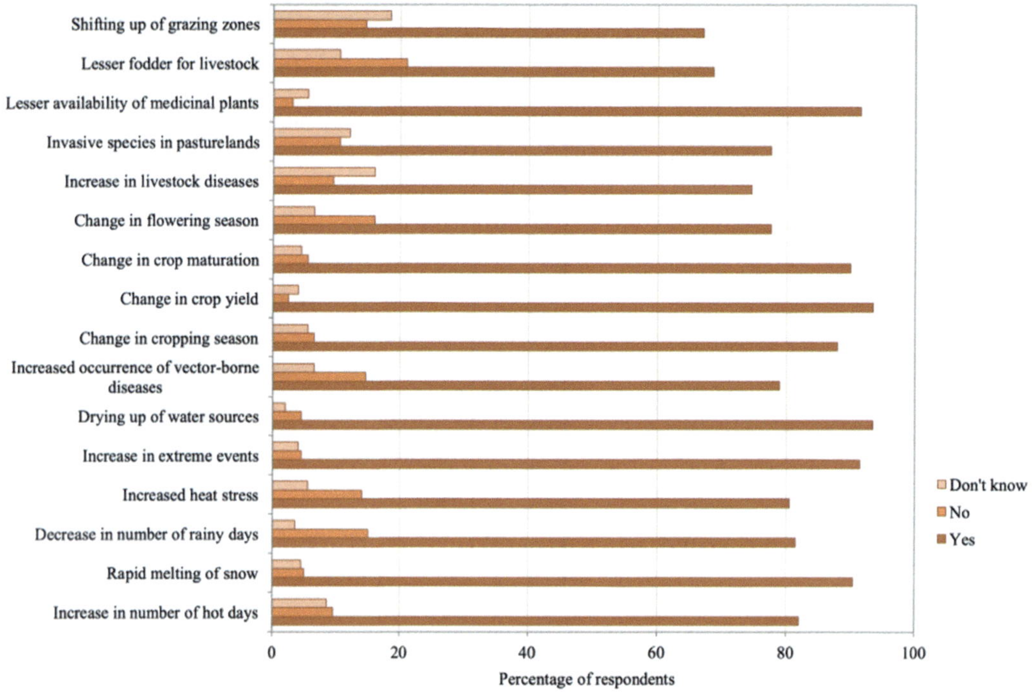

Fig. 18.9 Transhumant pastoralists' level of agreement towards different statements

in the study area. Major changes in cropping patterns include changed timing of agronomic practices (sowing, harvesting, irrigation frequency, etc.), use of HYV seeds and cultivating less water-intensive crops (beans, potatoes, colocasia, cabbage, carrots, maize, etc.) to cope up with ongoing changes in climate. More recently, there has been a shift to horticulture (apple farming), supported by incentives given by the government. Similar kind of alterations in cropping patterns have also been observed in similar to other regions in Uttarakhand (Sati 2015; Rautela and Karki 2015; Shukla et al. 2019) and other Himalayan states of Jammu and Kashmir (Batool et al. 2019) and Himachal Pradesh (Basannagari and Kala 2013; Ndungu

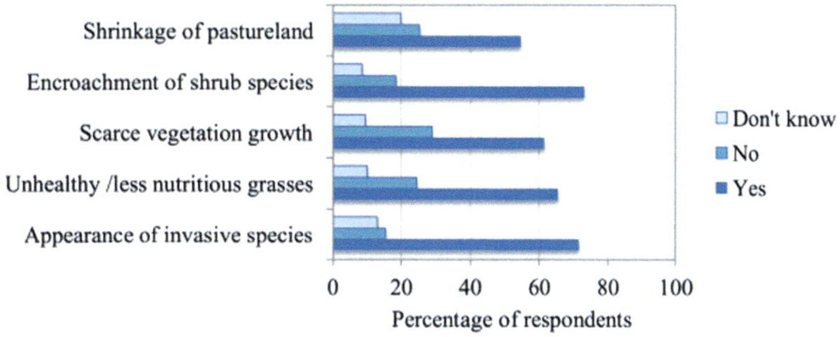

Fig. 18.10 Transhumant pastoralists' level of agreement towards pasture productivity

and Bhardwaj 2015; Loria and Bhardwaj 2016). The observed results regarding overall increase in mean annual temperature and warmer winters are consistent with recent studies in Kailash Sacred Landscape region of India by Zomer et al. (2014), North-west Himalayan region over the last century (Bhutiyani et al. 2007, 2009; Bhutiyani 2016), and upto 0.6 °C/decade between 1980 and 2009 in western Nepal regions (Kattel and Yao 2013). Additionally, the warming trend of temperature and declining trend of precipitation in the study area is found to be consistent with previous studies in Uttarakhand (Bhutiyani et al. 2007; Bhutiyani et al. 2009; Mishra 2014; Dobhal et al. 2015; Mishra 2017; Mal et al. 2019), in Kashmir valley, Jammu and Kashmir (Shafiq et al. 2018) and Nepal (Qi et al. 2013; Kattel and Yao 2013; DHM 2017). Another study by Negi et al. (2012) in Uttarakhand reveals that precipitation and temperature trends affect the discharge, volume, and availability of water which in turn affects the farmer's communities. In Ladakh, agriculture is being affected due to reduced snow and ice as farmers are experiencing an acute shortage of water for irrigating the crops (Clouse et al. 2017). In the study area, increased incidences of extreme events (landslides, flash-floods, forest fires, etc.) were also perceived by the respondents. The 2013 flood in Uttarakhand led to wide-scale misery and havoc in the lives of people all across the state, killing more than 5000 people (Rautela 2013; Awasthi et al. 2014), including people in the study area.

18.4 Conclusion

The adverse changes in climatic variables, especially temperature, precipitation and snow cover in Himalayan regions have altered the socio-economic dynamics of the marginalized local communities. At the same time, ongoing globalization, rapid urbanization, low levels of economic development, and other social transition processes have significant implications on the lives and livelihoods of mountain communities. The combined effects of socio-economic change and changing precipitation patterns, rise in temperatures and less snowfall availability have resulted in water scarcity, crop loss, less nutritive pastures, fodder scarcity, change in cropping patterns, emergence of insects and pests, and low livestock production. This has further led to economic uncertainties, food insecurities, abandonment of land holdings, accelerated outmigration and societal conflicts over resources, affecting various socio-economic aspects of mountain pastoral communities.

Across Himalayan regions, mountain livelihoods are constantly adapting to change, diversifying and evolving. People are combining farm with non-farm activities (such as daily wage labour, tourism services and labor migration) to gain more economic benefits (Gioli et al. 2019) and further to adapt to socio-economic and environmental changes being experienced by them. Over time, environmental and socio-economic changes (a particular stressor was the

Indo-China war of 1962) have resulted in a decline of traditional practices of transhumance and sedentary agriculture among the pastoral community in Darma valley. People have diversified their livelihoods by engaging in government wage labor schemes, tourism services (trekking and homestay), transport business, and collection and selling of medicinal plants to name a few. Considering the potential impacts of climate change on mountain livelihoods, there is an urgent need to mitigate and adapt to these environmental changes. Local community's perception of climate change can be crucial for policy makers to better plan and implement mitigation and adaptation strategies at regional, national and global levels as perception-based studies can substantially complement the observed climate change evidences for remote data-deficient locations like Darma valley in Uttarakhand. Also, in the face of the current climate change scenario, building the capacity for the affected pastoral communities to adapt and strengthen the socio-ecological system through effective sustainable adaptation measures demands attention. However, whether these benefits are realized depends on how well these mitigation and adaptation measures are proactively adopted, implemented and managed.

Acknowledgements The authors are thankful to Dr. Suraj Mal for his valuable suggestions and inputs. We are also thankful to Antonia Gollasch for extending her support with satellite data. Our heartiest gratitude lies with the people of Darma valley who provided their great support during field data collection.

References

Adger N, Dessai S, Goulden M, Hulme M, Lorenzoni I, Nelson R, Naess O, Wolf J, Wreford A (2009) Are there social limits to adaptation to climate change? Clim Change 93:335–354

Afzal J, Ahmed M, Begum I (2008) Vision for development of rangelands in Pakistan: a policy perspective. Q Sci Vis 14(1):53–58

Aryal S, Cockfield G, Masareni TK (2014) Vulnerability of Himalayan transhumant communities to climate change. Clim Change 125:193–208

Aryal S, Cockfield G, Maraseni TN (2016) Perceived changes in climatic variables and impacts on the transhumance system in the Himalayas. Clim Dev 8 (5):435–446

Awasthi IC, Mehta GS, Mamgain RP (2014) Uttarakhand disaster: lessons and way forward, Occasional-6. Gri Institute of Development Studies, Lucknow, India

Banerjee S (2009) Shift from transhumance and subtle livelihood patterns of the Bhotia community and its impact on Tibetan sheep population in Sikkim (India). World Appl Sci J 7(12):1540–1546

Banerjee S, Gerlitz JY, Hoermann B (2011) Labour migration as a response strategy to water hazards in the Hindu Kush Himalayas. ICIMOD, Kathmandu

Basannagari B, Kala CP (2013) Climate change and apple farming in Indian Himalayas: a study of local perceptions and responses. PLoS One 8(10):e77976

Batool N, Shah SA, Dar SN, Skinder S (2019) Rainfall variability and dynamics of cropping pattern in Kashmir Himalayas: A case study of climate change and agriculture. SN Appl Sci 1:1–9

Bergmann C (2016) The Himalayan border region: trade, identity and mobility in Kumaon, India. In: Advances in Asian human-environment research. Switzerland, Springer, Cham

Bergmann C, Gerwin M, Nüsser M, Sax WS (2008) Living in a high mountain border region. The case of the 'Bhotiyas' of the Indo-Chinese border region. J Mt Sci 5(3):209–217

Bergmann C, Gerwin M, Sax WS, Nüsser M (2011) Politics of scale in a high mountain border region: being mobile among the Bhotiyas of the Kumaon Himalaya, India. Nomadic Peoples 15:104–129

Berhanu W, Beyene F (2015) Climate variability and household adaptation strategies in Southern Ethiopia. Sustainability 7:6353–6375

Bhambri R, Bolch T, Chaujar RK (2011a) Mapping of debris-covered glaciers in the Garhwal Himalayas using ASTER DEMs and thermal data. Int J Remote Sens 32:8095–8119

Bhambri R, Bolch T, Chaujar RK, Kulshreshtha SC (2011b) Glacier changes in the Garhwal Himalaya, India, from 1968 to 2006 based on remote sensing. J Glaciol 57:543–556

Bhambri R, Bolch T, Chaujar RK (2012) Frontal recession of Gangotri Glacier, Garhwal Himalayas, from 1965 to 2006, measured through high resolution remote sensing data. Curr Sci 102:489–494

Bhasin V (2011) Pastoralists of Himalayas. J Human Ecol 33(3):147–177

Bhutiyani MR (2016) Spatial and temporal variability of climate change in high-altitude regions of NW Himalaya. In: Singh RB, Schickhoff U, Mal S (eds) Climate change, glacier response, and vegetation dynamics in the Himalaya. Springer, Cham, pp 87–101

Bhutiyani MR, Kale VS, Pawar NJ (2007) Long-term trends in maximum, minimum and mean annual air temperatures across the Northwestern Himalaya during the twentieth century. Clim Change 85:159–177

Bhutiyani MR, Kale VS, Pawar NJ (2009) Climate change and precipitation variations in the Northwestern Himalaya: 1866–2006. Int J Climatol 30:535–548

Bolch T, Shea JM, Liu S, Azam FM, Gao Y, Gruber S et al (2019) Status and change of the cryosphere in the Extended Hindu Kush Himalaya Region. In: Wester P, Mishra A, Mukherji A, Shrestha AB (eds) The Hindu Kush Himalaya assessment: mountains, climate change, sustainability and people. Springer, Cham, pp 209–255

Chaudhary P, Bawa KS (2011) Local perceptions of climate change validated by scientific evidence in the Himalayas. Biol Lett 7:767–770

Chaudhary P, Rai S, Wangdi S, Mao A, Rehman N, Chettri S, Bawa KS (2011) Consistency of local perceptions of climate change in the Kangchenjunga Himalaya landscape. Curr Sci 101(4, 25):504–513

Clouse C, Anderson N, Shippling T (2017) Ladakh's artificial glaciers: climate-adaptive design for water scarcity. Clim Dev 9(5):428–438

DHM (Department of Hydrology and Meteorology) (2017) Observed climate trend analysis in the districts and physiographic regions of Nepal (1971–2014), Government of Nepal, Kathmandu

District Census Handbook of Pithoragarh (2011) Village and town wise primary census abstract (PCA). Series 6, Part XII-B, Census of India, Uttarakhand

Dobhal DP, Pratap B (2015) Variable response of glaciers to climate change in Uttarakhand Himalaya, India. In: Joshi R, Kumar K, Palni LMS (eds) Dynamics of climate change and water resources of Northwestern Himalaya, Springer, Cham, pp 141–150

Dong S, Wen L, Zhu L, Li X (2010) Implication of coupled natural and human systems in sustainable rangeland ecosystem management in HKH region. Front Earth Sci China 4(1):42–50

Ehlers E, Kreutzmann H (2000) High mountain ecology and economy: potential and constraints. In: Ehlers E, Kreutzmann H (eds) High mountain pastoralism in Northern Pakistan. Erdkundliches Wissen 132:9–36

Eriksson M, Jianchu X, Shrestha AB, Vaidya RA, Nepal S, Sandström K (2009) The changing Himalayas: impact of climate change on water resources and livelihoods in the Greater Himalayas. ICIMOD, Kathmandu, Nepal

Gentle P, Maraseni TN (2012) Climate change, poverty, and livelihoods: adaptation practices by rural mountain communities in Nepal. Environ Sci Policy 21:24–34

Gentle P, Thwaites R (2016) Transhumant pastoralism in the context of socioeconomic and climate change in the mountains of Nepal. Mt Res Dev 36(2):173–182

Gioli G, Thapa G, Khan F, Dasgupta P, Nathan D, Chhetri N, Adhikari L, Mohanty S, Aurino E, Scott L (2019) Understanding and tackling poverty and vulnerability in mountain livelihoods in the Hindu Kush Himalaya. In: Wester P, Mishra A, Mukherji A, Shrestha AB (eds) The Hindu Kush Himalaya assessment. Springer, Cham, pp 421–425

Hall DK, Riggs GA, Salomonson VV (1995) Development of methods for mapping global snow cover using moderate resolution imaging spectroradiometer data. Remote Sens Environ 54:27–140

Hoermann B, Banerjee S, Kollmair M (2010) Labour migration as a responses strategy to water hazards in the Hindu Kush Himalayas—understanding a livelihood strategy in the context of socioeconomic and environmental change. ICIMOD, Kathmandu, Nepal

Hoermann B, Kollmair M (2009) Labour migration and remittances in the Hindu Kush-Himalayan region. ICIMOD, Kathmandu

Intigrinova T (2005) Transhumance in transition: consequences of socio-economic reform. A case study of Khoito Gol, Buryatia. Inner Asia 7(1):87–105

IPCC (2014) Impacts, adaptation, and vulnerability, part B: regional aspects. In: Contribution of working group II to the fifth assessment report of the intergovernmental panel on climate change. Cambridge University Press, Cambridge, United Kingdom, New York

IPCC (ed) (2018) Global warming of 1.5 °C. An IPCC special report. IPCC, Geneva

Joshi S, Jasra WA, Ismail M (2013a) Herders' perceptions of and responses to climate change in Northern Pakistan. Environ Manage 52:639–648

Joshi L, Shrestha RM, Jasra AW (2013b) Rangeland ecosystem services in the Hindu Kush Himalayan Region. In: Wu N, Rawat GS, Joshin S (eds) High-altitude Rangelands and their Interfaces in the Hindu Kush Himalayas, Kathmandu, ICIMOD, pp 157–175

Kattel DB, Yao Y (2013) Recent temperature trends at mountain stations on the southern slope of the central Himalayas. J Earth Syst Sci 122:215–227

Kelkar U, Narula K, Sharma V, Chandna U (2008) Vulnerability and adaptation to climate variability and water stress in Uttarakhand State, India. Global Environ Change 18:564–574

Kerven C, Steimann B, Dear C, Ashley L (2012) Researching the future of pastoralism in central Asia's mountains: examining development orthodoxies. Mt Res Dev 32(3):368–377

Loria N, Bhardwaj SK (2016) Farmers' response and adaptation strategies to climate change in low-hills of Himachal Pradesh in India. Nat Environ Pollut Technol 15(3):895–901

Luo T, Pan Y, Ouyang H, Shi P, Luo J, Yu Z, Lu Q (2004) Leaf area index and net primary productivity along subtropical to alpine gradients in the Tibetan Plateau. Glob Ecol Biogeogr 13:345–358

Macchi M, Gurung AM, Hoermann B (2015) Community perceptions and responses to climate variability and change in the Himalayas. Clim Dev 7(5):414–425

Mader TL (2003) Environmental stress in confined beef cattle. J Anim Sci 81(2):110–119

Maikhuri RK, Rao KS, Semwal RL (2001) Changing scenario of Himalayan agroecosystems: loss of agrobiodiversity, an indicator of environmental change in Central Himalaya, India. Environmentalist 21(1): 23–39

Mal S, Singh RB (2014) Changing glacial lakes and associated outburst floods risks in Nanda Devi Biosphere Reserve, Indian Himalaya: Proc. IAHS 364: 255–260

Mal S, Mehta M, Singh RB, Schickhoff U, Bisht MPS (2019) Recession and morphological changes of the debris-covered Milam glacier in Gori Ganga Valley, Central Himalaya, India, derived from satellite data. Front Environ Sci 7:42
Mishra A (2014) Changing climate of Uttarakhand, India. J Geol Geosci 3(4):163
Mishra A (2017) Changing temperature and rainfall patterns of Uttarakhand. Int J Environ Sci Nat Resour 7(4):90–95
Mueller V, Gray C, Kosec K (2014) Heat stress increases long-term human migration in rural Pakistan. Nat Clim Change 4:182–185
Nainwal HC, Banerjee A, Shankar R, Semwal P, Sharma T (2016) Shrinkage of Satopanth and Bhagirath Kharak glaciers, India, from 1936 to 2013. Ann Glaciol 57:131–139
Namgay K, Millar J, Black R, Samdup T (2014) Changes in Transhumant Agropastoralism in Bhutan: a disappearing livelihood? Human Ecol 42:779–792
Ndungu C, Bhardwaj SK (2015) Assessment of people's perceptions and adaptations to climate change and variability in mid-hills of Himachal Pradesh, India. Int J Curr Microbiol Appl Sci 4:47–60
Negi CS (2007) Declining transhumance and subtle changes in livelihood patterns and biodiversity in the Kumaon Himalaya. Mt Res Dev 27(2):114–118
Negi G, Samal P, Kuniyal J, Kothyari B, Sharma R, Dhyani P (2012) Impact of climate change on the western Himalayan mountain ecosystems: an overview. Tropical Ecol 53(3):345–356
Negi VS, Maikhuri RK, Pharswan D (2017) Climate change in the Western Himalaya: People's perception and adaptive strategies. J Mt Sci 14(2):403–416
Nüsser M (2006) Ressourcennutzung und nachhaltige Entwicklungim Kumaon-Himalaya (Indien). Geographische Rundschau 58:14–22
O'Brien KL, Leichenko RM (2000) Double exposure: assessing the impacts of climate change within the context of economic globalization. Global Environ Change 10:221–232
Ogra MV, Badola R (2015) Gender and climate change in the Indian Himalayas: global threats, local vulnerabilities, and livelihood diversification at the Nanda Devi Biosphere Reserve. Earth Syst Dyn 6:505–523
Qi W, Zhang Y, Gao J, Yang X, Liu L, Khanal NR (2013) Climate change on the southern slope of Mt. Qomolangma (Everest) Region in Nepal since 1971. J Geog Sci 23:595–611
Rautela P (2013) Lessons learnt from the deluge of Kedarnath, Uttarakhand, India. Asian J Environ Disaster Manage 5(2):43–51
Rautela P, Karki B (2015) Impact of climate change on life and livelihood of indigenous people of higher Himalayan Uttarakhand, India. Am J Environ Protect 3(4):112–124
Ren YY, Ren GY, Sun XB, Shrestha AB, You QL, Zhan YJ (2017) Observed changes in surface air temperature and precipitation in the Hindu Kush Himalayan region during 1901–2014. Adv Clim Change Res 8(3):148–156
Sati VP (2015) Climate change and socio-ecological transformation in high mountains: an empirical study of Garhwal Himalaya. Change Adapt Socio Ecol Syst 2:45–56
Schickhoff U, Mal S (2020) Current changes in alpine ecosystems of Asia. In: Goldstein MI, DellaSala DA (eds) Encyclopedia of the world's biomes, Vol. 1. Elsevier, Amsterdam, pp 589–598
Schickhoff U, Bobrowski M, Böhner J, Bürzle B, Chaudhary RP, Gerlitz L, Heyken H, Lange J, Müller M, Scholten T, Schwab N, Wedegärtner R (2015) Do Himalayan treelines respond to recent climate change? An evaluation of sensitivity indicators. Earth Syst Dyn 6:245–265
Shafiq MU, Rasool R, Ahmed P, Dimri AP (2018) Temperature and precipitation trends in Kashmir valley, North Western Himalayas. Theoret Appl Climatol 135:293–304
Sharma E, Chettri N, Tse-ring K, Shrestha AB, Jing F, Mool P, Eriksson M (2009) Climate change impacts and vulnerability in the Eastern Himalayas. ICIMOD, Kathmandu, Nepal
Shrestha UB, Gautam S, Bawa KS (2012) Widespread climate change in the Himalayas and associated changes in local ecosystems. PLoS One 7(5):e36741
Shukla R, Agarwal A, Gornott C, Sachdeva K, Joshi P (2019) Farmer typology to understand differentiated climate change adaptation in Himalaya. Sci Rep 9:20375
Siddiqui T, Bhagat RB, Banerjee S, Liu C, Sijapati B, Memon R, Thinley P, Ito M, Nemat O, Arif G (2019) Migration in the Hindu Kush Himalaya: Drivers, consequences, and governance. In: Wester P, Mishra A, Mukherji A, Shrestha AB (eds) The Hindu Kush Himalaya assessment. Springer, Cham, pp 517–544
Sujakhu NM, Ranjitkar S, Niraula RR, Pokharel BK, Schmidt-Vogt D, Xu J (2016) Farmers' perceptions of and adaptations to changing climate in the Melamchi Valley of Nepal. Mt Res Dev 36(1):15–30
Tiwari PC, Joshi B (2016) Gender processes in rural out-migration and socio-economic development in the Himalaya. Migr Dev 5(2):330–350
Vasquez EA, James JJ, Monaco TA, Cummings DC (2010) Invasive plants on rangelands: a global threat. Rangelands 32:3–5
Vedwan N, Rhoades RE (2001) Climate change in the Western Himalayas of India: a study of local perception and response. Clim Res 19:109–117
Vidya ARK, Mahajan PK, Negi YS, Bhardwaj SK (2015) Trend analysis of weather parameters and people perception in Kullu district of western Himalayan region. Environ Ecol Res 3(1):24–33
Weber EU (2010) What shapes perceptions of climate change? Wiley Interdisc Rev Clim Change 1:332–342
Wilkes A (2008) Towards mainstreaming climate change in grassland management policies and practices on the

Tibetan Plateau. Working paper No. 67. World Agroforestry Centre, ICRAF-China, Beijing

Williams CA, Albertson JD (2006) Dynamical effects of the statistical structure of annual rainfall on dryland vegetation. Glob Change Biol 12(5):777–792

Wu N, Yan Z (2002) Climate variability and social vulnerability on the Tibetan Plateau. Dilemma on the road of pastoral reform. Erdkunde 56:2–14

Wu N, Ismail. M., Joshi S, Yi SL, Shreshtha RM, Jasra AW (2014) Livelihood diversification as an adaptation approach to change in the pastoral Hindu-Kush Himalayan region. J Mt Sci 11(5):1342–1355

Xu J, Grumbine RE, Shrestha A (2009) The melting Himalayas: cascading effects of climate change on water, biodiversity, and livelihoods. Conserv Biol 23:520–530

Yi SL, Ning W, Peng L, Qian W, Fusun S, Geng S et al (2007) Changes in livestock migration patterns in a Tibetan-style agro-pastoral system. Mt Res Dev 27 (2):138–145

Zomer RJ, Trabucco A, Metzger MJ, Wang M, Oli KP, Xu J (2014) Projected climate change impacts on spatial distribution of bioclimatic zones and ecoregions within the Kailash sacred landscape of China, India. Nepal. Clim Change 125(3–4):445–460

Zomer RJ, Wang M, Trabucco A, Xu JC (2016) Projected climate change impact on hydrology, bioclimatic conditions and terrestrial ecosystems in the Asian highlands. (ICRAF Working Paper 222). World Agroforestry Centre East and Central Asia, Kunming, China

Current Crisis and Future Woes: The Case of Climate Change in the Drakensberg Mountains Region of Southern Africa and Its Socio-economic Impacts in the Region

19

Geoffrey Mukwada

Abstract

The Drakensberg, the largest mountain range of southern Africa, is currently facing a crisis from its changing climate, with the certainty of future woes looming. This paper investigates how the climate of the Drakensberg mountain region has changed between 1960 and 2016 and the implications that this change has for the social-ecological systems of the region and southern Africa as a whole. The paper uses temperature and precipitation data to conduct time series and spatial analyses and assesses how maximum temperature and precipitation varied during this period. Anomalies for maximum temperature and standardized precipitation index (SPI) were calculated and used to evaluate the changes that have occurred to these variables, as well as implications for local habitats, water supply and the socio-economic status of southern Africa. The results indicate that between 1960 and 2016 the average maximum temperatures of the Drakensberg mountains increased by more than 2 °C, while changes in precipitation were statistically insignificant ($p = 0.899$). However, the frequency of extreme and severe droughts has increased. As a conclusion, the paper recommends the development of efficient climate adaptation strategies by policy makers, to ensure that the current crisis and imminent woes are averted.

Keywords

Drakensberg mountains · Drought · Ecosystem services · Social-ecological systems · Standardized precipitation index

19.1 Introduction

African mountains are under threat from climate change. Their vulnerability is influenced by altitude. Climate modelling has shown that climate change will affect the production of climate sensitive crops in mountain regions in some parts of southern Africa. For instance, the climatic suitability of coffee in Eastern Highlands of Zimbabwe, especially Chipinge, Chimanimani and Mutare districts, is likely to decrease (Chemura et al. 2016). Similarly, the effects of climate change have been equally conspicuous in the East African Mountains, particularly Mt Meru and Mt Kilimanjaro (Hemp 2005, 2009; Munish and Sawere 2014). Hemp (2009) reported of varnishing glaciers and loss of forest cover on Mt Kilimanjaro. Compared to the glacier cover in 1912, Mt Kilimanjaro lost 82% of its ice cap in 92 years (Hemp 2005). Kaser et al. (2004)

G. Mukwada (✉)
Afromontane Research Unit, Department of Geography, University of the Free State, Bloemfontein, South Africa

U. Schickhoff et al. (eds.), *Mountain Landscapes in Transition*,
Sustainable Development Goals Series, https://doi.org/10.1007/978-3-030-70238-0_19

attribute this glacial retreat to a complex combination of changes in air temperature, air humidity, precipitation, cloudiness and incoming shortwave radiation, rather than warming alone. Mt Kilimanjaro has also lost a third of its forest cover, partly due to climate induced fires (Hemp 2009). While most of the fires in this mountain area have started by humans, their effect would be less devastating if the climate had not become drier (Hemp 2009). In East African Mountains, despite an increase in rainfall that has been reported in some catchments (Munishi and Sawere 2014), climate change has altered hydrologic systems and worsened water and food insecurity in local communities. Overall, there has been a decrease in regional precipitation and an increase in drought frequency in these mountain areas (Munishi and Sawere 2014). In Mt Kilimanjaro, climate change has even been reported to have triggered speciation in Orthoptera species (Voje et al. 2009). Research in the Atlas Mountains suggests that climate change could lead to a temporal shift in both the vegetation cover and rainfall patterns (Simonneaux et al. 2015). Anticipated heavier rainfall on poorly protected soils, resulting from intensification of human activity and a shift in the vegetation patterns, will contribute to large increases in rates of soil erosion (Simonneaux et al. 2015). The vulnerability of mountains to climate change arises from the fragile nature and high biodiversity value of mountain biota (Carbutt and Edwards 2015).

The objective of this paper is to assess climate change in the Drakensberg mountains of southern Africa and evaluate the impact of this change on social-ecological systems of the communities living in these mountains, as well as the implications for livelihoods in southern Africa as a whole. Using Climate Research Unit Time Series 4.01 monthly observation data for the Drakensberg mountains, the paper assesses how climate change has affected precipitation and temperature trends, as well as the anomalies of these variables between 1960 and 2016. The Drakensberg mountains sprawl over Lesotho and the eastern part of South Africa (Fig. 19.1), where they cover large portions of the Eastern Cape, Free

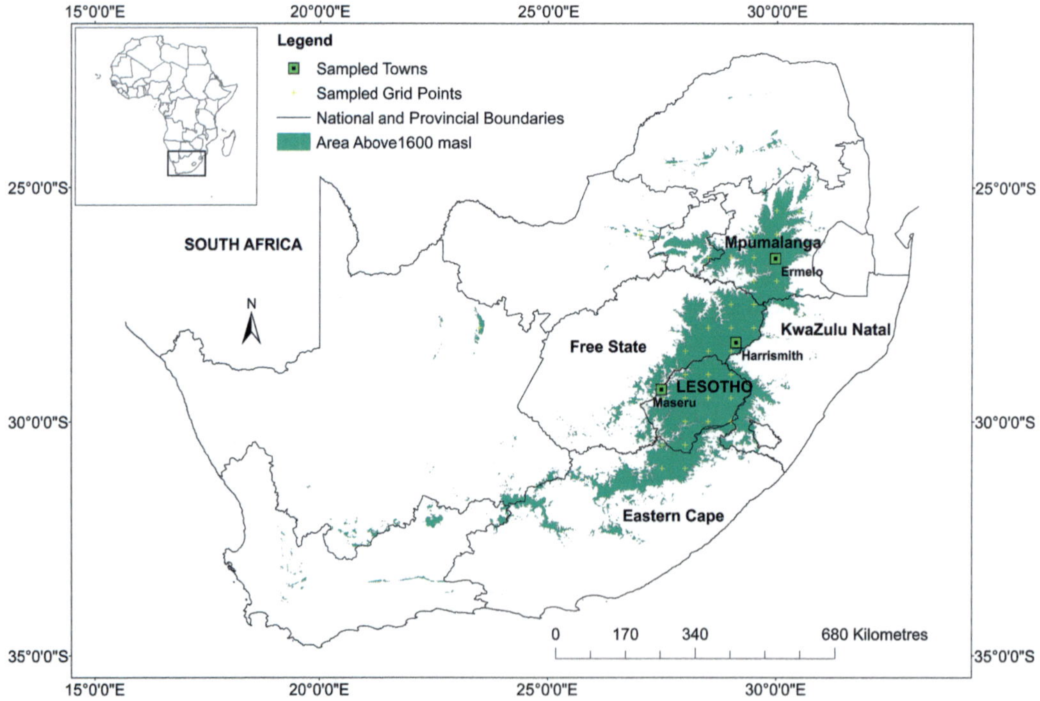

Fig. 19.1 South African main mountain belt and selected grid points

State, KwaZulu Natal and Mpumalanga provinces of South Africa. Understanding how the climate of this region is changing will not only help policy makers in these two countries to plan better for mitigation but will also have a variety of applications in hydrological, ecological, social and agricultural studies and related fields.

This study is based on two key interrelated approaches. The first involves the analysis of trends for precipitation and maximum temperature for the period between 1960 and 2016. Trends for three locations, including (from north to south) Ermelo, Harrismith and Maseru (Fig. 19.1), were compared in order to determine how the climate of the region has been changing through time. The second approach involves the spatial analysis of severe and extreme climate events, including drought and wet conditions in order to identify areas that are most vulnerable to climate change.

19.2 Study Area

The Drakensberg mountains are a chain of mountains stretching for a distance of over 1000 kms and are roughly divided into six units, including from the north to the south the Mpumalanga, Enkangala, Northern, Southern, Maluti and Eastern Cape Drakensberg units in that order. Drakensberg literally translates to "Dragon Mountains", in Afrikaans. The Drakensberg are also locally known as Ukahlamba, in isiZulu, depicting the spear-like resemblance of their morphological appearance that is characterized by jutting summits. The Sotho people call them the "Maluti", meaning mountains. The Drakensberg Mountains are the highest in southern Africa and they lie at the interface between the sub-continent's drier, colder, more seasonal interior and its perennially productive subtropical coastal belt (Stewart and Mitchell 2018). Some prominent scenic landforms found along the Drakensberg include the Cathedral Peak and Sentinel Peak, Champagne Castle, uThukela Falls and Monks Cowl. The Monks Cowl is located between the Champagne Castle and the Cathedral Peak and earned its name from its hood like shape.

The Drakensberg mountains experience cool summers and severe cold winters, with summer temperatures ranging between 10.8 and 23 °C at the summit (Carbutt et al. 2013), while winter temperatures fall below 0 °C, though both vary spatially according to altitude, latitude and continentality. The Drakensberg mountains are characterized by low and variable rainfall (Grundling et al. 2013). The Drakensberg Escarpment receives relatively higher amounts of precipitation compared to the rest of the region, with annual precipitation exceeding 2000 mm in some places (Nel and Sumner 2008). Mukwada et al (2016: 384) referred to the Drakensberg as "water factories of the region" due to their higher levels of precipitation and the prevalence of wetlands, which feed water into the most viable catchments that Lesotho, South Africa and Namibia depend on for water supply, mainly through two water transfer schemes, the Tugela Vaal Scheme and the Lesotho Highlands. These schemes are the primary source of water for agriculture, industry and commerce in South Africa.

Phytogeographically, the Drakensberg Alpine Centre is the only centre of plant endemism in southern Africa which is characterized by an alpine environment, making the Drakensberg mountains a region of rare assemblage of unique floristic and ecosystem diversity, depicting complex vegetation formations and habitats. Pumeza (2015) notes that the Drakensberg comprises three distinct altitudinal zones, namely montane, sub-alpine and alpine zones. The montane zone is generally found between 1280 and 1829 m above sea level (masl). This zone is generally dominated by *Protea* savannah and Afromontane forest, especially in sheltered gorges, with the former being the most dominant on north-facing slopes and the latter on south-facing slopes of the gorges. The sub-alpine zone ranges between 1829 and 2865 masl. This zone is dominated by fynbos and Themeda-Festuca grassland (Pumeza 2015). The uppermost zone is the alpine zone, which is largely found between 2865 and 3353 masl. *Erica* and *Helichrysum* are the most dominant species in this zone (Pumeza 2015).

Due to its species richness and spectacular landscapes, as well as occurrence of endemic species and other natural resources, the Drakensberg mountains are associated with numerous livelihood activities that are undertaken by local communities, including mining, agriculture, tourism (Magi and Nzama 2009; Mukwada et al. 2016; Mutana and Mukwada 2019) and pastoralism (Morris 2017a). Morris (2017a: 219) notes that "livestock, which graze largely on rangeland, are an indispensable part of the national and household economies and cultural life in Lesotho as well as providing many useful goods and services such as draught, transport, meat, milk, dung, hides and skins, for the Basotho people". Ninety per cent of people living in rural communities around the Mariepskop Mountain in the Mpumalanga Drakensberg region depend on firewood and water from the mountain, while 56% of them need better access to its resources (Ngwenya et al. 2019). Some ecosystem services relate to the land use activities practiced in the region (Fig. 19.2). These activities are central to the livelihoods and are therefore critical for sustaining the social-ecological systems of local communities.

Due to high levels of poverty, communities living in the Drakensberg mountains depend directly on resources drawn from the mountains, including wild foods such as vegetables and fruits, as well as phytomedicines.

19.3 Database and Methodology

Climate Research Unit Time Series (CRU-TS) 4.01 data were downloaded from the Climate Explorer Website. The CRU-TS data set "was developed and has been subsequently updated, improved and maintained with support from a number of funders, principally the UK's Natural Environment Research Council (NERC) and the US Department of Energy" (https://crudata.uea.ac.uk/cru/data/hrg/cru_ts_4.01/Release_Notes_CRU_TS4.01.txt).

The data comprised of both temperature and precipitation data for the period between 1960 and 2016. The resolution of the data was 0.5° × 0.5° and covered the mountain belt that was delineated from a digital elevation model (DEM), comprising all areas whose altitude exceeds 1 600 m above sea level (masl). Figure 19.1 shows the delineated belt and the grid points that were included in the analysis. Time series and spatial analyses of the data were undertaken. Standardized precipitation index (SPI) values were calculated from the rainfall

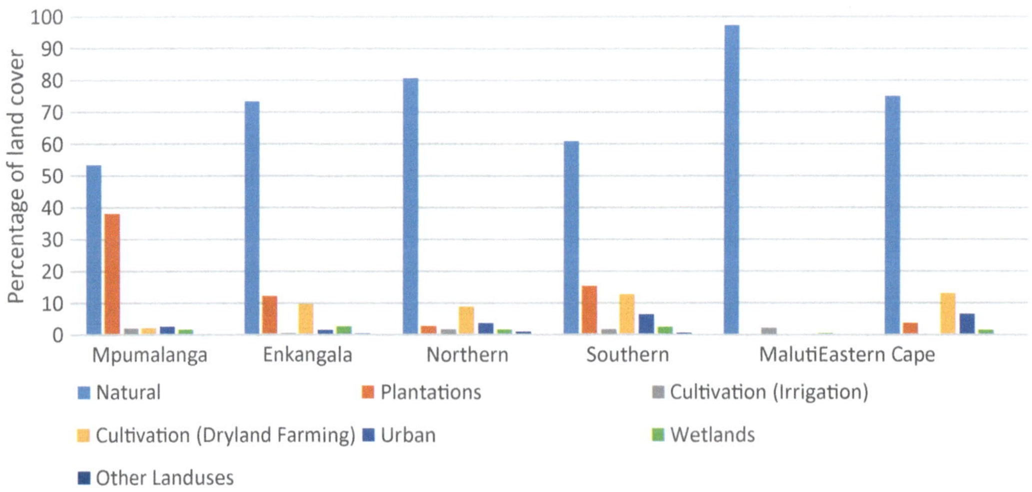

Fig. 19.2 Land use activities in the Drakensberg mountains. Data sourced from https://water.cer.org.za/areas/mpumalanga-drakensberg

data using the drought indices calculator, an online tool. Originally developed by Vicente-Serrano, the SPI allows for the monitoring of dry and wet periods over a wide spectrum of temporal scales due to the advantage that it is temporarily and spatially comparable, independent of geographical and topographical differences based on precipitation anomalies. This makes assessments of drought (and also wet conditions) possible even in the absence of other hydrometeorological measurements (Manatsa et al. 2010). The SPI is the number of standard deviations that an observed precipitation value would deviate from the long-term average if the precipitation was normally distributed (Türkeş and Tatli 2009; Mbiriri et al. 2018a). However, one important limitation of this study is the unavailability of long-term station data, which necessitated use of modelled data. Another limitation relates to the resolution of the data (0.5° × 0.5°), which was not high enough to allow micro-level analysis in a mountain environment where micro-level topographic variations are obvious.

The SPIs that were calculated were for the October–March period. This provided an overall picture about adequacy of precipitation throughout the entire summer season, the main season in which the mountain region receives most of its precipitation. The SPI values were then classified using McKee et al (1993) classification, and the years that were characterized by severe and extremely wet and dry conditions were identified, while their frequencies were calculated. Table 19.1 provides a summary of the classification that was developed by using McKee et al. (1993).

19.4 Results

The results from the empirical data used in this study confirm that the climate of the Drakensberg is changing. As shown in Figs. 19.3, 19.4 and 19.5, maximum temperatures increased significantly between 1960 and 2016. The trends of temperature change for all the three locations whose trends were analysed in this study, namely Ermelo, Harrismith and Maseru, are generally uniform, suggesting that the cause of the change could be the same. At all the three locations, temperatures increased by more than 2 °C between 1960 and 2016. This suggests that warming has been a universal phenomenon in the mountain areas.

However, as shown in Fig. 19.6, there has not been any significant changes in the amount of precipitation received between 1960 and 2016. Nevertheless, the analysis revealed that average precipitation for the whole region is highly variable, occasionally leading to frequent droughts.

When correlation analysis was performed to determine if there is any relationship between temperature changes and average precipitation recorded at the 34 grid points located in the mountain region, the results indicate that at all the locations, temperature negatively correlates with precipitation. This applies for all the three towns that were sampled for this study, namely Ermelo, Harrismith and Maseru, with correlation coefficients of −0.30, −0.24 and −0.27, respectively. However, the relatively low correlation coefficient values suggest that there could be other factors that account for the variability of precipitation in the region, apart from temperature.

Figure 19.7 shows the temporal variation of severe and extreme wet years in the Drakensberg region. As shown in Fig. 19.7, the frequency of these events has generally decreased since the early 1990s, with notable gaps for the period between 2002 and 2009 and the post 2012 period.

The highest frequencies of extreme wet years have been evident in the extreme northern and southern parts of the mountain region, as shown in Fig. 19.8, especially in the northern parts of the Eastern Cape Province and central regions of Mpumalanga Province. On the other hand, severe wet years have been mostly associated with the middle belt of the mountains, especially in the eastern parts of the Free State Province and the northern parts of Lesotho (Fig. 19.9).

When compared to severe wet and extreme wet events, severe and extreme drought events seem to have increased since 2000 (Fig. 19.10).

The most severe and widespread drought that was recorded in the Drakensberg occurred during

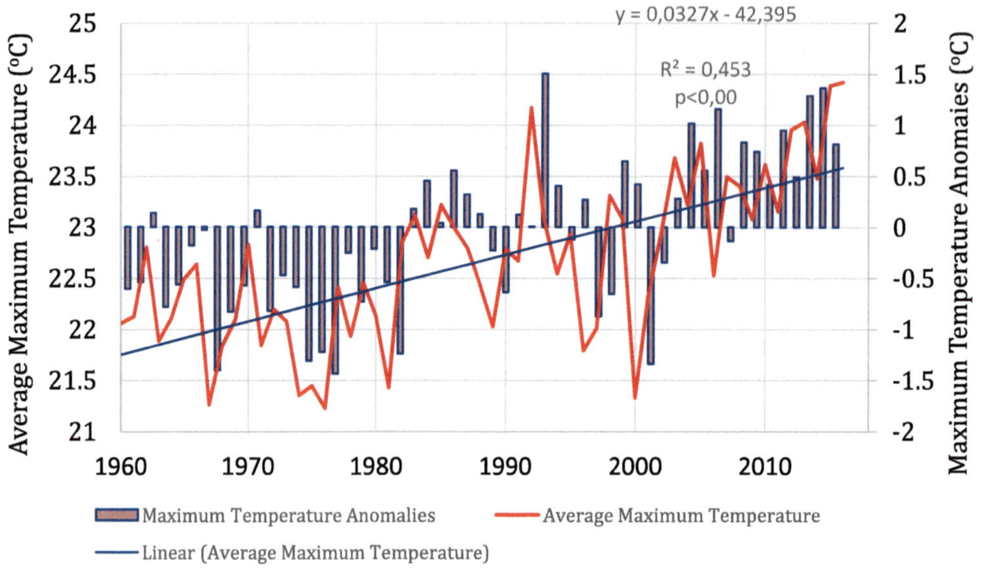

Fig. 19.3 Changes of maximum temperature at Ermelo between 1960 and 2016 and maximum temperature anomalies between 1960 and 2014. Data sourced from the Climate Research Unit Time Series 4.1

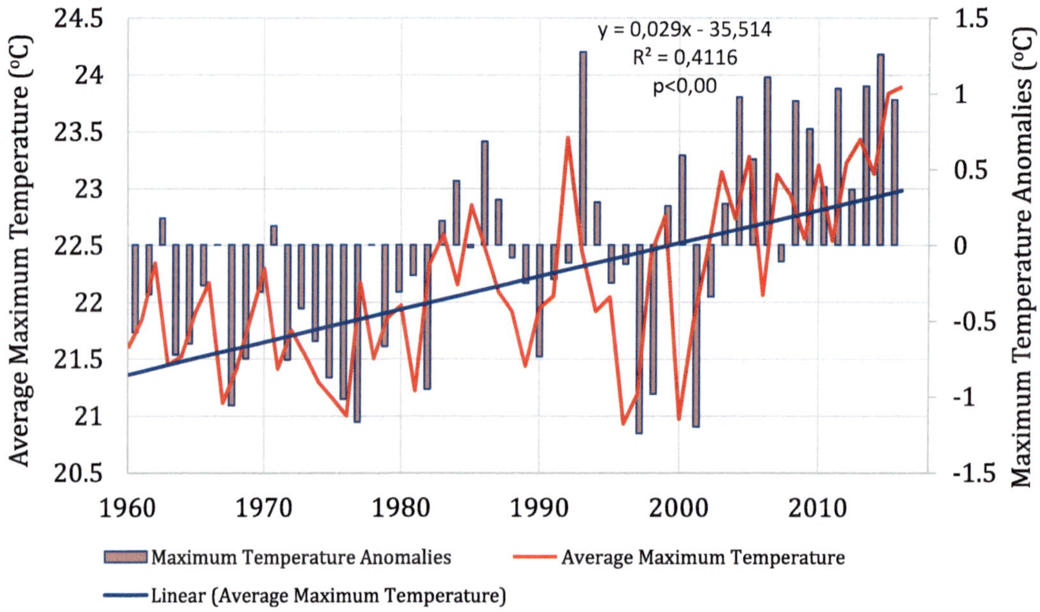

Fig. 19.4 Changes of maximum temperature at Harrismith between 1960 and 2016 and maximum temperature anomalies between 1960 and 2014. Data sourced from the Climate Research Unit Time Series 4.1

the 2015–2016 regions. Between 1960 and 2016, the southern parts of Lesotho experienced the highest frequency of extreme drought, as indicated in Fig. 19.11, whereas the central parts of Lesotho and western regions of the Mpumalanga Province recorded the highest frequency of severe droughts compared to the rest of the mountain belt (Fig. 19.12).

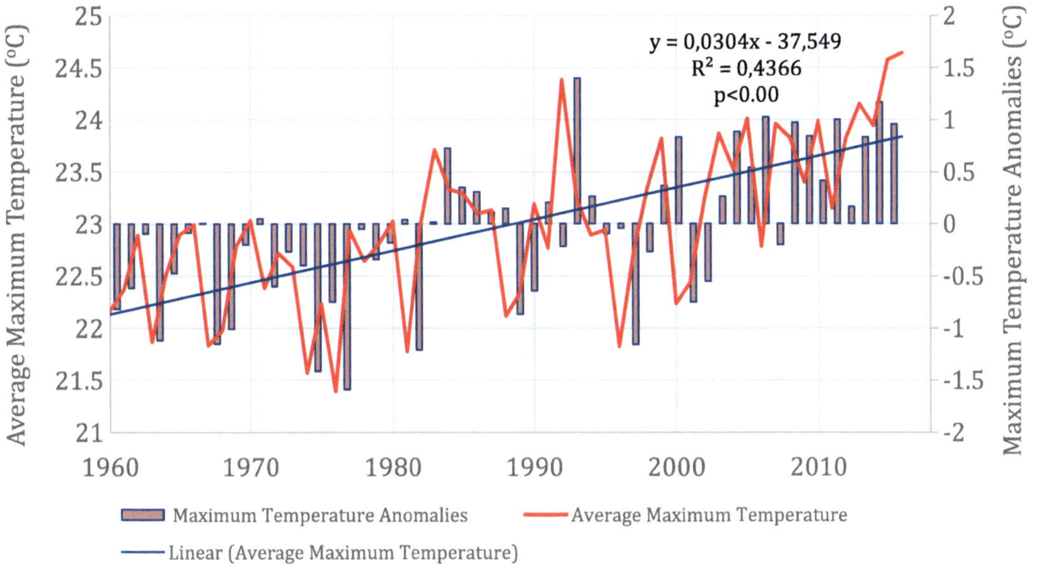

Fig. 19.5 Changes of maximum temperature at Maseru between 1960 and 2016 and maximum temperature anomalies between 1960 and 2014. Data sourced from the Climate Research Unit Time Series 4.1

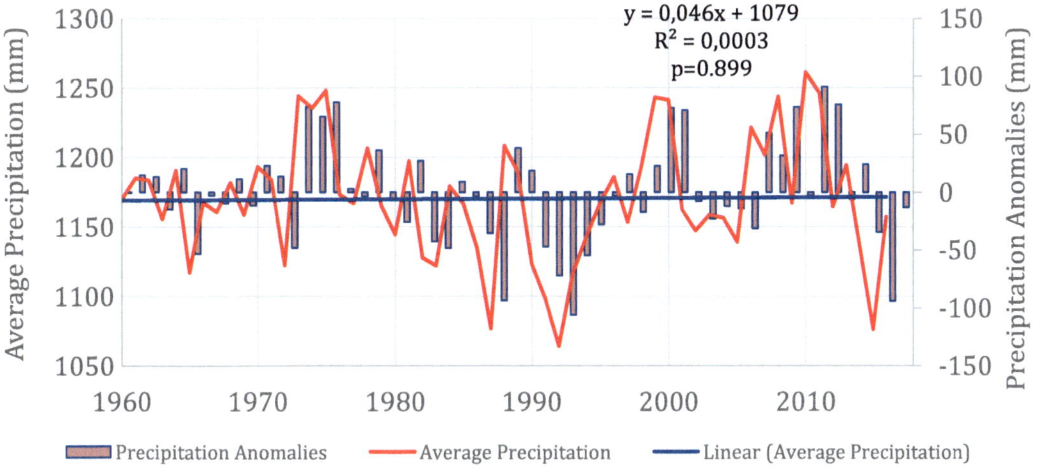

Fig. 19.6 Average amount of precipitation recorded at 34 mountain grid points between 1960 and 2016. Data sourced from the Climate Research Unit Time Series 4.1

It is noteworthy that in some cases the areas that have experienced extreme and severe wet conditions are the same as those that are affected by severe and extreme droughts. As shown in Fig. 19.13, during the 1975–1976, drought Lesotho experienced the most extreme wet conditions, as denoted by the low SPI values.

However, during the 1995–1996 period, Mpumalanga Province (Fig. 19.14) recorded the wettest conditions. Figures 19.15 and 19.16 provide typical examples of years when the lowest SPI values were recorded in regions that normally receive the highest amounts of precipitation, while Fig. 19.17 shows the areas that were affected by the 2015–2016 drought. As shown in Fig. 19.17, the 2015–2016 drought was not only widespread but also quite severe.

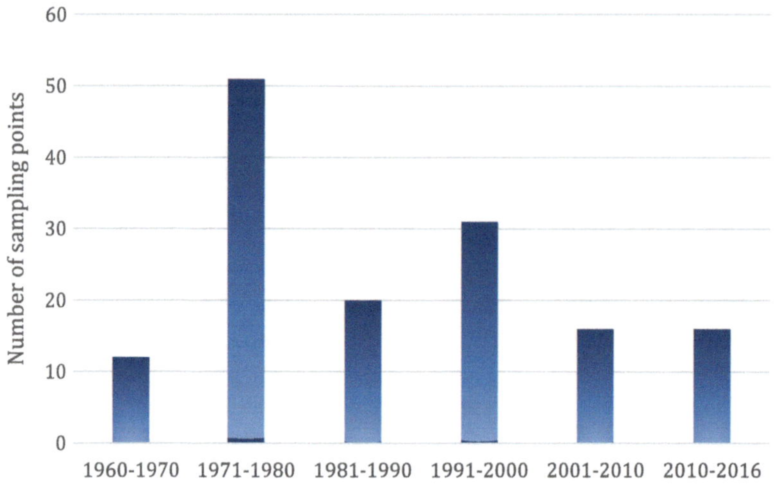

Fig. 19.7 Temporal variation of severe and extreme wet years in the Drakensberg mountains between 1960 and 2016

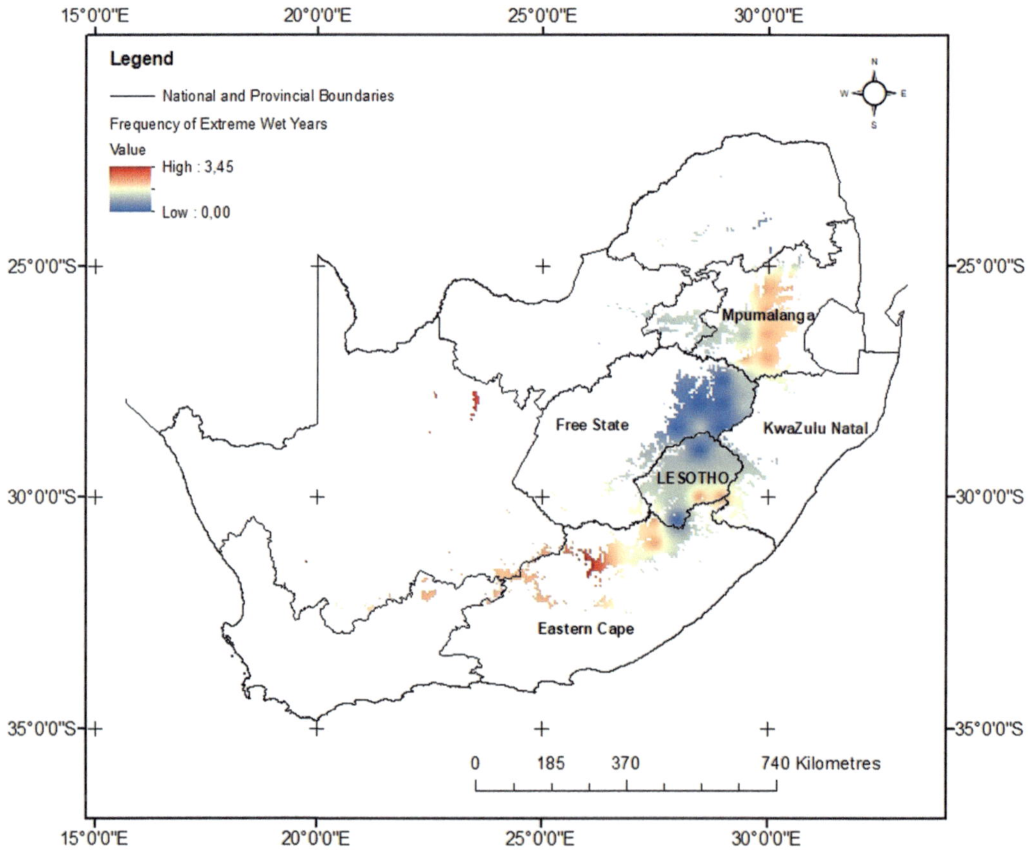

Fig. 19.8 Spatial variability of the frequency of extreme wet years between 1960 and 2016

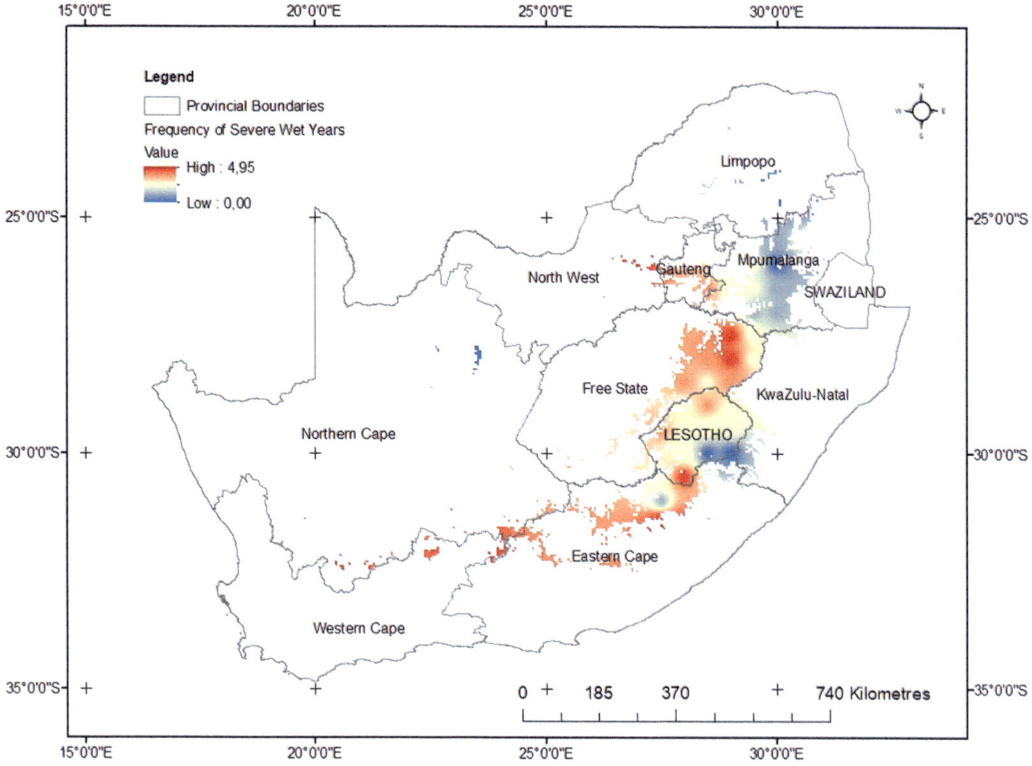

Fig. 19.9 Spatial variability of the frequency of severe wet years between 1960 and 2016

Fig. 19.10 Temporal variation of severe and extreme drought years in the Drakensberg mountains between 1960 and 2016

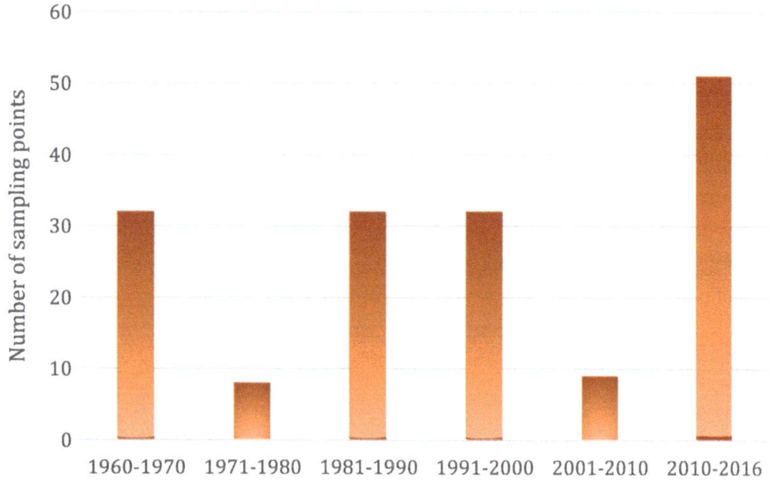

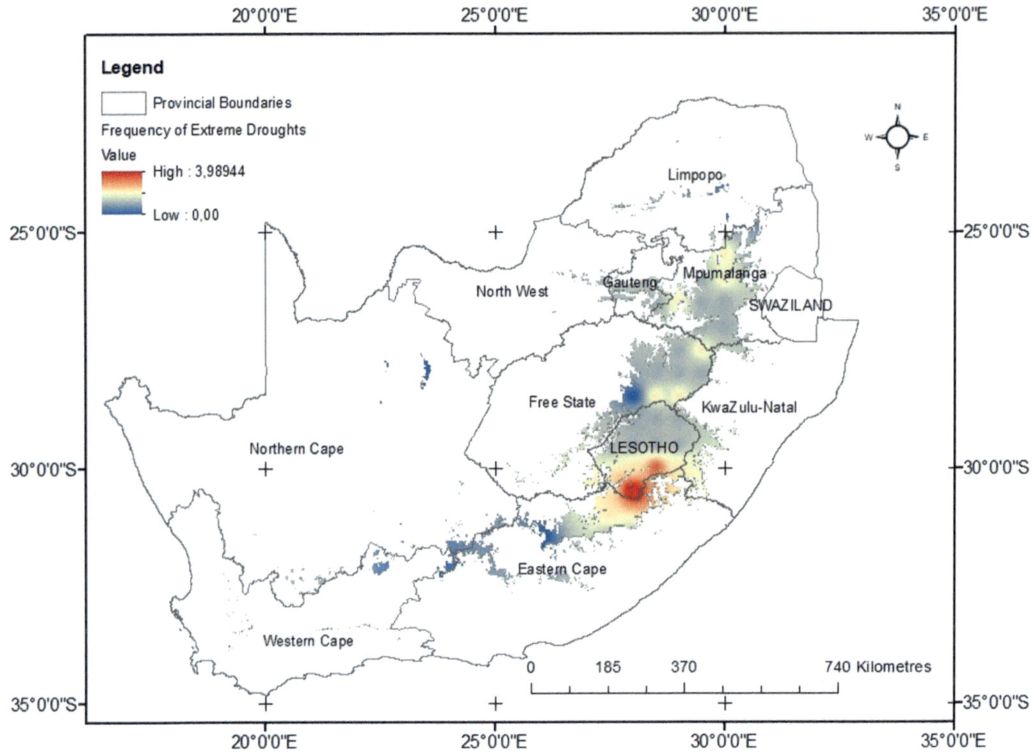

Fig. 19.11 Frequency of extreme drought years in South African mountain regions between 1960 and 2016

19.5 Discussion and Future Implication

The results of this study indicate that even though the amounts of precipitation received in the Drakeneberg mountains have been variable, changes in the amount of precipitation received in the Drakensberg mountains between 1960 and 2016 are not statistically significant. However, maximum temperature has risen significantly, as shown in Figs. 19.3, 19.4 and 19.5. The swing of temperature anomalies from the negative values that prevailed before 1980 to the positive ones in the post 2000 phase confirms a rise in temperature.

From the results presented in the foregoing section, there are some inferences that can be drawn about the relationship between temperature and precipitation. Between 1960 and 2016, most of the precipitation was received in summer when temperatures are generally high, suggesting that the dominant type of rainfall received in the mountain region is of convectional type. However, an increase in temperature has a negative effect on precipitation and has the capacity to trigger other undesirable environmental conditions which reduce chances of precipitation occurring. Results from this study indicate that the years that recorded the highest maximum temperatures also recorded the lowest average amounts of precipitation, leading to drought. Examples are 1992 and 2015, both of which were drought years. This can be explained within the context of changes in relative humidity. Excessively high temperatures reduce relative humidity and lower chances of precipitation, since the ambient air can never fully reach its dew point temperature. Beniston (2003) has reported similar rapid and systematic changes in climatic parameters in the Alps, in particular temperature

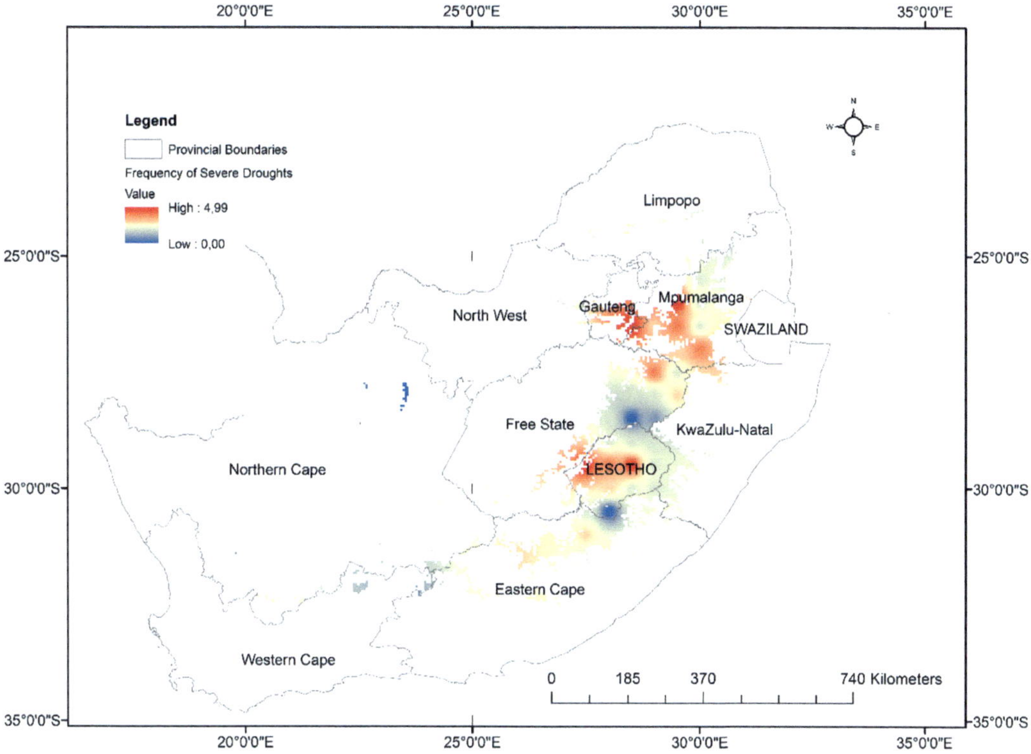

Fig. 19.12 Frequency of severe droughts years in South African mountain regions between 1960 and 2016

and precipitation, which have varied drastically over very short distances and led to enhanced direct runoff and erosion. Beniston (2003) also noted systematic variations in other environmental conditions such as radiation, as well as soil types, all of which have been shown to exhibit an elevation dependency on temperature trends and anomalies.

The results discussed above suggest serious implications for both the environment and the social-ecological systems of the Drakensberg region. Even if precipitation does not change, a warming climate is likely to put additional water stress on water resources in the future (Hulme et al. 2001), posing a threat to vegetation health (Mukwada and Manatsa 2018) and ecosystem services in general. The results of this study also confirm findings from earlier studies conducted in the region. However, in the Drakensberg mountains, climate change is not only a threat to economic activities undertaken by the mountains and downstream communities, but will also an onslaught on the livelihoods of these communities. It has already been established that the role of mountains in global water resources could be significantly altered by climate change (Viviroli et al. 2011).

The impact of a warming climate in the Drakensberg mountain ecosystems could extend to the hydrological, ecological and societal systems of these mountains (Mbiriri et al. 2018b) and enhance uncertainties associated with extreme weather events. Rosenzweig et al. (2001) note that even relatively small changes in mean temperature can result in disproportionately large changes in the frequency of extreme weather events such as spells of very high temperature, torrential rains and droughts. From the

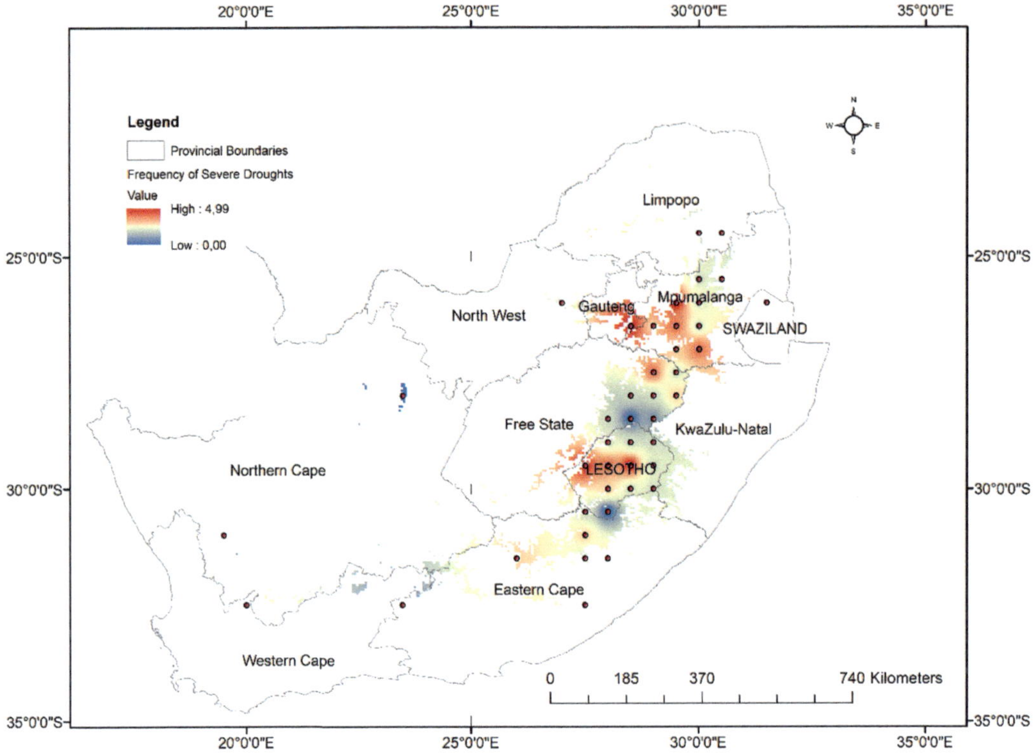

Fig. 19.13 Frequency of extreme wet years in the Drakensberg mountains

results above, it is evident that maximum temperature has risen significantly at all the three locations included in this study, signifying universal warming in the Drakensberg region. This has serious implications for both the region and South Africa as a whole. The remaining part of this paper highlights the implications for two categories of impacts, namely those associated with habitat changes and those that relate to socio-economic impacts.

In the Drakensberg mountains, a number of biophysical changes are likely to arise due to climate change. These include direct impacts such as loss of biodiversity and indirect impacts like soil erosion and siltation. Bishop et al. (2017: 106) note that "organisms tend to have larger thermal tolerance ranges in environments known to be more variable and this is due to greater variation in lower thermal tolerance limits". In mountain regions, climate change is likely to undermine the survival of many organisms whose thermal tolerance limits are already restricted. Plant and animal communities are likely to be affected by changes in thermal temperature limits. For example, since low temperatures limit the distribution of C4 grasses in the Drakensberg mountains, a warmer climate and longer growing seasons will enable C4 grasses to colonize higher and previously cooler slopes (Angelo and Daehler 2015), thus altering grazing dynamics. Adjorlolo et al. (2015: 49) note that "increases in temperature should favour C4 species, and C3 species could possibly move up to higher altitudes or into south-facing cooler and moister slopes in response to increasing temperatures". The elevational range of shrubs is likely to increase (Morris 2017b), hence leading to bush encroachment and increase pressure on remaining pastures (Morris 2017b).

However, only a few studies have investigated temporal variations of elevational diversity of species (Bishop et al. 2014). Climate change is one factor that can alter elevational diversity. Due to

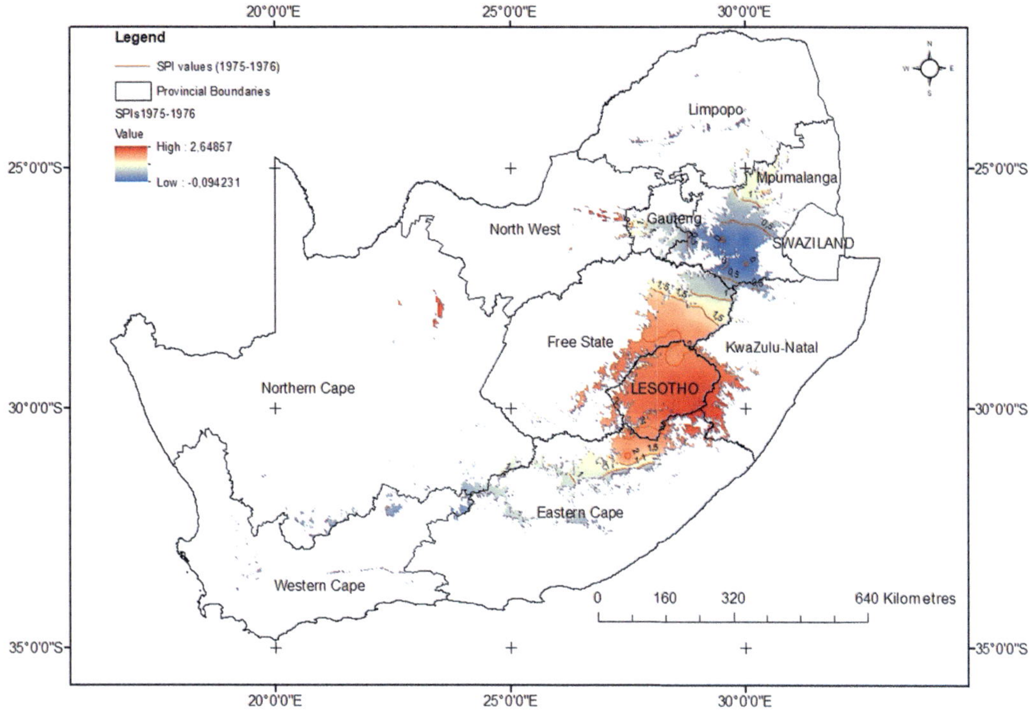

Fig. 19.14 Variability of SPI values during the 1975–1976 wet season

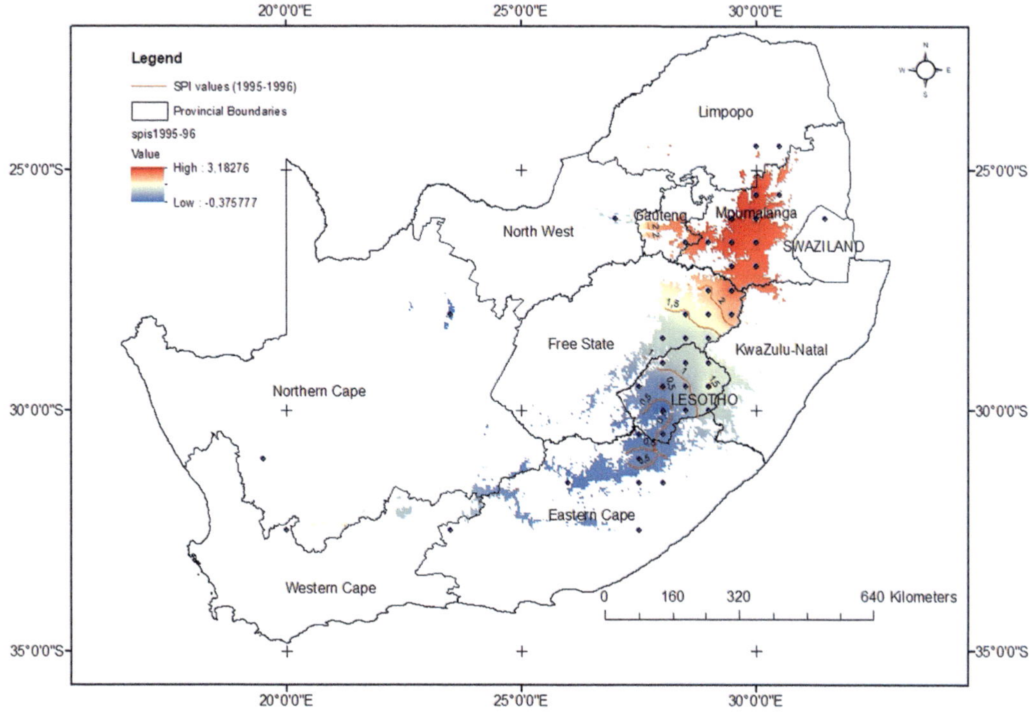

Fig. 19.15 Variability of SPI values during the 1995–1996 wet season

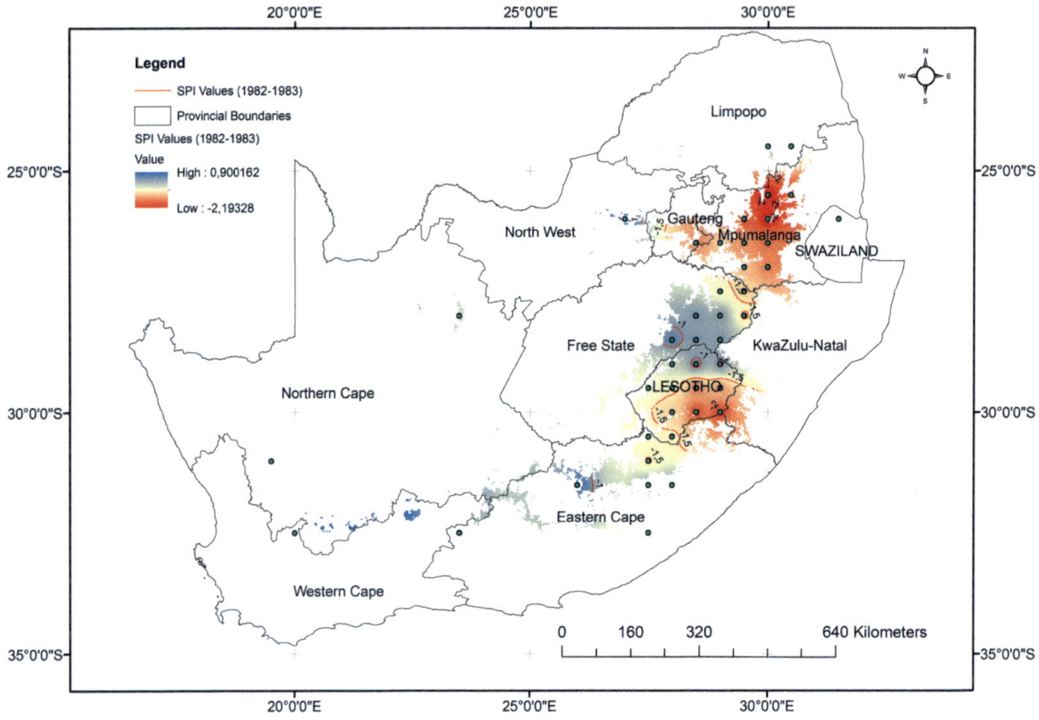

Fig. 19.16 Variability of SPI values during the 1982–1983 drought season

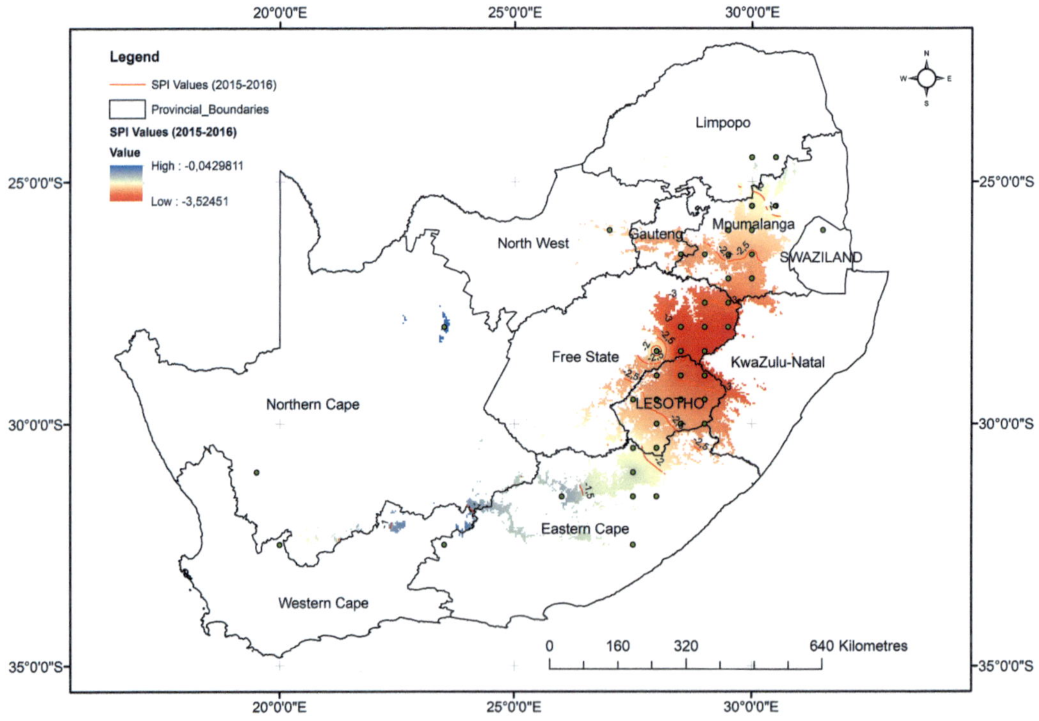

Fig. 19.17 Variability of SPI values during the 2015–2016 drought season

Fig. 19.18 **a** A Basotho vendor selling artefacts at a traffic circle near Katse Dam in Lesotho where tourists frequently visit. **b** On display are artefacts such as brooms, hats, mats, walking sticks (knobkerries) and other accoutrements sold by on a pavement by vendors in Phuthaditjhaba, in Qwaqwa, the former homeland of the Sotho people in South Africa

climate change, species distributions are expected to shift along elevational gradients as well as along latitudinal gradients (Wilson et al. 2005; Colwell et al. 2008). Changes in species distribution will have many negative effects on social-ecological systems in mountain communities. Some rare plant species, especially endemic species, might become locally extinct and unavailable to mountain communities that depend on them for food, pasture or medicines. Many Basotho people, the indigenous people of the Maluti Drakensberg region, depend on phytomedicines for their health. A number of forb species are used for medicinal and horticultural purposes and are openly traded by venders (see Figs. 19.18 and 19.19). Species such as the Arum lily (*Scilla natalensis*), *Helichrysum spp*., *Ledeburia spp*., Pineapple lily (*Eucomis spp*) and *Watsonia spp*. are used by the Basotho for medicinal purposes (Zondo, 2016). Similarly, rare species such as *Brunsvigia spp*., Arum lily and pineapple lily are highly sought after because of their ornamental worth. Many other plants, including grasses and sedges, are used by locals as thatch or raw materials for brooms, mats and hats. This demonstrates the range of goods and ecosystem services that local people derive from the Drakensberg mountains. These rare and highly sought after species are threatened with extinction due to climate change.

Changes within plant communities could put pressure on pastures and worsen overgrazing, as some palatable grass species are heavily foraged by livestock. Previous studies have revealed that sustained overgrazing contributes to accelerated erosion and alter the composition, structure and overall productivity of mountain rangelands (Morris 2017a). Also, overgrazing enhances surface runoff and accelerated erosion (Fig. 19.20), which will contribute to high siltation and reduction of surface water resources in downstream communities. Overgrazing can pave way for invasion by species that are not palatable to livestock. The most prominent plant invader is the black wattle (*Acacia mearnsii*), an exotic tree species which communities in the Maluti Drakensberg are already struggling to bring under control. Silver wattle (*Acacia dealbata*), Green wattle (*Acacia decurrens*) and Tree of heaven (*Ailanthus altissima*) are some of the invasive species that have been declared noxious weeds because of the threat they pose to wildlife conservation in the Golden Gate Highlands National Park (SanParks 2012). Some invasive species such as the black wattle and *Leucosidea sericea*, a native species, have a high affinity for water. Their proliferation might lead to widespread soil desiccation and undermine the survival of other species in the environment.

Considering that the Drakensberg mountains are a critical water source for South Africa, 50% of whose GDP depends on this water source, it is

Fig. 19.19 A stall of herbal medicines located in front of a clothing shop in Phuthaditjhaba

Fig. 19.20 Overgrazed land around Kase Dam in Lesotho. In the foreground is degraded land largely affected by sheetwash erosion. In the middle ground is cultivated land. As shown in the photo, steeply inclined land is terraced to make it possible for farmers to grow crops. Unfortunately, due to land shortage, cultivation is often practiced in unsuitable areas, including those close to water bodies where siltation is slowly taking its toll

evident that climate change will have serious socio-economic implications for both South Africa and the southern African region as a whole. Water shortages due to increased evapotranspiration caused by a warming climate will affect a number of sectors of the economy, leading to a fall in GDP, reduced employment and worsening of poverty in the region. As explained below, within both the Sequ-Orange Basin and the Gauteng Region, sectors such as industry, agriculture and tourism which depend directly on water from the Drakensberg mountain region will all be negatively affected.

The Gauteng Region is South Africa's biggest economic hub and directly and indirectly supports livelihoods of many people across the whole of southern Africa. Commodities produced in this region are traded in the whole of the Southern Africa Development Community (SADC) Region, embracing Botswana, Mozambique, Zimbabwe, Zambia, Malawi, Democratic Republic of Congo, Namibia, Madagascar, Tanzania and South Africa itself. In recent years, the food security of millions of people in this region has been dependent on South Africa because of recurrence of drought in the region. In the Drakensberg mountain region, where the prominence of tourism is steadily increasing, local water shortages and land degradation are likely to worsen by climate change. In recent years, the region has been affected by a spate of serious droughts. As shown in Fig. 19.11, the areas that have been affected by severe and extremely severe drought include the central and southern parts of Lesotho and western parts of Mpumalanga.

The effect of shortage of water will be most felt in human settlements that depend on water from the Drakensberg mountains. Critical shortages of water have already been reported in some communities in the Drakensberg mountains. For instance, communities in Qwaqwa, which is located in the eastern part of the Free State Province, are currently facing severe water shortages. Shortage of water is likely to become the biggest threat to the sustainability of human settlements. With persisting positive temperature anomalies, shortage of water for both economic and domestic use could worsen in the future as a result of high recurrence of drought caused by a warming climate, leading to economic depression and the worsening of the livelihoods millions of people in southern Africa. The ensuing hydropolitical problems will fuel social conflicts in a region that has already been ravaged by civil wars. This will exacerbate the prevailing refugee and xenophobic crisis. Water shortages will be most critical in towns and cities that are found in karoo or semi-arid environments, for instance Bethlehem and Bloemfontein, both of which are situated within the Sequ-Orange Basin. Nearly 70% of Bloemfontein's water supply is derived from the Drakensberg mountains through the Lesotho Highland Water Project. Similarly, Durban, an industrial and tourist hub situated along the eastern coast of the country, depends on water from the Drakensberg mountains. The Thukela River, which drains into the Indian Ocean, is an important source of freshwater for this city and is part of the water transfer scheme linking Lesotho and South Africa. Loss of revenue from water exports to South Africa will undermine Lesotho's GDP and worsen poverty in the country. Currently, plans for Phase 2 of the Lesotho Highland Water Project, which is scheduled for completion by 2020, are underway to augment water exports to South Africa at a cost of R20 billion.

19.6 Conclusion

The results of this study indicate that the climate of the Drakensberg mountain region is becoming warmer, while droughts are becoming more frequent due to reduced rainfall reliability. These changes have already started to create environmental challenges which will impact negatively on habitats, ecosystem services and the livelihoods of both mountain and downstream communities in the future. Thus, it can be concluded that the effects of climate change in the Drakensberg mountains, just like those in other parts of southern Africa, are similar to those experienced or anticipated in mountain regions in other parts of the African continent. Patterns and trends of climate change experienced or anticipated in the Drakensberg closely resemble those reported in other mountain regions elsewhere in Africa.

In order to address the impact of climate change induced uncertainties in the Drakeneberg mountains and their implications for the social-ecological systems of southern Africa, more research is needed on the adaptation measures that can be developed to ensure the continual supply of ecosystem services that support the

Table 19.1 Classification used in the categorization of SPI values

SPI value	Category	Probability (%)
≥2.0	Extreme wet	1.7
1.50–1.99	Severe wet	2.7
1.99–1.49	Moderate wet	9.1
0.5–0.99	Mild wet	16.5
0.49 to −0.49	Normal	40.0
−0.50 to −0.99	Mild drought	16.5
−1.00 to −1.49	Moderate drought	9.1
−1.50 to −1.99	Severe drought	2.7
≤−2.00	Extreme drought	1.7

Adapted from McKee et al. (1993)

livelihoods of both mountain and downstream communities. The uncertainties arising from impacts of climate change in the Drakensberg make the future of southern Africa bleak. Pangs of future woes have already started manifesting themselves as the climate change crisis deepens. Policy makers in southern Africa, particularly in Lesotho, South Africa and Namibia, need to redesign climate adaptation policies to ensure secure and sustainable livelihoods of these communities and avert socio-political upheavals in the future.

References

Adjorlolo C, Mutanga O Cho MA (2015) Predicting C3 and C4 grass nutrient variability using in situ canopy reflectance and partial least squares regression.Int J Remote Sens 36(6):1743–1761

Angelo CL, Daehler CC (2015) Temperature is the major driver of distribution patterns for C4 and C3 BEP grasses along tropical elevation gradients in Hawaii, and comparison with worldwide patterns. Botany 93:9–22

Beniston M (2003) Climatic change in mountain regions: a review of possible impacts. In: Climate variability and change in high elevation regions: Past, present and future. Springer, Dordrecht, pp 5–31

Bishop TR, Robertson MP, van Rensburg BJ, Parr CL (2014) Elevation–diversity patterns through space and time: ant communities of the Maloti-Drakensberg Mountains of southern Africa. J Biogeogr 41 (12):2256–2268

Bishop TR, Robertson MP, Van Rensburg BJ, Parr CL (2017) Coping with the cold: minimum temperatures and thermal tolerances dominate the ecology of mountain ants. Ecol Entomol 42(2):105–114

Bishop TR, Robertson MP, van Rensburg BJ, Parr CL (2014) Elevation–diversity patterns through space and time: ant communities of the Maloti-Drakensberg Mountains of southern Africa. Int J Remote Sens 41 (12):2256–2268

Carbutt C, Edwards TJ (2015) Plant–soil interactions in lower–upper montane systems and their implications in a warming world: a case study from the Maloti-Drakensberg Park, southern Africa. Biodiversity 16 (4):262–277

Carbutt C, Edwards TJ, Fynn RW, Beckett RP (2013) Evidence for temperature limitation of nitrogen mineralisation in the Drakensberg Alpine Centre. South African J Botany 88:447–454

Chemura A, Kutywayo D, Chidoko P, Mahoya C (2016) Bioclimatic modelling of current and projected climatic suitability of coffee (Coffea arabica) production in Zimbabwe. Reg Environ Change 16(2):473–485

Colwell RK, Brehm G, Cardelus CL, Gilman AC, Longino JT (2008) Global warming, elevational range shifts, and lowland biotic attrition in the wet tropics. Science 322:258–261

Grundling P, Grootjans AP, Price JP, Ellery WN (2013) Development and persistence of an African mire: How the oldest South African fen has survived ina marginal climate. Catena 110:176–183. https://doi.org/10.1016/j.catena.2013.06.004

Hemp A (2005) Climate change-driven forest fires marginalize the impact of ice cap wasting on Kilimanjaro. Glob Change Biol 11(7):1013–1023

Hemp A (2009) Climate change and its impact on the forests of Kilimanjaro. Afr J Ecol 47:3–10

Kaser G, Hardy DR, Mölg T, Bradley RS, Hyera TM (2004) Modern glacier retreat on Kilimanjaro as evidence of climate change: observations and facts. Int J Climatol J Royal Meteorol Soc 24(3):329–339

Magi L, Nzama TA (2009) Tourism strategies and local community responses around the World Heritage Sites in KwaZulu-Natal. S Afr Geogr J 91(2):94–102

Manatsa D, Mukwada G, Siziba E, Chinyanganya T (2010) Analysis of multidimensional aspects of agricultural droughts in Zimbabwe using the Standardized

Precipitation Index (SPI). Theoret Appl Climatol 102 (3–4):287–305
Mbiriri M, Mukwada G, Manatsa D (2018a) Influence of altitude on the spatiotemporal variations of meteorological droughts in mountain regions of the Free State Province, South Africa (1960–2013). Adv Meteorolhttps://doi.org/10.1155/2018/5206151
Mbiriri M, Mukwada G, Manatsa D (2018b) About surface temperature and their shifts in the Free State Province, South Africa (1960–2013). Appl Geogr 97:142–151
McKee TB, Doesken NJ, Kleist J (1993) The relationship of drought frequency and duration to time scales. In: Proceedings of the eighth conference on applied climatology, 17–22 Jan 1993, Anaheim, California
Morris C (2017a) Historical vegetation–environment patterns for assessing the impact of climatic change in the mountains of Lesotho African. J Range Forage Sci 34(1):45–51
Morris C (2017b) Vegetation gradients around cattleposts in the eastern mountains of Lesotho African. J Range Forage Sci 34(4):219–225
Mukwada G, Manatsa D (2018) Spatiotemporal analysis of the effect of climate change on vegetation health in the Drakensberg Mountain Region of South Africa. Environ Monit Assess 190(6):358
Mukwada G, Le Roux A, Hlalele D, Lombard C (2016) The afromontane research unit (ARU) in South Africa. Mt Res Dev 36(3):384–387
Munishi LK, Sawere PC (2014) Climate change and decline in water resources in Kikuletwa Catchment, Pangani, Northern Tanzania. African J Environ Sci Technol 8(1):58–65
Mutana S, Mukwada G (2019) Are policies and guidelines shaping tourism sustainability in South Africa? Critical success factors for tourism sustainability governance in the Drakensberg Region. Tourism Hospitality Res 1467358419841100
Nel W, Sumner P (2008) Rainfall and temperature attributes on the Lesotho-Drakensberg escarpment edge, southern Africa. Geografiska Annaler Ser Phys Geogr 90(1):97–108
Ngwenya SJ, Torquebiau E, Ferguson JWH (2019) Mountains as a critical source of ecosystem services: the case of the Drakensberg, South Africa. Environ Dev Sustain 21(2):1035–1052
Pumeza C (2015) Plant diversity and morphology in seasonally snow-abundant niches of the Drakensberg Alpine Centre, Lesotho. Doctoral dissertation, Faculty of Science, University of the Witwatersrand, Johannesburg
Rosenzweig C, Iglesias A, Yang XB, Epstein PR, Chivian E (2001) Climate change and extreme weather events; implications for food production, plant diseases, and pests. Global Change Hum Health 2(2):90–104
SANParks (2012) Golden gate highlands national park management plan, SANParks, Pretoria
Simonneaux V, Cheggour A, Deschamps C, Mouillot F, Cerdan O, Le Bissonnais Y (2015) Land use and climate change effects on soil erosion in a semi-arid mountainous watershed (High Atlas, Morocco). J Arid Environ 122:64–75
Stewart BA, Mitchell PJ (2018) Late Quaternary palaeoclimates and human-environment dynamics of the Maloti-Drakensberg region, southern Africa. Quatern Sci Rev 196:1–20
Türkeş M, Tatlı H (2009) Use of the standardized precipitation index (SPI) and a modified SPI for shaping the drought probabilities over Turkey. Int J Climatol 29(15):2270–2282
Viviroli D, Archer DR, Buytaert W, Fowler HJ, Greenwood GB, Hamlet AF, Huang Y, Koboltschnig G, Litaor MI, López-Moreno JI, LorentzS (2011). Climate change and mountain water resources: overview and recommendations for research, management and policy. Hydrol Earth Syst Sci 15(2):471–504
Voje KL, Hemp C, FLAGSTAD Ø, SÆTRE GP, Stenseth NC (2009) Climatic change as an engine for speciation in flightless Orthoptera species inhabiting African mountains. Mole Ecol 18(1):93–108
Wilson RJ, Guti_errez D, Guti_errez J, Mart_ınez D, Agudo R, Monserrat VJ (2005) Changes to the elevational limits and extent of species ranges associated with climate change. Ecol Lett 8:1138–1146
Zondo SA (2016) Value chain analysis in the proposed Witsehoek community conservation area of the eastern Free State region of South Africa. M.Sc. dissertation, University of the Free State

Part II
Response Processes of Mountain Environments to Land Use Change

Introduction

This part presents a compilation of contributed chapters from researchers in a variety of fields, who address effects of land use changes in mountain environments from different perspectives and in different mountain regions of Asia, Europe, and North America. The collection of case studies allows for developing a comprehensive view on current and historical land use changes in mountains of the Global South and North, ranging from effects of changing agriculture and pastoralism over forest/water resources management and urbanization processes to landscape management and biodiversity conservation. The studies showcased in this part are based on fieldwork, complemented by remote sensing and modelling approaches. The common ground of the following case studies is the dynamics of mountain environments, triggered by the alteration of complex resource utilization strategies in space and time and accelerated by impacts of globalization. Over the past decades, modified land use intensities and land use systems have had tremendous effects on mountain landscapes. Geoecological settings and economic, political and socio-cultural aspects in mountain regions are interwoven in a very complex manner. In this context, changing land use systems point to changing interrelationships between state policies, local power relations, cultural adaptations and external interferences at the interface between the ecosystem and the anthroposystem. As obvious from the following case studies, effects of changing land use systems often contradict the UN Sustainable Development Goals. Thus, it is imperative, in particular in mountain regions, to redirect land use modifications towards promoting sustainable use of terrestrial ecosystems, reversing land degradation, and halting biodiversity loss.

The first four case studies address land use/land cover changes in the Himalayan mountain system. Respective changes in the western Himalaya (S. Rani and S. Sreekesh) and in the foothills of the eastern Himalaya (V. P. Sati; J. Debnath et al.) are stongly linked to driving forces such as demographic changes, cropland expansion, political/administrative interventions, rural-urban migration, and shift to non-agrarian activities. Heterogeneous land use patterns in each study area and diverging trajectories of change suggest that caution is advisable in attempting to generalize results within a mountain system. Another Himalayan case study (M. M. Anees et al.) addresses the impact of urbanization processes on land use/land cover changes and infrastructure development with focus on the Indian Himalayan regions, highlighting the role of rural to urban migration as major driver. The need for a revised agriculture and land use policy in supporting crop diversification, agricultural productivity, and climate resilience is stressed in a case study in the Central Highlands of Sri Lanka (M. Peiris and A. D. N. Gunarathne). Montane and alpine grasslands responses to changing grazing regimes are analysed in the next two case studies in the Qilian Mountains (NE Tibetan Plateau) and in the mountains of NW Greece. Degradation of grazinglands, reflected in vegetation structure, plant species composition, and soil physical and chemical properties, is a consequence of overuse of grazing resources in the Qilian Mountains in recent decades (A. Baranova and U. Schickhoff), while the NW Greece case study (M. Vrahnakis and Y. Kazoglou) highlights the chance of improved rangeland management and biodiversity conservation associated with changes in grazing management schemes. The significance of maintaining traditional grazing and forest management practices for biodiversity and cultural landscape conservation becomes particularly evident in the case study in the mountains of NW Spain (I. J. Diaz-Maroto) where sustainable land management faces the challenge of land abandonment. Providing paleoperspectives is often the key for understanding present-day landscape patterns as obvious from the case study in the northern European Alps (A. Friedmann et al.). Pastoral use of alpine grazinglands has played a major role in the history of vegetation, and sheep grazing continues to be highly significant for recent vegetation dynamics on the Zugspitzplatt. The difficulties of disentangling land use and climate change effects on treeline dynamics are highlighted in the case study from Norway (A. Bryn and K. Potthoff) which identified land use changes as the main driver for treelines shifting towards higher elevations while climate warming is facilitating this process. Future water demand exceeding water supply is a prominent feature of change in mountain regions. The case study in the Wasatch Mountains in Utah (M. A. Baker and C. G. Flint) explores the dynamics of this water tower system and the challenges on the road to a mutual agreement on water management and planning. Several research and development institutions have emerged in recent decades as important drivers in generating awareness and increasing attention to sustainable mountain development. The International Centre for Integrated Mountain Development (ICIMOD), supporting regional and transboundary cooperation in the HKH region, is an outstanding example in this respect. The final chapter (N. Chettri et al.) presents a retrospect of the achievements of ICIMOD in transboundary landscape management since its establishment in 1983, focusing on socio-economic, ecological and environmental dimensions. Concluding Part II of ‘Mountain landscapes in transition’, it represents an encouraging example of facilitating cooperation and policy coherence among different countries aiming at sustainable mountain development.

Assessment and Prediction of Land Use/Land Cover Changes of Beas Basin Using a Modeling Approach

20

Seema Rani and Sreedharan Sreekesh

Abstract

Ever-increasing human population is leading to modification in land use/land cover (LULC) conditions around the world. Changes in LULC are often happening in an undesirable direction. It is imperative to monitor the patterns of LULC change for appropriate planning and management of an area. Remote sensing and GIS offer an opportunity to analyze these changes and associated driving factors. The present work is an effort to analyze the LULC changes in the upper Beas river basin, Western Indian Himalaya. Markov chain (MC) model is used to analyze the LULC changes over time and space. The LULC of 1991 and 2000 was used to optimize the MC model, while the image of 2010 was used for validation of an MC model using ground control points. Then, LULC maps of the years 2000 and 2018 were applied to foreshadow the possible future LULC of the year 2036. It is predicted that the area under built-up and cultivated land will increase at the expense of forest land. The findings indicate the influence of anthropogenic factors on the growth of settlements and agricultural land. The LULC modeling approach provides useful inputs to the planners of the area, and these data are also useful in hydrological modeling for effective water resource planning.

Keywords

LULC prediction · Transition matrix · Markov chain analysis · Himalayas

20.1 Introduction

Land use/land cover (LULC) is a dynamic operation that has taken place as a result of the interaction of human beings and the environment over a period of time. In recent decades, concern about changes in LULC has been increasing around the globe. According to Di Gregorio and Jansen (2000), "land use is characterized by the arrangements, activities, and inputs, people undertake in a certain land cover type to produce, change or maintain it such as built-up area, etc. On the other hand, land cover is the observed (bio) physical cover on the Earth's surface such as forest, grassland, barren, water bodies, snow, etc.". Changes in LULC are the major issue and challenges for the sustainable development of any area. Increasing human population results in

S. Rani
Department of Geography, Institute of Science, Banaras Hindu University, Varanasi, Uttar Pradesh, India

S. Sreekesh (✉)
Centre for the Study of Regional Development, Jawaharlal Nehru University, New Delhi, India
e-mail: sreekesh@mail.jnu.ac.in

U. Schickhoff et al. (eds.), *Mountain Landscapes in Transition*,
Sustainable Development Goals Series, https://doi.org/10.1007/978-3-030-70238-0_20

a rise of demand for food and energy. Consequently, human activities such as cultivation, construction of hydro projects, and deforestation become intense over time and lead to changes in LULC. Hence, modeling of changes in LULC is an important focus of studies in recent times to mitigate the consequences of their modification (Lambin et al. 2001).

Modification of LULC indicates the utilization of land in an area, and it is also considered as one of the important elements that affect the ecological system (Saleem et al. 2018). Different studies have shown changes in the LULC at different spatial scales (Osgouei and Kaya 2017; Gashaw et al. 2017; Palmate et al. 2017; Wang et al. 2018; Lu et al. 2019). Studies have reported the LULC changes across India (Singh et al. 2015; Maithani 2015; Mishra and Rai 2016). Human actions are not always leading to alterations in the LULC, sometimes physical factors such as forest fire, landslide, and flash flood also initiate modifications upon the landscape. Still, the alterations are largely initiated by anthropogenic factors, followed by natural factors (Niemelä et al. 2000; Srivastava et al. 2012). Anthropogenic factors include the deforestation, expansion of agriculture, urbanization, and losses of biodiversity. Urban development has been the most significant driver in LULC changes in recent decades (Remondi et al. 2016). LULC changes influence the relationship among the diverse parts of the ecosystem and sometimes result in destructive impacts such as soil erosion, landslide, floods, and so on. Changes in land-use practices may lead to considerable effect on the hydrology of a basin (Zheng et al. 2012; Shaw et al. 2014) and water quality (Wang et al. 2014; Goldshleger et al. 2015). Therefore, spatio-temporal analysis of LULC of an area is imperative to address the mitigation and adaptation options in the context of climate change, ecosystem balance, and natural resources management (Tallis and Polasky 2009; Nelson et al. 2009; Turner and Annamalai 2012; Bagan and Yamagata 2012). Understanding the impact of LULC changes is essential for the management of the watershed.

Good quality of data is required for monitoring and effective planning of land and water resources. Satellite remote sensing is providing an opportunity for the management of natural resources because of high spatial and temporal resolution (Chandra et al. 2003). It provides synoptic and continuous data for LULC change and related studies (Szabó et al. 2012). Information provided by satellite images is helpful in the prediction of LULC change and dynamics associated with it. Many researchers have used different types of satellite images to understand the LULC changes (Mishra and Rai 2016; Palmate et al. 2017; Anand et al. 2018; Lu et al. 2019). Various statistical and spatial models are available to understand and predict the probable LULC pattern of an area (Costanza and Ruth 1998) such as logistic regression model (Hu and Lo 2007), Markov chain (MC) (Kamusoko et al. 2009), cellular automata (CA) (Han et al. 2009; Mitsova et al. 2011), and machine learning algorithms (Huang et al. 2010). All mentioned methods are widely applied to study urban dynamics. Fewer studies used them at basin level, particularly in the Himalayas. Therefore, this study is an attempt to apply the given model to predict LULC at the basin level. The objectives of the study are: (i) to understand the spatio-temporal changes in the LULC of the study area during 1991–2018, (ii) to check the reliability of the Land Change Modeler Ecological Sustainability (LCMES) model (MC) for the LULC prediction in the basin, and (iii) to predict the LULC changes for the year 2036 using the model.

20.2 Study Area

The area of the present study is the upper Beas river basin up to the Pandoh dam (Fig. 20.1), which is one of the main tributaries of the Indus river. It originates at the Beas Kund in the Kullu district of the state of Himachal Pradesh at 4085 m. The area and length of the Beas river up to the Pandoh dam are 5300 km^2 and 116 km, respectively, of which 14% is under permafrost conditions (BBMB 1988). The altitude of the basin varies from 840 to 6582 m asl. The population and its density in the area are 437,903 and 80 persons/km^2, respectively (Census of India 2011). The area contains nearly 6.38% population of the state of Himachal Pradesh. There are many important towns in the basin, namely, Manali, Kullu, Kasol, Bhuntar, and Banjar. The density of population is high in the towns. Besides tourism, horticultural plantations such as apple are the main sources of livelihood of the people.

20.3 Data and Methods

Satellite images of the study area were taken from the global visualization viewer of the United States Geological Survey (USGS) (Table 20.1) (NASA Landsat Program 2019) because these images have high spatial and temporal resolutions. Toposheets of 1964–65 and 2005–06 at 1:50,000 were obtained from the Survey of India (SoI), Government of India, for preparing the base map of the upper Beas basin.

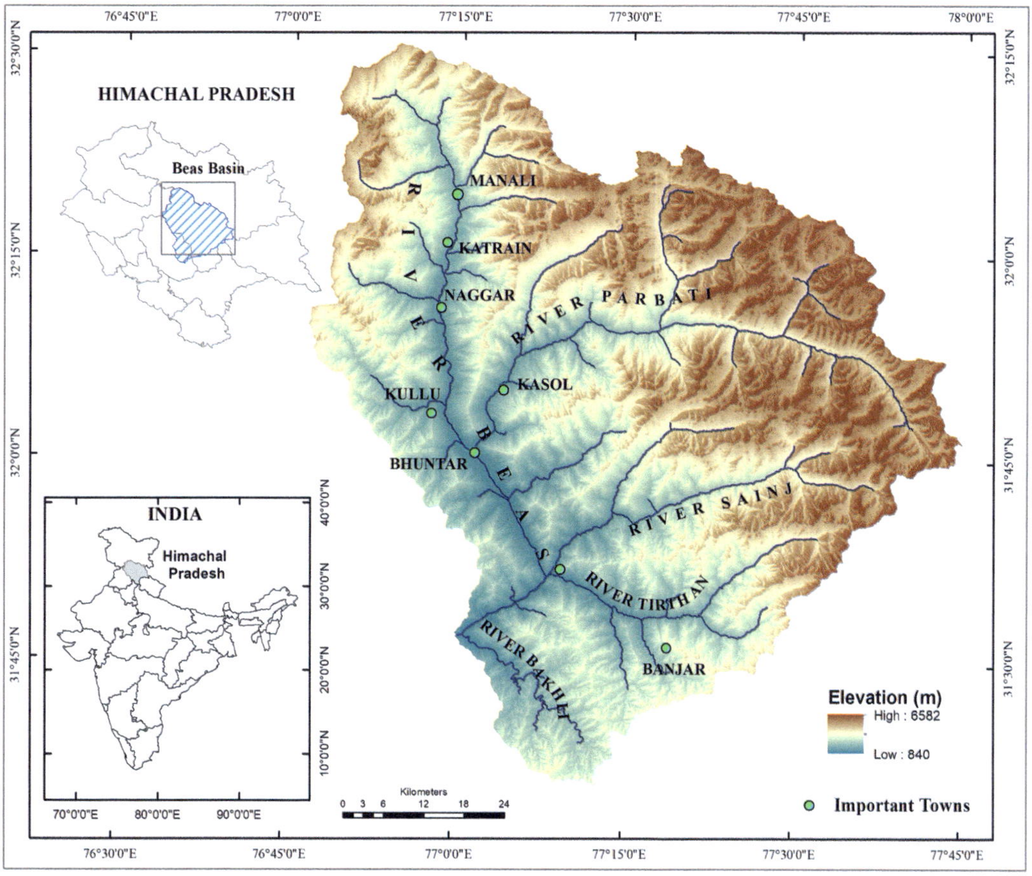

Fig. 20.1 Location and extent of the upper Beas river basin in India. The background 30 m DEM is derived from Cartosat 1 satellite image

Table 20.1 Satellite images used in the study

Satellite	Sensor	Path/Row	Date of acquisition	Number of bands
Landsat 5	TM	147/038	11/16/1991	7
Landsat 7	ETM + SLC on	147/038	10/15/2000	8
Landsat 5	TM	147/038	10/3/2010	7
Landsat 8	OLI_TIRS	147/038	10/25/2018	11

TM Thematic Mapper, *ETM + SLC* On-Enhanced Thematic Mapper Plus, *OLI_TIRS* Operational Land Imager, and Thermal Infrared Sensor

Cartosat-1 digital elevation model (DEM) data (30 m) was taken from the Bhuvan Web site for analyzing the topographical details of the area to develop a decision tree (Cartosat-1 DEM 2014).

20.3.1 Land Use/Land Cover Mapping

A methodological framework for LULC mapping of the study area is presented in Fig. 20.2. Firstly, the basin boundary was demarcated from the DEM in ArcGIS 10.1. The study area from the Landsat images was clipped by the demarcated basin boundary. The top-of-atmosphere (TOA) reflectance of the satellite images was estimated in ENVI 5.2. Dark object subtraction (DOS) atmospheric correction was applied to the images for removing the dark pixel values. It is based on the assumption that dark pixel has no reflection of light. It searches the darkest pixel value in each band and subtracts this value from the value of each band pixel.

Following the LULC classification scheme of the National Remote Sensing Centre (NRSC) and Indian Space Research Organization (ISRO) (2011), seven LULC types were identified and used in the study. They are built-up area, cultivated land, forest, grassland, barren/unculturable/wasteland (BUW), water bodies, and snow. To classify the LULC, a decision tree classification method was developed for the study area in ENVI 5.2. Decision tree[1] classification method is based on certain rules or algorithm that can be used either in one image or multiple images.

It consists of a number of binary conclusions that help to categorize each pixel. Segmentation of the image is not completely based on a single decision in the tree. Rather, each decision rule categorizes the image into two potential categories. Training samples for all LULC classes were prepared with the help of toposheets and field knowledge. Decision rules were derived from spectral separability analysis of the spectral profile of the training samples. Descriptive statistics were computed from the spectral profile of training samples of all LULC classes. Band ratios and their threshold, most suited for identification of different LULC types, were placed in conjunction with the area-specific knowledge of the study area. For better classification results, along with spectral properties, some spectral indices such as a normalized difference snow index (NDSI), water ratio index (WRI), normalized difference built-up index (NDBI), and normalized difference vegetation index (NDVI) were also used.

Quantitative data of LULC changes of images between 1991 and 2018 were compiled. The change in the area and the rate of change in each LULC classes of the study area were, thus, computed as follows:

$$A_{\mathrm{a}} = T_{\mathrm{a}(y2)} - T_{\mathrm{a}(y1)}$$

$$A_{\mathrm{r}} = \frac{A_{\mathrm{a}}}{T_{\mathrm{a}(y1)}} \times 100$$

where y_1 and y_2 are base and current time (year) of the land cover map.

[1]https://perso.univ-lemans.fr/.../06%20Decision%20Tree%20Classification/Decision_Tree.pdf.

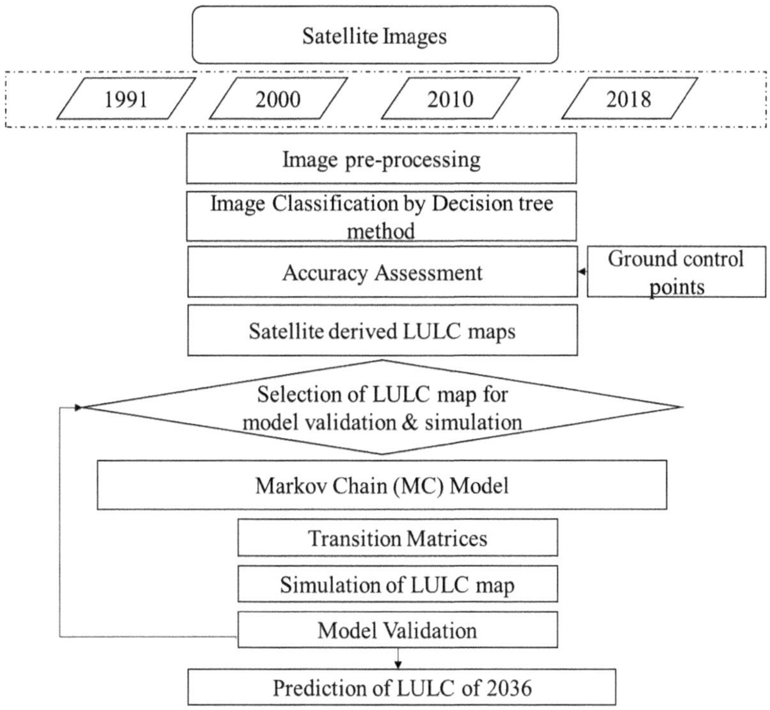

Fig. 20.2 Framework for preparing LULC map of the study area

T_a = Total area; A_a = absolute change in area; A_r = relative change in area.

The LULC transition matrix was also prepared to determine the spatial conversion from one class to another during the period of analysis. Through simple random sampling, samples were chosen for the field survey. Points above 2400 m and far from the road network were considered inaccessible and excluded from the sample of points to be visited. Garmin GPS etrex 10 was used for the field survey. Photographs were clicked on all the GPS points noted during the field study. Kappa (K) coefficient was used for accuracy assessment of the classified map (Cohen 1960). The accuracy of classified maps was also evaluated by an error matrix. "Accuracies of each category in error matrix are plainly described along with the errors of inclusion (commission errors) and exclusion (omission errors) present in the classification" (Congalton 1991), which are calculated as follows:

$$\text{Overall accuracy} = \frac{\sum_{k=1}^{q} n_{kk}}{n} \times 100$$

$$\text{Producer's accuracy} = \frac{n_{ii}}{n_{i+}} \times 100$$

$$\text{User's accuracy} = \frac{n_{ii}}{n_{i+}} \times 100$$

where q is a number of classes; n_{ii} is the total number of correct pixels in a category, n_{i+} is the total number of pixels of that category as derived from the reference data. The overall accuracy of the classified maps lies between 85 and 87%. Kappa coefficients of the mentioned years varied from 0.82 to 0.86 (Table 20.2).

Table 20.2 Accuracy level (in percentage) of classified maps of the basin

LULC classes	1991		2000		2010		2018	
	User's accuracy	Producer's accuracy	User's accuracy	Producer's accuracy	User's accuracy	Producer's accuracy	User's accuracy	Producer's accuracy
Built-up area	93	68	95	72	100	82	96	71
Cultivated land	80	95	82	94	75	84	63	71
Forest	89	90	81	100	88	85	82	78
Grassland	69	78	93	93	80	88	80	98
Barren/unculturable/wasteland (BUW)	71	91	78	88	78	93	78	87
Water bodies	100	82	100	89	100	86	100	91
Snow	100	79	96	84	94	88	96	93
Overall accuracy	85		88		87		85	
Kappa coefficient	0.82		0.86		0.84		0.82	

20.3.2 Prediction of Land Use/Land Cover

Future LULC conditions of the area are predicted, using the MC model in IDRISI software environment. Steps involve reclassification and conversion of format of classified maps, computation of transition probability matrix of the classified maps, and estimation of final suitability of transition for the images. The number of years used for the prediction is equal to the number of years (gap) between the base and current classified map (Palmate et al. 2017). Prediction of LULC includes analysis of changes, transition potentials, and prediction (Eastman 2012). The percentages of the area of two classified images of the years 1991 and 2000 are computed to observe the changes in LULC classes. Then, the earlier and later classified images were used in the Land Change Modeler (LCM) change analysis module to provide gain, loss, and net change in the area of LULC classes. Multilayer perceptron (MLP) method was applied to compute the transition potentials in LULC classes to run the transition submodel (Eastman 2012). MLP method was selected because it can run multiple transitions, up to 9, per submodel. The transition submodel consists of a single or group of land cover transitions in the LULC classes with controlling (driving) factors (elevation and slope in the present study). These controlling factors are helpful to understand the process of past change. Finally, the transition probability of future LULC changes is modeled through MC analysis. Using the rates of change of the LULC images and the transition potential model, future LULC map of a specified year can be predicted. As per the prediction theory in the MC model, the prediction period is equal to the interval between the base and current image used for the simulation (Wang et al. 2018). Classified images of 1991 and 2000 were used for predicting the LULC of 2010, which is validated with the classified image of 2010 using the Kappa (K) coefficient. Similarly, the output from the transition potentials of classified LULC maps of 2000 and 2018 was utilized for predicting the LULC of the year 2036.

20.4 Results and Discussion

20.4.1 Present Scenario of LULC

Land cover of the basin is dominated by forest and snow (Table 20.3). The coverage of snow in the area varies with seasons. Currently, built-up area and cultivated land together constitute approximately 8.5% of the total basin area. Built-up area in the basin remained less than 1% of the total basin area during the last three decades.

On the other hand, cultivated land has been found 4.98% and 7.72% in 1991 and 2018, respectively. Forest area varied between 28% (2018) and 34% (1991). Grassland covers about 15% of the basin area. The extent of BUW has increased during this season due to the least snow-covered area. It remained higher than 20% during the study period. It varies with variations in the extent of snow as debris is also included in this category. Water bodies showed an area below 0.23% of the basin area.

20.4.2 Changes in LULC

The analysis of changes in LULC of the basin has shown a minor reduction in the forest by about 7 km^2 during 1991–2000 (Table 20.4). Cultivated land and snow-covered area have increased by 14 and 53% while grassland decreased by 2% during 1991–2000. On the other hand, cultivated land, grassland, and snow have increased by 20, 55, and 52% during 2000–2010. The major reduction was noted in the area of forest (73 km^2) and BUW (723 km^2) during the period. In the last eight years (2010–18), the major reduction was found in a forest (234 km^2) followed by grassland (103 km^2). In contrast, the cultivated land of the study area has increased by 50 km^2. Two main trends of LULC were observed during 1991–2018: the increase of built-up area (28 km^2) and cultivated land (148 km^2) and a decrease in the forest (314 km^2). Waterbody showed a slight increase of about 6 km^2 during the study period. An increase in snow cover is also observed during

Table 20.3 Land cover conditions of the study area

LULC classes	1991		2000		2010		2018	
	km^2	%	km^2	%	km^2	%	km^2	%
Built-up area	14	0.26	19	0.34	25	0.46	42	0.78
Cultivated land	268	4.98	305	5.67	365	6.78	416	7.72
Forest	1831	34.03	1824	33.89	1751	32.53	1517	28.19
Grassland	605	11.24	594	11.03	921	17.11	818	15.20
Barren/unculturable/wasteland (BUW)	1262	23.44	1976	36.71	1253	23.27	1101	20.45
Water bodies	7	0.12	5	0.1	4	0.08	12	0.23
Snow	1396	25.93	659	12.25	1000	18.57	1476	27.42

Table 20.4 Absolute and relative changes in LULC classes in the study area

LULC classes	1991–2000		2000–2010		2010–2018		1991–2018	
	$A_a(km^2)$	$A_r(\%)$	$A_a(km^2)$	$A_r(\%)$	$A_a(km^2)$	$A_r(\%)$	$A_a(km^2)$	$A_r(\%)$
Built-up area	5	33	6	33	17	94	28	201
Cultivated land	37	14	60	20	50	17	148	55
Forest	−7	0	−73	−4	−234	−13	−314	−17
Grassland	−11	−2	327	55	−103	−17	214	35
Barren/unculturable/wasteland (BUW)	714	57	−723	−37	−152	−8	−161	−13
Waterbodies	−1	−22	−1	−23	8	157	6	84
Snow	−736	−53	340	52	476	72	80	6

the period, but it is difficult to conclude any trend based on results of two-years' images as it requires high temporal resolution and continuous monitoring.

Continuous reduction of forest along the valleys during 1991–2018 is a matter of concern because it may increase the slope instability in the area. Reduction of forests in hilly area can be due to natural factors such as landslides and forest fire and also due to developmental activities such as construction of roads and other infrastructure facilities. Degradation of forested areas at an increasing rate in the Himalaya is also reported by a previous study (Munsi et al 2012).

Spatial distribution of all land cover classes indicates that main changes have occurred in and around the lower reaches of the valley, namely Kullu, Parbati, and Sainj (Fig. 20.3). It is also to be noted that the cultivated area has marginally engrossed to higher altitudes. These may hold severe implications on the ecological system of the basin (Munsi et al 2012). The transition matrix of the land cover classes of 1991 and 2018 is prepared to assess the conversion among different land cover classes. During 1991–2018, about 17% area of cultivated land has been converted into grassland and 3% into BUW (Table 20.5). About 4% area of forest has been converted into cultivated land, 10% area into grassland, and 7% into BUW. From grassland, about 12% area has been converted into cultivated land, 6% area into a forest, and 14% into BUW. Approximately 21% of BUW has converted into the snow as the snow cover area is higher in 2018.

A major rise in settlement and the cultivated area was found after 2000, probably due to a rise in population and tourism activities. A past study

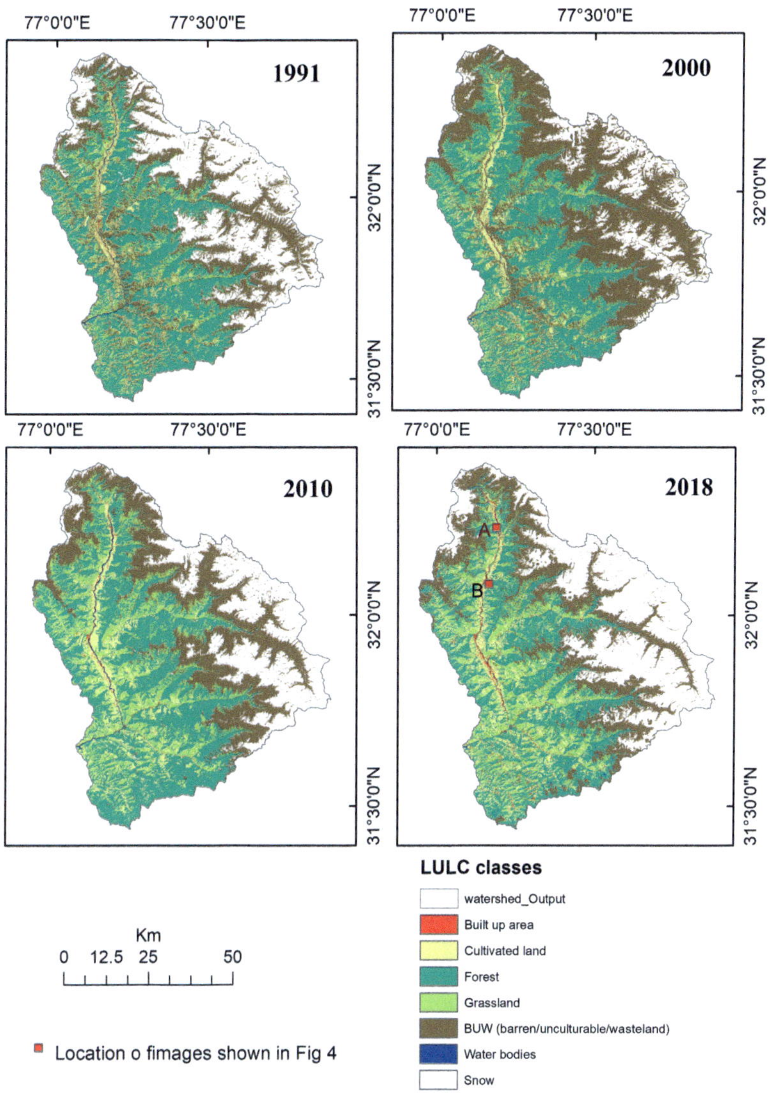

Fig. 20.3 Distribution of LULC classes of the study area

has also attributed the changes in LULC of the area to the rising population (Singh 1998). The density of settlements has increased over the time (Fig. 20.4). As per the data from the Census of India (2011), Manali, Kullu, Samshi, Bhuntar, and Banjar are important towns in the area. Among them, Manali and Bhuntar have been experiencing a population growth rate of 252 and 62%, respectively, between 1981 and 2011.

The former town has recorded a substantial population growth rate of about 157% during the period 1991–2001. It was noted as the highest among all the towns in the state of Himachal Pradesh indicating that Manali has attracted a significant migrant population during this period. Bhuntar, during the same decade, has also shown a population growth of around 43%. Subsequently, anthropogenic activities have increased in the region. The upper Beas basin is located mainly in Kullu district (leaving Ani and Hermand block), which has registered a decadal population growth 15% during 2001–2011, compared to the growth rate of about 13% in the state. The main source of income of the area is agriculture and tourism. It holds the 4th position among the districts of the state in terms of

Table 20.5 Transition matrix (in percentage) of the land cover classes of study area

		1991						
	LULC classes	Built-up area	Cultivated land	Forest	Grassland	Barren/unculturable/wasteland (BUW)	Waterbodies	Snow
2018	Built-up area	27	5	0	0	1	21	0
	Cultivated land	54	68	4	12	6	4	0
	Forest	7	6	78	6	1	29	1
	Grassland	3	17	10	67	15	1	0
	Barren/unculturable/wasteland (BUW)	4	3	7	14	56	6	13
	Waterbodies	5	0	0	0	0	39	0
	Snow	0	0	0	0	21	0	86

Fig. 20.4 Settlements in the Kullu valley of the basin

cultivators. It results in the expansion of cultivated land over other land categories. The main changes observed in the LULC in the transition probability matrix were from cultivated land to built-up area, from forest to cultivated land and from grassland to cultivated land.

20.4.3 Predicted LULC

Predicted LULC map of the basin is validated to check the model's uncertainty and set the model for future land cover map prediction. LULC maps for the years 1991 and 2000 were used to predict the land cover conditions of 2010 in the study area using an MC model. Then, predicted maps were compared with the LULC of 2010 derived from the satellite image to understand the level of uncertainty in the prediction of different LULC classes.

Results indicate that some classes were predicted well, except grassland, BUW, and snow (Table 20.6 and Fig. 20.5). Grassland and snow are underpredicted, though the underpredicted area is about 6% of the entire watershed area. Further, BUW was overpredicted, i.e., around

13% of the study area because it contains a major portion of snow debris that covered a larger area in 2000 due to a reduction in snow coverage in the basin. The model's transition probability matrix assumes an increase in BUW in the predicted map. Likewise, snow cover was underestimated because of the snow-covered less area in 2000. Kappa coefficient of predicted and observed LULC map of 2010 is about 0.82, which indicates the overall reliability in the predicted map. It shows the limitation of the application of the MC model in a larger area, which has low and dispersed urban growth like hilly areas. It also emphasis on the importance of usage of continuous high-resolution satellite images for larger hilly terrain for minimizing the uncertainty in the prediction of LULC conditions. Such predicted map is useful for evaluating the effect of future LULC on the discharge conditions of a Himalayan watershed because the uncertainty in the predicted area is a small fraction of the entire basin area and would make a minor or no alteration in the streamflow in the region and can give better estimation at the outlet.

Transition probability matrix for predicting the LULC conditions 2036 of the study area is to indicate an increase in built-up area, cultivated land, snow cover, and decrease in the forest, BUW (Table 20.7).

Changes in BUW due to snow are common in a Himalayan watershed that is dominated by the snow because snow cover varies with the seasons in the basin (Rani and Sreekesh 2016). Hence, it is not logical to accurately predict the BUW and snow based on two time period images, which is also a limitation of this model because it is unable to consider multiple period satellite images for prediction.

Predicted LULC of 2036 shown LULC transitions during 2018–2036 (Figs. 20.5 and 20.6). Built-up area and cultivated land will increase to 1.12 and 8.447% of the basin area by 2036, particularly along with the valley areas. Built-up area and cultivated land shown a rise of 43 and 9%, whereas the area under forest has shown a reduction of 4%. BUW area will reduce to 12% of the basin area. While snow covers shown an increase to 35% of the basin. However, the previous study has shown a reduction in snow cover area over the years in the basin (Rani and Sreekesh 2016). Singh (1998) and Munsi et al. (2012) have shown concern of reducing the biodiversity of the region due to deforestation. A long-term database might provide a potential tool for further related studies, planning, and implementation for land and water resource management, thus leading to sustainable development.

Table 20.6 Validation of predicted LULC map (2010) of the study area

LULC classes	2010		Difference b/w observed and predicted area (km^2)	Area (%)
	Area (km^2)			
	Observed	Predicted		
Built-up area	14.4	18.4	4	0.1
Cultivated land	348.2	310.9	−37.3	−0.7
Forest	1832.9	1817.3	−15.6	−0.3
Grassland	924.7	597.2	−327.6	−6.1
Barren/unculturable/wasteland (BUW)	1230.8	1957.8	727	13.5
Waterbodies	15	5.6	−9.4	−0.2
Snow	999.7	658.6	−341.1	−6.3

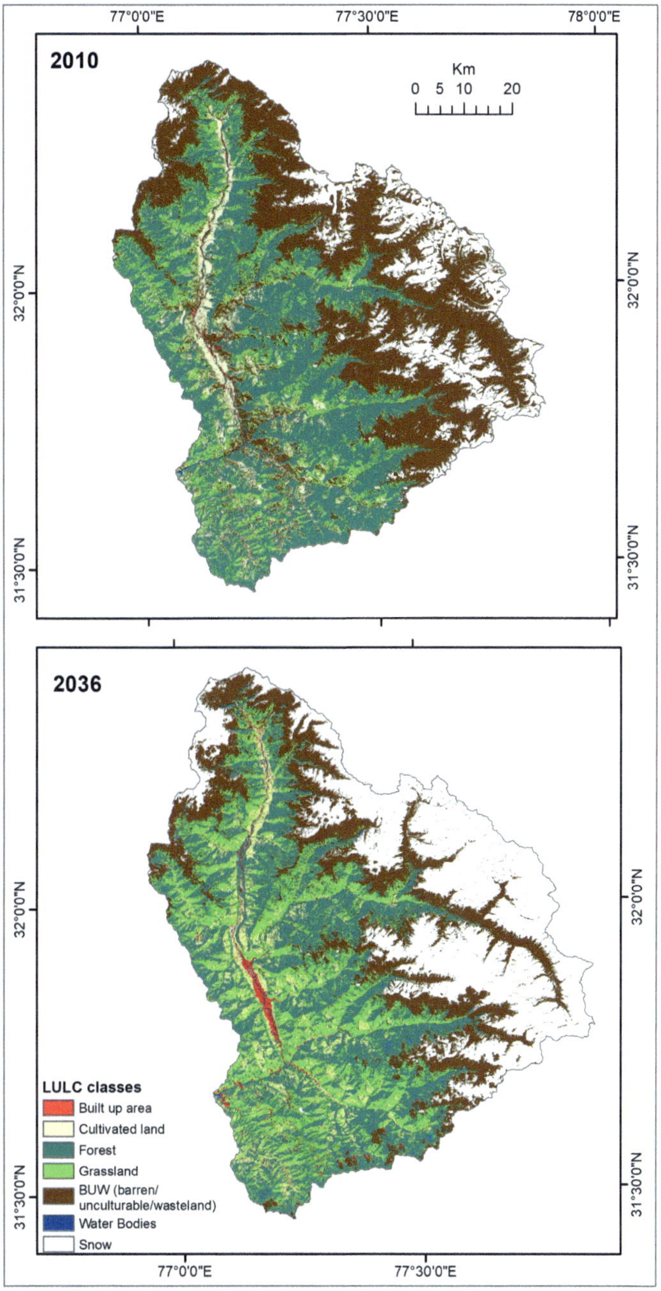

Fig. 20.5 Predicted LULC conditions of the study area during 2010 and 2036

Table 20.7 Transition probability matrix on the bases of LULC maps of 2000 and 2018

LULC classes	Built-up area	Cultivated land	Forest	Grassland	BUW—barren/unculturable/wasteland	Waterbodies	Snow
Built-up area	0.43	0.20	0.13	0.12	0.07	0.04	0.00
Cultivated land	0.04	0.70	0.05	0.18	0.03	0.00	0.00
Forest	0.00	0.04	0.79	0.11	0.05	0.00	0.00
Grassland	0.00	0.13	0.05	0.66	0.15	0.00	0.00
BUW—barren/unculturable/wasteland	0.01	0.02	0.10	0.09	0.55	0.00	0.32
Waterbodies	0.06	0.01	0.48	0.01	0.01	0.43	0.00
Snow	0.00	0.00	0.00	0.00	0.02	0.00	0.97

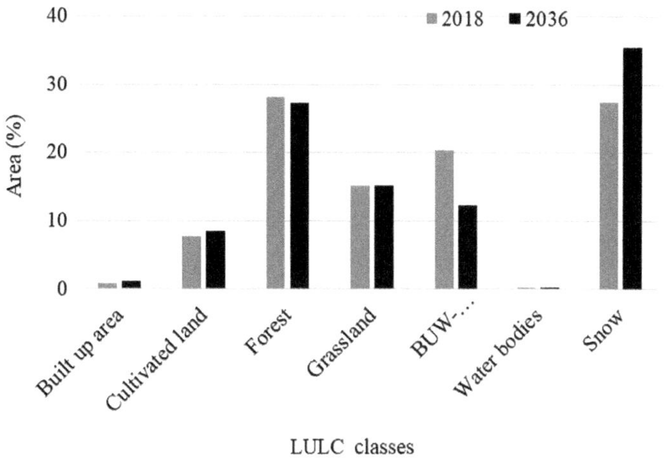

Fig. 20.6 Changes in LULC of the study area during 2018–2036

20.5 Conclusion

Concern about emerging LULC changes in the Himalayas has increased over the time due to the occurrence of disasters such as floods and landslides. Anthropogenic activities have increased over the time in the region due to the rising population and increased demand for food and energy. Hence, studies are making efforts to predict future LULC changes for effective land and water resource planning and management. Hence, the present work is an effort to understand the existing LULC conditions of the upper Beas basin and predict their changes by 2036. Digitally analyzed satellite images were used in the LCMES module of IDRISI by MC analysis. The LULC during 2018 has shown that built-up area and cultivated land together constitute about 8.5% of the total basin area. Forest and grassland cover an area of 28 and 15%. Snow cover is about 27% of the basin area. The LULC analysis showed an increase of built-up area and cultivated land and a decrease in forest cover. The transition matrix indicates the conversion of cultivated land into built-up area along the valleys and conversion of forest into cultivated land in the lower altitudes. An increase in built-up area and cultivated land may be attributed to rising anthropogenic activities such as tourism and expansion of horticulture along the hill slopes, respectively. Reduction in the forest area is a continuous process in the basin in the last three decades. It may be attributed to the expansion of cultivated land (including plantation) along the lower elevations and natural factors (landslide or forest fire).

Results of LULC prediction indicated the reliability of the MC model and deployability in the Himalayan environment for the wider applications, such as land and water resource management. As per the predicted LULC outputs, built-up area and cultivated land together will cover about 9% of the basin area by 2036, particularly along with the valley areas. The built-up area and cultivated land will increase by 43% and 9%, whereas the area under forest has shown a reduction of 4%. Expansion of concrete surface along the river banks makes the population vulnerable to flood-related hazards. The reduction in the forest area in the region is relatively less. Nevertheless, continuous reduction in the forest is a matter of concern because it may harm the ecological balance. Decreasing forest along with an increase in built-up area can lead to slope instability, which may make the area prone to landslides and other associated risks. Forest cover in the area may be increased by plantation, but it will not increase the biodiversity of the region, which is very important for maintaining the ecological balance and overall sustainability in the area.

Acknowledgements The authors acknowledge the United States Geological Survey (USGS) for providing the remote sensing data. The first author is thankful to the University Grant Commission (UGC), Ministry of Human Resource Development, Government of India, for providing the senior research fellowship for the study.

References

Anand J, Gosain AK, Khosa R (2018) Prediction of land use changes based on land change modeler and attribution of changes in the water balance of Ganga basin to land use change using the SWAT model. Sci Total Environ 644:503–519

Bagan H, Yamagata Y (2012) Landsat analysis of urban growth: How Tokyo became the world's largest megacity during the last 40 years. Remote Sens Environ 127:210–222

Bhakra Beas Management Board (BBMB) (1988) Snow hydrology studies in India with particular reference to the Satluj and Beas catchments. In: Proceeding of workshop on snow hydrology. Manali, India 23–26 Nov, pp 1–14

Cartosat-1 DEM (2014) https://bhuvan-noeda.nrsc.gov.in/data/download/index.php. Accessed 20 Dec 2018

Census of India (2011) Provisional population totals paper 1 of 2011: Himachal Pradesh. https://www.censusindia.gov.in/2011-prov-results/prov_data_products_himachal.html. Accessed 20 May 2018

Chandra G, Defourny P, Surendra S (2003) Land cover characterization and mapping of continental Southeast Asia using multi-resolution satellite sensor data. Int J Remote Sens 24(21):4181–4196

Cohen J (1960) A coefficient of agreement for nominal scales. Educ Psychol Meas 20(1):37–46

Congalton RG (1991) A review of assessing the accuracy of classifications of remotely sensed data. Remote Sens Environ 37(1):35–46

Costanza R, Ruth M (1998) Using dynamic modeling to scope environmental problems and build consensus. Environ Manage 22(2):183–195

Di Gregorio A, Jansen LJM (2000) Land Cover Classification System (LCCS). Classification concepts and user manual. Food and Agriculture Organization of the United Nations (FAO). Rome, Italy

Eastman JR (2012) IDRSI Selva manual. www.eng.usm.my/file/27/.../IDRISI+Manual.pdf. Accessed on 20 Jan 2018

Gashaw T, Tulu T, Argaw M, Worqlul AW (2017) Evaluation and prediction of land use/land cover changes in the Andassa watershed, Blue Nile Basin, Ethiopia. Environ Syst Res 6(1):6–17

Goldshleger N, Maor A, Garzuzi J, Asaf L (2015) Influence of land use on the quality of runoff along Israel's coastal strip (demonstrated in the cities of Herzliya and Ra' anana). Hydrol Process 29(6):1289–1300

Han J, Hayashi Y, Cao X, Imura H (2009) Application of an integrated system dynamics and cellular automata model for urban growth assessment: a case study of Shanghai, China. Landscape Urban Plan 91(3):133–141

Hu Z, Lo CP (2007) Modeling urban growth in Atlanta using logistic regression. Comput Environ Urban 31 (6):667–688

Huang GB, Ding X, Zhou H (2010) Optimization method based extreme learning machine for classification. Neurocomputing 74(1):155–163

Kamusoko C, Aniya M, Adi B, Manjoro M (2009) Rural sustainability under threat in Zimbabwe-simulation of future land use/cover changes in the Bindura district based on the Markov-cellular automata model. Appl Geogr 29(3):435–447

Lambin EF, Turner BL, Geist HJ, Agbola SB, Angelsen A, Bruce JW, Coomes OT, Dirzo R, Fischer G, Folke C (2001) The causes of land-use and land-cover change: moving beyond the myths. Glob Environ Chang 11(4):261–269

Lu Y, Wu P, Ma X, Li X (2019) Detection and prediction of land use/land cover change using spatiotemporal data fusion and the Cellular Automata–Markov model. Environ Monit Assess 191:68. https://doi.org/10.1007/s10661-019-7200-2

Maithani S (2015) Neural networks-based simulation of land cover scenarios in Doon valley, India. Geocarto Int 30(2):163–185

Mishra NK, Rai PK (2016) A remote sensing aided multi-layer perceptron-Markov chain analysis for land use and land cover change prediction in Patna district (Bihar), India. Arab J Geosci 9:249

Mitsova D, Shuster W, Wang X (2011) A cellular automata model of land cover change to integrate urban growth with open space conservation. Landscape Urban Plan 99(2):141–153

Munsi M, Areendran G, Joshi PK (2012) Modeling spatio-temporal change patterns of forest cover: a case study from the Himalayan foothills (India). Reg Environ Change 12:619–632

NASA Landsat Program (2019) Landsat TMLT51470381991320ISP00, 11/16/1991; Landsat LE71470382000289SGS00, 10/15/2000; Landsat LT51470382010276KHC00, 10/3/2010; Landsat LC08147038201810252018103101, 10/25/2018, L1T, Terrain Corrected, USGS. L1T, Terrain Corrected, USGS. http//glovis.usgs.gov, https://earthdata.nasa.gov/

Nelson E, Mendoza G, Regetz J, Polasky S, Tallis H, Cameron D et al (2009) Modeling multiple ecosystem services, biodiversity conservation, commodity production, and tradeoffs at landscape scales. Front Ecol Environ 7(1):4–11

Niemelä J, Kotze J, Ashworth A, Brandmayr P, Desender K, New T, Penev L, Samways M, Spence J (2000) The search for common anthropogenic impacts on biodiversity: a global network. J Insect Conserv 4(1):3–9

National Remote Sensing Centre (NRSC) and Indian Space Research Organization (ISRO) (2011) "Manual on preparation of geo spatial layers using high resolution (Cartosat-1Pan+LISS-IV mx) Orthorectified satellite imagery". Space based information support for decentralized planning (SIS-DP), remote sensing and GIS applications area National Remote Sensing Centre, Indian Space Research Organisation (ISRO), Department of Space, Government of India, Hyderabad

Osgouei PE, Kaya S (2017) Analysis of land cover/use changes using Landsat 5TM data and indices. Environ Monit Assess 189(4):136

Palmate SS, Pandey A, Mishra SK (2017) Modelling spatiotemporal land dynamics for a trans-boundary river basin using integrated Cellular Automata and Markov Chain approach. Appl Geogr 82:11–23

Rani S, Sreekesh S (2016) An analysis of pattern of changes in snow cover in the upper Beas river basin, Western Himalaya. In: Raju N (ed) Geostatistical and geospatial approaches for the characterization of natural resources in the environment. Springer, Germany, pp 899–903

Remondi F, Burlando P, Vollmer D (2016) Exploring the hydrological impact of increasing urbanisation on a tropical river catchment of the metropolitan Jakarta, Indonesia. Sustain Cities Soc 20:210–221

Saleem A, Corner R, Awange J (2018) On the possibility of using corona and landsat data for evaluating and mapping long-term lulc: case study of iraqi kurdistan. Appl Geogr 90:145–154

Shaw SB, Marrs J, Bhattarai N, Quackenbush L (2014) Longitudinal study of the impacts of land cover change on hydrologic response in four mesoscale watersheds in New York State, USA. J. Hydrol 519:12–22

Singh SK, Mustak SK, Srivastava PK, Szabó S, Islam T (2015) Predicting spatial and decadal LULC changes through cellular automata Markov chain models using earth observation datasets and geo-information. Environ Process 2:61–78

Singh RB (1998) Land use/cover changes, extreme events and ecohydrological responses in the Himalayan region. Hydrol Process 12:2043–2055

Srivastava PK, Han D, Gupta M, Mukherjee S (2012) Integrated framework for monitoring groundwater pollution using a geographical information system and multivariate analysis. Hydrol Sci J 57(7):1453–1472

Szabó S, Csorba P, Szilassi P (2012) Tools for landscape ecological planning—scale, and aggregation sensitivity of the contagion type landscape metric indices. Carpathian J Earth Environ Sci 7:127–136

Tallis H, Polasky S (2009) Mapping and valuing ecosystem services as an approach for conservation and natural-resource management. Ann Ny Acad Sci 1162 (1):265–283

Turner AG, Annamalai H (2012) Climate change and the South Asian summer monsoon. Nat Clim Change 2 (8):587–595

Wang G, Xu Z, Zhang S (2014) The influence of land use patterns on water quality at multiple spatial scales in a river system. Hydrol Process 28(20):5259–5272

Wang R, Hou H, Murayama Y (2018) Scenario-base simulation of Tianjin City using a cellular automata–Markov model. Sustainability 10:2633. https://doi.org/10.3390/su10082633

Zheng J, Yu X, Deng W, Wang H, Wang Y (2012) Sensitivity of land-use change to Streamflow in Chaobai river basin. J Hydrol Eng 18(4):457–464

21 Dynamics of Land-Use/Cover Change in Mizoram, Eastern Extension of Himalaya

Vishwambhar Prasad Sati

Abstract

Land-use/cover change has become a global phenomenon and a major concern, partly due to large-scale anthropogenic activities and because of its impact on the natural environment. This paper assesses land-use/cover changes in Mizoram state, which is located in the eastern extension of the Himalaya. Land-use data were obtained from the secondary sources—Forest Survey of India and land-use statistics, Ministry of Agriculture, Government of India, during the two consecutive periods—2005 and 2010. Data on changes in cropping pattern were also gathered from the State Agricultural Department for the same period. A case study of 16 villages of Mizoram state was conducted and land-use data were obtained through the household level survey. Land-use pattern in Mizoram is very peculiar. Forest covers about 75.58% land (2010), while forest area, notified by the forest department is 86% (2017). In the meantime, land under net area sown is only 4.5%. Out of the total sown area, 47.5% area is under shifting cultivation. Land cover changes have been noticed during the corresponding years. Area under *Jhum* and wet paddy cultivation has largely decreased, while net area sown was increased slightly (1.1%). Land under forest cover has decreased about 2% during the period of production and productivity of a number of crops have decreased. This study shows that increase and decrease in area, production and yields of various crops do not present any specific trend rather it is heterogeneous and the various factors are affecting these changes.

Keywords

Land-use pattern · Land cover change · Cropping pattern · Shifting cultivation · Mizoram

21.1 Introduction

Land cover may be defined as the biophysical earth surface, while land-use is often shaped by human, socioeconomic and political influences on the land (Nagendra et al. 2003). Land-use refers to uses of land under various categories such as forest land, cultivable land, wasteland, land under settlements and other miscellaneous land while land cover changes denote the changes in various categories of land during a certain period of time (Sati 2014). The land-use/cover changes (LUCC) are mainly anthropogenic and are being increasingly recognized as critical

V. P. Sati (✉)
Department of Geography and Resource Management, Mizoram University, Aizawl 796004, Mizoram, India

U. Schickhoff et al. (eds.), *Mountain Landscapes in Transition*,
Sustainable Development Goals Series, https://doi.org/10.1007/978-3-030-70238-0_21

factors influencing global environmental changes (Helmut et al. 2002; Nagendra et al. 2004). The LUCC is primarily and largely confined to tropical countries (Meyers 1994). However, the conversion of forest to other agricultural land-use, mainly shifting agriculture is common in the Northeast India (Singh et al. 1984; Rai et al. 1994; Ramakrishna et al. 1994; Schweik et al. 1997; Sen et al. 2002), as it continues to be the main source of livelihood for majority of people.

Land cover change is a dynamic process and influenced by various factors. In Mizoram, change in land cover is common, because of many driving forces, mainly due to shifting cultivation, which dominates the farming systems. Shifting cultivation leads to drastic changes under area sown from year to year, because of its cyclic nature, and therefore, it has tremendous impact on changes in all the land-use categories. The hilly terrain and extensive forestland (75.5%) characterizes land-use pattern in Mizoram. Forestland that includes grasses and bushes covers about 86% geographical area (FSI 2017). Net area sown is only 4.5% and out of which, 56.8% land is under shifting cultivation. Shifting cultivation is mainly carried out in gentle to steep slopes, while valley regions constitutes very small area where permanent agriculture is practiced which is known as wet rice cultivation (WRC). The other land-use patterns are permanent pasture and other grazing lands, land under misc. tree crops and groves, cultivable wasteland, fallow land other than current fallow and current fallow land. Settlements are scattered, located mainly along the hills and ridges. The people prefer residing on the top of the hills and it is also an indicator of their social status. Traditional subsistence cereal crops dominate the cropping pattern, carried out mainly under the shifting cultivation and it is the main occupation of the people. About 54% people are engaged in practicing shifting cultivation. Area and production of paddy are higher in comparison with the other crops and it grows both in the hilly terrain under shifting cultivation and in the small valley patches as WRC. Pulses, oilseeds, fruits and vegetables, the other crops, grow under shifting cultivation, whereas the proportion of arable land is less under these crops. Further, area under WRC is proportionately very less. Even being as the main occupation, overall output from both shifting and WRC is subsistence, does not substantiate the food requirement of the rural marginal farmers, and therefore, they struggle for sustaining livelihoods. Further, often poverty and malnutrition are observed common. This paper assesses the land-use/cover changes and examines the major factors affecting it. It broadly analyzed forest and agricultural land-use/cover change of Mizoram state in general and case study of villages in particular.

21.2 Materials and Methods

21.2.1 The Study Area

Mizoram, a state of republic of India is located in the extreme northeast part, bordered with Myanmar in the east and south and Bangladesh in the west. Its border with the states of northeast India includes Assam and Manipur in the north and northeast, and Tripura delimits its small boundary from the northwest. It has eight districts and 26 administrative blocks with Aizawl as the state capital city. Its total population is 10.92 lakh and out of it, 2.9 lakh people (26.6%) live in Aizawl city (2011). About 98% people are Christian. Population distribution is sparse and the settlements are mostly located in the highlands; therefore, Mizoram is called the 'land of highlanders.' Literacy rate stands for 91.85% while sex ratio is 975 and density of population is only 52 persons/km^2. Mizoram is located in the eastern extension of the Himalaya. It is an integral part of the Indo-Myanmar Arc. Arakan-Yoma and Patkoi Hills are located in the Northeast India and the Mizoram hills constitute their parts. The average altitude of these hills varies from 500 m to 800 m AMSL and the maximum elevation is 2,157 m, which is found in the Blue Mountain (Phawngpui) (ISFR 2011a). The hills of Mizoram are highly fragile. Land degradation in the forms of soil erosion and landslides is common. Its total geographical area is 21,087 km^{2}and shares 0.64% of the country's geographical area. Lying to the south of the Brahmaputra River, it forms a part of the Northeast hill states bio-geographical zone. A landlocked state, Mizoram extends between 21° 58' to 24° 35' N and 92° 15' to 93 ° 29' E (Fig. 21.1).

21.2.2 Methodology

This study was conducted using both qualitative and quantitative approaches. Data were collected from the primary and secondary sources. Primary data on land-use pattern were collected through conducting a study of 16 villages of eight districts of Mizoram state and household (HH) level survey was carried out (2014). Out of total 2010 HHs, 1527 HHs (76%) were selected for inclusive HH level survey. A structured questionnaire was constructed on the various issues and the local people were interviewed; land-use and cropping pattern data from the 16 case study villages were obtained. Land-use data were also obtained from the secondary sources, mainly from the land-use statistics, Ministry of Agriculture, Government of India, 2005 and 2010; land cover changes were illustrated and presented in the tables and figures. Data were also calculated using descriptive statistics.

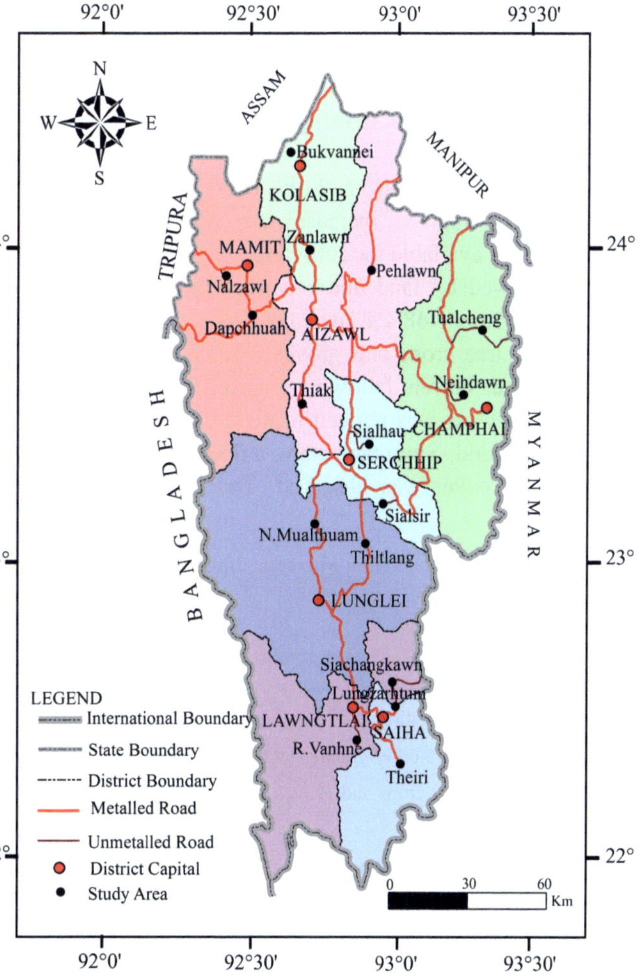

Fig. 21.1 Location map of Mizoram state

21.3 Results

21.3.1 Land-Use Pattern

Forest land and shifting cultivation largely characterize the land-use pattern in Mizoram state. Forest covers about 75.58% land, excludes the extensive grasslands, while total forest area is 86% (2017). It is followed by fallow land (current fallows and fallow land other than current fallow) which occupies 10.95%. The land, which is not available for cultivation, ranks third (6.31%). Land under misc. tree crops and groves have 2.18% area. The other land-uses are permanent pasture and other grazing lands (0.24%) and cultivable wasteland (0.24%). Meanwhile, net area sown obtains only 4.5% geographical areas (Table 21.1).

21.3.2 Land Cover Change

Change in land cover from 2005 to 2010 was assessed. The highest increase was obtained under land not available for cultivation (731.3%). It was followed by land under current fallows, i.e., 66.7%. About 48.4% increase was noticed under misc. tree crops and groves land. Fallow land other than current fallow increased by 9.6% and 1.1% increase was noticed in net area sown during the period. Simultaneously, a decrease in various land cover was observed. The highest decrease was noticed under cultivable wasteland (96.1%) followed by permanent pasture and other grazing lands (78.3%). Forest land decreased by about 2% (Table 21.2). During the recent period (2011–2017), forests cover has decreased by 6%.

21.3.3 Forest Land-Use/Cover Change

Mizoram state enjoys rich biodiversity. It forms a part of the Indo-Burma Global Biodiversity Hotspot. Natural vegetation comprises from tropical evergreen in the lower altitudes to semi-evergreen on the upper slopes (Champion and Seth 1968). Diversity in forests is found according to altitude, rainfall and dominant species composition (Singh et al. 2002). One of the major sources of livelihoods, forest resources in Mizoram state dominate in the natural resources potential. Tropical evergreen rainforest and semi-evergreen forest characterize forest diversity, as most of the part of the state falls under tropical regime. Meanwhile, montane and temperate forests are also found from 900 m to 2000 m, respectively. Bamboo forests are found almost in all the altitudinal zones but restricted up to 1500 m. The forestland in the state was noticed 19,117 km^2, which represents 90.68% of the state's geographical area (ISFR 2011b). Out of the total forest cover, protected area occupies

Table 21.1 Land-use pattern in Mizoram: 2005 and 2010

Land-use	2005		2010	
	Area in '000 ha	Percentage	Area in '000 ha	Percentage
Forest	1626	77.10	1594	75.58
Not available for cultivation	16	0.76	133	6.31
Permanent pasture and other grazing lands	23	1.09	5	0.24
Land under misc. tree crops and groves	31	1.47	46	2.18
Cultivable wasteland	127	6.02	5	0.24
Fallow land other than current fallow	156	7.40	171	8.11
Current fallows	36	1.71	60	2.84
Net area sown	94	4.46	95	4.50
Total	2109	100	2109	100

Source Land-use statistics, Ministry of Agriculture, Government of India, 2005 and 2010

Table 21.2 Land covers change: 2005–2010

Land-use	Land cover change (ha)	Percentage of land cover change
Forest	−32	−2
Not available for cultivation	117	731.3
Permanent pasture and other grazing lands	−18	−78.3
Land under misc. tree crops and groves	15	48.4
Cultivable wasteland	−122	−96.1
Fallow land other than current fallow	15	9.6
Current fallows	24	66.7
Net area sown	1	1.1

Source Analyzed by author

1,240.75 km^2, represents 5.88%. In terms of forest canopy density classes, the state has 134 km^2 areas under the dense forests, 6,086 km^2 area under moderately dense forests and 12,897 km^2 areas under open forests.

Table 21.3 shows altitudinal distribution of forests and their characteristics in Mizoram state. About 42.4% forest area lies below 500 m altitude. It is followed by area under 500-1,000 m altitude that occupies 41.5% of the total forest cover. The area, which lies between 1,000 m and 2,000 m altitude, forest area is 16%, whereas 0.02% forestland lies above 2000 m altitude. It shows that the highest geographical area of the state lies below 500 m altitude; therefore, forest cover decreases along with increase in altitude.

21.3.4 Agricultural Land-Use/Cover Change

Agricultural practices comprise both shifting cultivation and WRC. Shifting cultivation is practiced largely in the hilly slopes and spread from mountains to lowland ecosystems and from tropical forests to grasslands (Spencer 1966). It is the main occupation of the rural people. Singh and Ramakrishnan (1982) observed that shifting cultivators comprise of 82% of the rural main workers and few urban main workers also involved in shifting cultivation. Maithani (2005) observed that shifting cultivation is widely practiced in Mizoram. It is the main occupation of the populace and a major source of economy. There has been an ambiguity in terms of reporting area under shifting cultivation. A report of the Inter-Ministerial National Task Force on Rehabilitation of Shifting Cultivation Areas (GBPIHED 2008) reveals that annual area under shifting cultivation in Mizoram was 63,000 ha. Kumar (2012) reported that during 2004–2005, area under shifting cultivation was 64,536 ha, whereas Pachuau (2009) in his study of Mizoram denotes that 40,969 ha land (50% of the total cropped area) was under shifting cultivation, during the same period (2004–05). Figure 21.2 shows area under WRC and Shifting cultivation.

Table 21.3 Forest cover according to altitude (Km^2)

Altitude (M)	VDF	%	MDF	%	OF	%	Total	%
<500	15	11.2	1971	32.4	6129	47.5	8115	42.4
500–1000	56	41.8	2872	47.2	5001	38.8	7929	41.5
1000–2000	62	46.3	1241	20.4	1765	13.7	3068	16.0
>2000	1	0.7	2	0.03	2	0.02	5	0.02
Total	134	100	6086	100	12897	100	19117	100

Source ISFR (2011). Abr. VDF: Very dense forests, MDF: Moderate dense forests, OF: Open forests. (Based on SRTM, Digital Elevation Model)

Agriculture land-use varies from paddy crop that grows under both shifting and WRC to cultivating various other crops largely in the *Jhumland*. Paddy crop grows twice in a year in both Rabi and Kharif seasons only under WRC, in the valleys and flat land, where irrigation facilities are adequate. Mizoram state has suitable agro-climatic conditions, as average annual temperature is 23° and minimum and maximum temperature is 15° and 30°, respectively. Therefore, it supports cultivating paddy crop twice in a year along with varieties of pulses, oilseeds, vegetables and fruits, vertically and horizontally. On account of arable land under various crops, paddy crop under shifting cultivation dominates, as it covers 36,841 ha area, which is 56.8% of the total arable land. WRC follows it with covering 10,363 ha arable land (16% of the total area sown). The third largest crop is maize, which obtains 8,551 ha arable land (13.2%). Pulses (6%) and oilseeds (4.2%) follow it. Maize, pulses and oilseeds also grow twice in a year in both seasons. Arable land of other crops—potato, sugarcane and tapioca—varies from 0.3 to 2.2% and the crops are. Table 21.4 reveals agricultural land-use in 2005 and 2010 and shows that shifting cultivation obtained about 56.8% area in 2010.

Change in cropping pattern (area, production and yield) was noticed during the years 2005–2010 (Table 21.5). Except sugarcane, which area was increased to 3.7%, area under all other crops decreased significantly. Area under potato has decreased by 70.1% and oilseeds (-53.3%) and pulses (42.9%) follow it. Area of paddy under *Jhum* was decreased by 8.1% and under WRC; it decreased by 36.7%. Average area under all crops has decreased by 23.1%. Simultaneously, production and yield of crops decreased (average 47.3 and 36.8%, respectively).

21.4 Case Study

A case study of 16 villages of eight districts of Mizoram state was carried out and HH level survey was conducted. Out of 2010 HHs of the sixteen villages, 1527 HHs (76%) was surveyed. Data on land-use pattern were gathered and discussed to understand the present trend. Shifting cultivation is the main occupation of the people of these villages as about 25.6% people are engaged in this practice. It is followed by people, who are engaged in WRC (25.4%). The third largest category of people is involved in daily wages. Discussion on the occupational structure revealed that out of the total population, above 50% people are engaged in practicing agriculture.

Table 21.6 shows land-use pattern in the sixteen case study villages and descriptive statistics is shown in Table 21.7. Total land under different categories, except settlement, is 1,821.64 ha, out of which 38.2% land is under shifting

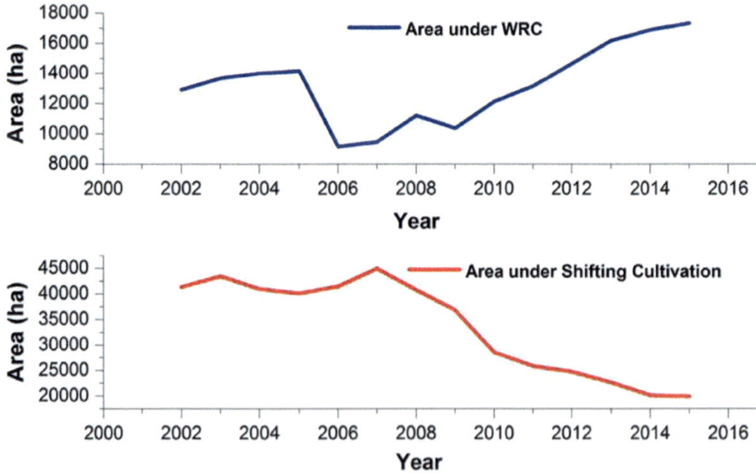

Fig. 21.2 Area under WRC and shifting cultivation

Table 21.4 Agricultural land-use

Crops	2005					2010				
	Area (ha)	%	Prod (MT)	%	Yield (MT/ha)	Area (ha)	%	Prod (MT)	%	Yield (MT/ha)
Paddy (Jhum)	40,100	47.5	63,100	32.2	1574	36,841	56.8	43,985	42.5	1194
Paddy (WRC)	16,360	19.3	44,640	22.7	2729	10,363	16	22,147	21.4	2137
Maize	11,742	13.9	22,703	11.6	1933	8551	13.2	11,510	11.1	1346
Pulses	6861	8.1	8663	4.4	1263	3920	6	6479	6.3	1653
Tapioca	300	0.3	1222	0.6	4073	193	0.3	1397	1.3	7238
Oilseeds	5870	6.9	5560	2.8	947	2741	4.2	2988	2.9	1090
Cotton	308	0.7	241	0.1	782	201	0.3	128	0.1	637
Tobacco	511	0.6	364	0.2	712	371	0.6	249	0.2	671
Sugarcane	1383	1.6	45,953	23.4	33,227	1434	2.2	12,368	12	8623
Potato	953	1.1	3891	2	4083	285	0.4	2235	2.2	7853
Total	84,388	100	196,337	100	5132.3	64,900	100	103,486	100	3244.2

Source Data were collected from the State Agricultural Department Reports: 2005 and 2010, Mizoram and by author; Abr.: Prod = Production
Note Figures in parentheses are percentage of cropped area

Table 21.5 Changes in cropping pattern in percentage (2005–2010)

Crops	Area (ha)	Production (MT)	Yield (MT/ha)
Paddy (Jhum)	−8.1	−30.3	−24.1
Paddy (Permanent)	−36.7	−50.4	−21.7
Maize	−27.2	−49.3	−30.4
Pulses	−42.9	−25.2	30.9
Tapioca	−35.7	14.3	77.7
Oilseeds	−53.3	−46.3	15.1
Cotton	−34.7	−46.9	−18.5
Tobacco	−27.4	−31.6	−5.8
Sugarcane	3.7	−73.1	−74
Potato	−70.1	−42.6	92.3
Total	**−23.1**	**−47.3**	**−36.8**

Source By author

cultivation. The second highest area is under cultivable wasteland which is 32.9%. Percentage share of wasteland is 17.9 and only 11% land is under permanent farming, which is mainly comprised of WRC. Area of crops grow under shifting cultivation is quite higher than permanent cultivation, the study revealed.

Table 21.8 shows a comparison of area, production and yield of crops between 2011 and 2014. There are total 11 crops grow in these villages and out of these crops, paddy crop obtains the highest area (64% of the total cropped area of sixteen village in 2011) production (96%) and yield (1361.1 per ha). Banana crop follows the figure with 10% of the total area. Ginger is the third largest crop (4.7% area) and the area under various other corps ranges from 1.7 to 3.8%. On account of production, ginger has 3.8% which follows the production of paddy. Production of other crops varies between 0.003 and

Table 21.6 Land-use pattern in 16 case study villages

Land category	Area (ha)	Percentage
Land under shifting cultivation	692.7	38.2
Permanent agricultural land	198.24	11
Wasteland	323.8	17.9
Cultivable wasteland	597.9	32.9
Total land	1812.64	100

Source by author

Table 21.7 Statistical description of land-use pattern (area in ha) in study villages

Variables	Minimum	Maximum	Mean	Std. deviation
Total land (n = 16)	4.00	572.20	1.01E2	155.33
Land under shifting cultivation (n = 15)	0.40	408.00	46.18	103.39
Wasteland (n = 8)	0.80	98.40	24.78	36.52
Land under WRC (n = 15)	0.60	86.80	21.58	26.47
Cultivable wasteland (n = 16)	2.00	332.00	37.36	79.50

Source Data collected from the primary sources and calculated by the author

0.3%. In productivity, ginger follows paddy with 671.1 kg/ha. The other crops have comparatively low productivity of crops. Area, production and productivity vary from 2011 to 2014. In a nutshell, the figure shows that paddy crop dominates in all categories, i.e., area, production and productivity, as paddy is the staple food and the other crops substantiate it.

Table 21.9 depicts change in cropping pattern that includes area, production and yield in percent. Total 11 crops comprises cropping pattern in the 16 case study villages. Change in area under all 11 crops was noticed acceding between 2011 and 2014. It was multi-times under lemon crop (highest) and 60.4% under paddy crops. As a whole, 281.5% area was increased under all crops. On account of change in production, rubber and ginger crops received decreasing trends during these two consecutive periods. On the other hand, the production of other crops, mainly of fruit crops, was increased multifold (average 32.8%). However, per ha yield of other crops was decreased, e.g., paddy, palm oil, rubber, mango and ginger (average -29.9%). Concisely, area and production of crops were increased and yield of crops was decreased, simultaneously.

21.5 Discussion and Conclusions

The result of the above-cited study shows that a substantial change in land-use/cover has taken place both at state level and villages during the last decade. At the state level, area, production and yield of crops have decreased substantially, whereas in the villages, although, area has increased, but production and yield of crops have decreased. Although paddy rice is the staple food yet, during the past, its production and yield have also decreased largely. There was an increase in area under forest cover till 2005; however, during the last decade, about 6% forest area has decreased. A long lasting debate is going on about decrease in forest area. The government officials are arguing that decrease in forest cover is due to practices of shifting cultivation; however, a study conducted by the author on shifting cultivation shows that area under shifting cultivation has decreased about 58% during the last three decades (Sati 2017). Therefore, the question is unanswered. It is true that shifting cultivation is causing forest depletion and degrading the spectacular landscape.

Table 21.8 Area, production and per ha yield of crops

Crops	2011					2014				
	Area (Ha)	%	P (kg)	%	Yield (kg/ha)	Area (Ha)	%	P (kg)	%	Yield (kg/ha)
Paddy	170.5	64	288,861	96.6	1694.2	273.5	26.9	372,256	93.7	1361.1
Oil Palm	8	3	65	0.02	8.1	72.3	7.1	277.5	0.1	3.8
Betel Nut	9.5	3.5	10	0.003	1.1	63.9	6.3	2100.04	0.5	32.9
Rubber	6	2.3	5	0.007	0.8	70.5	6.9	2	0.09	0.02
Mango	5.5	2.1	65.02	0.04	11.8	75.14	7.4	700	0.2	9.3
Lemon	6	2.3	25.01	0.03	4.1	170.63	16.8	2158.37	0.51	12.6
Pineapple	4.5	1.7	266.66	0.09	59.3	45.45	4.5	2966.65	0.7	65.3
Orange	7	2.6	23	0.01	3.3	65	6.4	5716.19	1.4	87.9
Ginger	12.5	4.7	8388.2	2.8	671.1	35	3.5	1180	0.3	33.7
Grapes	10	3.8	800	0.3	80	28	2.8	3628	0.9	129.6
Banana	26.7	10	474.33	0.1	17.8	116.07	11.4	6192.49	1.6	53.4
Total	266.2	100	298,983	100	1123.1	951.59	100	394,799	100	414.9

Source By author

Table 21.9 Changes (percent) in cropping pattern—area, production and yield

Crops	Chang in area (ha)	Change in production (MT)	Change in yield (kg/ha)
Paddy	60.4	28.9	−19.7
Oil Palm	803.8	326.9	-53.1
Betel Nut	572.6	2090	2890.9
Rubber	1075	−60	−97.5
Mango	1266	976.6	−21.2
Lemon	2743	8530.3	207.3
Pineapple	910	1012.5	10.1
Orange	828.6	24,753	2563.6
Ginger	180	−85.9	−94.9
Grapes	180	353.5	62
Banana	334.7	1205.5	200
Total	281.5	32.8	−29.9

Source Data collected from the primary sources and calculated by the author (2011–2014)

Among the drivers of land-use/cover change, New Land-use Policy (NLUP) of the state government dominates which has affected shifting cultivation by decrease in its area and production. NLUP promotes settled agriculture; and henceforth, the area under shifting cultivation received a substantial decrease in Mizoram state. This has resulted in increase in net area sown under WRC. NLUP of Mizoram government is pushing hard to eradicate *Jhum* cultivation and it is the major driving force (Tiwari 1991, Raman 2014). Among the other driving forces that have changed cropping and land-use patterns are low output from the cropped land and rural-urban migration. Mizoram state has very sparse distribution of population and the villages are remotely located. Further, migration from rural areas to urban centers mainly to Aizawl city has changed the land-use pattern. Area under oil palm plantation is increasing in Mizoram. So far 101,000 hectares of land has been identified for oil palm cultivation. Agricultural land-use has also been influenced by high literacy and improving education. The young generation of rural farmers is now involved in tertiary activities, as they are getting substantial wages through providing services and as a result, many of them have left practicing shifting cultivation.

Dynamics of land-use/cover changes in Mizoram state have illustrated broadly. Two categories of land-use mainly forest land-use/cover change and agricultural land-use/cover change were discussed using both primary and secondary data. Change in all types of land-use was observed. The highest increase was obtained under land not available for cultivation. This was followed by land under current fallows. An increase under misc. tree crops, groves, fallow land other than current fallows and net area sown was noticed. Meanwhile, a decrease under cultivable wasteland, permanent pasture, other grazing lands and forest land was noticed. On account of change in agricultural land-use, area of paddy crop under shifting cultivation is the highest, followed by WRC, maize, pulses and oilseeds. Rice and maize are the major crops in Mizoram. Changes in cropping pattern were noticed decreased except sugarcane. In case study, villages and area under all crops have increased, whereas yield of crops has decreased. This study revealed that there has not been any clear cut trend of observed land-use/cover change. Shifting cultivation is the major driving factor of change in land cover, and due to it, forestland was noticed decreased.

References

Champion HG, Seth SK (1968) A revised survey of the forest types of India. Manager of Publications, Government of India, Delhi

FSI (Forest Survey of India) (2017) India State of forest report. Dehradun, p 189

ISFR (2011 a&b) India State Forest Report, Forest Survey of India, Dehradun

Kumar G (2012) Dynamics of development and planning: Mizoram a comprehensive regional analysis. Kalpaz Publications, Delhi

Maithani BP (2005) Shifting cultivation in north-east India policy issues and options. Published and printed by Krishna for Mittal Publications, Mohan Garden, New Delhi, India

Meyers M (1994) Journal of Communication, Volume 44, Issue 2, June 1994, Pages 47–63. https://doi.org/10.1111/j.1460-2466.1994.tb00676.x

Nagendra H, Southworth J, Tucker C (2003) Accessibility as a determinant of landscape transformation in western honduras: linking pattern and process. Lands Ecol 18:141–158

Nagendra H, Munroe DK, Southworth J (2004) From pattern to process: landscape fragmentation and the analysis of land-use/land cover change. Agric. Ecosyst Environ 101:111–115

Nduwamungu J, Bloesch U, Munish PTK, Hagedorn F, Lulu K (n.d.) Recent land cover and use change in Miombo woodlands of Eastern Tanzania. Unpublished research report found at (www.adansonia-consulting.ch/document/Article.Jean.Land_cover_use_changes_in_miombo-NEW11.pdf) site visited on 12/03.2012

Pachuau R (2009) Mizoram: a study in comprehensive geography. Northern Book Centre, New Delhi

Rai SC, Sharma E, Sundriyal RC (1994) Conservation in the Sikkim Himalaya: traditional knowledge and land-use of the Malay Watershed. Environ Conserv 21:30–34

Ramakrishna PS, Purohit AN, Saxena KG, Rao KS (1994) Himalayan 'environment and sustainable development. Indian National Science Academy Diamond Jubilee Publication, New Delhi

Raman TRS (May 2014) Newspaper article, The Hindu

Sati VP (2017) A sustainable livelihood approach to poverty reduction: an empirical analysis of Mizoram, the eastern extension of the Himalaya. Springer Publications, Cham, Switzerland

Sati VP (2014) Land-use/cover changes in the kewer gadhera sub-watershed, Central Himalaya. In: Grover VI (eds) Impact of global changes on mountains, CRC Press, Taylor & Francis Group, pp 298–311

Schweik CMK, Adhikari KN (1997) Pandit. Land cover change and forest institutions: a comparison of two subbasins in the Southern Shivalik Hills of Nepal' mountain research and development 17:99–116

Sen KK, Semwal RL, Rana U, Nautiyal S, Maikhuri RK, Saxena KS, Rao KG (2002) Patterns and implications of land-use/cover change: a case study in Pranmati Watershed. (Garhwal Himalaya, India). Mt Res Dev 22:56–62

Singh JS, Pandey U, Tiwari AK (1984) Man and forests: a central Himalayan case study. Ambio 13:80–87

Singh KK, Das MM, Samanta AK, Kundu SS, Sharma SD (2002). Evaluation of certain feed resources for carbohydrate and protein fractions and in situ digestion characteristics. Indian J Anim Sci 72 (9): 794–797

Singh J, Ramakrishnan PS (1982) Structure and function of a sub-tropical humid forest of Meghalaya I. Vegetation, biomass and its nutrients. Proc Indian Acad Sci (Plant Sci.) 91, 241. https://doi.org/10.1007/BF03167128
Spencer JE (1966) Shifting cultivation in southeastern Asia. University of California Press, Berkeley
Tiwari DN (1991) Shifting cultivation in India. Indian Forester 117:91–104

Changing Scenario of Tropical Forests Due to Shifting Cultivation in the Indo-Burma Bio-Geographical Hotspot: A Study on Three Major Hill Ranges of Tripura, North-East India

22

Jatan Debnath, Nibedita Das (Pan), Amal Debnath, and Istak Ahmed

Abstract

The present study analysed the effects of shifting cultivation and alteration of forest areas in three hill ranges of Tripura, which belong to the Indo-Burma bio-geographical realm of India. Landsat imagery along with the modern geoinformatics tool was used for this study. NDVI was applied to evaluate the vegetation scenario and justify the land use/land cover maps. These maps of three hill ranges indicate that the areas under shifting cultivation and degraded forest have increased significantly while areas under dense forest, open forest and water body have declined sharply. The vegetation index indicates the absence of high NDVI value in 2015 in all these hill ranges. The findings of the present study affirm that the overwhelming non-traditional shifting cultivation has arrested the secondary succession of forest ecosystem enormously and helped for weed invasion in the study sites which resulted in obstacle to regeneration of tropical tree species.

Keywords

Hill ranges of Tripura • Tropical forest • Hotspot • Non-traditional shifting cultivation • Weed invasion

22.1 Introduction

Tropical forests are often referred as one of the most productive terrestrial habitats in the globe. Their immense biodiversity produces a variety of natural resource which helps to sustain the basic needs of local communities (Mishra 1968). They are extensively harvested in the Indo-Burma bio-geographical realm of the tropics as primary source of food for the rural inhabitants. But these natural heavens are suffering from over-exploitation and degradation.

Shifting cultivation was a sustainable use of forest ecosystem when plenty of forests were available to these cultivators (Myers 1990), but due to the modification from traditional to non-traditional jhuming (Gupta 2000), vegetation do not get enough time to re-establish (Yadav et al. 2013). Thus, the retardation of natural regeneration of trees, i.e. degradation which is regarded as one of the most critical factors affecting climate change, biodiversity, ecosystem and environmental conditions (Skole et al. 1994), results to manipulation of land use pattern and conversion to various other land use/land cover (LULC) categories (Subramani and Vishnumanoj 2014).

J. Debnath · N. Das (Pan) (✉) · I. Ahmed
Department of Geography and Disaster Management, Tripura University, Tripura 799022, India

A. Debnath
Department of Forestry and Biodiversity, Tripura University, Tripura 799022, India

U. Schickhoff et al. (eds.), *Mountain Landscapes in Transition*,
Sustainable Development Goals Series, https://doi.org/10.1007/978-3-030-70238-0_22

Therefore, in this region, LULC cover change has received much attention in current years (Achard et al. 2002).

The north-eastern region of India is well-known for rich biodiversity and becomes a priority for leading conservation agencies of the world. This region is identified as a geographic gateway for much of the living organisms of India. It is characterised by large scale variation in flora and fauna and consequently identified as Indo-Burma hotspot, one of the 35 recognised global biodiversity hotspots (Myers 1988; Tripathi et al. 2016). The richness of the avifauna reflects the diversity of habitats along with wide altitudinal range of this region. The region is famous as a 'cradle of flowering plants' due to its diversified angiosperm and many important cultivated plant species (Takhtajan 1969; Agarwal 1996; Tripathi et al. 2016). However, according to Forest Survey of India Report (2017), the forest areas of this region are gradually reducing for some anthropogenic activities. Increasing settlement due to rising population, illegal tree felling, shifting cultivation (locally known as jhuming), etc., are largely responsible for this reduction. Almost 0.45 million families in this region annually cultivate 10,000 km^2 forests while about 44,000 km^2 forest areas are affected by the jhum cultivation (Anonymous 2009; Tripathi et al. 2016). On the other hand, reduced jhum cycle from 25–30 years (earlier) to 4–5 years (present) has accelerated the degradation process of the natural ecosystem (Grogan et al. 2012). According to the State Forest Report (2011), among the eight north-eastern states the highest change was recorded in Manipur, whereas Tripura occupied seventh position in this respect unlike Sikkim (no change observed). Being a small state, such type of change has attracted the attention of the conservators. Typically, the hill ranges are dominated by the forest cover and much more affected by the shifting cultivation activity from the earlier periods. About 223 km^2 of the forests (3.76% of the total forest area of the state) is cleared annually for jhuming which is higher than the national average of 2.26% (Gupta 2000). Thus, the natural ecosystem of the state is degrading day by day due to the change of LULC categories, but the measures have not yet been taken due to the lack of relevant research works.

In modern times, remote sensing becomes a prime tool for monitoring the vegetation coverage and land use change over a period of time. Normalised difference vegetation index (NDVI) is the most widely used tool for recognising vegetation growth, vegetation variation, its heterogeneity (Tuker et al. 1985) and thus consistent spatio-temporal comparison and inter-annual variation in vegetation (Kinthada et al. 2014; Jung and Chang 2015) can be computed.

The main purpose of this study was to view the spatio-temporal changes of the LULC and relation of those changes with anthropogenic indices (shifting cultivation) within three hill ranges, namely Longtarai, Sakhantang and Jampui (Fig. 22.1). The NDVI status at different altitudinal level with different biophysical condition was also considered.

22.2 Study Area

Tripura with undulating hilly terrain is one of the smallest states of the Indo-Burma bio-geographical region of India. Geographically, it lies between 22° 56′ to 24° 32′ N latitude and 91° 09′ to 92° 20′ E longitude covering an area of 10,491.69 km^2 among which about 60% is covered by forest with diversified flora and fauna. The mean annual rainfall is 2024.4 mm while temperature ranges from 6 to 39 °C. Altitude ranges from 12.5 m in the west to 939 m in the east. Most of the forests are concentrated within north–south trending five parallel hill ranges and few in the southern plains. The study region is mainly occupied by the tropical evergreen forest, tropical moist deciduous forest, sal forest and tropical moist deciduous mixed forest. Moreover, there are also found some patches of grasslands, swamp vegetation, bamboo, riverine forest and garjan forest throughout the state. These forests are highly disturbed by shifting cultivation (jhuming) and monoculture practice. Shifting cultivation is regarded as the main reason for

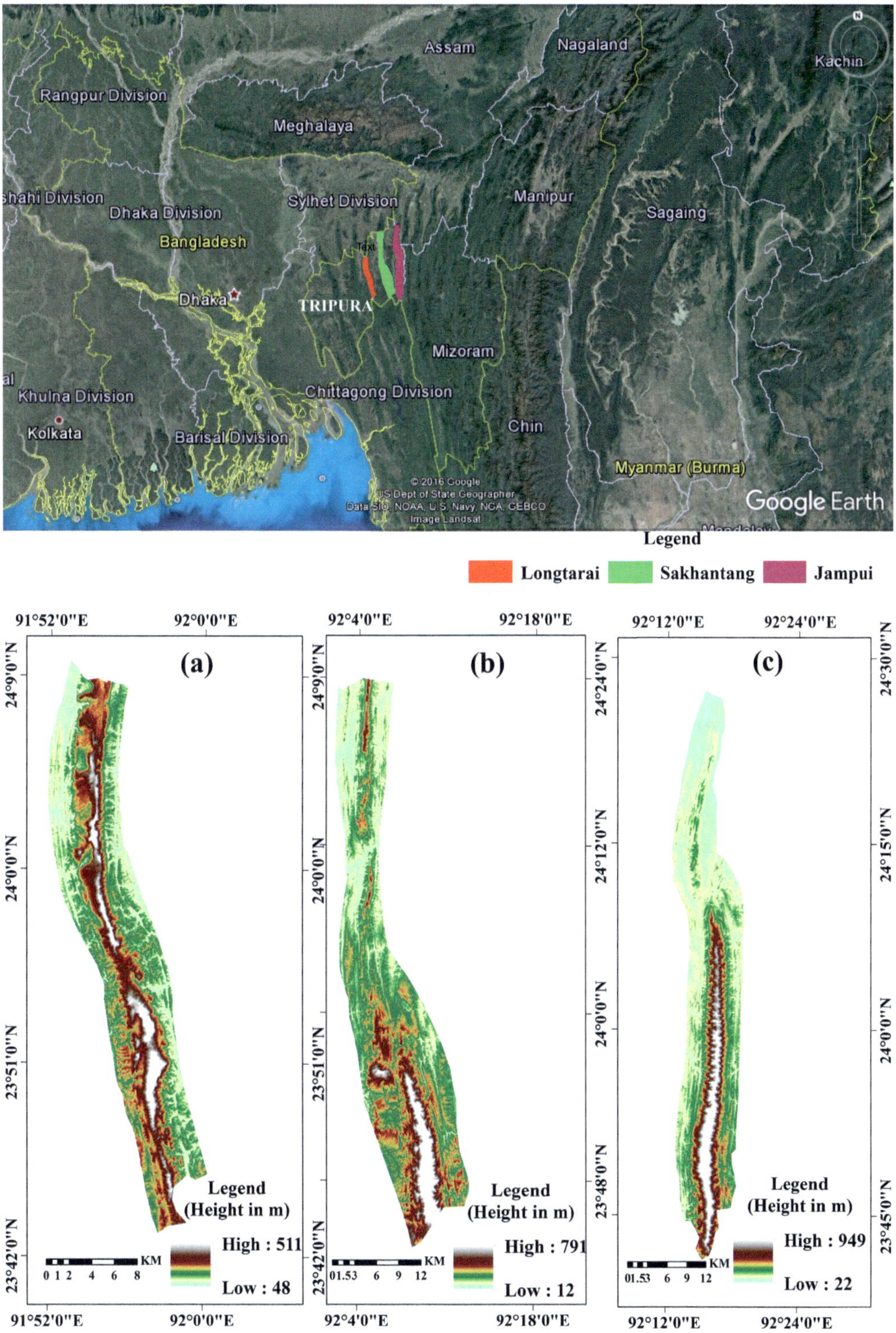

Fig. 22.1 Location map of the study area: **a** Longtarai hill range, **b** Sakhantang hill range, **c** Jampui hill range

land cover change. This cultivation is mainly practised by the indigenous tribes of Tripura. Generally, the plot selection for jhum cultivation starts during the months of November and December and slash and burn during mid-February to early March, before the rain begins. The weeding starts from June and continues till August and harvesting in the next year during the month from June to October. But increasing population from 6,39,029 to 36,73,917 (474.92% growth) during the census years 1951–2011 converted the traditional jhum cultivation into non-traditional jhum cultivation. Botanically, the state is of great interest because of its hot summer and favourable environmental condition which supports the growth of tropical aliens. As a result, many exotic plants have vigorously propagated and established in this region, thus, altered the native vegetation and become an obstacle to the natural regeneration of tree species.

22.3 Methodology

22.3.1 Datasets and Pre-processing

The satellite images used in this study were derived from the United States Geological Survey (USGS) using earth explorer (https://earthexplorer.usgs.gov). Three satellite images, i.e. Landsat TM 1989, 2005 and Landsat OLI 2015 were used for analysing the LULC change and NDVI. Each of the datasets was of 30 m resolution and acquired following the path 136 and row 044 (Table 22.1). The datasets of the month of February and March for each study year were collected as it is the commencing month of jhum in north-east India. The researchers have used Aster Global Digital Elevation Model (GDEM) data collected from USGS to extract topographic information and derived altitude-wise jhum areas in the hill ranges.

In order to obtain essential information from the satellite data, image pre-processing, i.e. image enhancement, geometric rectification, radiometric correction and mosaicing were done, as well as authenticated and corrected data was assembled for further analysis (Jensen 1996; Iqbal and Khan 2014). WGS 1984, UTM Zone 46 N was assigned to all the datasets separately using the project transformation system. Though very small patches of cloud cover were observed in the upper part of the datasets (beyond Tripura state), but it had not covered the study area, and hence cloud removal technique was not applied. Image enhancement technique was used to all the three datasets. Image geometric rectification was employed to reduce the geometric distortion of the images and the adjoining images were brought into registration as well. Image mosaicing was done for the selected datasets. In order to reduce error during overlay and change detection analysis, all the images were set into a standard projection (Kardoulas et al. 1996) and used radiometric correction to avoid radiometric errors or distortions; in this way, the reliability of the pixels' brightness value was increased (Xie et al. 2008). Shadow removal was performed on Landsat datasets and ancillary DEM using the topographic correction technique of ArcGIS.

Table 22.1 Details about the datasets

Satellite sensor	Date of satellite acquisition	Path/Row	Cloud cover	Scene Id
Landsat TM	18-03-1989	136/044	0	LT51360441989077BKT00
Landsat TM	18-03-1989	136/043	4	LT51360431989077BKT00
Landsat TM	26-02-2005	136/044	0	LT51360442005057BKT00
Landsat Tm	26-02-2005	136/043	3	LT51360432005057BKT00
Landsat OLI	26-03-2015	136/044	0	LC81360442015085LGN00
Landsat OLI	26-03-2015	136/043	4	LC81360432015085LGN00

22.3.2 Classification Scheme

The Arc GIS 10.1 software was used to process the Landsat images by using the pixel-based supervised image classification. Five land use/land cover classes like (a) shifting cultivation (SC), (b) dense forest (DF), (c) open forest (OF), (d) degraded forest (DGF) and (e) water body (WB) were selected following the USGS LULC classification system (Anderson et al. 1976). Maximum likelihood classification algorithm was used as it has been recognised as one of the most proficient parametric methods for LULC classification (Rawat and Kumar 2015). Ground truth verification was conducted using Global Positioning System (GPS) device and the local people were also consulted. The survey was done for two months (June and July 2015) and almost 40–50 points were verified for each class of the classified LULC maps. The areas under each class were calculated and analysed statistically for better illustration of the data.

22.3.3 Detection of Altitude-Wise Shifting Cultivation Area

Shifting cultivation is the ancient traditional farming system, mainly practised by the aboriginals all over the tropical and sub-tropical regions of the world (Sharma 1976; Sati and Rinawma 2014; Datta et al. 2014; Deka and Sarmah 2010), known as 'jhum' in north-eastern part of India (Dasgupta 1986; Gupta 2000; Das and Das 2014; Reimangam 2017; Debnath et al. 2017).The areas under shifting cultivation were identified from the false colour combination (FCC) of the imagery using the on-screen visual interpretation technique. The pre-burnt jhum plots have higher reflectance in both the red and NIR band, whereas the burnt jhum plots have low reflectance in the red as well as NIR band (Pebam 2018). The current year jhum have normally black (freshly burnt area), greenish blue tone, whereas jhum plots with second year crop have either light blue-green or blue tinge tone (Thong et al. 2019). Here, the areas of densely vegetated steep slopes with sufficient rain water are highly favoured for shifting cultivation. In order to estimate the altitude-wise area, under shifting cultivation within the study sites contours at 200 m interval were generated on Aster DEM of 30 m resolution using Arc Map 10.1. The overlay tool was applied for extraction of the altitude-wise jhum areas. This type of study assists to identify the areas where ecological balance is under threat due to shifting cultivation and could be a most effective tool for taking a proper management strategy.

22.3.4 Accuracy Assessment

The level of relationship between the remotely sensed data and ground reference information were mainly scrutinised with the help of classification accuracy (Congalton 1991). Confusion matrix (error matrix) is a most proficient method to express the classification accuracy in the LULC study (Lillesand and Kiefer 2007). Following Burnicki (2011) and Gass et al. (2013), error matrix table of the classified maps was prepared by taking random points on the basis of ground truth verification. From the error matrix, table omission error percentage, commission error percentage, accuracy of producer's, user's accuracy, overall accuracy and Kappa coefficient (K_{hat}) were calculated for each classified map using the following formula:

$$\text{Overall accuracy: } \frac{\sum_{i=1}^{r} x_{ii}}{N} \times 100 \qquad (22.1)$$

$$\text{Kappa accuracy: } \frac{N\sum_{i=1}^{r} x_{ii} - \sum_{i=1}^{r} (x_{i+} * x_{+i})}{N^2 - \sum_{i=1}^{r} (x_i + * x_{+i})} \qquad (22.2)$$

Source: Congalton (1991).

where r = number of rows in the matrix, x_{ii} = total number of correctly classified pixels in row i and column i, x_{i+} and x_{+i} = marginal totals of row i and column i, respectively, and N = total number of pixels in the matrix table.

However, the omission percentages were extracted to know the number of pixels which could not be classified into proper classes, whereas the commission percentage deliberate about the number of pixels which belong to the other class but confusedly added with another class. Moreover, user's accuracy measures the proportion of each class which was classified correctly in the maps as the actual landscape present on the ground, based on the training class. Whereas, producer's accuracy expresses how the proportion of LULC classes are correctly classified on the basis of the training pixels (Debnath et al. 2017). The overall accuracy was calculated as dividing the total accurately classified pixels by the total number of pixels in the confusion/error matrix table (Congalton 1991), while kappa coefficient (K_{hat}) was recognised as the discrete multivariate technique applied during the accuracy assessment of the Landsat images (Cohen 1960). According to Monserud and Leemans (1992), the value of Kappa coefficient of <0.4 symbolises poor conformity, 0.4–0.55 fair conformity, 0.55–0.70 good agreement, 0.70–0.85 very good agreement and >0.85 represents excellent conformity of the classified LULC maps.

22.3.5 Change Detection

Post classification comparison method is the most useful change detection technique (Rawat and Kumar 2015) which evaluates and detects the LULC change from one class to the other by comparing multi-dated images (Sinha et al. 2015; Iqbal et al. 2014). The LULC maps of the year 1989, 2005 and 2015 were produced after preprocessing the Landsat images.

Moreover, the identification of the trend, whether positive or negative change, is calculated using the following formula:

$$\frac{\text{Total area of type } i \text{ change}}{\text{Total positive or negative changed area of type } i \text{ of previous date}} \times 100 \tag{22.3}$$

The resulted LULC maps were overlaid and compared on pixel by pixel basis. 'From–To' change map/conversion map was prepared using this simple pixel by pixel combination of images from two temporal datasets.

22.3.6 Normalised Difference Vegetation Index

Recently, NDVI has been developed as an extensively used indicator for detecting land cover (Musa and Jiya 2011; Jung and Chang 2015). The NDVI was calculated using the following formula:

$$\text{NDVI} = \frac{\text{NIR} - \text{red}}{\text{NIR} + \text{red}} \tag{22.4}$$

where red corresponds to band 3 (0.63–0.69 μm) in Landsat TM and in Landsat 8 OLI band 4 (0.64–0.67 μm) and NIR corresponds to Landsat TM band 4 (0.76–0.90) and Landsat 8 OLI band 5 (0.85–0.88 μm). The ArcGIS 10.1 software was used to estimate the NDVI classes and each raster file had a 30 m cell size. The NDVI value always ranges between −1 to +1 and whenever the sensor gets some chlorophyll content, it reflects positive vegetation index value. In this study, the researchers have developed a vegetation index range on the basis of the works of Lakshmi Kumar et al. (2013) and Sahebjalal and Kazem (2013). It indicates 0–0.2 for low vegetation, 0.2–0.4 medium and >0.4 for high vegetation.

22.4 Results and Discussion

22.4.1 Accuracy Assessment

The classified images of all the hill ranges indicate overall accuracy of >88 with kappa coefficient of >0.8 (Table 22.2). According to Anderson et al. (1976), the classified map of >85%accuracy can be processed for further analysis.

Table 22.2 Error matrix of the classified LULC maps of three hill ranges

	Longtarai						Sakhantang						Jampui					
	1989		2005		2015		1989		2005		2015		1989		2005		2015	
Class	PA (%)	UA (%)	PA (%)	UA (%)	PA (%)	UA (%)	PA (%)	UA (%)	PA (%)	UA (%)	PA (%)	UA (%)	PA (%)	UA (%)	PA (%)	UA (%)	PA (%)	UA (%)
SC	100	100	100	100	77	94	97.6	95	100	100	98	78	100	98	100	100	100	98
DF	100	97.4	98	100	100	100	77.5	91	86.5	100	100	100	32	100	100	100	100	100
OF	100	96.8	92	80	100	86	97.4	97	100	95	100	89	100	100	100	96	100	100
DGF	96	100	94	90.9	88	82	91.7	97	98.3	93	50	90	100	3.2	97	100	100	100
WB	100	100	88	100	93	100	93.6	76	100	100	76	100	96	100	100	100	98	100
OA	98%		95%		91%		91%		97%		89%		88%		99%		99%	
KS	0.97		0.93		0.91		0.9		0.97		0.88		0.86		0.99		0.98	

Note: *DF* Dense Forest; *SC* Shifting Cultivation; *OF* Open Forest; *DGF* Degraded Forest; *WB* Water Body; *PA* Producer's Accuracy; *UA* User's Accuracy; *OA* Overall Accuracy; *KS* Kappa Statistics

22.4.2 Image Classification

The comparative analysis indicates that the LULC classes varied considerably in different years (Fig. 22.2 and Table 22.3). The Longtarai, Sakhantang and Jampui hill ranges share largest portion of the state's forest and so the land cover was classified into dense, open and degraded forest. In case of all the hill ranges, a significant loss in dense forest and water body was noticed, but expansion of shifting cultivation and degraded forest was prominent from the year 1989 to 2015 (Table 22.4).

22.4.3 Change Detection

The overlaid classified images of 1989 and 2015 provided information about 'From-To' change over three hill ranges (Fig. 22.3 and Table 22.5). The conversion maps indicated the preponderance of degraded forest (50–74%) in all the study areas (Table 22.3). In the Longtarai Range, changes were mainly observed from open (56.06%) and dense forests (54.95%) to degraded forest; in the Sakhantang Range from dense forest (59.39%) and shifting cultivation (59.234%) to degraded forest; in the Jampui Range from dense forest (75.20%) and shifting cultivation (71.95%) to degraded forest which was highest among all the ranges. The results also indicate that the present jhum lands will gradually be transformed into a degraded ecosystem.

The shrinkage of water body became another challenging issue. These modifications of LULC had eventually created a negative impact on the socio-economic life of the indigenous people.

22.4.4 Normalised Difference Vegetation Index (NDVI)

The results obtained from the NDVI analysis specified that the grid values within the study areas range between +0.94 and –0.79 which indicates significant LULC change during three selected study years. For better representation of changes in vegetation index (VI), the DN values were categorised into low, medium and high-density vegetation index (Fig. 22.4). The NDVI analysis indicates that high vegetal cover (>0.4) is absent in all the ranges and significant decrease in medium (0.2–0.4) category occurs in 2015. The low-density vegetation (<0.2) was increased enormously in each of the selected hill ranges (Tables 22.6 and 22.7). It was mainly due to the presence of unhealthy degraded forests, especially shrubs, bushes and weeds, and as a result, the DN values remained within low-density range.

Moreover, the significant interference of shifting cultivation had cleared the natural

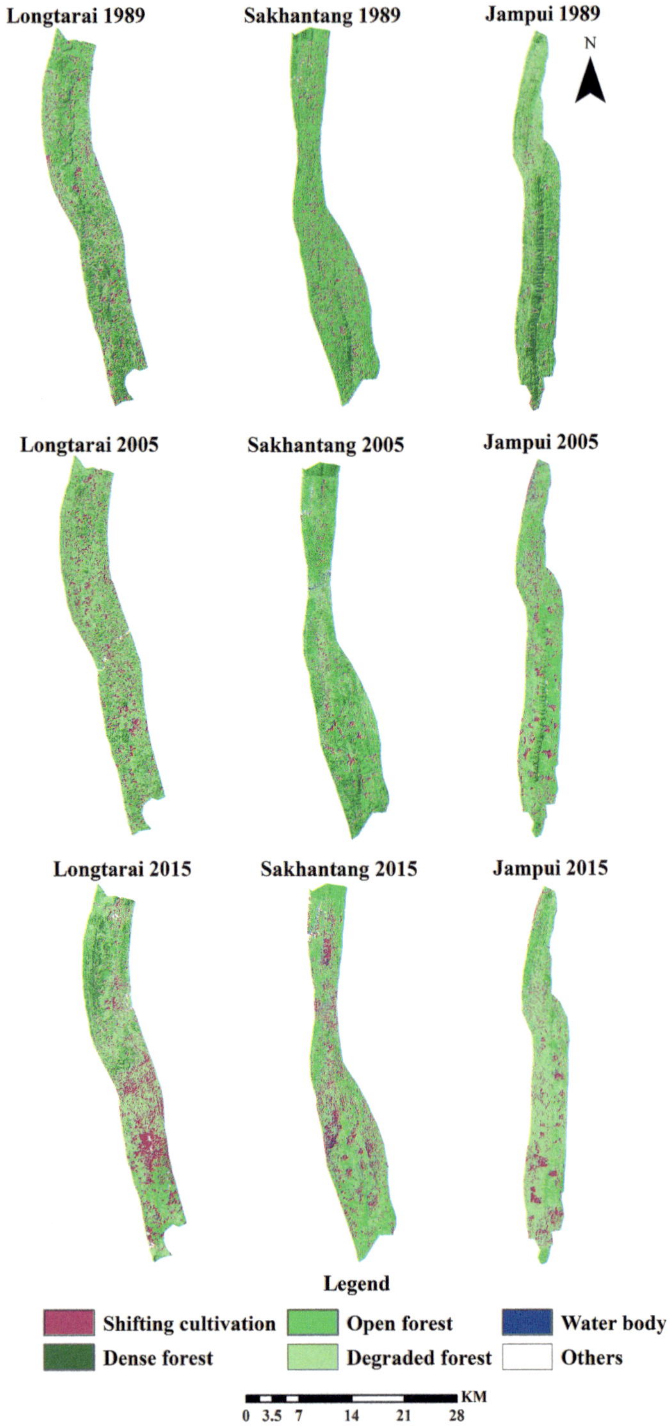

Fig. 22.2 Land use/land cover map of the three selected hill ranges of Tripura

Table 22.3 Area under land use/land cover classes during 1989–2005–2015

LULC	Longtarai			Sakhantang			Jampui		
	Area in km^2 (%)			Area in km^2 (%)			Area in km^2 (%)		
	1989	2005	2015	1989	2005	2015	1989	2005	2015
SC	21.11 (7.67)	25.71 (9.34)	41.93 (15.24)	35.77 (6.01)	37.12 (6.23)	79.57 (13.34)	33.23 (5.40)	47.36 (7.69)	50.13 (8.14)
DF	58.44 (21.24)	35.74 (12.99)	24.62 (8.95)	93.9 (15.77)	84.85 (14.25)	7.98 (1.34)	83.06 (13.49)	21.28 (3.45)	18.27 (2.97)
OF	60.72 (22.06)	38.9 (14.13)	45.38 (16.49)	364.93 (61.27)	299.65 (50.31)	210.79 (35.33)	158.3 (25.71)	275.75 (44.78)	91.60 (14.87)
DGF	130.68 (47.52)	170.63 (62.00)	158.94 (57.76)	98.01 (16.46)	169.95 (28.54)	293.99 (49.28)	334.42 (54.31)	263.64 (42.81)	448.75 (72.88)
WB	2.32 (0.84)	2.01 (0.73)	1.5 (0.54)	1.4 (0.23)	1.84 (0.31)	1.11 (0.19)	4.73 (0.77)	4.93 (0.80)	3.21 (0.52)
Others	1.82 (0.66)	2.2 (0.80)	2.82 (1.02)	1.55 (0.26)	2.15 (0.36)	3.12 (0.52)	2.02 (0.33)	2.8 (0.45)	3.8 (0.62)
Total	275.19 (100)	275.19 (100)	273.27 (100)	594.01 (100)	594.01 (100)	594.01 (100)	615.76 (100)	615.76 (100)	615.76 (100)

Note: *DF* Dense Forest; *SC* Shifting Cultivation; *OF* Open Forest; *DGF* Degraded Forest; *WB* Water Body

Table 22.4 Areal change of different land use/land cover classes during 1989–2005–2015

LULC	Longtarai			Sakhantang			Jampui		
	Area in km^2 (%)			Area in km^2 (%)			Area in km^2 (%)		
	1989–2005	2005–2015	1989–2015	1989–2005	2005–2015	1989–2015	1989–2005	2005–2015	1989–2015
SC	4.6	16.22	20.82	1.35	42.45	43.8	14.13	2.77	16.9
	(21.79)	(63.09)	(98.63)	(3.77)	(114.36)	(122.45)	(42.52)	(5.85)	(50.86)
DF	−22.7	−11.12	−33.82	−9.05	−76.87	−85.92	−61.78	−3.01	−64.79
	(−38.84)	(−31.11)	(−57.87)	(−9.64)	(−90.59)	(−91.50)	(−74.38)	(−14.14)	(−78.00)
OF	−21.82	6.48	−15.34	−65.28	−88.86	−154.14	117.45	−184.15	−66.7
	(−35.93)	(16.66)	(−25.26)	(−17.89)	(−29.65)	(−42.24)	(74.19)	(−66.78)	(−42.13)
DGF	39.85	−11.69	28.16	71.94	124.04	195.98	−70.78	185.11	114.33
	(30.47)	(−6.85)	(21.53)	(73.4)	(72.99)	(199.99)	(−21.16)	(70.21)	(34.19)
WB	−0.31	−0.51	−0.82	−0.44	−0.73	−0.29	−0.2	−1.72	−1.52
	(−13.36)	(−25.37)	(−35.34)	(−31.43)	(−39.67)	(−20.71)	(−4.23)	(−34.85)	(–32.14)
Others	0.38	0.62	1	0.6	0.97	1.57	0.78	1	1.78
	(20.88)	(28.18)	(54.94)	(38.71)	(45.12)	(101.29)	(38.61)	(35.71)	(88.12)

Note: *DF* Dense Forest; *SC* Shifting Cultivation; *OF* Open Forest; *DGF* Degraded Forest; *WB* Water Body

healthy forests year after year, but due to the growing population and adaptation of the non-traditional way of jhuming, these areas became highly degraded. This non-traditional way had reduced the 20–25 years' duration of fallow period of jhum plots to 4–5 years' or even less than that. As a result, although the medium density vegetation index was still present, but high-density vegetal cover was totally missing over the entire hill ranges under study.

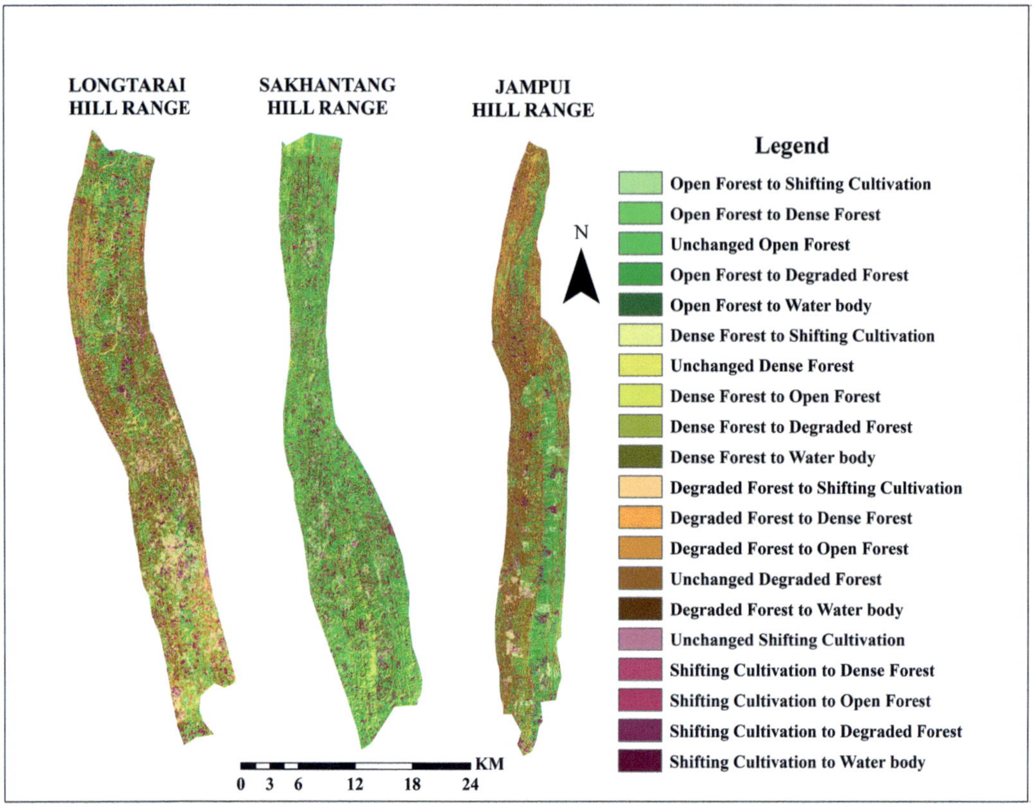

Fig. 22.3 'From-To' change map (1989–2015) of the three selected hill ranges of Tripura

22.5 Altitude-Wise Shifting Cultivation Area

In Tripura, shifting cultivation has been modified to '*hill slope cultivation*' where the entire hill slope is cultivated plot-wise year after year, from higher to lower altitude (Fig. 22.5). The Longtarai hill range was classified into three altitudinal zones like <200 m, 200–400 m and 400–600 m, whereas the Sakhantang and Jampui hill ranges were divided into four zones of <200 m, 200–400 m, 400–600 m and >600 m. It was observed that in all the hill ranges fairly flat and gentle slopes below 200 m altitude were greatly affected by this practice (Table 22.8).

According to Dupin et al. (2009) in the steep slopes the problems of reduced soil fertility, lesser crop productivity and increased soil erosion are greater than the gentle slopes. In spite of that the higher altitude zones of the study area were occupied by shifting cultivation (Table 22.8).

Due to the commencement of Wildlife (Protection) Act, 1972, the Wildlife (Protection) Amendment Act, 1991 and Forest (Conservation) Act, 1980 area under Reserved Forest were increased in comparison with the unreserved forests. This has recurred jhuming in a particular plot over a short period. Therefore, the secondary succession of forest community was detained and plots became highly invaded by invasive weeds like *Chromolaenaodorata, Ageratum conyzoides, Mikania micrantha, Lantana camara,* etc. Since there were fewer forests to burn, therefore, ash, used to increase the fertility of soil, was reduced significantly. Ultimately declined soil fertility, low productivity and enhanced soil erosion, as well as detained secondary succession by

Table 22.5 Area under 'From-To' change

'From-To' (1989–2015)	Longtarai		Sakhantang		Jampui	
	Area (km^2)	Area (%)	Area (km^2)	Area (%)	Area (km^2)	Area (%)
DF- SC	6.74	12.81	6.37	7.05	5.9	7.56
UC_DF	7.11	13.51	5.07	5.61	2.72	3.48
DF-OF	9.64	18.32	53.66	59.39	10.34	13.24
DF-DGF	28.91	54.95	20.92	23.15	58.72	75.2
DF-WB	0.21	0.4	4.33	4.79	0.4	0.51
UC_SC	3.21	19.55	7.9	24.26	4.5	14.61
SC-DF	0.84	5.12	0.01	0.03	0.32	1.04
SC-OF	7.8	47.5	5.32	16.33	3.76	12.21
SC-DGF	4.56	27.77	19.29	59.23	22.16	71.95
SC-WB	0.01	0.06	0.05	0.15	0.06	0.19
OF-SC	6.18	10.9	41.13	11.69	14.18	9.35
OF-DF	5.36	9.45	2.82	0.8	0.77	0.51
UC_OF	13.37	23.57	125.46	35.67	36.94	24.35
OF-DGF	31.8	56.06	179.15	50.93	99.73	65.74
OF-WB	0.01	0.018	3.17	0.9	0.09	0.06
DGF-SC	22.51	17.99	18.53	20.15	24.14	7.26
DGF-DF	9.58	7.66	0.16	0.17	14.49	4.36
DGF-OF	16.73	13.37	15.65	17.02	35.86	10.78
UC_DGF	76.2	60.9	57.28	62.29	256.5	77.1
DGF-WB	0.11	0.09	0.34	0.37	1.7	0.51

Note: *DF* Dense Forest, *SC* Shifting Cultivation, *UC* Unchanged, *OF* Open Forest, *DGF* Degraded Forest, *WB* Water Body

invasive weeds (which extract soil nutrients) might create massive obstacle for natural regeneration of trees. As a result, the need of virgin forest has forced the indigenous people to move towards higher altitudes.

On the other hand, during pre- and post-independence period, a huge number of immigrants entered Tripura from Bangladesh, which eventually pushed the aboriginals towards these hill ranges. Consequently, they were forced to settle in the hill ranges and practised jhuming rigorously by clearing natural vegetation. One of the main causes of such increase in shifting cultivation was probably the improvement in their economic status by selling jhum crops in the nearby markets.

22.6 Conclusion and Remarks

The study on LULC of Longtarai, Sakhantang and Jampui hill ranges for 27 years' depicted significant change in land cover as reflected by vegetation index. The non-traditional jhuming as well as increasing population and over-exploitation of forest resources by the forest dwellers have worsened the status of natural forests. The results indicate that the present jhum lands will gradually be transformed into a degraded ecosystem. Biodiversity of the hill ranges has declined significantly due to the subsequent loss of secondary succession. Various noxious weeds as well as grasses like

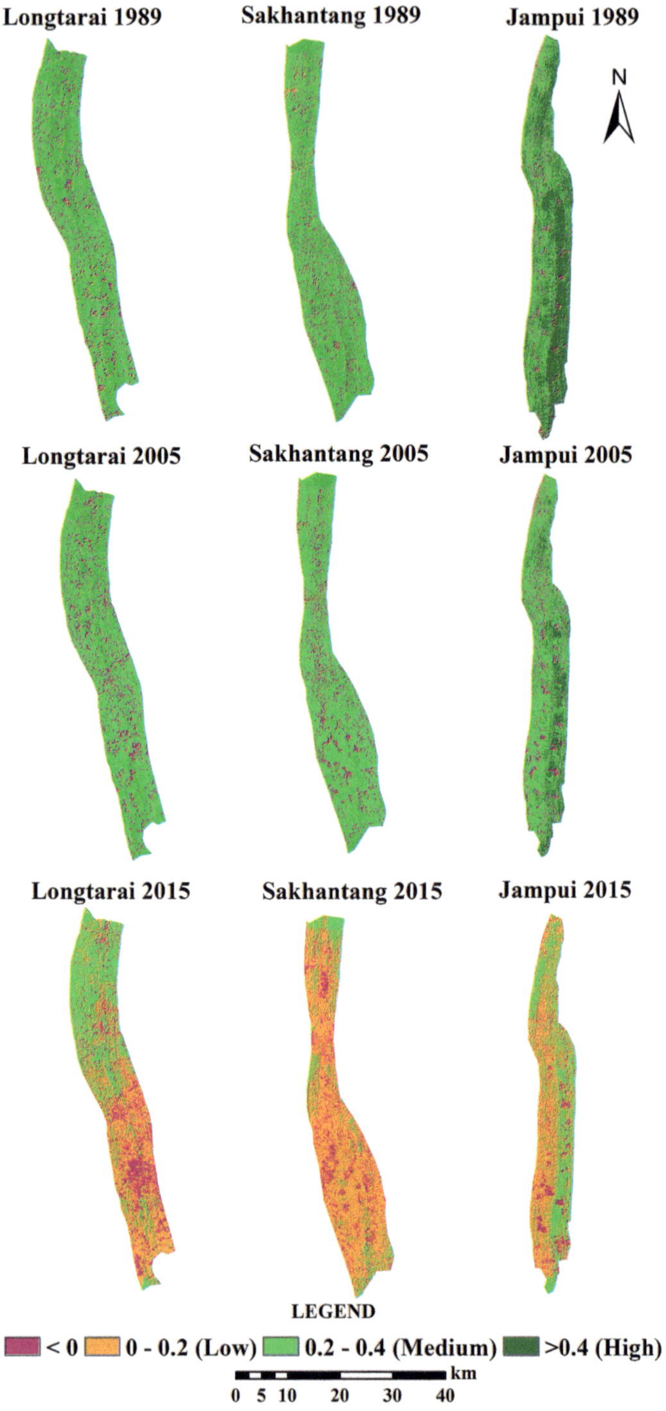

Fig. 22.4 Normalised Difference Vegetation Index (NDVI) map of the three hill ranges

Table 22.6 NDVI class-wise area under three hill ranges (1989, 2005 and 2015)

NDVI class	Longtarai			Sakhantang			Jampui		
	Area in km^2 (%)			Area in km^2 (%)			Area in km^2 (%)		
	1989	2005	2015	1989	2005	2015	1989	2005	2015
0–0.2 (Low)	12.88	12.7	144.77	26.77	47.89	471.4	109.58	36.66	428.29
	(4.75)	(4.69)	(55.74)	(4.52)	(8.09)	(82.73)	(18.82)	(5.99)	(72.15)
0.2–0.4 (Medium)	170.06	173.01	114.97	343.65	436.49	98.41	334.52	444.64	165.28
	(62.73)	(63.93)	(44.26)	(57.99)	(73.74)	(17.27)	(57.44)	(72.69)	(27.85)
>0.4 (High)	88.14	84.93	Absent	222.16	107.56	Absent	138.28	130.4	Absent
	(32.51)	(31.38)		(37.49)	(18.17)		(23.74)	(21.32)	
Total	271.08	270.64	259.74	592.58	591.94	569.81	582.38	611.7	593.57
	(100)	(100)	(100)	(100)	(100)	(100)	(100)	(100)	(100)

Table 22.7 NDVI class-wise areal change (1989–2005–2015)

NDVI class	Longtarai		Sakhantang		Jampui	
	Area in km^2 (%)		Area in km^2 (%)		Area in km^2 (%)	
	1989–2005	2005–2015	1989–2005	2005–2015	1989–2005	2005–2015
0–0.2 (Low)	−0.06	51.04	3.57	74.64	−12.82	66.16
	(−0.46)	(401.92)	(13.36)	(155.86)	(−11.7)	(180.47)
0.2–0.4 (Medium)	1.19	−19.66	15.75	−56.47	15.25	−44.84
	(0.7)	(−11.36)	(4.58)	(−12.94)	(4.56)	(−10.09)
>0.4 (High)	−1.13	−31.38	−19.32	−18.17	−2.43	−21.32
	(−1.28)	(−36.95)	(−8.7)	(−16.89)	(−1.75)	(−16.35)

Chrysopogon Aciculatus, Imperata Cylindrica, etc., were grown extensively in the abandoned jhum plots. On the other hand, soil erosion hazard was reflected through the sedimentation scenario of the major rivers like Gumti, Manu, Deo, Dhalai and Juri.

In case of all the hill ranges, a significant loss in dense forest and water body was noticed, but expansion of shifting cultivation and degraded forest became prominent from the year 1989 to 2015.

Although the Government and the Forest Department of the state have provided rehabilitation programme for these jhumias, but still the cultivation is going on without following any protection measures like contour bunding. Here, communication gap between these jhumias and the local government still exists. Therefore, the government should implement such programmes which will deal with the socio-economic development of the jhumias along with sustainable protection of the forests for future generation.

Fig. 22.5 Altitude-wise area under shifting cultivation in three studied hill ranges [Contours were generated on DEM using ArcGIS]

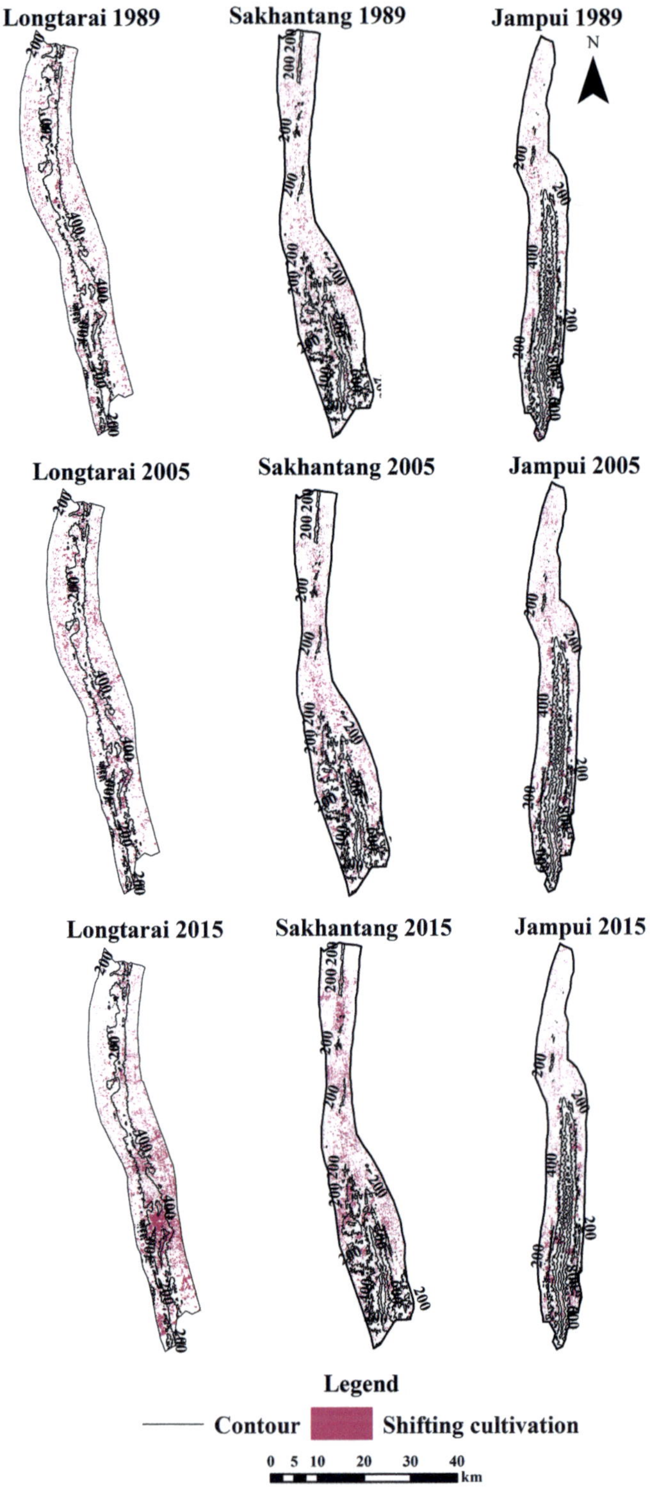

Table 22.8 Altitude-wise area under shifting cultivation during 1989, 2005, and 2015

Longtarai						
	1989		2005		2015	
Elevation (m)	Area (km^2)	Area (%)	Area (km^2)	Area (%)	Area (km^2)	Area (%)
<200	13.2	62.56	15.69	60.36	24.68	58.85
200–400	5.42	25.69	6.06	23.33	10.54	25.14
400–600	2.48	11.75	4.22	16.31	8.01	16.01
Sakhantang						
<200	19.4	54.23	17.9	48.23	37.67	47.48
200–400	10.2	28.51	9.74	26.25	26.38	33.25
400–600	4.66	13.03	6.12	16.5	8.16	10.29
>600	1.51	4.23	3.34	9.02	4.5	8.98
Jampui						
<200	18.3	55.07	25.15	53.11	24.57	49.01
200–400	8.87	26.77	11.63	24.56	16.32	32.56
400–600	5.03	15.13	8.59	18.15	6.11	12.2
>600	1.01	3.03	1.98	4.18	3.12	6.23

References

Achard F, Eva HD, Stibig HJ, Mayaux P, Gallego J, Richards T, Malingreau JP (2002) Determination of deforestation rates of the world's humid tropical forests. Sci 297:999–1002

Agarwal KC (1996) Biodiversity. Agra Botanical Publishers, India

Anderson JM, Hardy EE, Roach JT, Witmert RE (1976) A land use classification system for use with remote sensing data. U.S. Geological Survey Professional Paper, No 964, Washington D. C.: Government Printing Office Anonymous (2009) Report of the inter-ministerial national task force on rehabilitation of shifting cultivation areas. Report submitted to the Ministry of Environment and Forests, Government of India, p 95

Anonymous (2009) Report of the inter-ministerial national task force on rehabilitation of shifting cultivation areas. Report submitted to the Ministry of Environment and Forests, Government of India, p 95

Burnicki AC (2011) Spatio-temporal errors in land cover change analysis: implications for accuracy assessment. Int J Remote Sens 32(22):7487–512

Cohen J (1960) A coefficient of agreement for nominal scales. Educ Psychol Meas 20(1):37–46

Congalton RG (1991) A review of assessing the accuracy of classifications of remotely sensed data. Remote Sens Environ 37(1):35–46

Dasgupta M (1986) Jhumias of Tripura. Econ Polit Wkly XXI (44 and 45).

Das S, Das M (2014) Shifting cultivation in Tripura—a critical analysis. J Agric Life Sci 1(1):48–54

Datta J, Gangadharappa NR, Biradar GS (2014) Livelihood status of tribal people practicing shifting (Jhum) cultivation in Tripura state of North-East India. Trop Agric Res 25(3)316–326

Debnath J, Das (Pan) N, Ahmed I, Bhowmik M (2017) Channel migration and its impact on land use/land cover using RS and GIS: a study on Khowai River of Tripura, North-East India. Egypt J Remote Sens Space Sci 20:197–210

Deka PK, Sarmah D (2010) Shifting cultivation and its effects in regarding of perspective in Northern India. Int J Com Bus Manag 3(2):157–165

Dupin B, de Rouw A, Phantahvong KB, Valentin C (2009) Assessment of tillage erosion rates on steep slopes in northern Laos. Soil Tillage Res 103:19–126

Forest Survey of India (2017) Forest and tree resources in States and Union territories. pp 296–300. Retrieved from: https://fsi.nic.in/isfr2017/tripura-isfr-2017.pdf

Gass L, Norman LM, Villarreal ML, Tolle C, Coe M, Jamwal P(2013) A test of methods to measure vegetation change adjacent to gabions in Sonora, Mexico using Landsat imagery. Presented at the Santa Cruz River Researchers' Day, Tucson, Arizona.

Grogan P, Lalnunmawia F, Tripathi SK (2012) Shifting cultivation in steeply sloped regions: a review of management options and research priorities for Mizoram state, Northeast India. Agroforest Syst 84:163–177

Gupta AK (2000) Shifting cultivation and conservation of biological diversity in Tripura, Northeast India,. Hum Ecol 28(4):605–629

Iqbal MF, Khan IA (2014) Spatio-temporal land use land cover change analysis and erosion risk mapping of Azad Jammu and Kashmir, Pakistan. Egypt J Remote Sens Space Sci 17(2):209–229

Jensen JR (1996) Introductory digital image processing: A remote sensing perspective. Prentice-Hall, Upper Saddle, NJ. p, p 307

Jung M, Chang E (2015) NDVI-based land cover change detection using harmonic analysis. Int J Remote Sens 36(4):1097–1113

Kardoulas NG, Bird AC, Lawan AI (1996) Photogramm Eng Remote Sens 62(10):1173–1177

Kinthada NR, Gurram MK, Eadara A, Velagala VR (2014) Land use/Land cover and NDVI analysis for monitoring the health of micro-watersheds of Sarada River Basin, Visakhapatnam District, India. J GeolGeosci 3(2):1–8

Lakshmi Kumar TV, Rao KK, Barbos H, Jothi EP (2013) Studies on spatial pattern of NDVI over India and its relationship with rainfall, air temperature, soil moisture adequacy and ENSO. Geofizika 30:1–17

Lillesand TM, Kiefer RW (2007) Remote sensing and image interpretation, 5th edn. Wiley, New York. p, p 820

Mishra R (1968) Ecology work book. Oxford & IBH Co. New Delhi, p 244

Monserud RA, Leemans R (1992) Comparing global vegetation maps with the Kappa statistics. Ecol Model 62(4):275–293

Musa HD, Jiya SN (2011) An assessment of mining activities impact on Vegetation in Bukuru Jos Plateau, State Nigeria using Normalized Difference Vegetation Index (NDVI). J Sustain Dev 4(6):150–159

Myers N (1990) The World's forests and human population: the environmental interface. In: Churchill E (ed) Human demography and natural resources. The Population Council, New York, USA

Myers N (1988) Threatened biotas: "hot spots" in tropical forests. Environmentalist 8:187–208

Pebam R (2018) A novel approach to understand the spatial and temporal pattern of shifting cultivation fields using GIS techniques in Longding Division of Arunachal Pradesh, India. Int J Eng Res App 8 (10):61–67

Rawat JS, Kumar M (2015) Monitoring land use/cover change using remote sensing and GIS techniques: a case study of Hawalbagh Block, District Almora, Uttarakhand, India. Egypt J Remote Sens Space Sci 18:77–84

Reimeingam M (2017) Shifting cultivation in Manipur: land. Labour Environ J Rural Dev 36(1):97–119

Sahebjalal E, Kazem D (2013) Analysis of land use-land covers changes using Normalized Difference Vegetation Index (NDVI) differencing and classification methods. Afr J Agric Res 8(37):4614–4622

Sati VP, Rinawma P (2014) Practices of shifting cultivation and its implications in Mizoram, North-East India: a review of existing research. Nat Environ 19 (2):179–187

SFR (2011) India state of forest report 2011. Government of India Publication, Dehradun, Forest Survey of India

Sharma TC (1976) The pre-historic background of shifting cultivation. J Shifting Cultivation in Northeast India, Meghalaya.

Sinha S, Sharma LK, Nathawat MS (2015) Improved land-use/land cover classification of semi-arid deciduous forest landscape using thermal remote sensing. Egypt J Remote Sens Space Sci 18:217–233

Skole DL, Chomentowski WH, Salas WA, Nobre AD (1994) Physical and human dimension of deforestation in Amazonia. BioSci 44(5):314–328

Subramani T, Vishnumanoj V (2014) Land use and land cover change detection and urban sprawl analysis of Panamarathupatti Lake, Salem. Int J Eng Res Appl 4 (6):217–227

Takhtajan A (1969) flowering plants-origin and dispersal. Oliver & Boyd Ltd., Edinburgh

Thonga P, Sahoo UK, Pebamb R, Thangjama U (2019) Spatial and temporal dynamics of shifting cultivation in Manipur, Northeast India based on time-series satellite data. Remote Sens Appl Soci Environ 14:126–137

Tripathi S, Roy A, Kushwaha D, Lalnunmawia F, Lalnundanga, et al. (2016) Perspectives of forest biodiversity conservation in Northeast India. J BiodiversBiopros Dev 3(2):157. https://doi.org/10.4172/2376-0214.1000157

Tucker CJ, Townshend RG, Goff TE (1985) African landcover classification using satellite data. Sci 227 (4685):369–375

Xie Y, Sha Z, Yu M (2008) Remote sensing imagery in vegetation mapping: a review. J Plant Ecol 1(1):9–23

Yadav PK, Sarma K, Mishra AK (2013) Geospatial modelling to assess geomorphological risk for relentless shifting cultivation in Garo hills of Meghalaya. North-East India. Int J Environ 2(1):91–104

23 Urbanization in Himalaya—An Interregional Perspective to Land Use and Urban Growth Dynamics

Mangalasseril Mohammad Anees,
Richa Sharma, and Pawan Kumar Joshi

Abstract

Himalaya is one of the most tectonically unstable mountain ranges known for continuous uplift and being highly vulnerable. Human population growth and densification in and around habitation have initiated urbanization, a new phenomenon which is adding to its vulnerability. Urbanization is taking a toll of Himalaya through dual pathways, directly by altering the land use and land cover to meet these requirements and indirectly in the form of infrastructure constructions to meet their communication, commuting and energy demands. This chapter gives an account of urbanization in Indian Himalaya which is adding threat to this fragile system.

Keywords

Himalaya · Western Himalaya · Eastern Himalaya · Urbanization · Heterogeneity

M. M. Anees · P. K. Joshi (✉)
School of Environmental Sciences, Jawaharlal Nehru University, New Delhi 110067, India

R. Sharma
Public Health Foundation of India (PHFI), Gurgoan 122 002, India

P. K. Joshi
Special Center for Disaster Research, Jawaharlal Nehru University, New Delhi 110067, India

23.1 Introduction

Himalaya, a name derived from Sanskrit language and translating to—'the abode of snow', is one of the youngest and tallest mountain ranges in the world. It is located between the Tibetan plateau and the fertile Indo-Gangetic plains of northern India, the mountains span across 2500 km from east to west forming the northern border of Indian subcontinent while varying from 200 to 400 km in width (north to south). The varied topography of the region with altitudinal, micro-climatic and aspect variation endows rich biodiversity and a mosaic of ecosystems with diverse ecological gradients such as tropical forest, subtropical forest, temperate broadleaf deciduous forest, temperate coniferous forest, grasslands and shrubland (Wester et al. 2019). Having a large network of protected areas, with vast tracts of green cover, these mountains are one of global biodiversity hotspots and hold a special position as a giant carbon sink.

Himalaya is famously known as the 'third pole' owing to its vast expanse of ice and snow cover outside the polar regions. Himalaya is also called the 'water tower' being the source of a number of big and small perennial rivers that ensure freshwater supply to the subcontinent (Savoskul and Smakhtin 2013); simultaneously, the silt brought down by the rivers enriches the soil, creating the fertile Indo-Gangetic plains. But Himalaya is also one of the most tectonically

U. Schickhoff et al. (eds.), *Mountain Landscapes in Transition*,
Sustainable Development Goals Series, https://doi.org/10.1007/978-3-030-70238-0_23

unstable, and a continuous uplift has made these mountain ranges highly vulnerable to large-scale tectonic movements (Valdiya and Bartarya 1991). This process is still continuing and has resulted in high loss disasters in the recent past. In addition to exhibiting the stupendous topography and coupled vulnerability, the mountain ranges also play very important role in influencing the climate over the Indian subcontinent (Wester et al. 2019) and is the source of water for all three major river systems in the Asian continent.

The vast watersheds and integrated river systems of the Himalaya are continuously stressed by rapidly growing human population and their energy demands met through energy generated from various mega hydropower projects flocking all along the flow of these rivers. The water dependence of countries across the mountains has shaped their economies around agriculture and employ the greatest share of population (Shrestha et al. 2015). As a result, any degradation to the ecosystems of the Himalaya will bring a significant threat to the livelihood and food security of millions of people living downstream. The physical geography of the Himalaya seems to be demystified with large-scale historic geological surveys and recent technological advancements to map the various geographical features across it. However, the growth of urban population in the fragile ecosystem has largely gone unnoticed.

Undoubtedly, over the years, this sensitive ecosystem has become highly susceptible to impacts of global environmental change, including climate change, and has caught the attention of national and international climate scientists. However, very little is explored to understand the local-scale effects of the rapid urbanization taking place in these mountains. Dotting the Himalaya is a mixed group of most populated and fastest developing countries in the world. This brings an additional challenge to counter the loss of biodiversity and natural resources including the issue of water sustainability. Urbanization is taking a toll of Himalaya through dual pathways, directly by altering the land-use practices (e.g. agriculture and horticulture) and land cover (e.g. deforestation) to meet the requirements of local urbanization, and indirectly in the form of infrastructure constructions to meet the communication, commuting and energy demands of urbanization within the Himalaya as well as in far off lowland settlements (Pandit and Grumbine 2012). Thus, Himalayan landscape is equally affected by population growth far away from its formidable peaks. Many of the towns, meant and planned for a small population in the mountains, have outgrown their capacity due to natural growth, industrialization and raising benefits from tourism industry. However, the paucity of information on growth of these cities, similarities and dissimilarities of nature and the vulnerabilities they face is a major hindrance in coordinating action plans to devise adaptation practices and putting them on loci of sustainability. Only major urban centres such as Srinagar (Nengroo et al. 2017), Dehradun (Diksha and Kumar 2017), and Kathmandu (Bahadur and Murayama 2012; Rimal et al. 2017, 2018) are fairly studied by geographers and environmental scientist while leaving a void to understand other centres located within Himalaya. Interregional and altitudinal variation in the Himalayan cities can provide the clues to understand the spatial variation and its drivers and contribute in developing region-specific policies to bring countermeasures. Due to its major share in population, the western region of Himalaya (WH) is often the focus of policy and discussion with respect to urbanization, side-lining the growth in the eastern Himalaya (EH). There is a major variation in demographic, climatic and topographic conditions in the WH and EH influencing urbanization trends while at the same time altitudinal variation paints an obvious but understudied trend. The next section looks into the details of such variations and discusses research studies focussed to understand the effects of urbanization on this pristine landscape.

23.2 Interregional Variation in Urbanization

In spite of being a ubiquitous phenomenon, urbanization is often considered a conundrum in lowlands only. Himalaya is home to cities of various sizes, though smaller in population and area occupied as compared to cities in the plains, but is not deterred by the challenges brought in by the topography. In fact, cities have modified their structures (construction capabilities on steeper slopes, construction of all-weather roads etc.) to utilize the *advantages* provided by the topography and grown into major tourist and religious centres catering majorly to population from the lowlands. The Indian Himalayan Region (IHR) covering an area of 530,000 km^2 is divided across ten states in the Indian Territory and also contributes the highest urban population in Himalayan region, prompting the focus of this chapter. The next sections look deeper into the regional variations of urbanization across the length and breadth of the Himalaya.

23.2.1 West Versus East

The states of Jammu and Kashmir, Himachal Pradesh and Uttarakhand constitute the Western part of Himalaya while the states of Sikkim, Assam, Tripura, Mizoram, Manipur, Nagaland, Arunachal Pradesh and hill region of Assam and West Bengal constitute the eastern part (ENVIS-CHE 2015). A total of 109 districts cover the IHR and population data provides an interesting outlook on urban spread across the range. Figure 23.1 represents the major settlements in Himalaya in terms of urban population per unit (sqkm) area of the district.

The districts with high percentages ($\geq 10\%$) of urban area are Srinagar (14.9%), Bandipora (14.4%), Kulgam (12.2%) and Jammu (10.8%), all in the WH, in the state of Jammu and Kashmir. Interestingly, major area of these districts lies in the valley of Kashmir, favouring the expansion of urban growth. The EH districts, on the other hand, are sparsely covered by urban area and only few cities accommodate the major share of urban population. In terms of urban population density, Srinagar (616/km^2), Dehradun (305/km^2), Hardwar (294/km^2), and Udham Singh Nagar (231/km^2) located in the WH are the only districts with more than 200 persons per km^2 (Census of India 2011). In general, the lower Himalayan districts possess greater percentages of urban area and higher urban population densities. The WH has higher percentage of urban area and higher urban population density in comparison to EH, where except East District of Sikkim (128/km^2) all other districts have urban population density of less than 50/km^2. Overall, the average density of cities in WH is 4534/km^2 and in EH 3475/km^2. Class I cities (see Table 23.2 for definition of classes) in WH have an average density of 9060/km^2 while EH Class I cities have 9395/km^2 with the highest densities in Kashipur (Uttarakhand) and Darjeeling (West Bengal) in WH and EH, respectively. But a major difference is seen in Class II cities in which WH cities accommodate double the number of people (10,409/km^2) as compared to EH cities (5059/km^2).

Consistent with the general understanding that mountains would accommodate a far lesser population than the plains, Himalaya records a low population size overall and even lower urban population percentage when compared to the plains. But globally, they remain the most populated mountain systems in the world (ENVIS-CHE 2015), and thus, studying this geography of human settlement becomes extremely important. The current population numbers and the growth rates in the Himalaya are far higher than in past decades. Figure 23.1 represents the distribution of urban population across districts of IHR. On average, the urban population of WH districts is 0.15 million, while that of EH districts are 0.07 million, highlighting the skewed distribution across Himalaya. However, of the total, 26 districts with above 0.1 million populations, both WH and EH, share equal number of districts (13 each). The WH districts of Srinagar, Dehradun and Jammu top the list with an urban population of 1.2, 0.94 and 0.76 million, respectively (Table 23.1). All the districts have flourished on

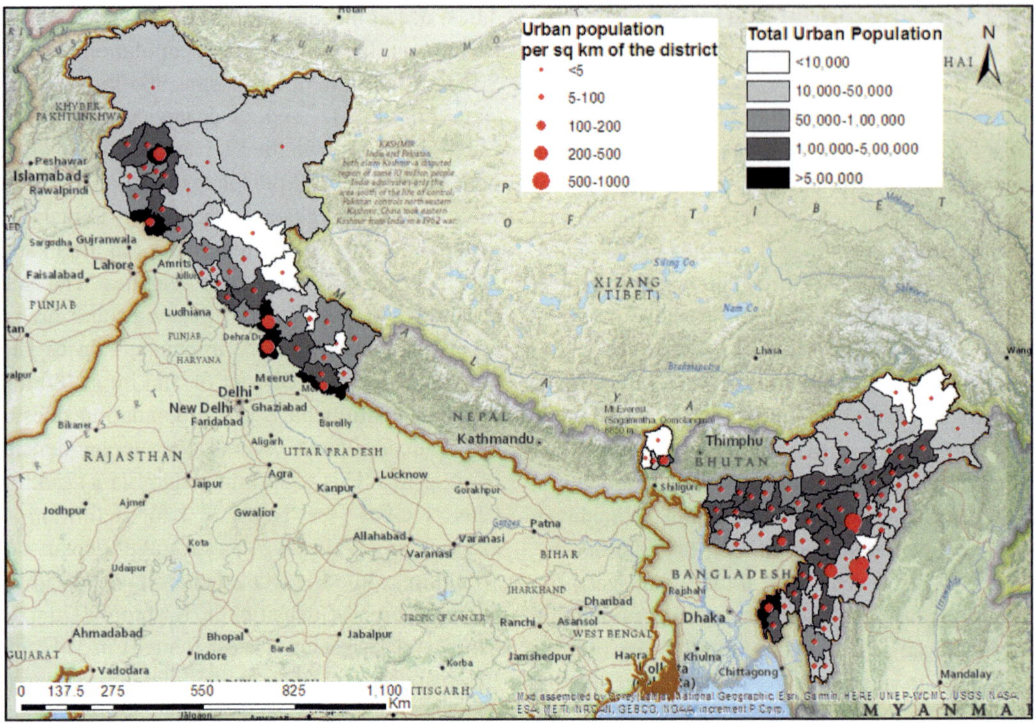

Fig. 23.1 Regions with urban population per square km and percentage of urban area across Himalaya

Table 23.1 List of districts with highest and lowest urban population in WH and EH (Census of India 2011)

WH			EH		
Highest					
State	District	Population	State	District	Population
J&K	Srinagar	1,219,516	West Bengal	Darjeeling	727,963
Uttarakhand	Dehradun	941,941	Tripura	West Tripura	677,638
J&K	Jammu	765,013	Meghalaya	East Khasi Hills	366,481
Uttarakhand	Hardwar	693,094	Manipur	Imphal West	322,879
Uttarakhand	Udham Singh Nagar	586,760	Mizoram	Aizawl	314,754
Lowest					
Himachal Pradesh	Kinnaur	0	A. Pradesh	Anjaw	982
Himachal Pradesh	Lahul & Spiti	0	A. Pradesh	Kurung Kumey	2345
Uttarakhand	Bageshwar	9079	A. Pradesh	Dibang Valley	2384
Uttarakhand	Rudraprayag	9925	Sikkim	North District	4644
J&K	Ramban	11,811	Sikkim	West District	5248

the foothills of Himalaya and thus advantaged by road connectivity, location of capital centre(s) of the respective states (Jammu—winter state capital, Srinagar—summer state capital of Jammu and Kashmir; Dehradun—state capital of Uttarakhand) and relatively expansion—favouring topography. In the EH, the highest urban population is observed in Darjeeling (0.72 million), the only hill district of West Bengal. It also serves as an important tourist destination in the eastern India and thus harbour a large migrant population as well (Zurik et al. 2005). Darjeeling is followed by West Tripura (0.67 million) and East Khasi hills (0.36 million) that inhabit the high populations and also serve as state capitals of Tripura and Meghalaya, respectively.

The least populated districts in IHR predominantly lie in the EH. However, two districts (Lahul and Spiti and Kinnaur—nil urban population) of WH also fall under five least populated districts of the range. The skewed distribution of urban population between WH and EH could be majorly attributed to the higher percentage of tribal population in north-eastern India that has higher dependence on agriculture, lack accessibility and hence tourism opportunities.

Comparing the data on cities, the WH has 297 cities as compared to 244 cities in EH. Table 23.2 describes the distribution of cities according to different population classes and shows that majority of cities in IHR have less than 20,000 of population. One of the reasons for this could be attributed to the rugged and complex topography and lack of transportation networks on their expansion.

Figure 23.2 represents the growth of number of cities in WH and EH between 2001 and 2011. Maximum increase in number of towns and cities between 2001 and 2011 were witnessed by Jammu and Kashmir where the count increased from 69 to 122. This was followed by Uttarakhand (from 76 to 116) and Manipur (from 29 to 52). The WH states are expected to have gained urban population due to rural–urban migration mainly due lack of employment opportunities and in search of better facilities. Uttarakhand has also observed greater tourism opportunities with advancement of roads and increased reach from tourist in the plains during summers. This growth of incoming tourist in Jammu and Kashmir and Uttarakhand is also highlighted in recent report of 'India Tourism Statistics 2018'. EH states except Meghalaya and Arunachal Pradesh record minimal growth of incoming tourist. Singh (2018) states that the urbanization in Manipur is mostly driven by rural–urban migration as well as illegal migration from neighbouring countries. However, Singh and Golson (2019) state reclassification of urban centres as one of the reasons for increased number of towns in the state.

However, based on the Census data (Envistat 2019), urbanization in EH states is seen to be more rapid than those in the WH. As indicated by Fig. 23.3, the average rate of urban population growth in WH and EH states is 3% per year and 6% per year, respectively. While in terms of urban area, WH states witnessed 18% increase, EH observed 53% (excluding Sikkim and Arunachal Pradesh) as per the Envistat Report 2019. This indicates the greater development of greenfield for urban expansion in WH, whereas EH expands the existing urban population at a much higher rate.

Table 23.2 Number of cities according to size-class distribution in EH and WH (Census of India 2011)

Class	Population size	WH	EH
Class I	>100,000	10	08
Class II	50,000–99,999	10	07
Class III	20,000–49,999	41	34
Class IV	10,000–19,999	70	72
Class V	5000–9999	98	85
Class VI	<5000	68	38

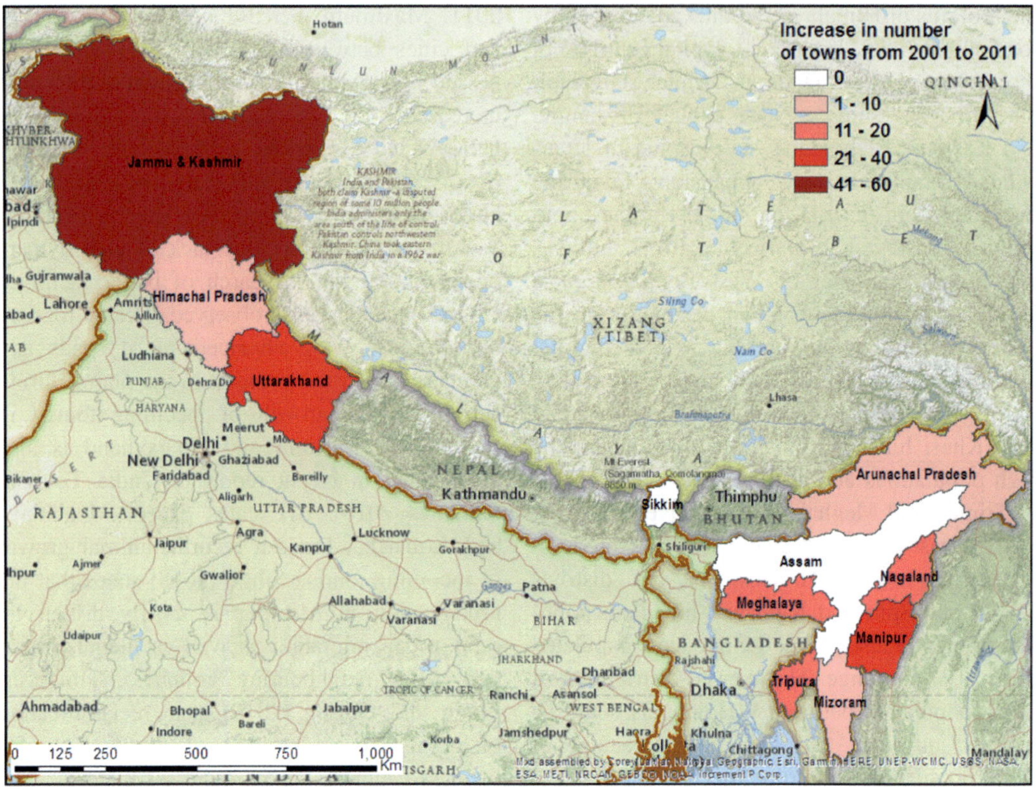

Fig. 23.2 Growth in number of towns from 2001 to 2011 (state-wise). Note—Only hill districts of Assam were considered

Decadal growth rate of Class I–VI cities (2001–2011) in EH (59.77%) is higher as compared WH (46.67%). Considering the growth in Class I cities (population > 0.1million), Agartala, state capital of Tripura, in EH shows the highest urban population growth (110.53%), while Rudrapur in Uttarakhand has the highest (58.84%) in WH. Table 23.3 lists the fastest-growing towns and cities in each class size. Overall, EH towns also report a higher number of towns (19) with greater than 100% growth rate as compared to WH towns (11). Thus, looking beyond the figures of usually projected 'total urban population', urban growth rate implies a rapid growth in towns and major urban agglomerations of the EH as compared to WH.

Not many research studies have been carried out in the IHR addressing the urban expansion and many of those attempted have been targeted in the WH region at variable scales. This *scholarly deficit* is compensated with a strong correlation of research interest to larger size and population of the cities/regions which invariably are located in western side, whereas EH with a higher growth rate is underrepresented (Mcduiera and Chettri 2018). A simple overview of published research articles listed in 'Web of Science' using keywords pertaining to urbanization in IHR provides a mix of limited results addressing both extent and impacts (discussed in Sect. 23.3) of urbanization. Studies attribute population growth, unplanned urban expansion and intense resource extraction as the key drivers of land-use land cover (LULC) changes. Remote sensing (RS) and geographical information system (GIS) remain the most preferred tool to understand the spatial changes in these cities owing to its spatial and temporal suitability. Most of the studies in IHR have focussed on understanding urban areas growth in terms of LULC

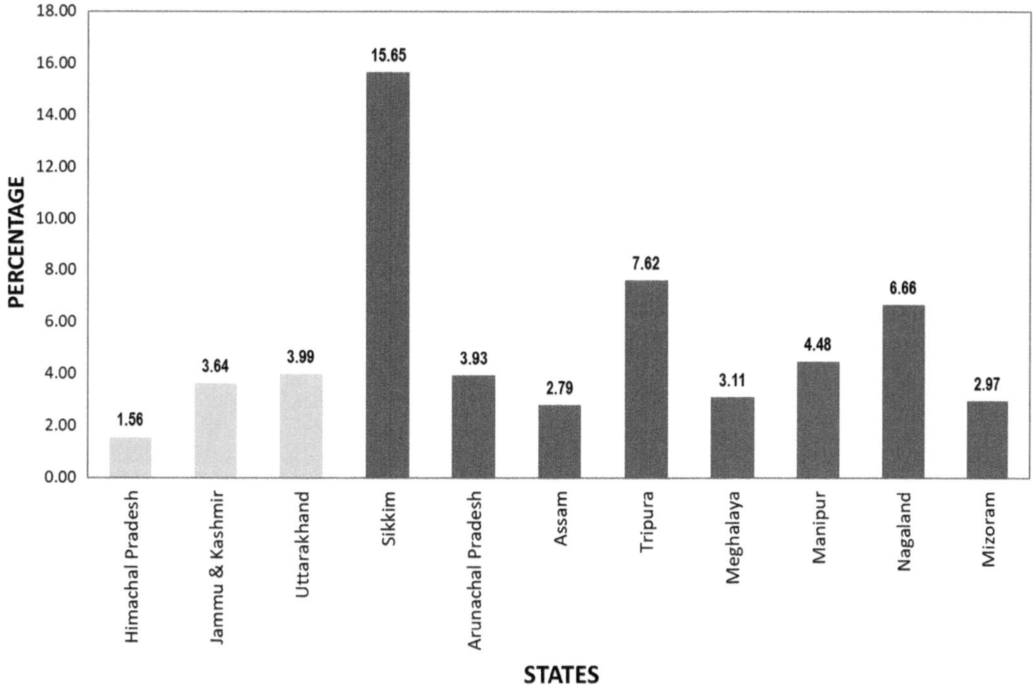

Fig. 23.3 Urban population decadal growth rate (% per year) from 2001–2011 (Envistat 2019). Grey for WH and Black for EH

Table 23.3 Number of fastest-growing towns and cities according to size-class distribution in EH and WH (Census of India 2011)

Class	WH	Growth rate (%)	EH	Growth rate (%)
Class I	Rudrapur (Uttarakhand)	58.84	Gangtok (Sikkim)	241.6
Class II	Jaspur (Uttarakhand)	29.76	Itanagar (Arunachal Pradesh)	69.9
Class III	Doda (J&K)	1808.56	Madanriting (Meghalaya)	78.91
Class IV	Tral (J&K)	3909.88	Namchi (Sikkim)	1145.1
Class V	Rudraprayag (Uttarakhand)	313.9	Jorethang (Sikkim)	203.6
Class VI	Gulmarg (J&K)	123.8	Luwangsangbam (Manipur)	3458

over a city while only few have coordinated studies over multiple cities or region covering a larger area. Figure 23.4 shows the distribution of major studies addressing urbanization in Himalaya. However, the comparison of urbanization trend in different cities is difficult owing to varied temporal periods the studies have considered.

In the WH, urbanization has been studied mostly as part to address the LULC in the region and only recently does a trend of focussing on

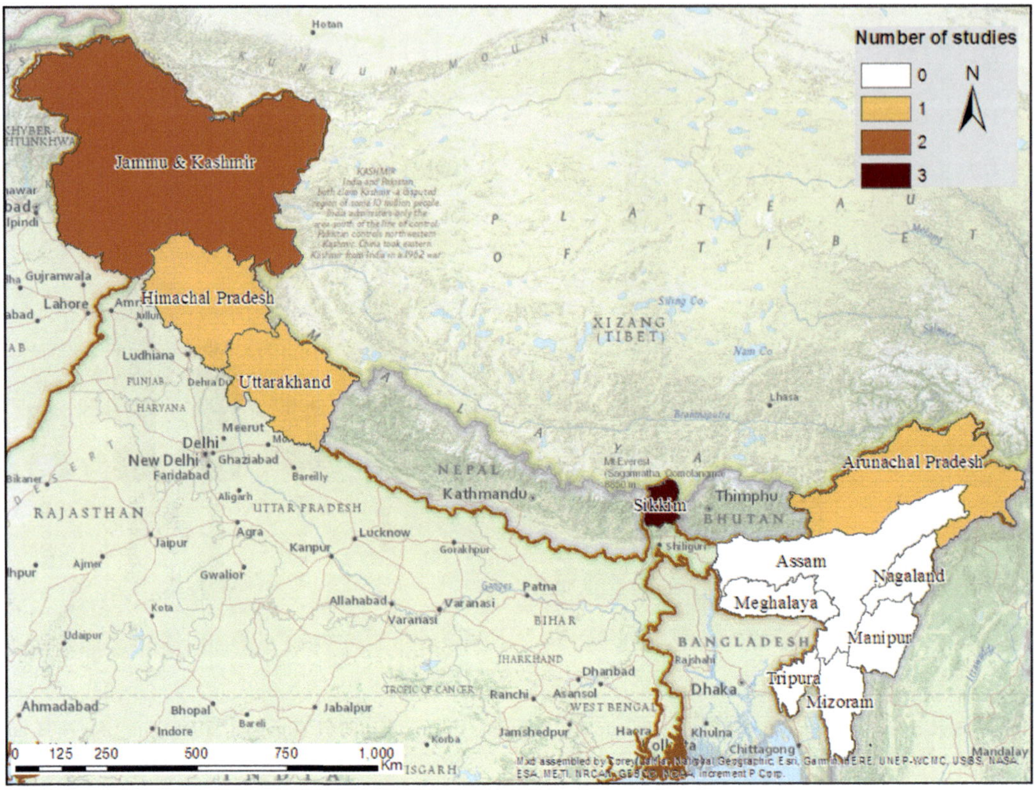

Fig. 23.4 Number of major studies for different Himalayan states on urban growth

urban areas viz. cities/towns appear. Cities such as Leh, Srinagar, Shimla and Dehradun are assessed for their LULC and report a trend of rapid urbanization over the past decades. Table 23.4 lists the towns/cities/urban settlements in the WH and EH which have undergone tremendous change over the past 50 years. Leh, the highest city (3500 m above sea level) in India and Himalaya and a Class III city has grown over from a trade node to an important military base and tourist attraction city. The city has seen rapid urban growth due to inward migration and tourist inflow leading to permanent changes in city structure as locals convert their land from agricultural activity to housing for tourist. Between 1969 and 2017, the city has grown its built-up area from mere 0.36 ha to 1.96 km^2 (196 ha). The urbanization trend in this once low-density city has led to densification of settlements and consistent loss of agricultural land. The city, although disconnected from Indian mainland by road for approximately 7 months of the year, is equally challenged by urban problems of water scarcity, water pollution and increased vulnerability to natural hazards (Dame et al. 2019). Addressing the growth in major cities of WH (Srinagar, Shimla and Dehradun), Diksha and Kumar (2017) provide a comprehensive detailed account of spatio-temporal urban sprawl. Srinagar, the administrative headquarters of the Jammu and Kashmir, is the largest city across Himalaya. The city has grown for mere 9.36 km^2 in 1972 to 142.19 km^2 in 2015 with major growth phase lying in the recent past (1992–2015). This extent of urban expansion is incomparable to any city in the entire Himalaya. However, the city has also observed reduced urban density primarily as a result of continuous conflicts in the region. Shimla, the state capital of Himachal Pradesh and one of the most visited tourist destinations in Himalaya, has undergone comparatively minimal urban growth. With just

Table 23.4 List of areas in WH with significant change in built-up areas

Western Himalaya	Period	Change (%)	Source
Leh (J & K)	1969–2017	444	Dame et al. (2019)
Upper Garwhal (Uttarakhand)	1990–2006	31	Raman and Punia (2012)
Kumaon (Uttarakhand)	1990–2014	955	Chakraborty et al. (2016)
Pithoragrah (Uttarakhand)	1976–2006	203	Munsi and Malaviya (2010)
Srinagar (J & K)	1972–2015	1419	Diksha and Kumar (2017)
Shimla (Himachal Pradesh)	1972–2015	162	Diksha and Kumar (2017)
Dehradun (Uttarakhand)	1972–2015		Diksha and Kumar (2017)
Kullu (Himachal Pradesh)	1989–2016	91	Vaidya et al. (2018)
Rudraprayag district (Uttarakhand)	1976–2014	328	Batar et al. (2017)

3.04 km^2 of net increase between 1972 and 2015, the city is by scale far smaller than others mainly due to the limitations brought in by the topography. However, the population of city exceeds many of that in the Himalaya. Dehradun, capital of Uttarakhand state, is located in the Doon valley providing favourable conditions for urban expansion and supporting road networks. Between 1972 and 2015, the city has added 45.99 km^2 of built-up area (Diksha and Kumar 2017).

Evidences of modifying Himalayan landscape due to increase in human settlements have also been recorded from remote parts of WH. Garhwal Himalaya located in the north-western part of the state of Uttarakhand (Raman and Punia 2012) is one such example. Primarily a watershed area with rugged terrain, the rise in built-up area has played an important role in expansion of agricultural land and fragmentation of forest. Another study of Garwhal Himalaya focussed on Rudraprayag district observes an expansion of urban area from mere 2.78 to 11.91 km^2 between 1976 and 2014 (Batar et al. 2017). While in Kumaon Himalaya of the Uttarakhand, an extensive study undertaken by Chakraborty et al. (2016) highlighted the increase in built-up area in the districts of Nainital (1.1–9.0 km^2) and Udham Singh Nagar (2.1–24.9 km^2). Overall, Kumaon division of the state witnessed a rise in built-up area from 3.80 to 40.10 km^2 between 1990 and 2014. Such accounts of LULC revel the footprints of urbanization in remote areas even though away from the pull of megacities. Other cases of drastic urbanization rate in remote areas include Pithoragarh district, bordering Nepal and China, of Uttarakhand state. Munsi and Malaviya (2010) report an increase from 2.82 to 8.55 km^2 in a span of three decades. With the highest rate of LULC conversion, urban areas are predicted to gain most from agricultural land. A study on LULC change in Kullu valley of Himachal Pradesh, more known for its tourist inflow than agricultural output, reveals a growing threat from industrialization and urbanization to agricultural land. This is reported from both increase in existing urban population (and associated area) and upcoming small towns being converted from rural land. Between 1989 and 2016, the built-up area increased from 5.66 to 10.8 km^2, with a loss of 15.74 km^2 of agricultural land (Vaidya et al. 2018).

The EH remains far less explored in terms of scientific studies in comparison to WH. The only coordinated systematic study is undertaken by Diksha and Kumar (2017), but restricted to the two capital cities (Gangtok and Itanagar). In comparison to the WH capital cities, Gangtok, capital of eastern Himalayan state Sikkim bordering China and Bhutan, marks an insignificant growth between 1972 and 2015. Higher growth rate is observed in between 1972 and 1991 as compared to later period up to 2015. Majorly, topography has led to scattered development of urban areas. Another study in and around Gangtok city provides a similar picture of slow urban expansion from approximately 25.41 to 34.93 km^2 between 1990 and 2010. The study also highlights the projected urban sprawl in different parts of the city along with expansion of step-cultivation at the expense of forest land (Mukhopadhyay et al. 2014). The capital city of Arunachal Pradesh, a very sparsely populated state, Itanagar, has undergone modest growth with just total addition of 7.19 km^2 to the initial 3.12 km^2 of urban land between 1972 and 2015.

Other isolated studies across EH (Table 23.5) provide a glimpse of *sprouting* urbanization in regions with majorly rural background. A district-level analysis of LULC in Assam (Kamrup district) reveals doubling (559–1128 km^2) of urban settlements over a span of three decades along with increased forest fragmentation and loss of dense forest cover (Kumar 2017). Dynamics of LULC in Rani Khola watershed of Sikkim Himalaya, close to the Gangtok city, between 1988 and 2017 highlight the increasing footprint of urbanization. In a largely rural setup, the built up is seen to consistently rise (7.18–12.59 km^2) along the watershed area and gaining mostly from agricultural and forest land (Mishra et al. 2019). A detailed study of LULC changes and its impacts on ecosystem vulnerability of Darjeeling district observed a moderate growth from 91.41 to 124.93 km^2 between 1977 and 2012 (Mor 2012). However, demographic records highlight the rapid growth of Darjeeling district, highest in EH. Contrasting to studies utilizing remote sensing and GIS techniques, ethnographic detailing by researcher in the small city of Namchi, located in Sikkim, gives a similar picture of booming urbanization. Having the second-highest urban population in the state, Namchi has undergone rapid expansion to accommodate nearly 12 times in past two decades and provides a perfect example of rural–urban transformation (Mcduie-ra and Chettri 2018). Thus, in EH, overall the expansion of urban area seems insignificant, although population growth rate in EH cities, exceeds those in WH. This indicates that EH cities continue to densify without much expansion in area, the restriction brought in by topography and lack of transportation networks playing a major role. However, with the lack of information on other regions of EH, discussion on urbanization remains limited and undervalued.

23.2.2 Altitudinal Variation

Altitudinal variation across IHR districts varies greatly, not just among each other, but also within each district. Figure 23.5a and b shows the location of cities across Class I–VI in different altitudinal zones. Across Himalaya, cities are located from foothills to a maximum height of 3500 m. However, a clear distinction is observed in location of towns with rising altitude as number of towns decreases. About 63% of the cities are situated below the altitude of 1000 m, with more than half of these cities below 300 m altitude. This indicates the preference of urban settlements in the lower Himalaya with favourable accessibility and land to expand without facing challenges of terrain. However, there is also a distinction between EH and WH cities. A higher proportion of cities (74%) is located below the altitude of 1000 m in EH as compared to WH (55%). This can be related to the known lack of road networks and higher coverage of forest, making accessibility and expansion of cities more challenging in EH as compared to WH. Population pressure exerted from the plains of northern India could also be another reason for resultant urbanization at higher altitudes in WH. Across Himalaya, about 32% of cities are located between an altitude of 1000–2000 m. This

Table 23.5 List of areas in EH with significant change in built-up areas

Eastern Himalaya	Period	Change (%)	Source
Kamrup district (Assam)	1977–2010	102	Kumar (2017)
Gangtok (Sikkim)	1972–2015	676	Diksha and Kumar (2017)
Gangtok (Sikkim)	1990–2010	37	Mukhopadhyay et al. (2014)
Itanagar (Arunachal Pradesh)	1972–2015	230	Diksha and Kumar (2017)
Rani Khola (Sikkim)	1988–2017	75	Mishra et al. (2019)
Darjeeling district (West Bengal)	1977–2012	37	Mor (2012)

proportion is higher in WH (39%) as compared to EH (24%). Cities in this range hold a greater significance as they reflect the true mountainous cities and their growing presence at this height range can be concerning given the vulnerability of Himalaya to various natural disasters. However, cities above an altitude of 2000 m are considerably less across Himalaya. Only 5% of the cities are located beyond this height and share is even lesser in EH (2%) as compared to WH (7%). In the WH, such cities are primarily a mixture of tourist and religious centres which cater to increasing population from plains and other parts of Himalaya. In EH, Darjeeling is the only touristic city catering to visitor population.

23.3 Urbanization Impacts on Himalayan Systems

The rapidly changing demography and land use in the Himalaya have generated far-reaching effects in and around the mountains. A globally known eco-sensitive region and biodiversity hotspot, these mountains harbour rare endangered species which are constantly pulled into the loop of urbanization-climate change nexus. The inefficiently and unsustainably *planned* urban growth gives rise to a number of activities which are adversely impacting the environment and humans alike. Given the intimate dependence of communities with mountains, these issues are primarily related to air and water quality, water availability, flooding, local climate, biodiversity, land and soil, as well as socio-economic impacts. The literature available on impacts of urbanization and other human-driven activities provide a glimpse of deterioration across these mountains and call for immediate policy intervention. Following are the impacts on subsystems of the Himalaya:

23.3.1 Air

Air pollution in the Himalayan cities is increasing over the years (Ghosh 2007), including cities of Manali (Dasgupta et al. 2017), Mussorie (Sundriyal et al. 2018), Srinagar (Sheikh and Najar 2018), Rishikesh (Deep et al. 2018), Guwahati and Haridwar (Deb 2019). The severity of increased pollution level in Himalaya also arises due the topographical and climatic factors which naturally lead to trapping of pollutants by temperature inversion (Wester et al. 2019). A number of air pollution sources have been reported across Himalaya, however, the tourism-related vehicular pollution and the construction dust have been particularly found to be the major source in this diesel fuel-usage dominated mountainous environment (Dasgupta et al. 2017). Tourism being an important driver of development in many mountainous cities has in turn aggravated the traffic mismanagement and resultant air pollution. A considerable impact of diesel-run vehicles, especially higher sulphur levels along the Manali-

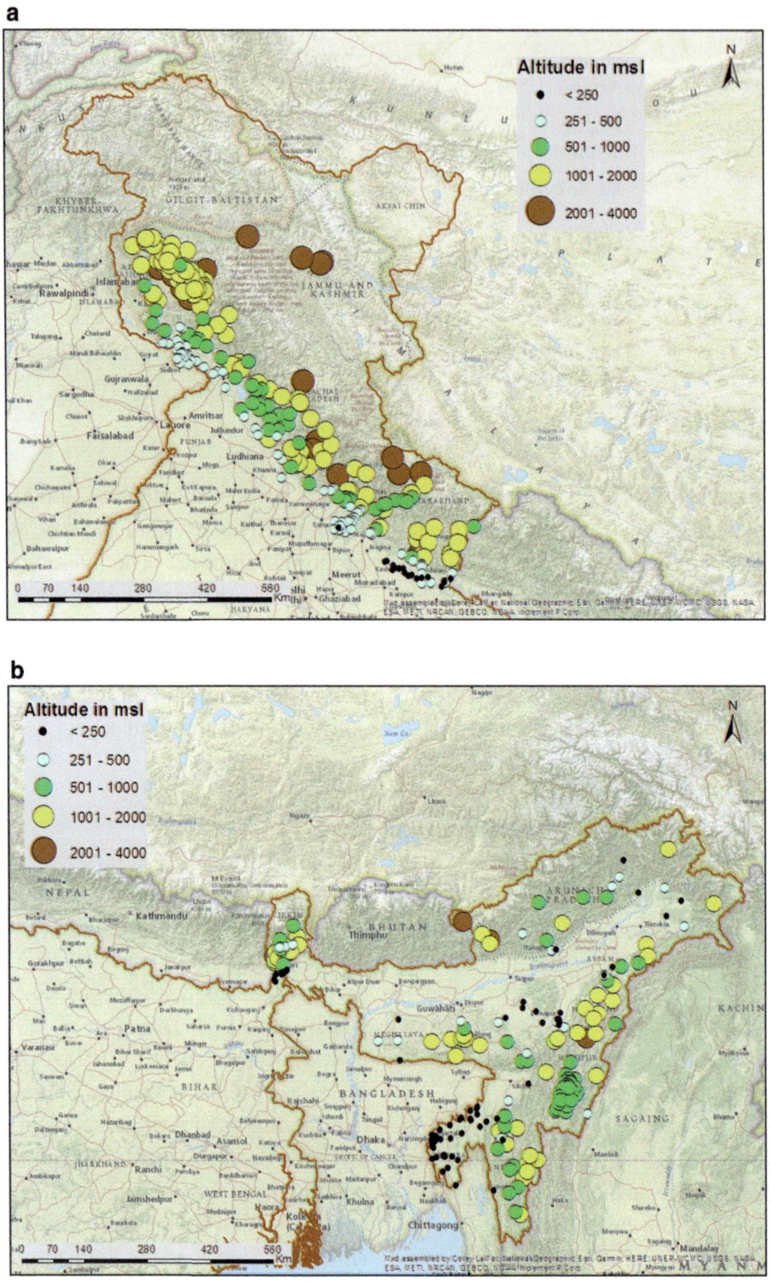

Fig. 23.5 **a** and **b** shows the location of all Class I-VI cities in WH and EH. The concentration of cities around capital cities and valleys is a visible trend indicating the growth of larger number of smaller cities around the main city. For example, a large number of cities can be observed to concentrate around Agartala, Imphal, Shillong (state capitals) and Darjeeling. While in WH, cities are more randomly spread. Considerable concentration can be observed around the cities of Jammu and Srinagar. Higher altitudinal gain of WH (from plains to high altitudes) as compared to EH can be seen as one of the factors for WH cities to concentrate around diagonal stretch from north to south

Leh Highway, is reported in roadside soils (Dasgupta et al. 2017). These remote and low vehicular load destinations point to growing human presence along the road network. One of the most visited mountain cities in the WH, Mussoorie, is reported to be far exceeding the national prescribed air pollutant limits. A strong relation is observed between the inflows of tourist in summer months and increased particulate matter. Increase in black carbon resultant of biomass burning in winter season is also another contributing factor effecting air quality (Madan and Rawat 2000; Sundriyal et al. 2018). Sharma et al. (2013) reported the increase in O_3 and black carbon in Kullu valley in summer months, with strong influence of meteorological conditions prevailing in the mountains. EH also has reported cases of growing air pollution. A study on atmospheric aerosols by Adak et al. (2014) in Darjeeling city highlights the significant contribution of local anthropogenic sources and biomass burning. The study indicated higher presence of ultrafine aerosol particles in the premonsoon season, coinciding with the influx of tourist. Reports of air quality from remote areas, such as Barapani, Meghalaya, with insignificant local sources and ample rainfall around the year show a dominant contribution from long-range transport of aerosols from Indo-Gangetic Planes (IGP). Pollutants from such transport include $PM_{2.5}$ and polycyclic aromatic hydrocarbons (PAHs) (Rajput et al. 2013). A review of literature on air pollution in the Hindu Khush Himalaya by Bonasoni et al. (2019) summarizes the increased levels of anthropogenic ozone and black carbon over the Himalaya. The impacts of local and regional sources of air pollution are also understood to have concerning impacts on glaciers and resultant hydrological cycle. Impacts of air pollution are expected to disrupt and lead to northward shift of monsoon rain belt, which will result in increased rainfall over northern India (Lau et al. 2010). Air quality is one of the regulatory ecosystem services valued in the Himalaya (Mondal and Zhang 2018), especially as perceived by the people in lowlands, a decrease in its quality can seriously hamper the economic gains and drive tourism out of the mountains. A major role is also played by the IGP south of the Himalaya, and thus, emphasis should be on developing policies which are integrated across the region. However, air quality monitoring stations across Himalaya remain a grey area in any efforts to cut down on emissions (Wester et al. 2019).

23.3.2 Water

Impacts of urbanization on water sources are one of the most critically perceived threats on Himalaya. Any degree of loss in their function as water towers will have a threatening chain of events across and beyond these mountains. On the one hand, glacier melt as result of climate change will lead to loss of volume and disrupt the water cycle, leading to flooding and other disasters. While, on the other, the increasing load of sewage and solid waste is rendering them unusable for human use and sets a pattern of countermeasure unsustainable infrastructural development harming the mountain environment. Himalayan landscape is also dotted by the development of large number of dams—an inseparable infrastructure feature of urbanization serving cities as well as rural areas far from its reach. Different hydrometeorological drivers along the length of Himalaya have given rise to multiple flooding events over the year (Elalem and Pal 2015). But, unplanned urbanization has played a contributing factor in elevating the arisen losses. Especially, the urban settlements in the valley and parts of lower Himalaya have faced disastrous consequences, while events of flash flood in high altitude cities are also on the rise. Rawat et al. (2017) studied the increasing flood vulnerability in foothills of Himalaya as result of urban growth. Results show that fast urbanizing areas of the Ramnagar town, Uttarakhand, are under high to extreme risk of flooding. Addition of new settlements in flood hazard zone and rising of river bed due to sedimentation are the main reasons for increasing proximity of settlements to river. A large number of studies have also highlighted the changes in the quality of water bodies (surface and groundwater) and the disasters arising from them across Himalaya. Impacts of tourism are observed to add vulnerability to groundwater through pollution and degradation of watersheds along the trekking trails in Ladakh region (Geneletti and Dawa 2009). Mountains also bear a greater threat of groundwater pollution due to toxic minerals as the terrain allows easy sediment transportation due to mineral extraction activities (Xu et al. 2007). A study in Lidder valley, Jammu and Kashmir, reported a decline in water quality correlated with increase in touristic inflow. Tourist data analysis reveals an exponential rise in the tourist visiting this valley after 2003, especially Pahalgam (Rashid and Romshoo 2013). The massive amount of solid waste generated from these cities is often dumped along the mountain slopes or in water bodies resulting in leaching into the soil and contamination of lakes and springs (Bashir and Goswami 2016; Qadir and Singh 2019). Release of untreated sewage from the urban areas is another reason for the deteriorating quality of the water bodies in hilly areas (Rashid and Romshoo 2013). In addition, encroachment of water bodies, construction in flood-prone areas and the increased run-off from impervious urban built-up, is further depleting the blue cover accompanied with frequent and more intense flooding in these hill cities. This coupled with climate change impacts, often results in water scarcity in Himalayan cities, with situation worsening during peak summer periods which coincide with peak tourism season. Urbanization led hydrological changes are resulting in long-term decreasing trend of stream discharge, drying of springs and dwindling capacity of urban lakes (Rashid and Romshoo 2013). In the recent times, Shimla, one of the most famous tourist destinations, has faced major water crisis and led to decline in number of tourist visiting it.

23.3.3 Land and Soil

Although the area of land occupied by cities in Himalaya is relatively limited as compared to vast landscape, urbanization generates far-reaching additional problems. Encroachment of fertile land and forest are the most the visible impacts but is aggravated by haphazard growth of unauthorized real estate growth, followed by construction of roads, solid waste mismanagement among others (Ghosh 2007). The solid waste dumping, emissions from diesel-run vehicles and the large-scale construction cause land degradation and contamination of the soil. All the three contributing factors are amplified by unplanned urban growth. In Mussoorie, increase in generated solid waste-23 metric tons per day was observed during the peak tourist seasons (2015), while up to 18 metric tons are generated on average around the year. The composition of waste shows that majority (66%) of the waste generated was bio-degradable followed by plastics (15%) (Sundriyal et al. 2018). In north-western Himalaya along the Manali-Leh Highway, various pollutants from diesel-run vehicles have been contaminating the soils along the roads (Dasgupta et al. 2017). In Lidder valley of Kashmir Himalaya, the solid waste generated by tourist and associated accommodation facilities accounted for 83% of the total annual contribution in between touristic months of June–August, 2011 as reported by Centre of Research for Development (CORD). Rashid and Romshoo (2013) noted this to be intrinsically linked to pollution of water bodies in the region as proper disposal sites for solid waste management are grossly lacking prompting disposal on river banks and forest areas. In Ladak, the growth of recent camping and trekking tourism has opened remote and poorly accessible areas without infrastructure and awareness (both by visitors and locals) to serious threat of land degradation (Geneletti and Dawa 2009). The study recognized waste dumping sites, trekking trials, camping and off-road driving as important stressors in the region leading to land degradation. Improper solid waste management along the trails and susceptibility to soil erosion due to climatic conditions was also found to the increase vulnerability to groundwater pollution. Geologically, Himalaya is known to be highly susceptibility to erosion owing to its seismotectonic position, meandering rivers and weak geological composition. Especially, Kashmir valley has been affected by land degradation as a result of vegetation loss and growing urban settlements. Many of the watersheds in the valley have experienced increased silt load due to sedimentation and eutrophication (Zaz and Romshoo 2012). Another conjunction to erosion susceptibility in the region is instability of slopes along roads built in Himalaya. The rapid pace of road development to facilitate easy transportation in Himalaya has led to destabilization of many slopes and results in large-scale landslides (Siddque and Pradhan 2018).

23.3.4 Biodiversity

Phytogeographically, the Himalaya represents two distinct biodiversity hotspots of the world, straddling between two biogeographic realms. The mountains harbour rich biological resources with high number of endemic flora and fauna. This has also been acknowledged with large number of protected areas across Himalaya (Chettri et al. 2010; Nautiyal and Kaechele 2009). Humans have been estimated to utilize nearly 25% of the earth's primary productivity (Vitousek et al. 1997) and resource competition is known to be more intensive in the Himalaya due to higher dependency of local community on natural resources. But, being the most populous mountains in the world, urban growth also intensively and rapidly changes the land cover resulting in degradation and disruption of critical ecosystem services provided by them (Guneralp and Seto 2013). Increasing demand of resources by growing population is also presenting obvious challenges in maintaining a resilient ecosystem. Climate change has been identified as one of the pressers to biodiversity loss in the region. However, impacts of urbanization in the region have not well been documented and would need establishment of concrete pathways (Wester et al. 2019). Pandit et al. (2014) delineated the existing drivers of ecological transformation and emphasis on the greater potential of urbanization in this process. Migration and tourism are known to be complementary add-ons to available opportunities and increased accessibility, respectively. Despite the difficulty imposed by terrain, road development has been one of the main elements assisting urbanization and leading to fragmentation of forest landscapes. Away from the cities, which are already established and growing continuously, biodiversity in the Himalaya is also impacted by the hydropower plants located in riverine ecosystems (Pandit and Grumbine 2012), reliant on a trade-off between gained energy and lost undervalued critical biodiversity. Some studies have attempted to bridge the missing link relating direct link between urbanization and biodiversity loss. The rippling effects of urbanization in the watersheds of Himalaya are also reported to threaten nearly 49 species of butterflies in the Bhimtal region of Nainital district, Uttarakhand (Smetacek 2012). Joshi et al. (2011) utilized the potential of geospatial modelling tools to report the barriers and shifts in ecological corridors serving as migratory routes in two important protected areas of Uttarakhand. The results indicate the impact of 'interspersed mosaic of forested and non-forested areas' in the form of increased human–wildlife conflicts. Within cities, biodiversity plays an important role to maintain the ecological balance skewed towards low hemeroby. But cities also have accumulated impact of invasive species, both purposefully introduced and accidentally brought in. Mehraj et al. (2018) recorded 325 alien species around Srinagar city mainly introduced for floriculture, followed by agriculture in the area. The favourable conditions provided by the urban landscape (like higher temperature and nutrient content) have led to naturalization of many species and pose a threat to local species already facing the wrath of climate change and anthropogenic disturbances.

23.4 Drivers of Urbanization in Himalaya

The trend of growing urbanization is spreading across the range of Himalaya with increasing size, number and complexity of urban settlements. This rapid and rampant urbanization across the Himalaya is resulting in a range of significant socio-economic changes. Urbanization in these mountainous ranges is primarily characterized by two pathways: urban expansion and urban intensification (Tiwari et al. 2018). The three observed modes of urban expansion in the Himalaya are; *low-density continuous* sprawl, *leapfrog* sprawl and *ribbon* sprawl. Of these, the leapfrog and ribbon sprawl are more commonly encountered, as complex terrain and topography constrain the low-density continuous urban sprawl. There are four major drivers of urbanization observed in the mountain range contributing to the growing Himalayan cities via the three sprawl pathways (Fig. 23.6).

Out-migration from rural areas in Himalayan states has been found to be the most prominent cause for urban growth in the mountain range. Declining agricultural productivity and decreasing size of landholdings is making agriculture unsustainable livelihood option and resulting in increasing rural to urban migration (Hunzai et al. 2011).

This migration is resulting in a gradual shift in livelihood practices from primary resource development to secondary and tertiary sectors. Rural to urban migration is of dual concern for the respective state governments, as this shifts the strain onto the existing urban settlements and intensifies the urban transformation of land in the sensitive ecosystem of Himalaya. For instance, the declining population in districts of Almora and Pauri Garhwal has been recognized as a serious problem by the Uttarakhand Government, and it has set up the *Rural Development and Migration Commission* in August 2017, with an aim to examine all aspects of the problem and inform government to efficiently address the problem and ensure sustainable rural development. Most of this rural to urban migration in Uttarakhand is into the districts of Dehradun, Udham Singh Nagar, Nainital and Haridwar (Government of Uttarakhand 2018). Climate change is also making agriculture an unsustainable livelihood option, further facilitating rural to urban migration in these areas. Such migration activities increase the urban density as well as cause urban expansion.

In addition, the blooming tourism, both domestic and international, is also driving the urbanization in the region (Tiwari et al. 2018). Tourism in form of pilgrimage for IHR has

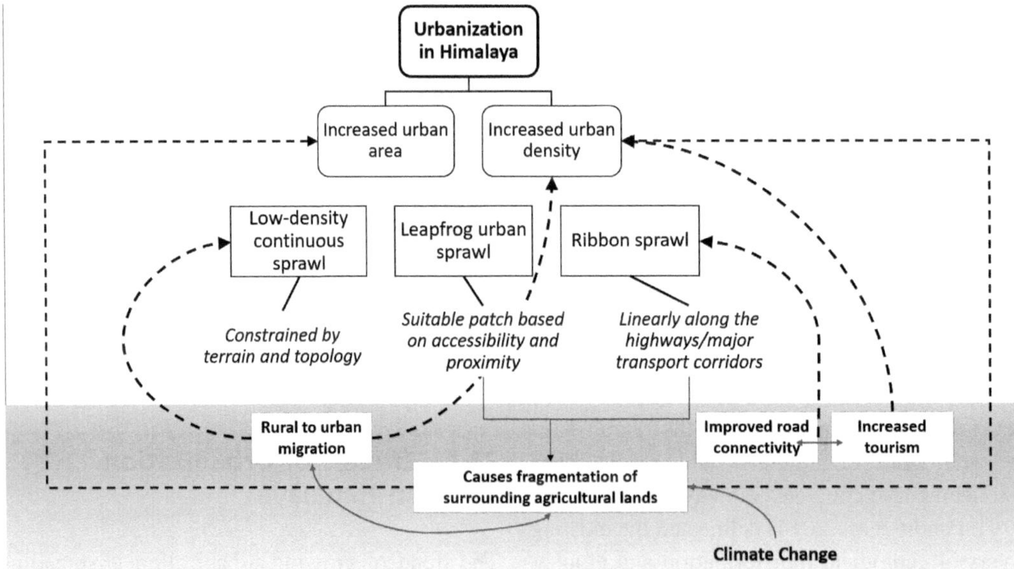

Fig. 23.6 Process of urbanization in Himalaya, its two broad pathways, three modes (with characteristics) and drivers (in grey portion). Dotted lines represent how drivers contribute to pathways and modes (from drivers)

existed for very long. With the arrival of the British in the nineteenth century, summer resorts and 'hill stations' were established. And the current publicity and marketing of new tourist sites are bringing in huge number of tourists throughout the year, facilitating the consumption-based urbanization, referred as tourism urbanization.

Improved road connectivity is another important driver of urbanization in IHR. The highway and other major transport corridors (district roads) drive the ribbon sprawl urbanization in the IHR, for instance, as reported by Kuchay and Bhat (2014) in the case of Srinagar city. Over the past few years, the governments (both state and central) have been working towards improving the connectivity in the Himalaya. Figure 23.7 indicates the rapidly increasing length roads in two of the Himalayan states; Uttarakhand and Meghalaya (ENVIS 2015). Construction of roads and highways promotes urban growth along their length and results in fragmentation of agricultural and forest lands which has grave ecological consequences.

These factors of rural to urban migration, tourism urbanization and improved road connectivity are often coupled with poor enforcement of land-use policies. This is resulting in unregulated and unplanned urban development in the region that is causing severe damage to the ecosystem.

23.5 Policy Gaps and Suggestions

A number of government policies in the Himalayan region have been focussing on the conservation of natural resources, quite often bundled in policies of forest, wildlife and environment. Consistent with the mixed form of governance, multiple stakeholders in the Himalayan region share various responsibilities in the conservation of valuable resources providing essential ecosystem services. The due attention required for sustaining them has been attempted through mission like 'National Mission on Sustaining Himalayan Ecosystem' (NMSHE), established in 2008 as one of the eight missions under 'National Action Plan on Climate Change' (NAPCC). The NMSHE commits to contribute towards sustainable development of the Himalaya by engaging up-to-date

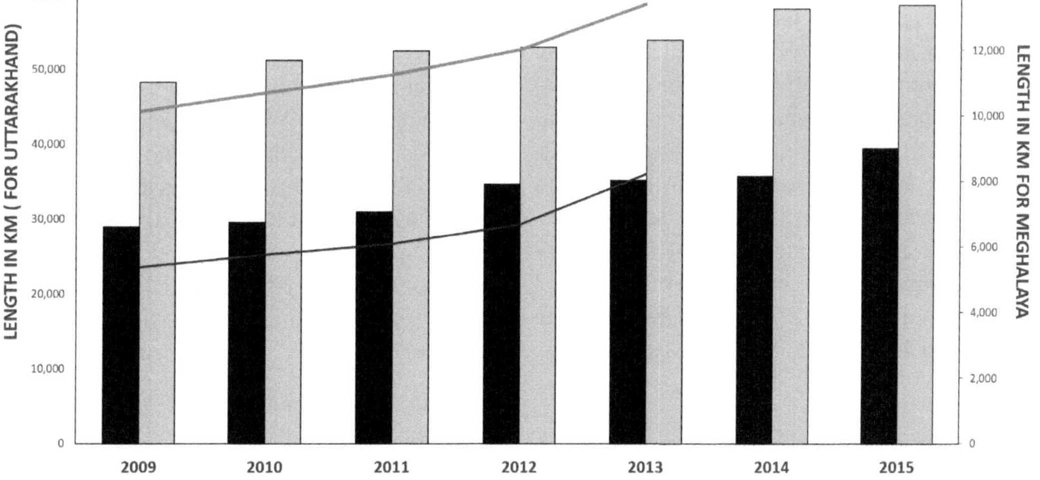

Fig. 23.7 Surface and total road lengths in Uttarakhand (UK) and Meghalaya over the years (2009–2015)

management and policy measures. While to tap the existing adaptation knowledge and bridge the science-policy-practice through technological advancements, the Govt. of India launched the 'National Mission on Himalayan Studies' in 2015. Programmes like 'North-Eastern Region Urban Development Programme (NERUDP)' have attempted to consider the specific needs of north-eastern states. Specific policy like this has been directed to provide required support to five capital cities of the eastern in sectors of—water supply, sewerage and solid waste management. National Tourism Policy of India (2018) tends to promote adventure tourism in Himalayan states. At the same time, the urban development policies of the country are guiding the urban growth and development in India. With tourism being an important driver of urbanization in Himalaya, there is a need for integrated policies that simultaneously focusses on the promoting eco-tourism and managing sustainable urban growth in the hill cities of Himalaya. In the latest development, Niti Aayog, the central planning branch of Govt. of India, made a coordinated effort by setting up five working groups in thematic areas relevant to sustainable development in the mountains. Though, discussed in brief, nodes of problems related to unsustainable urban growth find a place but with minimal commitments to tackle it. However, in order to make these policies more effective, it is important to understand and address the drivers of change of which urbanization is the most prominent one. There is a need to develop a dedicated action plan for Himalaya that is inclusive of the biological (flora and fauna), environmental (air, water, land), social (cultures, communities) as well as economical (energy, industry, tourism and urbanization) dimensions. To achieve this, policies need to combine direct regulation, provide economic incentives for protection and ensure community participation for framing more informed policies (Badola et al. 2010).

23.6 Conclusion

In the urban race, Himalayan urbanization is relatively a new aspirant with most of the patterns and process revealed in recent decade. But, contrary to the attention received to physical geography and climate change science in Himalaya, urban growth (and its immediate effects) remains largely ignored. Himalaya remains the most populated mountain chains in the world, majorly in rural landscape, but with a large unrecognized potential for urban transformation. This potential finds its drivers from the lack of employment opportunities in the region, harsh weather condition impeding regular income from agricultural sources, lack of proper health, education and infrastructure facilities, and complimented by the flourishing tourism industry, increased road connectivity, accessibility to better facilities, etc. But Himalaya is by nature very fragile ecosystem and is also tectonically alive, further threatened by the global climate change with frequent extreme events such as cloudbursts, landslides and flash floods (e.g. Srinagar, Leh, Uttarakhand, Sikkim). Developing cities in such high-risk area would require that cities not only develop sustainably but should also plan for climate resilience to face future climate change impacts.

Diversity in culture and nature in the Himalaya has led to differential urbanization patterns that need to be understood using a different lens which equally recognize the importance of valuable ecosystem services, ill-effects of mountain-side development and aspirations of a developing nation. WH with comparatively earlier exposure to urbanization trend and increased accessibility from road infrastructure in difficult terrain has outgrown its urban space as compared to EH, which still remains under large forest cover but with very high potential to grow. This rapid pace of urbanization in this vulnerable mountain backdrop requires special attention which counters the haphazard growth in them with systematic plan and bylaws. The implementation of bylaws and new policies needs to be micromanaged with due concern for localized differential conditions in the foothills and hills. EH urbanization, on the other hand, is unique with majority of urban population settled in state capitals, thereof with higher population densities. EH, at its current rate, has less number of cities in almost all population size classes. This provides the opportunity to implement sustainable development models with focus on appropriate resource conservation and management of urban services which are considerate of rising population density. The urbanization trend in different altitudinal ranges is largely concentrated in lower ranges of Himalaya. Accessibility factors and topographic reasons could be reasons for larger number of settlement in lower altitudes. WH concentrates larger number of cities in higher elevation (as compared to EH) and could lead to higher population pressure on biodiversity as technologies advance and land prices decrease. Discouraging advancement of roads and capping the population of cities at higher altitudes should find a place in policy implementation as pristine environments at these heights face the threat of human expansion.

Immediate assessment of identified impacts over entire Himalaya is necessary to develop countermeasures which are suitable for mountain landscape and not a mere replication of tools and techniques applied in the plains. In these mountains, known vulnerability and unknown value of ecosystem services provided especially make any impact assessed reasonably higher than that in the plains. Following points can be considered while planning mitigation options for various impacts in the Himalaya:

(a) Topography and climatic factors present a natural challenge to mitigation of air pollutants from mountainous cities, and thus, vehicular pollution coinciding with peak touristic season should be considered in these months. For these months, strictly adopting a sustainable way of transportation in, around and out of the city would reduce the accumulation of pollutants.
(b) The nature and dynamics of rivers in Himalaya demand exclusion of human settlements from immediate vicinity of the floodplains as this can benefit both ways. Recent disasters highlight the vulnerability of urban population to changing water cycle and lack of infrastructure to support such proximal habitation. On the other hand, proximity and mismanagement of sewage waste are continuously leading to increase in pollutant load, with impacts running over to downstream dependents as well.
(c) With limited feasible land for development, mountain cities should consider the load of increasing tourist inflow with time and ensure infrastructure which cater to floating population. Unlike in plains, factors of erosion can play a prominent role in land degradation and thus result in transport of both valuable soil and toxic pollutants. Land conservation with the assistance of remote sensing and GIS can benefit in assessing the susceptibility and management of such landscape.
(d) The two biodiversity hotspots in Himalaya represent a unique assembly of endemic and valued flora and fauna. Despite the higher number of protected areas, fragmentation of landscape brings in critical changes and shifts in ecological corridors. The inundation of forest landscape due to large number of hydropower projects will have a higher impact as compared to the plains due to higher biodiversity concentration in the region.

Urbanization in Himalaya has been found to driven by rural to urban migration, which in turn is accelerated by climate change impacts on agriculture making it an unsustainable livelihood option. The state of Uttarakhand has recognized this as a serious problem and has set up the Rural Development and Migration Commission towards understanding the underlying issues and to work towards the policy solution for the same. However, other states yet need to acknowledge the problem and examine the causal factors.

Others drivers of urbanization in IHR are tourism and increasing road connectivity. Though the two factors mentioned are often indicators of improvement in economy, but due to lack of appropriate land policies and ineffective enforcements, tourism and road connectivity are resulting in unplanned rampant urbanization instead of a more sustainable one. Hence, there is a need for land-use policies that are informed with evidences from economic sectors, social aspects as well as with evidences from environmental and ecological indicators. Above all, any such policy in this region should mainstream the climate change that is largely redefining the socio-economic behaviour and ecological conditions across the globe. This is particularly crucial for the Himalaya which has been found to be at higher risk due to climate change.

Overall, Himalayan cities have a special character in terms of their geography. The city development in these areas has its unique requirements, and hence, urban planning of Himalayan cities calls for more specialized focus that is inclusive of the needs of local communities and the Himalayan ecosystem at large. This has special implications for the urban development policies of the government. For instance, urban planners and authorities should take learnings from the climatologically and environmentally sustainable traditional ways of urbanizing in hill cities. Himalaya is highly sensitive ecosystems and even the mild disturbance could adversely impact their ecological dynamics.

References

Adak A, Chatterjee A, Singh AK, Sarkar C (2014) Atmospheric fine mode particulates at Eastern Himalaya, India: role of meteorology, long-range transport and local anthropogenic sources atmospheric fine mode particulates at Eastern Himalaya, India. Aerosol Air Qual Res 14(January):440–450. https://doi.org/10.4209/aaqr.2013.03.0090

Badola R, Hussain SA, Mishra BK, Konthoujam B, Thapliyal S, Dhakate PM (2010) An assessment of ecosystem services of Corbett Tiger Reserve, India. Environmentalist 30(4):320–329. https://doi.org/10.1007/s10669-010-9278-5

Bahadur R, Murayama Y (2012) Landscape and urban planning scenario based urban growth allocation in Kathmandu Valley, Nepal. Landsc Urban Plan 105(1):140–148. https://doi.org/10.1016/j.landurbplan.2011.12.007

Bashir S, Goswami S (2016) Tourism induced challenges in municipal solid waste management in hill towns: Case of Pahalgam. Procedia Environ Sci 35:77–89. https://doi.org/10.1016/j.proenv.2016.07.048

Batar AK, Watanabe T, Kumar A (2017) Assessment of land-use / land-cover change and forest fragmentation in the Garhwal Himalayan region of India. Environements 4(2):1–16. https://doi.org/10.3390/environments4020034

Bonasoni P, Cristofanelli P, Marinoni A, Vuillermoz E, Adhikari B (2019) Atmospheric pollution in the Hindu Kush—Himalaya Region. Mt Res Dev 32(4):468–479

Census of India (2011) Office of the registrar general & census commissioner of India, New Delhi, Government of India

Chakraborty A, Sachdeva K, Joshi PK (2016) Mapping long-term land use and land cover change in the central Himalayan region using a tree-based ensemble classification approach. Appl Geogr 74:136–150. https://doi.org/10.1016/j.apgeog.2016.07.008

Chettri N, Shakya B, Thapa R, Sharma E (2010) Status of a protected area system in the Hindu Kush-Himalayas: an analysis of PA coverage. Int J Biodivers Sci Manag 4(3):164–178. https://doi.org/10.3843/Biodiv.4.3

Dame J, Schmidt S, Müller J, Nüsser M (2019) Urbanisation and socio-ecological challenges in high mountain towns: insights from Leh (Ladakh), India. Landsc Urban Plan 189(May):189–199. https://doi.org/10.1016/j.landurbplan.2019.04.017

Dasgupta R, Crowley BE, Maynard JB (2017) Organic and inorganic pollutant concentrations suggest anthropogenic contamination of soils along the Manali-Leh Highway, Northwestern Himalaya, India. Arch Environ Contam Toxicol 72(4):505–518. https://doi.org/10.1007/s00244-017-0396-7

Deep A, Pandey C, Singh N, Nandan H., Dhaka S, Dimri AP, Purohit K (2018) Study of ambient air pollutants over Rishikesh at foothills of north-western IndianHimalaya. Indian j radio space phys 47(December):49–60. https://doi.org/10.13140/RG.2.2.17593.19047

Diksha, Kumar A (2017) Analysing urban sprawl and land consumption patterns in major capital cities in the Himalayan region using geoinformatics. Appl Geogr 89(October):112–123. https://doi.org/10.1016/j.apgeog.2017.10.010

Elalem S, Pal I (2015) Mapping the vulnerability hotspots over Hindu-Kush Himalaya region to flooding disasters. Weather Clim Extremes 8:46–58. https://doi.org/10.1016/j.wace.2014.12.001

ENVIS Centre on Himalayan Ecology (2015) State at a Glance: Uttarakhand 1(5)

EnviStats-India (2019) Environment statistics, central statistics office. Ministry of Statistics and Programme Implementation, Government of India, New Delhi

Geneletti D, Dawa D (2009) Environmental impact assessment of mountain tourism in developing regions: a study in Ladakh, Indian Himalaya. Environ Impact Assess Rev 29(4):229–242. https://doi.org/10.1016/j.eiar.2009.01.003
Ghosh (2007) Urbanization—a potential threat to the fragile Himalayan environment. Curr Sci 93(2):126–127
Government of Uttarakhand (2018) Rural development and migration comission. Available at: https://www.uttarakhandpalayanayog.com/
Guneralp B, Seto KC (2013) Futures of global urban expansion: uncertainties and implications for biodiversity conservation. Environ Res Lett 8:1–10. https://doi.org/10.1088/1748-9326/8/1/014025
Hunzai K, Gerlitz JY, Hoermann B (2011) Understanding mountain poverty in the Hindu Kush-Himalayas: regional report for Afghanistan, Bangladesh, Bhutan, China, India, Myanmar, Nepal, and Pakistan. Kathmandu: ICIMOD
India Tourism Statistics (2018) Ministry of tourism. Market Research Division, Government of India
Joshi PK, Yadav K, Sinha VSP (2011) Assessing impact of forest landscape dynamics on migratory corridors: a case study of two protected areas in Himalayan foothills. Biodivers Conserv 20:3393–3411. https://doi.org/10.1007/s10531-011-0123-z
Kuchay NA, Bhat MS (2014) Analysis and simulation of urban expansion of Srinagar City. Transaction 36 (1):109–121
Kumar D (2017) Monitoring and assessment of land use and land cover changes (1977–2010) in Kamrup District of Assam, India using remote sensing and GIS techniques. Appl Ecol Environ Res 15(3):221–239
Lau WKM, Kim M, Kim K, Lee W-S (2010) Enhanced surface warming and accelerated snow melt in the Himalayas and Tibetan Plateau induced by absorbing aerosols. Environ Res Lett Enhanced 5. https://doi.org/10.1088/1748-9326/5/2/025204
Madan S, Rawat L (2000) The impacts of tourism on the environment of Mussoorie, Garhwal Himalaya, India. Environmentalist 20:249–255
Mcduie-ra D, Chettri M (2018) Himalayan boom town: Rural—urban transformations in Namchi, Sikkim. Dev Chang 49(6):1471–1494. https://doi.org/10.1111/dech.12450
Mehraj G, Khuroo AA, Qureshi S, Muzafar I, Friedman CR, Rashid I (2018) Patterns of alien plant diversity in the urban landscapes of global biodiversity hotspots: a case study from the Himalayas. Biodivers Conserv 27(5):1055–1072. https://doi.org/10.1007/s10531-017-1478-6
Mishra KP, Rai A, Chand S (2019) Land use and land cover change detection using geospatial techniques in the Sikkim Himalaya, India. Egypt J Remote Sens Space Sci (xxxx):1–11. https://doi.org/10.1016/j.ejrs.2019.02.001
Mondal PP, Zhang Y (2018) Research progress on changes in land use and land cover in the Western Himalayas (India) and effects on ecosystem services. Sustainability 10:1–14. https://doi.org/10.3390/su10124504
Mor S (2012) Critical ecosystem modeling and analysis of Darjeeling district, West Bengal, India using geospatial techniques. Andhra University
Mukhopadhyay A, Mondal A, Mukherjee S, Khatua D (2014) Forest cover change prediction using hybrid methodology of geoinformatics and Markov chain model: a case study on sub-Himalayan town Gangtok, India. J Earth Syst Sci 123(6):1349–1360
Munsi M, Malaviya S (2010) A landscape approach for quantifying land-use and land-cover change (1976–2006) in middle Himalaya, pp 145–155. https://doi.org/10.1007/s10113-009-0101-0
Nautiyal S, Kaechele H (2009) Natural resource management in a protected area of the Indian Himalayas: a modeling approach for anthropogenic interactions on ecosystem. Environ Monit Assess 153:253–271. https://doi.org/10.1007/s10661-008-0353-z
Nengroo ZA, Bhat MS, Kuchay NA (2017) Measuring urban sprawl of Srinagar city, Jammu and Kashmir, India. J Urban Manag 6(2):45–55. https://doi.org/10.1016/j.jum.2017.08.001
Pandit MK, Grumbine RE (2012) Potential effects of ongoing and proposed hydropower development on terrestrial biological diversity in the Indian Himalaya. Conserv Biol 26(6):1061–1071. https://doi.org/10.1111/j.1523-1739.2012.01918.x
Pandit MK, Manish K, Koh LPIN (2014) Dancing on the roof of the world: ecological transformation of the Himalayan landscape. Bioscience 64(11):980–992. https://doi.org/10.1093/biosci/biu152
Qadir J, Singh P (2019) Land use/cover mapping and assessing the impact of solid waste on water quality of Dal Lake catchment using remote sensing and GIS (Srinagar, India). SN Appl Sci 1(1). https://doi.org/10.1007/s42452-018-0027-6
Rajput P, Sarin M, Kundu SS (2013) Atmospheric particulate matter (PM2.5), EC, OC, WSOC and PAHs from NE–Himalaya: abundances and chemical characteristics. Atmos Pollut Res 4(2):214–221. https://doi.org/10.5094/APR.2013.022
Raman R, Punia M (2012) Land use dynamics and landscape fragmentation in higher Garhwal Himalaya, India. Asian J Geoinfor 12(1)
Rashid I, Romshoo SA (2013) Impact of anthropogenic activities on water quality of Lidder River in Kashmir Himalayas. Environ Monit Assess 185:4705–4719. https://doi.org/10.1007/s10661-012-2898-0
Rawat PK, Pant CC, Bisht S (2017) Geospatial analysis of climate change and emerging flood disaster risk in fast urbanizing Himalayan foothill landscape, 5705. https://doi.org/10.1080/19475705.2016.1222314
Rimal B, Zhang L, Fu D, Kunwar R, Zhai Y (2017) Monitoring urban growth and the nepal earthquake 2015 for sustainability of Kathmandu Valley, Nepal. Land 6(42):1–23. https://doi.org/10.3390/land6020042
Rimal B, Zhang L, Keshtkar H, Haack BN, Rijal S, Zhang P (2018) Land use/land cover dynamics and modeling of urban land expansion by the integration

of cellular automata and Markov chain. Int J Geo-Inf 154(7):1–21. https://doi.org/10.3390/ijgi7040154

Savoskul OS, Smakhtin V (2013) Glacier systems and seasonal snow cover in six major Asian river basins: hydrological role under changing climate, 53p. International Water Management Institute (IWMI), Colombo, Sri Lanka (IWMI Research Report 150). https://doi.org/10.5337/2013.204

Sharma P, Kuniyal JC, Chand K, Guleria RP, Dhyani PP, Chauhan C (2013) Surface ozone concentration and its behaviour with aerosols in the northwestern Himalaya, India. Atmos Environ (October):44–53. https://doi.org/10.1016/j.atmosenv.2012.12.042

Sheikh M, Najar IA (2018) Preliminary study on air quality of Srinagar, (J&K), India. J Environ Sci Stud 1 (1):45. https://doi.org/10.20849/jess.v1i1.421

Shrestha AB, Agrawal NK, Alfthan B, Bajracharya SR, Maréchal J, van Oort B (eds) (2015) The Himalayan climate and water atlas: impact of climate change on water resources in five of Asia's major river basins. ICIMOD, GRID-Arendal and CICERO

Siddque T, Pradhan SP (2018) Stability and sensitivity analysis of Himalayan road cut debris slopes: an investigation along NH-58, India. Nat Hazards 93(2): 577–600. https://doi.org/10.1007/s11069-018-3317-9

Singh KJ (2018) Urbanization trends in Manipur: emerging problems and prospects. IOSR J Humanit Soc Sci 23(11):1–6

Singh KP, Golson N (2019) A Study in the trends and pattern of urbanisation of Manipur during last five decades (1951–2011). Transaction 41(1):99–110

Smetacek P (2012) Butterflies (Lepidoptera: Papilionoidea and Hesperoidea) and other protected fauna of Jones Estate, a dying watershed in the Kumaon Himalaya, Uttarakhand, India. J Threatened Taxa 4(9): 2857–2874

Sundriyal S, Shridhar V, Madhwal S, Pandey K, Sharma V (2018) Impacts of tourism development on the physical environment of Mussoorie, a hill station in the lower Himalayan range of India. J Mt Sci 15(10):2276–2291

Tiwari PC, Tiwari A, Joshi B (2018) Urban growth in Himalaya: understanding the process and options for sustainable development. J Urban Reg Stud Contemp India 4(2):15–27

Vaidya P, Bhardwaj SK, Sood S (2018) Land use and land cover changes in Kullu valley of Himachal Pradesh Land use and land cover changes in Kullu valley of Himachal Pradesh. Indian J Agric Sci 88(6): 94–98

Valdiya KS, Bartarya SK (1991) Hydrological studies of Springs in the catchment of Gaula River, Kumaon Lesser Himalaya, India. Mt Res Dev 11:17–25

Vitousek PM, Mooney HA, Lubchenco J, Melillo JM (1997) Human domination of Earth's ecosystems. Science 277:494–499. https://doi.org/10.1007/978-0-387-73412-5_1

Wester P, Mishra A, Mukherji A, Shrestha AB (2019) The Hindu Kush Himalaya assessment—mountains, climate change, sustainability and people. Springer Nature, Switzerland AG, Cham.

Xu J, Sharma R, Fang J, Xu Y (2007) Critical linkages between land-use transition and human health in the Himalayan region. Environ Int 34:239–247. https://doi.org/10.1016/j.envint.2007.08.004

Zaz SN, Romshoo SA (2012) Assessing the geoindicators of land degradation in the Kashmir Himalayan region, India. Nat Hazards 64:1219–1245. https://doi.org/10.1007/s11069-012-0293-3

Zurik D, Pacheco J, Shrestha B, Bajracharya B (2005) Atlas of the Himalaya. In: Pennington GM, Gurung H, Pradhan P, Dangol G (eds) (2005th ed.). International Centre for Integrated Mountain Development (ICIMOD), Kathmandu, Nepal

The Changing Landscape of the Plantation Sector in the Central Highlands of Sri Lanka

24

H. Mahendra P. Peiris and Nuwan Gunarathne

Abstract

The topography of Sri Lanka consists of the mountainous centre of the southern half of the island, referred to as the Central Highlands. A significant proportion of the land in this region is used for plantation agriculture. The plantation companies in this region have started to change their traditional land uses. Since these changes are in their early stage, how and why they are happening, their impact and implications are largely unknown. This chapter presents a successful pilot project undertaken by a tea plantation company aimed at improving land productivity and developing climate resilience in the Central Highlands of Sri Lanka. In addition to the initial results obtained, this chapter also discusses the challenges and way forward for projects of this nature.

Keywords

Central Highlands · Climate-smart agriculture · Plantations · Sri Lanka · Tea industry

H. Mahendra P.Peiris
Postgraduate Institute of Agriculture, University of Peradeniya, Peradeniya, Sri Lanka

Nuwan Gunarathne (✉)
University of Sri Jayewardenepura, Nugegoda, Sri Lanka
e-mail: nuwan@sjp.ac.lk

24.1 Introduction

The plantation sector has been largely responsible for the changing mountain landscape in the world. This is particularly evident in the tropical montane regions throughout the globe. For instance, from the 1970s in Sumatra (Indonesia), 12 million hectares (ha) of natural forest cover have been cleared for timber and large-scale crop plantations (Laumonier et al. 2010). In the Bungo district of Sumatra alone, the forest cover has been reduced from 75 to 30% from 1973 to 2005. This is mainly due to clearing of forests for monoculture plantations such as rubber and oil palm, which has increased from 3 to over 40% in the district (Ekadinata and Vincent 2011). More specifically, oil palm land use has increased from only 6,259 (1%) in 1988 to 88,355 (19%) ha. Globally, it has been estimated that about 11 million ha are under coffee cultivation almost entirely in tropical forest regions (Clay 2004). Similarly, tea cultivation spans over 2.3 million ha around the world (Clay 2004).

Further, the Western Ghats hill range in India,[1] a biodiversity hotspot covering nearly 160,000 km^2 in the west coast is also subjected to large-scale clearing due to the spread of plantations, particularly tea, coffee, teak, cardamom, and eucalyptus (Bali et al. 2007). It has been

[1]Western Ghats and Sri Lanka are designated as a single biogeographical unit which first identified by Wallace in 1876 (Wickramagamage 2017).

U. Schickhoff et al. (eds.), *Mountain Landscapes in Transition*,
Sustainable Development Goals Series, https://doi.org/10.1007/978-3-030-70238-0_24

estimated that between 1920 and 1990, the forest cover in the region had fallen by 40% (Menon and Bawa 1997). In the Western Ghats of Karnataka, areca, coffee, and rubber plantations occupy a land area of over 10,000 km^2 (Karanth et al. 2016). While most of these tropical plantations are located in key biodiversity areas of forest and grassland, they are usually cultivated as intensive monocultures with concerns over soil erosion and agrochemical inputs (Brockerhoff et al. 2008; Clay 2004).

This situation is more pronounced in the mountain landscape in Sri Lanka (5°55′ N—9° 51′, 79°41′ E—81°53′ E), an island nation in the Indian Ocean with an area of 65,610 km^2. In this island, the plantation companies own 190,000 ha of land in the Central Highlands (Sri Lanka Tea Board 2014). In a bid to improve profitability, some of these plantation companies have adopted various strategies including changes in land use in tea estates. Since these changes in the plantation sector are at an early stage, how and why they are happening, their impact and implications are largely unknown. This chapter presents a successful pilot project undertaken by a tea plantation company aimed at improving land productivity and developing climate resilience in the Central Highlands of Sri Lanka.

24.2 Study Area

The study area covered in this chapter is the Central Highlands of Sri Lanka which were formed by metamorphic rocks of the Precambrian Age, some of the few oldest rock types in existence in the earth. Recent geological studies have revealed and confirmed that the majority of this landscape, especially the upper montane zone, was created by some 1.8–2.2 billion-year-old-rock formations (Kroner and Williams 1993). This region covers six districts[2] of Sri Lanka (Table 24.1 and Fig. 24.1), where 26% of the total population live and cover 21% of the total land area of the country.

Like many other mountain regions of the world, the Central Highlands of Sri Lanka are important from a variety of perspectives. From an environmental point of view, this region is home to three UNESCO World Heritage listed natural forest ranges, which are the remaining areas of the submontane and montane rain forests of Sri Lanka. Accordingly, the International Union for Conservation of Nature (ICUN) has declared the natural ecosystems in the Central Highlands of Sri Lanka "a biodiversity superhot spot" in the world (International Union for Conservation of Nature (IUCN) 2017). This region is also considered the heart of Sri Lanka, since all major rivers that supply water to the entire island originate there (Wickramagamage 1998).

The Central Highlands are important from an economic point of view as it is home to two of the country's main agricultural exports—tea and rubber. The world famous pristine "Ceylon Tea" comes mostly from the tea plantations located in the Central Highlands with its salubrious climate. Moreover, other valuable resources such as minerals, gems, timber, and agricultural products also originate in this region adding value to the country's national economy. In addition, the country's major hydropower stations such as the River Mahaweli and Kelani Valley hydropower complexs are also located in this region.

24.3 Material and Methods

In this study, the researchers used an innovation action research approach where the given organization is considered a "client organization" for exploring land use change based on climate smart agriculture (Brown and McIntyre 1981; Kaplan 1998). This research approach is largely experimental, particularly in the early stages of operationalizing an innovative idea, as both the organization and the researchers endeavour to learn about the novel idea and how it can be successfully implemented (Kaplan 1998). In this study, one researcher was engaged with a tea

[2] In the administrative structure of Sri Lanka, districts represent the second-level administrative divisions and are included in a province. In Sri Lanka, there are twenty-five districts organized into nine provinces (Ministry of Internal and Home Affairs and Provincial Councils and Local Government 2019).

Table 24.1 Statistics of the Central Highland districts in Sri Lanka (year = 2017)

District	Province	Population '000	Land area '000 ha	Population density (per ha)	Land use (%)[a]											
					Forest	Home garden	Paddy land	Perennial crops	Major crops[d]	Field crops	Inland waters	Abandoned Land	Built up land	Chena	Other[b]	Total
Kandy	Central	1,468	1,940	0.757	21.4	25.67	8.31	1.33	17.76	5.97	2.18	4.92	0.96	11.21	0.29	100
Matale	Central	519	1,993	0.260	38.4	19.2	11.2	9.9	8.7	3.4	2.1	1.8	1.2	0.7	3.4	100
Nuwara Eliya	Central	763	1,741	0.438	32.69	8.2	3.52	0.01	29.8	9.08	2.77	0	2.15	9.16	2.62	100
Kegalle	Sabaragamuwa	884	1,693	0.522	4.08	34.97	6.33		37.03[c]	1.12	0.51	11.21	0.02	2.8	1.93	100
Ratnapura	Sabaragamuwa	1,163	3,275	0.355	24.2	21.9	6.1		28.7[c]	5.9	1.9	0.4	0.5	9.3	1.1	100
Badulla	Uva	873	2,861	0.305	21.23	0.54	10.85	5.94	11.22	4.38	2	0.39	4.58	10.31	28.56	100
Total		**5,670**	**13,503**	**0.420**												

Source Department of Census and Statistics (2018)

Note

[a]The land-use categorization is based on the categories used by the Department of Census and Statistics of Sri Lanka
[b]Other land-use categories include sacred places, roads, cemeteries, etc.
[c]Land-use data is available for perennial and major crops together for these two districts
[d]Major crops include tea, rubber, and coconut

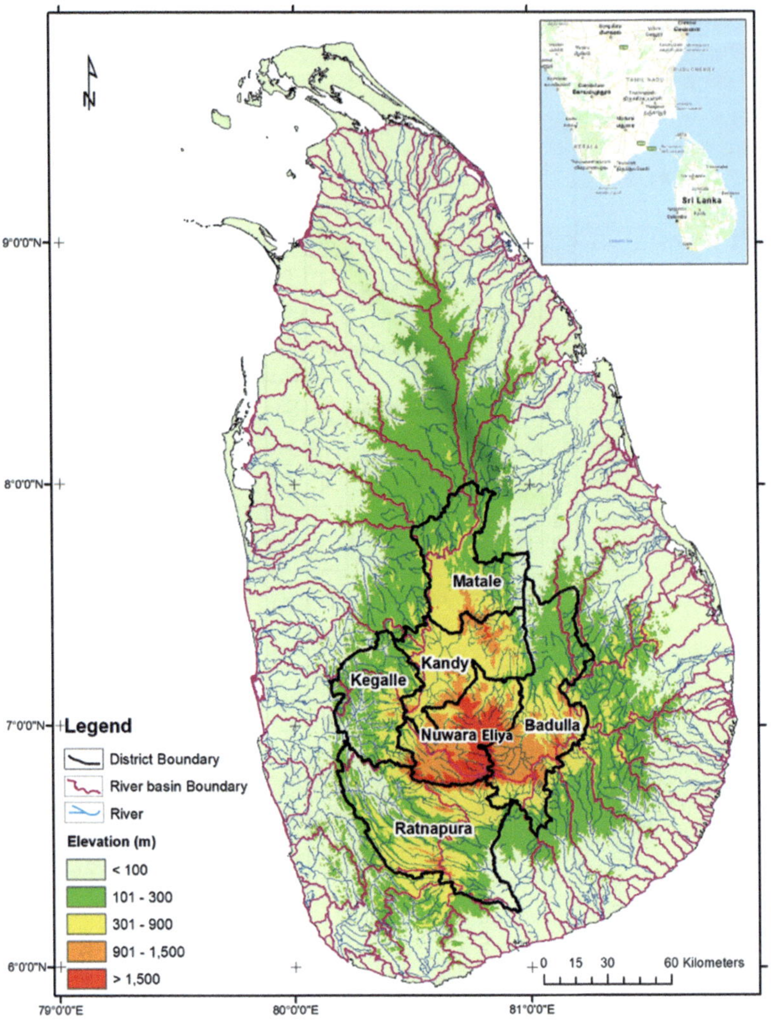

Fig. 24.1 Central Highlands of Sri Lanka (Provided by Author)

plantation company to spearhead a multidisciplinary pilot project aimed at improving business performance through minimized resource wastage and enhanced land productivity.

This innovative action research project on land-use change was carried out from April 2017 to October 2019 and is ongoing with a series of experiments at several estates belonging to the plantation company. The data for this study was collected through different methods including interviews, field visits, on-site verification and recording of data by physical observation and documentation at various stages, participation in meetings, examination of various company documents and scrutiny of external information (such as statistical reports, population data, Web sites and newspapers). In the various internal companies examined, annual reports, sustainability reports, monthly progress reports, and tea production data were studied. The range of data used in this study permitted extensive triangulation of the analysis of results.

24.4 Results

24.4.1 Historical Land-Use Change in the Central Highlands

The British commercialized the coffee industry in Sri Lanka to create the country's first largely successful export crop. Coffee was a major economic crop in Sri Lanka from 1830 to 1880. Unfortunately, this move led to the massive deforestation of pristine mountain forests that had evolved over millions of years and were also rich with great biodiversity (von Blanckenburg et al. 2004). During the British period, out of the forest cover of the Central Highlands, 176,455 and 80,163 ha within 1000–1500 m and above 1500 m (amsl), respectively, were sold to Europeans for coffee plantations (Wickramagamage 2017). It also resulted in forced colonization by a foreign community brought from South India for work in the plantation districts of Sri Lanka causing an indirect effect on the Central Highlands through population increase (von Blanckenburg et al. 2004; Wickramagamage 1998).

During the period from 1830 to 1880 over 222,773 ha (44%) of montane forests were cleared for coffee plantations in this region (Ferguson 1893). However, the entire coffee industry of Ceylon was devastated by the coffee leaf rust, a fungal disease, that spread in the 1880s. Subsequently coffee production dropped rapidly and almost all the areas under coffee plantations were soon converted into tea gardens. Since then, more than 120,000 ha of coffee land were converted into tea land (Sri Lanka Tea Board 2014). By 1920, tea cover claimed 162,000 ha, which peaked up to 238,000 ha in 1965.

There was a tremendous growth of population in the montane zone of Sri Lanka in parallel with the expansion of the plantations industry, especially with the plantation workers brought from South India by the British (von Blanckenburg et al. 2004). From zero level in the 1820s, the plantation worker population is reported to have increased to 55,000 in 1855 and risen to 100,000 in 1880. The 1948 plantation worker population of 800,000 further increased to 975,000 in 1964. In addition, there was a local population flow toward the Central Highlands with the development of the tea industry. The exponential growth in the population of the Central Highlands and the development of numerous new settlements exerted a heavy demand on the remaining land for purposes of both agriculture and settlements (Wickramagamage 1998).

These historical developments in relation to land-use change resulted in the environmental, social, and economic problems that this region is experiencing today. The depletion of forest cover that eroded biodiversity resulted in depleting the quality of the ecosystem of the region. Further, many natural disasters such as landslides, floods, and seasonal water scarcity in every part of the country resulted from land degradation. In addition, the decline in crop yields in terms of both quality and quantity led to massive losses to the national economy (Wickramagamage 1998).

24.4.2 Current Land Use in the Central Highlands

As presented in Table 24.1, the land in the Central Highlands is used for many purposes. Crop cultivation (i.e., perennial, major, and field crops) account for above 20% of land use in this region. While this land is used mainly by the regional plantation companies (RPCs), there are other players such as households, who use the land for cultivation of crops. Further, RPCs also use the land for purposes such as forestry, home gardening, inland waters and building and construction.

Figure 24.2 shows the overall land use by the plantation companies in the Central Highlands of Sri Lanka. Under a typical plantation company only 50–60% of the land is cultivated with tea and/or other commercial crops and the rest of the land is either forestry or uncultivated or used for other purposes.

This land use presented in Fig. 24.2 point to two major issues:

1. A large portion of the land is either unutilized or underutilized for commercial crops;
2. A single crop dominates the greater part of the land use.

The second issue typifies "monocrop culturing," an agricultural practice where one crop is planted in a farming system. Unlike in the temperate region with clear seasons, plantations in the tropics are blessed with plenty of free-flowing

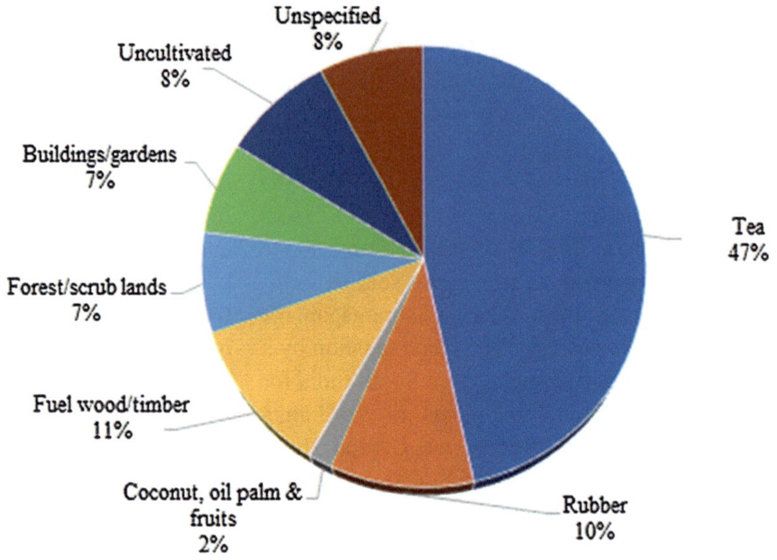

Fig. 24.2 Overall land use by the corporate tea plantation sector (Based on Samansiri et al. 2011)

natural resources such as sunshine, rainfall, and "living soils" that support the non-stop growth of vegetation, a high level of biodiversity, and a year-round flow of free ecosystem services. These environments are therefore capable of supporting year-round a variety of flora and fauna species simultaneously. However, in the traditional plantations, with the underutilization of land and monocrop culturing, a major portion of the said freely available natural resources are wasted and depleted. This not only results in a gross erosion and waste of natural resources within the plantation landscapes but also endangers the biodiversity and quality of other high-value free ecosystem services. Consequently, the degraded conditions demand greater quantities of synthetic inputs resulting in ever-increasing production costs for the plantation companies. This, in turn, reduces product quality while narrowing profit margins and creating environmental issues. These issues coupled with other factors such as the rising cost of production, labor shortages, market uncertainties, and exchange rate fluctuations have created major challenges for the plantation sector in the Central Highlands of Sri Lanka, where the tea plantation companies operate (Gunarathne and Peiris 2017; Sangakkara and Frossard 2016;).

The promotion and development of tea-based multicrop models provide a strong foundation and a sustainable method to overcome these issues faced by the plantation companies. Based on the climate-smart agriculture (CSA), it not only secures food and income for the country but also helps to withstand the changing climatic conditions while reducing greenhouse gas emissions in the long run (see Fig. 24.3 for a brief description of CSA).

24.4.3 Changing Use of the Mountain Landscape by Plantation Companies

Having identified the potential for improving land productivity, some plantation companies have resorted to crop diversification as a means of minimizing business risk. However, they still follow the traditional concept of monocrop systems. In fact, it should be understood that many of the problems in the plantations were caused by the waste of resources within such systems.

Globally, agricultural production has been seriously affected by climate change and this poses threats to food security. In response, the agriculture sector has followed adaptation and mitigation strategies in a sustainable way while contributing to food security in a bid to be climate-smart. More specifically, climate smart agriculture (CSA) is a concept to enhance the production capacity of agricultural systems through improved resource use efficiency to support food security, while building resilience against the negative effects of climate change through sustainable agricultural development strategies.

CSA focuses on transforming and reorienting agricultural systems to meet the challenges of climate change and thereby to support food security. It is focused on establishing and strengthening the impact of beneficial "*free ecosystem services*" generated within and around the agricultural systems in bringing down production cost and building resilience against the negative effects of climate change through sustainable agriculture development strategies.

Since CSA is an evolving concept and there is no blanket approach to how it should be practiced. These practices should be shaped by country/site specific contexts and varying capacities. However, in general CSA systems should aim to:

1. Sustainably improve the agricultural productivity and farmer income leading to food security and development.
2. Adapt and build resilience to climate change at multiple levels (from farms to nations); and finally sharing at the global scale.
3. Reduce and/or remove greenhouse gases emissions contributing to global warming at all possible levels while increasing biomass production through plant growth trapping carbon dioxide absorbed from the atmosphere (*carbon sinks*).

Fig. 24.3 Climate-smart agriculture (CSA) systems (Campbell et al. 2014; Food and Agriculture Organization (FAO) 2013; Lipper et al. 2014)

Ranging from in situ issues such as soil erosion, weed growth, seasonal water scarcity and declining productivity to ex situ hazards, viz. siltation of water courses, reservoirs and flooding of foothills and lowland areas followed by seasonal water deficiency are all caused by poor resource management in the Highlands. Nevertheless, a large plantation company, which manages about 15,000 ha, has been engaged in a successful pilot project since 2017 to explore the local forms of tropical montane agriculture systems such as the Kandyan Forest Gardens (KFGs).[3] This pilot project has focused on diversifying land use with three strategies:

- Expansion of productive vegetative cover within the plantation landscapes;
- Establishment/improvement of forest cover in uncultivated areas; and
- Improvement of the functional role of hydro catchments.

These three strategies followed by this company that changed land use are explained below:

[3]This is a traditional agro-forestry system in which the vegetation mostly consists of introduced exotic plants (e.g., nutmeg, clove, pepper, cardamom, vanilla, breadfruit) and mixed with native crop species (e.g., cinnamon, *kitul* palm, arecanut, and coconut) and numerous other fruit, vegetable, and medicinal plant species. These traditional, near forest agricultural systems established on mountain slopes are ecologically rich and self-sufficient, resilient to a wide range of fluctuations in climatic conditions due to the package of strong and healthy ecosystem services established within this system (Mohri, et al. 2013; Perera and Rajapakse 1991).

24.4.3.1 Expansion of Productive Vegetative Cover Within the Plantation Landscapes

This strategy focused on planting commercial crops to cover the ground so as to utilize all the available free-flowing natural resources. Its goal was to increase land productivity and profitability with the added income generated by the crops.

As discussed in the previous section, the warmth and moisture produced by the rich tropical conditions with year-round sunshine are ideal for continuous plant growth throughout the year. Thus, trying to keep the unutilized land of the plantation companies free of weed is an impossible task as it means a never-ending battle with natural forces. The control of weeds in the open land areas of the plantations requires heavy use of agrochemical inputs such as weedicides. Further, this exposed land increases the temperature of the environment due to the direct effect of sunshine on the ground while causing heavy moisture loss. In addition, straight rainfall on exposed land causes soil erosion and continuous weed growth on the ground.

In order to improve the productive vegetative cover while overcoming the above problems, two different approaches were adapted. The first was to increase the ground cover of the existing crop canopy in the marginal tea land by expanding the canopy cover of the tea bushes. This was achieved by enlarging the tea bush after the periodical pruning cycle.[4] The second was to introduce compatible crop species and develop mixed-crop intercrops with a view to minimizing monocrop culturing practices. This second approach is explained below:

Development of mixed-crop intercrops started with the identification of site-specific proper crop models which are compatible and also commercially viable. At this stage, attention was paid to the cultivation, harvesting, processing, transport, and marketing aspects in addition to considerations of crop duration and return on investment. Further, considerable attention was paid to proper land zoning.[5] These considerations at the initial stages of the planning process are expected to make this approach viable in the long run.

In selecting crops for the multicrop model, the following steps were followed:

- Step 1: A region-wide survey was conducted to identify and select suitable crop species that perform well under different agroclimatic regions by observing a variety of existing indigenous farming systems such as KFG and natural forest patches in each area.
- Step 2: After selecting the crop varieties, relevant crop research stations were consulted for their opinion on crop suitability and also to gather information on any improved varieties of the selected crop types recommended for particular regions for commercial cultivation.
- Step 3: Expert advice was also obtained from various sources to strengthen the viability of the model.

The following crops were identified for initial trial planting (Table 24.2):

As shown in Fig. 24.4, the plantation company has five tea plantation regions and the tree planting is currently ongoing in all these regions.

Coffee was identified as the most suitable intercrop species to plant in the vacant exposed land area in marginal tea lands and infills.

[4]This method is called Stripe-Spreading of Tea Bushes. It allows the tea bushes after pruning to grow up to 120 days and then, a radial spread of shoots is done using tight parallel stripes arranged along the tea rows instead of traditional practice of tipping the shoots after pruning (See Gunarathne and Peiris 2017 for more details).

[5]Land zoning is the identification and demarcation of different sections of a landscape to apply the most appropriate management strategies to optimize the land use. Proper land zoning enables effective decisions making to allocate the land for the most appropriate use. Landslide prevention, identification, and allocation of the best land area for highest profitable crops, planning of areas for timber and energy crops (e.g., fuel wood forestry), establishment of native forestry and non-timber agroforestry systems on hill tops, enhancement of hydro catchments and water way buffer zones are a few examples land zoning.

Table 24.2 Details of the crops in the multicrop model

Crop species	Botanical name	No. of trees planted/trialled	No. of trees planned for future planting
Coffee	*Coffee arabica*	5000	60,000
Cinnamon	*Cinnamomum verum*	20,000	50,000
Asian pears	*Pyrus pyrifolia*	2000	5000
Bibile sweet orange	*Citrus sinensis*	2500	5000
Local mandarin	*Citrus reticulata*	10,000	–
Lemon	*Citrus reticulata*	250	1000
Avocado	*Persea americana*	500	1000
Papaya	*Carica papaya*	100	–
Passion fruit	*Passiflora edulis*	250	1000
Ceylon gooseberry	*Dovyalis hebecarpa*	–	100
Local apple	*Malus domestica*	50	100
Improved mandarin	*Citrus × iyo*	–	2000
Cherrymoya	*Annona cherimola*	–	50
Guava	*Psidium guajava*	100	1000
Jakfruit	*Artocarpus heterophyllus*	5000	10,000
Kitul Palm	*Caryota urens*	100	500
African mahogany	*Khaya senegalensis*	500	50,000
Agarwood	*Aquilaria malaccensis*	200	1000
Giant bamboo	*Dendrocalamus hookeri*	1000	20,000

(Compilation by the authors based on internal company records)

Lak Perakum, a superior coffee variety ranked within the first ten best commercial coffee cultivars in the world, developed and released recently by the Department of Export Agriculture in Sri Lanka, was selected for this purpose. An initial evaluation showed that this particular coffee variety was 20–50 times more profitable than tea in terms of the present profitability levels of tea and coffee. This variety of coffee ensures an early return on investment within three years of planting depending on the agroclimatic region cultivated. Figure 24.5 presents how coffee has been cultivated as an intercrop species to infill exposed ground in marginal tea lands to make use of wasted resources and make the land generate income instead of spending on costly weed control.

Locally developed varieties of Asian pears and local mandarin had been tested for commercial cultivation since 2013 by the plantation company and proven successful. Commercial cultivation of two fruit species, Bibile sweet orange[6] and Ceylon gooseberry,[7] were also commenced under this project. In addition, Avocado Cherrymoya and varieties of mandarin recently released by the Department of Agriculture were selected as components of the crop models. In the meantime, short-term crops such as lemon, strawberry, passion fruit, and banana

[6]Bibile sweet orange (*Citrus sinensis*) is a juice type sweet orange variety grown in the Bibile region of Sri Lanka. Once a popular traditional crop, the cultivation of this orange variety was destroyed some time ago by a disease. However, presently plans are under way to revive the recultivation of the crop.

[7]Ceylon gooseberry (*Dovyalishebecarpa)* is endemic to Sri Lanka and India and was introduced into USA in 1920. It is a small spherical fruit characterized by a deep purple-red, sour juice, and small seeds enclosed in the pulp (Bochi et al. 2014). It is now cultivated in Hawaii, South America and Israel.

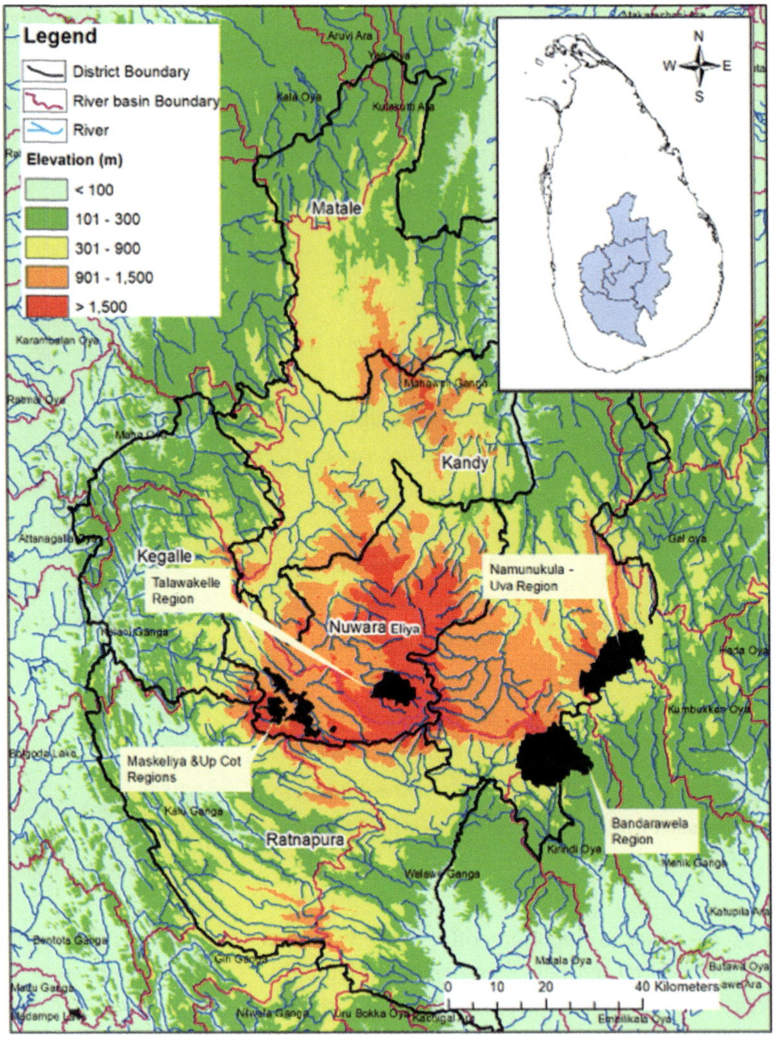

Fig. 24.4 Tea estates where the project is implemented (Provided by Author)

were also cultivated to obtain quick cash flows. Figure 24.6 shows the varieties of fruit cultivated in the vacant tea land.

24.4.3.2 Establishing/Improving Forest Cover in Uncultivated Areas

This second strategy focused on diversifying land use with forestry in the plantation company. The forestry in plantations was mainly confined to the Eucalyptus species-based timber and fuel wood cultivations grown as monocrops. Generally, these forestry cultivations are home to many exotic invasive plant species which have little or no value to the neighboring people. Since the local community does not derive any tangible benefits, these forestry blocks are periodically subjected to fires that destroy the entire understory vegetation. This, in turn, facilitates the further establishment of invasive plant cover over the ground.

Considering these facts, fast growing and high yielding species of giant bamboo, which are also fire-resistant, were selected for planting in the exposed land within the timber and fuel wood forestry blocks. Within the third year of planting,

Fig. 24.5 Coffee cultivated as an intercrop species (Provided by Author)

Fig. 24.6 Other intercrops planted [Ceylon Gooseberry (left) and Asian Pears (right)] (Provided by Author)

bamboo is capable of producing a continuous and regular supply of fuel wood to meet the energy demands of the factories of the plantation companies for about 40 years. In addition, fast-growing high-value timber species such as Agar wood and other species such as Balsa and African mahogany were also introduced. Moreover, another set of native plants indigenous to each region had been identified and planned for cultivation in the native forests in the Central Highlands.

24.4.3.3 Improvements in the Functional Role of Hydro Catchments

This third strategy focused on mitigating the adverse impact of climate change resulting in the increased severity of weather events such as intense rainfall within shorter time periods leading to flash floods. These heavy downpours cause increased surface runoff of rain water, severe soil erosion, and poor recharge of groundwater tables with consequent water scarcity in the highlands. Hence, special attention was paid to identifying effective agricultural practices to improve soil water holding capacity and nutrient retention ability. "Deep envelope forking[8]" was one such practice recognized to promote rainwater infiltration

[8]This is a periodic agricultural practice carried out in tea gardens to break up the compacted upper soil layer by inserting an agricultural fork to a depth of about 40–50 cm. This practice is performed to facilitate water infiltration and groundwater recharge, improve nutrient

into the soil and strengthen groundwater recharge. This method was applied in other important micro-hydro catchments such as spring areas of drinking water with a view to supporting continuous crop growth with natural and synthetic resources. Cascading mini-reservoirs were also identified as a rainwater harvesting measure while inland fisheries were selected as a component of this system to enhance food security for the plantation community.

24.5 Discussion

This pilot project has been in operation since 2017 in many different agroclimatic regions in the Central Highlands of Sri Lanka and shown early promise of success. Further, it marks a gradual change in conventional land use in this territory under a CSA approach. Plantations in similar land in other countries such as Indonesia, India, and Vietnam have followed somewhat similar land use changes by interchanging crop models under tea, rubber, and oil palm monocrops with coffee, cocoa, cardamom, pepper, and other spice crops (Brockerhoff et al. 2008; Mudappa and Raman 2012). However, these attempts are aimed at either increasing financial gains (i.e., the profitability of land area by filling the land by introducing a few crop plant species into the monocrop systems) or conserving biodiversity and ecosystem services. These indicate narrowly focused conventional approaches to improve land productivity or to conserve the environment. Hence, these approaches are not based on the concept in which this project was rooted. While the results of these approaches produce similar marginal benefits, they have failed to derive the full benefit of this project.

Although most of the results of this project have yet to be seen, this section describes the initial outcomes under the triple economic, environmental, and social bottom line. Although these benefits are described under each heading for the sake of logical arrangement, their interconnection should be appreciated.

24.5.1 Economic Gains

The initial trials such as deep envelop forking have produced impressive high performing results with quick improvement in crop yields in the tea fields in the range of 8% to 10%. Further, during the prolonged dry weather spells this method has produced positive results such as persistent crop yields compared to the rest of the tea plots that were wilting with declining yields due to moisture stress.

In the meantime, locally grown fresh Asian pears and mandarin crops have come into the market with a heavy consumer demand over their imported counterparts. This is an indication that expanding the area under these crops would save at least US$ 10 million of foreign exchange annually[9] for the country, which is currently spent on the importation of these fruits. Early-bearing fruit crops like lemon, passion fruit, and strawberry also have shown promising financial results. The newly introduced species of forestry such as balsa, African mahogany, and giant bamboo have also exhibited high growth in this territory proving their commercial viability under plantation conditions. However, the cultivation of many other crop trials is still in the early stages and under observation and need further evaluation. Once established with necessary site-specific refinements and expanded coverage, this would make a positive change toward a climate-resilient agricultural model for the tropical mountains while improving land productivity and profitability. The advances in ecosystem services should also bring down production costs for the plantations and other agricultural systems.

retention, aerate soil, and enhance subsoil biological activity.

[9]Sri Lanka imported 47,800 metric tons (MT) of pears, 8400 MT of oranges and 11,000 MT of mandarin in 2018 with a combine value of 34 million US$ (Ministry of Finance, 2018). This figure is estimated based on the assumption that the project of this nature can reduce these imports by one third.

24.5.2 Environmental Gains

The environmental benefits of this project include improvements in biodiversity within the region at all levels from within species (genetic) diversity to ecosystem diversity due to improvements in the total carrying capacity of the territory. Improved vegetative cover could reduce the ambient temperature and increase relative humidity, which would, in turn, increase frequent rain formation. Consequently, it would result in stress-free steady plant growth thus boosting crop production. Further, the increased efficiency in agricultural input use due to the minimization of the system losses can reduce pollution of the environment. Moreover, raising the ecosystem immunity of the agricultural systems prevents pest and disease outbreaks and the adverse impact of climate change.

24.6 Social Gains

From a social perspective, the benefits of this project accrue to the plantation community mainly in terms of additional work created,[10] generating in situ extra earning opportunities. In addition, non-cash benefits on offer such as a year-round steady supply of quality drinking water and intangible benefits such as saving on traveling expenses and time in search of earning opportunities etc. are bonus points for the community. This also reduces fluctuations in their monthly income, which is a main cause of worker migration out of the plantation sector. Further, multicrop models, inland fisheries, and climate-resilient farming conditions add to the diversity of their food supply, improve their nutrition status, and ensure food security. Particularly, the inland fisheries projects have drawn support from the local community as a source of fish protein and also as recreation destinations. Inhabitants of the lowlands outside the plantations too will have a dependable steady supply of water for their domestic and agricultural needs improving their crop yields and income. Moreover, they too will be offered seasonal work in the plantations especially during the crop harvesting periods thus boosting their income levels.

Increased soil moisture retention through effective agricultural practices ensures a steady year-round water supply for the immediate plantation community. As the Central Highlands are where the major rivers originate improved hydrology in the watershed would also support millions of people who live downstream and depend on the Central Highlands for water for drinking and agriculture. This could be monitored with frequent and regular stream flow measurements. Eco-tourism is another aspect that can be promoted in the mountain landscapes with their greenery, biodiversity, attractive scenery, and salubrious climate.

24.7 Challenges and the Way Forward

The benefits mentioned above can be affected by a variety of challenges, some controllable in the long-term and others not. However, a careful analysis of these challenges is paramount for the success of this project and for learning opportunities. This section describes these challenges encountered in steering this pilot project under two categories: technical and human-related.

24.7.1 Technical Challenges

Changing climatic conditions

These changes in the mountain landscape are spread over about 15,000 ha in the Central Highlands of Sri Lanka. This region is characterized by a variety of environmental factors such as soil types, land slope, rainfall patterns, wind exposure, relative humidity, ambient temperature, day–night temperature differences, and spread of sunshine intensity. Therefore, the combined effect of these climatic factors would

[10]Every work done in an estate is recorded and accounted as per the type of crop where the job was involved. Worker's income is calculated based on daily wage rate. Thus, measurement/calculation of the additional income generated by the project can be specifically done.

invariably vary from one point to another. Due to the changing climatic patterns experienced at present, it is very difficult to predict future climatic conditions in this region. They could be either favorable or unfavorable for some crops and multicrop models. This would affect the expected performance of these crops.

Absence of data on new varieties of crops

Some crops such as Lak Perakum coffee and the latest improved varieties of mandarin are crop cultivars recently released by the agriculture research stations in Sri Lanka. Further, Ceylon gooseberry, African mahogany, and balsa are novel introductions to the Central Highlands. Thus, there is a lack of evidence and data on the performance of these crops under the conditions prevailing in this montane region. This makes the long-term evaluation of the viability of these crops extremely difficult.

Market uncertainty

The next likely gray area relates to the marketing aspects of the produce of the new crops cultivated. Fluctuations in market prices and consumer demand, unpredictable competition and a host of other factors pose a high level of uncertainty for the financial viability of the project. Hence, further monitoring, continuous improvement, and site-specific refinements will be necessary to meet these future challenges.

24.7.2 Human-Related Challenges

Resistance to change

Some of the novel concepts used in this pilot project contradict conventional agricultural practices and other operations followed by the plantation industry. Even the recommendations and guidelines given by the plantation crop research stations have not yet been revised or updated according to CSA concepts. Hence, there is a conflict created between unchallenged routine practices followed in the plantation sector and the new practices introduced under this project. In addition, the personal interest and attention paid by corporate-level and estate-level management have greater impact on the success or failure of the project at field level. This is further aggravated by the general human resistance to change. Hence, these factors all converge to reduce the level of acceptance of CSA-based practices as introduced under this project in the plantation agriculture sector.

Opposition of agricultural input sellers

As CSA emphasizes, by improving the use efficiency of agricultural inputs such as fertilizer and pesticides in pursuance of the principles based on CSA, the use of such products in agriculture would greatly decrease. Consequently, sellers and dealers who engage in selling fertilizer and other agrochemicals are naturally inclined to resist these novel practices.

24.7.3 Way Forward

As stated in the introduction, the Central Highlands of Sri Lanka particularly bear characteristics similar to the Western Ghats hill range in India and other tropical forests in South-east Asia. This study offers several learning opportunities to do with the plantations such as rubber, tea, coffee, and oil palm in these mountain regions which aspire to change the current land use.

Availability of solutions in the close proximity

In the endeavour to identify strategies for changing land use, the environment itself can be a valuable source of viable clues. However, as highlighted in this study, one should be vigilant to identify these solutions "out there" with careful observation and an open mind. For instance, this study indicates that the main reason for many issues in tropical plantation models is the underutilization of the free flow of natural resources available in abundance. Hence, it is necessary to devise strategies to effectively utilize them for land use change rather than engaging in a never-ending battle to control them.

Need to capture the full spectrum of resource availability

One of the major differences between the tropical and temperate regions is the modes, intensity, and duration of the availability of major forms of resources such as sunshine, moisture, and biological activity. While there is a year-round rich flow of natural resources and ecosystem services available in the tropics, in the temperate regions there is a natural control mechanism on areas such as soil erosion, pests, and diseases due to seasonal factors. However, this key fact had been totally ignored by the conventional plantation agriculture systems in the tropics introduced by Western planters. This in turn has been the reason for many of the issues faced by the plantation industry at present. For instance, weed growth is promoted by the waste of natural resources, direct exposure of the soil without ground cover to intense rainfall and harsh sunlight leading to soil erosion and land degradation. Instead of spending on controlling troublesome weed growth, this situation could be simply turned into a state of income generation by placing productive plant (crop) cover to capture these resources that go unutilized.

Acquisition of knowledge from long-lasting traditional agricultural systems

The mountain landscape of this region is home to many traditional agricultural systems sustained locally for centuries or millennia and enriched by indigenous knowledge. When plantation companies devise strategies to change land use, these traditional systems offer a valuable source of knowledge in areas such as selecting matching components for the crop models, especially to identify compatible crop species/cultivars. This can save time and resources that could be otherwise devoted to trial and error studies.

Focus on applied climate smart agriculture solutions

The scale of agriculture operations can have a significant effect on the magnitude of climate resilience. Thus, it is necessary to practice CSA applications over a wide landscape to effectively withstand the effects of climate change. The plantation agriculture systems are capable of meeting this condition owing to the massive scale of operations over a great extent of land. That is, plantations have the ability to mitigate the effects of climate change by changing land use to implement climate smart agriculture solutions on a wide scale with the aim of establishing "climate-resilient regions" or "mini-ecological zones which are immune to the ill effects of climate change."

Need for policy level interventions

The study highlights the need for policy level support to change land use for higher productivity and better climate resilience, particularly in the developing countries. This is relevant to agriculture sectors such as tea, where the whole industry has recently witnessed diminishing productivity and adverse effects due to climate change. However, the plantation industry is not in a position to get out of this situation by itself. Hence, it urgently needs policy level interventions aimed at changing land use. These could include the promotion of land use-related research, revision of existing regulations, capacity building, funding, and awareness creation grounded on a national agriculture and land use policy.

References

Bali A, Kumar A, Krishnaswamy J (2007) The mammalian communities in coffee plantations around a protected area in the Western Ghats, India. Biol Conserv 139(1–2):93–102

Bochi VC, Barcia MT, Rodrigues D, Speroni CS, Giusti MM, Godoy HT (2014) Polyphenol extraction optimisation from Ceylon gooseberry (Dovyalis hebecarpa) pulp. Food Chem 164:347–354

Brockerhoff EG, Jactel H, Parrotta JA, Quine CP, Sayer J (2008) Plantation forests and biodiversity: oxymoron or opportunity? Biodivers Conserv 17(5):925–951

Brown S, McIntyre D (1981) An action-research approach to innovation in centralized educational systems. Eur J Sci Educ 3(3):243–258

Campbell BM, Thornton P, Zougmoré R, Van Asten P, Lipper L (2014) Sustainable intensification: what is its role in climate smart agriculture? Curr Opin Environ Sustain 8:39–43

Clay JW (2004) World agriculture and the environment: a commodity-by-commodity guide to impacts and practices. Island Press, Washington

Department of Census and Statistics (2018) District statistical handbook. https://www.statistics.gov.lk/DistrictStatHBook.asp. Accessed 06 July 2019

Ekadinata A, Vincent G (2011) Rubber agroforests in a changing landscape: analysis of land use/cover trajectories in Bungo district, Indonesia. For Trees Livelihoods 20(1):3–14

Food and Agriculture Organization (FAO) (2013) Climate-smart agriculture sourcebook. FAO, Rome

Ferguson J (1893) Ceylon in 1893. Observer Press, Colombo

Gunarathne ADN, Peiris HMP (2017) Assessing the impact of eco-innovations through sustainability indicators: the case of the commercial tea plantation industry in Sri Lanka. Asian J Sustain Social Respon 2 (1):41–58

International Union for Conservation of Nature (ICUN) (2017) Central highlands of Sri Lanka 2017 conservation outlook assessment. ICUN, Gland, Switzerland

Kaplan RS (1998) Innovation action research: creating new management theory and practice. J Man Acc Res 10:89–118

Karanth KK, Sankararaman V, Dalvi S, Srivathsa A, Parameshwaran R, Sharma S, Robbins P, Chhatre A (2016) Producing diversity: Agroforests sustain avian richness and abundance in India's Western Ghats. Front Ecol Evol 4:111

Kroner A, Williams IS (1993) Age of metamorphism in the high-grade rocks of Sri Lanka. J Geol 101(4):513–521

Laumonier Y, Uryu Y, Stüwe M, Budiman A, Setiabudi B, Hadian O (2010) Eco-floristic sectors and deforestation threats in Sumatra: identifying new conservation area network priorities for ecosystem-based land use planning. Biodivers Conserv 19:1153–1174

Lipper L, Thornton P, Campbell BM, Baedeker T, Braimoh A, Bwalya M, Caron P, Cattaneo A, Garrity D, Henry K, Hottle R (2014) Climate-smart agriculture for food security. Nat Clim Chang 4 (12):1068–1072

Menon S, Bawa KS (1997) Applications of geographic information systems, remote-sensing, and a landscape ecology approach to biodiversity conservation in the Western Ghats. Curr Sci 73(2):134–145

Ministry of Finance (2018) Annual report 2018 https://www.treasury.gov.lk/documents/10181/12870/2018/c65529f6-5fac-4f9b-8f38-758ab93676fd. Accessed 15 Sept 2019

Ministry of Internal and Home Affairs and Provincial Councils and Local Government (2019) District secretariats of Sri Lanka, https://www.moha.gov.lk/web/index.php?option=com_content&view=article&id=41&Itemid=173&lang=en. Accessed 17 July 2019

Mohri H, Lahoti S, Saito O, Mahalingam A, Gunatilleke N, Hitinayake G, Takeuchi K, Herath S (2013) Assessment of ecosystem services in homegarden systems in Indonesia, Sri Lanka, and Vietnam. Ecosyst Serv 5:124–136

Mudappa D, Raman TSR (2012) Beyond the borders: wildlife conservation in landscapes fragmented by plantation crops in India. NCF Working Paper 1. National conservation foundation. https://ncf-india.org/publications/565. Accessed 21 Sept 2019

Perera AH, Rajapakse RN (1991) A baseline study of Kandyan forest gardens of Sri Lanka: structure, composition and utilization. For Ecol Manage 45(1–4):269–280

Samansiri BAD, Rajasinghe JCK, Mahindapala KGJP (2011) Agronomic profile of the corporate sector tea plantations in Sri Lanka. Tea Research Institute, Sri Lanka

Sangakkara UR, Frossard E (2016) Characteristics of South Asian rural households and associated homegardens—a case study from Sri Lanka. J Trop Ecol 57 (4):765–777

Sri Lanka Tea Board (2014) History of Ceylon tea. http://www.srilankateaboard.lk/index.php/features/history-of-ceylon-tea. Accessed 12 Sept 2019

Von Blanckenburg F, Hewawasam T, Kubik PW (2004) Cosmogenic nuclide evidence for low weathering and denudation in the wet, tropical highlands of Sri Lanka. J Geophys Res Earth Surf 109(F3):F03008. https://doi.org/10.1029/2003JF000049

Wickramagamage P (1998) Large-scale deforestation for plantation agriculture in the hill country of Sri Lanka and its impact. Hydrol Process 12(13–14):2015–2028

Wickramagamage P (2017) Role of human agency in the transformation of the biogeography of Sri Lanka. Ceyl J Sci 46:19–31

25 Mountain Pastures of Qilian Shan Under Continuous Grazing: Main Environmental Gradients, Vegetation Composition and Soil Properties

Alina Baranova and Udo Schickhoff

Abstract

Degradation of mountain pastures in Qilian Mountains has increased in recent decades; soil erosion accelerated by extensive grazing is widespread. The aim of this study is to identify spatially differentiated and grazing-induced changes in vegetation patterns and associated changes in soil properties. The study area is located in the spring/autumn and summer pastures in the middle section of Qilian Mountains between 2600 and 3300 m a.s.l., representing montane/subalpine and alpine plant communities modified by continuous grazing with sheep, goat and yak. Quantitative and qualitative relevé data were collected for vegetation classification and analysing of gradual changes in vegetation patterns along altitudinal gradient. Vegetation was classified using hierarchical cluster analysis. Five vegetation groups were identified: (1) montane xerophytic shrubby grassland, (2) montane xerophytic grassland, (3) montane grassland—forest meadow, (4) grazing-modified alpine shrubby meadow, (5) alpine meadow. Direct gradient analysis was used to analyse variation in relationships between the vegetation and corresponding environmental variables. ANOVA was used to detect the differences between identified vegetation groups in given environmental conditions. The results showed distinct variation in soil pH, bulk density, OM, carbon, nitrogen and water content and soil minerals concentrations between the identified vegetation groups. Along the altitudinal gradient, increases in soil conductivity, carbon and nitrogen, organic matter and water content as well as decreases in soil pH and basic saturation were observed. Communities of degraded montane grassland with low concentration of soil OM, nitrogen and carbon were widespread on south-facing slopes at lower altitudes. Although all pastures were exposed to extensive grazing, montane grasslands seem to experience more severe degradation in terms of total vegetation cover, soil properties and mineral concentrations.

Keywords

Alpine meadow · Direct gradient analysis · Indicator species analysis (ISA) · Montane grasslands · Pasture degradation · Soil properties

25.1 Introduction

The Qilian Mountains are of prime functional significance for maintaining the ecological integrity of the adjacent Alxa highlands and the

A. Baranova (✉) · U. Schickhoff
CEN Center for Earth System Research and Sustainability, Institute of Geography, University of Hamburg, KlimaCampus, Hamburg, Germany
e-mail: alina.baranova@uni-hamburg.de

U. Schickhoff et al. (eds.), *Mountain Landscapes in Transition*,
Sustainable Development Goals Series, https://doi.org/10.1007/978-3-030-70238-0_25

hydrological stability of the HeiHe river lowlands and irrigation agriculture of the Hexi corridor (Zhao et al. 2011). Located on the northern edge of Tibetan Plateau, they represent both Mongolian and Tibetan floristic provinces (Kürschner et al. 2005; Froese 2012). *Picea crassifolia* forests play a major water protection role (Yang et al. 2005; Sun et al. 2016). At the same time, according to the modeling results of Liu et al. (2004), actual forest cover has been significantly reduced up to only 6% of the potential forest areas. Grasslands cover deforested slopes and are mostly used for animal grazing (Baranova et al. 2016).

Grasslands and shrublands in alpine and subalpine areas of the Qilian Mountains have been experiencing severe overgrazing in the recent past. Vegetation cover is very low during the growing season (Huang et al. 2011). The percentage of unpalatable and toxic plant species in grassland communities is increasing (Baranova et al. 2016). Total vegetation cover is comparatively low on the south-facing slopes, which are prone to erosion. Landslides and other types of soil erosion are often met in the vicinity of the herders summer camps in the alpine pastures (own observations).

Examining the environmental variables allows interpretations of ecologically regulating factors driving the vegetation patterns. In the Qilian Mountains, diverse spectrum of local ecological studies has been conducted, however most of them are published in Chinese. Main focus of the research lies in the field of hydrology and its responses to environmental changes (Yang et al. 2005; Li et al. 2009; Sun et al. 2016; Tian et al. 2017). Some studies were dealing with the response of forest stands to climate change (Deng et al. 2013; Yang et al. 2013). Detailed descriptions are available of variation in soil organic carbon and nitrogen as well as in other edaphic factors along the altitudinal gradient (Yuan and Hou 2015; Yang et al. 2018). Other investigations were focussing on the effect of grazing on plant composition and forage quality, species richness and soil properties (Chang et al. 2004; Baranova et al. 2016; Wang et al. 2017a; Baranova et al. 2019). Although some preliminary studies on the relations between vegetation structure, its dynamics and soil functioning were conducted (Wang et al. 2002; Yang et al. 2018), more extensive research covering unrepresented parts of the altitudinal gradient is necessary. In particular, the lack of qualitative vegetation analyses (Kürschner et al. 2005) represents a gap in environmental studies to be filled in the coming years in order to get a better understanding of the balance in fragile mountain ecosystems under the impact of biotic and abiotic site factors, including anthropogenic disturbances, grazing impact and climate change in the Qilian Mountains.

In the past, mountain rangelands were assumed to represent the ecosystems in equilibrium (Casimir 1992). Based on the modern theory of rangeland ecosystem functioning, both, equilibrium and non-equilibrium models, are to be found in the mountain regions along the altitudinal gradient (c.f. Hoppe et al. 2016; Wang et al. 2017b). Abiotic site factors and animal grazing both affect rangeland ecosystems; however, the effect of grazing is more pronounced in the humid areas, while in the arid conditions, unstable precipitation and its annual variations play a major role and overwhelm the impact of grazing (Behnke et al. 1993; Ellis and Swift 1988). Therefore, for the Qilian Mountains, we expect that grazing impact is more pronounced in the humid alpine zone, while at the lower elevations in presence of more arid conditions, vegetation dynamic is controlled by the moisture regime (von Wehrden et al. 2012). Soil responses to grazing could reveal similar patterns due to the plant–soil interactions (Wang et al. 2017b).

Therefore, we hypothesize that a) in the alpine zone the grazing impact exerts a more distinct influence on vegetation differentiation and underlying topsoil characteristics compared to other site factors; (b) in the montane-subalpine zone soil moisture (as a proxy for precipitation) and other related abiotic site factors would have a greater impact on the vegetation differentiation, while grazing effects would be less pronounced.

25.2 Methodology

25.2.1 Study Area

The Qilian Mountains are located in the middle part of the Heihe River Basin (97°24′–102°08′ E to 37°44′–42°42′ N), adjacent to the Hexi corridor on the north and to the Tibetan Plateau on the south (Fig. 25.1). The Qilian Mountains are covered by 43.61×10^4 ha of forests and 811.2×10^8 m^3 of glaciers which feed the headwaters of the Heihe, Shiyang and Shule rivers and support 4 million people living in the Hexi Corridor (Yang et al. 2005). The southern part of the Qilian Mountains is characterized by semi-arid cold and cold humid mountain climate. Temperature and precipitation show a distinct vertical gradient. The annual mean precipitation increases with elevation (from 250 to 700 mm), while annual mean temperature decreases with elevation (from 6.2 °C to −9.6 °C) (Zhao et al. 2006). A part of the study area belongs to the semi-arid zone with dominant winter drought, at higher elevations alpine conditions prevail (Nagy and Grabherr 2009).

Soils of the pastures in the study area were identified as haplic Leptosol, haplic cambic Regosol and Cambisol, with relatively shallow soil profile, rough texture (silt loam and silt) and intermediate organic matter content (Lieder 2013). The results of Friedrich (2015) on the analyses of the soil physical properties along the wider altitudinal range (2600–3700 m a.s.l.) suggest that investigated soil types refer to haplic Phaeozem and calcic Luvisol (Zech et al. 2014; Friedrich 2015); while Wang et al. (2002) reported chromic Luvisols and Cambisols. Permafrost soils and seasonally frozen soil horizons are widespread in the middle and high elevations.

In the Qilian Mountains, transhumance pastoral practice is in use (Yuan and Hou 2015): the herds of sheep, goats and yaks are kept close to the villages during winter (2400–2600 m a.s.l.), and in spring, the animals are moved upwards to graze on montane-subalpine pastures. In the beginning of June, herders move with their livestock to the summer camps in the alpine zone (above 3000 m a.s.l.). In autumn, the animals are brought back to the areas where they grazed in spring (2600–3000 m a.s.l.). In the Qilian Mountains, the growing season usually begins in the second half of May; the flowering of the mixed grasslands peaks in July and in the beginning of August. According to Wang et al. (2002) and Zhao et al. (2006), most common vegetation classes in the study area are subalpine and alpine shrubland, dominated by *Dasiphora fruticosa, Caragana jubata, Salix gilashanica* and *Spiraea spp.*; subalpine and alpine meadow (2400–3800 m a.s.l.), dominated by *Stipa purpurea, S. przewalskii, Carex lansuensis, Polygonum viviparum, P.bistorta, Dasiphora fruticosa* and *Caragana jubata*; between 2500 and 3600 m a.s.l. forest-steppe vegetation is common, dominated by *Picea crassifolia* and *Sabina przewalskii.*

25.2.2 Sampling Design

Vegetation sampling was conducted in the summer seasons of 2012 and 2013, following an adapted relevé method (Braun-Blanquet 1964; Kent 2012). We applied standard relevé size of 10 × 10 m for all plots, exceeding the requirement of minimal area size (Mueller-Dombois and Ellenberg 1974). On each relevé plot, we described species data according to the Braun-Blanquet cover-abundance scale (7 classes), including the complete list of vascular, bryophyte and lichen species. In order to identify the species, we used collections of the herbarium in the Academy of Water Conservation Forest of the Qilian Mountains (AWCFQ, Zhangye, China), together with local flora catalogues (Xiande et al. 2001; Anlin and Zongli 2009) and internet accessible databases (eFloras; Subject Database of China Plants; The Plant List; Plantarium). Nomenclature of the plant species follows eFloras (2008). For the remaining unknown specimens, we used additional expertise of the botanists in the Herbarium of the Komarov Botanical Institute of the Russian Academy of Sciences (St. Petersburg, Russia).

We conducted field sampling in the spring/autumn and summer pasture areas, covering the altitudinal range from 2650 to 3600 m a.s.l. (Fig. 25.1). Altogether 71 sample sites were randomly selected in different accessible slope exposures, representing the variety of habitat types. On each sampled plot, data on altitude, latitude, longitude and slope angle were obtained using Garmin GPS 60 (with accuracy of 4–6 m) and inclinometer Suunto MB-6 Nord. We collected biomass data on 1 × 1 m plots, placed in the centre of the relevé plot. We clipped the plant specimens on the ground level and measured wet biomass weight shortly after the sampling; we assessed dry biomass weight after oven-drying for 8–10 h at 65 °C. Grazing impact was visually estimated on each plot, on the scale from 3 to 14, using a developed set of environmental indicators (Baranova et al. 2016). On each relevé plot, we extracted soil samples from

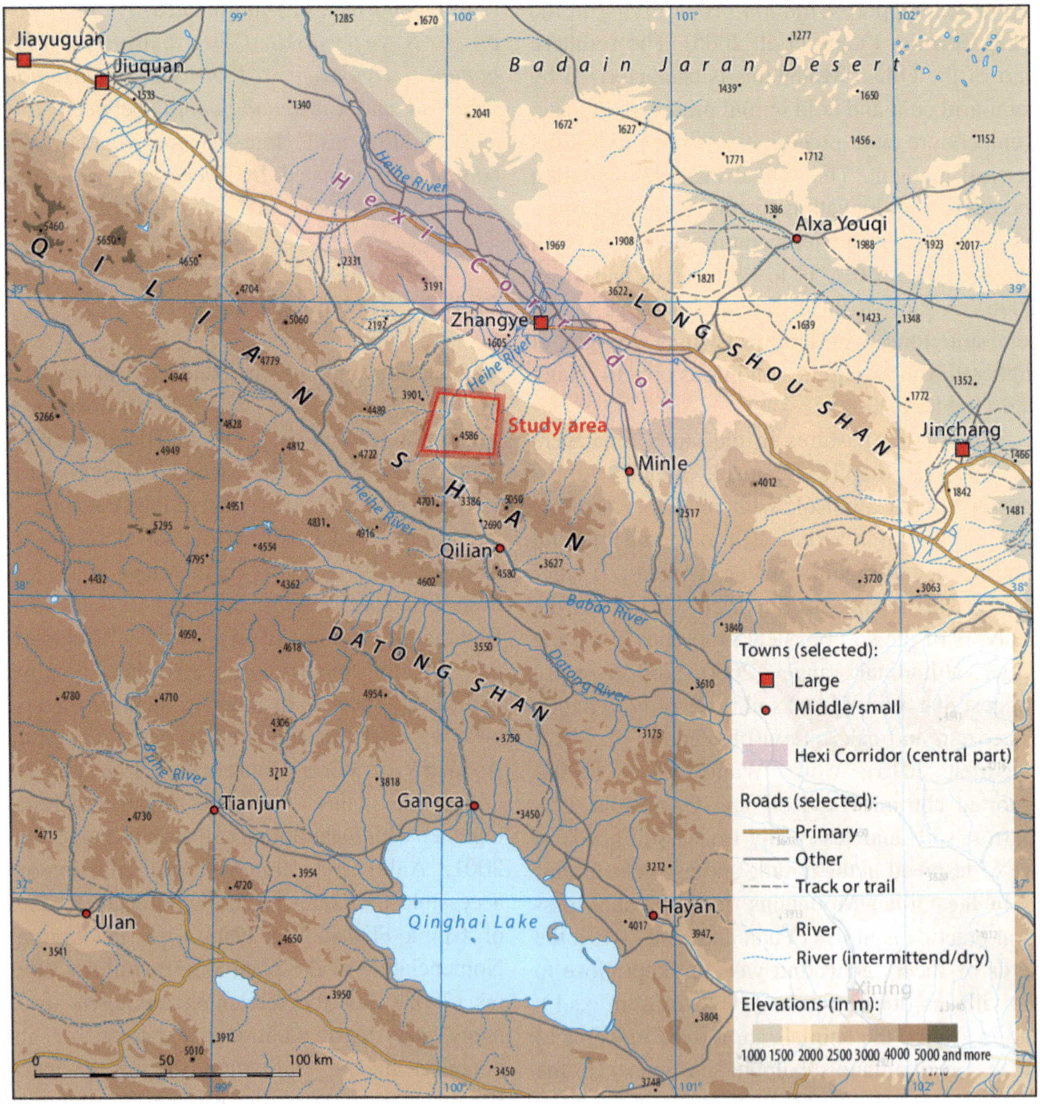

Fig. 25.1 Location of the study area (borders are marked with red square) in the Qilian Shan, Heihe River Basin, Gansu province, NW China

the uppermost mineral soil horizon using soil sampling rings (3 samples of 100 cm^3 per site; in 10–15 cm depth). We stored fresh soil samples in plastic bags and determined the weight at the same day with sampling. Dry soil weight was measured after oven-drying for 5–6 h at 105 °C. Due to misconduct during the sample preparation in Chinese field laboratory, only 63 soil samples were used in further analyses, associated with 63 corresponding relevés, excluding samples from alpine shrub thickets (3400–3600 m a.s.l.).

We performed soil analyses in the Laboratory of the Department of Physical Geography, University of Hamburg. Soil bulk density, organic matter content, water and skeleton content, total nitrogen and total carbon, carbon/nitrogen ratio, pH (in $CaCl_2$ and in H_2O), electroconductivity (EC), cation-exchange capacity (CEC), base saturation (BS) and concentration of the mineral protons were measured. Standard soil analyses followed DIN 19,684-1 (pH value in H_2O and in $CaCl_2$), DIN ISO 11,265 (conductivity), DIN ISO 11,465 (water content), and DIN EN 12,879 (organic matter). CEC, BS and mineral concentrations (in proton equivalent in µmol/g) were analysed according to Meiwes et al. (1984), using Inductively Coupled Plasma-related-Optical Emission Spectroscopy (ICP-OES) and ICP-OES-Software. The remaining analyses followed HFA (2009).

25.2.3 Statistical Analysis

We performed all statistical analyses using the R software and packages "*indicspecies*" (De Cáceres and Legendre 2009), "*mass*", "*pgirmess*", "*plyr*", "*vegan*" (Oksanen et al. 2018) and "*stats*" (Hothorn et al. 2006) (R version 3.4.1, Foundation for Statistical Computing, Vienna, R Core Team 2015).

In order to perform multivariate statistical analyses, we converted the Braun-Blanquet scale according to Wildi (2010) into percentage values; slope exposure degrees (0–360°) were recalculated into two independent variables "eastness" and "northness" after Zar (1999): Eastness = sin [(slope exposure in degrees × Pi)/180]; Northness = cos [(aspect in degrees × Pi)/180]. Log or square root transformation of the rest of environmental variables was performed when needed (Borcard et al. 2011). We applied several transformation techniques on species data to compare the results. To reduce the importance of observations with high values, we applied square root transformation of the species matrix using the function *decostand* (R package "*vegan*").

25.2.3.1 Classification

To identify vegetation patterns, we applied agglomerative clustering using a function *hclust* (Murtagh and Legendre 2014). In order to obtain a metric distance matrix of ecological resemblance, species cover-abundance values were subject to transformation using the Hellinger distance measure (Ruokolainen and Blanchet 2014). Hellinger distance measure gives less weight to species abundances and resolves the double-zero problem (Borcard et al. 2011; Oksanen et al. 2018). To verify the goodness of the clustering, we applied three-steps approach: with the first approach, we aimed to test the performance of the selected clustering methods (correlation between original distance and cophonetic matrixes); the second and third approach served to validate the results of the clustering based on the species composition and variation of environmental variables (Analysis of variance and Indicator Species Analyses (ISA), respectively) (Dufrene and Legendre 1997).

25.2.3.2 Ordination

We applied non-metric multidimensional scaling (NMDS) ordination technique, using function *metaMDS* in package "*vegan*". It is a favourable choice for representation of the objects in two- or three-dimensional space (Legendre and Legendre 2012) and often shows less deformed representation of the relationships among the objects than other ordination techniques could show on the same number of axes (Borcard et al. 2011). In order to assess the relevance of the NMDS and to observe the relationship between distance and cophenetic matrices, stress plot was performed. Stress plot value, obtained from two-dimensional

NMDS space, was still comparatively low (0.2215293), satisfying the condition of monotonicity and keeping the non-metric fit of R^2 close to 1 (Legendre and Legendre 2012).

25.2.3.3 Analysis of Variance

We applied ANOVA statistics, followed by the post-hoc test, to detect the differences between vegetation groups in environmental conditions. First, in order to check if statistical assumptions for ANOVA statistics are met, we tested normality and homogeneity of variance using Shapiro Wilks and Bartlett's test, respectively. If the data were meeting the criterion of normality and homogeneity, one-way ANOVA was applied to compare the variation for each variable, followed by a Tukey post-hoc comparison of means. If the criterion of normality and homogeneity was not met, we applied the Kruskal–Wallis test, followed by non-parametric post-hoc multiple comparisons ($p > 0.05$) after Siegel and Castellan (1998).

25.3 Results

25.3.1 Classification

Figure 25.2 presents the dendrogram of the cluster analysis: in vertical direction on the left side, the distance measure is shown; in horizontal direction, relevé plots are placed, grouped together according to greatest species similarity. The dendrogram illustrates two major patterns: on the left side vegetation of montane and subalpine zone is depicted, while on the right side—plant communities of alpine zone are located, corresponding to distinct vegetation groups 1, 2, 3, 4, and 5, respectively.

25.3.2 Diversity Indexes and Indicator Species Analysis (ISA)

The analysis of species constancy shows that only 23 species out of 176 have high constancy level (above 2.5). Most of the species are perennial. Among the plant functional types, 11 forb species, 6 graminoid species and 5 Fabaceae species were found as well as 1 additional tree species of Pinaceae (*Picea crassifolia*). Most species-rich (abundant) families were Rosaceae, Poaceae, Fabaceae, Cyperaceae and Asteraceae. In total, only 171 species were used in the ISA and 5 species were excluded from the analysis due to their presence in most of the plots (these were: *Achnatherum* sp., *Adenophora* sp., *Leymus* sp., *Melandrium apricum* and *Oxytropis imbricata).*

Species richness analysis showed on average 23 (SD $\pm$ 5) species per plot, with a maximum of 37 and a minimum of 2 species. The lowest average number was found in group 4–19 (SD $\pm$ 8) species pro plot, while group 1 contained the highest average—29 (SD $\pm$ 5) species per plot. Similar trend was observed in species diversity indices: Shannon entropy varied from 2.59 to 2.03 in the respective groups, Simpson diversity and Pielou evenness followed the same trend (N1, N2 and J: Table 25.1). When calculated with Hill's ratio instead (E1), Shannon diversity index picked at 0.47 in group 5 and had a minimum value of 0.45 in group 2.

According to the results of ISA presented in Table 25.2, strong indicator species in group 1 were *Iris lactea var. chinensis, Kobresia humilis, Poa attenuata* and *Artemisia austriaca.* Among them, *Iris lactea var. chinensis* revealed absolute fidelity for the group 1, whereas *Poa attenuata* and *Artemisia austriaca* showed absolute specificity. Group 2 presents the highest number of indicator species—11, among them strong indicators: *Dracocephalum heterophyllum, Heteropappus altaicus* and *Artemisia xerophytica.* These species serve as environmental indicators, corresponding to dry conditions of the study site. Altogether, indicator species of the group 2 represent a typical pattern of south-facing dry slopes, heavily affected by erosion processes, grazing and trampling. A weak indicator index value of the only indicator in group 3 and Silhouette plot, mentioned in the methods section, suggests that in clustering procedure, most of the relevé plots in this group were misinterpreted. Based on the additional analysis of synoptic tables, group 3 shares the same vegetation pattern as group 2, with a few site-specific species,

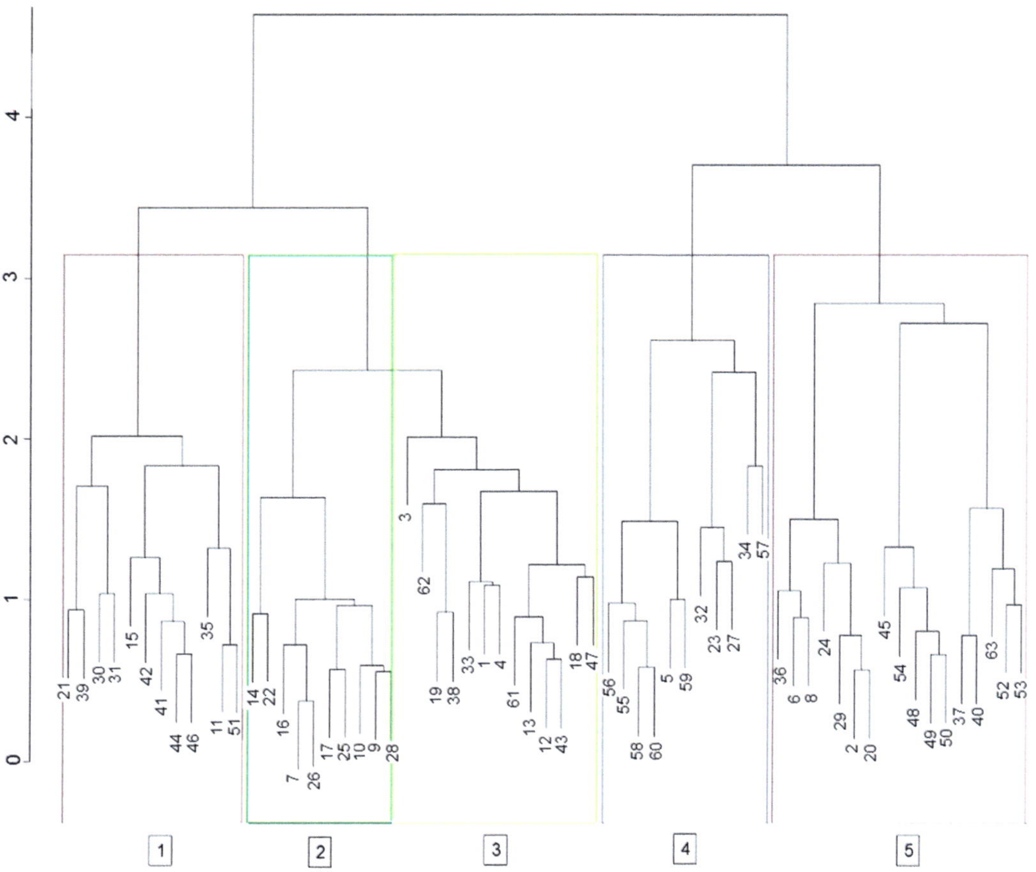

Fig. 25.2 Dendrogram of cluster analysis, based on advanced Ward's agglomerative clustering and Hellinger transformed species data. In colours, five vegetation groups are distinguished. The numbers refer as following: (1) montane xerophytic grassland, (2) montane xerophytic shrubby grassland, (3) montane mesophytic grassland, (4) grazing-modified alpine shrubby meadow and (5) alpine meadow

Table 25.1 Species richness, evenness and diversity indices of the five vegetation groups obtained in cluster analyses

Vegetation group	N0	SD	H	N1	N2	E1	E2	J
gr1	29.42	4.50	2.593	13.713	10.434	0.470	0.356	0.770
gr2	23.90	4.07	2.352	10.726	8.143	0.451	0.341	0.744
gr3	24.38	5.94	2.359	11.022	8.203	0.457	0.345	0.746
gr4	19.55	7.74	2.039	8.498	6.342	0.482	0.382	0.742
gr5	22.71	5.42	2.348	10.848	8.193	0.484	0.369	0.759

N0—species richness
H—Schannon entropy
N1—Schannon diversity number
N2—Simpson diversity number (inv)
J—Pielou evenness
E1 = N1/ N0—Schannon evenness (Hill's ratio)
E2 = N2/N0—Simpson evenness (Hill's ratio)

like *Stipa breviflora* and S. *krilovii* instead of *S. capillata.* In addition, companion species were identified: *Stellera chamaejasme, Agropyron cristatum* and *Leontopodium leontopodioides*.

In group 4 (Table 25.2), *Anemone obtusiloba* and *Ranunculus indivisus* had specificity values close to 1, explaining their occurrence only in alpine meadows. Other strong indicators in group 4 were *Kobresia pusilla, Sibbaldia procumbens* and *Phaeophyscia* sp. (lichen species). Group 5 contains 10 indicators, among them some with absolute specificity values: *Viola bifurca* and *Draba eriopoda* (an indicator of grazing disturbance). Other strong indicators in group 5, associated with heavily grazed alpine meadows, were *Plantago asiatica, Elymus* sp., *Poa* sp. and *Saussurea* sp.

Based on the results of agglomerative clustering (Fig. 25.2), supported by ISA and the outcome of synoptic tables, vegetation of spring/autumn and summer pastures in Qilian Mountains was classified into five main groups with the following rankless communities (further named as vegetation groups): (1) montane xerophytic shrubby grassland (*Iris lactea var. chinensis—Artemisia austriaca*; with dwarf-shrubs *Potentilla davurica, Dasiphora fruticosa)*; (2) montane xerophytic grassland (*Dracocephalum heterophyllum—Heteropappus altaicus)* (3) montane grassland—forest meadow (*Stipa krylovii—Potentilla multifida)*; (4) grazing-modified alpine shrubby meadow (*Anemone obtusiloba—Ranunculus indivisus (with dwarf-shrubs Potentilla bifurca, Caragana jubata);* (5) alpine meadow (*Anemone obtusiloba —Ranunculus indivisus).*

25.3.3 Ordination

According to the results of NMDS ordination, illustrated in Fig. 25.3a–d, moist alpine communities are determined by the increasing concentrations of soil nitrogen, carbon, organic matter and water content (Fig. 25.3: d), as well as by the increasing concentration of soil potassium, manganese and iron ions (Fig. 25.3: c). Xeric montane/subalpine communities show opposite trends.

Their differentiation is predetermined by higher pH and base saturation (Fig. 25.3d). We found concentrations of soil minerals not to be differentiating for these communities (Fig. 25.3 c). Altitude, north exposure, soil water content and concentration of iron showed the strongest correlation with the first NMDS axis (Table 25.3), which could be best characterized as an elevation/moisture gradient (Fig. 25.3a, c, d). The second NMDS axis could be interpreted as an inclination/woody gradient, where increasing tree, shrub, and moss cover, as well as increasing number of species per plot, are associated with more steep slopes (Fig. 25.3a, b). At the same time, increase in herb cover was higher on less inclined slopes and was related to high soil skeleton content and high concentration of potassium (Fig. 25.3c, d); most of the other soil minerals showed a negative correlation with the second NMDS axis (Table 3). Increase in carbon/nitrogen ratio was also associated with the inclination/woody gradient, while increase in concentrations of carbon and nitrogen was strongly related to the elevation/moisture gradient (Fig. 25.3d). Among measured environmental variables, soil electric conductivity, aspect, eastness and grazing impact showed no significant correlation with NMDS axes and did not appear within the ordination space (Table 25.3).

25.3.4 Vegetation Groups and Environmental Variables

Applying the ANOVA statistics and Kruskal–Wallis tests, the results showed significant differentiation between five vegetation groups, based on the following soil variables: water content, OM, pH, base saturation and soil bulk density (Fig. 25.4a–e), which were used as diagnostic variables for the five vegetation groups. Group 1 refers to S-facing shrublands and shrubby grasslands along the altitudinal gradient. Most of the diagnostic soil variables did not significantly differ within group 1, except for soil bulk density, which showed the second

Table 25.2 Indicator species analysis of five vegetation groups (without group combinations)

Vegetation groups/Indicator species	Indicator value			
	A	B	Stat	*p* value
1. Montane xerophytic grassland	#sps. 5			
Iris lactea var. chinensis	**0.3743**	**1**	**0.612**	**0.008****
Astragalus/Oxytropis sp.	0.7177	0.4167	0.547	0.023*
Kobresia humilis	**0.4828**	**0.5833**	**0.531**	**0.018***
Artemisia austriaca	**1**	**0.25**	**0.5**	**0.010****
Poa attenuata	**1**	**0.25**	**0.5**	**0.016***
Cerastium sp.-2	0.6813	0.3333	0.477	0.038*
2. Montane xerophytic shrubby grassland	#sps. 11			
Dracocephalum heterophyllum	**0.6738**	**0.7826**	**0.687**	**0.005***
Heteropappus altaicus	**0.6062**	**0.7261**	**0.651**	**0.002***
Artemisia xerophytica	**0.8481**	**0.4269**	**0.582**	**0.003***
Allium przewalskianum	**0.403**	**0.8261**	**0.568**	**0.041***
Oxytropis melanocalyx	0.9714	0.3176	0.54	0.035*
Allium cyaneum	0.727	0.4568	0.539	0.020*
Thalictrum cultratum	0.4758	0.6487	0.534	0.017*
Stipa capillata	0.5333	0.5454	0.516	0.050*
Potentilla acaulis	0.8794	0.3256	0.514	0.030*
Caragana opulens	0.5578	0.4678	0.472	0.041*
Chenopodium pamiricum	0.9118	0.2341	0.427	0.047*
3. Montane grassland - forest meadow	#sps.1			
Carex sp.-4	1.00000	0.2308	0.5	0.021*
4. Grazing-modified alpine shrubby meadow	#sps. 6			
Anemone obtusiloba	**0.9963**	**0.3636**	**0.602**	**0.002****
Phaeophyscia sp.	0.712	0.4545	0.569	0.013*
Kobresia pusilla	0.4881	0.6364	0.557	0.025*
Carex sp.-1	0.4879	0.5455	0.516	0.038*
Sibbaldia procumbens	0.7255	0.3636	0.514	0.015*
Ranunculus indivisus	**0.9221**	**0.2727**	**0.501**	**0.031***
5. Alpine meadow	#sps. 10			
Plantago asiatica	**0.7536**	**0.6471**	**0.698**	**0.001****
***Elymus* sp.**	**0.8065**	**0.5882**	**0.689**	**0.003****
Viola bifurca	**1**	**0.4118**	**0.642**	**0.002****
***Poa* sp.-1**	**0.989**	**0.4118**	**0.638**	**0.001****
***Saussurea* sp.**	**0.8559**	**0.4118**	**0.594**	**0.003****
Poa **sp.-2**	0.9706	0.2941	0.534	0.011*

(continued)

Table 25.2 (continued)

Vegetation groups/Indicator species	Indicator value			
	A	B	Stat	*p* value
Polygonum viviparum	0.465	0.5882	0.523	0.041*
Draba eriopoda	1	0.2353	0.485	0.024*
Cerastium caespitosum	0.9829	0.2353	0.481	0.049*
Parnassia oreophila	0.7073	0.2941	0.456	0.042*

List of species associated to each group. Indicator value components: A—specificity; B—fidelity. Only those species are shown, which indicator index value (stat) $\geq$ 0.4, with significance level (p) > 0.05). Significance codes: 0 '***'. 0.001 '**'. 0.01 '*'. 0.05 '.' (1), (2) and (3)—montane mesophytic grasslands, (4)—grazing-modified alpine shrubby meadow and (5)—alpine meadow

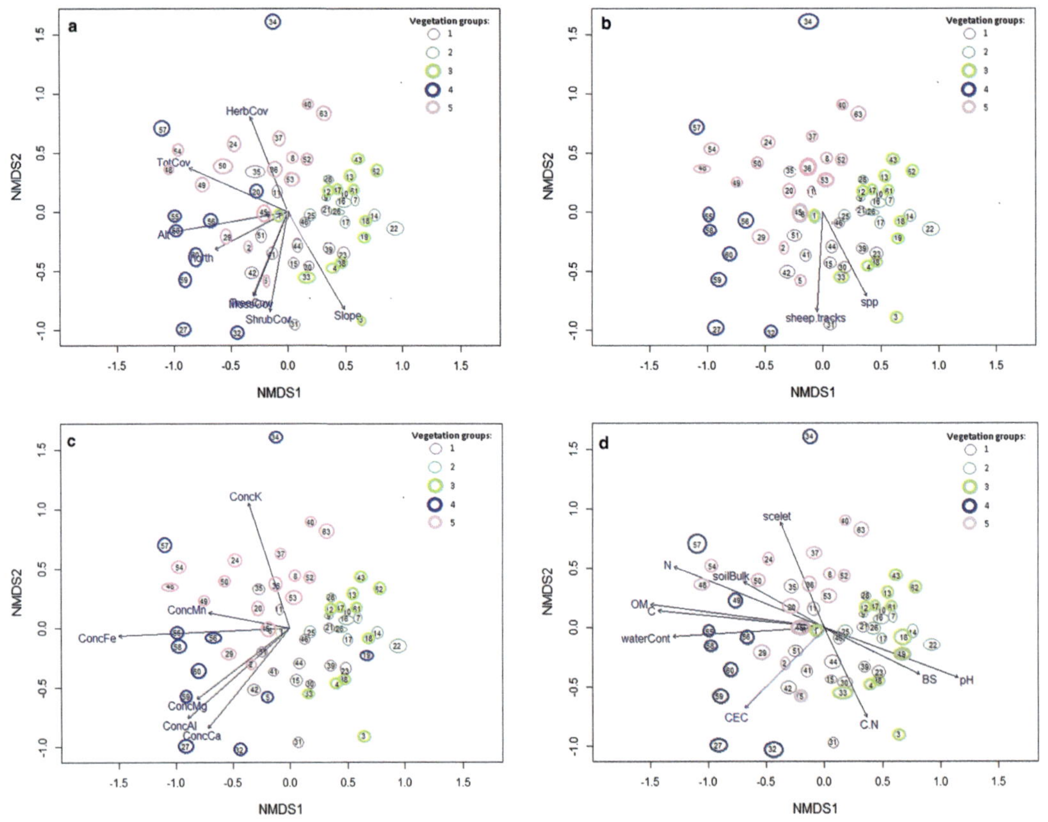

Fig. 25.3 Two-dimensional NMDS ordination of five vegetation groups against different environmental variables. Only vectors with significant correlation with NMDS axes are presented (p > 0.05); detailed numbers of Pearson's rank correlation coefficients are provided in Table 25.3. Description of vegetation groups is the same as on Fig. 25.2 and Table 25.2. **a**—Altitude (Alt), slope, northness (north) and total cover, moss cover, shrub cover, herb cover; **b**—Sheep tracks (e.g. indicator of grazing); **c**—Concentrations of soil nutrients: potassium (ConcK), iron (ConcFe), manganese (ConcMn), magnesium (ConcMg), aluminium (ConcAl), calcium (ConcCa); **d**—Soil properties: soil skeleton (skeleton), soil bulk density (soilBulk), nitrogen (N), organic matter (OM), carbon (C), water content (waterCont), carbon/nitrogen ration (C.N), cation-exchange capacity (CEC), base saturation (BS), pH

Table 25.3 Pearson's rank correlation coefficients of the variables and two axes of non-metric multidimensional scaling (NMDS), using monoMDS function

Variables	NMDS1	NMDS2	r2	Pr (>r)	Signif. level
Soil bulk density [g/cm^3]	**−0.87662**	**0.48118**	0.1458	0.008	**
Soil skeleton [%]	−0.39061	**0.92056**	0.2151	0.004	**
Water content [%]	**−0.99839**	−0.0568	0.3982	0.001	***
OM [%]	**−0.99183**	0.12756	0.5303	0.001	***
pH (CaCl2)	**0.93947**	−0.34263	0.3506	0.001	***
EC [µS/cm]	0.85651	0.51613	0.0028	0.919	n.s
C [%]	**−0.99534**	0.09638	0.4818	0.001	***
N [%]	**−0.92989**	0.36784	0.4516	0.001	***
C/N [%]	0.43416	**−0.90084**	0.1633	0.007	**
CEC [cmol/kg]	**−0.70667**	**−0.70754**	0.2832	0.001	***
BS [%]	**0.90274**	−0.43019	0.1925	0.003	**
Altitude	**−0.98633**	−0.16481	0.3274	0.001	***
Aspect	0.01168	0.99993	0.0045	0.869	n.s
Slope (grad)	0.51248	**−0.8587**	0.2792	0.001	***
Total cover [%]	**−0.91947**	0.39317	0.2713	0.001	***
Tree cover [%]	−0.39898	**−0.91696**	0.1799	0.007	**
Shrub cover [%]	−0.186	**−0.98255**	0.2259	0.001	***
Herb cover [%]	−0.37984	**0.92505**	0.2301	0.001	***
Moss cover [%]	−0.38759	**−0.92183**	0.1834	0.008	**
Northness	−0.89475	−0.44657	0.1569	0.005	**
Eastness	0.57618	−0.81732	0.0327	0.372	n.s
Sheep tracks [%]	−0.06189	**−0.99808**	0.1524	0.022	*
Grazing impact	0.21438	−0.97675	0.0141	0.642	n.s
Number of species	0.4681	**−0.88367**	0.1389	0.017	*
Al [µmol/g]	**−0.76531**	**−0.64367**	0.3224	0.001	***
Ca [µmol/g]	**−0.65124**	**−0.75887**	0.2871	0.001	***
K [µmol/g]	−0.32627	**0.94528**	0.2901	0.001	***
Mg [µmol/g]	**−0.80804**	**−0.58913**	0.2387	0.002	**
Na [µmol/g]	0.59186	**−0.80604**	0.0288	0.394	n.s
Fe [µmol/g]	**−0.99904**	−0.04378	0.5262	0.001	***
Mn [µmol/g]	**−0.98376**	0.1795	0.1226	0.019	*

Species data transformation: Wisconsin (sqrt) on Bray distances. Significance codes for Pr(>r): 0 '***' 0.001 '**' 0.01 '*' 0.05; *n.s.* not significant. Permutation: free. Number of permutations: 999; r2—squared correlation coefficient between the factor and two matrixes; Pr(>r) premutational significance test

highest value after the group 5—"the group of productive grasslands" (Fig. 25.4e). Also, it had the highest number of species per plot—29.50 (±4.80), suggesting that the shrub encroachment has a positive effect on the species diversity, reducing the grazing pressure.

Group 2, representing typical S-facing shrubby grasslands, showed higher pH, BS and

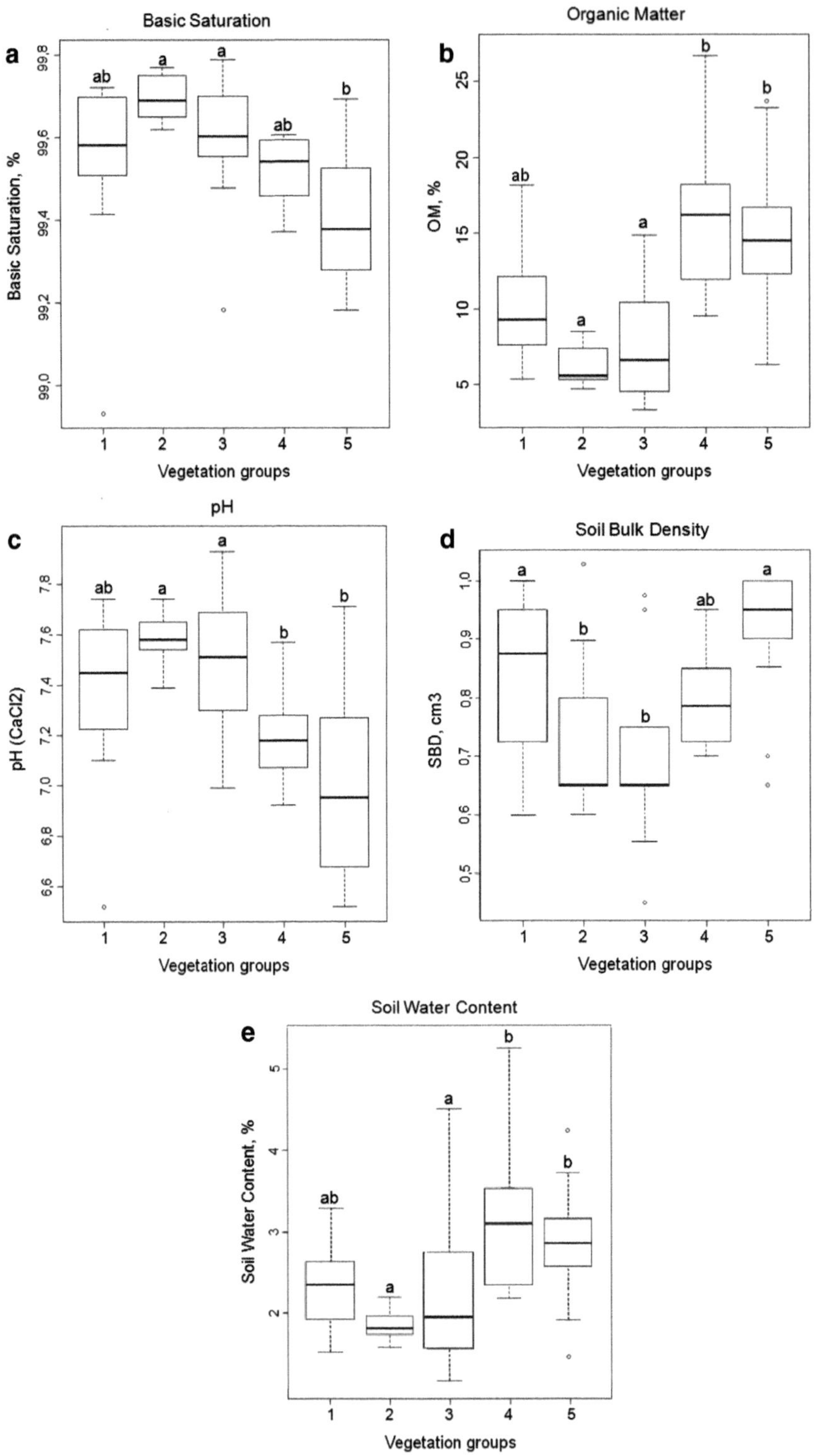

Fig. 25.4 Distribution of the diagnostic soil variables among five vegetation groups ($p > 0.05$): **a**—Base Saturation (%); **b**—Organic Matter (%); **c**—pH ($CaCl_2$); **d**—Soil Bulk Density (cm^3); **e**—Soil Water Content (%). Horizontal axes represent five vegetation groups, defined in Fig. 25.2 and Table 25.2

CEC values, in comparison with moist north-facing "productive grasslands"—groups 4 and 5 (Fig. 25.4a, b, c). By contrast, group 5, showed significantly higher water content, OM and SBD values, as well as higher content of soil carbon and nitrogen, which were considerably different in S-facing grasslands. In group 3, the variables were performing very similar to group 2, without any statistically significant differences among them.

Groups 4 and 5 are representing plant communities on more gentle, N-/NW - exposed slopes along the altitudinal gradient. Diagnostic soil variables did not significantly vary between them, although group 5 had the highest soil bulk density (0.91(±0.12)). At the same time, grazing impact was the lowest in group 5 (5.35(±2.34)). Mean concentrations of soil potassium and manganese were reaching maximum in group 5. In group 4, the highest mean concentration of aluminium and iron was observed (Fig. 25.5a, d). Here, mean calcium content reached a significant maximum of 521.55 μmol/g (Fig. 25.5e).

25.3.5 Physical Characteristics of the Soils

Most of the sampled soils present a pH range indicating neutral and slightly alkaline pH conditions (6.99–7.58); base saturation is often exceeding 99%; on our ordination space BS and pH vectors follow in the same direction (Fig. 25.3d). Among most of the sites, especially in group 1 (EC = 338.75, SD = 264.88), wide range of electrical conductivity values is observed. The mean values for each vegetation group vary in the range between 120 and 435 μS/cm, without any significant difference; current range of EC refers to relatively low salt content in the soil solution (AK Standortskartierung 2003). Our results reveal insignificant variation of the soil skeleton, electroconductivity, CEC and C/N ratio between identified vegetation groups, corresponding to a broad range of values within each group.

In general, the concentration of the organic matter in the investigated soil samples reveals high humus content. On dry, S-facing slopes the lowest OM content was observed (group 2, Fig. 25.3d). Vegetation groups referring to the alpine zone are associated with the highest OM content, even characterized as swampy conditions; therefore, the water storing capacity of these soils is estimated as very high. Variation of C/N ratio also corresponded to high humus content (10–15%) and was not significantly different between vegetation groups. Mineral composition of the soils is characterized by a very high calcium content, which showed significant variation among the identified vegetation groups (Fig. 25.5e). Increase of iron, calcium, aluminium, potassium and manganese mean concentrations was associated with the alpine zone. Concentrations of the iron in the uppermost soil horizon increased with altitude and were comparatively higher under grazing-modified alpine shrubby meadow.

The length of the vector formed by iron reveals the strongest differentiation impact on the vegetation composition along the altitudinal gradient (Fig. 25.3a, c).

25.4 Discussion

25.4.1 Species Diversity and Grazing Impact

The area of the Qilian Mountains has been grazed since prehistoric times (Rhode et al. 2007; Miehe et al. 2009); therefore, the composition of the plant communities indicates a high level of grazing-resistance (Milchunas and Lauenroth 1993; Suttie et al. 2005). At the same time, the percentage of unpalatable and toxic plant species has been increasing in recent decades (Chang et al. 2004; Baranova et al. 2016). Results of our study are consistent with previous findings, outlining the community-forming role of the unpalatable *Iris lactea var. chinensis* in montane xerophytic grassland, which had been expanding over the vast areas of the pastureland due to selective animal grazing. *Kobresia humilis,* occurring in montane grasslands in Qilian Mountains, similar in abundance to *K. pygmaea* in Tibet, has developed morphological adaptations to store the main

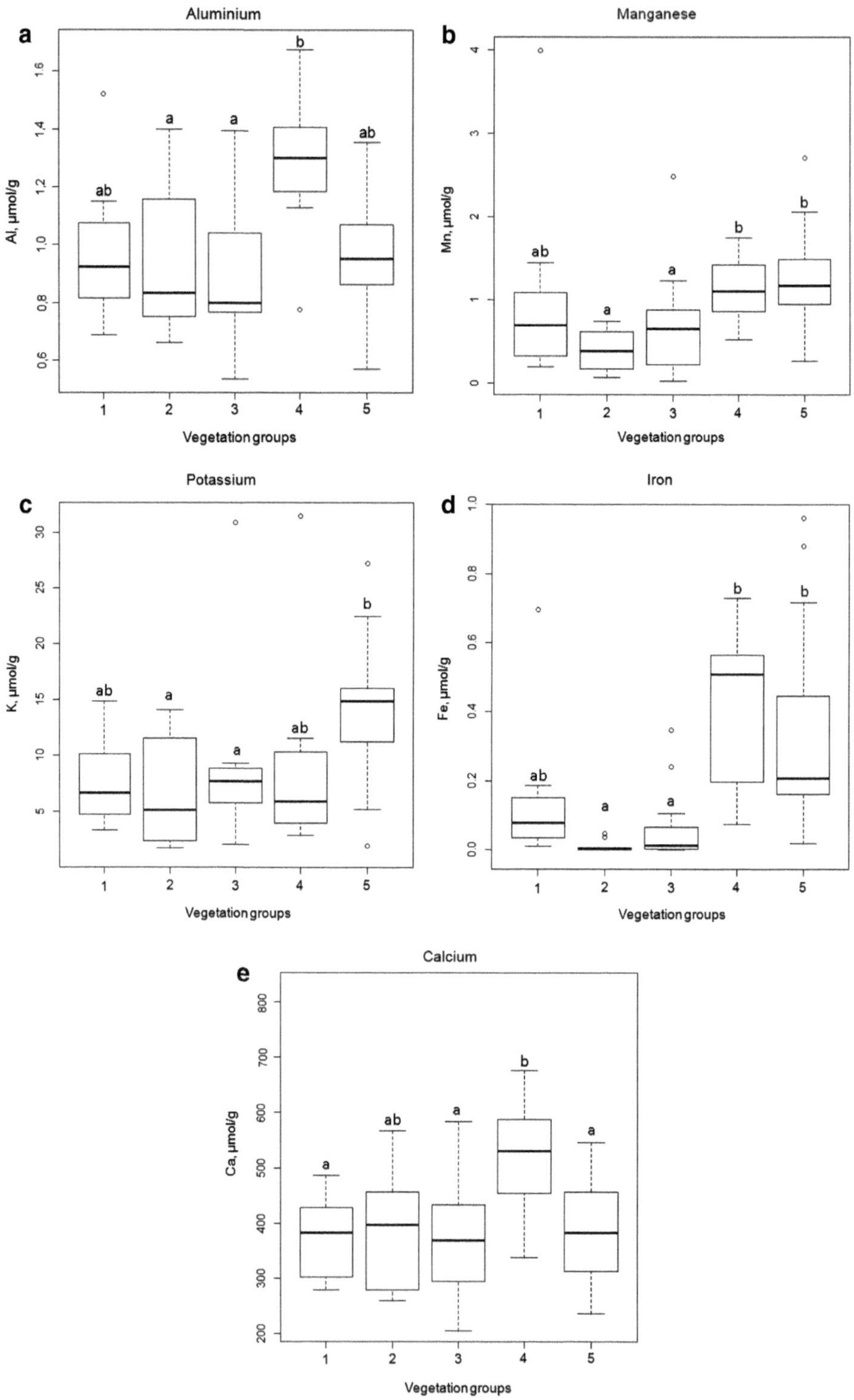

Fig. 25.5 Distribution of the soil minerals among five vegetation groups ($p > 0.05$): **a**—Aluminium [µmol/g]. Horizontal axis represents five vegetation groups, defined in Fig. 25.2 and Table 25.2. Distribution of the soil minerals among five vegetation groups ($p > 0.05$): **b**—Manganese [µmol/g]; **c**—Potassium [µmol/g]; **d**—Iron [µmol/g]; **e**—Calcium [µmol/g]

nutrients in belowground biomass, being an indicator of the long-term grazing by its means (Miehe et al. 2008; Etzold et al. 2016). In each of five vegetation groups, identified in our study, indicators of continuous grazing were found, suggesting a strong impact of intensive pasture utilization on the present composition of the plant communities.

With regard to altitudinal gradient, different studies show that species richness curve usually has a hump-shape on the middle elevations (Lomolino 2001; McCain and Grytnes 2010; Yang et al. 2018), which was not confirmed in our study, as variation of the species richness with increasing elevation was insignificant. However, the latter could be explained by the fact that altitudinal gradient was found to overlap the effect of grazing on plant species richness and diversity (Brinkmann et al. 2009). At the same time, plant species richness and diversity were shown to decrease under moderate to high levels of grazing intensity (Herrero-Jáuregui and Oesterheld 2018). Moreover, our results reveal a maximum species number at 2850 m a.s.l., while in the study of Yang et al. (2018), it peaked at 3177 m a.s.l. Although both studies are conducted in the Qilian Mountains, such a difference could be explained by the sampling design: the latter study covers only north-facing slopes, which usually have more gentle slopes, higher herb cover and therefore reveal greater species diversity.

Addressing the distribution of the plant species richness along the elevation gradient, Etzold et al. (2016) suggested to take into account the effect of the land use and its intensity. Recent metaanalysis of Herrero-Jáuregui and Oesterheld (2018) revealed larger negative response of species richness to grazing in arid and low productive rangelands than in humid and productive ones. By contrast, in our study, most of the diversity indices show the highest values in montane xerophytic grassland, and the lowest—in grazing-modified alpine shrubby meadow (Table 25.1), reflecting a higher negative impact of grazing on the plant species diversity in alpine communities in Qilian Mountains.

However, there are two restrictions, making an estimation of species diversity more vulnerable—species abundance distribution and sampling density (Chao et al. 2014). Therefore, in modern diversity measurements, classical approach of Shannon entropy and other diversity indexes usually correlating with each other is preferred to estimations using Hill numbers (i.e. effective number of species) and ratios, expressed in the same units (Borcard et al. 2011; Chao et al. 2014). It allows incorporating relative abundance and species richness (Chao et al. 2014). In our study, evenness indices calculated with the Hill numbers varied differently in comparison to Pielou evenness: the latter was in line with classical diversity indexes, while evenness based on the Hill numbers has shown the higher values in alpine, instead of montane, vegetation communities (Table 25.1).

Species constancy highly depends on the plot size (Dengler et al. 2009). In natural communities, usually a large number of species have relatively low abundances (Chao and Shen 2003). In our study, 87% of the species were found to have low constancy level below 2.5%. For that reason, square root transformation of the species data was used prior to constancy analysis in order to decrease the impact of the high-score species (Borcard et al. 2011). Nevertheless, the results of the cluster analysis as well as the performance of the species within NMDS ordination space have shown a high species heterogeneity within each of the identified vegetation groups.

25.4.2 Main Environmental Gradients

Our results show that the elevation/moisture gradient is responsible for the strongest change in species composition in alpine areas of the Qilian Mountains, which is in line with other altitudinal studies around the globe (Nagy and Garbher 2009; Etzold et al. 2016). North-facing slopes provide moist environments for grasslands and cause higher productivity compared to south-facing slopes on the same elevation. Within the montane zone, edaphic moisture was identified as an important driving factor for the vegetation differentiation (Zemmrich et al. 2010). In our study in the montane-subalpine, slope inclination and exposition have a distinguishable impact on vegetation distribution as well as on variation of the dry biomass along the altitudinal gradient (data not shown). N-, NW- and NE-facing slopes contain most productive and less disturbed plant communities with total cover close to 100%. North exposure was found to be an important factor, differentiating grasslands in alpine areas, while in montane/subalpine areas north-facing slopes are mostly covered by *Picea crassifolia* forests.

25.4.3 Soil Properties and Interactions Between Them

Soil water content is an important parameter, defining the vertical variation of the soil nutrients (Liu et al. 2017). Moreover, in arid and semi-arid areas with low-rainfall variation, cations of calcium, magnesium and sodium are forming an exchange complex, defining the value of cation-exchange capacity and influencing other soil properties (Brady and Weil 2014). In our study, cations of calcium play the major role in CEC, having mean calcium saturation of 86%; therefore, there is a strong correlation observed between calcium and CEC vectors (Fig. 25.3c, d). However, our results reveal no distinct correspondence between them: increasing CEC has high correlation with both NMDS axes; therefore, indicating that moisture gradient has the same impact on CEC values, as soil pH (Fig. 25.3d; Table 25.2).

Base saturation is mainly to be attributed to cations of Ca, K, Mg and Na. An increase in BS values indicates a tendency to neutrality or alkalinity in soil pH (Brady and Weil 2014). Our data reveal high values of base saturation in the range above 99%; thus, corresponding topsoils could be classified as highly elastic, with the high potential to intercept the soil disturbances (AK Standortskartierung 2003). Although in our study area, high values of base saturation could have been affected by the high values of calcium ions (Friedrich 2015).

Variation in soil bulk density is usually associated with the soil texture: sandy soils have higher values of SBD than silt loams or clays, which could be explained by the presence of micropores in clayey particles. In our study area, soil texture was defined as silt, silt loam and sandy loam (Lieder 2013; Tian et al. 2017), which corresponds to low values of soil bulk density with mean value of 0.80 g/cm^3, corresponding to uncultivated forest and grassland vegetation types (Brady and Weil 2014). Yang et al. (2018) reported about dependency of the soil bulk density on the soil depth as well as on the elevation: in deeper soil layers at high altitudes, soil bulk density was increasing, whereas on the upper-most soil horizon soil bulk density tends to decrease with altitude. The latter is only partly supported by our results: soil bulk density had the strongest correlation with elevation gradient, although minor correlation with inclination/woody gradient was also observed, revealing unequal distribution of the fine and grain-textured soils (Brady and Weil 2014).

25.4.4 Soil Organic Matter

It is known that soil organic matter develops at higher rates under grassland compared to forest or shrubland, which is related to the type of the root system and its decomposing ability (Brady and Weil 2014). Results of our study illustrate that the direction mentioned above: the direction of the carbon, nitrogen and organic matter accumulation follows elevation/moisture gradient, whereas the increase of total shrub and tree cover represents a separate inclination/woody gradient (Fig. 25.3a, d). Instead of carbon and nitrogen, C/N ratio was found to be correlated with inclination/woody gradient, which is supported by the previous studies, showing the increase in C/N ratio under the forest/shrubland cover (Lieder 2013; Friedrich 2015). These two independent gradients represent the driving environmental factors of the vegetation differentiation in montane, subalpine and alpine areas of Qilian Mountains, identified in our study. With regard to grazing intensity, no change in soil C/N ratio was observed between continuously grazed pastures and those excluded from grazing (Shrestha and Stahl 2008; Wang et al. 2017a, b). Similarly, in our findings, variation of C/N ratio was not predetermined neither by elevation nor by vegetation type or grazing pressure.

In our study, C/N ratio reveals comparatively narrow range of 9–20% and mean value of 12%, which corresponds to a sufficient supply of nitrogen and moderate level of humus content in the investigated soils (AK Standortskartierung 2003; Lieder 2013). However, in combination with low soil OM and high pH values, it corresponds to calcium-rich soils underlying the semi-arid grasslands (Zech et al. 2014; Brady and Weil 2014). At the same time, previous studies on the alpine meadows in the Qilian Mountains were reporting the presence of fertile black soils (Miao et al. 2015; Yuan and Hou 2015). Such contradiction explains the primary role of topographic factors: investigations mentioned above were hold in the outwash plains, whereas our sample plots mostly belong to the catchment areas with incomparably higher erosion rates and low soil water content. To complete the identification of the corresponding soil types according to FAO, detailed investigation of the soil profiles in varying geological units of corresponding pasture areas is necessary.

Some studies in mountain environment show that the total amount of the soil nitrogen and carbon content tends to increase in the topsoil with increasing elevation (Nagy and Grabbher 2009; Yang et al. 2018), which is also illustrated in our study (Fig. 25.2d). Depletion of soil carbon and nitrogen is often associated with increasing grazing pressure directly—by the reduction of the primary source of organic matter coming from the plant biomass, and indirectly—through the change in species composition and decrease of Fabaceae species (known for their N-fixator ability) due to selective grazing and decrease of species diversity (Wang et al. 2017a). However, some studies have shown high nitrogen content in heavily grazed pastures close to the camp or water place (Zemmrich et al. 2010; Hoppe et al. 2016; Wang et al. 2017b) due to direct depositing of the cattle dung. However, dung of the yaks is an essential source of fuel, which has been collected since prehistoric times in the highlands of Tibet (Rhode et al. 2007); therefore, even in extremely grazed spots, concentration of the soil nitrogen remains comparatively low (Miller 2005; Wang et al. 2017a).

25.4.5 Soil Nutrients

Among the essential mineral elements required for the plant growth, we have analysed variation of concentrations of potassium, calcium, magnesium, iron and manganese, and additional aluminium and sodium. Compared with results of Baranova et al. (2019), all the minerals were found in sufficient leaf concentrations for the plant growth, except for iron, which concentration during the growing season was exceeding toxicity levels (White and Greenwood 2013). Indirect ordination reveals a feasible increase of the iron concentration along the elevation gradient (Fig. 25.2c), which could be explained by decrease in redox potential and / or decrease of pH value (White and Greenwood 2013). The latter is supported by our results, where the pH values show a negative correlation with elevation/moisture gradient, pointing in the opposite direction to increasing iron concentration (Fig. 25.2c, d). Whereas increasing concentration of soil calcium showed an opposite trend to increasing soil pH, which is probably associated with leaching of calcium ions and longer accumulation of organic matter only possible at higher altitudes (Etzold et al. 2016).

In general, soil erosion has negative effect on availability of the soil nutrients, causing depletion of the soil organic matter from the uppermost soil horizon (Brady and Weil 2014). In our study area due to the high rates of the soil erosion on the S-, SW-facing slopes, decreased concentrations of the soil nutrients were observed; therefore, soil minerals were not playing an important role in the differentiation of the vegetation groups on the respective slope exposures at low elevations (Fig. 25.2c). Moreover, similar results were obtained by the investigation of the deeper soil layers (up to 90 cm), indicating a decrease of the soil mineral concentrations under intensified land use (Liu et al. 2017).

25.5 Conclusions

Due to the distinct spatial differentiation of the pastures in the mountain and alpine zones of the Qilian Shan, long-term grazing in these areas has resulted in diverse responses of the vegetation structure and soil properties. South-facing slopes are found to experience severe pasture degradation in terms of low percentage of the palatable plant species and low total vegetation cover and depleted soils (e.g. low mineral concentrations, high pH and low OM), At the same time, north-exposed slopes and gentle slopes in the alpine zone have moist soils with high concentrations of OM and soil minerals, and high vegetation cover, therefore seems to be more resistant to soil erosion triggered by animal grazing. In terms of vegetation composition, in both mountain and alpine zones, we have found different indicator species of—indicators of continuous grazing, suggesting a strong impact of intensive pasture utilization on the present plant community composition.

References

The Plant List. Version 10. Published on the Internet https://www.theplantlist.org/ [last accessed 20 Apr 2018]

AK Standortskartierung (2003) Forstliche Standortsaufnahme. 6. Aufl. München.

Anlin G, Zongli W (2009) Atlas of rangeland plants in Northern China. Agricultural Science and Technology Press, Beijing

Baranova A, Schickhoff U, Wang S, Jin M (2016) Mountain pastures of Qilian Mountains: plant communities, grazing impact and degradation status (Gansu province, NW China). Hacquetia 15(2):21–35

Baranova A, Oldeland J, Wang S, Schickhoff U (2019) Grazing impact on forage quality and macronutrient content of rangelands in Qilian Mountains NW China. J Mt Sci 16(1):43–53

Behnke RH, Scoones I, Kerven C (eds) (1993) Range ecology at disequilibrium, new models of natural variability and pastoral adaptation in African savannas. Overseas Development Institute and International Institute for Environment and Development, London, UK

Borcard D, Gillet F, Legendre P (2011) Numerical ecology with R. Springer, New York

Brady NC, Weil RR (2014) The nature and properties of soils, Fourteenth. Always learning, Dorling Kindersley, Noida, India

Braun-Blanquet J (1964) Pflanzensoziologie, 3rd edn. Wien, 819 p. [in German]

Brinkmann K, Patzelt A, Dickhoefer U, Schlecht E, Buerkert A (2009) Vegetation patterns and diversity along an altitudinal and a grazing gradient in the Jabal al Akhdar mountain range of northern Oman. J Arid Environ 73(11):1035–1045

eFloras (2008) Missouri botanical garden, St Louis, MO & Harvard University Herbaria, Cambridge, MA. Published on the Internet https://www.efloras.org [last accessed on 20 Apr 2015]

Casimir MJ (1992) The determinants of right to pasture: territorial organisation and ecological constraints. In: Casimir MJ, Rao A (eds) Mobility and territoriality: social and spatial boundaries among foragers, fishers, pastoralists and peripatetics: conference entitled "Territoriality among spatially mobile populations", pp 91–134

Chang X, Zhao W, Zhao A (2004) Species diversity of pasture community at different altitude levels in Qilian Mountains. Chin J Appl Ecol 15(9):1599–1603

Chao A, Shen TJ (2003) Nonparametric estimation of Shannon's diversity index when there are unseen species in sample. Environ Ecol Stat 10:429–443

Chao A, Gotelli NJ, Hsieh TC, Sander EL, Ma KH, Colwell RK, Ellison AM (2014) Rarefaction and extrapolation with Hill numbers: a framework for sampling and estimation in species diversity studies. Ecol Monogr 84(1):45–67

Subject Database of China Plant (Record number: ICP 09112257). Institute of botany, Chinese Academy of Sciences Beijing. Published on the Internet https://www.plant.csdb.cn/ [last accessed 15 Apr 2016] [in Chinese]

De Cáceres M, Legendre P (2009) Associations between species and groups of sites: indices and statistical inference. Ecology 90(12):3566–3574

Dengler J, Löbel S, Dolnik C (2009) Species constancy depends on plot size—a problem for vegetation classification and how it can be solved. J Veg Sci 20:754–766

Deng SF, Yang TB, Zeng B, Zhu XF, Xu HJ (2013) Vegetation cover variation in the Qilian Mountains and its response to climate change in 2000–2011. J Mt Sci 10(6):1050–1062

Dufrêne M, Legendre P (1997) Species assemblages and indicator species: the need for a flexible asymmetrical approach. Ecol Monogr 67(3):345–366

Ellis J, Swift D (1988) Stability of African pastoral ecosystems: alternate paradigms and implications for development. J Range Manag 41(6):450–459

Etzold J, Münzner F, Manthey M (2016) Sub-alpine and alpine grassland communities in the northeastern Greater Caucasus of Azerbaijan. Appl Veg Sci 19 (2):316–335

Friedrich I (2015) Pedologische Charakterisierung in der Quellregion des HeiHe River im Qilian Mountains (Provinz Gansu–China). Bachelor Thesis, Universität Hamburg

Froese M (2012) A phytogeographical analysis of the flora of the Qilian Mountains. Master Thesis, Universität Hamburg

Herrero-Jáuregui C, Oesterheld M (2018) Effects of grazing intensity on plant richness and diversity: a meta-analysis. Oikos 127(6):757–766

HFA (2009) Handbuch Forstliche Analytik. Göttingen. [in German]

Hoppe F, ZhusuiKyzy T, Usupbaev A, Schickhoff U (2016) Rangeland degradation assessment in Kyrgyzstan: vegetation and soils as indicators of grazing pressure in Naryn Oblast. J Mt Sci 13(9):1567–1583

Hothorn T, Hornik K, Zeileis A (2006) Unbiased recursive partitioning: a conditional inference framework. J Comput Graph Stat 15(3):651–674

Huang D, Yu L, Zhang Y, Zhao X (2011) Aboveground numerical characteristics of five natural grassland and their relationships to environmental factors in the northern slopes of the mountains Qilian. Northern J Agric Sci 20:174–180 [in Chinese]

Kent M (2012) Vegetation description and data analysis: a practical approach, 2nd edn. Wiley, Chichester, West Sussex, UK, Hoboken, NJ

Kürschner H, Herzschuh U, Wagner D (2005) Phytosociological studies in the north-eastern Tibetan Plateau (NW China) A first contribution to the subalpine scrub and alpine meadow vegetation. Botanische Jahrbücher Der Systematik 126(3):273–315

Legendre P, Legendre L (2012) Numerical ecology, Third English edition. Developments in environmental modelling, vol 24. Elsevier, Amsterdam

Li Z, Xu Z, Shao Q, Yang J (2009) Parameter estimation and uncertainty analysis of SWAT model in upper reaches of the Heihe river basin. Hydrol Process 23 (19):2744–2753

Lieder E (2013) Direction dependence of soil-water conductivity in high-altitude mountains environment in NW China. Master Thesis, Albert-Ludwigs-University, Freiburg, Germany

Liu Y, Zou S, Chen F (2004) Bioclimatic modeling the spatial distribution of mountain forests in the Qilian Mountains, Northwest of China, using down-scaled climatic models. In: IGARSS 20–24 Sept 2004. IEEE International Geoscience and Remote Sensing Symposium, pp 4625–2628

Liu X, Ma J, Ma ZW, Li LH (2017) Soil nutrient contents and stoichiometry as affected by land-use in an agro-pastoral region of northwest China. CATENA 150:146–153

Lomolino M (2001) Elevation gradients of species-density: historical and prospective views. Glob Ecol Biogeogr 10(1):3–13

McCain CM, Grytnes JA (2010) Elevational gradients in species richness. In: Fullerlove G (ed) Encyclopedia of life sciences, vol 26. Nature Publishing Group, London, p 456

Meiwes KJ, König N, Khanna PK, Prenzel J, Ulrich B (1984) Chemische Untersuchungsverfahren für Mineralböden, Auflagehumus und Wurzeln zur Charakterisierung und Bewertung der Versauerung in Waldböden. Ber. Forschungszentrum Waldökosysteme/-Wald- Sterben 7:1–67 [in German]

Miao F, Guo Z, Xue R, Wang X, Shen Y (2015) Effects of grazing and precipitation on herbage biomass, herbage nutritive value, and yak performance in an alpine meadow on the Qinghai-Tibetan Plateau. PLoS One 10(6)

Miehe G, Miehe S, Kaiser K, Liu J, Zhao X (2008) Status and dynamics of the Kobresia pygmaea ecosystem on the Tibetan Plateau. Ambio 37(4):272–279

Miehe G, Miehe S, Kaiser K, Reudenbach C, Behrendes L, Duo L, Schlütz F (2009) How old is pastoralism in Tibet? An ecological approach to the making of a Tibetan landscape. Palaeogeogr Palaeoclimatol Palaeoecol 276(1–4):130–147

Milchunas D, Lauenroth WK (1993) Quantitative effects of grazing on vegetation and soil over a global range of environments. Ecol Monogr 63:327–366

Miller DJ (2005) The Tibetan Steppe. In: Suttie JM, Reynolds SG, Batello C (eds) Grasslands of the world. FAO, Rome, Italy, pp 305–342

Mueller-Dombois D, Ellenberg H (1974) Aims and methods of vegetation ecology. Wiley, NY

Murtagh F, Legendre P (2014) Ward's Hierarchical agglomerative clustering method: which algorithms implement ward's criterion? J Classif 31:274–295

Nagy L, Grabherr G (2009) The biology of alpine habitats. The biology of habitats series. Oxford University Press, Oxford, New York, 376 p

Oksanen J, Blanchet FG, Friendly M, Kindt R, Legendre P et al (2018) Vegan: community ecology package. Ordination methods, diversity analysis and other functions for community and vegetation ecologists. Version 2.5-1. https://CRAN.R-project.org/package=vegan

Plantarium—illustrated online atlas of plants species of former USSR Republics and adjusted territories Manual of plants. Free assessable uncommercial public Project. Published on the Internet www.plantarium.ru [last accessed 22 Dec 2015] [in Russian]

R Core Team (2015) A language and environment for statistical computing. Foundation for Statistical Computing, Vienna, Austria. https://www.R-project.org/ [last accessed 9 Sept 2015]

Rhode D, Madsen DB, Brantingham JP, Dargye T (2007) Yaks, yak dung, and prehistoric human habitation of the Tibetan Plateau. In: Madsen DB, Chen F, Gao G (eds) Late quaternary climate change and human adaptation in Arid China, pp 205–224

Ruokolainen L, Blanchet G (2014) Introduction to ecological multivariate analysis. https://blogs.helsinki.fi/laruokol/files/2016/10/IEMA.pdf [last accessed 5 July 2018]

Shrestha G, Stahl PD (2008) Carbon accumulation and storage in semi-arid sagebrush steppe: effects of long-term grazing exclusion. Agr Ecosyst Environ 125 (1):173–181

Siegel S, Castellan JN (1998) Nonparametric statistics for the behavioral sciences. McGraw-Hill Inc., New York

Sun F, Lyu Y, Fu B, Hu J (2016) Hydrological services by mountain ecosystems in Qilian Mountain of China: a review. Chin Geogr Sci 26(2):174–187

Suttie JM, Reynolds SG, Batello C (eds) (2005) Grasslands of the world. In: Plant Production and Protection Series, vol 34. FAO, Rome, Italy. https://www.fao.org/docrep/008/y8344e/y8344e00.htm [last accessed 20 Apr 2018]

Tian J, Zhang B, He C, Yang L (2017) Variability in soil hydraulic conductivity and soil hydrological response under different land covers in the mountainous area of the Heihe River watershed, Northwest China. Land Degrad Dev 28(4):1437–1449

von Wehrden H, Hanspach J, Kaczensky P, Fischer J, Wesche K (2012) Global assessment of the non-equilibrium concept in rangelands. Ecol Appl 22 (2):393–399

Wang G, Zhou G, Yang L, Li Z (2002) Distribution, species diversity and life-form spectra of plant communities along an altitudinal gradient in the northern slopes of Qilianshan Mountains, Gansu, China. Plant Ecol 165:169–181

Wang TW, Zhang Z, Li ZB, Li P (2017a) Grazing management affects plant diversity and soil properties in a temperate steppe in northern China. CATENA 158:141–147

Wang Y, Heberling G, Görzen E, Miehe G, Seeber E, Wesche K (2017b) Combined effects of livestock grazing and abiotic environment on vegetation and soils of grasslands across Tibet. Appl Veg Sci 20 (3):327–339

White PJ, Greenwood DJ (2013) Properties and management of cationic elements for crop growth. In: Gregory PJ, Nortcliff S (eds) Soil conditions and plant growth. Blackwell Publishing Ltd., Oxford, pp 160–194

Wildi O (2010) Data Analysis in Vegetation Ecology. A John Wiley & Sons, Ltd.

Xiande L, Chunyun Z, Duolong D, Xiaochun M (2001) Catalogue of medicine plants of Qilian Mountains. Lanzhou Agricultural University Press, Lanzhou [in Chinese]

Yang YS, Zhang L, Li HQ, He HD, Wei YX, Luo J, Zhang GR, Huang YR, Li YN, Zhou HK (2018) Soil physicochemical properties and vegetation structure along an elevation gradient and implications for the response of alpine plant development to climate change on the northern slopes of the Qilian Mountains. J Mt Sci 15(5):1006–1019

Yang G, Xiao D, Zhou L, Tang C (2005) Hydrological effects of forest landscape patterns in the Qilian Mountains. Mt Res Dev 25(3):262–268

Yang B, He M, Melvin T, Zhao Y, Briffa K, Newsom L (2013) Climate control on tree growth at the upper and lower treelines: a case study in the Qilian Mountains, Tibetan Plateau. PLoS ONE 8(7):e69065

Yuan H, Hou F (2015) Grazing intensity and soil depth effects on soil properties in alpine meadow pastures of Qilian Mountain in northwest China. Acta Agriculturae Scandinavica, Section B—Soil Plant Sci 65 (3):222–232

Zar JH (1999) Biostatistical analysis. Prentice Hall, New Jersey

Zech W, Schad P, Hintermaier-Erhard G (2014) Böden der Welt: Ein Bildatlas, 2nd edn. SpringerLink Bücher, Springer Spektrum, Berlin, Heidelberg [in German]

Zemmrich A, Manthey M, Zerbe S, Oyunchimeg D (2010) Driving environmental factors and the role of grazing in grassland communities: a comparative study along an altitudinal gradient in Western Mongolia. J Arid Environ 74:1271–1280

Zhao C, Nan Z, Cheng G, Zhang J, Feng Z (2006) GIS-assisted modelling of the spatial distribution of Qinghai spruce (*Picea crassifolia*) in the Qilian Mountains, northwestern China based on biophysical parameters. Ecol Model 191:487–500

Zhao CY, Peng SZ, Feng ZD (2011) Spatial modelling of the variability of the soil moisture regime at the landscape scale in the Southern Qilian Mountains, China. Hydrol Earth Syst Sci Discuss 6:6335–6358

26 Mountain Habitats Dynamics Under Changing Grazing Management Schemes in Greece

Michael Vrahnakis and Yannis Kazoglou

Abstract

Dry and littoral grasslands, shrublands and traditional silvopastoral systems form a major land use on the mountains of Prespa National Park, northwestern Greece, an area with exceptional biodiversity located on the borders with Albania and North Macedonia. On altitudes ranging from 850 to 2330 m a.s.l. and geological substrates of granite, limestone and metamorphic origins, extensive grazing has been shaping the landscapes of the area, particularly affecting 19 out of the 51 habitat types recorded in the Park and its surrounding area according to the European Unions' Natura 2000 network typology (Directive 92/43/EEC). Over the last years, sheep and goat numbers are drastically reduced, numbers of beef cattle are increasing, and evidence of under- or over-grazing are observed at various sites mainly due to lack or mispositioning of basic infrastructure for grazers. These facts unquestionably affect the dynamics of grazing-dependent habitat types in the study area, including four priority ones, i.e. the "Pseudo-steppes with grasses and annuals of the Thero-Brachypodietea", the "Species-rich *Nardus* grasslands on siliceous substrates in mountain areas", the "Pannonic sand steppes" and the "Endemic forests with *Juniperus* spp.". The present contribution discusses (a) the management practices needed to maintain good conservation status of rangelands in the Park and its surrounding area, and (b) the survey and monitoring techniques to apply in order to identify effects on habitat types caused by the changing grazing schemes.

Keywords

Natura 2000 network · Habitat types · Extensive grazing · Mediterranean ecosystems · Ecological succession · Monitoring · Grassland management · Prespa National Park

26.1 Introduction

Extensive grazing has been practiced for millennia in the Mediterranean ecosystems and, thus, has contributed to the shaping of landscapes from the lower to the higher elevation zones (Papanastasis 1997; Grove and Rackham 2003; Blondel and Aronson 2004; Guarino et al. 2019) and to the creation of cultural elements that have inspired numerous writers from all over the world (e.g. Baumann 1999; Pritchett 2005; Papayannis 2008; Papayannis and Pritchard 2011; Besson 2017 and literature cited therein). Sheep, goats and beef cattle of various

M. Vrahnakis (✉) · Y. Kazoglou
Department of Forestry, Wood Sciences & Design, University of Thessaly, 11&13 V. Griva Str., 43100 Karditsa, Greece
e-mail: mvrahnak@uth.gr

U. Schickhoff et al. (eds.), *Mountain Landscapes in Transition*,
Sustainable Development Goals Series, https://doi.org/10.1007/978-3-030-70238-0_26

autochthonous and improved breeds are the main grazing animals that use the rangelands throughout the whole year in most Balkan countries, while grazers are provided with supplementary feeding mainly in areas with harsh winters, as well as during other periods of the year depending on the condition of the animals and the availability of forage of sufficient quality and quantity taken from grasslands, rangelands with phrygana, shrublands or woodlands. Although not always highly profitable, extensive stockbreeding was and still is a crucial sector of rural economy in Greece and a basic reason for many young or middle-aged people to remain in the family farming businesses and live in rural, often remote, areas located on mountainous sites or islands. For decades, the free-ranging sheep—with or without the supervision of shepherds, a fact primarily determined by the presence or absence of predators respectively—have been the main source of milk for the production of the well-known *feta* cheese, a product of Protected Designation of Origin with increasing demand from global markets.

Apart from its very important economic and social role, extensive stockbreeding also plays a significant environmental role, as a tool to halt ecological succession on habitats, that under no grazing or under grazing with low stocking densities, would evolve to shrubland or woodland, thus decreasing mosaic structure and habitat diversity in Mediterranean ecosystems. Habitat types of the European Union are classified under a common methodology for all 28 member states according to phyto-sociological characteristics; the list includes 233 habitat types (71 of which identified as priority habitat types) belonging to nine broad categories of terrestrial, coastal and marine ecosystems (European Commission 2013), while additional habitat types have been identified as important at national levels. Habitat types of European interest along with other important biodiversity elements—except birds—are the basic components of Sites of Community Interest, which form a major part of the "Natura 2000" network of protected sites according to the Habitats Directive (92/43/EEC). The Natura 2000 network is complemented by Special Protection Areas which have been designated as protected areas according to the Birds Directive (2009/147/EC). Grazing affects the vegetation of particular habitat types belonging to the categories of Temperate Heath and Scrub, Sclerophyllous Scrub (Mattoral), Natural and Semi-Natural Grassland Formations, Raised Bogs and Mires and Fens, and Forests. It is worth noting that in international literature grazing-dependent habitat types or habitat types used as pastures are defined as "semi-natural" (e.g. semi-natural grasslands) to indicate the man-induced management practice taking place through extensive stockbreeding. However, in the case of Mediterranean ecosystems, the term "semi-natural" is not readily accepted or used by many native (coming from Mediterranean countries) authors because of the long-term presence and influence of pastoral, often nomadic, stockbreeding on those habitats and ecosystems, a fact that, in their opinion and approach, defines grazing as a "natural" process; thus, such rangeland habitat types are often not necessarily named "semi-natural".

In Prespa National Park and its surrounding area (51,231 ha), grazing positively affects 19 out of the total 51 habitat types recorded in the four Natura 2000 sites of the total area and adjacent areas (Vrahnakis et al. 2011; Tsitoura et al. 2015), which all together constitute the geographic area of the Municipality of Prespa, NW Greece (Figs. 26.1 and 26.2). The present contribution aims at (a) discussing the management practices needed to maintain the rangelands of the Prespa area in good conservation status, and (b) presenting the survey and monitoring techniques that should be applied in order to identify the effects on habitat types caused by the changing grazing schemes observed over the last years in the study area.

Fig. 26.1 Sheep and goats grazing on rangelands of the habitat types 61A0 and 4090 of Mt Triklario (Vrahnakis 2015-08-06)

Fig. 26.2 The Prespa basin (foreground: Lake Mikri Prespa, background: Lake Megali Prespa & mountains of N. Macedonia) photographed from the peak of Mt Triklario, 6170 habitat type (Kazoglou, 2015-12-07)

26.2 Biodiversity in Prespa National Park, Greece

Prespa National Park and its surroundings within Greece is an area of remarkable biodiversity, the basic elements of which can be summarized in the following points:

- 51 habitat types, of which 33 are included in Annex I of the Habitats Directive with five (5) of them listed as priority habitat types, 16 are of national importance, and two (2) are mixed habitat types (Vrahnakis et al. 2011; Fotiadis et al. 2014; Tsitoura et al. 2015).

- 1847 plant taxa (Pavlides 1985; Pavlides 1997a, b; Fotiadis and Kazoglou 2010; Strid et al. 2017; Sakellarakis et al. 2019).
- 263 macrofungi species (Svetasheva et al. 2019; Kazoglou unpublished data).
- 279 bird species (62% of the species totally recorded in Greece), of which 143 regularly breed in the Park, including the Dalmatian pelican (*Pelecanus crispus*) with its globally largest colony established in the reedbeds of Lake Mikri Prespa, as well as species of grassland ecosystems, such as the eastern race Greylag geese (*Anser rubrirostris*), which breeds in reedbeds and feeds on littoral wet grasslands (Catsadorakis 1997; Portolou et al. 2009; Kazoglou and Bousbouras 2010) and the Corncrake (*Crex crex*), which was first recorded in June 2011 at highland dry grasslands in the eastern sector of the Park (Bonetti-Bousbouras-Gletsos-Kazoglou unpublished data, Society for the Protection of Prespa 2019a).
- 61 mammals (Catsadorakis 1995; Bousbouras and Kazoglou 2010; Papadatou et al. 2011; Society for the Protection of Prespa 2019b, Theodoroglou pers. comm.), of which 27 are bat species often feeding on sites used for livestock grazing (Galand et al. 2010; Vrahnakis et al. 2010; Papadatou et al. 2011, 2013).
- More than 163 butterfly species (Pamperis 2007).
- 24 fish species, of which nine (9) are endemic (Crivelli et al. 1997; Spirkovski et al. 2012; Society for the Protection of Prespa 2019c).
- 11 amphibian species and 22 reptile species (Bousbouras and Ioannidis 1997; Ioannidis and Bousbouras 1997; Bousbouras 2010).
- One (1) autochthonous bovine population belonging to the wider Greek shorthorn (*brachyceros*) or Balkan "Busha" groups (Georgoudis 1993; Grunenfelder 2006; Medugorac et al. 2009; Kazoglou et al. 2010) which – according to recent findings based on DNA profiling—seems to be genetically fairly distant from most similar populations in Greece or other Balkan countries, suggesting that it may form a different breed (Bizelis pers. comm.).

This significant biodiversity originates from the complex nature of the wider Prespa area in the three neighboring countries, declared a transboundary protected area in 2000, which includes (i) two large and ancient lakes, i.e. Mikri/Lesser Prespa Lake which is shared between Greece and Albania and Megali/Greater Prespa Lake (shared by North Macedonia, Albania and Greece), (ii) geological substrates of granite, limestone and metamorphic origin, (iii) a basin with different land uses and mosaic of habitats at altitudes ranging from 850 (Greater Prespa Lake) to 2601 m above sea level (Pelister peak, North Macedonia) connected to the Adriatic Sea through Ohrid Lake and River Drin, (iv) remoteness, and (v) relatively low impact to nature by human activities, that mainly consist of agriculture (bean cultivation in the Greek part, apple orchards in North Macedonia, cereals in Albania), livestock breeding (sheep, goats and beef cattle) mostly in the Greek and Albanian parts, fishing, forestry, few secondary sector activities and tourism (Hollis and Stevenson 1997; Catsadorakis and Malakou 1997; Perennou et al. 2009; Giannakis et al. 2010; Papadatou et al. 2011). Tourism is a fairly promising sector for the improvement of local economies and promotion of sustainable development on all three sides of transboundary Prespa provided that strong political decisions will prevail in the near future, especially after the signing of the "Prespa Agreement", June 17th 2018, between Greece and North Macedonia, which resolves many problems between the two countries and facilitates transboundary collaboration.

26.3 Livestock Grazing Effects on Rangeland Habitat Types

Rangelands in the study area (21,581.31 ha) form the largest land use category (42.12% of the total study area) as they include (a) grasslands (11,526.35 ha, 22.50% of the total study area), (b) shrublands (6841.45 ha, 13.35%), and (c) open-canopy wood pastures (3213.51 ha, 6.27%), followed by forests (32.62%), lakes (14.50%), farmland (8.23%), wetlands (2.07%)

and settlements (0.46%) (Tsitoura et al. 2015). From the total number of 51 habitat types recorded in the National Park and adjacent areas in the Greek part of Prespa, 41 are of low to high interest for livestock grazing (Tsitoura et al. 2015; Vrahnakis et al. 2018), while the conservation of 19 habitat types highly or moderately depends on grazing management (*the following list includes code and name of each habitat type, asterisks (*) indicate priority habitat types, (–) indicates habitat types of national importance not included in the EU Habitats Directive 92/43/EEC*):

(1) 4060 Alpine and Boreal heaths.
(2) 4090 Endemic oro-Mediterranean heaths with gorse.
(3) 5110 Stable xerothermophilous formations with *Buxus sempervirens* on rock slopes (*Berberidion* p.p.).
(4) 5130 *Juniperus communis* formations on heaths or calcareous grasslands.
(5) 5160 (–) South-eastern sub-mediterranean deciduous thickets (Schilbjak).
(6) 5210 Arborescent matorral with *Juniperus* spp.
(7) 5340 (–) Eastern Garrigues.
(8) 6170 Alpine and subalpine calcareous grasslands.
(9) 62A0 Eastern sub-mediteranean dry grasslands (*Scorzoneratalia villosae*).
(10) 6220 * Pseudo-steppe with grasses and annuals of the *Thero-Brachypodietea*.
(11) 6230 * Species-rich *Nardus* grasslands, on siliceous substrates in mountain areas (and submountain areas, in Continental Europe).
(12) 6260 * Pannonic sand steppes.
(13) 6290 (–) Mediterranean subnitrophilous grasslands.
(14) 6420 Mediterranean tall humid herb grasslands of the *Molinio-Holoschoenion*.
(15) 6430 Hydrophilous tall herb fringe communities of plains and of the montane to alpine levels.
(16) G645 *Greek hyper-Mediterranean humid grasslands*.
(17) 91M0 Pannonian-Balkanic turkey oak-sessile oak forests (in their open-canopy woodland forms combined with 62A0 grassland vegetation in the understorey).
(18) 9250 *Quercus trojana* woods.
(19) 9562 * Endemic forests with *Juniperus* spp.
(20) Sixteen of the habitat types listed above highly depend on livestock grazing to maintain good conservation status, while only three habitat types (i.e. 5160, open-canopy wood pastures of 91MO and 9250) are moderately grazing-dependent.

Alpine and Boreal heaths (4060) cover an area of 346 ha at elevations of 1890–2000 m and inclinations of 10–60% (Fig. 26.3). They mainly consist of *Vaccinium myrtillus*, *Juniperus communis* ssp. *nana* and *Chamaecytisus* spp., while forbs such as *Trifolium parnassi*, *Geum coccineum* and *Geum montanum* complement the floristic composition of the habitat type (Vrahnakis et al. 2011; Tsitoura et al. 2015). The intensity of livestock grazing has decreased over the last years due to the decrease of transhumant sheep and goat herds grazing on Mt Varnous (Kazoglou 2011) and to the fact that cattle do not seem to spend much time grazing on these heaths; thus, encroachment of tree species is possible at sites located close to beech forests, while fire also becomes a threat as dry plant biomass is accumulated on ungrazed sites (Vrahnakis et al. 2011).

Endemic oro-Mediterranean heaths with gorse (4090) are recorded in the southern part of the study area (Mt Triklario, 2361 ha, 1310–1570 m, 5–40% inclinations) on limestone soils of very low depths. The dominant plant taxa are *Anthyllis vulneraria*, *Prunus prostrata*, *Eryngium amethystinum* and *Stipa pennata*, while their floristic composition includes more than 50 rare and endemic plant taxa such as *Onosma visianii*, *Erodium guicciardii*, *Hypericum rumeliacum*, *Sideritis raeseri* subsp. *raeseri*, *Astragalus lacteus* and *Onobrychis alba* subsp. *calcarea* (Tsitoura et al. 2015). Moderate sheep and goat grazing along with cattle grazing is the driving conservation factor for these heaths as well as for the adjacent grasslands belonging to the 62A0 and 6170 habitat types, mainly as a

Fig. 26.3 Sheep grazing on heaths of the 4060 habitat type on Mt Varnous (2000 m). *Nardus* grasslands (6230*) and beech forests can be seen in the background (Kazoglou, 2014-08-14)

tool for the control of bushes at specific sites (Vrahnakis et al. 2011). Based on field research carried out in 2015 and 2016, above-ground forage production of these heaths was found equal to 3865 kg/ha (Kazoglou et al. 2019).

Stable xerothermophilous formations with Buxus sempervirens on rock slopes (Berberidion p.p.) (5110) are a nationally rare habitat type (present only at another two areas in the country), situated on limestone substrates of the western part of the Park (740 ha) at elevations of 880–1090 m and inclinations ranging from 10 to 90%. The main floristic components of these shrublands include *Buxus sempervirens*, *Prunus prostrata*, *Teucrium chamaedrys, Poa bulbosa*, *Sideritis montana*, while species of special interest are *Minuartia attica*, *Verbascum epixanthinum* and *Neotinea tridentata* (Tsitoura et al. 2015). Goat grazing and pruning, on which conservation of the habitat type depends (Vrahnakis et al. 2011), are decreasing over the last years and their continuation should clearly be promoted in the forthcoming management plan to avoid encroachment by broadleaved tree species and increase of woodland density.

Juniperus communis formations on heaths or calcareous grasslands (5130) is a habitat type first recorded by Fotiadis and Kazoglou in 2015 which covers a very small area of 9 ha in the southern part of the study area on Mt Malimadi (Kazoglou 2015), which the third record of the habitat type in Greece (Dafis et al. 2001). Its floristic composition in the study area consists of seven (7) grass species, three (3) legumes including *Astragalus angustifolius*, 16 broad-leaved species (forbs) including *Alyssum montanum*, *Eryngium amethystinum* and *Thymus longicaulis*, as well as two (2) perennial shrubs namely *Crataegus* sp. and *Juniperus communis* (Kazoglou 2015).

South-eastern sub-mediterranean deciduous thickets (Schilbjak) (5160) are mainly located on soils of granite origin in the eastern part of the Park at altitudes of 860–1520 m with mild slopes (0–10%), covering a total area of 848 ha. They are characterized by the presence of *Prunus cocomilia*, *Rosa canina*, *Rubus* spp., *Potentilla argentea* and *Cynosurus echinatus*, as well as *Linaria peloponnesiaca, Dianthus viscidus, Dianthus stenopetalus* and *Minuartia verna* and represent the last stage of succession before the appearance of forests and, thus, its conservation highly depends on grazing (Vrahnakis et al. 2011; Tsitoura et al. 2015). However, with the decreasing grazing pressure caused by browsers (in this case goats and to a lesser extent free-ranging horses i.e. approximately 80 individuals on Mt Varnous) and the low grazing pressure caused by cattle, it is expected that their physiognomy will deteriorate in the near future.

Arborescent matorral with Juniperus spp. (5210) is a habitat type first recorded in the study area by Fotiadis and Kazoglou in 2015 which covers an area of 85 Ha on Mt Malimadi i.e. the southern part of the study area. Its floristic composition consists of five (5) grasses, five (5) legumes including *Lotus corniculatus* and *Trifolium campestre*, 28 forbs including *Acinos alpinus, Alyssum montanum, Hieracium hoppeaneum* and *Potentilla inclinata*, and three (3) perennial shrubs namely *Crataegus* sp., *Juniperus communis* and *Juniperus oxycedrus* (Kazoglou 2015).

Eastern Garrigues (5340) is a typical Mediterranean habitat type covering a very small area (<1 ha,970 m) in the centre of the study area, which is very likely to evolve to *Quercus* woodlands of the 91M0 type due to succession and decreasing sheep and goat grazing pressure. Its characteristic species are *Cistus creticus, Teucrium capitatum, Eryngium campestre, Hippocrepis emerus subsp. emeroides* and *Hypericum rumeliacum* (Vrahnakis et al. 2011; Tsitoura et al. 2015).

Alpine and subalpine calcareous grasslands (6170) are located in the southern part of the study area (420 ha) at elevations ranging from 1510 to 1760 m and inclinations up to 80% (Fig. 26.2). *Astragalus angustifolius, Carex kitaibeliana, Sideritis raeseri, Eryngium amethystinum, Inula oculus-christi* and *Anthyllis vulneraria* ssp. *rubriflora* are the main floristic components of the habitat type (Vrahnakis et al. 2011), which also hosts plant taxa of special interest such as *Hypericum rumeliacum, Sideritis raeseri* subsp. *raeseri, Anthyllis vulneraria* subsp. *bulgarica, Astragalus lacteus, Onobrychis alba* subsp. *calcarea* and *Minuartia attica* (Tsitoura et al. 2015). These grasslands are very important to domestic grazers, especially to beef cattle and, secondarily, to sheep and goats. Grazing needs to be adjusted in space to attract herds on presently undergrazed sites and reduce grazing pressure on currently overgrazed sites.

Eastern sub-mediteranean dry grasslands (Scorzoneratalia villosae) (now coded 62A0, formerly listed as "*6210 Semi-natural dry grasslands and scrubland facies on calcareous substrates (Festuco-Brometalia) (* important orchid sites)*") is the largest, in terms of surface area, grassland resource of the study area with 9302 ha. These typical semi-natural grasslands are situated on all soil substrates at altitudes of 850–1940 m and inclinations of 0–65%. The dominant plant taxa are *Eryngium campestre, Festuca valesiaca, Achillea nobilis, Juniperus oxycedrus, Trifolium scabrum* and *Trifolium arvense*, while taxa of special interest include *Linaria peloponnesiaca, Dianthus deltoides, Erysimum microstylum, Hypericum rumeliacum, Phelypaea boissieri* (only on the peak of Mt Devas and occasionally on other sites of the same mountain, Fig. 26.4), *Erodium guiciardii, Dactylorhiza sambucina, Minuartia attica, Dianthus pinifolius, Dianthus viscidus, Dianthus stenopetalus, Minuartia verna, Campanula spatulata* subsp. *spatulata, Cynoglottis barrelieri* subsp. *serpentinicola* and *Viola orphanidis* (Tsitoura et al. 2015). On undergrazed sites, *Juniperus* spp. and broadleaved shrubs spread at the expense of 62A0 grasslands, an issue that requires proper grazing management and, possibly, other means (e.g. selective logging) to be resolved. Above-ground forage production in these grasslands was calculated at 3575 kg/ha (Kazoglou et al. 2019).

Pseudo-steppe with grasses and annuals of the Thero-Brachypodietea (6220*) are a priority habitat type covering a total area of 355 Ha at five localities of the Prespa Lakes watershed. The 6220* grasslands are very rich in species composition (e.g. *Bromus rubens, Arenaria leptoclados, Bromus hordeaceus, Dasypyrum villosum, Trifolium arvense, Logfia arvense, Poa bulbosa, Silene conica* as well as *Erysimum microstylum, Linaria peloponnesiaca, Dianthus pinifolius, Lilium candidum* and *Minuartia verna*) (Tsitoura et al. 2015), but at the same time threatened by (a) overgrazing at specific parts of its range mainly because of cattle grazing on loose and friable soils at high inclinations resulting in the creation of crevices and bare-soil

Fig. 26.4 The rare plant taxon *Phelypaea boissieri* on the dry grasslands of the 62A0 habitat type, Mt Devas, western sector of Prespa National Park (Kazoglou, 2010-05-22)

paths, or (b) undergazing especially at grassland fringes surrounded by expanding oak woods or at sites where sheep and goat grazing has decreased, which leads to the spreading of shrubs such as *Pyrus amygdaliformis*, *Juniperus oxycedrus* and *Quercus pubescens* (Vrahnakis et al. 2011).

Species-rich Nardus grasslands, on siliceous substrates in mountain areas (and submountain areas, in Continental Europe) (6230*) are situated at high altitudes (1730–2060 m) of the eastern sector of the National Park covering a total area of 1463 ha (Fig. 26.5). The dominant plant taxa are *Nardus stricta, Bellardiochloa variegata* and *Festuca* spp., but many rare and higly interesting species are also found in these grasslands such as *Cerastium banaticum* subsp. *speciosum, C. decalvans, C. rectum, Dianthus deltoides* subsp. *degenii, D. integer* subsp. *minutiflorus, D. myrtinervius, D. pinifolius* subsp. *lilacinus, D. stenopetalus, Herniaria parnassica, Bruckenthalia spiculifolia, Trifolium parnassi, Geum coccineum, G. montanum, Lilium carniolicum* subsp. *albanicum, Crocus cvijicii, C. pelistericus, C. sieberi* subsp. *sublimis, C. veluchensis, Festuca koritnicensis, Viola tricolor* subsp. *macedonica, V. eximia, V. orphanidis* and *Gentiana lutea* (Tsitoura et al. 2015), which was considered extinct (Kavadas 1956; Pavlides 1985, 1997a) until a few individuals and an important population were re-discovered by two naturalists in 2010 (Daikopoulos and Zografou pers. comm.) and the present authors in 2015–16 respectively (Fig. 26.6), on different localities of Mt Varnous (Strid et al. 2017). *Nardus* grasslands are highly significant for livestock grazing in late summer—autumn, but at the same time threatened by (a) reduction of grazing at specific sites which leads to the expansion of *Juniperus communis* ssp. *nana* and *Vaccinium myrtillus* shrubs, (b) overgrazing at other sites, especially in the lower altitudes of their range, which favours *Chamaecytisus* spp. and may lead to complete change of their floristic composition, and (c) opening of new or widening of existing dirt roads that cause direct habitat loss and erosion (Vrahnakis et al. 2011). Above-ground forage production in *Nardus* grasslands was calculated at 7900 kg/ha, which is the second highest productivity amongst the 11 habitat types examined by Kazoglou et al. (2019).

Pannonic sand steppes (6260*) are a habitat type reported for the first time in Greece and meet their southern-most occurrence on Prespa

Fig. 26.5 *Nardus* grasslands (habitat type 6230*) on their highest locality on Mt Varnous (~2100 m). with a small patch of 4060 heaths (Vrahnakis, 2010-05-23)

Fig. 26.6 The rare plant taxon *Gentiana lutea* on the dry grasslands of the 6230* habitat type, Mt Varnous, eastern sector of Prespa National Park (Kazoglou, 2015-10-20)

National Park; they are represented by six (6) vegetation units found on a sandy islet on the northern part of Lake Mikri Prespa and on sand dunes situated on the alluvial strip of land separating Lake Mikri from Lake Megali Prespa (Fotiadis et al. 2014) covering in total 167 ha (Vrahnakis et al. 2011). The dominant plant taxa of the habitat type are *Cruciata pedemontana, Rumex acetosella, Poa bulbosa, Trifolium arvense, Linaria genistifolia, Vulpia myurus, Filago arvensis, Silene conica, Erysimum microstylum, Bromus rubens, Avena sterilis, Eryngium campestre* and *Hypericum perforatum* (Tsitoura et al. 2015). Although not very productive or important for livestock grazing, their conservation largely depends on moderate grazing, which will inhibit the establishment of closed grass swards and dense moss carpets, but also avoid over-enrichment of soils with nutrients that might facilitate the establishment of plant species of other habitat types (Fotiadis et al. 2014). Above-ground forage production of Pannonic sand steppes was found equal to 2345 kg/Ha (Kazoglou et al. 2019).

Mediterranean subnitrophilous grasslands (6290) cover a total area of 231 Ha in the study area mostly at elevations of 850–1250 m and, often, on abandoned fields. They are dominated by annual grasses and various forbs including *Convolvulus arvensis, Hordeum murinum, Medicago sativa, Plantago lanceolata* and *Eryngium campestre* (Tsitoura et al. 2015). Their good conservation status depends on the continuation of moderate livestock grazing and other practices halting succession such as mowing followed by baling.

Mediterranean tall humid herb grasslands of the Molinio-Holoschoenion (6420) represent the semi-natural habitat type usually referred to as "wet grasslands" or "wet meadows" that occupy a total area of 120 Ha (Vrahnakis et al. 2011) in the littoral zone of Lake Mikri Prespa at sites with inclinations less than 1% and elevations of 853–855 m, on soils temporarily flooded following the fluctuations of the lake water level (less than 1 m over the last years) with maxima recorded in late spring and minima in late autumn. They are situated (a) between the reedbeds (habitat type 72A0) fringing the lake and drier habitats including farmland, and (b) adjacent to large sedge communities (72B0) and meadows of the 6450 habitat type. Their conservation directly depends on traditional farming activities taking place in the spring–summer-early autumn period, namely grazing, mechanical mowing or mowing with aftermath grazing which prevent encroachment of aggressive high emergent helophytes on the lower elevations and establishment of woody plants on the drier and higher parts of wet grasslands. Wet grasslands are very significant for biodiversity as they sustain high plant diversity (e.g. *Lythrum salicaria, Galium palustre, Alisma plantago-aquatica, Mentha aquatica, Lycopus europaeus, Carex* spp., *Sparganium erectum, Utricularia vulgaris* and *Salvinia natans*), offer key feeding sites for many bird species and spawning grounds for fish, and support plenty of invertebrates and amphibians at various parts of their life cycles (Crivelli et al. 1997; Kazoglou et al. 2004, 2008; Kazoglou 2007; Tsitoura et al. 2015). Above-ground forage production in these wet grasslands was calculated at 6370 kg/Ha, which is the third highest productivity amongst the 11 habitat types examined by Kazoglou et al. (2019).

Hydrophilous tall herb fringe communities of plains and of the montane to alpine levels (6430) are small wetlands situated on Mt Varnous at elevations of 1400–1950 m on a total area of 113. Dominant and characteristic plant taxa of these communities include *Geum coccineum, Deschampsia cespitosa, Doronicum austriacum, Myosotis sylvatica, Veratrum album, Cynosurus cristatus, Alchemilla acutiloba, Silene asterias, Viola orphanidis, Cirsium appendiculatum* and *Linaria peloponnesiaca* (Tsitoura et al. 2015). Grazing is very important for their conservation as it controls the growth of hydrophilous shrubs and trees, however, as these communities are used by cattle for drinking water and grazing (especially in summer—autumn), overgrazing should be avoided to minimize impacts caused by heavy trampling of the soils (Vrahnakis et al. 2011). Above-ground forage production in these communities was found equal to 8015 kg/ha, which is the highest productivity amongst the 11 habitat types examined by Kazoglou et al. (2019).

Greek hyper-Mediterranean humid grasslands (G645) appear beside the 6420 wet grasslands on the littoral zone of Lake Mikri Prespa or at higher elevations (855–1400 m), on flat sites, reaching a total area of 159 ha in the National Park. They include meadows with significant floristic elements such as *Narcissus poeticus* (at two localities) and *Limodorum abortivum* (Tsitoura et al. 2015), while pressures affecting their conservation status include undergrazing or overgrazing, expansion of agriculture and modifications of hydrological regime (Vrahnakis et al. 2011). Above-ground forage production in these meadows was calculated at 5170 kg/ha (Kazoglou et al. 2019).

Pannonian-Balkanic turkey oak- sessile oak forests (now coded 91M0, formerly listed as "*924A (–) Eastern white oak woods and balcanic thermophilous oak woods*") cover a great proportion of the study area non- or lightly-grazed forests (860–1370 m, 0–90% inclinations, 11,537 ha), but are presented here for their smaller part (~5%) that is characterized by open-canopy woodlands with canopy cover 25–60% and understorey vegetation of the dry grassland habitat type 62A0. Their floristic composition includes *Quercus pubescens*, *Q. frainetto*, *Q. petraea* ssp. *medwediewii*, *Q. cerris*, *Potentilla micrantha*, *Fraxinus ornus*, *Helleborus odorus* ssp. *cyclophylus* and *Luzula forsteri* (Tsitoura et al. 2015). These woodlands are readily grazed by domestic herbivores at all seasons of the year for their herbs, oak leaves and acorns, and are of great value for grazers in mid-late summer when their shade sustains greener herb layer compared to non-wooded pastures. Above-ground forage production in these open-canopy wood pastures was calculated at 2370 kg/ha (Kazoglou et al. 2019), while the same research revealed that the same parameter for closed-canopy woods of the 91M0 habitat type was found equal to 990 kg/ha (the lowest productivity amongst the 11 habitat types examined by Kazoglou et al. (2019)), indicating significant difference between the two forms of the same habitat type.

Quercus trojana woods (code 9250) cover an area of 177 ha, at elevations of 1040–1350 m and inclinations of 5–30%, mostly sited over limestone of the west-northwest aspects of Mts Devas and Vrondero (Vrahnakis et al. 2011), as well as Mt Malimadi (Tsitoura et al. 2015). They usually form open stands of small crown cover in the periphery of *Q. pubescens* woods. Besides *Q. trojana*, dominant plant taxa include *Thymus sibthorpii, Fraxinus ornus, Festuca valesiaca, Q. pubescens* and *Dactylis glomerata*, while taxa of special interest are *Trifolium pignantii, Helleborus odorus* subsp. *cyclophyllus, Lilium carniolicum* and *Dianthus cruentus*. The limited distribution of *Q. trojana* subsp. *trojana* in western Greece requires the conservation and recovery of its few clusters (Dafis et al. 2001). The major threat to these woods is their dynamics, which will gradually lead to the establishment and dominance of other species, such as *Q. pubescens*. *Quercus trojana* woods serve as wood pastures, as livestock grazing is a traditional activity that shapes the characteristic physiognomy of the woods. Tree management also includes the firewood and acorn collection, and the pruning of branches as fodder for livestock during the winter time when green forage is scarce. The absence of grazing and its associated man-imposed activities will lead to the loss of the habitat type, and its replacement by the dense forms of type "91M0 Pannonian-Balkanic turkey oak- sessile oak forests" (Vrahnakis et al. 2011).

The *Grecian juniper woods* (9562*) are considered extremely rare for the EU and are therefore considered a priority habitat type in accordance with Directive 92/43 /EEC, while according to the Natura 2000 typology the code 9562 corresponds only to the woods of Prespa (2192 ha). More than 300 plant species have been recorded in these woods until now; many of them are endemic or rare species, such as *Cynoglottis barrelieri* subsp. *serpentinicola*, *Lilium candidum*, *L. chalcedonicum* (Giannakis et al. 2010). The typical species of Grecian juniper woods are *Juniperus excelsa* and *J. foetidissima*, and the Prespa National Park is the western limit of the species natural range. According to Vrahnakis et al. (2011), Grecian Juniper forests are found in an intermediate succession stage; they tend to replace grasslands or *Buxus sempervirens* shrublands after termination of livestock grazing and logging and they are replaced by deciduous broad-leaved forests (various oaks, oriental hornbeam etc.) as observed in the southern part of the Lesser Prespa watershed. Thus, the habitat type highly depends on livestock grazing since animals can control the natural succession which takes place by sexual or asexual propagation, the expansion and finally domination of broadleaved wood species. Overall, the restoration of livestock grazing emerges as a great goal to avoid habitat loss of "9562 * Endemic forests with *Juniperus* spp." due to natural succession. Above-ground forage production of juniper wood pastures was calculated at 2135 kg/ha, which is the second lowest productivity amongst the 11 habitat types examined by Kazoglou et al. (2019).

26.4 Land Use—Grazing Management Scheme Changes

As clearly reported in the previous chapter, an important change affecting conservation management of grazing-dependent habitat types in the study area is related to the decreasing numbers of sheep and goats, as well as to the cessation of traditional activities that control the excessive expansion of woody species at the expense of grasslands, such as pruning and clearing. Besides our own knowledge of the study area over the last 22 years and unpublished data, substantial evidence on the increasing density of forests and their expansion over habitat types of earlier succession stages, namely rangelands, over the last decades, as well as gradual changes in the livestock capital of the area comes from various sources. Pyrovetsi and Karteris (1986) report that patches of rangelands dispersed among forest slopes in 1945 had changed to forests by 1984; Raus also mentions that forests in the area become denser at the expense of fringing (rangeland) openings (Phitos et al. 1995), while Catsadorakis and Malakou (1997) also mention that after 1990 forests became denser and expanded onto abandoned fields (most probably used as pastures by livestock). The same authors provide detailed data on livestock numbers in Prespa (which, in 1997, was smaller than today as it did not include the Community of Krystallopigi) from which a 57% reduction in livestock units was recorded from 1964 to 1993 (7652 and 3301 units respectively) and was mainly attributed to the decrease in sheep, local cattle breeds and equids numbers. Coming to more recent data on livestock numbers in the whole study area, Tsitoura et al. (2015) report decreasing numbers of sheep and goats from 2000 to 2015 (minus 10,760 heads i.e. −52%, minus 3661 heads i.e. −48%, respectively) and concurrent doubling of beef cattle (from 1033 to 2164 heads), trends which seem to continue until nowadays (2019) as some sheep and goat breeders sold their stock and abandoned the business, while a few others replaced their sheep and goats with beef cattle; at the same time transhumant sheep pastoralism is still practiced on Mt Varnous by only one stockbreeder (700 sheep) coming from Thessaly (Kazoglou, unpublished data) compared to the seven families (with more than 2000 heads) that used the same grasslands until 1997 (Kazoglou 2011). It is worth mentioning that the basic reason for the recent abandonment of sheep breeding lies with the very low prices for milk in 2017–2018, although it is the basic ingredient for the production of the famous *feta* cheese, a fact that—under thorough market conditions—should press for increasing the livestock and improve the breeding conditions. However, this is not the only problem of livestock breeding in remote areas, therefore many families either choose a relatively easier breeding sector, such as beef cattle, or decide to totally change their lives and professional activities, often starting by moving to urban areas.

The ongoing decrease in numbers of small ruminants and the simultaneous increase of cattle of improved breeds, along with the observed over-expansion of woody plants on grasslands of the study area, are of vital importance for the near-future management of rangelands. Increasing stocking density by heavy cattle, often on friable soils and rangelands situated on steep slopes, as well as decreasing browsing by goats on rangelands that contain woody plant species, in combination with lack of basic infrastructure (e.g. water troughs and sheds) that would favour the dispersal of livestock grazing, are the main issues to tackle, especially in an area with such high biodiversity.

26.5 Rangeland Conservation Projects in the Study Area

The high biodiversity of Prespa National Park and the related Natura 2000 sites has played a crucial role for the acquisition of EU, other international and national funds for the elaboration of necessary studies and researches, as well as for the implementation of successful nature conservation projects in the area since the early 90s. The most important studies and researches dealt with the institutional and management aspects of the National Park and the transboundary Prespa Park, threatened wildlife species and necessary monitoring activities. Some of the projects focused on rangeland ecosystems and were supported by the EU LIFE-Nature mechanism. One of them dealt with the restoration of wet meadows in the littoral zone of Lake Mikri Prespa, as major feeding grounds for bird species of EU concern. It was implemented in 2002–2007, had a budget of approx. 1.9 million euro, and was awarded by the European Commission as one of the five "Best of the Best" projects of those completed in 2007–8 (398 in total). The major achievements of the project, in brief, included (i) the construction of a new structure (sluice-gates-bridge) to control the water level of Lake Mikri Prespa so that wet meadows can be flooded in spring and its outflows to Lake Megali Prespa, (ii) management of littoral vegetation by means of water buffalo and cattle grazing, summer mowing and combination of the two practices, which resulted in the tripling of wet grasslands to 100 ha, (iii) systematic vegetation and bird monitoring activities, the latter proving —among other accomplishments (mentioned by Kazoglou (2011))—that the Glossy Ibis (*Plegadis falcinellus*) came back to breed in the area after 35 years of no breeding, and (iv) production of a management plan for the wet grassland and reedbed habitats of the lake, that set the basis for all management action after 2008.

Another LIFE-Nature project was implemented in 2013–2017 (budget approx. 1.05 million euro) and focused on the priority habitat type "9562* Grecian juniper woods", one of the most important in Prespa National Park due to its rarity and high biodiversity. The project actions tackled several pressures and threats to these woods, which stem from changes in management, such as reduced grazing and abandonment of traditional livestock and forestry practices (e.g. branching). The main pressures and threats dealt with the expansion of the most competitive woody broadleaved species, the limited regeneration of Grecian juniper (*Juniperus excelsa*) and foetid juniper (*J. foetidissima*) and the increased risk of fire. Project actions aimed at (a) eliminating the causes of degradation, and (b) contributing to the restoration of the habitat type conservation status. To achieve the objectives, concrete actions were implemented to control the broadleaved woody species, to improve the natural regeneration of junipers of interest and to facilitate extensive stockbreeding in the project area: (i) a series of forest clearing operations took place from spring 2015 to spring 2017, covering an area of 208.6 ha (Logotheti 2017), (ii) controlled grazing was applied for approximately 5 weeks with 650 goats and 19 weeks with 200 goats, (iii) artificial re-establishment of Grecian juniper juveniles in places where natural regeneration was very difficult, (iv) garbage collection and management, and (v) construction of a new open water storage tank and maintenance of an older one to promote extensive grazing and cause grazing dispersal. Early monitoring data collected in 2016 and 2017 indicated that the status of the habitat type has improved as (a) the typical species *J. excelsa* and *J. foetidissima* dominated the shrub- and tree-canopy layers; (b) the presence of other typical species, such as *Cynoglottis barrelieri* subsp. *serpentinicola*, *Silene graeca*, *Goniolimon dalmaticum*, *Thalictrum minus*, *Caucalis platycarpos* and *Anthyllis vulneraria* subsp. *rubriflora* had increased, and (c) the structure and functions of the habitat type had improved. The combination of forestry operations and re-introduction of grazing significantly reduced the pressure from broadleaved species and the threat of fire on the habitat type (Fotiadis et al. 2018).

Finally, important boost for comprehensive management of habitat types of Prespa National Park and adjacent areas that are associated to livestock grazing was given after the production of the Final Grazing Management Plan for the Municipality of Prespa (Tsitoura et al. 2015), a study that was funded by EU and national funds. According to the Plan, the study area was divided into 50 land parcels (Rangeland Units) where vegetation is represented by 51 habitat types; 41 of which are of interest for grazing, while 19 fulfill the criteria (institutional, ecological) for the management of farm animals as the conservation of these habitat types depends to a large or moderate extent on grazing. The Plan includes the spatial and temporal organization of grazing in full details (with proposed technical works up to 1.5 million euro), and it was designed on the basis of the equilibrium of stocking density to grazing capacity. For the first time, the designation of Rangeland Units on the basis of habitat types offers their institutional protection and appears as a great challenge for the grazing management of other areas to fulfill the principle of protection (Vrahnakis et al. 2018). According to that principle, Mediterranean rangelands must be managed by considering them as protected areas (Vrahnakis 2015; Guarino et al. 2019).

26.6 Conclusions—Proposed Actions for Rangelands

Extensive livestock grazing has substantially contributed to the conservation of important elements of biodiversity in the mountain habitats of Prespa National Park and adjacent areas. The changes in grazing management schemes observed over the last decades and especially over the last years cause some concern on the dynamics of mountain ecosystems, but at the same time act as a challenge for improved rangeland management. Based on previous experience gained by relevant survey and monitoring actions, as well as management plans and successfully implemented projects briefly presented in the present contribution, local authorities and conservationists have sufficient tools to achieve this objective. Actions to be undertaken are: (i) partial updating of the Grazing Management Plan (Tsitoura et al. 2015) to cover the needs of recent legislation (relevant Ministerial Decision of 2017), (ii) discussion with stockbreeders on their needs concerning technical works that will improve their daily breeding practices and will contribute to the even use of rangeland resources to avoid over- and undergrazing phenomena, (iii) prioritization of the technical works to be applied (e.g. water troughs, sheds, fenced parcels with traps for handling livestock, scrub clearing) taking into account policy initiatives, conservation aspects, budget restrictions, potential common use of infrastructure by more than one stockbreeder and other details depending on the nature of each site and the technical work to be set up, (iv) census of the present livestock capital in the study area and record of the problems of each stockbreeding business (e.g. with the use of simple-to-fill questionnaires), (v) training of the personnel of the Management Body of Prespa National Park on (a) the survey of Rangeland Units proposed in the Grazing Management Plan for the period after the establishment of the above-mentioned infrastructure, (b) the monitoring of habitat types within the Rangeland Units in order maintain good conservation status of rangelands, and (c) yearly record of livestock herds that use the area's rangelands extensively, (vi) training of farmers targeting at the improvement of their stockbreeding businesses and revenue, and (vii) based on all the above-mentioned experiences in the study area, compilation of measures to be proposed for implementation in the next Common Agricultural Policy of the EU, so that other remote mountainous areas gain from the Prespa case study.

References

Baumann H (1999) Greek Flora in myth, in art, in literature. Hellenic Society for the Protection of Nature, Athens (in Greek)

Besson F (2017) Mediterranean vegetation in Anglophone literature as a sign of man's relationship with the world. Athens J Mediterr Stud 3(1):7–20

Blondel J, Aronson J (2004) Biology and wildlife of the mediterranean region. Oxford Univ Press Inc., New York

Bousbouras D (2010) Catalogue of amphibians and reptiles of Prespa National Park. In: Giannakis N, Bousbouras D, Argyropoulos D, Kazoglou Y (eds) Prespa National Park management plan. Prefecture of Florina, Management Body of Prespa National Park, Florina (in Greek)

Bousbouras D, Ioannidis Y (1997) The distribution and habitat preferences of the amphibians of Prespa National Park. Hydrobiologia 351:127–133

Bousbouras D, Kazoglou Y (2010) Catalogue of mammals of Prespa National Park. In: Giannakis N, Bousbouras D, Argyropoulos D, Kazoglou Y (eds) Prespa National Park management plan. Prefecture of Florina, Management Body of Prespa National Park, Florina (in Greek)

Catsadorakis G (1995) The texts of the information centre of Prespa. Society for the Protection of Prespa, Agios Germanos (126 p, in Greek)

Catsadorakis G (1997) The importance of Prespa National Park for breeding and wintering birds. Hydrobiologia 351:157–174

Catsadorakis G, Malakou M (1997) Conservation and management issues of Prespa National Park. Hydrobiologia 351:175–196

Crivelli AJ, Catsadorakis G, Malakou M, Rosecchi E (1997) Fish and fisheries of the Prespa lakes. Hydrobiologia 351:107–125

Dafis S, Papastergiadou E, Lazaridou T, Tsiafouli M (2001) Technical guide for the identification, description and mapping of habitat types in Greece, Greek Biotope/Wetland Centre, Thessaloniki (in Greek)

European Commission (2013) Interpretation Manual of European Union Habitat Types – EUR 28. DG Environment, Nature ENV B.3, p 146

Fotiadis G, Kazoglou Y (2010) Floristic catalogue of Prespa National Park. In: Giannakis N, Bousbouras D, Argyropoulos D, Kazoglou Y (eds) Prespa National Park management plan. Prefecture of Florina, Management Body of Prespa National Park, Florina (in Greek)

Fotiadis G, Vrahnakis M, Kazoglou Y, Tsiripidis I (2014) Dry grassland types in the Prespa National Park (NW Greece), including the southern-most occurrence of the priority habitat type "Pannonic sand steppes" (code 6260). Hacquetia 13(1):171–189

Fotiadis G, Vrahnakis M, Kakouros P (2018) Monitoring of the habitat type "Grecian juniper forests" (*9562) in Prespa National Park, after conservation actions. Proceedings of the 9th Congress of the Hellenic Range and Pasture Society (HRPS), Ministry of Environment and Energy, HRPS, Athens, pp 335–342 (in Greek with English summary)

Galand N, Declercq S, Cheyrezy T, Puechmaille SJ, Deguines N, Grémillet X, Papadatou E, Kazoglou Y (2010) Bat survey on the subalpine grasslands of Mt Varnous (Florina, Greece): preliminary results. In: Sidiropoulou A, Mantzanas K., Ispikoudis I (eds) Rangeland science and quality of life, Proceedings of the 7th Congress of the Hellenic Rangeland and Pasture Society, Greek Ministry of Environment, Energy and Climate Change & Hellenic Rangeland and Pasture Society, Athens, pp 305–313

Georgoudis A (1993) The Greek shorthorn bovine breed in the Prespa area. Poster text, Society for the Protection of Prespa, Agios Germanos

Giannakis N, Bousbouras D, Argyropoulos D, Kazoglou Y, Nalpantidou M, Saliaris D, Kakouros P, Fotiadis G, Doanidis L, Vrahnakis M, Maglaras G, Georganda A, Lagadinou E, Korili S, Evaggelou C (2010) Prespa National Park management plan. Prefecture of Florina, Management Body of Prespa National Park, Florina (in Greek)

Grove AT, Rackham O (2003) The nature of mediterranean Europe: an ecological history. Yale University Press, New Heaven

Grunenfelder HP (2006) Prespa cattle: Identification and possible conservation measures. Monitoring Institute for Rare Breeds & Seeds in Europe, SAVE Foundation, St Gallen

Guarino R, Vrahnakis M, Rodriguez Rojo MP, Giouga L, Pasta S (2019) Mediterranean grasslands and shrublands. In: Dengler J, Török P (eds) Encyclopedia of the World's Biomes. Elsevier, Amsterdam (in press)

Hollis GE, Stevenson AC (1997) The physical basis of the Lake Mikri Prespa systems: geology, climate, hydrology and water quality. Hydrobiologia 351:1–19

Ioannidis Y, Bousbouras D (1997) The space utilization by the reptiles in Prespa National Park. Hydrobiologia 351:135–142

Kavadas D (1956) Illustrated botanological-phytological dictionary, vol. 1–9, Athens

Kazoglou Y (2007) Effects of water buffalo grazing on the wet meadows of Prespa National Park. PhD dissertation, School of Forestry and Natural Environment, Aristotle University of Thessaloniki, p 254 (in Greek with English summary)

Kazoglou Y (2011) Agriculture, stockbreeding and wetlands: the case of Prespa. In: Papayannis T, Pritchard D (eds) Culture and wetlands in the mediterranean: an evolving story. Med-INA, Athens, pp 146–158

Kazoglou Y (2015) Progress report (April–June 2015) on the project "Development of tools for rangeland management and support to livestock breeding in the rangeland ecosystems of the Prespa Municipality (Prespa National Park, Ladopotamos valley, Munic. District of Krystallopigi)". Forest Research Institute, Thesaloniki, p 55 (in Greek)

Kazoglou YE, Mesléard F, Papanastasis VP (2004) Water buffalo (*Bubalus bubalis*) grazing and summer cutting as methods of restoring wet meadows at Lake Mikri Prespa, Greece. Grassland Sci Europe 9:225–227

Kazoglou Y, Mesléard F, Papanastasis VP (2008) Wet meadow restoration at Lake Mikri Prespa, Greece: results of vegetation monitoring (2002–2007). Proceedings of the 6th European Congress on Ecological Restoration, Society for Ecological Restoration—Europe, Ghent

Kazoglou Y, Bousbouras D (2010) Catalogue of birds of Prespa National Park. In: Giannakis N, Bousbouras D, Argyropoulos D, Kazoglou Y (eds) Prespa National Park management plan. Prefecture of Florina, Management Body of Prespa National Park, Florina (in Greek)

Kazoglou Y, Traianopoulou I, Fotiadis G, Vrahnakis M, Yiakoulaki M (2019) Rangeland production and grazing capacity in protected and non-protected sites of the Prespa area, NW Greece. In: Proceedings of the 19th Congress of the Hellenic Forestry Society (in press, in Greek with English summary)

Kazoglou Y, Xega N, Logotheti A, Doleson F (2010) Rare bovine breeds in the transboundary Prespa Park. In: Sidiropoulou A, Mantzanas K, Ispikoudis Y (eds) Range science and quality of life. Proceedings of the 7th Congress of the Hellenic Range and Pasture Society (HRPS), Ministry of Environment and Energy, HRPS, Athens, pp 223–229 (in Greek with English summary)

Logotheti A (2017) 3rd phase of implementation of silviculture treatments (Concrete Action C1). LIFE12 NAT/GR539-JunEx, Society for the Protection of Prespa, Laimos, Greece

Medugorac I, Medugorac A, Russ I, Veit-Kensch CE, Taberlet P, Luntz B, Mix HM, Förster M (2009) Genetic diversity of European cattle breeds highlights the conservation value of traditional unselected breeds with high effective population size. Mol Ecol 18 (16):3394–3410

Pamperis L (2009) The butterflies of Greece. Pamperis Editions, KOAN Editions, Hellenic Society for the Protection of Nature, Athens (in Greek)

Papadatou E, Grémillet X, Bego F, Petkovski S, Stojkoska E, Avramoski O, Kazoglou Y (2011) Status survey and conservation action plan for the bats of Prespa. Society for the Protection of Prespa, Agios Germanos

Papadatou E, Puechmaille S, Grémillet X, Georgiakakis P, Galand N, Deguines N, Declercq S, Cheyrezy T, Kazoglou Y (2013) Bat diversity and activity at subalpine grasslands of Varnous and Triklarion Mountains (NW Greece). In: Vrahnakis M, Kyriazopoulos AP, Chouvardas D, Fotiadis G (eds) Dry Grasslands of Europe: grazing and ecosystems services, Proceedings of the 9th European Dry Grassland Meeting, Greek Ministry of Environment, Energy and Climate Change & Hellenic Rangeland and Pasture Society, Athens, pp 233–239

Papanastasis VP (1997) Livestock grazing in the Mediterranean ecosystems: a historical and policy perspective. In: Papanastasis VP, Peter D (eds) Ecological Basis of Livestock Grazing in Mediterranean Ecosystems, European Commission, Belgium

Papayannis T (2008) Action for Culture in mediterranean wetlands. Med-INA, Athens

Papayannis T, Pritchard D (2011) Culture and wetlands in the Mediterranean: an evolving story. Med-INA, Athens

Pavlides G (1985) Geobotanical study of Prespa National Forest, Florina, Part A: Ecology, Flora, Phytogeography. Aristotle University of Thessaloniki, Thessaloniki (308 pp + map, in Greek)

Pavlides G (1997a) The flora of Prespa National Park with emphasis on species of conservation interest. Hydrobiologia 351:35–40

Pavlides G (1997b) Aquatic and terrestrial vegetation of the Prespa area. Hydrobiologia 351:41–60

Perennou C, Gletsos M, Chauvelon P, Crivelli A, DeCoursey M, Dokulil M, Grillas P, Grovel R, Sandoz A (2009) Development of a transboundary monitoring system for the Prespa Park Area. In: Gletsos M, Kazoglou Y, Perennou C (eds) Aghios Germanos, vol 1 (381 p + Annexes)

Phitos D, Strid A, Snogerup S, Greuter W (eds) (1995) The red data book of rare and threatened plants of Greece. WWF, Athens

Portolou D, Bourdakis S, Vlachos C, Kasritis T, Dimalexis T (eds) (2009) The important bird areas of Greece: priority sites for the biodiversity conservation. Hellenic Ornithological Society, Athens (in Greek)

Pritchett K (2005) Sacred spaces: the mediterranean in poems by Antonio Colinas. Hispania 88:278–283

Pyrovetsi M, Karteris M (1986) Forty years of land cover/use changes in Prespa National Park, Greece. J Environ Manage 23:173–183

Sakellarakis F-N, Manolopoulos A, Fotiadis G, Kazoglou Y, Vrahnakis M, Swinkels C, Utermann S, Strid A, Bergmeier E (2019) Flora of Prespa area database, NW Greece. In: Book of abstracts of the 16th Congress of the Hellenic Botanical Society, 10–13 October 2019, Athens

Society for the Protection of Prespa (SPP) (2019a) Data on the birds of Prespa presented in the SPP website. https://www.prespawaterbirds.gr. Accessed Sept 2019

Society for the Protection of Prespa (SPP) (2019b) Data on the mammals of Prespa presented in the SPP website. https://www.spp.gr/thilastika_updated.10.2015.pdf. Accessed Sept 2019

Society for the Protection of Prespa (SPP) (2019c) Data on the fish species of Prespa presented in the SPP website. spp.gr/psaria_update.10.2015.pdf. Accessed Sept 2019

Spirkovski Z, Ilik-Boeva D, Taleski T, Palluqi A, Kapedani E (2012) The fish of Prespa. UNDP, Skopje

Strid A, Bergmeier E, Sakellarakis F-N, Kazoglou Y, Vrahnakis M, Fotiadis G (2017) Additions to the flora of the Prespa National Park, Greece. Phytologia Balcanica 23(2):207–269

Svetasheva T, Vrachnakis M, Kazoglou Y, Konstantinidis G (2019) Combined list of 262 macrofungi species recorded from the Prespa National Park (Greece) based on results of 2014, 2016, 2017 expeditions. Tula State Lev Tolstoy Pedagogical University, Tula, Russia (unpublished report)

Tsitoura P, Vrahnakis M, Kazoglou Y, Fotiadis G, Chouvardas D, Bousbouras D, Kotsios L, Papaporfyriou P, Spyridis A, Tsiripidis I, Koutalou V, Nasiakou S, Georgaki D, Zagalikis G, Keskilidou K., Kigkas N (2015) Final grazing management plan for the municipality of Prespa. Region of Western Macedonia, Management Body of Prespa National Park. Agios Germanos (204 p + Appendices, in Greek)

Vrahnakis M (2015) Rangeland science. Hellenic Academic Libraries Link, Kalippos, Athens

Vrahnakis M, Papadatou E, Kazoglou Y (2010) Grasslands as bat-lands: evidence from Mount Varnous, Greece. Euro Dry Grassland Group Bull 8:7–9

Vrahnakis M, Fotiadis G, Kazoglou Y (2011) Habitat types of Prespa National Park: record, evaluation and geographical presentation 2011. Society for the Protection of Prespa – Technological Educational Institute of Larisa, Agios Germanos (104 p + Appendices, in Greek)

Vrahnakis M, Kazoglou Y, Fotiadis G, Chouvardas D, Papaporfyriou P, Nasiakou S, Kotsios L, Ambas V (2018) Incorporating Natura 2000 habitat types into pasture management plans. In: Parisi Z, Kakouros P (eds) New challenges for Greek range science and management. Proceedings of the 9th Congress of the Hellenic Range and Pasture Society (HRPS), Ministry of Environment and Energy, HRPS, Athens, pp 237–244 (in Greek with English summary)

27 Landscape Dynamics in the Northwestern Mountains of the Iberian Peninsula: Case Study Ancares-Courel Mountain Range

Ignacio J. Diaz-Maroto

Abstract

The interaction between people and the environment plays a key role in the landscape, above all in those landscapes subject to human pressure in which socio-ecological relationships control biodiversity. The diversity maintenance of the cultural landscapes is often greater than in the natural landscapes, subject on the heterogeneity created by nature and human activities. Traditional land-use in the NW Mountains of the Iberian Peninsula has shaped a system managed by seasonal cycles and spatial models of human action. Agriculture, cattle grazing, unsuitable forestry management, forest fires, and afforestation with fast-growing species, have led to a reduction of the forests. In the last times, globalization has prompted complex changes in rural areas. The land abandonment has limited the open spaces and has induced an increase of the forests with effects for both ecological system and biodiversity. This means reduction of open habitats, difficulty to the agropastoral activities, and increasing wildfires. Moreover, an increase in forest species is likely to the loss of open-habitat and ecotone species. The sustainability depends on the conservation of traditional uses: extensive grazing and suitable forest management. The environmental, cultural and economic integration of agropastoral and forestry activities seems vital to ensure the cultural landscapes. Our aim is to study the evolution and historical background of this landscape to establish measures for their conservation and recovery. The main focus will be on native broadleaf forests, intensively exploited since ancestral times.

Keywords

Land-use · Sustainability · Cultural landscape · Biodiversity · Native forests

27.1 Introduction

The interplay between society and the environment perform a key role in the pattern of the landscapes, mainly in those subject to human influence where socio-ecological relationships and feedback tools govern biodiversity (Farina 2000). The conservation of biodiversity in cultural landscapes is often greater than in the remnants of natural landscapes, depending on the heterogeneity created by natural forces and anthropogenic actions (Burel and Baudry 2001; Farina 2000).

I. J. Diaz-Maroto (✉)
Department of Agroforestry Engineering, Higher Polytechnic School of Engineering, University of Santiago de Compostela, Campus Terra s/n, 27002 Lugo, Spain
e-mail: ignacio.diazmaroto@usc.es

U. Schickhoff et al. (eds.), *Mountain Landscapes in Transition*,
Sustainable Development Goals Series, https://doi.org/10.1007/978-3-030-70238-0_27

Traditional practices of land-use in the Eastern Mountains of Galicia, Atlantic region of the Iberian Peninsula, have shaped a structured system controlled by seasonal cycles and spatial patterns of human activities (Diaz-Maroto and Vila-Lameiro 2008). Agriculture and food production, cattle grazing, unsuitable management of forests, the confiscation of forests owned by the Church, frequent forest fires, and the contemporary afforestation by fast-growing species, have led to a reduction of the forest area, within a mosaic of highly integrated and structured landscape. However, during the last decades, socioeconomic globalization has induced deep changes in rural areas (Ahern 1994). The depopulation of the land has reduced the extension of the open spaces and has generated an expansion of the forests (Diaz-Maroto and Vila-Lameiro 2008), with repercussions for the ecological regime and biodiversity. This landscape homogenization means reduction of open habitats (crops, grasslands, and scrublands), problematic in the maintenance of agropastoral activities, and increasing risk of wildfires because to reduction in fragmentation. Moreover, from the biological perspective, an increase in forest species and core habitat specialists is expected to the detriment of open-habitat and ecotone species (Pons et al. 2003).

Sustainability depends on the maintenance of traditional uses: extensive agropastoral activity and suitable forest management. Extensive grazing, allowing an adequate number of livestock and avoiding their concentration in small areas, prevents soil erosion and vegetation impoverishment, increases mosaic diversity, and maintains open-habitat patches. Selective felling appears to be an environmentally integrated and viable economic activity that reduces wildfires by promoting landscape fragmentation. To guarantee the sustainability of these activities, measures such as clear cutting and prescribed burning to create open habitats, improvement of forest access, increment of public awareness about mountainous areas, and agro-environmental measures need to be adopted (Debussche et al. 1999).

The environmental, cultural and economic integration of agroforestry activities seems fundamental (Farina 2000) to ensure the conservation of this landscape mosaic. The approach to the problem is to study the evolution and historical background of these forests to establish measures for their conservation and recovery. The main focus will be on natural broadleaf forests, which have been intensively exploited since ancestral times.

27.2 The Exploitation System in the Eastern Mountains of Galicia: Dynamics and Historical Background

27.2.1 Study Area: Ancares-Courel Mountain Range

The study area is located in Galicia, northwestern Spain. The Ancares-Courel Mountains is a transitional area from typical Atlantic to Mediterranean flora (Diaz-Maroto and Vila-Lameiro 2007). It extends at the western end of the Cantabrian Mountains, and it is an area of sloped land with elevations ranging between 250 and 1935 m. The climate is characterized by rainfall fluctuating between 700 mm in the lower areas and nearly 2500 mm in the summits. The annual temperature regime is very extreme, with long winters in the highlands. Annual average temperatures range from 4.6 to 14.0 °C in the most protected stations (Ramil et al. 2013) (Fig. 27.1).

The climax vegetation that should currently occupy this region would be the native broadleaf forests, characterized by diverse species of oaks (Buide et al. 1998). According to different studies, these forests were established in the study area between five and seven thousand years ago. Historical factors, site conditions and requirements of each species gave rise to different type of forests and floristic composition (Peterken and Game 1984).

In the northwestern Iberian Peninsula there is confirmation of the presence of *Quercus* species since the Cretaceous Era, and their diversification in the Tertiary period (Allen et al. 1996). The

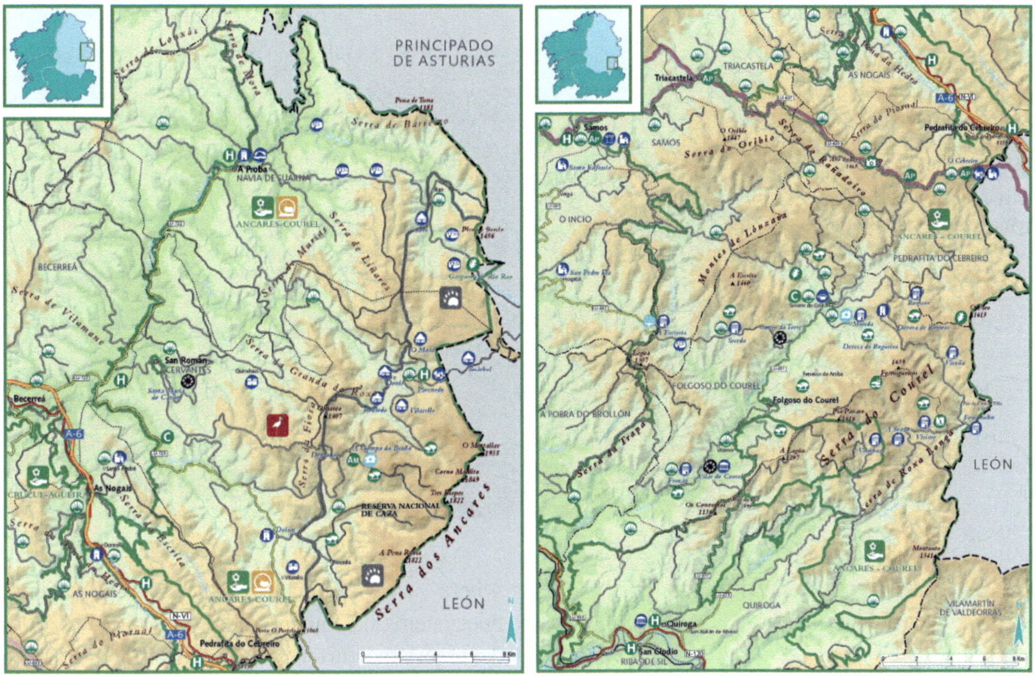

Fig. 27.1 Study area: Ancares-Courel Mountains (Galicia, NW Spain) (*Source* www.turgalicia.es)

palynological data indicate that deciduous forests were dominant in this area between 9000 and 11,000 years ago, after the last glacial phase, in the Quaternary period (Santos et al. 2000; Sobrino et al. 2001). The decline of broadleaved forests began in prehistoric times and coincided with the expansion of human activity and the establishment of crops and pastures. The wood felling and the burning were already used to clear forests for agricultural production, and many forested areas became scrubs and pastures for livestock (Bauer 1980).

27.2.2 Agricultural-Silvicultural-Pastoral System

The exploitation system in the Eastern Galician Mountains is an agricultural-silvicultural-pastoral organization adjusted to the heterogeneity of the highland environment, founded on cereal crops and extensive grazing with orchard crops and fruit trees (Diaz-Maroto and Vila-Lameiro 2008). This pattern of social-economic model has shaped the particular cultural landscape in this area (Sobrino et al. 2001). Agricultural lands are located at the bottom of the valleys and the villages on middle slopes with sunny orientation. Orchards and fruit trees placed between the houses and other farm construction surround the villages and give way to chestnut stands "*soutos*" and rye fields (Manuel and Gil 2001). Shrubs dominate the steep slopes where even periodic burning is performed to regenerate mountain pastures. The meadows are located in the low-lying areas near to the rivers. On the shady slopes, far from the villages and where the humidity is higher, the anthropogenic pressure is very low, being refuge from the best examples of primitive forests, mixed forests with abundance of different species of Atlantic oaks (Manuel and Gil 2001; Diaz-Maroto and Vila-Lameiro 2007) (Fig. 27.2).

Fig. 27.2 The exploitation system in the Eastern Galician Mountains; Agricultural-silvicultural-pastoral organization based on the maintenance of traditional uses: extensive agropastoral activity and suitable forest management (Diaz-Maroto, 2015-08-20)

27.2.3 Landscape Dynamics and Forecast

The landscapes are dynamic systems continuously influenced by human activities. Throughout history, the intensity of these activities has been increasing, which has brought about greater anthropic pressure on the landscape (Gökyer 2013). As a result of human pressure, negative impacts on the landscape and the species that inhabit it occur. These negative effects are particularly severe in the most vulnerable landscapes. In these landscapes, where fragmentation has increased, habitats have been more damaged (Jaeger 2000).

Current landscapes have been shaped by powerful anthropogenic forces always present in space and time (Sanderson and Harris 2000). Landscapes are influenced by climate changes, land-use changes and human activities that change mosaic structure, shape and size of fragments in a landscape. All these changes have different could be seen different spatial sizes and periodicities (Farina 2000).

Landscape dynamics in the study area has always been marked by progression-regression periods of the autochthonous forests. Across of the time, the regression periods have been more wide and intensive, except in recent decades, where an important expansion of natural ecosystems has taken place (Santos et al. 2000). To assess the current conditions of the landscape, as well as the future management forecast, historical process must be known. Suchlike knowledge based on the historical information can be used now and in the future as a complementary tool to develop conservation strategies (Allen et al. 1996). Depending on the natural and cultural influences the changes are seen over time in the landscape structure. In this case, landscape structure and relationship between ecosystems are changed. As a result of change studies, functions and conditions in the mosaic of different sized and shaped patches can be revealed (Wu and Hobbs 2002).

There are two main factors of landscape change. These are the natural processes and human actions. Both natural conditions and human activities are changed over time. Complex transformations can be developed in the landscape structure linked to natural environment and human needs (Farina 2000). Landscapes change naturally as they are the expression of the interaction between the environment and human's activities. At the same form, the natural conditions and the human needs change in time and are controlled by different but interactive aspects (Antrop 1998).

27.2.4 Chorological and Historical Information

The chorological information allows us to analyze the land-use changes caused by human activity. The importance of these data has been highlighted by different authors (Chocarro et al. 1990; Van Eetvelde and Antrop 2004). The review and analysis of chorological information enables the origin of vegetal formations and the natural changes that happen in their distribution to be defined (Marris 2007). These data were essentially useful for analyzing changes related to the natural distribution of the vegetal ecosystems under study, i.e. the native broadleaved forests, dominated for different species of *Quercus* genus.

Information about the land modifications because to anthropological management was obtained from several sources. As our research focuses on the latter, the historical data was completed with more recent information included in the Annual Agrarian Statistics (from 2000 to present), National Forest Inventories (NFIs), Forest Plan of Galicia of 1992, and the first revision of the Forest Plan of 2018. All databases used were indispensable for to achieve our objective because each provides a different type of information. For example, the Forest Plan of Galicia and the revision data showed that the increase in the area covered by the natural broadleaved forests during the last few decades is likely the result of the natural dynamic, but may also be influenced by other factors such as the depopulation and abandonment of agricultural land (Fig. 27.3).

27.3 Evolution and Historical Transformation in the Landscape of the Galician Eastern Mountains

27.3.1 First Signs of Deforestation: Beginning of Agriculture

The changes of landscape in the Galician Eastern Mountains include a prolonged period of dominance by oak forests (Bauer 1980). In the study area, intensive exploitation and timber extraction have been carried out for centuries because of the abundance of high quality wood, such as oak, beech and chestnut. However, Forestry Administration afforestation policies have recommended fast-growing species use such as pine and eucalyptus until recent times (Rico 1995). This fact played a very important role in shaping

Fig. 27.3 Meadows near to the rivers and abandoned agricultural land colonized by scrubland and incipient birch regeneration (Ancares-Courel Mountains) (Diaz-Maroto, 2016-05-06)

of the landscape, but new changes in the productive system have been a fundamental factor (Crecente et al. 2002). Moreover, the rural needs have changed during the last centuries, including a decrease in agriculture and extensive cattle farming and much less use of wood and firewood (Chocarro et al. 1990).

Alterations in the forest landscape were not significant until agricultural activities became generalized, as reflected by pollen analyses (Santos et al. 2000; Sobrino et al. 2001). Agriculture began in Galicia around 5500 years ago, during the megalithic culture (Torras et al. 1980). The first farmers were nomads who seeded crops after removing existing vegetation by fire. They always cultivated grain in the same place until production declined, and then left the land. The introduction of livestock speeded soil degradation (Pons et al. 2003).

27.3.2 Historical Development of Native Broadleaved Forests

During the Middle Ages before the mentioned socioeconomic transformations several factors resulted in an to an important reduction in the forest area. This decrease of the forests extended until the second half of the nineteenth century, followed by continuous increase of forest cover (Manuel and Gil 2001). As a result of the afforestation, the increase in forest area was very important after the Spanish Civil War and towards the end of the twentieth century, also due to the abandonment of agricultural land and the growth of the natural regeneration of autochthonous species (Poyatos et al. 2003). These practices were driven by private owners and the public administration (Bauer 1980; Rico 1995; Crecente et al. 2002) (Fig. 27.4).

Artificial afforestation has played a fundamental role in the current evolution of the forest landscape in the study area (Chocarro et al. 1990; Roura-Pascual et al. 2005). The most important consequence of this transformation has been the increase in the dimension of forests in general, and particularly the native broadleaved forests (Rico 1995; Diaz-Maroto and Vila-Lameiro 2008). The application of forest management practices to improve and preserve these formations would allow the recovery of the cultural landscape of the Mountains of Eastern Galicia as part of our Natural Heritage (Diaz-Marota and Vila-Lameiro 2007).

Fig. 27.4 Native broadleaf forest in an optimal condition of conservation in the Eastern Mountains of Galicia (Diaz-Maroto, 2017-06-14)

27.3.3 Future Perspectives: Effect for the Native Forests Conservation

The Forest Plan of Galicia of 1992 projected that by 2032, productive broadleaved forests with the exception of eucalyptus would occupy 410,000 ha. The change proposed was more qualitative than quantitative and only a slight increase in the area covered by broadleaved forests was planned and it has almost been reached. The aim was to develop productive forests of chestnut and other broadleaved species for saw timber and sheet wood (Diaz-Maroto and Vila-Lameiro 2008). Today, the area occupied by deciduous broadleaved forests in Galicia has significantly increased. According to the published last data, IV National Forest Inventory, the forests of native broadleaves occupy 31% of the forest area, 441,289 ha, with the area covered by *Quercus robur* L. being 246,446 ha, 17.4% of the forest cover (Galicia Government 2018). Most oak forests lack of management because of the limited economic interest of their use. However, they have a high environmental relevance being habitats of Community interest (Decree 92/43/EU) as part of the Natura 2000 Network (Skliar et al. 2019).

27.4 Conclusions

From prehistoric times, the presence of abundant broadleaved forests in the northwestern of the Iberian Peninsula has been associated with significant transformations in land-use as result of anthropogenic influence. The processes implicated in the reduction of natural broadleaved forests have been very long and complex. They have involved wars, invasions, felling for shipbuilding and steel industries, change to agricultural use, construction of railways, charcoal making, and the recent massive afforestation with pines and eucalyptus.

These facts gave rise to a progressive decrease in the area occupied by these formations until the middle of the nineteenth century, without to have done any proper action to favor the natural regeneration. In the second half of the nineteenth century, there was an abrupt change in this trend, and an increase in the area occupied by the broadleaved species occurs.

Rural demands are continually subjected to change, including modifications in agriculture, extensive cattle farming, firewood production, and exploitation of timber for the naval, iron and railway industries, among other. Afforestation has played a fundamental role in shaping the forest landscape, particularly in relation to fast growing species, involving critical changes in the productive system. The changes that have taken place in recent decades explain the distribution of current cultural landscape as a common system of rural management (agriculture-forestry-grazing) adapted to the environmental diversity of the mountains.

This structural pattern is associated with a predominance of small owners and a tendency for agrarian practices to be abandoned. The property is often inherited by town or city dwellers with no interest in agrarian or forest practices. The absence of interest drives the emergence of unproductive land with the consequent risk of forest fires occurrence.

The application of silvicultural practices with the aim of improving and restoring native broadleaved forests would enable the recovery of the cultural landscape as part of the natural and environmental heritage. This would minimize the effects of forest fires and of the increase of unproductive land and also maximize diversification of obtained forest products, improve biodiversity and stimulate the interest of the possible new owners.

References

Ahern J (1994) Greenways as ecological networks in rural areas. In: Cook EA, van Lier HN (eds) Landscape planning and ecological networks. Elsevier, Amsterdam, pp 159–177

Allen JRM, Huntley B, Watts WA (1996) The vegetation and climate of northwest Iberia over the last 14000 yr. J Quat Sci 11:125–147

Antrop M (1998) Landscape change: plan or chaos. Landsc Urban Plan 41:155–161

Bauer E (1980) Los Montes de España en la Historia. Ministerio de Agricultura, Madrid

Buide ML, Sánchez JM, Guitián J (1998) Ecological characteristics of the flora of the Northwest Iberian Peninsula. Plant Ecol 135:1–8

Burel F, Baudry J (2001) Ecologie du Paysage: Concepts. Méthodes et Applications. Éditions Tec & Doc-Lavoisier, París

Crecente R, Alvarez CJ, Fra U (2002) Economic, social and environmental impact of land consolidation in Galicia. Land Use Pol 19:135–147

Chocarro C, Fanlo R, Fillat F, Marin P (1990) Historical evolution of natural resource use in the Central Pyrenees of Spain. Mt Res Dev 10:257–265

Debussche M, Lepart J, Dervieux A (1999) Mediterranean landscape changes: evidence from old postcards. Global Ecol Biogeogr 8:3–15

Diaz-Maroto IJ, Vila-Lameiro P (2007) Deciduous and semi-deciduous oak forests (*Quercus robur*, *Q. petraea* and *Q. pyrenaica*) floristic composition in the Northwest Iberian Peninsula. Biologia 62:163–172

Diaz-Maroto IJ, Vila-Lameiro P (2008) Historical evolution and land-use changes in natural broadleaved forests in the north-west Iberian Peninsula. Scand J For Res 23:371–379

Farina A (2000) Landscape ecology in action. Springer, Dordrecht

Galicia Government (2018) First review of the forestry plan of Galicia: diagnosis report of the forests and the Galician forest sector. Ministry of Rural Environment, Santiago de Compostela

Gökyer E (2013) Understanding landscape structure using landscape metrics. In: Özyavuz M (ed) Advances in landscape architecture. IntechOpen, London, pp 663–676

Jaeger JA (2000) Landscape division, splitting index, and effective mesh size: new measures of landscape fragmentation. Landsc Ecol 15:115–130

Manuel CM, Gil L (2001) La Transformación Histórica del Paisaje Forestal en Galicia. Ministerio de Medio Ambiente, Madrid

Marris E (2007) Linnaeus at 300: the species and the specious. Nature 7133:250–253

Peterken GF, Game M (1984) Historical factors affecting the number and distribution of vascular plant-species in the woodlands of central Lincolnshire. J Ecol 72:155–182

Pons P, Lambert B, Rigolot E, Prodon R (2003) The effects of grassland management using fire on habitat occupancy and conservation of birds in a mosaic landscape. Biodivers Conserv 12:1843–1860

Poyatos R, Latron J, Llorens P (2003) Land use and land cover change after agricultural abandonment–The case of a Mediterranean mountain area (Catalan Pre-Pyrenees). Mt Res Dev 23:362–368

Ramil P, Rodríguez MA, López H, Ferreiro da Costa J, Muñoz C (2013) Loss of European dry heaths in NW Spain: a case study. Diversity 5:557–580

Rico E (1995) Política forestal y repoblaciones en Galicia (1941–1971). Universidad de Santiago de Compostela, Spain

Roura-Pascual N, Pons P, Etienne M, Lambert B (2005) Transformation of a rural landscape in the eastern Pyrenees between 1953–2000. Mt Res Dev. 25:252–261

Sanderson J, Harris LD (ed) (2000) Landscape ecology a top-down approach. Lewis Publishers, by CRC Press LLC, Boca Raton, FL

Santos L, Romani JRV, Jalut G (2000) History of vegetation during the Holocene in the Courel and Queixa Sierras, Galicia, northwest Iberian Peninsula. J Quat Sci 15:621–632

Skliar V, Kovalenko I, Skliar I, Sherstiuk M (2019) Vitality structure and its dynamics in the process of natural reforestation of *Quercus robur* L. AgroLife Sci J 8:233–241

Sobrino CM, Ramil-Rego P, Guitián MR (2001) Vegetation in the mountains of northwest Iberia during the last glacial-interglacial transition. Veg Hist Archaeobot 10:7–21

Torras M, Díaz-Fierros F, Vázquez J (1980) Sobre el comienzo de la agricultura en Galicia. Gallaecia 6:51–59

Van Eetvelde V, Antrop M (2004) Analyzing structural and functional changes of traditional landscapes–two examples from southern France. Landsc Urban Plan 67:75–95

Wu J, Hobbs R (2002) Key issues and research priorities in landscape ecology: an idiosyncratic synthesis. Landsc Ecol 17:355–365

History of Vegetation and Land-Use Change in the Northern Calcareous Alps (Germany/Austria)

28

Arne Friedmann, Philipp Stojakowits, and Oliver Korch

Abstract

A review of the vegetation history of the Northern Calcareous Alps of Germany and Austria for the montane, subalpine and alpine vegetation belts above 900 m a.s.l. is presented. Natural vegetation changes are recorded until the end of the Atlantic period. Human impact started locally in the valleys, but soon spread into higher altitude. First alpine pastoral farming is documented since the Bronze Age. Furthermore, a monitoring study of alpine and subnival vegetation dynamics as well as a grazing experiment on the Zugspitzplatt (Wetterstein Mountains) is outlined. The Zugspitzplatt is Germany's highest investigation site with a long altitudinal gradient from 2000 to 2700 m providing excellent conditions. Along with the specific site climate, pedogenetic processes, and soil conditions show considerable variation in the investigation area. The vegetation on these sites is highly influenced by anthropo-zoogenic impact.

Keywords

Pollen analysis · Human impact · Vegetation monitoring · Grazing

28.1 Introduction

To understand the dynamics of the present vegetation it is important to know the local to regional vegetation history of an area. Therefore, palynological studies represent a useful tool for reconstructing past environments and for measuring the anthropogenic influences on high mountain ecosystems in the past.

In this study, first an outline of the Late Glacial and postglacial vegetation history of the Northern Calcareous Alps of Bavaria and Austria is presented. There are numerous studies of lower montane sites, but studies in the higher montane, subalpine and alpine mountain zone are relatively rare (Friedmann and Stojakowits 2017; Adamski and Friedmann 2019). Here we reconstruct the vegetation and land use history by palynological investigations of peat bogs and lake sediments at sites located above 900 m a.s.l. (Fig. 28.1). All studies comprise high-resolution pollen analysis and well dated profiles.

Secondly, we present a case study of recent vegetation dynamics on the Zugspitzplatt (Wetterstein Mountains, Fig. 28.5) to emphasize the current anthropo-zoogenic influences of the alpine

A. Friedmann (✉) · P. Stojakowits · O. Korch
Institute of Geography, AG Biogeography, University of Augsburg, Alter Postweg 118, 86159 Augsburg, Germany
e-mail: friedmann@geo.uni-augsburg.de

U. Schickhoff et al. (eds.), *Mountain Landscapes in Transition*, Sustainable Development Goals Series, https://doi.org/10.1007/978-3-030-70238-0_28

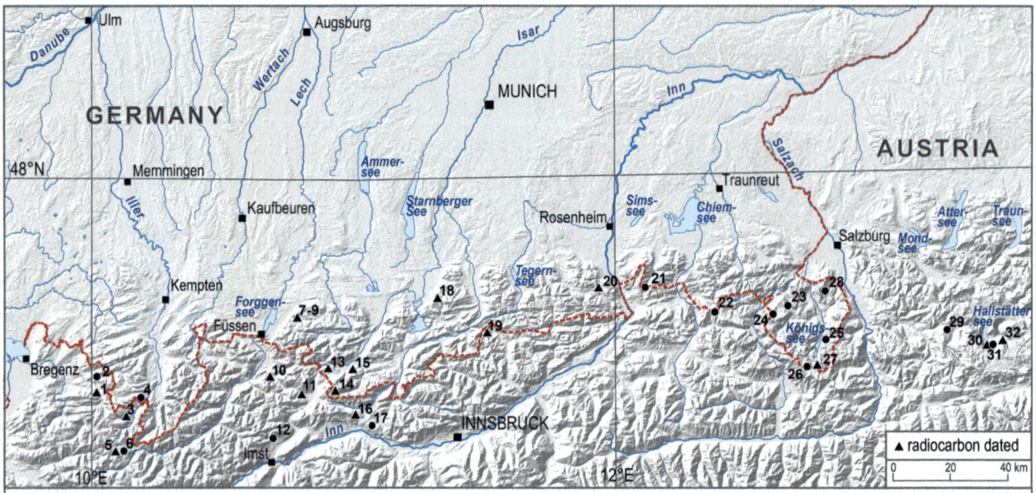

1 Moor am Krähenberg, 940 m a.s.l. (Kral 1993)
2 Kojenmoos, 1220 m a.s.l. (Kral 1993)
3 Moor am Gasthof Bergblick (I, III + IV), 1050 m a.s.l. (Dieffenbach-Fries 1981)
4 Fünfblänkenmoor, 1180 m a.s.l. (Grosse-Brauckmann 1998)
5 Kalbelesee, 1650 m a.s.l. (Walde & Oeggl 2003)
6 Körbersee, 1670 m a.s.l. (Walde & Oeggl 2004)
7 Moor bei der Sennalpe, 1350 m a.s.l. (Obidowicz & Schober 1985)
8 Sattlermoos, 1120 m a.s.l. (Bludau 1985)
9 Wasserfilz, 1075 m a.s.l. (Bludau 1985)
10 Heiterwanger Moor, 980 m a.s.l. (Kral 1989, Walde 2010)
11 Lermoos, 970 m a.s.l. (Kral 1989, Oeggl 2004)
12 Moor nordöstlich vom Göfelesee, 1485 m a.s.l. (Kral 1989)
13 Stanglmoos, 1115 m a.s.l. (Weber 1999)
14 Zugspitzplatt, 2030 + 2290 m a.s.l. (Grüger & Jerz 2010, Stojakowits et al. 2019)
15 Kreuzjoch, 1700 m a.s.l. (Weber 1999)
16 Katzenloch, 1220 m a.s.l. (Wahlmüller 1985)
17 Seefelder See, 1200 m a.s.l. (Wahlmüller 1985)
18 Rothmoos, 1030 m a.s.l. (Bludau & Görres 1993)
19 Halslkopfmoor, 1370 m a.s.l. (Adamski & Friedmann 2019)
20 Oberes Arzmoos, 1115 m a.s.l. (Peters 2010)
21 Moor auf der Ackeralm, 1356 m a.s.l. (Schmeidl 1980)
22 Winkelmoos, 1160 m a.s.l. (Mayer 1963)
23 Schwimmend Moos, 1370 m a.s.l. (Mayer 1966)
24 Reiteralpe, 1600 to 1800 m a.s.l. (Kral 1990)
25 Priesbergmoos, 1380 m a.s.l. (Mayer 1966)
26 Funtensee, 1600 m a.s.l. (Müller et al. 1985)
27 Baumgartl, 1720 m a.s.l. (Mayer 1965, 1966)
28 Stöhrhaus, 1895 m a.s.l. (Kral 1987)
29 Moosklausalm, 1130, m a.s.l. (Draxler 1977)
30 Gjaidalm, 1715 m a.s.l. (Kral 1971, Draxler 1977)
31 Hirzkarsee, 1800 m a.s.l. (Schmidt 1978)
32 Hirschgrube, 1690 m a.s.l. (Drescher-Schneider 2014)

Fig. 28.1 Overview of high altitude (>900 m) palynological study sites in the northern Calcareous Alps of Germany and Austria

vegetation belt in the Northern Calcareous Alps. Particularly, we focus on results of long-term vegetation monitoring, phytodiversity, species turnover and present a grazing exclusion experiment.

28.2 Study Area

The Northern Calcareous Alps in our context are comprised of the Allgäu Alps and Bavarian Alps in Germany and the Bregenzer Wald, the North Tyrolean Alps and the Dachstein Mountains in northern Austria (Fig. 28.1). These mountain ranges consist mainly of limestone and dolomite. The northern Pre-Alps can also be made up of Flysch and Molasse. The highest peaks in the study area are Mädelegabel (2644 m a.s.l.), Zugspitze (2962 m), Watzmann (2713 m) and Hoher Dachstein (2995 m).

Due to the dominating limestone rocks, intensive physical weathering and the harsh climate eutric Leptosols and eutric Regosols as well as folic Histosols are the most widely spread soil types (Grashey-Jansen et al. 2014). As a consequence many karst features can be found within the area and the nearly complete absence of surface runoff can even lead to edaphic aridity.

The climate of the Northern Calcareous Alps can be described as cool and humid. Suboceanic conditions prevail in the west and subcontinental conditions in the east. Based upon data (1983–2012) from the Deutscher Wetterdienst (2013) for the stations Zugspitze (2964 m) and Garmisch-Partenkirchen (719 m) the annual mean temperature ranges from 6.5 °C (700 m), 0,72 °C (2000 m) to −2.84 °C (2700 m a.s.l.). For the same period the mean rainfall recorded at Garmisch was 1364 mm, at the Zugspitze it was 2057 mm. Generally, precipitation decreases from the west to the east in the Northern Calcareous Alps and is highest on the windward slope of the northerly exposed peaks.

28.3 Late and Postglacial Vegetation History of the Northern Calcareous Alps

In this chapter the vegetation development in the Northern Calcareous Alps for the Late Glacial and Holocene is explained by example of the reference site Seefelder See (Wahlmüller 1985, Figs. 28.2 and 28.3) with special reference to other important palynological investigations (Fig. 28.1).

After the Würmian ice retreat a sparse herbaceous pioneer vegetation with steppe character evolved during the Oldest Dryas (*c.* 18,000–14,700 cal. BP) in ice-free parts of the montane zone. In these pioneer communities Poaceae, *Artemisia*, Chenopodiaceae, *Helianthemum*, and *Thalictrum* frequently occurred (e.g. Draxler 1977; Wahlmüller 1985). At the beginning of the Bølling interstadial (*c.* 14,700–13,800 cal. BP) a shrub phase with *Juniperus*, *Hippophaë*, *Salix* and dwarf birches initiated the following reforestation with tree birches (Fig. 28.2). Further to the east, as shown in the Chiemgau Alps (Schmeidl 1980) and the Dachstein area (Draxler 1977), birches were less important. At the end of the Bølling *Pinus sylvestris* reimmigrated into a mosaic of small open birch stands and meadows, whereas isolated valleys with their surrounding mountain slopes were colonized by *Betula* and *Pinus* during the Allerød interstadial first (Dieffenbach-Fries 1981; Stojakowits et al. 2014). In the Allerød (*c.* 13,800–12,880 cal. BP) open pine forests dominated in the montane zone. According to Schmeidl (1980), Wahlmüller (1985) and Walde (2010) *Pinus cembra* was also present. Furthermore, the existence of *Pinus mugo* above the tree line ('krummholz zone') is assumed (Schmeidl 1980). The climatic deterioration of the Younger Dryas (*c.* 12,880–11,590 cal. BP) led to an opening of forests mainly consisting of *Pinus* and subordinated *Betula*. At higher altitudes, the timberline was

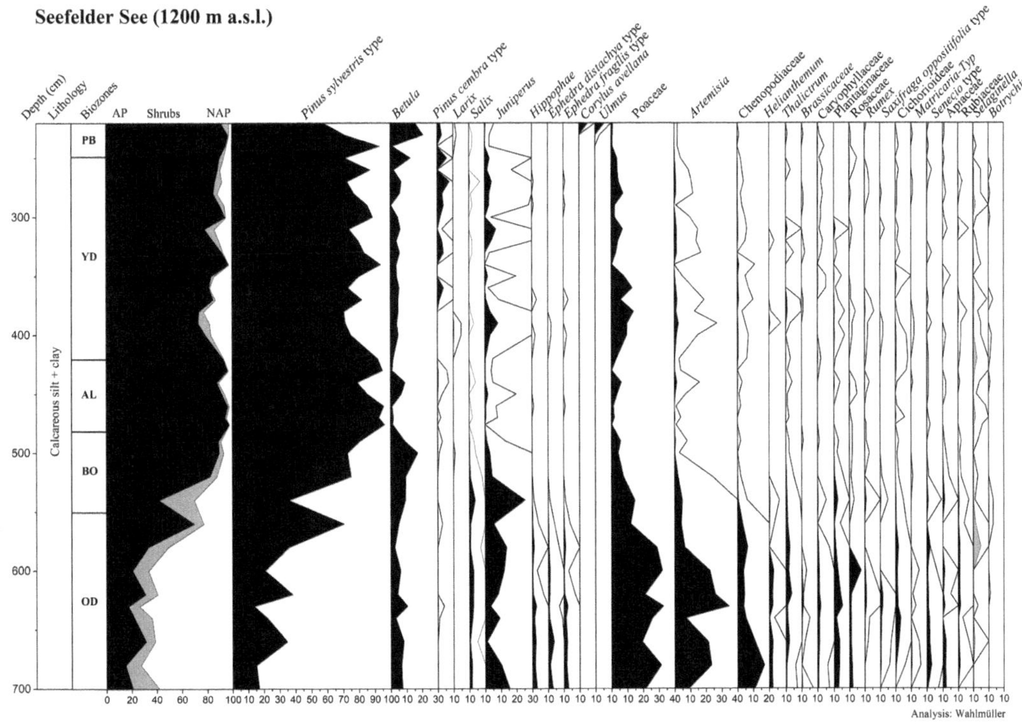

Fig. 28.2 Pollen diagram of Seefelder See-Late Glacial section (redrawn after Wahlmüller 1985) *Data source* EPD—https://www.europeanpollendatabase.net/

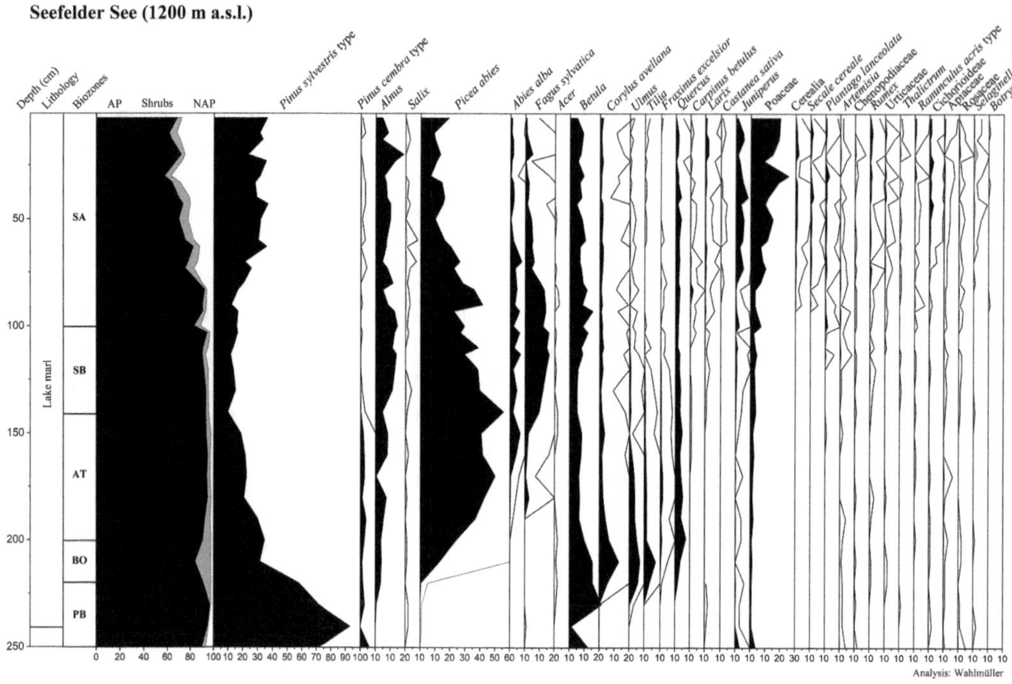

Fig. 28.3 Pollen diagram of Seefelder See-Holocene section (redrawn after Wahlmüller 1985) *Data source* EPD—https://www.europeanpollendatabase.net/

depressed up to 200 m (Kral 1979; Burga and Perret 1998; Stojakowits et al. 2014). The depressed timberline is indicated by increased findings of *Pinus cembra* (Wahlmüller 1985, Fig. 28.2). Generally, non-arboreal pollen and especially heliophilous elements like *Artemisia* increased. The Younger Dryas can be subdivided into two parts: a moister period rich in Poaceae and Cyperaceae preceding a drier one with more *Juniperus* in the pollen content.

As a consequence of the rapid warming at the Late Glacial-Holocene transition the pine woodlands became denser and the tree line rose again through the Preboreal (*c.* 11,590–10,300 cal. BP). High charcoal values point to the occurrence of frequent fires in these light forests (Adamski and Friedmann 2019). Since the Preboreal, *Larix decidua* and *Pinus cembra* occupied the timberline ecotone in many areas (e.g. Mayer 1966; Kral 1979, 1989). At the end of this period thermophilous trees like *Corylus avellana* and *Ulmus glabra* reached the Northern Calcareous Alps as well as *Picea abies* (Fig. 28.3), which expanded in the montane zone and subalpine zone from east to west (Kral 1979) reaching the western part of the Bavarian Alps in the Boreal. *Quercus*, *Tilia*, and *Acer* immigrated with some delay. In the Boreal (*c.* 10,300–8500 cal. BP) thermophilous forests developed in the lower montane zone at the expense of *Pinus*. At higher altitudes *Picea abies* expanded. This change in forest composition towards more deciduous trees in the lower montane zone lead to the reduction of wildfires.

During the Atlantic Period (*c.* 8500–6250 cal. BP) *Picea* dominated the forests of the higher montane and subalpine zones and displaced *Corylus*. *Abies alba* and *Fagus sylvatica* immigrated and gained importance after the 8200 cal. BP-cold event (Rohling and Pälike 2005) leading to mixed montane forests of *Picea, Abies*, and *Fagus* in the northernmost Calcareous Alps during the late Atlantic period (Friedmann and Stojakowits 2017; Adamski and Friedmann 2019). Depending on site conditions and altitude, these mixed mountain forests were dominated by either one of the three species. This was the last great forest restructuring without the influence of humans (Fig. 28.3).

In the Subboreal (c. 6250–2900 cal. BP) high montane beech forests composed of *Fagus sylvatica* with some *Acer pseudoplatanus* and *Abies alba* spread. High montane and subalpine forests consisting of *Abies alba* and *Picea abies* also established, gaining ground in the inner Calcareous Alps and rising timberlines. At the beginning of the Subboreal a widespread elm decline is witnessed in several profiles of the Northern Calcareous Alps (Kral 1979; Wahlmüller 1985; Oeggl 2013) probably by multiple causes including climatic deterioration. First local forest clearings and grazing indicators are recorded in several higher elevation pollen records dating to the younger Neolithic and the Bronze age (Bludau 1985; Wahlmüller 1985; Walde 2010; Oeggl and Nicolussi 2009; Friedmann and Stojakowits 2017; Adamski and Friedmann 2019). This lead to a depression of the timberline (Fig. 28.4). Alpine farming during the Bronze Age in the Dachstein area (Drescher-Schneider 2014) coincides with the salt mining activities in the surroundings and the climate optimum in that time (Gilck and Poschlod 2019).

During Subatlantic times (*c*. 2900–0 cal. BP) human activities become more widespread and changes in forest composition are now mainly caused by man. Late Iron Age and Roman high alpine pastoral land-use is documented in many investigated pollen diagrams, very often continuously until modern times. At the same time charcoal values increase again indicating the use of fire by man to clear land (Adamski and Friedmann 2019). Timberlines were anthropogenically lowered again.

In the High Middle Ages many more forest stands of the montane zone were cleared in order to create pasture ground (e.g. Mayer 1966; Kral 1987). Locally, the clearances extended up to the tree line (Mayer 1966). Cultural indicators are recorded in most reviewed pollen diagrams. In the Modern Era also higher isolated sites of the subalpine and alpine zone were affected by large-scale grazing (Stojakowits et al. 2019, Fig. 28.4). In Modern times charcoal values increase again further because of charcoal burning and wildfires.

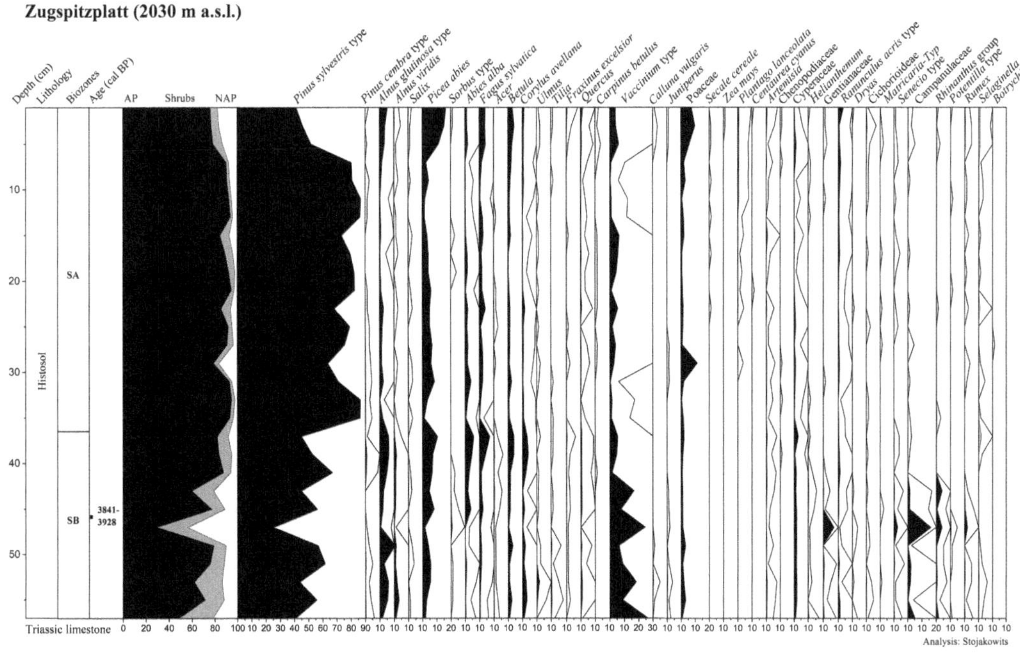

Fig. 28.4 Pollen diagram of Zugspitzplatt (modified according to Stojakowits et al. 2019)

From the Late Glacial up to the Atlantic period, the Northern Calcareous Alps were almost completely covered with forests up to the tree line. Only flood plains, bogs, rocky terrain, avalanche tracks, and the areas beyond the tree limit were naturally treeless. The natural vegetation has thereafter been substantially altered and modified by human impact (pasture, alpine dairy and forestry) since thousands of years leading to anthropogenic substitutional communities (current actual vegetation, managed forests and grasslands).

28.4 Recent Vegetation Dynamics on the Zugspitzplatt

About 4500 species of vascular plants are recorded in the entire Alps (Aeschimann et al. 2004). Of these about 650 species can be found in the alpine and nival vegetation belt thus representing the characteristic alpine flora. The flora and vegetation of the Alps is marked by an adaptation to altitude and extreme environmental conditions and can be divided along an altitudinal gradient with decreasing temperature into different belts (Körner 2003).

The Zugspitzplatt is part of the Wetterstein Mountains in southern Bavaria (Germany) which culminates in the Zugspitze peak (2962 m). The investigated area consists of almost the entire Zugspitzplatt between 2000 and 2700 m with a total surface area of approximately 8 km^2 (Fig. 28.5). It includes parts of the upper subalpine vegetation belt, the alpine vegetation belt and, a vast area of scree, rock and snowbed communities, mainly located in the subnival vegetation belt (Fig. 28.5).

The following chapter presents results of recent vegetation mapping and monitoring programmes and is based on vegetation mapping of over 300 rélevés.

The **subalpine belt** is restricted to the lowest parts of the Zugspitzplatt in the SE and reaching an altitude of hardly 2100 m a.s.l. (Fig. 28.5). Within this zone, a patchy vegetation of *Pinus*

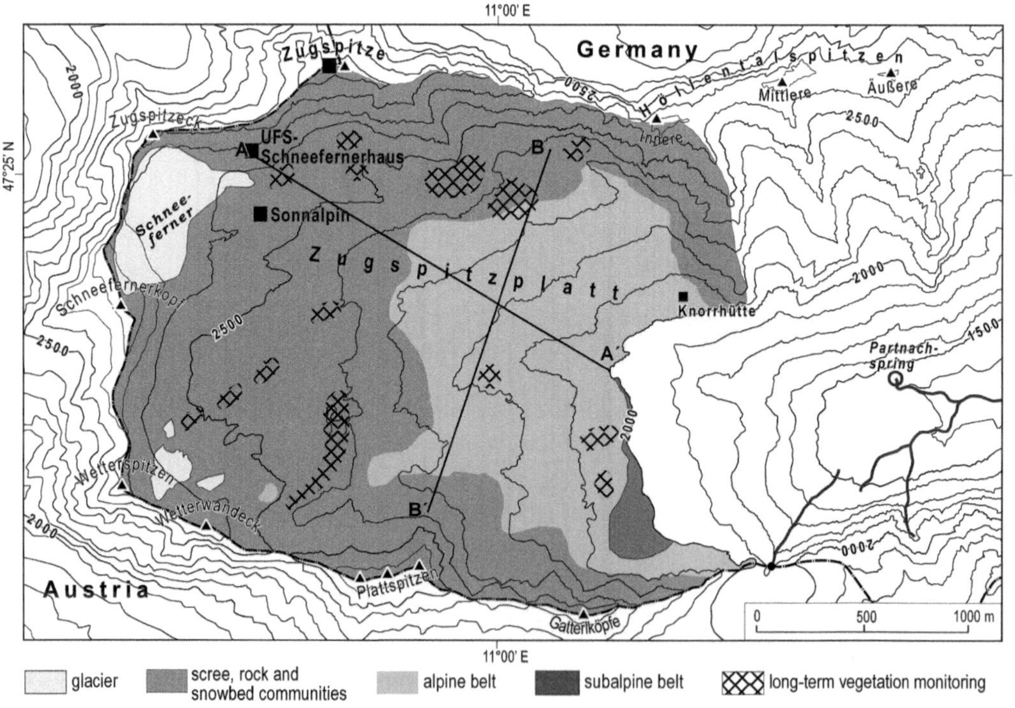

Fig. 28.5 Location and vegetation belts of the Zugspitzplatt in the Wetterstein Mountains (Bavarian Alps, Germany)

mugo-krummholz, grassland communities as well dwarf-shrub heath is typical. The latter two formations can also be found on the adjacent lower alpine zone what underlines the ecotone character of this area. Species-rich grasslands mark the ecotone between the subalpine and the alpine belt (Friedmann and Korch 2010; Korch 2014). These areas share many elements with the *Seslerio-Caricetum sempervirentis* (Br.-Bl. in Br.-Bl. et Jenny 1926). The *Caricetum firmae* is the typical association of the **alpine belt** and the most common grassland-community on the Zugspitzplatt. It is reaching an altitude of >2500 m on climatically favored slopes on the southwestern Platt. It prevails also on consolidated scree as on karst-bedrock, but overall species-richness decreases with increasing altitude. While the *Caricetum firmae* must be considered as a mostly stable community in the upper part of the alpine belt, further development leading to more complex associations like the *Seslerio-Caricetum sempervirentis* is possible in the lower parts.

Directly linked to acidified slopes, either by advancing pedogenetic development or on former krummholz sites, the Geo montani-*Nardetum strictae* (Lüdi 1928 nom. mutat. propos.) is the first of two plant communities within the investigation area of anthropogenic-zoogenic origin (Korch and Friedmann 2012, 2016). Promoted by grazing sheep *Nardus stricta* invades other communities with suitable conditions which additionally shows big tolerance to trampling. The community is restricted to Cambisols and former *Pinus mugo*-krummholz on the lower part of the Zugspitzplatt. Also of anthropogenic-zoogenic origin is the *Alchemillo-Poetum supinae* with a distribution all over the alpine belt. It is typically found in places where grazing livestock prevails. Of low diversity and dominated by graminoids and other species with a high nutrient requirement (e.g. *Taraxacum alpinum, Urtica dioica*), it is adapted to herbivory.

Above the 2400 m contour line in the **subnival belt** the general vegetation cover rapidly diminishes giving way to associations adapted to harsh climatic conditions as well as to scree slopes, unprotected bedrock, and snowbeds. Among these the *Salicetum retuso-reticulatae* (Br.-Bl. in Br.-Bl. et Jenny 1926) takes an interim position between alpine meadows and scree communities (Zöttl 1950). On the Zugspitzplatt it has only a small distribution and is dominated by the dwarf willows *Salix retusa* and *Salix serpillifolia*. The typical association of the snowbeds is the *Arabidetum caeruleae* (Br.-Bl. 1918). Beyond the snowbeds the *Arabidetum caeruleae* is widely found on the ski slopes of the upper Zugspitzplatt. Due to slope preparation and artificial accumulation and compacting of snow these sites become snow-free significantly later than the surrounding environment. The steeper scree slopes with moving scree are dominated by the *Thlaspietum rotundifolii*. On southerly exposed sites, stagnant scree is covered by the *Leontodontetum montani* (Korch 2014; Korch and Friedmann 2016).

The vegetation dynamics on the Zugspitzplatt are driven by the following factors (Korch 2014; Korch and Friedmann 2016).

28.4.1 Site Conditions

The spatial vegetation patterns are strongly influenced by geomorphologic processes as well as the given and changing soil formations. The autochthonous Wetterstein-limestone strongly influences soil developing processes leading to a basophilic vegetation. Besides, some sites show signs of soil acidification due to late glacial and actual aeolian deposition of mica from the Central Alps (Küfmann 2003; Grashey-Jansen et al. 2014). Azonal acidophilic plant communities as the *Geo-montani Nardetum strictae* on cambisols are directly linked to this phenomenon.

28.4.2 (Site) Climate

Different exposition to solar radiation and wind as well as the duration of annual snow cover directly influence the occurrence of plant communities. Boundary layer climate investigations on the Zugspitzplatt have shown that neighboring sites often strongly differ in their site climate leading to a completely altered vegetation. For example, the *Arabidetum caeruleae*, which is linked to snowbeds with an annual snow cover up to ten months, can be found neighboring *Caricetum firmae*-meadows indicating stronger insolation and thus a shorter snow cover.

Although meteorological records show an increase of temperature and precipitation for the whole area, long-term changes of vegetation patterns such as an upward shift of vegetation belts are not yet possible to prove due to missing long-term vegetation monitoring over decades.

28.4.3 Grazing

At least since the sixteenth century, there are up to (nowadays) 400 sheep grazing at the Zugspitzplatt during the summer months. Along with wildlife this leads to an area-wide alteration of the flora and vegetation. Some of the recorded plant communities as the *Geo-montani Nardetum strictae* and the *Alchemillo-Poetum supinae* are even directly linked to grazing and trampling-activity within the study site.

28.4.4 Anthropogenic Disturbance

Since the beginning of the 1930s when the cog-wheel railway to the Zugspitzplatt and the Hotel Schneefernerhaus where built the Zugspitzplatt has been made accessible to mass tourism. Today the area around the buildings and the ski-resort show no or severely influenced vegetation patches. A good example for the latter are the frequently found species of the *Arabidetum caeruleae* on ski-slopes. The anthropogenically prolonged snow cover on the slopes gives them an advantage over plants of other communities. On the more remote sites and during the summer months trampling by hiking-tourists causes damages and alteration to plant life mainly along hiking-paths.

28.4.5 Time

The already mentioned factors and processes proceed at different time scales. Time must be considered as a superordinate factor linking them. In the course of time, different connections, intensifications, and interactions between the named elements can be observed and presumed.

28.4.6 Long-Term Monitoring on Permanent Sample Plots

Since 2010 a long-term monitoring programme on permanent sample plots has been established within the investigated area on the Zugspitzplatt (Fig. 28.5). Figure 28.6 shows the results for selected plots on 8 selected sample plots for the period 2010–2018. The columns show the development of species magnitude and species richness on these plots. Unlike the exclosure experiment (3.4.7) no clear trend can be read out of the results so far. The reason for this is probably the great complexity of the interactions and influences of the already described driving factors of vegetation dynamics on these sites. Another problem is the difficulty of mapping the plots at the same stage of phenology in the course of the different years.

28.4.7 Exclosure Grazing Experiment 2016 and 2017

In 2016 and 2017 an exclosure experiment was carried out on the southwestern part of the Zugspitzplatt. As grazing was identified as one of the major factors influencing vegetation dynamics within the investigated area, the aim of this experiment was to figure out how the suspension of grazing affects the flora and vegetation within the investigated sites.

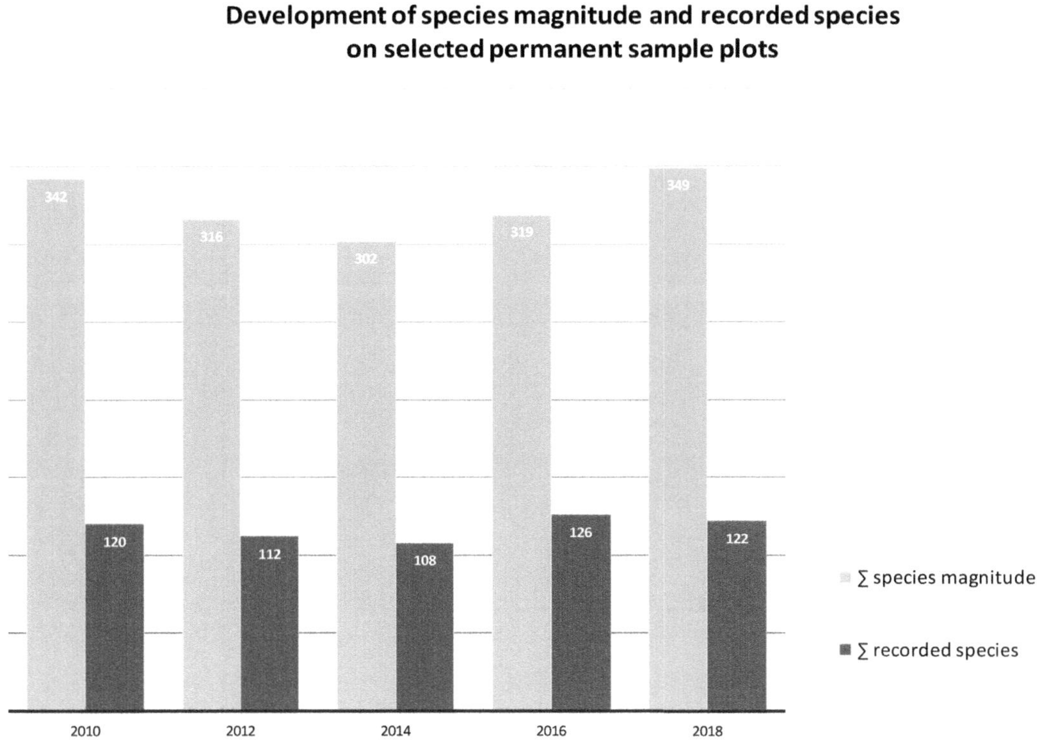

Fig. 28.6 Long-term monitoring results for species magnitude and species richness in 8 selected sample plots of the Zugspitzplatt for the period 2010–2018

The results of the two exclosures are illustrated in Figs. 28.7 and 28.8. The first column shows species magnitude and species richness for either site on August 17th 2016. At this stage at the end of the grazing season, both sites had been completely grazed. The second column shows the recorded parameters on July 05th 2017. At this date shortly before annual grazing sets in, the electric fences were installed. The third and fourth columns finally plot the recorded data for August 17th 2017 for the fenced sites and for nearby grazed reference sites, respectively.

The results clearly show, that both, species magnitude as well as recorded species numbers constantly increased prior to the beginning of the grazing season 2017 and within the fenced areas during grazing season 2017. In contrast, the situation on both reference sites 2017 is similar to the situation of the later fenced sites in 2016.

As a conclusion, the suspension of sheep grazing clearly leads to a short-term increase of phytomass and species richness on the Zugspitzplatt. However, it must be doubted that these results are also valid on a long-term scale, especially referring to species richness (Erschbamer et al. 2015; Mayer and Erschbamer 2017). The permanent absence of grazing would probably allow the quick expansion of competitive species leading to less niches for less competitive, specialized species. In the long range it can be expected, that species richness would decrease with a complete ceasing of sheep grazing.

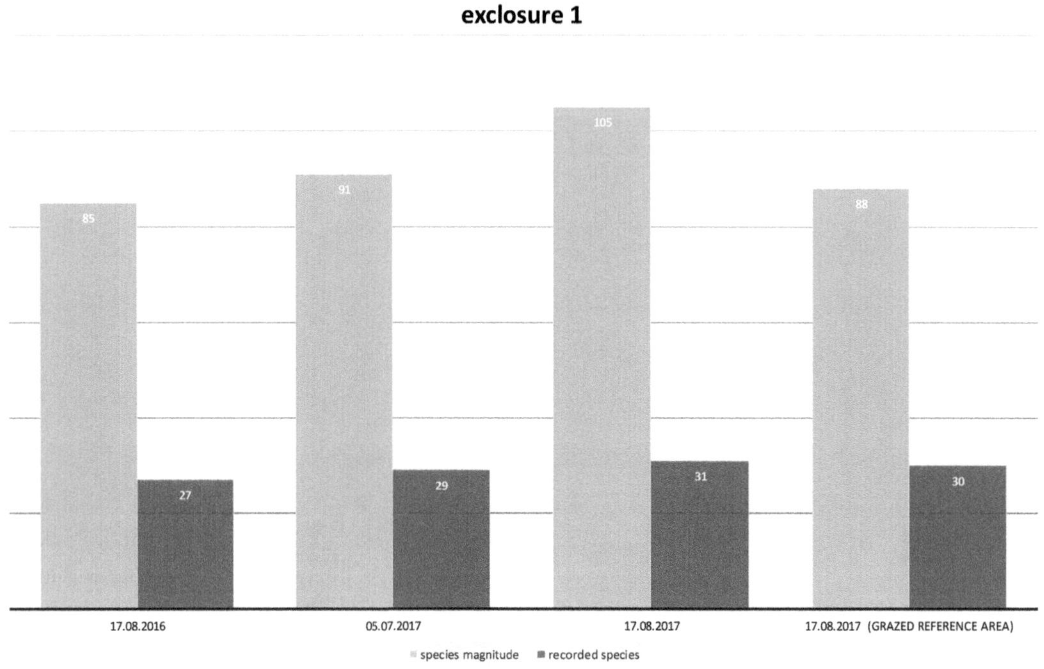

Fig. 28.7 Results of the grazing exclosure experiment 1 for the period 2016–2017: see text for explanation

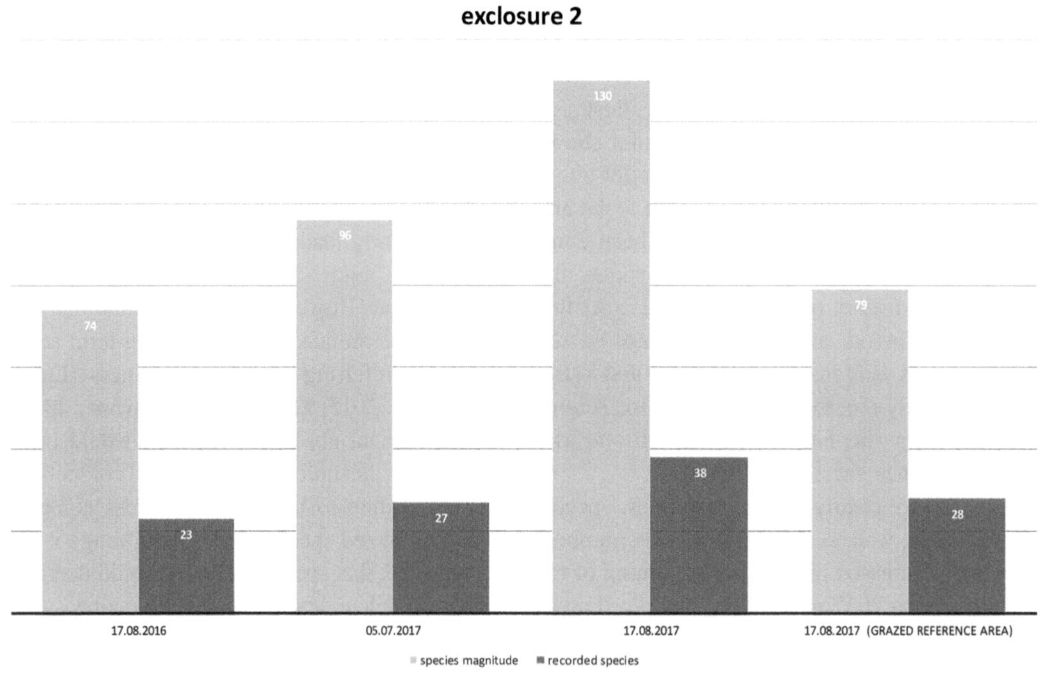

Fig. 28.8 Results of the grazing exclosure experiment 2 for the period 2016–2017: see text for explanation

Acknowledgements We would like to thank the Bavarian State Ministry of the Environment and Consumer Protection (former Bavarian State Ministry of the Environment and Public Health) for funding the projects "Ecological analysis of the subalpine to subnival vegetation zones on the Zugspitzplatt (HÖHENZUG)" and "Vegetation dynamics on the Zugspitzplatt" which are part of the collaborative program "Consequences of climatic change in the Alps—analysis by altitudinal gradients" Klimagrad (2009–2013) and Klimagrad 2 (2015–2018). We also would like to thank the Environmental Research Station Schneefernerhaus (UFS) for the logistic support, the Bayerische Zugspitzbahn Bergbahn AG (BZB) for providing free cable car access to the Zugspitzplatt 2010–2017 and the Weidegenossenschaft Partenkirchen for the support of the grazing pattern investigations.

References

Adamski S, Friedmann A (2019) Holozäne Vegetations- und Feuergeschichte des Halslkopfmoores (NO-Karwendel, Österreich). Innsbrucker Geograph Studien 41:45–67

Aeschimann D, Lauber K, Moser DM and Theurillat J-P (2004) Flora alpina: Ein Atlas sämtlicher 4500 Gefäßpflanzen der Alpen. 3 vol., Haupt Verlag, Bern, 2670 p

Bludau W (1985) Zur Paläoökologie des Ammergebirges im Spät- und Postglazial. Schäuble, Rheinfelden

Bludau W, Görres M (1993) Untersuchungen zur Siedlungstätigkeit des Menschen im süddeutschen Gebirge am Beispiel eines ombrogenen Moores – Pollenanalytische und geochemische Ergebnisse. Telma 23:213–236

Burga C, Perret R (1998) Vegetation und Klima der Schweiz seit dem jüngeren Eiszeitalter. Ott, Thun

Deutscher Wetterdienst (2013) Weather request and distribution system. https://werdis.dwd.de

Draxler I (1977) Pollenanalytische Untersuchungen von Mooren zur spät- und postglazialen Vegetationsgeschichte im Einzugsgebiet der Traun. Jahrb. Geol. B.-A. 120:131–163

Drescher-Schneider R (2014) Pollenanalysen zur Frage der Klimaveränderungen und des menschlichen Einflusses im Dachsteingebiet und im Salzkammergut. Gmunder Geo-Studien 5:57–63

Dieffenbach-Fries H (1981) Zur spät- und postglazialen Vegetationsentwicklung bei Oberstdorf (Oberallgäu) und im Kleinwalsertal (Vorarlberg). Pollen- und makrofossilanalytische Untersuchungen an drei Mooren der montanen Stufe. Dissertation, Technische Hochschule Darmstadt

Erschbamer B, Mayer R, Mallaun M, Unterluggauer P (2015) Alpine Pflanzengesellschaften unter dem Einfluss von Sukzession und Klimawandel. Ber Reinhold-Tüxen-Ges. 27:187–200

Friedmann A, Korch O (2010) Die Vegetation des Zugspitzplatts (Wettersteingebirge, Bayerische Alpen): Aktueller Zustand und Dynamik. Ber Reinhold-Tüxen-Ges. 22:114–128

Friedmann A, Stojakowits P (2017) Zur spät- und postglazialen Vegetationsgeschichte des Allgäu mit Alpenanteil. In: Lechterbeck J, Fischer E (eds) Kontrapunkte. Festschrift für Manfred Rösch. Universitätsforschungen zur prähistorischen Archäologie, vol 300, pp 51–63

Gilck F, Poschlod P (2019) The origin of alpine farming: A review of archaeological, linguistic and archaeobotanical studies in the Alps. The Holocene 29:1503–1511

Grashey-Jansen S, Korch O, Beck C, Friedmann A, Bernhard R, Dubitzky C (2014) Aeolian influenced soil sites in consideration of atmospheric circulation types—a case study in the alpine zone of the Zugspitzplatt (Northern Calcareous Alps, Germany). J Geol Agric Environ Sci 2(4):11–19

Grosse-Brauckmann G (1998) Das Fünfblänkenmoor am Engenkopf, ein bemerkenswertes ombrosoligenes Moor in einem Karstgebiet des südlichen Allgäus. Carolinea 56:29–62

Grüger E, Jerz H (2010) Untersuchung einer Doline auf dem Zugspitzplatt. E&G Quaternary Sci J 59:66–75

Korch O (2014) Untersuchungen zu Flora und Vegetation des Zugspitzplatts (Wettersteingebirge, Bayerische Alpen) – Rezente Vegetationsdynamik unter besonderer Berücksichtigung klimatischer und anthropozoogener Prozesse. - Dissertation, Universität Augsburg

Korch O, Friedmann A (2012) Phytodiversität und Dynamik der Flora und Vegetation des Zugspitzplatts. Jahrb Ver Schutz Bergwelt 76(77):217–234

Korch O, Friedmann A (2016) Vegetation und Vegetationsdynamik auf dem Zugspitzplatt (Bayerische Alpen): Natur- und Kulturlandschaft im hochalpinen Raum als Produkt natürlicher, anthropogener und zoogener Einflüsse. Polarforschung 86(1):35–45

Körner C (2003) Alpine plant life. Springer, Berlin, p 349

Kral F (1971) Pollenanalytische Untersuchungen zur Waldgeschichte des Dachsteinmassivs. Inst. Waldbau, Hochschule f. Bodenkultur, Wien, Veröff

Kral F (1979) Spät- und postglaziale Waldgeschichte der Alpen aufgrund der bisherigen Pollenanalysen. Inst. Waldbau, Hochschule f. Bodenkultur, Wien, Veröff

Kral F (1987) Ein pollenanalytischer Beitrag zur Waldgeschichte des Salzburger Untersberges. Jahrbuch Verein Z Schutz D Bergwelt 52:93–105

Kral F (1989) Pollenanalytische Untersuchungen im Fernpaßgebiet (Tirol): Zur Frage des Reliktcharakters der Bergsturz-Kiefernwälder. Verh. Zool.-Bot.-Ges. Österreich 126:127–138

Kral F (1990) Ein pollenanalytischer Beitrag zur natürlichen und anthropogenen Waldentwicklung in den Berchtesgadener Alpen. Forschungsberichte Nationalpark Berchtesgaden 20:7–20

Kral F (1993) Zum Aufbau von Fichten-Tannen-Buchenwäldern im jüngeren Postglazial (Bregenzerwald und Obersteiermark). Verh. Zool.-Bot.-Ges. Österreich 130:171–188

Küfmann C (2003) Soil types and eolian dust in high-mountainous karst of the Northern Calcareous Alps (Zugspitzplatt, Wetterstein Mountains, Germany). CATENA 53:211–227

Mayer H (1963) Tannenreiche Wälder am mittleren Nordabfall der Ostalpen. BLV Verlagsgesellschaft, München

Mayer H (1965) Zur Waldgeschichte des Steinernen Meeres. Jahrbuch Verein z Schutz d Alpenpflanzen und –tiere 30:1–20

Mayer H (1966) Waldgeschichte des Berchtesgadener Landes (Salzburger Kalkalpen). Beihefte Z. Forstwiss Centralblatt 22:1–42

Mayer R, Erschbamer B (2017) Long-term effects of grazing on subalpine and alpine grasslands in the Central Alps, Austria. Basic Appl Ecol 24:9–18

Müller JR, Schmidt R, Schmid AM, Froh J (1985) Die postglaziale Entwicklungsgeschichte des Funtensees (palynologische, sedimentologische und paläolimnologische Untersuchungen eines Bohrkerns). Forschungsberichte Nationalpark Berchtesgaden 7:67–96

Obidowicz A, Schober H (1985) Moorkundliche und vegetationsgeschichtliche Untersuchungen des Sennalpenmoores im Trauchgauer Flysch (Ammergebirge). Berichte Bayer Botan Ges 56:147–165

Oeggl K (2004) Palynologische Untersuchungen zur vor- und frühgeschichtlichen Erschließung des Lermooser Beckens in Tirol. Ber Reinhold-Tüxen-Ges 16:75–86

Oeggl K (2013) Vom Ulmensterben zur Waldverwüstung: anthropogene Vegetationsveränderungen in den Alpen seit dem Neolithikum. Ber Reinhold-Tüxen-Ges 25:95–107

Oeggl K, Nicolussi K (2009) Prähistorische Besiedlung von zentralen Alpentälern in Bezug zur Klimaentwicklung. In Schmid R, Matulla C, Psenner R (eds) Klimawandel in Österreich. Die letzten 20 000 Jahre … und ein Blick voraus. Innsbruck university press. alpine space-man and environment, vol 6, pp 77–86

Peters M (2010) Vergleichende Untersuchung zur Landschafts-, Vegetations- und Siedlungsgeschichte zwischen Donau und Alpen in Südbayern während der letzten 15.000 Jahre. Habilitationsschrift, Universität Augsburg

Rohling E, Pälike H (2005) Centennial-scale climate cooling with a sudden cold event around 8200 years ago. Nature 434:975–979

Schmeidl H (1980) Zur spät- und postglazialen Vegetations- und Waldentwicklung in der montanen Stufe des Kartenblattes Aschau i. Chiemgau. In: Ganns O: Geologische Karte von Bayern 1:25000. Erläuterungen zum Blatt Nr. 8239 Aschau i. Chiemgau. Bayerisches Geologisches Landesamt, München, pp 116–132

Schmidt R (1978) Postglaziale Vegetationsentwicklung und Klimaoszillationen im Pollenbild des Profiles Hirzkarsee/Dachstein 1800 m NN (O.Ö.). Linzer Biol Beitr 10:161–169

Stojakowits P, Friedmann A, Bull A (2014) Die spätglaziale Vegetationsgeschichte im oberen Illergebiet. E&G Quaternary Sci J 63:130–142

Stojakowits P, Korch O, Grashey-Jansen S, Friedmann A (2019) Contributions to the European Pollen database: 45. Zugspitzplatt, Wetterstein Mountains (Germany). Grana 58:396–398

Wahlmüller N (1985) Beiträge zur Vegetationsgeschichte Tirols V: Nordtiroler Kalkalpen. Ber Naturwiss-Medizin Verein Innsbruck 72:101–144

Walde C (2010) Palynologische Untersuchungen zur Kulturlandschaftsgeschichte in Westtirol. Dissertation, Universität Innsbruck

Walde C, Oeggl K (2003) Blütenstaub enthüllt dreitausendjährige Siedlungsgeschichte im Tannberggebiet. Walserheimat 73:162–175

Walde C, Oeggl K (2004) Neue Ergebnisse zur Siedlungsgeschichte am Tannberg. Die Pollenanalysen Aus Dem Körbersee. Walserheimat 75:309–317

Weber K (1999) Vegetations- und Klimageschichte im Werdenfelser Land. Augsburger Geogr Hefte 13:1–127

Zöttl H (1950) Die Vegetationsentwicklung auf Felsschutt in der alpinen und subalpinen Stufe des Wettersteingebirges. Dissertation, LMU München, Munich, 201 p

Assessing the Impact of Climate Change Versus Land Use on Tree- and Forest Line Dynamics in Norway

29

Anders Bryn and Kerstin Potthoff

Abstract

Alpine tree- and forest lines (TFL) are dynamic, influenced by a variety of processes. In Norway, where TFLs are dominated by deciduous mountain birch, the lines are generally on the move to higher altitudes. Contemporary land use changes interfere the interpretation of the consequences of recent high latitude warming for TFL dynamics due to possible combined effects. We have assessed all available long-term studies (>30 years) (n = 8) of TFL dynamics focusing on separating the effects of climate change from those of changed land use in Norway. Most of the studies emphasize the importance of changed land use on TFL dynamics. However, a deeper understanding of the relevance of land use changes and climate changes for TLF dynamics and of possible interacting effects is hampered by a lack of spatially representative TFL change data and site-specific climate and land use change data.

Keywords

Betula pubescens ssp. *czerepanovii* · Climate warming · Mountain birch · Range shift · Time-lag · Timberline · Treeline

29.1 Introduction

Tree- and forest lines (TFLs) in Norway are generally expanding into new elevational ranges (Bryn and Potthoff 2018), although with great variation in speed and altitude (Normark 2012; Wehn et al. 2012). The same pattern is evident from Europe (Cudlín et al. 2017), whereas globally TFLs are mainly either stable or rising (Harsch et al. 2009). However, site-specific natural disturbances, such as caterpillar outbreaks, extreme drought, top-breaks, snow avalanches and fungi attacks can lower TFLs, and create short-term local downwellings (Aas 1969; Rannow 2013; Volden 2018). If these, or other disturbance events become more frequent with a changing climate, TLFs could be lowered on a more permanent basis.

TFLs in Norway are dominated by mountain birch (*Betula pubescens* ssp. *czerepanovii*). Mountain birch is a small deciduous tree, adapted to oceanic climates and frequently browsed by free ranging domestic animals (Karlsson et al. 2005; Speed et al. 2010). Birch treelines are commonly defined as the highest elevational occurrence of upright trees above a certain height

A. Bryn
Natural History Museum, University of Oslo, Oslo, Norway

Department of Geography, University of Bergen, Bergen, Norway

K. Potthoff (✉)
School of Landscape Architecture, Norwegian University of Life Sciences, Ås, Norway
e-mail: kerstin.potthoff@nmbu.no

U. Schickhoff et al. (eds.), *Mountain Landscapes in Transition*,
Sustainable Development Goals Series, https://doi.org/10.1007/978-3-030-70238-0_29

(usually 2, 2.5 or 3 m) (Bryn and Potthoff 2018) (Fig. 29.1). Forest lines are defined as the uppermost patches, tongues or continuous stretches of trees; however, definitions may vary among studies with respect to for example distance between trees, canopy cover, and patch size (Bryn and Potthoff 2018). Some studies also focus on timberlines (Holtmeier 2009; Odland 2015), defined here as the upper occurrences of closed forest. The mountain birch alpine forest line form, sensu Harsch and Bader (2011, termed treeline in their publication), varies from site-to-site and includes diffuse, abrupt and island forms (Fig. 29.2). Although birch dominated treelines respond quicker to external forces than forest lines (Bryn and Potthoff 2018), both lines constitute physiognomic height limits of the same species correlated with similar environmental factors (i.e. climate, disturbances and edaphic and topographic conditions). It is therefore meaningful to treat both lines within one study.

There is increasing scientific attention towards the contribution of climate change to range expansion of TFLs into alpine regions (Gatti et al. 2019; Kullman and Öberg 2009; Körner 2012; Sigdel et al. 2018). A number of studies emphasise the potential for future TFL range expansions, as climate gets warmer and/or wetter (Bobrowski et al. 2017; de Wit et al. 2014; Karlsen et al. 2017), as well as climatic consequences thereof (Rydsaa et al. 2017). Since the uppermost regional TFLs of Norway are well correlated with climate variables such as summer temperature and growth season (Aas and Faarlund 2000; Odland 1996), and the same climate variables have changed during the last century (Tveito 2014), this focus is justified for Norwegian TFL research. Some studies also point to the importance of changed winter temperature and/or snow cover for TFL dynamics (Hagedorn et al. 2014; Harsch et al. 2009). In Norway, winter temperature has increased and the snow cover extent has decreased during the last decades (Rizzi et al. 2018).

However, in many populated mountain areas, especially along the Scandes mountain chain of

Fig. 29.1 Mountain birch treeline at 1342 m a.s.l. in Visdalen, southcentral Norway (Bryn, 2018-06-23)

Fig. 29.2 Abrupt mountain birch forest line at 1259 m a.s.l. in Haverdalen, southeast Norway (Bryn, 2015-08-12)

the Nordic region, land use changes have appeared simultaneously with the last centuries of warming (Bryn 2008; Hofgaard 1997). The background for large-scale effects of land use changes on TFL dynamics is the long-term use of all Norwegian mountain regions. Agriculture got established in South Norway around 3500 BC (Myhre 2004). Through centuries of varied agricultural activities such as cultivation, domestic grazing, forest logging, heath burning and outfield fodder collection, the forest cover gradually decreased (Aas and Faarlund 1995; Bjune 2005; Bryn and Daugstad 2001). From the sixteenth century on, the Sami people started to herd semi-domesticated reindeer in the north, middle and eastern parts of Norway (Hansen and Olsen 2004), and at the same time large-scale mining industry, salt and tar production and timber sales increased throughout the country (Jacobsen and Follum 2008). All these activities slowly confined the forest distribution, lowered TFLs and restructured the tree-species dominance in different regions of Norway (Aas and Faarlund 1995; Bryn et al. 2013).

During the twentieth century, most of the above-mentioned activities gradually declined, and today the use of outfield resources in Norway is at a historically low level (Almås 2004; Bryn and Daugstad 2001; Christensen 2002). Outfield domestic grazing was reduced or concentrated, mountain forest felling was reduced to a negligible amount and outfield scything, heath burning and fodder collection ceased (Almås 2004; Aune et al. 2018; Austrheim et al. 2011; Bryn and Daugstad 2001; Måren 2009; Potthoff 2007). In particular, the abandonment of mountain summer farms, reduced by around 98% from c.1850 to 2017 (Reinton 1955; data received from Statistics Norway), has led to a fundamental reduction of land use intensity at TFL elevations (Bryn and Daugstad 2001; Olsson et al. 2004; Potthoff 2009, 2017). Although domestic grazing in outfields has declined considerably in most parts of Norway, with the notable exception of semi-domestic reindeer grazing in northern parts, it is still considered as a major restricting factor for natural forest regeneration and expansion in Norway (Aune et al. 2011; Hofgaard et al. 2010;

Potthoff 2009; Speed et al. 2010; Wehn et al. 2012). Past and contemporary land use and land use changes therefore interfere the interpretation of climate driven TFL dynamics, leading to possible combined or interacting effects of climate and land use changes on the TFL dynamics.

The main aim of this chapter is to address the impact of climate change versus land use change on TFLs dynamics in Norway. We have assessed all available studies focusing on separating these effects. The studies cover time-scales from 30–100 years. Specifically, the assessment aims at answering the following questions:

1. What is the knowledge status regarding causes of TFL dynamics regulated by climate change versus land use change?
2. To which degree are the assessed studies able to separate the effects of climate change from land use change?
3. How can science move forward to close potential knowledge gaps identified through the assessment of existing studies?

29.2 Material and Methods

To identify all available and relevant studies focusing on TLF change in Norway, we used two search engines: the international WEB of Science (https://apps.webofknowledge.com) and the Norwegian Oria (https://bibsys-almaprimo.hosted.exlibrisgroup.com/primo-explore/search?vid=BIBSYS&lang=en_US). Whereas WEB of Science provides published peer-reviewed journal articles, the Oria database also contains scientific reports as well as master's and doctoral theses. The search terms in WEB of Science included: Norway/Norwegian AND treeline/forest line/timber line in all combinations with land use, climate, birch, regrowth, reforestation, expansion, and range (latest search 12 June 2019). The search terms in Oria included: skoggrense/tregrense AND arealbruk, klima, gjengroing, beite and seter/sæter in all combinations (latest search 12 June 2019).

We extracted all publications that (1) provided information on elevational TFL dynamics in Norway, and (2) that attempted to separate the effects of climate changes from those of land use changes. We excluded palaeobotanical studies based on micro- and/or macro fossils, since the registered TFL elevation dynamics in such studies have higher uncertainty (Bjune 2005; Paus and Haugeland 2017). We also excluded studies which did not include climatic TFLs (i.e. studies with study sites at mountains lower than the potential climatic TFLs) (Bryn and Potthoff 2018). We did not assess short-term TFL monitoring studies or experiments, since the long-term population dynamics and time-lag of range expansion exceeds that of the studies' designs (Máliš et al. 2016; Rannow 2013). However, we have included relevant findings of the latter studies in the discussion.

The search provided 8 studies focusing on TFL changes with a particular focus on separating land use and climate change effects on TFL dynamics (Table 29.1). The studies were assessed systematically, following two schemes. First, we identified relevant descriptors of the studies, such as temporal coverage, methods for registering elevational TFLs changes, empirical TFL change results and location of the study (Table 29.1). Second, we assessed the separation of the effects of climate and land use change focusing on study design, data sampling, register data, analysis of data and conclusions (Table 29.2).

A critical question regards the study design: How is the contribution of land use changes versus climate changes for temporal TFL dynamics decomposed methodologically (Table 29.2)? We have categorized the studies according to three main methods:

(1) Studies that have included both climate and land use change data enabling a statistical separation of the processes' contribution (Category: Statistically);
(2) Studies that have included both types of data, but where statistical methods have not been implemented to separate the contribution of each process (Category: By comparison);

Table 29.1 Description of the studies

Topic		Aschwanden (2002)	Bryn (2008)	Engum (2006)	Hofgaard (1997)	Normark (2012)	Rössler et al. (2008)	Volden (2018)	Wehn et al. (2012)
Time period	Time period covered	c.1900–2000	1959–2001	1973–1993	–	1913–2011	1960s/1970s–1990s	1938–2017	1960–2002
Methods	Methods used to map TFL changes	IV/MC	API	API	PM	RM	API	RM	API
	Data used to interpret the cause of TFL changes	TR/RD	TH/RD/PV	RD	–	RD	TR/RD/IV	RD	RD
Results	Lines considered	FL	FL	TL/FL[a]	TL/FL	TL/FL	FL[b]	TL/FL	TiL[c]
	Direction of TL change	–	–	–	–	A	–	A	–
	Direction of FL change	A	A	S	–	A	A	A	–
	Direction of TiL change			A	–	–	–	–	A
	Average/annual TL change (m)	–	–	–	–	74/0.76	–	55/0.70	–
	Average/annual FL change (m)	–	32/0.76	0/0	–	26 (not sig.)/0.27	–	48/0.61	–
	Average/annual TiL change (m)			50–60/2.5–3	–	–	–	–	4.24/0.1
	Degree to which change is consequence of land use change?	I	L	S	L	N/S	L	I	L
	Degree to which is change a consequence of climate change	I	N/S	L	N/S	L	N/S	I	N/S
Location		N	E	W	W	N	W, E	W	E
Height-demand defining a tree		>2 m	>2.5 m	>1 m	>2 m	>2 m	–(>50 cm)	>2.5 m	Not provided

API aerial photo interpretation; *RM* remapping; *IV* interview (results reported); *MC* map comparison; *PM* plot mapping; *TR* tree-ring; *TH* tree height; *RD* register data; *PV* potential vegetation map; *FL* forest line; *TL* treeline; *TiL* timberline; *A* advance; *S* stability; *L* large; *I* intermediate; *N/S* non/small; *N* North Norway; *E* East Norway; *W* West Norway

[a]Engum's (2006) TL is according to our definition FL and his FL is according to our definition TiL

[b]Rössler et al. (2008) use 'treeline' in their publication

[c]Wehn et al. (2012) use 'forest line' in their publication

Table 29.2 Assessment of separation climate versus land use change

Evaluation topic	Evaluation criteria	Aschwanden (2002)	Bryn (2008)	Engum (2006)	Hofgaard (1997)	Normark (2012)	Rössler et al. (2008)	Volden (2018)	Wehn et al. (2012)
Study design	Is the study designed to document vertical TFL changes?	Y	Y	Y	N	Y	Y	Y	Y
	Is the study designed for evaluation of the contribution of land use and climate change effects?	Y	Y	P	N	P	Y	P	Y
	How is the contribution of land use versus climate change methodologically decomposed?	I	Cm	Cm	I	Cm	Cm	I	S
Data sampling	What is the number of methods used for TFL change detection?	2	1	1	–	1	1	1	1
	What is the sampling size regarding TFL change detection?	>30	>100	>100	–	30–100	>100	30–100	>100
Register data	Is the land use change data site-specific, local or regional?	SP	L	L	R	L	L	–	SP
	Is the climate data local, county-wise or regional?	R	L/R	R	R	R	R	L	R
Analysis of data	Is the analysis of TFL change causes based on qualitative or quantitative analysis?	Ql	Qn	Ql	Ql	Qn	Ql	Qn	Qn
Conclusions	What geographic domain is the conclusion representative for in Norway—local, regional or national?	L	L/R	L	L	R	R	L/R	L/R

Y yes; *N* no; *P* partially; *S* statistically; *Cm* by comparison; *I* by indication; *SP* side-specific; *L* local; *R* regional; *N* national; *Ql* qualitative; *Qn* quantitative

(3) Studies that have data on only one of the processes, and where for example correlations with that dataset are indicative for which of the two processes have regulated the TFL changes (Category: By indication).

29.3 Results and Discussion

29.3.1 Empirical Evidence of TFL Change in Norway

The eight studies assessed cover mainly areas in West and East Norway while two are located in North Norway (Table 29.1). The most northern and southwestern parts of Norway as well as mid-Norway are not represented. The conclusions drawn from most of the studies are representative at a local scale (Table 29.2). However, the studies of Bryn (2008), Wehn et al. (2012) and Volden (2018) either cover rather large areas (above c. 150 km^2) or have their examined sites distributed within large areas. We consider such study designs to represent more than just the local scale, but they do not represent regional scale variation (i.e. reflecting longer environmental gradients as the coastal to inland gradient).

Five studies present empirical TFL and timberline change data (Table 29.1). All these studies report average advance although the specific numbers differ. Engum's (2006) stable forest line and Wehn et al.'s (2012) timberline decline in one of their study sites are exceptions. Annual average TL changes range between 0.70 and 0.76 m, FL changes between 0.27 and 0.76 m and timberline changes between 0.1 and 2.5–3 m. Ranges of TFL change are comparable to those reported for *Betula pubescens* ssp. *czerepanovii* in the Swedish mountains (0.74 m, treeline) but also to changes of TFLs being made up of other tree species in different European mountain areas (Ameztegui et al. 2016; Cudlín et al. 2017; Kullman and Öberg 2009). However, average data conceal rather large variability. The site-specific range varies from −0.86 to 1.55 m/year in Wehn et al.'s (2012) to −0.86 to 2.62 m/year (TL) and −0.54 and 2.14 m/year (FL) in Volden's (2018) single locations. Site-specific factors as disturbances and edaphic and topographic conditions and interactions among different factors are the reason for this large local variability (Bryn and Potthoff 2018). A typical pattern of advance along the FL is the infilling of open areas between existing forest patches (Bryn 2008; Potthoff 2017; Rössler et al. 2008) also observed in other European mountain areas (Ameztegui et al. 2016; Gehrig-Fasel et al. 2007).

Little evidence exists to compare the dynamics of TLs and FLs. The 2 studies that report both type of changes confirm that TLs are more dynamic than FLs (Bryn and Potthoff 2018; Normark 2012; Volden 2018). Similar indications of expansion divergence among TFLs have been observed in Swedish mountains (Kullman 2010). These divergences reflect idiosyncratic response to environmental change. Thus, although TFLs are physiognomic height limits of the same species correlated with similar environmental factors, population processes such as dispersal, germination, establishment and growth and survival vary between TFLs. Moreover, environmental factors, feedbacks and interactions operate on different spatiotemporal scales. Whereas TL expansion mirrors occupation of new favourable microsites, FLs reflect the optimum of a larger set of variables including abiotic conditions, biotic interactions as well as soil-, and snow-vegetation-atmosphere interactions (Rydsaa et al. 2017). FL advance involves all processes needed to support an entire ecosystem at new locations. This means that TLs will respond more quickly to environmental change than FLs.

Although the different lines respond with temporal differences to environmental change, they are as mentioned above nevertheless related. Volden (2018) shows a strong correlation between the shrub line in 1938 and the treeline in 2018 ($r = 0.722$, $p < 0.01$) and the treeline in 1938 and the forest line in 2018 ($r = 0.680$, $p < 0.01$). The relationship of shrub lines and treelines is however, not very surprising, since studies of northern latitude shrub growth indicate that shrubs are limited by climatic variables (i.e. summer temperature and soil–water availability in the growing season) (Myers-Smith et al. 2015; Myers-Smith and Hik 2018).

29.3.2 Temporal Dynamics and Time-Lags

Maximum time period covered by the change data is c. 100 year (Table 29.1). Studies using remapping methods are among those covering the largest time periods while aerial photographs of mountain areas in a scale appropriate for mapping FL change are commonly available since the early 1960s. However, even a time period of 100 years is rather short taking into consideration that TFL responses to climate change in a short-term perspective may be strongly influenced by processes resulting in distributional time-lags (Kharuk et al. 2010; Rannow 2013). These distributional time-lags lead to a non-equilibrium among TFLs and the current climate system (Davis and Gedalof 2018; Rydsaa et al. 2017), and are probably an important source of error in TFL studies that correlate TFLs with contemporary climate variables (Rannow 2013). While the forest line tends to lag behind present-day conditions by decades or even centuries, the treeline responds faster, usually within a few decades after a change in climate (Bryn 2008; de Wit et al. 2014; Hofgaard et al. 2013; Kullman and Öberg 2009; Körner 2012). The distribution of the species itself however, may respond within one or only a few years (Kullman 2007). In addition to differences in time-lags between TL and FLs, the duration of time-lags varies among tree species and regions (Kullman and Öberg 2009; Körner 2012, 178).

An important cause for delayed responses to climate change is disturbances. While any kind of disturbance may result in short-term local TLF retractions (Aas 1969; Rannow 2013; Volden 2018), disturbance regimes (i.e. spatial and temporal dynamics of disturbances over a longer period) are important for extensive time-lags. For example, certain types of human land use, such as life stock grazing, have been operating as long-term disturbance regimes in Norway for centuries (Bryn and Daugstad 2001; Ross et al. 2016; Väisänen et al. 2014) and are still influencing current TFL positions (Speed et al. 2010; Wehn et al. 2012) (see Introduction). Differences in duration and intensity of these disturbance regimes will result in local and regional variation among empirical TFLs and, thereby, response time after land use abandonment will differ.

Besides disturbances, unfavourable site conditions and negative biotic interactions, the same population processes causing idiosyncratic responses of TFL to environmental change may also retain a potential range advance or result in retraction of TFLs: dispersal (e.g. low seed production), germination (e.g. lack of gaps in the vegetation cover), establishment and growth (e.g. lack of nutrients) and survival (e.g. interspecific competition) (Alatalo and Ferrarini 2017; Donato 2013; Lenoir et al. 2010; Máliš et al. 2016; Svenning et al. 2014) (Fig. 29.3; see also Holtmeier and Broll 2007 for other ways to conceptualize processes causing time-lags). Stability of treeline forming species have been shown for *Betula pubescens* ssp. *czerepanovii* in Northern Sweden and Finland caused by unfavourable site conditions and disturbances hampering TLF expansion (Holtmeier et al. 2003; Van Bogaert et al. 2011). Individuals of *Picea abies* in continental Sweden have established at sites with particular good environmental conditions at very high locations and endured for millennia due to vegetative in-situ survival (Öberg and Kullman 2011).

29.3.3 Climate Versus Land Use Change

Most of the assessed studies conclude that TFL changes to a large or intermediate degree are the consequence of land use changes while climate changes only to a limited degree contribute to TFL dynamics (Table 29.1). Normark (2012) reports large importance of climate change; however, both historical and contemporary land use intensity may have been too low to significantly impact TFL elevation (Bryn and Potthoff

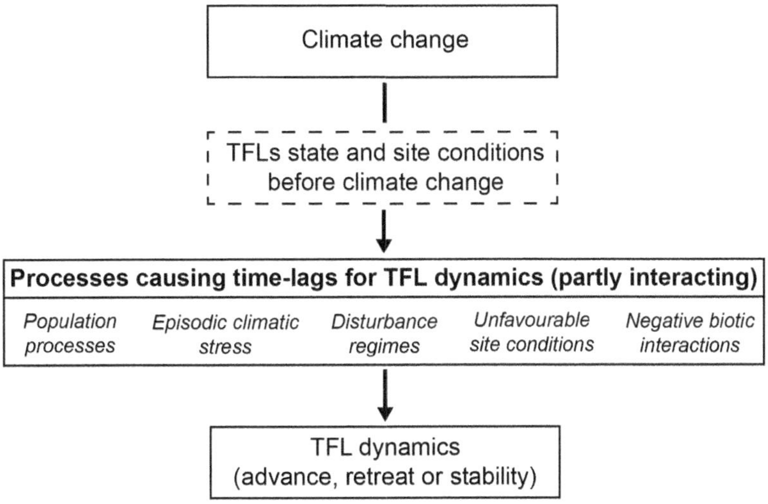

Fig. 29.3 Processes causing distributional time-lags after climate change. Some of the processes interact. Episodic stress, for example summer drought events, can delay upslope TFL dynamics independently of ongoing climate changes

2018). Moreover, Troms county, Normark's study area, is among the areas in Norway with the best outfield grazing resources (Bjørklund et al. 2012), meaning that low grazing intensity may have less impact on the TFLs than in areas with more scarce grazing resources. A reason for Engum's (2006) deviant results may be uncertainties connected to the recording of changes in grazing intensity (Engum 2006; Rößler 2005). Thus, contemporary TFL changes are most likely the result of combined effects of climate and land use change. Increased temperatures facilitate TFL advance; however, land use changes are so far, according to the assessed studies, of overriding importance.

All studies except the one by Hofgaard (1997) are designed for, or at least partially designed, to evaluate the contribution of land use and climate change to TFL change. All studies draw conclusions regarding to which degree the two factors impact TFL changes (Table 29.2). However, none of the studies is able to quantify the relative importance of the factors irrespective of sample size, data resolution (site-specific, local or regional) or methods used to decompose and analyse causes of change. Wehn et al. (2012) tested the significance of changes in temperature and precipitation for their considered time period. Due to non-significant changes in summer temperatures they did not include climate change data in modelling of forest line change.

29.4 How to Move Forward?

The assessment has revealed three related domains of research challenges that hamper our understanding of the effect of climate change versus land use change on TFL dynamics in Norway:

(1) All studies draw conclusions about the effects of climate and land use change on long-term TFL dynamics (> 30 years); however, they are not providing the degree of contribution from each of the two processes.
(2) None of the studies is representative for large parts of Norway or Norway as a whole.
(3) Empirical data for TLs and FLs are rarely available for the same study area.

29.4.1 The Degree of the Contribution of Land Use Versus Climate Change

Knowledge about the numerical contribution of land use and climate change is of fundamental importance for modelling future TFL dynamics. Site specific and high resolution data on both climate and land use change are needed to study the impact of climate change versus land use change on TFLs (Ameztegui et al. 2016). In Norway, the number of long-term weather stations providing climate data from mountain areas is rather restricted. However, the access to long-term spatially explicit historical climate data is improving. Presently, relevant climate data is provided back to 1957 in 1 × 1 km resolution (Lussana et al. 2018, 2016). Still, higher resolution data are needed for improved understanding of the contribution of climate change to TLF dynamics. For land use and land use change, the resolution of official data is even coarser, but goes further back in time. Agricultural census data are available since 1907 but are aggregated for municipalities (Statistics Norway 2019). Farm-specific information may go much further back in time (Arkivverket 2019); however, even farm-specific information, as number of animals kept on a farm, does not necessarily reflect land use intensity along TLFs (Bryn and Potthoff 2018). With regard to data availability, the lack of spatiotemporal data on previous natural disturbances presents an additional challenge.

Lack of high-resolution climate and land use change data could be tackled in different ways. Although the site-specific climate varies among locations, climate change can be assumed to be regionally fairly similar. When climate is changing, it is highly unlikely that it will change fundamentally different among comparable sites within the same region. Thus, with regard to long-term climate beyond 1957, robust estimates of change can probably be generated within regions, and implemented locally. This will not enable a precise modelling of the climate that regulates TFLs, but it can be used for studies of TFL dynamics through time within regions. A common practice to get access to site-specific land use change data is carrying out interviews with local people or others knowledgeable of land use history (Potthoff 2004; Wehn et al. 2011). However, the number of people with local knowledge of outfield resource use before c.1950 is declining. Alternatively, land use intensity can be estimated by distance to summer farms (Volden 2018) or modelled as grazing pressure (Wehn et al. 2012). Uncertainty of such efforts however is probably high. Grazing intensity is not only a result of distance but, for example, modified by topography. Modelling change in grazing pressure is partly based on site-specific historical data which may be imprecise and difficult to obtain. Finally, distance between past TFLs and potential upper climatic lines can be used as a proxy for past land use (Ameztegui et al. 2016).

Different statistical methods can be used to distinguish between the contribution of climate change and land use change (Ameztegui et al. 2016; Cudlín et al. 2017; de Wit et al. 2014; Gehrig-Fasel et al. 2007; Kulakowski et al. 2011; Wehn et al. 2012). Training spatial models is another approach (e.g. Guida et al. 2019 (MaxEnt for upward shifts of climate sensitive desert tree species); Macias-Fauria and Johnson 2013 (Random Forests for forest lines)). Although none of these studies includes explicit land use data, both methods can be used to estimate variable importance in the models (Halvorsen et al. 2015; Hastie et al. 2009). Lastly, finding previously mapped pristine climatic TFLs, unspoiled by any forms of land use (e.g. protected by fences, topography or rivers), would allow a straight forward comparison with neighboring TFLs influenced by land use. Such locations however, are difficult to find in Norway, and we do not know if any such sites have been mapped previously (Bryn and Potthoff 2018).

29.4.2 Spatial Representativeness

Both land use changes and climate changes vary throughout the country (Almås 2004; Tveito 2014), as well as a number of ecological background variables (Bakkestuen et al. 2008). However, since climate change and land use change may only differ slightly within study areas at local or even at regional scale, TLF changes will most likely mainly reflect micro-site-specific variation and change rather than broader scale climate and land use changes. Thus, studies covering larger areas are needed, preferable including important environmental gradients throughout the country. To obtain necessary empirical data on long-term TLF change, efforts to remap previous TLF studies should be increased (e.g. Fries 1921; Norman 1894; for further references see Bryn and Potthoff 2018). When planning remapping, special attention should be paid to reduce the spatial biases within Norway, in addition to prioritizing studies covering large areas.

29.4.3 Idiosyncrasy of Time-Lags

While birch FL dynamics are frequently treated in studies from the Nordic region (see references in Holtmeier 2009; Körner 2012; Wielgolaski 2005), very few studies report the dynamics of birch TL (Bryn and Potthoff 2018). This includes studies investigating the relevance of land use vs. climate change for TFL dynamics. Volden (2018) indicates that knowledge about previous treelines, shrub lines and average annual change can help to predict future TFLs. However, this means that more empirical data on long-term shrub line changes—to our knowledge not existing for Norway, and TL changes are needed. Moreover, knowledge about line specific time-lags is of importance.

To improve the knowledge of processes causing idiosyncrasy, field experiments should be set up to investigate differences among shrub lines, TLs and FLs, including potential line-specific responses to climate and land use change. Moreover, model experiments targeting dynamics of TFLs should pay more attention to the transition zones including boreal and alpine vegetation and not only focus on one of the vegetation zones (de Wit et al. 2014; Rydsaa et al. 2017).

29.4.4 Other Topics Important to Be Considered in Future TFL Research

TFL studies indicate that interactions and feedbacks between different processes may play an important role in the future development of TFL dynamics (de Wit et al. 2014; Rydsaa et al. 2017). According to Rydsaa et al. (2017), vegetation-climate feedbacks such as reduced albedo following expanding TFLs, might increase the local temperature, and thus trigger further TFL expansion. This expansion however, will interact with land use such as domestic and semi-domestic grazing (Engelkraut et al. 2018; Speed et al. 2010). Therefore, although not considered specifically by the assessed studies, it is likely that some of these interactions and feedbacks have played a role in the processes investigated by the assessed studies. Such effects should therefore probably be included in future TFL studies.

29.5 Conclusion

The assessed studies of mountain birch dominated TFLs in Norway document an ongoing range shift towards higher elevations. The rate of range shift is comparable with TFL dynamics documented by other studies in Europe. However, the rate of range shift is highly variable, both between studies and among sites within each study. The studies document that changes in land use is one of the main drivers for long-term (30–100 years) TFL dynamics; however, increased temperatures have most likely facilitated TFL advance. The assessed studies are probably not representative for the variation in TFL dynamics in Norway and do not provide the degree of contribution from each of the two processes, knowledge needed to model future TFL dynamics. To achieve a more complete understanding of the ongoing elevational range shift, more large-scale regional remapping projects and more site-specific data about land use change and climate change are needed.

References

Aas B (1969) Climatically raised birch lines in south-eastern Norway 1918–1968. Nor Geogr Tidsskr 23:119–130

Aas B, Faarlund T (1995) Skoggrenseutviklingen i Norge, særlig i det 20. århundre. AmS-Varia 24:89–100

Aas B, Faarlund T (2000) Forest limits and the subalpine birch belt in North Europe with a focus on Norway. AmS-Varia 37:103–147
Alatalo JM, Ferrarini A (2017) Braking effect of climate and topography on global change-induced upslope forest expansion. Int J Biometeorol 61:541–548
Almås R (ed) (2004) Norwegian agricultural history. Tapir Akademisk Forlag, Trondheim
Ameztegui A, Coll L, Brotons L, Ninot JM (2016) Land-use legacies rather than climate change are driving the recent upward shift of the mountain tree line in the Pyrenees. Glob Ecol Biogeogr 25:263–273
Arkivverket (2019) Kilder til gårdshistorie. https://www.arkivverket.no/opplysninger-om-eiendom/kilder-til-gardshistorie. Accessed 24 Jun 2019
Aschwanden S (2002) Changes in the tree- and forest-limits of Nordic mountain birch in Narvik municipality during the 20t century. Master's thesis. Department of Geography, Norwegian University of Science and Technology, Trondheim
Aune S, Bryn A, Hovstad KA (2018) Loss of semi-natural grassland in a boreal landscape: impacts of agricultural intensification and abandonment. J Land Use Sci 13:375–390
Aune S, Hofgaard A, Söderström L (2011) Contrasting climate- and land-use-driven tree encroachment patterns of subarctic tundra in northern Norway and the Kola Peninsula. Can J For Res 41:437–449
Austrheim G, Solberg EJ, Mysterud A (2011) Spatio-temporal variation in large herbivore pressure in Norway during 1949–1999: has decreasing grazing by livestock been countered by increased browsing by cervids? Wildl Biol 17:286–298
Bakkestuen V, Erikstad L, Halvorsen R (2008) Step-less models for regional environmental variation in Norway. J Biogeogr 35:1906–1922
Bjune AE (2005) Holocene vegetation history and tree-line changes on a north–south transect crossing major climate gradients in southern Norway—evidence from pollen and plant macrofossils in lake sediments. Rev Palaeobot Palynol 133:249–275
Bjørklund, PK, Rekdal, Y, Strand, G-H (2012) Arealregnskap for utmark. Arealstatistikk for Troms. Ressursoversikt 05/2012. Skog og landskap, Ås
Bobrowski M, Gerlitz L, Schickhoff U (2017) Modelling the potential distribution of *Betula utilis* in the Himalaya. Glob Ecol Cons 11:69–83
Bryn A (2008) Recent forest limit changes in south-east Norway: effects of climate change or regrowth after abandoned utilisation? Nor J Geogr 62:251–270
Bryn A, Daugstad K (2001) Summer farming in the subalpine birch forest. In Wielgolaski FE (ed) Nordic Mountain Birch ecosystems. Man and the biosphere series. UNESCO & Parthenon Publishing Group, Paris & New York, pp 307–315
Bryn A, Dourojeanni P, Hemsing LØ, O'Donnell S (2013) A high-resolution GIS null model of potential forest expansion following land use changes in Norway. Scand J For Res 28:81–98
Bryn A, Potthoff K (2018) Elevational treeline and forest line dynamics in Norwegian mountain areas—a review. Landsc Ecol 33:1225–1245
Christensen AL (2002) Det norske landskapet. Om landskap og landskapforståelse i kulturhistorisk perspektiv, Pax, Oslo
Cudlín P, Klopčič M, Tognetti R, Máliš F, Alados CL, Bebi P, Grunewald K, Zhiyanski M, Andonowski V, La Porta N, Bratanova-Doncheva S, Kachaunova E, Edwards-Jonášová M, Ninot JM, Rigling A, Hofgaard A, Hlásny T, Skalák P, Wielgolaski FE (2017) Drivers of treeline shift in different European mountains. Clim Res 73:135–150
Davis EL, Gedalof Ze (2018) Limited prospects for future alpine treeline advance in the Canadian Rocky Mountains. Glob Chang Biol 24:4489–4504
de Wit HA, Bryn A, Hofgaard A, Karstensen J, Kvalevåg MM, Peters GP (2014) Climate warming feedback from mountain birch forest expansion: reduced albedo dominates carbon uptake. Glob Chang Biol 20:2344–2355
Donato DC (2013) Limits to upward movement of subalpine forests in a warming climate. Proc Natl Acad Sci USA 110:7971–7972
Engelkraut D, Aronsson K-Å, Allard A, Åkerholm M, Stark S, Olofsson J (2018) Multiple feedbacks contribute to a centennial legacy of reindeer on tundra vegetation. Ecosystems 21:1545–1563
Engum H-C (2006) Alpine tre- og skoggrenseendringer. Indikator på klimaforandringer eller endret arealbruk? Master's thesis. Department of Geography, University of Bergen, Bergen
Fries TCE (1921) Björkskogsgränsens höjdläge inom Tromsö amt. Tidsskr Skogbruk 29:48–72
Gatti RC, Callaghan T, Velichevskaya A, Dudko A, Fabbio L, Battipaglia G, Liang J (2019) Accelerating upward treeline shift in the Altai Mountains under last-century climate change. Sci Rep 9:1–13
Gehrig-Fasel J, Guisan A, Zimmermann NE (2007) Tree line shifts in the Swiss Alps: climate change or land abandonment? J Veg Sci 18:571–582
Guida RJ, Abella SR, Robers CL, Norman CM, Smith WJ Jr (2019) Assessing historical and future habitat models for four conservation-priority Mojave Desert species. J Biogeogr. https://doi.org/10.1111/jbi.13645
Hagedorn F, Shiyatov SG, Mazepa VS, Devi NM, Grigor'ev AA, Bartysh AA, Fomin VV, Kapralov DS, Terent'ev M, Bugman H, Rigling A, Moiseev PA (2014) Treeline advances along the Urals mountain range—driven by improved winter conditions? Glob Chang Biol 20:3530–3543
Halvorsen R, Mazzoni S, Bryn A, Bakkestuen V (2015) Opportunities for improved distribution modelling practice via a strict maximum likelihood interpretation of MaxEnt. Ecography 38:172–183
Hansen LI, Olsen B (2004) Samenes historie fram til 1750. Cappelen Akademisk Forlag, Oslo
Harsch MA, Bader MY (2011) Treeline form—a potential key to understanding treeline dynamics. Glob Ecol Biogeogr 20:582–596

Harsch MA, Hulme PE, McGlone MS, Duncan RP (2009) Are treelines advancing? A global meta-analysis of treeline response to climate warming. Ecol Lett 12:1040–1049

Hastie T, Tibshirani R, Friedman J (2009) Elements of statistical learning. Data mining, inference, and prediction. Springer, New York

Hofgaard A (1997) Inter-relationships between treeline position, species diversity, land use and climate change in the central Scandes Mountains of Norway. Glob Ecol Biogeogr Lett 6:419–429

Hofgaard A, Løkken JO, Dalen L, Hytteborn H (2010) Comparing warming and grazing effects on birch growth in an alpine environment—a 10-year experiment. Plant Ecol Divers 3:19–27

Hofgaard A, Tømmervik H, Rees G, Hanssen F (2013) Latitudinal forest advance in northernmost Norway since the early 20th century. J Biogeogr 40:938–949

Holtmeier F-K (2009) Mountain timberlines. Ecology, patchiness, and dynamics. Springer, Berlin

Holtmeier F-K, Broll G (2007) Treeline advance—driving processes and adverse factors. Landsc Online 1:1–33

Holtmeier F-K, Broll G, Müterthies A, Anschlag K (2003) Regeneration of trees in the treeline ecotone: Northern Finnish Lapland. Fennia 181:103–128

Jacobsen H, Follum J-R (2008) Kulturminner i Norge. Spor etter mennesker gjennom 10000 år. Tun forlag & Skogbrukets kursinstitutt, Oslo

Karlsen SR, Tømmervik H, Johansen B, Riseth JÅ (2017) Future forest distribution on Finnmarksvidda, North Norway. Clim Res 73:125–133

Karlsson PS, Weih M, Borg C (2005) Mountain birch growth in relation to climate and herbivores. In: Wielgolaski FE (ed) Plant ecology, herbivory, and human impact on Nordic Mountain Birch Forests. Springer, Berlin, Heidelberg, pp 71–86

Kharuk VI, Im ST, Dvinskakya ML, Ranson KJ (2010) Climate-induced mountain tree-line evolution in southern Siberia. Scand J For Res 25:446–554

Kulakowski D, Bebi P, Rixen C (2011) The interacting effects of land use change, climate change and suppression of natural disturbances on landscape forest structure in the Swiss Alps. Oikos 120:216–225

Kullman L (2007) Tree line population monitoring of *Pinus sylvestris* in the Swedisch Scandes, 1973–2005: implications for tree line theory and climate change ecology. J Ecol 95:41–52

Kullman L (2010) A richer, greener and smaller alpine world: review and projection of warming-induced plant cover change in the Swedish Scandes. Ambio 39:159–169

Kullman L, Öberg L (2009) Post-Litte Ice Age tree line rise and climate warming in the Swedish Scandes: a landscape ecological perspective. J Ecol 97:415–429

Körner C (2012) Alpine treelines. Functional ecology of the global high elevation tree limits. Springer, Basel

Lenoir J, Gégout J-C, Guisan A, Vittoz P, Wohlgemuth T, Zimmermann NE, Dullinger S, Pauli H, Willner W, Svenning J-C (2010) Going against the flow: potential mechanisms for unexpected downslope range shifts in a warming climate. Ecography 33:295–303

Lussana C, Saloranta T, Skaugen T, Magnusson J, Tveito OE, Andersen J (2018) seNorge2 daily precipitation, an observational gridded dataset over Norway from 1957 to the present day. Earth Syst Sci Data 10:235–249

Lussana C, Tveito OE, Uboldi F (2016) seNorge v2.0: an observational gridded dataset of temperature for Norway. MET report 14–2016. Norwegian Meteorological Institute, Oslo

Macias-Fauria M, Johnson EA (2013) Warming-induced upslope advance of subalpine forest is severely limited by geomorphic processes. PNAS 110:8117–8122

Máliš F, Kopecký M, Petřík P, Vladovič J, Merganič J, Vida T (2016) Life stage, not climate change, explains observed tree range shifts. Glob Chang Biol 22:1904–1914

Myers-Smith IH, Elmendorf SC, Beck PSA, Wilmking M, Hallinger M, Blok D, Tape KD, Rayback SA, Macias-Fauria M, Forbes BC, Speed JDM, Boulanger-Lapointe N, Rixen C, Lévesque E, Schmidt NM, Baittinger C, Trant AJ, Hermanutz L, Collier LS, Dawes MA, Lantz TC, Weijers S, Jørgensen RH, Buchwal A, Buras A, Naito AT, Ravolainen V, Schaepman-Strub G, Wheeler JA, Wipf S, Guay KC, Hik DS, Vellend M (2015) Climate sensitivity of shrub growth across the tundra biome. Nat Clim Chang 5: 887–891

Myers-Smith IH, Hik DS (2018) Climate warming as a driver of tundra shrubline advance. J Ecol 106:547–560

Myhre B (2004) Agriculture, landscape and society. In: Almås R (ed) Norwegian agricultural history. Tapir, Trondheim, pp 14–20

Måren IE (2009) Effects of management on heathland vegetation in Western Norway. PhD Thesis. Department of Biology and Bergen Museum, University of Bergen, Bergen

Norman JM (1894) Norges Arktiske Flora I. Speciel Plantegoegrafi, Oscar Andersens Bogtrykkeri, Kristiania

Normark K (2012) The shift in forest and tree limits in Troms County—with a main focus on temperatures and herbivores. Examensarbete i äment biologi. Fakulteten för skogsvetenskap, Institutionen för vilt, fist och miljö, Sveriges landbruksuniversitet, Umeå

Öberg L, Kullman L (2011) Ancient subalpine clonal spruces (*Picea abies*): Sources of postglacial vegetation history in the Swedish Scandes. Arctic 64:183–196

Odland A (1996) Differences in the vertical distribution pattern of *Betula pubescens* in Norway and its ecological significance. In: Frenzel B (ed) Holocene treeline oscillations, dendrochronology and paleoclimate. Gustav Fischer Verlag, Stuttgart, pp 43–59

Odland A (2015) Effect of latitude and montain height on the timberline (*Betula pubescens* ssp. *czerpanovii*) elevation along the central Scandinavian mountain range. Fennia 193:260–270

Olsson EGA, Hanssen SK, Rønningen K (2004) Different conservation values of biological diversity? A case study from the Jotunheimen mountain range, Norway. Nor J Geogr 58:204–212

Paus A, Haugeland V (2017) Early- to mid-Holocene forest-line and climate dynamics in southern Scandes mountains inferred from contrasting megafossil and pollen data. Holocene 27:361–383

Potthoff K (2004) Change in mountain summer farming practices: a case study from Stølsheimen, Western Norway. Nor J Geogr 56:158–170

Potthoff K (2007) Persistence of alpine grass-dominated vegetation on abandoned mountain summer farms in Western Norway. Nor J Geogr 61:192–206

Potthoff K (2009) Grazing history affects the tree-line ecotone: a case study from Hardanger, Western Norway. Fennia 187:81–98

Potthoff K (2017) Spatiotemporal patterns of birch regrowth in a Western Norwegian treeline ecotone. Landsc Res 42:63–77

Rannow S (2013) Do shifting forest limits in south-west Norway keep up with climate change? Scand J For Res 28:574–580

Reinton L (1955) Sæterbruket i Noreg I. Sætertypar og driftsformer. H. Aschehoug & Co., Oslo, Norway

Rizzi J, Nilsen IB, Stagge JH, Gisnås K, Tallaksen LM (2018) Five decades of warming: impacts on snow cover in Norway. Hydrol Res 49:670–688

Ross LC, Austrheim G, Asheim L-J, Bjarnason G, Feilberg J, Fosaa AM, Hester AJ, Holand Ø, Jónsdóttir IS, Mortensen LE, Mysterud A, Olsen E, Skonhoft A, Speed JDM, Steinheim G, Thompson DBA, Thórhallsdóttir AG (2016) Sheep grazing in the North Atlantic region: A long-term perspective on environmental sustainability. Ambio 45:551–566

Rydsaa JH, Stordal F, Bryn A, Tallaksen LM (2017) Effects of shrub and tree cover increase on the near-surface atmosphere in northern Fennoscandia. Biogeosciences 14:4209–4227

Rößler O (2005) Die alpine Baumgrenze in Zentralnorwegen unter dem Einfluss von Klima- und Landnutzungswandel. Eine sozialgeographische, landschaftsökologische und dendroökologische Synthese. Master's thesis. Institute for Biology and Environmental Sciences, Carl von Ossietzky Universität Oldenburg, Oldenburg

Rössler O, Bräuning A, Löffler J (2008) Dynamics and driving forces of treeline fluctuation and regeneration in central Norway during the past decades. Erdkunde 62:117–128

Sigdel SR, Wang Y, Camarero JJ, Zhu H, Liang E, Peñuelas J (2018) Moisture-mediated responsiveness of treeline shifts to global warming in the Himalayas. Glob Chang Biol 24:5549–5559

Speed JDM, Austrheim G, Hester AJ, Mysterud A (2010) Experimental evidence for herbivore limitation of the treeline. Ecology 91:3414–3420

Statistics Norway (2019) Historiske landbruksteljingar (1907–1999). https://www.ssb.no/a/histstat/landbrukstellinger.html. Accessed 24 June 2019

Svenning J-C, Gravel D, Holt RD, Schurr FM, Thuiller W, Münkemüller T, Schiffers KH, Dullinger S, Edwards TC Jr, Hickler T, Higgins SI, Nabel JEMS, Pagel J, Normand S (2014) The influence of interspecific interactions on species range expansion rates. Ecography 37:1198–1209

Tveito OE (2014) Klimaendringer og betydning for skogbruket. MET report no. 25. Norwegian Meteorological Institute, [Oslo]

Van Bogaert R, Haneca K, Hoogesteger J, Jonasson C, De Dapper M, Callaghan TV (2011) A century of tree line changes in sub-Arctic Sweden shows local and regional variability and only a minor influence of 20th century climate warming. J Biogeogr 38:907–921

Volden IKF (2018) Dynamics of tree- and forest lines over time. A case study from Lærdal, Western Norway. Master's thesis. Department of Biosciences & Natural History Museum, University of Oslo, Oslo

Väisänen M, Ylänne H, Kaarlejärvi E, Sjögersten S, Olofsson J, Crout N, Stark S (2014) Consequences of warming on tundra carbon balance determined by reindeer grazing history. Nat Clim Chang 4:384–388

Wehn S, Olsson G, Hanssen S (2012) Forest line changes after 1960 in a Norwegian mountain region—implications for the future. Nor J Geogr 66:2–10

Wehn S, Pedersen B, Hanssen SK (2011) A comparison of influences of cattle, goat, sheep and reindeer on vegetation changes in mountain cultural landscapes in Norway. Landsc Urban Plan 102:177–187

Wielgolaski FE (ed) (2005) Plant ecology, herbivory and human impact in Nordic mountain birch forests. Springer, Berlin

Social-Ecological-Technical Misalignments Threaten Mountain Water Tower Resilience in Utah, USA

30

Michelle A. Baker and Courtney G. Flint

Abstract

The essential "water tower" role played by mountains is compromised by climate change and human development. Misalignments in various socio-ecological-technical dimensions threaten adaptive capacity and resilience in mountain water-dependent regions. Interdisciplinary research in Utah's Wasatch Mountains reveals a complex set of mid-elevation dynamics and stakeholder perspectives complicating water resource planning at local and state levels. Rapid urban development and population growth in the region point to water demand exceeding supply in the near future. Climate change is already influencing snowpack levels, snow water equivalent, and phase changes in mountain precipitation. Winter forest evapotranspiration rates present unexpected water loss with warming air temperatures. Mountain water quality is deteriorating due to up-slope nitrogen deposition as well as mid-elevation grazing, fire, and residential development. Multiple data sources point to diverse and conflicting stakeholder perspectives throughout the region suggesting considerable work to be done to find common ground for water management and planning in this dynamic water tower system. We explore the opportunities and constraints related to a range of adaptation pathways being considered and attempted at local, regional, and state government scales, including water reuse, water transfers and pipelines, new reservoirs, water banking, and water conservation promotion.

Keywords

Climate adaptation · Water resources · Urbanization · Social science · Vulnerability

30.1 Introduction

It is well known that mountains disproportionally supply runoff to adjacent lowlands, providing an important water provisioning service as "water towers" (Viviroli et al. 2007). Indeed, more than 50% of mountainous regions world-wide have an essential or supportive role as water supply (Viviroli et al. 2007). These regions are particularly vulnerable when reliance on mountain-derived runoff coincides with population growth and precipitation declines (Viviroli et al. 2011). It is less clear how well aligned this water

M. A. Baker
Department of Biology and Ecology Center, Utah State University, Logan, UT, USA

C. G. Flint (✉)
Department of Sociology, Social Work, and Anthropology, Utah State University, Logan, UT, USA
e-mail: courtney.flint@usu.edu

U. Schickhoff et al. (eds.), *Mountain Landscapes in Transition*,
Sustainable Development Goals Series, https://doi.org/10.1007/978-3-030-70238-0_30

provisioning service is with broader global change drivers and complex social-ecological-technical system (SETS) dynamics (Grabowski et al. 2017; Markolf et al. 2018).

The Intermountain West region of the United States is bound by the Sierra Nevada Mountains in the west and Wasatch/Rocky Mountains in the east. Human settlement in the region currently consists of mixed agricultural and urban land uses concentrated in arid and semi-arid valleys where water is available as snowmelt-driven runoff from surrounding mountain water towers (Powell 1875; Gollehon and Quinby 2000). Humans developed elaborate infrastructure (e.g., dams, canals, wells, and field drainage), water rights laws, institutions, and corporate entities (Reisner 1993; McCool 1995; Huffaker et al. 2000) to distribute water necessary for settlement. Thus, the modern waterscape of the Intermountain West results from the interplay of natural and human processes.

Rapid climatic and demographic changes in recent decades pose new challenges to water availability, water quality, and water demand in the Intermountain West (Anderson and Woosley 2005). These changes are likely to accelerate in the coming decade, with human population in the six-state Intermountain West expected to increase 50–90% by 2030 (GOPB 2010; US Census Bureau 2010). At the same time, climate change threatens to alter the amount and timing of precipitation throughout the region (Udall and Bates 2007; Barnett et al. 2008). Potential outcomes threaten the resilience of the region's mountain water towers; these include reduced snowpack (Mote et al. 2018), earlier snowmelt (Stewart et al. 2005), and more precipitation falling as rain rather than snow (Knowles et al. 2006).

Adaptations to these demographic and climatic changes will be shaped by complex technical, legal, and behavioral constraints that characterize the Intermountain West SETS. Potential responses include increasing efficiency of water transmission and use (Huffaker and Whittlesey 2003; Peterson and Ding 2005), changing reservoir operations (Webb et al. 1999), shifting water management objectives (Gosnell et al. 2007), and/or reallocating water rights from agricultural to other users (Young and Brozovic 2019). Clearly, changes in water use behaviors will have major ramifications for how water will flow through the landscape (Green and Hamilton 2000).

In this chapter, we present the Wasatch Front as a case study of potential social-ecological-technical misalignments that threaten resilience of mountain water towers.

The Wasatch Front lies at the eastern boundary of the Intermountain West in northern Utah, USA and encompasses three catchments that flow into the Great Salt Lake. From north to south, these include the Bear River, Logan metropolitan area; the Weber River, Ogden-Clearfield metropolitan area; and the Provo-Jordan, Salt Lake City and Provo-Orem metropolitan areas (Fig. 30.1). The Wasatch Front is home to 85% of the state's population of 3 million people. Elevation ranges from 1200 to 4000 m, and precipitation is elevation-dependent ranging from 400 to 1400 mm, with the majority as snow during winter and early spring (Ehleringer et al. 1992). Anglo settlement of the Wasatch Front began in 1847 with the arrival of the Mormon pioneers, who settled near the base of the foothills along perennial streams that then were modified to make "the desert bloom like a rose" (Kay 1995).

The current water system in the Wasatch Front is a highly regulated network of dams, diversions, and pipelines. Water demand is estimated at 533,000 acre-feet (Null 2018)—and not all of this water is supplied by Wasatch Mountain water towers—nearly 20% (101,900 acre-feet) is diverted from the Colorado River Basin through an intermountain aqueduct called the Central Utah Project (U.S. Bureau of Reclamation, https://www.doi.gov/cupcao/Overview). Despite irrigation return, discharge of treated wastewater to surface water networks, and other return flows, consumptive water use in the Wasatch Front has decreased water input to the Great Salt Lake, resulting in a decreased volume of 48%. This is significant because the Great Salt Lake is the 8th largest saline lake in the world with an estimated economic value of US$ 1.32 billion/year (Wurtsbaugh et al. 2017).

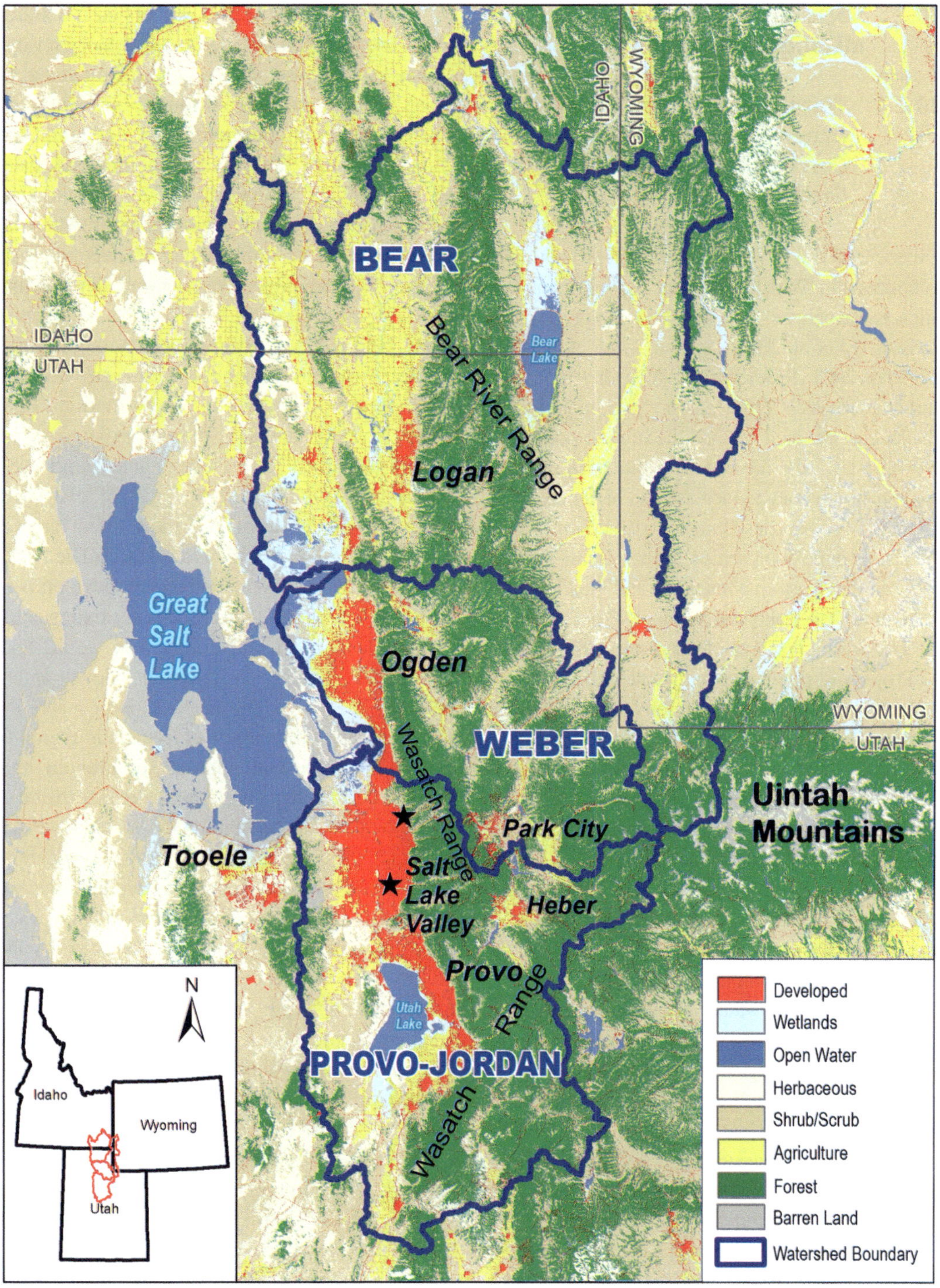

Fig. 30.1 Map showing the location of the Wasatch Front metropolitan region, from Logan, Utah in the north, south to Provo/Orem (red)

30.2 Socio-Ecological-Technical Misalignments

30.2.1 Water Supply and Demand—Urban-Mountain Corridor Population Growth

Experienced and projected population growth in the Intermountain West region and its water resource impacts have been a critical focal point in recent decades (Li et al. 2016; Jackson-Smith et al. 2006; Doremus 2001; Reisner 1993; Worster 1985). Utah and the Wasatch Front are no exception to this challenging nexus. Utah had the highest population growth rate of the fifty US states in 2018 due to a combination of natural increase and high rates of in-migration (US Census Bureau 2019). With 85% of the state's population settled in the narrow corridor between the Wasatch Mountains and the Great Salt Lake and Utah Lake, the Wasatch Front is one of the US West's hotspots of urban growth.

Projecting population and water demand illuminates an alarming collision course for the Wasatch Front. Edwards et al. (2017) showed an additional water demand in 2060 for five Wasatch Front counties to be about 200% higher than today (at the current per-person water use rate). They highlight the challenges in reaching the water conservation goals that would offset that added demand, including consequences of out-of-basin water transfers on local water availability, environmental costs of ecosystem degradation from water development projects, the cost of water transportation infrastructure, and factors limiting adoption of conservation practices of urban residents and agricultural producers. Nonetheless, Edwards et al. (2017) demonstrated that agricultural and urban irrigation are the largest water users in Utah, and have the greatest potential for low-cost water efficiency options if mechanisms are implemented to offset current barriers. An important complication on future water demand is that Utah's water is fully allocated in the urban corridor with new uses only possible by modifying existing water rights (Utah Division of Water Rights 2019).

The population trajectory of the Wasatch Front has a corresponding land use footprint that relates to water use and the need for planning. Li et al. (2016) modeled land use scenarios for Cache County in northern Utah, including the current trend of land-use regulations and management plans, a smart growth scenario of managed growth, and water-smart growth with full and moderate implementation depending on degree of development restrictions. The current trend of urban growth is projected to disperse urban land use and development throughout the county, whereas planned smart growth and water-smart growth models show more concentrated development near current existing urban centers and away from water-related land uses. Based on the study by Li et al. (2016), even with smart growth development, most urban growth would occur on prime farmland. The fully implemented water-smart growth development scenario would save over 80% of existing agricultural land from conversion, but would take substantial changes to current development policies.

Aggregate projections of population growth can mask important internal differences that make related water supply and demand difficult to anticipate. Residents of Wasatch Front municipalities are an increasingly diverse mix, with correspondingly varying water attitudes and behaviors. Flint et al. (2017) showed that water concerns varied significantly across locations and factors such as gender, religion, race and ethnicity, income, age, and education of Northern Utah residents. These differences suggest the value of understanding demographic and social context in order to better understand water attitudes and support for effective water governance (Flint et al. 2017). Furthermore, nontraditional housing arrangements are becoming more common in the US (Barnett et al. 2019). Through a recent Northern Utah survey, Barnett et al. (2019) found urban renters and multiunit dwellers to be less likely to have authority over key household water decisions, such as purchasing efficient appliances or outdoor landscaping and irrigation choices. Furthermore, those with nonwhite ethnic or racial identities indicated stronger water conservation-oriented attitudes than whites. These findings suggest the need to appreciate a more complex demographic mosaic in the future and to carefully design water conservation promotion strategies to reach an increasingly diverse population.

30.2.2 Regional Sensitivity of Intermountain West Water Cycle to Climate Change

Mountain water towers are expected to be very sensitive to climate warming (IPCC 2007; Viviroli et al. 2011). Globally, it is well documented that mountain regions of the world are warming more rapidly than lower elevations because of less albedo, related surface feedbacks, and changes to the energy balance (Pepin et al. 2015). Warmer air temperatures are associated with a decline in spring snowpack. For example, Mote et al. showed that April 1 snow water equivalent has declined between 15 and 30% since 1955 in the Western US (Mote et al. 2018), with the volume of that lost water roughly the same as Lake Mead, the largest reservoir in the region.

Such observations may not be so simply explained. Oversimplification of the mountain water cycle, and water cycles writ large (e.g. Abbott et al. 1999), creates a misalignment in understanding vulnerability of Intermountain West water towers to climate change. For example, in the Wasatch and other mountains in Utah, Gillies et al. (2012) showed that mountain snowpack decline may be better explained by a shift in phase of precipitation, with a 9% decline in the proportion of winter precipitation occurring as snow since 1950. This was associated with a modest decrease in depth of snowpack that was confirmed with associated satellite observations of both snow cover and surface albedo. All this while the total amount of winter precipitation increased over the period of observation. In the Intermountain West, such reductions in snowpack are even stronger at lower (<2000 m) elevation (e.g. Gillies et al. 2012; Tennant et al. 2015), reflecting some combination of phase change and earlier melt.

Wasatch Mountain water towers differ from other significant water tower regions (e.g. Himalayans and Alps) because there is less alpine area. As such, forest vegetation plays an important role in how these water towers function hydrologically. Tennant et al. (2017) found that variability in snow depth was inversely related to forest vegetation height, likely because of differences in ablation, interception, redistribution, and shading. Further, ecohydrologists assume that the dominant trees (evergreen conifers including pines, firs, spruce) are dormant during the winter when terrain is snow covered. Surprisingly, this is not the case in the Wasatch Front. Even with 50–100 cm of snow on the ground, both white fir and subalpine fir were capable of transpiration on days in the shoulder season when air temperature was above 0 °C (Chan and Bowling 2017). Such activity represents an unexpected and unaccounted for water loss from the Wasatch Front.

Mountain aquifers remain poorly understood (Viviroli et al. 2007), and aquifers associated with the Wasatch Front are no different. Mountain block recharge of groundwater is significant, with most subsurface residence times between 5 and 15 years (Manning and Solomon 2005). Given the importance of mountain block recharge, local aquifers on the Wasatch Front are vulnerable to loss of recharge from melting snowpack (Meixner et al. 2016). Currently, recharge exceeds groundwater withdrawals along the Wasatch Front (Meixner et al. 2016); however, if withdrawals increase in response to surface water overallocation, water scarcity could be exacerbated.

30.2.3 Mountain Water Quality

Mountain water towers, like all water sources, are susceptible to water quality declines due to human activity, as our research in the Wasatch Front revealed. While point sources of pollutants are readily identified and managed, diffuse sources are less so. Relative to their lowland counterparts, mountain water towers are particularly vulnerable to atmospheric deposition of anthropogenic contaminants, either as wet or dry deposition. In a synoptic sampling of snow-associated dust in several mountain ranges in the Intermountain West, Dastrup et al. (2018) found that Wasatch Mountain sites had dust chemistry associated with human activities, including the heavy metals copper, lead, and antimony that could leach into soil and surface waters during snowmelt. Such contaminants are sourced from the urban Wasatch Front and/or produced by mining and smelting activities that occur in the desert west of Salt Lake City (Dastrup et al. 2018).

Nitrogen deposition is another potential threat to mountain water quality. A well-studied example of this phenomenon is along the Colorado Front Range in the Rocky Mountains (Williams et al. 1996) where atmospherically derived nitrogen supplied to high elevation lakes has increased nitrate concentrations and caused rapid shifts in the lake diatom community (Baron et al. 2000). Along the Wasatch Front, we observed similar rates of nitrogen deposition as the Colorado Front Range, ~4 kg/ha/y (Hall et al. 2016a). We hypothesized that the total quantity and sources of atmospherically derived N would differ between agricultural valleys in the northern Wasatch compared to the urban central Wasatch, but deposition rates did not differ, and isotopic composition of the nitrogen was not able to distinguish sources. In both sites the nitrogen was derived from a combination of sources, including transportation, combustion, and volatilization (Hall et al. 2016a).

While the above examples highlight the potential for airsheds to impact mountain water quality, once water is transported down valley to urban settings, the potential for water quality degradation increases as a result of more proximal point and non-point sources. Stormwater, in particular, is identified as a source of pollution to urban stream networks (Walsh et al. 2005). In an effort to understand stormwater impacts on Wasatch Front water quality, Hall et al. (2016b) and Gabor et al. (2017) instrumented a subwatershed of the Jordan River Utah from the mountain headwaters through the urban environment. They found that while stormwater had episodic influence on stream water quality, an unexpected persistent source of pollution was from perennial springs that discharged contaminated groundwater, high in nitrogen and other constituents (Hall et al. 2016b; Gabor et al. 2017). Isotopic composition of water and nitrate suggested that the urban groundwater intercepted nitrate from leaking sewage infrastructure. These studies highlight that groundwater-surface water exchange should be recognized in managing mountain water quality.

Wildfire is an additional threat to mountain water security. In the western US, recent increases in burned area and subsequent erosion are could double sediment yield in over a third of watersheds by 2050 (Murphy et al. 2018). Not only does increased sediment load impact habitat for fisheries, but sedimentation in water storage reservoirs could accelerate climate-associated reductions in water storage for tens of millions of people in the western US (Murphy et al. 2018).

30.2.4 Stakeholder Perceptions of Mountain Water

Water stakeholders typically hold a diverse range of values related to water (Witt et al. 2019) and water is simultaneously a key part of social and political relationships (Krause and Strang 2016). Given this complex human-water relationship, water resource decision makers must balance myriad ecological, social, and economic interests (Viviroli et al. 2011). In the Wasatch Front, when 88 representatives of key professional stakeholder groups representing natural resource management, business, city and county planning, emergency management, and recreational and environmental groups were asked in interviews about the key sustainability issues for the mountain region, fewer than half (44%) mentioned water. Furthermore, perspectives diverged on the relationship between water and growth and development in the region. While some said, “We are loving [the mountains] to death” and

"There needs to be some limits on development" in order to protect natural ecosystems, others emphasized the need to manage resources in order to enable continued future economic growth across urban, agricultural, and recreation sectors.

In the Wasatch Mountain region, attitudes and concerns about water vary socially and geographically (Flint et al. 2017, 2016; Barnett et al. 2019). Public intercept interviews in two northern Utah cities revealed that water recreation, water supply, water quality, and water as essential for life, were themes mentioned by over 75% of participants when they were asked how they value and relate to water in their landscape (Flint et al. 2016). However, many questioned whether there was anything they or their community could or should do about water in the future.

In household surveys conducted in 2014 across 23 Wasatch Front neighborhoods in three valleys, concerns about water shortage, poor water quality, and climate change impacts on water supply were highest in urban Salt Lake Valley, whereas water concerns in less urbanized and more politically conservative areas were significantly lower (Flint et al. 2017). Age, gender, income, race, education, religion, and recreational activity levels were important differentiating variables with regard to water concerns (Flint et al. 2017). Older, nonwhite, middle income, non-Latter-Day Saint, and active recreationist respondents were more concerned about water shortages. Based on the same survey, a third (33.7%) of all respondents indicated they thought there was enough water to meet the current needs of their valley, whereas only 14% thought there would be enough water to meet future needs of their valley and these perspectives also varied significantly across northern Utah valleys (Jackson-Smith and Flint 2014). Just over half of respondents (54.2%) indicated they believe there is more their household could do to reduce indoor water use, while just over a third (37.8%) believed they could do more to reduce outdoor water use (by far the greater household water use). Varying perspectives on household capacity to conserve water was found to be related to housing ownership and type, with lower values on water reduction capacity found among renters and those from multi-unit dwellings (Barnett et al. 2019). As for support for state water policies and strategies, more survey respondents supported using state funds to help replace aging water system infrastructure in cities (72.4%) and to build new reservoirs and other storage projects (61.2%), while considerably fewer supported using state funds to construct pipelines to bring water to urban areas from other regions (35.9%) or facilitating transfers of water from agriculture to urban users (27.5%) (Jackson-Smith and Flint 2014). These survey findings suggest there is a lack of consensus around the water issues facing the Wasatch Mountain region in the future.

Diverse perspectives on water can reflect different patterns of exposure or vulnerability to risks and suggest that decision makers should consider how stakeholder perspectives are tied to support and opposition for water management strategies (Flint et al. 2017). Yet, local leaders often diverge from their constituents on what the critical water issues are for the future of the region, with residents more concerned than local leaders on water shortages and high cost of water (despite Utah water prices being notably low), and local leaders more concerned about deteriorating water infrastructure (Haeffner et al. 2018). These attitudinal misalignments present a challenging context for adapting to evolving water conditions in the Wasatch Mountain region.

30.3 Adapting Mountain Water Solutions to Scale

There is a growing recognition of the need to build in capacity of socio-ecological-technical systems to sustainably adapt to changing internal and external conditions such as population growth, land development, and climate change (Markolf et al. 2018). One first step to building this capacity is to identify misalignments in our understanding of this complex SETS as described above (Fig. 30.2). The SETS approach to resilience recognizes system interdependence and connectivity, with an aim to avoid lock-in of past social and technological decisions that constrain system adaptation. This approach lends itself to transdisciplinary efforts and co-production of knowledge among researchers, practitioners, decision makers, and citizens (Gleick 2018; Markolf et al. 2018). Indeed, Grabowski et al. (2017, p. 7) state "The ASCE [American Society of Civil Engineers] has identified interdisciplinary coordination as a key to infrastructure planning and management" and "agency coordination without public engagement around qualitatively different goals will not evolve systems."

In planning for Utah's water future, the governor commissioned an advisory team to recommend adaptation strategies to enhance water sustainability. "The State Water Strategy reflects a shifting paradigm that recognizes the unique and diverse climates, topography, water uses, and management challenges throughout our state. This shifting paradigm is evident in the development of regional water conservation goals and the effort to set up Basin Councils that draw on the expertise of local water users and stakeholders. These efforts sharpen the focus on developing locally-relevant solutions and strategies that are appropriate at the regional and watershed scale and recognize that many localized challenges will require localized solutions and strategies to be truly effective." (Evan Curtis, Personal Communication, 23 July 2019). Looking at the Wasatch mountain water system through a SETS lens reveals that some of the state's adaptation strategies (Governor's Water Strategy Advisory Team 2017) may be maladaptive. One strategy ripe for lock-in is a proposed reservoir and pipeline on the Bear River in northern Utah aimed at developing 220,000 acre-

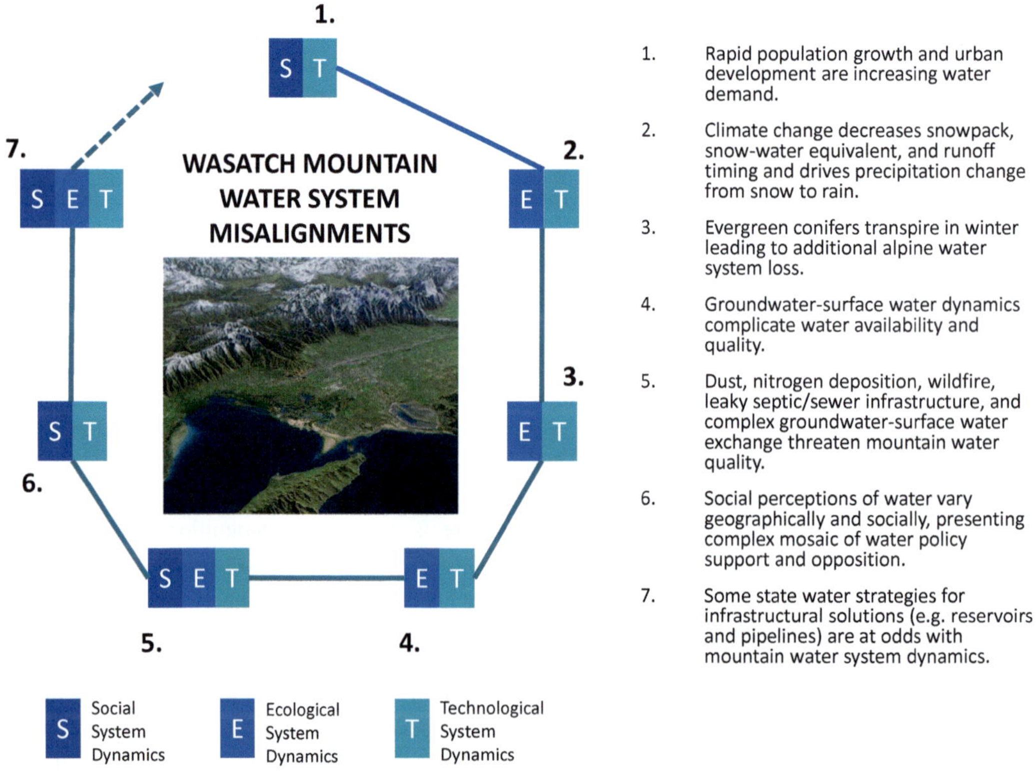

Fig. 30.2 Misalignments in social, ecological, and technical dynamics pose potential threats to sustainability of Wasatch Mountain water towers. Image from Google Earth (Ehleringer, personal communication). Figure adapted from Markolf et al. (2018)

feet for supply to the Wasatch Front at a cost of at least US $1.5 billion (Bowen Collins 2014). Fostering water conservation is another strategy being pursued locally and state-wide. While conservation is not a maladaptive strategy, as our research showed, citizens were more apt to believe that more could be done to conserve water indoors than outdoors, despite that outdoor use is far greater, and more water could potentially be conserved outdoors, and at lower cost (Edwards et al. 2017).

Clearly new information, data, and ideas are needed in order to enhance resilience of Utah's mountain water towers in the face of water resource transitions due to changing physical conditions, demographics, social preferences, and maladaptive conditions (Gleick 2018). Building a resilient water system relies on anticipatory governance—i.e. adopting a longer-term policy vision and bringing experts and citizens to work together to monitor changing conditions and manage systems, rather than waiting for crises to reveal themselves (Boyd et al. 2015). Democratically setting goals, matching the scale of process with the scale of intervention, designing system changes for multiple cross-cutting benefits, building in multi-functionality and redundancies, learning and sharing knowledge at multiple levels, and letting systems evolve modularly are considerations that support resilience (Grabowski et al. 2017; Biggs et al. 2012; Viviroli et al. 2011).

30.4 Conclusion

Our case study in northern Utah highlights that in trying to *understand* the complexity of mountain water systems, collaborative learning *within the science community* is key. i.e. if we don't link hydrology to vegetation systems, we miss shoulder season transpiration. If we don't link demography and urban development with water engineering and management, we don't anticipate spatial patterns of demand, etc. If we don't link study of groundwater with surface water, we miss the complex interactions that impact supply and quality. If we don't link study of quality with supply, we miss impacts such as algal blooms and flow of nutrients that compromise water uses. If we don't link biophysical water science with understanding perspectives of stakeholders, we lack appreciation for where adaptive efforts are possible or challenged. Finally, if we don't work on collaborative learning *among/between science and decision makers and local societies*, what's happening in the mountain water system (or what is likely to happen) is not well aligned with anticipatory and management decisions where it matters most.

References

Abbott BW, Bishop K, Zarnetske JP, Minaudo C, Chapin III FS, Krause S, Hannah DM, Conner L, Ellison D, Godsey SE, Plont S, Marcais J, Kolbe T, Huebner A, Frei RJ, Hampton T, Gu S, Buhman M, Sayedi SS, Ursache O, Chapin M, Henderson KD, Pinay G (1999) Human domination of the global water cycle absent from depictions and perceptions. Nat Geosci 5. https://doi.org/10.1038/s41561-019-0374-y

Anderson MT, Woosley LH (2005) Water availability for the western United States- key scientific challenges. Circular 1261. United States Geological Survey, Reston

Barnett MJ, Jackson-Smith D, Endter-Wada J (2019) Implications of nontraditional housing arrangements for urban water management in the United States Intermountain West. Soc Nat Resour 32(5):508–529

Barnett TP, Pierce DW, Hidalgo HG, Bonfils C, Santer BD, Das T, Bala G, Wood AW, Nozawa T, Miran AA, Cayan DR, Detlinger MD (2008) Human-induced changes in the hydrology of the western United States. Science 319:1080–1083

Baron JS, Rueth HM, Wolfe AM, Nydick KR, Allstott EJ, Minear R, Moraska B (2000) Ecosystem responses to nitrogen deposition in the Colorado Front Range. Ecosystems 3:352–368

Biggs R, Schlüter M, Biggs D, Bohensky EL, BurnSilver S, Gundill G, Dakos V, Daw TM, Evans SL, Kotschy K, Leitch AM, Meek C, Quinlan A, Raudsepp-Hearne C, Robards MD, Schoon ML, Schultz L, West PC (2012) Toward principles for enhancing the resilience of ecosystem services. Annu Rev Environ Resour 37:421–448. https://doi.org/10.1146/annurev-environ-051211-123836

Bowen Collins for Utah Division of Water Resources (2014) Volume I of II—Bear River Pipeline Concept Report—Final. Consultant Job 233-09-01

Boyd E, Nykvist B, Borgström S, Stacewicz IA (2015) Anticipatory governance for socio-ecological resilience. Ambio 44(Suppl. 1):S149–S161

Chan AM, Bowling DR (2017) Assessing the thermal dissipation sap flux density method for monitoring cold season water transport in seasonally snow-covered forests. Tree Physiol 36:984–995

Dastrup DB, Carling GT, Collins SA, Nelson ST, Fernandez DP, Tingey DG, Hahnenberger M, Aandurd ZY (2018) Aeolian dust chemistry and bacterial communities in snow are unique to airshed locations across northern Utah, USA. Atmos Environ 193:251–261

Doremus H (2001) Water, population growth, and endangered species in the West. Univ Colorado Law Rev 72:361–414

Edwards EC, Bosworth RC, Adams P, Baji V, Burrows A, Gerdes C, Jones M (2017) Economic insight from Utah's water efficiency supply curve. Water 9:214. https://doi.org/10.3390/w9030214

Ehleringer JR, Arnow LA, Arnow T, McNulty IB, Negus NC (1992) Red Butte Canyon Research Natural Area: history, flora, geology, climate, and ecology. Great Basin Naturalist 5:95–121

Flint CG, Dai X, Jackson-Smith D, Endter-Wada J, Yeo SK, Hale R, Dolan MK (2017) Social and geographic contexts of water concerns in Utah. Soc Nat Resour 30(8):885–902

Flint CG, Mascher C, Oldroyd Z, Valle PA, Wynn E, Cannon Q, Brown A, Unger B (2016) Public intercept interviews and surveys for gathering place-based perceptions: observations from community water research in Utah. J Rural Soc Sci 31(3):105–125

Gabor RS, Hall SJ, Eiriksson D, Jameel Y, Stout T, Barnes ML, Tennant H, Bowen GJ, Neilson BT, Brooks PD (2017) Persistent urban influence on surface water quality via impacted groundwater. Environ Sci Technol 51:9477–9487

Gillies RR, Wang S-Y, Boothe MR (2012) Observational and synoptic analyses of winter precipitation regime change over Utah. J Clim 25:4679–4698

Gleick PH (2018) Transitions to freshwater sustainability. PNAS 115(36): 8863–8871. https://doi.org/10.1073/pnas.1808893115

Gollehon N, Quinby H (2000) Irrigation in the American West: Area, water, and economic activity. Int J Water Resour Dev 16:187–195

GOPB (2010) 2010 Economic Report to the Governor. State of Utah, Governor's Office of Planning and Budget, Salt Lake City

Gosnell H, Haggerty JH, Byorth PA (2007) Ranchland ownership change and new approaches to water resource management in southwestern Montana. J Am Water Resour Assoc 43:990–1003

Governor's Water Strategy Advisory Team (2017) Recommended State Water Strategy [for Utah] https://www.envisionutah.org/images/FINAL_Recommended_State_Water_Strategy_7.14.17_5b15d.pdf. Envision Utah, Salt Lake City, Utah

Grabowski ZJ, Matsler AM, Thiel C, McPhillips L, Hum R, Bradshaw A, Miller T, Redman C (2017) Infrastructures as socio-eco-technical systems: five considerations for interdisciplinary dialog. J Infrastruct Syst 23:4. https://doi.org/10.1061/(ASCE)IS.1943-555X.0000383

Green GP, Hamilton JR (2000) Water allocation, transfers, and conservation: links between policy and hydrology. Water Resour Dev 16:197–208

Haeffner M, Jackson-Smith D, Flint CG (2018) Social position influencing the water perception gap between local leaders and constituents in a socio-hydrological system. Water Resource Res 54 https://doi.org/10.1002/2017WR021456

Hall SJ, Ogata EM, Weintraub SR, Baker MA, Ehleringer JR, Czimczik CI, Bowling DR (2016a) Convergence in nitrogen deposition and cryptic isotopic variation across urban and agricultural valleys in northern Utah. J Geophys Res-Biogeosci 121:2340–2355

Hall SJ, Weintraub SR, Eiriksson D, Brooks PD, Baker MA, Bowen GJ, Bowling DR (2016b) Stream nitrogen inputs reflect groundwater across a snowmelt-dominated montane to urban watershed. Environ Sci Technol 50:1137–1146

Huffaker R, Whittlesey N (2003) A theoretical analysis of economic incentive policies encouraging agricultural water conservation. Int J Water Resour Dev 19:37–53

Huffaker R, Whittlesey N, Hamilton JR (2000) The role of prior appropriation in allocating water resources in the 21st century. Water Resour Dev 16:265–273

IPCC [Intergovernmental Panel on Climate Change] (2007) Climate Change 2007: Synthesis Report. Contribution of Working Groups I, II, and III to the Fourth Assessment Report of the Intergovernmental Panel on Climate Change [Core Writing Team, Pachauri RK and Reisinger A (eds)]. IPCC, Geneva Switzerland, 104 p

Jackson-Smith D, Jensen E, Jennings B (2006) Changing land use in the rural Intermountain West. In: Kandel WA, Brown DL (eds) Population change and rural society. The Springer Series on Demographic Methods and Population Analysis, vol 16. Springer, Dordrecht, pp 253–276

Jackson-Smith D, Flint C (2014) Utah's Water Future - 2014 Household Survey. HydroShare http://www.hydroshare.org/resource/72ab49b468bc427fa2024b5b716d3103

Kay J (1995) Mormons and mountains. In: Wyckoff W (ed) The mountainous west: explorations in historical geography. University of Nebraska Press, Lincoln, pp 368–395

Knowles N, Dettinger MD, Cayan DR (2006) Trends in snowfall versus rainfall in the western United States. J Clim 19:4545–4559

Krause F, Strang V (2016) Thinking relationships through water. Soc Nat Resource 29:633–638. https://doi.org/10.1080/08941920.2016.1151714

Li E, Li S, Endter-Wada J (2016) Water-smart growth planning: linking water and land in the arid urbanizing American West. J Environ Plann Manage 60(6):1056–1072

Manning AH, Solomon DK (2005) An integrated environmental tracer approach to characterizing groundwater circulation in a mountain block. Water Resour Res 41:W12412. https://doi.org/10.1029/2005WR004178

Markolf SA, Chester MV, Eisenberg DA, Iwaniec DM, Davidson CI, Zimmerman R, Miller TR, Ruddell BL, Chang H (2018) Interdependent infrastructure as linked social, ecological, and technical systems (SETSs) to address lock-in and enhance resilience. Earth's Future 6:1638–1659

McCool DC (1995) The waters of Zion: the law, policy, and politics of water in Utah. University of Utah Press, Salt Lake City

Meixner T, Manning AH, Stonestrom DA, Allen DM, Ajami H, Blasch KW, Brookfield AE, Castro CL, Clark JF, Gochis DJ, Flint AF, Neff KL, Niraula R, Rodell M, Scanlon BR, Singha K, Walvoord MA (2016) Implications of projected climate change for groundwater recharge in the western United States. J Hydrol 534:124–138

Mote PW, Li S, Lettenmaier DP, Xiao M, Engel R (2018) Dramatic declines in snowpack in the western US. NPJ Clim Atmospheric Sci 1:2. https://doi.org/10.1038/s41612-018-0012-1

Murphy BP, Yocom LL, Belmont P (2018) Beyond the 1984 perspective: Narrow focus on modern wildfire trends underestimates future rishs to water security. Earth's Future 6:1492–1497

Null, SE (2018) Economic water demand functions to value urban water scarcity along Utah's Wasatch Front. Technical Report Aquatic Habitat, Climate, and Water Analysis Laboratory. HydroShare. https://doi.org/10.4211/hs.a6921eef1cbf4968b271d972bd997ab3

Pepin N, Bradley RS, Diaz HF, Baraer M, Caceres EB, Forsythe N, Fowler H, Greenwood G, Hashmi MZ, Liu XD, Miller JR, Ning L, Ohmura A, Papazzi E, Rangwala I, Schoner W, Severskiy I, Shahgedanova M, Wang MB, Williamson SN, Yang DQ (2015) Elevation-dependent warming in mountain regions of the world. Nat Clim Chang. https://doi.org/10.1038/NCLIMATE2563

Peterson JM, Ding Y (2005) Economic adjustments to groundwater depletion in the high plains: do water-saving irrigation systems save water? Am J Agric Econ 87:147–159

Powell JW (1875) The exploration of the Colorado River and its Canyons. Dover Press, New York

Reisner M (1993) Cadillac desert. Penguin Books, New York

Stewart IT, Cayan DR, Dettinger MD (2005) Changes toward earlier streamflow timing across western North America. J Clim 18:1136–1155

Tennant CJ, Crosby BT, Godsey SE (2015) Elevation-dependent responses of streamflow to climate warming. Hydrol Process 29:991–1001

Tennant CJ, Harpold AA, Lohse KA, Godsey SE, Crosby BT, Larson LG, Brooks PD, Van Kirk RW, Glenn NF (2017) Regional sensitivities of seasonal snowpack to elevation, aspect, and vegetation cover in western North America. Water Resour Res 53:6908–6926

Udall B, Bates G (2007) Climate and hydrologic yrends in the western US: a review of recent peer-reviewed research. Intermountain West Clim Summary 2:2–8

US Census Bureau (2010) Population Estimates. US Department of Commerce, Washington DC

US Census Bureau (2019) Population Estimates. US Department of Commerce, Washington DC

Utah Division of Water Rights (2019) Ground water policy. https://waterrights.utah.gov/gisinfo/maps/agwpol.pdf. Accessed 23 July 2019

Viviroli D, Durr HH, Messerli B, Meybeck M, Weingartner R (2007) Mountains of the world, water towers for humanity: typology, mapping, and global significance. Water Resour Res 43:W07447. https://doi.org/10.1029/2006WR005653

Viviroli D, Archer DR, Buytaert W, Fowler HJ, Greenwood GB, Hamlet AF, Huang Y, Koboltschnig G, Litaor MI, López-Moreno JI, Lorentz S, Schädler B, Schreier H, Schwaiger K, Vuille M, Woods R (2011) Climate change and mountain water resources: overview and recommendations for research, management, and policy. Hydrol Earth Syst Sci 15:471–504. https://doi.org/10.5194/hess-15-471-2011

Walsh CJ, Roy AH, Feminella JW, Cottingham PD, Groffman PM, Morgan RP (2005) The urban stream syndrome: current knowledge and the search for a cure. J N Am Benthol Soc 24:706–723

Webb RH, Schmidt JC, Marzolf GR, Valdez RA (1999) The controlled flood in Grand Canyon. Geophysical Monograph 110. American Geophysical Union, Washington, DC

Williams MW, Baron JS, Caine N, Sommerfeld R, Sanford R (1996) Nitrogen saturation in the Rocky Mountains. Environ Sci Technol 30:640–646

Witt K, Ross H, Shaw S, Jones N, Rissik D, Pinner B (2019) How do local people value rural waterways? A study in the upper catchments of south east Queensland's rivers. Soc Nat Resour 32(6):638–656. https://doi.org/10.1080/08941920.2019.1578910

Worster D (1985) Rivers of empire: water, aridity, and the growth of the American West. Pantheon, New York

Wurtsbaugh WA, Miller C, Null SE, DeRose J, Wilcock P, Hahnenberger M, Howe F, Moore J (2017) Devline of the world's saline lakes. Nat Geosci. https://doi.org/10.1038/NGEO3052

Young R, Brozovic N (2019) Agricultural water transfers in the Western United States. Daugherty Water for Food Global Institute and Mammoth Trading. University of Nebraska, Lincoln

31 Changing Paradigm in Transboundary Landscape Management: A Retrospect from the Hindu Kush Himalaya

Nakul Chettri, Srijana Joshi, Bandana Shakya, Sunita Chaudhary, Lipy Adhikari, Nabin Bhattarai, Eklabya Sharma, and David J. Molden

Abstract

The Hindu Kush Himalaya (HKH), the highest mountain biome, also referred as the third pole or the water tower of Asia, is an important repository of biological and cultural diversities and source of varied ecosystems services to 240 million people living within and 1.9 billion in the mountains and downstream. The region has been in spotlight for being part of the 36 'Global Biodiversity Hotspot' and 'Crises Ecoregions' as well as climate change hotspot. However, there is still knowledge gap on understanding the dynamics of changing landscapes and climate and its linkage to people, mostly challenged by poverty. International Centre for Integrated Mountain Development (ICIMOD), an intergovernmental regional knowledge and enabling centre, has been instrumental in developing better understanding on the dynamics of these fragile ecosystems and support its regional member countries through science-based integrated approaches. Since its inception, ICIMOD has been engaged in developing knowledge and supporting policies for mountain development focusing on socio-economic, ecological and environmental dimensions. In this chapter, we present the retrospect of our interventions in science, policy and practice in transboundary landscape management through regional cooperation mostly focused on biodiversity conservation and community development perspectives.

Keywords

Climate change · Ecosystem degradation · Nature-based solutions · Participatory approach · Regional cooperation

31.1 Introduction

The Hindu Kush Himalaya (HKH), the highest mountain ecosystem in the world with world's ten highest mountain peaks, considered as a *third pole* and *water tower* of Asia, is a global asset (Wester et al. 2019). Stretched over 3500 km (km) and covering more than four million square kilometers, the HKH includes all of Bhutan and Nepal and parts of Afghanistan, Bangladesh, China, India, Myanmar and Pakistan. It is one of the most diverse ecosystems among the global mountain biomes with extreme variations in vegetation, climate and ecosystems, resulted from altitudinal and latitudinal gradients (Xu et al. 2009; Molden et al. 2017). Owing to these enabling ecological conditions, the region is among the richest with two of its member

N. Chettri (✉) · S. Joshi · B. Shakya · S. Chaudhary · L. Adhikari · N. Bhattarai · E. Sharma · D. J. Molden
International Centre for Integrated Mountain Development, Kathmandu, Nepal
e-mail: Nakul.Chettri@icimod.org

U. Schickhoff et al. (eds.), *Mountain Landscapes in Transition*, Sustainable Development Goals Series, https://doi.org/10.1007/978-3-030-70238-0_31

countries—India and China, being 'Mega Diversity Countries' (Brooks et al. 2006). Blessed with diverse plants and animals of global significance and unique ecosystems, the HKH is also among the Global 200 Ecoregions (Olson and Dinerstein 2002). This rich biodiversity has also nurtured culture and traditions of more than 1000 ethnic groups (Turin 2005). In addition, the region is source of ten major river systems as the source of a wide range of ecosystem services supporting 240 million people in the region and benefit some 1.9 billion people in the mountains and downstream river basin areas (Wester et al. 2019).

However, in the course of human civilization, the HKH lost more than 70% of its original ecosystems, which resulted the HKH to be considered as parts of 'Crisis Ecoregions' and 'Biodiversity Hotspots' (Brooks et al. 2006; Mittermeier et al. 2011). Environmental degradation has been identified as a major threat to the functioning of HKH ecosystems and flow of ecosystem services (Chettri et al. 2010; Xu et al. 2019). Among others—climate change, habitat change, overexploitation, pollution and invasive alien species—are the major drivers of changes (Chettri and Sharma 2016; Wang et al. 2019). Adapting to and mitigating the effects of these changes and sustaining ecosystem services in the context of a burgeoning human population and climate change is a major challenge in the HKH as elsewhere (Molden 2020). Hence, it is important to improve scientific understanding of ecosystem structure and functioning and drivers of change as a basis for formulating comprehensive ecosystem management approaches and strategies that link to human well-being and poverty alleviation (Sharma et al. 2019). But, despite the significance of the HKH biodiversity, there have been little coordinated efforts to understand the drivers of biodiversity loss or their impact on conservation and economic development (Xu et al. 2019).

The concept of transboundary conservation planning is getting more traction to address the complexity through transdisciplinary approach at landscape level (Liu et al. 2020). With the changing global discourse in biodiversity conservation and sustainable development, the HKH also witnessed this changing paradigm (Sharma et al. 2010). The key characteristics of this approach include 'participation' of stakeholders including communities from the planning to the management and monitoring levels, 'transdisciplinarity' by engaging multiple stakeholders from various sectors through participatory processes, 'multifunctionality' to address the multiple objectives of a landscape, 'complexity' of social-ecological systems that include various land-use types within a landscape, and 'sustainability' for sustained provision of ecosystem services (Freeman et al. 2015). To ensure effective management of such areas, cooperation across international boundaries is required (Vasilijević and Pezold 2011; Lambertucci et al. 2014). In this chapter, we present how International Centre for Integrated Mountain Development (ICIMOD), as an intergovernmental organization, is playing role to translate the concept of transboundary landscape into a reality to enhance the resilience of HKH mountain landscape.

31.2 ICIMOD and Changing Paradigm

The International Centre for Integrated Mountain Development (ICIMOD), established in 1983 with an eight-country charter, serves as a regional intergovernmental centre for cooperation on the sustainable development of the HKH. Over the years, ICIMOD has been a global player in advocating mountains perspectives and agendas since its existence. Being an active mountain partnership member and observers for numerous multilateral environmental agreements such as Convention on Biological Diversity (CBD) and the United Nations Framework Convention on Climate Change (UNFCCC), ICIMOD has been playing an important bridging role for customizing global agendas to local and regional levels and vice versa. Catering to its member countries based on contemporary challenges and opportunities, ICIMOD advanced through different phases and priorities (Molden and Sharma

2013). At present, ICIMOD is in its fourth medium term action plan (2018–2022) with six integrated programmes supported by four thematic areas (ICIMOD 2017). Among the six programmes, transboundary landscapes is an important integrated and multidisciplinary regional programme that focusses on scientific understanding of mountain ecosystems and to develop people-centred interventions to resource conservation for sustainable and equitable development (see Molden et al. 2017). At ICIMOD, we work across HKH countries to help attain common goals related to sustainable development, by bringing together different groups within programmatic transboundary approaches covering scales from households, springshed, landscapes to river basins.

ICIMOD supports regional and transboundary cooperation to meet challenges of climate change, disaster risks and sustainable development in the HKH through integrating transboundary landscape and river basin programmatic focus. Actions to sustain the HKH have the potential to directly improve the lives of more than one-third of the world's population. However, facilitating cooperation and policy coherence among the countries sharing HKH resources is a persistent challenge in a region with highly variable priorities regarding development. ICIMOD's interventions focus on knowledge development, human resources development, technology transfer, policy outreach and innovations through demonstrations. In terms of changing paradigm in conservation and development perspectives in the HKH, ICIMOD passed through three broad phases—(a) pre-2002, (b) 2002 to 2008 and (c) post-2008. These phases reflect contextualization, developing an enabling environment for transboundary cooperation and planning and implementation respectively. Here we present a concise narrative for each of the three phases.

31.2.1 Contextualization (Pre-2002)

In the past, the classical approach of biodiversity conservation started with emphasizing conservation of flagship species (Yonzon 1989; Wikramanayake et al. 1998). Research and management interventions were made to save flagship species such as tigers, rhinoceros, elephants and snow leopards. considering them as climax species with an assumption that if they are conserved, the ecosystem and habitat will automatically be saved. Within few years, the realization to protect habitats evolved and concept of protected area came in. The first protected area recorded from the HKH is Pidaung Wildlife Sanctuary established in 1918 in Myanmar (Chettri et al. 2008). However, in many instances, the protected areas were declared but with restrictions to human intervention which is key to effective conservation. As the approach could not get the desired result, the United Nations Conference on Environment and Development (UNCED) in 1992 placed a premium on people's participation and promotion of this conceptual shift in both natural resources management and biodiversity conservation (UN 1992). Thus, the protected area concept further evolved with adding buffer zones and connectivity corridors considering wide habitat range used by flagship species. Since late 1990, the concept of landscape approach was introduced (Sherpa et al. 2004; Gurung 2005). These development witnessed significant conceptual evolution in biodiversity conservation, from 'People exclusionary' and 'species-focused' to 'People-centred community-based' approaches (Sharma et al. 2010).

During the course of evolution, ICIMOD was instrumental for documenting biodiversity of the HKH through collective minds of experts for the first time at HKH level (Pei 1995). With a regional mandate, ICIMOD also facilitated transboundary dialogue and cooperation for conservation between Nepal and China in the Everest complex (Sherpa et al. 2003), Bhutan, Nepal, India, in the Kangchenjunga complex (Rastogi et al. 1997) and China, India and Myanmar in the Far-Eastern Himalaya (Guangwei 2002). In 1998, an international meeting on Ecoregional Cooperation for Biodiversity Conservation in the Himalayas was co-organized by UNDP, WWF and ICIMOD followed by a workshop on 'A Biodiversity Vision for the Eastern Himalaya' in 1999 (WWF and ICIMOD 2001). These engagements were instrumental to rationalize and contextualized the need for wider consultations, comprehensive planning and facilitation for regional cooperation at the landscape level.

31.2.2 Developing Enabling Environment (2002–2008)

In 2002, ICIMOD initiated a transboundary initiative in Kangchenjunga complex by building on conservation priorities identified in previous years (Sharma and Chettri 2005). A participatory planning approach was adopted involving local community, conservation practitioners, academicians and policymakers to identify possible corridors (Chettri et al. 2007). The process also led to development of a regional cooperation framework for implementation of the Convention on Biological Diversity (CBD) in the Kangchenjunga Landscape (Sharma et al. 2007). During this time, the 'ecosystem approach' that was adopted by the Conference of the Parties (COP) of the CBD in 1995 (Secretariat of the CBD 2004), was endorsed and called various actions including regional or transboundary collaboration (Sharma and Acharya 2004). Subsequently, the Sacred Himalayan Landscape—covering an area of 29,021 km^2 was conceived (MFSC 2006).

In 2007, with the release of the IPCC's Fourth Assessment Report (AR4), the HKH was identified as a 'data deficient' region (Rosenzweig et al. 2008). Realizing the significance of the HKH and potential impacts of climate change on biodiversity and human well-being, ICIMOD brought more than 80 global experts in Kathmandu, Nepal and conceived the 'trans-Himalayan transect' and 'Transboundary Landscape' framework in the HKH in 2008 (Chettri et al. 2009). In this framework, four north–south transects and six transboundary complexes, i.e. Hindukush Karakoram-Pamir, Kailash, Everest, Kangchenjunga, Far-Eastern Himalaya and Cherrapunjee-Chittagong were proposed to achieve transboundary collaboration for research, and monitoring to develop social-ecological resilience (Fig. 31.1). Simultaneously, ICIMOD produced a series of comprehensive reports on climate change vulnerability highlighting trends and potential impacts on biodiversity and ecosystems in the Eastern Himalaya (Sharma et al. 2009; Chettri et al. 2010; Tse-ring et al. 2010). These interventions strongly recommended the need for regional cooperation through landscape approach to address the twin challenges of biodiversity loss and climate change.

31.2.3 Planning and Implementation of Landscape Framework (2009–2020)

Since 2009, ICIMOD along with partners initiated systematic planning processes considering four steps—diagnose (determine the need for transboundary conservation), design (match the process to situation), take action (formulate and implement action) and evaluate (learn and adapt) as planning tool (Erg et al. 2012, Fig. 31.2). While planning, the three pillars of sustainable development (social, environmental and economic) perspectives, three elements of programme management (science, policy and practice interface) and scales (local, national and regional) were also considered. During extensive consultative processes, participation of all key stakeholders (local communities, civil societies, conservation practitioners, academicians and policymakers) were ensured (Chettri et al. 2007). Subsequently, four out of six transboundary landscapes went through the four steps planning process and achieved various targets (Fig. 31.2), including Feasibility Assessment Reports, Conservation and Development Strategies and Regional Cooperation Frameworks (Fig. 31.2).

As indicated in Fig. 31.2, four different landscapes followed the four steps of planning and implementation but at different timeframe. The KSL started feasibility phase in 2009 and implementation phase since 2012 and at present

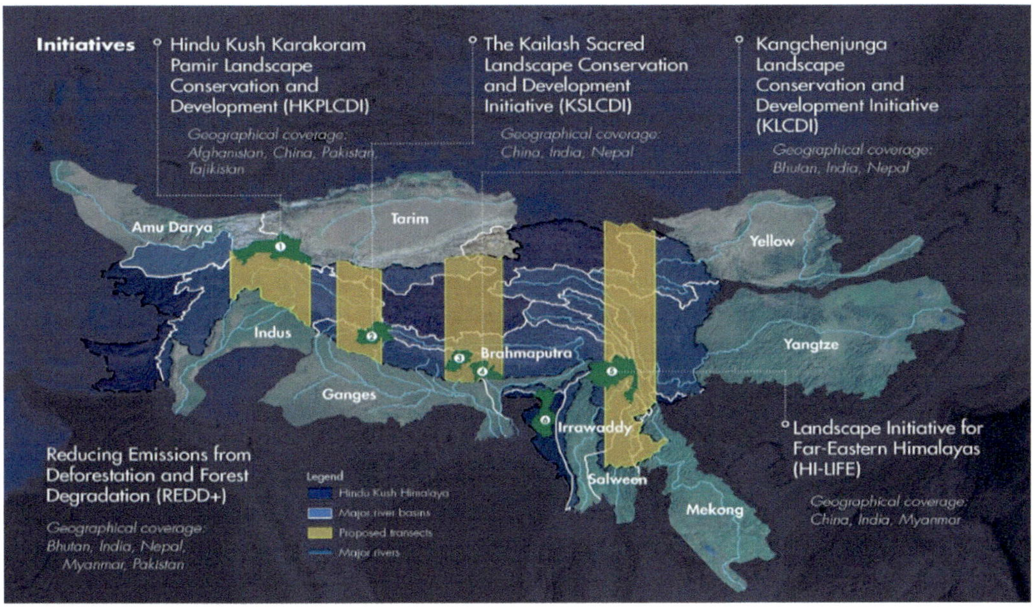

Fig. 31.1 Trans-Himalayan Transects and Transboundary Landscapes in the HKH

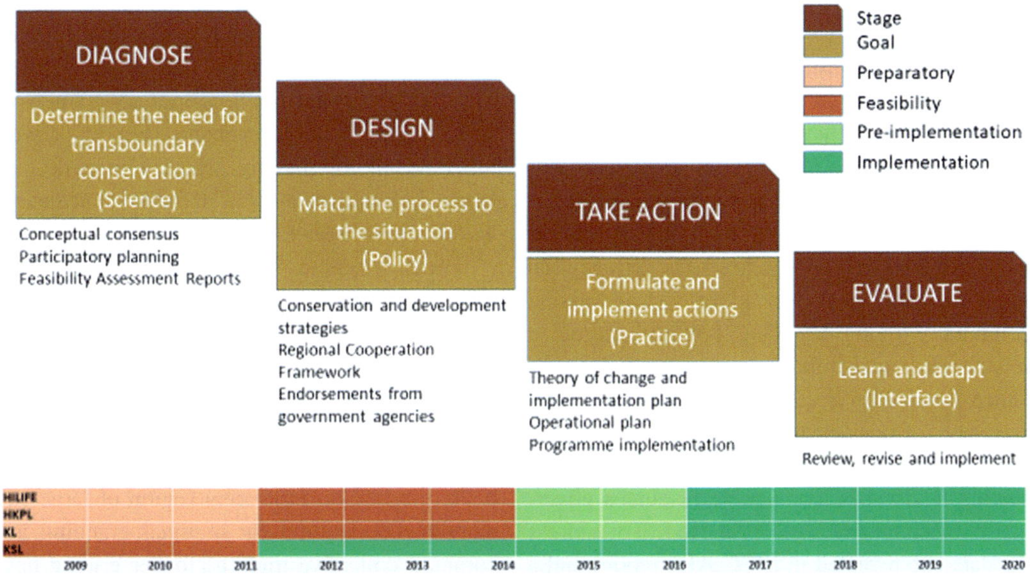

Fig. 31.2 Systematic planning process adopted in the transboundary landscape

it is in its second phase of implementation (Zomer and Oli 2011; Pandey et al. 2016; Kotru et al. 2020). Whereas, other three landscapes started with preparatory phase in 2009, feasibility phase in 2012, pre-implementation during 2015 and 2016 and finally implementation phase since 2017 (ICIMOD 2012, 2018a, 2019a; Ning et al. 2014; ICIMOD, WCD, GBPNIHESD & RECAST 2017a, b; Gurung et al. 2019). However, in all the four landscapes, objectives were similar for each of the four planning steps. First, the consensus building among the key government authorities followed by extensive consultations to develop feasibility assessment reports. In the second step, once the issues were identified and need for landscape level interventions were realized and agreed, conservation and development strategies and regional cooperation frameworks with 20 year visions were prepared, discussed and endorsed. This led to the third phase of development of theory of change, operational plans and implementation (see Fig. 31.2). All these steps followed an extensive participatory approach, engaging all stakeholders and each steps emphasized on science-based determination of transboundary landscape, dialogue and documentation for policy support, participatory planning and implementation to translate into practice and with the learning during the implementation, the interface between three were strengthened at local, national and regional levels (Fig. 31.2).

Guided by conservation and development strategies and regional cooperation frameworks, the transboundary landscape programme started implementation considering five major common thematic areas agreed along with three subsidiary conditions to strengthen integrated landscape management (Fig. 31.3). The emphasis was given for developing social-ecological resilience through interventions on human and ecosystem well-being as central themes. The priority on these thematic areas was given for addressing community living in acute poverty and/or in remote areas and improving ecosystem for sustained flow of services and conservation of biodiversity. Considering different institutional mechanism and governing structure of countries sharing landscapes, compatible resources governance mechanisms were proposed. To address the data gap revealed in IPCC AR4 report, and a dynamic nature of ecosystems, long-term monitoring for science-based decision making was considered as fourth thematic area. As regional cooperation is the main theme for transboundary landscape, cooperation among the member countries sharing landscapes was the fifth component with aspiration of designing common research protocol, data sharing platform and capacity building based on countries strength (Fig. 31.3).

31.3 Implementation for Resilience Building Through Nature-Based Solutions (NbS)

Improving resilience in the HKH is multidimensional and complex. The geo-political sensitivity, variance in investment capacities for research and development, challenges towards fighting poverty, inequality and food and nutrition insecurity, and limited access to water and energy along with degrading ecosystems make the regional cooperation important for improving resilience (Mishra et al. 2019; Xu et al. 2019). Since the inception of transboundary landscape programme in 2008, ICIMOD, as a regional knowledge and enabling centre, working with its capable partnership network across the region, has always put balancing conservation and development dilemma as a main thrust in landscape management (Chettri and Sharma 2016). The integrated and multidisciplinary framework with five thematic components (Fig. 31.3) were instrumental for making progress towards resilience building in the HKH focused on nature-based solutions (NbS).

The programme invested immensely on capitalizing natures' bounty for social-economic development through value chain development in mountain niche products such as *allo*, a unique local-level enterprise based on nettle plant for fibre and garment enterprises in KSL (Adhikari et al. 2018; Shrestha et al. 2018). Likewise, KSL being the highest producing region of large cardamom, cooperation for regional branding and common collective trade for better pricing have been initiated (ICIMOD 2019b). Considering the rich natural and traditional heritages, advocacy on yak-based enterprises along with nature-based tourism were promoted at regional scale to benefit local communities (Dorji et al. 2019; Lama et al. 2019). With the success of conservation interventions and shrinking natural ecosystems, human wildlife conflict was identified as an

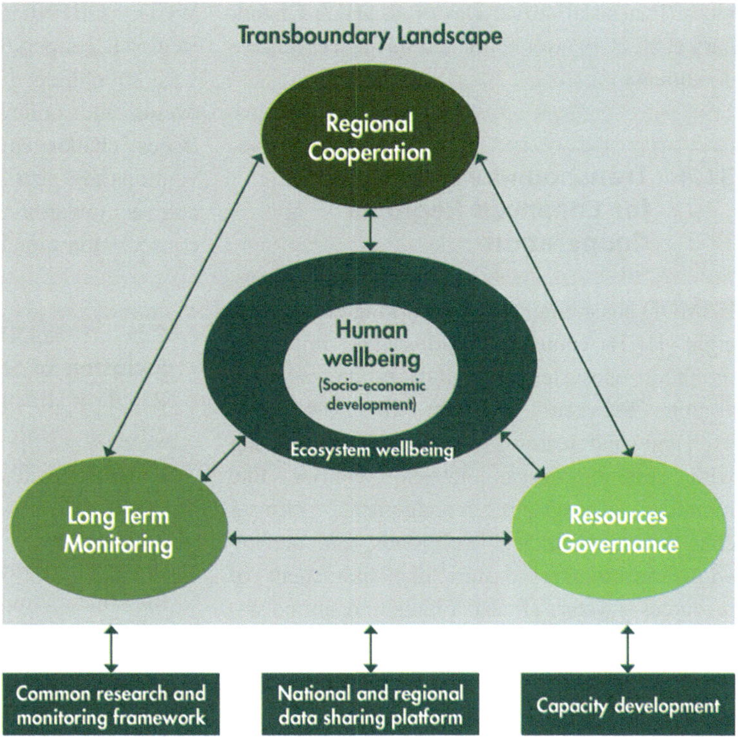

Fig. 31.3 Thematic areas agreed to consider for implementation of landscape approach in the HKH

important regional issue. Attempt has been made to address this with facilitating regional policy dialogue for collective mitigation strategies.

Likewise, ecosystem management interventions were made based on ecosystem management framework (Yi et al. 2017). Considering the framework, conservation corridors for connecting habitats and protected areas (Chettri et al. 2007; Uddin et al. 2019), valuation of ecosystem services from diverse ecosystems (Pant et al. 2012; Chaudhary et al. 2019) and piloting of springshed management (Kotru et al. 2017), are some of the highlight interventions across the landscape. Similarly, through REDD + programme, ICIMOD complemented with income-generating activities for community forest user groups in Nepal, Myanmar and Mizoram state of India, contributing to local economy, environmental sustainability and strengthen the rights of indigenous and local communities (Yamasaki and Bhattarai 2020). Efforts have also been made to strengthen gender and social inclusion in planning and implementation of programmes and institutionalized through networks with key stakeholders including private sector for better governance (Adhikari et al. 2018; Shrestha et al. 2018). Some of the effective strategies used for inclusive planning and implementations are (a) making participation of 30% women mandatory in all events, (b) inclusion of gender action plan in programme implementation, (c) gender focused natural resources planning etc. (Khadka and Verma 2012; Molden et al. 2014a; Yi et al. 2017). To sustain the programme and develop ownership, regional programme steering committees were established and made operational (Kotru et al. 2020).

As part of long-term monitoring, a comprehensive long-term social ecological monitoring framework was prepared and implemented (Chettri et al. 2015; Sinha et al. 2018; Negi et al. 2019). The knowledge on biodiversity (Kandel et al. 2016, 2018, 2019; Basnet et al. 2019), agrobiodiversity (Aryal et al. 2017, 2020), traditional practices (Uprety et al. 2016; Aryal et al. 2018; Ghimire et al. 2018) and ecosystems services (Pant et al. 2012; Das et al. 2017; Chaudhary et al. 2019) were enriched through extensive documentations.

31.4 Transboundary Landscapes for Enhanced Regional Cooperation

ICIMOD has a unique niche working between its eight HKH countries sharing the important mountain ecosystem. ICIMOD promotes cooperation between countries by knowledge exchange and learning across borders, dealing with common issues across countries like mountain ecosystem management, climate change and enhancing livelihoods; and dealing with transboundary issues like movement of resources—water, air and plants and animals—across borders. An ideal goal for the transboundary landscapes programme is to set up joint management of the landscapes for long-term sustainability, but we recognize that this goal will take time given the present political realities. However, significant progress has been made as the programme could brought together communities, professionals and managers across countries to work on common objectives, which has built trust. Some examples on such interventions are on regional tourism (Lama et al. 2019), large cardamom (ICIMOD 2019b), and REDD + (Box 1). Such enabling condition has also produced joint scientific outputs on status of biodiversity (Kandel et al. 2016, 2018, 2019; Basnet et al. 2019), non-timber forest products and agrobiodiversity (Uprety et al 2016; Aryal et al. 2018), climate change impacts (Zomer et al. 2014), land use mapping (Uddin et al. 2015) etc. As part of the regional cooperation, ICIMOD has promoted exchange of yak genetic material across countries (Manandhar 2020), and the countries sharing the landscape are discussing on mitigating human wildlife conflict and extending corridors (Gurung et al. 2019; Sharma et al. 2020). In addition, the regional cooperation frameworks as agreements to work together on the landscapes between countries have already been an important policy achievement (ICIMOD, WCD, GBPNIHESD & RECAST 2017b). Regional cooperation in each of the landscapes was strengthened through numerous regional events and policy dialogue on transboundary issues, challenges and opportunities. The interventions are also supplementing the global discourses towards achieving global biodiversity conservation targets (Desai et al. 2011).

Box 1: REDD + in policy influence-inclusion of SRAP in NRS

REDD + Himalaya project, launched in 2013 in Bhutan, India, Myanmar and Nepal was an important intervention to capacitate countries on the REDD + mechanism and support in readiness phase. During the programme implementation, a manual on State REDD + Action Plan (SRAP) was developed by ICIMOD for implementing REDD + at a sub-national level. Since the inception of REDD + Himalaya project in India, it provided the technical support for developing National REDD Strategy (NRS). States are required to develop their SRAPs as indicated in the NRS. Since, India covers a large geographical area with regional differences in forest ecosystems and drivers of D&D, policies and measures developed at the state level can help to achieve the goals set forth by the NRS. Thus, there is a coherent derivation of policy from the wider NRS to the more area-specific SRAPs. These plans highlight the relevance of the REDD + mechanism as a multi-sectorial approach and its potential for effective climate change mitigation and adaptation. All the interventions were designed through multi-stakeholders approach and in close consultation with local stakeholders. The interventions support local communities to participate in REDD + activities for the implementation of interventions. In other words, SRAPs are the most effective medium for operationalizing the NRS.

Indian Council of Forestry Research and Education (ICFRE) and ICIMOD developed SRAPs for the two states in India (Viz. Mizoram & Uttarakhand). Mizoram has become the first state to endorse the SRAP and come up with REDD Cell at the state level. Uttarakhand became the second state to develop the SRAP. SRAPs for 2 more states are under progress.

31.5 Emerging Challenges

The HKH, while being sensitive to climate change, is also subjected to a wide range of natural and anthropogenic drivers of change, such as habitat change, land use and land cover change, overexploitation of natural resources, fragmentation, expansion of invasive alien species, urbanization and pollution (Wang et al. 2019). These factors continually affect fragile ecosystems and have implications on the social-ecological resilience (Xu et al. 2019). Among these factors, climate change and land use land cover change (LULCC) are considered as leading causes of biodiversity loss and ecosystem degradation in the HKH (Chettri et al. 2010; Chettri and Sharma 2016).

31.5.1 Climate Change Impacts

During the past few decades, our understanding on climate change and its potential risks to mountain ecosystem has increased substantially (Singh et al. 2017; Wester et al. 2019; Chettri et al. 2020). The HKH is characterized by diverse climate due to diversity in topography, monsoon influence and ecosystems. Though with paucity of long-term studies, it was observed that the HKH witnessed changes in climate over the period with evidence of change in phenology and species range shift altering ecosystem functions (Xu et al. 2019). During 1901–2014, annual mean surface air temperature significantly increased in the HKH at a rate of about 0.104 °C per decade showing significant upward trend (Ren et al. 2017). However, there were exceptions with deviations from the general pattern. In the Karakoram region, decreasing (most notably) summer temperature have been measured (Forsythe et al. 2017). The intense precipitation also showed increasing trend in annual intense precipitation amount, days and intensity with 5.28 mm per decade, 0.14 day per decade and 0.39 mm/day per decade respectively (Zhan et al. 2017). The elevation dependent warming has also been prominent in the HKH with higher warming with the increasing elevation. Higher warming is projected during winter and the projected warming differs by more than 1 °C between the eastern and western HKH, with relatively higher values during winter, and the highest warming is projected to be over the central Himalaya for the far-future period with the RCP8.5 scenario (Sanjay et al. 2017). The projections made by the study for the near-future and far-future periods for HKH are relatively higher than the seasonal global means. These changes have indicated that rapidly changing climatic conditions could significantly thwart efforts for ecosystem resilience at national and regional scales (Chettri et al. 2020). There has been a wide range of interpretation from observed and people's perceptions impacting ecosystems and biodiversity at different scales (Chaudhary and Bawa 2011; Shrestha et al. 2012; Wangchuk and Wangdi 2018). Climate change is reshaping global biodiversity as species respond to changing temperatures, precipitation and other climatic conditions (Antão et al. 2020) and the trend is not different in the HKH (Salick et al. 2019). However, there is still a major gap in understanding the cross-linkages among areas of research, for example, linking social-ecological knowledge on resilience contributing to evolutionary adaptation. Although numerous important contributions have emerged in recent years, synthesis of such practices and its consequences has not yet been achieved.

31.5.2 Land Use Land Cover Change and Impacts

The HKH comprises approximately 39% grassland, 20% forest, 15% shrub land, 5% agricultural land and the remaining 21% is made up of barren land, rock outcrops, built-up areas, snow cover and waterbodies (Chettri et al. 2008). These land use and land cover types have witnessed significant changes over the last century. Over 70% of the natural ecosystems have been changed in the region (Myers et al. 2000). The rangelands, where substantial population depends on pastoralist economy, have witnessed degradation, mainly caused by climate warming and human disturbance (Ali et al. 2019; Duan et al. 2019). Declining land cover of alpine meadows and swamps (Niu et al. 2019), and conversion of alpine meadows to woody shrubs suggesting regime shift (Brandt et al. 2013), have been apparent. Natural wetlands and lakes, that occupy 10% of the HKH, are also experiencing transformation driven by land use changes, including over-extraction of resources, pollution and environmental changes (Chettri et al. 2015; Chaudhary et al. 2017). Similarly, there has been visible changes in forests across the HKH (Uddin et al 2019). Road expansion, human settlement expansion and natural disasters together with climatic influences have been identified as the major drivers of forest cover change and fragmentation (Uddin et al. 2015; Murthy et al. 2016). The trend of change basically shows the decline of healthy forests. About 16% decline in forest cover in Sikkim between 1990 and 2013 (Kanade and John 2018), 9% decline in forest cover (1990–2009) and 7% decline in forest cover (1976–2011) in Nepal and India part of Kailash sacred landscape, respectively (Uddin et al. 2015; Singh et al. 2018), and 6.5% decline in forest cover in Hindu Kush mountain range of Pakistan (Ullah et al. 2016) have been reported. The forest ecosystems in the Indo-Burma Biodiversity Hotspots in the Far-Eastern Himalayan region have also witnessed significant changes with 77.1% in 1950 to 50.6% in 2016 and it is projected to be 48.2% in 2027 (Reddy et a. 2019). It is projected that with the present trend of deforestation, the HKH face significant loss of its natural forests (Xu et al. 2019).

With regard to agro-ecosystems—that form the primary basis of livelihoods and income for 90% of rural mountain communities in the HKH, the changes have been in two directions—either conversions of natural ecosystems to agriculture or the conversion of arable lands into the built-up areas or plantations (Tulachan 2001). For example, in central Himalaya, between 1963 and 1993, agricultural land increased by 35% in protected forests and 5% in reserved forests, and that agricultural expansion was conspicuous at elevations between 1800 and 2600 m (Semwal et al. 2004). Likewise, in high mountain areas in Pakistan, the forested area decreased by 30% between 1968 and 2007, mostly due to agricultural expansion (Qasim et al. 2013). In Chittagong Hill tract in Bangladesh, a model predicted huge increment of built-up area with shrinkage of cultivated lands, forests and grassland areas between 2010 and 2040 (Hasan et al. 2019). Similar simulation model indicated significant reduction of arable land area for Yunnan, China, and increase of construction area by 2020 (Zhang et al. 2018), elaborating the negative consequences for agro-ecosystems and food production. As these remaining natural ecosystems are both sources of ecosystem services (Kandel et al. 2016; Chaudhary et al. 2018; Karki et al 2018), and habitats for globally significant species (Chettri et al. 2010, 2015), losing them could have detrimental consequences to the people living in the HKH and downstream (Xu et al. 2019).

31.5.3 Upstream–Downstream Linkages

The transboundary landscape programme interventions are in the high mountain watersheds of the water towers of Asia—the headwaters of the Indus, Ganges, Brahmaputra, Mekong, Salween and Ayerwaddy Rivers. Changes in land use and climate change have an impact on river flows serving nearly 2 billion people. ICIMOD's transboundary landscape and cryosphere and river basin programmes are working together to understand these changes and to enhance ecosystem services valuable to downstream people (Molden et al. 2014b). While water is important for downstream, so is the energy generated from hydropower, the forest products both timber and non-timber and that mountains offer a retreat for people from the plains. ICIMOD's role then is make sure that downstream beneficiaries are well aware of mountain concerns, which in fact are everyone's concerns (Mukherji et al. 2015). ICIMOD interacts with policy makers in HKH capital cities and in the plains to ensure that beneficial policies are made for everyone. For example, ICIMOD working with India's Niti Ayog has emphasized mountains as sources of ecosystem services for downstream and need collaborative efforts to maintain ecosystems through springshed management (ICIMOD 2018b). In Pakistan, ICIMOD has made sure the issues of mountain agriculture are addressed in its food security policy (Rasul and Hussain 2015). Likewise in Bhutan, early warning system addressing gender in upstream–downstream linkages were explored (Shrestha et al. 2016) and in China farmers' perceptions of the effectiveness of policies implemented to deal with drought (Pradhan et al. 2017).

31.6 Lessons and Opportunities

Since the inception of transboundary landscape programme in 2008, there has been significant progress on making the concept a reality. Over the period of 12 years, four of the six identified landscapes are now operational. All four landscapes went through a thorough process of planning, identifying challenges, delineating the areas, translating the challenges into opportunities and developing long-term programme. The implementation of the programme substantially contributed towards social-ecological resilience by providing demonstrated livelihood opportunities based on local biodiversity. Ecosystem management interventions have been initiated focusing on restoration for degraded ecosystems, advocated connectivity corridors between the protected areas and contiguous habitat across borders, established database with extensive documentation and monitoring mechanism in place, strengthen governance by focusing on equity and gender along with institutional development and facilitated regional cooperation for common objectives of conservation and development. The outstanding progress made by the programme was recognized by an award from the United States of America (Box 2). However, this progress was not limited to what has been achieved in last 12 years (Gurung et al. 2019). ICIMOD advocated regional cooperation and integrated approach for resources management since its establishment in 1983. Since then, the progress witnessed changing paradigm from sectoral interventions to integrated landscape with multidisciplinary to transdisciplinary for transformative change and two-way programmatic orientations from local to global scale and *vice versa* (Fig. 31.4), and progressive investment towards developing long-term multidisciplinary and multisectoral partnerships over the years, at all scales of programmatic interventions.

Box 2: Global outstanding achievement award 2018

ICIMOD's 'Transboundary Landscapes Regional Programme' received the 'Renewable Natural Resources Foundation's (RNRF) Outstanding Achievement Award' in 2018. The award was an acknowledgement to ICIMOD's continued efforts towards promoting the use of the landscape approach in the HKH. The Outstanding Achievement Award is determined through a global competition. The then Regional Programme Manager, Dr. Rajan Kotru, received the award on November 29, 2018 in Maryland, USA.

Considering the prevailing challenges from climatic change and its potential impacts on ecosystems and livelihood, the transboundary programme has scaled out its interventions to

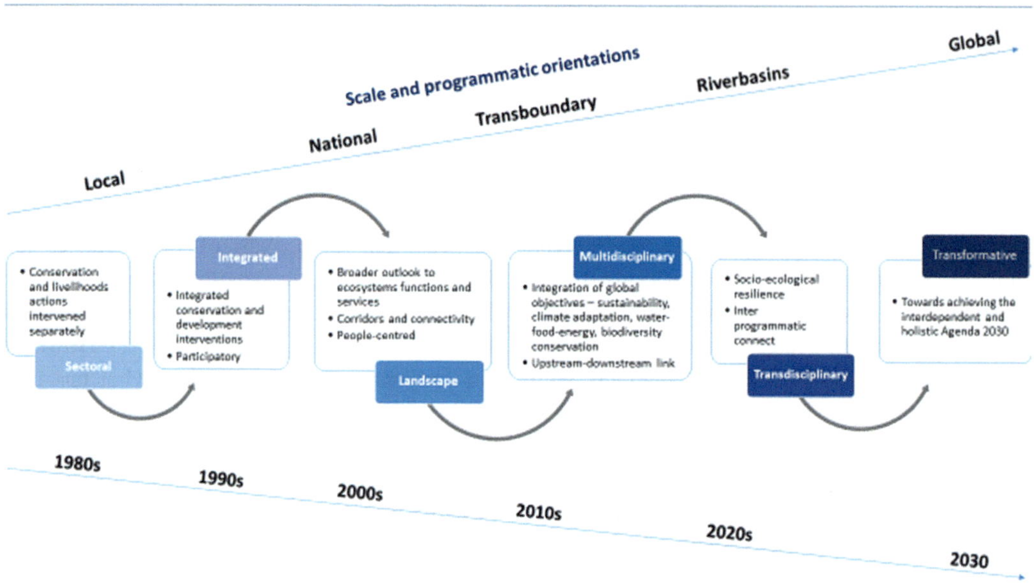

Fig. 31.4 Schematic presentation of changing programmatic paradigm and orientations to facilitate sustainable transformation of HKH mountain landscapes

make synergistic upstream and downstream linkages (Molden et al. 2017). Efforts have been made to replicate the good practices of community-based disaster management, an award winning practice of the river basin programme (Box 3), in the landscapes planning. Restoration of ecosystems are now considered through springshed management framework (Pradhan et al. 2015). Efforts have also been made to promote nature-based solutions (NbS) in the landscapes complementing to adaptation and resilience building strategy suggested by UNFCCC (Lafortezza et al. 2018). ICIMOD put forward the *HKH Call to Action* which outlines six urgent actions to address the issues and sustain the mountain environments and improve the livelihoods of the HKH region (Wester et al. 2020). One of the actions focuses to enhance ecosystem resilience for sustained flow of services by halting biodiversity loss and land degradation, and sustainably managing forests and other ecosystems in the HKH through promoting transboundary cooperation for landscapes and river basins. This will further strengthen the programme to look at the nexus between social, ecological and environmental dimensions from household to river basin levels.

Box 3: Cooperation across the borders to floods risk reduction in the Himalaya

The HKH is highly prone to natural hazards and disasters, which has been further exacerbated by climate change and its impacts. Floods and flash floods are the major climate-induced natural hazards threatening lives and livelihoods of billions in the region and beyond. Floods in small rivers and its tributaries, in particular, strike with little or no warning but with disastrous causalities. The lack of timely information especially to the vulnerable communities downstream has been the major reason for significant causalities by floods in the region. In response, ICIMOD developed and launched community-based flood early warning system (CBFEWS) in 2010. It is a low cost, people-centred integrated system of tools and plans to detect and respond to flood emergencies managed by communities to provide real-time flood information. The disseminated information by the upstream communities to the vulnerable communities downstream through a network of communities and government bodies provides enough time to prepare and respond to floods on the ground.

CBFEWS is an outstanding example of cooperation across the borders to climate-induced disasters. The community cooperation in upstream–downstream, and the synergies between communities and local partners, experts, private sectors and the government line agencies are the key factors for its successful implementation. First piloted in India, the CBFEWS has been successfully scaled out in other countries like Afghanistan, Nepal and Pakistan. The benefits of the approach reported have been enormous. It has enabled the flood vulnerable individuals, communities and organizations to prepare and act to reduce harm and loss of lives and property across the region. As such, CBFEWS can save lives and livelihoods of millions, empowered communities especially marginalized and built resilience to tackle to floods disasters across the border countries in the region. Acknowledging the impacts, CBFEWS has been awarded the 'Momentum for Change; 2014 Lighthouse Activity Award' in the Information and Communication Technology (ICT) category by UNFCCC.

31.7 Conclusion

ICIMOD has been promoting transboundary landscape management perspective with shared goal of healthy landscapes that balance conservation and development. The integrated landscape management approach is flexible enough to be applied beneficially to a range of geographies, cultures and types of actors, institutions and livelihood needs. So far, the programme has contributed to (a) creating common management of shared ecosystems and biodiversity given the plethora of local and transboundary-scale issues; (b) enhanced collaboration for sciences and cross learning opportunities from good practices; (c) mainstream standardized frameworks, research protocols for long-term research and monitoring to inform policy and national development strategies for transboundary cooperation. Since its implementation, there has been growing buy-in from participating national governments along with co-financing on piloted interventions in the fields of tourism, commerce and protected area management, among others. To successfully bring together policy, science and practice in transboundary landscapes, our experiences show that the following elements are necessary:

- Establishing common understanding and priority setting on conservation and development needs at scale across countries;
- Developing a jointly owned transboundary platform for collaborative planning, knowledge generation and effective implementation that uses a systematic program cycle; and
- Customizing local and national learning to global agendas, and vice versa.

References

Adhikari L, Shrestha AJ, Dorji T, Lemke E, Subedee BR (2018) Transforming the lives of mountain women through the Himalayan Nettle value chain: a case study from Darchula, Far West Nepal. Mt Res Dev 38 (1):4–13

Ali K, Bajracharya RM, Chapagain NR, Raut N, Kumar B, Begum F, Khan MZ, Ali M, Ahmed A (2019) Analyzing land cover change using remote sensing and GIS : a case study of Gilgit. Int J Econ Environ Geol 10(1):100–105

Antão LH, Bates AE, Blowes SA, Waldock C, Supp SR, Magurran AE, Dornelas M, Schipper AM (2020) Temperature-related biodiversity change across temperate marine and terrestrial systems. Nat Ecol Evol. https://doi.org/10.1038/s41559-020-1185-7

Aryal K, Poudel S, Chaudhary RP, Chettri N, Chaudhary P, Ning W, Kotru R (2018) Diversity and use of wild and non-cultivated edible plants in the Western Himalaya. J Ethnobiol Ethnomed. 14:10

Aryal K, Poudel S, Chaudhary P, Chaudhary RP, Ghimire KH, Shrestha DS, Joshi BK (2020) Agro-morphological diversity of high altitude bean landraces in the Kailash Sacred Landscape of Nepal. J Nepal Agri Res Council 6:1–13

Aryal K, Poudel S, Chaudhary RP, Chettri N, Chaudhary P, Ning W, Shaoliang Y, Kotru R (2017) Conservation and management practices of traditional crop genetic diversity by the farmers: a case from Kailash Sacred Landscape Nepal. J Agri Environ 18:15–28

Basnet D, Kandel P, Chettri N, Yang Y, Lodhi M S, Htun N Z, Uddin K, Sharma E (2019) Biodiversity research trends and gaps from the confluence of three global biodiversity hotspots in the far-eastern Himalaya. Int J Ecol 1323419

Brandt JS, Haynes M, Kuemmerle T, Waller DM, Radeloff VC (2013) Regime shift on the roof of the world: Alpine meadows converting to shrublands in the southern Himalayas. Biol Conser 158:116–127

Brooks TM, Mittermeier RA, da Fonseca GA, Gerlach J, Hoffmann M, Lamoreux JF, Rodrigues AS (2006) Global biodiversity conservation priorities. Science 313(5783):58–61

Chaudhary P, Bawa KS (2011) Local perceptions of climate change validated by scientific evidence in the Himalayas. Biol Lett 7(5):767–770

Chaudhary S, McGregor A, Houston D, Chettri N (2018) Environmental justice and ecosystem services: a disaggregated analysis of community access to forest benefits in Nepal. Ecosyst Serv 29:99–115

Chaudhary S, McGregor A, Houston D, Chettri N (2019) Spiritual enrichment or ecological protection? A multi-scale analysis of cultural ecosystem services at the Mai Pokhari, a Ramsar site of Nepal. Ecosyst Serv 39:100972

Chaudhary S, Tshering D, Phuntsho T, Uddin K, Shakya B, Chettri N (2017) Impact of land cover change on a mountain ecosystem and its services: case study from the Phobjikha valley Bhutan. Ecosyst Heal Sustain 3:1–12

Chettri N, Bubb P, Kotru R, Rawat G, Ghate R, Murthy MSR, Wallrapp C, Pauli H, Shrestha AB, Mool PK, Chaudhary D, Chaudhary R P, Mathur PK, Peili S, Ning W, Sharma E (2015) Long-term environmental and socio-ecological monitoring in transboundary landscapes. An interdisciplinary implementation framework. ICIMOD Working Paper 2015/2. ICIMOD, Kathmandu

Chettri N, Shakya B, Thapa R, Sharma E (2008) Status of a protected area system in the Hindu Kush-Himalayas: an analysis of PA coverage. Int J Biodivers Sci Manag 4:164–178

Chettri N, Sharma E (2006) Prospective for developing a transboundary conservation landscape in the Eastern Himalayas. In: McNeeley TMM, Smith A, Whittaker O, Wikramanayake E (eds) Conservation biology in Asia. Society of Conservation Biology, Kathmandu, Nepal, pp 21–44

Chettri N, Sharma E (2016) Reconciling the mountain biodiversity conservation and human wellbeing: drivers of biodiversity loss and new approaches in the Hindu-Kush Himalayas. Proc Indian Natl Sci Acad 82:53–73

Chettri N, Sharma E, Shakya B, Bajracharya B (2007) Developing forested conservation corridors in the Kangchenjunga Landscape Eastern Himalaya. Mt Res Dev 27(3):211–214

Chettri N, Sharma E, Shakya B, Thapa R, Bajracharya B, Uddin K, Oli K, Choudhury D (2010) Biodiversity in the Eastern Himalayas: status, trends and vulnerability to climate change; Climate change impact and vulnerability in the Eastern Himalayas—Technical report 2. ICIMOD, Kathmandu

Chettri N, Sharma E, Thapa R (2009) Long term monitoring using transect and landscape approaches within Hindu Kush Himalayas. In: Proceedings of the international mountain biodiversity Conference, Kathmandu, 16–18 November 2008, pp 201–208. ICIMOD, Kathmandu

Chettri N, Shrestha A B, Sharma E (2020) Climate change trends and ecological resilience in the Hindu Kush Himalaya. In: Dimri AP, Bookhagen B, Stoffel M, Yasunari T, Petra van Steenbergen (eds) Himalayan weather and climate and their impact on the environment. Springer Nature, Switzerland, pp 525–552

Cui X, Graf HF, Langmann B, Chen W, Huang R (2006) Climate impacts of anthropogenic land use changes on the Tibetan Plateau. Glob Planet Change 54:33–56

Das S, Rai RK, Bhatta LD, Somanathan E, Kotru R, Khadayat MS, .Rawal RS, Negi GCS (2017) Valuation of ecosystem services in the Kailash Sacred Landscape. ICIMOD, Kathmandu

Desai BH, Oli KP, Yang Y, Chettri N, Sharma E (2011) Implementation of the convention on biological diversity: a retrospective analysis in the Hindu Kush-Himalayan countries. ICIMOD, Kathmandu

Dorji T, Gaira K S, Rabgay T, Pandey A, Pant, B, Chettri, N (2019) Protecting a Himalayan icon: The need for transboundary cooperation to secure the future of yak in the Kangchenjunga Landscape. Issue Brief Kathmandu: ICIMOD

Duan C, Shi P, Song M, Zhang X, Zong N, Zhou C (2019) Land use and land cover change in the Kailash Sacred Landscape of China. Sustainability 11(6):1788

Erg B, Vasilijević M, McKinney M (2012) Initiating effective transboundary conservation: a practitioner's guideline based on the experience from the Dinaric Arc. Gland, Switzerland and Belgrade, Serbia: IUCN Programme Office for South-Eastern Europe. ix+98p

Forsythe N, Fowler HJ, Li X-F, Blenkinsop S, Pritchard D (2017) Karakoram temperature and glacial melt driven by regional atmospheric circulation variability. Nat Clim Change 7:664–670

Freeman OE, Duguma LA, Minang PA (2015) Operationalizing the integrated landscape approach in practice. Ecol Soc 20(1):24. https://doi.org/10.5751/ES-07175-200124

Ghimire K, Adhikari M, Uprety Y, Chaudhary R (2018) Ethnomedicinal use of plants by the highland communities of Kailash Sacred Landscape Far-West Nepal. Acad J Med Plants 6(11):365–378

Guangwei C (2002) Biodiversity in the eastern Himalayas: conservation through dialogue. Summary reports of workshops on biodiversity conservation in the Hindu Kush-Himalayan ecoregion. ICIMOD, Kathmandu, Nepal

Gurung J, Chettri N, Sharma E, Ning W, Chaudhary RP, Badola HK, Wangchuk S, Uprety Y, Gaira KS, Bidha N, Phuntsho K, Uddin K, Shah GM (2019) Evolution of a transboundary landscape approach in the Hindu Kush Himalaya: key learnings from the Kangchenjunga Landscape. Global Ecol Conserv 17: e00599

Gurung PC (2005) Terai arch landscape: a new paradigm in conservation and sustainable development. In: Harmone D, Worboys GL (eds) Managing mountain protected areas: challenges and responses for the 21st century. Andromeda Editrice, Italy, pp 80–86

Hasan S, Sarmin N, Miah M (2019) Assessment of scenario-based land use changes in the Chittagong Hill Tracts of Bangladesh. Environ Dev 1:100463. https://doi.org/10.1016/j.envdev.2019.100463

ICIMOD (2012) Towards developing the Brahmaputra-Salween Landscape—Report on the experts regional consultation for transboundary biodiversity management and climate change adaptation. ICIMOD Working Paper 2012/4. ICIMOD, Kathmandu, Nepal

ICIMOD (2017) Medium term action plan 2018–2022. ICIMOD, Kathmandu, Nepal

ICIMOD (2018a) Regional workshop on planning transboundary technical collaboration for landscape management. ICIMOD Workshop Report 2018. ICIMOD, Kathmandu, Nepal

ICIMOD (2018b) Knowledge forum: climate resilient development in Himalayan and downstream regions. ICIMOD Proceedings 2018/7. ICIMOD, Kathmandu, Nepal

ICIMOD (2019a) Harmonizing conservation and development along the silk road. Facilitating a Network of Protected Areas in the Hindu Kush, Karakoram, and Pamir. ICIMOD Proceedings 2019/1. ICIMOD, Kathmandu, Nepal

ICIMOD (2019b) Proceedings of the regional workshop on 'exploring opportunities for transboundary collaboration on large Cardamom value chain in the Kangchenjunga Landscape. 22–23 May 2019, Phungling, Taplejung, Nepal. ICIMOD, Kathmandu, Nepal

ICIMOD and NCD 2018. Proceedings of the policy dialogue: shared natural and cultural heritage for sustainable tourism in Kanchenjunga Transboundary Landscape. ICIMOD, Proceedings 2018/01. ICIMOD, Kathmandu, Nepal

ICIMOD, WCD, GBPNIHESD & RECAST (2017a) Kangchenjunga landscape feasibility assessment report. ICIMOD Working Paper 2017/9. ICIMOD, Kathmandu, Nepal

ICIMOD, WCD, GBPNIHESD & RECAST (2017b) Kangchenjunga Landscape conservation and development strategy and regional cooperation framework. ICIMOD Working Paper 2017/2. ICIMOD, Kathmandu, Nepal

Kanade R, John R (2018) Topographical in fluence on recent deforestation and degradation in the Sikkim Himalaya in India; Implications for conservation of East Himalayan broadleaf forest. Appl Geogr 92:85–93

Kandel P, Chettri N, Chaudhary RP, Badola HK, Gaira KS, Wangchuk S, Bidha N, Uprety Y, Sharma E (2019) Plant diversity of the Kangchenjunga landscape, Eastern Himalaya. Plant Diversity 41:153–165

Kandel P, Gurung J, Chettri N, Ning W, Sharma E (2016) Biodiversity research trends and gap analysis from a transboundary landscape, Eastern Himalayas. J Asia-Pacific Biod 9:1–10

Kandel P, Thapa I, Chettri N, Pradhan R, Sharma E (2018) Birds of the Kangchenjunga Landscape, the Eastern Himalaya: status, threats and implications for conservation. Avian Res 9:9

Karki S, Thandar AM, Uddin K, Tun S, Aye WM, Aryal K, Kandel P, Chettri N (2018) Impact of land use land cover change on ecosystem services: a comparative analysis on observed data and people's perception in Inle Lake, Myanmar. Environ Syst Res 7:25

Khadka M, Verma R (2012) Gender and biodiversity management in the Greater Himalayas: towards equitable mountain development. ICIMOD, Kathmandu, Nepal

Kotru R, Chaudhari S, Lemke E, Mueller M, Chettri R, Basnet S, Shaoliang Y (2017) Kailash Sacred Landscape conservation and development initiative (2012–2017) annual progress report 2016. ICIMOD, Kathmandu, p 101

Kotru R, Pradhan N, Shakya B, Amatya S (eds) (2020) Beyond boundaries: contouring transboundary landscapes in the Hindu Kush Himalaya. ICIMOD, Kathmandu

Lafortezza R, Chen J, Van Den Bosch CK, Randrup TB (2018) Nature-based solutions for resilient landscapes and cities. Environ Res 165:431–441

Lama AK, Kandel P, Chaudhary S, Dema K, Uprety Y, Gaira K, Pandey A, Dorji T, Chettri N (2019) Transboundary ecotourism in the Kangchenjunga Landscape: opportunities for sustainable development through regional cooperation. ICIMOD Working Paper 2019/4. ICIMOD, Kathmandu, Nepal

Lambertucci SA, Alarcón PA, Hiraldo F, Sanchez-Zapata JA, Blanco G, Donázar JA (2014) Apex scavenger movements call for transboundary conservation policies. Biol Conserv 170:145–150

Liu J, Yong DL, Choi CY, Gibson L (2020) Transboundary frontiers: an emerging priority for biodiversity conservation. Trends Ecol Evol. https://doi.org/10.1016/j.tree.2020.03.004

Manandhar A (2020) Yak across borders: Bhutan gifts breeding bulls to Nepal and India for gene pool improvement. ICIMOD, Kathmandu, Nepal

MFSC (2006) Sacred Himalayan Landscape Strategic Plan 2006–2016. Ministry of forests and soil conservation, Government of Nepal, Kathmandu, Nepal, Nepal

Mishra A, Appadurai AN, Choudhury D, Regmi BR, Kelkar U, Alam M, Chaudhary P, Mu SS, Ahmaed AU, Lotia H, Fu C, Namgyel T, Sharma U (2019). Adaptation to climate change in the Hindu Kush Himalaya: stronger action urgently needed. In: Wester P, Mishra A, Mukherji A, Shrestha A (eds) The Hindu Kush Himalaya assessment. Springer, Cham

Mittermeier RA, Turner WR, Larsen FW, Brooks TM, Gascon C (2011) Global biodiversity conservation: the critical role of hotspots. Biodiversity hotspots. Springer, Berlin, Heidelberg, pp 3–22

Molden D, Sharma E (2013) ICIMOD's strategy for delivering high-quality research and achieving impact for sustainable mountain development. Mt Res Dev 33 (2):179–183

Molden D, Sharma E, Shrestha AB, Chettri N, Shrestha N (2017) Advancing regional and transboundary cooperation in the conflict-prone Hindu Kush Himalaya. Mt Res Dev 37(4):503–508

Molden D, Verma R, Sharma E (2014a) Gender equality as a key strategy for achieving equitable and sustainable development in mountains: the case of the Hindu Kush-Himalayas. Mt Res Dev 34(3):297–300

Molden DJ, Vaidya RA, Shrestha AB, Rasul G, Shrestha MS (2014b) Water infrastructure for the Hindu Kush Himalayas. Int J Water Res Dev 30 (1):60–77

Molden D (2020) Scarcity of water or scarcity of management? Int J Water Resour Dev 36(2–3):258–268

Mukherji M, Molden D, Nepal S, Rasul G, Wagnon P (2015) Himalayan waters at the crossroads: issues and challenges. Int J Water Res Dev 31(2):151–160

Murthy MSR, Das P, Behera MD (2016) Road accessibility, population proximity and temperature increase are major drivers of forest cover change in the Hindu Kush Himalayan Region. Curr Sci 111:1599–1602

Myers N, Mittermeier RA, Mittermeier CG, Da Fonseca GA, Kent J (2000) Biodiversity hotspots for conservation priorities. Nature 403(6772):853

Negi VS, Pathak R, Rawal RS, Bhatt ID, Sharma S (2019) Long-term ecological monitoring on forest ecosystems in Indian Himalayan Region: criteria and indicator approach. Ecol Indicators 102:374–381

Ning W, Ismail M, Joshi S, Qamar FM, Phuntsho K, Weikang Y, Khan B, Shaoliang Y, Kotru R, Sharma E (2014) Understanding the transboundary Karakoram-Pamir landscape. Feasibility and Baseline Studies #1. ICIMOD, Kathmandu, Nepal

Niu Y, Zhu H, Yang S, Ma S, Zhou J, Chu B, Hua R, Hua L (2019) Overgrazing leads to soil cracking that later triggers the severe degradation of alpine meadows on the Tibetan Plateau. Land Degrad Dev 30 (10):1243–1257

Olson D M, Dinerstein E (2002) The global 200: priority ecoregions for global conservation. Ann Missouri Bot Gard 199–224

Pandey A, Kotru R, Pradhan N (2016) Kailash Sacred Landscape: bridging cultural heritage, conservation and development through a transboundary landscape approach. In: Asian Sacred natural sites, pp 167–180. Routledge

Pant KP, Rasul G, Chettri N, Rai KR, Sharma E (2012) Value of forest ecosystem services: a quantitative estimation from Eastern Nepal, Kangchenjunga Landscape. ICIMOD Working Paper 2012/5. ICIMOD, Kathmandu

Pei S (1995) Banking on biodiversity. Report on the regional consultations on biodiversity assessment in the Hindu Kush-Himalaya. ICIMOD, Kathmandu, Nepal

Pradhan N, Kotru R, Mukherji A (2015) An integrated Springshed management approach linking science, policy, and practice: collaborative applied research in the Kailash Sacred Landscape (India and Nepal). ICIMOD, Kathmandu, Nepal

Pradhan NS, Fu Y, Zhang L, Yang Y (2017) Farmers' perception of effective drought policy implementation: A case study of 2009–2010 drought in Yunnan province, China. Land Use Policy 67:48–56

Qasim M, Hubacek K, Termansen M, Fleskens L (2013) Modelling land use change across elevation gradients in district Swat, Pakistan. Reg Environ Change 13:567–581

Rastogi A., Shengji P, Amatya D (1997) Regional consultation on conservation of the Kanchanjunga mountain ecosystem. World Wildlife Fund (WWF) Nepal Programme and International Centre for Integrated Mountain Development

Rasul G, Hussain A (2015) Sustainable food security in the mountains of Pakistan: towards a policy framework. Ecol Food Nutr 54(6):625–643

Reddy CS, Pasha SV, Satish KV, Unnikrishnan A, Chavan SB, Jha CS, Diwakar PG, Dadhwal VK (2019) Quantifying and predicting multi-decadal forest cover changes in Myanmar: a biodiversity hotspot under threat. Biod Conserv 28(5):1129–1149

Ren YY, Ren GY, Sun XB et al (2017) Observed changes in surface air temperature and precipitation in the Hindu Kush Himalayan region over the last 100-plus years. Adv Clim Change Res 8(3):148–156

Rosenzweig C, Karoly D, Vicarelli M, Neofotis P, Wu Q, Casassa G, Tryjanowski P (2008) Attributing physical and biological impacts to anthropogenic climate change. Nature 453(7193):353–357

Salick J, Fang Z, Hart R (2019) Rapid changes in eastern Himalayan alpine flora with climate change. Am J Bot 106(4):520–530

Sanjay J, Krishman R, Shrestha AB, Rajbhandari R, Ren GU (2017) Downscaled climate change projections for the Hindu Kush Himalayan region using CORDEX South Asia regional climate models. Adv Clim Change Res. 8(3):185–198

Secretariat of the CBD (2004) The ecosystem approach (CBD guidelines). Secretariat of the convention on biological diversity, Montreal, Canada

Semwal RL, Nautiyal S, Sen K, Rana U, Maikhuri R, Rao K, Saxena K (2004) Patterns and ecological implications of agricultural land-use changes: a case study from central Himalaya, India. Agri Ecosyst Environ 102(1):81–92

Sharma E, Chettri N (2005) ICIMOD's transboundary biodiversity management initiative in the Hindu Kush-Himalayas. Mt Res Dev 25(3):280–283

Sharma E, Chettri N, Gurung J, Shakya B (2007) The landscape approach in biodiversity conservation. A regional cooperation framework for implementation of the convention on biological diversity in the Kangchenjunga Landscape. ICIMOD, Kathmandu, Nepal

Sharma E, Chettri N, Oli KP (2010) Mountain biodiversity conservation and management: a paradigm shift in policies and practices in the Hindu Kush-Himalayas. Ecol Res 25:905–923

Sharma E, Chettri N, Tse-ring K, Shrestha AB, Jing F, Mool P, Eriksson M (2009) Climate change impacts and vulnerability in the Eastern Himalayas. ICIMOD, Kathmandu, Nepal

Sharma E, Molden D, Rahman A, Khatiwada YR, Zhang L, Singh SP, Yao T, Wester P (2019) Introduction to the Hindu Kush Himalaya assessment. In: Wester P, Mishra A, Mukherji A, Shrestha A (eds) The Hindu Kush Himalaya assessment. Springer, Cham, pp 1–16

Sharma E, Acharya N (2004) Summary report on mountain biodiversity in the convention on biological diversity (CBD). Mt Res Dev 24:263–265

Sharma P, Chettri N, Uddin K, Wangchuk K, Joshi R, Tandin T, Pandey A, Gaira K, Basnet K, Wangdi S, Dorji T, Wangchuk N, Chitale VS, Uprety V, Sharma (2020) Mapping human wildlife conflict hotspots in a transboundary landscape, Eastern Himalaya. Global Ecology and Conservation. Under Review

Sherpa LN, Peniston B, Lama W, Richard C (2003) Hands around everest: transboundary cooperation for conservation and sustainable livelihoods. ICIMOD, Kathmandu, Nepal

Sherpa MN, Wangchuk S, Wikramanayake ED (2004) Creating biological corridors for conservation and development: a case study from Bhutan. In: Harmone D, Worboys GL (eds) Managing mountain protected areas: challenges and responses for the 21st century. Andromeda Editrice, Italy, pp 128–134

Shrestha AJ, Adhikari L, Amatya R, Subedee BR, Dorji T (2018) Allo value chain in Darchula, Nepal: Process documentation. ICIMOD Working Paper 2018/9. ICIMOD, Kathmandu

Shrestha UB, Gautam S, Bawa KS (2012) Widespread climate change in the Himalayas and associated changes in local ecosystems. PLoS ONE 7(5): e36741. https://doi.org/10.1371/journal.pone.0036741

Shrestha M, Goodrich C, Udas P, Rai D, Gurung M, Khadgi V (2016) Flood early warning systems in Bhutan: a gendered perspective. ICIMOD, Kathmandu, Nepal

Singh SP, Bassignana-Khadka I, Karki BS, Sharma E (2011) Climate change in the Hindu Kush-Himalayas: the state of current knowledge. ICIMOD, Kathmandu

Singh G, Sarkar MS, Pandey A, Lingwal S, Rai ID, Adhikari BS, Rawat GS, Rawal RS (2018) Quantifying four decades of changes in land use and land cover in India's Kailash Sacred Landscape: suggested option for priority based patch level future forest conservation. J Indian Soc Remote Sens 46:1625–1635

Sinha S, Badola HK, Chhetri B, Gaira KS, Lepch J, Dhyani PP (2018) Effect of altitude and climate in shaping the forest compositions of Singalila National Park in Khangchendzonga Landscape, Eastern Himalaya, India. J Asia-Pacific Biod 11:267–275

Tse-ring K, Chettri N, Sharma E, Shrestha AB (2010) Climate change vulnerability of mountain ecosystems in the Eastern Himalayas—synthesis report. ICIMOD, Kathmandu, Nepal

Tulachan P (2001) Mountain agriculture in the Hindu Kush Himalaya. Mt Res Dev 21:260–267

Turin M (2005) Language endangerment and linguistic rights in the Himalayas: a case study from Nepal. Mt Res Dev 25(1):4–9

Uddin K, Chaudhary S, Chettri N, Kotru R, Murthy M, Chaudhary RP, Ning W, Shrestha SM, Gautam SK (2015) The changing land cover and fragmenting forest on the roof of the world: a case study in Nepal's Kailash Sacred Landscape. Landsc Urban Plan 141:1–10

Uddin K, Chettri N, Yang Y, Lodhi M S, Htun N Z, Sharma E (2019) Integrating geospatial tools and species for conservation planning in a data-poor region of the Far Eastern Himalayas. Geol Ecol Landsc 1–16

Ullah S, Farooq M, Shafique M, Siyab M A, Kareem F, Dees M (2016) Spatial assessment of forest cover and land-use changes in the Hindu-Kush mountain ranges of northern Pakistan. J Mt Sci 13:1229–1237. https://doi.org/10.1007/s11629-015-3456-3

UN (1992) Report of the United Nations Conference on Environment and Development. Rio de Janeiro 3–14 June 1992. A/CONF.151/26, vol 1

Uprety Y, Poudel RC, Gurung J, Chettri N, Chaudhary RP (2016) Traditional use and management of NTFPs in Kangchenjunga Landscape: implications for conservation and livelihoods. J Ethnobiol Ethnomed 12 (1):19

Vasilijević M, Pezold T (2011) Crossing borders for nature. European examples of transboundary conservation. IUCN Programme Office for South-Eastern Europe, Belgrade, Serbia

Wang Y, Wu N, Kunze C, Long R, Perlik M (2019) Drivers of change to mountain sustainability in the Hindu Kush Himalaya. In: Wester P; Mishra A, Mukherji A, Shrestha AB (eds) The Hindu Kush Himalaya assessment. Springer, Berlin. https://doi.org/10.1007/978-3-319-92288-1_5

Wangchuk K, Wangdi J (2018) Signs of climate warming through the eyes of yak herders in northern Bhutan. Mt Res Dev 38(1):45–52

Wester P, Mishra A, Mukherji A, Shrestha A B (eds) (2019) The Hindu Kush Himalaya assessment—mountains, climate change, sustainability and people. Springer Nature Switzerland AG, Cham

Wester P, Rathore, B., Vasily L, Sharma, E and D Molden (2020) The Hindu Kush Himalaya call to action: sustaining mountain environments and improving livelihoods in the Hindu Kush Himalaya. Mt Res Dev (in press)

Wikramanayake ED, Dinerstein E, Robinson JG, Karanth U, Rabinowitz A, Olson D, Mathew T, Hedao P, Conner M, Hemley G, Bolze D (1998) An ecology-based method for defining priorities for large mammal conservation: the tiger as case study. Conserv Biol 12(4):865–878

WWF, ICIMOD (2001) Ecoregion-based conservation in the Eastern Himalaya: identifying important areas for biodiversity conservation. WWF Nepal, Kathmandu

Xu J, Badola R, Chettri N, Chaudhary RP, Zomer R, Pokhrel B, Sunita P, Rebecca P, Hussain SA (2019) Sustaining biodiversity and ecosystem services in the Hindu Kush Himalaya. In: Wester P, Mishra A, Mukherji A, Shrestha AB (eds) The Hindu Kush Himalaya assessment. Springer, Berlin, pp 127–165

Xu J, Grumbine RE, Shrestha A, Ericksson M, Yang X, Wang Y, Wilkes A (2009) The melting Himalayas: cascading effects of climate change on water, biodiversity, and livelihoods. Conserv Biol 23(3):520–530

Yamasaki Y, Bhattarai N (2020) Benefiting from the REDD+ Himalaya programme: success stories from Bhutan, India, Myanmar, and Nepal. ICIMOD, Kathmandu, Nepal

Yi S, Rawat GS, Wu N, Bubb P, Chettri N, Kotru R, Sharma E, Bhatta LD, Bisht N, Aryal K, Gurung J (2017) Framework for integrated ecosystem management in the Hindu Kush Himalaya. ICIMOD, Kathmandu, Nepal

Yonzon PB (1989) Ecology and conservation of the red panda in the Nepal-Himalayas. Doctoral dissertation, University of Maine

Zhan YJ, Ren GY, Shrestha AB et al (2017) Change in extreme precipitation events over the Hindu Kush Himalayan region during 1961–2012. Adv Clim Change Res 8(3):166–175

Zhang H, Liao X, Zhai T (2018) Evaluation of ecosystem service based on scenario simulation of land use in Yunnan Province. Phys Chem Earth, Parts a/B/C 104:58–65

Zomer R, Oli KP (2011) Kailash sacred landscape conservation initiative: feasibility assessment report. ICIMOD, Kathmandu, Nepal

Zomer RJ, Trabucco A, Metzger MJ, Wang M, Oli KP, Xu J (2014) Projected climate change impacts on spatial distribution of bioclimatic zones and ecoregions within the Kailash Sacred Landscape of China, India, Nepal. Clim Change 125(3–4):445–460

Correction to: Mountain Landscapes in Transition

Udo Schickhoff, R.B. Singh,
and Suraj Mal

Correction to:
U. Schickhoff et al. (eds.), *Mountain Landscapes in Transition*, Sustainable Development Goals Series, https://doi.org/10.1007/978-3-030-70238-0

The original version of the book was inadvertently published with incorrect references in Acknowledgement for chapters 15 and 16 and in chapter 3 page 163, for co-author "R. Karki" was mentioned "Deceased" this information has been updated. Correction to the previously published version has been updated with changes.

The updated versions of these chapters can be found at
https://doi.org/10.1007/978-3-030-70238-0_3
https://doi.org/10.1007/978-3-030-70238-0_16
https://doi.org/10.1007/978-3-030-70238-0_15

U. Schickhoff et al. (eds.), *Mountain Landscapes in Transition*,
Sustainable Development Goals Series, https://doi.org/10.1007/978-3-030-70238-0_32

GPSR Compliance

The European Union's (EU) General Product Safety Regulation (GPSR) is a set of rules that requires consumer products to be safe and our obligations to ensure this.

If you have any concerns about our products, you can contact us on ProductSafety@springernature.com

In case Publisher is established outside the EU, the EU authorized representative is:

Springer Nature Customer Service Center GmbH
Europaplatz 3
69115 Heidelberg, Germany

Batch number: 10732805

Printed by Printforce, the Netherlands